Representation Theory of Semisimple Groups

PRINCETON LANDMARKS
IN MATHEMATICS AND PHYSICS

Non-standard Analysis, *by Abraham Robinson*

General Theory of Relativity, *by P.A.M. Dirac*

Angular Momentum in Quantum Mechanics, *by A. R. Edmonds*

Mathematical Foundations of Quantum Mechanics,
by John von Neumann

Introduction to Mathematical Logic, *by Alonzo Church*

Convex Analysis, *by R. Tyrrell Rockafellar*

Riemannian Geometry, *by Luther Pfahler Eisenhart*

The Classical Groups: Their Invariants and Representations,
by Hermann Weyl

Topology from the Differentiable Viewpoint, *by John W. Milnor*

Algebraic Theory of Numbers, *by Hermann Weyl*

Continuous Geometry, *by John von Neumann*

Linear Programming and Extensions, *by George B. Dantzig*

Operator Techniques in Atomic Spectroscopy, *by Brian R. Judd*

The Topology of Fibre Bundles, *by Norman Steenrod*

Mathematical Methods of Statistics, *by Harald Cramér*

Theory of Lie Groups, *by Claude Chevalley*

Homological Algebra, *by Henri Cartan and Samuel Eilenberg*

PCT, Spin and Statistics, and Ali That
by Raymond F. Streater and Arthur S. Wightman

Stable and Random Motions in Dynamical Systems: With Special
Emphasis on Celestial Mechanics, *by Jürgen Moser*

Representation Theory of Semisimple Groups: An Overview Based on Examples,
by Anthony W. Knapp

Representation Theory of Semisimple Groups

AN OVERVIEW BASED ON EXAMPLES

With a new preface by the author

ANTHONY W. KNAPP

PRINCETON UNIVERSITY PRESS
PRINCETON AND OXFORD

Published by Princeton University Press, 41 William Street,
Princeton, New Jersey 08540
In the United Kingdom: Princeton University Press, 3 Market Place,
Woodstock, Oxfordshire OX20 1SY

Originally published as part of the Princeton Mathematical Series, No. 36

Third printing, and first Princeton Landmarks in Mathematics edition,
with a new preface by the author, 2001

Library of Congress Cataloging-in-Publication Data

Knapp, Anthony W.
Representation theory of semisimple groups, an overview based on examples / with a new preface by the author Anthony W. Knapp.
p. cm. — (Princeton landmarks in mathematics)
Originally published: 1986.
Includes bibliographical references and index.
ISBN 0-691-09089-0 (pbk. : alk. paper)
1. Semisimple Lie groups. 2. Representations of groups. I. Series.
QA387 .K58 2001
512′.55—dc21 2001027843

British Library Cataloging-in-Publication Data is available

This book has been composed in Lasercomp Times Roman

Printed on acid-free paper. ∞

www.pup.princeton.edu

Printed in the United States of America

P

Representation Theory of Semisimple Groups

AN OVERVIEW BASED ON EXAMPLES

With a new preface by the author

ANTHONY W. KNAPP

PRINCETON UNIVERSITY PRESS
PRINCETON AND OXFORD

Published by Princeton University Press, 41 William Street,
Princeton, New Jersey 08540
In the United Kingdom: Princeton University Press, 3 Market Place,
Woodstock, Oxfordshire OX20 1SY

Originally published as part of the Princeton Mathematical Series, No. 36

Third printing, and first Princeton Landmarks in Mathematics edition,
with a new preface by the author, 2001

Library of Congress Cataloging-in-Publication Data

Knapp, Anthony W.
Representation theory of semisimple groups, an overview based on examples / with a new preface by the author Anthony W. Knapp.
p. cm. — (Princeton landmarks in mathematics)
Originally published: 1986.
Includes bibliographical references and index.
ISBN 0-691-09089-0 (pbk. : alk. paper)
1. Semisimple Lie groups. 2. Representations of groups. I. Series.
QA387 .K58 2001
512′.55—dc21 2001027843

British Library Cataloging-in-Publication Data is available

This book has been composed in Lasercomp Times Roman

Printed on acid-free paper. ∞

www.pup.princeton.edu

Printed in the United States of America

P

To Susan

and

Sarah and William

for their patience

Contents

Preface to the Princeton Landmarks in Mathematics Edition

I am pleased that Princeton University Press has decided to reprint in its Landmarks in Mathematics series *Representation Theory of Semisimple Groups: An Overview Based on Examples*. The original hardback edition of the book has been out of print for two years, and the book continues to be in demand.

The subject matter is at least as important today as it was at the time of the book's original publication in 1986. Two of the fields of application—automorphic forms and analysis of semisimple symmetric spaces—have undergone remarkable advances, and the theory in the book has been indispensable for both. Newer fields, such as Kac-Moody algebras and quantum groups, show promise of using more and more of this theory. And attempts at solving the key problem in Chapter XVI—that of finding all the irreducible unitary representations for all semisimple groups—have led to new approaches and new problems in the subjects of algebraic groups and geometric group actions.

Even with all these advances, the approach taken in the hardback edition continues to be an appropriate one for learning the subject. None of the text has been changed in the Landmarks edition, and thus it remains true to this approach.

A.W.K.
April 2001

Preface to the Princeton Landmarks in Mathematics Edition

[illegible]

[illegible]

[illegible]

Preface

The intention with this book is to give a survey of the representation theory of semisimple Lie groups, including results and techniques, in a way that reflects the spirit of the subject, corresponds more to a person's natural learning process, and stops at the end of a single volume.

Our approach is based on examples and has unusual ground rules. Although we insist (at least ultimately) on precisely stated theorems, we allow proofs that handle only an example. This is especially so when the example captures the idea for the general case. In fact, we prefer such a proof when the difference between the special case and the general case is merely a matter of technique and a presentation of the technique would not contribute to the goals of the book. The reader will be confronted with a first instance of this style of proof with Proposition 1.2. In some cases later on, when the style of a proof is atypical of the subject matter of the book, we omit the proof altogether.

Another aspect of the ground rules is that we feel no compulsion to state results in maximum generality. Even when the effect is to break with tradition, we are willing to define a concept narrowly. This is especially so with concepts for which one traditionally makes a wider definition and then proves as a theorem that the narrower definition gives all examples. Thus, for instance, a semisimple Lie group for us has a built-in Cartan involution, whereas traditionally one proves the existence of a Cartan involution as a theorem; since the involution is apparent in examples, we take it as part of the definition.

An essential companion to this style of writing is a careful guide to further reading for people who are interested. The section of Notes and its accompanying References are for just this purpose—so that a reader can selectively go more deeply into an aspect of the subject at will.

Twice we depart somewhat from our ground rules and proceed in a more thorough fashion. The first time is in Chapter IV with the Cartan-Weyl theory for compact Lie groups. The theory is applied often, and its general techniques are used frequently. The second time is in Chapter VIII and Appendix B with admissible representations. The heart of this theory consists of two brilliant papers by Harish-Chandra [1960] on the role of differential equations, a fundamental contribution by Langlands

[1973] on the classification of irreducible admissible representations, and a striking application of the theory by Casselman [1975]. The original papers are unpublished manuscripts, although Harish-Chandra's have been included in his collected works and parts of all the papers have been incorporated into the books by Warner [1972b] and by Borel and Wallach [1980] and into the paper by Casselman and Miličić [1982]. Since the original papers are not otherwise widely accessible, since they have been simplified somewhat by several people, and since their content is so important, we have chosen to go into some detail about them.

The finite-dimensional representation theory of semisimple groups is due chiefly to E. Cartan and H. Weyl. The infinite-dimensional theory began with Bargmann's treatment of SL(2, ℝ) in 1947 and then was dominated for many years by Harish-Chandra in the United States and by Gelfand and Naimark in the Soviet Union. Although functional analysts such as Godement, Mackey, Mautner, and Segal made early contributions to the foundations of the subject, it was Harish-Chandra, Gelfand, and Naimark who set the tone for research by using deeper structural properties of the groups to get at explicit results in representation theory. The early work by these three leaders established the explicit determination of the Plancherel formula and the explicit description of the unitary dual as important initial goals. This attitude of requiring explicit results ultimately forced a more concrete approach to the subject than was possible with abstract functional analysis, and the same attitude continues today. More recently this attitude has been refined to insist that significant results not only be explicit but also be applicable to all semisimple groups. A group-by-group analysis is rarely sufficient now: It usually does not give the required amount of insight into the subject. To be true to the field, this book attempts to communicate such attitudes and approaches, along with the results.

Bruhat's 1956 thesis was the first major advance in the field by another author that was consistent with the attitudes and approaches of the three leaders. Toward 1960 other mathematicians began to make significant contributions to parts of the theory beyond the foundations, but the goals and attitudes remained.

Beginning with Cartan and Weyl and lasting even beyond 1960, there was a continual argument among experts about whether the subject should be approached through analysis or through algebra. Some today still take one side or the other. It is clear from history, though, that it is best to use both analysis and algebra; insight comes from each. This book reflects that philosophy. To present both viewpoints for compact groups, for example, we begin with Cartan's algebraic approach and switch abruptly to Weyl's analytic approach in the middle. The reader will notice other instances of this philosophy in later chapters.

The author's introduction to this subject came from a course taught by S. Helgason at M.I.T. in 1967, a seminar run with C. J. Earle, W.H.J. Fuchs, S. Halperin, O. S. Rothaus, and H.-C. Wang in 1968, a course from Harish-Chandra in the fall of 1968, and conversations with E. M. Stein beginning in 1968. Some of these first insights are reproduced in this book. More of the book comes from lectures and courses given by the author over a period of fifteen years. There are a few new theorems and many new proofs.

All of this material came together for a course at Université Paris VII in Spring 1982, and the notes given for that course constituted a preliminary edition of the present book.

Prerequisites for the book are a one-semester course in Lie groups, some measure theory, some knowledge of one complex variable, and a few things about Hilbert spaces. For the one-semester course in Lie groups, knowledge of the first four chapters of the book by Chevalley [1947] and some supplementary material on Lie algebras are appropriate; a summary of this material constitutes Appendix A. In addition to these prerequisites, existence and uniqueness of Haar measure are assumed, as is the definition of a complex manifold; references are provided for this material. Other theorems are sometimes cited in the text; they are not intended as part of the prerequisites, and references are given.

Beginning at a certain point in one's mathematical career—corresponding roughly to the second or third year of graduate school in the United States and to the troisième cycle in France—one rarely learns a field of mathematics by studying it from start to finish. Later courses may be given as logical progressions through a subject, but the alert instructor recognizes that the students who master the mathematics do not do so by mastering the logical progressions. Instead the mastery comes through studying examples, through grasping patterns, through getting a feeling for how to approach aspects of the subject, and through other intangibles. Yet our advanced mathematics books seldom reflect this reality.

The subject of semisimple Lie groups is especially troublesome in this respect. It has a reputation for being both beautiful and difficult, and many mathematicians seem to want to know something about it. But it seems impossible to penetrate. A thorough logical-progression approach might require ten thousand pages.

Thus the need and the opportunity are present to try a different approach. The intention is that an approach to representation theory through examples be a response to that need and opportunity.

A.W.K.
August 1984

Acknowledgments

It is difficult to see how the writing of this book could have been finished without the help of four people who gave instruction to the author, provided missing proofs, and solved various problems of exposition. The author is truly grateful to these four—R. A. Herb, R. P. Langlands, D. A. Vogan, and N. R. Wallach—for all their help.

The author appreciates also the contributions of J.-L. Clerc, K. Lai, H. Schlichtkrull, Erik Thomas, and E. van den Ban, who read extensive portions of the manuscript and offered criticisms and corrections.

Other people who provided substantive help in large or small ways were J. Arthur, M. W. Baldoni-Silva, Y. Benoist, B. E. Blank, M. Duflo, J. P. Gourdot, J.A.C. Kolk, P. J. Sally, W. Schmid, and J. Vargas Soria. Their contributions gave continual encouragement to the author during the course of the writing.

Financial support for the writing came from Université Paris VII and the Guggenheim Foundation. Some published research of the author that is reproduced in this book was supported by the National Science Foundation and the Institute for Advanced Study in Princeton, New Jersey.

Representation Theory of Semisimple Groups

CHAPTER I

Scope of the Theory

§1. The Classical Groups

A **linear connected reductive group** is a closed connected group of real or complex matrices that is stable under conjugate transpose. A **linear connected semisimple group** is a linear connected reductive group with finite center. To avoid having cumbersome statements of theorems, we may or may not make an exception for the trivial one-element group in these definitions.

We shall denote such a group typically by G. Since G is a closed subgroup of a Lie group, it is a Lie group, by (A.107). We let $\mathfrak{g}$ be its Lie algebra, regarded as a real Lie algebra of real or complex matrices. Inverse conjugate transpose, which we denote by Θ, is an automorphism of G called the Cartan involution. Note that $\Theta^2 = 1$. Let

$$K = \{g \in G \mid \Theta g = g\};$$

this is a subgroup of G that will be observed presently to be a maximal compact subgroup of G.

The differential θ of Θ at 1 is an automorphism of $\mathfrak{g}$, given as negative conjugate transpose. Let $\mathfrak{k}$ and $\mathfrak{p}$ be the eigenspaces in $\mathfrak{g}$ under θ for eigenvalues $+1$ and -1, respectively. Since $\theta^2 = 1$, we have a **Cartan decomposition** for $\mathfrak{g}$ given by

$$\mathfrak{g} = \mathfrak{k} \oplus \mathfrak{p}.$$

Here $\mathfrak{k}$ consists of the skew-Hermitian members of $\mathfrak{g}$, and $\mathfrak{p}$ consists of the Hermitian members. Since θ is an automorphism, we have the relations

$$[\mathfrak{k}, \mathfrak{k}] \subseteq \mathfrak{k}, \qquad [\mathfrak{k}, \mathfrak{p}] \subseteq \mathfrak{p}, \qquad [\mathfrak{p}, \mathfrak{p}] \subseteq \mathfrak{k}.$$

In particular, $\mathfrak{k}$ is a Lie subalgebra of $\mathfrak{g}$.

Before coming to the examples, we give two propositions about such groups and their Lie algebras, deferring the proofs to §2. The decomposition $G = K \exp \mathfrak{p}$ in the second one is called the **Cartan decomposition** of G.

Proposition 1.1. If G is linear connected semisimple, then $\mathfrak{g}$ is semisimple. More generally, if G is linear connected reductive, then

$$\mathfrak{g} = Z_{\mathfrak{g}} \oplus [\mathfrak{g}, \mathfrak{g}],$$

as a direct sum of ideals. Here $Z_{\mathfrak{g}}$ denotes the center of $\mathfrak{g}$, and the commutator ideal $[\mathfrak{g}, \mathfrak{g}]$ is semisimple.

Proposition 1.2. If G is linear connected reductive, then K is compact connected and is a maximal compact subgroup of G. Its Lie algebra is $\mathfrak{k}$. Moreover, the map of $K \times \mathfrak{p}$ into G given by $(k, X) \to k \exp X$ is a diffeomorphism onto G.

The **classical groups** are certain linear connected semisimple groups that are intimately connected with classical geometry. We shall list them along with a few closely related reductive groups. We divide the list into three sections—complex groups, compact groups, and real noncompact groups.

Complex groups:

$$\mathrm{GL}(n, \mathbb{C}) = \{\text{nonsingular } n\text{-by-}n \text{ complex matrices}\}$$

$$\mathrm{SL}(n, \mathbb{C}) = \{g \in \mathrm{GL}(n, \mathbb{C}) \mid \det g = 1\}$$

$$\mathrm{SO}(n, \mathbb{C}) = \{g \in \mathrm{SL}(n, \mathbb{C}) \mid gg^{\mathrm{tr}} = 1\}$$

$$\mathrm{Sp}(n, \mathbb{C}) = \left\{g \in \mathrm{SL}(2n, \mathbb{C}) \mid g^{\mathrm{tr}}Jg = J \text{ for } J = \begin{pmatrix} 0 & I_n \\ -I_n & 0 \end{pmatrix}\right\}.$$

These are called the complex general linear group, complex special linear group, complex special orthogonal group, and complex symplectic group, respectively. Each of these is given by polynomial conditions imposed on entries of members of $\mathrm{GL}(n, \mathbb{C})$ and hence is a closed subgroup of $\mathrm{GL}(n, \mathbb{C})$. They are thus Lie groups. Their Lie algebras are

$$\mathfrak{gl}(n, \mathbb{C}) = \{n\text{-by-}n \text{ complex matrices}\}$$

$$\mathfrak{sl}(n, \mathbb{C}) = \{X \in \mathfrak{gl}(n, \mathbb{C}) \mid \mathrm{Tr}\, X = 0\}$$

$$\mathfrak{so}(n, \mathbb{C}) = \{X \in \mathfrak{sl}(n, \mathbb{C}) \mid X + X^{\mathrm{tr}} = 0\}$$

$$\mathfrak{sp}(n, \mathbb{C}) = \{X \in \mathfrak{sl}(2n, \mathbb{C}) \mid X^{\mathrm{tr}}J + JX = 0\}.$$

Each group is stable under Θ. Note that θ is not complex linear.

Connectedness requires a little argument. For $\mathrm{SL}(n, \mathbb{C})$ we can reason as follows, the other groups being handled in an analogous manner. $\mathrm{SL}(n, \mathbb{C})$ acts transitively on the column vectors in $\mathbb{C}^n - \{0\}$ by matrix multiplication, and the subgroup that leaves fixed the last standard basis vector is

$$\begin{pmatrix} \mathrm{SL}(n-1, \mathbb{C}) & 0 \\ \mathbb{C}^{n-1} & 1 \end{pmatrix} = \mathrm{SL}(n-1, \mathbb{C}) \ltimes \mathbb{C}^{n-1}.$$

By (A.108) we obtain a homeomorphism of $\mathrm{SL}(n, \mathbb{C})/(\mathrm{SL}(n-1, \mathbb{C}) \ltimes \mathbb{C}^{n-1})$ with $\mathbb{C}^n - \{0\}$. Then we can argue inductively, using the fact (A.56.2) that H connected and G/H connected implies G connected.

All the groups but $\mathrm{GL}(n, \mathbb{C})$ have finite center and thus are semisimple.

For these groups the Lie algebra is actually a Lie algebra over $\mathbb{C}$. This property is a reflection of the fact that the exponential map can be used to form charts on G that make G into a complex manifold in such a way that multiplication is holomorphic. (Cf. (A.97).)

Compact groups:

$$\mathrm{SO}(n) = \{g \in \mathrm{SL}(n, \mathbb{C}) \mid g^{\mathrm{tr}} g = 1 \text{ and } g \text{ has real entries}\}$$

$$\mathrm{U}(n) = \{g \in \mathrm{GL}(n, \mathbb{C}) \mid \bar{g}^{\mathrm{tr}} g = 1\}$$

$$\mathrm{SU}(n) = \{g \in \mathrm{U}(n) \mid \det g = 1\}$$

$$\mathrm{Sp}(n) = \left\{g \in \mathrm{U}(2n) \mid g^{\mathrm{tr}} J g = J \text{ for } J = \begin{pmatrix} 0 & I_n \\ -I_n & 0 \end{pmatrix}\right\}.$$

The first three are called the special orthogonal group, the unitary group, and the special unitary group, respectively. Each of these is bounded as a subset of $\mathbb{C}^{n^2}$ (or $\mathbb{C}^{4n^2}$ in the case of $\mathrm{Sp}(n)$), as well as closed, and hence is compact by the Heine-Borel Theorem. Their Lie algebras are

$$\mathfrak{so}(n) = \{X \in \mathfrak{sl}(n, \mathbb{C}) \mid X^{\mathrm{tr}} + X = 0 \text{ and } X \text{ has real entries}\}$$

$$\mathfrak{u}(n) = \{X \in \mathfrak{gl}(n, \mathbb{C}) \mid \bar{X}^{\mathrm{tr}} + X = 0\}$$

$$\mathfrak{su}(n) = \{X \in \mathfrak{u}(n) \mid \mathrm{Tr}\, X = 0\}$$

$$\mathfrak{sp}(n) = \{X \in \mathfrak{u}(2n) \mid X^{\mathrm{tr}} J + JX = 0\}.$$

Each group is fixed elementwise by Θ. In fact,

$$\mathrm{U}(n) \text{ is } K \text{ for } \mathrm{GL}(n, \mathbb{C})$$

$$\mathrm{SU}(n) \text{ is } K \text{ for } \mathrm{SL}(n, \mathbb{C})$$

$$\mathrm{SO}(n) \text{ is } K \text{ for } \mathrm{SO}(n, \mathbb{C})$$

$$\mathrm{Sp}(n) \text{ is } K \text{ for } \mathrm{Sp}(n, \mathbb{C}).$$

By Proposition 1.2 each of these groups is connected. Thus each of the groups is linear connected reductive. Of these, $\mathrm{SU}(n)$ is semisimple for $n \geq 2$, $\mathrm{SO}(n)$ is semisimple for $n \geq 3$, and $\mathrm{Sp}(n)$ is semisimple for $n \geq 1$.

The group Sp(n) has another realization—as the n-by-n unitary group for the quaternions $\mathbb{H}$. This realization will be established in the Problems at the end of the chapter.

Real noncompact groups: We list G, $\mathfrak{g}$, K, and $\mathfrak{k}$ for the noncomplex noncompact classical groups.

G	$\mathfrak{g}$	K	$\mathfrak{k}$
$SL(n, \mathbb{R})$	$\mathfrak{sl}(n, \mathbb{R})$	$SO(n)$	$\mathfrak{so}(n)$
$SL(n, \mathbb{H})$	$\mathfrak{sl}(n, \mathbb{H})$	$Sp(n)$	$\mathfrak{sp}(n)$
$SO_0(m, n)$	$\mathfrak{so}(m, n)$	$SO(m) \times SO(n)$	$\mathfrak{so}(m) \oplus \mathfrak{so}(n)$
$SU(m, n)$	$\mathfrak{su}(m, n)$	$S(U(m) \times U(n))$	$\mathfrak{s}(\mathfrak{u}(m) \oplus \mathfrak{u}(n))$
$Sp(m, n)$	$\mathfrak{sp}(m, n)$	$Sp(m) \times Sp(n)$	$\mathfrak{sp}(m) \oplus \mathfrak{sp}(n)$
$Sp(n, \mathbb{R})$	$\mathfrak{sp}(n, \mathbb{R})$	$U(n)$	$\mathfrak{u}(n)$
$SO^*(2n)$	$\mathfrak{so}^*(2n)$	$U(n)$	$\mathfrak{u}(n)$.

SL(n, $\mathbb{R}$) and SL(n, $\mathbb{H}$) refer to matrices of determinant one with real and quaternion entries, respectively. $SO_0(m, n)$, SU(m, n), and Sp(m, n) are the linear isometry groups for the Hermitian form

$$|z_1|^2 + \cdots + |z_m|^2 - |z_{m+1}|^2 - \cdots - |z_{m+n}|^2$$

defined over $\mathbb{R}$, $\mathbb{C}$, and $\mathbb{H}$, respectively, with the subscript "0" referring to the identity component. The group $K = S(U(m) \times U(n))$ for SU(m, n) is the subgroup of $U(m) \times U(n)$ of matrices of determinant one. To stick strictly to our definition of "linear connected reductive group," we ought to define SL(n, $\mathbb{H}$) and Sp(m, n) as groups of complex matrices; such definitions by means of complex matrices will be given in the Problems at the end of the chapter. The group Sp(n, $\mathbb{R}$) is the subgroup of real matrices in Sp(n, $\mathbb{C}$) and can be conjugated by a unitary matrix so as to become

$$\{g \in SU(n, n) \mid g^{tr} J g = J\};$$

then $SO^*(2n)$ is the analogous group

$$\left\{g \in SU(n, n) \,\middle|\, g^{tr}\begin{pmatrix} 0 & I \\ I & 0 \end{pmatrix} g = \begin{pmatrix} 0 & I \\ I & 0 \end{pmatrix}\right\}.$$

We omit all verifications of connectedness for these groups.

Direct product: The direct product $G \times H$ of two linear connected reductive groups G and H is linear connected reductive when realized as $\begin{pmatrix} G & 0 \\ 0 & H \end{pmatrix}$. If G and H are semisimple, so is $G \times H$.

§2. Cartan Decomposition

Our definitions are arranged so as to impose the extra structure of the Cartan involution (Θ or θ) on our groups and Lie algebras. (By contrast, in the general theory one would prove the existence of Θ or θ from some weaker definition of reductive or semisimple group.) We should check that Θ is determined up to isomorphism by G, and we give a result in this direction in Proposition 1.4. But first we shall introduce a useful tool, the trace form for $\mathfrak{g}$, and we shall use it to give quick proofs of Propositions 1.1. and 1.2.

The **trace form** B_0 for $\mathfrak{g}$ is defined by

$$B_0(X, Y) = \operatorname{Tr}(XY).$$

This is a complex-valued symmetric bilinear form on $\mathfrak{g} \times \mathfrak{g}$. With respect to B_0, each ad X acts by skew transformations:

$$B_0((\operatorname{ad} X)Y, Z) = -B_0(Y, (\operatorname{ad} X)Z),$$

as we see by expanding both sides. The real part Re B_0 of the trace form is a real-valued symmetric bilinear form on $\mathfrak{g} \times \mathfrak{g}$ such that each ad X acts by skew transformations. Both B_0 and Re B_0 are nondegenerate on $\mathfrak{g} \times \mathfrak{g}$ because

$$B_0(X, \theta X) < 0 \qquad \text{if} \quad X \neq 0. \tag{1.1}$$

Inequality (1.1) shows that Re B_0 is negative definite on $\mathfrak{k}$ and positive definite on $\mathfrak{p}$. In addition, $\mathfrak{k}$ and $\mathfrak{p}$ are orthogonal with respect to Re B_0. [To see that $\operatorname{Re}\operatorname{Tr}(XY) = 0$ for X skew-Hermitian and Y Hermitian, conjugate X and Y by a unitary matrix to make Y diagonal with real entries, and then compute the trace.]

Applying (1.1), we can define an inner product on the real vector space $\mathfrak{g}$ by

$$\langle X, Y \rangle = -\operatorname{Re} B_0(X, \theta Y). \tag{1.2}$$

From the previous paragraph, $\mathfrak{k}$ and $\mathfrak{p}$ are orthogonal in this inner product. This inner product allows us to define adjoints of linear transformations from $\mathfrak{g}$ into itself, and we check that

$$\operatorname{ad} \theta X = -(\operatorname{ad} X)^*. \tag{1.3}$$

The **Killing form** B for $\mathfrak{g}$ is given in (A.20) by

$$B(X, Y) = \operatorname{Tr}(\operatorname{ad} X \operatorname{ad} Y),$$

and (1.3) shows that

$$-B(X, \theta X) = \operatorname{Tr}((\operatorname{ad} X)(\operatorname{ad} X)^*) > 0 \tag{1.4}$$

unless ad $X = 0$.

Proof of Proposition 1.1. If G has finite center, then $\mathfrak{g}$ has 0 center and $X \neq 0$ implies ad $X \neq 0$. Thus in this case, (1.4) shows that the Killing form for G is nondegenerate, and $\mathfrak{g}$ is semisimple by Cartan's criterion for semisimplicity (A.23).

For general G we can conclude in the same way that $\mathfrak{g}/Z_{\mathfrak{g}}$ is semisimple. Let $Z_{\mathfrak{g}}^{\perp}$ be the "orthogonal complement" of $Z_{\mathfrak{g}}$ with respect to the nondegenerate form Re B_0. Since $Z_{\mathfrak{g}}$ is a θ-stable ideal, so is $Z_{\mathfrak{g}}^{\perp}$. Then the restriction of Re B_0 to $Z_{\mathfrak{g}}^{\perp} \times Z_{\mathfrak{g}}^{\perp}$ is nondegenerate, and it follows from (A.19) that $\mathfrak{g} = Z_{\mathfrak{g}} \oplus Z_{\mathfrak{g}}^{\perp}$. Since $\mathfrak{g}/Z_{\mathfrak{g}}$ is semisimple, $Z_{\mathfrak{g}}^{\perp}$ is semisimple. But then $\mathfrak{g} = Z_{\mathfrak{g}} \oplus Z_{\mathfrak{g}}^{\perp}$ shows $Z_{\mathfrak{g}}^{\perp} = [\mathfrak{g}, \mathfrak{g}]$. Hence $[\mathfrak{g}, \mathfrak{g}]$ is semisimple and $\mathfrak{g} = Z_{\mathfrak{g}} \oplus [\mathfrak{g}, \mathfrak{g}]$.

Proof of Proposition 1.2 for the classical groups. Since $K = G \cap \mathrm{U}(n)$, K is a closed subgroup of a compact group and so is compact. Since Θ has differential θ, K has Lie algebra $\mathfrak{k}$. To obtain the decomposition $G = K \exp \mathfrak{p}$, let g be in G and write $g = k \exp X$ for the Polar Decomposition of g as a member of $\mathrm{GL}(n, \mathbb{C})$; here k is in $\mathrm{U}(n)$ and $\exp X$ is positive definite Hermitian, written as the exponential of a unique Hermitian matrix X. Then $\Theta g = k \exp(-X)$ and hence $(\Theta g)^{-1} g = \exp 2X$ is in G. If we can prove that X is in $\mathfrak{g}$, then $\exp X$ and $k = g(\exp X)^{-1}$ are in G, and the map $(k, X) \to k \exp X$ consequently carries $K \times \mathfrak{p}$ onto G. The map is one-one by the Polar Decomposition Theorem, and it is smooth because exp and multiplication are smooth. Its inverse is built from the smooth maps

$$g \to (\theta g)^{-1} g = \exp 2X, \qquad \exp 2X \to 2X \to X,$$

and

$$(g, X) \to g\,(\exp X)^{-1} = k$$

and so is smooth. (The verification that the Jacobian determinant of $X \to \exp X$ is nowhere 0 for X Hermitian is omitted.)

Thus we want to deduce that X is in $\mathfrak{g}$ from the fact that $\exp X$ is in G. Each classical group is the connected component of the identity of the zero locus of some set of real-valued polynomials in the real and imaginary parts of the matrix entries. (For $\mathrm{GL}(n)$ we must first imbed the group in $\mathrm{GL}(n+1)$, putting $(\det g)^{-1}$ in the lower right entry.) Let us conjugate matters so that $\exp X$ is diagonal, say $X = \mathrm{diag}(a_1, \ldots, a_n)$ with each a_j real. Since $\exp X$ and its integral powers are in G, the transformed polynomials vanish at

$$(\exp X)^k = \mathrm{diag}(e^{ka_1}, \ldots, e^{ka_n})$$

for every integer k. Therefore the transformed polynomials vanish at

$$\mathrm{diag}(e^{ta_1}, \ldots, e^{ta_n}) \tag{1.5}$$

for all real t. Since (1.5) is connected, $\exp tX$ is contained in G for all real t. Therefore X is in $\mathfrak{g}$ by (A.103.5).

Finally we want to prove that K is maximal among compact subgroups of G. If K_1 is a compact subgroup properly containing K, then the decomposition $G = K \exp \mathfrak{p}$ shows that K_1 contains an element of $\exp \mathfrak{p}$ other than the identity. The eigenvalues of this element are positive and not all 1. Raising the element to powers, we see that the powers cannot all lie in a bounded set. Thus the existence of K_1 leads to a contradiction, and we conclude K is maximal compact.

Corollary 1.3. If G is linear connected reductive, then the center Z_G of G satisfies

$$Z_G = (Z_G \cap K) \exp(\mathfrak{p} \cap Z_{\mathfrak{g}}).$$

Remarks. If Z_G is discrete, it follows from this result that Z_G is finite. This fact explains a bit the condition "finite center" in the definition of linear connected semisimple groups.

Proof. For z in Z_G, write $z = k \exp X$ as in Proposition 1.2. Then $(\Theta z)^{-1} z = \exp 2X$ is in Z_G. By the finite-dimensional Spectral Theorem, X is a polynomial in $\exp 2X$. Since $\exp 2X$ commutes with G, so does X. Thus X is in $Z_{\mathfrak{g}}$. Since $\exp X$ is then in Z_G, k is in Z_G.

Proposition 1.4. If G_1 and G_2 are linear connected reductive groups whose Lie algebras have Cartan decompositions $\mathfrak{g}_1 = \mathfrak{k}_1 \oplus \mathfrak{p}_1$ and $\mathfrak{g}_2 = \mathfrak{k}_2 \oplus \mathfrak{p}_2$, if G_1 and G_2 have compact center, and if $\varphi: \mathfrak{g}_1 \to \mathfrak{g}_2$ is an isomorphism, then there is another isomorphism $\varphi': \mathfrak{g}_1 \to \mathfrak{g}_2$ such that $\varphi'(\mathfrak{k}_1) = \mathfrak{k}_2$ and $\varphi'(\mathfrak{p}_1) = \mathfrak{p}_2$.

Proof in semisimple case. Let θ_1 and θ_2 be the Cartan involutions of $\mathfrak{g}_1$ and $\mathfrak{g}_2$, and let B_1 and B_2 be the Killing forms for $\mathfrak{g}_1$ and $\mathfrak{g}_2$. Since we are assuming G_1 and G_2 are semisimple, Proposition 1.1 says $\mathfrak{g}_1$ and $\mathfrak{g}_2$ are semisimple. Therefore the forms

$$\langle X, Y \rangle_1 = -B_1(X, \theta_1 Y) \qquad \text{and} \qquad \langle X, Y \rangle_2 = -B_2(X, \theta_2 Y),$$

which are seen to be positive semidefinite from (1.4), are positive definite.

Let ψ be the automorphism $\varphi^{-1}\theta_2\varphi\theta_1$ of $\mathfrak{g}_1$. For $X \neq 0$ in $\mathfrak{g}_1$, we have

$$\begin{aligned} \langle \psi(X), X \rangle_1 &= -B_1(\varphi^{-1}\theta_2\varphi\theta_1 X, \theta_1 X) = -B_2(\theta_2\varphi\theta_1 X, \varphi\theta_1 X) \\ &= \langle \varphi\theta_1 X, \varphi\theta_1 X \rangle_2 > 0. \end{aligned}$$

Also
$$\begin{aligned} \langle \psi(X), Y \rangle_1 &= -B_1(\varphi^{-1}\theta_2\varphi\theta_1 X, \theta_1 Y) \\ &= -B_1(X, \theta_1\varphi^{-1}\theta_2\varphi\theta_1 Y) = \langle X, \psi(Y) \rangle_1. \end{aligned}$$

Hence ψ is a positive definite symmetric transformation on $\mathfrak{g}_1$. We shall show that $\varphi' = \varphi\psi^{1/2}$ has the required properties, where $\psi^{1/2}$ is the positive definite square root of ψ.

First we show $\psi^{1/2}$ is an automorphism of $\mathfrak{g}_1$. Let $\{X_i\}$ be a basis of eigenvectors, say with $\psi(X_i) = c_i X_i$ and $c_i > 0$, and define structural constants c_{ij}^k for $\mathfrak{g}_1$ by $[X_i, X_j] = \sum_k c_{ij}^k X_k$. Since ψ is an automorphism of $\mathfrak{g}_1$, we have

$$\sum_k c_{ij}^k c_k X_k = \psi[X_i, X_j] = [\psi X_i, \psi X_j] = c_i c_j [X_i, X_j] = c_i c_j \sum_k c_{ij}^k X_k,$$

and thus $c_i c_j c_{ij}^k = c_k c_{ij}^k$ for all i, j, k. This equation implies that

$$c_i^{1/2} c_j^{1/2} c_{ij}^k = c_k^{1/2} c_{ij}^k \qquad \text{for all } i, j, k. \tag{1.6}$$

Since $\psi^{1/2}(X_i) = c_i^{1/2} X_i$, we have

$$\psi^{1/2}[X_i, X_j] = \psi^{1/2}\left(\sum_k c_{ij}^k X_k\right) = \sum_k c_{ij}^k c_k^{1/2} X_k$$

and

$$[\psi^{1/2} X_i, \psi^{1/2} X_j] = c_i^{1/2} c_j^{1/2} \sum_k c_{ij}^k X_k,$$

and (1.6) proves $\psi^{1/2}$ is an automorphism of $\mathfrak{g}_1$. Thus $\varphi' = \varphi\psi^{1/2}$ is an isomorphism of $\mathfrak{g}_1$ onto $\mathfrak{g}_2$.

By the finite-dimensional Spectral Theorem, we can choose a polynomial P such that $\psi^{-1/2} = P(\psi)$ and $\psi^{1/2} = P(\psi^{-1})$. From $\psi\theta_1 = \theta_1\psi^{-1}$, we deduce $\psi^2\theta_1 = \psi(\theta_1\psi^{-1}) = \theta_1\psi^{-2}$ and in general $\psi^n\theta_1 = \theta_1\psi^{-n}$. Using P, we obtain $\psi^{-1/2}\theta_1 = \theta_1\psi^{1/2}$. Therefore

$$\theta_2\varphi' = \varphi(\varphi^{-1}\theta_2\varphi\theta_1)\theta_1\psi^{1/2} = \varphi\psi\psi^{-1/2}\theta_1 = \varphi'\theta_1. \tag{1.7}$$

Applying both sides of (1.7) to X in $\mathfrak{k}_1$, we see that $\varphi'(X)$ is in $\mathfrak{k}_2$. Applying both sides of (1.7) to X in $\mathfrak{p}_1$, we see that $\varphi'(X)$ is in $\mathfrak{p}_2$. Thus φ' is an isomorphism of $\mathfrak{g}_1$ onto $\mathfrak{g}_2$ that carries $\mathfrak{k}_1$ to $\mathfrak{k}_2$ and $\mathfrak{p}_1$ to $\mathfrak{p}_2$. This completes the proof.

Remarks in general case. In Proposition 5.5, we shall see that the analytic subgroup G_{ss} of G with Lie algebra $[\mathfrak{g}, \mathfrak{g}]$ is closed and that G is the commuting product $G = (Z_G)_0 G_{ss}$. Consequently Proposition 5.5 reduces the general case to the semisimple case.

§3. Representations

Let G be a topological group. A **representation** of G on a complex Hilbert space $V \neq 0$ is a homomorphism Φ of G into the group of bounded linear operators on V with bounded inverses, such that the resulting map of $G \times V$ into V is continuous. To have this continuity property, it is enough to have strong continuity—that $g \to \Phi(g)v$ is continuous from G to V at $g = 1$ for each v in V—and a uniform bound for $\|\Phi(g)\|$ in some neighborhood of 1.

[This is proved by writing

$$\begin{aligned}|\Phi(g)v - \Phi(g_0)v_0| &\le \|\Phi(g_0)\|\,|\Phi(g_0^{-1}g)v - v_0| \\ &\le \|\Phi(g_0)\|\,|\Phi(g_0^{-1}g)(v - v_0)| + \|\Phi(g_0)\|\,|\Phi(g_0^{-1}g)v_0 - v_0| \\ &\le \|\Phi(g_0)\|\,\|\Phi(g_0^{-1}g)\|\,|v - v_0| + \|\Phi(g_0)\|\,|\Phi(g_0^{-1}g)v_0 - v_0|.\end{aligned}$$

Under our assumptions both terms on the right are small if g is close enough to g_0 and v is close enough to v_0.]

An **invariant subspace** for such a Φ is a vector subspace U such that $\Phi(g)U \subseteq U$ for all g in G. The representation is **irreducible** if it has no closed invariant subspaces other than 0 and V.

Such a representation Φ is **unitary** if $\Phi(g)$ is unitary (i.e., $\Phi(g)^*\Phi(g) = \Phi(g)\Phi(g)^* = I$) for all g in G. For a unitary representation the orthogonal complement $U^\perp$ of a closed invariant subspace U is a closed invariant subspace because

$$\langle \Phi(g)u^\perp, u\rangle = \langle u^\perp, \Phi(g)^{-1}u\rangle \in \langle u^\perp, U\rangle = 0 \qquad \text{for } u^\perp \in U^\perp,\ u \in U.$$

Two representations of G, Φ on V and Φ' on V', are **equivalent** if there is a bounded linear $E: V \to V'$ with a bounded inverse such that $\Phi'(g)E = E\Phi(g)$ for all g in G. If Φ and Φ' are unitary, they are **unitarily equivalent** if they are equivalent via an operator E that is unitary.

Examples.

(1) $G =$ a classical group as in §1, $\Phi(g) = g$. With any inner product on the underlying vector space, this is a representation. It can be unitary in the cases where G is compact but not where G is noncompact.

(2) $G = \mathrm{SO}(n)$, $V =$ all polynomials in n real variables with complex coefficients and all terms of degree N, and

$$\Phi(g)P\begin{pmatrix}x_1\\ \vdots\\ x_n\end{pmatrix} = P\left(g^{-1}\begin{pmatrix}x_1\\ \vdots\\ x_n\end{pmatrix}\right).$$

With any inner product on V, this is a representation. It is unitary if V has the inner product

$$(P, Q) = \int_{S^{n-1}} P(x)\overline{Q(x)}\,dx.$$

(3) $G = \mathrm{SL}(2, \mathbb{R})$, $V = L^2(\mathbb{R}^2)$, and

$$\Phi(g)f(x) = f(g^{-1}x).$$

With the usual inner product on V, this is a unitary representation.

(4) G = a locally compact group, $V = L^2(G, d_l x)$ taken with respect to a left-invariant measure $d_l x$, and $\Phi(g)f(x) = f(g^{-1}x)$. This is called the **left regular representation** of G. The **right regular representation** is given by $\Phi'(g)f(x) = f(xg)$ on $L^2(G, d_r x)$. The continuity property is a standard fact in measure theory.

The first example raises the question: Why study representations for a group that is already represented as matrices? A hint of an answer is in the other three examples: Representations are often forced on us by some outside area of mathematics, and they may not be the ones we are familiar with. Studying such representations may involve constructing other representations or using some classification scheme or working with the representation directly, and we need techniques for this study.

Proposition 1.5 (Schur's Lemma). A unitary representation Φ of a topological group G on a Hilbert space V is irreducible if and only if the only bounded linear operators on V commuting with all $\Phi(g)$, $g \in G$, are the scalar operators.

Proof. If V is reducible and U is a nontrivial closed invariant subspace, then the orthogonal projection on U is a nonscalar bounded linear operator on V commuting with all $\Phi(g)$.

Conversely let L be a nonscalar bounded linear operator on V commuting with all $\Phi(g)$. Then L^* commutes with all $\Phi(g)$ since Φ is unitary, and so do the self-adjoint operators $A = \frac{1}{2}(L + L^*)$ and $B = \dfrac{1}{2i}(L - L^*)$. Since $L = A + iB$, at least one of A and B is not scalar, say A for definiteness. The Spectral Theorem, applied to A, then produces at least one nonscalar orthogonal projection E that is a function of A. Since E is a function of A, E commutes with all $\Phi(g)$.

A handy device in working with a representation Φ of a locally compact group G on V is the system of operators $\Phi(f)$ as f varies over some class of scalar-valued functions on G. Let us suppose G has a two-sided Haar measure dx. For unitary Φ, $\Phi(f)$ is defined for f in $L^1(G)$; for general Φ we assume also that f has compact support. The operator $\Phi(f)$ can be defined either in terms of integration in V as

$$\Phi(f)v = \int_G f(x)\Phi(x)v\,dx \tag{1.8a}$$

or by means of inner products as

$$(\Phi(f)v, v') = \int_G f(x)(\Phi(x)v, v')\,dx. \tag{1.8b}$$

Note that $\Phi(f)$ leaves stable any closed invariant subspace of V.

If Φ is unitary, then

$$\|\Phi(f)\| \leq |f|_1. \tag{1.9}$$

Also

$$\Phi(f)^* = \Phi(f^*) \tag{1.10}$$

where $f^*(x) = \overline{f(x^{-1})}$, and

$$\Phi(f)\Phi(h) = \Phi(f * h), \tag{1.11}$$

where $f * h(x) = \int_G f(xy^{-1})h(y)\,dy$.

Let $\mathfrak{g}$ be a real Lie Algebra. A **representation** of $\mathfrak{g}$ on a complex vector space $V \neq 0$ is a homomorphism φ of $\mathfrak{g}$ into the Lie algebra of all linear transformations of V into itself. In particular, φ is to satisfy

$$\varphi[X, Y] = \varphi(X)\varphi(Y) - \varphi(Y)\varphi(X) \qquad \text{for all } X, Y \in \mathfrak{g}. \tag{1.12}$$

An **invariant subspace** for such a φ is a vector subspace U such that $\varphi(X)U \subseteq U$ for all X in $\mathfrak{g}$. The representation is **irreducible** if it has no invariant subspaces other than 0 and V. Two representations of $\mathfrak{g}$, φ on V and φ' on V', are **equivalent** if there is a linear invertible $E: V \to V'$ such that $\varphi'(X)E = E\varphi(X)$ for all X in $\mathfrak{g}$.

If G is a Lie group with Lie algebra $\mathfrak{g}$ and if V is finite-dimensional, then there is a correspondence between representations of G and representations of $\mathfrak{g}$. If Φ is a representation of G on V, then Φ is automatically smooth from G into the general linear group on V (see Corollary 3.16), and the differential $d\Phi$ at the identity is a representation of $\mathfrak{g}$ on V. Moreover, $d\Phi$ uniquely determines Φ. In the reverse direction, if φ is a representation of $\mathfrak{g}$ on V and if G is connected and simply connected, then there exists a representation Φ of G with $d\Phi = \varphi$.

Under this correspondence when V is finite-dimensional, invariant subspaces correspond, and Φ is irreducible if and only if $\varphi = d\Phi$ is irreducible. Moreover, equivalences on the group level correspond to equivalences on the Lie algebra level.

Matters are not so simple when V is infinite-dimensional. We return to this point in Chapter III.

For finite-dimensional representations of G, there is a natural definition of **tensor product:** The tensor product of two finite-dimensional Hilbert spaces is well defined, and the tensor product representation $\Phi = \Phi_1 \otimes \Phi_2$ is to satisfy

$$\Phi(g)(v_1 \otimes v_2) = \Phi_1(g)v_1 \otimes \Phi_2(g)v_2.$$

(With this definition we can even allow one of the two Hilbert spaces to be infinite-dimensional.) The tensor product $\varphi = \varphi_1 \otimes \varphi_2$ of two representations of a Lie algebra is defined as if it came from the differential of the tensor product of two representations of a Lie group; thus the formula is

$$\varphi(X)(v_1 \otimes v_2) = \varphi_1(X)v_1 \otimes v_2 + v_1 \otimes \varphi_2(X)v_2.$$

§4. Concrete Problems in Representation Theory

The three main problems in representations theory for a particular group G are these:

(1) Classify explicitly the irreducible unitary representations of G. Give classifications for any other interesting manageable classes of irreducible representations.
(2) Give an explicit analysis of $L^2(G)$.
(3) Analyze other representations of G, particularly unitary ones, that arise naturally in mathematics.

We shall be interested in these problems—and techniques for handling them—for linear connected reductive groups, especially for the classical groups. Since these groups are so concrete, it is reasonable to insist on detailed answers to the questions. At this writing, the irreducible unitary representations of such groups have not been classified; however, some related classes of irreducible representations are classified and well understood. The analysis of $L^2(G)$ is complete, in the form of an explicit Plancherel formula for a generalized Fourier transform on the group. In addition, some information is known about other naturally arising representations of G, although we shall not pursue such results here.

When G is compact, there is an abstract theory that gives some information without the restriction that G be a Lie group. Among other things, it shows what the answer is to (2) if the answer to (1) is known. This abstract theory will be presented in the next section. When the compact group G is a connected Lie group., there is also a concrete theory, which will be given in Chapter IV, and the theory includes a complete solution of (1).

There is a certain amount of abstract theory that one can do for general locally compact groups. However, the results that one obtains beyond those in the previous section will not have sufficient bearing on our concrete groups to justify the effort involved.

§5. Abstract Theory for Compact Groups

Let G be a compact topological group, and let dx denote Haar measure on G normalized so that $\int_G dx = 1$.

Proposition 1.6. If Φ is a representation of G on a finite-dimensional V, then V admits a Hermitian inner product such that Φ is unitary.

Proof. Let $(\cdot\,,\cdot)$ be any Hermitian inner product on V, and define

$$\langle u, v\rangle = \int_G (\Phi(x)u, \Phi(x)v)\,dx.$$

It is straightforward to see that $\langle\cdot,\cdot\rangle$ has the required properties.

Corollary 1.7. If Φ is a representation of G on a finite-dimensional V, then Φ is the direct sum of irreducible representations. (That is, $V = V_1 \oplus \ldots \oplus V_k$, with each V_j an invariant subspace on which Φ acts irreducibly.)

Proof. Form $\langle\cdot,\cdot\rangle$ as in Proposition 1.6. Find an invariant subspace $U \neq 0$ of minimal dimension and take its orthogonal complement $U^\perp$. Then $U^\perp$ is invariant. Repeating the argument with $U^\perp$ and iterating, we obtain the required decomposition.

Proposition 1.8 (Schur's Lemma). Suppose Φ and Φ' are irreducible representations of G on finite-dimensional spaces V and V', respectively. If $L: V \to V'$ is a linear map such that $\Phi'(g)L = L\Phi(g)$ for all g in G, then L is one-one onto or $L = 0$.

Proof. We see easily that ker L and image L are invariant subspaces of V and V', respectively, and then the only possibilities are the ones listed.

Corollary 1.9. Suppose Φ is an irreducible unitary representation of G on a finite-dimensional V. If $L: V \to V$ is a linear map such that $\Phi(g)L = L\Phi(g)$ for all g in G, then L is scalar.

Remark. This is a special case of Proposition 1.5, but there is a much easier proof here because V is finite-dimensional.

Proof. Let λ be an eigenvalue of L. Then $L - \lambda I$ is not one-one onto but does commute with $\Phi(g)$ for all g in G. By Proposition 1.8, $L - \lambda I = 0$.

Corollary 1.10 (Schur orthogonality relations).

(a) Let Φ and Φ' be inequivalent irreducible unitary representations on finite-dimensional spaces V and V', respectively. Then

$$\int_G (\Phi(x)u, v)\overline{(\Phi'(x)u', v')}\,dx = 0 \qquad \text{for all } u, v \in V \text{ and } u', v' \in V'.$$

(b) Let Φ be an irreducible unitary representation on a finite-dimensional V. Then

$$\int_G (\Phi(x)u_1, v_1)\overline{(\Phi(x)u_2, v_2)}\,dx = \frac{(u_1, u_2)\overline{(v_1, v_2)}}{\dim V} \qquad \text{for } u_1, v_1, u_2, v_2 \in V.$$

Proof. For (a), let $l: V' \to V$ be linear and form

$$L = \int_G \Phi(x) l \Phi'(x^{-1})\, dx.$$

(This integration can be regarded as occurring for matrix-valued functions and is to be handled entry-by-entry.) Then it follows that $\Phi(y)L\Phi'(y^{-1}) = L$, so that $\Phi(y)L = L\Phi'(y)$ for all y in G. By Proposition 1.8, $L = 0$. Thus $(Lv', v) = 0$. Choose $l(w') = (w', u')u$, and (a) results.

For (b), we proceed in the same way, starting from $l: V \to V$ and obtain $L = \lambda I$ from Corollary 1.9. Taking the trace of both sides, we find

$$\lambda \dim V = \operatorname{Tr} L = \operatorname{Tr} l,$$

so that $\lambda = \operatorname{Tr} l/\dim V$. Thus

$$(Lv_2, v_1) = \frac{\operatorname{Tr} l}{\dim V} \overline{(v_1, v_2)}.$$

Choosing $l(w) = (w, u_2)u_1$, we obtain (b).

We can interpret Corollary 1.10 as follows. Let $\{\Phi^{(\alpha)}\}$ be a maximal set of mutually inequivalent irreducible unitary representations of G. For each $\Phi^{(\alpha)}$, choose an orthonormal basis for the underlying vector space and let $\Phi_{ij}^{(\alpha)}(x)$ be the matrix of $\Phi^{(\alpha)}(x)$ in this basis. Then the functions $\{\Phi_{ij}^{(\alpha)}(x)\}_{i,j,\alpha}$ form an orthogonal set in $L^2(G)$. In fact, if $d^{(\alpha)}$ denotes the **degree** of $\Phi^{(\alpha)}$ (i.e., the dimension of the underlying vector space), then $\{(d^{(\alpha)})^{1/2}\Phi_{ij}^{(\alpha)}(x)\}_{i,j,\alpha}$ is an orthonormal set in $L^2(G)$. The Peter-Weyl Theorem, which follows, uses the maximality of the system $\{\Phi^{(\alpha)}\}$ to conclude that this orthonormal set is an orthonormal basis.

If Φ is a unitary representation of G, a **matrix coefficient** of Φ is any function on G of the form $(\Phi(x)u, v)$. If Φ acts on a finite-dimensional space, its **character** is the function

$$\chi_\Phi(x) = \operatorname{Tr} \Phi(x) = \sum_i (\Phi(x)u_i, u_i), \tag{1.13}$$

where $\{u_i\}$ is an orthonormal basis.

Lemma 1.11. Characters of finite-dimensional irreducible unitary representations Φ and Φ' satisfy

$$\chi_\Phi(x) = \overline{\chi_\Phi(x^{-1})}$$

$$\chi_\Phi * \chi_{\Phi'} = 0 \qquad \text{if } \Phi \text{ is not equivalent with } \Phi'$$

$$d_\Phi \chi_\Phi * \chi_\Phi = \chi_\Phi.$$

Proof. The first relation follows by summing for $i = j$ the relation $(\Phi(x)u_i, u_j) = \overline{(\Phi(x^{-1})u_j, u_i)}$. The other two relations are routine consequences of Schur orthogonality (Corollary 1.10).

Theorem 1.12 (Peter-Weyl Theorem).

(a) The linear span of all matrix coefficients for all finite-dimensional irreducible unitary representations of G is dense in $L^2(G)$.

(b) If $\{\Phi^{(\alpha)}\}$ is a maximal set of mutually inequivalent finite-dimensional irreducible unitary representations of G and $\{(d^{(\alpha)})^{1/2}\Phi_{ij}^{(\alpha)}(x)\}_{i,j,\alpha}$ is a corresponding orthonormal set of matrix coefficients, then $\{(d^{(\alpha)})^{1/2}\Phi_{ij}^{(\alpha)}(x)\}_{i,j,\alpha}$ is an orthonormal basis of $L^2(G)$.

(c) Every irreducible unitary representation of G is finite-dimensional.

(d) Let Φ be a unitary representation of G on a Hilbert space V. Then V is the orthogonal sum of finite-dimensional irreducible invariant subspaces.

(e) Let Φ be a unitary representation of G on a Hilbert space V. For each irreducible unitary representation τ of G, let E_τ be the orthogonal projection on the closure of the sum of all irreducible invariant subspaces of V that are equivalent with τ. Then E_τ is given by $d_\tau\Phi(\overline{\chi_\tau})$, where d_τ is the degree of τ and χ_τ is the character of τ. Moreover, if τ and τ' are inequivalent, the $E_\tau E_{\tau'} = E_{\tau'} E_\tau = 0$. Finally every v in V satisfies

$$v = \sum_\tau E_\tau v,$$

with the sum taken over a set of representatives τ of all equivalence classes of irreducible unitary representations of G.

Remarks. The heart of the proof is for (a). For compact groups of matrices, conclusion (a) is an immediate consequence of the Stone-Weierstrass Theorem: In fact, the linear span of all matrix coefficients forms an algebra (by use of tensor products) of continuous functions closed under conjugation (by use of the complex conjugate representation), containing the constants (by use of the trivial representation), and separating points (by use of the given representation of the group). This algebra is then uniformly dense in the space of all continuous complex-valued functions on G, and it follows that it is dense (in the L^2 norm) in $L^2(G)$. In our applications our compact groups will always be groups of matrices, but not automatically; in fact, we shall need the full strength of the Peter-Weyl Theorem to realize our groups as matrix groups. Theorem 1.15 below will give the details.

Proof.

(a) If $h(x) = (\Phi(x)u, v)$ is a matrix coefficient, then the following functions of x are also matrix coefficients for the same representation:

$$\overline{h(x^{-1})} = (\Phi(x)v, u)$$

$$h(gx) = (\Phi(x)u, \Phi(g^{-1})v)$$

$$h(xg) = (\Phi(x)\Phi(g)u, v).$$

Thus the closure U in $L^2(G)$ of the linear span of all matrix coefficients of all finite-dimensional irreducible unitary representations is stable under $h(x) \to \overline{h(x^{-1})}$ and under left and right translation. Arguing by contradiction, suppose $U \neq L^2(G)$. Then $U^\perp \neq 0$ and $U^\perp$ is closed under $h(x) \to \overline{h(x^{-1})}$ and under left and right translation.

We first prove there is a nonzero continuous function in $U^\perp$. Thus let $H \neq 0$ be in $U^\perp$. For each open neighborhood N of 1, we use the left regular representation L to define

$$F_N(x) = L\left(\frac{1}{|N|} I_N\right) H(x) = \frac{1}{|N|} \int_G I_N(y) H(y^{-1}x)\, dy,$$

where I_N is the characteristic function of N and $|N|$ is the Haar measure of N. Since F_N is the convolution of two L^2 functions, it is continuous. As N shrinks to $\{1\}$, the functions F_N tend to H in L^2; hence some F_N is not 0. Finally F_N is obtained as $L(f)v$ for a suitable f in $L^1(G)$ and a suitable v in $U^\perp$ and thus is in $U^\perp$.

Thus $U^\perp$ contains a nonzero continuous function. Using translations and scalar multiplication, find a continuous F_1 in $U^\perp$ with $F_1(1)$ real and nonzero. Set

$$F_2(x) = \int_G F_1(yxy^{-1})\, dy.$$

Then F_2 is continuous and is in $U^\perp$, $F_2(gxg^{-1}) = F_2(x)$ for all g in G, and $F_2(1) = F_1(1)$ is real and nonzero. Finally put

$$F(x) = F_2(x) + \overline{F_2(x^{-1})}.$$

Then F is continuous and is in $U^\perp$, $F(gxg^{-1}) = F(x)$ for all g in G, $F(1) = 2F_2(1)$ is real and nonzero, and $F(x) = \overline{F(x^{-1})}$. In particular, F is not the 0 function in $L^2(G)$.

Form the function $k(x, y) = F(x^{-1}y)$ and the integral operator

$$Tf(x) = \int_G k(x, y) f(y)\, dy = \int_G F(x^{-1}y) f(y)\, dy, \qquad f \in L^2(G).$$

Then $k(x, y) = \overline{k(y, x)}$ and $\int_{G \times G} |k(x, y)|^2\, dx\, dy = \int_G |F(x)|^2\, dx < \infty$, and hence T is a Hilbert-Schmidt operator from $L^2(G)$ into itself. Also T is not 0 since $F \neq 0$. Such an operator has a real nonzero eigenvalue λ, and the corresponding eigenspace $V_\lambda \subseteq L^2(G)$ is finite-dimensional.

The subspace V_λ is invariant under the left regular representation L because f in V_λ implies

$$\begin{aligned} TL(g)f(x) &= \int_G F(x^{-1}y) f(g^{-1}y)\, dy = \int_G F(x^{-1}gy) f(y)\, dy \\ &= Tf(g^{-1}x) = \lambda f(g^{-1}x) = \lambda L(g) f(x). \end{aligned}$$

Since V_λ is finite-dimensional, it contains (by dimensionality) an irreducible invariant subspace $W_\lambda \neq 0$.

Let $f_1, \ldots, f_n$ be an orthonormal basis of W_λ. The matrix coefficients for W_λ are

$$h_{ij}(x) = (L(x)f_j, f_i) = \int_G f_j(x^{-1}y)\overline{f_i(y)}\,dy$$

and by definition are in U. Since F is in $U^\perp$, we have

$$\begin{aligned}
0 &= \int_G F(x)\overline{h_{ii}(x)}\,dx = \int_G \int_G F(x)\overline{f_i(x^{-1}y)}f_i(y)\,dy\,dx \\
&= \int_G \int_G F(x)\overline{f_i(x^{-1}y)}f_i(y)\,dx\,dy = \int_G \int_G F(yx^{-1})\overline{f_i(x)}f_i(y)\,dx\,dy \\
&= \int_G \left[\int_G F(x^{-1}y)f_i(y)\,dy\right]\overline{f_i(x)}\,dx \qquad \text{since } F(gxg^{-1}) = F(x) \\
&= \int_G Tf_i(x)\overline{f_i(x)}\,dx \\
&= \lambda \int_G |f_i(x)|^2\,dx,
\end{aligned}$$

in contradiction to the fact that $W_\lambda \neq 0$. We conclude that $U^\perp = 0$ and $U = L^2(G)$.

(b) The linear span of the functions in question is the linear span considered in (a). Then (a) and general Hilbert space theory imply (b).

(c) This will follow from (d).

(d) By Zorn's Lemma choose a maximal orthogonal set of finite-dimensional irreducible invariant subspaces. Let U be the closure of the sum. Arguing by contradiction, we suppose U is not all of V. Then $U^\perp$ is a nonzero closed invariant subspace. Fix $v \neq 0$ in $U^\perp$, and form $\Phi(I_N)v$, where I_N is the characteristic function of an open neighborhood N of 1. Then $\Phi(I_N)v$ is in $U^\perp$ for every N. As N shrinks to $\{1\}$, $\frac{1}{|N|}\Phi(I_N)v$ tends to v; hence some $\Phi(I_N)v$ is not 0. Fix such an N.

If h is a linear combination of matrix coefficients, then h lies in a finite-dimensional subspace S of $L^2(G)$ that is invariant under left translation. Let $h_1, \ldots, h_n$ be a basis of this space S. Then

$$\begin{aligned}
\Phi(g)\Phi(h)v = \Phi(g)\int_G h(x)\Phi(x)v\,dx &= \int_G h(x)\Phi(gx)v\,dx \\
&= \int_G h(g^{-1}x)\Phi(x)v\,dx = \sum_{j=1}^n c_j \int_G h_j(x)\Phi(x)v\,dx,
\end{aligned}$$

and hence the finite-dimensional subspace $\sum_j \mathbb{C}\Phi(h_j)v$ is an invariant subspace for Φ. Consequently we will obtain a contradiction if we show that $\Phi(h)v \neq 0$ for some linear combination h of matrix coefficients.

To do this, choose h by (a) so that

$$|I_N - h|_1 \leq |I_N - h|_2 \leq \tfrac{1}{2}|\Phi(I_N)v|/|v| \tag{1.14}$$

Then
$$|\Phi(I_N)v - \Phi(h)v| = |\Phi(I_N - h)v| \leq |I_N - h|_1|v|$$
$$\leq |I_N - h|_2|v| \leq \tfrac{1}{2}|\Phi(I_N)v|$$

by (1.9) and (1.14). Hence

$$|\Phi(h)v| \geq |\Phi(I_N)v| - |\Phi(I_N)v - \Phi(h)v|$$
$$\geq \tfrac{1}{2}|\Phi(I_N)v| > 0.$$

Thus h has the required property. This proves (d) and thus also (c).

(e) Put $E'_\tau = d_\tau\Phi(\overline{\chi_\tau})$. By (1.10) and Lemma 1.11,

$$E'^*_\tau = d_\tau\Phi(\chi_\tau(x^{-1})) = d_\tau\Phi(\overline{\chi_\tau}) = E'_\tau$$
$$E'_\tau E'_{\tau'} = d_\tau d_{\tau'}\Phi(\overline{\chi_\tau} * \overline{\chi_{\tau'}}) = 0 \qquad \text{for } \tau \not\cong \tau'$$
$$E'^2_\tau = d^2_\tau\Phi(\overline{\chi_\tau} * \overline{\chi_\tau}) = d_\tau\Phi(\overline{\chi_\tau}) = E'_\tau$$

Thus E'_τ is an orthogonal projection, and $E'_\tau E'_{\tau'} = E'_{\tau'}E'_\tau = 0$ for τ and τ' inequivalent.

Let U be an irreducible finite-dimensional subspace of V on which $\Phi|_U$ is equivalent with τ, and let $u_1, \ldots, u_n$ be an orthonormal basis of U. If

$$\Phi_{ij}(x) = (\Phi(x)u_j, u_i),$$

then
$$\chi_\tau(x) = \sum_{i=1}^{n} \Phi_{ii}(x) \qquad \text{and} \qquad \Phi(x)u_j = \sum_{i=1}^{n} \Phi_{ij}(x)u_i.$$

Hence Schur orthogonality gives

$$E'_\tau u_j = d_\tau \int \overline{\chi_\tau(x)}\Phi(x)u_j\,dx = d_\tau \int \sum_{i,k} \overline{\Phi_{kk}(x)}\Phi_{ij}(x)u_i\,dx = u_j.$$

Thus E'_τ is the identity on every irreducible subspace of type τ.

For u in a space of type τ', we have $E'_\tau u = E'_\tau E'_{\tau'}u = 0$. Now let us apply E'_τ to a decomposition as in (d). All terms are then annihilated except the ones of type τ, since E'_τ vanishes on spaces of type τ' with τ' not equivalent with τ. Consequently $E'_\tau = E_\tau$ and $v = \sum_\tau E_\tau v$ for all v in V. This completes the proof of the theorem.

Part (b) of the Peter-Weyl Theorem, together with abstract Hilbert space theory, implies the **Parseval-Plancherel formula:**

$$\int_G |f(x)|^2\,dx = \sum_\Phi d_\Phi \sum_{i,j} |\Phi(f)_{ij}|^2. \tag{1.15}$$

Here Φ runs over a complete set of representatives of all equivalence classes of irreducible unitary representations of G. The inside sum on the right in (1.15) is independent of basis and gives the Hilbert-Schmidt norm squared of $\Phi(f)$, denoted $\|\Phi(f)\|_{\mathrm{HS}}^2$. Thus we have

$$\int_G |f(x)|^2\,dx = \sum_{\Phi} d_\Phi \|\Phi(f)\|_{\mathrm{HS}}^2. \tag{1.16}$$

The proof of (e) in the Peter-Weyl Theorem contains information even in the case that Φ is the right regular representation on $L^2(G)$. If τ is an irreducible unitary representation and $u_1, \ldots, u_{d_\tau}$ is an orthonormal basis of the space on which τ operates, then the span of a row of matrix coefficients

$$(\tau(x)u_j, u_i), \qquad i \text{ fixed and } 1 \le j \le d_\tau,$$

is an invariant subspace of $L^2(G)$ of type τ. By Schur orthogonality the different spaces, as j varies, are orthogonal. In the decomposition of (d) of the theorem, as made specific in (b), these d_τ spaces are the only ones of type τ, because the proof of (e) shows that their span is the image of E_τ. Thus in the case of $L^2(G)$, image E_τ has dimension d_τ^2, with τ occurring d_τ linearly independent times.

The conclusion of (e) in the Peter-Weyl Theorem says that the number of occurrences of τ in a decomposition (d) is independent of the decomposition. The number is obtained as the quotient (dim image E_τ)/d_τ. We write $[\Phi:\tau]$ for this quantity, calling it the **multiplicity** of τ in Φ.

Lemma 1.13. Let Φ and τ be unitary representations of G on spaces V^Φ and V^τ, respectively, and suppose τ is irreducible. Then

$$[\Phi:\tau] = \dim \operatorname{Hom}_G(V^\Phi, V^\tau) = \dim \operatorname{Hom}_G(V^\tau, V^\Phi),$$

where the subscripts "G" refer to linear maps respecting the indicated actions by G.

Proof. By Schur's Lemma and the Peter-Weyl Theorem, any member of $\operatorname{Hom}_G(V^\Phi, V^\tau)$ annihilates $(E_\tau V^\Phi)^\perp$. Thus write $E_\tau V^\Phi$ as the orthogonal sum of irreducible subspaces V_α, by Theorem 1.12d. Each V_α is equivalent with V^τ, by Theorem 1.12e. Thus for each V_α the space of G-maps from V_α to V^τ is at least one-dimensional. It is at most one-dimensional by Schur's Lemma. Then it follows that

$$[\Phi:\tau] = \dim \operatorname{Hom}_G(V^\Phi, V^\tau).$$

Taking adjoints, we obtain

$$\dim \operatorname{Hom}_G(V^\Phi, V^\tau) = \dim \operatorname{Hom}_G(V^\tau, V^\Phi).$$

The lemma follows.

Let H be a closed subgroup of G, and let σ be a unitary representation of H on a space V^σ. The **induced representation**

$$\Phi = \operatorname{ind}_H^G \sigma$$

operates in the space

$$V^\Phi = \left\{ V^\sigma\text{-valued } L^2 \text{ functions on } G \,\middle|\, \begin{array}{l} f(gh) = \sigma(h)^{-1} f(g) \text{ for all} \\ h \in H \text{ and almost every } g \in G \end{array} \right\}$$

by

$$\Phi(g)f(x) = f(g^{-1}x).$$

The result is a unitary representation of G. (To get around technical measure-theory problems, one usually works with the continuous members of V^Φ and defines V^Φ as the completion in the L^2 norm.)

Theorem 1.14 (Frobenius Reciprocity Theorem). Let H be a closed subgroup of G, let σ be an irreducible unitary representation of H on V^σ, let τ be an irreducible unitary representation of G on V^τ, and let $\Phi = \operatorname{ind}_H^G \sigma$ act on V^Φ. Then

$$[\operatorname{ind}_H^G \sigma : \tau] = [\tau|_H : \sigma].$$

Proof. We form the spaces of linear maps

$$\operatorname{Hom}_G(V^\tau, V^\Phi) \tag{1.17}$$

and

$$\operatorname{Hom}_H(V^\tau, V^\sigma). \tag{1.18}$$

Here the subscripts "G" and "H" refer to linear maps respecting the indicated actions. In the first case, V^Φ is contained in $L^2(G, V^\sigma)$, and $L^2(G, V^\sigma)$ is simply the direct sum of d_σ copies of $L^2(G)$. Therefore τ occurs exactly $d_\sigma d_\tau$ times in $L^2(G, V^\sigma)$ and at most that many times in V^Φ. By Schur's Lemma we then know that the image of any member of (1.17) lies in the subspace of continuous members of V^Φ.

If e denotes evaluation at 1 in G, it therefore makes sense to form eA whenever A is in (1.17). For v in V^τ we have

$$\begin{aligned} \sigma(h)(eAv) &= \sigma(h)[(Av)(1)] = (Av)(h^{-1}) \\ &= (\Phi(h)(Av))(1) = (A\tau(h)v)(1) = eA\tau(h)v. \end{aligned}$$

Thus eA is in (1.18), and e carries (1.17) into (1.18).

We shall show e is an isomorphism. Since Lemma 1.13 implies that the dimension of (1.17) is $[\operatorname{ind}_H^G \sigma : \tau]$ and the dimension of (1.18) is $[\tau|_H : \sigma]$, the theorem will follow.

To see that e is one-one, suppose $eAv = 0$ for all v in V^τ. Then $(Av)(1)=0$ for all v. Apply this conclusion to $v = \tau(g)^{-1}v'$. Then

$$0 = (Av)(1) = (A\tau(g)^{-1}v')(1) = (\Phi(g)^{-1}Av')(1) = (Av')(g),$$

and $Av' = 0$. Since v' is arbitrary, $A = 0$. Thus e is one-one.

To see that e is onto, let a be in (1.18). Define

$$Av(g) = a(\tau(g)^{-1}v) \qquad \text{for } v \in V^\tau,\ g \in G.$$

Then

$$Av(gh) = a(\tau(h)^{-1}\tau(g)^{-1}v) = \sigma(h)^{-1}a\tau(g)^{-1}v = \sigma(h)^{-1}Av(g),$$

so that Av is in V^Φ. In fact, A is in (1.17) because

$$(\Phi(g_0)Av)(g) = Av(g_0^{-1}g) = a(\tau(g)^{-1}(\tau(g_0)v)) = A(\tau(g_0)v)(g)$$

implies $\Phi(g_0)A = A\tau(g_0)$. Finally A maps onto a because

$$eAv = Av(1) = a(\tau(1)v) = av$$

implies $eA = a$. Thus e is onto. This completes the proof.

§6. Application of the Abstract Theory to Lie Groups

We can use the Peter-Weyl Theorem to see how compact connected Lie groups fit with our definitions in §1.

Theorem 1.15. Any compact connected Lie group G can be realized as a linear connected reductive Lie group.

Proof. Since a finite-dimensional representation of a compact group is unitary for a suitable Hermitian inner product, it is enough to produce a one-one finite-dimensional representation of G. It follows from (a) in the Peter-Weyl Theorem that for each $x \neq 1$ in G there is a finite-dimensional representation Φ_x of G such that $\Phi_x(x) \neq 1$. Pick $x_1 \neq 1$ in G. Then $G_1 = \ker \Phi_{x_1}$ is a closed proper subgroup of G, hence of lower dimension than G. If $(G_1)_0 \neq 1$, choose $x_2 \neq 1$ in $(G_1)_0$, and form Φ_{x_2}. Then $G_2 = \ker(\Phi_{x_1} \oplus \Phi_{x_2})$ is a closed subgroup of G_1, and its identity component is a proper subgroup of $(G_1)_0$. Continuing in this way and using the finite-dimensionality of G, we see that we can find a finite-dimensional representation Φ_0 of G such that $\ker \Phi_0$ is 0-dimensional. Then $\ker \Phi_0$ is finite since G is compact and is a Lie group. Say $\ker \Phi_0 = \{y_1, \ldots, y_n\}$. Then

$$\Phi = \Phi_0 \oplus \sum_{j=1}^{n} \oplus \Phi_{y_j}$$

is a one-one finite-dimensional representation of G.

§7. Problems

1. Let $\mathfrak{g}$ be a Lie algebra over $\mathbb{R}$.
   (a) Let $\mathfrak{g}^{\mathbb{C}}$ be the vector space complexification $\mathfrak{g}^{\mathbb{C}} = \mathfrak{g} \otimes_{\mathbb{R}} \mathbb{C}$, and identify $\mathfrak{g}$ with $\mathfrak{g} \otimes 1$. Show that the bracket operation on $\mathfrak{g}$ extends consistently and uniquely to $\mathfrak{g}^{\mathbb{C}}$ so that $\mathfrak{g}^{\mathbb{C}}$ becomes a complex Lie algebra.
   (b) Let $\mathfrak{g}$ be finite-dimensional, and let B and $B^{\mathbb{C}}$ be the Killing forms of $\mathfrak{g}$ and $\mathfrak{g}^{\mathbb{C}}$. Show that $B^{\mathbb{C}}|_{\mathfrak{g}\times\mathfrak{g}} = B$.
   (c) For $\mathfrak{g}$ finite-dimensional, prove that $\mathfrak{g}$ is semisimple if and only if $\mathfrak{g}^{\mathbb{C}}$ is semisimple.
   (d) For $\mathfrak{g}$ finite-dimensional, let $(\mathfrak{g}^{\mathbb{C}})_{\mathbb{R}}$ be $\mathfrak{g}^{\mathbb{C}}$ regarded as a real Lie algebra of twice the dimension of $\mathfrak{g}$, and let B' be its Killing form. Show that $B'(X, Y) = 2 \operatorname{Re} B^{\mathbb{C}}(X, Y)$.

2. In $\mathfrak{sl}(n, \mathbb{R})$ the Killing form and the trace form are multiples of one another. Identify the multiple.

3. Suppose that the Lie algebra $\mathfrak{g}$ of a linear connected semisimple group is actually simple and is closed under multiplication by i (so that $\mathfrak{g}$ is a complex Lie algebra).
   (a) Show that $\mathfrak{g}^{\mathbb{C}}$ is not a simple Lie algebra.
   (b) Show that the Killing form of $\mathfrak{g}$ (regarded as a complex Lie algebra) is a multiple of the trace form.

4. (a) Show that $\mathfrak{so}(3)$ is isomorphic with the vector product Lie algebra in $\mathbb{R}^3$, in which there is a basis $\mathbf{i}, \mathbf{j}, \mathbf{k}$ with $[\mathbf{i}, \mathbf{j}] = \mathbf{k}$, $[\mathbf{j}, \mathbf{k}] = \mathbf{i}$, and $[\mathbf{k}, \mathbf{i}] = \mathbf{j}$.
   (b) Under the isomorphism in (a), show that the Killing form B on $\mathfrak{so}(3)$ gets identified with a multiple of the dot product in $\mathbb{R}^3$.

5. Let $GL(n, \mathbb{R})$ be the Lie group of nonsingular n-by-n real matrices, $n \geq 1$.
   (a) Show that $GL(n, \mathbb{R})$ has two components.
   (b) Verify that the identity component of $GL(n, \mathbb{R})$ is a linear connected reductive group.

6. Find the cardinality of the center of $SU(n)$, $SO(n)$, $Sp(n)$, $SL(n, \mathbb{C})$, $SO(n, \mathbb{C})$, and $Sp(n, \mathbb{C})$.

7. For $G = U(n)$, identify $[\mathfrak{g}, \mathfrak{g}]$. Let G_{ss} be the analytic subgroup of $U(n)$ with Lie algebra $[\mathfrak{g}, \mathfrak{g}]$. Is G isomorphic with the direct product of Z_G and G_{ss}. Why or why not?

8. Go over the proof of Proposition 1.2, dropping the assumption of connectedness of G, in order to obtain a suitable "Cartan decomposition" of the full (disconnected) group $SO(p, q)$. Use the result and the

known connectedness of SO(n) to compute the number of components of SO(p, q) as a function of (p, q).

9. Let Φ be the unitary representation of SO(n) on the space V of polynomials in n real variables homogeneous of degree N, given by $\Phi(g)P(x) = P(g^{-1}x)$.
(a) Show that the operator $L: V \to V$ given by

$$L(P) = |x|^2 \left(\frac{\partial^2}{\partial x_1^2} + \ldots + \frac{\partial^2}{\partial x_n^2} \right) P$$

commutes with $\Phi(g)$ for all g in SO(n).
(b) Conclude that Φ is reducible if $n \geq 2$ and $N \geq 2$.

10. Let Φ be the unitary representation of SL(2, $\mathbb{R}$) on $V = L^2(\mathbb{R}^2)$ given by $\Phi(g)f(x) = f(g^{-1}x)$.
(a) Show for each $r > 0$ that the operator $L_r: V \to V$ given by $L_r f(x) = f(rx)$ commutes with $\Phi(g)$ for all g in SL(2, $\mathbb{R}$).
(b) Conclude that Φ is reducible.

11. Prove that any compact topological group with a one-one representation by matrices is a Lie group.

12. Let Φ be the left regular representation of $\mathbb{R}$ on $L^2(\mathbb{R})$. Define the Fourier transform on $L^2(\mathbb{R})$ in the usual way by a limiting process from

$$\hat{f}(y) = \int_{-\infty}^{\infty} e^{-2\pi ixy} f(x)\, dx.$$

(a) Fix m in $L^\infty(\mathbb{R})$. Show that the equation $(Tf)\hat{}(y) = m(y)\hat{f}(y)$ defines T as a bounded linear operator on $L^2(\mathbb{R})$ that commutes with $\Phi(x)$ for all x in $\mathbb{R}$.
(b) Prove that there is no nonzero closed invariant subspace H of $L^2(\mathbb{R})$ on which Φ is an irreducible representation.

13. Let $\mathfrak{g}$ be the Lie algebra

$$\left\{ \begin{pmatrix} 0 & x & z \\ 0 & 0 & y \\ 0 & 0 & 0 \end{pmatrix}, \text{ real entries} \right\}.$$

Let V be the complex vector space of complex-valued functions on $\mathbb{R}$ of the form $e^{-\pi s^2}P(s)$, where P is a polynomial, and let $\hbar$ be a positive constant.

(a) Show that the linear mappings $i\dfrac{d}{ds}$ and "multiplication by $-i\hbar s$" carry V into itself.

(b) Define $\varphi\begin{pmatrix}0&1&0\\0&0&0\\0&0&0\end{pmatrix} = i\dfrac{d}{ds}$ and $\varphi\begin{pmatrix}0&0&0\\0&0&1\\0&0&0\end{pmatrix}$ = multiplication by $-i\hbar s$. How should $\varphi\begin{pmatrix}0&0&1\\0&0&0\\0&0&0\end{pmatrix}$ be defined so that the linear extension of φ to $\mathfrak{g}$ is a representation of $\mathfrak{g}$ on V.

(c) With φ defined as in (b), prove that φ is irreducible.

14. (a) Show that $\text{ind}_{SO(n-1)}^{SO(n)} 1$ can be identified with the representation Φ of $SO(n)$ on $L^2(S^{n-1})$ given by $\Phi(g)f(x) = f(g^{-1}x)$.
(b) What is the multiplicity of the standard n-dimensional representation of $SO(n)$ in this representation Φ?

Problems 15 to 21 show how to relate certain groups and Lie algebras of matrices having quaternion entries to suitable groups and Lie algebras of complex matrices. The division algebra $\mathbb{H}$ of quaternions is a four-dimensional vector space over $\mathbb{R}$ with basis 1, i, j, k and with relations $i^2 = j^2 = k^2 = -1$, $ij = k$, $jk = i$, $ki = j$, $ji = -k$, $kj = -i$, $ik = -j$. Conjugation $h \to \bar{h}$ is the $\mathbb{R}$-linear map that is 1 on 1 and is -1 on i, j, and k. The norm is $|h| = (h\bar{h})^{1/2}$.

15. Identify $\mathbb{C}$ with $\{h \in H \mid h = \alpha + \beta i\}$, and let $\mathbb{H}^n$ denote the set of n-component column vectors with entries in $\mathbb{H}$. For v in $\mathbb{H}^n$, write $v = a + ib + jc + kd$ with a, b, c, d in $\mathbb{R}^n$, and define functions z_1 and z_2 from $\mathbb{H}^n$ into $\mathbb{C}^n$ by $z_1(v) = a + ib$ and $z_2(v) = c - id$, so that $v = z_1(v) + jz_2(v)$. Show that the $\mathbb{R}$-isomorphism $v \to \begin{pmatrix}z_1(v)\\z_2(v)\end{pmatrix}$ of $\mathbb{H}^n$ into $\mathbb{C}^{2n}$ is a $\mathbb{C}$-isomorphism if $\mathbb{H}^n$ is regarded as a right vector space over $\mathbb{C}$.

16. In analogy with Problem 15, define z_1 and z_2 from n-by-n matrices of quaternions to n-by-n complex matrices so that $M = z_1(M) + jz_2(M)$. Under the isomorphism of Problem 15, prove that left multiplication by M on $\mathbb{H}^n$ corresponds to left multiplication by $\begin{pmatrix}z_1(M) & -\overline{z_2(M)}\\ z_2(M) & \overline{z_1(M)}\end{pmatrix}$ on $\mathbb{C}^{2n}$.

17. Let $M \to Z(M)$ be the correspondence of n-by-n quaternion matrices with $2n$-by-$2n$ complex matrices given in Problem 16.
(a) Prove that $Z(MN) = Z(M)Z(N)$.
(b) Prove that $Z(M^*) = Z(M)^*$, where M^* denotes the conjugate transpose, conjugation being taken entry-by-entry.

18. (a) Interpret Problem 17a as implying an isomorphism of real Lie algebras

$$\mathfrak{gl}(n, \mathbb{H}) = \{n\text{-by-}n \text{ matrices over } \mathbb{H} \text{ with usual bracket}\}$$

with

$$\mathfrak{u}^*(2n) = \left\{ \begin{pmatrix} z_1 & -\overline{z_2} \\ z_2 & \overline{z_1} \end{pmatrix} \middle|\, z_1 \text{ and } z_2 \text{ are in } \mathfrak{gl}\,(n, \mathbb{C}) \right\}$$

and an isomorphism of topological groups

$$GL(n, \mathbb{H}) = \{\text{invertible } n\text{-by-}n \text{ matrices over } \mathbb{H}\}$$

with

$$U^*(2n) = GL(2n, \mathbb{C}) \cap \left\{ \begin{pmatrix} z_1 & -\overline{z_2} \\ z_2 & \overline{z_1} \end{pmatrix} \right\}.$$

(b) Interpret Problems 17b as implying that the decomposition of $\mathfrak{gl}(n, \mathbb{H})$ into a skew Hermitian part and a Hermitian part is consistent with the Cartan decomposition of $\mathfrak{u}^*(2n)$.

19. (a) Under the correspondence of Problem 16, show that $\operatorname{Re} \operatorname{Tr}(M) = 0$ in $\mathfrak{gl}(n, \mathbb{H})$ corresponds to $\operatorname{Tr}(Z(M)) = 0$ in $\mathfrak{u}^*(2n)$. [Remark: Then we can use these conditions to define Lie algebras $\mathfrak{sl}(n, \mathbb{H})$ and $\mathfrak{su}^*(2n)$.]

(b) Show that $\det M = 1$ in $GL(n, \mathbb{H})$ corresponds to $\det Z(M) = 1$ in $U^*(2n)$. [Remarks: Then we can use these conditions to define subgroups $SL(n, \mathbb{H})$ and $SU^*(2n)$. For determinants defined over $\mathbb{H}$, see the bibliographical notes.]

20. For v and v' in $\mathbb{H}^n$, define $\langle v, v' \rangle = \sum_{l=1}^n v_l \overline{v_l'}$.

(a) Show that $\langle Mv, v' \rangle = \langle v, M^* v' \rangle$ for any n-by-n matrix of quaternions M.

(b) Under the isomorphism of Problem 16, show that the space of skew Hermitian matrices (i.e., $M + M^* = 0$) corresponds to $\mathfrak{sp}(n)$.

(c) Under the isomorphism of Problem 16, show that the space of unitary matrices M (i.e., $\langle Mv, Mv' \rangle = \langle v, v' \rangle$ for all v and v') corresponds to $Sp(n)$.

21. Let $p + q = n$. For v and v' in $\mathbb{H}^n$, define

$$\langle v, v' \rangle = \sum_{l=1}^{p} v_l \overline{v_l'} - \sum_{l=p+1}^{n} v_l \overline{v_l'}.$$

Carry out the steps of Problem 20 for this Hermitian form in order to identify a Lie algebra of complex matrices isomorphic to $\mathfrak{sp}(p, q)$ and to identify a Lie group of complex matrices isomorphic to $Sp(p, q)$.

CHAPTER II

Representations of SU(2), SL(2, $\mathbb{R}$) and SL(2, $\mathbb{C}$)

§1. The Unitary Trick

An important class of finite-dimensional representations of SL(2, $\mathbb{C}$) is obtained as follows. Fix an integer $n \geq 0$, and let V_n be the complex vector space of polynomials in two complex variables z_1 and z_2 homogeneous of degree n. Define a representation Φ_n by

$$\Phi_n\begin{pmatrix} a & b \\ c & d \end{pmatrix} P\begin{pmatrix} z_1 \\ z_2 \end{pmatrix} = P\left(\begin{pmatrix} a & b \\ c & d \end{pmatrix}^{-1}\begin{pmatrix} z_1 \\ z_2 \end{pmatrix}\right) \tag{2.1}$$

Then $\dim V_n = n + 1$, and $\Phi_n : \mathrm{SL}(2, \mathbb{C}) \to \mathrm{GL}(V_n)$ is holomorphic. In §2 we shall see that Φ_n is irreducible. It will turn out that there are no other irreducible finite-dimensional holomorphic representations of SL(2, $\mathbb{C}$), up to equivalence.

If we restrict Φ_n to either of the subgroups SL(2, $\mathbb{R}$) or SU(2) of SL(2, $\mathbb{C}$), we obtain a representation of the subgroup. The representation of the subgroup is itself irreducible. In fact, let φ_n be the representation of $\mathfrak{sl}(2, \mathbb{C})$ given by $\varphi_n = d\Phi_n$. Since Φ_n is holomorphic, φ_n is complex linear. Now

$$\mathfrak{sl}(2, \mathbb{C}) = \mathfrak{sl}(2, \mathbb{R}) \oplus i\mathfrak{sl}(2, \mathbb{R}) = \mathfrak{su}(2) \oplus i\mathfrak{su}(2), \tag{2.2}$$

and consequently φ_n is determined by its restriction to $\mathfrak{sl}(2, \mathbb{R})$ or to $\mathfrak{su}(2)$. Thus invariant subspaces for $\mathfrak{sl}(2, \mathbb{R})$ or $\mathfrak{su}(2)$ are invariant for $\mathfrak{sl}(2, \mathbb{C})$, hence for SL(2, $\mathbb{C}$). In other words, the irreducibility of Φ_n for SL(2, $\mathbb{C}$) implies the irreducibility of Φ_n for SL(2, $\mathbb{R}$) and SU(2). This is a special case of Weyl's **unitary trick** given in the following proposition.

Proposition 2.1. Let V be a finite-dimensional complex vector space. Then a representation of any of the following kinds on V leads, via (2.2), to a representation of each of the other kinds. Under this correspondence, invariant subspaces and equivalences are preserved:

(a) a smooth representation of SL(2, $\mathbb{R}$) on V
(b) a smooth representation of SU(2) on V
(c) a holomorphic representation of SL(2, $\mathbb{C}$) on V
(d) a representation of $\mathfrak{sl}(2, \mathbb{R})$ on V

(e) a representation of $\mathfrak{su}(2)$ on V

(f) a complex-linear representation of $\mathfrak{sl}(2, \mathbb{C})$ on V.

Remark. It will be shown in Corollary 3.16 that finite-dimensional representations of Lie groups are automatically smooth. Thus the word "smooth" can be dropped in (a) and (b).

Proof. We can pass from (c) to (a) or (b) by restriction and from (a) or (b) to (d) or (e) by taking differentials. Formula (2.2) allows us to pass from (d) or (e) to (f). Finally Proposition 1.2 shows that $SL(2, \mathbb{C})$ is topologically the product of $SU(2)$ and a Euclidean space. Since

$$SU(2) = \left\{ \begin{pmatrix} \alpha & \beta \\ -\bar{\beta} & \bar{\alpha} \end{pmatrix} \middle| \alpha \in \mathbb{C}, \beta \in \mathbb{C}, |\alpha|^2 + |\beta|^2 = 1 \right\}$$

is homeomorphic with the 3-sphere, $SU(2)$ is simply connected. Thus $SL(2, \mathbb{C})$ is simply connected. If φ is given as in (f), then (A.113) asserts the existence of a representation Φ of $SL(2, \mathbb{C})$ with differential φ. Since φ is complex-linear, Φ is holomorphic. Thus we can pass from (f) to (c). If we follow the steps all the way around, starting from (c), we end up with the original representation, since a differential uniquely determines a homomorphism of Lie groups. Thus invariant subspaces and equivalences are preserved.

The name "unitary trick" comes from the fact that item (b) in the list is a representation of a compact group, which can be made unitary by Proposition 1.6. Then Corollary 1.7 yields the following result.

Corollary 2.2. Every finite-dimensional representation of $SL(2, \mathbb{R})$ or of $\mathfrak{sl}(2, \mathbb{R})$ is the direct sum of irreducible representations. The same thing is true of holomorphic finite-dimensional representations of $SL(2, \mathbb{C})$ and of complex-linear representations of $\mathfrak{sl}(2, \mathbb{C})$, with each summand respectively holomorphic or complex-linear.

Corollary 2.3. Every finite-dimensional unitary representation of $SL(2, \mathbb{R})$ is trivial.

Proof. Let $\mathfrak{sl}(2, \mathbb{R}) = \mathfrak{k} \oplus \mathfrak{p}$ be the Cartan decomposition of $\mathfrak{sl}(2, \mathbb{R})$. Then $\mathfrak{su}(2) = \mathfrak{k} \oplus i\mathfrak{p}$. For a given finite-dimensional representation of $SL(2, \mathbb{R})$, form the associated family of representations as in Proposition 2.1. In a suitable inner product, $SU(2)$ acts by unitary operators, by Proposition 1.6, and thus $\mathfrak{su}(2)$ acts by skew-Hermitian operators. Hence $i\mathfrak{p}$ acts by diagonable operators with imaginary eigenvalues, and $\mathfrak{p}$ acts by diagonable operators with real eigenvalues. If the representation of $SL(2, \mathbb{R})$ is unitary in some inner product, then $\mathfrak{p}$ acts with imaginary eigenvalues. Hence $\mathfrak{p}$ acts as 0. Since $[\mathfrak{p}, \mathfrak{p}] = \mathfrak{k}$, all of $\mathfrak{sl}(2, \mathbb{R})$ acts as 0.

§2. Irreducible Finite-Dimensional Complex-Linear Representations of $\mathfrak{sl}(2, \mathbb{C})$

Corollary 2.2 reduces the study of several classes of finite-dimensional representations to the study of irreducible ones, and Proposition 2.1 says that it is enough to study the irreducible complex-linear representations of $\mathfrak{sl}(2, \mathbb{C})$.

We shall make repeated use of the basis $\{h, e, f\}$ of $\mathfrak{sl}(2, \mathbb{C})$ over $\mathbb{C}$ given by

$$h = \begin{pmatrix} 1 & 0 \\ 0 & -1 \end{pmatrix}, \qquad e = \begin{pmatrix} 0 & 1 \\ 0 & 0 \end{pmatrix}, \qquad f = \begin{pmatrix} 0 & 0 \\ 1 & 0 \end{pmatrix}.$$

These elements satisfy the bracket relations

$$[h, e] = 2e, \qquad [h, f] = -2f, \qquad [e, f] = h. \tag{2.3}$$

Theorem 2.4. For each integer $m \geq 1$ there exists up to equivalence a unique irreducible complex-linear representation π of $\mathfrak{sl}(2, \mathbb{C})$ on a space V of dimension m. In V there is a basis $\{v_0, \ldots, v_{m-1}\}$ such that (with $n = m - 1$)

(1) $\pi(h)v_i = (n - 2i)v_i$
(2) $\pi(e)v_0 = 0$
(3) $\pi(f)v_i = v_{i+1}$ with $v_{n+1} = 0$
(4) $\pi(e)v_i = i(n - i + 1)v_{i-1}$ with $v_{-1} = 0$.

Moreover, the representation π can be realized as the differential of the representation Φ_n in (2.1).

Remark. Property (1) gives the eigenvalues of $\pi(h)$. Notice for $\mathfrak{sl}(2, \mathbb{C})$ that the smallest eigenvalue is the negative of the largest. The eigenvalues of $\pi(h)$ will generalize in Chapter IV to "weights," and the weight corresponding to n will be the "highest weight" of π.

Proof of uniqueness. Let π be a complex-linear irreducible representation of $\mathfrak{sl}(2, \mathbb{C})$ on V with $\dim V = m$. Let $v \neq 0$ be an eigenvector for $\pi(h)$, say with $\pi(h)v = \lambda v$. Then $\pi(e)v, \pi(e)^2 v, \ldots$ are also eigenvectors because

$$\begin{aligned} \pi(h)\pi(e)v &= \pi(e)\pi(h)v + \pi([h, e])v && \text{by (1.6)} \\ &= \pi(e)\lambda v + 2\pi(e)v && \text{by (2.3)} \\ &= (\lambda + 2)\pi(v). \end{aligned}$$

Since $\lambda, \lambda + 2, \lambda + 4, \ldots$ are distinct, these eigenvectors are independent (or 0). By finite-dimensionality we can find v_0 in V with (λ redefined and)

(a) $v_0 \neq 0$
(b) $\pi(h)v_0 = \lambda v_0$
(c) $\pi(e)v_0 = 0$.

Define $v_i = \pi(f)^i v_0$. Then $\pi(h)v_i = (\lambda - 2i)v_i$, by the same argument as above, and so there is a minimum integer n with $\pi(f)^{n+1}v_0 = 0$. Then $v_0, \ldots, v_n$ are independent and

(1) $\pi(h)v_i = (\lambda - 2i)v_i$
(2) $\pi(e)v_0 = 0$
(3) $\pi(f)v_i = v_{i+1}$ with $v_{n+1} = 0$.

We claim $V = \text{span}\{v_0, \ldots, v_n\}$. It is enough to show $\text{span}\{v_0, \ldots, v_n\}$ is stable under $\pi(e)$. In fact, we show

(4) $\pi(e)v_i = i(\lambda - i + 1)v_{i-1}$ with $v_{-1} = 0$.

We proceed by induction for (4), the case $i = 0$ being (2). Assume (4) for case i. To prove case $i + 1$, we write

$$\begin{aligned}\pi(e)v_{i+1} &= \pi(e)\pi(f)v_i = \pi([e,f])v_i + \pi(f)\pi(e)v_i \\ &= \pi(h)v_i + \pi(f)\pi(e)v_i \\ &= (\lambda - 2i)v_i + \pi(f)(i(\lambda - i + 1))v_{i-1} \\ &= (i + 1)(\lambda - i)v_i,\end{aligned}$$

and the induction is complete.

To finish the proof of uniqueness, we show $\lambda = n$. We have

$$\operatorname{Tr} \pi(h) = \operatorname{Tr}(\pi(e)\pi(f) - \pi(f)\pi(e)) = 0.$$

Thus $\sum_{i=0}^{n} (\lambda - 2i) = 0$, and we find $\lambda = n$.

Proof of existence. From the proof of uniqueness and from Corollary 2.2, it follows that $\pi(h)$ cannot have eigenvalue $n = m - 1$ in a reducible complex-linear representation of dimension m. Form the differential $d\Phi_n$ of the representation Φ_n of (2.1). Here $d\Phi_n$ is complex-linear and has dimension m. Also

$$d\Phi_n(h)(z_2^n) = \frac{d}{dt}\Phi_n\begin{pmatrix} e^t & 0 \\ 0 & e^{-t}\end{pmatrix}(z_2^n)\bigg|_{t=0} = \frac{d}{dt}e^{nt}z_2^n\bigg|_{t=0} = nz_2^n,$$

so that $d\Phi_n(h)$ does have n as an eigenvalue. Consequently $d\Phi_n$ is irreducible, and existence follows.

§3. Finite-Dimensional Representations of $\mathfrak{sl}(2, \mathbb{C})$

The previous section dealt only with complex-linear finite-dimensional representations of $\mathfrak{sl}(2, \mathbb{C})$. But $\mathfrak{sl}(2, \mathbb{C})$ has other finite-dimensional representations, e.g., $\Phi\begin{pmatrix} a & b \\ c & -a\end{pmatrix} = \begin{pmatrix} \bar{a} & \bar{b} \\ \bar{c} & -\bar{a}\end{pmatrix}$. The relevant fact for handling them is the following proposition.

Proposition 2.5. For the complex Lie algebra $\mathfrak{g} = \mathfrak{sl}(2, \mathbb{C})$, the complexification $\mathfrak{g}^{\mathbb{C}} = \mathfrak{g} \oplus i\mathfrak{g}$ is $\mathbb{C}$-isomorphic as a Lie algebra to $\mathfrak{g} \oplus \mathfrak{g}$.

Proof. Let J denote multiplication by $\sqrt{-1}$ in $\mathfrak{g}$, and define an $\mathbb{R}$-linear map $L: \mathfrak{g}^{\mathbb{C}} \to \mathfrak{g} \oplus \mathfrak{g}$ by

$$L(X + iY) = (X + JY, X - JY).$$

Then it is easy to see that L is one-one and preserves brackets. Hence L is an $\mathbb{R}$-isomorphism. Moreover, L satisfies

$$L(i(X + iY)) = (J(X + JY), -J(X - JY)).$$

This equation exhibits L as a $\mathbb{C}$-isomorphism of $\mathfrak{g}^{\mathbb{C}}$ with $\mathfrak{g} \oplus \bar{\mathfrak{g}}$, where $\bar{\mathfrak{g}}$ is the same real Lie algebra as $\mathfrak{g}$ but where the multiplication by $\sqrt{-1}$ is defined as multiplication by $-i$. Then complex conjugation of matrices exhibits $\bar{\mathfrak{g}}$ and $\mathfrak{g}$ as $\mathbb{C}$-isomorphic, and the result follows.

We can handle finite-dimensional representations of $SL(2, \mathbb{C})$ by means of Proposition 2.5 and a suitable unitary trick. The finite-dimensional representations of $SL(2, \mathbb{C})$ and of $\mathfrak{sl}(2, \mathbb{C})$ correspond to one another since $SL(2, \mathbb{C})$ is simply connected. (We are implicitly using the automatic smoothness of the representations of the group, as proved in Chapter III.) A finite-dimensional representation of $\mathfrak{sl}(2, \mathbb{C})$ yields by Proposition 2.5 a complex-linear representation of $\mathfrak{sl}(2, \mathbb{C}) \oplus \mathfrak{sl}(2, \mathbb{C})$, which is fully reducible because $\mathfrak{sl}(2, \mathbb{C}) \oplus \mathfrak{sl}(2, \mathbb{C})$ is the complexification of $\mathfrak{su}(2) \oplus \mathfrak{su}(2)$. An irreducible unitary representation of $SU(2) \times SU(2)$ has to be a tensor product of irreducible representations of the factors, by an easy application of the Peter-Weyl Theorem. Thus an irreducible complex-linear finite-dimensional representation of $\mathfrak{sl}(2, \mathbb{C}) \oplus \mathfrak{sl}(2, \mathbb{C})$ has to be a tensor product and must be given, according to Theorem 2.4, by a pair of nonnegative integers.

Consequently the finite-dimensional representations of $\mathfrak{sl}(2, \mathbb{C})$ are direct sums of irreducible representations, and the irreducible representations are parametrized by pairs of nonnegative integers. The representation is complex-linear exactly when the second integer is 0 and is complex conjugate linear exactly when the first integer is 0. A global realization of the irreducible representation $\Phi_{m,n}$ of $SL(2, \mathbb{C})$ with parameters (m, n) is in the vector space of polynomials in z_1, z_2, $\bar{z}_1$, $\bar{z}_2$ that are homogeneous of degree m in (z_1, z_2) and homogeneous of degree n in $(\bar{z}_1, \bar{z}_2)$, with action

$$\Phi_{m,n}\begin{pmatrix} a & b \\ c & d \end{pmatrix} P\begin{pmatrix} z_1 \\ z_2 \end{pmatrix} = P\left(\begin{pmatrix} a & b \\ c & d \end{pmatrix}^{-1}\begin{pmatrix} z_1 \\ z_2 \end{pmatrix}\right). \tag{2.4}$$

§4. Irreducible Unitary Representations of SL(2, ℂ)

The **principal series,** or **unitary principal series,** of SL(2, ℂ) is a family of representations in $L^2(\mathbb{C})$ that is indexed by pairs (k, iv) with $k \in \mathbb{Z}$ and $v \in \mathbb{R}$. The representation $\mathcal{P}^{k,iv}$ is given by

$$\mathcal{P}^{k,iv}\begin{pmatrix} a & b \\ c & d \end{pmatrix} f(z) = |-bz + d|^{-2-iv}\left(\frac{-bz+d}{|-bz+d|}\right)^{-k} f\left(\frac{az-c}{-bz+d}\right)$$

for $z \in \mathbb{C}$ and $f \in L^2(\mathbb{C})$.

Proposition 2.6. For each pair (k, iv), $\mathcal{P}^{k,iv}$ is an irreducible unitary representation of SL(2, ℂ). Moreover, $\mathcal{P}^{k,iv}$ is unitarily equivalent with $\mathcal{P}^{-k,-iv}$.

Remarks. The proof of irreducibility may be approached in several ways. This time we use the Euclidean Fourier transform. In Chapter VII we shall indicate a more elementary argument.

Proof. A straightforward change of variables shows that $\mathcal{P}^{k,iv}(g)$ is isometric for each g, and it is easy to check that $g \to \mathcal{P}^{k,iv}(g)$ is a homomorphism. For f in $C_{\mathrm{com}}^{\infty}(\mathbb{C})$, the continuity at 1 of $g \to \mathcal{P}^{k,iv}(g)f$ as a map of G into $L^2(\mathbb{C})$ follows by dominated convergence. Since $C_{\mathrm{com}}^{\infty}(\mathbb{C})$ is dense in $L^2(\mathbb{C})$, the strong continuity of $g \to \mathcal{P}^{k,iv}(g)$ follows. Thus $\mathcal{P}^{k,iv}$ is a unitary representation.

For the irreducibility we shall show that actually $\mathcal{P}^{k,iv}$ is irreducible when restricted to the lower triangular group. On that subgroup, the representation is given by

$$\mathcal{P}^{k,iv}\begin{pmatrix} 1 & 0 \\ z_0 & 1 \end{pmatrix} f(z) = f(z - z_0) \tag{2.5}$$

$$\mathcal{P}^{k,iv}\begin{pmatrix} a & 0 \\ 0 & a^{-1} \end{pmatrix} f(z) = |a|^{2+iv}\left(\frac{a}{|a|}\right)^{k} f(a^2 z). \tag{2.6}$$

Motivated by Proposition 1.5, let B be a bounded linear operator on $L^2(\mathbb{C})$ commuting with $\mathcal{P}^{k,iv}(g)$ for all g. Since B commutes with (2.5), B is given on the Fourier transform side by multiplication by a bounded measurable function m:

$$\widehat{Bf}(\zeta) = m(\zeta)\hat{f}(\zeta) \qquad \text{for } f \in L^2(\mathbb{C}). \tag{2.7}$$

Here the Fourier transform is given by

$$\hat{f}(\zeta) = \int_{\mathbb{C}} e^{-2\pi i z \cdot \zeta} f(z)\, dz,$$

where $z \cdot \zeta$ denotes $x\xi + y\eta$ if $z = x + iy$ and $\zeta = \xi + i\eta$. Since

$$\mathscr{P}^{k,iv}\begin{pmatrix} a & 0 \\ 0 & a^{-1} \end{pmatrix} B = B\mathscr{P}^{k,iv}\begin{pmatrix} a & 0 \\ 0 & a^{-1} \end{pmatrix},$$

(2.6) gives $(Bf)(a^2 z) = B(f(a^2 \cdot))(z)$ for all $a \neq 0$. Multiplying by $e^{-2\pi i z \cdot \zeta}$ and integrating and using (2.7), we obtain from the left side

$$\begin{aligned} \int_{\mathbb{C}} e^{-2\pi i z \cdot \zeta} Bf(a^2 z)\, dz &= |a|^{-4} \int_{\mathbb{C}} e^{-2\pi i z \cdot a^{-2}\zeta} Bf(z)\, dz \\ &= |a|^{-4}\widehat{Bf}(a^{-2}\zeta) \\ &= |a|^{-4} m(a^{-2}\zeta)\hat{f}(a^{-2}\zeta). \end{aligned} \tag{2.8}$$

From the right side we get

$$\begin{aligned} m(\zeta) f(a^2 \cdot)\hat{}\,(\zeta) &= m(\zeta) \int_{\mathbb{C}} e^{-2\pi i z \cdot \zeta} f(a^2 z)\, dz \\ &= |a|^{-4} m(\zeta) \int_{\mathbb{C}} e^{-2\pi i z \cdot a^{-2}\zeta} f(z)\, dz \\ &= |a|^{-4} m(\zeta)\hat{f}(a^{-2}\zeta). \end{aligned} \tag{2.9}$$

The equality of (2.8) and (2.9) for all f means that

$$m(a^{-2}\zeta) = m(\zeta) \tag{2.10}$$

for almost all ζ for each a. By Fubini's Theorem, (2.10) holds for some $\zeta = \zeta_0 \neq 0$ for almost every a. Since $a^{-2}\zeta_0$ sweeps out $\mathbb{C} - \{0\}$, m is constant almost everywhere. Thus B is scalar, and $\mathscr{P}^{k,iv}$ is irreducible.

Further Fourier analysis, which we omit, will establish the equivalences. An elementary proof will be given in Chapter VII.

The full **nonunitary principal series** of SL(2, ℂ) is a family of representations indexed by pairs (k, w) with $k \in \mathbb{Z}$ and $w = u + iv \in \mathbb{C}$. The representation $\mathscr{P}^{k,w}$ is given by

$$\mathscr{P}^{k,w}\begin{pmatrix} a & b \\ c & d \end{pmatrix} f(z) = |-bz + d|^{-2-w}\left(\frac{-bz + d}{|-bz + d|}\right)^{-k} f\left(\frac{az - c}{-bz + d}\right), \tag{2.11}$$

and the Hilbert space is L^2 of $\mathbb{C}$ with respect to the measure $(1 + |z|^2)^{\operatorname{Re} w}\, dx\, dy$. Although $\mathscr{P}^{k,w}$ is a representation for every value of w, it is not unitary for this inner product unless w is imaginary.

However, for $k = 0$ and w real, it becomes unitary for $0 < w < 2$ with respect to the inner product

$$\langle f, g\rangle = \int_{\mathbb{C}} \int_{\mathbb{C}} \frac{f(z)\overline{g(\zeta)}\, dz\, d\zeta}{|z - \zeta|^{2-w}}, \tag{2.12}$$

and the resulting representations are called **complementary series.** This topic will be taken up further in Chapter XVI.

Up to equivalence the trivial representation, the unitary principal series, and the complementary series are the only irreducible unitary representations of SL(2, ℂ). The only equivalences among members of this list are those in Proposition 2.6. This completeness will be proved in Chapter XVI.

The nonunitary principal series contains all the irreducible finite-dimensional representations of SL(2, ℂ) as subrepresentations. To see this, we first rewrite $\Phi_{m,n}$ in (2.4) by replacing z_1 by 1 and z_2 by z. The polynomials are no longer homogeneous but now are of degree $\leq m$ in z and $\leq n$ in $\bar{z}$. The action becomes

$$\Phi_{m,n}\begin{pmatrix} a & b \\ c & d \end{pmatrix} P(z) = (-bz + d)^m(-\bar{b}\bar{z} + \bar{d})^n P\left(\frac{az - c}{-bz + d}\right). \tag{2.13}$$

Comparing (2.11) and (2.13), we see that

$$\Phi_{m,n} \subseteq \mathscr{P}^{n-m,\ -2-m-n}. \tag{2.14}$$

§5. Irreducible Unitary Representations of SL(2, ℝ)

We begin with a list of some unitary representations of SL(2, ℝ):

(1) Discrete series $\mathscr{D}_n^+$ and $\mathscr{D}_n^-$, $n \geq 2$ an integer. The Hilbert space for $\mathscr{D}_n^+$ is

$$\left\{f \text{ analytic for } \operatorname{Im} z > 0 \,\Big|\, \|f\|^2 = \iint\limits_{\operatorname{Im} z > 0} |f(z)|^2 y^{n-2}\, dx\, dy < \infty\right\},$$

and the action is

$$\mathscr{D}_n^+\begin{pmatrix} a & b \\ c & d \end{pmatrix} f(z) = (-bz + d)^{-n} f\left(\frac{az - c}{-bz + d}\right).$$

The space for $\mathscr{D}_n^+$ is not 0 because $(z + i)^{-n}$ is in it. The representation $\mathscr{D}_n^-$ is obtained by using complex conjugates. All these representations are unitary, and we verify below that they are irreducible. In addition, these representations are **square-integrable** in the sense that some nonzero matrix coefficient is in $L^2(G)$. We shall verify that

$$\int_G |(\mathscr{D}_n^+(g)(z + i)^{-n}, (z + i)^{-n})|^2\, dg < \infty$$

after Proposition 5.28, using another realization of $\mathscr{D}_n^+$ given in §6. Then it will follow (from Proposition 9.6) that every matrix coefficient is in $L^2(G)$.

(2) Principal series $\mathscr{P}^{+,iv}$ and $\mathscr{P}^{-,iv}$, $v \in \mathbb{R}$. The Hilbert space for each of these is $L^2(\mathbb{R})$, and the action is

$$\mathscr{P}^{\pm,iv}\begin{pmatrix} a & b \\ c & d \end{pmatrix} f(x)$$
$$= \begin{cases} |-bx+d|^{-1-iv} f((ax-c)/(-bx+d)) & \text{if } +. \\ \operatorname{sgn}(-bx+d)|-bx+d|^{-1-iv} f((ax-c)/(-bx+d)) & \text{if } -. \end{cases}$$

These representations are all unitary, and we verify below, using Fourier analysis, that all but $\mathscr{P}^{-,0}$ are irreducible. Unitary equivalences $\mathscr{P}^{+,iv} \cong \mathscr{P}^{+,-iv}$ and $\mathscr{P}^{-,iv} \cong \mathscr{P}^{-,-iv}$ are implemented by operators that will be constructed in Chapter VII. All these representations are induced from the upper triangular subgroup in a sense that is to be made precise in Chapter VII.

(3) Complementary series $\mathscr{C}^u$, $0 < u < 1$. Here the Hilbert space is

$$\left\{ f : \mathbb{R} \to \mathbb{C} \,\middle|\, \|f\|^2 = \int_{-\infty}^{\infty} \int_{-\infty}^{\infty} \frac{f(x)\overline{f(y)}\,dx\,dy}{|x-y|^{1-u}} < \infty \right\}$$

and the action is

$$\mathscr{C}^u \begin{pmatrix} a & b \\ c & d \end{pmatrix} f(x) = |-bx+d|^{-1-u} f\left(\frac{ax-c}{-bx+d}\right).$$

These representations are irreducible unitary and will be discussed further in Chapter XVI. As with SL(2, ℂ) they arise from certain nonunitary principal series (defined below) by redefining the inner product.

(4) Others. There is the trivial representation, and there are two "limits of discrete series," $\mathscr{D}_1^+$ and $\mathscr{D}_1^-$. The group action with $\mathscr{D}_1^+$ and $\mathscr{D}_1^-$ is like that in the discrete series, but the norm is given by

$$\|f\|^2 = \sup_{y>0} \int_{-\infty}^{\infty} |f(x+iy)|^2\,dx.$$

The reduction of $\mathscr{P}^{-,0}$ is given by

$$\mathscr{P}^{-,0} \cong \mathscr{D}_1^+ \oplus \mathscr{D}_1^-.$$

The representations $\mathscr{D}_1^+$ and $\mathscr{D}_1^-$ are not square integrable.

Proposition 2.7. For $n \geq 1$, $\mathscr{D}_n^+$ and $\mathscr{D}_n^-$ are irreducible unitary representations of SL(2, ℝ). For all pairs $(\pm, iv)$, $\mathscr{P}^{\pm,iv}$ is a unitary representation, and it is irreducible except for the case of $(-, 0)$. Moreover, $\mathscr{P}^{+,iv}$ is unitarily equivalent with $\mathscr{P}^{+,-iv}$, and $\mathscr{P}^{-,iv}$ is unitarily equivalent with $\mathscr{P}^{-,-iv}$.

Proof of irreducibility of $\mathscr{D}_n^+$. If U is a nonzero closed invariant subspace, then we can find f in U with $f(i) \neq 0$, and the average

$$\frac{1}{2\pi} \int_0^{2\pi} e^{-in\theta} \mathscr{D}_n^+ \begin{pmatrix} \cos\theta & \sin\theta \\ -\sin\theta & \cos\theta \end{pmatrix} f\,d\theta \tag{2.15}$$

will be in U, also. We can evaluate (2.15) at z as

$$= \frac{1}{2\pi}\int_0^{2\pi} e^{-in\theta}\mathscr{D}_n^+\begin{pmatrix}\cos\theta & \sin\theta\\ -\sin\theta & \cos\theta\end{pmatrix} f(z)\,d\theta$$

$$= \frac{1}{2\pi}\int_0^{2\pi} e^{-in\theta}(-z\sin\theta+\cos\theta)^{-n} f\left(\frac{z\cos\theta+\sin\theta}{-z\sin\theta+\cos\theta}\right)d\theta$$

$$= \frac{1}{2\pi i}\oint_{|\zeta|=1}\zeta^{-n}\left(-\frac{z}{2i}(\zeta-\zeta^{-1})+\frac{1}{2}(\zeta+\zeta^{-1})\right)^{-n}$$

$$\times f\left(\frac{\frac{z}{2}(\zeta+\zeta^{-1})+\frac{1}{2i}(\zeta-\zeta^{-1})}{\frac{-z}{2i}(\zeta-\zeta^{-1})+\frac{1}{2}(\zeta+\zeta^{-1})}\right)\frac{d\zeta}{\zeta}\qquad (\text{with } \zeta=e^{i\theta})$$

$$= \frac{1}{2\pi i}\oint_{|\zeta|=1}(2i)^n(z+i+\zeta^2(-z+i))^{-n} f\left(\frac{i(z+i)+\zeta^2(iz+1)}{z+i+\zeta^2(-z+i)}\right)\frac{d\zeta}{\zeta}.$$

As a function of ζ the integrand is analytic for $|\zeta|\le 1$ except at $\zeta=0$, where it has a simple pole. By the Cauchy Integral Formula the integral is

$$= (2i)^n f(i)(z+i)^{-n}.$$

Thus $(z+i)^{-n}$ is in U. If U is not the whole space, then $(z+i)^{-n}$ is in $U^{\perp}$ similarly, and we have a contradiction. Hence $\mathscr{D}_n^+$ is irreducible.

Proof of irreducibility of $\mathscr{P}^{\pm,iv}$ except for $\mathscr{P}^{-,0}$. We proceed as in Proposition 2.6. Let B be a bounded linear operator on $L^2(\mathbb{R})$ commuting with $\mathscr{P}^{\pm,iv}(g)$ for all g. An argument completely analogous to that in Proposition 2.6 shows that

$$\widehat{Bf}(\xi) = m(\xi)\hat{f}(\xi)\qquad \text{for } f\in L^2(\mathbb{R}),$$

where m is a linear combination of 1 and signum(ξ). (We still get (2.10), but $a^{-2}\xi_0$ gives only a half-line; hence signum(ξ) is not immediately excluded as a possibility for m.) To complete the proof, we shall show that the subspace U of $L^2(\mathbb{R})$ of functions whose Fourier transforms vanish on the right half-line is not stable under $\mathscr{P}^{\pm,iv}$ (except in the case of $\mathscr{P}^{-,0}$).

This subspace U is the space of boundary values of analytic functions $F(z)$ in the upper half-plane with $\sup_{y>0}\int_{-\infty}^{\infty}|F(x+iy)|^2\,dx<\infty$. Then $(x+i)^{-1}$ is in U, being the boundary value of $F(z)=(z+i)^{-1}$. Suppose that

$$\mathscr{P}^{\pm,iv}\begin{pmatrix}0 & -1\\ 1 & 0\end{pmatrix}(x+i)^{-1} = \begin{cases}|x|^{-1-iv}i^{-1}x(x+i)^{-1} & \text{if } +\\ (\operatorname{sgn} x)|x|^{-1-iv}i^{-1}x(x+i)^{-1} & \text{if } -\end{cases}$$

is in U. Then

$$\begin{cases}(\operatorname{sgn} x)\,|x|^{-iv} & \text{if } +\\ |x|^{-iv} & \text{if } -\end{cases}$$

is the nontangential boundary value of an analytic function F in the upper half-plane. Denote the principal branch of log z in Im $z > 0$ by Log z and form $F(z) - e^{-iv \operatorname{Log} z}$. This has boundary value 0 on the right half-line but not on the left half-line (since we have excluded $(\pm, iv) = (-, 0)$), and this behavior cannot occur for an analytic function, by a theorem of Privalov. Thus U is not stable under $\mathscr{P}^{\pm, iv}$, and the irreducibility follows.

The representations listed in (1), (2), (3), and (4) above are the only irreducible unitary representations of SL(2, ℝ), up to equivalence. The only equivalences among them are those listed in Proposition 2.7. This completeness will be proved in Chapter XVI.

The full **nonunitary principal series** of SL(2, ℝ) is a family of representations indexed by pairs $(\pm, w)$ with $w = u + iv$ in ℂ. The representation $\mathscr{P}^{\pm, w}$ is given by

$$\mathscr{P}^{\pm, w}\begin{pmatrix} a & b \\ c & d \end{pmatrix} f(x)$$
$$= \begin{cases} |-bx + d|^{-1-w} f((ax - c)/(-bx + d)) & \text{if } + \\ \operatorname{sgn}(-bx + d)|-bx + d|^{-1-w} f((ax - c)/(-bx + d)) & \text{if } -, \end{cases} \tag{2.16}$$

and the Hilbert space is L^2 of ℝ with respect to $(1 + x^2)^{\operatorname{Re} w}\, dx$. Again $\mathscr{P}^{\pm, w}$ is not unitary except for w imaginary. However, $\mathscr{P}^{+, u}$ can be renormed for $0 < u < 1$ so as to become unitary, and we obtain the complementary series $\mathscr{C}^u$.

The nonunitary principal series contains all the irreducible finite-dimensional representations of SL(2, ℝ) as subrepresentations. To see this, we rewrite Φ_n in (2.1) by replacing z_1 by 1 and z_2 by x. The polynomials are no longer homogeneous but are of degree $\le n$ in x. The action becomes

$$\Phi_n\begin{pmatrix} a & b \\ c & d \end{pmatrix} P(x) = (-bx + d)^n P\left(\frac{ax - c}{-bx + d}\right). \tag{2.17}$$

Comparing (2.16) and (2.17), we see that

$$\Phi_n \subseteq \begin{cases} \mathscr{P}^{+, -(n+1)} & \text{if } n \text{ even} \\ \mathscr{P}^{-, -(n+1)} & \text{if } n \text{ odd.} \end{cases} \tag{2.18}$$

We can see some additional reducibility in $\mathscr{P}^{\pm, w}$ on a formal level by specializing the parameter w and by passing from z in the upper half-plane for discrete series to x on the real axis. We obtain continuous (not bicontinuous) inclusions

$$\mathscr{D}_n^+ \oplus \mathscr{D}_n^- \subseteq \begin{cases} \mathscr{P}^{+, n-1} & \text{if } n \text{ even} \\ \mathscr{P}^{-, n-1} & \text{if } n \text{ odd.} \end{cases} \tag{2.19}$$

There is no other reducibility for $\mathscr{P}^{\pm, w}$. The quotient by a finite-dimensional Φ is essentially the sum of two $\mathscr{D}$'s, and vice versa. All of these facts about reducibility are a little easier to see rigorously in a different realization of the representations, which we take up in the next section.

§6. Use of SU(1, 1)

A number of facts about representations of SL(2, $\mathbb{R}$) are easier to see when the restriction of the representation to K is more transparent. For this purpose it is more convenient to use the group SU(1, 1), which is conjugate to SL(2, $\mathbb{R}$) within SL(2, $\mathbb{C}$):

$$\begin{pmatrix} 1 & i \\ i & 1 \end{pmatrix} \mathrm{SU}(1, 1) \begin{pmatrix} 1 & i \\ i & 1 \end{pmatrix}^{-1} = \mathrm{SL}(2, \mathbb{R}). \tag{2.20}$$

Here
$$\mathrm{SU}(1, 1) = \left\{ \begin{pmatrix} \alpha & \beta \\ \bar{\beta} & \bar{\alpha} \end{pmatrix} \middle| \, |\alpha|^2 - |\beta|^2 = 1 \right\}$$

acts by linear fractional transformations on the unit disc in the same way that SL(2, $\mathbb{R}$) acts on the upper half plane.

We use z for an upper half plane variable and ζ for a disc variable. The two are related by

$$z = \frac{\zeta + i}{i\zeta + 1} \qquad \text{and} \qquad \zeta = \frac{z - i}{-iz + 1}.$$

To change a representation from SL(2, $\mathbb{R}$) acting on functions $f(z)$ to SU(1, 1) acting on functions $F(\zeta)$, we use a change of variables and an extra factor

$$F(\zeta) = Tf(\zeta) = m\left(\frac{1}{\sqrt{2}}(i\zeta + 1)\right) f(z(\zeta))$$

$$f(z) = T^{-1}F(z) = m\left(\frac{1}{\sqrt{2}}(-iz + 1)\right) F(\zeta(z)).$$

Here the extra factor m is a homomorphism of $\mathbb{C}^\times$ into $\mathbb{C}^\times$ that is at our disposal and can depend on the representation; m is called a **multiplier.** If R is a representation of SL(2, $\mathbb{R}$) acting on functions $f(z)$, the corresponding representation of SU(1, 1) (which we denote also by R) acting on functions $F(\zeta)$ is

$$R\begin{pmatrix} \alpha & \beta \\ \bar{\beta} & \bar{\alpha} \end{pmatrix} F(\zeta) = T\left(R\left[\begin{pmatrix} 1 & i \\ i & 1 \end{pmatrix} \begin{pmatrix} \alpha & \beta \\ \bar{\beta} & \bar{\alpha} \end{pmatrix} \begin{pmatrix} 1 & i \\ i & 1 \end{pmatrix}^{-1} \right] (T^{-1}F) \right)(\zeta).$$

Assuming that R is given on SL(2, $\mathbb{R}$) by

$$R\begin{pmatrix} a & b \\ c & d \end{pmatrix} f(z) = s(-bz + d) f\left(\frac{az - c}{-bz + d}\right),$$

we find that R is given on SU(1, 1) by

$$R\begin{pmatrix}\alpha & \beta\\ \bar\beta & \bar\alpha\end{pmatrix}F(\zeta) = \frac{m(-\beta\zeta+\bar\alpha)}{m((i\alpha-\beta)\zeta+(-i\bar\beta+\bar\alpha))}\, s\left(\frac{(i\alpha-\beta)\zeta+(-i\bar\beta+\bar\alpha)}{i\zeta+1}\right)$$

$$\times\, m(i\zeta+1)F\left(\frac{\alpha\zeta-\bar\beta}{-\beta\zeta+\bar\alpha}\right). \tag{2.21}$$

For $\mathscr{D}_n^+$ we choose $m = s$. Then $\mathscr{D}_n^+$, as a representation of SU(1, 1), acts on analytic functions on the disc by

$$\mathscr{D}_n^+\begin{pmatrix}\alpha & \beta\\ \bar\beta & \bar\alpha\end{pmatrix}F(\zeta) = (-\beta\zeta+\bar\alpha)^{-n}F\left(\frac{\alpha\zeta-\bar\beta}{-\beta\zeta+\bar\alpha}\right),$$

and the norm, except for a constant factor, is given by

$$\|F\|^2 = \begin{cases}\displaystyle\int_{|\zeta|<1}|F(\zeta)|^2(1-|\zeta|^2)^{n-2}\,d\zeta & \text{for } n\ge 2\\ \displaystyle\sup_{0\le r<1}\int_0^{2\pi}|F(re^{i\theta})|^2\,d\theta & \text{for } n=1.\end{cases}$$

Under our conjugation (2.20), the maximal compact subgroup becomes

$$K = \left\{\begin{pmatrix}e^{i\theta} & 0\\ 0 & e^{-i\theta}\end{pmatrix}\right\}.$$

It is clear that the functions $\{\zeta^N, N\ge 0\}$ form an orthogonal basis for the representation space of $\mathscr{D}_n^+$, and each ζ^N is an eigenfunction of $\mathscr{D}_n^+(K)$:

$$\mathscr{D}_n^+\begin{pmatrix}e^{i\theta} & 0\\ 0 & e^{-i\theta}\end{pmatrix}\zeta^N = e^{(n+2N)i\theta}\zeta^N.$$

It is instructive to re-examine the proof of irreducibility of $\mathscr{D}_n^+$ in this realization.

For $\mathscr{P}^{\pm,w}$, we specialize to $z = x$ real and to $\zeta = e^{i\psi}$ on the unit circle. In (2.21), s is evaluated at a real point, hence need be defined only on $\mathbb{R}^\times$. We can take m to be any extension of s to $\mathbb{C}^\times$. In any event the norm becomes a multiple of the L^2 norm on the circle. For $\mathscr{P}^{+,w}$, we can choose $m = |\cdot|^{-1-w}$ and then the formula is

$$\mathscr{P}^{+,w}\begin{pmatrix}\alpha & \beta\\ \bar\beta & \bar\alpha\end{pmatrix}F(e^{i\psi}) = |-\beta e^{i\psi}+\bar\alpha|^{-1-w}F\left(\frac{\alpha e^{i\psi}-\bar\beta}{-\beta e^{i\psi}+\bar\alpha}\right).$$

The functions $e^{iN\psi}$ form an orthogonal basis of the representation space, and

$$\mathscr{P}^{+,w}\begin{pmatrix}e^{i\theta} & 0\\ 0 & e^{-i\theta}\end{pmatrix}e^{iN\psi} = e^{2Ni\theta}e^{iN\psi}.$$

Differentiating the representation formally along the one-parameter group $\begin{pmatrix} \cosh t & \sinh t \\ \sinh t & \cosh t \end{pmatrix}$, in a way that will be made precise starting in Chapter III, we can see the irreducibility of $\mathscr{P}^{+,iv}$. (A closed invariant subspace must be generated by exponentials, by the Peter-Weyl Theorem applied to $\mathscr{P}^{+,iv}(K)$, and the differentiated action from $\begin{pmatrix} \cosh t & \sinh t \\ \sinh t & \cosh t \end{pmatrix}$ allows us to deduce the presence of all the exponentials from the presence of one of them.)

A different way of realizing $\mathscr{P}^{+,w}$ comes by choosing

$$m(z) = |z|^{-1-w}\left(\frac{z}{|z|}\right)^{-2N}.$$

Then the formula for $\mathscr{P}^{+,w}$ is

$$\mathscr{P}^{+,w}\begin{pmatrix} \alpha & \beta \\ \bar{\beta} & \bar{\alpha} \end{pmatrix} F(e^{i\psi}) = |-\beta e^{i\psi} + \bar{\alpha}|^{-1-w+2N}(-\beta e^{i\psi} + \bar{\alpha})^{-2N} F\left(\frac{\alpha e^{i\psi} - \bar{\beta}}{-\beta e^{i\psi} + \bar{\alpha}}\right).$$

Taking $w = 2N - 1$, we see that the F's that extend to analytic functions on the closed disc form a non-closed invariant subspace that is equivalent with the non-closed subspace of functions for $\mathscr{D}_{2N}^{+}$ that are analytic on the closed disc. From this fact we easily deduce the continuous inclusion $\mathscr{D}_{2N}^{+} \subseteq \mathscr{P}^{+,2N-1}$.

Similar remarks apply to realizing $\mathscr{P}^{-,w}$ and to deducing the continuous inclusion of $\mathscr{D}_{2N+1}^{+}$ in $\mathscr{P}^{-,2N}$.

§7. Plancherel Formula

For a compact group, (1.10) gives the Plancherel formula as

$$\int_G |f(x)|^2\,dx = \sum_{\Phi} d_{\Phi} \|\Phi(f)\|_{\mathrm{HS}}^2, \qquad f \in L^2(G).$$

From Theorem 2.4 and the unitary trick, it follows that the representations Φ_n of (2.1) are the only irreducible unitary representations of SU(2), up to equivalence. Thus the Plancherel formula for SU(2) reads

$$\int_{\mathrm{SU}(2)} |f(x)|^2\,dx = \sum_{n=0}^{\infty} (n+1)\|\Phi_n(f)\|_{\mathrm{HS}}^2, \qquad f \in L^2(\mathrm{SU}(2)). \tag{2.22}$$

In Chapter XI we shall prove a Fourier inversion formula for SU(2):

$$h(1) = \sum_{n=0}^{\infty} (n+1)\,\mathrm{Tr}(\Phi_n(h)), \qquad h \in C^\infty(\mathrm{SU}(2)). \tag{2.23}$$

The Plancherel formula follows from the Fourier inversion formula by taking $h = f^* * f$ and doing a passage to the limit, since $\|A\|_{HS}^2 = \mathrm{Tr}(A^*A)$.

The groups SL(2, ℂ) and SL(2, ℝ) also each have a Plancherel formula and a Fourier inversion formula for $C_{\mathrm{com}}^\infty(G)$, but no longer involving only a discrete sum. We give only the Fourier inversion formulas, since they again immediately imply corresponding Plancherel formulas. For SL(2, ℂ), the formula is

$$h(1) = \sum_{n=-\infty}^{\infty} \int_{-\infty}^{\infty} \mathrm{Tr}(\mathscr{P}^{n,iv}(h))(n^2 + v^2)\,dv, \qquad h \in C_{\mathrm{com}}^\infty(\mathrm{SL}(2,\mathbb{C})) \quad (2.24)$$

for a suitable normalization of Haar measure. For SL(2, ℝ), the formula is

$$h(1) = \int_{-\infty}^{\infty} \mathrm{Tr}(\mathscr{P}^{+,iv}(h))v \tanh\left(\frac{\pi v}{2}\right) dv + \int_{-\infty}^{\infty} \mathrm{Tr}(\mathscr{P}^{-,iv}(h))v \coth\left(\frac{\pi v}{2}\right) dv$$
$$+ \sum_{n=2}^{\infty} 4(n-1)\,\mathrm{Tr}(\mathscr{D}_n^+(h) + \mathscr{D}_n^-(h)), \qquad h \in C_{\mathrm{com}}^\infty(\mathrm{SL}(2,\mathbb{R})) \quad (2.25)$$

for a suitable normalization of Haar measure. These formulas are quite a bit more subtle than (2.23); the very existence of the indicated traces is not immediately evident. Proofs of the formulas will be given in Chapter XI, §§2–3.

§8. Problems

1. (a) Prove that $\mathfrak{su}(2)$ is isomorphic to $\mathfrak{so}(3)$.
   (b) Prove that $\mathfrak{su}(2)$ is simple as a Lie algebra.
   (c) Prove that $\mathfrak{su}(2)$ is not isomorphic to $\mathfrak{sl}(2,\mathbb{R})$.
2. Prove that $\mathfrak{sl}(2,\mathbb{R})$ is isomorphic to $\mathfrak{so}(2,1)$. More specifically let $C(u, v)$ be the bilinear form on $\mathbb{R}^3$ given by

$$C\left(\begin{pmatrix} u_1 \\ u_2 \\ u_3 \end{pmatrix}, \begin{pmatrix} v_1 \\ v_2 \\ v_3 \end{pmatrix}\right) = u_1v_1 + u_2v_2 - u_3v_3,$$

   and realize $\mathfrak{so}(2,1)$ concretely as the Lie algebra of linear maps of $\mathbb{R}^3$ into itself that are skew-symmetric relative to C:

$$\mathfrak{so}(2,1) = \{L \in \mathrm{End}(\mathbb{R}^3) \mid C(Lu, v) + C(u, Lv) = 0, \qquad \text{for all } u, v \in \mathbb{R}^3\}.$$

   Exhibit this $\mathfrak{so}(2,1)$ concretely as isomorphic with $\mathfrak{sl}(2,\mathbb{R})$.
3. This problem leads by steps to a direct proof, without structure theory, that $\mathfrak{sl}(2,\mathbb{C})$ is the only three-dimensional simple Lie algebra $\mathfrak{g}$ over ℂ, up to isomorphism.
   (a) Show from $[\mathfrak{g}, \mathfrak{g}] = \mathfrak{g}$ that $\mathrm{Tr}(\mathrm{ad}\, X) = 0$ for all X.

(b) Using Engel's Theorem (A.17), choose X_0 such that $\operatorname{ad} X_0$ is not nilpotent. Show from (a) and linear algebra that $\operatorname{ad} X_0$ is diagonable.

(c) Show that a suitable multiple X of X_0 is a member of a basis $\{X, Y, Z\}$ of $\mathfrak{g}$ in which $\operatorname{ad} X$ has the matrix realization

$$\operatorname{ad} X = \begin{array}{c} \\ X \\ Y \\ Z \end{array}\!\!\begin{array}{c} \begin{matrix} X & Y & Z \end{matrix} \\ \begin{pmatrix} 0 & 0 & 0 \\ 0 & 2 & 0 \\ 0 & 0 & -2 \end{pmatrix} \end{array}.$$

(d) Writing $[Y, Z]$ in terms of the basis $\{X, Y, Z\}$ and applying the Jacobi identity to $(\operatorname{ad} X)[Y, Z]$, show that $X \leftrightarrow h$, $Y \leftrightarrow ce$ leads to an isomorphism of $\mathfrak{g}$ with $\mathfrak{sl}(2, \mathbb{C})$.

4. Following steps as in Problem 3, prove that $\mathfrak{sl}(2, \mathbb{R})$ and $\mathfrak{so}(3)$ are the only three-dimensional simple Lie algebras $\mathfrak{g}$ over $\mathbb{R}$, up to isomorphism. [Hints: (a) is still okay. Define X_0 as in (b), and show that the nonzero eigenvalues of $\operatorname{ad} X_0$ are two in number, are distinct, have sum 0, and have product real. For (c), find a basis $\{X, Y, Z\}$ with $X = c_0 X_0$ such that $\operatorname{ad} X$ has matrix realization either

$$\operatorname{ad} X = \begin{array}{c} \\ X \\ Y \\ Z \end{array}\!\!\begin{array}{c} \begin{matrix} X & Y & Z \end{matrix} \\ \begin{pmatrix} 0 & 0 & 0 \\ 0 & 2 & 0 \\ 0 & 0 & -2 \end{pmatrix} \end{array} \quad \text{or} \quad \operatorname{ad} X = \begin{array}{c} \\ X \\ Y \\ Z \end{array}\!\!\begin{array}{c} \begin{matrix} X & Y & Z \end{matrix} \\ \begin{pmatrix} 0 & 0 & 0 \\ 0 & 0 & -1 \\ 0 & 1 & 0 \end{pmatrix} \end{array}.$$

For (d), show that the first case leads to an isomorphism with $\mathfrak{sl}(2, \mathbb{R})$ and the second case leads to an isomorphism with $\mathfrak{so}(3)$.]

5. (a) Exhibit a continuous homomorphism of SU(2) onto SO(3) with kernel $\{\pm I\}$. [Hint: The map

$$z = \frac{x_1 + i x_2}{1 - x_3}$$

carries the unit sphere S^2 in $\mathbb{R}^3$ one-one onto $\mathbb{C} \cup \{\infty\}$, on which SU(2) acts by linear fractional transformations:

$$w = g(z) = \frac{\alpha z + \beta}{-\bar{\beta} z + \bar{\alpha}} \qquad \text{for } g = \begin{pmatrix} \alpha & \beta \\ -\bar{\beta} & \bar{\alpha} \end{pmatrix} \text{ in SU(2)}.$$

Reinterpret these linear fractional transformations back on S^2 as the restrictions to S^2 of rotations of $\mathbb{R}^3$.]

(b) Which of the representations Φ_n of SU(2) given by (2.1) yield well-defined representations of SO(3)? Give an explicit realization of an irreducible representation of SO(3) of dimension 5.

6. The representation π of Theorem 2.4 is realized concretely on a space of polynomials by the formula (2.1). In this realization what polynomial corresponds to v_i? (The answer should be stated up to a multiplicative constant independent of i.)
7. Prove that the discrete series representation $\mathscr{D}_n^+$ of SL(2, ℝ) remains irreducible when restricted to the upper triangular subgroup of SL(2, ℝ).
8. With the complementary series $\mathscr{C}^u$ of SL(2, ℝ), $0 < u < 1$, realized as in §5, show that every f in $L^1(\mathbb{R}) \cap L^2(\mathbb{R})$ has finite norm.
9. What subgroup of SU(1, 1) plays the role of the diagonal subgroup of SL(2, ℝ) when SU(1, 1) and SL(2, ℝ) are identified as in §6?

Problems 10 to 13 rederive some classical formulas for the Mellin transform in $\mathbb{R}^2$ and interpret the results as a decomposition of the representation in Problem 10 of Chapter I as an integeral of principal series representations of SL(2, ℝ).

10. Fix a real number v, and let H_v^+ be the Hilbert space of functions

$$\{F:\mathbb{R}^2 \to \mathbb{C} \mid F(tx, ty) = |t|^{-1-iv}F(x, y) \quad \text{for } t \in \mathbb{R}\}$$

of finite norm

$$\|F\|_v^2 = \frac{1}{2\pi}\int_0^{2\pi} |F(\cos\theta, \sin\theta)|^2\, d\theta.$$

Let SL(2, ℝ) act on $\mathbb{R}^2$ as left multiplication on column vectors, and let $P^{+,iv}$ be the unitary representation of SL(2, ℝ) on H_v^+ given by

$$P^{+,iv}(g)F(x, y) = F(g^{-1}(x, y)).$$

Prove that the map of H_v^+ to $L^2(\mathbb{R})$ given by $F \to f$ with $f(x) = F(1, x)$ is a unitary equivalence (except for a constant factor) of $P^{+,iv}$ and $\mathscr{P}^{+,iv}$.

11. Starting from the analogous Hilbert space H_v^- obtained from

$$\{F:\mathbb{R}^2 \to \mathbb{C} \mid F(tx, ty) = (\operatorname{sgn} t)|t|^{-1-iv}F(x, y) \quad \text{for } t \in \mathbb{R}\},$$

construct a unitary representation $P^{-,iv}$ of SL(2, ℝ) on H_v^-, and show it is unitarily equivalent with $\mathscr{P}^{-,iv}$.

12. For g in SL(2, ℝ) and F in $L^2(\mathbb{R}^2)$, define $\Phi(g)F(x, y) = F(g^{-1}(x, y))$. On the dense subspace $C_{\text{com}}^\infty(\mathbb{R}^2 - \{0\})$, define

$$\{F:\mathbb{R}^2\, E_vF(x, y) = \int_0^\infty F(tx, ty)t^{iv}\, dt$$

Prove that E_v carries this dense subspace in equivariant fashion onto a dense subspace of

$$H_v^+ \oplus H_v^- = \{F:\mathbb{R}^2 \to \mathbb{C} \mid F(tx, ty) = t^{-1-iv}F(x, y) \qquad \text{for } t > 0\}.$$

13. Applying the Fourier inversion formula and Plancherel formula to the function $h(s) = e^s F(e^s x, e^s y)$, $-\infty < s < \infty$, establish the formulas

$$F(x, y) = \frac{1}{2\pi} \int_{-\infty}^{\infty} (E_v F)(x, y)\, dv$$

and

$$\|F\|^2 = \int_{-\infty}^{\infty} \|E_v F\|_v^2\, dv$$

for F in $C_{\mathrm{com}}^{\infty}(\mathbb{R}^2 - \{0\})$.

CHAPTER III

C^∞ *Vectors and the Universal Enveloping Algebra*

§1. Universal Enveloping Algebra

For any representation φ of a Lie algebra $\mathfrak{g}$, one has to compose operators $\varphi(X)$, $X \in \mathfrak{g}$, in order to understand reducibility. In the process the composite operators satisfy certain identities because of (1.12), and these identities do not really depend on φ. There is a universal gadget, namely a certain associative algebra, in which these identities are valid and imply the identities for φ.

To be prepared to take advantage of the unitary trick we shall assume that our basic Lie algebra is a complex Lie algebra; in practice, it will be the complexification of a real Lie algebra that we want to study.

Thus let $\mathfrak{g}$ be a finite-dimensional Lie algebra over $\mathbb{C}$, and let

$$T(\mathfrak{g}) = \mathbb{C} \oplus \mathfrak{g} \oplus (\mathfrak{g} \otimes \mathfrak{g}) \oplus \ldots = \sum_{r=0}^{\infty} (\bigotimes^r \mathfrak{g})$$

be the tensor algebra of $\mathfrak{g}$. $T(\mathfrak{g})$ is an associative algebra with identity over $\mathbb{C}$, and there is a natural map of $\mathfrak{g}$ into $T(\mathfrak{g})$ given by identifying $\mathfrak{g}$ with the first-order terms. $T(\mathfrak{g})$ has the following universal mapping property: Whenever τ is a linear map of $\mathfrak{g}$ into a complex associative algebra A with identity, then there exists a unique associative algebra homomorphism $\bar{\tau}$ with $\bar{\tau}(1) = 1$ such that the diagram below commutes.

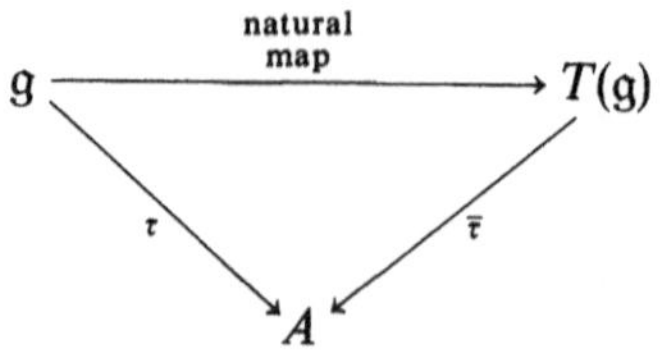

The tensor algebra is not the algebra we want to work with, because it does not incorporate the Lie algebra structure of $\mathfrak{g}$. Thus we want to force $XY - YX$ to be equal to $[X, Y]$ for X and Y in $\mathfrak{g}$. To this end, we let $U(\mathfrak{g})$ be the quotient of $T(\mathfrak{g})$ by the two-sided ideal generated by all $(X \otimes Y - Y \otimes X - [X, Y])$ for X and Y in $\mathfrak{g}$. $U(\mathfrak{g})$ is a complex associative algebra with identity and is called the **universal enveloping algebra** of $\mathfrak{g}$.

Let

$$U^N(\mathfrak{g}) = \text{image of } \sum_{r=0}^{N} (\bigotimes^r \mathfrak{g}) \subseteq U(\mathfrak{g}).$$

Then $U^N(\mathfrak{g})$ is a finite-dimensional subspace, and $U(\mathfrak{g}) = \bigcup_{N=0}^{\infty} U^N(\mathfrak{g})$.

Let $\sigma:\mathfrak{g} \to U(\mathfrak{g})$ be the canonical map obtained by imbedding $\mathfrak{g}$ in $T(\mathfrak{g})$ and passing to the quotient. Then

$$\sigma[X, Y] = \sigma(X)\sigma(Y) - \sigma(Y)\sigma(X)$$

for all X and Y. Relative to σ, $U(\mathfrak{g})$ has the universal mapping property given in the following proposition.

Proposition 3.1. Let σ be the canonical map of $\mathfrak{g}$ into $U(\mathfrak{g})$, let A be a complex associative algebra with identity, and let φ be a linear mapping of $\mathfrak{g}$ into A such that

$$\varphi[X, Y] = \varphi(X)\varphi(Y) - \varphi(Y)\varphi(X)$$

for all X and Y in $\mathfrak{g}$. Then there exists a unique algebra homomorphism $\varphi_0:U(\mathfrak{g}) \to A$ such that $\varphi_0\sigma = \varphi$ and $\varphi_0(1) = 1$.

Proof. Uniqueness follows from the fact that $U(\mathfrak{g})$ is generated by 1 and $\sigma(\mathfrak{g})$. For existence, let $\tilde{\varphi}:T(\mathfrak{g}) \to A$ be the unique algebra homomorphism of $T(\mathfrak{g})$ into A extending φ and having $\tilde{\varphi}(1) = 1$. For X and Y in $\mathfrak{g}$, we have

$$\tilde{\varphi}(X \otimes Y - Y \otimes X - [X, Y]) = \tilde{\varphi}(X)\tilde{\varphi}(Y) - \tilde{\varphi}(Y)\tilde{\varphi}(X) - \tilde{\varphi}[X, Y] = 0.$$

Hence $\tilde{\varphi}$ annihilates the ideal generated by these elements and therefore passes to the quotient $U(\mathfrak{g})$; the homomorphism on $U(\mathfrak{g})$ is the required φ_0.

Theorem 3.2 (Birkhoff-Witt Theorem). The canonical map $\sigma:\mathfrak{g} \to U(\mathfrak{g})$ is one-one. Dropping it from the notation, let $X_1, \ldots, X_n$ be a basis of $\mathfrak{g}$ over $\mathbb{C}$. Then the monomials $X_1^{j_1} \cdots X_n^{j_n}$, $j_k \geq 0$, form a basis of $U(\mathfrak{g})$ over $\mathbb{C}$.

We shall prove the theorem in several steps. The first two lemmas prove that the indicated monomials span $U(\mathfrak{g})$. The second one uses the notation

$$Y_i = \sigma(X_i), \qquad 1 \leq i \leq n, \tag{3.1}$$

$$Y_I = Y_{i_1} \cdots Y_{i_p} \qquad \text{if } I = (i_1, \ldots, i_p) \text{ is a } p\text{-tuple of integers from 1 to } n.$$

For the proof of the linear independence we shall assume that $\mathfrak{g}$ is the complexification of a real Lie algebra $\mathfrak{g}_{\mathbb{R}}$, although the proof can easily be adapted to handle the general case.

Lemma 3.3. Let $W_1, \ldots, W_p$ be in $\mathfrak{g}$ and let π be a permutation of $\{1, \ldots, p\}$. Then $\sigma(W_1) \cdots \sigma(W_p) - \sigma(W_{\pi(1)}) \cdots \sigma(W_{\pi(p)})$ is in $U^{p-1}(\mathfrak{g})$.

Proof. Because we can iterate matters, it is enough to handle the case that π is a consecutive transposition, say of j with $j+1$. Then

$$\sigma(W_j)\sigma(W_{j+1}) - \sigma(W_{j+1})\sigma(W_j) = \sigma[W_j, W_{j+1}],$$

and the lemma follows by multiplying through on the left by

$$\sigma(W_1) \cdots \sigma(W_{j-1})$$

and on the right by $\sigma(W_{j+2}) \cdots \sigma(W_p)$.

Lemma 3.4. The Y_I, for all increasing sequences I of $\leq p$ integers, generate $U^p(\mathfrak{g})$.

Proof. If we use *all* tuples with $\leq p$ members, we certainly have a set of generators. Lemma 3.3 then says inductively that the increasing sequences suffice.

Before proving the linear independence in Theorem 3.2, we digress to introduce an algebra of differential operators that will provide us with an analytic interpretation of $U(\mathfrak{g})$.

Let $G_{\mathbb{R}}$ be an analytic group, and let $\mathfrak{g}_{\mathbb{R}}$ be its Lie algebra. For X in $\mathfrak{g}_{\mathbb{R}}$, the left-invariant vector field $\tilde{X}$ is the endomorphism of the set $C^\infty(G_{\mathbb{R}})$ of *complex-valued* smooth functions on $G_{\mathbb{R}}$ given by

$$\tilde{X}f(x) = \frac{d}{dt} f(x \exp tX)\bigg|_{t=0}$$

(Cf. (A.90), (A.91), and (A.99). The use of complex-valued functions here is a departure from the notation of Appendix A.) The operator $\tilde{X}$ is a particular example of a **left-invariant differential operator** on $G_{\mathbb{R}}$, which is defined as any linear endomorphism D of $C^\infty(G_{\mathbb{R}})$ with the properties

(i) D commutes with all left translations by members of $G_{\mathbb{R}}$, and
(ii) for each $g \in G_{\mathbb{R}}$ there is a chart (φ, S) about g, say $\varphi = (x^1, \ldots, x^n)$, and there are functions $a_{k_1 \cdots k_n}$ in $C^\infty(S)$ such that

$$Df(x) = \sum\nolimits_{\text{bounded}} a_{k_1 \cdots k_n}(x) \frac{\partial^{k_1 + \cdots + k_n} f}{\partial x_1^{k_1} \cdots \partial x_n^{k_n}}(x) \qquad (3.2)$$

for all x in S and f in $C^\infty(G_{\mathbb{R}})$.

Such operators form a subalgebra $D(G_{\mathbb{R}})$ of $\operatorname{End}_{\mathbb{C}}(C^\infty(G_{\mathbb{R}}))$ with identity. Moreover, any such D has an expansion of the form (3.2) in any chart about x.

Now we can return to the proof of Theorem 3.2. We shall assume that $\mathfrak{g} = (\mathfrak{g}_{\mathbb{R}})^{\mathbb{C}}$ for some real Lie algebra $\mathfrak{g}_{\mathbb{R}}$, which is the only case that will be of interest to us. Matching the numbers of monomials in $U^p(\mathfrak{g})$ for different bases and taking into account Lemma 3.4, we see that we may assume $X_1, \ldots, X_n$ are all in $\mathfrak{g}_{\mathbb{R}}$. Let $G_{\mathbb{R}}$ be an analytic group with Lie algebra $\mathfrak{g}_{\mathbb{R}}$ (existence by (A.129)). The real-linear map $X \to \tilde{X}$ of $\mathfrak{g}_{\mathbb{R}}$ into $D(G_{\mathbb{R}})$ extends, via multiplication by i, to a complex-linear map of $\mathfrak{g} = (\mathfrak{g}_{\mathbb{R}})^{\mathbb{C}}$ into $D(G_{\mathbb{R}})$, and we then have

$$\widetilde{[X, Y]} = \tilde{X}\tilde{Y} - \tilde{Y}\tilde{X} \qquad \text{for } X, Y \text{ in } \mathfrak{g}.$$

By Proposition 3.1, we obtain a unique extension $D \to \tilde{D}$ to an algebra homomorphism of $U(\mathfrak{g})$ into $D(G_{\mathbb{R}})$ carrying 1 to 1. In terms of our notation (3.1), we have

$$\tilde{Y}_{(i_1, \ldots, i_p)} = \tilde{Y}_{i_1} \cdots \tilde{Y}_{i_p} = \tilde{X}_{i_1} \cdots \tilde{X}_{i_p}.$$

To prove the linear independence in Theorem 3.2, it is enough to show that the elements $\tilde{Y}_I$, I increasing, are linearly independent in $D(G_{\mathbb{R}})$. The latter independence follows from the following lemma.

Lemma 3.5. For each increasing tuple I there is a function h_I in $C^\infty(G_{\mathbb{R}})$ such that each increasing tuple J satisfies

$$\tilde{Y}_J h_I(1) = \begin{cases} 1 & \text{if } J = I \\ 0 & \text{if } J \text{ is increasing and } J \neq I. \end{cases}$$

Proof. If W is in $\mathfrak{g}_{\mathbb{R}}$ and f is in $C^\infty(G_{\mathbb{R}})$, then we know from (A.98) that

$$\tilde{W}^k f(x) = \frac{d^k}{ds^k} f(x \exp sW)\bigg|_{s=0}. \tag{3.3}$$

From (3.3) it follows that

$$\tilde{X}_1^{j_1} \cdots \tilde{X}_n^{j_n} f(1) = \frac{\partial^{j_1 + \cdots + j_n}}{\partial s_1^{j_1} \cdots \partial s_n^{j_n}} f(\exp s_1 X_1 \cdots \exp s_n X_n)\bigg|_{s_1 = \ldots = s_n = 0}. \tag{3.4}$$

In a small enough open neighborhood N of 1 in G, (A.104) shows that we get a well-defined chart (ψ, N) from

$$\psi((\exp s_1 X_1)(\exp s_2 X_2) \cdots (\exp s_n X_n)) = (s_1, s_2, \ldots, s_n) \tag{3.5}$$

We shall write $\psi = (s_1, \ldots, s_n)$.

Now let I be an increasing tuple with i_1 ones, i_2 twos, etc. Define f in $C^\infty(G_{\mathbb{R}})$ to have compact support in N and to be given by

$$f(x) = s_1(x)^{i_1} \cdots s_n(x)^{i_n}$$

in a neighborhood of 1. For $(s_1, \ldots, s_n)$ small enough, we then have

$$f(\exp s_1 X_1 \cdots \exp s_n X_n) = s_1^{i_1} \cdots s_n^{i_n}.$$

Applying (3.4) and supposing that J has j_1 ones, j_2 twos, etc., we see that

$$\tilde{Y}_J f(1) = \tilde{X}_1^{j_1} \cdots \tilde{X}_n^{j_n} f(1) \text{ is } \begin{cases} c \neq 0 & \text{if } j_1 = i_1, \ldots, j_n = i_n \\ = 0 & \text{if some } j_k \neq i_k. \end{cases}$$

Thus we can take $h_I = c^{-1}f$, and the lemma is proved. This completes also the proof of Theorem 3.2.

Theorem 3.6. If $G_{\mathbb{R}}$ is a Lie group with Lie algebra $\mathfrak{g}_{\mathbb{R}}$, then the natural algebra homomorphism of $U((\mathfrak{g}_{\mathbb{R}})^{\mathbb{C}})$ into the algebra $D(G_{\mathbb{R}})$ of left-invariant differential operators on $G_{\mathbb{R}}$ is an isomorphism onto.

Proof. Lemma 3.5 shows that the map is one-one. We prove the map is onto. In the local coordinate system (3.5) about 1, write

$$Df(x) = \sum_{\text{bounded}} a_{j_1 \cdots j_n}(x) \frac{\partial^{j_1 + \cdots + j_n}}{\partial_{s_1}^{j_1} \cdots \partial s_n^{j_n}} f(x).$$

For each n-tuple $(j_1, \ldots, j_n)$ in the sum, form a $(j_1 + \ldots + j_n)$-tuple $J_{(j_1, \ldots, j_n)}$ that has j_1 ones, followed by j_2 twos, etc. Then (3.4) shows that

$$Df(1) = \sum_{\text{bounded}} a_{j_1 \cdots j_n}(1) \tilde{Y}_{J(j_1, \ldots, j_n)} f(1).$$

Since D commutes with left translations,

$$D = \sum_{\text{finite}} a_{j_1 \cdots j_n}(1) \tilde{Y}_{J(j_1, \ldots, j_n)}.$$

Hence the map is onto.

§2. Actions on Universal Enveloping Algebra

Let G be an analytic group, and let $\mathfrak{g}$ be its Lie algebra. We define a representation ad of $\mathfrak{g}$ on $U^N(\mathfrak{g}^{\mathbb{C}})$ for each N by

$$(\text{ad } X)D = XD - DX.$$

(Lemma 3.3 shows that $(\text{ad } X)D$ remains in $U^N(\mathfrak{g}^{\mathbb{C}})$.)

On the group level, each $\text{Ad}(g)$ for g in G gives an automorphism of $\mathfrak{g}$, hence a complex-linear map $\text{Ad}(g)$ of $\mathfrak{g}^{\mathbb{C}}$ into $U(\mathfrak{g}^{\mathbb{C}})$ such that

$$\text{Ad}(g)[X, Y] = [\text{Ad}(g)X, \text{Ad}(g)Y] = (\text{Ad}(g)X)(\text{Ad}(g)Y) - (\text{Ad}(g)Y)(\text{Ad}(g)X).$$

By Proposition 3.1 (with $A = U(\mathfrak{g}^{\mathbb{C}})$), $\text{Ad}(g)$ extends to a homomorphism of $U(\mathfrak{g}^{\mathbb{C}})$ into itself carrying 1 to 1, and moreover

$$\text{Ad}(g_1 g_2) = \text{Ad}(g_1)\text{Ad}(g_2). \tag{3.6}$$

Therefore each $\text{Ad}(g)$ is an automorphism of $U(\mathfrak{g}^{\mathbb{C}})$. Because $\text{Ad}(g)$ leaves $U^1(\mathfrak{g}^{\mathbb{C}})$ stable, it leaves each $U^N(\mathfrak{g}^{\mathbb{C}})$ stable. Its smoothness on $U^1(\mathfrak{g}^{\mathbb{C}})$ implies its smoothness on $U^N(\mathfrak{g}^{\mathbb{C}})$. Thus Ad provides for all N a consistently defined family of smooth representations of G on $U^N(\mathfrak{g}^{\mathbb{C}})$.

Proposition 3.7

(a) The differential at 1 of Ad on $U^N(\mathfrak{g}^{\mathbb{C}})$ is ad.

(b) On each $U^N(\mathfrak{g}^{\mathbb{C}})$, $\mathrm{Ad}(\exp X) = e^{\mathrm{ad}\, X}$ for all X in $\mathfrak{g}$.

Proof. For (a) let $D = X_1^{k_1} \cdots X_n^{k_n}$ be a basis vector of $U^N(\mathfrak{g}^{\mathbb{C}})$. For X in $\mathfrak{g}$ we have

$$\mathrm{Ad}(\exp tX)D = (\mathrm{Ad}(\exp tX)X_1)^{k_1} \cdots (\mathrm{Ad}(\exp tX)X_n)^{k_n}$$

since each $\mathrm{Ad}(g)$ is an automorphism of $U(\mathfrak{g}^{\mathbb{C}})$. Differentiating both sides with respect to t and applying the product rule for differentiation, we obtain at $t = 0$

$$\begin{aligned} &\frac{d}{dt}\mathrm{Ad}(\exp tX)D\bigg|_{t=0} \\ &\quad = \sum_{i=1}^{n}\sum_{j=1}^{k_i} X_1^{k_1}\cdots X_{i-1}^{k_{i-1}}X_i^{j-1}\left(\frac{d}{dt}\mathrm{Ad}(\exp tX)X_i\right)_{t=0} X_i^{k_i-j}X_{i+1}^{k_{i+1}}\cdots X_n^{k_n} \\ &\quad = (\mathrm{ad}\, X)D. \end{aligned}$$

This proves (a). For (b), fix X and consider $\mathrm{Ad}(\exp tX)$ and $e^{\mathrm{ad}\, tX}$ as one-parameter groups in $GL(U^N(\mathfrak{g}^{\mathbb{C}}))$. These have the same differential at 0, namely ad X, and so are equal. Now take $t = 1$ to obtain (b).

The center $Z(\mathfrak{g}^{\mathbb{C}})$ of $U(\mathfrak{g}^{\mathbb{C}})$ will play an important role in the theory. We can characterize elements of $Z(\mathfrak{g}^{\mathbb{C}})$ in several ways.

Proposition 3.8. The following conditions on an element D of $U(\mathfrak{g}^{\mathbb{C}})$ are equivalent:

(a) D is in $Z(\mathfrak{g}^{\mathbb{C}})$

(b) $DX = XD$ for all X in $\mathfrak{g}$

(c) $e^{\mathrm{ad}\, X}D = D$ for all X in $\mathfrak{g}$

(d) $\mathrm{Ad}(g)D = D$ for all g in G.

Proof. First (a) implies (b) trivially, and (b) implies (a) since $\mathfrak{g}$ generates $U(\mathfrak{g}^{\mathbb{C}})$. If (b) holds, then $(\mathrm{ad}\, X)D = 0$, and (c) follows by summing the series for the exponential. Conversely if (c) holds, then replace X by tX in it and differentiate to obtain (b). Finally (c) follows from (d) by taking $g = \exp X$ and applying Proposition 3.7b, while (d) follows from (c) by Proposition 3.7b and (3.6).

§3. C^∞ Vectors

Let Φ be a representation of a Lie group G on a Hilbert space V, let $\mathfrak{g}$ be the Lie algebra of G, and let $U(\mathfrak{g}^{\mathbb{C}})$ be the universal enveloping algebra of $\mathfrak{g}^{\mathbb{C}}$. In this section we shall show how to associate to Φ a "representation" φ of $U(\mathfrak{g}^{\mathbb{C}})$ on the subspace of "C^∞ vectors" of V, and we shall establish some elementary properties of this representation. Here φ is to be an associative algebra homomorphism of $U(\mathfrak{g}^{\mathbb{C}})$ into

$$\mathrm{End}_{\mathbb{C}}(\text{space of } C^\infty \text{ vectors})$$

and is to carry 1 to 1. In particular, $\varphi|_{\mathfrak{g}}$ will be a representation of the Lie algebra $\mathfrak{g}$ on the space of C^∞ vectors.

We recall the definition of a C^∞ mapping of G into the possibly infinite-dimensional V. A function f from an open set S in $\mathbb{R}^n$ into a linear topological space E is **differentiable** at x_0 in S if there is a (necessarily unique) linear map $f'(x_0): \mathbb{R}^n \to E$ such that

$$\lim_{x \to x_0} \frac{f(x) - f(x_0) - f'(x_0)(x - x_0)}{|x - x_0|} = 0, \tag{3.7}$$

where $|\cdot|$ is any norm on $\mathbb{R}^n$. We call $f'(x_0)$ the **differential** of f at x_0. Now $\mathrm{End}(\mathbb{R}^n, E)$ is a linear topological space in a canonical way, since $\mathbb{R}^n$ is finite-dimensional, and if f is differentiable at each point of S, then $x \to f'(x)$ is a map of S into $\mathrm{End}(\mathbb{R}^n, E)$. We say that f is of class C^1 if $x \to f'(x)$ is continuous, of class C^2 if $x \to f'(x)$ is of class C^1, and so on. We say f is of **class** C^∞ if f is of class C^k for all $k \geq 1$.

Just as in the case of functions with finite-dimensional range, one proves that

(i) f is of class C^k if and only if all its partial derivatives up to and including order k exist and are continuous,

(ii) the chain rule holds for a composition $f \circ h$, where h is a mapping from an open subset of a Euclidean space into S, and

(iii) we can carry the definitions over to a C^∞ manifold as domain in the obvious way.

We say that v in V is a C^∞ **vector** for the representation Φ if $g \to \Phi(g)v$ is of class C^∞ on G. Evidently the C^∞ vectors form a vector subspace of V, and we denote this subspace by $C^\infty(\Phi)$.

Let v be in $C^\infty(\Phi)$, and let

$$f(X) = \Phi(\exp X)v$$

for X in $\mathfrak{g}$. Then f is of class C^∞. Put

$$\varphi(X)v = f'(0)(X). \tag{3.8a}$$

By the chain rule,

$$\lim_{t \to 0} \frac{\Phi(\exp tX)v - v}{t} = \varphi(X)v. \tag{3.8b}$$

In particular, the limit on the left side of (3.8b) exists for all X in $\mathfrak{g}$, and (3.8) shows that $\varphi(X)$ is a linear mapping of $C^\infty(\Phi)$ into V that depends linearly on X.

Proposition 3.9. Let Φ be a representation of the Lie group G on a Hilbert space V. For X in $\mathfrak{g}$, define a linear mapping from $C^\infty(\Phi)$ into V by

$$\varphi(X)v = \lim_{t\to 0} \frac{\Phi(\exp tX)v - v}{t}. \tag{3.9}$$

Then each $\varphi(X)$ leaves $C^\infty(\Phi)$ stable, and φ is a representation of $\mathfrak{g}$ on $C^\infty(\Phi)$. Consequently φ extends to a representation of $U(\mathfrak{g}^{\mathbb{C}})$ on $C^\infty(\Phi)$ with $\varphi(1) = 1$.

Notation. The representation of $U(\mathfrak{g}^{\mathbb{C}})$ will be called φ also.

Proof. Let v be in $C^\infty(\Phi)$, and let f be the map $g \to \Phi(g)v$. Applying $\Phi(g)$ to both sides of (3.9), we see that

$$\tilde{X}f(g) = \lim_{t\to 0} t^{-1}\{\Phi(g \exp tX)v - \Phi(g)v\} = \Phi(g)\varphi(X)v. \tag{3.10}$$

By assumption $g \to f(g)$ is of class C^∞, and it follows that $g \to \tilde{X}f(g)$ is of class C^∞. Then (3.10) says that $\varphi(X)v$ is in $C^\infty(\Phi)$.

Hence $\varphi(X)$ leaves $C^\infty(\Phi)$ stable. Consequently $X \to \varphi(X)$ extends to an algebra homomorphism, also denoted φ, of $T(\mathfrak{g})$ into $C^\infty(\Phi)$ such that $\varphi(1) = 1$. Thus Proposition 3.1 will finish the proof if we can show that

$$\varphi(X)\varphi(Y) - \varphi(Y)\varphi(X) - \varphi[X, Y] = 0 \tag{3.11}$$

for all X and Y in $\mathfrak{g}$.

Let

$$c(t) = \exp((-\operatorname{sgn} t)|t|^{1/2}X)\exp(-|t|^{1/2}Y)\exp((\operatorname{sgn} t)|t|^{1/2}X)\exp(|t|^{1/2}Y);$$

then $c(t)$ is a class C^1 curve in G with $c'(0) = [X, Y]$, by (A.105). Let v be in $C^\infty(\Phi)$. By the chain rule, the map $t \to \Phi(c(t))v$ has differential $f'(1)([X, Y])$ at $t = 0$, where $f(g) = \Phi(g)v$. That is,

$$\lim_{t\to 0} \frac{\Phi(c(t))v - v}{t} = \varphi([X, Y])v.$$

Hence
$$\lim_{t\to 0} \frac{\Phi(c(t^2))v - v}{t^2} = \varphi([X, Y])v.$$

This equation and the strong continuity of Φ imply that

$$\begin{aligned}\lim_{t\to 0} &\frac{\Phi(\exp tX \exp tY)v - \Phi(\exp tY \exp tX)v}{t^2} \\ &= \lim_{t\to 0} \Phi(\exp tY)\Phi(\exp tX)\frac{\Phi(c(t^2))v - v}{t^2} \\ &= \varphi([X, Y])v.\end{aligned} \tag{3.12}$$

Meanwhile the mapping $(s, t) \to \Phi(\exp sX \exp tY)v$ is of class C^∞, being the composition of the C^∞ maps

$$(s, t) \to (\exp sX, \exp tY), \qquad (g_1, g_2) \to g_1 g_2, \qquad \text{and} \qquad g \to \Phi(g)v.$$

Consequently for each w in V the map

$$(s, t) \to \langle \Phi(\exp sX \exp tY)v, w \rangle$$

is of class C^∞. But then $\dfrac{\partial}{\partial s}\dfrac{\partial}{\partial t}$ of this map at $s = t = 0$ can be computed as a double limit, instead of just an iterated limit, and the iterated limit is equal to the diagonal limit. Thus

$$\begin{aligned}
&\langle \varphi(X)\varphi(Y)v, w \rangle \\
&\quad = \frac{\partial}{\partial s}\frac{\partial}{\partial t} \langle \Phi(\exp sX \exp tY)v, w \rangle_{s=t=0} \\
&\quad = \lim_{t \to 0} \langle t^{-2}\{\Phi(\exp tX \exp tY) - \Phi(\exp tX) - \Phi(\exp tY) + I\}v, w \rangle.
\end{aligned}$$

Interchanging X and Y and subtracting, we find

$$\begin{aligned}
&\langle (\varphi(X)\varphi(Y) - \varphi(Y)\varphi(X))v, w \rangle \\
&\qquad = \lim_{t \to 0} \langle t^{-2}\{\Phi(\exp tX \exp tY) - \Phi(\exp tY \exp tX)\}v, w \rangle. \qquad (3.13)
\end{aligned}$$

Comparison of (3.12) and (3.13) gives (3.11). This completes the proof of the proposition.

Proposition 3.10. Let Φ be a unitary representation of G on a Hilbert space V. For the associated representation φ of $\mathfrak{g}$ on $C^\infty(\Phi)$, each $\varphi(X)$ for X in $\mathfrak{g}$ is skew-Hermitian.

Proof. If v and w are in $C^\infty(\Phi)$, then Φ unitary implies

$$\left\langle v, i\frac{\Phi(\exp tX)w - w}{t} \right\rangle = \left\langle i\frac{\Phi(\exp - tX)v - v}{-t}, w \right\rangle.$$

In the limit as $t \to 0$, (3.9) gives $\langle v, i\varphi(X)w \rangle = \langle i\varphi(X)v, w \rangle$. Hence $\varphi(X)$ is skew-Hermitian.

Now we show that $C^\infty(\Phi)$ is an invariant subspace (though not usually closed) under Φ.

Proposition 3.11. If Φ is a representation of G on a Hilbert space V, then $C^\infty(\Phi)$ is stable under each $\Phi(g)$. If φ denotes the corresponding representation of $U(\mathfrak{g}^{\mathbb{C}})$ on $C^\infty(\Phi)$, then

$$\Phi(g)\varphi(D)\Phi(g)^{-1} = \varphi(\mathrm{Ad}(g)D) \qquad (3.14)$$

for all D in $U(\mathfrak{g}^{\mathbb{C}})$ and g in G.

Proof. The map $g \to \Phi(g)\Phi(g_0)v$ is the composition of $g \to gg_0$ and $g \to \Phi(g)v$. Hence $C^\infty(\Phi)$ is stable under $\Phi(g_0)$. Now let X be in $\mathfrak{g}$, g be in G, and v be in $C^\infty(\Phi)$. Then

$$\frac{\Phi(\exp tX) - I}{t}\Phi(g)^{-1}v = \Phi(g)^{-1}\left(\frac{\Phi(g(\exp tX)g^{-1}) - I}{t}\right)v$$
$$= \Phi(g)^{-1}\left(\frac{\Phi(\exp \mathrm{Ad}(g)tX) - I}{t}\right)v.$$

Passing to the limit as $t \to 0$, we obtain $\varphi(X)\Phi(g)^{-1}v = \Phi(g)^{-1}\varphi(\mathrm{Ad}(g)X)v$. That is, $\Phi(g)\varphi(X)\Phi(g)^{-1} = \varphi(\mathrm{Ad}(g)X)$. The result for D follows from this equation and Proposition 3.1.

Corollary 3.12. If Φ is a representation of G on a Hilbert space V and if D is in the center $Z(\mathfrak{g}^{\mathbb{C}})$ of $U(\mathfrak{g}^{\mathbb{C}})$, then $\varphi(D)$ commutes with $\Phi(g)$ for all g in G.

Proof. Apply Propositions 3.8 and 3.11.

§4. Gårding Subspace

We are going to prove in this section that the C^∞ subspace is dense. We do so by showing that a manageable subspace of the C^∞ subspace, known as the Gårding subspace, is itself dense.

Lemma 3.13. If Φ is a representation of the Lie group G on a Hilbert space V and if $S \subseteq V$ is a vector subspace that is stable under all limit operators

$$\bar{\varphi}(X)v = \lim_{t\to 0}\frac{\Phi(\exp tX)v - v}{t}, \tag{3.15}$$

then $S \subseteq C^\infty(\Phi)$.

Proof. For v in S let f_v be the map $f_v(g) = \Phi(g)v$. By assumption all first partial derivatives of f_v exist at 1, the partial in the direction X being $\bar{\varphi}(X)v$. If $g\exp X \to X$ is a local coordinate system near g in G, then the first partial of f_v in the direction X exists at g and is $\Phi(g)\bar{\varphi}(X)v$, because we can simply apply $\Phi(g)$ to both sides of (3.15). For each X the map $g \to \Phi(g)\bar{\varphi}(X)v$ is continuous, and thus f_v has continuous first partials everywhere and must be of class C^1. The formula for the differential is then $f_v'(g)(X) = \Phi(g)\bar{\varphi}(X)v$.

Proceeding inductively, suppose each f_v for v in S is known to be of class C^k, $k \geq 1$. If v is given, then $\bar{\varphi}(X)v$ is in S and $g \to \Phi(g)\bar{\varphi}(X)v$ is of class C^k, by assumption. That is, $g \to f_v'(g)(X)$ is of class C^k for each X. Thus f_v has continuous partials of order $k+1$ and must be of class C^{k+1}. The induction is completed, and each f_v is of class C^∞.

Let Φ be a representation of the Lie group G on a Hilbert space V. The **Gårding subspace** for Φ is the vector subspace of V spanned by all vectors of the form

$$\Phi(f)v = \int_G f(g)\Phi(g)v\,dg$$

for v in V and f in $C^\infty_{\text{com}}(G)$; here dg is a left-invariant Haar measure on G.

Proposition 3.14. If Φ is a representation of the Lie group G on a Hilbert space V, then the Gårding subspace is stable under the limit operators $\bar{\varphi}(X)$ of (3.15), with $\bar{\varphi}(X)$ given by

$$\varphi(X)(\Phi(f)v) = -\Phi(X(f)v), \tag{3.16}$$

where $X(f)$ refers to the action of the **right-invariant vector field** on G whose value at 1 is X: $Xf(x) = \frac{d}{dt} f((\exp tX)^{-1}x)\Big|_{t=0}$. Consequently the Gårding subspace is contained in $C^\infty(\Phi)$.

Proof. If $t \neq 0$, a change of variables gives

$$\begin{aligned}\left(\frac{\Phi(\exp tX) - I}{t}\right)\Phi(f)v &= t^{-1}\int_G f(g)\{\Phi(\exp tX)\Phi(g) - \Phi(g)\}v\,dg \\ &= -\int_G \frac{f((\exp - tX)g) - f(g)}{-t}\Phi(g)v\,dg.\end{aligned}$$

As $t \to 0$, we have dominated convergence on the right side. Thus (3.16) follows. Hence each $\bar{\varphi}(X)$ leaves the Gårding subspace stable, and Lemma 3.13 shows the Gårding subspace is contained in $C^\infty(\Phi)$.

Theorem 3.15. If Φ is a representation of the Lie group G on a Hilbert space V, then the Gårding subspace is dense in V. Consequently $C^\infty(\Phi)$ is dense in V.

Proof. Let v in V and $\varepsilon > 0$ be given. Since Φ is strongly continuous, the set

$$S = \{g \in G \mid |\Phi(g)v - v| < \varepsilon\}$$

is open. Thus we can choose a function $f \geq 0$ in $C^\infty_{\text{com}}(G)$ supported in S such that $\int_G f(g)\,dg = 1$. Then

$$\begin{aligned}|\Phi(f)v - v| &= \left|\int_G f(g)[\Phi(g)v - v]\,dg\right| \\ &\leq \int_S f(g)|\Phi(g)v - v|\,dg \\ &\leq \varepsilon \int_G f(g)\,dg = \varepsilon.\end{aligned}$$

Thus the Gårding subspace is dense. By Proposition 3.14, $C^\infty(\Phi)$ is dense.

Corollary 3.16. If Φ is a representation of the Lie group G on a finite-dimensional space V, then Φ is smooth as a mapping of G into $\mathrm{GL}(V)$.

Proof. If follows from Theorem 3.15 that $g \to \Phi(g)v$ is of class C^∞ for all v in V. Hence $g \to \Phi(g)$ is smooth.

§5. Problems

1. Let $\mathfrak{g}$ be a real finite-dimensional Lie algebra, and let $\mathfrak{h}$ be a Lie subalgebra. Prove that the complex associative subalgebra of $U(\mathfrak{g}^{\mathbb{C}})$ generated by 1 and $\mathfrak{h}$ is isomorphic with $U(\mathfrak{h}^{\mathbb{C}})$.
2. Let the real finite-dimensional Lie algebra $\mathfrak{g}$ be a direct sum of subspaces $\mathfrak{a}$ and $\mathfrak{b}$ that happen to be subalgebras: $\mathfrak{g} = \mathfrak{a} \oplus \mathfrak{b}$. Prove that the multiplication mapping yields a canonical isomorphism of the vector space $U(\mathfrak{a}^{\mathbb{C}}) \otimes_{\mathbb{C}} U(\mathfrak{b}^{\mathbb{C}})$ onto $U(\mathfrak{g}^{\mathbb{C}})$.
3. Let $\mathfrak{g}$ be $\mathfrak{sl}(2, \mathbb{R})$, and regard $\Omega = \frac{1}{2}h^2 + ef + fe$ as a member of $U(\mathfrak{g}^{\mathbb{C}})$.
   (a) Calculate $\varphi_n(\Omega)$ directly from the formulas of Theorem 2.4, where φ_n is the differential of the representation Φ_n of (2.1).
   (b) Use Proposition 3.8 to show Ω is in $Z(\mathfrak{g}^{\mathbb{C}})$.
   (c) How can $\varphi_n(\Omega)$ be calculated more easily than in (a) if account is taken of (b) and of the irreducibility of φ_n?
   (d) Characterize the (possibly reducible) finite-dimensional representations φ of $\mathfrak{g}$ with $\varphi(\mathfrak{g})v = 0$ for some $v \neq 0$ as the representations for which $\varphi(\Omega)$ is singular.
4. Prove that $U(\mathfrak{g}^{\mathbb{C}})$ has no zero divisors.

Problems 5 and 6 establish properties of the left regular representation Φ of $\mathbb{R}$ on $L^2(\mathbb{R})$.

5. (a) Show that every member of $C^\infty_{\mathrm{com}}(\mathbb{R})$ is a C^∞ vector and that the Lie algebra acts on $C^\infty_{\mathrm{com}}(\mathbb{R})$ by imaginary multiples of the first derivative operator.
   (b) If f is a C^∞ function on $\mathbb{R}$ that is a C^∞ vector for Φ, show that all derivatives $f^{(n)}(t)$ are in $L^2(\mathbb{R})$.
   (c) Show that $(1 + t^2)^{-1}$ is a C^∞ vector for Φ.
   (d) Show that the Gårding subspace for Φ is $C^\infty_{\mathrm{com}}(\mathbb{R}) * L^2(\mathbb{R})$.
6. Let V be the subspace of members of $C^\infty_{\mathrm{com}}(\mathbb{R})$ with support in $[0, 1]$, and regard V as a subspace of C^∞ vectors for Φ. Show that V is invariant under the Lie algebra action but that the closure of V in $L^2(\mathbb{R})$ is not invariant under the group action.

Problems 7 and 8 give further relations between $U(\mathfrak{g}^{\mathbb{C}})$ and left-invariant derivatives.

7. Let G be an analytic group with Lie algebra $\mathfrak{g}$, and fix a basis $X_1, \ldots, X_n$ of $\mathfrak{g}$. Compute in $U(\mathfrak{g}^{\mathbb{C}})$ that

$$(t_1 X_1 + \ldots + t_n X_n)^r = \sum_{|m|=r} \frac{t_1^{m_1} \cdots t_n^{m_n}}{m_1! \cdots m_n!} \sum_{\sigma \in S_r} X_{i_{\sigma(1)}} \cdots X_{i_{\sigma(r)}} \quad (3.17)$$

where S_r denotes the symmetric group on r letters and where the indices on the right side are given by $i_1 = i_2 = \cdots = i_{m_1} = 1$, $i_{m_1+1} = i_{m_1+2} = \cdots = i_{m_1+m_2} = 2$, etc. With $X(t)$ defined as $t_1 X_1 + \ldots + t_n X_n$ and with $X(m)$ defined as the coefficient of $t_1^{m_1} \cdots t_n^{m_n}$ on the right side of (3.17), (3.17) becomes

$$(X(t))^r = \sum_{|m|=r} t^m X(m).$$

8. Using Taylor's Theorem (A.98), prove that

$$(X(m)f)(x) = \frac{1}{m_1! \cdots m_n!} \frac{\partial^{|m|}}{\partial t_1^{m_1} \cdots \partial t_n^{m_n}} f(x \exp X(t))\bigg|_{t=0}.$$

Problems 9 to 16 deal with symmetric algebras and their relationship to universal enveloping algebras. The key to the relationship is the symmetrization mapping.

9. Let V be a finite-dimensional complex vector space. The **symmetric algebra** $\mathscr{S}(V)$ of V is the quotient $T(V)/I$, where $T(V)$ is the tensor algebra and I is the two-sided ideal generated by all $X \otimes Y - Y \otimes X$ for X and Y in V. Establish the following universal mapping property of $\mathscr{S}(V)$: Any linear map of V into a complex commutative associative algebra A with identity extends uniquely to an algebra homomorphism of $\mathscr{S}(V)$ into A such that 1 maps to 1.

10. Prove that the symmetric algebra $\mathscr{S}(V)$ is the direct sum of subspaces $\mathscr{S}^r(V)$, where $\mathscr{S}^r(V)$ is the image of $\bigotimes^r V$ under the quotient mapping, and that the product $\mathscr{S}^p(V)\mathscr{S}^q(V)$ is contained in $\mathscr{S}^{p+q}(V)$.

11. Prove that $\mathscr{S}^r(V)$ has the following universal property: If U is a complex vector space and if $\varphi: V \times \ldots \times V \to U$ is a symmetric r-linear map, then there is a unique linear map $\tilde{\varphi}: \mathscr{S}^p(V) \to U$ such that

$$\tilde{\varphi}(v_1 \cdots v_r) = \varphi(v_1, \ldots, v_r)$$

for all r-tuples of members v_i of V.

12. Let $\mathfrak{g}$ be a finite-dimensional real Lie algebra. Applying the results of Problems 10 and 11 to the symmetric r-linear map $\lambda_r: \mathfrak{g}^{\mathbb{C}} \times \ldots \times \mathfrak{g}^{\mathbb{C}} \to U(\mathfrak{g}^{\mathbb{C}})$ given by

$$\lambda_r(X_1, \ldots, X_r) = \frac{1}{r!} \sum_{\sigma \in S_r} X_{\sigma(1)} \cdots X_{\sigma(r)}, \qquad (S_r = \text{symmetric group})$$

construct a canonical $\mathbb{C}$-linear map $\lambda:\mathscr{S}(\mathfrak{g}^{\mathbb{C}}) \to U(\mathfrak{g}^{\mathbb{C}})$ such that

$$\lambda(X_1 \cdots X_r) = \frac{1}{r!} \sum_{\sigma \in S_r} X_{\sigma(1)} \cdots X_{\sigma(r)}.$$

Prove that λ is a linear isomorphism of $\mathscr{S}(\mathfrak{g}^{\mathbb{C}})$ onto $U(\mathfrak{g}^{\mathbb{C}})$. [The map λ is called the **symmetrization mapping.**]

13. Let $X_1, \ldots, X_n$ be a basis of $\mathfrak{g}$. For an n-tuple $m = (m_1, \ldots, m_n)$, define $X(m)$ as in Problem 7. Prove that

$$\lambda(X_1^{m_1} \cdots X_n^{m_n}) = \frac{m_1! \cdots m_n!}{(|m|)!} X(m).$$

[Remark: Therefore Problem 8 gives an analytic interpretation of the symmetrization mapping.]

14. Let G be an analytic group with Lie algebra $\mathfrak{g}$. Fix a basis $X_1, \ldots, X_n$ of $\mathfrak{g}$, so that $\{X_1^{k_1} \cdots X_n^{k_n}\}$ is a basis over $\mathbb{C}$ of $\mathscr{S}(\mathfrak{g}^{\mathbb{C}})$. Reading off the coefficients of elements of $\mathscr{S}(\mathfrak{g}^{\mathbb{C}})$ relative to this basis allows us to identify $\mathscr{S}(\mathfrak{g}^{\mathbb{C}})$ with the polynomial algebra $\mathbb{C}[t_1, \ldots, t_n]$. With this identification in force, prove that

$$[\lambda(P)(f)](x) = P\left(\frac{\partial}{\partial t_1}, \ldots, \frac{\partial}{\partial t_n}\right) f(x \exp(t_1 X_1 + \ldots + t_n X_n))\bigg|_{t=0}$$

for all $x \in G$, $f \in C^\infty(G)$, and $P \in \mathbb{C}[t_1, \ldots, t_n]$. [Hint: Let $X(t)$ be as in Problem 6. Form a finite Taylor series expansion (with remainder) for $f(x \exp X(t))$ by means of (A.98), and use the results of Problems 7 and 13.]

15. For x in G, use Problem 9 to extend $\operatorname{Ad}(x)$ from $\mathfrak{g}^{\mathbb{C}}$ to an algebra automorphism of $\mathscr{S}(\mathfrak{g}^{\mathbb{C}})$. Prove that

$$\lambda \operatorname{Ad}(x) = \operatorname{Ad}(x)\lambda$$

as mappings of $\mathscr{S}(\mathfrak{g}^{\mathbb{C}})$ into $U(\mathfrak{g}^{\mathbb{C}})$.

16. Prove that the restriction of λ to the $\operatorname{Ad}(G)$-invariant members of $\mathscr{S}(\mathfrak{g}^{\mathbb{C}})$ is a linear isomorphism onto the center $Z(\mathfrak{g}^{\mathbb{C}})$ of $U(\mathfrak{g}^{\mathbb{C}})$.

CHAPTER IV

Representations of Compact Lie Groups

§1. Examples of Root Space Decompositions

The Cartan-Weyl theory analyzes representations of a compact connected Lie group G by decomposing them under an abelian subgroup in such a way that much information about the action by the full Lie algebra $\mathfrak{g}$ is transparent. The theory applies to any representation and in particular to Ad as a representation of G on $\mathfrak{g}^{\mathbb{C}}$, and the information that it gives about Ad is relevant for the general case.

In this section we shall write down explicitly this decomposition of the Ad representation, or of its differential ad, for the classical compact groups. For an exercise we shall determine which classical compact groups have Lie algebras that are simple.

Example 1. $G = \mathrm{SU}(n)$, $\mathfrak{g} = \mathfrak{su}(n)$, $\mathfrak{g}^{\mathbb{C}} = \mathfrak{sl}(n, \mathbb{C})$.

Let
$$\mathfrak{h} = \text{diagonal matrices in } \mathfrak{g}$$
$$\mathfrak{h}_{\mathbb{R}} = i\mathfrak{h}$$
$$\mathfrak{h}^{\mathbb{C}} = \mathfrak{h} \oplus i\mathfrak{h} = \mathfrak{h}_{\mathbb{R}} \oplus i\mathfrak{h}_{\mathbb{R}}.$$

Define a matrix E_{ij} to be 1 in the $(i, j)^{\text{th}}$ place and 0 elsewhere. Define a linear functional e_i in the dual space $(\mathfrak{h}^{\mathbb{C}})'$ by

$$e_i\begin{pmatrix} h_1 & & \\ & \ddots & \\ & & h_n \end{pmatrix} = h_i.$$

For each H in $\mathfrak{h}^{\mathbb{C}}$, ad H is diagonalized by the basis of $\mathfrak{g}^{\mathbb{C}}$ consisting of members of $\mathfrak{h}^{\mathbb{C}}$ and the E_{ij} for $i \neq j$. We have

$$(\operatorname{ad} H)E_{ij} = [H, E_{ij}] = (e_i(H) - e_j(H))E_{ij}.$$

So E_{ij} is a simultaneous eigenvector for all ad H, with eigenvalue $e_i(H) - e_j(H)$. In its dependence on H, the eigenvalue is linear. So the eigenvalue is a linear functional on $\mathfrak{h}^{\mathbb{C}}$, namely $e_i - e_j$. The $(e_i - e_j)$'s, for $i \neq j$, are called **roots.** The set of roots is denoted Δ. We have

$$\mathfrak{g}^{\mathbb{C}} = \mathfrak{h}^{\mathbb{C}} \oplus \sum_{i \neq j} \mathbb{C}E_{ij},$$

which we can rewrite as

$$\mathfrak{g}^{\mathbb{C}} = \mathfrak{h}^{\mathbb{C}} \oplus \sum_{i \neq j} \mathfrak{g}_{e_i - e_j}, \tag{4.1}$$

where

$$\mathfrak{g}_{e_i - e_j} = \{X \in \mathfrak{g}^{\mathbb{C}} \mid (\operatorname{ad} H)X = (e_i - e_j)(H)X \qquad \text{for all } H \in \mathfrak{h}^{\mathbb{C}}\}.$$

The decomposition (4.1) is called a **root-space decomposition.** Notice that the roots span $(\mathfrak{h}^{\mathbb{C}})'$ over $\mathbb{C}$.

The bracket relations are easy, relative to (4.1). If α and β are roots, we can compute $[E_{ij}, E_{i'j'}]$ and see that

$$[\mathfrak{g}_\alpha, \mathfrak{g}_\beta] \begin{cases} = \mathfrak{g}_{\alpha+\beta} & \text{if } \alpha + \beta \text{ is a root} \\ = 0 & \text{if } \alpha + \beta \text{ is not a root or } 0 \\ \subseteq \mathfrak{h}^{\mathbb{C}} & \text{if } \alpha + \beta = 0. \end{cases} \tag{4.2}$$

In the last case, the exact formula is

$$[E_{ij}, E_{ji}] = E_{ii} - E_{jj} \in \mathfrak{h}^{\mathbb{C}}.$$

All the roots are real on $\mathfrak{h}_{\mathbb{R}}$. We introduce an ordering on the roots as follows. The elements $H_i = E_{ii} - E_{nn}$, $1 \le i \le n-1$, form a basis of $\mathfrak{h}_{\mathbb{R}}$. If f is in $(\mathfrak{h}^{\mathbb{C}})'$, we say f is **positive** if f is real on $\mathfrak{h}_{\mathbb{R}}$ and if

$$\begin{aligned} & f(H_1) > 0 \\ \textit{or} \quad & f(H_1) = 0 \quad \text{and} \quad f(H_2) > 0 \\ \textit{or} \quad & f(H_1) = f(H_2) = 0 \quad \text{and} \quad f(H_3) > 0 \\ \textit{or} \ldots \textit{or}\ & f(H_1) = \ldots = f(H_{n-2}) = 0 \quad \text{and} \quad f(H_{n-1}) > 0. \end{aligned}$$

If f is not 0 but is real on each H_i, then exactly one of f and $-f$ is positive. The sum of positive elements is positive. Let us say $f > g$ if $f - g$ is positive. Then

$$\begin{aligned} e_1 - e_n > e_2 - e_n > \ldots > e_{n-1} - e_n > e_1 - e_{n-1} > e_1 - e_{n-2} > \ldots \\ > e_1 - e_2 > e_2 - e_{n-1} > \ldots > e_2 - e_3 > e_3 - e_{n-1} > \ldots \\ > e_3 - e_4 > \ldots > e_{n-2} - e_{n-1} > 0, \end{aligned}$$

and afterward we have the negatives.

We shall prove that $\mathfrak{g}^{\mathbb{C}}$ is simple over $\mathbb{C}$ (and hence $\mathfrak{g}$ is simple over $\mathbb{R}$) for $n \ge 2$. Let $\mathfrak{a} \subseteq \mathfrak{g}^{\mathbb{C}}$ be an ideal, and first suppose $\mathfrak{a} \subseteq \mathfrak{h}^{\mathbb{C}}$. Let $H \neq 0$ be in $\mathfrak{a}$. Since the roots span $(\mathfrak{h}^{\mathbb{C}})'$, we can find a root α with $\alpha(H) \neq 0$. If X is in $\mathfrak{g}_\alpha$ and $X \neq 0$, then

$$\alpha(H)X = [H, X] \in [\mathfrak{a}, \mathfrak{g}^{\mathbb{C}}] \subseteq \mathfrak{a} \subseteq \mathfrak{h}^{\mathbb{C}},$$

and so X is in $\mathfrak{h}^{\mathbb{C}}$, contradiction. Hence $\mathfrak{a} \subseteq \mathfrak{h}^{\mathbb{C}}$ implies $\mathfrak{a} = 0$.

Now suppose $\mathfrak{a} \not\subseteq \mathfrak{h}^{\mathbb{C}}$. Let $X = H + \sum X_\alpha$ be in $\mathfrak{a}$ with each X_α in $\mathfrak{g}_\alpha$ and some $X_\alpha \neq 0$. For the moment, assume there is some $\alpha < 0$ with $X_\alpha \neq 0$, and let β be the smallest such α. Say $X_\beta = cE_{ij}$ with $i > j$, $c \neq 0$. Form

$$[E_{1i}, [X, E_{jn}]]. \tag{4.3}$$

We claim (4.3) is a nonzero multiple of E_{1n}. {In fact, if $i < n$, then $[E_{ij}, E_{jn}] = aE_{in}$ with $a \neq 0$. Next, $[E_{1i}, E_{in}] = bE_{1n}$ with $b \neq 0$. So (4.3) has a nonzero component in $\mathfrak{g}_{e_1 - e_n}$. The other components of (4.3) must correspond to larger roots than $e_1 - e_n$ if they are nonzero, but $e_1 - e_n$ is the largest root. Hence the claim follows if $i < n$. If $i = n$, then (4.3) is

$$= [E_{1n}, [cE_{nj} + \ldots, E_{jn}]] = c[H_j, E_{1n}] + \ldots = -cE_{1n}.\}$$

In any case we conclude E_{1n} is in $\mathfrak{a}$. Now

$$E_{kl} = c'[E_{k1}, [E_{1n}, E_{nl}]]$$

(with obvious changes if $k = 1$ or $l = n$) shows E_{kl} is in $\mathfrak{a}$, and

$$[E_{kl}, E_{lk}] = E_{kk} - E_{ll}$$

shows a spanning set for $\mathfrak{h}^{\mathbb{C}}$ is in $\mathfrak{a}$. Hence $\mathfrak{a} = \mathfrak{g}^{\mathbb{C}}$.

Thus $\mathfrak{a} \not\subseteq \mathfrak{h}^{\mathbb{C}}$ implies $\mathfrak{a} = \mathfrak{g}^{\mathbb{C}}$ if there is some $\alpha < 0$ with $X_\alpha \neq 0$ above. Similarly if there is some $\alpha > 0$ with $X_\alpha \neq 0$, let β be the largest such α, say $\alpha = e_i - e_j$ with $i < j$. Form $[E_{ni}, [X, E_{j1}]]$ and argue with E_{n1} similarly to get $\mathfrak{a} = \mathfrak{g}^{\mathbb{C}}$. Thus $\mathfrak{g}^{\mathbb{C}}$ is simple over $\mathbb{C}$. This completes the first example.

We can abstract these properties. The complexification $\mathfrak{g}^{\mathbb{C}}$ will be simple whenever we can arrange that

(1) $\mathfrak{h}$ is an abelian subalgebra of $\mathfrak{g}$ such that $\mathfrak{g}^{\mathbb{C}}$ has a simultaneous eigenspace decomposition relative to ad $\mathfrak{h}^{\mathbb{C}}$ and
  (a) the 0 eigenspace is $\mathfrak{h}^{\mathbb{C}}$,
  (b) the other eigenspaces are one-dimensional,
  (c) with the set Δ of roots defined as before, (4.2) holds,
  (d) the roots are all real on $\mathfrak{h}_{\mathbb{R}} = i\mathfrak{h}$.
(2) the roots span $(\mathfrak{h}^{\mathbb{C}})'$. If α is a root, so is $-\alpha$.
(3) $\sum_{\alpha \in \Delta} [\mathfrak{g}_\alpha, \mathfrak{g}_{-\alpha}] = \mathfrak{h}^{\mathbb{C}}$.
(4) each root $\beta < 0$ relative to an ordering of the kind for $\mathfrak{sl}(n, \mathbb{C})$ has the following property: There exists a sequence of roots $\alpha_1, \ldots, \alpha_k$ such that each partial sum from the left of $\beta + \alpha_1 + \ldots + \alpha_k$ is a root or 0 and the full sum is the largest root. If a partial sum $\beta + \ldots + \alpha_j$ is 0, then the member $[E_{\alpha_j}, E_{-\alpha_j}]$ of $\mathfrak{h}^{\mathbb{C}}$ is such that $\alpha_{j+1}([E_{\alpha_j}, E_{-\alpha_j}]) \neq 0$.

We shall say what these constructs are in the case of the other compact classical groups.

Example 2. $G = SO(2n+1)$, $\mathfrak{g} = \mathfrak{so}(2n+1)$, $\mathfrak{g}^{\mathbb{C}} = \mathfrak{so}(2n+1, \mathbb{C})$.

Here $\mathfrak{g}^{\mathbb{C}}$ is simple for $n \geq 1$. However, for $n = 1$, we have an isomorphism $\mathfrak{so}(3) \cong \mathfrak{su}(2)$. The constructs used in proving that $\mathfrak{g}^{\mathbb{C}}$ is simple are as follows:

$$\mathfrak{h}^{\mathbb{C}} = \{H \in \mathfrak{so}(2n+1, \mathbb{C}) \,|\, H = \text{matrix below}\}$$

$$H = \begin{pmatrix} \begin{pmatrix} 0 & ih_1 \\ -ih_1 & 0 \end{pmatrix} & & & & \\ & \begin{pmatrix} 0 & ih_2 \\ -ih_2 & 0 \end{pmatrix} & & & \\ & & \ddots & & \\ & & & \begin{pmatrix} 0 & ih_n \\ -ih_n & 0 \end{pmatrix} & \\ & & & & 0 \end{pmatrix}$$

$$e_j \,(\text{above } H) = h_j, \qquad 1 \leq j \leq n$$

$$\mathfrak{h} = \mathfrak{h}^{\mathbb{C}} \cap \mathfrak{so}(2n+1) = \{H \in \mathfrak{h}^{\mathbb{C}} \,|\, \text{entries are real}\}$$

$$\Delta = \{\pm e_i \pm e_j \text{ with } i \neq j\} \cup \{\pm e_k\}.$$

In the obvious ordering the largest root is $e_1 + e_2$. The root space decomposition is

$$\mathfrak{g}^{\mathbb{C}} = \mathfrak{h}^{\mathbb{C}} \oplus \sum_{\alpha \in \Delta} \mathfrak{g}_\alpha, \qquad \mathfrak{g}_\alpha = \mathbb{C}E_\alpha.$$

To define E_α, first let $i < j$ and $\alpha = \pm e_i \pm e_j$. Then E_α is 0 except in the sixteen entries corresponding to the ith and j^{th} pairs of indices, where it is

$$E_\alpha = \begin{matrix} & i & j \\ & \begin{pmatrix} 0 & X_\alpha \\ -X_\alpha^{\text{tr}} & 0 \end{pmatrix} \end{matrix}$$

with

$$X_{e_i - e_j} = \begin{pmatrix} 1 & i \\ -i & 1 \end{pmatrix}, \qquad X_{e_i + e_j} = \begin{pmatrix} 1 & -i \\ -i & -1 \end{pmatrix}$$

$$X_{-e_i + e_j} = \begin{pmatrix} 1 & -i \\ i & 1 \end{pmatrix} \qquad X_{-e_i - e_j} = \begin{pmatrix} 1 & i \\ i & -1 \end{pmatrix}.$$

To define E_α for $\alpha = \pm e_k$, write

$$E_\alpha = \begin{matrix} & \text{pair } k & \text{entry } 2n+1 \\ & \begin{pmatrix} 0 & X_\alpha \\ -X_\alpha^{\text{tr}} & 0 \end{pmatrix} \end{matrix}$$

with 0's elsewhere and with

$$X_{e_k} = \begin{pmatrix} & 1 \\ -i & \end{pmatrix} \quad \text{and} \quad X_{-e_k} = \begin{pmatrix} 1 \\ i \end{pmatrix}.$$

Example 3. $G = \mathrm{Sp}(n)$, $\mathfrak{g} = \mathfrak{sp}(n)$, $\mathfrak{g}^{\mathbb{C}} = \mathfrak{sp}(n, \mathbb{C})$.

Here $\mathfrak{g}^{\mathbb{C}}$ is simple for $n \geq 1$. However, for $n = 1$ and $n = 2$ we have isomorphisms $\mathfrak{sp}(1) \cong \mathfrak{su}(2)$ and $\mathfrak{sp}(2) \cong \mathfrak{so}(5)$. The constructs used in proving $\mathfrak{g}^{\mathbb{C}}$ is simple are

$$\mathfrak{h}^{\mathbb{C}} = \left\{ H = \begin{pmatrix} h_1 & & & & & \\ & \ddots & & & & \\ & & h_n & & & \\ & & & -h_1 & & \\ & & & & \ddots & \\ & & & & & -h_n \end{pmatrix} \right\}$$

$$e_j \,(\text{above } H) = h_j, \qquad 1 \leq j \leq n$$

$$\mathfrak{h} = \mathfrak{h}^{\mathbb{C}} \cap \mathfrak{sp}(n) = \{ H \in \mathfrak{h}^{\mathbb{C}} \,|\, \text{entries are imaginary} \}$$

$$\Delta = \{ \pm e_i \pm e_j \text{ with } i \neq j \} \cup \{ \pm 2e_k \}$$

$$E_{e_i - e_j} = E_{i,j} - E_{j+n,\, i+n} \qquad E_{2e_k} = E_{k,\, k+n}$$

$$E_{e_i + e_j} = E_{i,\, j+n} + E_{j,\, i+n} \qquad E_{-2e_k} = E_{k+n,\, k}$$

$$E_{-e_i - e_j} = E_{i+n,\, j} + E_{j+n,\, i}$$

Example 4. $G = \mathrm{SO}(2n)$, $\mathfrak{g} = \mathfrak{so}(2n)$, $\mathfrak{g}^{\mathbb{C}} = \mathfrak{so}(2n, \mathbb{C})$.

Here $\mathfrak{g}^{\mathbb{C}}$ is simple for $n \geq 3$. However, for $n = 3$ we have an isomorphism $\mathfrak{so}(6) \cong \mathfrak{su}(4)$. The constructs used in proving $\mathfrak{g}^{\mathbb{C}}$ is simple are

$\mathfrak{h}^{\mathbb{C}}$ as with $\mathfrak{so}(2n + 1)$ but with the last row and column deleted

$e_j(H) = h_j, \qquad 1 \leq j \leq n, \qquad$ as with $\mathfrak{so}(2n + 1)$

$\mathfrak{h} = \mathfrak{h}^{\mathbb{C}} \cap \mathfrak{so}(2n) = \{ H \in \mathfrak{h}^{\mathbb{C}} \,|\, \text{entries are real} \}$

$\Delta = \{ \pm e_i \pm e_j \text{ with } i \neq j \}$

E_α as for $\mathfrak{so}(2n + 1)$ when $\alpha = \pm e_i \pm e_j$.

The root-space decomposition in our examples gives a great deal of insight into the specific Lie algebras and the corresponding Lie groups. Actually such a decomposition exists for general compact connected semisimple Lie groups and is the point of departure for studying the representation theory of these groups. We shall study the decomposition in §§2–4 and apply it to representation theory starting in §5.

§2. Roots

Recall from Theorem 1.15 that any compact connected Lie group can be realized as a compact linear connected reductive group. It will be handy to think of a particular realization of such a group, and thus we shall repeatedly refer to G as a "linear group."

Thus let G be a compact linear connected reductive group, and let $\mathfrak{g}$ be its Lie algebra. Since $G \subseteq \mathrm{U}(n)$ for some n, $\mathfrak{g} \cap i\mathfrak{g} = 0$. Thus the complexification $\mathfrak{g}^{\mathbb{C}}$ is given by $\mathfrak{g}^{\mathbb{C}} = \mathfrak{g} \oplus i\mathfrak{g}$ in terms of matrices.

A **Cartan subalgebra** $\mathfrak{h}$ of $\mathfrak{g}$ is a maximal abelian subspace of $\mathfrak{g}$. Such subalgebras exist, by dimensional considerations, and contain the center $Z_{\mathfrak{g}}$. A **Cartan subgroup** T of G is a maximal connected abelian subgroup; such subgroups are closed and are exactly the analytic subgroups of G corresponding to Cartan subalgebras.

Fix $\mathfrak{h}$ and T. Define $\mathfrak{h}_{\mathbb{R}} = i\mathfrak{h}$ and $\mathfrak{h}^{\mathbb{C}} = \mathfrak{h} \oplus i\mathfrak{h} = \mathfrak{h}_{\mathbb{R}} \oplus i\mathfrak{h}_{\mathbb{R}}$. The trace form $B_0(X, Y) = \mathrm{Tr}(XY)$ is complex bilinear on $\mathfrak{g}^{\mathbb{C}} \times \mathfrak{g}^{\mathbb{C}}$, and its restriction to $\mathfrak{h}_{\mathbb{R}} \times \mathfrak{h}_{\mathbb{R}}$ is real-valued and positive definite.

Let Φ be a representation of G on a finite-dimensional complex vector space V. By Proposition 1.6 we may assume that Φ is unitary. By Corollary 3.16, Φ is smooth; let $\varphi: \mathfrak{g} \to \mathrm{End}\, V$ be its differential. When we need to, we can extend φ to $U(\mathfrak{g}^{\mathbb{C}})$, calling the extension φ. Each $\varphi(H)$, for H in $\mathfrak{h}$, is skew-Hermitian by Corollary 3.10 and hence is diagonable with imaginary eigenvalues. Let $H_1, \ldots, H_l$ be a basis of $\mathfrak{h}$. These matrices commute and hence so do the $\varphi(H_j)$, by (3.11). Therefore we can find a simultaneous eigenspace decomposition of V under all the $\varphi(H_j)$. Since φ is linear on $\mathfrak{h}$ and thus on $\mathfrak{h}^{\mathbb{C}}$, this decomposition is a simultaneous eigenspace decomposition for all of $\varphi(\mathfrak{h}^{\mathbb{C}})$, and each eigenvalue is linear. A typical eigenvalue is $\lambda(H)$, $H \in \mathfrak{h}^{\mathbb{C}}$, and the eigenspace is V_λ:

$$V_\lambda = \{v \in V \mid \varphi(H)v = \lambda(H)v \qquad \text{for all } H \in \mathfrak{h}^{\mathbb{C}}\}.$$

These eigenvalues, which are certain linear functionals on $\mathfrak{h}^{\mathbb{C}}$ that are real on $\mathfrak{h}_{\mathbb{R}}$, are called the **weights** of the representation, and the spaces V_λ are called **weight spaces.** There are only finitely many weights, and we have an orthogonal direct sum **weight space decomposition**

$$V = \sum_{\text{weights}} V_\lambda. \tag{4.4}$$

For any linear functional λ on $\mathfrak{h}^{\mathbb{C}}$ that is imaginary on $\mathfrak{h}$, we let H_λ be the element of $\mathfrak{h}_{\mathbb{R}}$ such that $\lambda(H) = B_0(H, H_\lambda)$ for all H in $\mathfrak{h}_{\mathbb{R}}$.

Example 1. For $G = \mathrm{SU}(2)$, let us take $\mathfrak{h} = \left\{\begin{pmatrix} i\theta & 0 \\ 0 & -i\theta \end{pmatrix}\right\}$ and put $e_1\begin{pmatrix} w & 0 \\ 0 & -w \end{pmatrix} = w$. If Φ_n is the representation in (2.1) on homogeneous polynomials

$$a_n z_1^n + a_{n-1} z_1^{n-1} z_2 + \ldots + a_0 z_2^n,$$

then the weight spaces for $\varphi_n \begin{pmatrix} w & 0 \\ 0 & -w \end{pmatrix}$ are $\mathbb{C}z_1^n, \mathbb{C}z_1^{n-1}z_2, \ldots, \mathbb{C}z_2^n$ with respective weights $-ne_1, -(n-2)e_1, \ldots, ne_1$. A special feature of this example is that every weight space has dimension one.

Example 2. For $G = \mathrm{SU}(n)$ operating on $\mathfrak{sl}(n, \mathbb{C})$ by conjugation (that is, $\Phi = \mathrm{Ad}$), $\mathfrak{g}$ operates with $\varphi = \mathrm{ad}$, and we are in the case of the first example of §1. With $\mathfrak{h}$ and the e_j as in that example, the weights are 0 and $e_i - e_j$ for $i \neq j$.

Example 2 generalizes to a representation of any compact G by conjugation (that is, $\Phi = \mathrm{Ad}$) on $\mathfrak{g}^{\mathbb{C}}$. Again $\varphi = \mathrm{ad}$. In this case the nonzero weights are called **roots.** Let $\Delta = \Delta(\mathfrak{h}^{\mathbb{C}} : \mathfrak{g}^{\mathbb{C}})$ be the set of roots. The weight space decomposition, here called a **root space decomposition,** is written

$$\mathfrak{g}^{\mathbb{C}} = \mathfrak{g}_0 \oplus \sum_{\alpha \in \Delta} \mathfrak{g}_\alpha.$$

For now we shall concentrate on the adjoint representation and roots, deferring most of the general theory of representations and weights to §5. Further examples of representations will be given in §5.

Proposition 4.1 Roots and weights have the following properties:

(a) All roots and weights are real-valued on $\mathfrak{h}_{\mathbb{R}}$.

(b) If λ is a weight of φ with weight space V_λ and if α is a root or 0, then

$$\varphi(\mathfrak{g}_\alpha) V_\lambda \subseteq \begin{cases} V_{\lambda+\alpha} & \text{if } \lambda + \alpha \text{ is a weight} \\ 0 & \text{if not.} \end{cases}$$

(c) The root space decomposition of $\mathfrak{g}^{\mathbb{C}}$ is

$$\mathfrak{g}^{\mathbb{C}} = \mathfrak{h}^{\mathbb{C}} \oplus \sum_{\alpha \in \Delta} \mathfrak{g}_\alpha.$$

(d) If α is a root, so is $-\alpha$. In fact, if E_α is in $\mathfrak{g}_\alpha$, then θE_α is in $\mathfrak{g}_{-\alpha}$, where θ is the Cartan involution of $\mathfrak{g}^{\mathbb{C}}$ ($=1$ on $\mathfrak{g}$ and -1 on $i\mathfrak{g}$).

(e) If α is a root and E_α and $E_{-\alpha}$ are members of $\mathfrak{g}_\alpha$ and $\mathfrak{g}_{-\alpha}$, respectively, then $[E_\alpha, E_{-\alpha}] = B_0(E_\alpha, E_{-\alpha}) H_\alpha$.

(f) Every root vanishes on $Z_{\mathfrak{g}}^{\mathbb{C}}$ and is therefore determined by its values on $(\mathfrak{h} \cap [\mathfrak{g}, \mathfrak{g}])^{\mathbb{C}}$. In the sense of being defined on $(\mathfrak{h} \cap [\mathfrak{g}, \mathfrak{g}])^{\mathbb{C}}$, the roots span the dual of $(\mathfrak{h} \cap [\mathfrak{g}, \mathfrak{g}])^{\mathbb{C}}$.

(g) If α is a root, then $\dim \mathfrak{g}_\alpha = 1$ and $n\alpha$ is not a root for any integer $n \geq 2$.

Proof. Result (a) follows from the fact that $\varphi(H)$ has only imaginary eigenvalues if φ is a weight and H is in $\mathfrak{h}$. For (b) let E_α be in $\mathfrak{g}_\alpha$, H be in $\mathfrak{h}^{\mathbb{C}}$, and X be in V_λ. Then $\varphi(E_\alpha)X$ is in $V_{\lambda+\alpha}$ because

$$\begin{aligned}\varphi(H)\varphi(E_\alpha)X &= \varphi(E_\alpha)\varphi(H)X + \varphi([H, E_\alpha])X \\ &= \lambda(H)\varphi(E_\alpha)X + \alpha(H)\varphi(E_\alpha)X.\end{aligned}$$

Result (c) says that $\mathfrak{g}_0 = \mathfrak{h}^{\mathbb{C}}$, which follows from the fact that $\mathfrak{h}$ is maximal abelian in $\mathfrak{g}$. In (d) let H be in $\mathfrak{h}_{\mathbb{R}}$. Then $\alpha(H)$ is real, and so

$$[H, \theta E_\alpha] = \theta[\theta H, E_\alpha] = \theta(\alpha(\theta H)E_\alpha) = \theta(-\alpha(H)E_\alpha) = -\alpha(H)\theta E_\alpha.$$

Hence θE_α is in $\mathfrak{g}_{-\alpha}$.

In (e), both $[E_\alpha, E_{-\alpha}]$ and $B_0(E_\alpha, E_{-\alpha})H_\alpha$ are in $\mathfrak{h}^{\mathbb{C}}$, by (b) and (c). If H is in $\mathfrak{h}^{\mathbb{C}}$, then

$$\begin{aligned}B_0(H, [E_\alpha, E_{-\alpha}]) &= B_0([H, E_\alpha], E_{-\alpha}) = \alpha(H)B_0(E_\alpha, E_{-\alpha}) \\ &= B_0(H, H_\alpha)B_0(E_\alpha, E_{-\alpha}) = B_0(H, B_0(E_\alpha, E_{-\alpha})H_\alpha).\end{aligned}$$

Since B_0 is nondegenerate on $\mathfrak{h}^{\mathbb{C}} \times \mathfrak{h}^{\mathbb{C}}$, (e) follows.

For (f), let H be in $Z_{\mathfrak{g}}^{\mathbb{C}}$ and let α be a root. For E_α in $\mathfrak{g}_\alpha$, we have

$$0 = [H, E_\alpha] = \alpha(H)E_\alpha.$$

Hence α vanishes on $Z_{\mathfrak{g}}^{\mathbb{C}}$. Conversely suppose H in $\mathfrak{h}^{\mathbb{C}}$ satisfies $\alpha(H) = 0$ for all roots α. Then $[H, \mathfrak{g}_\alpha] = 0$ for all α, and also $[H, \mathfrak{g}_0] = 0$. Hence $[H, \mathfrak{g}^{\mathbb{C}}] = 0$ and H is in $Z_{\mathfrak{g}}^{\mathbb{C}}$. Then (f) follows because $\mathfrak{g} = Z_{\mathfrak{g}} \oplus [\mathfrak{g}, \mathfrak{g}]$ by Proposition 1.1 and because $Z_{\mathfrak{g}} \subseteq \mathfrak{h}$.

For (g) let $E_\alpha \neq 0$ be in $\mathfrak{g}_\alpha$ and define

$$\mathfrak{g}^* = \mathbb{C}E_\alpha + \mathbb{C}H_\alpha + \sum_{k=1}^{\infty} \mathfrak{g}_{-k\alpha}.$$

By (d), θE_α is in $\mathfrak{g}_{-\alpha}$. Since $B_0(E_\alpha, \theta E_\alpha) = -\mathrm{Tr}(E_\alpha \bar{E}_\alpha^{\mathrm{tr}}) < 0$, we can choose $E_{-\alpha}$ in $\mathfrak{g}_{-\alpha}$ with $B_0(E_\alpha, E_{-\alpha}) = 1$. By (e), $[E_\alpha, E_{-\alpha}] = H_\alpha$. Now $(\mathrm{ad}\, E_\alpha)\mathfrak{g}^* \subseteq \mathfrak{g}^*$ since $[E_\alpha, E_\alpha] = 0$, $[E_\alpha, H_\alpha] = cE_\alpha$, $[E_\alpha, \mathfrak{g}_{-\alpha}] = \mathbb{C}H_\alpha$, etc. Also $(\mathrm{ad}\, E_{-\alpha})\mathfrak{g}^* \subseteq \mathfrak{g}^*$ since $E_{-\alpha}$ is in $\mathfrak{g}_{-\alpha}$. Consequently

$$\mathrm{ad}\, H_\alpha = \mathrm{ad}\, E_\alpha\, \mathrm{ad}\, E_{-\alpha} - \mathrm{ad}\, E_{-\alpha}\, \mathrm{ad}\, E_\alpha$$

is a commutator leaving $\mathfrak{g}^*$ stable. Therefore

$$\begin{aligned}0 = \mathrm{Tr}(\mathrm{ad}\, H_\alpha|_{\mathfrak{g}^*}) &= \alpha(H_\alpha) + 0 + \sum_{k=1}^{\infty} -k\alpha(H_\alpha) \dim \mathfrak{g}_{-k\alpha} \\ &= \alpha(H_\alpha)\left\{1 - \sum_{k=1}^{\infty} k \dim \mathfrak{g}_{-k\alpha}\right\}.\end{aligned}$$

Since B_0 is an inner product on $\mathfrak{h}_{\mathbb{R}}$, $\alpha(H_\alpha)$ is not 0. Thus $\dim \mathfrak{g}_{-\alpha} = 1$ and $\dim \mathfrak{g}_{-n\alpha} = 0$ for $n \geq 2$. To finish (g), replace α by $-\alpha$.

Let α be a root. We consider the subspace

$$\mathbb{C}H_\alpha + \mathfrak{g}_\alpha + \mathfrak{g}_{-\alpha} \tag{4.5}$$

in more detail. By (g) of the proposition, $\dim \mathfrak{g}_\alpha = \dim \mathfrak{g}_{-\alpha} = 1$. Let us choose E_α and $E_{-\alpha}$ in $\mathfrak{g}_\alpha$ and $\mathfrak{g}_{-\alpha}$, respectively, with $B_0(E_\alpha, E_{-\alpha}) = 1$; the proof showed how to do this. Then we have

$$[H_\alpha, E_\alpha] = \alpha(H_\alpha)E_\alpha, \qquad [H_\alpha, E_{-\alpha}] = -\alpha(H_\alpha)E_{-\alpha},$$

and

$$[E_\alpha, E_{-\alpha}] = H_\alpha.$$

Put

$$H'_\alpha = \frac{2}{\alpha(H_\alpha)} H_\alpha, \qquad E'_\alpha = \frac{2}{\alpha(H_\alpha)} E_\alpha, \qquad E'_{-\alpha} = E_{-\alpha}. \tag{4.6}$$

Then

$$[H'_\alpha, E'_\alpha] = 2E'_\alpha, \qquad [H'_\alpha, E'_{-\alpha}] = -2E'_{-\alpha}, \qquad [E'_\alpha, E'_{-\alpha}] = H'_\alpha.$$

Hence the map $H'_\alpha \to h$, $E'_\alpha \to e$, $E'_{-\alpha} \to f$ exhibits (4.5) as isomorphic to $\mathfrak{sl}(2, \mathbb{C})$. Moreover, (d) of the proposition shows that (4.5) is θ-stable, and the members of (4.5) fixed by θ form a Lie subalgebra isomorphic to $\mathfrak{su}(2)$.

We see that $\mathfrak{g}^\mathbb{C}$ consists of a number of copies of $\mathfrak{sl}(2, \mathbb{C})$ pieced together appropriately (one for each pair $(\alpha, -\alpha)$ of roots), and the piecing together is done compatibly with θ. The detailed structure of $\mathfrak{g}^\mathbb{C}$ comes from understanding how this piecing together occurs. To investigate this, we study the action of the subalgebra (4.5) on all of $\mathfrak{g}^\mathbb{C}$; that is, we study a certain complex-linear representation of $\mathfrak{sl}(2, \mathbb{C})$.

Let α be in Δ and let β be in $\Delta \cup \{0\}$. The **α-string containing β** is the set of all members of $\Delta \cup \{0\}$ of the form $\beta + n\alpha$ with n in $\mathbb{Z}$. We define an inner product $\langle \cdot, \cdot \rangle$ on $\mathfrak{h}'_\mathbb{R}$ by

$$\langle \varphi, \psi \rangle = B_0(H_\varphi, H_\psi) = \varphi(H_\psi) = \psi(H_\varphi) \qquad \text{for } \varphi, \psi \in \mathfrak{h}'_\mathbb{R}.$$

The corresponding norm is denoted $|\cdot|$.

Example. Let $G = SU(n)$, $\mathfrak{g} = \mathfrak{su}(n)$, $\mathfrak{g}^\mathbb{C} = \mathfrak{sl}(n, \mathbb{C})$. In the notation of §1, the roots are $e_i - e_j$, $i \neq j$. Let k be an index different from i and j, and let $\beta = e_i - e_j$ and $\alpha = e_j - e_k$. Then the α-string through β is $\beta, \beta + \alpha$. Moreover, H_β is $E_{ii} - E_{jj}$ since we must have

$$e_i(H) - e_j(H) = \beta(H) = B_0(H, H_\beta).$$

Consequently in the inner product $\langle \cdot, \cdot \rangle$, the e_j behave as if they are orthogonal and normalized.

Proposition 4.2.

(a) If α is in Δ and β is in $\Delta \cup \{0\}$, then the α-string containing β has the form $\beta + n\alpha$ for $-p \le n \le q$ and $p \ge 0$, $q \ge 0$. There are no gaps. Furthermore,

$$p - q = \frac{2\langle \beta, \alpha \rangle}{|\alpha|^2} \tag{4.7}$$

and $2\langle \beta, \alpha \rangle / |\alpha|^2$ is in $\mathbb{Z}$.

(b) If α and β are in $\Delta \cup \{0\}$ and $\alpha + \beta \ne 0$, then $[\mathfrak{g}_\alpha, \mathfrak{g}_\beta] = \mathfrak{g}_{\alpha+\beta}$.

Proof.

(a) By Proposition 4.1g, we may assume $\beta + n\alpha \ne 0$ for all n. Let $\mathfrak{g}^* = \sum_n \mathfrak{g}_{\beta + n\alpha}$, and regard H'_α, E'_α, and $E'_{-\alpha}$ in (4.6) as spanning $\mathfrak{sl} = \mathfrak{sl}(2, \mathbb{C})$. Proposition 4.1b shows that $\mathfrak{g}^*$ is stable under $\operatorname{ad}(\mathfrak{sl})$. The operator $\operatorname{ad} H'_\alpha$ is diagonable on $\mathfrak{g}^*$, and Proposition 4.1g shows the eigenvalues $(\beta + n\alpha)(H'_\alpha)$ have multiplicity one. Here

$$(\beta + n\alpha)(H'_\alpha) = \frac{2}{|\alpha|^2}(\beta + n\alpha)(H_\alpha) = \frac{2}{|\alpha|^2}(\langle \beta, \alpha \rangle + n\langle \alpha, \alpha \rangle) = \frac{2\langle \beta, \alpha \rangle}{|\alpha|^2} + 2n.$$

Thus the eigenvalues of $\operatorname{ad} H'_\alpha$ are the sum of a fixed number and an even integer. It follows from Corollary 2.2 and Theorem 2.4 that $\operatorname{ad}(\mathfrak{sl})$ acts irreducibly on $\mathfrak{g}^*$ and that the n's that occur are consecutive, with no gaps, say $-p \le n \le q$. Since the largest eigenvalue must be the negative of the smallest, we have

$$\frac{2\langle \beta, \alpha \rangle}{|\alpha|^2} + 2q = -\left(\frac{2\langle \beta, \alpha \rangle}{|\alpha|^2} - 2p\right).$$

Then (4.7) follows, and also $2\langle \beta, \alpha \rangle / |\alpha|^2$ is in $\mathbb{Z}$.

(b) We saw in the proof of (a) that $\mathfrak{sl}$ acts irreducibly on $\mathfrak{g}^*$. It is clear that the vectors $E_{\beta + n\alpha}$, apart from scalars, are the vectors v_i of Theorem 2.4, with v_0 corresponding to $E_{\beta + q\alpha}$. Since the kernel of $\pi(e)$ in Theorem 2.4 is only $\mathbb{C}v_0$, it follows that $[E_\alpha, E_{\beta + n\alpha}] \ne 0$ if $-p \le n < q$. This proves (b).

Relative to the inner product $\langle \cdot, \cdot \rangle$ on $\mathfrak{h}'_{\mathbb{R}}$, we introduce the **root reflection**

$$s_\alpha(\varphi) = \varphi - \frac{2\langle \varphi, \alpha \rangle}{|\alpha|^2}\alpha \qquad \text{for } \alpha \in \Delta,\ \varphi \in \mathfrak{h}'_{\mathbb{R}}.$$

This is -1 on α and is $+1$ on the orthogonal complement of α.

Proposition 4.3. For α in Δ, the root reflection s_α carries Δ into itself.

Proof. If β is in Δ, then Proposition 4.2a gives

$$s_\alpha\beta = \beta - \frac{2\langle \beta, \alpha \rangle}{|\alpha|^2}\alpha = \beta - (p - q)\alpha = \beta + (q - p)\alpha.$$

Since $-p \leq q - p \leq q$, $\beta + (q - p)\alpha$ is in the α-string through β. Hence $s_\alpha \beta$ is a root or 0. Since s_α is an orthogonal transformation on $\mathfrak{h}'_{\mathbb{R}}$, $s_\alpha \beta$ is not 0. Thus s_α leaves Δ stable.

An **abstract root system** in a finite-dimensional real inner product space V is a finite set Δ of nonzero elements of V such that

(i) Δ spans V,

(ii) the orthogonal transformations $s_\alpha(\varphi) = \varphi - \dfrac{2\langle \varphi, \alpha \rangle}{|\alpha|^2} \alpha$, for $\alpha \in \Delta$, leave Δ stable, and

(iii) $2\langle \beta, \alpha \rangle / |\alpha|^2$ is in $\mathbb{Z}$ for all α and β in Δ.

An abstract root system is **reduced** if α in Δ implies 2α not in Δ. An abstract root system is **irreducible** if Δ admits no nontrivial disjoint decomposition $\Delta = \Delta' \cup \Delta''$ with every member of Δ' orthogonal to every member of Δ''.

Proposition 4.4. The root system for a compact connected linear semisimple group G with respect to a Cartan subalgebra $\mathfrak{h}$ forms a reduced abstract root system in $\mathfrak{h}'_{\mathbb{R}}$. The system is irreducible if and only if the Lie algebra $\mathfrak{g}$ of G is simple, if and only if $\mathfrak{g}^{\mathbb{C}}$ is simple.

Remarks. If G is merely reductive, we obtain an abstract root system in the subspace of $\mathfrak{h}'_{\mathbb{R}}$ spanned by Δ, namely $(\mathfrak{h}_{\mathbb{R}} \cap i[\mathfrak{g}, \mathfrak{g}])'$.

Proof of abstract root system properties. The three properties of an abstract root system follow from Proposition 4.1f, Proposition 4.3, and Proposition 4.2a, respectively. The system is reduced, according to Proposition 4.1g.

Proof that Δ irreducible imples $\mathfrak{g}^{\mathbb{C}}$ simple. Suppose that $\mathfrak{g}^{\mathbb{C}} = \mathfrak{g}' \oplus \mathfrak{g}''$ exhibits $\mathfrak{g}^{\mathbb{C}}$ as semisimple, not simple. Let α be a root, and write $E_\alpha = E'_\alpha + E''_\alpha$ correspondingly. For H in $\mathfrak{h}^{\mathbb{C}}$, we have

$$0 = [H, E_\alpha] - \alpha(H)E_\alpha = ([H, E'_\alpha] - \alpha(H)E'_\alpha) + ([H, E''_\alpha] - \alpha(H)E''_\alpha).$$

Since $\mathfrak{g}'$ and $\mathfrak{g}''$ are ideals and are disjoint, the two terms on the right are separately 0. Thus E'_α and E''_α are both in $\mathfrak{g}_\alpha$. Since $\dim \mathfrak{g}_\alpha = 1$, $E'_\alpha = 0$ or $E''_\alpha = 0$. Thus $\mathfrak{g}_\alpha \subseteq \mathfrak{g}'$ or $\mathfrak{g}_\alpha \subseteq \mathfrak{g}''$. Define

$$\Delta' = \{\alpha \in \Delta \mid \mathfrak{g}_\alpha \subseteq \mathfrak{g}'\}$$
$$\Delta'' = \{\alpha \in \Delta \mid \mathfrak{g}_\alpha \subseteq \mathfrak{g}''\},$$

so that $\Delta = \Delta' \cup \Delta''$ disjointly. Now with obvious notation, we have

$$\begin{aligned} \alpha'(H_{\alpha''})E_{\alpha'} &= [H_{\alpha''}, E_{\alpha'}] \subseteq [H_{\alpha''}, \mathfrak{g}'] \\ &= [[E_{\alpha''}, E_{-\alpha''}], \mathfrak{g}'] \subseteq [\mathfrak{g}'', \mathfrak{g}'] = 0, \end{aligned}$$

and thus $\alpha'(H_{\alpha''}) = 0$. Hence Δ' and Δ'' are mutually orthogonal.

Proof that $\mathfrak{g}$ *simple implies* Δ *irreducible.* Suppose that $\Delta = \Delta' \cup \Delta''$ exhibits Δ as reducible. Define

$$\mathfrak{g}' = \left\{\theta\text{-fixed elements in } \sum_{\alpha \in \Delta'} \{\mathbb{C}H_\alpha + \mathfrak{g}_\alpha + \mathfrak{g}_{-\alpha}\}\right\}$$

$$\mathfrak{g}'' = \left\{\theta\text{-fixed elements in } \sum_{\alpha \in \Delta''} \{\mathbb{C}H_\alpha + \mathfrak{g}_\alpha + \mathfrak{g}_{-\alpha}\}\right\}.$$

Proposition 4.1d shows that

$$(\mathfrak{g}')^{\mathbb{C}} = \sum_{\alpha \in \Delta'} \{\mathbb{C}H_\alpha + \mathfrak{g}_\alpha + \mathfrak{g}_{-\alpha}\}$$

$$(\mathfrak{g}'')^{\mathbb{C}} = \sum_{\alpha \in \Delta''} \{\mathbb{C}H_\alpha + \mathfrak{g}_\alpha + \mathfrak{g}_{-\alpha}\}.$$

Thus $\mathfrak{g}'$ and $\mathfrak{g}''$ are subspaces of $\mathfrak{g}$, and $\mathfrak{g} = \mathfrak{g}' \oplus \mathfrak{g}''$ as vector spaces. To complete the proof, it is enough to show that $(\mathfrak{g}')^{\mathbb{C}}$ and $(\mathfrak{g}'')^{\mathbb{C}}$ are ideals in $\mathfrak{g}^{\mathbb{C}}$. For α' in Δ' and α'' in Δ'',

$$[H_{\alpha'}, E_{\alpha''}] = \alpha''(H_{\alpha'})E_{\alpha''} = 0.$$

Also if $[\mathfrak{g}_{\alpha'}, \mathfrak{g}_{\alpha''}] \neq 0$, then $\alpha' + \alpha''$ is a root that is not osthogonal to every member of Δ' and is not orthogonal to every member of Δ''. Hence $[\mathfrak{g}_{\alpha'}, \mathfrak{g}_{\alpha''}] = 0$, and it follows that $(\mathfrak{g}')^{\mathbb{C}}$ and $(\mathfrak{g}'')^{\mathbb{C}}$ are ideals.

Examples of §1.

(1) $G = SU(n)$, $\mathfrak{g} = \mathfrak{su}(n)$, $\mathfrak{g}^{\mathbb{C}} = \mathfrak{sl}(n, \mathbb{C})$.

$$\Delta = \{e_i - e_j, i \neq j\}$$
$$s_{e_i - e_j} = \text{transposition of } i \text{ and } j.$$

(2) $G = SO(2n+1)$, $\mathfrak{g} = \mathfrak{so}(2n+1)$, $\mathfrak{g}^{\mathbb{C}} = \mathfrak{so}(2n+1, \mathbb{C})$.

$$\Delta = \{\pm e_i \pm e_j, i \neq j\} \cup \{\pm e_k\}$$

$$s_{e_i - e_j} = \text{transposition of } i \text{ and } j$$

$$s_{e_k} = s_{-e_k} = \text{change of } k^{\text{th}} \text{ sign}$$

$$s_{e_i + e_j} = s_{-e_i - e_j} = \text{transposition of } i \text{ and } j \text{ and change of both signs.}$$

(3) $G = Sp(n)$, $\mathfrak{g} = \mathfrak{sp}(n)$, $\mathfrak{g}^{\mathbb{C}} = \mathfrak{sp}(n, \mathbb{C})$.

$$\Delta = \{\pm e_i \pm e_j, i \neq j\} \cup \{\pm 2e_k\}.$$

Root reflections are as in $SO(2n+1)$, with $s_{2e_k} = s_{e_k}$.

(4) $G = SO(2n)$, $\mathfrak{g} = \mathfrak{so}(2n)$, $\mathfrak{g}^{\mathbb{C}} = \mathfrak{so}(2n, \mathbb{C})$.

$$\Delta = \{\pm e_i \pm e_j, i \neq j\}.$$

Root reflections are as in SO($2n + 1$). For SO(4), notice that

$$\Delta = \{\pm(e_1 + e_2)\} \cup \{\pm(e_1 - e_2)\}$$

exhibits Δ as reducible.

§3. Abstract Root Systems and Positivity

Abstract root systems arise in the theory in other situations besides the one in Proposition 4.4. (We shall see our first additional one in §5.3.) We therefore proceed temporarily in this greater generality. Let Δ be an abstract root system in a finite-dimensional real inner-product space V. The inner product will be written $\langle \cdot, \cdot \rangle$, and the corresponding norm will be written $|\cdot|$. If α is in Δ and β is in $\Delta \cup \{0\}$, we define the **α-string containing β** to be the set of all members of $\Delta \cup \{0\}$ of the form $\beta + n\alpha$ with n in $\mathbb{Z}$.

Proposition 4.5. Any abstract root system Δ has the following properties:

(a) If α is in Δ, so is $-\alpha$.
(b) If α is in Δ, then the only possible members of Δ proportional to α are $\pm\alpha$, or $\pm\alpha$ and $\pm 2\alpha$, or $\pm\alpha$ and $\pm\frac{1}{2}\alpha$.
(c) If α is in Δ and β is in $\Delta \cup \{0\}$, then $2\langle \beta, \alpha \rangle / |\alpha|^2$ equals 0 or ± 1 or ± 2 or ± 3 or ± 4, and ± 4 can occur only if $\beta = \pm 2\alpha$.
(d) If α and β are nonproportional members of Δ with $|\alpha| \leq |\beta|$, then $2\langle \alpha, \beta \rangle / |\beta|^2 = 0$, 1, or -1.
(e) If α and β are in Δ and $\langle \alpha, \beta \rangle > 0$, then $\alpha - \beta$ is in $\Delta \cup \{0\}$. If α and β are in Δ and $\langle \alpha, \beta \rangle < 0$, then $\alpha + \beta$ is in $\Delta \cup \{0\}$.
(f) If α and β are in Δ and neither $\alpha + \beta$ nor $\alpha - \beta$ is in $\Delta \cup \{0\}$, then $\langle \alpha, \beta \rangle = 0$.
(g) If α is in Δ and β is in $\Delta \cup \{0\}$, then the α-string containing β has the form $\beta + n\alpha$ for $-p \leq n \leq q$ and $p \geq 0$, $q \geq 0$. There are no gaps. Furthermore

$$p - q = \frac{2\langle \beta, \alpha \rangle}{|\alpha|^2},$$

and the α-string containing β contains at most four roots.
(h) If Δ is reduced, then Δ has these additional properties:
  (i) If α is in Δ, then the only members of Δ proportional to α are $\pm\alpha$.
  (ii) If α is in Δ and β is in $\Delta \cup \{0\}$, then $2\langle \beta, \alpha \rangle / |\alpha|^2$ equals 0 or ± 1 or ± 2 or ± 3.

Remark. The proof will not use the assumption that Δ spans V.

Proof.

(a) $s_\alpha(\alpha) = -\alpha$.

(b) and (h, i) Let $\beta = c\alpha$ and assume $|c| < 1$ without loss of generality. Then $2\langle\beta, \alpha\rangle/|\alpha|^2 = 2c$. For $2c$ to be in $\mathbb{Z}$, we must have $c = 0$ or $c = \pm\frac{1}{2}$. If Δ is reduced, $c = \pm\frac{1}{2}$ cannot occur.

(c) and (h, ii) By the Schwarz inequality, we have

$$\left|\frac{2\langle\alpha, \beta\rangle}{|\alpha|^2}\frac{2\langle\alpha, \beta\rangle}{|\beta|^2}\right| \leq 4$$

with equality only if $\beta = c\alpha$, a case handled by (b). If strict inequality holds, then $2\langle\alpha, \beta\rangle/|\alpha|^2$ and $2\langle\alpha, \beta\rangle/|\beta|^2$ are two integers whose product is ≤ 3. The result follows in either case.

(d) We have an inequality of integers

$$\left|\frac{2\langle\alpha, \beta\rangle}{|\alpha|^2}\right| \geq \left|\frac{2\langle\alpha, \beta\rangle}{|\beta|^2}\right|,$$

and (c) shows that the product of the two sides is ≤ 3. So the smaller side is 0 or 1.

(e) We may assume α and β are not proportional. For the first statement, suppose $|\alpha| \leq |\beta|$. Then $s_\beta(\alpha) = \alpha - \dfrac{2\langle\alpha, \beta\rangle}{|\beta|^2}\beta$ must be $\alpha - \beta$, by (d). So $\alpha - \beta$ is in Δ. If $|\beta| \leq |\alpha|$ instead, we find that $s_\alpha(\beta) = \beta - \alpha$ is in Δ, and we apply (a). For the second statement, we apply the first statement to (a).

(f) This is immediate from (e).

(g) Let $-p$ and q be the smallest and largest values of n with $\beta + n\alpha$ in $\Delta \cup \{0\}$. If the string is not an interval, we can find r and s with $r < s - 1$ such that $\beta + r\alpha$ is in $\Delta \cup \{0\}$, $\beta + (r+1)\alpha$ and $\beta + (s-1)\alpha$ are not in $\Delta \cup \{0\}$, and $\beta + s\alpha$ is in $\Delta \cup \{0\}$. By (e),

$$\langle\beta + r\alpha, \alpha\rangle \geq 0 \qquad \text{and} \qquad \langle\beta + s\alpha, \alpha\rangle \leq 0.$$

Subtracting, we obtain $(r - s)|\alpha|^2 \geq 0$ and thus $r \geq s$, contradiction. So the string is an interval. Next,

$$s_\alpha(\beta + n\alpha) = \beta + n\alpha - \frac{2\langle\beta + n\alpha, \alpha\rangle}{|\alpha|^2}\alpha = \beta - \left(n + \frac{2\langle\beta, \alpha\rangle}{|\alpha|^2}\right)\alpha,$$

and thus $-p \leq n \leq q$ implies $-q \leq n + \dfrac{2\langle\beta, \alpha\rangle}{|\alpha|^2} \leq p$. Taking $n = q$ and then $n = -p$, we obtain

$$\frac{2\langle\beta, \alpha\rangle}{|\alpha|^2} \leq p - q \text{ and then } p - q \leq \frac{2\langle\beta, \alpha\rangle}{|\alpha|^2}.$$

Thus $2\langle\beta, \alpha\rangle/|\alpha|^2 = p - q$. Finally, to investigate the length of the string, we may assume $q = 0$. The length of the string is then $p + 1$, with $p = 2\langle\beta, \alpha\rangle/|\alpha|^2$. The conclusion that the string has at most four roots now follows from (c) and (b).

We now introduce a notion of positive system for Δ. The definition does not use the assumption that Δ spans V, and we shall note the extent to which this assumption is used in Proposition 4.6.

Fix a basis $\varphi_1, \ldots, \varphi_l$ of V, and define an ordering on V as follows. We say that φ is **positive** (written $\varphi > 0$) if $\varphi = \sum_{i=1}^{l} x_i\varphi_i$ with $x_1 = \ldots = x_k = 0$ and $x_{k+1} > 0$ for some $k \geq 0$. This notion has these properties:

(i) each φ is > 0 or is $= 0$ or has $-\varphi > 0$, uniquely.
(ii) if $\varphi > 0$ and $\psi > 0$, then $\varphi + \psi > 0$ and $c\varphi > 0$ for $c > 0$.

We say $\varphi > \psi$ or $\psi < \varphi$ if $\varphi - \psi > 0$.

A root α is **simple** if $\alpha > 0$ and if α does not decompose as $\alpha = \beta_1 + \beta_2$ with β_1 and β_2 both positive roots.

Proposition 4.6. With $l = \dim V$, there are l simple roots $\alpha_1, \ldots, \alpha_l$, and they are linearly independent. If β is a root and is written as $\beta = \sum x_j\alpha_j$, then all the x_j have the same sign (with 0 allowed to be positive or negative), and all the x_j are integers.

Remark. If we drop the assumption that Δ spans V, then all the conclusions are still valid on the subspace of V spanned by the roots.

Before coming to the proof, we shall indicate choices of positive roots for the examples of §1 and we shall list the simple roots. The set of positive roots will be denoted Δ^+, and the set of simple roots will be denoted Π. The proof of Proposition 4.6 will then be preceded by a lemma.

Examples of §1.

(1) $G = SU(n)$, $\mathfrak{g} = \mathfrak{su}(n)$, $\mathfrak{g}^{\mathbb{C}} = \mathfrak{sl}(n, \mathbb{C})$.

$$\Delta^+ = \{e_i - e_j, i < j\}$$
$$\Pi = \{e_1 - e_2, e_2 - e_3, \ldots, e_{n-1} - e_n\}.$$

(2) $G = SO(2n+1)$, $\mathfrak{g} = \mathfrak{so}(2n+1)$, $\mathfrak{g}^{\mathbb{C}} = \mathfrak{so}(2n+1, \mathbb{C})$.

$$\Delta^+ = \{e_i \pm e_j, i < j\} \cup \{e_k\}$$
$$\Pi = \{e_1 - e_2, e_2 - e_3, \ldots, e_{n-1} - e_n, e_n\}.$$

(3) $G = Sp(n)$, $\mathfrak{g} = \mathfrak{sp}(n)$, $\mathfrak{g}^{\mathbb{C}} = \mathfrak{sp}(n, \mathbb{C})$.

$$\Delta^+ = \{e_i \pm e_j, i < j\} \cup \{2e_k\}$$
$$\Pi = \{e_1 - e_2, e_2 - e_3, \ldots, e_{n-1} - e_n, 2e_n\}.$$

(4) $G = \mathrm{SO}(2n)$, $\mathfrak{g} = \mathfrak{so}(2n)$, $\mathfrak{g}^{\mathbb{C}} = \mathfrak{so}(2n, \mathbb{C})$.

$$\Delta^+ = \{e_i \pm e_j, i < j\}$$

$$\Pi = \{e_1 - e_2, e_2 - e_3, \ldots, e_{n-1} - e_n, e_{n-1} + e_n\}.$$

Lemma 4.7. If α and β are distinct simple roots, then $\langle \alpha, \beta \rangle \leq 0$.

Proof. Suppose $\langle \alpha, \beta \rangle > 0$. By Proposition 4.5e, $\alpha - \beta$ is in Δ. Then either $\alpha = (\alpha - \beta) + \beta$ exhibits α as a nontrivial sum of positive roots or $\beta = (\beta - \alpha) + \alpha$ exhibits β as a nontrivial sum of positive roots. In either case we have a contradiction.

Proof of Proposition 4.6. Let $\beta > 0$ be in Δ. If β is not simple, write $\beta = \beta_1 + \beta_2$ with $\beta_1, \beta_2 > 0$ in Δ. Then decompose β_1 and/or β_2, and then decompose each of their components if possible. Continue in this way. We can list the decompositions as tuples (β, β_1, component of β_1, etc.) with each entry a component of the previous entry. The claim is that no tuple has more than $|\Delta^+|$ entries, and therefore the decomposition process must stop. In fact, otherwise some tuple would have the same $\gamma > 0$ in it twice, and we would have $\gamma = \gamma + \alpha$ with α a nonempty sum of positive roots, contradicting the properties of an ordering. Thus β is exhibited as $\beta = x_1\alpha_1 + \ldots + x_m\alpha_m$ with all x_j positive integers or 0 and with all α_j simple. Thus the simple roots span Δ in the fashion asserted.

Finally we prove the independence. Renumbering the α_j's, suppose

$$x_1\alpha_1 + \ldots + x_s\alpha_s - x_{s+1}\alpha_{s+1} - \ldots - x_m\alpha_m = 0$$

with all $x_j \geq 0$ in $\mathbb{R}$. Put $\beta = x_1\alpha_1 + \ldots + x_s\alpha_s$. Then

$$0 \leq \langle \beta, \beta \rangle = \left\langle \sum_{j=1}^{s} x_j\alpha_j, \sum_{k=s+1}^{m} x_k\alpha_k \right\rangle = \sum_{j,k} x_j x_k \langle \alpha_j, \alpha_k \rangle \leq 0,$$

the last inequality holding by Lemma 4.7. So $\langle \beta, \beta \rangle = 0$, $\beta = 0$, and all the x_j's equal 0 since a positive combination of positive roots cannot be 0.

Fix an ordering of V, and let $\alpha_1, \ldots, \alpha_l$ be the simple roots. The l-by-l matrix (A_{ij}) given by

$$A_{ij} = \frac{2\langle \alpha_i, \alpha_j \rangle}{|\alpha_i|^2}$$

is called the **Cartan matrix.** The Cartan matrix has the following properties:

(1) $A_{ij} \in \mathbb{Z}$
(2) $A_{ii} = 2$

(3) $A_{ij} \leq 0$ and $A_{ij} = 0, -1, -2$, or -3 if $i \neq j$
(4) For $i \neq j$, $A_{ij}A_{ji} < 4$
(5) $A_{ij} = 0$ if and only if $A_{ji} = 0$
(6) $\det(A_{ij})$ is a positive integer.

Proof of (6). The determinant is an integer by (1). Next, the matrix (A_{ij}) is the product of a diagonal matrix with diagonal entries $2/|\alpha_i|^2$ and a matrix with $(i, j)^{\text{th}}$ entry $\langle \alpha_i, \alpha_j \rangle$. Each of these has positive determinant, the latter because the matrix $(\langle \varphi_i, \varphi_j \rangle)$ is positive definite symmetric if $\{\varphi_i\}$ is any basis.

From a Cartan matrix we define a **Dynkin diagram.** Associate with each simple root α_i a point in the plane with a weight proportional to $|\alpha_i|^2$. Connect any two points α_i and α_j $(i \neq j)$ by $A_{ij}A_{ji}$ straight lines. In principle, the Dynkin diagram might be as in Figure 4.1, although it turns out that the diagram in the figure is too complicated to be a Dynkin diagram.

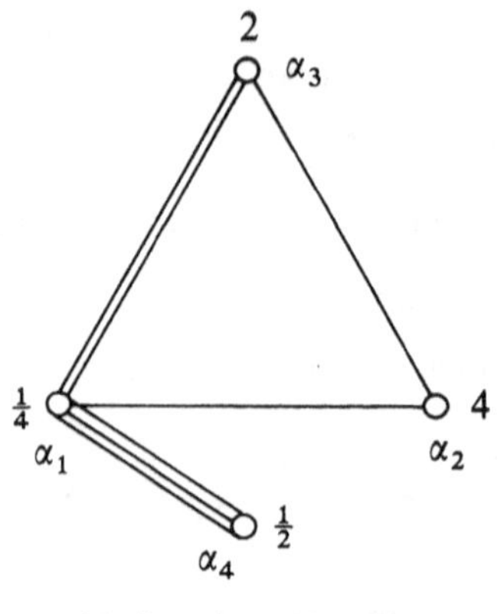

FIGURE 4.1. Conceivable Dynkin diagram

One can show that the Dynkin diagram for a compact group does not depend on the two choices we have made—Cartan subalgebra and ordering—and that distinct g's give distinct Dynkin diagrams. A different realization of G as a linear group can change the trace form B_0, and the result will be at most a change in the proportionality constants on each connected component. We list the diagrams for the examples of §1, together with their names, in Figure 4.2.

A connected Dynkin diagram corresponds to an irreducible root system. One can show that there are only five other possible connected Dynkin diagrams of reduced root systems beyond those in Figure 4.2 and

(1) $\mathfrak{g}^{\mathbb{C}} = \mathfrak{sl}(n, \mathbb{C})$

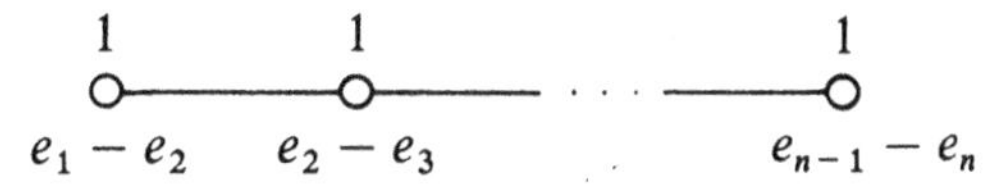

(A_{n-1})

(2) $\mathfrak{g}^{\mathbb{C}} = \mathfrak{so}(2n + 1, \mathbb{C})$

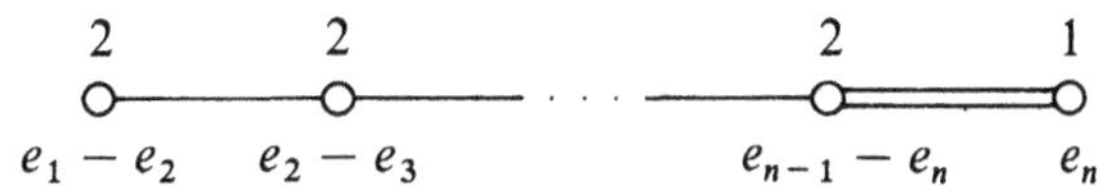

(B_n)

(3) $\mathfrak{g}^{\mathbb{C}} = \mathfrak{sp}(n, \mathbb{C})$

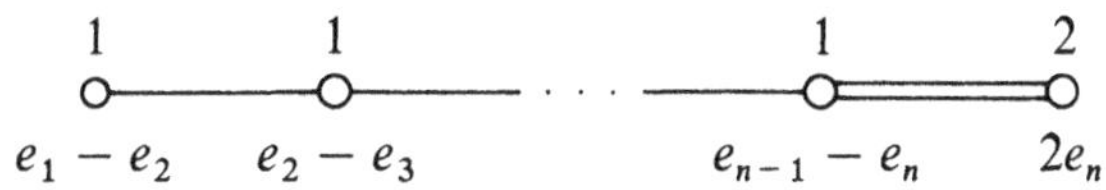

(C_n)

(4) $\mathfrak{g}^{\mathbb{C}} = \mathfrak{so}(2n, \mathbb{C})$

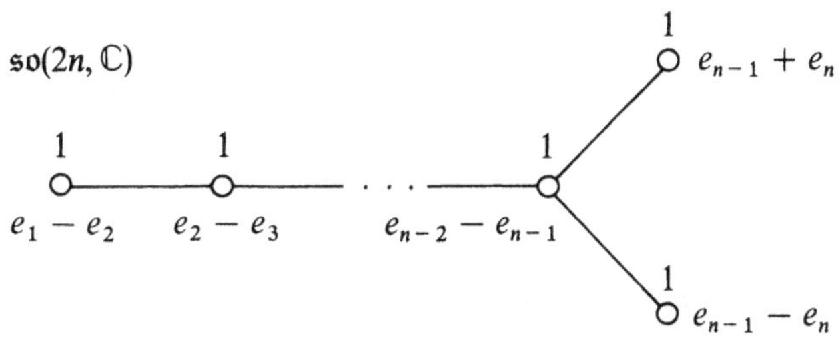

(D_n)

FIGURE 4.2. Dynkin diagrams for compact classical groups

that each corresponds to some compact semisimple group. These diagrams are known as E_6, E_7, E_8, F_4, and G_2.

There is only one family of irreducible nonreduced root systems. The set of roots is the union of that for B_n and C_n, and the system is denoted $(BC)_n$:

$$\Delta = \{\pm e_i \pm e_j, i \neq j\} \cup \{\pm e_k\} \cup \{\pm 2e_k\}.$$

§4. Weyl Group, Algebraically

In this section we shall define the Weyl group associated to a reduced abstract root system, and we shall develop its properties.

Let Δ be a reduced abstract root system in a finite-dimensional real inner product space V. The group W generated by the s_α for $\alpha \in \Delta$ is called the **Weyl group** of Δ. We quickly see two properties of this group:

(1) W is a finite group of orthogonal transformations of V. [In fact, $w\Delta = \Delta$. If w fixes Δ, then w fixes a spanning set of V and so fixes V.]
(2) For any orthogonal transformation r of V, $s_{r\alpha} = rs_\alpha r^{-1}$. In particular, if r is in W and $r\alpha = \beta$, then $s_\beta = rs_\alpha r^{-1}$. [In fact, we just compute that

$$s_{r\alpha}(r\varphi) = r\varphi - \frac{2\langle r\varphi, r\alpha\rangle}{|r\alpha|^2} r\alpha = r\varphi - \frac{2\langle \varphi, \alpha\rangle}{|\alpha|^2} r\alpha = r(s_\alpha \varphi).\Big]$$

Examples of §1.

(1) $\mathfrak{g} = \mathfrak{sl}(n, \mathbb{C})$. $W = \{\text{permutations of } \{e_1, \ldots, e_n\}\}$. $|W| = n!$
(2, 3) $\mathfrak{g} = \mathfrak{so}(2n+1, \mathbb{C})$ or $\mathfrak{sp}(n, \mathbb{C})$. $W = \{\text{permutations and sign changes of } \{e_1, \ldots, e_n\}\}$. $|W| = 2^n n!$
(4) $\mathfrak{g} = \mathfrak{so}(2n, \mathbb{C})$. $W = \{\text{permutations and even sign changes of } \{e_1, \ldots, e_n\}\}$. $|W| = 2^{n-1} n!$

In some ordering, let Δ^+ be the set of positive roots. Then Δ^+ determines a set Π of simple roots. In turn, Π can be used to pick out Δ^+ from Δ, since $\alpha > 0$ means $\alpha = \sum c_i\alpha_i$ with all $c_i \geq 0$.

Now suppose that Π is any basis of l elements α_i such that every expansion of a member α of Δ as $\alpha = \sum c_i\alpha_i$ has all nonzero c_i of the same sign. We call Π a **simple system.** We can then define a set Δ^+, and the claim is that Δ^+ is the set of positive roots for some ordering. In fact, we can use the dual basis to $\{H_{\alpha_i}\}$ to get such an ordering. Thus we have an abstract characterization of the possible Π's that can arise as sets of simple roots, namely all simple systems. For the remainder of this section, positivity will be taken relative to some specified simple system.

Lemma 4.8. Let $\Pi = \{\alpha_1, \ldots, \alpha_l\}$ be a simple system, and let $\alpha > 0$ be in Δ. Then

$$s_{\alpha_i}(\alpha) \text{ is } \begin{cases} = -\alpha_i & \text{if } \alpha = \alpha_i \\ > 0 & \text{if } \alpha \neq \alpha_i. \end{cases}$$

Proof. We have

$$\begin{aligned} s_{\alpha_i}(\textstyle\sum c_j\alpha_j) &= -c_i\alpha_i + \sum_{j\neq i} c_j\left(\alpha_j - \frac{2\langle\alpha_i, \alpha_j\rangle}{|\alpha_i|^2}\alpha_i\right) \\ &= \sum_{j\neq i} c_j\alpha_j + \left(\sum_{j\neq i}\left(-\frac{2\langle\alpha_i, \alpha_j\rangle}{|\alpha_i|^2}\right) - c_i\right)\alpha_i. \end{aligned}$$

If at least one c_j, $j \neq i$, is > 0, then $s_{\alpha_i}(\sum c_k\alpha_k)$ has that same coefficient for α_j and so must be > 0. Otherwise $\sum c_k\alpha_k = \alpha_i$, and we know that $s_{\alpha_i}(\alpha_i) = -\alpha_i$.

Proposition 4.9. Let $\Pi = \{\alpha_1, \ldots, \alpha_l\}$ be a simple system. Then W is generated by the root reflections s_{α_i} for α_i in Π. If α is any root, then there exist $\alpha_j \in \Pi$ and $s \in W$ such that $s\alpha_j = \alpha$.

Proof. Let $W' \subseteq W$ be the group generated by the s_{α_i}. We prove that any $\alpha > 0$ is of the form $s\alpha_j$, $s \in W'$. Writing $\alpha = \sum c_j\alpha_j$, we proceed by induction on $\sum c_j$, which we call level(α). The case of level one is the case of $\alpha = \alpha_i$ in Π, and we can take $s = 1$. Assume the assertion for level $<$ level(α). Write $\alpha = \sum c_i\alpha_i$. Since

$$0 < \langle\alpha, \alpha\rangle = \langle\alpha, \textstyle\sum c_i\alpha_i\rangle = \sum c_i\langle\alpha, \alpha_i\rangle,$$

we must have $\langle\alpha, \alpha_i\rangle > 0$ for some $i = i_0$. Then $\beta = s_{\alpha_{i_0}}(\alpha)$ is > 0 by Lemma 4.8 and has

$$\beta = \sum_{j\neq i_0} c_j\alpha_j + \left(c_{i_0} - \frac{2\langle\alpha, \alpha_{i_0}\rangle}{|\alpha_{i_0}|^2}\right)\alpha_{i_0}.$$

Since $\langle\alpha, \alpha_{i_0}\rangle > 0$, level($\beta$) $<$ level(α). By inductive hypothesis, $\beta = s'\alpha_j$ with s' in W'. Then $\alpha = s_{\alpha_{i_0}}\beta = s_{\alpha_{i_0}}s'\alpha_j$ with $s_{\alpha_{i_0}}s'$ in W'. This completes the induction. If $\alpha < 0$, then we can write $-\alpha = s\alpha_j$, and it follows that $\alpha = ss_{\alpha_j}\alpha_j$.

Finally we show that each s_α is in W'. Write $\alpha = s\alpha_j$ with $s \in W'$. Then property (2) shows that $s_\alpha = ss_{\alpha_j}s^{-1}$, which is in W'. Hence $W \subseteq W'$ and $W = W'$.

Theorem 4.10. If Π and Π' are two simple systems of roots, then there exists one and only one element s in W with $s\Pi = \Pi'$.

Proof of existence. Let Δ^+ and Δ'^+ be the sets of positive roots in question. We have $|\Delta^+| = |\Delta'^+| = q = \frac{1}{2}|\Delta|$. Also $\Delta^+ = \Delta'^+$ if and only if $\Pi = \Pi'$, and $\Delta^+ \neq \Delta'^+$ implies $\Pi \not\subseteq \Delta'^+$ and $\Pi' \not\subseteq \Delta^+$. Let $r = |\Delta^+ \cap \Delta'^+|$. We induct downward on r, the case $r = q$ being handled by using $s = 1$. Let $r < q$. Choose $\alpha_i \in \Pi$ with $\alpha_i \notin \Delta'^+$, so that $-\alpha_i \in \Delta'^+$. If β is in $\Delta^+ \cap \Delta'^+$, $s_{\alpha_i}\beta$ is in Δ^+ by Lemma 4.8. Thus $s_{\alpha_i}\beta$ is in $\Delta^+ \cap s_{\alpha_i}\Delta'^+$. Also $\alpha_i = s_{\alpha_i}(-\alpha_i)$ is in $\Delta^+ \cap s_{\alpha_i}\Delta'^+$. Hence $|\Delta^+ \cap s_{\alpha_i}\Delta'^+| \geq r + 1$. Now $s_{\alpha_i}\Delta'^+$ corresponds to the simple system $s_{\alpha_i}\Pi'$, and by inductive hypothesis we can find $t \in W$ with $t\Pi = s_{\alpha_i}\Pi'$. Then $s_{\alpha_i}t\Pi = \Pi'$, and the induction is complete.

Proof of uniqueness. We may assume $s\Pi = \Pi$, and we are to prove $s = 1$. Assuming the contrary, write $s = s_{i_m} \cdots s_{i_1}$. (Here $s_j = s_{\alpha_j}$.) We prove by induction on m that $s\Pi = \Pi$ implies $s = 1$. If $m = 1$, then $s = s_{i_1}$ and $s\alpha_{i_1} < 0$. If $m = 2$, we obtain $s_{i_2}\Pi = s_{i_1}\Pi$, whence $-\alpha_{i_2}$ is in $s_{i_1}\Pi$ and so $-\alpha_{i_2} = -\alpha_{i_1}$ by Lemma 4.8; hence $s = 1$. Thus assume inductively that

$$t\Pi = \Pi \quad \text{with} \quad t = s_{j_r} \cdots s_{j_1}, \quad r < m, \quad \text{implies} \quad t = 1, \tag{4.8}$$

and let $s = s_{i_m} \cdots s_{i_1}$ satisfy $s\Pi = \Pi$ with $m > 2$.

Put $s' = s_{i_{m-1}} \cdots s_{i_1}$, so that $s = s_{i_m}s'$. Then $s' \neq 1$ by (4.8) for $t = s_{i_m}$. Also $s'\alpha_j < 0$ for some j by (4.8) applied to $t = s'$. The latter fact, together with

$$s_{i_m}s'\alpha_j = s\alpha_j > 0,$$

says $-\alpha_{i_m} = s'\alpha_j$, by Lemma 4.8. Also if $\beta > 0$ and $s'\beta < 0$, then $s'\beta = -\alpha_{i_m} = s'\alpha_j$, so that $\beta = \alpha_j$. Thus s' satisfies

(i) $s'\alpha_j = -\alpha_{i_m}$
(ii) $s'\beta > 0$ for every $\beta > 0$ other than α_j.

Now $s_{i_{m-1}} \cdots s_{i_1}\alpha_j = -\alpha_{i_m} < 0$. Choose k so that $t = s_{i_{k-1}} \cdots s_{i_1}$ satisfies $t\alpha_j > 0$ and $s_{i_k}t\alpha_j < 0$. Then $t\alpha_j = \alpha_{i_k}$. By property (2), $ts_jt^{-1} = s_{i_k}$. Hence $ts_j = s_{i_k}t$.

Put $t' = s_{i_{m-1}} \cdots s_{i_{k+1}}$, so that $s' = t's_{i_k}t = t'ts_j$. Then $t't = s's_j$. Now $\alpha > 0$ and $\alpha \neq \alpha_j$ imply $s_j\alpha = \beta > 0$ with $\beta \neq \alpha_j$. Thus

$$t't\alpha = s's_j\alpha = s'\beta > 0 \qquad \text{by (ii)}$$

$$t't\alpha_j = s'(-\alpha_j) = \alpha_{i_m} > 0 \qquad \text{by (i).}$$

Hence $t't\Pi = \Pi$. Now $t't$ is a product of $m - 2$ s_i's. By inductive hypothesis, $t't = 1$. Then $s's_j = 1$, $s' = s_j$, and $s = s_{i_m}s' = s_{i_m}s_j$, in contradiction to the fact that (4.8) has been proved for $r = 2$. This completes the proof.

Fix a simple system Π and its associated positive system Δ^+. In view of Proposition 4.9, any w in W can be written as $w = s_{i_k} \cdots s_{i_1}$ with $\alpha_{i_k}, \ldots,$

α_{i_1} in Π. Let $l(w)$ be the smallest number of factors needed to represent w in this fashion; $l(w)$ is called the **length** of w.

Proposition 4.11. Fix a simple system Π. For w in W, $l(w)$ is the number of roots $\alpha > 0$ such that $w\alpha < 0$.

Proof. Write $w = s_{i_k} \cdots s_{i_1}$ with $k = l(w)$. First we observe that there are at most k roots $\alpha > 0$ such that $w\alpha < 0$. [In fact, if $s\alpha$ is <0 for j positive roots, then $s_i s\alpha$ is <0 for either $j + 1$ or $j - 1$ positive roots, since s_i either adjoins $\alpha = s^{-1}\alpha_i$ or removes $\alpha = -s^{-1}\alpha_i$, by Lemma 4.8. Thus the extreme case is that each factor of w makes one more positive root get sent into a negative root.]

To get the desired equality , we now show that if w sends exactly k positive roots into negative roots, then w can be expressed as $w = s_{i_k} \cdots s_{i_1}$. We do so by induction on k. For $k = 0$, this follows from the uniqueness in Theorem 4.10. Assume the assertion for $k - 1$. If w sends k positive roots into negative roots, it must send some simple root α_j into a negative root. Set $w' = ws_{\alpha_j}$. Then w' sends $k - 1$ positive roots into negative roots, and the induction goes through.

Proposition 4.12 (Chevalley's Lemma). Fix v in V and let $W_0 = \{w \in W \mid wv = v\}$. Then W_0 is generated by the s_α's such that $\langle v, \alpha \rangle = 0$.

Proof. Choose an ordering from an orthogonal basis with v first. Then $\langle v, \beta \rangle > 0$ implies $\beta > 0$. Choose $w \in W_0$ with $l(w)$ as small as possible so that w is not a product of elements s_α with $\langle v, \alpha \rangle = 0$. Let γ be a simple root such that $w\gamma < 0$. Then $\langle v, \gamma \rangle \geq 0$ since γ is positive. If $\langle v, \gamma \rangle > 0$, then

$$\langle v, w\gamma \rangle = \langle wv, w\gamma \rangle = \langle v, \gamma \rangle > 0,$$

contradiction. Hence $\langle v, \gamma \rangle = 0$. That is, s_γ is in W_0. But then ws_γ is in W_0 and $l(ws_\gamma) < l(w)$. By assumption ws_γ is a product of the required reflections, and therefore so is w.

§5. Weights and Integral Forms

Now let us return to the situation of §2. Thus we work with a compact linear connected reductive group G, its Lie algebra $\mathfrak{g}$, and the complexified Lie algebra $\mathfrak{g}^{\mathbb{C}} = \mathfrak{g} \oplus i\mathfrak{g}$. We fix a Cartan subalgebra $\mathfrak{h} \subseteq \mathfrak{g}$. Weights relative to $\mathfrak{h}$ were defined in §2. We begin with some more complicated examples of representations and weights. Further examples appear in the Problems at the end of the chapter.

Example 1. Polynomial representations of $SO(2n + 1)$.

Let $m = 2n + 1$, and let $SO(2n + 1)$ act on complex-valued polynomials on $\mathbb{R}^m$ of degree $\leq N$ by $\Phi(g)P(x) = P(g^{-1}x)$. We shall use notation for $SO(2n + 1)$ as in Example 2 of §1. Let H_1 be the member of the Cartan subalgebra $\mathfrak{h}$ equal to $\begin{pmatrix} 0 & 1 \\ -1 & 0 \end{pmatrix}$ in the first 2-by-2 block and 0 elsewhere. With φ denoting the Lie algebra representation, we have

$$\varphi(H_1)P\begin{pmatrix} x_1 \\ \vdots \\ x_m \end{pmatrix} = \frac{d}{dt} P \begin{pmatrix} x_1 \cos t - x_2 \sin t \\ x_1 \sin t + x_2 \cos t \\ x_3 \\ \vdots \\ x_m \end{pmatrix}_{t=0} = -x_2 \frac{\partial P}{\partial x_1}(x) + x_1 \frac{\partial P}{\partial x_2}(x). \tag{4.9}$$

For $P(x) = (x_1 + ix_2)^k$, $\varphi(H_1)$ thus acts as the scalar ik. The other 2-by-2 blocks of $\mathfrak{h}$ annihilate this P, and it follows that $(x_1 + ix_2)^k$ is a weight vector of weight $-ke_1$. Similarly $(x_1 - ix_2)^k$ is a weight vector of weight $+ke_1$.

Replacing P in (4.9) by $(x_{2j-1} \pm ix_{2j})Q$ and making the obvious adjustments in the computation, we obtain

$$\varphi(H)((x_{2j-1} \pm ix_{2j})Q) = (x_{2j-1} \pm ix_{2j})(\varphi(H) \mp e_j(H))Q \qquad \text{for } H \in \mathfrak{h}.$$

Since $x_{2j-1} + ix_{2j}$ and $x_{2j-1} - ix_{2j}$ together generate x_{2j-1} and x_{2j} and since $\varphi(H)$ acts as 0 on x_{2n+1}^k, this equation tells us how to compute $\varphi(H)$ on any monomial, hence on any polynomial.

It is clear that the subspace V_N of polynomials homogeneous of degree N is an invariant subspace under Φ. This invariant subspace is spanned by the weight vectors

$$(x_1 + ix_2)^{k_1}(x_1 - ix_2)^{l_1}(x_3 + ix_4)^{k_2} \cdots (x_{2n-1} - ix_{2n})^{l_n} x_{2n+1}^{k_0},$$

where

$$\sum_{j=0}^{n} k_j + \sum_{j=1}^{n} l_j = N. \tag{4.10}$$

Hence the weights of V_N are all expressions

$$\sum_{j=1}^{n} (l_j - k_j)e_j \tag{4.11}$$

subject to (4.10). The dimension of V_N is $\begin{pmatrix} N + m - 1 \\ N \end{pmatrix}$.

The representation Φ is still reducible on V_N if $N \geq 2$. To see this, we observe that

$$L(P) = |x|^2 \left(\frac{\partial^2 P}{\partial x_1^2} + \ldots + \frac{\partial^2 P}{\partial x_m^2} \right)$$

is linear, carries V_N into itself, and is not scalar since L annihilates $(x_1 + ix_2)^N$ but not x_1^N. By Schur's Lemma, Φ is reducible on V_N. The subspace $\mathscr{H}_N$ of harmonic polynomials homogeneous of degree N is an invariant subspace. To calculate $\dim \mathscr{H}_N$, one shows that $\text{image}(L) = |x|^2 V_{N-2}$. Then

$$\dim \mathscr{H}_N = \dim V_N - \dim V_{N-2} = \frac{(N+m-3)!\,(2N+m-2)}{N!\,(m-2)!}. \tag{4.12}$$

In §10, we shall see from (4.12) and (4.11) that Φ is irreducible on $\mathscr{H}_N$.

Example 2. Polynomial representations of SO(2n).

Let $m = 2n$, and let SO(2n) act on complex-valued polynomials on $\mathbb{R}^m$ of degree $\leq N$ by $\Phi(g)P(x) = P(g^{-1}x)$. With notation as in Example 4 of §1, we can go through the same reasoning as with SO(2n + 1), dropping x_{2n+1} when it appears. The subspace V_N of polynomials homogeneous of degree N is invariant, and the weights are as in (4.11) and (4.10), except that $k_0 = 0$. The subspace $\mathscr{H}_N$ of harmonic elements in V_N is invariant, and its dimension is given by (4.12). As with SO(2n + 1), we shall see that Φ is irreducible on $\mathscr{H}_N$.

Example 3. Alternating tensor representations of SO(2n + 1).

Let $m = 2n + 1$, and let $\varepsilon_1, \ldots, \varepsilon_m$ be the standard basis of $\mathbb{R}^m$. The action of SO(2n + 1) by multiplication on column vectors in $\mathbb{R}^m$ leads to a representation Φ of SO(2n + 1) on $V = \bigwedge^N \mathbb{C}^m$. The element H_1 of $\mathfrak{h}$ in Example 1 acts on $\varepsilon_1 + i\varepsilon_2$ by the scalar $-i$ and on $\varepsilon_1 - i\varepsilon_2$ by the scalar $+i$. Thus $\varepsilon_1 + i\varepsilon_2$ and $\varepsilon_1 - i\varepsilon_2$ are weight vectors in $\mathbb{R}^m$ of respective weights $+e_1$ and $-e_1$. Also ε_{2n+1} has weight 0. Then the product rule for differentiation allows us to compute the weights in $\bigwedge^N \mathbb{C}^m$, seeing that they are all expressions

$$\pm e_{j_1} \pm e_{j_2} \pm \ldots \pm e_{j_r} \text{ with } j_1 < j_2 < \ldots < j_r \text{ and } \begin{cases} r \leq N & \text{if } N \leq n \\ r \leq m - N & \text{if } N > n. \end{cases} \tag{4.13}$$

The complex dimension of $\bigwedge^N \mathbb{C}^m$ is $\binom{m}{N}$, and we shall see in §10 that Φ is irreducible.

Example 4. Alternating tensor representations of SO(2n).

The same considerations apply as in Example 3, but with $m = 2n$. The weights for the representation Φ on $\bigwedge^N \mathbb{C}^m$ are as in (4.13), except that $N - r$ must be even if $N \leq n$ and $m - N - r$ must be even if $N > n$. The dimension is still $\binom{m}{N}$, and here Φ will be seen to be irreducible unless $m = 2N$.

Now let us discuss properties of weights in the general case. Let T be the analytic subgroup of G corresponding to the Cartan subalgebra $\mathfrak{h} \subseteq \mathfrak{g}$; recall that T is closed. Let $\mathfrak{h}_{\mathbb{R}} = i\mathfrak{h}$ and $\mathfrak{h}^{\mathbb{C}} = \mathfrak{h} \oplus i\mathfrak{h}$. The set of roots is denoted Δ.

Proposition 4.13. If λ is in $(\mathfrak{h}^{\mathbb{C}})'$, then the following conditions on λ are equivalent:

(i) Whenever $H \in \mathfrak{h}$ satisfies $\exp H = 1$, then $\lambda(H)$ is in $2\pi i\mathbb{Z}$.
(ii) There is a character ξ_λ of T with $\xi_\lambda(\exp H) = e^{\lambda(H)}$ for all H in $\mathfrak{h}$.

Every weight of a finite-dimensional representation has these properties. If λ has these properties, then λ is real-valued on $\mathfrak{h}_{\mathbb{R}}$.

Terminology. A linear functional λ satisfying (i) and (ii) is said to be **analytically integral.**

Proof. If $\mathbb{R}^n$ denotes the universal covering group of T, then $\exp:\mathfrak{h} \to \mathbb{R}^n$ is an isomorphism and $\tilde{\xi}_\lambda(\exp H) = e^{\lambda(H)}$ is a well-defined homomorphism of $\mathbb{R}^n$ into $\mathbb{C}^\times$. Then $\tilde{\xi}_\lambda$ descends to T if and only if (i) holds, and so (i) and (ii) are equivalent. If $\tilde{\xi}_\lambda$ descends to ξ_λ on T, then ξ_λ has compact image in $\mathbb{C}^\times$, hence image in the unit circle, and it follows that λ is real-valued on $\mathfrak{h}_{\mathbb{R}}$. Every weight satisfies (i) and (ii) since it arises from the action of T in the representation space.

Proposition 4.14. If λ in $(\mathfrak{h}^{\mathbb{C}})'$ is analytically integral, then λ satisfies the condition

$$\frac{2\langle \lambda, \alpha \rangle}{|\alpha|^2} \text{ is in } \mathbb{Z} \text{ for each } \alpha \in \Delta. \tag{4.14}$$

Terminology. A linear functional λ satisfying (4.14) is said to be **algebraically integral.**

Proof. If α is in Δ, define H'_α, E'_α, and $E'_{-\alpha}$ as in (4.6). Since SU(2) is simply connected and since

$$\mathfrak{su}(2) \cong \mathfrak{g} \cap (\mathbb{C}H'_\alpha + \mathbb{C}E'_\alpha + \mathbb{C}E'_{-\alpha}),$$

the analytic subgroup with Lie algebra $\mathfrak{g} \cap (\mathbb{C}H'_\alpha + \mathbb{C}E'_\alpha + \mathbb{C}E'_{-\alpha})$ is a homomorphic image of SU(2). Since h in $\mathfrak{sl}(2, \mathbb{C})$ corresponds to H'_α, exp $2\pi i H'_\alpha$ is the identity in G. Since λ is analytically integral, $\lambda(2\pi i H'_\alpha)$ is in $2\pi i\mathbb{Z}$. This means that $2\langle \lambda, \alpha \rangle/|\alpha|^2$ is in $\mathbb{Z}$.

Proposition 4.15. Fix a simple system of roots $\{\alpha_1, \ldots, \alpha_l\}$. Then λ in $(\mathfrak{h}^{\mathbb{C}})'$ is algebraically integral if and only if $2\langle \lambda, \alpha_i \rangle/|\alpha_i|^2$ is in $\mathbb{Z}$ for each simple root α_i.

Proof. If λ is algebraically integral, then $2\langle\lambda, \alpha_i\rangle/|\alpha_i|^2$ is in $\mathbb{Z}$ for each i by definition. Conversely if $2\langle\lambda, \alpha_i\rangle/|\alpha_i|^2$ is an integer for each i, let $\alpha = \sum c_i\alpha_i$ be a positive root. We prove by induction on level(α) $= \sum c_i$ that $2\langle\lambda, \alpha\rangle/|\alpha|^2$ is an integer. Level 1 is the given case. Assume the assertion for level < level(α). Choose α_i with $\langle\alpha, \alpha_i\rangle > 0$, so that $\beta = s_{\alpha_i}\alpha$ has level < level(α). Then

$$\frac{2\langle\lambda, \alpha\rangle}{|\alpha|^2} = \frac{2\langle s_{\alpha_i}\lambda, \beta\rangle}{|\beta|^2} = \frac{2\langle\lambda, \beta\rangle}{|\beta|^2} - \frac{2\langle\lambda, \alpha_i\rangle}{|\alpha_i|^2}\frac{2\langle\alpha_i, \beta\rangle}{|\beta|^2},$$

and the right side is an integer by inductive hypothesis. The proposition follows.

Since all roots are weights, our propositions tell us that we have inclusions

$$\begin{aligned}\mathbb{Z}\text{-combinations of roots} &\subseteq \text{analytically integral forms}\\ &\subseteq \text{algebraically integral forms.}\end{aligned} \tag{4.15}$$

Each of these three sets may be regarded as an additive group in $\mathfrak{h}'_{\mathbb{R}}$.

Let us specialize to the case that G is semisimple. Then Propositions 4.6 and 4.15 show that the right member of (4.15) is a **lattice** in $\mathfrak{h}'_{\mathbb{R}}$, i.e., a discrete subgroup with compact quotient. Proposition 4.6 shows that the left member of (4.15) spans $\mathfrak{h}'_{\mathbb{R}}$ over $\mathbb{R}$ and hence is a sublattice. Thus (4.15) provides us with an inclusion relation for three lattices. Matters are controlled somewhat by the following result.

Proposition 4.16. If G is semisimple, then the index of the lattice of $\mathbb{Z}$-combinations of roots in the lattice of algebraically integral forms is exactly the determinant of the Cartan matrix.

Lemma 4.17. Let F be a free abelian group of rank l, and let A be a subgroup of rank l. Then it is possible to choose $\mathbb{Z}$ bases $\{t_i\}$ of F and $\{u_i\}$ of A such that $u_i = \delta_i t_i$ (with $\delta_i \in \mathbb{Z}$) for all i and such that δ_i divides δ_j if $i < j$.

Proof omitted.

Lemma 4.18. Let F be a free abelian group of rank l, and let A be a subgroup of rank l. Let $\{t_i\}$ and $\{u_i\}$ be $\mathbb{Z}$ bases of F and A, respectively, and suppose that $u_j = \sum_{i=1}^{l} d_{ij}t_i$. Then F/A has order $|\det(d_{ij})|$.

Proof. A change of basis in the t's or in the u's corresponds to multiplying (d_{ij}) by an integer matrix of determinant ± 1. In view of Lemma 4.17, we may therefore assume (d_{ij}) is diagonal. Then the result is obvious.

Proof of Proposition 4.16. Fix a simple system $\{\alpha_1, \ldots, \alpha_l\}$ and define $\{\lambda_1, \ldots, \lambda_l\}$ by $2\langle\lambda_i, \alpha_j\rangle/|\alpha_j|^2 = \delta_{ij}$. The $\{\lambda_i\}$ form a $\mathbb{Z}$ basis of the lattice of algebraically integral forms, by Proposition 4.15, and the $\{\alpha_i\}$ form a $\mathbb{Z}$

basis of the lattice generated by the roots, by Proposition 4.6. Write

$$\alpha_j = \sum_{k=1}^{l} d_{kj}\lambda_k$$

and apply $2\langle\alpha_i, \cdot\rangle/|\alpha_i|^2$ to both sides. Then we see that $d_{ij} = 2\langle\alpha_i, \alpha_j\rangle/|\alpha_i|^2$. Since the determinant of the Cartan matrix is known to be positive, the result follows from Lemma 4.18.

§6. Centralizers of Tori

The analytic theory for compact connected Lie groups begins with the deep fundamental fact that the exponential mapping is onto. We shall prove this result for the classical compact groups, giving only references for the proof in the general case. Our first application of this result will be to relate the analytically integral forms for a compact connected group with those for a covering group.

A **torus** (as a Lie group) is the product of circle groups, and a compact connected abelian Lie group is necessarily a torus (A.119).

Theorem 4.19. For any compact connected Lie group G, the exponential mapping carries $\mathfrak{g}$ onto G.

Proof for SU(n). If $x \in \mathrm{SU}(n)$ is given, we can write $x = gtg^{-1}$ with g in U(n) and t diagonal in SU(n), by the finite-dimensional Spectral Theorem. We can then multiply g by a scalar to make it have determinant 1. As a result, we see that x is conjugate within SU(n) to a member of the Cartan subgroup (maximal torus) T. Since exp carries $\mathfrak{h}$ onto T, it carries $\mathfrak{g}$ onto G.

Proof for SO(n). If $x \in \mathrm{SO}(n)$ is given, we can write $x = gtg^{-1}$ with g in U(n) and t complex diagonal. The eigenvalues of t must occur in conjugate pairs, since x is real, and we can conjugate t by a matrix in U(n) to get it to an element s in the maximal torus T of SO(n) (cf. §1). It follows from the Rational Canonical Form that x is conjugate to s by a real matrix, which we can take to be in O(n). If the conjugating matrix has determinant -1, we conjugate by the diagonal matrix diag$(1, 1, \ldots, 1, -1)$ to see that x is conjugate by a member of SO(n) to a member s' of T. Since exp carries $\mathfrak{h}$ onto T, it carries $\mathfrak{g}$ onto G.

Proof for Sp(n). If $x \in \mathrm{Sp}(n)$ is given, then x is in U($2n$) and x satisfies $x^{\mathrm{tr}}Jx = J$. Hence x satisfies $JxJ^{-1} = \bar{x}$. It follows that if v is an eigenvector for x with eigenvalue λ, then $J\bar{v}$ is an eigenvector for x with eigenvalue $\bar{\lambda}$.

Let us observe that $\{v, J\bar{v}\}$ always spans a two-dimensional space. [In fact, if $J\bar{v} = cv$, then conjugation gives $Jv = \bar{c}\bar{v}$. Hence

$$-\bar{v} = J^2\bar{v} = J(cv) = c\bar{c}\bar{v},$$

and $|c|^2 = -1$, contradiction.] Thus we can find an orthonormal basis of eigenvectors of x such that if the $(n+j)^{\text{th}}$ is $\begin{pmatrix} v_1 \\ v_2 \end{pmatrix}$, then the jth is $\begin{pmatrix} \bar{v}_2 \\ -\bar{v}_1 \end{pmatrix}$. Taking these vectors as the columns of a matrix g, we see that g is unitary and satisfies $Jg = \bar{g}J$. Hence g is in $\mathrm{Sp}(n)$, the condition $\det g = 1$ being forced. Then $g^{-1}xg$ is diagonal, with the $(n+j)^{\text{th}}$ entry the conjugate of the jth. So every member of $\mathrm{Sp}(n)$ is conjugate to a member of the maximal torus T, and then exp must carry $\mathfrak{g}$ onto G.

Lemma 4.20. Let A be a compact abelian Lie group such that A/A_0 is cyclic, where A_0 denotes the identity component of A. Then A has an element whose powers are dense in A.

Proof. Since A_0 is a torus, we can choose a_0 in A_0 such that the powers of a_0 are dense in A_0. Let $N = |A/A_0|$, and let b be a representative in A of a generating coset of A/A_0. Since b^N is in A_0, we can find c in A_0 with $b^Nc^N = a_0$. Then the closure of the powers of bc is a subgroup containing A_0 and a representative of each coset of A/A_0, hence is all of A.

Theorem 4.21. Let G be a compact connected Lie group, and let S be a torus (subgroup) of G. If g in G centralizes S, then there is a torus S' in G containing both S and g.

Proof. Let A be the closure of $\bigcup_{n=-\infty}^{\infty} g^nS$. Then the identity component A_0 is a torus. Since A_0 is open in A, $\bigcup_{n=-\infty}^{\infty} g^nA_0 = A$. By compactness some power of g is in A_0. If N denotes the smallest positive such power, then A/A_0 is cyclic of order N. Applying Lemma 4.20, we can find a in A whose powers are dense in A. By Theorem 4.19, we can write $a = \exp X$ for some X in $\mathfrak{g}$. Then the closure of $\{\exp tX, -\infty < t < \infty\}$ is a torus S' containing A, hence containing both S and g.

Corollary 4.22. In a compact connected Lie group, the centralizer of a torus (subgroup) is connected.

Proof. Theorem 4.21 shows the centralizer is the union of the tori containing the given torus.

Corollary 4.23. A maximal torus in a compact connected Lie group is equal to its own centralizer.

Proof. Apply Theorem 4.21.

Corollary 4.24. The center of a compact connected Lie group is contained in every maximal torus.

Proof. Apply Corollary 4.23.

Corollary 4.25. If G is a compact connected Lie group and $\tilde{G}$ is a finite covering group, then the index of the group of analytically integral forms for G in the group of analytically integral forms for $\tilde{G}$ equals the index of G in $\tilde{G}$.

Proof. This follows by combining Corollary 4.24 and Proposition 4.13.

Theorem 4.26 (Weyl's Theorem). If G is a compact linear connected semisimple Lie group, then the universal covering group of G is compact.

Lemma 4.27. If G is a compact connected Lie group, then its fundamental group is finitely generated.

Proof. By (A.115) let $G = \tilde{G}/Z$, where $\tilde{G}$ is the universal covering group of G and Z is a discrete subgroup of the center of $\tilde{G}$. Here Z is isomorphic to the fundamental group of G, by (A.61). Let $e:\tilde{G} \to G$ be the covering homomorphism. About each point x in G, choose a connected simply connected open neighborhood N_x and a connected simply connected neighborhood N'_x with closure in N_x. Extract a finite subcover of G of the N'_x, say $N'_{x_1}, \ldots, N'_{x_n}$. Then associate a component M_{x_j} of $e^{-1}(N'_{x_j}) \subseteq \tilde{G}$ to each N'_{x_j}. The result is that $U = \bigcup_{j=1}^{n} M_{x_j}$ is an open set in $\tilde{G}$ such that $\bar{U}$ is compact and such that $\tilde{G} = ZU$. By enlarging U, we may suppose that 1 is in U and that $U = U^{-1}$.

The set $\bar{U}\bar{U}^{-1}$ is compact in $\tilde{G}$ and is covered by the open sets zU, $z \in Z$. Thus we can find $z_1, \ldots, z_k$ in Z such that

$$\bar{U}\bar{U}^{-1} \subseteq \bigcup_{j=1}^{k} z_j U. \tag{4.16}$$

Let Z_1 be the subgroup of Z generated by $z_1, \ldots, z_k$, and let E be the image of $\bar{U}$ in $\tilde{G}/Z_1$. Then E contains the identity and $E = E^{-1}$, and (4.16) shows that $EE^{-1} \subseteq E$. Thus E is a subgroup of $\tilde{G}/Z_1$. Since it contains the image of U, it is open, and thus $E = \tilde{G}/Z_1$ by connectedness. Since $\bar{U}$ is compact, E is compact. Consequently E is a finite-sheeted covering group of G. That is, G has a finite-sheeted covering group whose fundamental group Z_1 is finitely generated. The lemma follows.

Proof of Theorem 4.26. Let $G = \tilde{G}/Z$, where $\tilde{G}$ is the universal covering group of G and Z is a discrete subgroup of the center of $\tilde{G}$. Here Z is a finitely generated abelian group, by Lemma 4.27. If Z is finite, we are done. Otherwise Z has an infinite cyclic direct summand, and we can find a subgroup Z_1 of Z such that Z_1 has finite index in Z greater than the determinant of the Cartan matrix. Then $\tilde{G}/Z_1$ is a compact covering group of G with a number of sheets exceeding the determinant of the Cartan matrix. By Corollary 4.25 the index of the lattice of analytically integral forms for G in the corresponding lattice for $\tilde{G}/Z_1$ exceeds the determinant of the Cartan matrix. By (4.10) and Proposition 4.16, $\tilde{G}/Z_1$ cannot be a linear group. But this conclusion contradicts Theorem 1.15. Hence Theorem 4.26 follows.

§7. Theorem of the Highest Weight

We continue with the notation G, $\mathfrak{g}$, $\mathfrak{g}^{\mathbb{C}}$, $\mathfrak{h}$, $\mathfrak{h}_{\mathbb{R}}$, $\mathfrak{h}^{\mathbb{C}}$, Δ of §§2 and 5. Introduce an ordering for $\mathfrak{h}'_{\mathbb{R}}$ as in §3, and let Δ^+ and Π be the corresponding sets of positive roots and simple roots, respectively. A linear functional λ on $\mathfrak{h}^{\mathbb{C}}$ that is real on $\mathfrak{h}_{\mathbb{R}}$ is said to be **dominant** if $2\langle\lambda, \alpha\rangle/|\alpha|^2 \geq 0$ for every $\alpha \in \Delta^+$.

Theorem 4.28 (Theorem of the highest weight). Let G be a compact linear connected reductive group. Apart from equivalence, the irreducible representations Φ of G stand in one-one correspondence with the dominant, analytically integral linear functionals λ on $\mathfrak{h}^{\mathbb{C}}$, the correspondence being that λ is the **highest weight** (largest weight in the ordering) of Φ_λ. The highest weight λ of Φ_λ has these properties:

(a) λ depends only on the simple system Π and not on the particular ordering
(b) the weight space V_λ for λ is one-dimensional
(c) each E_α, for α in Δ^+, annihilates the members of V_λ, and the members of V_λ are the only vectors with this property
(d) every weight for Φ_λ is of the form $\lambda - \sum_{i=1}^{l} n_i\alpha_i$ with the n_i integers ≥ 0 and the α_i in Π.

Furthermore, if G is semisimple and simply connected, then every algebraically integral linear functional on $\mathfrak{h}^{\mathbb{C}}$ is analytically integral, so that the correspondence in this case is between irreducible representations of G and dominant, algebraically integral, linear functionals on $\mathfrak{h}^{\mathbb{C}}$.

Examples of §5.

(1) $G = \mathrm{SO}(2n+1)$, $\mathscr{H}_N$ = harmonic polynomials homogeneous of degree N. We can use the positive system Δ^+ of §3, namely

$$\Delta^+ = \{e_i \pm e_j, i < j\} \cup \{e_k\}.$$

Referring to (4.11) and (4.10), we see that the highest weight is Ne_1. We do not yet know that the representation is irreducible on $\mathscr{H}_N$, but the proof of Theorem 4.28 will show for now that a highest weight vector generates an irreducible subspace. Theorem 4.48 will tell us the dimension of this irreducible subspace, and we will be able to compare it with (4.12) to conclude that the representation is irreducible on the full space $\mathscr{H}_N$.

(2) $G = SO(2n)$, $\mathscr{H}_N$ = harmonic polynomials homogeneous of degree N. We can use the positive system of §3, namely

$$\Delta^+ = \{e_i \pm e_j, i < j\}.$$

Then the highest weight is Ne_1.

(3) $G = SO(2n+1)$, $V = \bigwedge^N \mathbb{C}^{2n+1}$. Here $V = 0$ if $N > 2n+1$. If $N \leq 2n+1$, the highest weight is

$$\begin{aligned} &e_1 + \ldots + e_N && \text{if } N \leq n \\ &e_1 + \ldots + e_{2n+1-N} && \text{if } N > n. \end{aligned}$$

After we prove the irreducibility in §10, it will follow from Theorem 4.28 that $\bigwedge^N \mathbb{C}^{2n+1}$ and $\bigwedge^{2n+1-N} \mathbb{C}^{2n+1}$ give equivalent representations.

(4) $G = SO(2n)$, $V = \bigwedge^N \mathbb{C}^{2n}$. Here $V = 0$ if $N > 2n$. If $N \leq 2n$, the highest weight is

$$\begin{aligned} &e_1 + \ldots + e_N && \text{if } N \leq n \\ &e_1 + \ldots + e_{2n-N} && \text{if } N > n. \end{aligned}$$

After we prove the irreducibility for $N \neq n$ in §10, it will follow from Theorem 4.28 that $\bigwedge^N \mathbb{C}^{2n}$ and $\bigwedge^{2n-N} \mathbb{C}^{2n}$ give equivalent representations.

Proof of existence of the correspondence. Let Φ be an irreducible representation of G on a space V. Then Φ can be made unitary, and the theory of §2 and §5 applies. Let φ be the corresponding representation of $\mathfrak{g}$, and extend φ to a representation of $U(\mathfrak{g}^{\mathbb{C}})$ with $\varphi(1) = 1$. The representation φ has weights, and we let λ be the highest. Then λ is analytically integral, by Proposition 4.13.

If α is in Δ^+, then $\lambda + \alpha$ exceeds λ and cannot be a weight. Thus $E_\alpha \in \mathfrak{g}_\alpha$ and $v \in V_\lambda$ imply $\varphi(E_\alpha)v = 0$, by Proposition 4.1b. This proves the first part of (c).

Since Φ is irreducible, so is φ. Thus $\varphi(U(\mathfrak{g}^{\mathbb{C}}))v = V$ for each $v \neq 0$ in V. Let $\gamma_1, \ldots, \gamma_n$ be an enumeration of Δ^+, and let $H_1, \ldots, H_l$ be a basis of $\mathfrak{h}^{\mathbb{C}}$ over $\mathbb{C}$. By the Birkhoff-Witt Theorem, the monomials

$$E_{-\gamma_1}^{i_1} \cdots E_{-\gamma_n}^{i_n} H_1^{j_1} \cdots H_l^{j_l} E_{\gamma_1}^{k_1} \cdots E_{\gamma_n}^{k_n} \tag{4.17}$$

form a basis of $U(\mathfrak{g}^{\mathbb{C}})$. Let us apply φ of each of these monomials to v in V_λ. The E_γ's give 0, the H's multiply by constants, and the $E_{-\gamma}$'s push the weight down. Consequently the only members of V_λ that can be obtained by applying φ of (4.17) to v are the vectors $\mathbb{C}v$. Thus V_λ is one-dimensional, and (b) is proved.

The effect of φ of (4.17) applied to v in V_λ is to give a weight vector with weight

$$\lambda - \sum_{j=1}^{n} i_j \gamma_j, \tag{4.18}$$

and these vectors span V. Thus the weights (4.18) are the only weights of φ, and (d) follows from Proposition 4.6. Also (d) implies (a).

To prove the second half of (c), let $v \notin V_\lambda$ satisfy $\varphi(E_\alpha)\, v = 0$ for all $\alpha \in \Delta^+$. Subtracting the component in V_λ, we may assume v has 0 component in V_λ. Let λ_0 be the largest weight such that v has a nonzero component in V_{λ_0}, and let v' be the component. Then $\varphi(E_\alpha)v' = 0$ for all $\alpha \in \Delta^+$, and $\varphi(\mathfrak{h}^{\mathbb{C}})v' \subseteq \mathbb{C}v'$. Applying φ of (4.17), we see that

$$V = \sum \mathbb{C}\varphi(E_{-\gamma_1})^{i_1} \cdots \varphi(E_{-\gamma_n})^{i_n} v'.$$

Every weight of vectors on the right is strictly lower than λ, and we have a contradiction to the fact that λ occurs as a weight.

Finally we prove that λ is dominant. Let α be in Δ^+, and form H'_α, E'_α, and $E'_{-\alpha}$ as in (4.6). These vectors span a subalgebra of $\mathfrak{g}^{\mathbb{C}}$ isomorphic to $\mathfrak{sl}(2, \mathbb{C})$, and the isomorphism carries H'_α to h. For $v \neq 0$ in V_λ, the subspace of V spanned by all

$$\varphi(E'_{-\alpha})^p \varphi(H'_\alpha)^q \varphi(E'_\alpha)^r v$$

is stable under $\mathfrak{sl}(2, \mathbb{C})$, and (c) shows it is the same as the span of all $\varphi(E'_{-\alpha})^p v$. On these vectors, $\varphi(H'_\alpha)$ acts with eigenvalue

$$(\lambda - p\alpha)(H'_\alpha) = \frac{2\langle\lambda, \alpha\rangle}{|\alpha|^2} - 2p,$$

and the largest eigenvalue of $\varphi(H'_\alpha)$ is therefore $2\langle\lambda, \alpha\rangle/|\alpha|^2$. By Corollary 2.2 and Theorem 2.4 the largest eigenvalue for h in any representation of $\mathfrak{sl}(2, \mathbb{C})$ is ≥ 0. Thus $2\langle\lambda, \alpha\rangle/|\alpha|^2$ is ≥ 0, and λ is dominant.

Proof that correspondence is one-one. Let Φ and Φ' be irreducible on V and V', respectively, both with highest weight λ, and let φ and φ' be the corresponding representations of $U(\mathfrak{g}^{\mathbb{C}})$. Let v_0 and v'_0 be nonzero highest weight vectors. Form $\Phi \oplus \Phi'$ on $V \oplus V'$. We claim that

$$S = (\varphi \oplus \varphi')(U(\mathfrak{g}^{\mathbb{C}}))(v_0 \oplus v'_0)$$

is an irreducible subspace of $V \oplus V'$.

Certainly S is invariant. Write S as the direct sum $S_1 \oplus \ldots \oplus S_M$ of irreducibles, and let

$$v_0 \oplus v'_0 = v_1 + \ldots + v_M \tag{4.19}$$

be the corresponding decomposition of $v_0 \oplus v'_0$. We apply the formula

$$(\varphi \oplus \varphi')(H)(v_0 \oplus v'_0) = \varphi(H)v_0 \oplus \varphi'(H)v'_0 = \lambda(H)(v_0 \oplus v'_0) \qquad \text{for } H \in \mathfrak{h}^{\mathbb{C}}$$

to both sides of (4.19) and use the invariance of each S_j to conclude that each v_j is a weight vector in $V \oplus V'$ of weight λ. But S is given as φ of (4.17) applied to $v_0 \oplus v'_0$, and S can have only a one-dimensional subspace of vectors of weight λ because

$$(\varphi \oplus \varphi')(E_\alpha)(v_0 \oplus v_0') = \varphi(E_\alpha)v_0 + \varphi'(E_\alpha)v_0' = 0 \qquad \text{for } \alpha \in \Delta^+.$$

Thus the number m of irreducible constituents of S is 1, and $\varphi \oplus \varphi'$ is irreducible on S.

The projection of S to V commutes with the representations and is not identically 0. By Schur's Lemma, $\varphi \oplus \varphi'|_S$ is equivalent with φ. Similarly it is equivalent with φ'. Hence φ and φ' are equivalent.

Steps that remain. We still have to prove that every dominant, analytically integral λ arises as the highest weight of an irreducible representation and that if G is semisimple and simply connected, then algebraically integral implies analytically integral. We shall prove these facts in this section and the next, proceeding as follows.

First suppose G is semisimple and simply connected. For λ algebraically integral we shall introduce a simple system Π that makes λ dominant (Proposition 4.30), and then we shall give an algebraic construction of an irreducible representation of $\mathfrak{g}$, hence of G, with highest weight λ (Theorem 4.37). This will establish all our assertions about the semisimple, simply connected case.

Now suppose G is merely reductive. We shall note below (Proposition 4.32) that G is the product of commuting *closed* subgroups $G = (Z_G)_0 G_{ss}$, where $(Z_G)_0$ is the identity component of the center and G_{ss} is the analytic subgroup with Lie algebra $[\mathfrak{g}, \mathfrak{g}]$; then G_{ss} is linear semisimple, by Proposition 1.1. The universal covering group $\tilde{G}$ of G is of the form $\mathbb{R}^n \times \tilde{G}_{ss}$, where $\tilde{G}_{ss}$ is the universal covering group of G_{ss}, and $\tilde{G}_{ss}$ is compact by Theorem 4.26 and linear by Theorem 1.15. If λ is dominant and analytically integral for G, then λ is algebraically integral (Proposition 4.14), and the previous paragraph implies there is an irreducible unitary representation $\tilde{\Phi}$ of $\tilde{G}$ with highest weight λ. ($\tilde{\Phi}$ is scalar on $\mathbb{R}^n$ and is obtained as above on $\tilde{G}_{ss}$.) By Corollary 4.24, $Z_{\tilde{G}} \subseteq \exp \mathfrak{h}$ as subgroups of $\tilde{G}$. Since λ is analytically integral for G, ξ_λ on $\tilde{G}$ is trivial on the discrete subgroup Z of $Z_{\tilde{G}}$ such that $G \cong \tilde{G}/Z$. By Schur's Lemma, $\tilde{\Phi}$ is scalar on $Z_{\tilde{G}}$, and its scalar values must agree with those of ξ_λ since λ is a weight. Thus $\tilde{\Phi}$ is trivial on Z, and $\tilde{\Phi}$ descends to a representation Φ of G with the required properties.

In short, the proof of Theorem 4.28 is completed by Propositions 4.30 and 4.32, together with Theorem 4.37.

If we combine Theorem 4.28 and Corollary 4.25, then we obtain the following corollary.

Corollary 4.29. Let G be a compact linear connected semisimple Lie group. Then the order of the fundamental group of G is the index of the lattice of analytically integral forms on $\mathfrak{h}^{\mathbb{C}}$ in the lattice of algebraically integral linear forms.

Examples. It follows immediately from Corollary 4.29 that SU(n) is simply connected for $n \geq 2$, SO(n) has fundamental group of order 2 for $n \geq 3$, and Sp(n) is simply connected for $n \geq 1$. The double covering group of SO(n) is known as the **spin group.**

Now let us return to the proof of Theorem 4.28. We begin with two propositions we left unproved.

Proposition 4.30. If λ is in $\mathfrak{h}'_{\mathbb{R}}$, then there exists a simple system Π for which λ is dominant.

Proof. We may assume $\lambda \neq 0$. Choose $\varphi_1 = \lambda$, extend to an orthogonal basis $\varphi_1, \ldots, \varphi_l$ of $\mathfrak{h}'_{\mathbb{R}}$, obtain an ordering as in §3, and let Π be the set of simple roots. Then Π has the required property.

Corollary 4.31. If λ is in $\mathfrak{h}'_{\mathbb{R}}$ and if a positive system Δ^+ is specified, then λ is conjugate via W to a dominant element.

Proof. This follows from Proposition 4.30 and Theorem 4.10.

Proposition 4.32. If G is a compact linear connected reductive group, then the analytic subgroup G_{ss} with Lie algebra $[\mathfrak{g}, \mathfrak{g}]$ is compact, and G is the commuting product of two closed subgroups $G = (Z_G)_0 G_{ss}$.

Proof. By Proposition 1.1. $[\mathfrak{g}, \mathfrak{g}]$ is semisimple. Thus $\mathrm{Ad}(G_{ss}) = \mathrm{Ad}(G)$, and this is compact, being the continuous image of G under Ad. Then $\mathrm{Ad}(G_{ss})$ is a compact group with a semisimple Lie algebra, and Weyl's Theorem shows G_{ss} is compact. The rest is clear.

§8. Verma Modules

In this section we complete the proof of Theorem 4.28: Under the assumption that G is compact semisimple and λ is dominant and algebraically integral, we shall give an algebraic construction of an irreducible finite-dimensional representation of $\mathfrak{g}$ with highest weight λ.

By means of Proposition 3.1, we can identify representations of $\mathfrak{g}$ with left $U(\mathfrak{g}^{\mathbb{C}})$ modules, and it is customary to drop the name of the representation when working in this fashion. The idea is to consider all $U(\mathfrak{g}^{\mathbb{C}})$ modules that possess a vector that behaves like a highest weight vector with weight λ. Among these we shall see that there is one (called a Verma module) with a universal mapping property. A suitable quotient of the Verma module will give us our irreducible representation, and the main step will be to prove that it is finite-dimensional.

The notation is as follows: G is compact semisimple with Lie algebra $\mathfrak{g}$ and Cartan subalgebra $\mathfrak{h}$. We form $\mathfrak{g}^{\mathbb{C}}$, $U(\mathfrak{g}^{\mathbb{C}})$, $\mathfrak{h}_{\mathbb{R}} = i\mathfrak{h}$, $\mathfrak{h}^{\mathbb{C}}$, Δ, and W as in previous sections. Then we fix an ordering on $\mathfrak{h}'_{\mathbb{R}}$, and this determines the positive system Δ^+ and the simple system $\Pi = \{\alpha_1, \ldots, \alpha_l\}$. Define

$$\mathfrak{n}^+ = \sum_{\alpha \in \Delta^+} \mathfrak{g}_\alpha$$

$$\mathfrak{n}^- = \sum_{\alpha \in \Delta^+} \mathfrak{g}_{-\alpha}$$

$$\mathfrak{b} = \mathfrak{h} \oplus \mathfrak{n}^+$$

$$\delta = \tfrac{1}{2} \sum_{\alpha \in \Delta^+} \alpha.$$

Then $\mathfrak{n}^+$, $\mathfrak{n}^-$, and $\mathfrak{b}$ are Lie subalgebras of $\mathfrak{g}^{\mathbb{C}}$, by Proposition 4.1b, and $\mathfrak{g}^{\mathbb{C}} = \mathfrak{b} \oplus \mathfrak{n}^-$ as a direct sum of vector spaces.

In order to make the notation more symmetric later (with respect to the Weyl group W), we shall shift our parameters from the expected ones by the functional δ defined above. The relevant property of δ that leads to the symmetry is as follows.

Proposition 4.33. The element δ in $\mathfrak{h}'_{\mathbb{R}}$ satisfies $s_{\alpha_i}(\delta) = \delta - \alpha_i$ and $2\langle \delta, \alpha_i \rangle / |\alpha_i|^2 = 1$ for $1 \leq i \leq l$.

Proof. By Lemma 4.8, s_{α_i} permutes the positive roots other than α_i and sends α_i into $-\alpha_i$. Therefore

$$s_{\alpha_i}(2\delta) = s_{\alpha_i}(2\delta - \alpha_i) + s_{\alpha_i}(\alpha_i) = (2\delta - \alpha_i) - \alpha_i = 2(\delta - \alpha_i),$$

and $s_{\alpha_i}(\delta) = \delta - \alpha_i$. Using the definition of s_{α_i}, we then see that

$$2\langle \delta, \alpha_i \rangle / |\alpha_i|^2 = 1.$$

Let the complex vector space $V \neq 0$ be a left $U(\mathfrak{g}^{\mathbb{C}})$ module. If μ is in $(\mathfrak{h}^{\mathbb{C}})'$, we say μ is a **weight** if

$$V_\mu = \{v \in V \mid Hv = \mu(H)v \text{ for all } H \in \mathfrak{h}^{\mathbb{C}}\}$$

is not 0, and then V_μ is the **weight space** for μ. We call $\dim V_\mu$ the multiplicity of μ. A vector in V_μ is said to be of **weight** μ. The sum $\sum V_\mu$ is necessarily a direct sum. As in Proposition 4.1b, we have

$$\mathfrak{g}_\alpha(V_\mu) \subseteq V_{\mu+\alpha} \tag{4.20}$$

if α is in Δ and μ is in $(\mathfrak{h}^{\mathbb{C}})'$. Moreover, (4.20) and the root space decomposition of $\mathfrak{g}^{\mathbb{C}}$ show that

$$\mathfrak{g}^{\mathbb{C}}\Big(\sum_{\mu \in (\mathfrak{h}^{\mathbb{C}})'} V_\mu\Big) \subseteq \sum_{\mu \in (\mathfrak{h}^{\mathbb{C}})'} V_\mu. \tag{4.21}$$

A **highest weight vector** for V is by definition a weight vector $v \neq 0$ with $\mathfrak{n}^+ v = 0$. (Notice that $\mathfrak{n}^+ v$ will be 0 as soon as $E_\alpha v = 0$ for all α in Π, by induction on the level of the root.) A **highest weight module** is a $U(\mathfrak{g}^{\mathbb{C}})$ module generated by a highest weight vector. "Verma modules," to be defined below, will be universal highest weight modules.

Proposition 4.34. Let M be a highest weight module for $U(\mathfrak{g}^{\mathbb{C}})$, and let v be a highest weight vector generating M. Suppose v is of weight λ. Then

(a) $M = U(\mathfrak{n}^-)v$

(b) $M = \sum_{\mu \in (\mathfrak{h}^{\mathbb{C}})'} M_\mu$ with each M_μ finite-dimensional and with $\dim(M_\lambda) = 1$

(c) every weight of M is of the form $\lambda - \sum_{i=1}^{l} n_i \alpha_i$ with the α_i's in Π and with the n_i's integers ≥ 0.

Proof.

(a) We have $\mathfrak{g}^{\mathbb{C}} = \mathfrak{n}^- \oplus \mathfrak{h}^{\mathbb{C}} \oplus \mathfrak{n}^+$. By the Birkhoff-Witt Theorem, $U(\mathfrak{g}^{\mathbb{C}}) = U(\mathfrak{n}^-)U(\mathfrak{h}^{\mathbb{C}})U(\mathfrak{n}^+)$. On the vector v, $U(\mathfrak{n}^+)$ and $U(\mathfrak{h}^{\mathbb{C}})$ act to give multiples of v. Thus $U(\mathfrak{g}^{\mathbb{C}}) = U(\mathfrak{n}^-)v$. Since v generates M, $M = U(\mathfrak{g}^{\mathbb{C}})v = U(\mathfrak{n}^-)v$.

(b, c) By (4.21), $\sum M_\mu$ is $U(\mathfrak{g}^{\mathbb{C}})$-stable, and it contains v. Since $M = U(\mathfrak{g}^{\mathbb{C}})v$, $M = \sum M_\mu$. By (a), $M = U(\mathfrak{n}^-)v$, and (4.20) shows that

$$E_{-\gamma_1}^{j_1} \cdots E_{-\gamma_n}^{j_n} v \qquad (\text{all } \gamma_j \in \Delta^+) \tag{4.22}$$

is a weight vector with weight $\mu = \lambda - j_1\gamma_1 - \ldots - j_n\gamma_n$, from which (c) follows. The number of expressions (4.22) leading to this μ is finite, and so $\dim M_\mu < \infty$. The number of expressions (4.22) leading to λ is 1, from v itself, and so $\dim M_\lambda = 1$.

Before defining Verma modules, we recall some facts about tensor products of associative algebras. Let M_1 and M_2 be complex vector spaces, and let A and B be complex associative algebras with identity. Suppose M_1 is a right B module and M_2 is a left B module, and suppose M_1 is also a left A module in such a way that $(am_1)b = a(m_1 b)$. We define

$$M_1 \otimes_B M_2 = \frac{M_1 \otimes_{\mathbb{C}} M_2}{\text{subspace generated by all } m_1 b \otimes m_2 - m_1 \otimes bm_2},$$

and we let A act on the quotient by $a(m_1 \otimes m_2) = (am_1) \otimes m_2$. Then $M_1 \otimes_B M_2$ is a left A module, and it has the following universal mapping property: Whenever $\varphi: M_1 \times M_2 \to E$ is a bilinear map into a complex vector space E such that $\varphi(m_1 b, m_2) = \varphi(m_1, bm_2)$, then there exists a unique linear map $\tilde{\varphi}: M_1 \otimes_B M_2 \to E$ such that $\varphi(m_1, m_2) = \tilde{\varphi}(m_1 \otimes m_2)$.

Now let λ be in $(\mathfrak{h}^{\mathbb{C}})'$, and make $\mathbb{C}$ into a left $U(\mathfrak{b})$ module $\mathbb{C}_{\lambda-\delta}$ by defining

$$\begin{aligned} Hz &= (\lambda - \delta)(H)z && \text{for } H \in \mathfrak{h}^{\mathbb{C}}, z \in \mathbb{C} \\ Xz &= 0 && \text{for } X \in \mathfrak{n}^+. \end{aligned} \tag{4.23}$$

(Equation (4.23) defines a one-dimensional representation of $\mathfrak{b}$, and thus $\mathbb{C}_{\lambda-\delta}$ becomes a left $U(\mathfrak{b})$ module.) The algebra $U(\mathfrak{g}^{\mathbb{C}})$ itself is a right $U(\mathfrak{b})$ module and a left $U(\mathfrak{g}^{\mathbb{C}})$ module under multiplication, and we define the **Verma module** $V(\lambda)$ to be the left $U(\mathfrak{g}^{\mathbb{C}})$ module

$$V(\lambda) = U(\mathfrak{g}^{\mathbb{C}}) \otimes_{U(\mathfrak{b})} \mathbb{C}_{\lambda-\delta}.$$

Proposition 4.35. Let λ be in $(\mathfrak{h}^{\mathbb{C}})'$.

(a) $V(\lambda)$ is a highest weight module under $U(\mathfrak{g}^{\mathbb{C}})$ and is generated by $1 \otimes 1$ (the **canonical generator**), which is of weight $\lambda - \delta$.
(b) The map of $U(\mathfrak{n}^-)$ into $V(\lambda)$ given by $u \to u(1 \otimes 1)$ is one-one onto.
(c) If M is any highest weight module under $U(\mathfrak{g}^{\mathbb{C}})$ generated by a highest weight vector $v \neq 0$ of weight $\lambda - \delta$, then there exists one and only one $U(\mathfrak{g}^{\mathbb{C}})$ homomorphism φ of $V(\lambda)$ into M such that $\varphi(1 \otimes 1) = v$. The map φ is onto. Also φ is one-one if and only if $u \neq 0$ in $U(\mathfrak{n}^-)$ implies $u(v) \neq 0$ in M.

Proof.
(a) Clearly $V(\lambda) = U(\mathfrak{g}^{\mathbb{C}})(1 \otimes 1)$. Also

$$\begin{aligned} H(1 \otimes 1) &= H \otimes 1 = 1 \otimes H(1) = (\lambda - \delta)(H)(1 \otimes 1) && \text{for } H \in \mathfrak{h}^{\mathbb{C}} \\ X(1 \otimes 1) &= X \otimes 1 = 1 \otimes X(1) = 0 && \text{for } X \in \mathfrak{n}^+, \end{aligned}$$

and so $1 \otimes 1$ is a highest weight vector of weight $\lambda - \delta$.

(b) By the Birkhoff-Witt Theorem, we have $U(\mathfrak{g}^{\mathbb{C}}) \cong U(\mathfrak{n}^-) \otimes_{\mathbb{C}} U(\mathfrak{b})$, and this isomorphism is clearly an isomorphism of left $U(\mathfrak{n}^-)$ modules. Then we obtain a chain of canonical left $U(\mathfrak{n}^-)$ module isomorphisms

$$\begin{aligned} V(\lambda) &= U(\mathfrak{g}^{\mathbb{C}}) \otimes_{U(\mathfrak{b})} \mathbb{C} \cong (U(\mathfrak{n}^-) \otimes_{\mathbb{C}} U(\mathfrak{b})) \otimes_{U(\mathfrak{b})} \mathbb{C} \\ &\cong U(\mathfrak{n}^-) \otimes_{\mathbb{C}} (U(\mathfrak{b}) \otimes_{U(\mathfrak{b})} \mathbb{C}) \cong U(\mathfrak{n}^-) \otimes_{\mathbb{C}} \mathbb{C} \cong U(\mathfrak{n}^-), \end{aligned}$$

and (b) follows.

(c) We consider the bilinear map of $U(\mathfrak{g}^{\mathbb{C}}) \times \mathbb{C}_{\lambda-\delta} \to M$ given by $(u, z) \to u(zv)$. In terms of the action of $U(\mathfrak{b})$ on $\mathbb{C}_{\lambda-\delta}$, we have

$$(u, b(z)) \to u(b(z)v) = zu((b \cdot 1)v)$$

and

$$(ub, z) \to ub(zv) = zub(v) = zu((b \cdot 1)v)$$

by a check for b in $\mathfrak{h}^{\mathbb{C}}$ and then b in $\mathfrak{n}^+$. By the universal mapping property, there exists one and only one linear map

$$\varphi: U(\mathfrak{g}^{\mathbb{C}}) \otimes_{U(\mathfrak{b})} \mathbb{C}_{\lambda-\delta} \to M$$

such that $u(zv) = \varphi(u \otimes z)$ for all $u \in U(\mathfrak{g}^{\mathbb{C}})$ and $z \in \mathbb{C}$, i.e., such that $u(v) = \varphi(u(1 \otimes 1))$. This condition says that φ is a $U(\mathfrak{g}^{\mathbb{C}})$ homomorphism and that $1 \otimes 1$ maps to v. Hence existence and uniqueness follow. Clearly φ is onto.

If $u(v) = 0$ with $u \neq 0$, then $\varphi(u(1 \otimes 1)) = 0$ while $u(1 \otimes 1) \neq 0$, by (b). Hence φ is not one-one. Conversely if φ is not one-one, then Proposition 4.34a implies there exists u in $U(\mathfrak{n}^-)$ with $u \neq 0$ and $\varphi(u \otimes 1) = 0$. Then

$$u(v) = u(\varphi(1 \otimes 1)) = \varphi(u(1 \otimes 1)) = \varphi(u \otimes 1) = 0.$$

This completes the proof.

Proposition 4.36. Let λ be in $(\mathfrak{h}^{\mathbb{C}})'$, and let $V(\lambda)_+ = \sum_{\mu \neq \lambda - \delta} V(\lambda)_\mu$. Then every proper $U(\mathfrak{g}^{\mathbb{C}})$ submodule of $V(\lambda)$ is contained in $V(\lambda)_+$. Consequently the sum K of all proper $U(\mathfrak{g}^{\mathbb{C}})$ submodules is a proper $U(\mathfrak{g}^{\mathbb{C}})$ submodule, and $L(\lambda) = V(\lambda)/K$ is an irreducible $U(\mathfrak{g}^{\mathbb{C}})$ module. Moreover, $L(\lambda)$ is a highest weight module with highest weight $\lambda - \delta$.

Proof. If N is a $U(\mathfrak{h}^{\mathbb{C}})$ submodule, then $N = \sum_\mu (N \cap V(\lambda)_\mu)$. Since $V(\lambda)_{\lambda-\delta}$ is one-dimensional and generates $V(\lambda)$ (by Proposition 4.34), the $\lambda - \delta$ term must be 0 in the sum for N if N is proper. Thus $N \subseteq V(\lambda)_+$. Hence K is proper and $L(\lambda) = V(\lambda)/K$ is irreducible. The image of $1 \otimes 1$ in $L(\lambda)$ is not 0 and is annihilated by $\mathfrak{n}^+$ and is acted upon by $\mathfrak{h}^{\mathbb{C}}$ according to $\lambda - \delta$. Thus $L(\lambda)$ has the other required properties.

Theorem 4.37. Suppose λ in $(\mathfrak{h}^{\mathbb{C}})'$ is real-valued on $\mathfrak{h}_{\mathbb{R}}$ and is dominant and algebraically integral. Then the irreducible highest weight module $L(\lambda + \delta)$ is an irreducible finite-dimensional representation of $\mathfrak{g}$ with highest weight λ.

The proof will be preceded by three lemmas. The first two handle the special case of $\mathfrak{sl}(2, \mathbb{C})$, and the third shows how to imbed the $\mathfrak{sl}(2, \mathbb{C})$ result in the proof of the general case.

Lemma 4.38. In $U(\mathfrak{sl}(2, \mathbb{C}))$, $[e, f^n] = nf^{n-1}(h - (n-1))$.

Proof. Let

$$Lf = \text{left by } f \text{ in } U(\mathfrak{sl}(2,\mathbb{C}))$$
$$Rf = \text{right by } f$$
$$\operatorname{ad} f = Lf - Rf.$$

Then $Rf = Lf - \operatorname{ad} f$, and the terms on the right commute. By the binomial theorem,

$$(Rf)^n e = \sum_{j=0}^{n} \binom{n}{j} (Lf)^{n-j}(-\operatorname{ad} f)^j e$$

$$= (Lf)^n e + n(Lf)^{n-1}(-\operatorname{ad} f)e + \frac{n(n-1)}{2}(Lf)^{n-2}(-\operatorname{ad} f)^2 e$$

since $(\operatorname{ad} f)^3 e = 0$, and this expression is

$$= (Lf)^n e + nf^{n-1}h + \frac{n(n-1)}{2} f^{n-2}(-2f)$$

$$= (Lf)^n e + nf^{n-1}(h - (n-1)).$$

Thus

$$[e, f^n] = (Rf)^n e - (Lf)^n e = nf^{n-1}(h - (n-1)).$$

Lemma 4.39. For $\mathfrak{sl}(2, \mathbb{C})$, let $\mathfrak{h}^{\mathbb{C}} = \mathbb{C}h$, $\mathfrak{n}^+ = \mathbb{C}e$, $\mathfrak{n}^- = \mathbb{C}f$. If λ is in $(\mathfrak{h}^{\mathbb{C}})'$, then $V(\lambda)$ is reducible if and only if $\lambda(h)$ is a positive integer. In this case the space of vectors annihilated by $\mathfrak{n}^+$ is two-dimensional with basis $v_{\lambda-\delta}$ and $f^{\lambda(h)}v_{\lambda-\delta}$, where $v_{\lambda-\delta}$ denotes the canonical generator of $V(\lambda)$. These two vectors are highest weight vectors, and $f^{\lambda(h)}v_{\lambda-\delta}$ generates an isomorphic copy of $V(-\lambda)$. Moreover, $V(\lambda)/V(-\lambda)$ is a finite-dimensional (irreducible) representation with highest weight $\lambda - \delta$ and dimension $\lambda(h)$.

Proof. Every nonzero $U(\mathfrak{sl}(2, \mathbb{C}))$ submodule has a highest weight vector. (Just keep applying e.) Thus to deal with reducibility, it is enough to classify highest weight vectors. By Proposition 4.35b, $V(\lambda)$ has $\{f^n v_{\lambda-\delta}\}$ as a basis of weight vectors, and we want to know for which $n \geq 0$ we have $ef^n v_{\lambda-\delta} = 0$. Since $f^n e v_{\lambda-\delta} = 0$, Lemma 4.38 shows we want

$$nf^{n-1}(h - (n-1))v_{\lambda-\delta} = 0,$$

hence

$$n[(\lambda - \delta)(h) - (n-1)]f^{n-1}v_{\lambda-\delta} = 0.$$

Thus $n = 0$ or $n - 1 = (\lambda - \delta)(h) = \lambda(h) - 1$, so that $n = \lambda(h)$.

The weight of $f^{\lambda(h)}v_{\lambda-\delta}$ is $\lambda - \delta - \lambda(h)\alpha$, where α is the positive root. Since

$$(\lambda - \delta - \lambda(h)\alpha)(h) = \lambda(h) - 1 - 2\lambda(h) = (-\lambda - \delta)(h),$$

the $U(\mathfrak{sl}(2, \mathbb{C}))$ submodule generated by $f^{\lambda(h)}v_{\lambda-\delta}$ is a highest weight module with highest weight $-\lambda - \delta$. By Proposition 4.35c, it is isomorphic to $V(-\lambda)$.

Now $V(\lambda)/V(-\lambda)$ has to be irreducible, and its highest weight is $\lambda - \delta$. Its dimension is $\lambda(h)$ since $\{f^j v_{\lambda-\delta} | 0 \leq j \leq \lambda(h) - 1\}$ descends to a basis for it.

Lemma 4.40. For general $\mathfrak{g}$ let λ be in $(\mathfrak{h}^{\mathbb{C}})'$, let α be a simple root, and suppose that $m = 2\langle\lambda, \alpha\rangle/|\alpha|^2$ is a positive integer. Let $v_{\lambda-\delta}$ be the canonical generator of $V(\lambda)$ and let M be the $U(\mathfrak{g}^{\mathbb{C}})$ submodule generated by $(E_{-\alpha})^m v_{\lambda-\delta}$, where $E_{-\alpha}$ is nonzero in $\mathfrak{g}_{-\alpha}$. Then M is isomorphic to $V(s_\alpha\lambda)$.

Proof. The vector $v = (E_{-\alpha})^m v_{\lambda-\delta}$ is not 0, by Proposition 4.35b. Since $s_\alpha\lambda = \lambda - m\alpha$, v is in $V(\lambda)_{\lambda-\delta-m\alpha} = V(\lambda)_{s_\alpha\lambda-\delta}$. Thus the result will follow from Proposition 4.35c if we show that $E_\beta v = 0$ for every simple β. For $\beta \neq \alpha$, $[E_\beta, E_{-\alpha}] = 0$ since $\beta - \alpha$ is not a root. Thus

$$E_\beta v = E_\beta(E_{-\alpha})^m v_{\lambda-\delta} = (E_{-\alpha})^m E_\beta v_{\lambda-\delta} = 0.$$

For $\beta = \alpha$, we may normalize matters so that $[E_\alpha, E_{-\alpha}] = 2|\alpha|^{-2}H_\alpha$, and then, as in Lemma 4.39,

$$\begin{aligned} E_\alpha(E_{-\alpha})^m v_{\lambda-\delta} &= [E_\alpha, E_{-\alpha}^m]v_{\lambda-\delta} \\ &= m(E_{-\alpha})^{m-1}(2|\alpha|^{-2}H_\alpha - (m-1))v_{\lambda-\delta} \\ &= m\left(\frac{2\langle\lambda-\delta, \alpha\rangle}{|\alpha|^2} - (m-1)\right)E_{-\alpha}^{m-1}v_{\lambda-\delta} \\ &= 0 \end{aligned}$$

by Proposition 4.33.

Proof of Theorem 4.37. Let $v_\lambda \neq 0$ be a highest weight vector in $L(\lambda + \delta)$, with weight λ. We proceed in three steps.

(1) We show: For every simple root α, $E_{-\alpha}^n v_\lambda = 0$ for n sufficiently large. In fact, for $n = \dfrac{2\langle\lambda+\delta, \alpha\rangle}{|\alpha|^2} > 0$ (Proposition 4.33), $E_{-\alpha}^n(1 \otimes 1)$ in $V(\lambda + \delta)$ lies in a proper $U(\mathfrak{g}^{\mathbb{C}})$ submodule, according to Lemma 4.40, and hence is in K. Thus $E_{-\alpha}^n v_\lambda = 0$ in $L(\lambda + \delta)$.

(2) We show: The set of weights is stable under the Weyl group W. In fact, let α be a simple root, set $v^{(i)} = E_{-\alpha}^i v_\lambda$, and let n be the largest integer such that $v^{(n)} \neq 0$ (existence by (1)). Then $\mathbb{C}v^{(0)} + \ldots + \mathbb{C}v^{(n)}$ is stable under $\mathbb{C}H_\alpha + \mathbb{C}E_\alpha + \mathbb{C}E_{-\alpha} = \mathfrak{sl}_\alpha(2, \mathbb{C})$. The sum of all finite-dimensional irreducible $U(\mathfrak{sl}_\alpha(2, \mathbb{C}))$ submodules contains $v^{(0)} = v_\lambda$ by Corollary 2.2, and we claim it is $\mathfrak{g}$ stable. [In fact, if W is a finite-dimensional $U(\mathfrak{sl}_\alpha(2, \mathbb{C}))$ submodule, then

$$\mathfrak{g}W = \{\textstyle\sum Xw | X \in \mathfrak{g} \text{ and } w \in W\}$$

is finite-dimensional and for $S \in \mathfrak{sl}_\alpha(2, \mathbb{C})$ and $X \in \mathfrak{g}$ we have

$$SXw = XSw + [S, X]w = Xw' + [S, X]w \in \mathfrak{g}W.$$

So $\mathfrak{g}W$ is $\mathfrak{sl}_\alpha(2, \mathbb{C})$ stable and splits into irreducibles, by Corollary 2.2. The claim follows.] Thus the sum of all finite-dimensional irreducible $U(\mathfrak{sl}_\alpha(2, \mathbb{C}))$ submodules is all of $L(\lambda + \delta)$.

Let μ be a weight, and let $t \neq 0$ be in V_μ. We have just shown that t lies in a finite sum of finite-dimensional irreducible $U(\mathfrak{sl}_\alpha(2, \mathbb{C}))$ submodules, say as $t = \sum_{i \in I} t_i$, $t_i \in T_i$, $t_i \neq 0$. Then

$$\sum \mu(H_\alpha)t_i = \mu(H_\alpha)t = H_\alpha t = \sum H_\alpha t_i,$$

and so

$$\frac{2H_\alpha}{|\alpha|^2} t_i = \frac{2\langle \mu, \alpha\rangle}{|\alpha|^2} t_i$$

for each $i \in I$. If $\langle \mu, \alpha\rangle > 0$, we know that $(E_{-\alpha})^{2\langle \mu,\alpha\rangle/|\alpha|^2} t_i \neq 0$ from Theorem 2.4. Hence $(E_{-\alpha})^{2\langle \mu,\alpha\rangle/|\alpha|^2} t \neq 0$. This says

$$\mu - \frac{2\langle \mu, \alpha\rangle}{|\alpha|^2}\alpha = s_\alpha \mu$$

is a weight. If $\langle \mu, \alpha\rangle < 0$ instead, we know that $(E_\alpha)^{-2\langle \mu,\alpha\rangle/|\alpha|^2} t_i \neq 0$ from Theorem 2.4. Hence $(E_\alpha)^{-2\langle \mu,\alpha\rangle/|\alpha|^2} t \neq 0$, and so

$$\mu - \frac{2\langle \mu, \alpha\rangle}{|\alpha|^2}\alpha = s_\alpha \mu$$

is a weight. If $\langle \mu, \alpha\rangle = 0$, then $s_\alpha \mu = \mu$. In any case $s_\alpha \mu$ is a weight. So the set of weights is stable under each s_α, α simple, and Proposition 4.9 shows the set of weights is stable under W.

(3) We show: The set of weights of $L(\lambda + \delta)$ is finite, and $L(\lambda + \delta)$ is finite-dimensional. In fact, Corollary 4.31 shows that any linear functional on $\mathfrak{h}_{\mathbb{R}}$ is W-conjugate to a dominant one. Since (2) says the set of weights is stable under W, the number of weights is at most $|W|$ times the number of dominant weights, which are of the form $\lambda - \sum_{i=1}^{l} n_i\alpha_i$ by Proposition 4.34c. Each such dominant form must satisfy

$$\langle \lambda, \delta\rangle > \sum_{i=1}^{l} n_i\langle \alpha_i, \delta\rangle,$$

and Proposition 4.33 shows $\sum n_i$ is bounded; thus the number of dominant weights is finite. Then $L(\lambda + \delta)$ is finite-dimensional by Proposition 4.34b.

§9. Weyl Group, Analytically

The representation theory of compact connected Lie groups can be continued in the algebraic fashion of the previous sections, and one can obtain a character formula for the irreducible representations. But also one can develop much of the theory analytically, working with a maximal torus and with global properties of the group. As a compromise to illustrate both approaches, we shall switch now to the analytic approach, using it to complete the theory.

In this section we give the analytic definition of the Weyl group, showing it coincides with the algebraically defined Weyl group of §4, and then we study conjugacy classes in G. Thus let G, $\mathfrak{g}$, $\mathfrak{g}^{\mathbb{C}}$, $\mathfrak{h}$, $\mathfrak{h}^{\mathbb{C}}$, T, and Δ be as in §§2 and 5. The Weyl group $W(T:G)$ is defined as the quotient of normalizer by centralizer

$$W(T:G) = N_G(T)/Z_G(T).$$

The group $W(T:G)$ acts by automorphisms of T, hence by linear transformations on $\mathfrak{h}$ and the associated spaces $\mathfrak{h}_{\mathbb{R}} = i\mathfrak{h}$, $\mathfrak{h}^{\mathbb{C}}$, $\mathfrak{h}'_{\mathbb{R}}$, and $(\mathfrak{h}^{\mathbb{C}})'$, and only 1 acts as the identity. In the definition of $W(T:G)$, we can replace $Z_G(T)$ by T, according to Corollary 4.23.

Theorem 4.41. For a compact connected Lie group G, the analytically defined Weyl group $W(T:G)$, when considered as acting on $\mathfrak{h}'_{\mathbb{R}}$, coincides with the algebraically defined Weyl group of the root system Δ.

Proof. We may regard G as contained in some $U(n)$. In view of Proposition 4.32, we may assume G is semisimple.

To show that $W \subseteq W(T:G)$, it is enough to show that α in Δ implies s_α in $W(T:G)$. [First, we give the argument for SU(2): T is diagonal and s_α is given by $\begin{pmatrix} 0 & 1 \\ -1 & 0 \end{pmatrix}$. With $E_\alpha = \begin{pmatrix} 0 & 1 \\ 0 & 0 \end{pmatrix}$, we have $\begin{pmatrix} 0 & 1 \\ -1 & 0 \end{pmatrix} = \exp \frac{\pi}{2}(E_\alpha + \theta E_\alpha)$. The general case amounts to imbedding this argument in G.] Thus let E_α be in $\mathfrak{g}_\alpha$ and write $E_\alpha = X_\alpha + iY_\alpha$ with X_α and Y_α in $\mathfrak{g}$. Then $\theta E_\alpha = X_\alpha - iY_\alpha$ is in $\mathfrak{g}_{-\alpha}$. For H in $\mathfrak{h}^{\mathbb{C}}$, we have

$$[X_\alpha, H] = -\tfrac{1}{2}[H, E_\alpha + \theta E_\alpha] = -\tfrac{1}{2}\alpha(H)(E_\alpha - \theta E_\alpha) = -i\alpha(H)Y_\alpha. \quad (4.24)$$

Also Proposition 4.1e gives

$$\begin{aligned}[X_\alpha, Y_\alpha] &= \frac{1}{4i}[E_\alpha + \theta E_\alpha, E_\alpha - \theta E_\alpha] = -\frac{1}{2i}[E_\alpha, \theta E_\alpha] \\ &= -\frac{1}{2i}B(E_\alpha, \theta E_\alpha)H_\alpha. \end{aligned} \quad (4.25)$$

Since $B(E_\alpha, \theta E_\alpha) < 0$, we can define a real number t by

$$t = \frac{\sqrt{2}\pi}{|\alpha|\sqrt{-B(E_\alpha, \theta E_\alpha)}}.$$

Since X_α is in $\mathfrak{g}$, $g = \exp tX_\alpha$ is in G. We compute $\mathrm{Ad}(g)H$ for H in $\mathfrak{h}_{\mathbb{R}}$. We have

$$\mathrm{Ad}(g)H = e^{\mathrm{ad}\, tX_\alpha}H = \sum_{k=0}^{\infty} \frac{t^k}{k!}(\mathrm{ad}\, X_\alpha)^k H. \tag{4.26}$$

If $\alpha(H) = 0$, then (4.24) shows the series (4.26) collapses to H. If $H = H_\alpha$, we obtain

$$t^2(\mathrm{ad}\, X_\alpha)^2 H_\alpha = \tfrac{1}{2}|\alpha|^2 B(E_\alpha, \theta E_\alpha)t^2 H_\alpha = -\pi^2 H_\alpha$$

from (4.24) and (4.25). Then (4.26) shows that

$$\begin{aligned}
\mathrm{Ad}(g)H_\alpha &= \sum_{m=0}^{\infty} \frac{t^{2m}}{(2m)!}(\mathrm{ad}\, X_\alpha)^{2m}H_\alpha + \sum_{m=0}^{\infty} \frac{t^{2m+1}}{(2m+1)!}(\mathrm{ad}\, X_\alpha)^{2m+1}H_\alpha \\
&= \sum_{m=0}^{\infty} \frac{(-1)^m\pi^{2m}}{(2m)!} H_\alpha + t\sum_{m=0}^{\infty} \frac{(-1)^m\pi^{2m}}{(2m+1)!}[X_\alpha, H_\alpha] \\
&= (\cos \pi)H_\alpha + \pi^{-1}t(\sin \pi)[X_\alpha, H_\alpha] \\
&= -H_\alpha.
\end{aligned}$$

Thus every H in $\mathfrak{h}_{\mathbb{R}}$ satisfies

$$\mathrm{Ad}(g)H = H - \frac{2\alpha(H)}{|\alpha|^2}H_\alpha,$$

and $\mathrm{Ad}(g)$ normalizes $\mathfrak{h}_{\mathbb{R}}$ and operates as s_α on $\mathfrak{h}'_{\mathbb{R}}$.

Thus $W \subseteq W(T{:}G)$. Next let us observe that $W(T{:}G)$ permutes the roots. In fact, let g be in $N_G(T) = N_G(\mathfrak{h})$, let α be in Δ, and let E_α be in $\mathfrak{g}_\alpha$. Then

$$\begin{aligned}
(g\alpha)(H)\, \mathrm{Ad}(g)E_\alpha = \alpha(\mathrm{Ad}(g)^{-1}H)\, \mathrm{Ad}(g)\, E_\alpha &= \mathrm{Ad}(g)(\alpha(\mathrm{Ad}(g)^{-1}H)E_\alpha) \\
&= \mathrm{Ad}(g)[\mathrm{Ad}(g)^{-1}H, E_\alpha] = [H, \mathrm{Ad}(g)E_\alpha]
\end{aligned}$$

shows $g\alpha$ is in Δ and $\mathrm{Ad}(g)E_\alpha$ is a root vector for $g\alpha$. Thus $W(T{:}G)$ permutes the roots.

Fix a simple system Π for Δ, and let g be given in $W(T{:}G)$. It follows from the previous paragraph that $g\Pi$ is another simple system for Δ. By Theorem 4.10, choose w in W with $wg\Pi = \Pi$. We show wg fixes $\mathfrak{h}'_{\mathbb{R}}$. Then wg will be in $Z_G(T)$, and wg will represent the identity in $W(T{:}G)$. Hence $g = w^{-1}$ and g will be exhibited as in W, so that $W = W(T{:}G)$.

Thus let $wg\Pi = \Pi$. Let Δ^+ be the positive system corresponding to Π, and define $\delta = \frac{1}{2}\sum_{\alpha\in\Delta^+}\alpha$. Then $wg\delta = \delta$ and so $\mathrm{Ad}(wg)H_\delta = H_\delta$. If S denotes the closure of $\{\exp rH_\delta \mid r \in \mathbb{R}\}$, then S is a torus, and wg is in its centralizer. Let $\mathfrak{s}$ be the Lie algebra of S. We claim that $Z_{\mathfrak{g}}(\mathfrak{s}) = \mathfrak{h}$. If so, it follows from Corollary 4.23 that $Z_G(S) = T$; hence wg is in T, and the proof is complete.

Thus we compute

$$\begin{aligned} Z_{\mathfrak{g}}(\mathfrak{s}) &= \mathfrak{g} \cap Z_{\mathfrak{g}^{\mathbb{C}}}(\mathfrak{s}) = \mathfrak{g} \cap Z_{\mathfrak{g}^{\mathbb{C}}}(H_\delta) \\ &= \mathfrak{g} \cap \{X = H + \textstyle\sum X_\alpha \mid [H_\delta, X] = 0\} \\ &= \mathfrak{g} \cap \{X = H + \textstyle\sum X_\alpha \mid \sum \langle \delta, \alpha \rangle X_\alpha = 0\} \\ &= \mathfrak{g} \cap \mathfrak{h}^{\mathbb{C}} \qquad (4.27) \\ &= \mathfrak{h}, \end{aligned}$$

as required; here step (4.27) follows from Proposition 4.33.

Theorem 4.42. For a compact connected Lie group G, any two Cartan subalgebras of $\mathfrak{g}$ are conjugate via $\mathrm{Ad}(G)$.

Proof. Let $\mathfrak{h}$ and $\mathfrak{h}_1$ be Cartan subalgebras with T and T_1 the corresponding Cartan subgroups. Choose X in $\mathfrak{h}_1$ such that $Z_{\mathfrak{g}}(X) = \mathfrak{h}_1$, and choose Y in $\mathfrak{h}$ such that $Z_{\mathfrak{g}}(Y) = \mathfrak{h}$; here we can take X and Y to be the respective elements iH_δ, by (4.27) and Proposition 4.33. Regard G as contained in some $\mathrm{U}(n)$ and let B_0 be the trace form; then $\langle \cdot, \cdot \rangle = -\mathrm{Re}\, B_0(\cdot, \cdot)$ is an $\mathrm{Ad}(G)$ invariant inner product on $\mathfrak{g}$. Since G is compact, we can choose g_0 in G such that $\langle \mathrm{Ad}(g)X, Y \rangle$ is minimized for $g = g_0$. For any Z in $\mathfrak{g}$, $t \to \langle \mathrm{Ad}(\exp tZ)\, \mathrm{Ad}(g_0)X, Y \rangle$ is then a smooth function of t that is minimized for $t = 0$. Differentiating and setting $t = 0$, we obtain

$$0 = \langle (\mathrm{ad}\, Z)\, \mathrm{Ad}(g_0)X, Y \rangle = \langle [Z, \mathrm{Ad}(g_0)X], Y \rangle = \langle Z, [\mathrm{Ad}(g_0)X, Y] \rangle.$$

Since Z is arbitrary, $[\mathrm{Ad}(g_0)X, Y] = 0$. Thus $\mathrm{Ad}(g_0)X$ is in $Z_{\mathfrak{g}}(Y) = \mathfrak{h}$. Since $\mathfrak{h}$ is abelian, this means

$$\mathfrak{h} \subseteq Z_{\mathfrak{g}}(\mathrm{Ad}(g_0)X) = \mathrm{Ad}(g_0) Z_{\mathfrak{g}}(X) = \mathrm{Ad}(g_0)\mathfrak{h}_1.$$

Equality must hold since $\mathfrak{h}$ is maximal abelian. Thus $\mathfrak{h} = \mathrm{Ad}(g_0)\mathfrak{h}_1$.

Corollary 4.43. For a compact connected Lie group, any two maximal tori are conjugate.

Theorem 4.44. For a compact connected Lie group G, every element of G is conjugate to a member of the maximal torus T, and two elements of T are conjugate within G if and only if they are conjugate via W. Thus the conjugacy classes in G are parameterized by T/W. This correspondence respects the topologies in that a continuous complex-valued function f on T extends to a continuous function F on G invariant under conjugation if and only if f is invariant under W.

Proof. Theorem 4.19 shows that each element of G is contained in a torus, hence a maximal torus, and Corollary 4.43 shows that this maximal torus is conjugate to T. Hence every conjugacy class meets T.

Now suppose that s and t are in T and g is in G and $gtg^{-1} = s$. We show there is an element g_0 of $N_G(T)$ with $g_0 t g_0^{-1} = s$. In fact, consider the centralizer $Z_G(s)$. This is a closed subgroup of G with Lie algebra

$$Z_{\mathfrak{g}}(s) = \{X \in \mathfrak{g} \mid \mathrm{Ad}(s)X = X\}.$$

The identity component $(Z_G(s))_0$ is a group to which we can apply Theorem 4.42. Both $\mathfrak{h}$ and $\mathrm{Ad}(g)\mathfrak{h}$ are in $Z_{\mathfrak{g}}(s)$, and they are maximal abelian; hence there exists z in $(Z_G(s))_0$ with

$$\mathfrak{h} = \mathrm{Ad}(zg)\mathfrak{h}.$$

Then $g_0 = zg$ is in $N_G(T)$ and $(zg)t(zg)^{-1} = s$.

Thus the conjugacy classes in G are given by $T/W(T:G)$, and it is a simple matter to check that continuous functions correspond. Theorem 4.41 says that $W = W(T:G)$, and the proof is therefore complete.

§10. Weyl Character Formula

We retain the notation of §9. The space T/W is a compact Hausdorff space, and Theorem 4.44 shows that we obtain a positive linear functional l on $C(T/W)$ as follows: For f in $C(T/W)$, define F in $C(G)$ by $F(gtg^{-1}) = f(t)$ and let $l(f) = \int_G F(x)\,dx$, where dx is normalized Haar measure. Then it follows that there is a unique W-invariant measure $d\mu(t)$ on T such that

$$\int_G F(x)\,dx = \int_T \left[\int_G F(gtg^{-1})\,dg\right] d\mu(t) \tag{4.28}$$

for all F in $C(G)$. To find $d\mu(t)$ explicitly, it is a question of evaluating a certain Jacobian determinant. The result is as follows.

Theorem 4.45 (Weyl integration formula). Let G be compact semisimple and simply connected. Fix a positive system Δ^+ of Δ, and let δ be half the sum of the positive roots. Then the measure $d\mu(t)$ in (4.28) is given by $d\mu(t) = |W|^{-1}|D(t)|^2\,dt$, where

$$D(t) = \xi_\delta(t) \prod_{\alpha\in\Delta^+} (1 - \xi_{-\alpha}(t)) \tag{4.29}$$

with ξ_λ denoting the character of T with differential λ. Here dx and dt are normalized Haar measures on G and T, respectively.

Remarks.

(1) The characters $\xi_{-\alpha}$ are well defined since roots are analytically integral. Since G is simply connected, Theorem 4.28 shows that algebraically integral implies analytically integral, and δ is algebraically integral by Proposition 4.33; hence ξ_δ is well defined.

(2) Formally $D(t)$ is given by

$$D(t) = \prod_{\alpha \in \Delta^+} (\xi_{\alpha/2}(t) - \xi_{-\alpha/2}(t)),$$

but the individual linear functionals $\alpha/2$ need not be integral.

(3) The hypothesis on G can be weakened: $|D(t)|^2$ is defined for any compact connected Lie group, even though $D(t)$ may not be, and the formula for $d\mu(t)$ is valid in general. $D(t)$ itself is well defined as soon as δ is analytically integral.

Proof of Theorem 4.45 omitted.

Theorem 4.46 (Weyl character formula). Let G be compact semisimple and simply connected. Fix a positive system Δ^+ of Δ, and let δ be half the sum of the positive roots. If τ_λ is an irreducible finite-dimensional representation of G with highest weight λ, then the character χ_λ of τ_λ is given on T by

$$\chi_\lambda(t) = D(t)^{-1} \sum_{w \in W} (\det w) \xi_{w(\lambda+\delta)}(t),$$

where $\det w$ is computed in the linear action of W on $\mathfrak{h}_{\mathbb{R}}$.

We give the proof below. First we take note of the special case $\lambda = 0$, for which τ_λ is trivial and χ_λ is identically one. The character formula then yields the following corollary.

Corollary 4.47 (Weyl denominator formula). Let G be compact semisimple and simply connected. Then the function $D(t)$ in (4.29) is given by

$$D(t) = \sum_{w \in W} (\det w) \xi_{w\delta}(t). \tag{4.30}$$

Remark. For $G = \mathrm{SU}(n)$, let $t = \mathrm{diag}(t_1, \ldots, t_n)$ with $\prod t_i = 1$. Then W is the symmetric group on n letters, and $\det w$ is the sign of the permutation w. The right side of (4.30) can be seen to equal the determinant of the Vandermonde matrix with $(i, j)^{\text{th}}$ entry $(t_{n+1-j})^i$. The left side is the classical value for this determinant, namely $\prod_{i<j} (t_i - t_j)$.

Proof of Theorem 4.46. Let $R(T)$ be the ring of $\mathbb{Z}$-linear combinations of characters of T. Theorem 4.41 shows that W acts on $R(T)$ by the definition

$$w\xi_\lambda(t) = \xi_\lambda(w^{-1}t) = \xi_{w\lambda}(t).$$

We say that a member $f(t)$ of $R(T)$ is **even** under W if $wf = f$ for all w in W; $f(t)$ is **odd** under W if $wf = (\det w)f$ for all w in W. Then we can make the following observations:

(i) If χ is the character of a representation of G, then $\chi|_T$ is in $R(T)$ and is even. [In fact, $\chi|_T = \sum \xi_\mu$ with the sum taken over all weights μ repeated according to their multiplicities. Thus $\chi|_T$ is in $R(T)$. Since χ is invariant under conjugation, Theorem 4.44 say $\chi|_T$ is even under W.]

(ii) $D(t)$ is in $R(T)$ and is odd under W. [In fact, we see $D(t)$ is in $R(T)$ by expanding the right side of (4.29). To check that $D(t)$ is odd under W, we need only check that $s_\alpha D = -D$ for α simple, by Proposition 4.9, and this is an immediate consequence of Lemma 4.8.]

(iii) The product of two even members of $R(T)$ is even, and the product of two odd members of $R(T)$ is even. The product of an even member of $R(T)$ by an odd member of $R(T)$ is odd.

(iv) Every μ in $\mathfrak{h}'_{\mathbb{R}}$ is conjugate via W to a dominant element. [This follows from Corollary 4.31.]

(v) If μ in $\mathfrak{h}'_{\mathbb{R}}$ has $\langle \mu, \alpha \rangle \neq 0$ for all α in Δ, then $w\mu \neq \mu$ for all $w \neq 1$ in W. [This follows from Chevalley's Lemma (Proposition 4.12).]

Now let χ_λ be the character of τ_λ. By (i), (ii), and (iii), $\chi_\lambda(t)D(t)$ is in $R(T)$ and is odd under W. Let us write

$$\chi_\lambda(t)D(t) = \sum_\mu n_\mu \xi_\mu. \tag{4.31}$$

Applying w to both sides, multiplying by det w, and summing, we obtain

$$|W|\chi_\lambda(t)D(t) = \sum_{w,\nu} n_\nu(\det w)\xi_{w\nu}.$$

Thus

$$\sum_\mu |W| n_\mu \xi_\mu = \sum_\mu \sum_{\substack{w,\nu \\ w\nu=\mu}} n_\nu(\det w)\xi_\mu = \sum_\mu \left[\sum_{w \in W} n_{w^{-1}\mu} \det w \right] \xi_\mu,$$

and

$$|W| n_\mu = \sum_{w \in W} n_{w^{-1}\mu} \det w. \tag{4.32}$$

In a given orbit $W\mu$, let us arrange notation temporarily so that $|n_{w^{-1}\mu}|$ is a maximum for $w = 1$. In order for the right side of (4.32) to be as large as the left, all terms must be equal. Thus

$$n_{w^{-1}\mu} \det w = n_\mu \tag{4.33}$$

for all w in W. Consequently this relation holds even if μ does not have the special property.

If some w_0 in W satisfies $w_0^{-1}\mu = \mu$, then (v) says $s_\alpha\mu = \mu$ for some α in Δ. Then (4.33) says

$$-n_\mu = n_{s_\alpha\mu} \det s_\alpha = n_\mu,$$

and hence $n_\mu = 0$. Thus the sum (4.31) extends only over μ with $\langle \mu, \alpha \rangle \neq 0$ for all α in Δ.

Such a μ can be dominant only with respect to one positive system, namely $\{\alpha | \langle \mu, \alpha \rangle > 0\}$, and hence there is at most one w in W with $w\mu$

dominant. By (iv) there is at least one such w. Thus we can regroup (4.31) as

$$\chi_\lambda(t)D(t) = \sum_{\substack{\mu \text{ with } \langle \mu, \alpha \rangle > 0 \\ \text{for all } \alpha > 0 \text{ in } \Delta}} \sum_{w \in W} n_{w\mu} \xi_{w\mu}$$

and apply (4.33) to rewrite it as

$$\chi_\lambda(t)D(t) = \sum_{\substack{\mu \text{ with } \langle \mu, \alpha \rangle > 0 \\ \text{for all } \alpha > 0 \text{ in } \Delta}} n_\mu \left(\sum_{w \in W} (\det w) \xi_{w\mu} \right). \tag{4.34}$$

Since τ_λ is irreducible, Lemma 1.10 implies that $\int_G |\chi_\lambda(x)|^2 \, dx = 1$. By (4.28) and Theorem 4.45, we therefore have

$$\frac{1}{|W|} \int_T |\chi_\lambda(t)D(t)|^2 \, dt = 1.$$

Substituting from (4.34), we find that

$$\sum_{\substack{\mu \text{ with } \langle \mu, \alpha \rangle > 0 \\ \text{for all } \alpha > 0 \text{ in } \Delta}} |n_\mu|^2 = 1.$$

Thus

$$\chi_\lambda(t)D(t) = \pm \sum_{w \in W} (\det w) \xi_{w\mu} \tag{4.35}$$

for a suitable sign and for a suitable μ with $\langle \mu, \alpha \rangle > 0$ for all α in Δ^+. Comparing the coefficients of $\xi_{\lambda+\delta}$ on the two sides of (4.35), we see that the sign is plus and that $\mu = \lambda + \delta$. This completes the proof.

Theorem 4.48 (Weyl dimension formula). Let G be compact semisimple and simply connected. Fix a positive system Δ^+ of Δ, and let δ be half the sum of the positive roots. If τ_λ is an irreducible finite-dimensional representation of G with highest weight λ, then the degree d_λ of τ_λ is given by

$$d_\lambda = \frac{\prod_{\alpha \in \Delta^+} \langle \lambda + \delta, \alpha \rangle}{\prod_{\alpha \in \Delta^+} \langle \delta, \alpha \rangle}.$$

Proof. The degree can be evaluated as the value of the character at the identity. However, the Weyl character formula yields 0/0 in this case. Thus we shall proceed by using l'Hospital's Rule. First we define

$$\partial(\varphi) f(H) = H_\varphi f(H) = -i \frac{d}{ds} f(H + isH_\varphi) \Big|_{s=0}, \qquad H \in \mathfrak{h}, \varphi \in \mathfrak{h}'_{\mathbb{R}}.$$

Then
$$\partial(\varphi)\xi_\mu(\exp H) = -i \frac{d}{ds} e^{\mu(H + isH_\varphi)} \Big|_{s=0} = \langle \mu, \varphi \rangle \xi_\mu(\exp H).$$

Consider any derivative $\partial(\varphi_1)\cdots\partial(\varphi_n)$ of order less than the number of positive roots, and apply it to the Weyl denominator $D(t)$. We are then considering

$$\partial(\varphi_1)\cdots\partial(\varphi_n)\left\{\xi_\delta(\exp H)\prod_{\alpha\in\Delta^+}(1-\xi_{-\alpha}(\exp H))\right\}.$$

Each $\partial(\varphi_j)$ operates by the product rule and differentiates one factor, leaving the others alone. Thus each term in the product rule expansion has some undifferentiated $1-\xi_{-\alpha}(\exp H)$ and gives 0 at $H=0$. Since we know the limit of $\chi_\lambda(\exp H)$ exists as $H\to 0$, we can apply l'Hospital's Rule, using $\prod_{\alpha\in\Delta^+}\partial(\alpha)$ in both numerator and denominator. For the numerator we have

$$\begin{aligned}
&\prod_{\alpha\in\Delta^+}\partial(\alpha)\left(\sum_{w\in W}(\det w)\xi_{w(\lambda+\delta)}(\exp H)\right)\\
&\quad=\sum_{w\in W}(\det w)\left(\prod_{\alpha\in\Delta^+}\langle w(\lambda+\delta),\alpha\rangle\right)\xi_{w(\lambda+\delta)}(\exp H)\\
&\quad=\sum_{w\in W}(\det w^{-1})\left(\prod_{\alpha\in\Delta^+}\langle \lambda+\delta, w^{-1}\alpha\rangle\right)\xi_{w(\lambda+\delta)}(\exp H)\\
&\quad=\sum_{w\in W}\left(\prod_{\alpha\in\Delta^+}\langle \lambda+\delta,\alpha\rangle\right)\xi_{w(\lambda+\delta)}(\exp H)\qquad\text{by Proposition 4.11}\\
&\quad=\left(\prod_{\alpha\in\Delta^+}\langle \lambda+\delta,\alpha\rangle\right)\sum_{w\in W}\xi_{w(\lambda+\delta)}(\exp H).
\end{aligned}$$

Dividing this expression and the one obtained from $\lambda=0$, which is the one obtained from $D(t)$ according to Corollary 4.47, we get the desired formula upon letting $H\to 0$.

Examples of §5 and §7. We can use Theorem 4.48 to prove irreducibility of some of our representations of SO(m). (The hypothesis "simply connected" is clearly unnecessary, as we can pass to the universal covering group.) For SO($2n+1$), we have

$$\delta=(n-\tfrac{1}{2})e_1+(n-\tfrac{3}{2})e_2+\ldots+\tfrac{1}{2}e_n.$$

For $\lambda=Ne_1$, the expression for d_λ in Theorem 4.48 works out easily to

$$\frac{(N+2n-2)!(2N+2n-1)}{N!(2n-1)!},$$

which agrees with (4.12); thus SO($2n+1$) acts irreducibly on $\mathscr{H}_N$. For $\lambda=e_1+\ldots+e_N$ and $N\le n$, the expression for d_λ works out after some calculations to $\binom{2n+1}{N}$; thus SO($2n+1$) acts irreducibly on $\bigwedge^N\mathbb{C}^{2n+1}$ and $\bigwedge^{2n+1-N}\mathbb{C}^{2n+1}$, and these representations are equivalent. We can proceed similarly with SO($2n$), using

$$\delta = (n-1)e_1 + (n-2)e_2 + \ldots + e_{n-1},$$

and we find that $\mathscr{H}_N$ is irreducible, that $\wedge^N \mathbb{C}^{2n}$ is irreducible if $N < n$, and that $\wedge^N \mathbb{C}^{2n}$ and $\wedge^{2n-N} \mathbb{C}^{2n}$ are equivalent.

§11. Problems

1. (a) Write down the Cartan matrices for the classical Dynkin diagrams.
   (b) Find the determinants of these matrices by suitable inductive arguments.

Problems 2 to 5 deal with particular representations of SU(n).

2. Let Φ be the representation of SU(n) on the space V_N of polynomials in $z_1, \ldots, z_n$ homogeneous of degree N, given by $\Phi(g)P(z) = P(g^{-1}z)$. Using the notation of Example 1 of §1, find all weights, the highest weight, and the dimension of the representation. Use the Weyl dimension formula to prove the representation is irreducible.
3. Repeat Problem 2 for the space of polynomials in $\bar{z}_1, \ldots, \bar{z}_n$, with Φ still given by $\Phi(g)P(z) = P(g^{-1}z)$.
4. Allow SU(n) to act on the full space of polynomials in $z_1, \ldots, z_n$, $\bar{z}_1, \ldots, \bar{z}_n$ that are homogeneous of degree N.
   (a) Show for each pair (p, q) with $p + q = N$ that the subspace $V_{p,q}$ of polynomials with p z-type factors and q $\bar{z}$-type factors is invariant. What are the weights of $V_{p,q}$?
   (b) The Laplacian in these coordinates is a multiple of $\sum \dfrac{\partial^2}{\partial z_j \, \partial \bar{z}_j}$. Using the mapping property of the Laplacian given for SO(m) in Examples 1 and 2 of §5, prove that the Laplacian here carries $V_{p,q}$ onto $V_{p-1,q-1}$.
   (c) Prove that the subspace $H_{p,q}$ of harmonic polynomials in $V_{p,q}$ is an invariant subspace. Compute its dimension and highest weight. Prove $H_{p,q}$ is irreducible.
   (d) Prove that every irreducible unitary representation of SU(3) has a concrete realization as some $H_{p,q}$ but that the corresponding statement fails for SU(4).
5. Define alternating tensor representations for SU(n) in analogy with the ones in Examples 3 and 4 of §5. Find their highest weights, and prove the representations are irreducible. Are there any equivalences among them?

Problems 6 to 9 derive the Kostant multiplicity formula for the multiplicity of a weight.

6. Let G be a compact connected Lie group, and let notation be as in Chapter IV. Let Q^+ be the set of all sums of simple roots ($\sum n_i\alpha_i$ with each n_i an integer ≥ 0). Define an additive group $\mathbb{Z}\langle\mathfrak{h}'_{\mathbb{R}}\rangle$ to be given as the set

$$\mathbb{Z}\langle\mathfrak{h}'_{\mathbb{R}}\rangle = \{f:\mathfrak{h}'_{\mathbb{R}} \to \mathbb{Z} \mid \text{support } (f) \subseteq \text{finite union of sets } \nu - Q^+, \nu \in \mathfrak{h}'_{\mathbb{R}}\},$$

with addition defined pointwise.

   (a) Show how to identify the ring $\mathbb{Z}[\exp \mathfrak{h}'_{\mathbb{R}}]$ of all finite integral combinations of exponentials of members of $\mathfrak{h}'_{\mathbb{R}}$ with a subset of $\mathbb{Z}\langle\mathfrak{h}'_{\mathbb{R}}\rangle$.

   (b) Let e^λ be the member of $\mathbb{Z}\langle\mathfrak{h}'_{\mathbb{R}}\rangle$ that is 1 at λ and 0 elsewhere, and write a general element of $\mathbb{Z}\langle\mathfrak{h}'_{\mathbb{R}}\rangle$ formally as $\sum c_\lambda e^\lambda$ with λ in $\mathfrak{h}'_{\mathbb{R}}$ and c_λ in $\mathbb{Z}$. Define a multiplication in $\mathbb{Z}\langle\mathfrak{h}'_{\mathbb{R}}\rangle$ by

$$(\textstyle\sum c_\lambda e^\lambda)(\sum c'_\mu e^\mu) = \displaystyle\sum_\nu \Bigg(\sum_{\lambda+\mu=\nu} c_\lambda c'_\mu\Bigg) e^\nu.$$

   Show that this multiplication is well defined in $\mathbb{Z}\langle\mathfrak{h}'_{\mathbb{R}}\rangle$ and that $\mathbb{Z}\langle\mathfrak{h}'_{\mathbb{R}}\rangle$ becomes a commutative ring with identity e^0 whose multiplication is consistent with that in $\mathbb{Z}[\exp \mathfrak{h}'_{\mathbb{R}}]$.

   (c) For H in the positive Weyl chamber of $\mathfrak{h}_{\mathbb{R}}$ define the mapping "evaluation at H" on the subset of members $\sum c_\lambda e^\lambda$ of $\mathbb{Z}\langle\mathfrak{h}'_{\mathbb{R}}\rangle$ for which $\sum |c_\lambda| e^{\lambda(H)} < \infty$ by $\sum c_\lambda e^\lambda \to \sum c_\lambda e^{\lambda(H)}$. Prove that if evaluation at H is defined on two elements, then it is defined on their product and behaves multiplicatively.

7. For γ in $\mathfrak{h}'_{\mathbb{R}}$, the value of the **Kostant partition function** $\mathscr{P}(\gamma)$ is defined as the number of nonnegative integer tuples $\{n_\alpha, \alpha \in \Delta^+\}$ such that $\gamma = \sum n_\alpha \alpha$. Let

$$d = e^\delta \prod_{\alpha\in\Delta^+} (1 - e^{-\alpha}) \qquad (\text{in } \mathbb{Z}[\exp \mathfrak{h}'_{\mathbb{R}}])$$

$$K = \sum_{\gamma\in Q^+} \mathscr{P}(\gamma) e^{-\gamma}. \qquad (\text{in } \mathbb{Z}\langle\mathfrak{h}'_{\mathbb{R}}\rangle)$$

   Prove that $Ke^{-\delta}d = 1$ in the ring $\mathbb{Z}\langle\mathfrak{h}'_{\mathbb{R}}\rangle$; thus d^{-1} exists. [Hint: Show that $(1 - e^{-\alpha})(1 + e^{-\alpha} + e^{-2\alpha} + \ldots) = 1$ for all α in Δ^+, and take the product over all α.]

8. Let λ be a highest weight for G. Motivated by the Weyl character formula, define the formal character of an irreducible representation of highest weight λ to be the member of $\mathbb{Z}\langle\mathfrak{h}'_{\mathbb{R}}\rangle$ given by

$$\text{char}(\lambda) = d^{-1} \sum_{w\in W} (\det w) e^{w(\lambda+\delta)}.$$

Use the result of Problem 7 to prove that the coefficient of e^{μ} in char(λ) is $\sum_{w \in W} (\det w)\mathscr{P}(w(\lambda + \delta) - (\mu + \delta))$.

9. Combine the result of Problem 8, an application of the evaluation homomorphism of Problem 6c, and a suitable linear independence argument to prove that the multiplicity of the weight μ in an irreducible representation of G with highest weight λ is

$$\sum_{w \in W} (\det w)\mathscr{P}(w(\lambda + \delta) - (\mu + \delta)).$$

Problems 10 to 17 deal with decomposing tensor products into irreducible representations.

10. Let G be SU(2), and let Φ_n be as in (2.1). Prove that $\Phi_m \otimes \Phi_n$ decomposes as the sum of all Φ_k with k going from $|m - n|$ to $m + n$ in steps of 2; each such Φ_k has multiplicity one. [Hint: Write the character of Φ_m as a sum of exponentials of weights, write the character of Φ_n as in the Weyl character formula, multiply, and rearrange.]
11. Let G be a compact connected Lie group, and let notation be as in Chapter IV. Let Φ_λ and $\Phi_{\lambda'}$ be irreducible representations of G with highest weights λ and λ', respectively. Prove that the weights of $\Phi_\lambda \otimes \Phi_{\lambda'}$ are all sums $\mu + \mu'$, where μ is a weight of Φ_λ and μ' is a weight of $\Phi_{\lambda'}$. How is the multiplicity of $\mu + \mu'$ related to multiplicities in Φ_λ and $\Phi_{\lambda'}$?
12. Let v_λ and $v_{\lambda'}$ be highest weight vectors in Φ_λ and $\Phi_{\lambda'}$, respectively. Prove that $v_\lambda \otimes v_{\lambda'}$ is a highest weight vector in $\Phi_\lambda \otimes \Phi_{\lambda'}$. Conclude that $\Phi_{\lambda + \lambda'}$ occurs exactly once in $\Phi_\lambda \otimes \Phi_{\lambda'}$.
13. Let λ'' be any highest weight in $\Phi_\lambda \otimes \Phi_{\lambda'}$, i.e., the highest weight of some irreducible constituent. Prove that λ'' is of the form $\lambda'' = \lambda + \mu'$ for some weight μ' of $\Phi_{\lambda'}$. [Hint: Write the λ'' weight vector v as $v = \sum_{\mu + \mu' = \lambda''} (v_\mu \otimes v_{\mu'})$, and choose $\mu = \mu_0$ as large as possible so that v_μ in this sum is not 0. Prove that v_{μ_0} is highest in Φ_λ.]
14. Prove that if all weights of Φ_λ have multiplicity one, then each irreducible constituent of $\Phi_\lambda \otimes \Phi_{\lambda'}$ has multiplicity one.
15. If λ is an integral form and if there exists $w_0 \neq 1$ in W fixing λ, prove that $\sum_{w \in W} (\det w)\xi_{w\lambda} = 0$. [Hint: Use Chevalley Lemma. Cf. Step (v) in the proof of Theorem 4.46.]
16. Let $m_\lambda(\mu)$ be the multiplicity of the weight μ in Φ_λ, and define sgn μ by

$$\operatorname{sgn} \mu = \begin{cases} 0 & \text{if some } w \neq 1 \text{ in } W \text{ fixes } \mu \\ \det w & \text{otherwise, where } w \text{ is chosen in } W \text{ to make } w\mu \\ & \text{dominant.} \end{cases}$$

With $\check{\mu}$ denoting the result of applying an element of W to make μ dominant and $\check{}$ indicating "made dominant," prove that

$$\chi_\lambda \chi_{\lambda'} = \sum_{\lambda'' = \text{weight of } \Phi_\lambda} m_\lambda(\lambda'') \operatorname{sgn}(\lambda'' + \lambda' + \delta) \chi_{(\lambda'' + \lambda' + \delta)^\vee - \delta}.$$

[Hint: Write $\chi_\lambda = \sum m_\lambda(\lambda'')\xi_{\lambda''}$, write $\chi_{\lambda'}$ as in the Weyl character formula, and multiply. Change $\xi_{\lambda'' + w(\lambda' + \delta)}$ to $\xi_{w(\lambda'' + \lambda' + \delta)}$ by using the fact that $m_\lambda(\lambda'') = m_\lambda(w\lambda'')$. Eliminate unnecessary terms by means of Problem 15, and examine the remaining terms ξ_ν for which ν is dominant.]

17. Let $-\mu$ be the lowest weight of Φ_λ. Deduce from Problem 16 that if $\lambda' - \mu$ is dominant, then $\Phi_{\lambda' - \mu}$ occurs in $\Phi_\lambda \otimes \Phi_{\lambda'}$ with multiplicity one.

CHAPTER V

Structure Theory for Noncompact Groups

§1. Cartan Decomposition and the Unitary Trick

In Chapter I we defined linear connected reductive groups and linear connected semisimple groups. The reductive groups are the ones that arise in practice, and we shall see in this section that they differ from semisimple groups only trivially.

For semisimple groups themselves, our first examples were $SL(2, \mathbb{R})$ and $SU(2)$, imbedded as subgroups of $SL(2, \mathbb{C})$ with Lie algebras $\mathfrak{k} \oplus \mathfrak{p}$ and $\mathfrak{k} \oplus i\mathfrak{p}$; the relationships among these subgroups and Lie algebras leads to the unitary trick of Chapter II. We shall generalize these relationships in this section and deduce a unitary trick for the generalization. For $SL(2, \mathbb{R})$, we have $\mathfrak{k} \cap i\mathfrak{p} = 0$, and this equation allows us to regard $\mathfrak{g}^{\mathbb{C}}$ as the obvious set of matrices. We shall make the assumption $\mathfrak{k} \cap i\mathfrak{p} = 0$ in our generalization; the assumption is not really a restriction, since we can always imbed $G \subseteq GL(n, \mathbb{C})$ in $GL(2n, \mathbb{R})$, and there the condition $\mathfrak{k} \cap i\mathfrak{p} = 0$ will automatically be satisfied.

Lemma 5.1. If U is a dense analytic subgroup of an analytic group U_1 and if $\mathfrak{u}$ and $\mathfrak{u}_1$ are the respective Lie algebras, then $[\mathfrak{u}_1, \mathfrak{u}] \subseteq \mathfrak{u}$.

Proof. If X is in $\mathfrak{u}$ and z is in U, then $\text{Ad}(z)X$ is in $\mathfrak{u}$. Passing to the limit, we see that X in $\mathfrak{u}$ and z in U_1 imply $\text{Ad}(z)X$ is in $\mathfrak{u}$, since $\mathfrak{u}$ is closed. Differentiating along one-parameter subgroups of U_1, we obtain $[\mathfrak{u}_1, \mathfrak{u}] \subseteq \mathfrak{u}$.

Lemma 5.2. If $\mathfrak{g}$ is a real semisimple Lie algebra, then every derivation of $\mathfrak{g}$ is inner (i.e., is of the form $\text{ad } X$ for some X in $\mathfrak{g}$).

Proof. Let D be a derivation of $\mathfrak{g}$. Since the Killing form B of $\mathfrak{g}$ is nondegenerate, we can find X in $\mathfrak{g}$ with $\text{Tr}(D \text{ ad } Y) = B(X, Y)$ for all Y in $\mathfrak{g}$. The derivation property

$$[DY, Z] = D([Y, Z]) - [Y, DZ]$$

can be rewritten

$$\text{ad } DY = [D, \text{ad } Y]. \tag{5.1}$$

Therefore

$$\begin{aligned} B(DY, Z) &= \mathrm{Tr}(\mathrm{ad}\, DY\, \mathrm{ad}\, Z) \\ &= \mathrm{Tr}([D, \mathrm{ad}\, Y]\, \mathrm{ad}\, Z) && \text{by (5.1)} \\ &= \mathrm{Tr}(D\, \mathrm{ad}[Y, Z]) && \text{by expanding both sides} \\ &= B(X, [Y, Z]) && \text{by definition of } X \\ &= B([X, Y], Z), \end{aligned}$$

and $D = \mathrm{ad}\, X$ because B is nondegenerate.

Proposition 5.3. Let G be a linear connected reductive group with compact center, let $\mathfrak{g} = \mathfrak{k} \oplus \mathfrak{p}$ be the Cartan decomposition of its Lie algebra, and suppose $\mathfrak{k} \cap i\mathfrak{p} = 0$. Then the analytic group U of matrices with Lie algebra $\mathfrak{k} \oplus i\mathfrak{p}$ is compact, and so is the analytic subgroup U' with Lie algebra $([\mathfrak{g}, \mathfrak{g}] \cap \mathfrak{k}) \oplus i\mathfrak{p}$.

Remarks. The group U is called the **compact form** of G, and $\mathfrak{u} = \mathfrak{k} \oplus i\mathfrak{p}$ is called the **compact form** of $\mathfrak{g}$. For example, the compact forms of $SL(n, \mathbb{R})$ and $SU(m, n)$ are $SU(n)$ and $SU(m + n)$, respectively.

Proof. First suppose G is semisimple. Then $\mathfrak{g}^{\mathbb{C}}$ is semisimple and equals $(\mathfrak{k} \oplus i\mathfrak{p})^{\mathbb{C}}$. Hence $\mathfrak{k} \oplus i\mathfrak{p}$ is semisimple. Let U_1 be the closure of U. Then U_1 is compact, being a closed subgroup of a unitary group, and its Lie algebra $\mathfrak{u}_1$ normalizes $\mathfrak{u} = \mathfrak{k} \oplus i\mathfrak{p}$, by Lemma 5.1. Lemma 5.2 then says that to each X in $\mathfrak{u}_1$ corresponds $f(X)$ in $\mathfrak{u}$ with $\mathrm{ad}\, X = \mathrm{ad}\, f(X)$ on $\mathfrak{u}$, and $f(X)$ is unique since $\mathfrak{u}$ has zero center. Then $\mathfrak{u}_1 = \mathfrak{u} \oplus \ker f$ and $\ker f = Z_{\mathfrak{u}_1}$. Hence $\mathfrak{u} = [\mathfrak{u}_1, \mathfrak{u}_1]$, and Proposition 4.32 shows U is compact.

In the general case, let $\mathfrak{k}' = [\mathfrak{g}, \mathfrak{g}] \cap \mathfrak{k}$. Since G has compact center, $\mathfrak{k} = Z_{\mathfrak{g}} \oplus \mathfrak{k}'$, and $[\mathfrak{g}, \mathfrak{g}] = \mathfrak{k}' \oplus \mathfrak{p}$. The argument in the previous paragraph shows that the analytic group U' with Lie algebra $\mathfrak{k}' \oplus i\mathfrak{p}$ is compact. Since $U = Z_G U'$, U is compact, too.

Lemma 5.4. Let G be an analytic group with a semisimple Lie algebra $\mathfrak{g}$, and let Φ be a representation of G on a finite-dimensional V. Then $\det \Phi(g) = 1$ for all g in G.

Proof. Since $[\mathfrak{g}, \mathfrak{g}] = \mathfrak{g}$, the commutators $h = g_1 g_2 g_1^{-1} g_2^{-1}$ generate G. For each such h, $\det \Phi(h) = 1$, and so $\det \Phi(G) = 1$.

Proposition 5.5. Let G be a linear connected reductive group with compact center, let $\mathfrak{g} = \mathfrak{k} \oplus \mathfrak{p}$ be the Cartan decomposition of its Lie algebra, and suppose $\mathfrak{k} \cap i\mathfrak{p} = 0$. Then the analytic subgroup G_{ss} with Lie algebra $[\mathfrak{g}, \mathfrak{g}]$ is closed (hence linear connected semisimple), and G is the commuting product $G = (Z_G)_0 G_{ss}$.

Remarks. So linear reductive groups are only trivially more general than linear semisimple groups. In particular, Proposition 5.5 allows us to complete the proof of Proposition 1.4 just by inspection.

Proof. Form $\tilde{G} = (Z_G)_0 \times G_{ss}$. Multiplication of the two factors within G gives us a homomorphism m of $\tilde{G}$ into G whose differential at 1 is an isomorphism. So $m: \tilde{G} \to G$ is a covering map, and $G = (Z_G)_0 G_{ss}$. Since G_{ss} is closed in $\tilde{G}$, we will have proved the result if we show $m: \tilde{G} \to G$ is a finite cover, i.e., that ker m is finite. Here ker m is a discrete subgroup of $Z_{\tilde{G}} = (Z_G)_0 \times Z_{G_{ss}}$, and Z_G is compact by assumption. If we prove $Z_{G_{ss}}$ is finite, then $Z_{\tilde{G}}$ is compact and ker m has to be finite. So we are to prove $Z_{G_{ss}}$ is finite.

We are given $G \subseteq \mathrm{GL}(V)$ with $V = C^m$ for some m. Let U be as in Proposition 5.3. Then the invariant subspaces for G, $\mathfrak{g}$, $\mathfrak{u}$, and U within V coincide. Since U is compact, $V = \sum V_j$ with each V_j irreducible under G. By Schur's Lemma (proved for G just as in Corollary 1.9), any z in $Z_{G_{ss}} \subseteq Z_G$ has $z|_{V_j} = cI$. Moreover, Lemma 5.4 shows $\det(z|_{V_j}) = 1$, so that $c^{\dim V_j} = 1$. Thus $Z_{G_{ss}}$ is finite.

Proposition 5.6. Let G be a linear connected reductive group, let $\mathfrak{g} = \mathfrak{k} \oplus \mathfrak{p}$ be the Cartan decomposition of its Lie algebra, and suppose $\mathfrak{k} \cap i\mathfrak{p} = 0$. Then the analytic group $G^{\mathbb{C}}$ of matrices with Lie algebra $(\mathfrak{k} \oplus \mathfrak{p})^{\mathbb{C}}$ is closed and hence is a linear connected reductive group.

Proof in semisimple case. Let $G_1^{\mathbb{C}}$ be the closure of $G^{\mathbb{C}}$, and let $(\mathfrak{g}^{\mathbb{C}})_1$ be its Lie algebra. Arguing as in the first half of the proof of Proposition 5.3, we see that $(\mathfrak{g}^{\mathbb{C}})_1 = \mathfrak{g}^{\mathbb{C}} \oplus Z_{\mathfrak{g}_1^{\mathbb{C}}}$. The group $G_1^{\mathbb{C}}$ is linear connected reductive, and this decomposition shows that its semisimple part is $G^{\mathbb{C}}$. We cannot apply Proposition 5.5 immediately, because that $\mathfrak{k}$ and $i\mathfrak{p}$ parts of $(\mathfrak{g}^{\mathbb{C}})_1$ are not disjoint, but they become disjoint when we imbed our underlying $\mathrm{GL}(n, \mathbb{C})$ as a closed subgroup of $\mathrm{GL}(2n, \mathbb{R})$. Then Proposition 5.5 in this setting implies that $G^{\mathbb{C}}$ is closed. Hence it is closed in its original setting.

Proposition 5.7 (Weyl's unitary trick). Let G be a linear connected semisimple group, let $\mathfrak{g} = \mathfrak{k} \oplus \mathfrak{p}$ be the Cartan decomposition of its Lie algebra, and suppose $\mathfrak{k} \cap i\mathfrak{p} = 0$. Let U and $G^{\mathbb{C}}$ be the analytic groups of matrices with Lie algebras $\mathfrak{u} = \mathfrak{k} \oplus i\mathfrak{p}$ and $\mathfrak{g}^{\mathbb{C}} = (\mathfrak{k} \oplus \mathfrak{p})^{\mathbb{C}}$, and suppose U is simply connected. If V is any finite-dimensional complex vector space, then a representation of any of the following kinds on V leads, via the formula

$$\mathfrak{g}^{\mathbb{C}} = \mathfrak{g} \oplus i\mathfrak{g} = \mathfrak{u} \oplus i\mathfrak{u},$$

to a representation of each of the other kinds. Under this correspondence, invariant subspaces and equivalences are preserved:

(a) a representation of G on V
(b) a representation of U on V
(c) a holomorphic representation of $G^{\mathbb{C}}$ on V
(d) a representation of $\mathfrak{g}$ on V
(e) a representation of $\mathfrak{u}$ on V
(f) a complex-linear representation of $\mathfrak{g}^{\mathbb{C}}$ on V.

Proof. The proof proceeds as in Proposition 2.1, since U is compact (Proposition 5.3) and $G^{\mathbb{C}}$ is closed (Proposition 5.6). The group U is the maximal compact subgroup of $G^{\mathbb{C}}$, and $G^{\mathbb{C}}$ is topologically the product of U and a Euclidean space, by Proposition 1.2. Hence $G^{\mathbb{C}}$ is simply connected, and the proof goes through.

Example. Let $G = \mathrm{Ad}(SL(3, \mathbb{R}))$, realized through the adjoint representation of $SL(3, \mathbb{R})$ as a certain group of 8-by-8 matrices. Then we can form U and $G^{\mathbb{C}}$ contained in $GL(8, \mathbb{C})$. Here U is not simply connected, and the correspondence in the proposition breaks down. In fact, $SL(3, \mathbb{R}) \cong \mathrm{Ad}(SL(3, \mathbb{R}))$ since $SL(3, \mathbb{R})$ has trivial center; thus G has a three-dimensional nontrivial representation. But $U \cong SU(3)/\mathbb{Z}_3$ does not have a three-dimensional nontrivial representation.

Corollary 5.8. If G is a linear connected semisimple group with Lie algebra $\mathfrak{g}$, then every finite-dimensional representation of G or of $\mathfrak{g}$ is the direct sum of irreducible representations.

Proof. If $\mathfrak{k} \cap i\mathfrak{p} = 0$ and U is simply connected, the result follows from Proposition 5.7. We can handle $\mathfrak{k} \cap i\mathfrak{p} \neq 0$ by passing from $GL(n, \mathbb{C})$ to $GL(2n, \mathbb{R})$. If U is not simply connected, Weyl's Theorem shows that a finite covering is simply connected. This covering is also a matrix group and can be taken to consist of unitary matrices. Then by passing to the matrix group corresponding to $\mathfrak{g}$, we obtain a group with G as homomorphic image, and Proposition 5.7 applies to this group.

§2. Iwasawa Decomposition

Example 1. Let $G = SL(m, \mathbb{C})$. As usual $K = SU(m)$. Let A be the group of diagonal matrices with positive entries on the diagonal, and let N be the group of upper triangular matrices with 1 in every diagonal entry. Then every element of G is uniquely the product of a member of K, a member of A, and a member of N, and $G = KAN$ is called the Iwasawa decomposition of G. There is a simple geometric interpretation of this decomposition, as follows: Given g, we want to write $g = kan$, hence $k^{-1}g = an$. Let $u_1, \ldots, u_m$ be the standard orthonormal basis of $\mathbb{C}^m$, and form the basis $gu_1, \ldots, gu_m$. We apply the Gram-Schmidt orthogonalization procedure to obtain an orthonormal basis $v_1, \ldots, v_m$ with the properties that

$$\operatorname{span}\{gu_1, \ldots, gu_j\} = \operatorname{span}\{v_1, \ldots, v_j\}$$

and
$$v_j \in \mathbb{R}^+(gu_j) + \operatorname{span}\{v_1, \ldots, v_{j-1}\}, \qquad 1 \leq j \leq m.$$

If k^{-1} is the matrix that carries the column vector v_j to u_j for each j, then it is easy to see that k is in $\mathrm{SU}(m)$ and $k^{-1}g$ is in AN. This handles existence of the decomposition. Uniqueness follows after verifying that $K \cap AN = \{1\}$. Multiplication from $K \times A \times N$ into G is certainly smooth, and the inverse is smooth because, as is seen inductively, v_j depends smoothly on g. There is a corresponding but less subtle decomposition of the Lie algebra $\mathfrak{g} = \mathfrak{sl}(m, \mathbb{C})$. We have $\mathfrak{k} = \mathfrak{su}(m)$. Let $\mathfrak{a}$ consist of those members of $\mathfrak{g}$ with real entries on the diagonal and with 0's off the diagonal, and let $\mathfrak{n}$ consist of those members of $\mathfrak{g}$ that are 0 on and below the diagonal. Then we have a (vector space) direct sum formula $\mathfrak{g} = \mathfrak{k} \oplus \mathfrak{a} \oplus \mathfrak{n}$, which is called the Iwasawa decomposition of $\mathfrak{g}$.

Example 2. Let $G = \mathrm{SL}(m, \mathbb{R})$, $\mathfrak{g} = \mathfrak{sl}(m, \mathbb{R})$. Here $K = \mathrm{SO}(m)$ and $\mathfrak{k} = \mathfrak{so}(m)$, A and $\mathfrak{a}$ are as in Example 1, and N and $\mathfrak{n}$ are the real matrices of the same types as in Example 1. Again we have $\mathfrak{g} = \mathfrak{k} \oplus \mathfrak{a} \oplus \mathfrak{n}$ and $G = KAN$, with the global decomposition interpreted by means of the Gram-Schmidt procedure.

We turn to the general case, where we shall derive a Lie algebra decomposition and use it to obtain a group decomposition. Let G be a linear connected semisimple group with Lie algebra $\mathfrak{g} = \mathfrak{k} \oplus \mathfrak{p}$, maximal compact subgroup K, and trace form B_0, as in §§1.1–1.2. Let $\mathfrak{a}$ be a maximal abelian subspace of $\mathfrak{p}$ (existence by finite-dimensionality). With respect to the inner product $\langle X, Y\rangle = -\operatorname{Re} B_0(X, \theta Y)$ on the real vector space $\mathfrak{g}$, each member of $\operatorname{ad}\mathfrak{p}$ is symmetric. Thus the members of $\operatorname{ad}\mathfrak{a}$ form a commuting family of symmetric transformations on $\mathfrak{p}$, and they can be simultaneously diagonalized, with real eigenvalues. For each linear functional λ on $\mathfrak{a}$, let

$$\mathfrak{g}_\lambda = \{X \in \mathfrak{g} \mid [H, X] = \lambda(H)X \text{ for all } H \in \mathfrak{a}\}.$$

If $\lambda \not\equiv 0$ and $\mathfrak{g}_\lambda \neq 0$, we say λ is a **restricted root** of $\mathfrak{g}$ or a root of $(\mathfrak{g}, \mathfrak{a})$, with **restricted root space** $\mathfrak{g}_\lambda$. The set of restricted roots will be denoted either Σ or $\Delta(\mathfrak{a}:\mathfrak{g})$, depending on circumstances. By the same reasoning as with roots for compact groups, we obtain the following proposition.

Proposition 5.9. Restricted roots and their root spaces have the following properties:

(a) $\mathfrak{g} = \mathfrak{g}_0 \oplus \sum_{\lambda \in \Sigma} \mathfrak{g}_\lambda$
(b) $[\mathfrak{g}_\lambda, \mathfrak{g}_\mu] \subseteq \mathfrak{g}_{\lambda+\mu}$
(c) $\theta\mathfrak{g}_\lambda = \mathfrak{g}_{-\lambda}$ and hence $\lambda \in \Sigma$ implies $-\lambda \in \Sigma$
(d) $\mathfrak{g}_\lambda$ and $\mathfrak{g}_\mu$ are orthogonal with respect to $\langle \cdot, \cdot \rangle$ if $\lambda \neq \mu$
(e) $\mathfrak{g}_0 = \mathfrak{a} \oplus \mathfrak{m}$, where $\mathfrak{m} = Z_{\mathfrak{k}}(\mathfrak{a})$, and the sum is an orthogonal sum.

Proof of (e). Part (c) shows $\theta\mathfrak{g}_0 = \mathfrak{g}_0$, and so $\mathfrak{g}_0 = (\mathfrak{k} \cap \mathfrak{g}_0) \oplus (\mathfrak{p} \cap \mathfrak{g}_0)$. We must have $\mathfrak{a} = \mathfrak{p} \cap \mathfrak{g}_0$ since $\mathfrak{a} \subseteq \mathfrak{p} \cap \mathfrak{g}_0$ and $\mathfrak{a}$ is maximal abelian in $\mathfrak{p}$, and it is clear that $\mathfrak{k} \cap \mathfrak{g}_0 = Z_{\mathfrak{k}}(\mathfrak{a})$.

Later we shall see that Σ is an abstract root system. But for now, let us proceed without knowing this fact. We introduce an ordering in $\mathfrak{a}'$ just as in §4.3, and this ordering singles out a set Σ^+ of positive restricted roots. Define

$$\mathfrak{n} = \sum_{\lambda \in \Sigma^+} \mathfrak{g}_\lambda.$$

By Proposition 5.9, $\mathfrak{n}$ is a Lie subalgebra of $\mathfrak{g}$.

Proposition 5.10. For G linear connected semisimple, $\mathfrak{g}$ is a direct sum $\mathfrak{g} = \mathfrak{k} \oplus \mathfrak{a} \oplus \mathfrak{n}$. Here $\mathfrak{a}$ is abelian, $\mathfrak{n}$ is nilpotent, $\mathfrak{a} \oplus \mathfrak{n}$ is solvable, and $[\mathfrak{a} \oplus \mathfrak{n}, \mathfrak{a} \oplus \mathfrak{n}] = \mathfrak{n}$.

Proof. We constructed $\mathfrak{a}$ to be abelian, and $\mathfrak{n}$ is nilpotent by use of Proposition 5.9b. Since $[\mathfrak{a}, \mathfrak{g}_\lambda] = \mathfrak{g}_\lambda$ for each $\lambda \neq 0$, we see that $[\mathfrak{a}, \mathfrak{n}] = \mathfrak{n}$ and $\mathfrak{a} \oplus \mathfrak{n}$ is a solvable subalgebra with $[\mathfrak{a} \oplus \mathfrak{n}, \mathfrak{a} \oplus \mathfrak{n}] = \mathfrak{n}$. Every member of $\mathfrak{k}$ has nonzero projection either in $\mathfrak{m}$ or in $\sum_{\lambda \in \Sigma^+} \mathfrak{g}_{-\lambda}$, by Proposition 5.9, parts (c) and (e). Hence $\mathfrak{k} + \mathfrak{a} + \mathfrak{n}$ is a direct sum. The sum is all of $\mathfrak{g}$ because we can write

$$H + X_0 + \sum X_\lambda \qquad \text{with } H \in \mathfrak{a},\ X_0 \in \mathfrak{m},\ X_\lambda \in \mathfrak{g}_\lambda$$

as

$$= \left\{X_0 + \sum_{\lambda \in \Sigma^+} (X_{-\lambda} + \theta X_{-\lambda})\right\} + H + \left\{\sum_{\lambda \in \Sigma^+} (X_\lambda - \theta X_{-\lambda})\right\}$$

in $\mathfrak{k} \oplus \mathfrak{a} \oplus \mathfrak{n}$.

To pass to the decomposition at the group level, we use the following handy lemma.

Lemma 5.11. Let G be an analytic group with Lie algebra $\mathfrak{g}$, and suppose $\mathfrak{g}$ is a direct sum of subalgebras $\mathfrak{g} = \mathfrak{s} \oplus \mathfrak{t}$. If S and T denote the analytic subgroups corresponding to $\mathfrak{s}$ and $\mathfrak{t}$, then the multiplication map $\Phi(s, t) = st$ of $S \times T$ into G is everywhere regular.

Proof. The tangent space at (s_0, t_0) in $S \times T$ can be identified by left translation within S and within T with $\mathfrak{s} \oplus \mathfrak{t} = \mathfrak{g}$, and the tangent space at $s_0 t_0$ in G can be identified by left translation within G with $\mathfrak{g}$. With these identifications, we compute the differential $d\Phi$ at (s_0, t_0). Let X be in $\mathfrak{s}$ and Y be in $\mathfrak{t}$. Then

$$\Phi(s_0 \exp rX, t_0) = s_0(\exp rX)t_0 = s_0 t_0 \exp(\operatorname{Ad}(t_0^{-1})rX)$$

$$\Phi(s_0, t_0 \exp rY) = s_0 t_0 \exp rY,$$

from which it follows that

$$d\Phi(X) = \operatorname{Ad}(t_0^{-1})X$$

$$d\Phi(Y) = Y.$$

In matrix form, $d\Phi$ is block triangular, and thus

$$\det d\Phi = \frac{\det \operatorname{Ad}(t_0^{-1})}{\det \operatorname{Ad}_T(t_0^{-1})} = \frac{\det \operatorname{Ad}_T(t_0)}{\det \operatorname{Ad}(t_0)},$$

where Ad_T refers to the adjoint action of T on $\mathfrak{t}$. Since $\det d\Phi$ is exhibited by this formula as nonzero, Φ is regular.

Theorem 5.12 (Iwasawa decomposition). For G linear connected semisimple, let A and N be the analytic subgroups with Lie algebras $\mathfrak{a}$ and $\mathfrak{n}$. Then A, N, and AN are simply connected closed subgroups of G, and the multiplication map $K \times A \times N \to G$ given by $(k, a, n) \to kan$ is a diffeomorphism onto.

Most of proof. The group A is simply connected and closed, by Proposition 1.2. We omit the verifications that N and AN are simply connected and closed. Multiplication from $A \times N$ to AN is smooth and onto. It is one-one since each $\operatorname{Ad}(a)$ acts diagonably with positive eigenvalues, each $\operatorname{Ad}(n)$ acts with all eigenvalues 1, and $A \cap Z_G = \{1\}$. It is regular by Lemma 5.11 and thus is a diffeomorphism.

Next, multiplication from $K \times AN$ into G is smooth. To see it is onto, we note that the image is closed since K is compact and AN is closed, and the image is open by Lemma 5.11 and Proposition 5.10. Thus the image is all of G since G is connected. A consideration of eigenvalues shows that $K \cap AN = \{1\}$, so that multiplication is one-one. By Lemma 5.11 and Proposition 5.10 again, multiplication $K \times AN \to G$ is a diffeomorphism onto G. Thus multiplication $K \times A \times N \to G$ is a diffeomorphism onto G.

The decomposition of $\mathfrak{g}$ via $\mathfrak{a}$ into restricted root spaces is not merely analogous to the root space decomposition of a compact group but is actually deducible from the decomposition of the compact form $\mathfrak{u} = \mathfrak{k} \oplus i\mathfrak{p}$ into root spaces. Namely extend $\mathfrak{a}$ to a maximal abelian subspace of $\mathfrak{g}$, necessarily by adjoining a maximal abelian subspace $\mathfrak{b}$ of $\mathfrak{m}$. Then it is easy to see that $\mathfrak{h} = \mathfrak{b} \oplus i\mathfrak{a}$ is a Cartan subalgebra of $\mathfrak{u}$. Let Δ be the roots of $\mathfrak{u}^{\mathbb{C}}$ with respect to $\mathfrak{h}^{\mathbb{C}}$, and write

$$\mathfrak{g}^{\mathbb{C}} = \mathfrak{u}^{\mathbb{C}} = \mathfrak{h}^{\mathbb{C}} \oplus \sum_{\alpha \in \Delta} \mathfrak{u}_\alpha.$$

Then we have

$$\mathfrak{g}_\lambda = \mathfrak{g} \cap \left\{ \sum_{\alpha|_\mathfrak{a} = \lambda} \mathfrak{u}_\alpha \right\}$$

and

$$\mathfrak{m}^{\mathbb{C}} = \mathfrak{b}^{\mathbb{C}} \oplus \sum_{\substack{\alpha \in \Delta \\ \alpha|_\mathfrak{a} = 0}} \mathfrak{u}_\alpha. \tag{5.2}$$

That is, the restricted roots are the nonzero restrictions to $\mathfrak{a}$ of the roots, and $\mathfrak{m}$ arises from the roots that restrict to 0 on $\mathfrak{a}$.

The construction of the Iwasawa decomposition involves two choices beyond how G is imbedded as a linear group—namely, the maximal abelian subspace $\mathfrak{a}$ of $\mathfrak{p}$ and the positive system Σ^+ within Σ. We shall prove that all possible other results for these choices can be achieved by conjugation within G by a member of K. Theorem 5.13 demonstrates this fact for the choice of $\mathfrak{a}$.

Theorem 5.13. The space $\mathfrak{p}$ satisfies $\mathfrak{p} = \bigcup_{k \in K} \operatorname{Ad}(k)\mathfrak{a}$. In more detail, if $\mathfrak{a}$ and $\mathfrak{a}_1$ are two maximal abelian subspaces of $\mathfrak{p}$, then there is a member k of K such that $\operatorname{Ad}(k)\mathfrak{a} = \mathfrak{a}_1$.

Proof. Choose H in $\mathfrak{a}$ so that no root of $(\mathfrak{g}, \mathfrak{a})$ vanishes on H; this is possible since there are only finitely many restricted roots and their kernels are hyperplanes in $\mathfrak{a}'$. Choose H_1 in $\mathfrak{a}_1$ so that no root of $(\mathfrak{g}, \mathfrak{a}_1)$ vanishes on H_1. Then we can argue as in the proof of Theorem 4.42: If $k_0 \in K$ is chosen so as to minimize $\operatorname{Re} B_0(\operatorname{Ad}(k)H, H_1)$, then the same kind of argument as in Theorem 4.42 gives

$$\operatorname{Re} B_0([X, \operatorname{Ad}(k_0)H], H_1) = 0 \qquad \text{for all } X \in \mathfrak{k}$$

and hence

$$\operatorname{Re} B_0([\operatorname{Ad}(k_0)H, H_1], X) = 0 \qquad \text{for all } X \in \mathfrak{k}.$$

Since $[\operatorname{Ad}(k)H, H_1]$ is in $\mathfrak{k}$ and $B_0(X, X) < 0$ for nonzero X in $\mathfrak{k}$, we see that $[\operatorname{Ad}(k)H, H_1] = 0$. Since no root of $(\mathfrak{g}, \mathfrak{a}_1)$ vanishes on H_1, we conclude that $\operatorname{Ad}(k)H$ is in $\mathfrak{p} \cap \mathfrak{g}_0$ relative to $\mathfrak{a}_1$, and this means $\operatorname{Ad}(k)H$ is in $\mathfrak{a}_1$. So H is in $\operatorname{Ad}(k)^{-1}\mathfrak{a}_1$. Since $\operatorname{Ad}(k)^{-1}\mathfrak{a}_1$ is abelian and no root of $(\mathfrak{g}, \mathfrak{a})$

vanishes on H, $\mathrm{Ad}(k)^{-1}\mathfrak{a}_1$ is in $\mathfrak{p} \cap \mathfrak{g}_0$ relative to $\mathfrak{a}$. Thus $\mathrm{Ad}(k)^{-1}\mathfrak{a}_1 \subseteq \mathfrak{a}$ and $\mathrm{Ad}(k)\mathfrak{a} \subseteq \mathfrak{a}_1$. By maximality, $\mathrm{Ad}(k)\mathfrak{a} = \mathfrak{a}_1$.

Remarks. For $\mathfrak{sl}(n, \mathbb{C})$ the first conclusion of Proposition 5.13 is the assertion that every Hermitian matrix is conjugate via a unitary matrix to a diagonal matrix.

§3. Regular Elements, Weyl Chambers, and the Weyl Group

The other choice that enters the Iwasawa decomposition is the ordering on $\mathfrak{a}'$, or at least the selection of Σ^+. This choice determines $\mathfrak{n}$. We are going to analyze this choice in more detail than would be necessary just to show the conjugacy of the choices of $\mathfrak{n}$. We shall first make this notion more geometric by parametrizing the choices of Σ^+ by "Weyl chambers" in $\mathfrak{a}$, connected components of the "regular elements" of $\mathfrak{a}$. It will be apparent that these same considerations apply also to the analysis of roots for a compact group, with $\mathfrak{h}_{\mathbb{R}}$ there playing the role that is here played by $\mathfrak{a}$.

Afterward we show that Σ is an abstract root system, and then it follows that its Weyl group permutes the simple systems Σ^+ simply transitively. In more geometric terms, the Weyl group of Σ permutes the Weyl chambers of $\mathfrak{a}$ simply transitively. Finally we identify the Weyl group of Σ with an analytically defined Weyl group, and we conclude that any $\mathfrak{n}$ can be conjugated into any other $\mathfrak{n}$ by a member of K, while $\mathfrak{a}$ is conjugated into itself, and that there is an appropriate uniqueness statement.

The restriction of $\mathrm{Re}\, B_0(\cdot, \cdot)$ to $\mathfrak{a} \times \mathfrak{a}$ coincides with our form $\langle \cdot, \cdot \rangle$ and is an inner product on $\mathfrak{a}$. If μ is in $\mathfrak{a}'$, we define H_μ in $\mathfrak{a}$ by

$$\mu(H) = \mathrm{Re}\, B_0(H, H_\mu) \qquad \text{for } H \in \mathfrak{a},$$

and then we define an inner product $\langle \cdot, \cdot \rangle$ on $\mathfrak{a}'$ by

$$\langle \mu, \nu \rangle = \mu(H_\nu) = \nu(H_\mu) = \mathrm{Re}\, B_0(H_\mu, H_\nu).$$

A **regular element** H in $\mathfrak{a}$ is an element satisfying $\lambda(H) \neq 0$ for all $\lambda \in \Sigma$. The set $\mathfrak{a}_r$ of regular elements is nonempty and open, being the complement of the union of finitely many hyperplanes in $\mathfrak{a}$. (Note that regular elements played a role in the proof of Proposition 5.13.) The connected components of $\mathfrak{a}_r$ are called **Weyl chambers.** Every root has constant sign on each Weyl chamber, and it follows that each Weyl chamber is an open convex cone in $\mathfrak{a}$. Figure 5.1 shows the Weyl chambers for $G = SL(3, \mathbb{R})$.

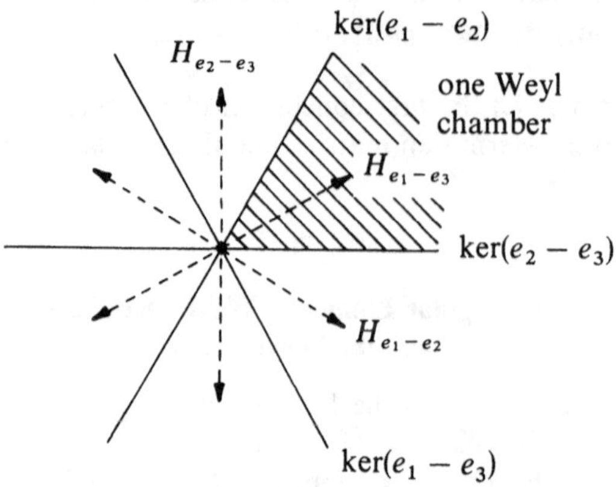

FIGURE 5.1. Weyl chambers for SL(3, ℝ)

Proposition 5.14. The map $\Sigma^+ \to \{H \in \mathfrak{a} \,|\, \lambda(H) > 0 \text{ for all } \lambda \in \Sigma^+\}$ of positive systems to subsets of $\mathfrak{a}$ is a one-one map of the set of all positive systems onto the set of all Weyl chambers in $\mathfrak{a}$.

Remarks.

(1) Hence the choices for $\mathfrak{n}$ are parametrized by the set of all Weyl chambers.

(2) It will be apparent from the proof that this relationship among **regular elements,** positive systems, and **Weyl chambers** is valid in the context of Chapter IV, as well.

Proof. Fix Σ^+ and let $\mathscr{C} = \{H \in \mathfrak{a} \,|\, \lambda(H) > 0 \text{ for all } \lambda \in \Sigma^+\}$. Then $\mathscr{C} \subseteq \mathfrak{a}_r$, and $\mathscr{C}$ is connected because it is convex. Clearly $\mathscr{C}$, if nonempty, is a maximal set on which each member of Σ has constant sign. Hence $\mathscr{C}$ is empty or $\mathscr{C}$ is a Weyl chamber; let us see that $\mathscr{C}$ is nonempty. Let $\varphi_1, \ldots, \varphi_l$ be a basis defining an ordering leading to Σ^+. Using the finiteness of Σ, we readily construct by induction (small) positive numbers $\varepsilon_2, \ldots, \varepsilon_l$ such that $\varphi_1 + \sum_{j=2}^{l} \varepsilon_j \varphi_j$ is in $\mathscr{C}$. Hence $\mathscr{C}$ is nonempty and is a Weyl chamber.

Since $\mathscr{C}$ is nonempty, any member of $\mathscr{C}$ can be used to test for positivity of members of Σ. Thus the map $\Sigma^+ \to \mathscr{C}$ is one-one.

Finally we show the map is onto. Let $\mathscr{C}$ be a Weyl chamber. Since $\mathscr{C}$ is nonempty, we can choose a basis $H_1, \ldots, H_l$ of $\mathfrak{a}$ with H_1 in $\mathscr{C}$. Let $\varphi_1, \ldots, \varphi_l$ denote the dual basis of $\mathfrak{a}'$, and let Σ^+ be the corresponding positive system. If $\lambda = \sum c_i \varphi_i$ is in Σ^+, then

$$0 \neq \lambda(H_1) = \sum c_i \varphi_i(H_1) = c_1$$

forces $c_1 > 0$ and $\lambda(H_1) > 0$. Thus every member of Σ^+ is positive on H_1, hence on $\mathscr{C}$. That is, $\mathscr{C}$ is the image of Σ^+.

Proposition 5.15. Let λ be a restricted root and let $E_\lambda \neq 0$ be in $\mathfrak{g}_\lambda$.

(a) $[E_\lambda, \theta E_\lambda] = \operatorname{Re} B_0(E_\lambda, \theta E_\lambda)H_\lambda$, and $\operatorname{Re} B_0(E_\lambda, \theta E_\lambda) < 0$.

(b) $\mathbb{R}H_\lambda + \mathbb{R}E_\lambda + \mathbb{R}\theta E_\lambda$ is a subalgebra of $\mathfrak{g}$ isomorphic to $\mathfrak{sl}(2, \mathbb{R})$, and under the isomorphism the vector $H'_\lambda = 2|\lambda|^{-2}H_\lambda$ corresponds to $h = \begin{pmatrix} 1 & 0 \\ 0 & -1 \end{pmatrix}$.

(c) If E_λ is normalized so that $\operatorname{Re} B_0(E_\lambda, \theta E_\lambda) = -2|\lambda|^{-2}$, then $k = \exp \frac{\pi}{2}(E_\lambda + \theta E_\lambda)$ is a member of the normalizer $N_K(\mathfrak{a})$, and $\operatorname{Ad}(k)$ acts as the reflection s_λ on $\mathfrak{a}'$.

(d) Σ is an abstract root system in $\mathfrak{a}'$.

Remarks.

(1) As a result of (d), we can parametrize the $\mathfrak{n}$'s in a second way—by using a fixed Weyl chamber as reference chamber and using the Weyl group of Σ to pass to the other chambers. The next step will be to use representatives within K of the Weyl group elements.

(2) The abstract root system Σ need not be reduced. For $SU(n, m)$ with $n > m$, it is of type $(BC)_m$. See Appendix C for more detail and further examples. Some of our results in Chapter IV about Weyl groups were proved only for reduced root systems. However, the only irreducible non-reduced root systems are those of type $(BC)_m$, where the Weyl group is of type C_m. Thus the results about Weyl groups do apply here with suitable interpretations.

Proof.

(a) We use Proposition 5.9. The vector $[E_\lambda, \theta E_\lambda]$ is in $[\mathfrak{g}_\lambda, \mathfrak{g}_{-\lambda}] \subseteq \mathfrak{g}_0$, and $\theta[E_\lambda, \theta E_\lambda] = [\theta E_\lambda, E_\lambda] = -[E_\lambda, \theta E_\lambda]$ shows $[E_\lambda, \theta E_\lambda]$ is in $\mathfrak{p}$. Thus $[E_\lambda, \theta E_\lambda]$ is in $\mathfrak{g}_0 \cap \mathfrak{p} = \mathfrak{a}$, and we can imitate the proof of Proposition 4.1e.

(b) Put

$$H'_\lambda = \frac{2}{|\lambda|^2} H_\lambda, \qquad E'_\lambda = \frac{2}{|\lambda|^2} E_\lambda, \qquad E'_{-\lambda} = \theta E_\lambda.$$

Then (a) shows

$$[H'_\lambda, E'_\lambda] = 2E'_\lambda, \qquad [H'_\lambda, E'_{-\lambda}] = -2E'_{-\lambda}, \qquad [E'_\lambda, E'_{-\lambda}] = H'_\lambda,$$

and (b) follows.

(c) We calculate that

$$\left(\operatorname{ad}\frac{\pi}{2}(E_\lambda + \theta E_\lambda)\right)\left(\frac{\pi}{2}(E_\lambda + \theta E_\lambda)\right) = -\tfrac{1}{2}\pi^2 H'_\lambda,$$

$$\left(\operatorname{ad}\frac{\pi}{2}(E_\lambda + \theta E_\lambda)\right) H'_\lambda = \pi(E_\lambda - \theta E_\lambda),$$

and

$$\left(\operatorname{ad}\frac{\pi}{2}(E_\lambda + \theta E_\lambda)\right)^2 H'_\lambda = \pi^2 H'_\lambda.$$

Then we can sum the infinite series for $\mathrm{Ad}(k)$ on H'_λ and $H_\lambda^\perp$ as in the proof of Theorem 4.40.

(d) Σ spans $\mathfrak{a}'$; in fact, $\mu(H) = 0$ for all $\mu \in \Sigma$ implies $[H, \mathfrak{g}_\mu] = 0$ for all μ, and hence H is in $Z_\mathfrak{g} = 0$. The subalgebra of (b) acts on $\mathfrak{g}$ by ad, and any member of $\mathfrak{g}_\mu$ is a weight vector whose weight on h is $2\langle\mu, \lambda\rangle/|\lambda|^2$. Corollary 2.2 and Theorem 2.4 show that the eigenvalues of h have to be integers, and thus $2\langle\mu, \lambda\rangle/|\lambda|^2$ is in $\mathbb{Z}$. Finally if μ and λ are in Σ, then we see that $s_\lambda(\mu)$ is in Σ as follows: Define k as in (c), and let H be in $\mathfrak{a}$ and X be in $\mathfrak{g}_\mu$. Then

$$\begin{aligned}[H, \mathrm{Ad}(k)X] &= \mathrm{Ad}(k)[\mathrm{Ad}(k)^{-1}H, X] = \mathrm{Ad}(k)[s_\lambda^{-1}(H), X] \\ &= \mu(s_\lambda^{-1}(H))\,\mathrm{Ad}(k)X = (s_\lambda\mu)(H)\,\mathrm{Ad}(k)X \end{aligned} \tag{5.3}$$

shows $\mathrm{Ad}(k)X$ is in $\mathfrak{g}_{s_\lambda\mu}$. Hence Σ is an abstract root system in $\mathfrak{a}'$.

We define the **Weyl group** $W(A:G)$ as the quotient of the normalizer by centralizer

$$W(A:G) = N_K(\mathfrak{a})/Z_K(\mathfrak{a}). \tag{5.4}$$

This is an effectively acting group of orthogonal transformations of $\mathfrak{a}$, hence also of $\mathfrak{a}'$. We write M for $Z_K(\mathfrak{a})$; this is a compact group with Lie algebra $Z_\mathfrak{k}(\mathfrak{a}) = \mathfrak{m}$. The theory of Chapter IV applies to the identity component M_0 and the Cartan subalgebra $\mathfrak{b}$ of (5.2), and (5.2) is the root space decomposition of $\mathfrak{m}^\mathbb{C}$.

Lemma 5.16. Let $\mathfrak{b}$ be a maximal abelian subspace of $\mathfrak{m}$. Every element of $N_K(\mathfrak{a})$ decomposes as the product zn, where n is in $N_K(\mathfrak{a} \oplus \mathfrak{b})$ and z is in $Z_K(\mathfrak{a})$.

Proof. Let k be in $N_K(\mathfrak{a})$ and consider $\mathrm{Ad}(k)\mathfrak{b}$. This is contained in $\mathfrak{k}$, and if H is in $\mathfrak{a}$ and X is in $\mathfrak{b}$ we have

$$[H, \mathrm{Ad}(k)X] = \mathrm{Ad}(k)[\mathrm{Ad}(k)^{-1}H, X] = 0$$

since $\mathrm{Ad}(k)^{-1}H$ is in $\mathfrak{a}$. Thus $\mathrm{Ad}(k)\mathfrak{b}$ is contained in $\mathfrak{m}$. Since $\mathrm{Ad}(k)\mathfrak{b}$ is clearly maximal abelian in $\mathfrak{m}$, Theorem 4.41 produces z in (the identity component of) $M = Z_K(\mathfrak{a})$ with $\mathrm{Ad}(z)^{-1}\mathrm{Ad}(k)\mathfrak{b} = \mathfrak{b}$. Then $n = z^{-1}k$ is in $N_K(\mathfrak{a})$ and in $N_K(\mathfrak{b})$, hence in $N_K(\mathfrak{a} \oplus \mathfrak{b})$.

Theorem 5.17. $W(A:G)$ coincides with the Weyl group of Σ.

Proof. Let us notice that $W(A:G)$ permutes the Weyl chambers. [In fact, the same computation as in (5.3) shows that $W(A:G)$ permutes the roots, hence leaves the set of regular elements stable, hence permutes the Weyl chambers.] Since Σ is an abstract root system (Proposition 5.15d), the Weyl group $W(\Sigma)$ of Σ is well defined. Proposition 5.15c shows that s_λ is in $W(A:G)$ if λ is in Σ, and hence $W(\Sigma) \subseteq W(A:G)$. By Theorem 4.10 and Proposition 5.14, $W(\Sigma)$ is simply transitive on the Weyl chambers. Thus we will be done if we show that no nontrivial element of $W(A:G)$ leaves a Weyl chamber stable.

We see easily from Proposition 5.9a that $N_K(\mathfrak{a})$ has Lie algebra $\mathfrak{m}$. Since $N_K(\mathfrak{a})$ and $Z_K(\mathfrak{a})$ have the same Lie algebra and are compact, $W(A:G)$ is finite. Thus let w leave the Weyl chamber $\mathscr{C}$ stable. Say $w^n = 1$. Choose H_0 in $\mathscr{C}$ and let $H_1 = \frac{1}{n}\sum_{k=0}^{n-1} w^k H_0$. Then H_1 is in $\mathscr{C}$ and is fixed by w.

Let k be a representative of w in K and write $k = zn$ as in Lemma 5.16. Here $\mathrm{Ad}(n)$ is in $N_K(\mathfrak{a} \oplus i\mathfrak{b})$ and $\mathrm{Ad}(n)H_1 = H_1$. Now $\mathfrak{a} \oplus i\mathfrak{b}$ is $\mathfrak{h}_{\mathbb{R}} = i\mathfrak{h}$ for a Cartan subalgebra $\mathfrak{h}$ of the compact form $\mathfrak{u}$ of $\mathfrak{g}$. In the situation of a compact group the analytic and algebraic Weyl groups coincide (Theorem 4.40). Hence $\mathrm{Ad}(n)$ is in the Weyl group of $\mathfrak{u}^{\mathbb{C}}$ with respect to $\mathfrak{h}^{\mathbb{C}}$, and Chevalley's Lemma (Proposition 4.12) shows it is the product of reflections leaving H_1 fixed. Each of these reflections is in a root of $\Delta(\mathfrak{b}^{\mathbb{C}}:\mathfrak{m}^{\mathbb{C}})$ and has a representative in $\mathrm{Ad}(M_0)$. Hence n is the product of members of M_0 and a member of $Z_K(\mathfrak{a} \oplus i\mathfrak{b})$. Consequently n centralizes $\mathfrak{a}$, k centralizes $\mathfrak{a}$, and $w = 1$. That is, $W(\Sigma) = W(A:G)$.

Corollary 5.18. For any two choices of Σ^+, the corresponding nilpotent subalgebras $\mathfrak{n}$ are conjugate via $\mathrm{Ad}(N_K(\mathfrak{a}))$, and the conjugating element is unique up to a member of $\mathrm{Ad}(M)$.

Concluding remarks.

(1) With suitable interpretations the results of §§2–3 are valid if G is merely linear connected reductive. [In that case $Z_{\mathfrak{g}} = (\mathfrak{k} \cap Z_{\mathfrak{g}}) \oplus (\mathfrak{p} \cap Z_{\mathfrak{g}})$, and $\mathfrak{k} \cap Z_{\mathfrak{g}}$ contributes to $\mathfrak{m}$. Any maximal abelian subspace of $\mathfrak{p}$ necessarily contains $\mathfrak{p} \cap Z_{\mathfrak{g}}$, so that we have $\mathfrak{p} \cap Z_{\mathfrak{g}} \subseteq \mathfrak{a}$. Two such maximal abelian subspaces of $\mathfrak{p}$ are still conjugate under $\mathrm{Ad}(K)$. It is no longer true that Σ spans $\mathfrak{a}'$, but Σ is an abstract root system in the subspace of $\mathfrak{a}$ that it spans. The Weyl groups $W(\Sigma)$ and $W(A:G)$ are still the same, and they leave $\mathfrak{p} \cap Z_{\mathfrak{g}}$ fixed.]

(2) The group $W(A:G)$ can equally well be defined as

$$W(A:G) = N_G(\mathfrak{a})/Z_G(\mathfrak{a}), \tag{5.5}$$

and it is not changed by this definition. [In fact, let $g = k \exp X$ be the Cartan decomposition of a member of $N_G(\mathfrak{a})$. Then $(\Theta g)^{-1}g = \exp 2X$ is in $N_G(\mathfrak{a})$, and so H in $\mathfrak{a}$ implies $\operatorname{Ad}(\exp 2X)H$ in $\mathfrak{a}$. Since $\operatorname{Ad}(\exp 2X)$ is positive definite on $\mathfrak{g}$, its Hermitian logarithm $\operatorname{ad} 2X$ is a polynomial in $\operatorname{Ad}(\exp 2X)$. Thus $(\operatorname{ad} 2X)(\mathfrak{a}) \subseteq \mathfrak{a}$, and we see easily that X is in $\mathfrak{a}$. So $N_G(\mathfrak{a}) = N_K(\mathfrak{a}) \times A$, and (5.5) is equivalent with (5.4) as a definition of $W(A:G)$.]

§4. Other Decompositions

Let G be a linear connected semisimple group. We shall discuss three other decompositions of G in this section:

(1) the KAK decomposition
(2) the Bruhat decomposition
(3) the decomposition of (almost all of) G into conjugacy classes.

Lemma 5.19. If k is in K and a and a' are in A with $kak^{-1} = a'$, then there exists k_0 in $N_K(\mathfrak{a})$ with $k_0 a k_0^{-1} = a'$.

Proof. The group $Z_G(a')$ is closed in G and is Θ-stable; hence its identity component is linear connected reductive. The Lie algebra of $Z_G(a')$ is

$$Z_{\mathfrak{g}}(a') = \{X \in \mathfrak{g} \,|\, \operatorname{Ad}(a')X = X\},$$

and $\mathfrak{a}$ and $\operatorname{Ad}(k)\mathfrak{a}$ are two maximal abelian subspaces of $Z_{\mathfrak{g}}(a') \cap \mathfrak{p}$. By Theorem 5.13 (applied in the reductive case), we can find k_1 in $K \cap Z_G(a')$ such that $\operatorname{Ad}(k_1)\operatorname{Ad}(k)\mathfrak{a} = \mathfrak{a}$. Then $k_0 = k_1 k$ is in $N_K(\mathfrak{a})$ and $k_0 a k_0^{-1} = k_1 a' k_1^{-1} = a'$.

Theorem 5.20 (KAK decomposition). Every element in G has a decomposition as $k_1 a k_2$ with k_1 and k_2 in K and a in A. In this decomposition, a is uniquely determined up to conjugation by a member of $W(A:G)$. If a is fixed as $\exp H$ with H regular, then k_1 is unique up to right multiplication by a member of M.

Example. In $SL(m, \mathbb{C})$ or $SL(m, \mathbb{R})$, existence of the decomposition follows when $g = g^*$ by the finite-dimensional Spectral Theorem. For other g, we first use the Polar Decomposition and then apply this diagonalization to the positive definite part. As we shall see, this style of proof handles general groups G, too.

Remark. It follows from the theorem that the A component can be taken to be the exponential of an element in the closed positive Weyl chamber of $\mathfrak{a}$. With this condition imposed, the A component is unique.

Proof. Existence of the decomposition follows by combining the decomposition $G = K \exp \mathfrak{p}$ and the conjugacy theorem $\mathfrak{p} = \bigcup_{k \in K} \mathrm{Ad}(k)\mathfrak{a}$ (Proposition 1.2 and Theorem 5.13). For uniqueness let us suppose

$$k_1' a' k_2' = k_1'' a k_2''.$$

If $k' = k_1''^{-1} k_1'$ and $k = k_2' k_2''^{-1}$, we then have $k'a'k = a$ and hence

$$(k'k)(k^{-1}a'k) = a.$$

The uniqueness in Proposition 1.2 then implies that $k'k = 1$ and $k^{-1}a'k = a$. Hence a' and a are conjugate via $N_K(\mathfrak{a})$, by Lemma 5.19.

Finally if $a = a'$ above and if $a = \exp H$ with H regular, we have seen that $k^{-1}ak = a$. The uniqueness in Proposition 1.2 implies that $\mathrm{Ad}(k)^{-1}H = H$. Since H is regular, the centralizer of H in $\mathfrak{g}$ is $\mathfrak{g}_0 = \mathfrak{a} \oplus \mathfrak{m}$. Hence the centralizer of H in $\mathfrak{p}$ is $\mathfrak{a}$. Thus the centralizer of $\mathrm{Ad}(k)^{-1}H$ in $\mathfrak{p}$ is $\mathrm{Ad}(k)^{-1}\mathfrak{a}$. But $\mathrm{Ad}(k)^{-1}H = H$ implies these centralizers are the same. Hence $\mathrm{Ad}(k)^{-1}\mathfrak{a} = \mathfrak{a}$. Thus k is in $N_K(\mathfrak{a})$. Since $\mathrm{Ad}(k)$ fixes the regular element H, Theorem 5.17 implies k is in M. Then $k'k = 1$ says k' is in M, and $k_1''^{-1}k_1' = k'$ says k_1' is in $k_1''M$.

The group $M = Z_K(\mathfrak{a})$ centralizes A, and it normalizes N (in fact, it normalizes each space $\mathfrak{g}_\lambda$) by the same computation as in (5.3). Hence MAN is a subgroup of G. It is a closed subgroup since M is compact and AN is closed.

Theorem 5.21 (Bruhat decomposition). The double cosets of $MAN\backslash G/MAN$ are parametrized in one-one onto fashion by $W(A:G)$, the double coset corresponding to w in $W(A:G)$ being $MAN\tilde{w}MAN$, where $\tilde{w}$ is any representative of w in $N_K(\mathfrak{a})$.

Sketch of proof of existence of decomposition for $\mathrm{SL}(m, \mathbb{R})$. To each x in $\mathrm{SL}(m, \mathbb{R})$ we attach a permutation σ of $\{1, \ldots, m\}$ as follows. Let $x_{i_1 \cdots i_k, j_1 \cdots j_k}$ denote the k-by-k determinant obtained by using only rows $i_1, \ldots, i_k$ and columns $j_1, \ldots, j_k$ of x. Now choose r as large as possible so that $x_{r1} \neq 0$, and define $\sigma(1) = r$. Inductively having defined $\sigma(j)$ for $1 \leq j \leq k-1$, choose s as large as possible but different from $\sigma(1), \ldots, \sigma(k-1)$ so that $x_{\sigma(1) \cdots \sigma(k-1), 1\, 2 \cdots k} \neq 0$ and define $\sigma(k) = s$. One checks that $b_1 x b_2$ gets the same permutation attached as x does if b_1 and b_2 are in the upper triangular group MAN.

Using this observation, one recursively can define a unique member n of N with the properties that $n_{ij} = 0$ if $\sigma^{-1}(i) < \sigma^{-1}(j)$ and $(nx)_{ij} = 0$ if $\sigma^{-1} > j$. Let $\tilde{w}$ have entries

$$\tilde{w}_{ij} = \begin{cases} 1 & \text{if } i = \sigma(j) \\ 0 & \text{if not,} \end{cases}$$

and change one of the 1's to a -1 if necessary to make $\tilde{w}$ have determinant 1. Then $\tilde{w}^{-1}nx = b$ is upper triangular, and hence $x = n^{-1}\tilde{w}b$ is the required decomposition.

Proof of uniqueness in general case. Suppose w_1 and w_2 are in $W(A:G)$, and suppose $\tilde{w}_1$ and $\tilde{w}_2$ are representatives. If b_1 and b_2 in MAN satisfy $b_1\tilde{w}_1 = \tilde{w}_2 b_2$, we argue as follows: Apply Ad of both sides to H in $\mathfrak{a}$, and project to $\mathfrak{a}$ along $\mathfrak{m} \oplus \sum_{\lambda\in\Sigma} \mathfrak{g}_\lambda$. Since $\text{Ad}(N_K(\mathfrak{a}))$ leaves $\mathfrak{m} \oplus \sum_{\lambda\in\Sigma} \mathfrak{g}_\lambda$ stable and since $\text{Ad}(N)$ carries $\mathfrak{a}$ to $\mathfrak{a} \oplus \mathfrak{n}$, with the same $\mathfrak{a}$ component, we obtain $\text{Ad}(\tilde{w}_1)H = \text{Ad}(\tilde{w}_2)H$. Since H is arbitrary, $w_1 = w_2$.

A θ-stable **Cartan subalgebra** of $\mathfrak{g}$ is a subalgebra $\mathfrak{h}$ of $\mathfrak{g}$ that is maximal among abelian θ-stable subalgebras of $\mathfrak{g}$. A Θ-stable **Cartan subgroup** H of G is the centralizer in G of a θ-stable Cartan subalgebra. These definitions are consistent with the case that G is compact, by Corollary 4.23.

In the compact case any element is conjugate to an element of a given Cartan subgroup, but this is not so in the noncompact case. Two things go wrong—the existence of nonconjugate Cartan subgroups and the existence of nilpotent (or unipotent) elements that are not conjugate to members of any Cartan subgroup. For example, in $\text{SL}(2, \mathbb{R})$ there are two nonconjugate Cartan subgroups

$$\left\{\begin{pmatrix} \cos\theta & \sin\theta \\ -\sin\theta & \cos\theta \end{pmatrix}\right\} \quad \text{and} \quad \left\{\pm\begin{pmatrix} e^t & 0 \\ 0 & e^{-t} \end{pmatrix}\right\}.$$

The elements x with $|\text{Tr}\, x| < 2$ are conjugate to members of the first group, and the elements x with $|\text{Tr}\, x| > 2$ are conjugate to members of the second group. This accounts for almost all the members of the group. But $\begin{pmatrix} 1 & a \\ 0 & 1 \end{pmatrix}$, with trace 2, is not conjugate to a member of either Cartan subgroup unless $a = 0$.

When working with Cartan subalgebras and subgroups, it is often convenient to refer to $\mathfrak{g}^{\mathbb{C}}$ and $G^{\mathbb{C}}$. As we saw in §1, the easy case for understanding $\mathfrak{g}^{\mathbb{C}}$ and $G^{\mathbb{C}}$ is that $\mathfrak{k} \cap i\mathfrak{p} = 0$, and we shall normally work with $\mathfrak{g}^{\mathbb{C}}$ and $G^{\mathbb{C}}$ as if this assumption is satisfied. However, there are important cases where it is not satisfied (e.g., $G = \text{SL}(2, \mathbb{C})$), and in such cases we need to interpret $\mathfrak{g}^{\mathbb{C}}$ and $G^{\mathbb{C}}$ in a more sophisticated fashion, say by regarding the inclusion $G \subseteq GL(n, \mathbb{C})$ as being given by $G \subseteq GL(2n, \mathbb{R})$. The exact way we force $\mathfrak{k} \cap i\mathfrak{p} = 0$ will not be important, and we shall normally make little or no mention of this point.

If $\mathfrak{h}$ is a θ-stable Cartan subalgebra of $\mathfrak{g}$, then we can write

$$\mathfrak{h} = (\mathfrak{h} \cap \mathfrak{k}) \oplus (\mathfrak{h} \cap \mathfrak{p}),$$

and $(\mathfrak{h} \cap \mathfrak{k}) \oplus i(\mathfrak{h} \cap \mathfrak{p})$ will be a Cartan subalgebra of the compact form $\mathfrak{u} = \mathfrak{k} \oplus i\mathfrak{p}$. By Theorem 4.41, we conclude that all θ-stable Cartan subalgebras of $\mathfrak{g}$ have the same dimension; this common dimension is called the (complex) **rank** of G or of $\mathfrak{g}$. The term **real rank** is used for $\dim \mathfrak{a}$, with $\mathfrak{a}$ as in the Iwasawa decomposition.

General examples (extreme cases).

(1) $\mathfrak{h} = \mathfrak{a} \oplus \mathfrak{b}$, where $\mathfrak{b}$ is a maximal abelian subspace of $\mathfrak{m}$. This has $\dim(\mathfrak{h} \cap \mathfrak{p})$ as large as possible.

(2) $\mathfrak{h} = Z_{\mathfrak{g}}(\mathfrak{b}_0)$, where $\mathfrak{b}_0$ is a maximal abelian subspace of $\mathfrak{k}$; it has $\dim(\mathfrak{h} \cap \mathfrak{k})$ as large as possible. [To see that $\mathfrak{h}$ is abelian, let X and Y be in $\mathfrak{h} \cap \mathfrak{p}$ and let Z be in $\mathfrak{b}_0$. Then $[X, Y]$ is in $\mathfrak{k}$ and commutes with $\mathfrak{b}_0$, by the Jacobi identity, so is in $\mathfrak{b}_0$. On the other hand, $[X, Y]$ is orthogonal to $\mathfrak{b}_0$ with respect to the real part of the trace form, and thus $[X, Y] = 0$.]

Specific examples.

(1) $\mathfrak{g} = \mathfrak{sl}(n, \mathbb{C})$. The diagonal subalgebra $\mathfrak{h}$ is the only θ-stable Cartan subalgebra, up to conjugacy. Here $\dim(\mathfrak{h} \cap \mathfrak{k}) = \dim(\mathfrak{h} \cap \mathfrak{p}) = n - 1$.

(2) $\mathfrak{g} = \mathfrak{sl}(n, \mathbb{R})$. For $0 \le j \le [n/2]$, define $\mathfrak{h}_j$ to be all

$$\begin{pmatrix} \begin{pmatrix} t_1 & \theta_1 \\ -\theta_1 & t_1 \end{pmatrix} & & & & & \\ & \ddots & & & & \\ & & \begin{pmatrix} t_j & \theta_j \\ -\theta_j & t_j \end{pmatrix} & & & \\ & & & s_{2j+1} & & \\ & & & & \ddots & \\ & & & & & s_n \end{pmatrix}$$

with trace 0. Then the $\mathfrak{h}_j$ form a complete set of nonconjugate θ-stable Cartan subalgebras. Here

$$\dim(\mathfrak{h}_j \cap \mathfrak{k}) = j \qquad \text{and} \qquad \dim(\mathfrak{h}_j \cap \mathfrak{p}) = n - j - 1.$$

The nonconjugacy may be seen by observing that the eigenvalues are different for the different $\mathfrak{h}_j$'s.

(3) $\mathfrak{g} = \mathfrak{su}(2, 2)$. Here the three subalgebras

$$\begin{pmatrix} i\theta_1 & & & \\ & i\theta_2 & & \\ & & i\theta_3 & \\ & & & -i(\theta_1+\theta_2+\theta_3) \end{pmatrix},$$

$$\begin{pmatrix} i\theta_1 & 0 & s & 0 \\ 0 & i\theta_2 & 0 & 0 \\ s & 0 & i\theta_1 & 0 \\ 0 & 0 & 0 & -i(2\theta_1+\theta_2) \end{pmatrix}, \quad \begin{pmatrix} i\theta & 0 & s & 0 \\ 0 & -i\theta & 0 & t \\ s & 0 & i\theta & 0 \\ 0 & t & 0 & -i\theta \end{pmatrix}$$

are a complete set of nonconjugate θ-stable Cartan subalgebras. In the respective cases we have $\dim(\mathfrak{h} \cap \mathfrak{k}) = 3, 2, 1$ and $\dim(\mathfrak{h} \cap \mathfrak{p}) = 0, 1, 2$.

(4) $\mathfrak{g} = \mathfrak{sp}(2, \mathbb{R})$. Here the four subalgebras

$$\begin{pmatrix} 0 & 0 & \theta_1 & 0 \\ 0 & 0 & 0 & \theta_2 \\ -\theta_1 & 0 & 0 & 0 \\ 0 & -\theta_2 & 0 & 0 \end{pmatrix}, \quad \begin{pmatrix} t & \theta & 0 & 0 \\ -\theta & t & 0 & 0 \\ 0 & 0 & -t & \theta \\ 0 & 0 & -\theta & -t \end{pmatrix},$$

$$\begin{pmatrix} 0 & 0 & \theta & 0 \\ 0 & t & 0 & 0 \\ -\theta & 0 & 0 & 0 \\ 0 & 0 & 0 & -t \end{pmatrix}, \quad \begin{pmatrix} s & & & \\ & t & & \\ & & -s & \\ & & & -t \end{pmatrix}$$

are a complete set of nonconjugate θ-stable Cartan subalgebras. In the respective cases we have $\dim(\mathfrak{h} \cap \mathfrak{k}) = 2, 1, 1, 0$ and $\dim(\mathfrak{h} \cap \mathfrak{p}) = 0, 1, 1, 2$.

In $SL(2, \mathbb{R})$ the elements with the nicest properties under conjugation are those with $|\operatorname{Tr} x| \neq 2$. We shall generalize this condition. Let l be the rank of G. We say x is **regular** if the characteristic polynomial of $\operatorname{Ad}(x)$ has 1 for a root with multiplicity l and no more. One can show that the set of regular elements of G is open and its complement is of lower dimension and Haar measure 0.

Theorem 5.22. There are only finitely many nonconjugate θ-stable Cartan subalgebras. Let $\mathfrak{h}_1, \ldots, \mathfrak{h}_r$ denote a complete set of representatives, and let $H_1, \ldots, H_r$ be the corresponding Cartan subgroups. Then

(a) there is just one $\mathfrak{h}_i$ for which $\mathfrak{h}_i \cap \mathfrak{p}$ is of maximum dimension, say for $i = r$. The intersection $\mathfrak{h}_r \cap \mathfrak{p}$ may be taken to be the Iwasawa $\mathfrak{a}$, and $\mathfrak{h}_i \cap \mathfrak{p}$ may be taken to be $\subseteq \mathfrak{a}$ for $1 \leq i \leq r$. Then the subsets Σ_i of Σ defined by

$$\Sigma_i = \{\lambda \in \Sigma \mid \lambda(\mathfrak{h}_i \cap \mathfrak{p}) = 0\}$$

are distinct.

(b) there is just one $\mathfrak{h}_i$ for which $\mathfrak{h}_i \cap \mathfrak{k}$ is of maximum dimension, say for $i = 1$, and H_1 is connected.

(c) each H_i is abelian and is contained in $\exp \mathfrak{h}_i^{\mathbb{C}}$.

(d) for each i with $1 \leq i \leq r$, the set

$$(H_i')^G = \{ghg^{-1} \mid g \in G, h \in H_i, h \text{ is regular}\}$$

is open in G, and every regular element of G is conjugate to some member of $(H_i')^G$ for one and only one i.

(e) all θ-stable Cartan subalgebras are conjugate if $\mathfrak{g}$ is complex.

Proof omitted.

Fix a θ-stable Cartan subalgebra $\mathfrak{h}$ of $\mathfrak{g}$; then

$$\mathfrak{h}_0 = (\mathfrak{h} \cap \mathfrak{k}) \oplus i(\mathfrak{h} \cap \mathfrak{p})$$

is a Cartan subalgebra of the compact form $\mathfrak{u} = \mathfrak{k} \oplus i\mathfrak{p}$ with $\mathfrak{h}_0^{\mathbb{C}} = \mathfrak{h}^{\mathbb{C}}$. Hence we can form roots of $\mathfrak{u}^{\mathbb{C}}$ with respect to $\mathfrak{h}_0^{\mathbb{C}}$ by the theory of Chapter IV, and we may think of these as the roots of $\mathfrak{g}^{\mathbb{C}}$ with respect to $\mathfrak{h}^{\mathbb{C}}$. Different choices of $\mathfrak{h}$ lead to isomorphic root systems, since the $\mathfrak{h}_0$'s are conjugate within $\mathfrak{u}$.

This root system, which we denote $\Delta(\mathfrak{h}^{\mathbb{C}}:\mathfrak{g}^{\mathbb{C}})$, has an algebraic Weyl group $W(\mathfrak{h}^{\mathbb{C}}:\mathfrak{g}^{\mathbb{C}})$ and an analytic Weyl group

$$W(H:G) = N_K(H)/Z_K(H).$$

Then we have the inclusion relation

$$W(H:G) \subseteq W(\mathfrak{h}^{\mathbb{C}}:\mathfrak{g}^{\mathbb{C}}). \tag{5.6}$$

[To see this, we assume (as we may) that $\mathfrak{k} \cap i\mathfrak{p} = 0$, and we write U, H_0, $H^{\mathbb{C}}$, and $G^{\mathbb{C}}$ for the analytic groups of matrices with Lie algebras $\mathfrak{u}$, $\mathfrak{h}_0$, $\mathfrak{h}^{\mathbb{C}}$, and $\mathfrak{g}^{\mathbb{C}}$. Then Theorem 4.41 and the same considerations as at the end of §3 show that

$$\begin{aligned} W(H:G) = N_K(H)/Z_K(H) &\cong N_G(H)/Z_G(H) \subseteq N_{G^{\mathbb{C}}}(H^{\mathbb{C}})/Z_{G^{\mathbb{C}}}(H^{\mathbb{C}}) \\ &\cong N_U(H_0)/Z_U(H_0) = W(\mathfrak{h}_0^{\mathbb{C}}:\mathfrak{u}^{\mathbb{C}}) = W(\mathfrak{h}^{\mathbb{C}}:\mathfrak{g}^{\mathbb{C}}).] \end{aligned}$$

In general, equality does not hold, as the following example shows.

Example. $G = \mathrm{SL}(2, \mathbb{R})$ has Cartan subgroups

$$H_1 = \left\{\begin{pmatrix} \cos\theta & \sin\theta \\ -\sin\theta & \cos\theta \end{pmatrix}\right\} \quad \text{and} \quad \left\{\pm\begin{pmatrix} e^t & 0 \\ 0 & e^{-t} \end{pmatrix}\right\}.$$

The abstract root system in either case is $\{\pm\alpha\}$, and its algebraic Weyl group is $\mathbb{Z}_2$. One can check that the analytic Weyl groups are

$$W(H_1:G) = \{1\} \qquad \text{and} \qquad W(H_2:G) = \mathbb{Z}_2.$$

Concluding remarks. Suitably interpreted, the three decompositions of this section apply also to linear connected reductive groups.

§5. Parabolic Subgroups

We continue with G as a linear connected semisimple group, but we shall change our notation somewhat in order to use the symbols A and N to denote a certain wider class of groups than arise in the Iwasawa decomposition. Thus let us write the Iwasawa decomposition as $G = KA_\mathfrak{p}N_\mathfrak{p}$ and take $M_\mathfrak{p} = Z_K(A_\mathfrak{p})$, so that the Bruhat decomposition is the double coset decomposition $M_\mathfrak{p}A_\mathfrak{p}N_\mathfrak{p}\backslash G/M_\mathfrak{p}A_\mathfrak{p}N_\mathfrak{p}$. Let $\mathfrak{m}_\mathfrak{p}$, $\mathfrak{a}_\mathfrak{p}$, and $\mathfrak{n}_\mathfrak{p}$ be the Lie algebras of $M_\mathfrak{p}$, $A_\mathfrak{p}$, and $N_\mathfrak{p}$.

A **parabolic subgroup** of G is a closed subgroup containing some conjugate of $M_\mathfrak{p}A_\mathfrak{p}N_\mathfrak{p}$; the conjugates of $M_\mathfrak{p}A_\mathfrak{p}N_\mathfrak{p}$ are called **minimal parabolic subgroups.** (The Iwasawa decomposition shows that it is enough to conjugate by K.)

Example. $G = \mathrm{SL}(n, \mathbb{R})$. Let $M_\mathfrak{p}A_\mathfrak{p}N_\mathfrak{p}$ be the usual upper triangular group. The parabolic subgroups S that contain $M_\mathfrak{p}A_\mathfrak{p}N_\mathfrak{p}$ are the block upper triangular subgroups, such as

$$\left(\begin{array}{cc|ccc|c}
\cdot & \cdot & \cdot & \cdot & \cdot & \cdot \\
\cdot & \cdot & \cdot & \cdot & \cdot & \cdot \\
\hline
0 & 0 & \cdot & \cdot & \cdot & \cdot \\
0 & 0 & \cdot & \cdot & \cdot & \cdot \\
0 & 0 & \cdot & \cdot & \cdot & \cdot \\
\hline
0 & 0 & 0 & 0 & 0 & \cdot
\end{array}\right)$$

in $\mathrm{SL}(6, \mathbb{R})$. Each has a decomposition $S = MAN$. The group MA is the block diagonal part, and N has the identity in each diagonal block and contains the entries that are strictly above and to the right of the blocks. The members of M have determinant ± 1 in each block along the diagonal, and the members of A are positive scalar matrices in each block along the diagonal.

For general G there is a corresponding decomposition $S = MAN$ of parabolic subgroups known as the **Langlands decomposition.** First of all, on the Lie algebra level there is a direct sum decomposition

$$\mathfrak{s} = \mathfrak{m} \oplus \mathfrak{a} \oplus \mathfrak{n}$$

that is uniquely defined by the following properties:

(i) $\mathfrak{m}$, $\mathfrak{a}$, and $\mathfrak{n}$ are mutually orthogonal with respect to the inner product $\langle X, Y \rangle = -\operatorname{Re} B_0(X, \theta Y)$
(ii) $\mathfrak{m} \oplus \mathfrak{a} = \mathfrak{s} \cap \theta\mathfrak{s}$
(iii) $\mathfrak{a} = \mathfrak{p} \cap Z_{\mathfrak{m} \oplus \mathfrak{a}}$.

This decomposition is easily shown to have the following additional properties:

(iv) $\mathfrak{m}$, $\mathfrak{a}$, and $\mathfrak{n}$ are Lie subalgebras with $\mathfrak{a}$ abelian, $\mathfrak{n}$ nilpotent, and $\mathfrak{m}$ and $\mathfrak{a}$ normalizing $\mathfrak{n}$
(v) $\mathfrak{m} \oplus \mathfrak{a} = Z_{\mathfrak{g}}(\mathfrak{a})$
(vi) ad $\mathfrak{a}$ acts as a commuting family of Hermitian operators on $\mathfrak{n}$, and $\mathfrak{n}$ decomposes as the orthogonal sum

$$\mathfrak{n} = \sum_{\mu \in \Gamma_S^+} \mathfrak{g}_\mu,$$

where Γ_S^+ is a finite subset of nonzero elements of $\mathfrak{a}'$ and

$$\mathfrak{g}_\mu = \{X \in \mathfrak{n} \mid [H, X] = \mu(H)X \text{ for all } H \in \mathfrak{a}\}.$$

If we define

$$\begin{aligned} A &= \text{analytic subgroup with Lie algebra } \mathfrak{a} \\ N &= \text{analytic subgroup with Lie algebra } \mathfrak{n} \\ M_0 &= \text{analytic subgroup with Lie algebra } \mathfrak{m} \\ M &= Z_K(\mathfrak{a})M_0, \end{aligned} \tag{5.7}$$

then one can show that

(i) $S = MAN$ and multiplication is a diffeomorphism of $M \times A \times N$ onto S
(ii) M_0 is linear connected reductive with compact center
(iii) $\Theta N \cap MAN = \{1\}$

Parabolic subgroups are classified by the following result.

Proposition 5.23. Fix $M_\mathfrak{p} A_\mathfrak{p} N_\mathfrak{p}$, let Σ^+ be the positive roots of $(\mathfrak{g}, \mathfrak{a}_\mathfrak{p})$, with positivity determined by $\mathfrak{n}_\mathfrak{p}$, and let Π be the simple roots in Σ^+. Then the parabolic subgroups $S = MAN$ containing $M_\mathfrak{p} A_\mathfrak{p} N_\mathfrak{p}$ are in one-one correspondence with the subsets Π_S of Π, the correspondence being that

$$\lambda \in \Pi \text{ is in } \Pi_S \text{ if and only if } \mathfrak{g}_{-\lambda} \subseteq \mathfrak{m}.$$

No two of these parabolic subgroups are conjugate within G.

Proof omitted. It is instructive to see how this result gives the block upper triangular subgroups of $SL(n, \mathbb{R})$; see the Problems at the end of the chapter. Presently we shall work out the case of $SU(2, 2)$ as a further illustration. It is always true that $\mathfrak{a}$ is the orthocomplement in $\mathfrak{a}_\mathfrak{p}$ of the elements H_λ for λ in Π_S. Then we have

$$\mathfrak{a} \subseteq \mathfrak{a}_\mathfrak{p}, \qquad \mathfrak{n} \subseteq \mathfrak{n}_\mathfrak{p}, \qquad \mathfrak{m} \supseteq \mathfrak{m}_\mathfrak{p}.$$

In the decomposition

$$\mathfrak{n} = \sum_{\mu \in \Gamma_S^+} \mathfrak{g}_\mu \tag{5.8}$$

of (vi) for $\mathfrak{s}$, the elements μ are the nonzero restrictions to $\mathfrak{a}$ of the members of Σ^+, and $\mathfrak{m}$ is built from $\mathfrak{m}_\mathfrak{p}$ and the members of Σ that restrict to 0 on $\mathfrak{a}$. Corresponding to (5.8) we have also

$$\theta\mathfrak{n} = \sum_{\mu \in \Gamma_S^+} \mathfrak{g}_{-\mu} \tag{5.9}$$

and thus

$$\mathfrak{g} = \mathfrak{a} \oplus \mathfrak{m} \oplus \sum_{\mu \in \Gamma_S^+} \mathfrak{g}_\mu \oplus \sum_{\mu \in \Gamma_S^+} \mathfrak{g}_{-\mu}, \tag{5.10}$$

which is a kind of **root space decomposition** for $\mathfrak{g}$ in which the roots of $(\mathfrak{g}, \mathfrak{a})$ are $\Gamma_S = \Gamma_S^+ \cup (-\Gamma_S^+)$. These roots do not always form an abstract root system, as for example in $SL(4, \mathbb{R})$ when $\Pi_S = \{e_1 - e_2\}$. But we can still define **regular elements** and **Weyl chambers** in the same way as in §3. The analog of Proposition 5.14 is still valid here, and each positive system in Γ_S (or each Weyl chamber) determines a group N_1 such that MAN_1 is a parabolic subgroup.

We shall have occasion later to work with all the groups MAN_1 at once. They are not all conjugate in G, in general. In fact, the analytically defined Weyl group

$$W(A{:}G) = N_K(\mathfrak{a})/Z_K(\mathfrak{a}) \cong N_G(\mathfrak{a})/Z_G(\mathfrak{a})$$

permutes the Weyl chambers, and no nontrivial element of $W(A{:}G)$ leaves a Weyl chamber stable. But $W(A{:}G)$ need not be transitive on the set of Weyl chambers, as for example in $SL(3, \mathbb{R})$ or $SL(4, \mathbb{R})$ when $\Pi_S = \{e_1 - e_2\}$. The tool for controlling $W(A{:}G)$ here is the following analog of Lemma 5.16.

Lemma 5.24. Let $\mathfrak{a}_M$ be a maximal abelian subspace of $\mathfrak{m} \cap \mathfrak{p}$. Every element of $N_K(\mathfrak{a})$ decomposes as the product zn, where n is in $N_K(\mathfrak{a} \oplus \mathfrak{a}_M)$ and z is in $Z_K(\mathfrak{a})$.

Proof. An easy adaptation of the proof of Lemma 5.16.

Let $S = MAN$ be any parabolic subgroup. Because of the Iwasawa decomposition, we have

$$G = KMAN. \tag{5.11}$$

It is clear that $K \cap MAN = K \cap M$ and therefore that the A and N components of (5.11) are well defined.

There is a natural construction of parabolic subgroups from θ-stable Cartan subalgebras. Namely let $\mathfrak{h}$ be a θ-stable Cartan subalgebra. Put

$$\mathfrak{a} = \mathfrak{h} \cap \mathfrak{p}$$

$$\mathfrak{m} = \text{orthocomplement of } \mathfrak{a} \text{ in } Z_{\mathfrak{g}}(\mathfrak{a}),$$

and form a "root space decomposition" relative to ad $\mathfrak{a}$ in the usual way. If we introduce an ordering in $\mathfrak{a}'$ in the usual way, we can define

$$\mathfrak{n} = \text{sum of root spaces for positive roots of } (\mathfrak{g}, \mathfrak{a}).$$

Then we obtain a parabolic subgroup MAN by means of the definitions (5.7). The parabolic subgroups constructed this way are characterized by the fact that $\mathfrak{m}$ has a Cartan subalgebra lying in $\mathfrak{k}$ (namely $\mathfrak{h} \cap \mathfrak{k}$); such parabolic subgroups are called **cuspidal** and are especially relevant for the Plancherel formula of G.

Example. $G = \mathrm{SU}(2, 2)$. We write down the parabolic subgroups containing a fixed minimal one $M_{\mathfrak{p}}A_{\mathfrak{p}}N_{\mathfrak{p}}$ and determine which ones are cuspidal. Here

$$\begin{matrix} & 2 & 2 & \\ \mathfrak{g} = & \Big(\begin{matrix} X_{11} \\ X_{12}^* \end{matrix} & \begin{matrix} X_{12} \\ X_{22} \end{matrix} \Big) & \begin{matrix} 2 \\ 2 \end{matrix} \end{matrix}$$

with X_{11} and X_{22} skew-Hermitian and the total trace equal to 0. We have

$$K = \left\{ \begin{pmatrix} X_{11} & 0 \\ 0 & X_{22} \end{pmatrix} \middle| X_{11} \in \mathrm{U}(2), X_{22} \in \mathrm{U}(2), (\det X_{11})(\det X_{22}) = 1 \right\}$$

and we can take

$$\mathfrak{a}_{\mathfrak{p}} = \left\{ \begin{pmatrix} 0 & 0 & s & 0 \\ 0 & 0 & 0 & t \\ s & 0 & 0 & 0 \\ 0 & t & 0 & 0 \end{pmatrix} \middle| s \text{ and } t \text{ in } \mathbb{R} \right\}.$$

Define linear functionals f_1 and f_2 on $\mathfrak{a}_{\mathfrak{p}}$ by saying that f_1 of the above matrix is s and f_2 of the matrix is t. Then

$$\Sigma = \{\pm f_1 \pm f_2, \pm 2f_1, \pm 2f_2\},$$

which is a root system of type C_2. Here $\pm f_1 \pm f_2$ have multiplicity 2, and the others have multiplicity 1. In the obvious ordering, $f_1 \pm f_2$ and $2f_1$ and $2f_2$ comprise Σ^+, and

$$\{af_1 + bf_2 | 0 < b < a\}$$

is the positive Weyl chamber of $\mathfrak{a}'$. With our definitions we can check that

$$N_\mathfrak{p} = \exp\left(\sum_{\lambda \in \Sigma^+} \mathfrak{g}_\lambda\right) \text{ with } \dim N_\mathfrak{p} = 6$$

$$A_\mathfrak{p} = \exp \mathfrak{a}_\mathfrak{p}$$

$$(M_\mathfrak{p})_0 = \{\mathrm{diag}(e^{i\theta}, e^{-i\theta}, e^{i\theta}, e^{-i\theta})\}$$

$$M_\mathfrak{p} = \{i^n t \mid 0 \le n \le 3 \text{ and } t \in (M_\mathfrak{p})_0\}.$$

Define γ to be the element of order 2 in $M_\mathfrak{p}$ given by

$$\gamma = \mathrm{diag}(1, -1, 1, -1).$$

Then $M_\mathfrak{p} = (M_\mathfrak{p})_0 \oplus \{1, \gamma\}$.

Proposition 5.23 says there are four parabolic subgroups S containing $M_\mathfrak{p}A_\mathfrak{p}N_\mathfrak{p}$. Two of these are

$$S_\phi = M_\mathfrak{p}A_\mathfrak{p}N_\mathfrak{p} \qquad \text{and} \qquad S_{\{f_1 - f_2,\, 2f_2\}} = G.$$

For $S_{\{f_1 - f_2\}}$ we have

$$\mathfrak{a}_{f_1 - f_2} = \{H \in \mathfrak{a}_\mathfrak{p} \mid (f_1 - f_2)(H) = 0\} \qquad (s = t \text{ in } \mathfrak{a}_\mathfrak{p})$$

$$A_{f_1 - f_2} = \exp \mathfrak{a}_{f_1 - f_2}$$

$$(M_{f_1 - f_2})_0 = \text{analytic subgroup of } G \text{ with Lie algebra}$$

$$\left\{\left.\begin{pmatrix} i\theta & w & x & z \\ -\bar{w} & -i\theta & \bar{z} & -x \\ x & z & i\theta & w \\ \bar{z} & -x & -\bar{w} & -i\theta \end{pmatrix}\right| x, \theta \in \mathbb{R} \text{ and } w, z \in C\right\}$$

$$M_{f_1 - f_2} = \{1, \gamma\} \ltimes (M_{f_1 - f_2})_0$$

$$N_{f_1 - f_2} = \exp(\mathfrak{g}_{2f_1} \oplus \mathfrak{g}_{f_1 + f_2} \oplus \mathfrak{g}_{2f_2}).$$

For $S_{\{2f_2\}}$ we have

$$\mathfrak{a}_{2f_2} = \{H \in \mathfrak{a}_\mathfrak{p} \mid 2f_2(H) = 0\} \qquad (t = 0 \text{ in } \mathfrak{a}_\mathfrak{p})$$

$$A_{2f_2} = \exp \mathfrak{a}_{2f_2}$$

$$M_{2f_2} = (M_\mathfrak{p})_0 \oplus \left\{\left.\begin{pmatrix} 1 & 0 & 0 & 0 \\ 0 & \alpha & 0 & \beta \\ 0 & 0 & 1 & 0 \\ 0 & \bar{\beta} & 0 & \bar{\alpha} \end{pmatrix}\right| \alpha, \beta \in C \text{ and } |\alpha|^2 - |\beta|^2 = 1\right\}$$

$$N_{2f_2} = \exp(\mathfrak{g}_{2f_1} \oplus \mathfrak{g}_{f_1 + f_2} \oplus \mathfrak{g}_{f_1 - f_2}).$$

The group $(M_\mathfrak{p})_0$ is a circle group, and we observe that

$$M_\mathfrak{p} = (M_\mathfrak{p})_0 \oplus \{1, \gamma\}$$
$$M_{f_1 - f_2} \cong \{1, \gamma\} \ltimes \mathrm{SL}(2, \mathbb{C})$$
$$M_{2f_2} \cong (M_\mathfrak{p})_0 \oplus \mathrm{SL}(2, \mathbb{R}).$$

Now SU(2, 2), SL(2, $\mathbb{R}$), and G itself have compact Cartan subgroups, but SL(2, $\mathbb{C}$) does not. Thus S_ϕ, $S_{\{2f_2\}}$, and G are cuspidal parabolic subgroups, but $S_{\{f_1 - f_2\}}$ is not cuspidal.

Concluding remarks.

(1) With suitable interpretations the discussion of parabolic subgroups is valid if G is merely linear connected reductive. We always have $\mathfrak{p} \cap Z_\mathfrak{g} \subseteq \mathfrak{a}$ and $\mathfrak{k} \cap Z_\mathfrak{g} \subseteq \mathfrak{m}$.

(2) A certain amount of the representation theory of semisimple groups depends on inductive constructions that pass from M to G. The fact that $\mathfrak{g}$ semisimple can lead to $\mathfrak{m}$ reductive is an important reason for studying reductive groups, not just semisimple groups. Unfortunately M need not be connected, and one should really work with a class of reductive groups that are not necessarily connected such that when one passes from G to M one stays in the class. Although we shall not pursue the point, a set of axioms suitable for this purpose is as follows: G can be realized as a group of matrices such that

(i) the identity component G_0 of G is linear connected reductive
(ii) G_0 has compact center
(iii) G has finitely many components
(iv) $G \subseteq (G_0)^\mathbb{C} Z(G)$, where $Z(G)$ denotes the centralizer of G in the total general linear group of matrices.

The various decomposition theorems for G and its parabolic subgroups can be framed in this context.

§6. Integral Formulas

Our group decompositions lead to a number of integration formulas that relate various Haar measures. We give here four basic integration formulas that are obtained by making calculations with differential forms. This use of differential forms will have limited other application for us, and we consequently omit the proofs. The four basic formulas have several consequences; some of them we shall state only qualitatively, so as to stress the principle rather than the result.

Proposition 5.25. For a connected Lie group G, right Haar measure $d_r x$ and left Haar measure $d_l x$ are related by

$$d_r x = c \det \mathrm{Ad}(x)\, d_l x, \tag{5.12}$$

where c is a constant. [Note $\det \mathrm{Ad}(x) > 0$ for all x.]

Consequence. Linear connected reductive groups and connected nilpotent Lie groups are unimodular (right Haar measure equals left Haar measure).

Proof. Proposition 1.2 shows that a linear connected reductive group is the product of some $\mathbb{R}^n$ by a linear connected reductive group with compact center; the second factor has no nontrivial homomorphisms into $\mathbb{R}^+$, and consequently $\det \mathrm{Ad}(x) = 1$ for all x in the group. For a connected nilpotent group, ad X is nilpotent and thus

$$\det \mathrm{Ad}(\exp X) = \det e^{\mathrm{ad}\, X} = e^{\mathrm{Tr}\, X} = 1$$

for all X in the Lie algebra; then $\det \mathrm{Ad}(x) = 1$ for all x in the group.

Proposition 5.26. Let G be an analytic group, and let S and T be closed subgroups such that $S \cap T$ is compact and such that the set of products ST exhausts G except possibly for a set of Haar measure 0. Then the left Haar measures on G, S, and T can be normalized so that

$$\int_G f(x)\, dx = \int_{S \times T} f(st) \frac{\det \mathrm{Ad}_T(t)}{\det \mathrm{Ad}_G(t)}\, ds\, dt \tag{5.13}$$

for all f in $C_{\mathrm{com}}(G)$.

Consequence 1. Under the above assumptions on G, S, and T, if G is unimodular, then

$$dx = d_l s\, d_r t, \tag{5.14}$$

where $d_l s$ is left Haar measure on S and $d_r t$ is right Haar measure on T.

Consequence 2. If G is linear connected reductive and $S = MAN$ is a parabolic subgroup, then we can relate the Haar measures of MAN and MA to the Haar measures of M, A, and N. Such formulas will have trivial dependence on M since $|\det \mathrm{Ad}(m)| = 1$ on $\mathfrak{m}$, $\mathfrak{a}$, and each $\mathfrak{g}_\mu$. But they will often involve the factor $\det \mathrm{Ad}(a)|_{\mathfrak{n}}$, which we calculate in a basis compatible with the direct sum $\sum \mathfrak{g}_\mu$ as

$$\det \mathrm{Ad}(a)|_{\mathfrak{n}} = e^{2\rho \log a}, \tag{5.15}$$

where

$$\rho = \rho_A = \tfrac{1}{2} \sum_{\mu \in \Gamma_S^+} (\dim \mathfrak{g}_\mu)\mu, \tag{5.16}$$

the sum taken over the N-positive roots of $(\mathfrak{g}, \mathfrak{a})$.

Consequence 3. If $G = KA_\mathfrak{p}N_\mathfrak{p}$ is the Iwasawa decomposition of a linear connected reductive group, then we can relate the Haar measure of G to the Haar measures of K and $A_\mathfrak{p}N_\mathfrak{p}$ as

$$dx = dk\, d_r(an); \tag{5.17}$$

hence we can relate the Haar measure of G to the Haar measures of K, $A_\mathfrak{p}$, and $N_\mathfrak{p}$.

Consequence 4. If G is linear connected reductive and $S = MAN$ is a parabolic subgroup, so that $G = KMAN$, then we can relate the Haar measure of G to the Haar measures of K and MAN as

$$dx = dk\, d_r(man); \tag{5.18}$$

hence we can relate the Haar measure of G to the Haar measures of K, M, A, and N.

Consequence 5. If $M_\mathfrak{p}A_\mathfrak{p}N_\mathfrak{p}$ is a minimal parabolic subgroup of a linear connected reductive group G and if $\Theta N_\mathfrak{p}$ is denoted $\bar{N}_\mathfrak{p}$, then

$\bar{N}_\mathfrak{p}M_\mathfrak{p}A_\mathfrak{p}N_\mathfrak{p}$ is open in G and its complement is of lower dimension and has Haar measure 0. (5.19)

Hence the Haar measures satisfy

$$dx = d\bar{n}\, d_r(man). \tag{5.20}$$

Sketch of proof. In the Bruhat decomposition of G, a double coset of $M_\mathfrak{p}A_\mathfrak{p}N_\mathfrak{p}$ is

$$M_\mathfrak{p}A_\mathfrak{p}N_\mathfrak{p}\tilde{w}M_\mathfrak{p}A_\mathfrak{p}N_\mathfrak{p} = N_\mathfrak{p}\tilde{w}M_\mathfrak{p}A_\mathfrak{p}N_\mathfrak{p} = \tilde{w}(\tilde{w}^{-1}N_\mathfrak{p}\tilde{w})M_\mathfrak{p}A_\mathfrak{p}N_\mathfrak{p}, \tag{5.21}$$

i.e., a translate of $(\tilde{w}^{-1}N_\mathfrak{p}\tilde{w})M_\mathfrak{p}A_\mathfrak{p}N_\mathfrak{p}$. We can compute the dimension of this set. On the Lie algebra level, we have

$$\dim \operatorname{Ad}(\tilde{w})^{-1}\mathfrak{n}_\mathfrak{p} + \dim(\mathfrak{m}_\mathfrak{p} \oplus \mathfrak{a}_\mathfrak{p} \oplus \mathfrak{n}_\mathfrak{p}) = \dim \mathfrak{g}.$$

Now $\operatorname{Ad}(\tilde{w})^{-1}\mathfrak{n}_\mathfrak{p}$ has zero intersection with $\mathfrak{m}_\mathfrak{p} \oplus \mathfrak{a}_\mathfrak{p} \oplus \mathfrak{n}_\mathfrak{p}$ if and only if $\operatorname{Ad}(\tilde{w})^{-1}\mathfrak{n}_\mathfrak{p} = \theta\mathfrak{n}_\mathfrak{p}$, which happens for exactly one Weyl group element w, by Theorem 5.17. When it happens, Lemma 5.11 shows that (5.21) is open in G. In the other cases, (5.21) has dimension less than that of G. Hence (5.19) follows, and (5.20) is an instance of (5.14).

Consequence 6. If MAN is a parabolic subgroup of a linear connected reductive group G and if ΘN is denoted $\bar{N}$, then

$\bar{N}MAN$ is open in G and its complement is of lower dimension and has Haar measure 0. (5.22)

Hence the Haar measures satisfy

$$dx = d\bar{n}\, d_r(man). \tag{5.23}$$

Proof. $\bar{N}MAN$ is open in G by Lemma 5.11. If $\tilde{w}$ is chosen so that (5.21) is open in G, then (5.21) is

$$= \tilde{w}\bar{N}_{\mathfrak{p}}M_{\mathfrak{p}}A_{\mathfrak{p}}N_{\mathfrak{p}} \subseteq \tilde{w}(MA\bar{N})(MAN) = \tilde{w}\bar{N}MAN.$$

Hence (5.22) follows, and (5.23) follows from (5.14).

Consequence 7. Let MAN be a parabolic subgroup of a linear connected reductive group G, let $\bar{N} = \Theta N$, and let ρ be as in (5.16). For g in G, decompose g according to $G = KMAN$ as

$$g = \kappa(g)\mu(g)\exp H(g)n. \tag{5.24}$$

Then Haar measures, when suitably normalized, satisfy

$$\int_K f(k)\, dk = \int_{\bar{N}} f(\kappa(\bar{n}))e^{-2\rho H(\bar{n})}\, d\bar{n} \tag{5.25}$$

for all f in $C(K)$ that are right invariant under $K \cap M$.

Remark. The technique of proof is an important one and will arise again.

Proof. Given f in $C(K)$ right invariant under $K \cap M$, extend f to a function F on G by

$$F(kman) = e^{-2\rho \log a} f(k); \tag{5.26}$$

the right invariance of f under $K \cap M$ makes F well defined. Next, fix a function $\varphi \geq 0$ in $C_{\text{com}}(MAN)$ with $\int_{MAN} \varphi(man) d_l(man) = 1$; by averaging over $K \cap M$, we may assume φ is left invariant under $K \cap M$. Extend φ to G by the definition

$$\varphi(kman) = \varphi(man);$$

the left invariance of φ under $K \cap M$ makes φ well defined. We have

$$\int_{MAN} \varphi(xman) d_l(man) = 1 \qquad \text{for all } x \text{ in } G.$$

Then

$$\begin{aligned}
\int_K f(k)\, dk &= \int_K f(k)\left[\int_{MAN} \varphi(kman) d_l(man)\right] dk \\
&= \int_{K \times MAN} f(k)\varphi(kman)e^{-2\rho \log a}\, dk\, d_r(man) && \text{by (5.12)} \\
&= \int_{K \times MAN} F(kman)\varphi(kman)\, dk\, d_r(man) && \text{by (5.26)} \\
&= \int_G F(x)\varphi(x)\, dx && \text{by (5.18).}
\end{aligned}$$

Also

$$\int_{\bar N} f(\kappa(\bar n))e^{-2\rho H(\bar n)}\,d\bar n = \int_{\bar N} F(\bar n)\left[\int_{MAN} \varphi(\bar n man)d_l(man)\right]d\bar n \qquad \text{by (5.26)}$$

$$= \int_{\bar N\times MAN} F(\bar n)e^{-2\rho\log a}\varphi(\bar n man)\,d\bar n\,d_r(man) \qquad \text{by (5.12)}$$

$$= \int_{\bar N\times MAN} F(\bar n man)\varphi(\bar n man)\,d\bar n\,d_r(man) \qquad \text{by (5.26)}$$

$$= \int_G F(x)\varphi(x)\,dx \qquad \text{by (5.23).}$$

The result (5.25) follows.

Proposition 5.27. Let G be linear connected reductive, let $\mathfrak{h}_1, \ldots, \mathfrak{h}_r$ be a maximal set of nonconjugate θ-stable Cartan subalgebras, and let $H_1, \ldots, H_r$ be the corresponding Cartan subgroups. Let the invariant measures on each H_j and G/H_j be normalized so that

$$\int_G f(x)\,dx = \int_{G/H_j}\left[\int_{H_j} f(gh)\,dh\right]d\dot g \qquad \text{for all } f \in C_{\text{com}}(G).$$

Then every F in $C_{\text{com}}(G)$ satisfies

$$\int_G F(x)\,dx = \sum_{j=1}^{r}\frac{1}{|W(H_j\colon G)|}\int_{(G/H_j)\times H_j} F(ghg^{-1})|D_{H_j}(h)|^2\,dh\,d\dot g,$$

where D_{H_j} denotes the **Weyl denominator**

$$D_{H_j}(h) = \xi_\delta(h)\prod_{\alpha\in\Delta^+(\mathfrak{h}_j^{\mathbb{C}}:\mathfrak{g}^{\mathbb{C}})}[1-\xi_\alpha(h)^{-1}] \qquad \text{for } h \in H_j.$$

Remarks. We shall refer to this integration formula as the **Weyl integration formula** for G. This formula is valid with no further assumptions since $|D_{H_j}|^2$ is always unambiguously defined. However, when we want D_{H_j} itself defined, we need to assume more. We assume that G has a realization in which $\mathfrak{k} \cap i\mathfrak{p} = 0$ and in which ξ_δ exists as a well-defined exponential on $\exp \mathfrak{h}_j^{\mathbb{C}} \subseteq G^{\mathbb{C}}$. Then ξ_δ is defined on H_j by restriction. We summarize this assumption on G by saying that G has a realization in which δ is **analytically integral.** For a particular realization of G, the existence of ξ_δ depends neither on the index j nor on the positive system of roots. The value of ξ_δ does not depend on the realization, as long as ξ_δ exists. By Proposition 4.33 and Theorem 4.28, δ is analytically integral if G is contained in $G^{\mathbb{C}}$ with $G^{\mathbb{C}}$ simply connected. For more information, see the Problems at the end of the chapter.

Proposition 5.28. Let G be linear connected reductive, and fix a positive system Σ^+ of restricted roots. For suitable normalizations of Haar measures, with $\mathfrak{a}^+$ as the positive Weyl chamber, and for f in $C_{\text{com}}(G)$,

$$\int_G f(x)\,dx = \int_{K\times\mathfrak{a}^+\times K}\left[\prod_{\lambda\in\Sigma^+}(\sinh\lambda(H))^{\dim\mathfrak{g}_\lambda}\right] f(k_1(\exp H)k_2)\,dk_1\,dH\,dk_2.$$

Example. In $G = \mathrm{SL}(2,\mathbb{R})$ let us show that the representation $\mathscr{D}_n^+$ is square integrable for $n \geq 2$. We use the SU(1, 1) picture and are to compute

$$\int_{\mathrm{SU}(1,1)} |(\mathscr{D}_n^+(g)1, 1)|^2\,dg \tag{5.27}$$

with notation as in §2.6. The integrand is left and right K-invariant since $\mathscr{D}_n^+(K)$ acts on 1 by scalars of modulus one. Hence Proposition 5.28 shows that (5.27), with a suitable normalization of Haar measure and with H taken as $\begin{pmatrix} 0 & t \\ t & 0\end{pmatrix}$, is equal to

$$\begin{aligned}
&\int_0^\infty \sinh 2t\left|\left(\mathscr{D}_n^+\begin{pmatrix}\cosh t & \sinh t\\ \sinh t & \cosh t\end{pmatrix}1, 1\right)\right|^2 dt\\
&= \int_0^\infty \sinh 2t\left|\int_{|\zeta|<1}(-(\sinh t)\zeta + \cosh t)^{-n}(1-|\zeta|^2)^{n-2}\,d\zeta\right|^2 dt\\
&= \int_0^\infty \sinh 2t\,(\cosh t)^{-2n}\left|2\pi\int_{r=0}^1 (1-r^2)^{n-2} r\,dr\right|^2 dt\\
&= \|1\|^2\int_0^\infty 2(\cosh t)^{-2n+1}\sinh t\,dt = \frac{\|1\|^2}{n-1},
\end{aligned}$$

which is finite for $n > 1$.

§7. Borel-Weil Theorem

The Borel-Weil Theorem gives an explicit realization of each irreducible representation of a compact connected Lie group. At the same time it provides a tidy application of some of the results of this chapter.

The structure theory that is involved amounts to imbedding our given compact connected group in a complexification, via Proposition 5.6, and then interpreting the Iwasawa decomposition of the complexification. Before stating the theorem, let us therefore establish the setting in the notation we have been using. Our basic compact connected Lie group will be denoted K here, and we regard K as a group of unitary matrices as usual. Its Lie algebra is $\mathfrak{k}$, and we write G for the (complex) analytic group $K^{\mathbb{C}}$ with Lie algebra $\mathfrak{g} = \mathfrak{k}^{\mathbb{C}} = \mathfrak{k} \oplus i\mathfrak{k}$. Here G is linear connected reductive, by Proposition 5.6, and $\mathfrak{g} = \mathfrak{k} \oplus i\mathfrak{k}$ is the Cartan decomposition of $\mathfrak{g}$.

Fix an Iwasawa decomposition $\mathfrak{g} = \mathfrak{k} \oplus \mathfrak{a} \oplus \mathfrak{n}$ with corresponding group decomposition $G = KAN$. Since $\mathfrak{p} = i\mathfrak{k}$, we have

$$\mathfrak{m} = Z_{\mathfrak{k}}(\mathfrak{a}) = i\mathfrak{a},$$

and $\mathfrak{m}$ is a Cartan subalgebra of $\mathfrak{k}$ with $\mathfrak{m}^{\mathbb{C}} = \mathfrak{a} \oplus \mathfrak{m}$. The roots in $\Delta = \Delta(\mathfrak{m}^{\mathbb{C}} : \mathfrak{k}^{\mathbb{C}})$ are just the complex-linear extensions to $\mathfrak{m}^{\mathbb{C}}$ of their restrictions to $\mathfrak{a}$, and it follows that the restricted root space decomposition of $(\mathfrak{g}, \mathfrak{a})$ coincides with the root space decomposition of $(\mathfrak{k}^{\mathbb{C}}, \mathfrak{m}^{\mathbb{C}})$:

$$\mathfrak{k}^{\mathbb{C}} = \mathfrak{m}^{\mathbb{C}} \oplus \sum_{\alpha \in \Delta} \mathfrak{k}_{\alpha} = \mathfrak{a} \oplus \mathfrak{m} \oplus \sum_{\alpha \in \Delta} \mathfrak{g}_{\alpha} = \mathfrak{g}. \tag{5.28}$$

Because of (5.28), the choice of $\mathfrak{n}$ determines a positive system Δ^+ of Δ.

With $M = Z_K(\mathfrak{a})$, the group $B = MA\bar{N}$ is a complex subgroup of G since its Lie algebra

$$\mathfrak{b} = \mathfrak{m} \oplus \mathfrak{a} \oplus \theta\mathfrak{n} = \mathfrak{m}^{\mathbb{C}} \oplus \sum_{\alpha \in \Delta^+} \mathfrak{k}_{-\alpha}$$

is complex. If λ is analytically integral on $(\mathfrak{m}^{\mathbb{C}})'$, let ξ_λ be the corresponding character of M. We can then extend ξ_λ to a holomorphic one-dimensional representation $\xi_\lambda : B \to \mathbb{C}^\times$ by defining

$$\xi_\lambda(x) = \begin{cases} \exp\{\lambda(\log x)\} & \text{for } x \in A \\ 1 & \text{for } x \in \bar{N}. \end{cases}$$

Theorem 5.29 (Borel-Weil Theorem). Let K be a compact connected Lie group, and let $G = K^{\mathbb{C}}$ and B be as above. If λ is dominant and analytically integral and if ξ_λ denotes the corresponding holomorphic one-dimensional representation of B, then a realization of an irreducible representation of K with highest weight λ is as follows: The space is

$$\Gamma(\lambda) = \left\{ F : G \to \mathbb{C} \;\middle|\; \begin{array}{l} \text{(i) } F \text{ is holomorphic} \\ \text{(ii) } F(xb) = \xi_\lambda(b)^{-1} F(x) \text{ for } x \in G,\, b \in B \end{array} \right\},$$

and K operates by the left regular representation.

Remarks. The proof will not give an alternate existence argument for the theorem of the highest weight. Rather it will use the theorem of the highest weight.

For the proof we proceed in several steps, isolating some of them as lemmas.

We make $\Gamma(\lambda)$ into an inner-product space, not obviously complete, by defining

$$(F_1, F_2) = \int_K F_1(k)\overline{F_2(k)}\, dk,$$

and then the left regular representation L of K operates isometrically.

[Only the 0 function has norm 0 since $\|F\| = 0$ implies $F = 0$ on K and since $G = KB$ then forces $F = 0$ on all of G.]

Lemma 5.30. Let τ be a finite-dimensional representation of K on a complex vector space V. Then τ extends to a holomorphic representation of G on V.

Proof if K is semisimple. By Weyl's Theorem and the fact that any compact connected Lie group may be taken to be a subgroup of a unitary group, we may assume K is simply connected. Then the result follows from Proposition 5.7.

Proof in general case. By Proposition 4.31, write $K = K_{ss}(Z_K)_0$ corresponding to $\mathfrak{k} = [\mathfrak{k}, \mathfrak{k}] \oplus Z_{\mathfrak{k}}$, with K_{ss} closed. Then $G = G_{ss}(Z_K)_0^{\mathbb{C}}$ by Proposition 5.5, where G_{ss} is closed and has Lie algebra $[\mathfrak{g}, \mathfrak{g}] = [\mathfrak{k}, \mathfrak{k}]^{\mathbb{C}}$ and where $(Z_K)_0^{\mathbb{C}}$ has Lie algebra $Z_{\mathfrak{k}} \oplus iZ_{\mathfrak{k}}$. The special case above allows us to extend τ_λ holomorphically from K_{ss} to G_{ss}. We can define τ_λ holomorphically also on the (Euclidean) universal covering of $(Z_K)_0^{\mathbb{C}}$, and the result descends to $(Z_K)_0^{\mathbb{C}}$ since $(Z_K)_0^{\mathbb{C}}$ is the product of $(Z_K)_0$ and a Euclidean group. Then we can put the results together to define τ_λ holomorphically on G since $G_{ss} \cap (Z_K)_0^{\mathbb{C}} \subseteq Z_{G_{ss}} \subseteq K_{ss} \subseteq K$, by Corollary 1.3.

Returning to Theorem 5.29, we apply the theorem of the highest weight to obtain an irreducible unitary representation Φ_λ of K with highest weight λ. Let us say Φ_λ acts in V and v_λ is a highest weight vector of norm one. We extend Φ_λ to a holomorphic representation of G by means of Lemma 5.30. For v in V we define ψ_v on G by

$$\psi_v(x) = (\Phi_\lambda(x)^{-1}v, v_\lambda).$$

Lemma 5.31. For each v in V, ψ_v is in $\Gamma(\lambda)$. Moreover, $L(k)\psi_v = \psi_{\Phi_\lambda(k)v}$, so that $\{\psi_v \mid v \in V\}$ is an irreducible subspace of $\Gamma(\lambda)$ under K equivalent with Φ_λ.

Proof. Let φ_λ be the differential of Φ_λ. Since Φ_λ is unitary on K, φ_λ is skew-Hermitian on $\mathfrak{k}$. Since φ_λ is complex-linear on $\mathfrak{g}$, $\varphi_\lambda(\theta X) = -\varphi_\lambda(X)^*$ for all X in $\mathfrak{g}$. Thus $\Phi_\lambda(\Theta x) = \Phi_\lambda(x^{-1})^*$ for x in G. Hence $b = ma\bar{n}$ in $MA\bar{N}$ implies

$$\begin{aligned}
\psi_v(xma\bar{n}) &= (\Phi_\lambda(ma\bar{n})^{-1}\Phi_\lambda(x)^{-1}v, v_\lambda) \\
&= (\Phi_\lambda(x)^{-1}v, \Phi_\lambda(ma^{-1}n)v_\lambda) && \text{for } n = \Theta\bar{n} \\
&= (\Phi_\lambda(x)^{-1}v, \Phi_\lambda(ma^{-1})v_\lambda) && \text{since } v_\lambda \text{ is a highest weight vector} \\
&= (\Phi_\lambda(x)^{-1}v, \xi_\lambda(m)\xi_\lambda(a)^{-1}v_\lambda) && \text{since } v_\lambda \text{ has weight } \lambda \\
&= \overline{\xi_\lambda(m)\xi_\lambda(a)^{-1}}(\Phi_\lambda(x)^{-1}v, v_\lambda) \\
&= \xi_\lambda(b)^{-1}\psi_v(x).
\end{aligned}$$

Clearly ψ_v is holomorphic, and thus ψ_v is in $\Gamma(\lambda)$. Finally

$$\begin{aligned}\psi_{\Phi_\lambda(k)v}(x) &= (\Phi_\lambda(x)^{-1}\Phi_\lambda(k)v, v_\lambda)\\ &= (\Phi_\lambda(k^{-1}x)^{-1}v, v_\lambda)\\ &= \psi_v(k^{-1}x) = L(k)\psi_v(x),\end{aligned}$$

and the lemma follows.

The idea now is to prove that $v \to \psi_v$ in Lemma 5.31 carries V *onto* $\Gamma(\lambda)$. This will complete the proof of the theorem. For this purpose we define $\psi_\lambda = \psi_{v_\lambda}$. The key lemma is the next one, which shows the fundamental role played by ψ_λ.

Lemma 5.32. If F is in $\Gamma(\lambda)$, then

$$\int_M F(mxm^{-1})\,dm = F(1)\psi_\lambda(x)$$

for all x in G. Here dm denotes normalized Haar measure on M.

Proof. The idea is to show near $x = 1$ that the left side is $F(1)$ times a power series in x that is independent of F. The power series is evaluated as the series for $\psi_\lambda(x)$ by putting $F = \psi_\lambda$. Then since both sides are holomorphic on G and equal in a neighborhood of 1, they are equal everywhere.

Thus for $X \in \mathfrak{g}$, let $\tilde{X}$ be the corresponding left-invariant vector field on G. Since F is real analytic, the Taylor series of F converges to F near the identity, by (A.100):

$$F(\exp X) = \sum_{n \geq 0} \frac{1}{n!}(\tilde{X}^n F)(1).$$

Conjugate by $m \in M$ and integrate to get

$$\int_M F(m(\exp X)m^{-1})\,dm = \sum_{n\geq 0} \frac{1}{n!}\left(\left\{\int_M \mathrm{Ad}(m)\tilde{X}^n\,dm\right\}F\right)(1). \quad (5.29)$$

Choose a basis $\{X_1, \ldots, X_s\}$ of $\mathfrak{g}$ over $\mathbb{C}$ consisting first of root vectors for positive roots, then members of $\mathfrak{m}^{\mathbb{C}}$, and finally root vectors for negative roots. Write $z_1X_1 + \ldots + z_sX_s$, and expand the expression in (5.29) into a sum of integrals of monomials. In this expansion, we can factor out the z_j's from their effect on F because $\tilde{X}$ is a complex-linear when applied to a holomorphic function. Using the Birkhoff-Witt Theorem, we can rewrite the expression as a linear combination of Ad(m) of monomials of the form $X_1^{j_1} \cdots X_s^{j_s}$. The integral of each monomial is an Ad(m)-invariant monomial. If the latter monomial has no factor $E_{-\gamma}$ for any $\gamma \in \Delta^+$, it cannot have a factor E_γ either, by the Ad(M)-invariance.

On the other hand, an Ad(M)-invariant monomial cannot have any factor $E_{-\gamma}$ for $\gamma \in \Delta^+$ since $\exp tE_{-\gamma}$ in $\bar{N}$ implies

$$\tilde{E}_{-\gamma}F(x) = \frac{d}{dt}F(x \exp tE_{-\gamma})\bigg|_{t=0} = 0.$$

So the Ad(M)-invariant monomials are in effect all in $U(\mathfrak{m}^{\mathbb{C}})$. Since $\exp \mathfrak{m}^{\mathbb{C}} = AM \subseteq B$, each member of $U(\mathfrak{m}^{\mathbb{C}})$ acts on F as a scalar depending only on λ:

$$\tilde{H}F(x) = \frac{d}{dt}F(x \exp tH)\bigg|_{t=0} = \frac{d}{dt}\xi_\lambda^{-1}(\exp tH)\bigg|_{t=0} F(x).$$

Hence an expression $\tilde{H}_1^{l_1} \cdots \tilde{H}_r^{l_r}F(1)$ is a scalar times $F(1)$, with the scalar depending on λ but not F. The lemma follows.

The lemma allows us to complete the proof of Theorem 5.29 in a few easy steps:

(1) $|F(1)| \le \|\psi_\lambda\|^{-1}\|F\|$ for all F in $\Gamma(\lambda)$. [In fact, we have

$$\begin{aligned}
\|F\|^2 &= \int_K |F(k)|^2\,dk = \int_K |F(mkm^{-1})|^2\,dk \qquad \text{for all } m \in M\\
&= \int_K\int_M |F(mkm^{-1})|^2\,dm\,dk\\
&\ge \int_K \left|\int_M F(mkm^{-1})\,dm\right|^2 dk \qquad \text{by the Schwarz inequality}\\
&= |F(1)|^2 \int_K |\psi_\lambda(k)|^2\,dk \qquad \text{by Lemma 5.32}\\
&= |F(1)|^2\|\psi_\lambda\|^2,
\end{aligned}$$

and the assertion follows.]

(2) To each compact set $E \subseteq G$ corresponds a constant $C_E < \infty$ such that

$$|F(x)| \le C_E\|F\|$$

for all F in $\Gamma(\lambda)$ and x in E. [In fact, this is a simple consequence of (1) since $\|L(k)F\| = \|F\|$ and since $|F(xb)| = |\xi_\lambda(b)|^{-1}|F(x)|$ for b in B.]

(3) $\Gamma(\lambda)$ is complete. [In fact, Cauchy sequences in $\Gamma(\lambda)$ converge uniformly on compact sets in G, by (2), and the limit function is holomorphic and satisfies the correct transformation law under B.]

(4) $\Gamma(\lambda)$ is irreducible, and hence the map $v \to \psi_v$ of Lemma 5.31 is onto $\Gamma(\lambda)$. [In fact, let $U \subseteq \Gamma(\lambda)$ be a nonzero closed invariant subspace. Let $F \neq 0$ be in U. After applying some $L(k)$, we may assume that $F(1) \neq 0$. Then

$$\int_M \overline{\xi_\lambda(m)}L(m)F\,dm$$

is in U, by (3). But Lemma 5.32 shows this is $F(1)\psi_\lambda$. Thus ψ_λ is in U. If $U^\perp \neq 0$ also, then ψ_λ is in $U^\perp$ also, which is a contradiction. Hence $U = 0$ or $U^\perp = 0$.]

§8. Problems

1. In $\mathrm{Sp}(2, \mathbb{R})$, take $\mathfrak{a}_\mathfrak{p}$ to be the diagonal subalgebra, writing a typical member of $\mathfrak{a}_\mathfrak{p}$ as $\mathrm{diag}(s, t, -s, -t)$. Let f_1 and f_2 of this typical member be s and t, respectively. Obtain the restricted root space decomposition, showing that $\mathfrak{m}_\mathfrak{p} = 0$, that the set of restricted roots is $\{\pm f_1 \pm f_2, \pm 2f_1, \pm 2f_2\}$, and that all restricted roots have multiplicity one.

2. (a) Prove that every element of $\mathrm{SL}(2, \mathbb{R})$ is conjugate to at least one matrix of the form

$$\begin{pmatrix} a & 0 \\ 0 & a^{-1} \end{pmatrix} \text{ with } a \neq 0, \quad \begin{pmatrix} 1 & t \\ 0 & 1 \end{pmatrix},$$
$$\begin{pmatrix} -1 & t \\ 0 & -1 \end{pmatrix}, \quad \begin{pmatrix} \cos\theta & \sin\theta \\ -\sin\theta & \cos\theta \end{pmatrix}.$$

(b) Prove that the exponential map from $\mathfrak{sl}(2, \mathbb{R})$ into $\mathrm{SL}(2, \mathbb{R})$ has range exactly

$$\{x \mid \mathrm{Tr}\, x > -2\} \cup \{-I\}.$$

3. In $\mathrm{SL}(n, \mathbb{R})$, take $M_\mathfrak{p} A_\mathfrak{p} N_\mathfrak{p}$ to be the upper triangular group.
   (a) Follow the prescription of Proposition 5.23 to see that the proposition leads to all possible block upper triangular subgroups of $\mathrm{SL}(n, \mathbb{R})$.
   (b) Give a direct proof for $\mathrm{SL}(n, \mathbb{R})$ that the only closed subgroups containing $M_\mathfrak{p} A_\mathfrak{p} N_\mathfrak{p}$ are the block upper triangular subgroups.
   (c) Give a direct proof for $\mathrm{SL}(n, \mathbb{R})$ that no two distinct block upper triangular subgroups are conjugate within $\mathrm{SL}(n, \mathbb{R})$.

4. The group $\mathrm{Sp}(2, \mathbb{R})$ has four nonconjugate Cartan subalgebras, listed explicitly in Example 4 of Cartan subalgebras in §4. For each of the four, construct the MA of an associated parabolic subgroup according to the recipe in §5. In two cases, A has dimension one; show that the corresponding M's are not isomorphic. [Note: This fact proves the nonconjugacy of the two Cartan subalgebras in question.]

5. Give an example of a two-dimensional linear connected reductive group G for which $\mathfrak{k} \cap i\mathfrak{p} = 0$ and B_0 is not real-valued.

6. Let H be a Θ-stable Cartan subgroup of G, and write $H = BA$ with $B \subseteq K$ and $A \subseteq \exp \mathfrak{p}$. Show that restriction to A defines a homomorphism of $W(H:G)$ into $W(A:G)$. Using Lemmas 5.16 and 5.24, show that this homomorphism is onto. Identify its kernel as $W(B:M)$.
7. Let $\psi: \bar{N}_\mathfrak{p} \to K/M_\mathfrak{p}$ be the composition of passage to the K component of the Iwasawa decomposition, followed by the quotient map to $K/M_\mathfrak{p}$. Using (5.22) and (5.25), show that ψ is a diffeomorphism onto an open dense set.

Problems 8 to 10 give further information about the Cartan decomposition of a linear connected reductive G with Lie algebra $\mathfrak{g}$.

8. Let $\mathfrak{p}_0$ be an ad $\mathfrak{k}$ invariant subspace of $\mathfrak{p}$, and let $\mathfrak{p}_0^\perp$ be its orthogonal complement with respect to the real part of the trace form B_0. Prove that $\operatorname{Re} B_0([\mathfrak{p}_0, \mathfrak{p}_0^\perp], \mathfrak{k}) = 0$ and conclude that $[\mathfrak{p}_0, \mathfrak{p}_0^\perp] = 0$.
9. If $\mathfrak{p}_0$ is an ad $\mathfrak{k}$ invariant subspace of $\mathfrak{p}$, prove that $[\mathfrak{p}_0, \mathfrak{p}] \oplus \mathfrak{p}_0$ is an ideal in $\mathfrak{g}$.
10. Under the additional assumption that $\mathfrak{g}$ is simple and G is noncompact, prove that
    (a) $[\mathfrak{p}, \mathfrak{p}] = \mathfrak{k}$
    (b) $\mathfrak{k}$ is a maximal Lie subalgebra of $\mathfrak{g}$
    (c) every finite-dimensional unitary representative of G is trivial.

Problems 11 to 13 concern linear forms that are analytically integral for a noncompact group. Let G be linear connected semisimple with $\mathfrak{k} \cap i\mathfrak{p} = 0$.

11. Using the example of $G = SL(3, \mathbb{R})$ that follows Proposition 5.7 and taking $\mathfrak{h}$ to be the diagonal subalgebra, show that G has one realization in which ξ_{e_1} exists on $\exp \mathfrak{h}^\mathbb{C}$ and another realization where ξ_{e_1} does not exist.
12. Prove that $Z_{G^\mathbb{C}} = Z_G$ if rank G = rank K.
13. Suppose rank G = rank K. Let $\mathfrak{h}$ be a Cartan subalgebra of $\mathfrak{g}$, and let $\lambda \in (\mathfrak{h}^\mathbb{C})'$ be algebraically integral. Prove that the question of whether ξ_λ exists on $\exp \mathfrak{h}^\mathbb{C}$ is independent of the realization of G.

Problems 14 to 17 deal with the Iwasawa decomposition of $SO(m, 1)$ and $SU(m, 1)$, $m \geq 2$. For each of these groups, take $\mathfrak{a}_\mathfrak{p} = \mathbb{R}(E_{1,m+1} + E_{m+1,1})$.

14. Verify that $\mathfrak{a}_\mathfrak{p}$ is maximal abelian in $\mathfrak{p}$.
15. Calculate the restricted root space decomposition for the Lie algebra of each group, showing that the restricted roots are of the form $\{\pm\alpha\}$ for $SO(m, 1)$ and $\{\pm\alpha, \pm 2\alpha\}$ for $SU(m, 1)$.
16. Find $N_\mathfrak{p}$ and $\bar{N}_\mathfrak{p}$ as explicit sets of matrices for each of the groups.

17. For each of the groups, calculate $H_{\mathfrak{p}}$ (= the log of the Iwasawa $A_{\mathfrak{p}}$ component), and find $e^{2\rho_{\mathfrak{p}}H_{\mathfrak{p}}(\bar{n})}$ for $\bar{n}$ in $\bar{N}_{\mathfrak{p}}$.

Problems 18 to 22 address the question of connectedness of $M_{\mathfrak{p}}$ under the assumption that G has real rank one.

18. Show that $W(A_{\mathfrak{p}}:G)$ has order 2.
19. Using the Bruhat decomposition, show that the image of ψ in Problem 7 is all of $K/M_{\mathfrak{p}}$ except for one point.
20. Using the results of Problems 7 and 19, identify $K/M_{\mathfrak{p}}$ topologically as the one-point compatification of $\bar{N}_{\mathfrak{p}}$, hence as a sphere of dimension $\dim \bar{N}_{\mathfrak{p}}$.
21. Conclude that $M_{\mathfrak{p}}$ is connected unless $\dim \bar{N}_{\mathfrak{p}} = 1$.
22. If $\dim \bar{N}_{\mathfrak{p}} = 1$, prove that $\mathfrak{n}_{\mathfrak{p}}$ is contained in a θ-stable ideal of $\mathfrak{g}$ that is isomorphic to $\mathfrak{sl}(2, \mathbb{R})$.

Problems 23 to 25 give properties of regular elements.

23. Let H be a Θ-stable Cartan subgroup. Prove that an element t of H is regular if and only if $\xi_\alpha(t) \neq 1$ for all roots α in $\Delta(\mathfrak{h}^{\mathbb{C}}:\mathfrak{g}^{\mathbb{C}})$.
24. Let H be a Θ-stable Cartan subalgebra, and let t be a regular element of H. Prove that $\{g \in G \mid gtg^{-1} \in H\}$ is generated by H and representatives of the members of the Weyl group $W(H:G)$. [Hint: Use Theorem 5.22d.]
25. Let H be a Θ-stable Cartan subgroup, and let H' be the set of regular elements. Prove that the map $\psi: G/H \times H' \to G$ given by $\psi(gH, t) = gtg^{-1}$ is everywhere smooth and regular and is everywhere $|W(H:G)|$ to 1. [Hint: Use Problem 24 and Proposition 5.27.]

CHAPTER VI

Holomorphic Discrete Series

§1. Holomorphic Discrete Series for SU(1, 1)

Recall from §2.6 that the discrete series representation $\mathscr{D}_n^-$ of SU(1, 1) acts in the space $\mathscr{V}_n$ of analytic functions in the unit disc with

$$\|f\|^2 = \int_{|z|<1} |f(z)|^2(1 - |z|^2)^{n-2}\, dx\, dy < \infty$$

by the formula

$$\mathscr{D}_n^- \begin{pmatrix} \alpha & \beta \\ \bar{\beta} & \bar{\alpha} \end{pmatrix} f(z) = (-\bar{\beta}z + \alpha)^{-n} f\left(\frac{\bar{\alpha}z - \beta}{-\bar{\beta}z + \alpha}\right).$$

The representation $\mathscr{D}_n^-$ is irreducible unitary for $n \geq 2$ and has a nonzero square integrable matrix coefficient.

Harish-Chandra found a different realization of this representation that suggests a generalization to a wider class of groups. To write down the different realization, let

$$G = \mathrm{SU}(1, 1) \qquad \text{and} \qquad B = \left\{\begin{pmatrix} a & 0 \\ c & a^{-1} \end{pmatrix}\right\} \subseteq \mathrm{SL}(2, \mathbb{C}).$$

Then we observe that

(i) every element of the set $GB \subseteq \mathrm{SL}(2, \mathbb{C})$ has a unique decomposition as a product

$$\begin{pmatrix} 1 & z \\ 0 & 1 \end{pmatrix}\begin{pmatrix} \gamma^{-1} & 0 \\ 0 & \gamma \end{pmatrix}\begin{pmatrix} 1 & 0 \\ \zeta & 1 \end{pmatrix}, \qquad \zeta \in \mathbb{C}, \gamma \in \mathbb{C}^\times, |z| < 1, \tag{6.1}$$

and every matrix (6.1) is in GB. [In fact, since $B = \left\{\begin{pmatrix} \gamma^{-1} & 0 \\ 0 & \gamma \end{pmatrix} \times \begin{pmatrix} 1 & 0 \\ \zeta & 1 \end{pmatrix}\right\}$ is a group, existence of the decomposition follows from

$$\begin{pmatrix} \alpha & \beta \\ \bar{\beta} & \bar{\alpha} \end{pmatrix} = \begin{pmatrix} 1 & \beta/\bar{\alpha} \\ 0 & 1 \end{pmatrix}\begin{pmatrix} \bar{\alpha}^{-1} & 0 \\ 0 & \bar{\alpha} \end{pmatrix}\begin{pmatrix} 1 & 0 \\ \bar{\beta}/\bar{\alpha} & 1 \end{pmatrix}. \tag{6.2}$$

The rest of the properties are easily checked.]

(ii) GB is an open subset of $SL(2, \mathbb{C})$, and its product complex structure obtained from (6.1) is the same as what it inherits from $SL(2, \mathbb{C})$. In particular, left translation by any member of G is a holomorphic automorphism of GB. [In fact, the relevant tool here is Lemma 5.11, which first is applied to G and the subgroup of B with real diagonal entries, then is applied to the factors of (6.1).]

To construct the representation, let $n \geq 2$ be an integer and let ξ_n be the one-dimensional holomorphic representation of B given by

$$\xi_n\begin{pmatrix} a & 0 \\ c & a^{-1} \end{pmatrix} = a^{-n}.$$

Let

$$V_n = \left\{ F: GB \to \mathbb{C} \;\middle|\; \begin{array}{l} \text{(i) } F \text{ is holomorphic} \\ \text{(ii) } F(xb) = \xi_n(b)^{-1}F(x) \text{ for } x \in GB, b \in B \\ \text{(iii) } \|F\|^2 = \int_G |F(g)|^2\, dg < \infty \end{array} \right\}$$

$$L(g)F(x) = F(g^{-1}x) \qquad \text{for } F \in V_n, g \in G, x \in GB.$$

We shall give a correspondence between $\mathscr{V}_n$ and V_n and show that it is a unitary equivalence if the Haar measure dg is normalized suitably. Given $F(z, \gamma, \zeta)$ in V_n, define $f(z)$ on the disc $\Omega = \{|z| < 1\}$ by

$$f(z) = F(z, 1, 0).$$

For the inverse, if $f(z)$ in $\mathscr{V}_n$ is given, put

$$F(z, \gamma, \zeta) = \gamma^{-n}f(z).$$

The F satisfies (i) and (ii) at least. To see that $F \leftrightarrow f$ is unitary, we use the fact that it is possible to normalize dg in such a way that for any reasonable $h_0(z)$ on Ω the function $h(g) = h_0(g(0))$ satisfies

$$\int_G h(g)\, dg = \int_\Omega h_0(z)(1 - |z|^2)^{-2}\, dx\, dy, \tag{6.3}$$

where $g(z) = \dfrac{\alpha z + \beta}{\bar{\beta}z + \bar{\alpha}}$ for $g = \begin{pmatrix} \alpha & \beta \\ \bar{\beta} & \bar{\alpha} \end{pmatrix}$. For F in V_n, (6.2) gives

$$F(g) = F(\beta/\bar{\alpha}, \bar{\alpha}, \bar{\beta}/\bar{\alpha}) = \bar{\alpha}^{-n}f(\beta/\bar{\alpha}) = \bar{\alpha}^{-n}f(g(0)).$$

Thus $|F(g)|^2 = |\alpha|^{-2n}|f(g(0))|^2$. Since $|\alpha|^2 - |\beta|^2 = 1$, $1 - |g(0)|^2 = |\alpha|^{-2}$ and

$$|F(g)|^2 = (1 - |g(0)|^2)^n|f(g(0))|^2.$$

Putting $h_0(z) = (1 - |z|^2)^n |f(z)|^2$ and $h(g) = |F(g)|^2$ in (6.3), we obtain

$$\|F\|^2 = \int_G |F(g)|^2 \, dg = \int_\Omega |f(z)|^2 (1 - |z|^2)^{n-2} \, dx \, dy = \|f\|^2,$$

and it follows that the correspondence is unitary.

To see that $F \leftrightarrow f$ commutes with the group actions, we compute that

$$L(g)F(z, \gamma, \zeta) = F\left(\frac{\bar{\alpha}z - \beta}{-\bar{\beta}z + \alpha}, \gamma(-\bar{\beta}z + \alpha), \zeta'\right).$$

Thus

$$\begin{aligned} L(g)F(z, 1, 0) &= (-\bar{\beta}z + \alpha)^{-n} F\left(\frac{\bar{\alpha}z - \beta}{-\bar{\beta}z + \alpha}, 1, 0\right) \\ &= (-\bar{\beta}z + \alpha)^{-n} f\left(\frac{\bar{\alpha}z - \beta}{-\bar{\beta}z + \alpha}\right) = \mathcal{D}_n^-(g)f(z), \end{aligned}$$

and $F \leftrightarrow f$ is a unitary equivalence.

§2. Classical Bounded Symmetric Domains

A **bounded symmetric domain** Ω is a bounded open subset of $\mathbb{C}^n$ of the form G/K, with G linear connected semisimple, such that G operates holomorphically on Ω. There are four infinite classes of classical examples, and we give three of them here. In each case it should be noted that the center of K has positive dimension.

Example 1. Let $m \leq n$ and $\Omega = \{Z \in M_{nm}(\mathbb{C}) \mid I_m - Z^*Z > 0\}$. Here M_{nm} refers to all n-by-m matrices, I_m is the identity of size m, and >0 means "is positive definite." Let

$$\begin{aligned} G &= \mathrm{SU}(n, m) \subseteq \mathrm{SL}(n + m, \mathbb{C}) \\ K &= \mathrm{S}(\mathrm{U}(n) \times \mathrm{U}(m)) \\ &= \left\{\begin{pmatrix} A & 0 \\ 0 & D \end{pmatrix} \middle| \, A \in \mathrm{U}(n), D \in \mathrm{U}(m), \det A \det D = 1\right\} \end{aligned}$$

The group G operates holomorphically on Ω by

$$g(Z) = (AZ + B)(CZ + D)^{-1} \quad \text{if} \quad g = \overset{\begin{matrix} n & m \end{matrix}}{\begin{pmatrix} A & B \\ C & D \end{pmatrix}} \begin{matrix} n \\ m \end{matrix}. \tag{6.4}$$

One can show that the action of G is transitive, and it is easy to see that the isotropy subgroup at $Z = 0$ is K.

Let us verify that $(CZ+D)^{-1}$ is defined in (6.4) and that $g(\Omega) \subseteq \Omega$. We write

$$\begin{aligned}(AZ+B)^*(AZ+B) &- (CZ+D)^*(CZ+D) \\ &= (Z^* \quad I_m)g^*\begin{pmatrix} I & 0 \\ 0 & -I \end{pmatrix} g \begin{pmatrix} Z \\ I_m \end{pmatrix} \\ &= (Z^* \quad I_m)\begin{pmatrix} I & 0 \\ 0 & -I \end{pmatrix}\begin{pmatrix} Z \\ I_m \end{pmatrix} \\ &= Z^*Z - I_m. \end{aligned} \tag{6.5}$$

Let $(CZ+D)v = 0$. Unless $v = 0$, we have

$$0 \le v^*(AZ+B)^*(AZ+B)v = v^*(Z^*Z - I_m)v < 0$$

by (6.5), a contradiction. So $(CZ+D)^{-1}$ exists, and then (6.5) gives

$$g(Z)^*g(Z) - I = (CZ+D)^{-1*}(Z^*Z - I_m)(CZ+D)^{-1} < 0.$$

Thus $g(\Omega) \subseteq \Omega$.

Example 2. Let $\Omega = \{Z \in M_{nn}(\mathbb{C}) \mid I_n - Z^*Z > 0, Z = -Z^{tr}\}$. Let

$$G = SO^*(2n) = \left\{ g \in SU(n,n) \,\middle|\, g^{tr}\begin{pmatrix} 0 & I \\ I & 0 \end{pmatrix} g = \begin{pmatrix} 0 & I \\ I & 0 \end{pmatrix} \right\}$$

$$K = \left\{ \begin{pmatrix} A & 0 \\ 0 & \bar{A} \end{pmatrix} \,\middle|\, A \in U(n) \right\}.$$

Then G acts in the same way as in Example 1, preserving Ω. The action is transitive with K the isotropy subgroup at $Z = 0$.

Example 3. Let $\Omega = \{Z \in M_{nn}(\mathbb{C}) \mid I_n - Z^*Z > 0, Z = Z^{tr}\}$. For G we use the alternate realization of $Sp(n, \mathbb{R})$ in Chapter I:

$$G = \left\{ g \in SU(n,n) \,\middle|\, g^{tr}\begin{pmatrix} 0 & I \\ -I & 0 \end{pmatrix} g = \begin{pmatrix} 0 & I \\ -I & 0 \end{pmatrix} \right\}$$

$$K = \left\{ \begin{pmatrix} A & 0 \\ 0 & \bar{A} \end{pmatrix} \,\middle|\, A \in U(n) \right\}.$$

Then G acts in the same way as in Example 1, preserving Ω. The action is transitive with K the isotropy subgroup at $Z = 0$.

§3. Harish-Chandra Decomposition

Let G be a linear connected reductive group. Although it is not critical to do so, we assume once more that $\mathfrak{k} \cap i\mathfrak{p} = 0$; this assumption allows us to regard $\mathfrak{g}^{\mathbb{C}}$ as the obvious matrices, since it makes $\mathfrak{g} + i\mathfrak{g}$ a direct sum. Let $\mathfrak{c}$ be the center of $\mathfrak{k}$. Our key assumption in this section is that $Z_{\mathfrak{g}}(\mathfrak{c}) = \mathfrak{k}$. This assumption is satisfied in the case of the three examples of §2. It is satisfied in the case also that G is compact, and in that case it is worth noting that the theory in this section and the next reduces exactly to the Borel-Weil Theorem of §5.7.

Let $\mathfrak{h}$ be a Cartan subalgebra of $\mathfrak{k}$. Then $\mathfrak{c} \subseteq \mathfrak{h}$, so that $Z_{\mathfrak{g}}(\mathfrak{h}) \subseteq Z_{\mathfrak{k}}(\mathfrak{h}) = \mathfrak{h}$ and $\mathfrak{h}$ is a Cartan subalgebra of $\mathfrak{g}$. If T denotes $Z_G(\mathfrak{h})$, then T is a torus with Lie algebra $\mathfrak{h}$, by Theorem 5.22b.

Let $\Delta = \Delta(\mathfrak{h}^{\mathbb{C}}:\mathfrak{g}^{\mathbb{C}})$. Since $\mathfrak{k}^{\mathbb{C}}$ brackets $\mathfrak{k}^{\mathbb{C}}$ and $\mathfrak{p}^{\mathbb{C}}$ into themselves, so does $\mathfrak{h}^{\mathbb{C}}$. The one-dimensionality of each root space therefore implies that any root space is either in $\mathfrak{k}^{\mathbb{C}}$ or in $\mathfrak{p}^{\mathbb{C}}$. We call each root **compact** or **noncompact,** accordingly, and we write Δ_K and Δ_n for the sets of compact and noncompact roots, respectively. Then

$$\mathfrak{k}^{\mathbb{C}} = \mathfrak{h}^{\mathbb{C}} \oplus \sum_{\alpha \in \Delta_K} \mathfrak{g}_\alpha$$

is the root space decomposition of $\mathfrak{k}^{\mathbb{C}}$ relative to $\mathfrak{h}^{\mathbb{C}}$, and we have

$$\mathfrak{p}^{\mathbb{C}} = \sum_{\alpha \in \Delta_n} \mathfrak{g}_\alpha.$$

Lemma 6.1. A root α is compact if and only if it vanishes on $\mathfrak{c}$.

Proof. If α is in Δ, then $\alpha(\mathfrak{c}) = 0$ if and only if $[\mathfrak{c}, \mathfrak{g}_\alpha] = 0$, if and only if $\mathfrak{g}_\alpha \subseteq Z_{\mathfrak{g}^{\mathbb{C}}}(\mathfrak{c})$, if and only if $\mathfrak{g}_\alpha \subseteq \mathfrak{k}^{\mathbb{C}}$, if and only if α is compact.

Choose a **good ordering** for $\mathfrak{h}_{\mathbb{R}} = i\mathfrak{h}$, in that every noncompact positive root is larger than every compact root. (For instance, determine an ordering by means of an orthogonal basis whose first members are a basis of $i\mathfrak{c}$. This ordering is good by Lemma 6.1.) The ordering determines sets of positive roots Δ^+, Δ_K^+, and Δ_n^+. Let

$$\mathfrak{p}^+ = \sum_{\alpha \in \Delta_n^+} \mathfrak{g}_\alpha \qquad \text{and} \qquad \mathfrak{p}^- = \sum_{\alpha \in \Delta_n^+} \mathfrak{g}_{-\alpha},$$

so that $\mathfrak{p}^{\mathbb{C}} = \mathfrak{p}^+ \oplus \mathfrak{p}^-$.

In our three examples of §2, we can choose $\mathfrak{h}$ diagonal, and $\mathfrak{k}^{\mathbb{C}}$ is block diagonal. For a suitable good ordering, $\mathfrak{p}^+$ has the block form $\begin{pmatrix} 0 & * \\ 0 & 0 \end{pmatrix}$, and $\mathfrak{p}^-$ has the block form $\begin{pmatrix} 0 & 0 \\ * & 0 \end{pmatrix}$.

Lemma 6.2. $[\mathfrak{k}^{\mathbb{C}}, \mathfrak{p}^+] \subseteq \mathfrak{p}^+$, $[\mathfrak{k}^{\mathbb{C}}, \mathfrak{p}^-] \subseteq \mathfrak{p}^-$, and $\mathfrak{p}^+$ and $\mathfrak{p}^-$ are abelian subspaces.

Proof. Let α, β, and $\alpha + \beta$ be in Δ with α compact and β noncompact. Then $[\mathfrak{g}_\alpha, \mathfrak{g}_\beta] \subseteq \mathfrak{g}_{\alpha+\beta}$, and β and $\alpha + \beta$ are both positive or both negative because the ordering is good. Summing on α and β, we obtain $[\mathfrak{k}^{\mathbb{C}}, \mathfrak{p}^+] \subseteq \mathfrak{p}^+$ and $[\mathfrak{k}^{\mathbb{C}}, \mathfrak{p}^-] \subseteq \mathfrak{p}^-$.

Next, if α and β are in Δ_n^+, then $\alpha + \beta$ cannot be a root. [In fact, $\alpha + \beta$ would have to be a compact root larger than α; there is no such root since the ordering is good.] Summing on α and β, we obtain $[\mathfrak{p}^+, \mathfrak{p}^+] = 0$. Similarly $[\mathfrak{p}^-, \mathfrak{p}^-] = 0$.

As in §5.1 we form $\mathfrak{g}^{\mathbb{C}} = \mathfrak{g} \oplus i\mathfrak{g}$ and the corresponding analytic group $G^{\mathbb{C}}$ of matrices; $G^{\mathbb{C}}$ is closed by Proposition 5.6. Define

$$\mathfrak{b} = \mathfrak{h}^{\mathbb{C}} \oplus \sum_{\alpha \in \Delta^+} \mathfrak{g}_{-\alpha}$$

and let P^+, $K^{\mathbb{C}}$, P^-, and B be the analytic subgroups of $G^{\mathbb{C}}$ with Lie algebras $\mathfrak{p}^+$, $\mathfrak{k}^{\mathbb{C}}$, $\mathfrak{p}^-$, and $\mathfrak{b}$.

Theorem 6.3 (Harish-Chandra decomposition). Let G be linear connected reductive, and suppose the center $\mathfrak{c}$ of $\mathfrak{k}$ has $Z_{\mathfrak{g}}(\mathfrak{c}) = \mathfrak{k}$. Then multiplication from $P^+ \times K^{\mathbb{C}} \times P^-$ into $G^{\mathbb{C}}$ is one-one, holomorphic, and regular (with image open in $G^{\mathbb{C}}$), GB is open in $G^{\mathbb{C}}$, and there exists a bounded open subset $\Omega \subseteq P^+$ such that

$$GB = GK^{\mathbb{C}}P^- = \Omega K^{\mathbb{C}}P^-.$$

Remarks. For g in G and ω in Ω, the theorem allows us to define $g(\omega)$ to be the P^+ component of the product $g\omega$; in this way G acts on Ω transitively with isotropy subgroup K at $\omega = \exp 0$. So G/K is realized as the complex manifold Ω, and G acts holomorphically. In the case of $SU(n, m)$, the Harish-Chandra decomposition is

$$\begin{pmatrix} A & B \\ C & D \end{pmatrix} = \begin{pmatrix} I & BD^{-1} \\ 0 & I \end{pmatrix} \begin{pmatrix} A - BD^{-1}C & 0 \\ 0 & D \end{pmatrix} \begin{pmatrix} I & 0 \\ D^{-1}C & I \end{pmatrix}$$

and the P^+ component of $\begin{pmatrix} A & B \\ C & D \end{pmatrix}\begin{pmatrix} I & Z \\ 0 & I \end{pmatrix}$ is just

$$\begin{pmatrix} I & (AZ + B)(CZ + D)^{-1} \\ 0 & I \end{pmatrix}.$$

Thus the Harish-Chandra action in the case of $SU(n, m)$ reproduces the classical action (6.4).

Proof. Since

$$\mathfrak{k}^{\mathbb{C}} \oplus \mathfrak{p}^+ = \mathfrak{h}^{\mathbb{C}} \oplus \sum_{\alpha \in \Delta_K \cup \Delta_n^+} \mathfrak{g}_\alpha \supseteq \mathfrak{h}^{\mathbb{C}} \oplus \sum_{\alpha \in \Delta^+} \mathfrak{g}_\alpha,$$

Proposition 5.23 shows that $P^+K^{\mathbb{C}}$ is a parabolic subgroup of $G^{\mathbb{C}}$. Moreover, $P^- = \Theta P^+$, and hence the map $P^+ \times K^{\mathbb{C}} \times P^- \to G^{\mathbb{C}}$ is one-one and regular (with image open in $G^{\mathbb{C}}$) as a consequence of facts about parabolic subgroups.

Let us define

$$\mathfrak{n}^+ = \sum_{\alpha\in\Delta^+} \mathfrak{g}_\alpha, \qquad \mathfrak{n}^- = \sum_{\alpha\in\Delta^+} \mathfrak{g}_{-\alpha}, \qquad \mathfrak{b}_K = \mathfrak{h}^{\mathbb{C}} \oplus \sum_{\alpha\in\Delta_K^+} \mathfrak{g}_{-\alpha}$$

$N^+, N^-, B_K, T_{\mathbb{R}}, T^{\mathbb{C}}$ = analytic subgroups with respective Lie algebras $\mathfrak{n}^+, \mathfrak{n}^-, \mathfrak{b}_K, \mathfrak{h}_{\mathbb{R}}, \mathfrak{h}^{\mathbb{C}}$.

We observe that

$$\mathfrak{g} \cap (\mathfrak{h}_{\mathbb{R}} \oplus \mathfrak{n}^-) = 0. \tag{6.6}$$

[In fact, let bar denote conjugation of $\mathfrak{g}^{\mathbb{C}}$ with respect to $\mathfrak{g}$. Since roots are imaginary on $\mathfrak{h}^{\mathbb{C}} \cap \mathfrak{g} = \mathfrak{h}$, we have $\bar{\mathfrak{g}}_\alpha = \mathfrak{g}_{-\alpha}$. Thus

$$\overline{H_{\mathbb{R}} + \sum_{\alpha>0} X_{-\alpha}} = -H_{\mathbb{R}} + \sum_{\alpha>0} \bar{X}_{-\alpha} \in -H_{\mathbb{R}} + \mathfrak{n}^+,$$

and (6.6) follows, since members of $\mathfrak{g}$ equal their conjugates.] Next, the real dimension of $\mathfrak{h}_{\mathbb{R}} \oplus \mathfrak{n}^-$ is half the real dimension of $\mathfrak{h}^{\mathbb{C}} \oplus \mathfrak{n}^+ \oplus \mathfrak{n}^- = \mathfrak{g}^{\mathbb{C}}$, and hence

$$\dim_{\mathbb{R}}(\mathfrak{g} \oplus \mathfrak{h}_{\mathbb{R}} \oplus \mathfrak{n}^-) = \dim_{\mathbb{R}} \mathfrak{g}^{\mathbb{C}}. \tag{6.7}$$

Combining (6.6) and (6.7), we see that

$$\mathfrak{g}^{\mathbb{C}} = \mathfrak{g} \oplus \mathfrak{h}_{\mathbb{R}} \oplus \mathfrak{n}^-.$$

Then Lemma 5.11 shows that $GT_{\mathbb{R}}N^- = GB$ is open in $G^{\mathbb{C}}$.

Let A be a particular A component for the Iwasawa decomposition of G to be specified shortly. We shall show below that this A satisfies

$$A \subseteq P^+K^{\mathbb{C}}P^- \tag{6.8}$$

and $\qquad$ P^+ components of members of A are bounded. $\qquad$ (6.9)

In any event, we have $G = KAK$. Since $\mathfrak{b} \subseteq \mathfrak{k}^{\mathbb{C}} \oplus \mathfrak{p}^-$, we have $B \subseteq K^{\mathbb{C}}P^-$. Thus (6.8) gives

$$GB \subseteq GK^{\mathbb{C}}P^- \subseteq KAKK^{\mathbb{C}}P^- \subseteq KP^+K^{\mathbb{C}}P^-K^{\mathbb{C}}P^- = P^+K^{\mathbb{C}}P^- \tag{6.10}$$

since $K^{\mathbb{C}}$ normalizes P^+ and P^-. From §5.7 we have $K^{\mathbb{C}} = KB_K$, and thus

$$GK^{\mathbb{C}}P^- = GKB_KP^- \subseteq GB_KP^- \subseteq GB. \tag{6.11}$$

Inclusions (6.10) and (6.11) together say

$$GB = GK^{\mathbb{C}}P^- \subseteq P^+K^{\mathbb{C}}P^-.$$

Thus $GB = GK^{\mathbb{C}}P^- = \Omega K^{\mathbb{C}}P^-$ for some open set Ω in P^+.

To see that Ω is bounded, we need to show the P^+ component of members of G remains bounded. Since $G = KAK$ with K normalizing P^+

and P^-, we see that we are to prove that $\|\log P^+(A)\|$ is bounded, and this is just (6.9). Thus the theorem comes down to proving (6.8) and (6.9), which we do in the two lemmas to follow.

Lemma 6.4. There exist positive noncompact roots $\gamma_1, \ldots, \gamma_s$ and normalizations of root vectors E_{γ_j} in $\mathfrak{g}_{\gamma_j}$, $1 \le j \le s$, such that

$$\mathfrak{a} = \sum_{j=1}^{s} \mathbb{R}(E_{\gamma_j} - \theta E_{\gamma_j})$$

is a maximal abelian subspace of $\mathfrak{p}$. (Here θ is 1 on $\mathfrak{k} \oplus i\mathfrak{p}$ and -1 on $i\mathfrak{k} \oplus \mathfrak{p}$.)

Proof in $\mathfrak{su}(n, m)$, $n \ge m$. Take

$$\mathfrak{a} = \begin{array}{c} \begin{array}{ccc} m & n-m & m \end{array} \\ \left(\begin{array}{c|c|c} 0 & 0 & \begin{matrix} & & t_1 \\ & \cdot^{\cdot^{\cdot}} & \\ t_m & & \end{matrix} \\ \hline 0 & 0 & 0 \\ \hline \begin{matrix} & & t_m \\ & \cdot^{\cdot^{\cdot}} & \\ t_1 & & \end{matrix} & 0 & 0 \end{array}\right) \begin{array}{l} m \\ n-m, \\ m \end{array} \end{array} \qquad \text{real entries.}$$

Lemma 6.5. With notation as in Lemma 6.4, if $Z = \sum_{j=1}^{s} t_j(E_{\gamma_j} - \theta E_{\gamma_j})$ is in $\mathfrak{a}$, then

$$\exp Z = \exp X_0 \exp H_0 \exp Y_0 \tag{6.12}$$

with

$$X_0 = \sum (\tanh t_j) E_{\gamma_j} \in \mathfrak{p}^+, \qquad Y_0 = -\sum (\tanh t_j)\theta E_{\gamma_j} \in \mathfrak{p}^-,$$
$$H_0 = -\sum (\log \cosh t_j)[E_{\gamma_j}, \theta E_{\gamma_j}] \in \mathfrak{h}_{\mathbb{R}} \subseteq \mathfrak{k}^{\mathbb{C}}. \tag{6.13}$$

Moreover, the P^+ components $\exp X_0$ of $\exp Z$ remain bounded.

Proof. For $SU(1, 1) \subseteq SL(2, \mathbb{C})$ the decomposition (6.12) is a special case of (6.2):

$$\begin{pmatrix} \cosh t & \sinh t \\ \sinh t & \cosh t \end{pmatrix} = \begin{pmatrix} 1 & \tanh t \\ 0 & 1 \end{pmatrix} \begin{pmatrix} (\cosh t)^{-1} & 0 \\ 0 & \cosh t \end{pmatrix} \begin{pmatrix} 1 & 0 \\ \tanh t & 1 \end{pmatrix}.$$

Here we are using $E_\gamma = \begin{pmatrix} 0 & 1 \\ 0 & 0 \end{pmatrix}$ and $\theta E_\gamma = \begin{pmatrix} 0 & 0 \\ -1 & 0 \end{pmatrix} \in \mathfrak{g}_{-\gamma}$. We can imbed this result in $G^{\mathbb{C}}$ for each j ($1 \le j \le s$) since the inclusion

$$\mathfrak{sl}(2, \mathbb{C}) \cong \mathbb{C}H_{\gamma_j} + \mathbb{C}E_{\gamma_j} + \mathbb{C}E_{-\gamma_j} \subseteq \mathfrak{g}^{\mathbb{C}}$$

induces a homomorphism $SL(2, \mathbb{C}) \to G^{\mathbb{C}}$, $SL(2, \mathbb{C})$ being simply connected. This imbedding handles each term separately in Z. Now the fact that $\mathfrak{a}$ is abelian means that E_{γ_i} and $E_{\pm\gamma_j}$ commute for $i \neq j$. Thus the contributions to X_0, Y_0, and H_0 for γ_i commute with those for γ_j, and the formula (6.12) follows for general Z.

Finally in the expression for X_0, the coefficients of each E_{γ_j} lie between -1 and $+1$ for all Z. Hence $\exp X_0$ remains bounded.

§4. Holomorphic Discrete Series

We continue with the notation of §3. The linear connected reductive group G is assumed to have $Z_{\mathfrak{g}}(\mathfrak{c}) = \mathfrak{k}$.

If λ is analytically integral in $(\mathfrak{h}^{\mathbb{C}})'$, let ξ_λ be the corresponding holomorphic one-dimensional representation of $T^{\mathbb{C}}$. Extend ξ_λ to a holomorphic one-dimensional representation of B by defining ξ_λ to be 1 on N^-. Now define

$$\Gamma(\lambda) = \left\{ F: GB \to \mathbb{C} \;\middle|\; \begin{array}{l} \text{(i) } F \text{ is holomorphic} \\ \text{(ii) } F(xb) = \xi_\lambda(b)^{-1} F(x) \text{ for } x \in GB,\ b \in B \end{array} \right\}$$

$$V_\lambda = \left\{ F \in \Gamma(\lambda) \;\middle|\; \|F\|^2 = \int_G |F(g)|^2 \, dg < \infty \right\}.$$

$$L(g)F(x) = F(g^{-1}x) \qquad \text{for } F \in \Gamma(\lambda),\ g \in G,\ x \in GB.$$

Then $L(g)$ preserves both $\Gamma(\lambda)$ and V_λ and acts in unitary fashion on V_λ. The inner product on V_λ is clearly definite but not clearly complete.

Theorem 6.6. Let G be linear connected reductive with $Z_{\mathfrak{g}}(\mathfrak{c}) = \mathfrak{k}$, and suppose that λ is analytically integral on $\mathfrak{h}^{\mathbb{C}}$ and is dominant with respect to Δ_K^+. Then V_λ is a Hilbert space and L is a (continuous) unitary representation on it. If furthermore

$$\langle \lambda + \delta, \alpha \rangle < 0 \qquad \text{for all } \alpha \in \Delta_n^+ \qquad \left(\text{with } \delta = \tfrac{1}{2} \sum_{\alpha \in \Delta^+} \alpha \right),$$

then V_λ is nonzero, L is irreducible, the matrix coefficients of L are square integrable, and the irreducible representation of K with highest weight λ occurs in $L|_K$.

Example of SU(1, 1). Here

$$\mathfrak{h} = \begin{pmatrix} i\theta & 0 \\ 0 & -i\theta \end{pmatrix}, \quad \mathfrak{h}^{\mathbb{C}} = \begin{pmatrix} a & 0 \\ 0 & -a \end{pmatrix}, \quad e_1 \begin{pmatrix} a & 0 \\ 0 & -a \end{pmatrix} = a, \quad \mathfrak{b} = \begin{pmatrix} a & 0 \\ c & -a \end{pmatrix},$$

and also $\lambda = -ne_1$, $\alpha = 2e_1$, $\delta = \frac{1}{2}\alpha = e_1$. The form λ is analytically integral if n is an integer, and it is automatically dominant for Δ_K^+ since there are no compact roots. The form $\lambda + \delta$ is $-(n-1)e_1$, and the condition $\langle \lambda + \delta, \alpha \rangle < 0$ means $-(n-1) < 0$, i.e., $n > 1$. Therefore $\xi_\lambda \begin{pmatrix} \gamma^{-1} & 0 \\ 0 & \gamma \end{pmatrix} = \gamma^n$ with $n \geq 2$. In the realization of $\mathscr{D}_n^-$ in §1, the constant function 1 plays a special role in that

$$\frac{1}{2\pi} \int_0^{2\pi} f(e^{2i\theta} z)\, d\theta = f(0),$$

i.e.,

$$\frac{1}{2\pi} \int_0^{2\pi} e^{-in\theta} \mathscr{D}_n^- \begin{pmatrix} e^{-i\theta} & 0 \\ 0 & e^{i\theta} \end{pmatrix} f(z)\, d\theta = f(0).$$

Here the constant function (of z) appears on the right. We translate this formula into Harish-Chandra's realization by means of the correspondence $F \leftrightarrow f$ of §1, where $F(z, \gamma, \zeta) = \gamma^{-n} f(z)$. If $\psi_n \leftrightarrow 1$, then ψ_n is given by $\psi_n(z, \gamma, \zeta) = \gamma^{-n}$, and we obtain

$$\frac{1}{2\pi} \int_0^{2\pi} F\left(\begin{pmatrix} e^{i\theta} & 0 \\ 0 & e^{-i\theta} \end{pmatrix} x \begin{pmatrix} e^{-i\theta} & 0 \\ 0 & e^{i\theta} \end{pmatrix} \right) d\theta = F(1)\psi_n(x) \qquad \text{for } x \in GB.$$

A generalization of this "constant function" ψ_n will play the same key role in the general case that ψ_n does here (and ψ_λ did in the Borel-Weil Theorem). See Lemma 6.8 below.

Returning to the general case of the theorem, let Φ_λ be an irreducible unitary representation of K with highest weight λ. To fix the notation, let us say that Φ_λ acts in a vector space V and that v_λ is a highest weight vector of norm one. We extend Φ_λ to a holomorphic representation of $K^{\mathbb{C}}$ by means of Lemma 5.30. In the product decomposition $P^+ K^{\mathbb{C}} P^- \subseteq G^{\mathbb{C}}$, let $\mu(\cdot)$ denote the middle component.

Lemma 6.7. For v in V the function ψ_v defined on GB by

$$\psi_v(x) = (\Phi_\lambda(\mu(x))^{-1} v, v_\lambda) \tag{6.14}$$

is in $\Gamma(\lambda)$ and satisfies

$$L(k)\psi_v = \psi_{\Phi_\lambda(k)v} \qquad \text{for } k \in K. \tag{6.15}$$

Remark. ψ_v is well defined because $GB \subseteq P^+ K^{\mathbb{C}} P^-$, by Theorem 6.3.

Proof. Formula (6.15) and the transformation law under B_K on the right are proved as in Lemma 5.31. It is trivial that $\psi_v(x)$ is right invariant under P^-. Passing to differentials and back to the group, we see therefore that ψ_v transforms correctly under B on the right. Since ψ_v is clearly holomorphic, the lemma follows.

Now define

$$\psi_\lambda(x) = \psi_{v_\lambda}(x) = (\Phi_\lambda(\mu(x))^{-1}v_\lambda, v_\lambda).$$

Lemma 6.8. For F in $\Gamma(\lambda)$ and x in GB,

$$\int_T F(txt^{-1})\,dt = F(1)\psi_\lambda(x).$$

Proof. The same as for Lemma 5.32, but with M replaced by T.

We can deduce the following consequences of Lemma 6.8, as in §5.7.

(1) $|F(1)| \le \|\psi_\lambda\|^{-1}\|F\|$ for all F in $\Gamma(\lambda)$, by the same proof as for (1) in §5.7.

(2) $V_\lambda \ne 0$ if and only if $\|\psi_\lambda\| < \infty$. [In fact, ψ_λ is in V_λ if $\|\psi_\lambda\| < \infty$. If $\|\psi_\lambda\| = \infty$, then (1) shows F and all functions $L(g)F$ vanish at 1. Hence $F = 0$.]

(3) To each compact set $E \subseteq GB$ corresponds a constant $C_E < \infty$ such that

$$|F(x)| \le C_E\|F\|$$

for all F in V_λ and x in E, by the same proof as for (2) in §5.7.

(4) V_λ is complete, by the same proof as for (3) in §5.7.

(5) If $\|\psi_\lambda\| < \infty$, then V_λ is irreducible and the map $v \to \psi_v$ of V into $\Gamma(\lambda)$ carries V into V_λ and exhibits the K representation Φ_λ as occurring in $L|_K$. [In fact, the irreducibility is proved as in (4) of §5.7. Then (6.15) says that $\psi_{\Phi_\lambda(k)v_\lambda}$ is in V_λ for all k; since these elements are ψ_v for a set of v in V that spans V, all ψ_v are in V_λ. Formula (6.15) shows Φ_λ occurs in $L|_K$.

(6) If $\|\psi_\lambda\| < \infty$, then $(L(x)\psi_\lambda, \psi_\lambda) = \psi_\lambda(x^{-1})\|\psi_\lambda\|^2$, so that $(L(x)\psi_\lambda, \psi_\lambda)$ is in $L^2(G)$ and every matrix coefficient of L on V_λ is square integrable. [In fact, application of Lemma 6.8 to $L(x)\psi_\lambda$ and use of the identity $\psi_\lambda(txt^{-1}) = \psi_\lambda(x)$ for $t \in T$ gives

$$\begin{aligned}(L(x)\psi_\lambda, \psi_\lambda) &= \int_G \psi_\lambda(x^{-1}y)\overline{\psi_\lambda(y)}\,dy \\ &= \int_G \psi_\lambda(x^{-1}tyt^{-1})\overline{\psi_\lambda(y)}\,dy \qquad \text{for all } t \in T \\ &= \int_{G\times T} \psi_\lambda(x^{-1}tyt^{-1})\overline{\psi_\lambda(y)}\,dt\,dy \\ &= \int_G \psi_\lambda(x^{-1})\psi_\lambda(y)\overline{\psi_\lambda(y)}\,dy \\ &= \psi_\lambda(x^{-1})\|\psi_\lambda\|^2.\end{aligned}$$

Once a single nonzero matrix coefficient of an irreducible unitary representation is square integrable, all matrix coefficients are square integrable, as we shall see in Proposition 9.6.]

Looking over these six facts and comparing them with the statement of Theorem 6.6, we see that we have reduced matters to proving the following:

Lemma 6.9. If $\langle \lambda + \delta, \alpha \rangle < 0$ for all $\alpha \in \Delta_n^+$, then $\int_G |\psi_\lambda(x)|^2 \, dx < \infty$.

We take up this step in the next section.

§5. Finiteness of an Integral

The proof of Theorem 6.6 has been reduced to proving Lemma 6.9. The idea in evaluating the integral in that lemma is to use the integral formula for the KAK decomposition (Proposition 5.28) and Schur orthogonality to simplify the integral to an integral over the positive Weyl chamber in $\mathfrak{a}$. The integrand will involve the character of the holomorphic extension of Φ_λ. When we substitute from the Weyl character formula and make an appropriate change of variables, some miraculous cancellation occurs that allows us to handle the integral.

We recall that $G = K(\exp \overline{\mathfrak{a}^+})K$, where $\mathfrak{a}^+$ is the positive Weyl chamber in $\mathfrak{a}$. Proposition 5.28 says that Haar measure dx on G can be normalized so as to be given relative to this decomposition as

$$dx = D_A(a)\, dk_1\, da\, dk_2, \qquad \text{where } D_A(a) = \prod_{\lambda \in \Sigma^+} (\sinh \lambda(\log a))^{\dim \mathfrak{g}_\lambda}. \tag{6.16}$$

Lemma 6.10. With Haar measure on G normalized as in (6.16),

$$\|\psi_\lambda\|^2 = d_\lambda^{-2} \int_{\exp \mathfrak{a}^+} \operatorname{Tr}(\Phi_\lambda(\mu(a)^{-2}))D_A(a)\, da, \tag{6.17}$$

where d_λ is the degree of Φ_λ.

Proof. Choose an orthonormal basis $v_1, \ldots, v_d$ in V with $v_1 = v_\lambda$. Since $\mu(kak') = k\mu(a)k'$, we have

$$\begin{aligned} \|\psi_\lambda\|^2 = \int_G |\psi_\lambda(x)|^2\, dx &= \int_{K \times \exp \mathfrak{a}^+ \times K} |(\Phi_\lambda(k\mu(a)k')^{-1}v_\lambda, v_\lambda)|^2 D_A(a)\, dk'\, da\, dk \\ &= d_\lambda^{-1} \int_{K \times \exp \mathfrak{a}^+} \|\Phi_\lambda(\mu(a))^{-1}\Phi_\lambda(k)v_\lambda\|^2 D_A(a)\, dk\, da \quad \text{by Corollary 1.9.} \end{aligned} \tag{6.18}$$

Write $\Phi_\lambda(k)v_j = \sum_i \Phi_{ij}(k)v_i$ and put $j = 1$. Then (6.18) is

$$\begin{aligned} &= d_\lambda^{-1} \int_{K \times \exp \mathfrak{a}^+} \sum_{i,l} (\Phi_\lambda(\mu(a))^{-1}v_i, \Phi_\lambda(\mu(a))^{-1}v_l)\Phi_{i1}(k)\overline{\Phi_{l1}(k)}D_A(a)\, dk\, da \\ &= d_\lambda^{-2} \int_{\exp \mathfrak{a}^+} \sum_i \|\Phi_\lambda(\mu(a))^{-1}v_i\|^2 D_A(a)\, da \qquad \text{by Corollary 1.9} \\ &= d_\lambda^{-2} \int_{\exp \mathfrak{a}^+} \sum_i (\Phi_\lambda(\mu(a))^{-2}v_i, v_i)D_A(a)\, da \end{aligned} \tag{6.19}$$

since $\mu(a)$ is in $T_{\mathbb{R}}$, by Lemma 6.5. The right side of (6.19) we recognize as the right side of (6.17), and the lemma is proved.

The application of Lemma 6.10 to prove Lemma 6.9 is rather complicated, requiring detailed relationships between Σ^+ and Δ. We shall do the evaluation in the case of SU(2, 2), which captures most features of the general case. Here we have

$$a = \begin{pmatrix} \cosh t_1 & 0 & 0 & \sinh t_1 \\ 0 & \cosh t_2 & \sinh t_2 & 0 \\ 0 & \sinh t_2 & \cosh t_2 & 0 \\ \sinh t_1 & 0 & 0 & \cosh t_1 \end{pmatrix}$$

and

$$\mu(a)^{-2} = \begin{pmatrix} (\cosh t_1)^2 & & & \\ & (\cosh t_2)^2 & & \\ & & (\cosh t_2)^{-2} & \\ & & & (\cosh t_1)^{-2} \end{pmatrix}. \tag{6.20}$$

In these coordinates

$$D_A(a) = \sinh 2t_1 \sinh 2t_2 \sinh^2(t_1 + t_2) \sinh^2(t_1 - t_2), \tag{6.21}$$

and the integration over $\exp \mathfrak{a}^+$ is taken over $0 < t_2 < t_1$.

The integrand contains $\operatorname{Tr} \Phi_\lambda(\mu(a))^{-2}$, and $\mu(a)$ is in $T^{\mathbb{C}}$. Thus the Weyl character formula is applicable. Allowing "a" to have a second meaning, let

$$\lambda = ae_1 + be_2 + ce_3.$$

Integrality means a, b, c are in $\mathbb{Z}$, and dominance with respect to

$$\Delta_K^+ = \{e_1 - e_2, e_3 - e_4\}$$

means $a \geq b$ and $c \geq 0$. If δ_K denotes the δ for Δ_K^+, we have

$$\delta_K = \tfrac{1}{2}(e_1 - e_2) + \tfrac{1}{2}(e_3 - e_4) = e_1 + e_3.$$

For the matrix $h = \operatorname{diag}(h_1, h_2, h_3, h_4)$ in $T^{\mathbb{C}}$ the Weyl character formula says

$$\begin{aligned} \operatorname{Tr} \Phi_\lambda(h) &= \frac{h_1^{a+1} h_2^b h_3^{c+1} - h_1^b h_2^{a+1} h_3^{c+1} - h_1^{a+1} h_2^b h_4^{c+1} + h_1^b h_2^{a+1} h_4^{c+1}}{h_1 h_3 (1 - h_2 h_1^{-1})(1 - h_4 h_3^{-1})} \\ &= \frac{\text{numerator}}{h_1 h_3 (1 - h_2 h_1^{-1})(1 - h_4 h_3^{-1})}, \text{ say.} \end{aligned} \tag{6.22}$$

Substituting for h from (6.20) and for $D_A(a)$ from (6.21), we find that the integral in Lemma 6.10 is

$$= \iint_{0<t_2<t_1} \frac{\text{numerator} \times \sinh 2t_1 \sinh 2t_2 \sinh^2(t_1+t_2)\sinh^2(t_1-t_2)\,dt_1\,dt_2}{\left(\dfrac{\cosh t_1}{\cosh t_2}\right)^2\left(1-\dfrac{\cosh^2 t_2}{\cosh^2 t_1}\right)^2}. \tag{6.23}$$

Since $\qquad \cosh^2 t_1 - \cosh^2 t_2 = \sinh(t_1+t_2)\sinh(t_1-t_2),$

the denominator in the integrand of (6.23) is

$$\begin{aligned} &= (\cosh t_1)^{-2}(\cosh t_2)^{-2}(\cosh^2 t_1 - \cosh^2 t_2)^2 \\ &= (\cosh t_1)^{-2}(\cosh t_2)^{-2}\sinh^2(t_1+t_2)\sinh^2(t_1-t_2). \end{aligned}$$

Therefore (6.23) is

$$= 4 \iint_{0<t_2<t_1} \text{numerator} \times \sinh t_1 \sinh t_2 \cosh^3 t_1 \cosh^3 t_2\,dt_1\,dt_2.$$

Each term of the numerator is of the form $\pm(\cosh t_1)^{N_1}(\cosh t_2)^{N_2}$, by (6.22) and (6.20), and we have

$$\begin{aligned} &\iint_{0<t_2<t_1} \sinh t_1 \sinh t_2 (\cosh t_1)^{N_1+3}(\cosh t_2)^{N_2+3}\,dt_1\,dt_2 \\ &\quad = \int_{t_2=0}^{\infty} \sinh t_2 (\cosh t_2)^{N_2+3}\left\{\frac{-(\cosh t_2)^{N_1+4}}{N_1+4}\right\}dt_2 \\ &\quad = \frac{1}{(N_1+4)(N_1+N_2+8)} \qquad \text{if } N_1+4<0 \text{ and } N_1+N_2+8<0. \end{aligned}$$

Thus we get a finite integral if all four terms of the numerator have $N_1+4<0$ and $N_1+N_2+8<0$. The four terms have, respectively,

$$\begin{aligned} N_1 &= 2(a+1) & N_2 &= 2(b-c-1) \\ N_1 &= 2b & N_2 &= 2(a-c) \\ N_1 &= 2(a-c) & N_2 &= 2b \\ N_1 &= 2(b-c-1) & N_2 &= 2(a+1). \end{aligned}$$

Since the Δ_K^+ dominance of λ means $a \geq b$ and $c \geq 0$, we see that the largest case for N_1+4 is the first one:

$$N_1 + 4 \leq (2a+2)+4 = 2a+6. \tag{6.24}$$

Also N_1+N_2+8 is the same in all cases:

$$N_1+N_2+8 = 2a+2b-2c+8. \tag{6.25}$$

We have not used the condition $\langle \lambda + \delta, \alpha \rangle < 0$ for $\alpha \in \Delta_n^+$. Here

$$\delta = \tfrac{3}{2}e_1 + \tfrac{1}{2}e_2 - \tfrac{1}{2}e_3 - \tfrac{3}{2}e_4 = 3e_1 + 2e_2 + e_3,$$

so that

$$\lambda + \delta = (a + 3)e_1 + (b + 2)e_2 + (c + 1)e_3.$$

The largest noncompact root is $e_1 - e_4$, and $\langle \lambda + \delta, e_1 - e_4 \rangle < 0$ says $a < -3$. For $a < -3$, (6.24) gives precisely

$$N_1 + 4 \leq 2a + 6 < 0$$

and (6.25) gives

$$N_1 + N_2 + 8 = 2a + 2b - 2c + 8 \leq 2a + 2a - 0 + 8 = 4a + 8 < 0.$$

Thus our integral is finite as asserted.

§6. Problems

1. Let G be $SO(n, 2)$.
   (a) Show that $G^{\mathbb{C}} \cong SO(n + 2, \mathbb{C})$.
   (b) Show that $Z_{\mathfrak{g}}(\mathfrak{c}) = \mathfrak{k}$.
   (c) The isomorphism in (a) identifies the root system of $SO(n, 2)$ as of type $B_{(n+1)/2}$ if n is odd and of type $D_{(n+2)/2}$ if n is even. (See §4.1 for notation.) Identify which roots are compact and which are noncompact.
   (d) Decide on some particular good ordering (in the sense of this chapter), and identify the positive roots.
2. Let G be linear connected reductive, and suppose $\mathfrak{g}$ is a simple Lie algebra. Under these conditions, prove the following:
   (a) Either the center $\mathfrak{c}$ of $\mathfrak{k}$ is 0 or else $Z_{\mathfrak{g}}(\mathfrak{c}) = \mathfrak{k}$. [Hint: Use Problem 10b of Chapter V.]
   (b) If $Z_{\mathfrak{g}}(\mathfrak{c}) = \mathfrak{k}$, then $Z_{\mathfrak{g}}(H) = \mathfrak{k}$ for every nonzero H in $\mathfrak{c}$.
   (c) If $Z_{\mathfrak{g}}(\mathfrak{c}) = \mathfrak{k}$ and α is a root that vanishes on some nonzero H in $\mathfrak{c}$, then α is compact.
   (d) The center $\mathfrak{c}$ of $\mathfrak{k}$ has dimension either 0 or 1.
3. Carry out the calculations of §5 for $Sp(2, \mathbb{R})$ in place of $SU(2, 2)$.

Problems 4 to 8 concern construction of an Iwasawa $\mathfrak{a}_\mathfrak{p}$ (cf. Lemma 6.4). In these problems it is assumed that G is linear connected reductive and that rank $G =$ rank K, i.e., that a maximal abelian subalgebra $\mathfrak{h}$ of $\mathfrak{k}$ is maximal abelian in $\mathfrak{g}$ and hence is a Cartan subalgebra. (A sufficient condition for this assumption to be satisfied is that $Z_{\mathfrak{g}}(\mathfrak{c}) = \mathfrak{k}$.) Under this assumption we can define compact and noncompact roots as in §3. Fix an ordering on $(i\mathfrak{h})'$ and define recursively

$$\Delta_0 = \Delta$$

$$\gamma_j = \text{largest noncompact member of } \Delta_{j-1} \qquad (j \geq 1)$$

$$\Delta_j = \{\alpha \in \Delta_{j-1} \mid \text{neither of } \alpha \pm \gamma_j \text{ is a root}\} \qquad (j \geq 1).$$

4. Show that Δ_j (if nonempty) is a reduced abstract root system in a subspace of $(i\mathfrak{h})'$.
5. For each root α, define $H'_\alpha = 2|\alpha|^{-2}H_\alpha$ as usual. Show for each noncompact α that it is possible to choose root vectors E'_α and $E'_{-\alpha}$ in $\mathfrak{g}_\alpha$ and $\mathfrak{g}_{-\alpha}$, respectively, so that $E'_\alpha + E'_{-\alpha}$ and $i(E'_\alpha - E'_{-\alpha})$ are in $\mathfrak{g}$ and $[E'_\alpha, E'_{-\alpha}] = H'_\alpha$.
6. Fix a choice of E'_α as in Problem 5. Show that $\sum \mathbb{R}(E'_{\gamma_j} + E'_{-\gamma_j})$ is an abelian subspace of $\mathfrak{p}$.
7. Show that $\sum \mathbb{R}(E'_{\gamma_j} + E'_{-\gamma_j})$ is actually maximal abelian in $\mathfrak{p}$. [Hint: Let $\sum_{\beta \in \Delta_n} E_\beta$ be a member of $\mathfrak{p}^{\mathbb{C}}$ commuting with all $E'_{\gamma_j} + E'_{-\gamma_j}$. Show that the terms E_β with $\beta \neq \pm\gamma_j$ must individually commute with $E'_{\gamma_j} + E'_{-\gamma_j}$. Then sort out the result.]
8. Show that the assumption rank $G =$ rank K is satisfied in $G = SO(2n, 1)$ for $n \geq 1$ despite the fact that $Z_{\mathfrak{g}}(\mathfrak{c}) = \mathfrak{k}$ fails for $n > 1$.

Problems 9 to 11 yield a realization of G/K, under the assumption $Z_{\mathfrak{g}}(\mathfrak{c}) = \mathfrak{k}$, as a particularly nice unbounded open subset Ω' of P^+. Let notation be as in the chapter.

9. In the special case that $G = SU(1, 1)$, let $u = \dfrac{1}{\sqrt{2}}\begin{pmatrix} 1 & i \\ i & 1 \end{pmatrix}$, $G' = SL(2, \mathbb{R})$, and $\Omega' = \left\{\begin{pmatrix} 1 & z \\ 0 & 1 \end{pmatrix} \middle| \operatorname{Im} z > 0\right\}$. From (2.20) it is known that $uGu^{-1} = G'$. Prove that $uGB = G'uB = \Omega' K^{\mathbb{C}} P^-$ and that G' acts on Ω' by the usual action of $SL(2, \mathbb{R})$ on the upper half plane. [Note: Because of this effect of u, u is called a **Cayley transform.**]
10. In the general case as in the chapter, let $\gamma_1, \ldots, \gamma_s$ be constructed as in Problems 4 to 8. For each j, construct an element u_j in $G^{\mathbb{C}}$ that behaves for the three-dimensional group corresponding to γ_j like the element u of Problem 9. Put $u = \prod_{j=1}^s u_j$.
    (a) Exhibit u as in $P^+K^{\mathbb{C}}P^-$. [Hint: Handle each u_j separately by a calculation in $SU(1, 1)$, and show that enough things commute so that a decomposition of u results.]
    (b) Let $\mathfrak{a}_{\mathfrak{p}}$ be the maximal abelian subspace of $\mathfrak{p}$ constructed in Problems 7 and 8, and let $A_{\mathfrak{p}} = \exp \mathfrak{a}_{\mathfrak{p}}$. Show that $uA_{\mathfrak{p}}u^{-1} \subseteq K^{\mathbb{C}}$.
    (c) Show for a particular ordering on $\mathfrak{a}'_{\mathfrak{p}}$ that $uN_{\mathfrak{p}}u^{-1} \subseteq P^+K^{\mathbb{C}}$.

(d) Writing $G = N_\mathfrak{p} A_\mathfrak{p} K$ by the Iwasawa decomposition, prove that $uGB \subseteq P^+ K^\mathbb{C} P^-$.

11. Let $g' = uGu^{-1}$. Prove that $G'uB = \Omega' K^\mathbb{C} P^-$ for some open subset Ω' of P^+. Prove also that the resulting action of G' on Ω' is holomorphic and transitive, and identify Ω' with G/K.

Problems 12 to 16 give further properties of linear connected reductive groups with rank G = rank K. Let $\mathfrak{h} \subseteq \mathfrak{k}$ be a Cartan subalgebra of $\mathfrak{g}$, and form roots, compact and noncompact, as in §3.

12. K acts on $\mathfrak{p}^\mathbb{C}$ via the representation Ad_G. Identify the weights as the noncompact roots, showing in particular that 0 is not a weight.
13. Show that the subalgebras of $\mathfrak{g}^\mathbb{C}$ containing $\mathfrak{k}^\mathbb{C}$ are of the form $\mathfrak{k}^\mathbb{C} \oplus \sum_{\alpha \in E} \mathfrak{g}_\alpha$, for some subset E of noncompact roots.
14. Suppose that $\mathfrak{k}^\mathbb{C} \oplus \sum_{\alpha \in E} \mathfrak{g}_\alpha$ is a subalgebra of $\mathfrak{g}^\mathbb{C}$. Prove that $\mathfrak{k}^\mathbb{C} \oplus \sum_{\alpha \in E} (\mathfrak{g}_\alpha \oplus \mathfrak{g}_{-\alpha})$ and $\mathfrak{k}^\mathbb{C} \oplus \sum_{\alpha \in (E \cap (-E))} \mathfrak{g}_\alpha$ are subalgebras of $\mathfrak{g}^\mathbb{C}$ that are the complexifications of subalgebras of $\mathfrak{g}$.
15. Suppose $\mathfrak{g}$ is a simple Lie algebra. Prove that the representation of K on $\mathfrak{p}^\mathbb{C}$ splits into at most two irreducible pieces. Prove, moreover, that when there are two pieces, there is a positive system Δ^+ of roots such that the irreducible pieces are

$$\sum_{\alpha \in \Delta_n^+} \mathfrak{g}_\alpha \qquad \text{and} \qquad \sum_{\alpha \in \Delta_n^+} \mathfrak{g}_{-\alpha}.$$

[Hint: Use Problem 10b in Chapter V.]

16. Suppose $\mathfrak{g}$ is a simple Lie algebra, and suppose the representation of K on $\mathfrak{p}^\mathbb{C}$ is reducible. Show that the center $\mathfrak{c}$ of $\mathfrak{k}$ is nonzero, that $Z_\mathfrak{g}(\mathfrak{c}) = \mathfrak{k}$, and that the irreducible pieces are $\mathfrak{p}^+$ and $\mathfrak{p}^-$. [Hint: Let Δ^+ be as in Problem 15. Show iH_{δ_n} is in $\mathfrak{c}$, where δ_n is half the sum of the positive noncompact roots. Then apply Problem 2a.]

CHAPTER VII

Induced Representations

§1. Three Pictures

The principal series representations of $SL(2, \mathbb{R})$, and also of $SL(2, \mathbb{C})$, can be realized as so-called "induced representations" in a way that immediately suggests generalization. In the case of $G = SL(2, \mathbb{R})$, let $M_\mathfrak{p} A_\mathfrak{p} N_\mathfrak{p}$ be the upper triangular group, and let $\mathscr{P}^{\pm, iv}$ be given. Define σ on $M_\mathfrak{p} = \{I, -I\}$ and ν on $\mathfrak{a}_\mathfrak{p}$ by

$$\sigma\begin{pmatrix} \varepsilon & 0 \\ 0 & \varepsilon \end{pmatrix} = \begin{cases} \varepsilon & \text{if } - \\ 1 & \text{if } + \end{cases}$$

$$\nu\begin{pmatrix} t & 0 \\ 0 & -t \end{pmatrix} = ivt.$$

Then $man \to e^{\nu \log a}\sigma(m)$ is a representation of $M_\mathfrak{p} A_\mathfrak{p} N_\mathfrak{p}$, and it is this representation that we induce to G.

To do so, we recall that the half-sum ρ of the positive restricted roots, given in (5.16), for this situation satisfies

$$\rho\begin{pmatrix} t & 0 \\ 0 & -t \end{pmatrix} = t,$$

and we form the space

$$\{F \in C(G) \,|\, F(xman) = e^{-(\nu+\rho)\log a}\sigma(m)^{-1}F(x)\}$$

with group operation

$$P^{\pm, iv}(g)F(x) = F(g^{-1}x).$$

The Iwasawa decomposition shows that F is determined by $F|_K$, and we put

$$\|F\|^2 = \int_K |F(k)|^2\, dk.$$

Then we complete the space to get a representation.

To see the equivalence of $P^{\pm, iv}$ with $\mathscr{P}^{\pm, iv}$, let F be as above and let y be in $\mathbb{R}$. Define LF to be essentially the restriction to $\bar{N}_\mathfrak{p} = \Theta N_\mathfrak{p}$:

$$(LF)(y) = F\begin{pmatrix} 1 & 0 \\ y & 1 \end{pmatrix}.$$

It is easy to check that $LP^{\pm,iv}(g) = \mathscr{P}^{\pm,iv}(g)L$. Moreover, L is onto a dense subspace of $L^2(\mathbb{R})$ because for f in $C_{\text{com}}(\mathbb{R})$ we can put

$$F\begin{pmatrix} a & b \\ c & d \end{pmatrix} = \begin{cases} |a|^{-(1+iv)}\sigma\begin{pmatrix} \operatorname{sgn} a & 0 \\ 0 & \operatorname{sgn} a \end{pmatrix} f(c/a) & \text{if } a \neq 0 \\ 0 & \text{if } a = 0 \end{cases}$$

and obtain F as above with $LF = f$. The proof that L preserves norms (if Haar measures are normalized suitably) uses the integral formula (5.25) and goes as follows:

$$\begin{aligned} \|F\|^2 = \int_K |F(k)|^2\,dk &= \int_{\bar{N}_{\mathfrak{p}}} |F(\kappa(\bar{n}))|^2 e^{-2\rho H(\bar{n})}\,d\bar{n} \\ &= \int_{\bar{N}_{\mathfrak{p}}} |F(\bar{n})|^2\,d\bar{n} = \|LF\|^2. \end{aligned} \tag{7.1}$$

Let us now generalize these matters. Let

G = linear connected reductive group with usual notation

$S = MAN$ = Langlands decomposition of a parabolic subgroup

$\bar{N} = \Theta N$

$\rho = \frac{1}{2}\sum_{\Gamma^+} (\dim \mu)\mu$, where $\Gamma^+ = \{\text{roots of } (\mathfrak{g}, \mathfrak{a}) \text{ positive for } N\}$

σ = irreducible unitary representation of M on a space V^σ

ν = member of $(\mathfrak{a}')^{\mathbb{C}}$.

We construct, out of the representation $\sigma \otimes \exp \nu \otimes 1$ of $S = MAN$, a representation of G denoted

$$U(S, \sigma, \nu, \cdot) = \operatorname{ind}_{MAN}^G(\sigma \otimes \exp \nu \otimes 1).$$

There will be three different "pictures," or realizations, of these representations, and the three pictures have different uses.

Induced picture. A dense subspace of the representation space is

$$\{F : G \to V^\sigma \text{ continuous} \mid F(xman) = e^{-(\nu+\rho)\log a}\sigma(m)^{-1}F(x)\}$$

with norm

$$\|F\|^2 = \int_K |F(k)|^2\,dk, \tag{7.2}$$

and G acts by

$$U(S, \sigma, \nu, g)F(x) = F(g^{-1}x).$$

The actual Hilbert space and representation is then obtained by completion. We shall see in the next section that U is actually a representation and is unitary if $\sigma \otimes \exp \nu \otimes 1$ is unitary (i.e., if ν is imaginary). The induced picture has the advantage that the group action is simple.

Compact picture. This is simply the restriction to K of the induced picture. Restriction is one-one since $G = KMAN$. More intrinsically, a dense subspace is

$$\{F: K \to V^\sigma \text{ continuous } | F(km) = \sigma(m)^{-1}F(k) \text{ for } k \in K, m \in K \cap M\}$$

with norm (7.2). If g decomposes under $G = KMAN$ as

$$g = \kappa(g)\mu(g)e^{H(g)}n, \tag{7.3a}$$

then the action is

$$U(S, \sigma, \nu, g)F(k) = e^{-(\nu+\rho)H(g^{-1}k)}\sigma(\mu(g^{-1}k))^{-1}F(\kappa(g^{-1}k)).$$

The action is more complicated than in the induced picture, but the space is independent of ν. Consequently the compact picture allows one to study the dependence of the representation on ν.

Noncompact picture. This is the restriction to $\bar{N}$ of the induced picture. The intuitive reason that restriction is one-one is that $\bar{N}MAN$ exhausts G except for a lower-dimensional set. More intrinsically, the representation space is $L^2(\bar{N}, e^{2\,\mathrm{Re}\,\nu H(\bar{n})}\,d\bar{n})$. If g decomposes under $G \doteq \bar{N}MAN$ as

$$g = \bar{n}(g)m(g)a(g)n, \tag{7.3b}$$

then the action is

$$U(S, \sigma, \nu, g)F(\bar{n}) = e^{-(\nu+\rho)\log a(g^{-1}\bar{n})}\sigma(m(g^{-1}\bar{n}))^{-1}F(\bar{n}(g^{-1}\bar{n})).$$

The proof that norms are preserved in passing from the induced picture to the noncompact picture proceeds exactly as in (7.1) above. The noncompact picture is useful for studying the induced representation by analytic methods.

§2. Elementary Properties

We list some properties of the induced representations of §1 that we shall use repeatedly without explicit reference.

(1) If ν is imaginary, so that $\sigma \otimes \exp \nu \otimes 1$ is unitary, then each $U(S, \sigma, \nu, g)$ is unitary.

Proof. As in the proof of (5.25), choose $\varphi \geq 0$ on G with $\int_{MAN} \varphi(xman)d_l(man) = 1$ for all x in G. For F continuous in the induced space, we have

$$\begin{aligned}
\|U(g)F\|^2 &= \int_K |F(g^{-1}k)|^2\,dk = \int_K |F(g^{-1}k)|^2\left[\int_{MAN} \varphi(kman)d_l(man)\right]dk \\
&= \int_{K\times MAN} |F(g^{-1}k)|^2\varphi(kman)e^{-2\rho\log a}\,dk\,d_r(man) \qquad \text{by (5.12)} \\
&= \int_{K\times MAN} |F(g^{-1}kman)|^2\varphi(kman)\,dk\,d_r(man) \\
&= \int_G |F(g^{-1}x)|^2\varphi(x)\,dx \qquad \text{by (5.18)} \\
&= \int_G |F(x)|^2\varphi(gx)\,dx \\
&= \int_{K\times MAN} |F(k)|^2 e^{-2\rho\log a}\varphi(gkman)\,dk\,d_r(man) \qquad \text{by (5.18)} \\
&= \int_K |F(k)|^2\left[\int_{MAN} \varphi(gkman)d_l(man)\right]dk = \|F\|^2.
\end{aligned}$$

(2) Let MAN be a parabolic subgroup. For f in $C(K)$ right invariant under $K \cap M$ and for g in G, the equality

$$\int_K e^{-2\rho H(g^{-1}k)} f(\kappa(g^{-1}k))\,dk = \int_K f(k)\,dk \tag{7.4}$$

holds. (The notation is as in (5.24) or (7.3a).)

Proof. This follows from (1) with σ trivial and $\nu = 0$.

(3) For any ν, $U(S, \sigma, \nu, g)$ is a (continuous) representation.

Proof. Let F be a continuous function in the induced space. Then

$$\begin{aligned}
\|U(g)F\|^2 &= \int_K e^{-2\,\mathrm{Re}\,\nu H(g^{-1}k)} e^{-2\rho H(g^{-1}k)} |F(\kappa(g^{-1}k))|^2\,dk \\
&\le \left\{\sup_{k\in K} e^{-2\,\mathrm{Re}\,\nu H(g^{-1}k)}\right\} \int_K e^{-2\rho H(g^{-1}k)} |F(\kappa(g^{-1}k))|^2\,dk \\
&= \left\{\sup_{k\in K} e^{-2\,\mathrm{Re}\,\nu H(g^{-1}k)}\right\} \|F\|^2 \qquad \text{by (7.4).}
\end{aligned}$$

Hence each $U(g)$ is bounded. It is clear that $U(g_1g_2) = U(g_1)U(g_2)$, and an easy argument with dominated convergence shows

$$\lim_{g\to 1} \|U(g)F - F\|^2 = 0$$

for F continuous. Hence U is a representation.

(4) **(Double induction formula)** Let MAN be a parabolic subgroup of G and let $M_*A_*N_*$ be a parabolic subgroup of M, so that $M_*(A_*A)(N_*N)$ is a parabolic subgroup of G. If σ is a unitary representation of M_* and

ν_* and ν are in $(\mathfrak{a}'_*)^{\mathbb{C}}$ and $(\mathfrak{a}')^{\mathbb{C}}$, respectively, then there is a canonical equivalence

$$\begin{aligned} &\operatorname{ind}_{MAN}^{G}(\operatorname{ind}_{M_*A_*N_*}^{M}(\sigma \otimes \exp \nu_* \otimes 1) \otimes \exp \nu \otimes 1) \\ &\quad \cong \operatorname{ind}_{M_*(A_*A)(N_*N)}^{G}(\sigma \otimes \exp(\nu_* + \nu) \otimes 1), \end{aligned} \tag{7.5}$$

and the equivalence preserves norms. Here $\operatorname{ind}_{M_*A_*N_*}^{M}(\sigma \otimes \exp \nu_* \otimes 1)$ may not be an irreducible unitary representation of M, but (7.5) is still valid if one makes the obvious definitions.

Remarks. The representation on the right side of (7.5) acts on functions on G with values in V^σ. The representation on the left side acts on functions on G whose values are V^σ-valued functions on M. The latter might be written as $F(g, m)$; the equivalence that maps the left side of (7.5) to the right side carries $F(\cdot, \cdot)$ to $F(\cdot, 1)$. It is not hard to check the asserted properties.

Proposition 7.1. Let $G = SL(n, \mathbb{C})$, and let MAN be the upper triangular subgroup. For any σ (necessarily one-dimensional) and for any imaginary ν, $U(MAN, \sigma, \nu)$ is irreducible.

Idea of proof. We proceed inductively, arguing in part as in Proposition 2.6. We sketch some details for $n = 3$. The representation $U(MAN, \sigma, \nu)$ can be regarded as induced from the parabolic subgroup

$$M^*A^*N^* = \left(\begin{array}{cc|c} \cdot & \cdot & \cdot \\ \cdot & \cdot & \cdot \\ \hline 0 & 0 & \cdot \end{array}\right)$$

because of the double induction formula. The group M^* is

$$M^* \cong \{g \in GL(2, \mathbb{C}) \mid |\det g| = 1\},$$

and the inducing representation is essentially a principal series representation $\mathscr{P}$ on M^* and is $\exp(\nu|_{\mathfrak{a}^*})$ on A^*.

We can thus view $U(MAN, \sigma, \nu)$ in the noncompact picture relative to $M^*A^*N^*$, i.e., as acting on square integrable functions on $\bar{N}^* \cong \mathbb{R}^2$ with values in the space $V^{\mathscr{P}}$ on which $\mathscr{P}$ acts. The point is to show that the restriction of $U(MAN, \sigma, \nu)$ to $M^*A^*\bar{N}^*$ is irreducible by arguing as in Proposition 2.6.

As in that proof, a bounded operator B commuting with $\bar{N}^*$ is given on the Fourier transform side by an $\operatorname{End}(V^{\mathscr{P}})$-valued L^∞ function b. We may regard the domain of b as the dual vector space to $\mathbb{R}^2 \cong \bar{N}^*$. Since B commutes with M^*A^* and since M^*A^* acts transitively on $\bar{N}^* - \{I\}$, b is determined by its value at any one point. Evaluating b at y_0, where y_0 is the linear functional that picks off the (3, 2) entry of $\bar{N}^*$, we find that

$b(y_0)$, which is an operator on $V^{\mathscr{P}}$, commutes with $\mathscr{P}$ on the lower triangular subgroup of M^*. The proof of Proposition 2.6 showed that $\mathscr{P}$ is irreducible on the lower triangular subgroup (of $SL(2, \mathbb{C})$), and hence $b(y_0)$ is scalar. Unwinding matters, we find that B is scalar. Hence $U(MAN, \sigma, \nu)$ is irreducible.

§3. Bruhat Theory

In $SL(n, \mathbb{C})$ even though the upper triangular group MAN is fairly small, the induced representations of Proposition 5.1 are irreducible. In $SL(2, \mathbb{R})$ in the analogous situation we know that $\mathscr{P}^{-,0}$ is reducible and the other unitary principal series representations are irreducible. We can ask what happens in general.

In this section we shall study the representations $U(S_\mathfrak{p}, \sigma, \nu)$ of G with $S_\mathfrak{p}$ minimal parabolic and with ν imaginary. These comprise the **principal series** of representations of G. The name **nonunitary principal series** will refer to $U(S_\mathfrak{p}, \sigma, \nu)$ for general ν. Since $M_\mathfrak{p}$ is compact, σ is in any case finite-dimensional.

Motivation. Suppose G is a finite group, π_1 and π_2 are irreducible representations of respective subgroups Γ_1 and Γ_2 on spaces V^{π_1} and V^{π_2}, and L is an **intertwining operator** between

$$U_1 = \operatorname{ind}_{\Gamma_1}^G \pi_1 \qquad \text{and} \qquad U_2 = \operatorname{ind}_{\Gamma_2}^G \pi_2,$$

i.e., a linear operator with $LU_1(g) = U_2(g)L$ for all g in G. We use integrals in place of sums even though the group is finite.

For $f: G \to V^{\pi_1}$, the function

$$(p_1 f)(x) = \int_{\Gamma_1} \pi_1(\gamma_1')f(x\gamma_1')\,d\gamma_1' \tag{7.6}$$

is in the space of U_1. Since $f \to L(p_1 f)(1)$ is a linear map from V^{π_1}-valued functions on G into V^{π_2}, we can express L as

$$L(p_1 f)(1) = \int_G l(y)f(y)\,dy \tag{7.7}$$

for a function $l: G \to \operatorname{Hom}(V^{\pi_1}, V^{\pi_2})$. For γ_1 in Γ_1, let f' denote the function $f'(x) = \pi_1(\gamma_1)f(x\gamma_1)$. Then

$$p_1 f'(x) = \int_{\Gamma_1} \pi_1(\gamma_1')\pi_1(\gamma_1)f(x\gamma_1'\gamma_1)\,d\gamma_1' = \int_{\Gamma_1} \pi_1(\gamma_1')f(x\gamma_1')\,d\gamma_1' = p_1 f(x).$$

Hence

$$\begin{aligned}\int_G l(y)f(y)\,dy &= L(p_1 f)(1) = L(p_1 f')(1) \\ &= \int_G l(y)\pi_1(\gamma_1)f(y\gamma_1)\,dy = \int_G l(y\gamma_1^{-1})\pi_1(\gamma_1)f(y)\,dy,\end{aligned}$$

and we obtain
$$l(y) = l(y\gamma_1^{-1})\pi_1(\gamma_1). \tag{7.8}$$
On the other hand, if γ_2 is in Γ_2, then the intertwining property of L gives

$$\begin{aligned}\pi_2(\gamma_2^{-1}) \int_G l(y)f(y)\,dy &= \pi_2(\gamma_2^{-1})L(p_1f)(1) = L(p_1f)(\gamma_2)\\ &= U_2(\gamma_2)^{-1}L(p_1f)(1) = L(U_1(\gamma_2)^{-1}p_1f)(1)\\ &= \int_G l(y)f(\gamma_2 y)\,dy = \int_G l(\gamma_2^{-1}y)f(y)\,dy\end{aligned}$$

and hence
$$\pi_2(\gamma_2^{-1})l(y) = l(\gamma_2^{-1}y). \tag{7.9}$$
Properties (7.8) and (7.9) may be combined as
$$l(\gamma_2 y\gamma_1) = \pi_2(\gamma_2)l(y)\pi_1(\gamma_1). \tag{7.10}$$

Property (7.10) says that l is determined by its values on a set of double coset representatives of $\Gamma_2\backslash G/\Gamma_1$. And it says more. We can rewrite (7.10) as
$$\pi_2(\gamma_2)l(\gamma_2^{-1}y\gamma_1) = l(y)\pi_1(\gamma_1).$$
That is, $l(y)$ is an intertwining operator between the representations
$$\gamma_1 \to \pi_1(\gamma_1) \qquad \text{and} \qquad \gamma_1 \to \pi_2(y\gamma_1 y^{-1}) \tag{7.11}$$
of the group $\Gamma_1 \cap y^{-1}\Gamma_2 y$. Thus L can be analyzed in terms of π_1 and π_2.

Let us return to the case of a linear connected semisimple group G. One would like to extend the above argument so as to handle the case of such a group G when $\Gamma_1 = \Gamma_2 = S_\mathfrak{p}$. Bruhat managed to overcome the serious technical problems involved and arrive at such a theory.

The theory uses smooth functions, initially of compact support, and distributions. The group $S_\mathfrak{p}$ plays the role of both Γ_1 and Γ_2. Although the inducing representation is $\sigma \otimes \exp \nu \otimes 1$, we have to take ρ into account, and the representation
$$\pi(man) = e^{(\rho+\nu)\log a}\sigma(m)$$
plays the role of π_1 and π_2 above. Let L be a bounded linear operator intertwining $U(S_\mathfrak{p}, \sigma, \nu)$ with itself.

The trick in bringing in distributions is to write
$$pf(x) = \int_{S_\mathfrak{p}} \pi(s')f(xs')\,d_r s'$$
and define a sesquilinear form C on $C^\infty_{\text{com}}(G) \times C^\infty_{\text{com}}(G)$ by
$$C(f_1, f_2) = (L(pf_1), pf_2),$$
where the form on the right side is the inner product for $U(S_\mathfrak{p}, \sigma, \nu)$. The Schwartz Kernel Theorem says that there is a distribution $d\mu(x, y)$ on

$G \times G$ with $C(f_1, f_2) = \int_{G\times G} f_1(x)\overline{f_2(y)}\, d\mu(x, y)$, and thus

$$(L(pf_1), pf_2) = \int_{G\times G} f_1(x)\overline{f_2(y)}\, d\mu(x, y).$$

This is the starting point of the argument and is analogous to (7.7).

The analysis leading to double cosets proceeds somewhat as in the case of finite groups. One has to prove that no distributions involving transverse derivatives to a double coset make any contribution. Matters finally come down to analyzing the analog of (7.10). By the Bruhat decomposition the double cosets $S_\mathfrak{p}\backslash G/S_\mathfrak{p}$ are parametrized by $W(A_\mathfrak{p}:G)$. For $\tilde{w}$ representing a member w of $W(A_\mathfrak{p}:G)$, we always have $S_\mathfrak{p} \cap \tilde{w}^{-1}S_\mathfrak{p}\tilde{w} \supseteq M_\mathfrak{p}A_\mathfrak{p}$, and $\sigma \otimes \exp \nu \otimes 1$ remains irreducible when restricted to $M_\mathfrak{p}A_\mathfrak{p}$. By Schur's Lemma the analog of (7.11) gives us a contribution of 0 or 1 for each w, and we get 1 only if

$$ma \to e^{\nu \log a}\sigma(m) \qquad \text{and} \qquad ma \to e^{\nu \log(\tilde{w}a\tilde{w}^{-1})}\sigma(\tilde{w}m\tilde{w}^{-1})$$

are equivalent. The theorem is thus as follows.

Theorem 7.2. For ν imaginary the dimension of the space of bounded linear operators L such that

$$LU(S_\mathfrak{p}, \sigma, \nu) = U(S_\mathfrak{p}, \sigma, \nu)L$$

is $\leq |W_{\sigma,\nu}|$, where

$$W_{\sigma,\nu} = \{[\tilde{w}] \in W(A_\mathfrak{p}:G) \mid \tilde{w}\sigma \cong \sigma \text{ and } \tilde{w}\nu = \nu\}.$$

Here $[\tilde{w}]$ denotes the class of $\tilde{w}$ in the Weyl group, and $\tilde{w}$ acts on σ and ν by

$$\begin{aligned} \tilde{w}\sigma(m) &= \sigma(\tilde{w}^{-1}m\tilde{w}), \qquad m \in M \\ \tilde{w}\nu(H) &= \nu(\operatorname{Ad}(\tilde{w})^{-1}H), \qquad H \in \mathfrak{a}_\mathfrak{p}. \end{aligned} \tag{7.12}$$

Remarks.

(1) The theorem implies for each σ that $U(S_\mathfrak{p}, \sigma, \nu)$ is irreducible for an open dense set of ν, in view of Proposition 1.4. The theorem is not sharp for the exceptional ν, as is shown by the case of $SL(n, \mathbb{C})$.

(2) We have to use $\tilde{w}$, rather than $[\tilde{w}]$ to define $\tilde{w}\sigma$. But then the equivalence class of $\tilde{w}\sigma$ depends only on $[\tilde{w}]$. Also $\tilde{w}\nu$ depends only on $[\tilde{w}]$.

§4. Formal Intertwining Operators

The Bruhat theory addresses also the more general question, for ν imaginary, of bounding the dimension of the space of L such that

$$LU(S_\mathfrak{p}, \sigma, \nu) = U(S_\mathfrak{p}, \sigma', \nu')L.$$

It shows that there can be no nonzero such L except when $\sigma' \cong \tilde{w}\sigma$ and $\nu' = \tilde{w}\nu$ for some $\tilde{w}$ in $N_K(A_\mathfrak{p})$.

For SL(2, $\mathbb{R}$), $W(A_\mathfrak{p}:G)$ has order 2, and $\tilde{w}\sigma$ always equals σ. Moreover, $\tilde{w}\nu$ equals ν or $-\nu$. Hence the above remarks prove another part of Proposition 2.7—that the only possible equivalences among unitary principal series $\mathscr{P}^{\pm,iv}$ are between $\mathscr{P}^{+,iv}$ and $\mathscr{P}^{+,-iv}$ and between $\mathscr{P}^{-,iv}$ and $\mathscr{P}^{-,-iv}$.

Conversely we do have equivalences $\mathscr{P}^{+,iv} \cong \mathscr{P}^{+,-iv}$ and $\mathscr{P}^{-,iv} \cong \mathscr{P}^{-,-iv}$. In the noncompact picture such equivalences from iv to $-iv$ are implemented by operators

$$
\begin{aligned}
f &\to \int_{-\infty}^{\infty} \frac{f(x-y)\,dy}{|y|^{1-iv}} && (+\text{ case}) \\
f &\to \int_{-\infty}^{\infty} \frac{f(x-y)\operatorname{sgn} y\,dy}{|y|^{1-iv}} && (-\text{ case}),
\end{aligned}
\tag{7.13}
$$

respectively, interpreted as limits of operators with iv replaced by z, as $\operatorname{Re} z \downarrow 0$. One can write down a corresponding expression in the induced picture, namely

$$A(\tilde{w}, \sigma, \nu)f(x) = \int_{\bar{N}_\mathfrak{p}} f(x\tilde{w}\bar{n})\,d\bar{n}, \tag{7.14}$$

where $\tilde{w}$ is $\begin{pmatrix} 0 & 1 \\ -1 & 0 \end{pmatrix}$ or its negative, σ corresponds to $\pm$, and ν corresponds to iv. (We shall verify this correspondence in §7.)

This expression suggests generalization, and that is the starting point for a systematic study of intertwining operators. Let us notice that the map $f(\cdot) \to f(\cdot\,\tilde{w}^{-1})$ for f in an induced space for SL(2, $\mathbb{R}$) carries the space for $U(S_\mathfrak{p}, \sigma, -\nu)$ to the space for $U(\Theta S_\mathfrak{p}, \sigma, \nu)$. Composing (7.14) and this map, we obtain an integral $\int_{\bar{N}_\mathfrak{p}} f(x\bar{n})\,d\bar{n}$ that intertwines $U(S_\mathfrak{p}, \sigma, \nu)$ with $U(\Theta S_\mathfrak{p}, \sigma, \nu)$. We study this integral first because it has a wider generalization.

Fix a parabolic subgroup $S = MAN$ of G and form the set Γ of roots of $(\mathfrak{g}, \mathfrak{a})$. The choice of N determines a positive system Γ^+ in Γ. If we change the positive system, we obtain a new parabolic subgroup $S' = MAN'$. Given σ and ν, we can write down a formal expression for an intertwining operator $A(S':S:\sigma:\nu)$ with

$$A(S':S:\sigma:\nu)U(S, \sigma, \nu) = U(S', \sigma, \nu)A(S':S:\sigma:\nu), \tag{7.15}$$

namely

$$A(S':S:\sigma:\nu)F(x) = \int_{\bar{N} \cap N'} F(x\bar{n})\,d\bar{n} \tag{7.16}$$

in the induced picture, where $\bar{N} = \Theta N$. Whenever we can make (7.16) rigorous, we call $A(S':S:\sigma:\nu)$ the **standard intertwining operator** satisfying (7.15).

It is clear that $A(S':S:\sigma:\nu)$ commutes with the left action of G. To check formally that it satisfies the intertwining property (7.15), we are to check that (7.16) transforms appropriately on the right side under S'. We can argue as follows:

(1) Behavior under M:

$$\begin{aligned} A(\text{——})F(xm) &= \int_{\bar{N} \cap N'} F(xm\bar{n})\,d\bar{n} = \int_{\bar{N} \cap N'} F(x(m\bar{n}m^{-1})m)\,d\bar{n} \\ &= \int_{\bar{N} \cap N'} F(x\bar{n}m)\,d\bar{n} \qquad \text{by a change of variables} \\ &= \sigma(m)^{-1} \int_{\bar{N} \cap N'} F(x\bar{n})\,d\bar{n} = \sigma(m)^{-1} A(\text{——})F(x). \end{aligned}$$

(2) Behavior under A:

$$\begin{aligned} A(\text{——})F(xa) &= \int_{\bar{N} \cap N'} F(xa\bar{n})\,d\bar{n} = \int_{\bar{N} \cap N'} F(x(a\bar{n}a^{-1})a)\,d\bar{n} \\ &= \det(\mathrm{Ad}(a)|_{\bar{\mathfrak{n}} \cap \mathfrak{n}'})^{-1} \int_{\bar{N} \cap N'} F(x\bar{n}a)\,d\bar{n} \\ &= e^{-(\nu+\rho_S)\log a} \det(\mathrm{Ad}(a)|_{\bar{\mathfrak{n}} \cap \mathfrak{n}'})^{-1} \int_{\bar{N} \cap N'} F(x\bar{n})\,d\bar{n} \\ &= e^{-(\nu+\rho_S)\log a} \det(\mathrm{Ad}(a)|_{\bar{\mathfrak{n}} \cap \mathfrak{n}'})^{-1} A(\text{——})F(x), \end{aligned}$$

and the argument is completed by an easy verification that

$$\det(\mathrm{Ad}(a)|_{\bar{\mathfrak{n}} \cap \mathfrak{n}'})^{-1} = e^{(\rho_S - \rho_{S'})\log a}.$$

(3) Behavior under N': Here we use a product decomposition

$$N' = (N' \cap \bar{N})(N' \cap N) \tag{7.17}$$

whose proof we omit, and moreover a Haar measure for N' is the product of the Haar measures of the factors. Then we can identify

$$N' \cap \bar{N} \leftrightarrow N'/(N' \cap N) \tag{7.18a}$$

and

$$d\bar{n} \text{ for } N' \cap \bar{N} \leftrightarrow \text{invariant measure } d\dot{n}' \text{ for } N'/(N' \cap N). \tag{7.18b}$$

For F in the representation space of $U(S, \sigma, \nu)$ and n' in N', $F(xn')$ is right invariant under $N' \cap N$ and thus depends only on the coset $\dot{n}'$ in $N'/(N' \cap N)$. Thus we conclude

$$A(\text{——})F(x) = \int_{N'/(N' \cap N)} F(x\dot{n}')\,d\dot{n}'$$

for a suitable normalization of the measure. If n_0' is in N', we thus have

$$\begin{aligned} A(\text{——})F(xn_0') &= \int_{N'/(N' \cap N)} F(xn_0'\dot{n}')\,d\dot{n}' \\ &= \int_{N'/(N' \cap N)} F(x\dot{n}')\,d\dot{n}' = A(\text{——})F(x). \end{aligned}$$

Therefore $A(S':S:\sigma:\nu)$ formally satisfies (7.15), carrying the one induced space into the other.

For w in $N_K(\mathfrak{a})$ and F in the space for $U(S, \sigma, \nu)$, let

$$R(w)F(x) = F(xw) \tag{7.19}$$

and

$$A_S(w, \sigma, \nu) = R(w)A(w^{-1}Sw:S:\sigma:\nu). \tag{7.20}$$

Then $A_S(w, \sigma, \nu)$ is given formally by

$$A_S(w, \sigma, \nu)F(x) = \int_{\bar{N} \cap w^{-1}Nw} F(xw\bar{n})\, d\bar{n}$$

and should be expected to satisfy

$$A_S(w, \sigma, \nu)U(S, \sigma, \nu) = U(S, w\sigma, w\nu)A_S(w, \sigma, \nu).$$

We call $A_S(w, \sigma, \nu)$ the **standard intertwining operator** from $U(S, \sigma, \nu)$ to $U(S, w\sigma, w\nu)$ whenever we can make matters rigorous.

The reason we refer to $A(S':S:\sigma:\nu)$ and $A_S(w, \sigma, \nu)$ as only formal is that their defining integrals need not converge. This is evident from (7.13), in which the kernel of the operator is not locally integrable.

The expectation, however, is that we always get an interesting operator satisfying (7.15) whenever we can make sense out of (7.16). The meaning that we shall ultimately attach to (7.16) is that the integral is convergent for Re ν sufficiently large and is to be interpreted by analytic continuation (in the parameter ν) for the remaining values of ν. Then also $A_S(w, \sigma, \nu)$ will be defined.

§5. Gindikin-Karpelevič Formula

The investigation of convergence of the integral (7.16) defining an intertwining operator begins with information about some special cases. First we give in Lemma 7.3 without proof an integral formula similar to the decomposition (7.17).

Lemma 7.3. Let MAN, MAN', and MAN'' be parabolic subgroups with the same MA, and suppose $\mathfrak{n}'' \cap \mathfrak{n} \subseteq \mathfrak{n}' \cap \mathfrak{n}$. If $\bar{N}$, $\bar{N}'$, and $\bar{N}''$ denote Θ of N, N', and N'' and if Haar measures are suitably normalized, then

$$\int_{\bar{N} \cap N''} f(\bar{n})\, d\bar{n} = \int_{\bar{N}' \cap N''} \left[\int_{\bar{N} \cap N'} f(\bar{n}'\bar{n})\, d\bar{n} \right] d\bar{n}' \tag{7.21}$$

for every nonnegative measurable function f.

Proposition 7.4 (Gindikin-Karpelevič formula). Let MAN, MAN', and MAN'' be parabolic subgroups with the same MA, and suppose $\mathfrak{n}'' \cap \mathfrak{n} \subseteq \mathfrak{n}' \cap \mathfrak{n}$. Define H and ρ relative to the two decompositions $G = KMAN$ and $G = KMAN'$, calling them H and ρ, H' and ρ'. If ν is real-valued and if Haar measures are normalized as in (7.21), then

$$\int_{\bar{N}\cap N''} e^{-(\nu+\rho)H(\bar{n})}\,d\bar{n} = \left[\int_{\bar{N}'\cap N''} e^{-(\nu+\rho')H'(\bar{n}')}\,d\bar{n}'\right]\left[\int_{\bar{N}\cap N'} e^{-(\nu+\rho)H(\bar{n})}\,d\bar{n}\right]. \tag{7.22}$$

This formula is valid also for a general ν if the integrals in question are convergent for Re ν.

Purpose. In the induced representation space for $U(S, 1, \nu)$, the function obtained by extending the constant function 1 on K to G is $f(kman) = e^{-(\nu+\rho)\log a}$, i.e., $f(x) = e^{-(\nu+\rho)H(x)}$. If we insert absolute value signs in the integral (7.16) in the general case, we find it is enough to study the integral for this special case. The left side of (7.22) is an intertwining operator on this special f evaluated at 1. The proposition gives a product formula for this integral, and we shall see below that iteration of the product formula reduces the integral to a situation where it can be evaluated.

Proof. Let $f(x) = e^{-(\nu+\rho)H(x)}$, and write $x = kman_0'$ relative to $G = KMAN'$. Then

$$\begin{aligned}
\int_{\bar{N}\cap N'} f(x\bar{n})\,d\bar{n} &= \int_{\bar{N}\cap N'} e^{-(\nu+\rho)H(x\bar{n})}\,d\bar{n} \\
&= \int_{N'/(N'\cap N)} e^{-(\nu+\rho)H(xn')}\,d\dot{n}' && \text{by (7.18)} \\
&= \int_{N'/(N'\cap N)} e^{-(\nu+\rho)H(an_0'n')}\,d\dot{n}' && \text{since } km \text{ drops out} \\
&= \int_{N'/(N'\cap N)} e^{-(\nu+\rho)H(an')}\,d\dot{n}' && \text{by translation} \\
&= \int_{\bar{N}\cap N'} e^{-(\nu+\rho)H(a\bar{n}a^{-1})}e^{-(\nu+\rho)\log a}\,d\bar{n} && \text{by (7.18)} \\
&= (\det \mathrm{Ad}(a)|_{\bar{\mathfrak{n}}\cap\mathfrak{n}'})^{-1}e^{-(\nu+\rho)\log a}\int_{\bar{N}\cap N'} e^{-(\nu+\rho)H(\bar{n})}\,d\bar{n} \\
&= e^{-(\nu+\rho')\log a}\int_{\bar{N}\cap N'} e^{-(\nu+\rho)H(\bar{n})}\,d\bar{n} \\
&= e^{-(\nu+\rho')H'(x)}\int_{\bar{N}\cap N'} e^{-(\nu+\rho)H(\bar{n})}\,d\bar{n}
\end{aligned}$$

Taking $x = \bar{n}'$, integrating over $\bar{N}'\cap N''$, and applying (7.21), we obtain (7.22), and the proof is complete.

Remarks. If we go over the proof carefully, we can obtain a better formula that will be useful later. Thus suppose $f(x) = f_0(H(x))e^{-\rho H(x)}$ is a nonnegative measurable function. Then

$$\int_{\bar{N}\cap N''} f_0(H(\bar{n}))e^{-\rho H(\bar{n})}\,d\bar{n}$$

$$= \int_{\bar{N}'\cap N''}\left[\int_{\bar{N}\cap N'} f_0(H'(\bar{n}') + H(\bar{n}))e^{-\rho' H'(\bar{n}')-\rho H(\bar{n})}\,d\bar{n}\right]d\bar{n}'. \tag{7.23}$$

Our application of Proposition 7.4 will be in the case of a minimal parabolic subgroup $M_\mathfrak{p}A_\mathfrak{p}N_\mathfrak{p}$. Let Σ^+ be the set of positive restricted roots. To each λ in Σ^+ for which $\frac{1}{2}\lambda$ is not in Σ^+ (in which case we say λ is **reduced**), we shall associate a certain analytic subgroup $G^{(\lambda)}$ of G. Namely let $\mathfrak{g}^{(\lambda)}$ be the subalgebra of $\mathfrak{g}$ generated by $\mathfrak{g}_\lambda$, $\mathfrak{g}_{2\lambda}$, $\mathfrak{g}_{-\lambda}$, and $\mathfrak{g}_{-2\lambda}$. This subalgebra is θ-stable, and we let $G^{(\lambda)}$ be the corresponding analytic subgroup. The style of argument in §5.1 shows $G^{(\lambda)}$ is closed; hence $G^{(\lambda)}$ is linear connected reductive. (Actually $G^{(\lambda)}$ is semisimple.)

Let $\mathfrak{k}^{(\lambda)} = \mathfrak{g}^{(\lambda)} \cap \mathfrak{k}$, $\mathfrak{p}^{(\lambda)} = \mathfrak{g}^{(\lambda)} \cap \mathfrak{p}$, $\mathfrak{a}^{(\lambda)} = \mathfrak{g}^{(\lambda)} \cap \mathfrak{a}_\mathfrak{p}$. Then $\mathfrak{g}^{(\lambda)} = \mathfrak{k}^{(\lambda)} \oplus \mathfrak{p}^{(\lambda)}$ is the Cartan decomposition of $\mathfrak{g}^{(\lambda)}$, and one can check that $\mathfrak{a}^{(\lambda)} = \mathbb{R}H_\lambda$, that $\mathfrak{a}^{(\lambda)}$ is maximal abelian in $\mathfrak{p}^{(\lambda)}$, and that $\mathfrak{m}^{(\lambda)} = Z_{\mathfrak{k}^{(\lambda)}}(\mathfrak{a}^{(\lambda)}) = \mathfrak{k}^{(\lambda)} \cap \mathfrak{m}$. The restricted root space decomposition of $\mathfrak{g}^{(\lambda)}$ is then

$$\mathfrak{g}^{(\lambda)} = \mathfrak{a}^{(\lambda)} \oplus \mathfrak{m}^{(\lambda)} \oplus (\mathfrak{g}_\lambda \oplus \mathfrak{g}_{2\lambda} \oplus \mathfrak{g}_{-\lambda} \oplus \mathfrak{g}_{-2\lambda}).$$

We can specify that λ and 2λ are to be considered as positive, and then $\mathfrak{n}^{(\lambda)} = \mathfrak{g}_\lambda \oplus \mathfrak{g}_{2\lambda}$ and $\bar{\mathfrak{n}}^{(\lambda)} = \mathfrak{g}_{-\lambda} \oplus \mathfrak{g}_{-2\lambda}$. In this case the Iwasawa decomposition of $G^{(\lambda)}$ is compatible with the Iwasawa decomposition of G, and in particular the H function of $G^{(\lambda)}$, say $H^{(\lambda)}$, is the restriction to $G^{(\lambda)}$ of the H function of G. The ρ function for G does not restrict to the ρ function $\rho^{(\lambda)}$ for $G^{(\lambda)}$ in general, but we do have

$$\rho\big|_{\mathfrak{a}^{(\lambda)}} = \rho^{(\lambda)} \qquad \text{if } \lambda \text{ is simple,} \tag{7.24}$$

by a calculation of the inner product of both sides with λ in the style of the proof of Proposition 4.32.

Corollary 7.5. Let $M_\mathfrak{p}A_\mathfrak{p}N_\mathfrak{p}$ be a minimal parabolic subgroup, and let w be in $N_K(\mathfrak{a}_\mathfrak{p})$. For ν in $(\mathfrak{a}_\mathfrak{p}')^\mathbb{C}$ and β in Σ^+, let $\nu_\beta = \dfrac{\langle \nu, \beta\rangle}{|\beta|^2}\beta$. Then

$$\int_{\bar{N}_\mathfrak{p} \cap w^{-1}N_\mathfrak{p}w} e^{-(\nu+\rho_\mathfrak{p})H_\mathfrak{p}(\bar{n})}\,d\bar{n} = \prod_{\substack{\beta\in\Sigma^+ \\ \frac{1}{2}\beta\notin\Sigma^+ \\ w\beta\notin\Sigma^+}} \int_{\bar{N}^{(\beta)}} e^{-(\nu_\beta+\rho^{(\beta)})H^{(\beta)}(\bar{n})}\,d\bar{n} \tag{7.25}$$

for any real-valued ν, and the formula remains valid for any general ν if the integrals in question are convergent for Re ν.

Remark. The more general formula that corresponds to (7.23) is

$$\int_{\bar{N}_\mathfrak{p} \cap w^{-1}N_\mathfrak{p}w} f_0(H(\bar{n}))e^{-\rho H(\bar{n})}\,d\bar{n}$$
$$= \int\cdots\int f_0\Big(\sum H^{(\beta)}(\bar{n}^{(\beta)})\Big)\Big[\prod e^{-\rho^{(\beta)}H^{(\beta)}(\bar{n}^{(\beta)})}\Big]\prod d\bar{n}^{(\beta)}, \tag{7.26}$$

and the β's that appear on the right are the reduced members of Σ^+ with $w\beta < 0$.

Proof. To simplify the notation, we use the same symbol for a Weyl group element as for a representative in K, and we drop the subscripts "$\mathfrak{p}$." The proof is by induction on the length of w, the case $w = 1$ being trivial. In the general case we write $w = s_\alpha w_1$ with α simple and $l(w) = l(w_1) + 1$. Define $\mathfrak{n}'' = \text{Ad}(w)^{-1}\mathfrak{n}$ and $\mathfrak{n}' = \text{Ad}(w_1)^{-1}\mathfrak{n}$. If

$$E' = \{\gamma \in \Sigma^+ \,|\, \gamma \text{ is reduced and } w_1\gamma > 0\}$$

$$E'' = \{\gamma \in \Sigma^+ \,|\, \gamma \text{ is reduced and } w\gamma > 0\},$$

then
$$E' = E'' \cup \{\beta\},$$
where
$$\beta = w_1^{-1}\alpha.$$

Since

$$\mathfrak{n}'' \cap \mathfrak{n} = \sum_{\gamma \in E''} \mathfrak{n}^{(\gamma)} \qquad \text{and} \qquad \mathfrak{n}' \cap \mathfrak{n} = \sum_{\gamma \in E'} \mathfrak{n}^{(\gamma)},$$

we have $\mathfrak{n}'' \cap \mathfrak{n} \subseteq \mathfrak{n}' \cap \mathfrak{n}$. Thus Proposition 7.4 gives

$$\int_{\bar{N} \cap w^{-1}Nw} e^{-(\nu+\rho)H(\bar{n})}\, d\bar{n}$$

$$= \left[\int_{\bar{N}' \cap N''} e^{-(\nu+\rho')H'(\bar{n}')}\, d\bar{n}'\right] \int_{\bar{N} \cap w_1^{-1}Nw_1} e^{-(\nu+\rho)H(\bar{n})}\, d\bar{n}.$$

To complete the induction, we need to show that

$$\int_{\bar{N}' \cap N''} e^{-(\nu+\rho')H'(\bar{n}')}\, d\bar{n}' = \int_{\bar{N}^{(\beta)}} e^{-(\nu_\beta+\rho^{(\beta)})H^{(\beta)}(\bar{n})}\, d\bar{n}. \tag{7.27}$$

We have

$$\begin{aligned} \bar{N}' \cap N'' &= w_1^{-1}\bar{N}w_1 \cap w_1^{-1}s_\alpha^{-1}Ns_\alpha w_1 = w_1^{-1}(\bar{N} \cap s_\alpha^{-1}Ns_\alpha)w_1 \\ &= w_1^{-1}\bar{N}^{(\alpha)}w_1 \end{aligned}$$

since α is simple, and $w_1^{-1}\bar{N}^{(\alpha)}w_1 = \bar{N}^{(\beta)}$. Thus the two sides of (7.27) match if it is shown that $\rho'|_{\mathfrak{a}(\beta)} = \rho^{(\beta)}$. In view of (7.24), it is enough to see that β is simple relative to the positive system defined by N', i.e., relative to $w_1^{-1}\Sigma^+$. Since α is simple relative to Σ^+, $\beta = w_1^{-1}\alpha$ is simple relative to $w_1^{-1}\Sigma^+$. This completes the proof.

Each of the integrals on the right side of (7.25) is attached to a real-rank-one group. These integrals can be explicitly calculated, or they can be evaluated indirectly by an argument using the hypergeometric equation. We give only the result.

Proposition 7.6. Suppose G has real rank one. Let β be the reduced positive restricted root, and let $p = \dim \mathfrak{g}_\beta$ and $q = \dim \mathfrak{g}_{2\beta}$. For ν in $(\mathfrak{a}_\mathfrak{p}')^{\mathbb{C}}$, let $\nu' = 2\langle \nu, \beta\rangle/|\beta|^2$. If Haar measure on $\bar{N}_\mathfrak{p}$ is normalized so that

$\int_{\bar{N}_\mathfrak{p}} e^{-2\rho H(\bar{n})}\, d\bar{n} = 1$, then

$$\int_{\bar{N}_\mathfrak{p}} e^{-(\nu+\rho)H(\bar{n})}\, d\bar{n} = \frac{\Gamma(p+q)}{\Gamma(\frac{1}{2}(p+q))} \frac{\Gamma(\frac{1}{2}\nu')}{\Gamma(\frac{1}{2}(\nu'+p))} \frac{\Gamma(\frac{1}{4}(\nu'+p))}{\Gamma(\frac{1}{4}(\nu'+p)+\frac{1}{2}q)}.$$

Corollary 7.7. For general G, for a minimal parabolic subgroup $M_\mathfrak{p}A_\mathfrak{p}N_\mathfrak{p}$, for $\rho_\mathfrak{p}$ the corresponding element ρ, and for w in $N_K(\mathfrak{a}_\mathfrak{p})$,

$$\int_{\bar{N}_\mathfrak{p} \cap w^{-1}N_\mathfrak{p}w} e^{-(\nu+\rho_\mathfrak{p})H_\mathfrak{p}(\bar{n})}\, d\bar{n}$$

is convergent if $\langle \operatorname{Re} \nu, \beta\rangle > 0$ for every β in Σ^+ for which $w\beta < 0$.

§6. Estimates on Intertwining Operators, Part I

The convergence just obtained allows us to prove convergence of intertwining operators in the minimal-parabolic case for suitable ν.

Proposition 7.8. Let *minimal* parabolic subgroups $S = MAN$ and $S' = MAN'$ be given, and suppose that ν satisfies $\langle \operatorname{Re} \nu, \beta\rangle > 0$ for every β in Σ that is positive for N and negative for N'.

(a) Then the integral (7.16) defining $A(S':S:\sigma:\nu)F(x)$ is convergent for every continuous F and for all x in G, and

$$A(S':S:\sigma:\nu)U(S, \sigma, \nu) = U(S', \sigma, \nu)A(S':S:\sigma:\nu) \tag{7.28}$$

on such F.

(b) Suppose also that w in $N_K(\mathfrak{a})$ is such that $N' = w^{-1}Nw$. Then the operator $A_S(w, \sigma, \nu)$ defined by (7.20) satisfies

$$A_S(w, \sigma, \nu)U(S, \sigma, \nu) = U(S, w\sigma, w\nu)A_S(w, \sigma, \nu) \tag{7.29}$$

on such F.

Remark. The condition on ν can be rephrased in terms of the element w in (b):

$$\langle \operatorname{Re} \nu, \beta\rangle > 0 \text{ for every } \beta \text{ in } \Sigma^+ \text{ for which } w\beta < 0. \tag{7.30}$$

Proof.

(a) Since F is bounded on K,

$$\int_{\bar{N} \cap N'} |F(x\bar{n})|\, d\bar{n} = \int_{\bar{N} \cap N'} e^{-(\operatorname{Re} \nu+\rho)H(x\bar{n})} |F(\kappa(x\bar{n}))|\, d\bar{n}$$
$$\leq C \int_{\bar{N} \cap N'} e^{-(\operatorname{Re} \nu+\rho)H(x\bar{n})}\, d\bar{n}.$$

The convergence therefore follows from Corollary 7.7. (The group N' is of the form $w^{-1}Nw$ for some w since the Weyl group $W(A:G)$ is transitive on the Weyl chambers.) The formal proof of the identity (7.15) in §4 now becomes a valid argument.

(b) We readily check by a change of variables that any w in $N_K(\mathfrak{a})$ has

$$A(S_2:S_1:\sigma:\nu) = R(w)^{-1}A(wS_2w^{-1}:wS_1w^{-1}:w\sigma:w\nu)R(w), \quad (7.31)$$

under the assumption that Haar measures have been normalized compatibly. Taking $S_1 = S$ and $S_2 = w^{-1}Sw$ and using the identity

$$R(w)U(w^{-1}Sw, \sigma, \nu) = U(S, w\sigma, w\nu)R(w),$$

we obtain (7.29).

Proposition 7.9. Let *minimal* parabolic subgroups $S = MAN$, $S' = MAN'$, and $S'' = MAN''$ be given, and suppose that $\mathfrak{n}'' \cap \mathfrak{n} \subseteq \mathfrak{n}' \cap \mathfrak{n}$ and that $\langle \operatorname{Re} \nu, \beta \rangle > 0$ for every β in Σ that is positive for N and negative for N''. Then

$$A(S'':S:\sigma:\nu) = A(S'':S':\sigma:\nu)A(S':S:\sigma:\nu). \quad (7.32)$$

Proof. This follows from Lemma 7.3 if we take into account the convergence in Proposition 7.8a.

Proposition 7.10. Let $S = MAN$ be a *minimal* parabolic subgroup, let w be in $N_K(\mathfrak{a})$, and suppose (7.30) holds. Then

(a) $$A_S(w, E\sigma E^{-1}, \nu) = EA_S(w, \sigma, \nu)E^{-1} \quad (7.33)$$

if E is a unitary operator on V^σ, and

(b) $$A_S(w_1w_2, \sigma, \nu) = A_S(w_1, w_2\sigma, w_2\nu)A_S(w_2, \sigma, \nu) \quad (7.34)$$

whenever w decomposes as $w = w_1w_2$ with $l(w) = l(w_1) + l(w_2)$.

Remarks. In (b), $l(\cdot)$ refers to the lengths of the corresponding members of the Weyl group. When $l(w) = l(w_1) + l(w_2)$, we say that $w = w_1w_2$ is a **minimal decomposition** of w. An equivalent formulation is: Whenever β is in Σ^+ and $w_1w_2\beta$ is in Σ^+, then $w_2\beta$ is in Σ^+. Note that if w is decomposed into a product of simple reflections in as short a fashion as possible, then the condition of minimality holds at every stage needed to write $A_S(w, \sigma, \nu)$ as a product of operators corresponding to simple reflections.

Proof. Conclusion (a) is routine, and (b) follows easily from Propositions 7.8b and 7.9 and formula (7.31).

If w in $N_K(\mathfrak{a})$ represents a member $[w]$ of $W(A_\mathfrak{p}:G)$, we can write $[w]$ as a minimal product of simple reflections, as in Proposition 4.11. Decomposing w and $A_S(w, \sigma, \nu)$ accordingly, by means of Proposition 7.10b, we see that $A_S(w, \sigma, \nu)$ is understood once it is understood in the case that w represents a simple reflection s_λ. In this case, the next result says that $A_S(w, \sigma, \nu)$ can be regarded as an intertwining operator for the group $G^{(\lambda)}M_\mathfrak{p}$, which has real rank one. (We skip over the problem that $G^{(\lambda)}M_\mathfrak{p}$

may be disconnected. See the end of §5.5.) The result, given as Proposition 7.11, is proved just by tracking down the definitions.

Proposition 7.11. Let $S = MAN$ be a *minimal* parabolic subgroup, and let w in $N_K(\mathfrak{a})$ represent a simple reflection s_λ in $W(A_\mathfrak{p}:G)$. Suppose that $\langle \operatorname{Re} \nu, \lambda \rangle > 0$. If F is in the space of $U(S, \sigma, \nu)$ in the induced picture and if F_k denotes the restriction to $G^{(\lambda)}M_\mathfrak{p}$ of the left translate of F by k in K, then F_k is in the space of $U(S^{(\lambda)}M_\mathfrak{p}, \sigma, \nu|_{\mathfrak{a}(\lambda)})$ and

$$A_S(w, \sigma, \nu)F(k) = (A_{S^{(\lambda)}M_\mathfrak{p}}(w, \sigma, \nu|_{\mathfrak{a}(\lambda)})F_k)(1). \tag{7.35}$$

§7. Analytic Continuation of Intertwining Operators, Part I

We now address, in the case of a *minimal* parabolic subgroup, the problem of extending the definitions of our intertwining operators by analytic continuation in the variable ν. Propositions 7.10b and 7.11 show that the critical case is the case that G has real rank one.

We begin with a computation of different forms of the intertwining operator $A_S(w, \sigma, \nu)$ in the real-rank-one case (S minimal) when w represents the (unique) nontrivial element of $W(A:G)$. If F is in the induced space for $U(S, \sigma, \nu)$ and if the notation is as in (7.3) but with $G = KAN$ since $M \subseteq K$, then the computation below is valid whenever one of the integrals in question is absolutely convergent:

$$\begin{aligned}
&\int_{\bar N} F(xw\bar n)\, d\bar n \\
&\quad = \int_{\bar N} e^{-(\rho+\nu)H(\bar n)} F(xw\kappa(\bar n))\, d\bar n && (7.36)\\
&\quad = \int_{\bar N} e^{-(\rho-\nu)\log a(\kappa(\bar n))} e^{-2\rho H(\bar n)} \sigma(m(\kappa(\bar n)))F(xw\kappa(\bar n))\, d\bar n \\
&\qquad\qquad \text{since } \bar n = kan \text{ implies } k = \bar n 1 a^{-1} n' \\
&\quad = \int_K e^{-(\rho-\nu)\log a(k)} \sigma(m(k))F(xwk)\, dk \qquad \text{by (5.25)} \\
&\quad = \int_K e^{-(\rho-\nu)\log a(w^{-1}k)} \sigma(m(w^{-1}k))F(xk)\, dk && (7.37)\\
&\quad = \int_{\bar N} e^{-(\rho-\nu)\log a(w^{-1}\kappa(\bar n))} \sigma(m(w^{-1}\kappa(\bar n)))F(x\kappa(\bar n)) e^{-2\rho H(\bar n)}\, d\bar n \\
&\quad = \int_{\bar N} e^{-(\rho-\nu)\log a(w^{-1}\bar n)} e^{-(\rho+\nu)H(\bar n)} \sigma(m(w^{-1}\bar n))F(x\kappa(\bar n))\, d\bar n && (7.38)\\
&\quad = \int_{\bar N} e^{-(\rho-\nu)\log a(w^{-1}\bar n)} \sigma(m(w^{-1}\bar n))F(x\bar n)\, d\bar n. && (7.39)
\end{aligned}$$

The left side of (7.36) is the formal expression for $A_S(w, \sigma, \nu)F(x)$. Formulas (7.37) and (7.39) show how to write this expression as a convolution on K and $\bar N$, respectively. Formula (7.39) in the case of $SL(2, \mathbb{R})$ proves the equivalence of (7.13) and (7.14).

In order to make statements about analytic dependence on ν, we need a space of functions that does not depend on ν; for this reason we use the compact picture. The analytic argument, however, is done essentially in the noncompact picture. The heart of the argument is seen by means of (7.39), and the argument is made rigorous by means of (7.38).

Theorem 7.12. For the real-rank-one minimal case, let $p = \dim \mathfrak{g}_\lambda$ and $q = \dim \mathfrak{g}_{2\lambda}$, so that $\rho = \frac{1}{2}(p + 2q)\lambda$. Suppose F is a C^∞ function in the space of the representation $U(S, \sigma, z\rho)$, realized in the compact picture, i.e., F is a C^∞ function on K satisfying $F(km) = \sigma(m)^{-1}F(k)$ for k in K, m in M. Then $A_S(w, \sigma, z\rho)F$, which is initially defined by a convergent integral for $\operatorname{Re} z > 0$, extends to a meromorphic function of z in all of $\mathbb{C}$ with at most simple poles at nonnegative integral multiples of $-(p + 2q)^{-1}$; except at the poles, the map $(z, F) \to A_S(w, \sigma, z\rho)F$ is continuous from $\mathbb{C} \times C^\infty$ into C^∞.

Remark. The proof will show also that at any pole z_0, $(z, F) \to (z - z_0)A_S(w, \sigma, z\rho)F$ is continuous.

Heart of proof for SL(2, $\mathbb{R}$). In concrete notation, formula (7.39) becomes (7.13), but with $i\nu$ replaced by z. Suppose in (7.13) that f is in $C^\infty_{\text{com}}(\mathbb{R})$. Fix x, replace the integral by $\int_a^b$ for suitable a and b, and expand $f(x - y)$ in a Taylor series in y with remainder term:

$$f(x - y) = c_0 + c_1 y + \ldots + c_n y^n + O(y^{n+1}).$$

Substitute in (7.13) and integrate each term. The terms $c_0, \ldots, c_n y^n$ contribute definite integrals of powers of y, or of sgn y times powers of y; these can be evaluated for large $\operatorname{Re} z$ and lead to functions of z with at most one pole each. The term $O(y^{n+1})$ contributes a definite integral convergent (hence analytic) for $\operatorname{Re} z > -n - 1$. Hence (7.13) is meromorphic as indicated. The coefficients c_j above depend continuously on x, and the joint continuity follows as asserted.

Further aspects of proof. The argument that is given above is incomplete in that we assumed f has compact support and in that it uses functions in the noncompact picture rather than the compact picture. We return to these points in a moment.

The above argument is easily generalized to G of real rank one. The main point is that the kernel of the operator (7.39) has a one-point singularity (at the identity of $\bar{N}$). In fact, G of real rank one forces $|W(A:G)| = 2$; hence all elements of G not in MAN are in $NwMAN$ for w representing the nontrivial element of $W(A:G)$. In particular this applies to $g = \bar{n} \in \bar{N}$ if $\bar{n} \neq 1$. Thus $\bar{n} \in NwMAN$, and $w^{-1}\bar{n} \in w^{-1}NwMAN = \bar{N}MAN$, so that the ingredients of (7.39) are well defined. In (7.39) we regard conjugation by elements of A as providing dilations of $\bar{N}$. The function $\sigma(m(w^{-1}\bar{n}))$ is homogeneous of degree 0, and $e^{-(\rho-\nu)\log a(w^{-1}n)}$ behaves like a radial function. If we expand $f(x\bar{n})$ in finite Taylor series with remainder and regroup terms according to their homogeneity under the dilations by A, then the argument for $SL(2, \mathbb{R})$ can be pushed through here.

Now let us adjust matters to work with the compact picture. With F given on K, write F as the sum of two functions, one supported within $\kappa(\bar{N})M = K - wM$ and the other supported away from M. The above argument applies to the first function, except that we use (7.38) in place of (7.39), and (7.37) easily shows that the second function leads to an expansion that is entire in ν.

Corollary 7.13. For general G let $S = MAN$ be a *minimal* parabolic subgroup. If F is a C^∞ function in the compact picture of $U(S, \sigma, \nu)$, then $A(S':S:\sigma:\nu)F$ and $A_S(w, \sigma, \nu)F$ extend to meromorphic functions of ν on $(\mathfrak{a}')^{\mathbb{C}}$, and formulas (7.28), (7.29), and (7.31) through (7.35) extend on such F to be identities of meromorphic functions.

Proof. This follows by combining the theorem with Propositions 7.10b and 7.11.

§8. Spherical Functions

For nonminimal parabolic subgroups, the intertwining operators are quite a bit more subtle. We can see this by referring to (7.39) in any situation where it applies. The group M is noncompact, and $\sigma(m(w^{-1}\bar{n}))$ depends on the asymptotic behavior of the matrix coefficients of σ at infinity; we can expect that the problem of analytic continuation of intertwining operators will similarly depend on this asymptotic behavior.

Our approach will be indirect. We will use detailed information about matrix coefficients to get a handle on the classification of representations, and we will use the first results about classification to treat the intertwining operators. Thus we begin an investigation of matrix coefficients, starting with our standard induced representations as yardsticks for the general case.

As we remarked in connection with Proposition 7.4, the function

$$\phi_\nu(x) = e^{-(\nu+\rho_{\mathfrak{p}})H(x)} \tag{7.40}$$

is in the representation space of $U(S_{\mathfrak{p}}, 1, \nu)$. It is to be understood that H in (7.3a) refers to the log of the A component of the Iwasawa decomposition. The function ϕ_ν in (7.40) arises by extending to G the constant function on K. The matrix coefficient of $U(S_{\mathfrak{p}}, 1, \nu)$ obtained by using this function twice is

$$(U(S_{\mathfrak{p}}, 1, \nu)\phi_\nu, \phi_\nu) = \int_K \phi_\nu(g^{-1}k)\overline{\phi_\nu(k)}\, dk = \int_K e^{-(\nu+\rho_{\mathfrak{p}})H(g^{-1}k)}\, dk.$$

Accordingly we define

$$\varphi_\nu^G(g) = \int_K e^{-(\nu+\rho_\mathfrak{p})H(g^{-1}k)}\,dk. \tag{7.41}$$

The function φ_ν^G on G is called a **spherical function.** Such functions will be used as basic measures of the size of matrix coefficients of representations. (We can and will use also more explicit measures of size, but these functions have nicer invariance properties than the more explicit measures that we give presently.) Elementary properties are

(i) $\varphi_\nu^G(1) = 1$ (7.42)
(ii) $\varphi_\nu^G(kxk') = \varphi_\nu^G(x)$ for all k, k' in K and x in G (7.43)
(iii) $|\varphi_\nu^G(x)| \leq \varphi_{\mathrm{Re}\,\nu}^G(x)$ (7.44)
(iv) $\int_K \varphi_\nu^G(xky)\,dk = \varphi_\nu^G(x)\varphi_\nu^G(y)$ for all x, y in G. (7.45)

Proof of (iv). We have

$$\begin{aligned}
&\int_K \varphi_\nu^G(xky)\,dk \\
&= \int_{K\times K} e^{-(\nu+\rho_\mathfrak{p})H(y^{-1}k^{-1}x^{-1}k')}\,dk\,dk' \\
&= \int_{K\times K} e^{-(\nu+\rho_\mathfrak{p})H(y^{-1}kx^{-1}k')}\,dk\,dk' \\
&= \int_{K\times K} e^{-(\nu+\rho_\mathfrak{p})H(y^{-1}k\kappa(x^{-1}k'))}e^{-(\nu+\rho_\mathfrak{p})H(x^{-1}k')}\,dk\,dk'.
\end{aligned}$$

Changing variables with $k \to k\kappa(x^{-1}k')^{-1}$, we obtain (iv) immediately.

Before listing some deeper properties of the φ_ν^G, we give more of an idea how these functions control the size of matrix coefficients. We say a vector Φ in a representation space for G is **K-finite** if its K translates span a finite-dimensional space. (If the restriction to K of the representation is unitary, the Peter-Weyl Theorem implies the K-finite vectors are dense.)

Proposition 7.14. Suppose $S = MAN$ and $S_\mathfrak{p} = M_\mathfrak{p}A_\mathfrak{p}N_\mathfrak{p}$ are parabolic subgroups of G with $S_\mathfrak{p}$ minimal and with $S \supseteq S_\mathfrak{p}$. Let $K_M = K \cap M$, and choose A_M and N_M as Iwasawa A and N components of M so that $A_\mathfrak{p} = AA_M$ and $N_\mathfrak{p} = NN_M$. Let σ be an irreducible unitary representation of M on a space V^σ, and suppose λ in $\mathfrak{a}_M'$ is such that

$$|(\sigma(m)\phi, \psi)_{V^\sigma}| \leq c_{\phi,\psi}\varphi_\lambda^M(m)$$

for all K_M-finite ϕ and ψ and all m in M. Put $U_\nu = U(S, \sigma, \nu)$. Then

$$|(U_\nu(g)\Phi, \Psi)| \leq C_{\Phi,\Psi}\varphi_{\lambda+\mathrm{Re}\,\nu}^G(g) \tag{7.46}$$

for all K-finite Φ and Ψ and all g in G.

Remarks. Note that the index $\lambda + \mathrm{Re}\,\nu$ of the spherical function (7.46) consists of two orthogonal parts λ and $\mathrm{Re}\,\nu$; there is no cancellation. If S is minimal, then M is compact and $\lambda = 0$; hence in this case (7.46) gives a trivial estimate similar to the one used in the proof of Proposition 7.8.

Proof. The images $\Phi(K)$ and $\Psi(K)$ in V^σ span finite-dimensional subspaces, and the vectors in these spaces are easily seen to be K_M-finite. Let $\phi_1, \ldots, \phi_r$ and $\psi_1, \ldots, \psi_s$ be orthonormal bases of these spaces. Then

$$
\begin{aligned}
&|(U_\nu(g)\Phi, \Psi)| \\
&= \left|\int_K (\Phi(g^{-1}k), \Psi(k))_{V^\sigma}\, dk\right| \\
&\leq \int_K e^{-(\operatorname{Re}\nu+\rho)H(g^{-1}k)}|(\sigma(\mu(g^{-1}k))^{-1}\Phi(\kappa(g^{-1}k)), \Psi(k))_{V^\sigma}|\, dk \\
&\qquad\qquad \text{with notation as in (7.3a)} \\
&= \int_K e^{-(\operatorname{Re}\nu+\rho)H(g^{-1}k)} \\
&\quad \times \left|(\sigma(\mu(g^{-1}k))^{-1} \sum_{i=1}^r (\Phi(\kappa(g^{-1}k)), \phi_i)\phi_i, \sum_{j=1}^s (\Psi(k), \psi_j)\psi_j)\right| dk \\
&\leq \sum_{i,j} \int_K e^{-(\operatorname{Re}\nu+\rho)H(g^{-1}k)} \\
&\quad \times |(\Phi(\kappa(g^{-1}k)), \phi_i)||\overline{(\Psi(k), \psi_j)}||(\sigma(\mu(g^{-1}k))^{-1}\phi_i, \psi_j)|\, dk \\
&\leq \sum_{i,j} (\sup_{k\in K} |\Phi(k)|)(\sup_{k\in K} |\Psi(k)|) \\
&\quad \times \int_K e^{-(\operatorname{Re}\nu+\rho)H(g^{-1}k)}|(\sigma(\mu(g^{-1}k))^{-1}\phi_i, \psi_j)|\, dk \\
&\leq \left\{\sum_{i,j}\left(\sup_{k\in K}|\Phi(k)|\right)\left(\sup_{k\in K}|\Psi(k)|\right)c_{\phi_i,\psi_j}\right\} \\
&\quad \times \int_{K\times K_M} e^{-(\operatorname{Re}\nu+\rho)H(g^{-1}k)}e^{-(\lambda+\rho_M)H_M(\mu(g^{-1}k)k_M)}\, dk_M\, dk \\
&= C_{\Phi,\Psi}\int_{K\times K_M} e^{-(\operatorname{Re}\nu+\rho)H(g^{-1}k)}e^{-(\lambda+\rho_M)H_M(\mu(g^{-1}k)k_M)}\, dk_M\, dk, \quad \text{say,} \\
&= C_{\Phi,\Psi}\int_{K\times K_M} e^{-(\operatorname{Re}\nu+\rho)H(g^{-1}k)}e^{-(\lambda+\rho_M)H_M(\mu(g^{-1}k))}\, dk_M\, dk \quad \text{under } kk_M \to k \\
&= C_{\Phi,\Psi}\int_K e^{-(\lambda+\operatorname{Re}\nu+\rho+\rho_M)H_{\mathfrak{p}}(g^{-1}k)}\, dk \\
&= C_{\Phi,\Psi}\varphi^G_{\lambda+\operatorname{Re}\nu}(g)
\end{aligned}
$$

since $\rho_{\mathfrak{p}} = \rho + \rho_M$ and since the H functions are appropriately consistent.

Our direct estimates of the functions φ^G_ν are given in the following proposition.

Proposition 7.15. The functions φ^G_ν satisfy

(a) $\varphi^G_{w\nu} = \varphi^G_\nu$ for all $w \in W(A_{\mathfrak{p}}:G)$
(b) $\varphi^G_\nu(a) \leq e^{\nu \log a}\varphi^G_0(a)$ for $a \in \exp \overline{\mathfrak{a}^+_{\mathfrak{p}}}$ if ν is real and Σ^+ dominant

(c) $\varphi_0^G(a) \le Ce^{-\rho_\mathfrak{p} \log a}(1 + \rho_\mathfrak{p} \log a)^d$ for $a \in \exp \overline{\mathfrak{a}_\mathfrak{p}^+}$, where d is a suitable constant ≥ 0.

Remarks. The three estimates together give an upper bound on all φ_ν^G for ν real, since $G = K(\exp \overline{\mathfrak{a}_\mathfrak{p}^+})K$. To get an idea of the size of φ_0^G, we can use (c) and the integration formula for $G = K(\exp \mathfrak{a}_\mathfrak{p}^+)K$ given in Proposition 5.28. Since

$$\prod_{\lambda \in \Sigma^+} (\sinh \lambda(H))^{\dim \mathfrak{g}_\lambda} \le e^{2\rho_\mathfrak{p}(H)} \tag{7.47}$$

we see that

$$\varphi_0^G \text{ is in } \overline{L^{2+\varepsilon}} \text{ for every } \varepsilon > 0. \tag{7.48}$$

We can give a sharper estimate. The real part of the trace form on $\mathfrak{g}$ is positive definite on $\mathfrak{p}$ and gives a Euclidean norm $\|\cdot\|$ on $\mathfrak{p}$. For x in G, decompose $x = k \exp X$ according to $G = K \exp \mathfrak{p}$ and define

$$\|x\| = \|X\|. \tag{7.49}$$

Then

$$\|k_1 x k_2\| = \|x\| \qquad \text{for } k_1, k_2 \in K, \tag{7.50}$$

and

$$\|a\| \le c_1(\rho_\mathfrak{p} \log a) \le c_2\|a\| \qquad \text{for } a \in \exp \overline{\mathfrak{a}_\mathfrak{p}^+}. \tag{7.51}$$

Then it is clear from (7.50) and (7.51) and Proposition 7.15c that

$$\frac{\varphi_0^G(x)}{(1 + \|x\|)^r} \text{ is in } L^2(G) \text{ for } r \text{ sufficiently large.} \tag{7.52}$$

Lemma 7.16.

(a) If ν is real and Σ^+ dominant, then

$$e^{\nu H(a\bar{n}a^{-1})} \le e^{\nu H(\bar{n})}$$

for all $\bar{n} \in \bar{N}_\mathfrak{p}$ and $a \in \exp \overline{\mathfrak{a}_\mathfrak{p}^+}$. Consequently $e^{\nu H(\bar{n})} \ge 1$ for all $\bar{n} \in \bar{N}_\mathfrak{p}$.

(b) If ν is real and Σ^+ dominant and if w_0 is the element of $W(A_\mathfrak{p}:G)$ with $\bar{N}_\mathfrak{p} = w_0^{-1} N_\mathfrak{p} w_0$, then

$$e^{-(\nu - w_0\nu)\log a} e^{\nu H(\bar{n})} \le e^{\nu H(a\bar{n}a^{-1})}$$

for all $\bar{n} \in \bar{N}_\mathfrak{p}$ and $a \in \exp \overline{\mathfrak{a}_\mathfrak{p}^+}$.

Proof of lemma deferred to §9.

Proof of Proposition 7.15a. Let us use "w" to denote both the Weyl group element and a representative. Put $\rho = \rho_\mathfrak{p}$. It is enough, by analyticity in ν, to prove the assertion under the assumption that $\langle \operatorname{Re} \nu, \beta \rangle > 0$ for all β in Σ^+ with $w\beta < 0$. Then $A_{S_\mathfrak{p}}(w, 1, \nu)\phi_\nu(x)$ is given by a convergent integral, according to Proposition 7.8, and is left invariant under K. Since the result is in the space for $U(S_\mathfrak{p}, 1, w\nu)$, we must have

$$\int_{\bar{N}_\mathfrak{p} \cap w^{-1}N_\mathfrak{p}w} e^{-(\nu+\rho)H(xw\bar{n})}\, d\bar{n} = c_w(\nu)e^{-(w\nu+\rho)H(x)}. \tag{7.53}$$

Taking $x = 1$, we see that

$$c_w(\nu) = \int_{\bar{N}_\mathfrak{p} \cap w^{-1}N_\mathfrak{p}w} e^{-(\nu+\rho)H(w\bar{n})}\, d\bar{n} = \int_{\bar{N}_\mathfrak{p} \cap w^{-1}N_\mathfrak{p}w} e^{-(\nu+\rho)H(\bar{n})}\, d\bar{n}$$

In (7.53), replace x by xk and integrate over k in K to get

$$\begin{aligned} c_w(\nu)\varphi^G_{w\nu}(x^{-1}) &= \int_K \int_{\bar{N}_\mathfrak{p} \cap w^{-1}N_\mathfrak{p}w} e^{-(\nu+\rho)H(xkw\bar{n})}\, d\bar{n}\, dk \\ &= \int_{\bar{N}_\mathfrak{p} \cap w^{-1}N_\mathfrak{p}w} e^{-(\nu+\rho)H(w\bar{n})} \left[\int_K e^{-(\nu+\rho)H(xk\kappa(w\bar{n}))}\, dk\right] d\bar{n} \\ &= \int_{\bar{N}_\mathfrak{p} \cap w^{-1}N_\mathfrak{p}w} e^{-(\nu+\rho)H(\bar{n})} \left[\int_K e^{-(\nu+\rho)H(xk)}\, dk\right] d\bar{n} \\ &= c_w(\nu)\varphi^G_\nu(x^{-1}). \end{aligned}$$

The result follows since $c_w(\nu)$ is not identically 0 in ν (e.g., for ν real).

Proof of Proposition 7.15b. With $\rho = \rho_\mathfrak{p}$, we have

$$\begin{aligned} \varphi^G_\nu(a) &= \int_K e^{-(\nu+\rho)H(a^{-1}k)}\, dk = \int_{\bar{N}_\mathfrak{p}} e^{-(\nu+\rho)H(a^{-1}\kappa(\bar{n}))} e^{-2\rho H(\bar{n})}\, d\bar{n} \quad \text{by (5.25)} \\ &= \int_{\bar{N}_\mathfrak{p}} e^{-(\nu+\rho)H(a^{-1}\bar{n})} e^{(\nu+\rho)H(\bar{n})} e^{-2\rho H(\bar{n})}\, d\bar{n} \\ &= \int_{\bar{N}_\mathfrak{p}} e^{-(\nu+\rho)H(a^{-1}\bar{n}a)} e^{(\nu+\rho)H(\bar{n})} e^{(\nu+\rho)\log a} e^{-2\rho H(\bar{n})}\, d\bar{n} \\ &= e^{(\nu-\rho)\log a} \int_{\bar{N}_\mathfrak{p}} e^{-(\nu+\rho)H(\bar{n})} e^{(\nu-\rho)H(a\bar{n}a^{-1})}\, d\bar{n}, \end{aligned} \tag{7.54}$$

after the change of variables $\bar{n} \to a\bar{n}a^{-1}$. By Lemma 7.16, (7.54) is

$$\le e^{(\nu-\rho)\log a} \int_{\bar{N}_\mathfrak{p}} e^{-\rho H(\bar{n})} e^{-\rho H(a\bar{n}a^{-1})}\, d\bar{n},$$

and this equals $e^{\nu \log a}\varphi^G_0(\mathrm{a})$ by (7.54) with $\nu = 0$.

Proof of Proposition 7.15c. For $\rho = \rho_\mathfrak{p}$, we shall prove in the next section (following Proposition 7.19) that

$$\int_{\bar{N}_\mathfrak{p}} e^{-\rho H(\bar{n})}\{1 + \rho H(\bar{n})\}^{-1-\varepsilon}\, d\bar{n} < \infty \tag{7.55}$$

for every $\varepsilon > 0$ if $\dim A_\mathfrak{p} = 1$. The integrand is well defined since $\rho H(\bar{n}) \ge 0$, by Lemma 7.16. Assuming (7.55) for now, we shall complete the proof of Proposition 7.15c.

First we prove that if Σ^+ has d reduced members λ, then

$$\int_{\bar{N}_\mathfrak{p}} e^{-\rho H(\bar{n})}\{1 + \rho H(\bar{n})\}^{-d-\varepsilon}\, d\bar{n} < \infty \tag{7.56}$$

for every $\varepsilon > 0$. We use (7.26) with $w^{-1}N_p w = \bar{N}_p$ and with $f_0(H) = \{1 + \rho(H)\}^{-d(1+\varepsilon)}$. (Again $f_0 \geq 0$ by Lemma 7.16.) The left side of (7.56) is the left side of (7.26), and the right side of (7.26) becomes

$$\int \cdots \int \left\{ \sum_{j=1}^{d} (d^{-1} + \rho H(\bar{n}_j)) \right\}^{-d(1+\varepsilon)} \prod e^{-\rho_j H(\bar{n}_j)} \, d\bar{n}_1 \cdots d\bar{n}_d \tag{7.57}$$

if we denote the various $\bar{n}^{(\beta)}$ by $\bar{n}_j$, $1 \leq j \leq d$. By Lemma 7.16, $d^{-1} + \rho H(\bar{n}_j) \geq d^{-1} > 0$. Hence the arithmetic mean/geometric mean inequality gives

$$\left\{ \prod_{j=1}^{d} (d^{-1} + \rho H(\bar{n}_j)) \right\}^{1/d} \leq \frac{1}{d} \sum_{j=1}^{d} (d^{-1} + \rho H(\bar{n}_j)) \leq \sum_{j=1}^{d} (d^{-1} + \rho H(\bar{n}_j))$$

and therefore

$$\left\{ \sum_{j=1}^{d} (d^{-1} + \rho H(\bar{n}_j)) \right\}^{-d(1+\varepsilon)} \leq \prod_{j=1}^{d} (d^{-1} + \rho H(\bar{n}_j))^{-(1+\varepsilon)}.$$

When we substitute this inequality into (7.57), the resulting integral on the right side separates into a product of integrals

$$\int_{\bar{N}_j} (d^{-1} + \rho H(\bar{n}_j))^{-(1+\varepsilon)} e^{-\rho_j H(\bar{n}_j)} \, d\bar{n}_j.$$

If $\rho(H_{\alpha_j}) = c_j \rho_j(H_{\alpha_j})$, then $\rho H(\bar{n}_j) = c_j \rho_j H(\bar{n}_j)$, and each factor integral on the right side is thus of the form (7.55). Therefore (7.56) follows from (7.55).

Now let $\bar{N}_r = \{\bar{n} \mid \rho H(\bar{n}) \leq 2^r\}$ for $r \geq 0$. By (7.56),

$$\int_{\bar{N}_{r+1} - \bar{N}_r} e^{-\rho H(\bar{n})} \{1 + \rho H(\bar{n})\}^{-d-\varepsilon} \, d\bar{n} \leq C < \infty \qquad \text{for } r \geq 0,$$

with C independent of r. Therefore

$$\begin{aligned} (2^{r+2})^{-d-\varepsilon} \int_{\bar{N}_{r+1} - \bar{N}_r} e^{-\rho H(\bar{n})} \, d\bar{n} &\leq (1 + 2^{r+1})^{-d-\varepsilon} \int_{\bar{N}_{r+1} - \bar{N}_r} e^{-\rho H(\bar{n})} \, d\bar{n} \\ &\leq \int_{\bar{N}_{r+1} - \bar{N}_r} e^{-\rho H(\bar{n})} \{1 + \rho H(\bar{n})\}^{-d-\varepsilon} \, d\bar{n} \\ &\leq C, \end{aligned}$$

so that

$$\int_{N_{r+1} - \bar{N}_r} e^{-\rho H(\bar{n})} \, d\bar{n} \leq 4^{d+\varepsilon} C 2^{r(d+\varepsilon)}. \tag{7.58}$$

Also (7.56) gives

$$\int_{\bar{N}_0} e^{-\rho H(\bar{n})} \{1 + \rho H(\bar{n})\}^{-d-\varepsilon} \, d\bar{n} \leq C.$$

Since $2^{-d-\varepsilon} \leq \{1 + \rho H(\bar{n})\}^{-d-\varepsilon}$ on $\bar{N}_0$, we have

$$\int_{\bar{N}_0} e^{-\rho H(\bar{n})} \, d\bar{n} \leq 2^{d+\varepsilon} C. \tag{7.59}$$

Putting (7.58) and (7.59) together, we obtain

$$\int_{\bar{N}_r} e^{-\rho H(\bar{n})}\,d\bar{n} = \int_{\bar{N}_0} + \sum_{k=1}^{r} \int_{\bar{N}_k - \bar{N}_{k-1}} \leq 2^{d+\varepsilon}C + \sum_{k=1}^{r} 4^{d+\varepsilon}C2^{(k-1)(d+\varepsilon)}$$

$$\leq 4^{d+\varepsilon}\,C\,\frac{2^{r(d+\varepsilon)}-1}{2^{d+\varepsilon}-1} \leq C'2^{r(d+\varepsilon)}, \qquad (7.60)$$

where $C' = 4^{d+\varepsilon}C(2^{d+\varepsilon}-1)^{-1}$.

Now let us estimate $\varphi_0^G(a)$. In view of (7.54), we are to show that

$$\int_{\bar{N}_\mathfrak{p}} e^{-\rho H(\bar{n}) - \rho H(a\bar{n}a^{-1})}\,d\bar{n} \leq \text{Const}(1 + \rho \log a)^{d+\varepsilon} \qquad (7.61)$$

for a in $\exp \overline{\mathfrak{a}_\mathfrak{p}^+}$. First suppose $\rho \log a > \frac{1}{4}$. Choose an integer $r > 0$ so that $2^{r-1} < 4\rho \log a \leq 2^r$, and write the left side of (7.61) as $\int_{\bar{N}_r} + \int_{\bar{N}_\mathfrak{p} - \bar{N}_r}$. Since $e^{-\rho H(a\bar{n}a^{-1})} \leq 1$ by Lemma 7.16a, (7.60) gives

$$\int_{\bar{N}_r} e^{-\rho H(\bar{n}) - \rho H(a\bar{n}a^{-1})}\,d\bar{n} \leq \int_{\bar{N}_r} e^{-\rho H(\bar{n})}\,d\bar{n} \leq C'2^{r(d+\varepsilon)}$$

$$\leq C'2^{d+\varepsilon}(4\rho \log a)^{d+\varepsilon}. \qquad (7.62)$$

On $\bar{N}_\mathfrak{p} - \bar{N}_r$, we have $\rho H(\bar{n}) \geq 2^r \geq 4\rho \log a$. Hence $\rho H(\bar{n}) - 4\rho \log a \geq 0$, and Lemma 7.16b thus gives

$$e^{-\rho H(a\bar{n}a^{-1})} \leq \{e^{(\rho H(\bar{n}) - 4\rho \log a) + \rho H(\bar{n})}\}^{-1/2} \leq e^{-(1/2)\rho H(\bar{n})}, \qquad \bar{n} \in \bar{N}_\mathfrak{p} - \bar{N}_r$$

Therefore

$$\int_{\bar{N}_\mathfrak{p} - \bar{N}_r} e^{-\rho H(\bar{n}) - \rho H(a\bar{n}a^{-1})}\,d\bar{n} \leq \int_{\bar{N}_\mathfrak{p} - \bar{N}_r} e^{-\rho H(\bar{n})}e^{-(1/2)\rho H(\bar{n})}\,d\bar{n}$$

$$\leq C'' \text{ by Corollary 7.7.} \qquad (7.63)$$

Inequalities (7.62) and (7.63) prove (7.61) for the case $\rho \log a > \frac{1}{4}$. In the case $\rho \log a \leq \frac{1}{4}$, Lemma 7.16a gives $e^{-\rho H(\bar{n})} \leq e^{-\rho H(a\bar{n}a^{-1})}$, and thus

$$\int_{\bar{N}_\mathfrak{p}} e^{-\rho H(\bar{n}) - \rho H(a\bar{n}a^{-1})}\,d\bar{n} \leq \int_{\bar{N}_\mathfrak{p}} e^{-2\rho H(a\bar{n}a^{-1})}\,d\bar{n}$$

$$= e^{2\rho \log a} \int_{\bar{N}_\mathfrak{p}} e^{-2\rho H(\bar{n})}\,d\bar{n} \qquad \text{under } a\bar{n}a^{-1} \to \bar{n}$$

$$\leq e^{1/2}C'''.$$

Hence (7.61) follows for $\rho \log a \leq \frac{1}{4}$, and the proof is complete.

§9. Finite-Dimensional Representations and the *H* Function

To complete the discussion of spherical functions in §8, we still have to prove Lemma 7.16 and formula (7.55). Both these results come from an interpretation of the *H* function in terms of finite-dimensional representations.

Let G be linear connected semisimple, and let $S_\mathfrak{p} = MAN$ be a *minimal* parabolic subgroup. If $\mathfrak{b}$ is a maximal abelian subspace of $\mathfrak{m}$, we have seen that $\mathfrak{a} \oplus \mathfrak{b}$ is a Cartan subalgebra of $\mathfrak{g}$. Hence we can form the set of roots $\Delta = \Delta((\mathfrak{a} \oplus i\mathfrak{b})^{\mathbb{C}}:\mathfrak{g}^{\mathbb{C}})$. The restricted roots (members of Σ) are obtained as the nonzero restrictions of members of Δ to $\mathfrak{a}$. Let us extend the ordering on $\mathfrak{a}'$ to an ordering for $\mathfrak{a}' \oplus (i\mathfrak{b})'$ by adjoining to our basis for $\mathfrak{a}'$ a basis for $(i\mathfrak{b})'$. Then the nonzero restrictions to $\mathfrak{a}$ of members of Δ^+ are in Σ^+, and $\mathfrak{n}^{\mathbb{C}}$ is contained in $\sum_{\alpha \in \Delta^+} \mathfrak{g}_\alpha$.

It will be technically simpler in part of our discussion to assume that $\mathfrak{k} \cap i\mathfrak{p} = 0$ and that $G \subseteq G^{\mathbb{C}}$ with $G^{\mathbb{C}}$ simply connected. We can always pass to a covering group of G to achieve this. [In fact, realize G as real matrices and let U be the compact form of G. Since U is compact semisimple, its universal covering group $\tilde{U}$ is compact, hence linear. Then we can pass to $\tilde{U}^{\mathbb{C}}$ and to the analytic subgroup of $\tilde{U}^{\mathbb{C}}$ with Lie algebra $\mathfrak{g}$, which will be the required covering group of G.]

Proposition 7.17. Let $\Phi_{\tilde{\nu}}$ be a finite-dimensional irreducible holomorphic representation of $G^{\mathbb{C}}$ with highest weight $\tilde{\nu}$ relative to $(\mathfrak{a} \oplus i\mathfrak{b})^{\mathbb{C}}$. Introduce a Hermitian inner product such that the compact form U acts by unitary operators. Let $\nu = \tilde{\nu}|_{\mathfrak{a}}$, and let u_ν be any unit vector in the sum of the weight spaces for weights that restrict to ν on $\mathfrak{a}$. Then the H function of G relative to $S_\mathfrak{p}$ satisfies

$$e^{2\nu H(x)} = \|\Phi_{\tilde{\nu}}(x)u_\nu\|^2 \tag{7.64}$$

for all x in G. Moreover,

$$2\langle \nu, \lambda\rangle/|\lambda|^2 \text{ is in } \mathbb{Z} \text{ for all } \lambda \in \Sigma. \tag{7.65}$$

Proof. Write $x = kan$, and let $\varphi = d\Phi_{\tilde{\nu}}$. For λ in Σ^+ and X in $\mathfrak{g}_\lambda$, the restriction to $\mathfrak{a}$ of any weight contributing to $\varphi(X)u_\nu$ is $\nu + \lambda$. Hence $\varphi(X)u_\nu = 0$. Thus $\varphi(\mathfrak{n})$ annihilates u_ν, and it follows that $\Phi_{\tilde{\nu}}(N)$ fixes u_ν. Hence $\Phi_{\tilde{\nu}}(n)u_\nu = u_\nu$. Clearly $\Phi_{\tilde{\nu}}(a)u_\nu = e^{\nu \log a}u_\nu$. Since $\Phi_{\tilde{\nu}}(k)$ is unitary, we conclude

$$\|\Phi_{\tilde{\nu}}(x)u_\nu\|^2 = \|\Phi_{\tilde{\nu}}(a)u_\nu\|^2 = e^{2\nu \log a}\|u_\nu\|^2 = e^{2\nu \log a},$$

and (7.64) follows.

For (7.65), let X_λ be in $\mathfrak{g}_\lambda$. Then X_λ, θX_λ, and $[X_\lambda, \theta X_\lambda]$ together span a copy of $\mathfrak{sl}(2, \mathbb{R})$, and (7.65) follows by the standard argument from Theorem 2.4 and Corollary 2.3.

Corollary 7.18. Under the assumptions on $\tilde{\nu}$ and ν in Proposition 7.17,

(a) $e^{\nu H(a\bar{n}a^{-1})} \leq e^{\nu H(\bar{n})}$ for all $\bar{n} \in \bar{N}$ and $a \in \exp \overline{\mathfrak{a}^+}$.
(b) $e^{\nu H(\bar{n})} \geq 1$ for all $\bar{n} \in \bar{N}$.
(c) $e^{-(\nu - w_0\nu)\log a}e^{\nu H(\bar{n})} \leq e^{\nu H(a\bar{n}a^{-1})}$ for $a \in \exp \overline{\mathfrak{a}^+}$, where w_0 is the element of $W(A:G)$ with $\bar{N} = w_0^{-1}Nw_0$.

Proof. With $\Phi_{\tilde{\nu}}$ as in Proposition 7.17, let us lump together weight spaces whose weights agree on $\mathfrak{a}$. The $\mathfrak{a}$ weights that occur are then $\nu_1, \ldots, \nu_r$, say, with $\nu = \nu_1$ and with each ν_j equal to the difference of ν and a sum of members of Σ^+, by the theory of the highest weight. Each member of $\theta\mathfrak{n}$ pushes the $\mathfrak{a}$ weights down. By (A.127) we can write $\bar{n} = \exp X$ with $X \in \theta\mathfrak{n}$, and then we see that $\Phi_{\tilde{\nu}}(\bar{n})u_\nu$ is of the form

$$\Phi_{\tilde{\nu}}(\bar{n})u_\nu = u_\nu + \sum_{j=2}^{r} P_j(\bar{n}) \tag{7.66}$$

with $P_j(\bar{n})$ in the weight space for the $\mathfrak{a}$ weight ν_j.

The weight spaces for distinct $\mathfrak{a}$ weights are orthogonal, and then (b) follows immediately from (7.66). Formula (7.66) gives also

$$\Phi_{\tilde{\nu}}(a\bar{n}a^{-1})u_\nu = u_\nu + \sum_{j=2}^{r} e^{(\nu_j - \nu)\log a} P_j(\bar{n}), \tag{7.67}$$

and (a) follows by comparison of (7.66) and (7.67). [We can obtain (b) alternatively now by letting $a \to \infty$ in (a).]

The lowest $\mathfrak{a}$ weight is $w_0\nu$, and thus

$$e^{(w_0\nu - \nu)\log a} \leq e^{(\nu_j - \nu)\log a}$$

for $1 \leq j \leq r$. Taking the norm squared of both sides of (7.67) and using this inequality, we obtain (c).

Proposition 7.19. Suppose that $\mathfrak{k} \cap i\mathfrak{p} = 0$ and that $G \subseteq G^{\mathbb{C}}$ with $G^{\mathbb{C}}$ simply connected. If ν in $\mathfrak{a}'$ is such that $2\langle \nu, \lambda\rangle/|\lambda|^2$ is in $4\mathbb{Z}$ and is ≥ 0 for every λ in Σ^+, then the extension $\tilde{\nu}$ of ν to $\mathfrak{a}' \oplus (i\mathfrak{b})'$ by 0 on $(i\mathfrak{b})'$ is dominant integral and has $\nu = \tilde{\nu}|_{\mathfrak{a}}$.

Remark. Then, of course, Proposition 7.17 and Corollary 7.18 apply to this ν. A sharper result will be given in Corollary 9.15.

Proof. For α in Δ, write $\alpha = \alpha_R + \alpha_I$ with $\alpha_R \in \mathfrak{a}'$ and $\alpha_I \in (i\mathfrak{b})'$. Then $\bar{\alpha} = \alpha_R - \alpha_I$ is in Δ also (with root vector obtained from E_α by applying the conjugation of $\mathfrak{g}^{\mathbb{C}}$ with respect to $\mathfrak{g}$). If $\alpha_R \neq 0$ (so that α_R is in Σ), then the equality $|\alpha|^2 = |\bar{\alpha}|^2$ implies that $2\langle \alpha, \bar{\alpha}\rangle/|\alpha|^2 = k$ for some integer $k > -2$. Hence

$$2(|\alpha_R|^2 - |\alpha_I|^2) = k|\alpha|^2.$$

Adding $2|\alpha|^2$ to both sides, we obtain $4|\alpha_R|^2 = (k+2)|\alpha|^2$. Hence $|\alpha|^{-2} = \frac{1}{4}(k+2)|\alpha_R|^{-2}$ and

$$\frac{2\langle \tilde{\nu}, \alpha\rangle}{|\alpha|^2} = \frac{1}{4}(k+2)\frac{2\langle \nu, \alpha_R\rangle}{|\alpha_R|^2}.$$

If $2\langle \nu, \alpha_R \rangle / |\alpha_R|^2$ is in $4\mathbb{Z}$, then $2\langle \tilde{\nu}, \alpha \rangle / |\alpha|^2$ is in $\mathbb{Z}$. If $\langle \nu, \alpha_R \rangle \geq 0$, then $\langle \tilde{\nu}, \alpha \rangle \geq 0$. The result follows.

Proof of Lemma 7.16. If ν is real and Σ^+ dominant, then ν is a nonnegative combination of members of $\mathfrak{a}'$ to which Proposition 7.19 applies. The conclusions of Corollary 7.18 then apply to each of these members of $\mathfrak{a}'$ and hence also to ν.

Lemma 7.20. If G has real rank one, then there exists a continuous function $Q(\bar{n}) \geq 0$ on $\bar{N}$ vanishing only at the identity such that $e^{8\rho H(\bar{n})}$ is bounded above and below by multiples of $1 + Q(\bar{n})$ and such that $Q(a\bar{n}a^{-1}) = e^{-16\rho \log a} Q(\bar{n})$ for a in A.

Proof. There is no loss of generality in assuming that $\mathfrak{k} \cap i\mathfrak{p} = 0$ and that $G \subseteq G^{\mathbb{C}}$ with $G^{\mathbb{C}}$ simply connected. We apply Proposition 7.19 to 4ρ and let Φ be the resulting irreducible finite-dimensional holomorphic representation of $G^{\mathbb{C}}$ with highest weight the extension of 4ρ. Let us refer to the proof of Corollary 7.18, denoting the $\mathfrak{a}$ weights by $\nu_1, \ldots, \nu_r$, with $\nu_1 = 4\rho$ the highest and with $\nu_r = w_0(4\rho) = -4\rho$ the lowest. In the notation of (7.66), we put $Q(\bar{n}) = |P_r(\bar{n})|^2$. It is clear that $Q(\bar{n}) \geq 0$ and that $Q(a\bar{n}a^{-1}) = e^{-16 \log a} Q(\bar{n})$.

Now let us address the vanishing of $Q(\bar{n})$. More generally let $Q(x)$ be the length squared of the projection of $\Phi(x)u_\nu$ on the $\mathfrak{a}$ weight space for the weight -4ρ, for x in G. That weight space is stabilized by $\Phi(MA)$. Also $n \in N$ implies $\Phi(n)\Phi(x)u_\nu$ and $\Phi(x)u_\nu$ have the same projection since n increases weights (by (A.127)) and -4ρ is the lowest. Thus

$$Q(x) \neq 0 \text{ implies } Q(man_1xn_2) \neq 0 \text{ for } m \in M, a \in A, n_1 \in N, n_2 \in N.$$

Let w represent the nontrivial element w_0 of $W(A:G)$. Since w_0 carries weight 4ρ to -4ρ, we have $Q(w) = 1 \neq 0$. Thus Q is nonvanishing on $MANwMAN$. Since the Bruhat decomposition gives

$$G = MAN \cup MANwMAN$$

(G being of real rank one) and since $MAN \cap \bar{N} = \{1\}$, $Q(\bar{n})$ can vanish only for $\bar{n} = 1$.

Let E be a "sphere" in $\bar{N}$, obtained by exponentiating a sphere about 0 in $\theta\mathfrak{n}$, and let

$$C = \max_{\bar{n} \in E} \{e^{8\rho H(\bar{n})} / Q(\bar{n})\}.$$

It is trivial that

$$1 + Q(\bar{n}) = |u_\nu|^2 + |P_r(\bar{n})|^2 \leq |u_\nu|^2 + \sum_{j=2}^{r} |P_j(\bar{n})|^2 = e^{8\rho H(\bar{n})}.$$

In the reverse direction, Corollary 7.18c gives, for $\bar{n} \in E$ and $a \in \exp \overline{\mathfrak{a}^+}$,

$$\begin{aligned} e^{8\rho H(a^{-1}\bar{n}a)} &\leq e^{8\rho H(\bar{n})} e^{16\rho \log a} \leq CQ(\bar{n}) e^{16\rho \log a} \\ &= CQ(a^{-1}\bar{n}a) \leq C(1 + Q(a^{-1}\bar{n}a)). \end{aligned}$$

The elements $a^{-1}\bar{n}a$ with $\bar{n} \in E$ and $a \in \exp \overline{\mathfrak{a}^+}$ exhaust the exterior of E, and it follows that $e^{8\rho H(\bar{n})}$ is bounded above and below by multiples of $1 + Q(\bar{n})$.

Lemma 7.21. If G has real rank one, then

$$\int_{\bar{N}} e^{-\rho H(\bar{n})} \{1 + \rho H(\bar{n})\}^{-1-\varepsilon} \, d\bar{n} < \infty$$

for every $\varepsilon > 0$.

Remark. This lemma is formula (7.55).

Proof. Let Q be as in Lemma 7.20, let C_0 be a "unit ball" in $\bar{N}$ large enough so that $Q(\bar{n}) \geq 1$ off C_0, and let E be the boundary of C_0. Fix H_0 in $\mathfrak{a}^+$, and coordinatize $\bar{N} - \{1\}$ by $\bar{n} \leftrightarrow (t, \omega) \in \mathbb{R} \times E$, $\bar{n} = a_t^{-1}\omega a_t$, where $a_t = \exp tH_0$. Put $C = a_1^{-1}C_0 a_1$. Then

$$\begin{aligned} &\int_{\bar{N}-C} Q(\bar{n})^{-1/8} \{\log Q(\bar{n})\}^{-1-\varepsilon} \, d\bar{n} \\ &\quad = \int_{t=1}^{\infty} \int_E Q(a_t^{-1}\omega a_t)^{-1/8} \{\log Q(a_t^{-1}\omega a_t)\}^{-1-\varepsilon} g(t, \omega) \, d\omega \, dt, \end{aligned} \tag{7.68}$$

where $g(t, \omega)$ is a certain Jacobian determinant. It is clear that g satisfies

$$\int_{-\infty}^{t} \int_E g(s, \omega) \, d\omega \, ds = \int_{a_t^{-1}C_0 a_t} d\bar{n} = e^{2t\rho(H_0)} |C_0|,$$

where $|C_0|$ denotes the measure of C_0, and hence

$$\int_E g(t, \omega) \, d\omega = 2\rho(H_0) |C_0| e^{2t\rho(H_0)}$$

by passage to the derivative. The homogeneity of Q thus means that the right side of (7.68) is

$$\begin{aligned} &= \int_{t=1}^{\infty} e^{-2t\rho(H_0)} \int_{\omega \in E} Q(\omega)^{-1/8} \{16t\rho(H_0) + \log Q(\omega)\}^{-1-\varepsilon} g(t, \omega) \, d\omega \, dt \\ &\leq \int_{t=1}^{\infty} e^{-2t\rho(H_0)} \{16t\rho(H_0)\}^{-1-\varepsilon} \int_{\omega \in E} g(t, \omega) \, d\omega \, dt \\ &= 2\rho(H_0) |C_0| \{16\rho(H_0)\}^{-1-\varepsilon} \int_1^{\infty} t^{-1-\varepsilon} \, dt < \infty. \end{aligned}$$

Thus the left side of (7.68) is finite, and it follows that

$$\int_{\bar{N}-C} e^{-\rho H(\bar{n})} \{1 + \rho H(\bar{n})\}^{-1-\varepsilon} \, d\bar{n} < \infty.$$

Since C is compact, the integral is finite also on C. Thus Lemma 7.21 follows.

§10. Estimates on Intertwining Operators, Part II

The estimates on spherical functions enable us to prove convergence of intertwining operators in the general case. Let us suppose we are in the situation of Proposition 7.14. We have the usual data $S = MAN$, σ irreducible unitary on M, and ν on $\mathfrak{a}$. An estimate on the $(K \cap M)$-finite matrix coefficients of σ implies, according to the proposition, an estimate on the K-finite matrix coefficients of $U(S, \sigma, \nu)$. We consider now the formal intertwining operator

$$A(S':S:\sigma:\nu)f(x) = \int_{\bar{N} \cap N'} f(x\bar{n})\, d\bar{n}. \tag{7.69}$$

The theorem that follows generalizes Proposition 7.8a to the case of nonminimal parabolics.

Theorem 7.22. Suppose $S = MAN$ and $S_\mathfrak{p} = M_\mathfrak{p} A_\mathfrak{p} N_\mathfrak{p}$ are parabolic subgroups of G with $S_\mathfrak{p}$ minimal and with $S \supseteq S_\mathfrak{p}$. Let $K_M = K \cap M$, and choose A_M and N_M as Iwasawa A and N components of M so that $A_\mathfrak{p} = AA_M$ and $N_\mathfrak{p} = NN_M$. Let σ be an irreducible unitary representation of M on a space V^σ, and suppose λ in $\mathfrak{a}_M'$ is such that

$$|(\sigma(m)\phi, \psi)_{V^\sigma}| \le c_{\phi,\psi} \varphi_\lambda^M(m)$$

for all K_M-finite ϕ and ψ and all m in M. Put $U_\nu = U(S, \sigma, \nu)$, and let $S = MAN'$ be a parabolic subgroup with the same MA as for S. If $\langle \lambda + \operatorname{Re} \nu, \alpha \rangle > 0$ for every $\mathfrak{a}_\mathfrak{p}$ root α such that $\mathfrak{g}_\alpha \subseteq \theta\mathfrak{n} \cap \mathfrak{n}'$, then

$$\int_{\bar{N} \cap N'} (f(x\bar{n}), \psi)_{V^\sigma}\, d\bar{n} \tag{7.70}$$

exists as a Lebesgue integral for all K-finite f in the space for U_ν.

Remarks. We come back to the question of how to pass from (7.70) to (7.69) after giving the proof.

Proof. We have decompositions $G = KMAN$, $G = KA_\mathfrak{p}N_\mathfrak{p}$, and $M = K_M A_M N_M$, and we let H, $H_\mathfrak{p}$, H_M be the respective H functions and ρ, $\rho_\mathfrak{p}$, ρ_M be the respective ρ functions. Some relationships among these are

$$\rho_\mathfrak{p} = \rho + \rho_M \qquad \text{since } \mathfrak{n}_\mathfrak{p} = \mathfrak{n} \oplus \mathfrak{n}_M \tag{7.71a}$$

$$H_\mathfrak{p} = H + H_M \tag{7.71b}$$

$$(\rho + \operatorname{Re} \nu)H(x) = (\rho + \operatorname{Re} \nu)H_\mathfrak{p}(x) \quad \text{by (7.71b) since } (\rho + \operatorname{Re} \nu)(\mathfrak{a}_M) = 0 \tag{7.71c}$$

$$(\lambda + \rho_M)H_M(\mu(x)) = (\lambda + \rho_M)H_M(x) \quad \text{by (7.71b) since } (\lambda + \rho_M)(\mathfrak{a}) = 0. \tag{7.71d}$$

We are to consider the Lebesgue integral

$$\int_{\bar{N}\cap N'} |(f(x\bar{n}), \psi)_{V^\sigma}|\, d\bar{n} = \int_{\bar{N}\cap N'} |e^{-(\rho+\nu)H(x\bar{n})}(\sigma(\mu(x\bar{n})))^{-1}f(\kappa(x\bar{n})), \psi)_{V^\sigma}|\, d\bar{n}. \tag{7.72}$$

Since f is K-finite, the span of $f(K)$ in V^σ is finite-dimensional, and we can therefore let $\phi_1, \ldots, \phi_r$ be an orthonormal basis for a finite-dimensional subspace of V^σ containing the image. Then the right side of (7.72) is

$$= \int_{\bar{N}\cap N'} e^{-(\rho+\text{Re }\nu)H(x\bar{n})}\left|\sum_{i=1}^{r} (f(\kappa(x\bar{n})), \phi_i)_{V^\sigma}(\sigma(\mu(x\bar{n})))^{-1}\phi_i, \psi)_{V^\sigma}\right| d\bar{n}$$

$$\leq \sum_{i=1}^{r}\left\{\sup_{k\in K}|(f(k), \phi_i)|\right\}\int_{\bar{N}\cap N'} e^{-(\rho+\text{Re }\nu)H(x\bar{n})}|(\sigma(\mu(x\bar{n})))^{-1}\phi_i, \psi)_{V^\sigma}|\, d\bar{n}$$

$$\leq \left\{\sum_{i=1}^{r} c_{\phi_i,\psi}\sup_{k\in K}|(f(k), \phi_i)|\right\}$$
$$\times \int_{\bar{N}\cap N'}\int_{K_M} e^{-(\rho+\text{Re }\nu)H(x\bar{n})}e^{-(\lambda+\rho_M)H_M(\mu(x\bar{n})k_M)}\, dk_M\, d\bar{n}$$

$$= C\int_{\bar{N}\cap N'}\int_{K_M} e^{-(\rho+\text{Re }\nu)H(xk_M\bar{n})}e^{-(\lambda+\rho_M)H_M(\mu(xk_M\bar{n}))}\, dk_M\, d\bar{n}$$
$$\text{under } \bar{n} \to k_M\bar{n}k_M^{-1}$$

$$= C\int_{K_M}\left[\int_{\bar{N}\cap N'} e^{-(\rho_\mathfrak{p}+\lambda+\text{Re }\nu)H_\mathfrak{p}(xk_M\bar{n})}\, d\bar{n}\right] dk_M \qquad \text{by (7.71)}$$

$$= C\left[\int_{K_M} e^{-(\rho'_\mathfrak{p}+\lambda+\text{Re }\nu)H'_\mathfrak{p}(xk_M)}\, dk_M\right]\left[\int_{\bar{N}\cap N'} e^{-(\rho_\mathfrak{p}+\lambda+\text{Re }\nu)H_\mathfrak{p}(\bar{n})}\, d\bar{n}\right],$$

the last step holding by the chain of equalities in the proof of Proposition 7.4. The right side here is finite by Corollary 7.7, and the proof is complete.

Let us return to the question of how to interpret (7.69). We shall see in the next chapter that each irreducible representation of K_M occurs with finite multiplicity in $\sigma|_{K_M}$ and that the Frobenius reciprocity theorem implies that each irreducible representation of K occurs with finite multiplicity in $U_\nu|_K$. If f transforms according to some representation of K (on the left), so must any reasonable interpretation of (7.69). Thus we have no choice but to define

$$A(S':S:\sigma:\nu)f(k) = \sum_j\left[\int_{\bar{N}\cap N'} (f(k\bar{n}), \psi_j)_{V^\sigma}\, d\bar{n}\right]\psi_j,$$

with ψ_j ranging over an orthonormal basis of a subspace of V^σ containing all images in V^σ of members of the induced space transforming according to the given K type; the sum is finite. One can show this definition is appropriately consistent.

The next question is how to interpret $A(S':S:\sigma:\nu)$ as an intertwining operator if it is defined only on the (non G-invariant) subspace of K-finite vectors. The answer is that the Lie algebra $\mathfrak{g}$ does leave the space of K-finite vectors stable (see Proposition 8.5), and the operator intertwines relative to the representation of the Lie algebra determined by $U(S, \sigma, \nu)$.

These are not serious problems in working with $A(S':S:\sigma:\nu)$, and we shall tend to ignore them for now and handle them all at once in §8.3.

§11. Tempered Representations and Langlands Quotients

We say that a representation π of G is **tempered** if its K-finite matrix coefficients satisfy

$$|(\pi(g)\phi, \psi)| \leq c_{\phi,\psi}\varphi_0^G(g).$$

This definition has the following properties:

(1) If π is tempered, then the K-finite matrix coefficients of π are in $L^{2+\varepsilon}$ for every $\varepsilon > 0$. [This is by (7.48).]

(2) If $S = MAN$ is a parabolic subgroup, if σ is irreducible tempered unitary on M, and if ν is imaginary on $\mathfrak{a}$, then $U(S, \sigma, \nu)$ is tempered on G. [This is by Proposition 7.14.]

(3) If $S = MAN$ and $S' = MAN'$ are parabolic subgroups, if σ is irreducible tempered unitary on M, and if ν on $\mathfrak{a}$ has $\operatorname{Re}\nu$ in the open positive Weyl chamber (relative to S), then $A(S':S:\sigma:\nu)f$ is convergent for K-finite f. [This is by Theorem 7.22.]

In Chapter VIII we shall obtain a converse to (1) as a preliminary to the "Langlands classification." The converse is that an irreducible representation (with the property that it is "admissible" in the sense of the next chapter) whose K-finite matrix coefficients are in $L^{2+\varepsilon}(G)$ for every $\varepsilon > 0$ is necessarily tempered.

Lemma 7.23. Let $S = MAN$ be a parabolic subgroup, let σ be irreducible tempered unitary on M, and let ν on $\mathfrak{a}$ be such that $\operatorname{Re}\nu$ is in the open positive Weyl chamber. For f and g K-finite in the space of $U(S, \sigma, \nu)$ and for m in M.

$$\lim_{\substack{a\to\infty\\ S}} e^{-(\nu-\rho)\log a}(U(S, \sigma, \nu, ma)f, g) = (\sigma(m)A(\bar{S}:S:\sigma:\nu)f(1), g(1))_{V^\sigma}.$$

Remark. Here $\bar{S}$ denotes ΘS, and $\lim_{a\to\infty, S}$ means that $\log a$ tends to infinity in $\mathfrak{a}^+$ in such a way that every positive root, applied to $\log a$, tends to infinity.

Formal argument. We have

$$
\begin{aligned}
&(U(S, \sigma, \nu, ma)f, g) \\
&= \int_K e^{-(\nu+\rho)H(a^{-1}m^{-1}k)}(\sigma^{-1}(\mu(a^{-1}m^{-1}k))f(\kappa(a^{-1}m^{-1}k)), g(k))_{V^\sigma}\, dk \\
&= \int_{\bar N} e^{-(\nu+\rho)H(a^{-1}m^{-1}\kappa(\bar n))}(\sigma^{-1}(\mu(a^{-1}m^{-1}\kappa(\bar n)))f(\kappa(a^{-1}m^{-1}\kappa(\bar n))), g(\kappa(\bar n)))_{V^\sigma} \\
&\qquad \times e^{-2\rho H(\bar n)}\, d\bar n \qquad \text{by (5.25)} \\
&= \int_{\bar N} e^{-(\nu+\rho)H(a^{-1}m^{-1}\bar n)} e^{(\nu-\rho)H(\bar n)} \\
&\qquad \times (\sigma(\mu(\bar n))(\sigma^{-1}(\mu(a^{-1}m^{-1}\bar n))f(\kappa(a^{-1}m^{-1}\bar n)), g(\kappa(\bar n)))_{V^\sigma}\, d\bar n \\
&= e^{(\nu+\rho)\log a}\int_{\bar N} e^{-(\nu+\rho)H(a^{-1}m^{-1}\bar n m a)} e^{(\nu-\rho)H(\bar n)} \\
&\qquad \times (\sigma(\mu(\bar n))\sigma(m)\sigma^{-1}(\mu(a^{-1}m^{-1}\bar n m a))f(\kappa(a^{-1}m^{-1}\bar n m a)), g(\kappa(\bar n)))_{V^\sigma}\, d\bar n \\
&= e^{(\nu-\rho)\log a}\int_{\bar N} e^{-(\nu+\rho)H(\bar n)} e^{(\nu-\rho)H(am\bar n m^{-1}a^{-1})} \\
&\qquad \times (\sigma(\mu(am\bar n m^{-1}a^{-1}))\sigma(m)\sigma^{-1}(\mu(\bar n))f(\kappa(\bar n)), g(\kappa(am\bar n m^{-1}a^{-1})))_{V^\sigma}\, d\bar n
\end{aligned}
$$

As a tends to infinity in the sense of $\lim_{a\to\infty, S}$, $a(m\bar n m^{-1})a^{-1}$ tends to 1 for each $\bar n \in \bar N$ and $m \in M$. Interchanging limit and integral, we thus obtain

$$
\begin{aligned}
&\lim_{\substack{a\to\infty\\ S}} e^{-(\nu-\rho)\log a}(U(S, \sigma, \nu, ma)f, g) \\
&\qquad = \int_{\bar N} e^{-(\nu+\rho)H(\bar n)}(\sigma(m)\sigma(\mu(\bar n))^{-1}f(\kappa(\bar n)), g(1))_{V^\sigma}\, d\bar n \\
&\qquad = \int_{\bar N} (\sigma(m)f(\bar n), g(1))_{V^\sigma}\, d\bar n \\
&\qquad = (\sigma(m)A(\bar S:S:\sigma:\nu)f(1), g(1))_{V^\sigma},
\end{aligned}
$$

as required.

Argument for interchange of limit and integral. We use dominated convergence. In the case of a minimal parabolic ($S = S_{\mathfrak p}$), we may as well take $m = 1$, and we can drop the term with the inner product over V^σ, by the Schwarz inequality. The problem is to show that $e^{-(\operatorname{Re}\nu+\rho)H(\bar n)}e^{(\operatorname{Re}\nu-\rho)H(a\bar n a^{-1})}$ is dominated by a fixed integrable function as a tends to infinity. Since $\operatorname{Re}\nu$ is in the open positive Weyl chamber, we can write

$$\operatorname{Re}\nu = \varepsilon\rho + \lambda \text{ with } \lambda \text{ dominant and } 0 < \varepsilon \le 1.$$

Then our function is

$$
\begin{aligned}
&e^{-(\operatorname{Re}\nu+\rho)H(\bar n)}e^{(\operatorname{Re}\nu-\rho)H(a\bar n a^{-1})} \\
&\qquad = e^{-(1+\varepsilon)\rho H(\bar n)}\{e^{-(1-\varepsilon)\rho H(a\bar n a^{-1})}\}\{e^{-\lambda H(\bar n)}e^{\lambda H(a\bar n a^{-1})}\}.
\end{aligned}
$$

The two expressions in braces are each ≤ 1, by Lemma 7.16a, and $e^{-(1+\varepsilon)\rho H(\bar{n})}$ is integrable by Corollary 7.7. Hence we indeed have dominated convergence. For the argument in the general case, see the bibliographical notes.

Theorem 7.24. Let $S = MAN$ be a parabolic subgroup, let σ be irreducible tempered unitary on M, and let ν on $\mathfrak{a}$ be such that $\operatorname{Re}\nu$ is in the open positive Weyl chamber. Then $U(S, \sigma, \nu)$ has a unique irreducible quotient $J(S, \sigma, \nu)$, and $J(S, \sigma, \nu)$ is isomorphic with the image of the intertwining operator $A(\bar{S}:S:\sigma:\nu)$ on $U(S, \sigma, \nu)$.

Remarks. $J(S, \sigma, \nu)$ is called the **Langlands quotient.** To be completely rigorous, we should work only with K-finite vectors and with representations of the Lie algebra $\mathfrak{g}$. We will, however, continue to ignore this complication for now, returning to it in §8.3, where we shall reformulate the theorem as Theorem 7.24′. In any event, let us note that the image $A(\bar{S}:S:\sigma:\nu)(U(S, \sigma, \nu))$ is a subrepresentation of $U(\bar{S}, \sigma, \nu)$.

Proof. We are to show that (the space of K-finite vectors of) any proper closed invariant subspace of $U(S, \sigma, \nu)$ is contained in the kernel of $A(\bar{S}:S:\sigma:\nu)$. It is enough to show that if a K-finite f is not in the kernel of $A(\bar{S}:S:\sigma:\nu)$, then f is cyclic under the action of $U(S, \sigma, \nu)$, in the sense that the closure of the linear span of the translates of f is dense. Failure to be cyclic means that there is a K-finite function $g \not\equiv 0$ with $(U(S, \sigma, \nu, x)f, g) = 0$ for all x in G. Let us write U_ν for $U(S, \sigma, \nu)$.

Replacing x by $k_1 m a k_2$ with $k_1, k_2 \in K$, $m \in M$, and $a \in A$, we see that

$$(U_\nu(ma)U_\nu(k_2)f,\ U_\nu(k_1)^{-1}g) = 0,$$

since $U_\nu(k_1)^{-1}$ is trivially unitary. Applying Lemma 7.23, we obtain

$$(\sigma(m)A(\bar{S}:S:\sigma:\nu)U_\nu(k_2)f(1),\ U_\nu(k_1)^{-1}g(1))_{V^\sigma} = 0$$

for all $k_1, k_2 \in K$ and $m \in M$. Since σ is irreducible, we conclude either that

$$A(\bar{S}:S:\sigma:\nu)U_\nu(k_2)f(1) = 0 \qquad \text{for all } k_2 \in K \tag{7.73a}$$

or

$$U_\nu(k_1)^{-1}g(1) = 0 \qquad \text{for all } k_1 \in K. \tag{7.73b}$$

Since the intertwining operator commutes with $U_\nu(K)$, (7.73a) says that

$$0 = U_\nu(k_2)A(\bar{S}:S:\sigma:\nu)f(1) = A(\bar{S}:S:\sigma:\nu)f(k_2^{-1}),$$

i.e., f is in the kernel of $A(\bar{S}:S:\sigma:\nu)$, contrary to assumption. Thus (7.73b) must hold, and this means that $g = 0$. This completes the proof.

§12. Problems

1. Let MAN be any proper parabolic subgroup of G. Use Corollary 7.5 to prove that $e^{-\rho H(\bar{n})}$ is not integrable on $\bar{N}$.

Problems 2 to 6 work with the Gindikin-Karpelevič formula for SL(3, $\mathbb{R}$), using the upper triangular subgroup as $M_\mathfrak{p}A_\mathfrak{p}N_\mathfrak{p}$. Introduce parameters for $\bar{N}_p$ by writing

$$\bar{N}_\mathfrak{p} = \left\{\begin{pmatrix} 1 & 0 & 0 \\ x & 1 & 0 \\ z & y & 1 \end{pmatrix}\right\}.$$

2. Using Proposition 7.17, show that
$$e^{2\rho_\mathfrak{p} H_\mathfrak{p}(\bar{n})} = (1 + x^2 + z^2)(1 + y^2 + (z - xy)^2) \qquad \text{for } \bar{n} \in \bar{N}_\mathfrak{p}.$$
3. Introduce parameters that describe a general ν in $\mathfrak{a}_\mathfrak{p}'$, and calculate $e^{-(\nu + \rho_\mathfrak{p})H_\mathfrak{p}(\bar{n})}$ as in Problem 2.
4. Let
$$\mathfrak{n} = \left\{\begin{pmatrix} 0 & a & c \\ 0 & 0 & b \\ 0 & 0 & 0 \end{pmatrix}\right\}, \qquad \mathfrak{n}' = \left\{\begin{pmatrix} 0 & a & c \\ 0 & 0 & 0 \\ 0 & b & 0 \end{pmatrix}\right\}, \qquad \mathfrak{n}'' = \left\{\begin{pmatrix} 0 & 0 & 0 \\ a & 0 & 0 \\ c & b & 0 \end{pmatrix}\right\}.$$
Take Haar measure on $\bar{N} \cap N'' = \bar{N}_\mathfrak{p}$ to be $dx\,dy\,dz$, on $\bar{N}' \cap N''$ to be $dx\,dz$, and on $\bar{N} \cap N'$ to be dy. Verify (7.21) for these normalizations.
5. With notation as in Problem 4, write out the two sides of (7.22) in coordinates. [Hint: Unwind H' by using the reflection $s_{e_2 - e_3}$ to relate H' to H.]
6. With notation as in Problem 5, find a Euclidean change of variables that proves (7.22) directly. [Hint: Regard the part of the denominator containing y as a quadratic expression in y. Start by completing the square.]

Problems 7 to 9 show that induction is a much richer construct than is at first suggested in this chapter, even when it is considered just from a parabolic subgroup MAN to G. Let l be a representation of MAN on a Hilbert space V^l, and define

$$V^L = \{F : G \to V^l \text{ continuous} \mid F(xman) = e^{-\rho \log a} l(man)^{-1} F(x)\}.$$

Let L act by the left regular representation of G on V^L, use the $L^2(K)$ norm, and complete. Call this representation $L = \operatorname{ind}_{MAN}^G l$.

7. Show that induction carries MAN intertwining operators to G intertwining operators. That is, if $e : V^{l_1} \to V^{l_2}$ commutes with the action

of MAN, then the map E defined by $(EF)(x) = e(F(x))$ carries $V^{\mathrm{ind}\, l_1}$ to $V^{\mathrm{ind}\, l_2}$, commuting with the action of G.

8. Conclude from Problem 7 that induction preserves equivalences.
9. With the convention of Problem 7 that induction acts on maps, as well as representations, show that induction carries short exact sequences of representations on the MAN level to short exact sequences of representations on the G level.

Problems 10 to 14 deal with the formula (7.39) for the intertwining operator in the noncompact picture.

10. Calculate the kernel

$$e^{-(\rho-\nu)\log a(w^{-1}\bar{n})}\sigma(m(w^{-1}\bar{n}))$$

of the intertwining operator for the groups $SO(n, 1)$ and $SU(n, 1)$. [See Problems 14 to 17 in Chapter V for preliminary information.]

11. Show for all real-rank-one groups that the A part

$$f(\bar{n}) = e^{-(\rho-\nu)\log a(w^{-1}\bar{n})}$$

of the kernel satisfies
   (a) $f(\bar{n})$ is smooth away from $\bar{n} = 1$
   (b) $f(a_0\bar{n}a_0^{-1}) = e^{2(\nu-\rho)\log a_0}f(\bar{n})$ for a_0 in A
   (c) $f(m\bar{n}m^{-1}) = f(\bar{n})$ for m in M
   (d) $f(\bar{n})\,d\bar{n}$, in the case $\nu = 0$, is a measure on $\bar{N}$ invariant under "dilations" $\bar{n} \to a_0\bar{n}a_0^{-1}$ with a_0 in A.
12. Show for all real-rank-one groups that the M part $h(\bar{n}) = \sigma(m(w^{-1}\bar{n}))$ of the kernel satisfies
   (a) $h(\bar{n})$ is smooth away from $\bar{n} = 1$
   (b) $h(a_0\bar{n}a_0^{-1}) = h(\bar{n})$ for a_0 in A
   (c) $h(m_0\bar{n}m_0^{-1}) = \sigma(w^{-1}m_0w)h(\bar{n})\sigma(m_0^{-1})$ for m_0 in M.
13. Suppose that MAN is a maximal parabolic subgroup of G, that $|W(A:G)| = 2$, and that w represents the nontrivial element of $W(A:G)$.
   (a) Adjust the derivation of (7.39) so that it remains valid in this generality.
   (b) Which conclusions of Problems 11 and 12 remain valid?
14. For $SL(2n, \mathbb{R})$, let MAN be the block upper triangular subgroup built from two n-by-n blocks.
   (a) Show that $|W(A:G)| = 2$.
   (b) Under the identification of $\bar{N}$ with the lower left n-by-n block, show that the A part of the kernel is a power of the absolute value of the determinant.

CHAPTER VIII

Admissible Representations

§1. Motivation

Our motivation revolves around using the nonunitary principal series

$$U = U(S_{\mathfrak{p}}, \sigma, \nu) = \mathrm{ind}_{S_{\mathfrak{p}}}^{G}(\sigma \otimes \exp \nu \otimes 1)$$

to get a handle on general representations.

Let us avoid technicalities and work with the continuous members of the induced space V. Then we can define an evaluation mapping $e: V \to V^{\sigma}$ by

$$ef = f(1) \qquad \text{for } f \text{ in } V.$$

The map e is $N_{\mathfrak{p}}$-invariant in that

$$eU(n)f = U(n)f(1) = f(n^{-1}) = f(1) = ef,$$

and it has the following transformation law under a in $A_{\mathfrak{p}}$:

$$eU(a)f = U(a)f(1) = f(a^{-1}) = e^{(\nu + \rho_{\mathfrak{p}}) \log a} f(1) = e^{(\nu + \rho_{\mathfrak{p}}) \log a} ef.$$

If we compose e with a linear functional on V^{σ}, then we get a member of the dual space V' with these same properties.

Now suppose π is a general representation of G, say on a space V^{π}, and suppose there exists $l \neq 0$ in $(V^{\pi})'$ with

$$\begin{aligned} l(\pi(n)v) &= l(v) && \text{for } n \in N_{\mathfrak{p}}, v \in V^{\pi} \\ l(\pi(a)v) &= e^{(\nu + \rho_{\mathfrak{p}}) \log a} l(v) && \text{for } a \in A_{\mathfrak{p}}, v \in V^{\pi}. \end{aligned} \tag{8.1}$$

We map $v \to f_v$ with $f_v(x) = l(\pi(x)^{-1} v)$ and see that

$$f_v(xan) = e^{-(\nu + \rho_{\mathfrak{p}}) \log a} f_v(x)$$

$$f_{\pi(g)v}(x) = f_v(g^{-1} x).$$

Hence we have a nonzero G map from V^{π} into the space of a representation that it is natural to call

$$\mathrm{ind}_{A_{\mathfrak{p}} N_{\mathfrak{p}}}^{G}(\exp \nu \otimes 1).$$

By double induction

$$\operatorname{ind}_{A_\mathfrak{p}N_\mathfrak{p}}^G(\exp \nu \otimes 1) = \operatorname{ind}_{M_\mathfrak{p}A_\mathfrak{p}N_\mathfrak{p}}^G[(\operatorname{ind}_1^{M_\mathfrak{p}}(1)) \otimes \exp \nu \otimes 1],$$

Here $\operatorname{ind}_1^{M_\mathfrak{p}}(1) = L^2(M_\mathfrak{p})$ decomposes discretely by the Peter-Weyl Theorem, since $M_\mathfrak{p}$ is compact. Thus our map $v \to f_v$ gives us a nonzero G map into $U(S_\mathfrak{p}, \sigma, \nu)$ for some σ. If π is irreducible in a suitable sense, this map has to be one-one, and π is exhibited as a subrepresentation of $U(S_\mathfrak{p}, \sigma, \nu)$.

A source of such linear functionals l is the asymptotics of matrix coefficients. In the simplest case suppose that

$$\lim_{\substack{a \to \infty \\ S}} e^{-(\nu - \rho_\mathfrak{p}) \log a} \langle \pi(a)v, v' \rangle = c_{v,v'} \tag{8.2}$$

exists for all v and some v', and suppose $c_{v,v'}$ is not identically 0. Put $l(v) = c_{v,v'}$. Then

$$l(\pi(a_0)v) = e^{(\nu - \rho_\mathfrak{p}) \log a_0} l(v) \tag{8.3}$$

trivially, and

$$\begin{aligned} l(\pi(\bar{n})v) &= \lim e^{-(\nu - \rho_\mathfrak{p}) \log a} \langle \pi(a\bar{n}a^{-1})\pi(a)v, v' \rangle \\ &= \lim e^{-(\nu - \rho_\mathfrak{p}) \log a} \langle \pi(a)v, \pi^{\mathrm{tr}}(a\bar{n}a^{-1})v' \rangle \qquad \text{for } \bar{n} \in \bar{N}_\mathfrak{p} = \Theta N_\mathfrak{p}. \end{aligned}$$

Since $a\bar{n}a^{-1} \to 1$, we can expect in the best case that we can replace $\pi^{\mathrm{tr}}(a\bar{n}a^{-1})v'$ by v' and get $l(v)$. Thus we expect

$$l(\pi(\bar{n})v) = l(v) \qquad \text{for } \bar{n} \in \bar{N}_\mathfrak{p}, v \in V^\pi. \tag{8.4}$$

Properties (8.3) and (8.4) are an adequate substitute for (8.1); in fact, $\rho_\mathfrak{p}$ for $\bar{S}_\mathfrak{p} = \Theta S_\mathfrak{p}$ is the negative of $\rho_\mathfrak{p}$ for $S_\mathfrak{p}$, and thus l defines a nonzero G map into some $U(\bar{S}_\mathfrak{p}, \sigma, \nu)$. (We have to get $\bar{S}_\mathfrak{p}$ rather than $S_\mathfrak{p}$; see Lemma 7.23.)

In the general case we will not quite get (8.2), but we get something close enough: an asymptotic expression for the matrix coefficients in which several terms predominate. A linear functional that evaluates the contribution from one such leading term behaves like l above. The fact that we have some control over the asymptotics of matrix coefficients comes from differential equations. For an irreducible representation every element of the center of the universal enveloping algebra will act as a scalar on nice matrix coefficients. The resulting system of differential equations on G can be reduced to a system on $A_\mathfrak{p}$ if we assume some transformation law under K on both sides, and it will be seen that the system is large enough so that the theory of Appendix B applies.

We shall take up this motivation again, in §4, after making precise the class of representations we study and after obtaining some preliminary information about the center of the universal enveloping algebra.

§2. Admissible Representations

Let G be linear connected reductive. If π is a representation of G on a Hilbert space V and if v is in V, we have called v **K-finite** if $\pi(K)v$ spans a finite-dimensional space. When K acts by unitary operators (e.g., if π is unitary or if π is one of the induced representations of Chapter VII), $\pi|_K$ splits into the orthogonal sum of spaces on which $\pi|_K$ is irreducible, by the Peter-Weyl Theorem, and the multiplicity of each irreducible representation of K is well defined, by the remarks just before Lemma 1.13. Thus we can write

$$\pi|_K \cong \sum_{\tau \in \hat{K}} n_\tau \tau, \tag{8.5}$$

where $\hat{K}$ is the set of equivalence classes of irreducible representations of K and where each **multiplicity** n_τ is a nonnegative integer or is $+\infty$. Any vector in one of the irreducible K spaces is K-finite, of course, and thus the subspace of K-finite vectors is dense. The equivalence classes τ that occur in (8.5) with positive multiplicity are called the K **types** that occur in π.

For a holomorphic discrete series representation (Chapter VI), the K-finite vectors are functions that restrict to polynomials on the bounded symmetric domain and can be analyzed; it is not hard to see that each K type has finite multiplicity. This is a general phenomenon, according to the theorem below.

Theorem 8.1. For G linear connected reductive and π an irreducible unitary representation on V, the multiplicities n_τ of the K types in $\pi|_K$ given in (8.5) satisfy $n_\tau \leq \dim \tau$ for every τ in $\hat{K}$.

By way of preparation let $M_n(\mathbb{C})$ be the full n-by-n matrix algebra over $\mathbb{C}$ of dimension n^2, and let $\mathscr{S}_r$ be the symmetric group on r letters. For any r the expression

$$\sum_{\varepsilon \in \mathscr{S}_r} (\operatorname{sgn} \varepsilon) X_{\varepsilon(1)} X_{\varepsilon(2)} \cdots X_{\varepsilon(r)} \qquad \text{with } X_i \in M_n(\mathbb{C}), \qquad 1 \leq i \leq r,$$

is an alternating multilinear expression on $M_n(\mathbb{C})$ and hence vanishes when the X_i's are linearly dependent. Thus

$$\sum_{\varepsilon \in \mathscr{S}_r} (\operatorname{sgn} \varepsilon) X_{\varepsilon(1)} X_{\varepsilon(2)} \cdots X_{\varepsilon(r)} = 0 \tag{8.6}$$

identically if $r > n^2$. Let $r(n)$ be the least integer for which it vanishes identically for all X.

Lemma 8.2. $r(n) \geq r(n-1) + 2$.

Proof omitted.

Lemma 8.3. $\pi(C_{\text{com}}(G))$ is strongly dense in the space $L(V)$ of all bounded linear operators on V.

Proof omitted. (This is true of an irreducible unitary representation of a locally compact unimodular group and follows from the double commutant theorem.)

Proof of Theorem 8.1. We may assume G is semisimple since the center of G acts as scalars. The argument is then in four steps.

(1) Let τ be in $\hat{K}$ and let $n > \dim \tau$; then $n\tau$ has no **cyclic vector** (one whose translates under K span the whole space). In fact, write $n\tau = \tau \otimes \iota$, where ι is the trivial n-dimensional representation, and suppose w_0 in $V^\tau \otimes V^\iota$ is a cyclic vector. Say

$$w_0 = u_1 \otimes v_1 + \ldots + u_d \otimes v_d,$$

where $d = \dim \tau$, $\{u_i\}$ is a basis of V^τ, and v_i is in V^ι. Put $H = \operatorname{span}\{v_i\} \subsetneqq V^\iota$. Then

$$(\tau \otimes \iota)(K)w_0 \subseteq V^\tau \otimes H \subsetneqq V^\tau \otimes V^\iota.$$

(2) If Φ is an irreducible finite-dimensional representation of G (which we may take unitary under K) and if $\Phi|_K = \sum n_\tau \tau$, then $n_\tau \leq \dim \tau$. In fact, let $G = KAN$ be an Iwasawa decomposition and let v_0 be a simultaneous nonzero eigenvector under AN (either by Lie's Theorem (A.12) or by the theory of the highest weight). Since $G = KAN$ and since Φ is irreducible, v_0 is cyclic under K. Hence the projection of v_0 on $n_\tau \tau$ is cyclic for $n_\tau \tau$, and (1) implies $n_\tau \leq \dim \tau$.

(3) If τ is in $\hat{K}$, if d and χ denote the dimension and character of τ, and if ψ is the product $d\bar{\chi}$ (so that the K convolution $\psi *_K \psi$ equals ψ as in Lemma 1.11), then $\psi *_K C_{\text{com}}(G) *_K \psi$ is a subalgebra $\mathscr{A}$ of $C_{\text{com}}(G)$ as a convolution algebra, and (8.6) holds in $\mathscr{A}$ with $r = r(d^2)$.

In fact, let C be the space of all linear combinations of matrix coefficients of finite-dimensional representations of G. C is an algebra of functions on G (because of tensor products) closed under conjugation (because of complex conjugate representations), and it separates points since G is linear. By the Stone-Weierstrass Theorem any $f \neq 0$ in $C_{\text{com}}(G)$ is the uniform limit on compact sets of members of C, and we can thus find g in C with $\int_G f(x)g(x)\,dx \neq 0$. This means that $\Phi(f) \neq 0$ for a suitable finite-dimensional representation Φ of G. Since finite-dimensional representations of G are fully reducible, we may take Φ irreducible.

If (8.6) is false in $\mathscr{A}$ for $r = r(d^2)$, let f be a nonzero function in $\mathscr{A}$ of the form

$$f = \sum_{\varepsilon \in \mathscr{S}_r} (\operatorname{sgn} \varepsilon) f_{\varepsilon(1)} * \ldots * f_{\varepsilon(r)}, \qquad r = r(d^2),$$

and find Φ irreducible finite-dimensional with $\Phi(f) \neq 0$, by the previous paragraph. Then

$$0 \neq \Phi(f) = \sum_{\varepsilon \in \mathscr{S}_r} (\operatorname{sgn} \varepsilon)\Phi(f_{\varepsilon(1)}) \cdots \Phi(f_{\varepsilon(r)}), \qquad r = r(d^2). \tag{8.7}$$

Since $\mathscr{A} = \psi *_K C_{\text{com}}(G) *_K \psi$, we have

$$\Phi(\mathscr{A}) = \Phi(\psi)\Phi(C_{\text{com}}(G))\Phi(\psi). \tag{8.8}$$

Here $\Phi(\psi)$ is the orthogonal projection on the subspace of the representation space of Φ corresponding to the K type τ. This subspace has dimension $\leq d^2$, by (2), and (8.8) says that the members of $\Phi(\mathscr{A})$ can be regarded as linear transformations of this space. Since $r = r(d^2)$ in (8.7), the right side of (8.7) is 0, contradiction.

(4) If the given π decomposes as $\pi|_K = \sum n_\tau \tau$, then $n_\tau \leq \dim \tau$ for each τ. In fact, fix τ and let the notation be as in (3). The operator $\pi(\psi)$ is the orthogonal projection E on the space S of $n_\tau \tau$. Hence (3) implies that (8.6) holds in the algebra $E\pi(C_{\text{com}}(G))E$ with $r = r(d^2)$. By Lemma 8.3, $\pi(C_{\text{com}}(G))$ is strongly dense in $L(V)$, and hence $E\pi(C_{\text{com}}(G))E$ is strongly dense in $EL(V)E = L(S)$. Consequently (8.6) holds in the algebra $L(S)$ for $r = r(d^2)$. By Lemma 8.2, $\dim S \leq d^2$. Thus $n_\tau \leq \dim \tau$, and the theorem is proved.

We say that a representation π of a linear connected reductive group G on a Hilbert space V is **admissible** if $\pi(K)$ operates by unitary operators and if each τ in K occurs with only finite multiplicity in $\pi|_K$. Theorem 8.1 says that irreducible unitary representations are admissible, and Proposition 8.4 below says that the induced representations of Chapter VII are admissible. The useful notion of equivalence of admissible representations is not the obvious one and will be introduced later in this section; it is called "infinitesimal equivalence."

Proposition 8.4. If G is linear connected reductive, if $S = MAN$ is a parabolic subgroup, and if σ is irreducible unitary on M, then $U(S, \sigma, \nu)$ is admissible.

Proof (disregarding any disconnectedness of M). We know that $U(S, \sigma, \nu, k)$ is unitary for k in K. Theorem 8.1 shows that

$$\sigma|_{K \cap M} = \sum_{\omega \in (K \cap M)^\wedge} n_\omega \omega$$

with each n_ω finite. Now a look at the definitions shows that

$$U(S, \sigma, \nu)|_K = \operatorname{ind}_{K \cap M}^K \sigma, \tag{8.9}$$

where the right side has the same meaning as in Chapter I. Then (8.9) is

$$= \sum_{\omega \in (K \cap M)^\wedge} n_\omega \operatorname{ind}_{K \cap M}^K \omega,$$

If τ is in $\hat{K}$, then it follows that

$$[U(S, \sigma, \nu)|_K : \tau] = \sum_{\omega \in (K \cap M)^\wedge} n_\omega[\operatorname{ind}_{K \cap M}^K \omega : \tau]$$
$$= \sum_{\omega \in (K \cap M)^\wedge} n_\omega[\tau|_{K \cap M} : \omega],$$

the last step following by Frobenius reciprocity (Theorem 1.14). Since each n_ω is finite, the right side is clearly finite, and the result follows.

Remarks.

(1) If we use the estimate $n_\omega \leq \dim \omega$, which remains valid even for disconnected M, then we see from the proof that $[U(S, \sigma, \nu)|_K : \tau] \leq \dim \tau]$.

(2) For ν imaginary and in general position and for $S = S_\mathfrak{p}$ minimal, Theorem 7.2 says $U(S_\mathfrak{p}, \sigma, \nu)$ is irreducible. For such ν, Proposition 8.4 follows from Theorem 8.1. The decomposition of $U(S_\mathfrak{p}, \sigma, \nu)|_K$ is independent of ν, by (8.9), and thus we have an alternate proof of the proposition when $S = S_\mathfrak{p}$ is minimal.

Proposition 8.5. For an admissible representation π, every K-finite vector is a C^∞ vector, and the space of K-finite vectors is stable under $\pi(\mathfrak{g})$.

Proof. Let f be a K-finite function on K, let h be in $C_{\text{com}}^\infty(\exp \mathfrak{p})$, and put $F(k \exp X) = f(k)h(\exp X)$ for $k \in K$, $X \in \mathfrak{p}$. Then F is in $C_{\text{com}}^\infty(G)$, and $\pi(F)v$ is a C^∞ vector, by Proposition 3.14, for each v. It is also K-finite, since

$$\pi(k_0)\pi(F)v = \int_G F(x)\pi(k_0 x)v\,dx = \int_G F(k_0^{-1}x)\pi(x)v\,dx$$

and since F is left K-finite. Then an approximation argument like that in Theorem 3.15 shows that the C^∞ K-finite vectors are dense. Since the spaces for the various K types are orthogonal, it follows that the C^∞ vectors of each K type are dense in the space of all vectors of that K type. Admissibility means that the latter space is finite-dimensional, and it follows that every vector in it is C^∞. Hence every K-finite vector is C^∞.

To see that the space of K-finite vectors is stable under $\pi(\mathfrak{g})$, let v be K-finite and let V_v be the finite-dimensional space $\pi(U(\mathfrak{k}^{\mathbb{C}}))v$. If X is in $\mathfrak{k}$, Y is in $\mathfrak{g}$, and v' is in V_v, then

$$\pi(X)\pi(Y)v' = \pi(Y)\pi(X)v' + \pi[X, Y]v'$$

shows that the linear span of $\pi(\mathfrak{g})V_v$ is stable under $\pi(\mathfrak{k})$. Since this linear span is finite-dimensional, we can exponentiate and conclude that $\pi(\mathfrak{g})V_v$ is stable under $\pi(K)$. Therefore Y in $\mathfrak{g}$ implies that $\pi(Y)v$ lies in a finite-dimensional space stable under $\pi(K)$ and is thus K-finite. This completes the proof.

Consequently an admissible π leads to a representation of $\mathfrak{g}$ on the space of K-finite vectors. We say two admissible representations π and π' of G are **infinitesimally equivalent** if the associated representations of $\mathfrak{g}$ on K-finite vectors are algebraically equivalent (i.e., if there is a linear isomorphism commuting with the action of $\mathfrak{g}$). Infinitesimal equivalence does not in general imply equivalence by bounded operators. However, we shall observe ultimately that infinitesimally equivalent irreducible unitary representations are unitarily equivalent.

§3. Invariant Subspaces

The main results of this section will follow from knowing that the K-finite matrix coefficients of an admissible representation are real analytic functions. To get at this real analyticity, we introduce the Casimir element Ω of the universal enveloping algebra.

Thus let G be linear connected reductive, and let $C(X, Y) = \operatorname{Re}\operatorname{Tr}(XY)$ be the real part of the trace form. Choose a basis $X_1, \ldots, X_n$ of $\mathfrak{g}$, and let $g_{ij} = C(X_i, X_j)$. Since C is nondegenerate, (g_{ij}) is nonsingular; let (g^{ij}) denote the inverse matrix $(g_{ij})^{-1}$. Put $X^j = \sum g^{ij} X_i$, so that $X_i = \sum g_{ji} X^j$. The **Casimir element** of $U(\mathfrak{g}^{\mathbb{C}})$ is defined by

$$\Omega = \sum_{i,j} g_{ij} X^i X^j.$$

Proposition 8.6. The Casimir element Ω of $U(\mathfrak{g}^{\mathbb{C}})$ is independent of the basis. It satisfies $\operatorname{Ad}(x)\Omega = \Omega$ for all x in G and hence is in the center $Z(\mathfrak{g}^{\mathbb{C}})$ of $U(\mathfrak{g}^{\mathbb{C}})$.

Proof. Let $Y_j = \sum a_j^i X_i$ be another basis, and let $(b_j^i) = (a_j^i)^{-1}$, so that $\sum_k a_k^i b_j^k = \delta_j^i = \sum_k a_j^k b_k^i$. Define

$$h_{ij} = C(Y_i, Y_j) = \sum_{k,m} a_i^k a_j^m C(X_k, X_m) = \sum_{k,m} a_i^k g_{km} a_j^m.$$

If we put $H^{jn} = \sum b_{k'}^j g^{k'm'} b_{m'}^n$, then

$$\sum_j h_{ij} H^{jn} = \sum a_i^k g_{km} a_j^m b_{k'}^j g^{k'm'} b_{m'}^n = \delta_i^n,$$

so that $(H^{ij}) = (h_{ij})^{-1}$. Thus

$$\begin{aligned} Y^j &= \sum h^{ij} Y_i = \sum b_k^i g^{km} b_m^j Y_i = \sum b_k^i g^{km} b_m^j a_i^{i'} X_{i'} \\ &= \sum g^{km} b_m^j X_k = \sum g^{km} b_m^j g_{rk} X^r = \sum b_r^j X^r. \end{aligned}$$

Letting Ω_Y and Ω_X denote the two possible versions of Ω, we thus have

$$\Omega_Y = \sum h_{ij} Y^i Y^j = \sum a_i^k g_{km} a_j^m b_r^i b_s^j X^r X^s = \sum g_{rs} X^r X^s = \Omega_X.$$

Hence Ω is independent of the basis.

Now if x is in G, we have

$$\begin{aligned}\operatorname{Ad}(x)\Omega &= \sum_{i,j} C(X_i, X_j)\operatorname{Ad}(x)X^i \operatorname{Ad}(x)X^j \\ &= \sum C(\operatorname{Ad}(x)^{-1}X_i, \operatorname{Ad}(x)^{-1}X_j)\operatorname{Ad}(x)X^i \operatorname{Ad}(x)X^j && \text{by invariance of } C \\ &= \sum C(X_i, X_j)X^iX^j && \text{by the above independence} \\ &= \Omega.\end{aligned}$$

By Proposition 3.8, Ω is in $Z(\mathfrak{g}^{\mathbb{C}})$.

Theorem 8.7. If G is linear connected reductive and π is an admissible representation of G, then every matrix coefficient of π of the form $g \to (\pi(g)u, v)$ for u K-finite is a real analytic function on G.

Remark. We do not need to assume v is K-finite.

Proof. The idea is to show that $(\pi(g)u, v)$ is annihilated by an elliptic differential operator with real analytic coefficients. If X is in $\mathfrak{g}$ and we regard X as acting as a left-invariant vector field, then

$$\begin{aligned}X(\pi(g)u, v) &= \frac{d}{dt}(\pi(g \exp tX)u, v)\bigg|_{t=0} \\ &= \frac{d}{dt}(\pi(\exp tX)u, \pi(g)^*v)\bigg|_{t=0} = (\pi(g)\pi(X)u, v).\end{aligned}$$

Iterating this equality, we obtain the fundamental formula

$$\boxed{D(\pi(g)u, v) = (\pi(g)\pi(D)u, v)} \tag{8.10}$$

for every D in $U(\mathfrak{g}^{\mathbb{C}})$.

Without loss of generality, we may assume u is wholly in the space S_τ transforming according to a single K type τ. For an irreducible representation of K of type τ, the Casimir operator Ω_K of K must act as a scalar c_τ, by Proposition 8.6 and Schur's Lemma. Thus $\pi(\Omega_K) = c_\tau$ on all of S_τ. The Casimir operator Ω of G commutes with $\pi(K)$, by Proposition 8.6, and hence maps S_τ into itself. By admissibility S_τ is finite-dimensional. Thus we can find constants $c_1, \ldots, c_n$, not necessarily distinct, such that

$$\prod_{j=1}^{n} (\pi(\Omega) - c_j) = 0 \qquad \text{on } S_\tau. \tag{8.11}$$

Now we can proceed with the construction of the elliptic differential operator. The form $\langle X, Y\rangle = -C(X, \theta Y)$ is an inner product on $\mathfrak{g}$, by (1.2). If we let $\{X_i\}$ be an orthonormal basis of $\mathfrak{k}$ and $\{Y_i\}$ be an orthonormal basis of $\mathfrak{p}$, then our definitions give

$$\Omega = -\sum X_i^2 + \sum Y_i^2$$

$$\Omega_K = -\sum X_i^2.$$

Hence
$$\Omega - 2\Omega_K = \sum X_i^2 + \sum Y_i^2$$

is elliptic as a differential operator on G, as we see by looking at canonical coordinates of the first kind (see (A.95)). Thus also

$$D = \prod_{j=1}^{n} [(\Omega - 2\Omega_K) + (2c_\tau - c_j)]$$

is elliptic. Using (8.10), we obtain

$$\begin{aligned} D(\pi(g)u, v) &= \left(\pi(g) \prod_{j=1}^{n} [\pi(\Omega) - \pi(2\Omega_K) + (2c_\tau - c_j)]u, v\right) \\ &= \left(\pi(g) \prod_{j=1}^{n} (\pi(\Omega) - c_j)u, v\right) \qquad \text{since } \pi(\Omega_K) = c_\tau \text{ on } S_\tau \\ &= 0 \qquad \text{by (8.11).} \end{aligned}$$

Since $(\pi(g)u, v)$ is annihilated by the elliptic differential operator D and since D has real analytic coefficients by (A.96), it follows that $(\pi(g)u, v)$ is real analytic.

Corollary 8.8. If π and π' are infinitesimally equivalent admissible representations of G, then π and π' have the same sets of K-finite matrix coefficients.

Remark. Here we are assuming that both the u and the v in $(\pi(g)u, v)$ are K-finite.

Proof. Let π act on V. We can characterize the linear functionals $(\cdot, v)$ for v K-finite in V as follows. Namely let V_0 be the K-finite subspace and let V_0' be its vector space dual. Then there is a natural transpose action $\pi^{\mathrm{tr}}(K)$ on V_0', and we let $(V_0')_K$ be the K-finite vectors in V_0'. Then $(V_0')_K$ gives us the linear functionals $(\cdot, v)$ for v K-finite.

If u and v are K-finite, the matrix coefficient $(\pi(g)u, v)$ can be characterized, in view of Theorem 8.7 and formula (8.10), as the unique real analytic function on G whose derivatives at $g = 1$ are given by

$$D(\pi(g)u, v)_{g=1} = (\pi(D)u, v).$$

The previous paragraph shows that the right side here is of the form $\langle \pi(D)u, v' \rangle$ for some v' in $(V_0')_K$, v' not depending on D. Thus we have a characterization of the matrix coefficients that is invariant under infinitesimal equivalence, and the corollary follows.

If π is an admissible representation of G on V and if U is a closed G-invariant subspace, then it follows from Proposition 8.5 that the K-finite vectors in U form a $\mathfrak{g}$-invariant subspace dense in U. The theorem below gives a converse result. (For comparison, see the Problems in Chapter III, where a $\mathfrak{g}$-invariant subspace of C^∞ vectors is exhibited whose closure is not G-invariant.)

Theorem 8.9. If G is linear connected reductive and π is an admissible representation of G on V, then the closure in V of any $\mathfrak{g}$-invariant subspace of K-finite vectors is G-invariant.

Proof. Let U_0 be a $\mathfrak{g}$-invariant subspace of K-finite vectors, and let U be its closure. Since each $\pi(g)$, for g in G, is bounded, it suffices to prove that $\pi(G)U_0 \subseteq U$.

Fix u in U_0 and v in $U^\perp$. The function $(\pi(g)u, v)$ is real analytic on G, according to Theorem 8.7. For X sufficiently small in $\mathfrak{g}$, Taylor's Theorem (A.100) therefore gives

$$(\pi(\exp X)u, v) = \sum_{n=0}^{\infty} \frac{1}{n!} X^n(\pi(x)u, v)\Big|_{x=1},$$

and this, by (8.10), is

$$= \sum_{n=0}^{\infty} \frac{1}{n!} (\pi(X)^n u, v),$$

which is 0 since $\pi(X)^n u$ is in U_0. Thus $(\pi(g)u, v)$ vanishes in a neighborhood of the identity. Since $(\pi(g)u, v)$ is real analytic on G, it vanishes everywhere. Since u in U_0 and v in $U^\perp$ are arbitrary, we conclude $\pi(G)U_0 \subseteq U$. This proves the theorem.

Corollary 8.10. Let π be an admissible representation of G on V, and let V_0 be the subspace of K-finite vectors in V. Then the closed G-invariant subspaces U of V stand in one-one correspondence with the $\mathfrak{g}$-invariant subspaces U_0 of V_0, the correspondence being $U_0 = U \cap V_0$ and $U = \bar{U}_0$.

Proof. If U is closed and G-invariant, then its space U_0 of K-finite vectors is $U \cap V_0$ and is $\mathfrak{g}$-invariant by Proposition 8.5. Conversely if $U_0 \subseteq V_0$ is $\mathfrak{g}$-invariant, then its closure U in V is G-invariant by Theorem 8.9. To complete the proof we need to know these operations are inverse to one another—that $\overline{U \cap V_0} = U$. But $\overline{U \cap V_0} \subseteq U$ since U is closed, and $U \cap V_0$ is dense in U since the space U_0 of K-finite vectors of U is dense in U.

Corollary 8.11. Let π be an admissible representation of G on V, and let V_0 be the subspace of K-finite vectors in V. Then $\pi(G)$ has no nontrivial closed invariant subspaces in V if and only if $\pi(\mathfrak{g})$ has no nontrivial invariant subspaces in V_0.

Remarks. This is a special case of Corollary 8.10. If these equivalent conditions on π are satisfied, we say that π is **irreducible admissible.**

Corollary 8.12. If π and π' are irreducible admissible representations of G on V and V' with a nonzero K-finite matrix coefficient in common, then π and π' are infinitesimally equivalent.

Proof. Let V_0 and V'_0 be the respective subspaces of K-finite vectors in V and V'. Our assumption is that

$$(\pi(g)u, v) = (\pi'(g)u', v')$$

for some nonzero u, v in V_0 and u', v' in V'_0. Let V''_0 be the subspace $U(\mathfrak{g}^{\mathbb{C}})(\pi(g)u, v)$ of $C^\infty(G)$, with the action of $U(\mathfrak{g}^{\mathbb{C}})$ given by left-invariant differentiation. Corollary 8.11 implies that $\pi(U(\mathfrak{g}^{\mathbb{C}}))u = V_0$, and thus (8.10) gives

$$V''_0 = (\pi(g)\pi(U(\mathfrak{g}^{\mathbb{C}}))u, v) = (\pi(g)V_0, v).$$

Thus we can define $\varphi: V_0 \to V''_0$ by $\varphi(\pi(D)u) = (\pi(\cdot)\pi(D)u, v)$, and φ will map V_0 onto V''_0. The map φ is $U(\mathfrak{g}^{\mathbb{C}})$-equivariant since D and D' in $U(\mathfrak{g}^{\mathbb{C}})$ imply

$$\begin{aligned}\varphi(\pi(D)(\pi(D')u)) &= \varphi(\pi(DD')u) = (\pi(\cdot)\pi(DD')u, v) \\ &= (\pi(\cdot)\pi(D)\pi(D')u, v) = D(\pi(\cdot)\pi(D')u, v),\end{aligned}$$

by (8.10). Since π is irreducible, $\ker \varphi = 0$. Thus φ is a $U(\mathfrak{g}^{\mathbb{C}})$ isomorphism.

The fact that we have a matrix coefficient in common means that V''_0 is the same space if we start from π'. Thus we construct a $U(\mathfrak{g}^{\mathbb{C}})$ isomorphism $\varphi': V'_0 \to V''_0$ in the same fashion, and $\varphi^{-1}\varphi'$ is the required infinitesimal equivalence.

It is now an easy matter to make rigorous the results of §§7.10–7.11. In the interpretation of Theorem 7.22, the induced representation is admissible. Hence the sum in the formula

$$A(S':S:\sigma:\nu)f(k) = \sum_j \left[\int_{\bar{N} \cap N'} (f(k\bar{n}), \psi_j)_{V^\sigma}\, d\bar{n}\right]\psi_j$$

may be taken as a finite sum. For the ν's in the statement of the theorem, the dependence on ν of each of the integrals on the right in this formula is holomorphic. Moreover, the integral (7.69) is convergent in the strong sense (because of the boundedness of the coefficients of σ) if $\operatorname{Re}\nu$ is large enough so that

$$\int_{\bar{N} \cap N'} e^{-(\operatorname{Re}\nu+\rho)H(\bar{n})}\, d\bar{n}$$

is finite, i.e. if ν is in the smaller region

$$\langle \operatorname{Re} \nu - \rho_M, \beta \rangle > 0 \qquad \text{for every restricted root } \beta \text{ with } \mathfrak{g}_\beta \text{ in } \mathfrak{n} \text{ but not in } \mathfrak{n}'. \tag{8.12}$$

In the region (8.12), $A(S':S:\sigma:\nu)$ is an intertwining operator in the usual sense for representations of G. By differentiation it is an intertwining operator for $\mathfrak{g}$ on K-finite vectors, for ν in the region (8.12). The intertwining property relative to $\mathfrak{g}$ on K-finite vectors extends by analytic continuation to the larger region of ν's in Theorem 7.22.

Similarly Theorem 7.24 is to be interpreted as a statement on the Lie algebra level with K-finite vectors, with the precise statement as given below. If we want, we can pass to closed invariant subspaces, etc., by means of Corollary 8.10.

Theorem 7.24′. Let $S = MAN$ be a parabolic subgroup, let σ be irreducible tempered unitary on M, and let ν on $\mathfrak{a}$ be such that $\operatorname{Re} \nu$ is in the open positive Weyl chamber. Then $U(S, \sigma, \nu)$, as a representation of $\mathfrak{g}$ on K-finite vectors, has a unique irreducible quotient $J(S, \sigma, \nu)$, and $J(S, \sigma, \nu)$ is infinitesimally equivalent with the image of the intertwining operator $A(\bar{S}:S:\sigma:\nu)$ on the K-finite vectors of $U(S, \sigma, \nu)$.

Corollary 8.13. Let π be an irreducible admissible representation of G on V, and let V_0 be the subspace of K-finite vectors in V. If $L: V_0 \to V_0$ is a linear operator commuting with $\pi(\mathfrak{g})$, then L is scalar.

Proof. Since L commutes with $\pi(\mathfrak{k})$ and every vector in V_0 is K-finite, we can exponentiate and see that L commutes with $\pi(K)$. Therefore L maps the full space for each K type into itself. Since this space is assumed finite-dimensional, L has an eigenvector, say with eigenvalue c. Then $L - c$ commutes with $\pi(\mathfrak{g})$ and has a nonzero kernel. Since the kernel is $\mathfrak{g}$-invariant and $\pi(\mathfrak{g})$ is irreducible on V_0, $L - c = 0$. Thus L is scalar.

Corollary 8.14. If π is irreducible admissible, and in particular if π is irreducible unitary, then each member of the center $Z(\mathfrak{g}^{\mathbb{C}})$ of $U(\mathfrak{g}^{\mathbb{C}})$ acts as a scalar operator on the K-finite vectors of π.

In combination with formula (8.10), Corollary 8.14 implies that the K-finite matrix coefficients of an irreducible admissible representation are necessarily eigenfunctions of $Z(\mathfrak{g}^{\mathbb{C}})$. The idea for how to classify representations will be to rewrite the resulting system of differential equations on $\exp \mathfrak{a}_{\mathfrak{p}}$ and to apply the theory of Appendix B. In this way we will see that the matrix coefficients are almost completely determined by the action of $Z(\mathfrak{g}^{\mathbb{C}})$ and the K dependence, so much so that a classification theorem results.

§4. Framework for Studying Matrix Coefficients

Let τ_1 and τ_2 be finite-dimensional representations of K on spaces U_1 and U_2, respectively, and let $\tau = (\tau_1, \tau_2)$. A **τ-spherical function** F on G is a member of $C^\infty(G, \operatorname{Hom}_{\mathbb{C}}(U_2, U_1))$ such that

$$F(k_1 x k_2) = \tau_1(k_1)F(x)\tau_2(k_2) \qquad \text{for all } k_1, k_2 \in K, x \in G.$$

A source of such functions is as follows: Let π be an admissible representation of G and write $\pi|_K = \sum n_\omega \omega$. Fix two finite subsets $\mathscr{S}_1$ and $\mathscr{S}_2$ of $\hat{K}$ let τ_1 and τ_2 be the subrepresentations

$$\tau_1 = \sum_{\omega \in \mathscr{S}_1} n_\omega \omega \qquad \text{and} \qquad \tau_2 = \sum_{\omega \in \mathscr{S}_2} n_\omega \omega,$$

respectively, of $\pi|_K$, and let E_1 and E_2 be the respective orthogonal projections on the spaces U_1 and U_2 of these subrepresentations. Then $E_1 \pi(x) E_2$ is a τ-spherical function for $\tau = (\tau_1, \tau_2)$.

We are going to study a system of differential equations satisfied by some τ-spherical functions. These equations will be easier to analyze on the dense open subset $G^{(0)}$ of G given by $G^{(0)} = KA^+K$, where $A^+ = \exp \mathfrak{a}_\mathfrak{p}^+$. Thus let

$$C_\tau^\infty(G^{(0)}) = \left\{ F \in C^\infty(G^{(0)}, \operatorname{Hom}_{\mathbb{C}}(U_2, U_1)) \,\middle|\, \begin{array}{l} F(k_1 x k_2) = \tau_1(k_1)F(x)\tau_2(k_2) \\ \text{for } k_1, k_2 \in K \text{ and } x \in G^{(0)} \end{array} \right\}. \tag{8.13}$$

A function F in $C_\tau^\infty(G^{(0)})$ is the restriction to $G^{(0)}$ of a τ-spherical function if and only if it extends from $G^{(0)}$ to G as a C^∞ function.

Let F be in $C_\tau^\infty(G^{(0)})$. We shall compute the effect on F of a member of $U(\mathfrak{g}^{\mathbb{C}})$ in a way that takes advantage of the dependence of F on K on the left and the right. Fix a in A^+. Then an easy computation given below as Proposition 8.24 shows that

$$\mathfrak{g} = \operatorname{Ad}(a^{-1})\mathfrak{k} + \mathfrak{a}_\mathfrak{p} + \mathfrak{k}.$$

By the Birkhoff-Witt Theorem every member of $U(\mathfrak{g}^{\mathbb{C}})$ is a linear combination of terms

$$(\operatorname{Ad}(a^{-1})X)HY \tag{8.14}$$

with X and Y in $U(\mathfrak{k}^{\mathbb{C}})$ and H in $U(\mathfrak{a}_\mathfrak{p}^{\mathbb{C}})$. The particular linear combination typically depends on a.

Now let us compute $(\operatorname{Ad}(a^{-1})X)HYF(a)$ for our F in $C_\tau^\infty(G^{(0)})$. If X is in $\mathfrak{k}$ (i.e., is a first-order term), this is

$$= \frac{d}{dt} HYF(a \exp \operatorname{Ad}(a^{-1})tX)\bigg|_{t=0} = \frac{d}{dt} HYF((\exp tX)a)\bigg|_{t=0}.$$

For general X in $U(\mathfrak{k}^{\mathbb{C}})$, all the terms of a product occur on the left of a in their same order, while the H gives derivatives on the right of a, and the Y is to the right of that. Performing the differentiations, we therefore obtain

$$(\mathrm{Ad}(a^{-1})X)HYF(a) = \tau_1(X)(HF)(a)\tau_2(Y). \tag{8.15}$$

Combining matters, let u be in $U(\mathfrak{g}^{\mathbb{C}})$, and write u as a linear combination of terms (8.14), the coefficients varying with a. Using (8.15), we see that

$$uF(a) = D_\tau(u)(F|_{A^+})(a), \tag{8.16}$$

where $D_\tau(u)$ is a differential operator on A^+ with variable coefficients. Specifically the term (8.15) contributes a differentiation from the H and a coefficient that we view as

$$(\text{left by } \tau_1(X)) \times (\text{right by } \tau_2(Y)). \tag{8.17}$$

Viewed this way, this coefficient is in $\mathrm{End}_{\mathbb{C}}(\mathrm{Hom}_{\mathbb{C}}(U_2, U_1))$. When a variable coefficient multiplies (8.15), the total coefficient is a function from A^+ to $\mathrm{End}_{\mathbb{C}}(\mathrm{Hom}_{\mathbb{C}}(U_2, U_1))$. Motivated by the KAK decomposition of G and by (8.16), we call $D_\tau(u)$ the **τ-radial component** of u.

Although (8.16) gives us information about uF on A^+, it does not readily give information about uF on $G^{(0)}$ unless u satisfies an extra condition. In fact, for X in $\mathfrak{g}$, we have

$$\begin{aligned} XF(k_1xk_2) &= \frac{d}{dt} F(k_1xk_2 \exp tX)\bigg|_{t=0} \\ &= \frac{d}{dt} F(k_1x \exp(t\,\mathrm{Ad}(k_2)X)k_2)\bigg|_{t=0} \\ &= \tau_1(k_1)(\mathrm{Ad}(k_2)X)F(x)\tau_2(k_2). \end{aligned} \tag{8.18}$$

If we iterate this formula, we find it is valid for X in $U(\mathfrak{g}^{\mathbb{C}})$. Hence our u satisfies

$$uF(k_1ak_2) = \tau_1(k_1)(uF)(a)\tau_2(k_2) \tag{8.19}$$

if u is fixed by $\mathrm{Ad}(K)$, in particular if u is in the center $Z(\mathfrak{g}^{\mathbb{C}})$. In short, equation (8.19) says that ZF is in $C_\tau^\infty(G^{(0)})$ if Z is in $Z(\mathfrak{g}^{\mathbb{C}})$.

Setting $x = a$ in (8.13), we see that the values of a member of $C_\tau^\infty(G^{(0)})$ satisfy

$$\tau_1(m)F(a) = F(ma) = F(am) = F(a)\tau_2(m) \qquad \text{for } m \text{ in } M = M_{\mathfrak{p}}.$$

That is, $F(a)$ is in the subspace $\mathrm{Hom}_M(U_2, U_1)$ of $\mathrm{Hom}_{\mathbb{C}}(U_2, U_1)$. Taking into account (8.19), we see an indication that $D_\tau(Z)$, for Z in $Z(\mathfrak{g}^{\mathbb{C}})$, has as coefficients certain functions on A^+ with values in $\mathrm{End}_{\mathbb{C}}(\mathrm{Hom}_M(U_2, U_1))$.

Detailed information about $D_\tau(Z)$ will be important for us, because we will want to apply the theory of Appendix B. For the time being, we shall illustrate matters by doing the relevant computations for the Casimir operator of $SL(2, \mathbb{R})$:

$$\Omega = \tfrac{1}{2}h^2 + ef + fe = \tfrac{1}{2}h^2 - h + 2ef. \tag{8.20}$$

Lemma 8.15. For $G = SL(2, \mathbb{R})$ let $\alpha = e_1 - e_2$ be the positive restricted root, and let $\xi_\alpha(a) = e^{\alpha \log a}$ for $a \in A$. Let Y be the member of $\mathfrak{k}$ given by $Y = \frac{1}{2}(e + \theta e) = \frac{1}{2}(e - f)$. If a is in A^+, then

$$e = \frac{2\xi_\alpha(a)}{\xi_\alpha(a)^2 - 1}\{\xi_\alpha(a)Y - \mathrm{Ad}(a^{-1})Y\}. \tag{8.21}$$

Proof.
$$\begin{aligned}\mathrm{Ad}(a^{-1})Y &= \tfrac{1}{2}\mathrm{Ad}(a^{-1})e - \tfrac{1}{2}\mathrm{Ad}(a^{-1})f \\ &= \tfrac{1}{2}\xi_\alpha(a)^{-1}e - \tfrac{1}{2}\xi_{-\alpha}(a)^{-1}f \\ &= \tfrac{1}{2}\xi_\alpha(a)^{-1}e + \xi_\alpha(a)Y - \tfrac{1}{2}\xi_\alpha(a)e \\ &= \xi_\alpha(a)Y - \tfrac{1}{2}(\xi_\alpha(a) - \xi_\alpha(a)^{-1})e,\end{aligned}$$

and the lemma follows.

Proposition 8.16. For $G = SL(2, \mathbb{R})$, let F be in $C_\tau^\infty(G^{(0)})$. Then

$$\begin{aligned}D_\tau(\Omega)F(a) = \tfrac{1}{2}h^2F(a) &+ \frac{\xi^2+1}{\xi^2-1}hF(a) + \frac{8\xi^2}{(\xi^2-1)^2}(F(a)\tau_2(Y)^2 + \tau_1(Y)^2F(a)) \\ &- \frac{8\xi(\xi^2+1)}{(\xi^2-1)^2}\tau_1(Y)F(a)\tau_2(Y),\end{aligned}$$

where $\xi = \xi_\alpha(a)$.

Proof. By (8.20) and (8.21),

$$\begin{aligned}\Omega &= \tfrac{1}{2}h^2 - h + 2ef \\ &= \tfrac{1}{2}h^2 - h + \frac{4\xi}{\xi^2-1}\xi Yf - \frac{4\xi}{\xi^2-1}(\mathrm{Ad}(a^{-1})Y)f \\ &= \tfrac{1}{2}h^2 - h + \frac{4\xi^2}{\xi^2-1}fY + \frac{4\xi^2}{\xi^2-1}[Y, f] \\ &\quad - \frac{4\xi}{\xi^2-1}(\mathrm{Ad}(a^{-1})Y)(e - 2Y).\end{aligned}$$

Since $[Y, f] = \frac{1}{2}h$ and $f = e - 2Y$, this is

$$\begin{aligned}&= \tfrac{1}{2}h^2 - h + \frac{2\xi^2}{\xi^2-1}h + \frac{4\xi^2}{\xi^2-1}(e - 2Y)Y \\ &\quad - \frac{4\xi}{\xi^2-1}(\mathrm{Ad}(a)^{-1}Y)e + \frac{8\xi}{\xi^2-1}(\mathrm{Ad}(a)^{-1}Y)Y.\end{aligned}$$

If we substitute again for e from (8.21), we see this is

$$= \tfrac{1}{2}h^2 + \frac{\xi^2+1}{\xi^2-1}h + \frac{4\xi^2}{\xi^2-1}\left(\frac{2\xi^2}{\xi^2-1}Y^2 - \frac{2\xi}{\xi^2-1}(\mathrm{Ad}(a)^{-1}Y)Y - 2Y^2\right)$$
$$-\frac{8\xi^2}{(\xi^2-1)^2}\xi(\mathrm{Ad}(a)^{-1}Y)Y - (\mathrm{Ad}(a)^{-1}Y)^2) + \frac{8\xi}{\xi^2-1}(\mathrm{Ad}(a)^{-1}Y)Y$$
$$= \tfrac{1}{2}h^2 + \frac{\xi^2+1}{\xi^2-1}h + \frac{8\xi^2}{\xi^2-1}\left(\frac{\xi^2}{\xi^2-1} - 1\right)Y^2 + \frac{8\xi^2}{(\xi^2-1)^2}(\mathrm{Ad}(a)^{-1}Y)^2$$
$$+\frac{8\xi}{\xi^2-1}\left(-\frac{\xi^2}{\xi^2-1} - \frac{\xi^2}{\xi^2-1} + 1\right)(\mathrm{Ad}(a)^{-1}Y)Y$$
$$= \tfrac{1}{2}h^2 + \frac{\xi^2+1}{\xi^2-1}h + \frac{8\xi^2}{(\xi^2-1)^2}Y^2 + \frac{8\xi^2}{(\xi^2-1)^2}(\mathrm{Ad}(a)^{-1}Y)^2$$
$$-\frac{8\xi(\xi^2+1)}{(\xi^2-1)^2}(\mathrm{Ad}(a)^{-1}Y)Y.$$

Then Ω is in the correct form to apply (8.15) and (8.16), and the result follows.

We can be more explicit by setting $a_t = \begin{pmatrix} e^{t/2} & 0 \\ 0 & e^{-t/2} \end{pmatrix}$. Then $\xi = \xi_\alpha(a_t) = e^t$ and the differential operator h becomes $2\dfrac{d}{dt}$. Substituting, we obtain

$$\tfrac{1}{2}D_\tau(\Omega)F(a_t) = \frac{d^2F}{dt^2} + (\coth t)\frac{dF}{dt} + \frac{1}{(\sinh t)^2}(F(a_t)\tau_2(Y)^2 + \tau_1(Y)^2F(a_t))$$
$$-\frac{2\cosh t}{\sinh^2 t}\tau_1(Y)F(a_t)\tau_2(Y).$$

If F arises from the matrix coefficients of an irreducible admissible representation, then (8.10) and Corollary 8.14 show $\Omega F = cF$ for a scalar c. Consequently $F(a_t)$ satisfies the differential equation $D_\tau(\Omega)F(a_t) = cF(a_t)$. Section 5 of Appendix B shows how the theory of Appendix B applies to give a series expansion for $F(a_t)$ about $t = +\infty$ in powers of e^{-t}. The result is our asymptotic expansion of τ-spherical functions for $SL(2, \mathbb{R})$.

To proceed for general G, we need to use all of $Z(\mathfrak{g}^{\mathbb{C}})$, not just Ω, in order to have a large enough system of differential equations. We shall study $Z(\mathfrak{g}^{\mathbb{C}})$ in the next two sections and then return to the differential equations in §7.

§5. Harish-Chandra Homomorphism

Let $\mathfrak{g}^{\mathbb{C}}$ be the complexification of the Lie algebra of a linear connected reductive group G, and let $\mathfrak{h}^{\mathbb{C}}$ be a Cartan subalgebra. The goal of this section is an understanding of the center $Z(\mathfrak{g}^{\mathbb{C}})$ of $U(\mathfrak{g}^{\mathbb{C}})$.

For motivation we introduce an ordering and take φ to be an irreducible complex-linear representation of $\mathfrak{g}^{\mathbb{C}}$ with highest weight λ. Suppose v_0 is a nonzero highest weight vector. Schur's Lemma implies that $\varphi(Z)$ is scalar for Z in $Z(\mathfrak{g}^{\mathbb{C}})$, and we compute $\varphi(Z)$ by calculating its effect on v_0. Let the positive roots be $\alpha_1, \ldots, \alpha_k$, and let $H_1, \ldots, H_l$ be a basis of $\mathfrak{h}^{\mathbb{C}}$ over $\mathbb{C}$. Then the monomials

$$E_{-\alpha_1}^{q_1} \cdots E_{-\alpha_k}^{q_k} H_1^{m_1} \cdots H_l^{m_l} E_{\alpha_1}^{p_1} \cdots E_{\alpha_k}^{p_k} \tag{8.22}$$

are a basis of $U(\mathfrak{g}^{\mathbb{C}})$, by the Birkhoff-Witt Theorem. In an expression of Z in terms of this basis we shall see that there are only two possibilities for a term of the expansion:

(1) Some E_{α_j} is present, and the term gives 0 on v_0.
(2) No E_{α_j} is present, no $E_{-\alpha_j}$ is present, and the term acts on v_0 as a member of $U(\mathfrak{h}^{\mathbb{C}})$.

Thus in principle we can calculate $\varphi(Z)$ in terms of λ. The Harish-Chandra homomorphism below will make this calculation universally, without reference to λ.

To define the Harish-Chandra homomorphism, we fix a positive system Δ^+ for $\Delta(\mathfrak{h}^{\mathbb{C}}:\mathfrak{g}^{\mathbb{C}})$ and let

$$\mathscr{H} = U(\mathfrak{h}^{\mathbb{C}})$$
$$\mathscr{P} = \sum_{\alpha \in \Delta^+} U(\mathfrak{g}^{\mathbb{C}})E_\alpha.$$

Here $\mathscr{H}$ coincides with the symmetric algebra of $\mathfrak{h}^{\mathbb{C}}$ since $\mathfrak{h}^{\mathbb{C}}$ is abelian, and $\mathscr{P}$ depends on the choice of Δ^+. The Weyl group $W = W(\mathfrak{h}^{\mathbb{C}}:\mathfrak{g}^{\mathbb{C}})$ acts on $\mathfrak{h}^{\mathbb{C}}$ and hence on $\mathscr{H}$. Let $\delta = \frac{1}{2}\sum_{\alpha \in \Delta^+} \alpha$.

Lemma 8.17. $\mathscr{H} \cap \mathscr{P} = 0$ and $Z(\mathfrak{g}^{\mathbb{C}}) \subseteq \mathscr{H} \oplus \mathscr{P}$.

Proof. To see that $\mathscr{H} \cap \mathscr{P} = 0$, we note first that $\varphi_\lambda(\mathscr{P})v_0 = 0$ if φ_λ is a complex-linear representation of $\mathfrak{g}^{\mathbb{C}}$ with highest weight λ and if v_0 is a highest weight vector. Hence if X is in $\mathscr{H} \cap \mathscr{P}$, then X is a member of $\mathscr{H}$ with $\varphi_\lambda(X)v_0 = 0$ for all φ_λ, i.e., $\lambda(X) = 0$ for all dominant integral λ, with $\lambda(X)$ defined factor by factor. (This definition of $\lambda(X)$ comes by applying Proposition 3.1 to $\lambda: \mathfrak{h}^{\mathbb{C}} \to \mathbb{C}$ since $[\mathfrak{h}^{\mathbb{C}}, \mathfrak{h}^{\mathbb{C}}] = 0$.) We can regard X as a polynomial on the λ space, and then X vanishes at all dominant integral points λ. Since the dominant integral points exhaust the lattice points in a cone, $X = 0$. Hence $\mathscr{H} \cap \mathscr{P} = 0$.

For the inclusion $Z(\mathfrak{g}^{\mathbb{C}}) \subseteq \mathscr{H} \oplus \mathscr{P}$, let Z be in $Z(\mathfrak{g}^{\mathbb{C}})$, and expand Z in terms of the basis (8.22). Each monomial (8.22) is an eigenvector for ad H, if H is in $\mathfrak{h}^{\mathbb{C}}$, and the eigenvalue is

$$\sum_i p_i\alpha_i(H) - \sum_i q_i\alpha_i(H). \tag{8.23}$$

Also $(\operatorname{ad} H)Z = 0$. Then it follows that each nonzero term in the expansion of Z has (8.23) equal to 0. If some $E_{-\alpha_i}$ is present in a term, it follows that some E_{α_j} is present. That is, each term is in $\mathscr{P}$ or in $\mathscr{H}$. Hence $Z(\mathfrak{g}^{\mathbb{C}}) \subseteq \mathscr{H} \oplus \mathscr{P}$.

Let γ'_{Δ^+} be the projection of $Z(\mathfrak{g}^{\mathbb{C}}) \subseteq \mathscr{H} \oplus \mathscr{P}$ into the $\mathscr{H}$ factor. We have seen the relation of γ'_{Δ^+} to highest weights. Harish-Chandra found that a slight adjustment of γ'_{Δ^+} leads to a more symmetric formula. Namely let $\sigma_{\Delta^+}: \mathfrak{h}^{\mathbb{C}} \to \mathscr{H}$ be given by

$$\sigma_{\Delta^+}(H) = H - \delta(H)1, \qquad \left(\delta = \tfrac{1}{2} \sum_{\alpha \in \Delta^+} \alpha\right)$$

and extend σ_{Δ^+} to an algebra automorphism of $\mathscr{H}$ by Proposition 3.1. Define the **Harish-Chandra homomorphism** γ by

$$\gamma = \sigma_{\Delta^+} \circ \gamma'_{\Delta^+}, \tag{8.24}$$

as a mapping of $Z(\mathfrak{g}^{\mathbb{C}})$ into $\mathscr{H}$. The maps γ and γ'_{Δ^+} are related by

$$\gamma(Z)(\Lambda) = \gamma'_{\Delta^+}(Z)(\Lambda - \delta) \tag{8.25}$$

for Z in $Z(\mathfrak{g}^{\mathbb{C}})$ and Λ in $(\mathfrak{h}^{\mathbb{C}})'$.

Theorem 8.18. The Harish-Chandra homomorphism γ is an algebra isomorphism of $Z(\mathfrak{g}^{\mathbb{C}})$ onto the algebra

$$\mathscr{H}^W = \{\text{members of } \mathscr{H} \text{ fixed by } W\},$$

and it does not depend upon the choice of the positive system Δ^+.

Example. $\mathfrak{g}^{\mathbb{C}} = \mathfrak{sl}(2, \mathbb{C})$, and $\Omega = \frac{1}{2}h^2 + ef + fe$ in $Z(\mathfrak{g}^{\mathbb{C}})$.

Let us agree that e corresponds to the positive root α. Then $ef = fe + [e,f] = fe + h$ implies

$$\Omega = \tfrac{1}{2}h^2 + ef + fe = (\tfrac{1}{2}h^2 + h) + 2fe \in \mathscr{H} \oplus \mathscr{P}$$

Hence
$$\gamma'_{\Delta^+}(\Omega) = \tfrac{1}{2}h^2 + h.$$

Now $\delta(h) = \frac{1}{2}\alpha\begin{pmatrix} 1 & 0 \\ 0 & -1 \end{pmatrix} = 1$, and so

$$\sigma_{\Delta^+}(h) = h - 1.$$

Thus
$$\gamma(\Omega) = \tfrac{1}{2}(h-1)^2 + (h-1) = \tfrac{1}{2}h^2 - \tfrac{1}{2}.$$

For $\mathfrak{sl}(2, \mathbb{C})$ the nontrivial element of W maps h to $-h$, and $\mathscr{H}^W$ is the set of polynomials in h^2. The formula above for $\gamma(\Omega)$ is a W-invariant member of $\mathscr{H}$. If we use Theorem 8.18, we can then conclude that $Z(\mathfrak{g}^{\mathbb{C}}) = \mathbb{C}[\Omega]$ for $\mathfrak{sl}(2, \mathbb{C})$.

The proof of Theorem 8.18 will be broken into several steps. The hardest step is that γ is onto, and this will be done last. We give the proof that γ is onto just in the case of $\mathfrak{gl}(n, \mathbb{C})$, saving some comments about the general case for the bibliographical notes.

Proof that image(γ) $\subseteq \mathscr{H}^W$. In view of (8.25), we are to prove that

$$\gamma'_{\Delta^+}(Z)(w\Lambda - \delta) = \gamma'_{\Delta^+}(Z)(\Lambda - \delta), \tag{8.26}$$

and it is enough to handle w equal to a simple reflection s_α. Moreover, since we are dealing with an equality of polynomials we may assume that $\Lambda - \delta$ is dominant integral.

Consider the Verma module $V(\Lambda)$ of §4.8; this is a universal highest weight module for the weight $\Lambda - \delta$, and thus Z acts as $\gamma'_{\Delta^+}(Z)(\Lambda - \delta)$ on its canonical generator. By Proposition 4.34c, Z acts as the scalar $\gamma^+_{\Delta^+}(Z)(\Lambda - \delta)$ on all of $V(\Lambda)$. According to Lemma 4.39, we have an imbedding $V(s_\alpha\Lambda) \subseteq V(\Lambda)$. Thus Z acts as both $\gamma'_{\Delta^+}(Z)(\Lambda - \delta)$ and $\gamma'_{\Delta^+}(Z)(s_\alpha\Lambda - \delta)$ on $V(s_\alpha\Lambda)$, and (8.26) follows.

Proof that γ is independent of Δ^+. This is a simple consequence of the invariance just proved and of the interpretation of $\gamma'_{\Delta^+}(Z)(\lambda)$ in terms of finite-dimensional representations.

Proof that γ is multiplicative. Since σ_{Δ^+} is an algebra isomorphism, we need to see that

$$\gamma'_{\Delta^+}(Z_1Z_2) = \gamma'_{\Delta^+}(Z_1)\gamma'_{\Delta^+}(Z_2). \tag{8.27}$$

We have

$$Z_1Z_2 - \gamma'_{\Delta^+}(Z_1)\gamma'_{\Delta^+}(Z_2) = Z_1(Z_2 - \gamma'_{\Delta^+}(Z_2)) + \gamma'_{\Delta^+}(Z_2)(Z_1 - \gamma'_{\Delta^+}(Z_1)),$$

which is in $U(\mathfrak{g}^{\mathbb{C}})\mathscr{P} = \mathscr{P}$. With γ'_{Δ^+} extended to all of $\mathscr{H} \oplus \mathscr{P}$, we therefore have

$$0 = \gamma'_{\Delta^+}(Z_1Z_2 - \gamma'_{\Delta^+}(Z_1)\gamma'_{\Delta^+}(Z_2)) = \gamma'_{\Delta^+}(Z_1Z_2) - \gamma'_{\Delta^+}(Z_1)\gamma'_{\Delta^+}(Z_2),$$

and (8.27) follows.

Proof that γ is one-one. We may assume $\mathfrak{g}^{\mathbb{C}}$ is semisimple. If $\gamma(Z) = 0$, then $\gamma'_{\Delta^+}(Z) = 0$ also. For λ a highest weight, $\gamma'_{\Delta^+}(Z)(\lambda)$ is the scalar $\varphi_\lambda(Z)$. In view of the complete reducibility of finite-dimensional representations of G, we may therefore assume that $\pi(Z) = 0$ for every finite-dimensional representation. Formula (8.10) then shows that Z, as a left-invariant differential operator on G, annihilates every matrix coefficient of a finite-dimensional representation of G. This holds for the k-fold tensor product of the standard representation of G with itself, for all k, and in this case the matrix coefficients at x are the kth degree monomials in the entries of the matrix x. It holds also for the conjugate of any representation. Thus

Z, as a differential operator, annihilates all monomials in the real and imaginary parts of the coordinates, and $Z = 0$ as a differential operator. By Theorem 3.6, $Z = 0$ in $U(\mathfrak{g}^{\mathbb{C}})$. Hence γ is one-one.

Proof that γ is onto for $\mathfrak{g}^{\mathbb{C}} = \mathfrak{gl}(n, \mathbb{C})$. Let $\{H_i\}$ be the basis of $\mathfrak{h}^{\mathbb{C}}$ with $H_i = E_{ii}$. Then $\mathscr{H}^W$ is the algebra of symmetric polynomials in the H_i and is well known to be given by

$$\mathscr{H}^W = \mathbb{C}[p_1, \ldots, p_n],$$

where

$$p_1 = \sum_i H_i$$

$$p_2 = \sum_{i<j} H_i H_j$$

$$p_3 = \sum_{i<j<k} H_i H_j H_k$$

$$\vdots$$

$$p_n = H_1 H_2 \cdots H_n.$$

The polynomials $p_1, \ldots, p_n$ are the elementary symmetric polynomials and are given also by the formula

$$(X - H_1) \cdots (X - H_n) = X^n - p_1 X^{n-1} + p_2 X^{n-2} - \ldots + (-1)^n p_n, \tag{8.28}$$

where X is an indeterminate.

Let X_i run through the basis of vectors $E_{-\alpha}$, H_j, E_α of $\mathfrak{g}^{\mathbb{C}}$. A calculation like that in Proposition 8.6 shows that

$$Z_k = \sum_{i_1, \ldots, i_k} \operatorname{Tr}(X_{i_1} \cdots X_{i_k}) X^{i_1} \cdots X^{i_k} \tag{8.29}$$

is in $Z(\mathfrak{g}^{\mathbb{C}})$, where X^j is defined by the relations $g_{ij} = \operatorname{Tr}(X_i X_j)$, $(g^{ij}) = (g_{ij})^{-1}$, and $X^j = \sum g^{ij} X_i$. It is clear that

$$\gamma'_{\Delta^+}(Z_k) \equiv \sum_{i_1, \ldots, i_k} \operatorname{Tr}(H_{i_1} \cdots H_{i_k}) H^{i_1} \cdots H^{i_k} \bmod U^{k-1}(\mathfrak{g}^{\mathbb{C}})$$

and then that

$$\gamma(Z_k) \equiv \sum_{i_1, \ldots, i_k} \operatorname{Tr}(H_{i_1} \cdots H_{i_k}) H^{i_1} \cdots H^{i_k} \bmod U^{k-1}(\mathfrak{g}^{\mathbb{C}}). \tag{8.30}$$

If we have chosen each E_α and $E_{-\alpha}$ to be 1 in the nonzero entry, then g_{ij} is given in obvious notation by

$$\begin{array}{c} \quad E_\alpha \quad H \quad E_{-\alpha} \\ (g_{ij}) = \begin{pmatrix} 0 & 0 & I \\ 0 & I & 0 \\ I & 0 & 0 \end{pmatrix} \begin{array}{l} E_\alpha \\ H \\ E_{-\alpha} \end{array} = (g^{ij}) \end{array}$$

Hence $H^j = H_j$, and evaluation of the traces gives

$$\gamma(Z_k) \equiv \sum_i (H_i)^k \bmod U^{k-1}(\mathfrak{g}^{\mathbb{C}}).$$

Let $1 \leq k \leq n$. Since $\gamma(Z_k)$ and $\sum_i (H_i)^k$ are both symmetric polynomials, their difference is a polynomial in $p_1, \ldots, p_{k-1}$. Thus we see inductively that p_k will be in the image of γ if $\sum (H_i)^k$ requires p_k for its expansion in terms of elementary symmetric polynomials.

To see that $\sum (H_i)^k$ requires p_k, we let $\zeta_1, \ldots, \zeta_k$ be the kth roots of unity, and we introduce the homomorphism

$$H_j \to \begin{cases} \zeta_j & \text{for } 1 \leq j \leq k \\ 0 & \text{for } k+1 \leq j \leq n. \end{cases}$$

Under this homomorphism, the left side of (8.28) becomes $X^n - X^{n-k}$, and thus (8.28) gives

$$p_j(\zeta_1, \ldots, \zeta_k, 0, \ldots, 0) = \begin{cases} 0 & \text{for } j \neq k \\ (-1)^{k+1} & \text{for } j = k. \end{cases} \tag{8.31}$$

The homomorphism sends $\sum (H_i)^k$ into n, and (8.31) thus implies that $\sum (H_i)^k$ cannot be expressed in terms of $p_1, \ldots, p_{k-1}$. This completes the proof.

Theorem 8.18 implies that $Z(\mathfrak{g}^{\mathbb{C}})$ is large in a certain sense. For our application to matrix coefficients, the following result captures its size adequately.

Theorem 8.19. If $\mathfrak{g}^{\mathbb{C}}$ is reductive, then $\mathscr{H}^W$ is a polynomial algebra in $\dim_{\mathbb{C}} \mathfrak{h}^{\mathbb{C}}$ variables, and $\mathscr{H}$ is a free $\mathscr{H}^W$ module of rank $|W|$. One of the generators of $\mathscr{H}$ over $\mathscr{H}^W$ can be taken as 1.

Proof omitted.

Remarks. For $\mathfrak{g}^{\mathbb{C}} = \mathfrak{gl}(n, \mathbb{C})$, let H_j be one in the jth diagonal entry and 0 elsewhere. Then $\mathscr{H} = \mathbb{C}[H_1, \ldots, H_n]$, and $\mathscr{H}^W$ is the subset of symmetric elements (under permutation). The first conclusion of Theorem 8.19 in this case is the classical result that $\mathscr{H}^W$ is the full polynomial algebra on n generators; the generators can be taken to be the elementary symmetric polynomials.

§6. Infinitesimal Character

We continue with $\mathfrak{g}^{\mathbb{C}}$ a reductive Lie algebra over $\mathbb{C}$ (obtained from a group G) and $\mathfrak{h}^{\mathbb{C}}$ a Cartan subalgebra of $\mathfrak{g}^{\mathbb{C}}$. We shall classify homomorphisms of the center $Z(\mathfrak{g}^{\mathbb{C}})$ of $U(\mathfrak{g}^{\mathbb{C}})$ into $\mathbb{C}$. This classification will limit the possibilities for the scalar values of the action of members of $Z(\mathfrak{g}^{\mathbb{C}})$ on an irreducible admissible representation of G (cf. Corollary 8.14).

Abstract examples of such homomorphisms are obtained by composing the Harish-Chandra homomorphism with evaluation at Λ in $(\mathfrak{h}^{\mathbb{C}})'$:

$$\chi_\Lambda(Z) = \Lambda(\gamma(Z)) \qquad \text{for } Z \in Z(\mathfrak{g}^{\mathbb{C}}). \tag{8.32}$$

Here we have extended Λ to an algebra homomorphism of $\mathscr{H} = U(\mathfrak{h}^{\mathbb{C}})$ into $\mathbb{C}$ by Proposition 3.1. We shall prove below that all homomorphisms of $Z(\mathfrak{g}^{\mathbb{C}})$ into $\mathbb{C}$ are obtained this way.

The homomorphisms χ_Λ satisfy

$$\chi_{w\Lambda} = \chi_\Lambda \qquad \text{for } w \in W \tag{8.33}$$

because $\gamma(Z)$ is in $\mathscr{H}^W$:

$$\chi_{w\Lambda}(Z) = w\Lambda(\gamma(Z)) = \Lambda(\mathrm{Ad}(w^{-1})\gamma(Z)) = \Lambda(\gamma(Z)) = \chi_\Lambda(Z).$$

In the converse direction we have the following proposition.

Proposition 8.20. If $\chi_{\Lambda'} = \chi_\Lambda$, then $\Lambda' = w\Lambda$ for some w in W.

Proof. If not, then choose a polynomial p on $(\mathfrak{h}^{\mathbb{C}})'$ that is 1 on $W\Lambda$ and 0 on $W\Lambda'$. Then $\tilde{p} = |W|^{-1} \sum_{w \in W} wp$ has the same property and is W-invariant. By Theorem 8.18 we can choose Z in $Z(\mathfrak{g})$ with $\gamma(Z) = \tilde{p}$. Then $\chi_{\Lambda'}(Z) = \gamma(Z)(\Lambda') = \tilde{p}(\Lambda') = 0$ while $\chi_\Lambda(Z) = 1$. Hence $\chi_{\Lambda'} \neq \chi_\Lambda$.

Proposition 8.21. Every homomorphism from $Z(\mathfrak{g}^{\mathbb{C}})$ into $\mathbb{C}$ is of the form χ_Λ for some Λ in $(\mathfrak{h}^{\mathbb{C}})'$.

Proof. Let χ be a homomorphism of $Z(\mathfrak{g}^{\mathbb{C}})$ into $\mathbb{C}$. We may regard the domain of χ as $\mathscr{H}^W$, by Theorem 8.18. Form $\mathscr{H}\ker\chi$, which is an ideal in $\mathscr{H}$. To see $\mathscr{H}\ker\chi$ is proper, we apply Theorem 8.19 to write $\mathscr{H} = \sum_{j=1}^{|W|} \mathscr{H}^W x_j$ with $x_1 = 1$. Then

$$\mathscr{H}\ker\chi = \sum_{j=1}^{|W|} \mathscr{H}^W x_j(\ker\chi) \subseteq \ker\chi + \sum_{j=2}^{|W|} \mathscr{H}^W x_j,$$

and 1 is not in the right side. By Zorn's Lemma, extend $\mathscr{H}\ker\chi$ to a maximal ideal I in $\mathscr{H}$. The Hilbert Nullstellensatz says that I has a nonempty locus of common zeros in $(\mathfrak{h}^{\mathbb{C}})'$, and this locus must be a single point Λ by maximality. Then it follows that $\chi = \chi_\Lambda$, and the proposition is proved.

Now let G be linear connected reductive, and suppose that π is an admissible representation of G for which $\pi(Z)$ is scalar on the K-finite vectors for all Z in $Z(\mathfrak{g}^{\mathbb{C}})$. (For example, take π to be irreducible.) Then the map $\chi: Z(\mathfrak{g}^{\mathbb{C}}) \to \mathbb{C}$ given by $\pi(Z) = \chi(Z)I$ is a homomorphism of $Z(\mathfrak{g}^{\mathbb{C}})$ into $\mathbb{C}$ and is of the form χ_Λ for some Λ in $(\mathfrak{h}^{\mathbb{C}})'$, by Proposition 8.21. Thus we have

$$\pi(Z) = \chi_\Lambda(Z)I \qquad \text{for } Z \in Z(\mathfrak{g}^{\mathbb{C}}).$$

In this case we say π has **infinitesimal character** Λ (or χ_Λ, depending on the context). By Proposition 8.20, Λ is determined up to a member of the Weyl group W.

Example. If G is a compact connected Lie group and π is an irreducible representation of G with highest weight λ, then the infinitesimal character of π is $\lambda + \delta$, where δ is half the sum of the positive roots.

Proposition 8.22. Let MAN be a parabolic subgroup of G, let $\mathfrak{t}$ be a θ-stable Cartan subalgebra of $\mathfrak{m}$, and let σ be an irreducible unitary representation of M. If σ has infinitesimal character Λ_σ relative to $\mathfrak{t}^{\mathbb{C}}$ and if ν is in $(\mathfrak{a}')^{\mathbb{C}}$, then $U(MAN, \sigma, \nu)$ has infinitesimal character $\Lambda_\sigma + \nu$ relative to $(\mathfrak{a} \oplus \mathfrak{t})^{\mathbb{C}}$.

Proof. Let $\Delta = \Delta((\mathfrak{a} \oplus \mathfrak{t})^{\mathbb{C}} : \mathfrak{g}^{\mathbb{C}})$, so that the set Γ of roots of $(\mathfrak{g}, \mathfrak{a})$ is obtained by restriction from $(\mathfrak{a} \oplus \mathfrak{t})^{\mathbb{C}}$ to $\mathfrak{a}$. By extending an ordering from $\mathfrak{a}^{\mathbb{C}}$ to $(\mathfrak{a} \oplus \mathfrak{t})^{\mathbb{C}}$, we can arrange that Γ^+ arises by restriction from Δ^+. The members of Δ vanishing on $\mathfrak{a}$ provide us with the root system $\Delta_M = \Delta(\mathfrak{t}^{\mathbb{C}} : \mathfrak{m}^{\mathbb{C}})$, and we have $\Delta_M^+ = \Delta^+ \cap \Delta_M$. Then we have $\delta = \rho_A + \delta_M$ for the half-sums of positive roots.

If we write

$$\mathfrak{g} = \mathfrak{n} \oplus (\mathfrak{a} \oplus \mathfrak{m}) \oplus \bar{\mathfrak{n}},$$

then the same kind of argument as in Lemma 8.17 shows that

$$Z(\mathfrak{g}^{\mathbb{C}}) \subseteq \mathfrak{n}^{\mathbb{C}} U(\mathfrak{g}^{\mathbb{C}}) \oplus U((\mathfrak{a} \oplus \mathfrak{m})^{\mathbb{C}}),$$

and we let μ_{Γ^-} be the projection of $Z(\mathfrak{g}^{\mathbb{C}})$ into $U((\mathfrak{a} \oplus \mathfrak{m})^{\mathbb{C}})$. We readily check that Z in $Z(\mathfrak{g}^{\mathbb{C}})$ implies $\mu_{\Gamma^-}(Z)$ in $Z((\mathfrak{a} \oplus \mathfrak{m})^{\mathbb{C}})$ and hence

$$\gamma'_{\mathfrak{g},\Delta^-} = \gamma'_{\mathfrak{a}\oplus\mathfrak{m},\Delta_M^-} \circ \mu_{\Gamma^-}. \tag{8.34}$$

Let f be in the induced space, and write U for $U(MAN, \sigma, \nu)$. If X is in $\mathfrak{n}$, then

$$U(X)U(g)f(1) = \frac{d}{dt} U(g)f((\exp tX)^{-1})\Big|_{t=0} = 0$$

because $U(g)f$ is right invariant under N. Hence $\mathfrak{n}^{\mathbb{C}} U(\mathfrak{g}^{\mathbb{C}})$ annihilates f, and

$$U(Z)f(1) = U(\mu_{\Gamma^-}(Z))f(1) \qquad \text{for } Z \in Z(\mathfrak{g}^{\mathbb{C}}). \tag{8.35}$$

If $X = X_{\mathfrak{a}} + X_{\mathfrak{m}}$ is in $\mathfrak{a} \oplus \mathfrak{m}$, then

$$\begin{aligned} U(X)f(1) = \frac{d}{dt} f((\exp tX)^{-1})\Big|_{t=0} &= \frac{d}{dt}(e^{\nu+\rho_A} \otimes \sigma)(\exp tX)f(1)\Big|_{t=0} \\ &= \{(\nu + \rho_A)(X_{\mathfrak{a}}) + \sigma(X_{\mathfrak{m}})\}f(1). \end{aligned}$$

Hence

$$U(\mu_{\Gamma^-}(Z))f(1) = ((\nu + \rho_A) \otimes \sigma)(\mu_{\Gamma^-}(Z))f(1). \tag{8.36}$$

Thus we compute

$$\begin{aligned}
U(Z)f(1) &= U(\mu_{\Gamma}\text{-}(Z))f(1) && \text{by (8.35)}\\
&= ((\nu + \rho_A) \otimes \sigma)(\mu_{\Gamma}\text{-}(Z))f(1) && \text{by (8.36)}\\
&= (\nu + \rho_A + \Lambda_\sigma)(\gamma_{\mathfrak{a}\oplus\mathfrak{m}}(\mu_{\Gamma}\text{-}(Z)))f(1) && \text{by assumption}\\
&= (\nu + \rho_A + \Lambda_\sigma)(\sigma_{\Delta_{\bar{M}}} \circ \gamma'_{\mathfrak{g},\Delta}\text{-}(Z))f(1) && \text{by (8.34)}\\
&= (\nu + \rho_A + \Lambda_\sigma + \delta_M)(\gamma'_{\mathfrak{g},\Delta}\text{-}(Z))f(1)\\
&= (\nu + \Lambda_\sigma + \delta)(\gamma'_{\mathfrak{g},\Delta}\text{-}(Z))f(1)\\
&= (\Lambda_\sigma + \nu)(\sigma_{\Delta}\text{-} \circ \gamma'_{\mathfrak{g},\Delta}\text{-}(Z))f(1)\\
&= (\Lambda_\sigma + \nu)(\gamma_{\mathfrak{g}}(Z))f(1)\\
&= \chi_{\Lambda_\sigma+\nu}(Z)f(1),
\end{aligned}$$

which is the result asserted.

§7. Differential Equations Satisfied by Matrix Coefficients

Let us return to the framework of §4. The group G is to be linear connected reductive. We shall assume G has compact center; in the contrary case, Proposition 1.2 shows that G is the direct product of a Euclidean group and a group with compact center, and hence there is no useful gain in generality by considering noncompact center.

A choice of positive Weyl chamber $\mathfrak{a}_\mathfrak{p}^+$ fixes a positive system Σ^+ in the set Σ of restricted roots. It will be convenient to carry along a system of roots for which Σ is the set of nonzero restrictions to $\mathfrak{a}_\mathfrak{p}$. Thus let $\mathfrak{b}_\mathfrak{p}$ be a Cartan subalgebra of $\mathfrak{m}_\mathfrak{p}$, so that $\mathfrak{a}_\mathfrak{p} \oplus \mathfrak{b}_\mathfrak{p}$ is a Cartan subalgebra of $\mathfrak{g}$, and let Δ be the set of roots $\Delta((\mathfrak{a}_\mathfrak{p} \oplus \mathfrak{b}_\mathfrak{p})^{\mathbb{C}}:\mathfrak{g}^{\mathbb{C}})$. As in the proof of Proposition 8.22, we can define Δ^+ in such a way that the members of Σ^+ are the nonzero restrictions of the members of Δ^+.

Let $\tau = (\tau_1, \tau_2)$ be as in §4, and define $C_\tau^\infty(G^{(0)})$ as in §4. With $A^+ = \exp \mathfrak{a}_\mathfrak{p}^+$ and $M = M_\mathfrak{p}$, let

$$C_\tau^\infty(A^+) = C^\infty(A^+, \operatorname{Hom}_M(U_2, U_1)).$$

Proposition 8.23. The restriction map $F \to F|_{A^+}$ is an isomorphism of $C_\tau^\infty(G^{(0)})$ onto $C_\tau^\infty(A^+)$.

Proof. If F is in $C_\tau^\infty(G^{(0)})$, then it is clear that $F|_{A^+}$ is in $C_\tau^\infty(A^+)$. The restriction map is one-one since $G^{(0)} = KA^+K$. To see it is onto, let f be in $C_\tau^\infty(A^+)$ and define $F(k_1ak_2) = \tau_1(k_1)f(a)\tau_2(k_2)$. Then F is well defined on $G^{(0)}$ by Theorems 5.20 and 5.17, and it satisfies the transformation rule (8.13). Use of the Jacobian determinant in Proposition 5.28 shows F is smooth.

Lemma 8.24. Let α be a restricted root, and let $\xi_\alpha(a) = e^{\alpha \log a}$ for a in A. To X in the restricted root space $\mathfrak{g}_\alpha$, associate the member Y of $\mathfrak{k}$ given by $Y = \frac{1}{2}(X + \theta X)$. If a is in A^+, then

$$X = \frac{2\xi_\alpha(a)}{\xi_\alpha(a)^2 - 1}\{\xi_\alpha(a)Y - \operatorname{Ad}(a)^{-1}Y\}.$$

Proof. Essentially the same as for Lemma 8.14.

Proposition 8.25. If a is in A^+, then

$$\mathfrak{g} = \operatorname{Ad}(a^{-1})\mathfrak{k} + \mathfrak{a}_\mathfrak{p} + \mathfrak{k}. \tag{8.37}$$

Consequently every member u of $U(\mathfrak{g}^\mathbb{C})$ decomposes as a linear combination of terms.

$$(\operatorname{Ad}(a)^{-1}X)HY \tag{8.38}$$

with X and Y in $U(\mathfrak{k}^\mathbb{C})$ and H in $U(\mathfrak{a}_\mathfrak{p}^\mathbb{C})$.

Remark. One can observe from the proof that the decompositions of particular elements as in (8.37) and (8.38) can be taken in a special fashion with coefficients that are global real analytic functions of a in A^+. We shall use this observation in our applications without further comment.

Proof. We know that $\mathfrak{g} = \bar{\mathfrak{n}}_\mathfrak{p} \oplus \mathfrak{a}_\mathfrak{p} \oplus \mathfrak{m}_\mathfrak{p} \oplus \mathfrak{n}_\mathfrak{p}$. Lemma 8.24 shows that $\bar{\mathfrak{n}}_\mathfrak{p}$ and $\mathfrak{n}_\mathfrak{p}$ are contained in $\operatorname{Ad}(a^{-1})\mathfrak{k} + \mathfrak{k}$. Thus $\mathfrak{g}$ equals the sum on the right of (8.37). The statement about $U(\mathfrak{g}^\mathbb{C})$ follows from the Birkhoff-Witt Theorem.

Combining (8.15) and Proposition 8.25, we obtain from u in $U(\mathfrak{g}^\mathbb{C})$ a differential operator $D_\tau(u)$ on A^+, with coefficients that are smooth functions from A^+ to $\operatorname{Hom}_\mathbb{C}(\operatorname{Hom}_M(U_2, U_1), \operatorname{Hom}_\mathbb{C}(U_2, U_1))$, with the property that

$$uf(a) = D_\tau(u)(F|_{A^+})(a).$$

Since Proposition 8.23 allows $F|_{A^+}$ in this equation to be an arbitrary smooth function with values in $\operatorname{Hom}_M(U_2, U_1)$, the operator $D_\tau(u)$ does not depend on the particular expression of u as a linear combination of terms (8.38). We call $D_\tau(u)$ the **τ-radial component** of u. As we saw in §4, the coefficients of $D_\tau(Z)$ are in $C^\infty(A^+, \operatorname{End}_\mathbb{C}(\operatorname{Hom}_M(U_2, U_1)))$ for any element Z of the center $Z(\mathfrak{g}^\mathbb{C})$.

We shall now introduce coordinates. Let $\{\alpha_1, \ldots, \alpha_l\}$ be the set of simple restricted roots, and define $\iota: A \to (\mathbb{R}^+)^l$ by

$$\iota(\exp H) = (e^{-\alpha_1(H)}, \ldots, e^{-\alpha_l(H)}).$$

Then $\iota(A^+) = (0, 1)^l \subseteq D^l$, where D denotes the unit disc in $\mathbb{C}$. Under ι, the function $e^{-\alpha_j(\log a)}$ corresponds to the function z_j. If $\beta = \sum n_j \alpha_j$ is in Σ^+, then $e^{-\beta(\log a)}$ corresponds to $z_1^{n_1} \cdots z_l^{n_l}$.

Let $H^{(1)}, \ldots, H^{(l)}$ be the dual basis of $\mathfrak{a}_\mathfrak{p}$ with $\alpha_i(H^{(j)}) = \delta_i^j$. We shall show that the differential operator $H^{(j)}$ on A corresponds to $-z_j \dfrac{\partial}{\partial z_j}$. In fact, let F be in $C^\infty(A)$ and put $f = F \circ \iota^{-1}$. Then

$$\begin{aligned}
H^{(j)}F(a) &= H^{(j)}(f \circ \iota)(a) = \frac{d}{dt} f \circ \iota(a \exp tH^{(j)})\bigg|_{t=0} \\
&= \frac{d}{dt} f(e^{-\alpha_1 \log a}, \ldots, e^{-\alpha_j \log a} e^{-t\alpha_j(H^{(j)})}, \ldots, e^{-\alpha_l \log a})\bigg|_{t=0} \\
&= -e^{-\alpha_j \log a} f_j(e^{-\alpha_1 \log a}, \ldots, e^{-\alpha_l \log a}) \\
&= -z_j \frac{\partial f}{\partial z_j}(z).
\end{aligned}$$

Thus $H^{(j)}$ corresponds under ι to $-z_j \dfrac{\partial}{\partial z_j}$, and $U(\mathfrak{a}_\mathfrak{p}^\mathbb{C})$ corresponds to $\mathbb{C}\left[-z_1 \dfrac{\partial}{\partial z_1}, \ldots, -z_l \dfrac{\partial}{\partial z_l}\right]$.

We are going ultimately to apply the theory of differential equations of Appendix B, and we introduce notation consistent with that in §5 of the appendix. Our basic vector space will be

$$U = \operatorname{Hom}_M(U_2, U_1),$$

and we let

$$\begin{aligned}
\mathscr{H}_U &= \{\text{holomorphic functions from } D^l \text{ to } U\} \\
\mathscr{H}_{\operatorname{End} U} &= \{\text{holomorphic functions from } D^l \text{ to } \operatorname{End} U\} \\
\partial^k &= \left(\frac{\partial}{\partial z_1}\right)^{k_1} \cdots \left(\frac{\partial}{\partial z_l}\right)^{k_l} \qquad \text{for } k \text{ in } (\mathbb{Z}^+)^l \\
(z\partial)^k &= \left(z_1 \frac{\partial}{\partial z_1}\right)^{k_1} \cdots \left(z_l \frac{\partial}{\partial z_l}\right)^{k_l} \qquad \text{for } k \text{ in } (\mathbb{Z}^+)^l \\
\mathscr{D} &= \left\{ \sum_{\substack{k \in (\mathbb{Z}^+)^l \\ \text{finite sum}}} A_k \partial^k \;\middle|\; A_k \in \mathscr{H}_{\operatorname{End} U} \text{ for all } k \right\} \\
\mathscr{D}^* &= \left\{ \sum_{\substack{k \in (\mathbb{Z}^+)^l \\ \text{finite sum}}} A_k (z\partial)^k \;\middle|\; A_k \in \mathscr{H}_{\operatorname{End} U} \text{ for all } k \right\}.
\end{aligned}$$

We have just seen that the members of $U(\mathfrak{a}_\mathfrak{p}^\mathbb{C})$, when regarded as operators on $\mathscr{H}_U$ by means of ι, correspond to members of $\mathscr{D}^*$.

For Z in $Z(\mathfrak{g}^\mathbb{C})$ we shall show that $D_\tau(Z)$ corresponds to a member of $\mathscr{D}^*$, and we shall relate the main terms of the member of $\mathscr{D}^*$ to the

Harish-Chandra homomorphism. Let $\mathfrak{l} = \mathfrak{a}_\mathfrak{p} \oplus \mathfrak{m}_\mathfrak{p}$. If we write

$$\mathfrak{g} = \bar{\mathfrak{n}}_\mathfrak{p} \oplus \mathfrak{l} \oplus \mathfrak{n}_\mathfrak{p},$$

then the same kind of argument as in Lemma 8.17 shows that

$$Z(\mathfrak{g}^\mathbb{C}) \subseteq \bar{\mathfrak{n}}_\mathfrak{p}^\mathbb{C} U(\mathfrak{g}^\mathbb{C}) \oplus U(\mathfrak{l}^\mathbb{C}),$$

and we let μ'_{Σ^+} be the projection of $Z(\mathfrak{g}^\mathbb{C})$ into $U(\mathfrak{l}^\mathbb{C})$. As in (8.34), Z in $Z(\mathfrak{g}^\mathbb{C})$ implies $\mu'_{\Sigma^+}(Z)$ is in $Z(\mathfrak{l}^\mathbb{C})$, and μ'_{Σ^+} is related to the Harish-Chandra homomorphism by

$$\gamma'_{\Delta^+} = \gamma'_{\Delta^+_{M_I}} \circ \mu'_{\Sigma^+} \text{ on } Z(\mathfrak{g}^\mathbb{C}). \tag{8.39}$$

Since $Z(\mathfrak{l}^\mathbb{C}) = U(\mathfrak{a}_\mathfrak{p}^\mathbb{C})Z(\mathfrak{m}_\mathfrak{p}^\mathbb{C})$, each $\mu'_{\Sigma^+}(Z)$ for Z in $Z(\mathfrak{g}^\mathbb{C})$ decomposes into the sum of terms (8.38) without the use of variable coefficients. Thus $D_\tau(\mu'_{\Sigma^+}(Z))$ corresponds under ι to a linear combination of operators $(z\partial)^k$ with constant coefficients; the coefficients are in End U.

We can extend $\mu'_{\Sigma^+}: Z(\mathfrak{g}^\mathbb{C}) \to Z(\mathfrak{l}^\mathbb{C})$ to a map $\mu'_{\Sigma^+}: U(\mathfrak{g}^\mathbb{C}) \to U(\mathfrak{a}_\mathfrak{p}^\mathbb{C})U(\mathfrak{k}^\mathbb{C})$ as follows: We write $\mathfrak{g} = \bar{\mathfrak{n}}_\mathfrak{p} \oplus \mathfrak{a}_\mathfrak{p} \oplus \mathfrak{k}$ and apply the Birkhoff-Witt Theorem to obtain

$$U(\mathfrak{g}^\mathbb{C}) = (U(\mathfrak{a}_\mathfrak{p}^\mathbb{C})U(\mathfrak{k}^\mathbb{C})) \oplus \bar{\mathfrak{n}}_\mathfrak{p}^\mathbb{C} U(\mathfrak{g}^\mathbb{C}). \tag{8.40}$$

Projection to the first component then gives the required extension of μ'_{Σ^+} to all of $U(\mathfrak{g}^\mathbb{C})$.

Theorem 8.26. For each Z in $Z(\mathfrak{g}^\mathbb{C})$, the τ-radial component $D_\tau(Z)$ corresponds under ι to a member of $\mathscr{D}^*$. Moreover, $D_\tau(Z) - D_\tau(\mu'_{\Sigma^+}(Z))$ corresponds under ι to a member of $\mathscr{D}^*$ whose coefficient functions vanish at $z = 0$.

Remark. The main term $D_\tau(\mu'_{\Sigma^+}(Z))$ of $D_\tau(Z)$ we can view as having constant coefficients. This operator is easier to understand in the special case $\mathfrak{m}_\mathfrak{p} = 0$. In this case (which includes $SL(2, \mathbb{R})$), $\mathfrak{a}_\mathfrak{p}$ is a Cartan subalgebra of $\mathfrak{g}$ and Σ^+ coincides with Δ^+. The main term $D_\tau(\mu'_{\Sigma^+}(Z))$ simplifies to $\gamma'_{\Delta^+}(Z)$, which we can view as an operator with numerical coefficients.

Before coming to the proof of the theorem, we prove a lemma.

Lemma 8.27. Let Z be in $U^n(\mathfrak{g}^\mathbb{C})$. Then there exist X_j and Y_j in $U(\mathfrak{k}^\mathbb{C})$, H_j in $U^{n-1}(\mathfrak{a}_\mathfrak{p}^\mathbb{C})$, and scalar-valued holomorphic functions f_j on D^l such that $f_j(0) = 0$ and

$$Z = \mu'_{\Sigma^+}(Z) + \sum_{j=1}^{p} f_j(\iota(a))(\mathrm{Ad}(a^{-1})X_j)H_jY_j$$

for all a in A^+.

Proof. We proceed by induction on n, the case $n = 0$ being obvious since $\mu'_{\Sigma^+}(1) = 1$. Suppose Z is in $U^n(\mathfrak{g}^{\mathbb{C}})$ with $n \geq 1$. Then $Z - \mu'_{\Sigma^+}(Z)$ is in $\bar{\mathfrak{n}}_{\mathfrak{p}}^{\mathbb{C}} U^{n-1}(\mathfrak{g}^{\mathbb{C}})$, and we can find $A_1, \ldots, A_s, B_1, \ldots, B_s$ with A_j in $\mathfrak{g}_{-\beta_j}$ and β_j in Σ^+, with each B_j in $U^{n-1}(\mathfrak{g}^{\mathbb{C}})$, and with

$$Z - \mu'_{\Sigma^+}(Z) = \sum_{j=1}^{s} A_j B_j.$$

By Lemma 8.24, applied with $\alpha = -\beta_j$, there exist $C_1, \ldots, C_s$ in $\mathfrak{k}^{\mathbb{C}}$ and holomorphic functions $g_1, \ldots, g_s, h_1, \ldots, h_s$ on D^l vanishing at 0 such that

$$A_j$$

Hence

$$\begin{aligned} Z - \mu'_{\Sigma^+}(Z) &= \sum_{j=1}^{s} (g_j(\imath(a))C_j B_j + h_j(\imath(a))(\operatorname{Ad}(a^{-1})C_j)B_j) \\ &= \sum_{j=1}^{s} (g_j(\imath(a))[C_j, B_j] + g_j(\imath(a))B_j C_j + h_j(\imath(a))(\operatorname{Ad}(a^{-1})C_j)B_j). \end{aligned}$$

Since $[C_j, B_j]$ and B_j are in $U^{n-1}(\mathfrak{g}^{\mathbb{C}})$, we can apply the inductive hypothesis, and the result follows.

Proof of Theorem 8.26. By Lemma 8.27, write Z in $Z(\mathfrak{g}^{\mathbb{C}})$ as

$$Z = \mu'_{\Sigma^+}(Z) + \sum_{j=1}^{p} f_j(\imath(a))(\operatorname{Ad}(a^{-1})X_j)H_j Y_j \tag{8.41}$$

with $f_j(0) = 0$ for all j. For m in M, $\operatorname{Ad}(m)$ fixes Z and $\mu'_{\Sigma^+}(Z)$. If we apply $\operatorname{Ad}(m)$ to (8.41) and evaluate both sides on F in $C_\tau^\infty(A^+)$, we obtain, by (8.15) and then (8.16),

$$\begin{aligned} ZF(a) &= \mu'_{\Sigma^+}(Z)F(a) + \sum_{j=1}^{p} f_j(\imath(a))\tau_1(\operatorname{Ad}(m)X_j)H_j F(a)\tau_2(\operatorname{Ad}(m)Y_j) \\ &= \mu'_{\Sigma^+}(Z)F(a) + \sum_{j=1}^{p} f_j(\imath(a))\tau_1(m)\tau_1(X_j)H_j F(a)\tau_2(Y_j)\tau_2(m)^{-1}. \end{aligned}$$

We shall form averages over M. For any T in $U = \operatorname{Hom}_M(U_2, U_1)$, such as $T = H_j F(a)$, we can form

$$B_j(T) = \int_M \tau_1(m)\tau_1(X_j)T\tau_2(Y_j)\tau_2(m^{-1})\,dm.$$

Then $B_j(T)$ is in $\operatorname{Hom}_{\mathbb{C}}(U_2, U_1)$, and we check readily that $B_j(T)$ is in $\operatorname{Hom}_M(U_2, U_1)$. Hence B_j is in $\operatorname{End} U$. Averaging (8.42) over M and applying (8.16), we thus obtain

$$D_\tau(Z) = \mu'_{\Sigma^+}(Z) + \sum_{j=1}^{p} f_j(\imath(a))B_j H_j F(a),$$

which is of the prescribed form.

Theorem 8.28. Let I be an ideal of finite codimension in $Z(\mathfrak{g}^{\mathbb{C}})$, and fix τ. Then the system $\{D_\tau(Z)(F|_{A^+}) = 0,\ Z \in I\}$, referred to D^l via ι, has a simple singularity (along $D^l - (D^\times)^l$).

The proof will be preceded by three lemmas. We are to prove that the left ideal in $\mathscr{D}^*$ generated by all $D_\tau(Z)$ for Z in I has finite codimension in $\mathscr{D}^*$. Control over the size of I will come from Theorems 8.18 and 8.19.

Lemma 8.29. There exist finitely many elements $H_1, \ldots, H_r$ in $Z(\mathfrak{l}^{\mathbb{C}})$ with the following property: To each H in $Z(\mathfrak{l}^{\mathbb{C}})$ correspond members $Z_1, \ldots, Z_r$ of $Z(\mathfrak{g}^{\mathbb{C}})$ such that $H - \sum_{j=1}^r Z_j H_j$ is in $\bar{\mathfrak{n}}_{\mathfrak{p}}^{\mathbb{C}} U^{\deg H - 1}(\mathfrak{g}^{\mathbb{C}})$ and such that $\deg Z_j + \deg H_j \le \deg H$ for $1 \le j \le r$.

Proof. Let $\gamma_G = \sigma_{\Delta^+} \circ \gamma'_{\Delta^+}$ and $\gamma_L = \sigma_{\Delta_M^+} \circ \gamma'_{\Delta_M^+}$ be the Harish-Chandra homomorphisms for $Z(\mathfrak{g}^{\mathbb{C}})$ and $Z(\mathfrak{l}^{\mathbb{C}})$, and let W and W_M denote the Weyl groups for G and M. Introduce

$$\sigma_{\Sigma^+} = \sigma_{\Delta^+} \circ \sigma_{\Delta_M^+}^{-1} \tag{8.43}$$

as an automorphism of $U((\mathfrak{a} \oplus \mathfrak{b})^{\mathbb{C}})$. Since the half-sums of roots for Δ^+, Σ^+, and Δ_M^+ satisfy $\delta = \rho_{\mathfrak{p}} + \delta_M$, σ_{Σ^+} is the automorphism of $U((\mathfrak{a} \oplus \mathfrak{b})^{\mathbb{C}})$ that extends the linear map

$$\sigma_{\Sigma^+}(H) = H - \rho_{\mathfrak{p}}(H)1 \tag{8.44}$$

of $\mathfrak{a} \oplus \mathfrak{b}$ into $U((\mathfrak{a} \oplus \mathfrak{b})^{\mathbb{C}})$. The formula (8.44) makes it clear that σ_{Σ^+} commutes with the action of W_M on $U((\mathfrak{a} \oplus \mathfrak{b})^{\mathbb{C}})$.

By Theorems 8.18 and 8.19, $U((\mathfrak{a} \oplus \mathfrak{b})^{\mathbb{C}})$ is a finitely generated free module over $\gamma_G(Z(\mathfrak{g}^{\mathbb{C}}))$. Let $A_1, \ldots, A_r$ in $U((\mathfrak{a} \oplus \mathfrak{b})^{\mathbb{C}})$ be a basis. Given H in $Z(\mathfrak{l}^{\mathbb{C}})$, we can write

$$\sigma_{\Sigma^+}\gamma_M(H) = \sum_{j=1}^r \gamma_G(Z_j)A_j$$

for suitable Z_j in $Z(\mathfrak{g}^{\mathbb{C}})$. Averaging this equation over W_M and using the fact that $W_M \subseteq W$, we obtain

$$\sigma_{\Sigma^+}\gamma_M(H) = \sum_{j=1}^r \gamma_G(Z_j)A_j^{W_M}, \tag{8.45}$$

where $A_j^{W_M}$ denotes the averaged version of A_j. We apply

$$\gamma_M^{-1}\sigma_{\Sigma^+}^{-1} = \gamma'^{-1}_{\Delta_M^+}\sigma_{\Delta^+}^{-1}$$

to both sides of (8.45) and apply (8.39) to obtain

$$H = \sum_{j=1}^r \mu'_{\Sigma^+}(Z_j)\gamma_M^{-1}\sigma_{\Sigma^+}^{-1}(A_j^{W_M}). \tag{8.46}$$

The lemma follows immediately with $H_j = \gamma_M^{-1}\sigma_{\Sigma^+}^{-1}(A_j^{W_M})$.

Lemma 8.30. Fix $H_1, \ldots, H_r$ as in Lemma 8.29. For each u in $U(\mathfrak{g}^{\mathbb{C}})$ there exist $X_1, \ldots, X_p$, $Y_1, \ldots, Y_p$ in $U(\mathfrak{k}^{\mathbb{C}})$, $W_1, \ldots, W_p$ in $\sum Z(\mathfrak{g}^{\mathbb{C}})H_j$, and complex-valued holomorphic functions $f_1, \ldots, f_p$ on D^l such that

$$u = \sum_{j=1}^{r} f_j(\iota(a))(\mathrm{Ad}(a^{-1})X_j)W_jY_j \tag{8.47}$$

for all a in A^+.

Proof. We suppose that u is in $U^n(\mathfrak{g}^{\mathbb{C}})$, and we proceed inductively on n. For $n = 0$ and 1, we observe from Lemma 8.29 that $H_1, \ldots, H_r$ includes a 0th degree element and a spanning set for the 1st degree elements of $U(\mathfrak{a}_\mathfrak{p}^{\mathbb{C}})$. Then the result for $n = 0$ and $n = 1$ follows from Proposition 8.25. So suppose $n > 1$. With μ'_{Σ^+} extended to $U(\mathfrak{g}^{\mathbb{C}})$ via (8.40), we can write

$$u - \mu'_{\Sigma^+}(u) \in \bar{\mathfrak{n}}_\mathfrak{p}^{\mathbb{C}} U^{n-1}(\mathfrak{g}^{\mathbb{C}}) \qquad \text{and} \qquad \mu'_{\Sigma^+}(u) = \sum_{j=1}^{q} H'_jK_j$$

with H'_j in $U(\mathfrak{a}_\mathfrak{p}^{\mathbb{C}})$, K_j in $U(\mathfrak{k}^{\mathbb{C}})$, and $\deg H'_j + \deg K_j \leq n$. By Lemma 8.29, there are elements Z_{jk} in $Z(\mathfrak{g}^{\mathbb{C}})$ with

$$H'_j - \sum_{k=1}^{r} Z_{jk}H_k$$

in $\bar{\mathfrak{n}}_\mathfrak{p}^{\mathbb{C}} U^{\deg H'_j - 1}(\mathfrak{g}^{\mathbb{C}})$ and with $\deg Z_{jk} + \deg H_k \leq \deg H'_j$. Substituting, we see that

$$u - \sum_{j,k} Z_{jk}H_kK_j \text{ is in } \bar{\mathfrak{n}}_\mathfrak{p}^{\mathbb{C}} U^{n-1}(\mathfrak{g}^{\mathbb{C}}).$$

Thus we can write

$$u = \sum_{j,k} Z_{jk}H_kK_j + \sum_i L_iM_i \tag{8.48}$$

with L_i in $\mathfrak{g}_{-\beta_i}$ for some β_i in Σ^+ and with M_i in $U^{n-1}(\mathfrak{g}^{\mathbb{C}})$. By Lemma 8.24 there exist members N_i of $\mathfrak{k}^{\mathbb{C}}$ and holomorphic functions g_i, h_i on D^l such that

$$L_i = g_i(\iota(a))N_i + h_i(\iota(a))\,\mathrm{Ad}(a^{-1})N_i.$$

Substituting in (8.48), we obtain

$$\begin{aligned} u &= \sum_{j,k} Z_{jk}H_kK_j + \sum_i (g_i(\iota(a))N_iM_i + h_i(\iota(a))(\mathrm{Ad}(a^{-1})N_i)M_i) \\ &= \sum_{j,k} Z_{jk}H_kK_j \\ &\quad + \sum_i (g_i(\iota(a))M_iN_i + g_i(\iota(a))[N_i, M_i] + h_i(\iota(a))(\mathrm{Ad}(a^{-1})N_i)M_i). \end{aligned}$$

Since M_i and $[N_i, M_i]$ are in $U^{n-1}(\mathfrak{g}^{\mathbb{C}})$, the lemma follows by induction.

Lemma 8.31. Fix $H_1, \ldots, H_r$ as in Lemma 8.29, and let $D_1, \ldots, D_r$ be the corresponding operators in $\mathscr{D}^*$ under ι. Identify members of $D_\tau(Z(\mathfrak{g}^{\mathbb{C}}))$ with their corresponding operators in $\mathscr{D}^*$. Then each member D of $\mathscr{D}^*$ admits a finite expansion

$$D = \sum_{j,k} F_j D_k D_\tau(Z_{jk}) \tag{8.49}$$

with all F_j in $\mathscr{H}_{\mathrm{End}\,U}$ and all Z_{jk} in $Z(\mathfrak{g}^{\mathbb{C}})$.

Proof. Fix X in $U(\mathfrak{a}_{\mathfrak{p}}^{\mathbb{C}})$, and expand X as in (8.47). Write $W_j = \sum H_k Z_{jk}$, substitute, and commute the Z_{jk} past Y_j to get

$$X = \sum_{j,k} f_j(\iota(a))(\mathrm{Ad}(a^{-1})X_j)H_k Y_k Z_{jk}.$$

As in the proof of Theorem 8.26, apply $\mathrm{Ad}(m)$ and average over M, letting

$$B_j(T) = \int_M \tau_1(m)\tau_1(X_j)T\tau_2(Y_j)\tau_2(m^{-1})\,dm$$

and noting that $\mathrm{Ad}(m)X = X$. Then B_j is in End U. Since $H_k D_\tau(Z_{jk})F(a)$ is in U, we obtain

$$XF(a) = \sum_{j,k} f_j(\iota(a))B_j H_k D_\tau(Z_{jk})F(a).$$

Since the basis elements $H^{(1)}, \ldots, H^{(l)}$ correspond to $-z_1 \dfrac{\partial}{\partial z_1}, \ldots,$ $-z_l \dfrac{\partial}{\partial z_l}$ under ι, X corresponds to a linear combination of powers of $(z\partial)$ with numerical coefficients. Linear combinations of the latter, with coefficients in $\mathscr{H}_{\mathrm{End}\,U}$, give all of $\mathscr{D}^*$, and thus the lemma follows.

Proof of Theorem 8.28. Let $Q_1, \ldots, Q_s$ in $Z(\mathfrak{g}^{\mathbb{C}})$ be a basis for a vector space complement to the ideal I in $Z(\mathfrak{g}^{\mathbb{C}})$, and denote by $\mathscr{I}$ the left ideal in $\mathscr{D}^*$ generated by all $D_\tau(Z)$ for Z in I. With $D_1, \ldots, D_r$ as in Lemma 8.31, consider the finite set $\{D_k D_\tau(Q_m)\}$ of elements in $\mathscr{D}^*$. If D in $\mathscr{D}^*$ is given, expand D as in (8.49) and choose constants c_{jkm} such that

$$Z_{jk} - \sum_{m=1}^{s} c_{jkm} Q_m$$

is in I. Then

$$D - \sum_{j,k,m} c_{jkm} F_j D_k D_\tau(Q_m)$$

is in $\mathscr{I}$. Consequently the set $\{D_k D_\tau(Q_m)\mathscr{I}\}$ is a finite set of generators for $\mathscr{D}^*/\mathscr{I}$ as a left $\mathscr{H}_{\mathrm{End}\,U}$ module. Thus our system has a simple singularity.

Let us apply the theorems of this section. Suppose that π is an irreducible admissible representation. Corollary 8.14 and the results of §6 say that π has an infinitesimal character Λ:

$$\pi(Z) = \chi_\Lambda(Z)I \qquad \text{for } Z \in Z(\mathfrak{g}^{\mathbb{C}}).$$

Let F be a τ-spherical function built from matrix coefficients of π as at the start of §4. Formula (8.10) says that

$$ZF = \chi_\Lambda(Z)F \qquad \text{for } Z \in Z(\mathfrak{g}^{\mathbb{C}}),$$

hence that $\quad D_\tau(Z)(F|_{A^+}) = \chi_\Lambda(Z)(F|_{A^+}) \qquad \text{for } Z \in Z(\mathfrak{g}^{\mathbb{C}}).$

Thus $F|_{A^+}$ is annihilated by $D_\tau(Z)$ for all Z in the kernel I of χ_Λ. Since χ_Λ is a linear functional, its kernel has codimension one. Thus Theorems 8.26 and 8.28 apply, and these theorems say that we can apply the theory of Appendix B.

§8. Asymptotic Expansions and Leading Coefficients

We shall now combine Theorem 8.26, Theorem 8.28, and the theory of Appendix B. Let π be an admissible representation of G, and suppose π has an infinitesimal character, say Λ. (Two situations in which π has an infinitesimal character are if π is irreducible or if π is a standard induced representation, by Corollary 8.14 and Proposition 8.22, respectively.)

For the time being, fix a nonempty finite block of K-finite matrix coefficients of π, and construct $\tau = (\tau_1, \tau_2)$ as in §4. Say τ_1 and τ_2 act on U_1 and U_2, respectively. The τ-spherical function

$$F(x) = E_1\pi(x)E_2 \tag{8.50}$$

has values in $U = \operatorname{Hom}_{\mathbb{C}}(U_2, U_1)$ and, as noted at the end of §7, is an eigenfunction of $Z(\mathfrak{g}^{\mathbb{C}})$ with eigenvalue χ_Λ.

We introduced coordinates in §7, using a mapping ι to carry A^+ to $(0, 1)^l$. Theorem B.16, in combination with Theorems 8.26 and 8.28, immediately yields the following expansion for $F|_{A^+}$.

Theorem 8.32. Let F be the τ-spherical eigenfunction of $Z(\mathfrak{g}^{\mathbb{C}})$ in (8.50). Then $F \circ \iota^{-1}$, initially defined on $(0, 1)^l$, extends to a global multiple-valued holomorphic solution on $(D^\times)^l$ of the system corresponding to

$$D_\tau(Z)(F|_{A^+}) = \chi_\Lambda(Z)(F|_{A^+}), \qquad Z \in Z(\mathfrak{g}^{\mathbb{C}}).$$

Moreover, there exist an integer q_0 and a finite subset $\mathscr{F} \subseteq \mathbb{C}^l$ such that $F \circ \iota^{-1}$ is given on $(D^\times)^l$ by

$$F \circ \iota^{-1}(z) = \sum_{s \in \mathscr{F}} \sum_{0 \le |q| \le q_0} z^s (\log z)^q F_{s,q}(z), \tag{8.51}$$

with each $F_{s,q}(z)$ in $\mathscr{H}_U$.

Each function $F_{s,q}(z)$, being holomorphic on D^l, has a convergent multiple power series expansion. The convergence of the series is absolute and uniform on any compact subset of D^l. This implies that each $F_{s,q}$ corresponds to a series expansion on A^+ of the form

$$F_{s,q}(\iota(a)) = \sum_{k \in (\mathbb{Z}^+)^l} c'_{s,q,k} e^{-(k_1\alpha_1 + \cdots + k_l\alpha_l)(\log a)}, \qquad c'_{s,q,k} \in U,$$

with the convergence absolute and uniform as long as $\alpha_j(\log a) \geq \varepsilon_j > 0$ for $1 \leq j \leq l$. Since the sum in (8.51) is a finite sum, $F(a)$ itself has an analogous asymptotic expansion, except that there are factors present of the form

$$z^s(\log z)^q \leftrightarrow \alpha_1(\log a)^{q_1} \cdots \alpha_l(\log a)^{q_l} e^{-(s_1\alpha_1 + \cdots + s_l\alpha_l)(\log a)}.$$

Let us regroup matters essentially as in (B.33), letting $\rho = \rho_{\mathfrak{p}}$ denote the half-sum of the members of Σ^+ (counted with multiplicities) and writing

$$F(\exp H) = \sum_{\nu} F_{\nu-\rho}(\exp H) \tag{8.52a}$$

$$F_{\nu-\rho}(\exp H) = e^{-\rho(H)} \sum_{|q| \leq q_0} c_{\nu,q} \alpha(H)^q e^{\nu(H)}. \tag{8.52b}$$

Guided by Appendix B, we say $\nu - \rho$ is an **exponent** of F if $F_{\nu-\rho} \not\equiv 0$. (We isolate ρ in the terminology in order to make results like Theorem 8.33 below be more symmetric-looking). Two exponents are **integrally equivalent** if their difference is an integral combination of the simple roots $\alpha_1, \ldots, \alpha_l$ of Σ^+. A **leading exponent** $\nu - \rho$ is an exponent such that $\nu - \nu'$ is a nonnegative integral combination of simple roots whenever $\nu' - \rho$ is an exponent of F integrally equivalent with $\nu - \rho$. The corresponding term $F_{\nu-\rho}$ will be called a **leading term** of F. We shall write $\nu' \leq \nu$ when ν and ν' are integrally equivalent and $\nu - \nu'$ is a nonnegative integral combination of simple roots. By Proposition B.17 the set of leading exponents of F is finite, and every exponent $\nu' - \rho$ has $\nu' \leq \nu$ for some leading exponent $\nu - \rho$.

Theorem 8.33. Let F be the τ-spherical eigenfunction of $Z(\mathfrak{g}^{\mathbb{C}})$ in (8.50) with eigenvalue χ_Λ under $Z(\mathfrak{g}^{\mathbb{C}})$. Then any leading exponent $\nu - \rho$ of F has the property that $\nu = w\Lambda|_{\mathfrak{a}_{\mathfrak{p}}}$ for some w in the Weyl group $W = W((\mathfrak{a} \oplus \mathfrak{b})^{\mathbb{C}} : \mathfrak{g}^{\mathbb{C}})$.

Proof. Let $F_{\nu-\rho}$ be a leading term of F. We shall identify ν by means of Lemma B.22. According to that lemma and to Theorem 8.26, we have

$$D_\tau(\mu'_{\Sigma^+}(Z))F_{\nu-\rho} = 0 \qquad \text{for all } Z \text{ in } Z(\mathfrak{g}^{\mathbb{C}}) \text{ with } \chi_\Lambda(Z) = 0.$$

Say $F_{\nu-\rho}$ is given as in (8.52b). Since $\mu'_{\Sigma^+}(Z)$ is in $Z(\mathfrak{l}^{\mathbb{C}})$, we can write

$$\mu'_{\Sigma^+}(Z) = \sum_j H'_j Z'_j \qquad \text{with } H'_j \in U(\mathfrak{a}_{\mathfrak{p}}^{\mathbb{C}}) \text{ and } Z'_j \in U(\mathfrak{m}_{\mathfrak{p}}^{\mathbb{C}}).$$

Then $$D_\tau(\mu'_{\Sigma^+}(Z)) = \sum_j (\text{right by } \tau_2(Z'_j))H'_j.$$

We apply this operator to $F_{\nu-\rho}$, set the result equal to 0, and read off the coefficient of $\alpha(H)^q e^{(\nu-\rho)(H)}$ with q as large as possible. The result is

$$0 = c_{\nu,q} \sum_j (\nu - \rho)(H'_j)\tau_2(Z'_j). \tag{8.53}$$

Here $c_{\nu,q}$ is a nonzero member of $\operatorname{Hom}_M(U_2, U_1)$. Choose an irreducible representation π_λ of the identity component M_0 (with highest weight λ) such that $c_{\nu,q}$ is not the 0 map on the subspace of U_2 transforming by π_λ under M_0, and restrict (8.53) to that subspace. On that subspace, $\tau_2(Z'_j)$ is the scalar operator $\lambda(\gamma'_{\Delta_M^+}(Z'_j))$, and thus (8.53) implies

$$\sum_j (\nu - \rho)(H'_j)\lambda(\gamma'_{\Delta_M^+}(Z'_j)) = 0.$$

Hence

$$\begin{aligned} 0 &= \sum_j (\nu - \rho + \lambda)(H'_j)(\nu - \rho + \lambda)(\gamma'_{\Delta_M^+}(Z'_j)) \\ &= (\nu - \rho + \lambda)\gamma'_{\Delta_M^+}(\textstyle\sum H'_j Z'_j) \\ &= (\nu - \rho + \lambda)\gamma'_{\Delta_M^+}\mu'_{\Sigma^+}(Z) \\ &= (\nu - \rho + \lambda)\gamma'_{\Delta^+}(Z) \qquad \text{by (8.39)} \\ &= (\nu + \lambda + \delta_M)\gamma_G(Z) \qquad \text{by (8.25).} \end{aligned}$$

This means that $\chi_{\nu+\lambda+\delta_M}(Z) = 0$ for Z in $\ker \chi_\Lambda$, hence that $\chi_{\nu+\lambda+\delta_M} = \chi_\Lambda$. The theorem follows by applying Proposition 8.20.

Now let us allow $\tau = (\tau_1, \tau_2)$ to vary, yielding different blocks of matrix coefficients of our original admissible representation π. If $\nu - \rho$ is an exponent of the solution F that goes with some τ, we say $\nu - \rho$ is an **exponent** of π. A **leading exponent** $\nu - \rho$ of π is an exponent of π such that $\nu' \leq \nu$ whenever $\nu' - \rho$ is an exponent of π integrally equivalent with $\nu - \rho$. Theorem 8.33 implies that the set of leading exponents for each τ lies in a finite set independent of τ, and we therefore obtain an immediate corollary.

Corollary 8.34. Let π be an admissible representation of G with an infinitesimal character. Then the set of leading exponents of π is finite and nonempty. Moreover, if $\nu' - \rho$ is any exponent of π, then there is some leading exponent $\nu - \rho$ of π with $\nu' \leq \nu$.

Proposition 8.35. Let π be an admissible representation of G with an infinitesimal character. As τ varies, the powers q of $\alpha(H)$ that can appear in the expansion (8.52b) of leading terms have $|q|$ uniformly bounded.

Remark. Inspection of the proof will show that a bound for $|q|$ is the order of the Weyl group $W((\mathfrak{a} \oplus \mathfrak{b})^{\mathbb{C}} : \mathfrak{g}^{\mathbb{C}})$.

Proof. Let $H_1, \ldots, H_r$ be the elements of $Z(\mathfrak{l}^{\mathbb{C}})$ in Lemma 8.29. That lemma (or, more directly, line (8.46) of its proof) shows that every H in

$Z(\mathfrak{l}^{\mathbb{C}})$ admits an expansion

$$H = \sum_{j=1}^{r} H_j \mu'_{\Sigma^+}(Z_j) \qquad \text{with } Z_1, \dots, Z_r \text{ in } Z(\mathfrak{g}^{\mathbb{C}}). \tag{8.54}$$

Let τ be given, form the τ-spherical F, and let $F_{\nu-\rho}$ be a leading term, written as in (8.52b). For H in $U(\mathfrak{a}_{\mathfrak{p}}^{\mathbb{C}})$, write H as in (8.54), and apply H to $F_{\nu-\rho}$ to get

$$\begin{aligned} HF_{\nu-\rho} &= \sum_{j=1}^{r} D_\tau(H_j \mu'_{\Sigma^+}(Z_j))F_{\nu-\rho} \\ &= \sum_{j=1}^{r} D_\tau(H_j)D_\tau(\mu'_{\Sigma^+}(Z_j))F_{\nu-\rho} \qquad \text{since all objects in question commute} \\ &= \sum_{j=1}^{r} D_\tau(H_j)D_\tau(\mu'_{\Sigma^+}(Z_j) - \chi_\Lambda(Z_j))F_{\nu-\rho} + \sum_{j=1}^{r} \chi_\Lambda(Z_j)D_\tau(H_j)F_{\nu-\rho} \\ &= \sum_{j=1}^{r} \chi_\Lambda(Z_j)D_\tau(H_j)F_{\nu-\rho} \qquad \text{by Lemma B.22.} \end{aligned}$$

The right side lies in the vector space $\sum_{j=1}^{r} \mathbb{C}D_\tau(H_j)F_{\nu-\rho}$, whose dimension is $\le r$. Thus $\dim\{U(\mathfrak{a}_{\mathfrak{p}}^{\mathbb{C}})F_{\nu-\rho}\} \le r$. Assuming that some maximal tuple q in (8.52b) has $|q| > r$, we obtain a contradiction by noting that the $|q|$ functions

$$\prod_{j=1}^{l} (H^{(j)} - (\nu - \rho)(H^{(j)})1)^{k_j} F_{\nu-\rho},$$

parametrized by tuples $k \le q$, are linearly independent.

If we differentiate our τ-spherical F, as given in (8.52), by members of $U(\mathfrak{a}_{\mathfrak{p}}^{\mathbb{C}})$, the differentiated F is given by the term-by-term differentiated series. This fact is a consequence of familiar properties of power series, since $H^{(j)}$ corresponds to $-z_j \dfrac{\partial}{\partial z_j}$ under ι. There is a more general result concerning derivatives by members of $U(\mathfrak{g}^{\mathbb{C}})$ that we give as the next proposition.

Proposition 8.36. Let π be an admissible representation of G with an infinitesimal character, and let F be the τ-spherical function corresponding to a block of matrix coefficients. For each u in $U(\mathfrak{g}^{\mathbb{C}})$, $(uF)|_{A^+}$ is given by a series expansion (8.52) with coefficients in $\operatorname{Hom}_{\mathbb{C}}(U_2, U_1)$, the series obtained by formal manipulation of the series for $F|_{A^+}$.

Remarks. This result deals with left-invariant derivatives. There is an analogous result for right-invariant derivatives and even for mixtures of left-invariant and right-invariant derivatives.

Proof. We expand u as in Lemma 8.30, commute the members of $Z(\mathfrak{g}^{\mathbb{C}})$ to the right end, and apply the result to F. Then the result is apparent from (8.15) and properties of power series.

We conclude this section by noting what Theorem 8.32 implies about a single matrix coefficient. Let π be an admissible representation with an infinitesimal character, and let v and v' be K-finite vectors. We consider the matrix coefficient $(\pi(x)v, v')$ restricted to A^+. Since v and v' are K-finite, we can form $\tau = (\tau_1, \tau_2)$ and the associated τ-spherical

$$F(x) = E_1\pi(x)E_2$$

with E_1 and E_2 chosen large enough so that $E_2 v = v$ and $E_1 v' = v'$. Then our matrix coefficient is simply

$$(\pi(x)v, v') = (F(x)v, v').$$

Applying our asymptotic expansion for $F|_{A^+}$ term-by-term to v and taking the inner product with v', we obtain an asymptotic expansion for $(\pi(x)v, v')$ on A^+.

§9. First Application: Subrepresentation Theorem

Now we can return to the problem discussed in §1. Guided by the motivation in that section, we obtain the following Subrepresentation Theorem as an application of the results of §§7–8.

Theorem 8.37. Let π be an irreducible admissible representation of G. Then π is infinitesimally equivalent with a subrepresentation of some nonunitary principal series representation. More specifically, if $\nu - \rho$ is a leading exponent of π, then π is infinitesimally equivalent with a subrepresentation of $U(\bar{S}_{\mathfrak{p}}, \sigma, \nu)$ for some irreducible unitary representation σ of $M_{\mathfrak{p}}$.

Proof. By Corollary 8.14, π has an infinitesimal character. Corollary 8.34 says that π has a leading exponent, say $\nu - \rho$. The powers of $\alpha(H)$ that can appear in terms (8.52) corresponding to $\nu - \rho$ are bounded, in view of Proposition 8.35, and we let q be a maximal such power. Then we choose two K-finite vectors v_1 and v_2 such that the asymptotic expansion of $(\pi(a)v_1, v_2)$ along $A_{\mathfrak{p}}^+$ contains the term $\alpha(H)^q e^{\nu-\rho}$ with nonzero coefficient.

Define a linear functional l on the space of K-finite vectors of π as follows: $l(v)$ is the coefficient of $\alpha(H)^q e^{\nu-\rho}$ in the asymptotic expansion of $(\pi(a)v, v_2)$ along $A_{\mathfrak{p}}^+$. The construction of v_2 above ensures that $l \neq 0$. Since $\nu - \rho$ is a leading exponent, it follows easily from Lemma 8.24 that

$$l(Xv) = 0 \qquad \text{for all } X \in \bar{\mathfrak{n}}_{\mathfrak{p}} \text{ and all } K\text{-finite } v.$$

Since q is as large as possible, direct calculation shows that

$$l(Hv) = (\nu - \rho)(H)l(v) \qquad \text{for all } H \in \mathfrak{a}_\mathfrak{p} \text{ and all } K\text{-finite } v.$$

Define a linear mapping L of the space of K-finite vectors to the space of (K-finite) complex-valued functions on K by

$$L(v)(k) = l(\pi(k)^{-1}v) \qquad \text{for } k \in K.$$

We regard the space of complex-valued functions on K as the compact picture of $\operatorname{ind}_{A_\mathfrak{p}\bar{N}_\mathfrak{p}}^G(e^\nu \otimes 1)$, extending the functions to G by

$$f(ka\bar{n}) = e^{(-\nu+\rho)(\log a)}f(k).$$

(This formula takes into account that the ρ associated to $\bar{N}_\mathfrak{p}$ is $-\rho$.)

We shall check that L is $\mathfrak{g}$-equivariant, i.e., that $L(Xv) = X(L(v))$ for all X in $\mathfrak{g}$. We are to verify that

$$l(\pi(k)^{-1}Xv) = X(e^{(-\nu+\rho)H(x)}\pi(\kappa(x))^{-1}v)_{x=k}, \tag{8.55}$$

where H is the log of the $A_\mathfrak{p}$ component when $G = KA_\mathfrak{p}\bar{N}_\mathfrak{p}$. We do so for one value of k at a time. Put $X = \operatorname{Ad}(k)Y$. Then the left side of (8.55) is $l(Y\pi(k)^{-1}v)$. The right side is easily checked to be

$$\begin{array}{ll} 0 & \text{if } Y \in \bar{\mathfrak{n}}_\mathfrak{p} \\ (\nu - \rho)(Y)l(\pi(k)^{-1}v) & \text{if } Y \in \mathfrak{a}_\mathfrak{p} \\ l(Y\pi(k)^{-1}v) & \text{if } Y \in \mathfrak{k}. \end{array}$$

Hence the right side is always $l(Y\pi(k)^{-1}v)$, and thus (8.55) holds. Consequently L is $\mathfrak{g}$-equivariant.

As in §1, we can now use the compactness of $M_\mathfrak{p}$ to obtain a nonzero $\mathfrak{g}$-equivariant map into some $\operatorname{ind}_{S_\mathfrak{p}}^G(\sigma \otimes e^\nu \otimes 1)$. Since π is irreducible, this map must be one-one. This completes the proof.

§10. Second Application: Analytic Continuation of Intertwining Operators, Part II

Our second application of the theory of §§7–8 will be to the intertwining operators of Chapter VII. Recall from §7.4 that we have defined formal intertwining operators $A(S':S:\sigma:\nu)$ and $A_S(w, \sigma\ \nu)$ on the standard induced representation $U(S, \sigma, \nu)$. These operators map $U(S, \sigma, \nu)$ to $U(S', \sigma, \nu)$ and $U(S, w\sigma, w\nu)$, respectively. The sense in which the operators are formal is that their defining integrals are not convergent for many values of ν.

We found that when $S = S_\mathfrak{p}$ is a minimal parabolic subgroup, the operators $A(S'_\mathfrak{p}:S_\mathfrak{p}:\sigma:\nu)$ and $A_{S_\mathfrak{p}}(w, \sigma, \nu)$ have a nonempty region of convergence, have suitable meromorphic continuations in ν, and satisfy a number of identities. The definitive results are Corollary 7.13 and the propositions of §7.6.

For general S we saw in §8.3 that the convergence of the intertwining integrals presented no problem in the region (8.12) because of the boundedness of the matrix coefficients of σ. (Actually Theorem 7.24 gave a better estimate, but we shall not need the better estimate for our current purposes.) What we lack for general S is results about analytic continuation. The point is that we can obtain an analytic continuation by using the Subrepresentation Theorem.

Without paying too much attention to details, we shall show how to obtain this analytic continuation. In particular, if we write $S = MAN$, we shall not give the easy supplementary arguments that allow the whole theory, including the Subrepresentation Theorem, to apply to M even if M is disconnected.

We may suppose that there is a minimal parabolic subgroup $S_{\mathfrak{p}} = M_{\mathfrak{p}}A_{\mathfrak{p}}N_{\mathfrak{p}}$ with $S \supseteq S_{\mathfrak{p}}$. We let $K_M = K \cap M$, and we can choose Iwasawa A and N components of M so that $A_{\mathfrak{p}} = AA_M$ and $N_{\mathfrak{p}} = NN_M$. Then $S_M = M_{\mathfrak{p}}A_M N_M$ is a minimal parabolic subgroup of M.

Given an irreducible unitary representation ξ of M, we choose parameters σ in $\widehat{M}_{\mathfrak{p}}$ and ν_M in $(\mathfrak{a}_M')^{\mathbb{C}}$ by the Subrepresentation Theorem so that ξ is infinitesimally equivalent with a subrepresentation of

$$\omega = \operatorname{ind}_{S_M}^M(\sigma \otimes e^{\nu_M} \otimes 1).$$

Disregarding the fact that ω may not be unitary, we form the induced representation $U(S, \omega, \nu)$ of G. From a double induction formula just as in §7.2, we obtain a canonical equivalence of representations

$$U(S, \omega, \nu) \cong U(S_{\mathfrak{p}}, \sigma, \nu \oplus \nu_M). \tag{8.56}$$

Specifically we obtain (8.56) as follows: ω acts in a certain space V^ω of functions on M with values in V^σ, and $U(S, \omega, \nu)$ acts in a certain space of functions on G with values in V^ω. If we evaluate the values of the latter kind of function at the identity of M, we obtain functions on G with values in V^σ. These are just the members of the space for $U(S_{\mathfrak{p}}, \sigma, \nu \oplus \nu_M)$, and the evaluation mapping implements (8.56). The evaluation mapping between the spaces is actually unitary, and thus there is no problem with any continuity questions.

For the moment let us suppose ξ is a genuine (i.e., global) subrepresentation of ω. Then (8.56) says that $U(S, \xi, \nu)$ may be regarded as a genuine subrepresentation of $U(S_{\mathfrak{p}}, \sigma, \nu \oplus \nu_M)$. Under this identification an element $F(g)$, $g \in G$, in the representation space of $U(S, \xi, \nu)$ gets identified by the inclusion with a function $F(g)(m)$, $g \in G$ and $m \in M$, in the representation space of $U(S, \omega, \nu)$ and then by evaluation with the function $F(g)(1)$, $g \in G$, in the representation space of $U(S_{\mathfrak{p}}, \sigma, \nu \oplus \nu_M)$. Tracking down the intertwining integrals, we find that

$$A(S':S:\xi:\nu)F(x) = \int_{\bar{N} \cap N'} F(x\bar{n})\, d\bar{n}$$

gets identified with

$$\int_{\bar{N} \cap N'} F(x\bar{n})(1)\, d\bar{n} = \int_{(\bar{N}\bar{N}_M) \cap (N'N_M)} F(x\bar{n})(1)\, d\bar{n},$$

the two sets of integration being the same. The right side here is the operator $A(S'S_M:SS_M:\sigma:\nu \oplus \nu_M)F(x)(1)$ attached to the minimal parabolic subgroup $S_{\mathfrak{p}}$. We can make our spaces of F's be independent of ν by restricting x to be in K, just as in §7.7, and then Corollary 7.13 says that $A(S':S:\xi:\nu)F$ is meromorphic in a suitable sense, being the restriction to $(\mathfrak{a}')^{\mathbb{C}}$ of a meromorphic function on $(\mathfrak{a}_{\mathfrak{p}}')^{\mathbb{C}}$. (Our ν's do not all lie in the singular set of the function on $(\mathfrak{a}_{\mathfrak{p}}')^{\mathbb{C}}$, since we know we have a convergent integral for Re ν large.)

This argument presupposes that ξ is a global subrepresentation of ω, and the Subrepresentation Theorem gives us only an imbedding of the K_M-finite vectors. If we use only what the Subrepresentation Theorem gives us, the result is that we get a meromorphic continuation only on certain F's—K-finite F's in particular. The result is as follows.

Theorem 8.38. Let MAN be a general parabolic subgroup of G, and let ξ be an irreducible unitary representation of M. If F is a K-finite function in the compact picture of $U(S, \xi, \nu)$, then $A(S':S:\xi:\nu)F$ and $A_S(w, \xi, \nu)F$, which are defined by convergent integrals for Re ν in the region (8.12), extend to meromorphic functions of ν on $(\mathfrak{a}')^{\mathbb{C}}$, and the following formulas are valid on such functions F as identities of meromorphic functions in ν, provided Haar measures are normalized suitably.

(a) $A(S':S:\xi:\nu)U(S, \xi, \nu, X)F = U(S', \xi, \nu, X)A(S':S:\xi:\nu)F$ for all X in $\mathfrak{g}$ and all $S' = MAN'$,

(b) $A_S(w, \xi, \nu)U(S, \xi, \nu, X)F = U(S, w\xi, w\nu, X)A_S(w, \xi, \nu)F$ for all X in $\mathfrak{g}$ and all w in $N_K(\mathfrak{a})$,

(c) $A(S_2:S_1:\xi:\nu)F = R(w)^{-1}A(wS_2w^{-1}:wS_1w^{-1}:w\xi:w\nu)R(w)F$ for all w in $N_K(\mathfrak{a})$,

(d) $A(S'':S:\xi:\nu)F = A(S'':S':\xi:\nu)A(S':S:\xi:\nu)F$ for all $S'' = MAN''$ and $S' = MAN'$ with $\mathfrak{n}'' \cap \mathfrak{n} \subseteq \mathfrak{n}' \cap \mathfrak{n}$,

(e) $A_S(w, E\xi E^{-1}, \nu)F = EA_S(w, \xi, \nu)E^{-1}F$ if E is a unitary operator on V^ξ and w is in $N_K(\mathfrak{a})$,

(f) $A_S(w_1w_2, \xi, \nu)F = A_S(w_1, w_2\xi, w_2\nu)A_S(w_2, \xi, \nu)F$ whenever w_1 and w_2 are members of $N_K(\mathfrak{a})$ such that $w_1w_2\beta < 0$ for each $\mathfrak{n}$-positive root β of $(\mathfrak{g}, \mathfrak{a})$ with $w_2\beta < 0$.

There is also an obvious analog for our nonminimal S of the analytically continued version of Proposition 7.11, but we shall not bother to formulate it.

§11. Third Application: Control of K-Finite $Z(\mathfrak{g}^{\mathbb{C}})$-Finite Functions

Our third application will be to functions on G that transform within finite-dimensional spaces under the action of $Z(\mathfrak{g}^{\mathbb{C}})$ and the left and right actions of K. (K-finite matrix coefficients of irreducible representations provide examples.) We shall see that there are not many such functions. A sample consequence is that a K-finite $Z(\mathfrak{g}^{\mathbb{C}})$-finite function in $L^2(G)$ lies in the finite direct sum of closed subspaces of $L^2(G)$ invariant and irreducible under the right regular representation. Other consequences restrict the size of some other naturally arising invariant subspaces of $C^\infty(G)$ and allow us to analyze them. The full import of such results will not be evident until after the classification results at the end of this chapter and the main theorems on discrete series in later chapters.

Let $\mathscr{S}$ be a finite subset of $\hat{K}$ and let $\alpha_{\mathscr{S}} = \sum_{\omega\in\mathscr{S}} d_\omega \chi_\omega$, where d_ω and χ_ω are the degree and the character of ω. Convolution on K with $\alpha_{\mathscr{S}}$ is the orthogonal projection of $L^2(K)$ on the linear span of the matrix coefficients of the representations in $\mathscr{S}$. If f is in $C^\infty(G)$, then the K convolutions

$$\alpha_{\mathscr{S}} *_K f(x) = \int_K \alpha_{\mathscr{S}}(k) f(k^{-1}x)\, dk$$

and

$$f *_K \alpha_{\mathscr{S}}(x) = \int_K f(xk^{-1})\alpha_{\mathscr{S}}(k)\, dk$$

are left K-finite and right K-finite, respectively, on G. Conversely for any (two-sided) K-finite f in $C^\infty(G)$ we can choose the finite set $\mathscr{S}$ large enough so that

$$\alpha_{\mathscr{S}} *_K f = f = f *_K \alpha_{\mathscr{S}}.$$

Theorem 8.39. Let I be an ideal of finite codimension in $Z(\mathfrak{g}^{\mathbb{C}})$, and let $\mathscr{S}$ be a finite subset of $\hat{K}$. Then the subspace of functions f in $C^\infty(G)$ satisfying

$$\alpha_{\mathscr{S}} *_K f = f = f *_K \alpha_{\mathscr{S}} \tag{8.57}$$

and

$$Zf = 0 \qquad \text{for all } Z \in I \tag{8.58}$$

is finite-dimensional. All such functions are real analytic on G.

Remark. The theorem obviously remains valid if f is allowed to take its values in a fixed finite-dimensional space. In applications we shall use this vector-valued version.

Proof. By taking projections we may replace (8.57) by the stronger condition

$$(d_{\tau_1}\chi_{\tau_1}) *_K f = f = f *_K (d_{\tau_2}\chi_{\tau_2}), \tag{8.59}$$

where τ_1 and τ_2 are irreducible representations of K. Fix specific realizations of τ_1 and τ_2 on spaces U_1 and U_2. Fix f. For each L in $\operatorname{Hom}_{\mathbb{C}}(U_2, U_1)$, define

$$F_L(x) = d_{\tau_1} d_{\tau_2} \int_{K \times K} f(k_1^{-1} x k_2^{-1}) \tau_1(k_1) L \tau_2(k_2) dk_1\, dk_2. \tag{8.60}$$

Then F_L is τ-spherical for $\tau = (\tau_1, \tau_2)$, and F_L is annihilated by the members of the ideal I.

Let us choose bases for U_1 and U_2 and write (8.60) out as a matrix-valued function in those bases. The result is

$$F_L(x)_{ij} = \sum_{m,n} d_{\tau_1} d_{\tau_2} \int_{K \times K} f(k_1^{-1} x k_2^{-1}) \tau_1(k_1)_{im} L_{mn} \tau_2(k_2)_{nj}\, dk_1\, dk_2.$$

Defining $L^{(ij)}$ to be the member of $\operatorname{Hom}_{\mathbb{C}}(U_2, U_1)$ whose matrix is 1 in the $(i, j)^{\text{th}}$ entry and 0 elsewhere, we see that

$$\sum_{i,j} F_{L^{(ij)}}(x)_{ij} = (d_{\tau_1} \chi_{\tau_1}) *_K f *_K (d_{\tau_2} \chi_{\tau_2})(x) = f(x).$$

That is, any f satisfying (8.58) and (8.59) can be recovered from the τ-spherical functions F annihilated by I. For fixed τ, the space of τ-spherical functions F annihilated by I is finite-dimensional, by Theorem 8.28 and Corollary B.23, and hence the space of f's is finite-dimensional. The real analyticity is proved as in Theorem 8.7.

We shall give four corollaries of Theorem 8.39. The first one, which is immediate from the theorem, gives the flavor of all of them, but it alone is not too useful without further structure around.

Corollary 8.40. Let f be a K-finite $Z(\mathfrak{g}^{\mathbb{C}})$-finite function in $C^\infty(G)$, and let $U(\mathfrak{g}^{\mathbb{C}})$ act on $C^\infty(G)$ by left-invariant differentiation (i.e., the differential of the right regular representation). Then the $U(\mathfrak{g}^{\mathbb{C}})$ module $U(\mathfrak{g}^{\mathbb{C}})f$ is "admissible" in the sense that $(U(\mathfrak{g}^{\mathbb{C}})f) *_K \alpha_{\mathscr{S}}$ is finite-dimensional for each finite subset $\mathscr{S}$ of $\hat{K}$.

Corollary 8.41. Let f be a K-finite $Z(\mathfrak{g}^{\mathbb{C}})$-finite function in $C^\infty(G)$, and let $\mathscr{H}$ be the set of all h in $C^\infty_{\text{com}}(G)$ such that $h(kxk^{-1}) = h(x)$ for all k in K and x in G. Then there exists h in $\mathscr{H}$ such that $f * h = f$.

Remarks. Note that $f * h$ refers to convolution on G, not on K. The corollary remains valid, with no change in proof, for f with values in a finite-dimensional vector space. We can conclude from the corollary that $h_1 * f * h = f$ for suitable h_1 and h in $\mathscr{H}$ within $C^\infty_{\text{com}}(G)$, and the proof will show that the supports of h_1 and h can be taken to be arbitrarily small about the identity.

Proof. For any finite subset $\mathscr{S}$ of $\hat{K}$ and for h in $\mathscr{H}$, we readily check that

$$(f *_G h) *_K \alpha_{\mathscr{S}} = (f *_K \alpha_{\mathscr{S}}) *_G h. \tag{8.61}$$

Choose $\mathscr{S}$ large enough so that $\alpha_{\mathscr{S}} *_K f *_K \alpha_{\mathscr{S}} = f$, and let I be an ideal of finite codimension in $Z(\mathfrak{g}^{\mathbb{C}})$ annihilating f. Then (8.61) shows that $f * \mathscr{H}$ is a subspace of the set of functions in $C^\infty(G)$ annihilated by I and transforming under K on the left and the right according to $\mathscr{S}$. By Theorem 8.39, $f * \mathscr{H}$ is finite-dimensional. Since $\mathscr{H}$ contains functions of arbitrarily small support near the identity, f can be approximated uniformly on compact sets by members of $f * \mathscr{H}$. The finite-dimensionality means that f is in $f * \mathscr{H}$. Hence h exists as asserted.

Corollary 8.42. Let f be a K-finite $Z(\mathfrak{g}^{\mathbb{C}})$-finite L^2 function in $C^\infty(G)$. Within L^2 there exist finitely many orthogonal closed subspaces invariant and irreducible under the right regular representation R such that f is in the sum of these subspaces. Every left-invariant derivative of f is in $L^2(G)$.

Remark. After Proposition 9.6, we shall be able to interpret this corollary as characterizing the linear span of the K-finite matrix coefficients of discrete series as the K-finite $Z(\mathfrak{g}^{\mathbb{C}})$-finite functions in $L^2(G)$.

Proof. By Corollary 8.41, f is in the Gårding subspace of $L^2(G)$ and hence is a C^∞ vector for R. In particular, every left-invariant derivative of f is in $L^2(G)$. Let V_1 be the linear span of all $R(x)f$ for x in G, and let V be the closure of V_1. Fix a finite subset $\mathscr{S}$ of K and an ideal I of finite codimension on $Z(\mathfrak{g}^{\mathbb{C}})$ with $\alpha_{\mathscr{S}} *_K f = f$ and $If = 0$. Every member f_1 of V_1 then satisfies $\alpha_{\mathscr{S}} *_K f_1 = f_1$ and $If_1 = 0$.

Let ω be in $\hat{K}$, and let E_ω be the orthogonal projection defined by $E_\omega v = d_\omega \int_K \overline{\chi_\omega(k)} R(k)v\, dk$. Then $E_\omega(V) \subseteq V$. Since E_ω is bounded, $E_\omega(V)$ is contained in the closure of $E_\omega(V_1)$. Now every member f_1 of $E_\omega(V_1)$ satisfies $\alpha_{\mathscr{S}} *_K f_1 *_K (d_\omega \chi_\omega) = f_1$ and $If_1 = 0$, and thus Theorem 8.39 implies $E_\omega(V_1)$ is finite-dimensional. Consequently $E_\omega(V)$ is finite-dimensional, and V is admissible.

Let $d = \dim E_\omega(V)$. We shall prove that $E_\omega(V)$ is contained in the sum of $\leq d$ orthogonal closed irreducible $R(G)$-invariant subspaces of V. In fact, let f_1 be in $E_\omega(V)$. Let U_1 be the closed linear span generated by all $R(x)f_1$ for x in G. Let $\mathscr{I}_1$ be the left ideal in $U(\mathfrak{g}^{\mathbb{C}})$ of all D with $R(D)f_1 = 0$, and by Zorn's Lemma let $\mathscr{I}$ be a maximal left ideal containing $\mathscr{I}_1$. Then $R(\mathscr{I})f_1$ is a $U(\mathfrak{g}^{\mathbb{C}})$-invariant subspace of K-finite vectors in U_1 not containing f_1. Since V is admissible, we can conclude from Corollary 8.10 that the closure U_2 of $R(\mathscr{I})f_1$ is G-invariant and does not contain f_1. Then $V \cap U_2^\perp$ is G-invariant and irreducible, and f_1 has a nonzero component in $V \cap U_2^\perp$. Consequently $\dim E_\omega(U_2) < \dim E_\omega(V)$. Thus we have decomposed $V = U_2 \oplus (V \cap U_2^\perp)$ with $V \cap U_2^\perp$ irreducible and with $\dim E_\omega(U_2) < \dim E_\omega(V)$. An easy induction completes the argument that

$E_\omega(V)$ is contained in the sum of $\leq d$ orthogonal closed irreducible $R(G)$-invariant subspaces of V.

Finally let us return to f. Since f is in the finite sum of spaces $E_\omega(V)$ with ω in K, we can iterate the above argument to complete the proof of the corollary.

The final corollary uses the notation $\|x\|$, $x \in G$, defined in §7.8. Recall that to define $\|x\|$, we write $G = K \exp \mathfrak{p}$, introduce the usual norm on $\mathfrak{p}$, and set $\|x\| = \|X\|$ if $x = k \exp X$. For a in $A_\mathfrak{p}^+$, (7.51) shows $\|a\|$ is comparable in size with $\rho_\mathfrak{p}(\log a)$.

Corollary 8.43. Let π be an admissible representation of G on V, and let V_0 be the space of K-finite vectors. Fix a finite subset $\mathscr{S}$ of K, an ideal I of finite codimension in $Z(\mathfrak{g}^{\mathbb{C}})$, an integer $q \geq 0$, and a spherical function φ_ν with a real index ν. Suppose there is a vector space V_0' of K-finite $Z(\mathfrak{g}^{\mathbb{C}})$-finite functions on G such that

(i) $\alpha_{\mathscr{S}} *_K f = f$ for all $f \in V_0'$
(ii) $Zf = 0$ for all $f \in V_0'$ and $Z \in I$
(iii) $U(\mathfrak{g}^{\mathbb{C}})V_0' \subseteq V_0'$ under the action by left-invariant differentiation
(iv) $|f(x)| \leq c_f(1 + \|x\|)^q \varphi_\nu(x)$ for all $f \in V_0'$, with c_f constant.

If V_0' maps onto V_0 via some $U(\mathfrak{g}^{\mathbb{C}})$-equivariant map, then all K-finite matrix coefficients of π are dominated by multiples of $(1 + \|x\|)^q \varphi_\nu(x)$.

Remark. We shall use without proof the inequality

$$\|xy\| \leq \|x\| + \|y\| \qquad \text{for } x, y \in G. \tag{8.62}$$

Proof. Let R be the right regular representation of G on $C^\infty(G)$, and fix a finite subset $\mathscr{S}'$ of K. Define V_1' to be the space of all f satisfying (i) and (ii) such that $f *_K \alpha_{\mathscr{S}'} = f$. By Theorem 8.39, V_1' is finite-dimensional. Let V_2' be the subspace $V_0' *_K \alpha_{\mathscr{S}'}$ of $V_1' \cap V_0'$. For f in V_0' and y in G, we claim that $(R(y)f) *_K \alpha_{\mathscr{S}'}$, which is certainly a member of V_1', is actually in V_2'.

Assuming the contrary, choose a linear functional l vanishing on V_2' but not on $(R(y)f) *_K \alpha_{\mathscr{S}'}$. By (iii), $l((R(D)f) *_K \alpha_{\mathscr{S}'}) = 0$ for all D in $U(\mathfrak{g}^{\mathbb{C}})$. The interchange of limits needed to obtain

$$D(l((R(y)f) *_K \alpha_{\mathscr{S}'}))_{y=1} = l((R(D)f) *_K \alpha_{\mathscr{S}'})$$

is justified since V_1' is finite-dimensional. Thus $l((R(y)f) *_K \alpha_{\mathscr{S}'})$ has all derivatives of all orders vanishing at $y = 1$. Since this function of y is real analytic on G, it vanishes identically, contradiction. Thus $(R(y)f) *_K \alpha_{\mathscr{S}'}$ is actually in V_2'.

The most general K-finite matrix coefficient for π is of the form $y \to l(\pi(\alpha_{\mathscr{S}'})\pi(y)v_0)$ with v_0 in V_0, $\mathscr{S}'$ finite in K, and l in the dual of $\pi(\alpha_{\mathscr{S}'})V$. We may assume that $\pi(\alpha_{\mathscr{S}'})v_0 = v_0$. We shall apply the results of the previous paragraphs.

Let ψ be the given $U(\mathfrak{g}^{\mathbb{C}})$-equivariant map of V_0' onto V_0, and choose f in V_0' with $\psi(f) = v_0$. We may assume that $f *_K \alpha_{\mathscr{S}'} = f$, so that f is in V_2'. By restriction, ψ maps the finite-dimensional space V_2' onto $\pi(\alpha_{\mathscr{S}'})V_0 = \pi(\alpha_{\mathscr{S}'})V$. The image $\psi((R(y)f) *_K \alpha_{\mathscr{S}'})$ in $\pi(\alpha_{\mathscr{S}'})V$ depends on y in real analytic fashion, and its D derivative at $y = 1$ is $\psi((R(D))f *_K \alpha_{\mathscr{S}'})$, which equals $\pi(\alpha_{\mathscr{S}'})\pi(D)v_0$ by the $U(\mathfrak{g}^{\mathbb{C}})$ equivariance. The latter expression is the D derivative at $y = 1$ of the real analytic function $\pi(\alpha_{\mathscr{S}'})\pi(y)v_0$. Hence

$$\psi((R(y)f) *_K \alpha_{\mathscr{S}'}) = \pi(\alpha_{\mathscr{S}'})\pi(y)v_0$$

for all y in G.

We can norm the vector space V_2' by using the least possible constant in (iv) as norm. Since V_2' is finite-dimensional, the linear function $l \circ \psi$ on it is bounded, say with norm C. Then the above relation gives

$$\begin{aligned}
|l(\pi(\alpha_{\mathscr{S}'})\pi(y)v_0)| &\le C\|(R(y)f) *_K \alpha_{\mathscr{S}'}\| \\
&= C \sup_x \left\{(1 + \|x\|)^{-q}\varphi_\nu(x)^{-1} \left| \int_K f(xky)\alpha_{\mathscr{S}'}(k^{-1})\, dk \right| \right\} \\
&\le (C\|f\| \sup|\alpha|) \\
&\quad \times \sup_x \left\{(1 + \|x\|)^{-q}\varphi_\nu(x)^{-1} \int_K \varphi_\nu(xky)(1 + \|xky\|)^q\, dk\right\} \\
&\le (C\|f\| \sup|\alpha|) \sup_x \left\{(1 + \|x\|)^{-q}\varphi_\nu(x)^{-1} \right. \\
&\quad \left. \times \int_K \varphi_\nu(xky)(1 + \|x\| + \|ky\|)^q\, dk\right\} \qquad \text{by (8.62)} \\
&= (C\|f\| \sup|\alpha|) \sup_x \left\{(1 + \|x\|)^{-q}(1 + \|x\| + \|y\|)^q \right. \\
&\quad \left. \times \varphi_\nu(x)^{-1} \int_K \varphi_\nu(xky)\, dk\right\} \\
&= (C\|f\| \sup|\alpha|) \sup_x \left\{(1 + \|x\|)^{-q}(1 + \|x\| + \|y\|)^q \varphi_\nu(y)\right\} \\
&\qquad \text{by (7.45)} \\
&\le C'(1 + \|y\|)^q \varphi_\nu(y),
\end{aligned}$$

and the proof is complete.

§12. Asymptotic Expansions near the Walls

For our remaining applications, which are largely to classification results that are sharper than the Subrepresentation Theorem, the theory of §§7–8 is not quite good enough. The difficulty is that the uniformity of convergence in (8.52) breaks down as we approach the walls of the positive Weyl chamber, where some $\alpha_j(H) \to 0$ as $z_j \to 1$. Thus, in making estimates of the size of matrix coefficients, we cannot (as yet) disregard all terms but the leading ones, and consequently we can say rather little.

The point of this section will be to remedy this situation. We shall suppose that F is a globally defined τ-spherical function on G annihilated by an ideal I in $Z(\mathfrak{g}^{\mathbb{C}})$ of finite codimension, and we shall take advantage of the global nature of F, not just the behavior of F on the dense set $G^{(0)}$.

As a reminder that the theory of §§7–8 was built around a minimal parabolic subgroup, let us restore the subscripts "$\mathfrak{p}$" for that case. Our point of departure now is to rewrite the expansion (8.52) as occurring on $M_\mathfrak{p}A_\mathfrak{p}^+$, not just on $A_\mathfrak{p}^+$. The $M_\mathfrak{p}$ dependence comes simply from τ_1 or τ_2, since $M_\mathfrak{p} \subseteq K$. The expansion is then

$$F(m \exp H) = \sum_\nu F_{\nu-\rho_\mathfrak{p}}(m \exp H) \tag{8.63a}$$

with

$$F_{\nu-\rho_\mathfrak{p}}(m \exp H) = e^{-\rho_\mathfrak{p}(H)} \sum_{|q| \le q_0} c_{\nu,q}(m)\alpha(H)^q e^{\nu(H)}. \tag{8.63b}$$

Here $c_{\nu,q}(\cdot)$ is a real analytic function on $M_\mathfrak{p}$ with values in $\mathrm{Hom}_\mathbb{C}(U_2, U_1)$, and H is in $\mathfrak{a}_\mathfrak{p}^+$.

The expansion (8.63) runs into difficulty for H's satisfying $\alpha_j(H) < \varepsilon_j$ for some subset of simple roots of Σ^+. Let us renumber our simple roots so that the subset in question is $\alpha_{l'+1}, \ldots, \alpha_l$. Then we can build a parabolic subgroup $S = MAN$ of G, following the procedure of Proposition 5.23, such that $S \supseteq S_\mathfrak{p}$ and such that the simple roots in Σ^+ for which $\mathfrak{g}_{-\alpha}$ is in $\mathfrak{m}$ are exactly $\alpha_{l'+1}, \ldots, \alpha_l$. Specifically

$$\mathfrak{a}_M = \sum_{j=l'+1}^{l} \mathbb{R}H_{\alpha_j} \qquad \mathfrak{a}_\mathfrak{p} = \mathfrak{a} \oplus \mathfrak{a}_M,$$

$$\mathfrak{n} = \sum_{\substack{\beta\in\Sigma^+\\ \beta|_\mathfrak{a}\ne 0}} \mathfrak{g}_\beta, \qquad \mathfrak{n}_M = \sum_{\substack{\beta\in\Sigma^+\\ \beta|_\mathfrak{a}= 0}} \mathfrak{g}_\beta, \qquad \mathfrak{n}_\mathfrak{p} = \mathfrak{n} \oplus \mathfrak{n}_M,$$

$$\mathfrak{m} = \mathfrak{m}_\mathfrak{p} \oplus \mathfrak{a}_M \oplus \sum_{\beta\in\Sigma} \mathfrak{g}_\beta = \mathfrak{m}_\mathfrak{p} \oplus \mathfrak{a}_M \oplus \mathfrak{n}_M \oplus \bar{\mathfrak{n}}_M,$$

$$M = Z_K(\mathfrak{a})M_0, \qquad A_\mathfrak{p} = AA_M, \qquad N_\mathfrak{p} = NN_M.$$

Moreover, $M_\mathfrak{p}A_MN_M$ is a minimal parabolic subgroup of M, and $K_M = K \cap M$ is maximal compact in M.

The idea will be to redo a certain amount of §7, replacing objects associated to $S_{\mathfrak{p}}$ by objects associated to S and seeking an expansion (8.63) for m in M and H in $\mathfrak{a}^+$. To succeed at this effort, we shall restrict m to lie in a compact subset of M and we shall require that H be sufficiently large (depending on the compact subset). The new ingredient is the replacement for Proposition 8.25, as follows.

Proposition 8.44. Fix a compact subset C of M such that $K_M C K_M = C$. Then there exists a number $R = R(C) \geq 1$ such that

$$\mathfrak{g} = \mathrm{Ad}(ma)^{-1}\mathfrak{k} + \mathfrak{a} + \mathfrak{m} + \mathfrak{k} \tag{8.64}$$

whenever m is in C and $a = \exp H$ is in A^+ with $\exp \alpha_j(H) > R$ for $1 \leq j \leq l'$. Consequently for such ma, every member u of $U(\mathfrak{g}^{\mathbb{C}})$ decomposes as a linear combination of terms

$$(\mathrm{Ad}(ma)^{-1}X)HY \tag{8.65}$$

with X and Y in $U(\mathfrak{k}^{\mathbb{C}})$ and H in $U((\mathfrak{m} \oplus \mathfrak{a})^{\mathbb{C}})$.

Remark. Just as in Proposition 8.25, the proof will show that the decomposition (8.64) can be made in a canonical fashion, with coefficients that therefore vary in ma in real analytic fashion for all ma under consideration. Hence the coefficients in (8.65) can be chosen to be equally good. The effect will be that our individual differential operators are globally defined for all ma under consideration.

Proof. Let $\mathfrak{q} = \{X + \theta X \mid X \in \mathfrak{n}\}$; $\mathfrak{q}$ is the orthocomplement of $\mathfrak{k} \cap \mathfrak{m}$ in $\mathfrak{k}$ and is stable under $\mathrm{Ad}(K_M)$. We shall prove that

$$\mathfrak{g} = \mathrm{Ad}(ma)^{-1}\mathfrak{q} \oplus \mathfrak{a} \oplus \mathfrak{m} \oplus \mathfrak{q} \tag{8.66}$$

for the indicated elements ma. Since $\dim \mathfrak{q} = \dim \mathfrak{n}$ and $\mathfrak{g}$ is known to be the direct sum of $\bar{\mathfrak{n}}$, $\mathfrak{a}$, $\mathfrak{m}$, and $\mathfrak{n}$, the dimensions are right for (8.66). Moreover, if γ is a root of $(\mathfrak{g}, \mathfrak{a})$, then $\mathrm{Ad}(MA)$ leaves $\mathfrak{g}_\gamma$ stable and hence leaves $\mathfrak{g}_\gamma + \mathfrak{g}_{-\gamma}$ stable. Thus the only possible intersection of the spaces in (8.66) is if $\mathrm{Ad}(ma)^{-1}(\mathfrak{q} \cap (\mathfrak{g}_\gamma + \mathfrak{g}_{-\gamma}))$ meets $\mathfrak{q} \cap (\mathfrak{g}_\gamma + \mathfrak{g}_{-\gamma})$.

Introduce a norm in $\mathfrak{g}$ such that θ and $\mathrm{Ad}(K)$ act isometrically, and choose $R > 0$ so that

$$R^{-1}\|X\| \leq \|\mathrm{Ad}(m)^{-1}X\| \leq R\|X\|$$

for all X in $\mathfrak{g}$ and m in C. Suppose $a = \exp H$ in A is such that $\exp \alpha_j(H) > R$ for $1 \leq j \leq l'$. We are to prove that $X \in \mathfrak{g}_\gamma$ and $m \in C$ imply $\mathrm{Ad}(ma)^{-1}(X + \theta X)$ is not in $\mathfrak{k}$. We do so by showing that $\|\mathrm{Ad}(ma)^{-1}X\| \neq \|\mathrm{Ad}(ma)^{-1}\theta X\|$. (This is enough since $\mathrm{Ad}(ma)^{-1}X$ is in $\mathfrak{g}_\gamma$ and $\mathrm{Ad}(ma)^{-1}\theta X$ is in $\mathfrak{g}_{-\gamma}$.)

If $\gamma > 0$, then γ is the restriction to $\mathfrak{a}$ of some nonnegative integer combination of $\alpha_1, \ldots, \alpha_{l'}$, and hence $\exp \gamma(H) > R$. Then

$$\begin{aligned}\|\mathrm{Ad}(ma)^{-1}X_\gamma\| &= e^{-\gamma(H)}\|\mathrm{Ad}(m)^{-1}X_\gamma\| \le Re^{-\gamma(H)}\|X_\gamma\| \\ &= Re^{-\gamma(H)}\|\theta X_\gamma\| \le R^2e^{-\gamma(H)}\|\mathrm{Ad}(m)^{-1}\theta X_\gamma\| \\ &= R^2e^{-2\gamma(H)}\|\mathrm{Ad}(ma)^{-1}\theta X_\gamma\| \\ &< \|\mathrm{Ad}(ma)^{-1}\theta X_\gamma\|.\end{aligned}$$

Similarly $\|\mathrm{Ad}(ma)^{-1}X_\gamma\| > \|\mathrm{Ad}(ma)^{-1}\theta X_\gamma\|$ if $\gamma < 0$. This completes the proof.

We introduce coordinates now similar to those in §7. Namely we define ι on our subset of MA, mapping it into $M \times (\mathbb{R}^+)^{l'}$ by

$$\iota(m \exp H) = (m, e^{-\alpha_1(H)}, \ldots, e^{-\alpha_{l'}(H)}) = (m, e^{-\alpha(H)}),$$

so that $e^{-\alpha_j(\log a)}$ corresponds to z_j in the disc $D^{l'}$, with m carried along as a parameter. The dual basis of $\mathfrak{a}$ to $\alpha_1, \ldots, \alpha_{l'}$ consists of the first l' members of our earlier dual basis $\{H^{(j)}\}$ used in §7, since $\alpha_i(H^{(j)}) = 0$ for $1 \le j \le l'$ and $l' + 1 \le i \le l$ is consistent with the requirement that $H^{(j)}$ be in $\mathfrak{a}$. Then $H^{(j)}$ as a differential operator on A corresponds under ι to $-z_j \dfrac{\partial}{\partial z_j}$, $1 \le j \le l'$. We seek to apply the theory of Appendix B with $U = \mathrm{Hom}_{\mathbb{C}}(U_2, U_1)$ and with m as a real analytic parameter.

Let $\mathfrak{l} = \mathfrak{a} \oplus \mathfrak{m}$. With Γ^+ referring to the nonzero restrictions to $\mathfrak{a}$ of the members of Σ^+, we introduce

$$\mu'_{\Gamma^+}: U(\mathfrak{g}^{\mathbb{C}}) \to U(\mathfrak{l}^{\mathbb{C}})U(\mathfrak{k}^{\mathbb{C}})$$

in analogy with μ'_{Σ^+} as follows: We write $\mathfrak{g} = (\mathfrak{l} + \mathfrak{k}) \oplus \bar{\mathfrak{n}}$ and apply the Birkhoff-Witt Theorem to obtain

$$U(\mathfrak{g}^{\mathbb{C}}) = (U(\mathfrak{l}^{\mathbb{C}})U(\mathfrak{k}^{\mathbb{C}})) \oplus \bar{\mathfrak{n}}^{\mathbb{C}}U(\mathfrak{g}^{\mathbb{C}}).$$

Projection to the first component then defines μ'_{Γ^+}. As in §7, μ'_{Γ^+} carries $Z(\mathfrak{g}^{\mathbb{C}})$ into $Z(\mathfrak{l}^{\mathbb{C}})$ and is related to the Harish-Chandra homomorphism by

$$\gamma'_{\Delta^+} = \gamma'_{\Delta_M^+} \circ \mu'_{\Gamma^+} \text{ on } Z(\mathfrak{g}^{\mathbb{C}}).$$

It is time to skip through the proofs of §7, checking whether they remain valid. Three comments are in order:

(1) The averaging over M in the proofs of Theorem 8.26 and Lemma 8.31 is unnecessary in the present case since we are content to use $U = \mathrm{Hom}_{\mathbb{C}}(U_2, U_1)$.

(2) After our system of differential equations is referred by ι to $C \times (0, 1/R)^{l'} \subseteq M \times (\mathbb{R}^+)^{l'}$, we want the coefficient functions to extend

holomorphically to the product $C \times (D_{1/R})^{l'}$ of C with a polydisc of radius $1/R$ in each variable, with the coefficients vanishing on $C \times \{0\}$. To achieve this, what we need to see is that when a member X of $\mathfrak{g}_{-\gamma}$, $\gamma > 0$, is decomposed according to (8.66), the coefficient functions extend holomorphically to $C \times (D_{1/R})^{l'}$ and vanish on $C \times \{0\}$. This conclusion comes by going over the proof of Proposition 8.44, replacing the element a in A by an element $a^{\mathbb{C}} = \exp H^{\mathbb{C}}$ in $A^{\mathbb{C}}$ with $\operatorname{Re} \log \alpha_j(H^{\mathbb{C}}) > \log R$ for $1 \leq j \leq l'$, and then using a removable singularity argument.

(3) When we apply Theorem B.16 to our system of differential equations as referred to variables (m, z), we apply it first with the parameter m fixed, obtaining various powers z^s and $(\log z)^q$. The possible q's that can arise are bounded as m varies through the compact set C, by dimensional considerations. But we need to control the set of exponents s as functions of m. To do so, we use two tools: Our new version of Theorem 8.26 shows that the Euler system corresponding to our system of differential equations is independent of m, and Theorem B.19 shows that the leading exponents s are contained in a finite set defined in terms of the Euler system. Thus we can adjust notation and find a finite set $\mathcal{S}$ of integrally inequivalent exponents independently of m, such that all the needed exponents t are $\geq$ some s in $\mathcal{S}$. Corollary B.26 implies that the expansion of our functions of (m, z) as sums of products of $z^s(\log z)^q$ and holomorphic functions of z is unique for each m once $\mathcal{S}$ is fixed. Then it follows that the holomorphic functions depend in real analytic fashion on the parameter m.

Then with minor modifications the proofs go through. The result is as follows.

Theorem 8.45. Let F be a τ-spherical function on G annihilated by an ideal I of finite codimension in $Z(\mathfrak{g}^{\mathbb{C}})$. Fix a compact subset C of M with $K_M C K_M = C$, and define $R \geq 1$ as in Proposition 8.44. Then $F \circ \iota^{-1}$, initially defined on $C \times (0, 1/R)^{l'}$, extends to a global multiple-valued solution on $C \times (D_{1/R}^{\times})^{l'}$ that is real analytic in the first variable and holomorphic in the second variable. Moreover, there exist an integer q_0 and a finite subset $\mathcal{S} \subseteq \mathbb{C}^{l'}$ such that $F \circ \iota^{-1}$ is given on $C \times (D_{1/R}^{\times})^{l'}$ by

$$F \circ \iota^{-1}(m, z) = \sum_{s \in \mathcal{S}} \sum_{0 \leq |q| \leq q_0} z^s (\log z)^q F_{s,q}(m, z),$$

with each $F_{s,q}(m, z)$ real analytic for m in C and holomorphic for z in all of $(D_{1/R})^{l'}$. Consequently F has an expansion

$$F(m \exp H) = \sum_{\nu} F_{\nu - \rho_A}(m \exp H) \tag{8.67a}$$

$$F_{\nu - \rho_A}(m \exp H) = e^{-\rho_A(H)} \sum_{|q| \leq q_0} c_{\nu,q}(m) \alpha(H) e^{\nu(H)}, \tag{8.67b}$$

valid when m is in C and $\exp \alpha_j(H) > R$ for $1 \leq j \leq l'$. The coefficients $c_{\nu,q}(\cdot)$ are real analytic, have values in $\mathrm{Hom}_{\mathbb{C}}(U_2, U_1)$, and are $\tau|_{K \cap M}$ spherical. Every leading term $F_{\nu-\rho_A}$ satisfies

$$\mu'_{\Gamma^+}(Z)F_{\nu-\rho_A} = 0 \tag{8.68}$$

for all Z in the ideal I of $Z(\mathfrak{g}^{\mathbb{C}})$. Since $\mu'_{\Gamma^+}(I)$ has finite codimension in $Z(\mathfrak{l}^{\mathbb{C}})$, the theory of §§7–8 can be applied to any such leading term.

In the theorem if we replace C by a larger set, then the two expressions (8.67b) must be compatible term-by-term, by uniqueness (Corollary B.26). Consequently $c_{\nu,q}(\cdot)$ extends to a global real analytic function on M, and $F_{\nu-\rho_A}$, which is given in (8.67b) as a finite sum, extends to a globally defined function on MA, though the series (8.67a) has not been proved to be convergent globally.

Comparison of (8.67) for S general and S minimal gives information about leading terms. Let us specialize m to be an element a_M lying in a bounded subset of $\exp \mathfrak{a}_M^+$. For H in $\mathfrak{a}^+$ sufficiently large (i.e., with each $\alpha_j(H)$ sufficiently large for $1 \leq j \leq l'$), the element $a_M \exp H$ both is in $\mathfrak{a}_\mathfrak{p}^+$ (so that the expansion (8.67a) relative to $S_\mathfrak{p}$ is convergent) and is such that the expansion (8.67a) relative to S is convergent. Taking $a_M = \exp H_M$ and using the identity $\rho_\mathfrak{p} = \rho_A + \rho_{A_M}$, we can regroup the expansion (8.67a) for $S_\mathfrak{p}$ in obvious notation as follows:

$$\begin{aligned} F(a_M \exp H) &= \sum_{\tilde{\nu}} F^{S_\mathfrak{p}}_{\tilde{\nu}-\rho_\mathfrak{p}}(a_M \exp H) \\ &= e^{-\rho_A(H)} \sum_{\tilde{\nu}} \sum_{\tilde{q}} c^{S_\mathfrak{p}}_{\tilde{\nu},\tilde{q}}(1)\alpha(H_M + H)^{\tilde{q}} e^{(\tilde{\nu}-\rho_{A_M})(H_M)} e^{\tilde{\nu}(H)} \\ &= e^{-\rho_A(H)} \sum_{\nu} \left\{ \sum_{\tilde{\nu}|_\mathfrak{a} = \nu} \sum_{\tilde{q}} c^{S_\mathfrak{p}}_{\tilde{\nu},\tilde{q}}(1)\alpha(H_M + H)^{\tilde{q}} e^{(\tilde{\nu}-\rho_{A_M})(H_M)} \right\} e^{\nu(H)}, \end{aligned} \tag{8.69}$$

the last equality following since the $\tilde{q}$'s are uniformly bounded and since the series otherwise are effectively power series. Meanwhile the expansion relative to S gives

$$F(a_M \exp H) = e^{-\rho_A(H)} \sum_{\nu} \left\{ \sum_{|q| \leq q_0} c^S_{\nu,q}(a_M)\alpha(H)^q \right\} e^{\nu(H)}. \tag{8.70}$$

The uniqueness result of Corollary B.26 allows us to equate the respective expressions in braces in the two expansions and obtain the first conclusion in the following corollary.

Corollary 8.46. In the expansion (8.70) of F relative to S, each exponent $\nu - \rho_A$ is the restriction to $\mathfrak{a}$ of an exponent $\tilde{\nu} - \rho_\mathfrak{p}$ for the expansion relative to $S_\mathfrak{p}$. (Thus also $\nu = \tilde{\nu}|_\mathfrak{a}$.) Conversely if $\tilde{\nu} - \rho_\mathfrak{p}$ is an exponent for the expansion relative to $S_\mathfrak{p}$, then the restriction $\nu - \rho_A$ is an exponent for the expansion relative to S.

Proof. For the converse we are to show that if $\tilde{\nu} - \rho_{\mathfrak{p}}$ is an exponent, then the expression in braces in (8.69) is nonzero. To do so, we show how to interpret the expression in braces as a series of the sort handled by the uniqueness result Corollary B.26, and we show there is a nonzero coefficient.

Introduce coordinates $w_j \leftrightarrow e^{-\alpha_j(H_M)}$, $l' + 1 \le j \le l$, for the bounded subset of $\exp \mathfrak{a}_M^+$ under consideration. Each $\alpha_j|_{\mathfrak{a}_M}$ for $1 \le j \le l'$ can be expressed in terms of the $\alpha_j|_{\mathfrak{a}_M}$ for $l' + 1 \le j \le l$, with coefficients that can be negative. Thus the expression in braces in (8.69) expands out as a bounded number of powers of $\alpha(H)$, each multiplied by a series whose terms have nonnegative integer-tuple powers of $\log w$ and complex-tuple powers of w. We have to make sure that the powers of w that arise are limited so that our uniqueness result applies. First let us observe the following: Since $\tilde{\nu}$ is determined by its values on $\mathfrak{a}$ and $\mathfrak{a}_M$ together, the various powers of w in braces are distinct as $\tilde{\nu}$ varies with ν fixed.

Now let $\tilde{\nu}_1$ and $\tilde{\nu}_2$ both restrict to ν on $\mathfrak{a}$, and suppose $\tilde{\nu}_1$ and $\tilde{\nu}_2$ are integrally equivalent. Choose an integrally equivalent $\tilde{\nu}_0$ with $\tilde{\nu}_1 \le \tilde{\nu}_0$ and $\tilde{\nu}_2 \le \tilde{\nu}_0$. Writing $\tilde{\nu}_0 - \tilde{\nu}_1 = \sum_{j=1}^{l} m_j \alpha_j$, we have

$$(\tilde{\nu}_0 - \tilde{\nu}_1)(H) = \sum_{j=1}^{l'} m_j \alpha_j(H) \qquad \text{for } H \in \mathfrak{a}$$

since $\alpha_{l'+1}, \ldots, \alpha_l$ vanish on $\mathfrak{a}$. Since $\alpha_1, \ldots, \alpha_{l'}$ are linearly independent on $\mathfrak{a}$, the m_j for $1 \le j \le l'$ are determined by $(\tilde{\nu}_0 - \tilde{\nu}_1)|_{\mathfrak{a}}$, hence by $\tilde{\nu}_0|_{\mathfrak{a}} - \nu$. Thus in the expression

$$e^{(\tilde{\nu}_1 - \rho_{A_M})(H_M)} = e^{(\tilde{\nu}_0 - \rho_{A_M})(H_M)} \exp\left\{-\sum_{j=1}^{l'} m_j \alpha_j(H_M)\right\} \prod_{j=l'+1}^{l} (e^{-\alpha_j(H_M)})^{m_j}, \tag{8.71}$$

the last factor on the right is the only one that can change when we replace $\tilde{\nu}_1$ by $\tilde{\nu}_2$. That is, any $\tilde{\nu}_1$ related to $\tilde{\nu}_0$ gives rise to a power of w that is the sum of some fixed tuple and a nonnegative integer tuple. Since all $\tilde{\nu}$'s are related to $\tilde{\nu}_0$'s in a finite set, the powers of w are appropriately limited so that the uniqueness theorem applies.

Finally we need to see that some coefficient is nonzero in the expression in braces in (8.69) after the powers of $\alpha(H)$ are factored out and terms are collected. Fix $q = (q_1, \ldots, q_{l'})$ to be maximal among the restrictions to $\{1, \ldots, l'\}$ of all $\tilde{q}$ that occur, and consider the coefficient of $\alpha(H)^q$ within the expression in braces. It is easy to see that the term in this coefficient involving

$$\alpha_{l'+1}(H_M)^{q_{l'+1}} \cdots \alpha_l(H_M)^{q_l} e^{(\tilde{\nu} - \rho_{A_M})(H_M)}$$

comes only from $c_{\tilde{\nu},\tilde{q}}^{S_{\mathfrak{p}}}$ with $\tilde{q} = (q_1, \ldots, q_l)$, and thus we have a nonzero coefficient.

§13. Fourth Application: Asymptotic Size of Matrix Coefficients

We shall now use the theory of §§7–8, supplemented by the results of §12, to show that the leading exponents of an irreducible admissible representation give global pointwise control of the size of the K-finite matrix coefficients. Consequently we shall find that L^p integrability of K-finite matrix coefficients can be characterized in terms of the leading exponents. The special case $p = 2$ will give information about discrete series that we shall exploit in our classification results in §§14–15.

We have worked with the basis $H^{(1)}, \ldots, H^{(l)}$ of $\mathfrak{a}_\mathfrak{p}$ dual to $\alpha_1, \ldots, \alpha_l$, and we now introduce the members $\omega_1, \ldots, \omega_l$ of $\mathfrak{a}'_\mathfrak{p}$ that correspond to $H^{(1)}, \ldots, H^{(l)}$ under the inner product. These elements satisfy $\langle \alpha_i, \omega_j \rangle = \delta_{ij}$ for all i and j.

In §7.8 we used the decomposition $G = K \exp \mathfrak{p}$ and a norm on $\mathfrak{p}$ to define $\|x\|$ for all x in G. For a in $A_\mathfrak{p}^+$, (7.51) shows $\|a\|$ is comparable in size with $\rho_\mathfrak{p}(\log a)$. Thus $\|a\|$ is logarithmic in size compared with $\exp \rho_\mathfrak{p}(\log a)$.

Theorem 8.47. If ν_0 in $\mathfrak{a}'_\mathfrak{p}$ is real-valued and if π is an irreducible admissible representation of G, then the following are equivalent:

(a) Every leading exponent $\nu - \rho_\mathfrak{p}$ of π has $\langle \operatorname{Re} \nu, \omega_j \rangle \leq \langle \nu_0, \omega_j \rangle$ for $1 \leq j \leq l$.

(b) There is an integer $q \geq 0$ such that each K-finite matrix coefficient of π is dominated on $\overline{A_\mathfrak{p}^+}$ by a multiple of $e^{(\nu_0 - \rho_\mathfrak{p})(\log a)}(1 + \|a\|)^q$.

Remark. When (b) holds, it follows from Corollary 8.41 that similar pointwise bounds are valid for all mixed left invariant and right invariant derivatives of the K-finite matrix coefficients. The proof will show that this theorem is a result about asymptotic expansions of $Z(\mathfrak{g}^\mathbb{C})$-finite τ-spherical functions, not just matrix coefficients; we shall use this observation in §15.

Proof that (b) $\Rightarrow$ (a). From the estimate in (b) we obtain a similar estimate for the finite block of matrix coefficients comprising a τ-spherical function. We expand the τ-spherical function relative to $S_\mathfrak{p}$ as in (8.63). After the usual change of variables $z = e^{-t}$, Theorem B.25 applies immediately to the estimate from (b) and allows us to conclude (a). (To unravel the correspondence between $\mathfrak{a}_\mathfrak{p}^+$ and $(\mathbb{R}^+)^l$, we write $H \leftrightarrow \{t_j\}$, where $t_j = \alpha_j(H)$. Then $H = \sum \alpha_j(H) H^{(j)}$ implies

$$\nu(H) = \sum \alpha_j(H)\nu(H^{(j)}) = \sum t_j \langle \nu, \omega_j \rangle,$$

and hence the tuple $\{\langle \nu, \omega_j \rangle\}$ is the coordinate realization of ν.)

Proof that (a) ⇒ (b). It is enough to prove (b) for a τ-spherical block F of matrix coefficients. We shall obtain the bound (b) separately on each of 2^l subsets of $\overline{A_\mathfrak{p}^+}$. The first such subset is where $\alpha_j(\log a) \geq 1$ for $1 \leq j \leq l$. For a in this set, we consider the product of $e^{-(\nu_0-\rho_\mathfrak{p})\log a}(1 + \|a\|)^{-q_0}$ by the expansion (8.63) of F. Because of the assumption (a), we can regard the product as a finite linear combination of power series with bounded functions as coefficients. For a in our set, we are staying within a compact subset of the region of convergence for all the power series. Hence the linear combination results in a bounded function of a, for a in our set. This is conclusion (b) for such a.

Next we consider the l subsets where $\alpha_j(\log a) < 1$ for the index j. In particular, let us choose $j = l$. Relative to the set $\{l\}$ of indices, we can construct as in §12 a parabolic subgroup $S = MAN$ containing $S_\mathfrak{p}$, with M built from α_l. For a in our set $\{\alpha_l(\log a) < 1\}$, the expansion (8.67) relative to S converges as nicely as in the previous paragraph, as long as $\alpha_j(\log a) \geq r_l$ for $1 \leq j \leq l - 1$, by Corollary 8.46 and the assumption (a). Thus we get the required estimate for $\alpha_l(\log a) < 1$ as long as $\alpha_j(\log a) \geq r_l$ for $1 \leq j \leq l - 1$. Let r be the maximum of 1 and the largest r_i, $1 \leq i \leq l$. Then it follows that the estimate (b) is valid unless $\alpha_j(\log a) < r$ for at least two indices j.

Now we can repeat the construction of the previous paragraph, starting from a two-element set of indices, constructing a parabolic subgroup S with M built from two simple roots, and applying (8.67) and Corollary 8.46. The result is that the estimate (b) is valid unless $\alpha_j(\log a) < r'$ for at least three indices j. By an obvious induction we can continue this argument, finally obtaining (b) on all of $\overline{A_\mathfrak{p}^+}$ except where $\alpha_j(\log a)$ is less than some constant for all j. This exceptional set is bounded, and thus (b) holds everywhere.

Theorem 8.48. Fix p with $1 \leq p < \infty$, and let π be an irreducible admissible representation of G. Then the following are equivalent:

(a) Every leading exponent $\nu - \rho_\mathfrak{p}$ of π has

$$\mathrm{Re}\langle \nu - \rho_\mathfrak{p}, \omega_j \rangle < -\frac{2}{p} \langle \rho_\mathfrak{p}, \omega_j \rangle$$

for $1 \leq j \leq l$.

(b) Every K-finite matrix coefficient of π is in $L^p(G)$.

Remark. By combining Corollary 8.41 with the proof of the easy half (a) ⇒ (b) of this theorem, we see readily that (b) implies that all mixed left-invariant and right-invariant derivatives of the K-finite matrix coefficients are in $L^p(G)$.

The proof will use the formula of Proposition 5.28 for Haar measure relative to the $K\overline{A_{\mathfrak{p}}^{+}}K$ decomposition, in which we can write

$$dx = C\left[\prod_{\beta\in\Sigma^{+}} (\sinh \beta(H))^{\dim \mathfrak{g}_{\beta}}\right] dk_1\, dH\, dk_2 \qquad \text{with } k_1, k_2 \in K \text{ and } H \in \mathfrak{a}_{\mathfrak{p}}'. \tag{8.72}$$

It is clear that

$$\prod_{\beta\in\Sigma^{+}} (\sinh \beta(H))^{\dim \mathfrak{g}_{\beta}} \le e^{2\rho_{\mathfrak{p}}(H)} \qquad \text{for } H \in \mathfrak{a}_{\mathfrak{p}}^{+}. \tag{8.73}$$

Before proving the theorem, we give two lemmas that address bounds in the other direction.

Lemma 8.49. On the subset of $H \in \mathfrak{a}_{\mathfrak{p}}'$ where $\alpha_2(H) < 1, \ldots, \alpha_l(H) < 1$, and $\alpha_1(H) \ge r > 0$, there is an everywhere positive continuous function $f(H)$ depending only on $\alpha_2(H), \ldots, \alpha_l(H)$ such that

$$e^{\langle 2\rho_{\mathfrak{p}}, \omega_1\rangle \alpha_1(H)} f(H) \le \prod_{\beta\in\Sigma^{+}} (\sinh \beta(H))^{\dim \mathfrak{g}_{\beta}}. \tag{8.74}$$

Proof. Decompose each β in Σ^{+} as $\beta = \sum \langle \beta, \omega_j\rangle \alpha_j$. When $\langle \beta, \omega_1\rangle = 0$, $\beta(H)$ depends only on $\alpha_2(H), \ldots, \alpha_l(H)$, and we shall take

$$f(H) = c \prod_{\substack{\beta\in\Sigma^{+}\\ \langle\beta,\omega_1\rangle = 0}} (\sinh \beta(H))^{\dim \mathfrak{g}_{\beta}}, \tag{8.75}$$

where $c > 0$ is a constant to be specified. Then the right side of (8.74) is

$$\begin{aligned} &= c^{-1} f(H) \prod_{\substack{\beta\in\Sigma^{+}\\ \langle\beta,\omega_1\rangle > 0}} (\sinh \beta(H))^{\dim \mathfrak{g}_{\beta}} \\ &\ge c^{-1} f(H) \prod_{\substack{\beta\in\Sigma^{+}\\ \langle\beta,\omega_1\rangle > 0}} (\sinh(\langle \beta, \omega_1\rangle \alpha_1(H)))^{\dim \mathfrak{g}_{\beta}} \\ &\ge c^{-1} 2^{\text{power}} f(H) \prod_{\substack{\beta\in\Sigma^{+}\\ \langle\beta,\omega_1\rangle > 0}} (\sinh \alpha_1(H))^{\langle\beta,\omega_1\rangle \dim \mathfrak{g}_{\beta}}, \end{aligned} \tag{8.76}$$

since $\sinh nx \ge 2^{n-1}(\sinh x)^n$ for any integer $n \ge 1$ if $x > 0$. In (8.76) we can extend the product to be over all β in Σ^{+} and see that (8.76) is

$$= c^{-1} 2^{\text{power}} f(H) (\sinh \alpha_1(H))^{\langle 2\rho_{\mathfrak{p}}, \omega_1\rangle}.$$

Since $\sinh \alpha_1(H)$ is comparable with $\frac{1}{2}\exp \alpha_1(H)$ for $\alpha_1(H) \ge r > 0$, (8.74) follows from (8.75) for a suitable choice of c.

Lemma 8.50. In the one-variable case, let $\mathscr{F}$ be a finite subset of $\mathbb{C}$ and let $r > 0$. Suppose, for each s in $\mathscr{F}$ and each integer q with $0 \le q \le q_0$, that $f_{s,q}(z)$ is a $\mathbb{C}^N$-valued function analytic for $|z| \le r$. Let $f(z)$ be the function

$$f(z) = \sum_{s\in\mathscr{F}} \sum_{0\le q\le q_0} z^s (\log z)^q f_{s,q}(z)$$

and rearrange f as

$$f(z) = \sum_{s \in \mathbb{C}} \sum_{0 \leq q \leq q_0} c_{s,q} z^s (\log z)^q.$$

If the function $F(t) = f(e^{-t})$ is in $L^p((-\log r, \infty), dt)$ for some p with $1 \leq p < \infty$, then $\operatorname{Re} s > 0$ for each s in $\mathbb{C}$ with $c_{s,q} \neq 0$ for some q.

Proof. By applying linear functionals we may assume that each $c_{s,q}$ is scalar. Let p' be the index conjugate to p. Suppose there is some $s = s_0$ occurring in the sum in which $\operatorname{Re} s_0 \leq 0$. The function equal to $\exp(i \operatorname{Im} s_0 t)$ on (R, T) and 0 on $[T, \infty)$ is in $L^{p'}$ with norm $\leq T^{1/p'}$ if $R = -\log r$. By Holder's inequality

$$\left| \int_R^T \sum_{s,q} c_{s,q} e^{-(s - i \operatorname{Im} s_0)t} (-t)^q \, dt \right| \leq T^{1/p'} \left\| \sum_{s,q} c_{s,q} e^{-st} (-t)^q \right\|_p = c T^{1/p'}. \tag{8.77}$$

Successive integrations by parts show that

$$\int_R^T e^{at} t^q \, dt = \begin{cases} a^{-(q+1)}(P(T)e^{aT} + Q) & \text{if } a \neq 0 \\ \dfrac{1}{q+1}(T^{q+1} + Q) & \text{if } a = 0; \end{cases} \tag{8.78}$$

here P is a monic polynomial of degree q, and Q is independent of T. On the left side of (8.77) the sum is uniformly convergent, by our hypotheses. Hence we can apply Theorem B.25 to the inequality (8.77). If $\operatorname{Re} s_0 < 0$ and q is chosen as large as possible so that $c_{s_0,q} \neq 0$, then (8.78) shows that the coefficient of $e^{-(\operatorname{Re} s_0)T} T^q$ on the left side of (8.77) after the integration is a nonzero multiple of $c_{s_0,q}$, hence is nonzero, in contradiction with Theorem B.25. Thus $\operatorname{Re} s_0 = 0$, and (8.78) shows that the coefficient of $e^{0T} T^{q+1}$ on the left side of (8.77) after the integration is a nonzero multiple of $c_{s_0,q}$. Theorem B.25 thus says that $q + 1 \leq 1/p'$ whenever $c_{s_0,q} \neq 0$, and we conclude that $q < 0$, contradiction. Hence s_0 does not occur in the original sum.

Proof that (a) ⇒ (b) in Theorem 8.48. If (a) holds here, then the implication (a) ⇒ (b) of Theorem 8.47 shows that each K-finite matrix coefficient of π (and hence also any block of matrix coefficients of π comprising a τ-spherical function) is dominated on $\overline{A_\mathfrak{p}^+}$ by a multiple of

$$\exp(-(\varepsilon + 2p^{-1}\rho_\mathfrak{p}(\log a)))[1 + \rho_\mathfrak{p}(\log a)]^q$$

for some q and for some ε in $\mathfrak{a}_\mathfrak{p}'$ with $\langle \varepsilon, \omega_j \rangle > 0$ for $1 \leq j \leq l$. Taking into account the integration formula (8.72) relative to $KA_\mathfrak{p}K$, the fact that τ-spherical functions transform under K in unitary fashion on both sides, and the inequality (8.73), we see that the τ-spherical function is in $L^p(G)$ if

$$\int_{\mathfrak{a}^+} e^{-p\varepsilon(H)} \, dH \tag{8.79}$$

is finite. We can coordinatize $\mathfrak{a}^+$ by $x_j \leftrightarrow \alpha_j(H)$, with all $x_j \geq 0$, and then dH is a multiple of $dx_1 \cdots dx_l$. In these coordinates $\varepsilon(H)$ equals $\sum x_j \langle \varepsilon, \omega_j \rangle$, and thus (8.79) is indeed finite.

Proof that (b) ⇒ (a) in Theorem 8.48. Arguing by contradiction, suppose we have a τ-spherical block F of K-finite matrix coefficients in $L^p(G)$ and that (a) fails for some j, say $j = 1$. If we write down the condition for F to be in $L^p(G)$ in terms of the decomposition (8.73), we can discard the K integrations because F transforms on both sides in unitary fashion under K. We shall discard most of the region of integration in $\mathfrak{a}_\mathfrak{p}^+$, and of course the finite integral remains finite.

Specifically let us construct a parabolic subgroup $S \supseteq S_\mathfrak{p}$ by building M from $\alpha_2, \ldots, \alpha_l$. Fix a compact subset C of M such that $K_M C K_M = C$ and C contains all $\exp H_M$ with H_M in $\mathfrak{a}_M^+$ and $\alpha_j(H_M) < 1$ for $2 \leq j \leq l$. Define $R \geq 1$ as in Proposition 8.44, and let E be the subset of $\mathfrak{a}_\mathfrak{p}^+$ given as sums $H + H_M$ with

$$H \in \mathfrak{a} \text{ and } \alpha_1(H) > \log R, \qquad H_M \in \mathfrak{a}_M \text{ and } \alpha_j(H_M) < 1 \text{ for } 2 \leq j \leq l.$$

Then we certainly have

$$\int_{H \in E} |F(\exp H)|^p \left[\prod_{\beta \in \Sigma^+} (\sinh \beta(H))^{\dim \mathfrak{g}_\beta} \right] dH < \infty. \tag{8.80}$$

On E, $\alpha_1(H + H_M)$ satisfies

$$\alpha_1(H) - c_1 \leq \alpha_1(H + H_M).$$

Thus if we let E_M be the subset of H_M in $\mathfrak{a}_M^+$ with $\alpha_j(H_M) < 1$ for $2 \leq j \leq l$, then (8.80) and Lemma 8.49 imply that

$$\int_{E_M \times \{H \in \mathfrak{a}^+ \mid \alpha_1(H) > \log R\}} |F(\exp H \exp H_M)|^p f(H_M) e^{2\langle \rho_\mathfrak{p}, \omega_1 \rangle \alpha_1(H)} \, dH \, dH_M < \infty \tag{8.81}$$

for a certain positive continuous function f on E_M.

Theorem 8.45 allows us to expand F in a series of the form (8.67). By Fubini's Theorem the result of the integral in the $\mathfrak{a}^+$ variable is finite for almost all H_M in E_M. Fix such an H_M, and parametrize the $\mathfrak{a}^+$ variable by $t = \alpha_1(H)$. Then we have

$$\int_{t=R}^{\infty} \left| \sum_{\nu, q} c_{\nu,q}(\exp H_M) t^q e^{\langle \nu - \rho_A, \omega_1 \rangle t} \right|^p e^{2\langle \rho_\mathfrak{p}, \omega_1 \rangle t} \, dt < \infty,$$

hence also

$$\int_R^{\infty} \left| \sum_{\nu, q} c_{\nu,q}(\exp H_M) t^q e^{\langle \nu - \rho_A + 2p^{-1}\rho_\mathfrak{p}, \omega_1 \rangle t} \right|^p dt < \infty.$$

The series inside the absolute value signs here is of the kind considered in Lemma 8.50, apart from a change of variables. That lemma says that $c_{\nu,q}(\exp H_M) \neq 0$ implies

$$\mathrm{Re}\langle \nu - \rho_A + 2p^{-1}\rho_{\mathfrak{p}}, \omega_1 \rangle < 0. \tag{8.82}$$

Allowing H_M to vary, we see that (8.82) holds whenever $c_{\nu,q}(\cdot) \not\equiv 0$.

Finally any leading exponent $\tilde{\nu} - \rho_{\mathfrak{p}}$ in the expansion of F relative to $S_{\mathfrak{p}}$ gives rise by restriction to $\nu - \rho_A$ as above, according to the converse half of Corollary 8.46. Thus each leading exponent $\tilde{\nu} - \rho_{\mathfrak{p}}$ satisfies

$$\mathrm{Re}\langle \tilde{\nu} - \rho_{\mathfrak{p}} + 2p^{-1}\rho_{\mathfrak{p}}, \omega_j \rangle < 0$$

for $j = 1$, by (8.82). Our selection of j as 1 was arbitrary, and thus (a) follows.

§14. Fifth Application: Identification of Irreducible Tempered Representations

The theorems of §13 provide us with the last tools we need in order to obtain classification theorems that are sharper than the Subrepresentation Theorem. We shall prove two such classification theorems in this section and the next, as the fifth and sixth applications of asymptotic expansions of matrix coefficients. In this section, we shall treat discrete series and tempered representations: After giving some properties of discrete series, we shall relate irreducible tempered representations to discrete series. (Discrete series of a particular kind were the subject of Chapter VI; tempered representations made their first appearance in §7.11.) In the next section, we shall relate general irreducible admissible representations to irreducible tempered representations by a remarkably similar argument.

We continue to denote the simple restricted roots of $\mathfrak{g}$ by $\alpha_1, \ldots, \alpha_l$ and to write $\omega_1, \ldots, \omega_l$ for the members of $\mathfrak{a}_{\mathfrak{p}}'$ with $\langle \alpha_i, \omega_j \rangle = \delta_{ij}$ for all i and j.

Theorem 8.51. Let π be an irreducible admissible representation of G.

(a) Then π has all its K-finite matrix coefficients in $L^2(G)$ if and only if $\mathrm{Re}\langle \tilde{\nu}, \omega_j \rangle < 0$ for every j and for every leading exponent $\tilde{\nu} - \rho_{\mathfrak{p}}$ of π.

(b) Moreover, if π has all its K-finite matrix coefficients in $L^2(G)$, then π is infinitesimally equivalent with the action of the right regular representation of G on some irreducible closed subspace of $L^2(G)$.

Remarks. If the irreducible admissible representation π satisfies the equivalent conditions of (a), we say that π is in the **discrete series** of G. In Proposition 9.6 we shall see from an elementary argument that the natural converse of (b) is valid, too: Any irreducible subrepresentation of the right regular representation on $L^2(G)$ has all K-finite matrix coefficients in $L^2(G)$. We shall need temporarily to take this converse result for granted to obtain the later theorem in this section. (The elementary proof of this converse result does not rely on asymptotic expansions, and there is no circularity. The argument is postponed merely because it does not fit well with the theme of the present chapter.)

Proof. Result (a) is the special case $p = 2$ of Theorem 8.48. For (b), fix $v_0 \neq 0$ in the space V of K-finite vectors of π, and define a linear map $L: V \to L^2(G)$ by $L(v) = (\pi(\cdot)v, v_0)$. Then X in $\mathfrak{g}$ implies

$$L(\pi(X)v) = (\pi(\cdot)\pi(X)v, v_0) = R(X)(\pi(\cdot)v, v_0) = R(X)L(v)$$

by (8.10). Since π is irreducible, L is an infinitesimal equivalence between V and a subspace S_0 of the right K-finite members of $L^2(G)$. The members of S_0 are real analytic on G, and the closure S of S_0 is G-invariant and irreducible, just as in Theorem 8.9 and its corollaries.

Corollary 8.52. If π is a discrete series representation of G and if $r \geq 0$ is specified, then each K-finite matrix coefficient of π is dominated in absolute value by a multiple of $(1 + \|x\|)^{-r}\varphi_0^G(x)$. Here φ_0^G denotes the 0th spherical function.

Proof. It is enough to prove the domination on $\overline{A_\mathfrak{p}^+}$. Substituting the conclusion of Theorem 8.51a into Theorem 8.47, we see that the product of any K-finite matrix coefficient by $(1 + \|x\|)^r$ is dominated on A_p^+ in absolute value by a multiple of

$$e^{-\rho_\mathfrak{p}(\log a)}[e^{\nu_0(\log a)}(1 + \rho_\mathfrak{p}(\log a))^{q+r}]$$

for some real ν_0 with $\langle \nu_0, \omega_j \rangle < 0$ for all j. The expression in brackets above is bounded, and hence the K-finite matrix coefficients are dominated on $\overline{A_\mathfrak{p}^+}$ by a multiple of $\exp\{-\rho_\mathfrak{p}(\log a)\}$. Meanwhile, (7.54) gives

$$\varphi_0^G(a) = e^{-\rho_\mathfrak{p}(\log a)} \int_{\bar{N}_\mathfrak{p}} e^{-\rho_\mathfrak{p} H(\bar{n})} e^{-\rho_\mathfrak{p} H(a\bar{n}a^{-1})} \, d\bar{n},$$

and the right side here is

$$\geq e^{-\rho_\mathfrak{p}(\log a)} \int_{\bar{N}_\mathfrak{p}} e^{-2\rho_\mathfrak{p} H(\bar{n})} \, d\bar{n} \tag{8.83}$$

by Corollary 7.18a. Since (5.25) establishes the integrability of $\exp\{-2\rho_\mathfrak{p} H(\bar{n})\}$ on $\bar{N}_\mathfrak{p}$, the present corollary follows.

Remark. The same conclusion applies to derivatives of the K-finite matrix coefficients of π, by Corollary 8.41 and its accompanying remarks.

Theorem 8.53. For an irreducible admissible representation π of G, the following are equivalent:

(a) All K-finite matrix coefficients are in $L^{2+\varepsilon}(G)$ for every $\varepsilon > 0$.
(b) Every leading exponent $\tilde{\nu} - \rho_{\mathfrak{p}}$ satisfies $\operatorname{Re}\langle \tilde{\nu}, \omega_j \rangle \leq 0$ for $1 \leq j \leq l$.
(c) π is infinitesimally equivalent with a subrepresentation of a standard induced representation $U(S, \omega, \nu)$ for some parabolic subgroup $S = MAN$, a discrete series representation ω of M, and an imaginary parameter ν on $\mathfrak{a}$.
(d) Each K-finite matrix coefficient is dominated in absolute value by a multiple of the spherical function φ_0^G.

Remarks. If the irreducible admissible representation π satisfies these equivalent conditions, we say that π is **irreducible tempered.** By (c), *such a representation is infinitesimally equivalent with a unitary representation.* Derivatives of the K-finite matrix coefficients of such a representation satisfy (a) and (d), by a simple estimate starting from Corollary 8.41.

Proof of all but (b) $\Rightarrow$ (c).

(a) $\Rightarrow$ (b). This follows from Theorem 8.48 if we take $p = 2 + \varepsilon$ and let ε tend to 0.

(c) $\Rightarrow$ (d). Corollary 8.52 implies that the $(K \cap M)$-finite matrix coefficients of ω are dominated in absolute value by multiples of φ_0^M. Then (d) follows from Proposition 7.14.

(d) $\Rightarrow$ (a). This is immediate from (7.48).

The idea behind the implication (b) $\Rightarrow$ (c) is fairly simple. We start with a *suitable* leading exponent $\tilde{\nu} - \rho_{\mathfrak{p}}$ for π and imbed π infinitesimally as a subrepresentation of some nonunitary principal series $U(\bar{S}_{\mathfrak{p}}, \sigma, \tilde{\nu})$. Next we decompose $\tilde{\nu}$ into two pieces, one corresponding to directions in $\mathfrak{a}_{\mathfrak{p}}'$ where $\langle \operatorname{Re} \tilde{\nu}, \omega_j \rangle = 0$ and the other corresponding to directions where $\langle \operatorname{Re} \tilde{\nu}, \omega_j \rangle < 0$. From this decomposition we construct a parabolic subgroup $S = MAN \supseteq S_{\mathfrak{p}}$. By the double induction formula, we can identify $(U(\bar{S}_{\mathfrak{p}}, \sigma, \tilde{\nu})$ with $U(\bar{S}, \xi, \tilde{\nu}|_{\mathfrak{a}})$, where ξ is induced up to M. We take the smallest subrepresentation ξ_0 of ξ needed so that $U(\bar{S}, \xi_0, \tilde{\nu}|_{\mathfrak{a}})$ still contains π, and then $\tilde{\nu}|_{\mathfrak{a}}$ will be imaginary and ξ_0 will have $(K \cap M)$-finite matrix coefficients in $L^2(M)$. It will be easy to replace ξ_0 with an irreducible representation ω of M, which will then be in the discrete series of M by the converse of Theorem 8.51b, given in the remarks after the statement of the theorem.

The heart of the proof is to get good control over ξ_0, e.g., to see that its $(K \cap M)$-finite matrix coefficients are in $L^2(M)$. One tool for this purpose is the combination of (8.67) and Corollary 8.42. But that tool is not sufficient by itself, as we must use some special property of $\tilde{\nu}$. The representation π can have more than one leading exponent, and the one we use must in some sense control them all.

Unfortunately the detailed pursuit of this idea is somewhat tedious. Much of the proof consists of formalism with the double induction formula. We shall be a little sketchy in handling this formalism, e.g., by allowing ourselves sometimes to work with group actions even though rigor requires sticking to actions by the Lie algebra. In addition, we shall disregard the fact that M may be disconnected; in places where a detailed argument might involve carrying along actions by both $K \cap M$ and $\mathfrak{m}^{\mathbb{C}}$, we may treat just $\mathfrak{m}^{\mathbb{C}}$.

We begin with any leading exponent $\tilde{\nu} - \rho_{\mathfrak{p}}$ of π relative to a fixed $S_{\mathfrak{p}}$. (We shall specialize $\tilde{\nu}$ later so that it has the appropriate property of control.) Let

$$\mathscr{F} = \{j \mid 1 \leq j \leq l \text{ and } \operatorname{Re}\langle \tilde{\nu}, \omega_j \rangle < 0\}. \tag{8.84}$$

If we write

$$\operatorname{Re} \tilde{\nu} = \sum_{i=1}^{l} c_i \alpha_i \tag{8.85}$$

and take the inner product with ω_j, we see that $c_j < 0$ for $j \in \mathscr{F}$ and $c_j = 0$ for $j \notin \mathscr{F}$. Moreover, the set

$$\{\alpha_j \mid j \in \mathscr{F}\} \cup \{\omega_j \mid j \notin \mathscr{F}\} \tag{8.86}$$

is a basis of $\mathfrak{a}_{\mathfrak{p}}'$. [In fact, each subset in braces is linearly independent, and the two subsets are orthogonal to each other. Thus (8.86) is a linearly independent set of l elements and hence is a basis.] Expanding $\operatorname{Im} \tilde{\nu}$ in terms of this basis and using what we know about $\operatorname{Re} \tilde{\nu}$ from (8.84) and (8.85), we obtain

$$\tilde{\nu} = \sum_{j \notin \mathscr{F}} b_j \omega_j - \sum_{j \in \mathscr{F}} a_j \alpha_j, \tag{8.87a}$$

where

$$\begin{aligned} \operatorname{Re} b_j &= 0 \qquad \text{for } j \notin \mathscr{F} \\ \operatorname{Re} a_j &> 0 \qquad \text{for } j \in \mathscr{F}. \end{aligned} \tag{8.87b}$$

Next we build a parabolic subgroup $S = MAN \supseteq S_{\mathfrak{p}}$ as in §12 and Proposition 5.23, using the indices in $\mathscr{F}$ to define M. In particular, we take

$$\mathfrak{a} = \sum_{j \notin \mathscr{F}} \mathbb{R} H_{\omega_j} \qquad \text{and} \qquad \mathfrak{a}_M = \sum_{j \in \mathscr{F}} \mathbb{R} H_{\alpha_j}.$$

Let us write $\tilde{\nu} = \nu + \nu_M$ correspondingly, with $\nu = \tilde{\nu}|_{\mathfrak{a}}$ and $\nu_M = \tilde{\nu}|_{\mathfrak{a}_M}$.

By the Subrepresentation Theorem we have an infinitesimal imbedding

$$\pi \subseteq \operatorname{ind}_{S_{\mathfrak{p}}}^{G}(\sigma \otimes e^{\tilde{\nu}} \otimes 1) \tag{8.88}$$

for some σ. Recall how this imbedding comes about: For a maximal $\tilde{q}$ the linear functional l that reads off the coefficient of $\alpha(H)^{\tilde{q}}\exp(\tilde{\nu} - \rho_{\mathfrak{p}})(H)$ in the asymptotic expansion of $(\pi(\exp H)v, v_2)$ on $A_{\mathfrak{p}}^{+}$ gives us an imbedding into the compact picture of

$$\operatorname{ind}_{A_{\mathfrak{p}}\bar{N}_{\mathfrak{p}}}^{G}(e^{\tilde{\nu}} \otimes 1),$$

namely $L(v)(k) = l(\pi(k)^{-1}v)$. Then we find some σ so that we have (8.88).

It will be convenient to make the role of σ more explicit; a clue of how to do so is contained in the proof of Theorem 8.33. With $\tilde{\nu}$ fixed, choose a maximal power $\tilde{q}$ of $\alpha(H)$ that can occur in the term $F_{\tilde{\nu}-\rho_{\mathfrak{p}}}$ of the expansion of the τ-spherical blocks of matrix coefficients of π. Fix a τ-spherical block where $\tilde{q}$ does occur. The coefficient $c_{\tilde{\nu},\tilde{q}}$ is a nonzero member of $\operatorname{Hom}_{M_{\mathfrak{p}}}(U_2, U_1)$, and we can choose an irreducible representation π_λ of $(M_{\mathfrak{p}})_0$ (with highest weight λ) such that $c_{\tilde{\nu},\tilde{q}}$ is not the 0 map on the subspace of U_2 transforming by π_λ under $(M_{\mathfrak{p}})_0$. Then there exists a nonzero member of U_1 in the image of $c_{\tilde{\nu},\tilde{q}}$ transforming under $(M_{\mathfrak{p}})_0$ according to π_λ. Disregarding the disconnectedness of $M_{\mathfrak{p}}$, we see that there exists an $M_{\mathfrak{p}}$ irreducible subspace of type π_λ in U_1 in the image of $c_{\tilde{\nu},\tilde{q}}$. Let σ be the restriction of $\tau_1(M_{\mathfrak{p}})$ to this subspace, and let us denote this subspace by V^σ and the orthogonal projection to it by E_σ. Then $E_\sigma E_1 = E_\sigma$.

Given a vector v in the K-finite subspace V for π, we choose E_2 large enough so that $E_2 v = v$. From the asymptotic expansion of $E_1\pi(x)E_2$ along $A_{\mathfrak{p}}^{+}$, we obtain an asymptotic expansion of $E_\sigma\pi(x)v = E_\sigma(E_1\pi(x)E_2)v$, and the coefficients are independent of E_2 because of the uniqueness of such an expansion. Let us write the $(\tilde{\nu}, \tilde{q})$ term in the expansion of $E_\sigma\pi(\exp H)v$ as

$$\alpha(H)^{\tilde{q}} e^{(\tilde{\nu}-\rho_{\mathfrak{p}})(H)} E_\sigma c_{\tilde{\nu},\tilde{q}}^{S_{\mathfrak{p}}}(v).$$

Then the linear map $l: V \to V^\sigma$ defined by

$$l(v) = E_\sigma c_{\tilde{\nu},\tilde{q}}^{S_{\mathfrak{p}}}(v) \tag{8.89}$$

gives us a $\mathfrak{g}$-equivariant imbedding L directly into the compact picture of $\operatorname{ind}_{S_{\mathfrak{p}}}^{G}(\sigma \otimes e^{\tilde{\nu}} \otimes 1)$ if we set

$$L(v)(k) = l(\pi(k)^{-1}v). \tag{8.90}$$

(The proof of this assertion imitates the proof of Theorem 8.37.)

Now we bring in double induction, generalizing matters to allow the intermediate representation (the representation of M) to be nonunitary. With

$$\xi = \operatorname{ind}_{S_{\mathfrak{p}} \cap M}^{M}(\sigma \otimes e^{\nu_M} \otimes 1),$$

the relevant formula is

$$\mathrm{ind}_{S_p}^{G}(\sigma \otimes e^{\tilde{\nu}} \otimes 1) \cong \mathrm{ind}_S^G(\xi \otimes e^{\nu} \otimes 1). \tag{8.91}$$

Specifically the space for the right side of (8.91) consists of certain functions on G whose values are V^σ-valued functions on M, and the isomorphism from right to left is obtained by evaluation at the identity of M. In the reverse direction we start with a V^σ-valued function f on G and obtain a V^ξ-valued function f^* on G by taking

$$f^*(x)(m) = F(xm).$$

Regarding π as imbedded in the right side of (8.91), we let V^{ξ_0} be the subspace of values $f^*(1)(\cdot)$ in V^ξ, as f^* ranges through the subspace identified with V, and let ξ_0 be the restriction of ξ to V^{ξ_0}. (More precisely, we take the space of K-finite vectors in the right side of (8.91) obtained from π, evaluate them at the identity of G, and take the closure in V^ξ as V^{ξ_0}.) Then ξ_0 is an admissible representation of M, and we have an imbedding

$$\pi \subseteq \mathrm{ind}_S^G(\xi_0 \otimes e^{\nu} \otimes 1) \tag{8.92a}$$

obtained from l and from (8.91). Abusing notation, we write L for this imbedding also. Our main objective is to see that ξ_0 is essentially in the discrete series if $\tilde{\nu}$ is chosen properly.

We shall write down two maps analogous to L and relate all three in a diagram. For one of them we start with our K-finite blocks of matrix coefficients of π and form the asymptotic expansion of Theorem 8.45 relative to the nonminimal S. Again we interpret this as an expansion of $E_\sigma\pi(x)v$. Letting q be the part of $\tilde{q}$ off $\mathscr{F}$, we write the (ν, q) term in the expression of $E_\sigma\pi(m \exp H)v$ as

$$\alpha(H)^q e^{(\nu - \rho_A)(H)} E_\sigma c_{\nu,q}^S(m, v).$$

Let V^{coef} be the space of V^σ-valued functions $E_\sigma c_{\nu,q}^S(m, v)$ on M as v ranges through V, and let R be the right regular representation of M on smooth functions. Then

$$R(m)(E_\sigma c_{\nu,q}^S(\cdot, v)) = E_\sigma c_{\nu,q}^S(\cdot, \pi(m)v)$$

on the level of the universal enveloping algebra $U(\mathfrak{m}^{\mathbb{C}})$, though perhaps not on the level of the group M since $\pi(m)v$ need not be K-finite. This means that R gives us a representation of $U(\mathfrak{m}^{\mathbb{C}})$ on V^{coef}; we shall treat R as if we have a representation of M, re-examining this matter later.

We define a linear mapping $l^*: V \to V^{\mathrm{coef}}$ simply by

$$l^*(v)(\cdot) = E_\sigma c_{\nu,q}^S(\cdot, v).$$

Using the second remark right before Theorem 8.45, we see that

$$\begin{aligned} l^*(Xv) &= 0 && \text{for } X \in \bar{\mathfrak{n}} \\ l^*(Hv) &= (\nu - \rho_A)(H) l^*(v) && \text{for } H \in \mathfrak{a} \\ l^*(Xv) &= R(X) l^*(v) && \text{for } X \in \mathfrak{m}. \end{aligned}$$

The proof of Corollary 8.46 shows that l^* is not 0 if q is maximal among all restrictions of exponents of $\alpha(H)$ that are associated with $\tilde{\nu}$. By imitating the proof of Theorem 8.37, we can then check that the map L^* defined by

$$L^*(v)(k)(m) = l^*(\pi(k)^{-1}v)(m) = E_\sigma c^S_{\tilde{\nu},q}(m, \pi(k)^{-1}v)$$

is a $\mathfrak{g}$-equivariant map of π into

$$\operatorname{ind}_S^G\, ((R \text{ on } V^{\text{coef}}) \otimes e^{\nu} \otimes 1).$$

The functions $E_\sigma c_{\nu,q}(m)v$ have asymptotic expansions along A_M^+ (relative to $M \cap S_p$) that are given by equating the expressions in braces in (8.70) and (8.69). Picking off the coefficient corresponding to ν_M and to $q_M = \tilde{q}|_{\mathscr{F}}$ by a map $l_M: V^{\text{coef}} \to V^\sigma$, we obtain an $\mathfrak{m}$-equivariant map L_M given by

$$L_M(E_\sigma c_{\nu,q}(m)v)(k_M) = l_M(R(k_M^{-1})E_\sigma c_{\nu,q}(m)v).$$

The map L_M carries R into ξ in $\mathfrak{m}$-equivariant fashion. By allowing L_M to act on values of functions defined on G, we obtain a corresponding map $\operatorname{ind}(L_M)$ carrying

$$\operatorname{ind}_S^G(R \otimes e^{\nu} \otimes 1) \qquad \text{into} \qquad \operatorname{ind}_S^G(\xi \otimes e^{\nu} \otimes 1).$$

Thus we have a diagram

$$\begin{array}{ccccc} & & \operatorname{ind}_S^G(R \otimes e^{\nu} \otimes 1) & & \\ & \overset{L^*}{\nearrow} & & \overset{\operatorname{ind}(L_M)}{\searrow} & \\ \pi & \xrightarrow{\ L\ } & \operatorname{ind}_S^G(\xi_0 \otimes e^{\nu} \otimes 1) & \xrightarrow{\ \text{incl.}\ } & \operatorname{ind}_S^G(\xi \otimes e^{\nu} \otimes 1). \end{array} \tag{8.92b}$$

The point of all this formalism is that the diagram (8.92b) commutes, in fact, that it does so in a way that has the space V^{coef} for R mapping onto the space V^{coef} for ξ_0 in $\mathfrak{m}$-equivariant fashion. [We readily check that $L(v)$, regarded as a function on K whose values are V^σ-valued functions on K_M, is given by

$$L(v)(k)(k_M) = E_\sigma c^{S_p}_{\tilde{\nu},\tilde{q}}(\pi(k_M k^{-1})v),$$

whereas $(\operatorname{ind} L_M)(L^*v)(k)(k_M)$ is the (ν_M, q_M) coefficient of

$$E_\sigma c^S_{\tilde{\nu},q}(a_M k_M, \pi(k)^{-1}v).$$

These are equal, by the considerations in the proof of Corollary 8.46, provided q (the restriction of $\tilde{q}$) is maximal among all restrictions of exponents going with $\tilde{\nu}$.]

Within this formalism let us see that ξ_0 has an irreducible quotient ω, so that π imbeds in $\operatorname{ind}_S^G(\omega \otimes e^\nu \otimes 1)$ and R maps onto ω. Fix a nonzero finite-dimensional K-invariant subspace V_1 of the original V, regard V via L as contained in the space of the induced representation $\operatorname{ind}_S^G(\xi_0 \otimes e^\nu \otimes 1)$, and form the closure in V^{ξ_0} of the M translates of the set

$$\{f(1) \mid f \in V_1\} \subseteq V^{\xi_0}.$$

This subspace of V^{ξ_0} is M-invariant, and we let ξ_{00} be the restriction of $\xi_0(M)$ to it. The members of V_1 have their values in this subspace, and the irreducibility of π thus implies that

$$\pi \subseteq \operatorname{ind}_S^G(\xi_{00} \otimes e^\nu \otimes 1).$$

The definition of ξ_0 therefore forces $\xi_{00} = \xi_0$. Hence we can conclude ξ_0 is finitely generated. Taking the quotient by the M span of all but one generator, we see that ξ_0 has a singly generated quotient. Say the quotient is ξ_1 on V^{ξ_1} and a generator is v_1. We may assume that π imbeds in $\operatorname{ind}_S^G(\xi_1 \otimes e^\nu \otimes 1)$ since otherwise we did not need the last generator. By Zorn's Lemma, extend the left ideal of $U(\mathfrak{m}^{\mathbb{C}})$ that annihilates v_1 to a maximal left ideal $\mathscr{I}$. The subset $\mathscr{I}v_1$ is $U(\mathfrak{m}^{\mathbb{C}})$-invariant, and the quotient ω by the closure of $\mathscr{I}v_1$ is M-invariant and irreducible. We cannot have $\pi \subseteq \operatorname{ind}_S^G(\xi_1|_{\mathscr{I}v_1} \otimes e^\nu \otimes 1)$, and thus π imbeds in $\operatorname{ind}_S^G(\omega \otimes e^\nu \otimes 1)$.

Now suppose that we can show that the members of V^{coef} are in $L^2(M)$. They are known to be $(K \cap M)$-finite, and they satisfy the system of differential equations (8.68). By the same kind of argument as in Lemma 8.29 (using Theorems 8.18 and 8.19), $Z(\mathfrak{m}^{\mathbb{C}})$ is a finitely generated module over $\mu_{\mathfrak{l}^+}^+(Z(\mathfrak{g}^{\mathbb{C}}))$, and thus (8.68) implies that the members of V^{coef} are $Z(\mathfrak{m}^{\mathbb{C}})$-finite. So Corollary 8.42 applies, and the closed linear span generated by the right M translates of any member of V^{coef} is a finite direct sum of irreducible closed invariant subspaces of $L^2(M)$. In each of these invariant subspaces, R acts as a discrete series representation, by the converse of Theorem 8.52b. Since R maps onto ω and ω is irreducible, ω is a discrete series representation, as required.

Thus we are to prove that the members of V^{coef} are in $L^2(M)$ if $\tilde{\nu}$ is chosen suitably. Before doing so, let us return to the difficulty that V^{coef} seems to support only a representation of $U(\mathfrak{m}^{\mathbb{C}})$ rather than of M. The point here is that the members of V^{coef} are $Z(\mathfrak{m}^{\mathbb{C}})$-finite and $(K \cap M)$-finite, as we saw above. As a result we can work with V^{coef} in Corollary 8.42 without using the full M action. In place of an induced representation of G from

V^{coef}, we can work with an action of $U(\mathfrak{g}^{\mathbb{C}})$, and the whole argument will go through.

So much for the formalism. Let us now select $\tilde{\nu}$ so that the members of V^{coef} are in $L^2(M)$. Recall the definition (8.84) of $\mathscr{F}$ in terms of $\tilde{\nu}$. We write $\mathscr{F}(\tilde{\nu})$ for this set so that we can examine the dependence of the set on $\tilde{\nu}$. Among all leading exponents $\tilde{\nu} - \rho_{\mathfrak{p}}$ of π, choose $\tilde{\nu} = \tilde{\nu}_0$ so that $\mathscr{F}(\tilde{\nu}_0)$ is minimal under inclusion, i.e., so that M is as small as possible. By Corollary 8.46 any exponent of $c_{v_0,q}(m)$ is of the form $\tilde{\nu}'|_{\mathfrak{a}_M} - \rho_{A_M}$ with $\tilde{\nu}'|_{\mathfrak{a}} = \tilde{\nu}_0|_{\mathfrak{a}}$ and

$$\tilde{\nu}' = \tilde{\nu}_1 - \sum n_j \alpha_j \qquad (n_j \geq 0)$$

for some leading exponent $\tilde{\nu}_1$ of π. By Theorem 8.51a we are to show that

$$\operatorname{Re}\langle \tilde{\nu}', \omega_{j,M} \rangle < 0 \tag{8.93}$$

for each member $\omega_{j,M}$ of the basis of $\mathfrak{a}'_M$ dual to $\{\alpha_j \mid j \in \mathscr{F}(\tilde{\nu}_0)\}$.

For $j \notin \mathscr{F}(\tilde{\nu}_0)$, we have $\operatorname{Re}\langle \tilde{\nu}_0, \omega_j \rangle = 0$ and $\langle \tilde{\nu}' - \tilde{\nu}_0, \omega_j \rangle = 0$ since $\tilde{\nu}'|_{\mathfrak{a}} = \tilde{\nu}_0|_{\mathfrak{a}}$. Thus

$$\begin{aligned} 0 &= \operatorname{Re}\langle \tilde{\nu}' - \tilde{\nu}_0, \omega_j \rangle + \operatorname{Re}\langle \tilde{\nu}_0, \omega_j \rangle \\ &= \operatorname{Re}\langle \tilde{\nu}', \omega_j \rangle = \operatorname{Re}\langle \tilde{\nu}_1, \omega_j \rangle - n_j \leq -n_j \leq 0, \end{aligned}$$

the next-to-last inequality holding by our assumption (b). Consequently $\operatorname{Re}\langle \tilde{\nu}_1, \omega_j \rangle = 0$, and $\mathscr{F}(\tilde{\nu}_1)$ is a subset of $\mathscr{F}(\tilde{\nu}_0)$. Then $\mathscr{F}(\tilde{\nu}_1) = \mathscr{F}(\tilde{\nu}_0)$ by minimality.

Arguing by contradiction, suppose that (8.93) fails for an index $j \in \mathscr{F}(\tilde{\nu}_0)$. Since $\omega_{j,M} - \omega_j$ is orthogonal to $\mathfrak{a}'_M$ and $\tilde{\nu}'$ differs from $\tilde{\nu}_0$ by a member of $\mathfrak{a}'_M$, we then have

$$\begin{aligned} 0 &\leq \operatorname{Re}\langle \tilde{\nu}', \omega_{j,M} \rangle \\ &= \operatorname{Re}\langle \tilde{\nu}', \omega_{j,M} - \omega_j \rangle + \operatorname{Re}\langle \tilde{\nu}', \omega_j \rangle \\ &= \operatorname{Re}\langle \tilde{\nu}_0, \omega_{j,M} - \omega_j \rangle + \operatorname{Re}\langle \tilde{\nu}', \omega_j \rangle \\ &= \operatorname{Re}\langle \tilde{\nu}', \omega_j \rangle \qquad\qquad \text{since } \tilde{\nu}_0(\mathfrak{a}'_M) = 0 \\ &= \operatorname{Re}\langle \tilde{\nu}_1, \omega_j \rangle - n_j. \end{aligned}$$

Thus j, which is in $\mathscr{F}(\tilde{\nu}_0)$, cannot be in $\mathscr{F}(\tilde{\nu}_1)$, and we have a contradiction. This completes the proof of Theorem 8.53.

§15. Sixth Application: Langlands Classification of Irreducible Admissible Representations

We come now to the main result of this chapter, the Langlands classification of irreducible admissible representations. Recall from Theorem 7.24 and its reformulation in §8.3 that any standard representation $U(S, \omega, \nu)$

for which ω is irreducible tempered unitary and Re ν is in the open positive Weyl chamber has a unique irreducible quotient $J(S, \omega, \nu)$. The representation $J(S, \omega, \nu)$ is called the **Langlands quotient** of $U(S, \omega, \nu)$, and its K-finite vectors can be regarded as the image in $U(\bar{S}, \omega, \nu)$ of the K-finite vectors of $U(S, \omega, \nu)$ under the intertwining operator $A(\bar{S}:S:\omega:\nu)$.

Theorem 8.54. Fix a minimal parabolic subgroup $S_{\mathfrak{p}} = M_{\mathfrak{p}}A_{\mathfrak{p}}N_{\mathfrak{p}}$ of G. Then the equivalence classes of irreducible admissible representations of G stand in one-one correspondence with all triples $(S, [\omega], \nu)$ such that

- $S = MAN$ is a parabolic subgroup containing $S_{\mathfrak{p}}$,
- ω is an irreducible tempered unitary representation of M and $[\omega]$ is its equivalence class,
- ν is a member of $(\mathfrak{a}')^{\mathbb{C}}$ with Re ν in the open positive Weyl chamber.

The correspondence is that $(S, [\omega], \nu)$ corresponds to the class of $J(S, \omega, \nu)$.

Terminology. We refer to (S, ω, ν) as the **Langlands parameters** of the representation in question.

The hardest part of the proof is to exhibit each irreducible admissible representation π as a Langlands quotient. By an easy duality argument that we give later in this section, it is enough to show that π admits an infinitesimal imbedding

$$\pi \subseteq U(\bar{S}, \omega, \nu) \tag{8.94}$$

with (S, ω, ν) as in the theorem.

The formal part of the proof of (8.94) is remarkably similar to that of the implication (b) ⇒ (c) in Theorem 8.53. We start with a *suitable* leading exponent $\tilde{\nu} - \rho_{\mathfrak{p}}$ for π and imbed π infinitesimally as a subrepresentation of some nonunitary principal series $U(\bar{S}_{\mathfrak{p}}, \sigma, \tilde{\nu})$. Next we use Lemma 8.56 below to decompose $\tilde{\nu}$ into two pieces in somewhat the same fashion as in (8.87):

$$\tilde{\nu} = \sum_{j \notin \mathscr{F}} b_j\omega_j - \sum_{j \in \mathscr{F}} a_j\alpha_j \tag{8.95a}$$

as before but with the coefficients now satisfying

$$\begin{aligned} &\text{Re}\, b_j > 0 \qquad \text{for } j \notin \mathscr{F} \\ &\text{Re}\, a_j \geq 0 \qquad \text{for } j \in \mathscr{F}. \end{aligned} \tag{8.95b}$$

(Again the α_j are the simple roots, and the ω_j are dual to the α_j.) Incorporating the α_j for $j \in \mathscr{F}$ into M, we construct a parabolic subgroup $S = MAN \supseteq S_{\mathfrak{p}}$. As in §14, we identify $U(\bar{S}_{\mathfrak{p}}, \sigma, \tilde{\nu})$ with $U(\bar{S}, \xi, \tilde{\nu}|_{\mathfrak{a}})$ by the double induction formula, and we take the smallest subrepresentation ξ_0 of ξ so that $U(\bar{S}, \xi_0, \tilde{\nu}|_{\mathfrak{a}})$ still contains π. Then $\nu = \tilde{\nu}|_{\mathfrak{a}}$ will have real part in the positive Weyl chamber, and we shall prove that ξ_0 has $(K \cap M)$-finite matrix coefficients dominated by multiples of φ_0^M. Again we readily replace ξ_0 by an irreducible representation ω of M, and ω will be tempered and (8.94) will hold.

Let us now tend to the details of proving (8.94). We begin with a sequence of lemmas that give the decomposition (8.95) and establish, for a special choice of $\tilde{\nu}$, estimates analogous to (8.93). The first lemma capitalizes on the fact that $\overline{(\mathfrak{a}_\mathfrak{p}')^+}$ both is a closed convex set in a Hilbert space and also is a cone. In concrete terms, $\overline{(\mathfrak{a}_\mathfrak{p}')^+}$ consists of all nonnegative linear combinations of $\omega_1, \ldots, \omega_l$.

Lemma 8.55. Let v be in $\mathfrak{a}_\mathfrak{p}'$, and let v_0 be the unique point in $\overline{(\mathfrak{a}_\mathfrak{p}')^+}$ closest to v. Then

(a) $\langle v - v_0, w\rangle \leq 0$ for every w in $\overline{(\mathfrak{a}_\mathfrak{p}')^+}$, and
(b) $\langle v - v_0, v_0\rangle = 0$.

Proof. Existence and uniqueness of v_0 are well-known properties of closed convex sets in Hilbert space. Let w be in $\overline{(\mathfrak{a}_\mathfrak{p}')^+}$. For $0 \leq \varepsilon \leq 1$, $v_0 + \varepsilon(w - v_0)$ is in $(\mathfrak{a}_\mathfrak{p}')^+$ by convexity. For such ε,

$$|v_0 + \varepsilon(w - v_0) - v|^2 \geq |v_0 - v|^2$$

and hence

$$-2\varepsilon\langle w - v_0, v - v_0\rangle + \varepsilon^2|w - w_0|^2 > 0.$$

For small enough $\varepsilon > 0$, this inequality forces

$$\langle w - v_0, v - v_0\rangle \leq 0. \tag{8.96}$$

Since v_0 is in $\overline{(\mathfrak{a}_\mathfrak{p}')^+}$ and $\overline{(\mathfrak{a}_\mathfrak{p}')^+}$ is a cone, $2v_0$ and $\frac{1}{2}v_0$ are in $\overline{(\mathfrak{a}_\mathfrak{p}')^+}$. Putting $w = 2v_0$ and then $w = \frac{1}{2}v_0$ in (8.96), we obtain (b). Then (a) follows immediately from (8.96) and (b).

Recall from (8.86) that

$$\{\alpha_j\}_{j\in\mathscr{F}} \cup \{\omega_j\}_{j\notin\mathscr{F}} \tag{8.97}$$

is a basis of $\mathfrak{a}_\mathfrak{p}'$ for each subset $\mathscr{F}$ of $\{1, \ldots, l\}$.

Lemma 8.56. For each v in $\mathfrak{a}_\mathfrak{p}'$ there exists a unique subset $\mathscr{F}$ of $\{1, \ldots, l\}$ such that the decomposition

$$v = \sum_{j\notin\mathscr{F}} b_j\omega_j - \sum_{j\in\mathscr{F}} a_j\alpha_j \tag{8.98a}$$

of v relative to the basis (8.97) has

$$\begin{aligned} b_j &> 0 \qquad \text{for } j \notin \mathscr{F} \\ a_j &\geq 0 \qquad \text{for } j \in \mathscr{F}. \end{aligned} \tag{8.98b}$$

For this $\mathscr{F}$ the vector $\sum_{j \notin \mathscr{F}} b_j\omega_j$ is the vector v_0 of Lemma 8.55.

Notation. We write $\mathscr{F}(v)$ for this particular $\mathscr{F}$.

Remarks. Figure 8.1 illustrates this lemma for SL(3, $\mathbb{R}$).

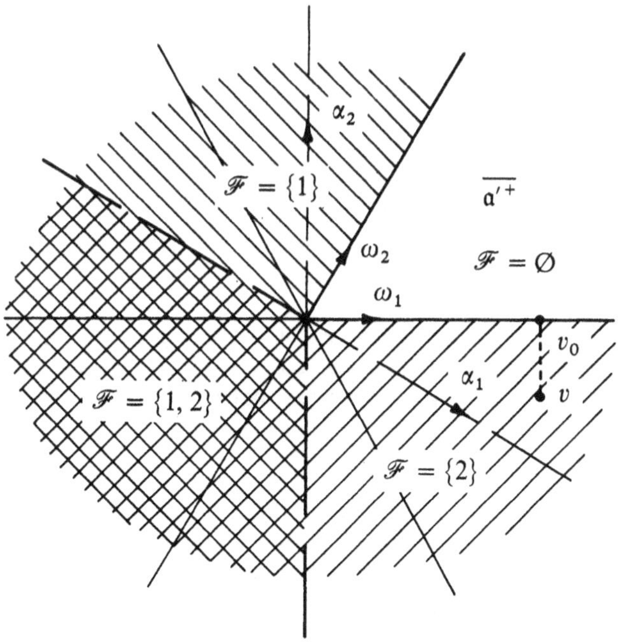

FIGURE 8.1. Decomposition of Lemma 8.56 for SL(3, $\mathbb{R}$)

Proof of existence. Let v_0 be as in Lemma 8.55, and define $\mathscr{F}$ to consist of those j's not needed for the expansion of v_0 in terms of the ω_j's:

$$v_0 = \sum_{j \notin \mathscr{F}} b_j\omega_j \qquad \text{with all } b_j > 0. \tag{8.99}$$

Define a_j, $1 \leq j \leq l$, by

$$v_0 - v = \sum_{\text{all } j} a_j\alpha_j. \tag{8.100}$$

Then Lemma 8.55a applied to $w = \omega_j$ gives $a_j \geq 0$ for all j. Using this fact, taking the inner product of (8.99) and (8.100), and applying Lemma 8.55b, we obtain

$$0 = \langle v_0 - v, v_0 \rangle = \sum_{j \notin \mathscr{F}} a_j b_j$$

with each term on the right ≥ 0. From this relation we conclude $a_j = 0$ for $j \notin \mathscr{F}$, and then (8.98) follows from (8.99) and (8.100).

Proof of uniqueness. Suppose $\mathscr{F}$ is a subset of $\{1, \ldots, l\}$ such that

$$v = \sum_{j \notin \mathscr{F}} b_j \omega_j - \sum_{j \in \mathscr{F}} a_j \alpha_j$$

with all $b_j > 0$ and all $a_j \geq 0$. Put $v_0' = \sum_{j \notin \mathscr{F}} b_j \omega_j$. Then v_0' is in $\overline{(\mathfrak{a}_\mathfrak{p}')^+}$, and we see directly that

$$\langle v - v_0', v_0' \rangle = 0. \tag{8.101}$$

If w is in $\overline{(\mathfrak{a}_\mathfrak{p}')^+}$, we find

$$\begin{aligned} |w - v|^2 &= |(w - v_0') + (v_0' - v)|^2 \\ &= |w - v_0'|^2 + 2\langle w - v_0', v_0' - v \rangle + |v_0' - v|^2 \\ &= |w - v_0'|^2 + 2\langle w, v_0' - v \rangle + |v_0' - v|^2 \qquad \text{by (8.101).} \end{aligned}$$

Here $v_0' - v = \sum_{j \in \mathscr{F}} a_j \alpha_j$ clearly has inner product ≥ 0 with any $w = \sum c_j \omega_j$ having all $c_j \geq 0$, and so $|w - v|^2 \geq |v_0' - v|^2$ with equality only if $w = v_0'$. Taking $w = v_0$, we see that $v_0' = v_0$. Then linear independence of the ω_j's forces $\mathscr{F}$ to be as in the existence half of the proof.

Lemma 8.56 is the tool needed to obtain (8.95). We now give four lemmas that prepare estimates analogous to (8.93). Lemmas 8.59 and 8.60 use the notations $v \succeq v'$ and $v' \preceq v$ to mean $v = v' + \sum c_j \alpha_j$ with all $c_j \geq 0$; the c_j are not assumed to be integers.

Lemma 8.57. $\langle \omega_i, \omega_j \rangle \geq 0$ for all i and j.

Proof. We apply Lemma 8.56 to $-\omega_i$, writing

$$-\omega_i = \sum_{j \notin \mathscr{F}} b_j \omega_j - \sum_{j \in \mathscr{F}} a_j \alpha_j$$

with all $b_j > 0$ and all $a_j \geq 0$. For $k \notin \mathscr{F}$, we take the inner product with α_k and obtain

$$0 \geq -\langle \omega_i, \alpha_k \rangle = b_k - \sum_{j \in \mathscr{F}} a_j \langle \alpha_j, \alpha_k \rangle. \tag{8.102}$$

Since $\langle \alpha_j, \alpha_k \rangle \leq 0$ for $j \in \mathscr{F}$ and $k \notin \mathscr{F}$, the right side of (8.102) is $\geq b_k \geq 0$.

Since the left side is 0, we conclude $b_k = 0$, i.e., $\mathcal{F}$ is all of $\{1, \ldots, l\}$. Hence $\omega_i = \sum a_k \alpha_k$ with all $a_k \geq 0$. Taking the inner product with ω_j, we obtain the result of the lemma.

Lemma 8.58. If j is in $\mathcal{F}$, then the expression

$$\omega_j = \sum_{k \in \mathcal{F}} c_k \alpha_k + \sum_{k \notin \mathcal{F}} d_k \omega_k \tag{8.103}$$

has all $c_k \geq 0$ and all $d_k \geq 0$.

Proof. Build the M of a parabolic subgroup in the usual way by incorporating those α_k into M for which k is in $\mathcal{F}$. Let P_M be the orthogonal projection of $\mathfrak{a}'_{\mathfrak{p}}$ on $\sum_{k \in \mathcal{F}} \mathbb{R}\alpha_k$. From (8.103) we obtain $P_M \omega_j = \sum_{k \in \mathcal{F}} c_k \alpha_k$. Hence $k \in \mathcal{F}$ implies

$$\langle P_M \omega_j, \alpha_k \rangle = \langle \omega_j, P_M \alpha_k \rangle = \langle \omega_j, \alpha_k \rangle = \delta_{jk}.$$

Since j is assumed to be in $\mathcal{F}$, this relation says that

$$P_M \omega_j = \omega_j^M, \tag{8.104}$$

i.e., $P_M \omega_j$ is the dual element to α_j for the group M. Lemma 8.57 for M thus says $c_k \geq 0$ for all $k \in \mathcal{F}$.

Finally, for $i \notin \mathcal{F}$, we take the inner product of (8.103) with α_i and obtain

$$0 = \langle \omega_j, \alpha_i \rangle = \sum_{k \in \mathcal{F}} c_k \langle \alpha_k, \alpha_i \rangle + d_i.$$

The right side is $\leq d_i$ since $\langle \alpha_k, \alpha_i \rangle \leq 0$. So all d_i are ≥ 0 for $i \notin \mathcal{F}$.

Lemma 8.59. If $v \succeq v'$, then $v_0 \succeq v'_0$ (in the notation of Lemma 8.55).

Proof. Lemma 8.56 shows that $v_0 \succeq v$. Thus

$$v_0 \succeq v \succeq v'. \tag{8.105}$$

Let $\mathcal{F} = \mathcal{F}(v')$. If j is not in $\mathcal{F}$, then $\langle v', \omega_j \rangle = \langle v'_0, \omega_j \rangle$. So (8.105) implies

$$\langle v_0, \omega_j \rangle \geq \langle v'_0, \omega_j \rangle \tag{8.106}$$

for $j \notin \mathcal{F}$. If j is in $\mathcal{F}$, then

$$\langle v_0, \alpha_j \rangle \geq 0 = \langle v'_0, \alpha_j \rangle. \tag{8.107}$$

Applying Lemma 8.58 and taking a suitable positive combination of (8.106) and (8.107), we obtain $\langle v_0, \omega_j \rangle \geq \langle v'_0, \omega_j \rangle$ for $j \in \mathcal{F}$. In combination with (8.106), this relation proves the lemma.

Lemma 8.60. Among all leading exponents $\tilde{v} - \rho_{\mathfrak{p}}$ of the irreducible admissible representation π, fix one for which $(\operatorname{Re} \tilde{v})_0$ is maximal with respect to $\succeq$. Put $\mathscr{F} = \mathscr{F}(\operatorname{Re} \tilde{\nu})$. Then

(a) $\operatorname{Re}\langle \tilde{\nu}|_{\mathfrak{a}}, \omega_j \rangle > 0$ for $j \notin \mathscr{F}$
(b) $\operatorname{Re} \langle \tilde{\nu}'|_{\mathfrak{a}_M}, \omega_j^M \rangle \leq 0$ for $j \in \mathscr{F}$ whenever $\tilde{\nu}' - \rho_{\mathfrak{p}}$ is an exponent of π with $\tilde{\nu}'|_{\mathfrak{a}} = \tilde{\nu}|_{\mathfrak{a}}$.

Proof. (a) is immediate from Lemma 8.57. For (b), we know from (8.104) that $\omega_j^M = P_M \omega_j$. Thus it is enough to prove

$$\operatorname{Re}\langle \tilde{\nu}'|_{\mathfrak{a}_M}, \omega_j \rangle \leq 0 \qquad \text{for } j \in \mathscr{F}. \tag{8.108}$$

Let us expand $\operatorname{Re} \tilde{\nu}'$ in terms of the basis $\{\omega_j\}_{j \notin \mathscr{F}} \cup \{\alpha_j\}_{j \in \mathscr{F}}$ as

$$\operatorname{Re} \tilde{\nu}' = \sum_{j \notin \mathscr{F}} b'_j \omega_j - \sum_{j \in \mathscr{F}} a'_j \alpha_j. \tag{8.109}$$

Since $\tilde{\nu}'|_{\mathfrak{a}} = \tilde{\nu}|_{\mathfrak{a}}$, each b'_j equals the corresponding coefficient b_j for $\operatorname{Re} \tilde{\nu}$ and hence is >0. To prove (8.108), we are to show $a'_j \geq 0$ for all $j \in \mathscr{F}$. Thus consider the set

$$\mathscr{F}_1 = \{j \in \mathscr{F} \mid a'_j < 0\}.$$

We write

$$\begin{aligned} \operatorname{Re} \tilde{\nu}' &= \sum_{j \notin \mathscr{F}} b_j \omega_j + \sum_{j \in \mathscr{F}_1} (-a'_j) \alpha_j - \sum_{j \in \mathscr{F} - \mathscr{F}_1} a'_j \alpha_j \\ &\succeq \sum_{j \notin \mathscr{F}} b_j \omega_j - \sum_{j \in \mathscr{F} - \mathscr{F}_1} a'_j \alpha_j. \end{aligned} \tag{8.110}$$

Since $\tilde{\nu}' - \rho_{\mathfrak{p}}$ is an exponent, there are integers $n_j \geq 0$ and there is a leading exponent $\tilde{\nu}^L - \rho_{\mathfrak{p}}$ such that $\tilde{\nu}' = \tilde{\nu}^L - \sum n_j \alpha_j$. Combining this relation with (8.110), we obtain

$$\operatorname{Re} \tilde{\nu}^L \succeq \sum_{j \notin \mathscr{F}} b_j \omega_j - \sum_{j \in \mathscr{F} - \mathscr{F}_1} a'_j \alpha_j. \tag{8.111}$$

By the uniqueness half of Lemma 8.56, we can identify $(\text{——})_0$ of the right side of (8.111) as the first term $\sum_{j \notin \mathscr{F}} b_j \omega_j$, which equals $(\operatorname{Re} \tilde{\nu})_0$. Thus Lemma 8.59 gives $(\operatorname{Re} \tilde{\nu}^L)_0 \succeq (\operatorname{Re} \tilde{\nu})_0$. By maximality we conclude

$$(\operatorname{Re} \tilde{\nu}^L)_0 = (\operatorname{Re} \tilde{\nu})_0. \tag{8.112}$$

Hence

$$\operatorname{Re} \tilde{\nu}^L = \sum_{j \notin \mathscr{F}} b_j \omega_j - \sum_{j \in \mathscr{F}} c_j \alpha_j \tag{8.113}$$

with $c_j \geq 0$ for $j \in \mathscr{F}$. Comparing (8.109) and (8.113), we see that

$$\sum_{j \in \mathscr{F}} (a'_j - c_j) \alpha_j \succeq 0,$$

hence in particular that $a'_j \geq c_j \geq 0$ for $j \in \mathscr{F}_1$. This relation forces $\mathscr{F}_1$ to be empty and completes the proof of (b).

Armed with these lemmas, we review §14 in order to adapt the proof of (b) $\Rightarrow$ (c) there to our present situation. We begin with a leading exponent $\tilde{\nu} - \rho_{\mathfrak{p}}$ of π. Although some of the argument will work for any such $\tilde{\nu}$ (as was the case in §14), we may as well assume from the outset that $(\operatorname{Re}\tilde{\nu})_0$ is maximal with respect to $\succeq$, so that we have Lemma 8.60 available. Then we construct $S = MAN$ from $\mathscr{F} = \mathscr{F}(\operatorname{Re}\tilde{\nu})$ just as in §14, and we obtain parameters for which

$$\pi \subseteq \operatorname{ind}_{S_{\mathfrak{p}}}^{G}(\sigma \otimes e^{\tilde{\nu}} \otimes 1) \cong \operatorname{ind}_{S}^{G}(\xi \otimes e^{\nu} \otimes 1)$$

just as in (8.88) and (8.91). Passing to $\xi_0 \subseteq \xi$ as in §14, we obtain

$$\pi \subseteq \operatorname{ind}_{S}^{G}(\xi_0 \otimes e^{\nu} \otimes 1)$$

as in (8.92a), and there is no change in the argument that the diagram (8.92b) commutes. As in §14, ξ_0 has an irreducible admissible quotient ω for which π imbeds in $\operatorname{ind}_{S}^{G}(\omega \otimes e^{\nu} \otimes 1)$ and R maps onto ω.

Every member of the space V^{coef} of $(K \cap M)$-finite vectors of R is $Z(\mathfrak{m}^{\mathbb{C}})$-finite, as in §14. The exponents of the members of V^{coef} are controlled by Lemma 8.60b, and the remark attached to Theorem 8.47 implies that the members of V^{coef} are dominated on $\overline{A_{\mathfrak{p}}^{+}}$ by multiples of $e^{-\rho_{\mathfrak{p}}(\log a)}(1 + \|a\|)^q$. By Theorem 8.53, ω is tempered (and hence also unitary). This completes the proof of (8.94).

Let us now give the simple duality argument that finishes off the existence part of Theorem 8.54. If f_1 and f_2 are functions in the compact pictures of $U(S, \omega, \nu)$ and $U(S, \omega, -\bar{\nu})$,respectively, and if it is assumed only that ω is unitary, then we can check that

$$\langle U(S, \omega, \nu, g)f_1, U(S, \omega, -\bar{\nu}, g)f_2\rangle_{L^2(K)} = \langle f_1, f_2\rangle_{L^2(K)}. \quad (8.114)$$

[In fact, we write out the left side as an integral over K and expand $g^{-1}k$ in terms of $G = KMAN$. Then ω and ν drop out, and the assertion reduces to the known unitarity of $U(S, 1, 0)$.]

For our parameters, we have $\operatorname{Re}\langle \nu, \beta\rangle > 0$ for all S positive $\mathfrak{a}$ roots β; hence $\operatorname{Re}\langle -\bar{\nu}, \beta\rangle < 0$ for all S positive β and $\operatorname{Re}\langle -\bar{\nu}, \beta\rangle > 0$ for all $\bar{S}$ positive β. By Theorem 7.24, $U(\bar{S}, \omega, -\bar{\nu})$ has a unique irreducible quotient. Then (8.114) (with $\bar{S}$ in place of S) shows that $U(\bar{S}, \omega, \nu)$ has a unique irreducible subrepresentation. In view of Theorem 7.24 and our construction above, $A(\bar{S}:S:\omega:\nu)U(S, \omega, \nu)$ and π are each irreducible subrepresentations of $U(\bar{S}, \omega, \nu)$. Thus they are equal, and existence is proved.

We turn to uniqueness. The proof of uniqueness rests on Lemma 7.23 and some relationships among the exponents of $U(S, \omega, \nu)$ given in the proposition below. The proposition is useful in its own right in that it allows the Langlands classification to be applied in an inductive fashion; an illustration of this principle appears in the Problems at the end of the chapter.

Proposition 8.61. Let $S_\mathfrak{p} = M_\mathfrak{p} A_\mathfrak{p} N_\mathfrak{p}$ be a minimal parabolic subgroup, let $S = MAN$ be a parabolic subgroup with $S \supseteq S_\mathfrak{p}$, let ω be an irreducible tempered representation of M, and let ν be a member of $(\mathfrak{a}')^\mathbb{C}$ with $\operatorname{Re} \nu$ in the open positive Weyl chamber. Then

(a) there exists an exponent $\tilde{\nu} - \rho_\mathfrak{p}$ of $J(S, \omega, \nu)$, hence of $U(S,.\omega, \nu)$, relative to $S_\mathfrak{p}$ such that $\tilde{\nu}|_\mathfrak{a} = \nu$.
(b) every exponent $\tilde{\lambda}$ of $U(S, \omega, \nu)$ relative to $S_\mathfrak{p}$ satisfies $\operatorname{Re} \tilde{\lambda} \preceq \operatorname{Re} \nu$. (Here $\operatorname{Re} \nu$ is extended to all of $\mathfrak{a}_\mathfrak{p}$ so as to be 0 on $\mathfrak{a}_M$.)
(c) strict inequality holds in (b) for every exponent of the subrepresentation $\ker A(\bar{S}{:}S{:}\omega{:}\nu)$ of $U(S, \omega, \nu)$.

Proof. (b) Since ω is tempered, its $(K \cap M)$-finite matrix coefficients are dominated by multiples of φ_0^M. By Proposition 7.14 the K-finite matrix coefficients of $U(S, \omega, \nu)$ are dominated by multiples of $\varphi_{\operatorname{Re}\nu}^G$. Then Proposition 7.15 shows that the K-finite matrix coefficients of $U(S, \omega, \nu)$ are dominated on $A_\mathfrak{p}^+$ by multiples of $e^{(\operatorname{Re}\nu - \rho_\mathfrak{p})(\log a)}(1 + \|a\|)^q$ for some q. Hence Theorem B.25 shows all exponents $\tilde{\lambda} - \rho_\mathfrak{p}$ satisfy $\operatorname{Re} \tilde{\lambda} \preceq \operatorname{Re} \nu$.

(a) In view of (b) and of Lemma 7.23, $\nu - \rho_A$ is an exponent of $J(S, \omega, \nu)$ relative to S. By Corollary 8.46, there exists an exponent $\tilde{\nu} - \rho_\mathfrak{p}$ of $J(S, \omega, \nu)$ relative to $S_\mathfrak{p}$ such that $\tilde{\nu}|_\mathfrak{a} = \nu$.

(c) The limit in Lemma 7.23 is always 0 when the two K-finite members f and g of the induced space are in $\ker A(\bar{S}{:}S{:}\omega{:}\nu)$. Form the asymptotic expansions of matrix coefficients of $\ker A(\bar{S}{:}S{:}\omega{:}\nu)$ relative to S. Applying Theorem B.25 with $q_0 = 0$, we see that there are no logarithms present in the terms whose exponents have real part $\operatorname{Re} \nu - \rho_A$. As a result we can move the limit in Lemma 7.23 under the sum and conclude that

$$\lim_{\substack{a \to \infty \\ S}} e^{(-\nu + \rho_A)(\log a)} \sum (\text{terms whose exponents have real part } \operatorname{Re} \nu - \rho_A)$$

is zero. Formula (B.45) implies that there are no such terms in the sum. That is, no exponents have real part $\operatorname{Re} \nu - \rho_A$.

Proof of uniqueness in Theorem 8.54. The proof consists in giving an intrinsic description of the parameters (S, ω, ν) of $J(S, \omega, \nu)$ from the asymptotics of the representation. For this purpose, fix $\tilde{\nu}$ as in Proposition 8.61a. Writing $\mathfrak{a}_\mathfrak{p} = \mathfrak{a} \oplus \mathfrak{a}_M$, we have

$$\operatorname{Re} \tilde{\nu} = \operatorname{Re} \nu - \sum_{\alpha_j \leftrightarrow M} a_j \alpha_j \tag{8.115}$$

for suitable real a_j. Since ω is tempered, $\langle \operatorname{Re} \tilde{\nu}|_{\mathfrak{a}_M}, \omega_j^M \rangle \leq 0$ for all j with $\alpha_j \leftrightarrow M$, by Theorem 8.53. Taking the inner product of ω_j^M with (8.115), we see that $a_j \geq 0$ for each j.

On the other hand, Re ν is assumed to be in the open positive Weyl chamber of $\mathfrak{a}'$, and thus

$$\operatorname{Re} \nu = \sum_{\alpha_j \not\leftrightarrow M} b_j \omega_j \tag{8.116}$$

with all $b_j > 0$. Combining (8.115) and (8.116) with the uniqueness in Lemma 8.56, we see that

$$(\operatorname{Re} \tilde{\nu})_0 = \operatorname{Re} \nu. \tag{8.117}$$

Proposition 8.61b says that every exponent $\tilde{\lambda} - \rho_{\mathfrak{p}}$ of $J(S, \omega, \nu)$ relative to $S_{\mathfrak{p}}$ satisfies Re $\tilde{\lambda} \preceq$ Re ν. Since Re ν is dominant, Lemma 8.59 gives

$$(\operatorname{Re} \tilde{\lambda})_0 \preceq (\operatorname{Re} \nu)_0 = \operatorname{Re} \nu. \tag{8.118}$$

Relations (8.117) and (8.118) together say that Re ν is the unique maximum of $(\operatorname{Re} \tilde{\lambda})_0$ for all exponents $\tilde{\lambda} - \rho_{\mathfrak{p}}$ of $J(S, \omega, \nu)$ relative to $S_{\mathfrak{p}}$. In short, Re ν has an intrinsic characterization in terms of $J(S, \omega, \nu)$.

If we expand Re ν in terms of the ω_j, then (8.116) shows that the j's needed for the expansion are exactly those for which α_j does not contribute to M. Thus Re ν determines S. Now suppose S is fixed, ν and ν' have the same real part, and $J(S, \omega, \nu)$ is infinitesimally equivalent with $J(S, \omega', \nu')$.

Let f and g be members of the quotient Hilbert space for $J(S, \omega, \nu)$. Choose any pre-image $\tilde{f}$ of f in the space for $U(S, \omega, \nu)$, and let $\tilde{g}$ be the pre-image of g orthogonal to ker $A(\bar{S}:S:\omega:\nu)$. Then

$$(f, g) = (\tilde{f}, \tilde{g})_{L^2(K)}$$

by the definition of the quotient inner product. Moreover, the orthogonality of the various K spaces implies that $\tilde{g}$ is K-finite if g is K-finite. For f and g K-finite, we form

$$\begin{aligned} &\lim_{\substack{a \to \infty \\ S}} e^{(\nu - \rho_A)(\log a)}(J(S, \omega, \nu, ma)f, g) \\ &\qquad = \lim_{\substack{a \to \infty \\ S}} e^{(\nu - \rho_A)(\log a)}(U(S, \omega, \nu, ma)\tilde{f}, \tilde{g})_{L^2(K)}. \end{aligned}$$

By Lemma 7.23 the limit on the right side exists and is

$$= (\omega(m)A(\bar{S}:S:\omega:\nu)\tilde{f}(1), \tilde{g}(1))_{V^\omega}. \tag{8.119a}$$

Let f' and g' be the K-finite members of the space for $J(S, \omega', \nu')$ that correspond to f and g under the infinitesimal equivalence, and form $\tilde{f}'$

and $\tilde{g}'$. Then we get also

$$\lim_{\substack{a\to\infty\\ S}} e^{(\nu'-\rho_A)(\log a)}(J(S,\omega,\nu,ma)f,g) = (\omega'(m)A(\bar{S}:S:\omega':\nu')\tilde{f}'(1),\tilde{g}'(1))_{V^{\omega'}}. \tag{8.119b}$$

We know this limit is not 0 identically as f and g vary. Dividing (8.119a) and (8.119b), we see that $e^{(\nu-\nu')(\log a)}$ tends to a limit. Since ν and ν' have the same real part, $\nu = \nu'$.

Finally if we put $\nu' = \nu$ in (8.119b), then we conclude that the right sides of (8.119a) and (8.119b) are equal. Thus ω and ω' have a nonzero $(K \cap M)$-finite matrix coefficient in common. By Corollary 8.12, ω and ω' are infinitesimally equivalent.

This completes the proof of uniqueness for the Langlands classification, apart from one final remark: We proved only that ω and ω' are infinitesimally equivalent; it will follow from Corollary 9.2 that ω and ω' are unitarily equivalent.

§16. Problems

1. Let π be the holomorphic discrete series representation of Chapter VI with parameter λ. Prove that the infinitesimal character of π is $\lambda + \delta$.

Problems 2 to 8 identify the irreducible tempered and irreducible admissible representations of SL(2, $\mathbb{R}$) under the assumption that the discrete series is known. Take as known that the representations $\mathcal{D}_n^+$ and $\mathcal{D}_n^-$, $n \geq 2$, exhaust the discrete series.

2. Let $\rho\begin{pmatrix}1 & 0\\ 0 & -1\end{pmatrix} = 1$. Show that the infinitesimal character of $\mathcal{P}^{\pm,w}$ is $w\rho$. Using (2.19), show that the infinitesimal character of $\mathcal{D}_n^{\pm}$ is $(n-1)\rho$.
3. Calculate the value of the Casimir operator $\Omega = \frac{1}{2}h^2 + ef + fe$ on $\mathcal{P}^{\pm,w}$ and $\mathcal{D}_n^{\pm}$ from the result of Problem 2 by using the Harish-Chandra homomorphism.
4. Using Theorem 8.53 and results of §2.5, prove that $\mathcal{D}_n^+$ for $n \geq 1$, $\mathcal{D}_n^-$ for $n \geq 1$, $\mathcal{P}^{+,iv}$ for v real, and $\mathcal{P}^{-,iv}$ for v nonzero real exhaust the irreducible tempered representations of SL(2, $\mathbb{R}$). Show that $v \to -v$ gives the only equivalences among these representations by taking into account the infinitesimal character, the value of the representation on $\begin{pmatrix}-1 & 0\\ 0 & -1\end{pmatrix}$, and the existence/nonexistence of a square-integrable matrix coefficient.

5. Adapt the proof of Corollary 8.42 to show that any irreducible subrepresentation of $\mathscr{P}^{\pm,w}$ has an irreducible (admissible) quotient.
6. Fix w with $\operatorname{Re} w > 0$. Taking into account Proposition 8.61 and the values of infinitesimal characters, prove that the only irreducible subquotients (quotients of subrepresentations) of $\mathscr{P}^{\pm,w}$ other than $J(S_{\mathfrak{p}}, \pm, w\rho)$ are suitable discrete series $\mathscr{D}_n^{\pm}$, $n \geq 2$.
7. Using the results of Problems 2, 5, and 6, prove for $\operatorname{Re} w > 0$ that $\mathscr{P}^{+,w}$ is irreducible unless w is an odd integer and that $\mathscr{P}^{-,w}$ is irreducible unless w is an even integer.
8. Using Theorem 8.54, Problem 7, and results of §2.5, prove that the following exhaust the nontempered irreducible admissible representations of $SL(2, \mathbb{R})$: $\mathscr{P}^{+,w}$ with $\operatorname{Re} w > 0$ and $w \notin 2\mathbb{Z} + 1$, $\mathscr{P}^{-,w}$ with $\operatorname{Re} w > 0$ and $w \notin 2\mathbb{Z}$, and Φ_n (as in (2.1)) with $n \geq 0$ and $n \in \mathbb{Z}$. Prove that no two of these representations are infinitesimally equivalent.

Problems 9 to 14 identify the irreducible tempered and irreducible admissible representations of $SL(2, \mathbb{C})$. Take as known that $SL(2, \mathbb{C})$ has no discrete series.

9. Use Theorem 8.53 and Proposition 2.6 to prove that the representations $\mathscr{P}^{k,iv}$ exhaust the irreducible tempered representations.
10. Identify $\mathfrak{sl}(2, \mathbb{C})^{\mathbb{C}}$ with $\mathfrak{sl}(2, \mathbb{C}) \oplus \mathfrak{sl}(2, \mathbb{C})$ by
$$H_1 + iH_2 \to (H_1 + JH_2, J(H_1 - JH_2))$$
as in Proposition 2.5. Let $\mathfrak{h}$ be the diagonal subalgebra of $\mathfrak{sl}(2, \mathbb{C})$, and let $\alpha\begin{pmatrix} t & 0 \\ 0 & -t \end{pmatrix} = 2t$ on $\mathfrak{h}$. Prove that $\Delta(\mathfrak{h}^{\mathbb{C}} : \mathfrak{g}^{\mathbb{C}})$ consists of $\pm\,(\alpha, 0)$ and $\pm(0, \alpha)$, and identify $W(\mathfrak{h}^{\mathbb{C}} : \mathfrak{g}^{\mathbb{C}})$.
11. Under the identification in Problem 10, show that the infinitesimal character of $\mathscr{P}^{k,w}$ is
$$\frac{k}{4}((\alpha, 0) - (0, \alpha)) + \frac{w}{4}((\alpha, 0) + (0, \alpha))$$
$$= \frac{1}{4}(w + k)(\alpha, 0) + \frac{1}{4}(w - k)(0, \alpha).$$
12. Observe from Problem 9 and Theorem 8.54 that the only irreducible admissible representations are $J(\mathscr{P}^{l,z})$ with $\operatorname{Re} z > 0$ and $\mathscr{P}^{l,z}$ with $\operatorname{Re} z = 0$. Proceeding as in Problem 6 and taking into account Problem 11, prove that $\mathscr{P}^{k,w}$ for $\operatorname{Re} w > 0$ can have a nontrivial such irreducible subquotient $J(\mathscr{P}^{l,z})$ or $\mathscr{P}^{l,z}$ only if $(z, l) = \pm(k, w)$.

13. Taking into account Problem 12 and a suitable analog of Problem 5, prove that $\mathscr{P}^{k,w}$ for $\operatorname{Re} w > 0$ is irreducible unless w is an integer and $|k| < w$. By looking at the value of the inducing representation on $\begin{pmatrix} -1 & 0 \\ 0 & -1 \end{pmatrix}$, prove that a further necessary condition for reducibility is that $k \equiv w \bmod 2$.

14. Using (2.14) and Problem 13, prove that the following exhaust the nontempered irreducible admissible representations of $SL(2, \mathbb{C})$: $\Phi_{m,n}$ for nonnegative integral m and n, and $\mathscr{P}^{k,w}$ for $\operatorname{Re} w > 0$ when w is not an integer with $|k| < w$ and $k \equiv w \bmod 2$. Prove that no two of these representations are infinitesimally equivalent.

Problems 15 to 19 apply the same circle of ideas to some other groups. Assume that G has real rank one, that $\mathfrak{b}_\mathfrak{p}$ is a maximal abelian subspace of $\mathfrak{m}_\mathfrak{p}$, that there is some root α in $\Delta((\mathfrak{a}_\mathfrak{p} \oplus \mathfrak{b}_\mathfrak{p})^\mathbb{C} : \mathfrak{g}^\mathbb{C})$ vanishing on $\mathfrak{b}_\mathfrak{p}$, and that every discrete series of G has algebraically integral infinitesimal character. Take for granted that the infinitesimal character λ of any $U(S_\mathfrak{p}, \sigma, 0)$ has $2\langle \lambda, \beta \rangle / |\beta|^2$ in $\frac{1}{2}\mathbb{Z}$ for every root β, i.e., λ is integral or half-integral.

15. Prove that the discrete series and the irreducible constituents of the unitary principal series exhaust the irreducible tempered representations.

16. Prove that if $U(S_\mathfrak{p}, \sigma, z\rho_\mathfrak{p})$ and $U(S_\mathfrak{p}, \sigma', z'\rho_\mathfrak{p})$ match on Z_G and have the same infinitesimal character with $\operatorname{Re} z > 0$ and $\operatorname{Im} z \neq 0$ and $\operatorname{Re} z' \geq 0$, then $z = z'$ and $\sigma \cong \sigma'$. [Hint: Divide into two cases, depending on whether $M_\mathfrak{p}$ is connected. Handle the disconnected case by using Problem 22 of Chapter V. When $M_\mathfrak{p}$ is connected, write each infinitesimal character as the sum of one that is real on $\mathfrak{a}_\mathfrak{p} \oplus i\mathfrak{b}_\mathfrak{p}$ and one that is imaginary there. Take a Weyl group element carrying one infinitesimal character into the other, and consider its effect on the imaginary part.]

17. Prove from Theorem 8.54 and Problems 15 and 16 that $U(S_\mathfrak{p}, \sigma, z\rho_\mathfrak{p})$ is irrreducible if $\operatorname{Re} z > 0$ and $\operatorname{Im} z \neq 0$.

18. Regard α as a positive multiple of $\rho_\mathfrak{p}$. Suppose $U(S_\mathfrak{p}, \sigma, c\alpha)$ is reducible with $c > 0$. Proceeding as in Problems 5 and 6, show that its infinitesimal character $\lambda + c\alpha$ (with λ integral or half-integral) must match some $\lambda' + d\alpha$ with λ' integral or half-integral and with $0 \leq d < c$.

19. Forming an equality $w(\lambda' + d\alpha) = \lambda + c\alpha$ from Problem 18 and taking the inner product with α and $w\alpha$, prove that c and d are in $\frac{1}{6}\mathbb{Z}$. Conclude that $U(S_\mathfrak{p}, \sigma, z\rho_\mathfrak{p})$ for $\operatorname{Re} z > 0$ is irreducible except for z

in a discrete subset of $\mathbb{R}$. [Note: This estimate can be sharpened with additional work. The only reducibility occurs at points where the infinitesimal character is integral.]

Problems 20 to 25 apply the theory of asymptotic expansions to spherical functions. The spherical function φ_ν^G is the τ-spherical function obtained with τ trivial from the representation $U(S_\mathfrak{p}, 1, \nu)$.

20. Going over the proof of Theorem 8.33, prove that the leading exponents $\nu' - \rho_\mathfrak{p}$ of φ_ν^G have $\nu' + \delta_M = w(\nu + \delta_M)$ for some w in $W((\mathfrak{a}_\mathfrak{p} \oplus \mathfrak{b}_\mathfrak{p})^\mathbb{C}:\mathfrak{g}^\mathbb{C})$. Conclude that $\nu' = s\nu$ for some s in $W(A_\mathfrak{p}:G)$, at least for ν imaginary.
21. Call ν nonsingular if $\langle \nu, \beta \rangle \neq 0$ for every restricted root β. Prove for ν nonsingular imaginary that the coefficient $c_{\nu',q}$ is 0 if $\nu' - \rho_\mathfrak{p}$ is a leading exponent and $q > 0$.
22. Define $\mathbf{c}(w, \nu)$ for ν nonsingular imaginary to be the coefficient of $\exp(w\nu - \rho_\mathfrak{p})$ in the expansion of φ_ν^G. Prove that $\mathbf{c}(w, \nu) = \mathbf{c}(1, w\nu)$ from the relation $\varphi_\nu^G = \varphi_{w\nu}^G$. [Remark: Then the leading terms of φ_ν^G are expressed in terms of $\mathbf{c}(\nu) = \mathbf{c}(1, \nu)$, which is called the **Harish-Chandra c function.**]
23. Prove that $\mathbf{c}(\nu)$ is real analytic for ν nonsingular imaginary and that its holomorphic extension yields a valid asymptotic expansion on a neighborhood in $(\mathfrak{a}_\mathfrak{p}')^\mathbb{C}$ of the nonsingular imaginary set.
24. Using Lemma 7.23, give an integral formula for $\mathbf{c}(\nu)$ valid in the subset of the set in Problem 23 where Re ν is in the positive Weyl chamber.
25. Using facts about intertwining operators, prove that $\mathbf{c}(\nu)$ extends to a meromorphic function on $(\mathfrak{a}_\mathfrak{p}')^\mathbb{C}$.

Problems 26 to 27 illustrate some unexpected failure of full reducibility in certain contexts with infinite-dimensional representations. Let G be SU(1, 1).

26. Let χ_ν be the eigenvalue of the Casimir operator Ω on the spherical function φ_ν^G. Fix $\nu_0 \neq 0$ and define
$$\varphi(x) = \frac{d}{d\nu}\,\varphi_\nu^G(x)\bigg|_{\nu=\nu_0}$$
Prove that $(\Omega - \chi_{\nu_0})^2$ annihilates φ but $\Omega - \chi_{\nu_0}$ does not annihilate φ.
27. Let $\{h, e, f\}$ be the usual basis of $\mathfrak{sl}(2, \mathbb{C})$, and regard $\mathcal{D}_1^+$ and $\mathcal{D}_1^-$ as representations of SU(1, 1). Let V^+ and V^- be the respective spaces of K-finite vectors, and let $V = V^+ \oplus V^-$. Let $v_1^+, v_3^+, \ldots$ be an

orthonormal basis of weight vectors of V^+ under h, and define $v_1^-, v_3^-, \ldots$ similarly. For X in $\mathfrak{sl}(2, \mathbb{C})$, define

$$\mathscr{D}(X)v_j^- = \mathscr{D}_1^-(X)v_j^-$$

$$\mathscr{D}(h)v_j^+ = \mathscr{D}_1^+(h)v_j^+$$

$$\mathscr{D}(e)v_j^+ = \mathscr{D}_1^+(e)v_j^+$$

$$\mathscr{D}(f)v_j^+ = \begin{cases} \mathscr{D}_1^+(f)v_j^+ & \text{if } j > 1 \\ cv_1^- & \text{if } j = 1. \end{cases}$$

Show that $\mathscr{D}$ defines a representation of $\mathfrak{su}(1, 1)$ for any choice of c. Show that $\mathscr{D}_1^-$ is a subrepresentation and that the quotient is $\mathscr{D}_1^+$. Show that $\mathscr{D}_1^+$ is a subrepresentation only for $c = 0$.

CHAPTER IX

Construction of Discrete Series

§1. Infinitesimally Unitary Representations

The Langlands classification and the results about irreducible tempered representations in Chapter VIII make it clear that an understanding of discrete series is an important step in understanding all irreducible admissible representations. By way of preparation for a general discussion of discrete series, we obtain in these first two sections some useful results about admissible representations of linear connected reductive groups.

An admissible representation π on a Hilbert space V is said to be **infinitesimally unitary** if its space V_0 of K-finite vectors admits an inner product with respect to which $\pi(\mathfrak{g})$ acts by skew-Hermitian operators. If π is unitary, then π is infinitesimally unitary, by Proposition 3.10. Two inequivalent irreducible unitary representations cannot lead to infinitesimally equivalent unitary representations, by Corollary 9.2 below.

Proposition 9.1. Let π be an irreducible admissible representation on a Hilbert space V, and let V_0 be the space of K-finite vectors. Up to a constant, there exists at most one sesquilinear form $\langle\cdot,\cdot\rangle$ on $V_0 \times V_0$ such that $\langle\pi(X)u, v\rangle = -\langle u, \pi(X)v\rangle$ for all X in $\mathfrak{g}$ and all u and v in V_0.

Proof. The idea is to set up matters so as to apply Schur's Lemma as stated in Corollary 8.12. Thus let $\langle\cdot,\cdot\rangle_1$ and $\langle\cdot,\cdot\rangle_2$ be two such forms, with $\langle\cdot,\cdot\rangle_1 \not\equiv 0$, say. The subspaces $\{u \mid \langle u,\cdot\rangle_1 = 0\}$ and $\{v \mid \langle\cdot, v\rangle_1 = 0\}$ of V_0 are both 0 since π is irreducible, and thus $\langle\cdot,\cdot\rangle_1$ is nondegenerate in both variables. One checks readily that $\pi(K)$ acts by unitary transformations for both $\langle\cdot,\cdot\rangle_1$ and $\langle\cdot,\cdot\rangle_2$ and that distinct K types lead to orthogonal subspaces in each. Then it follows from finite-dimensional linear algebra that there exists a linear mapping $B\colon V_0 \to V_0$ preserving each K type such that

$$\langle u, v\rangle_2 = \langle Bu, v\rangle_1$$

for all u and v in V_0. With B^* defined K space by K space relative to $\langle\cdot,\cdot\rangle_1$, we then have

$$\begin{aligned} -\langle u, \pi(X)B^*v\rangle_1 &= \langle \pi(X)u, B^*v\rangle_1 = \langle B\pi(X)u, v\rangle_1 \\ &= \langle \pi(X)u, v\rangle_2 = -\langle u, \pi(X)v\rangle_2 \\ &= -\langle Bu, \pi(X)v\rangle_1 = -\langle u, B^*\pi(X)v\rangle_1 \end{aligned}$$

for all X in $\mathfrak{g}$. By nondegeneracy B^* commutes with $\pi(\mathfrak{g})$, and then Corollary 8.13 says B^* is scalar. Hence B is scalar.

Corollary 9.2. Any two irreducible unitary representations that are infinitesimally equivalent are unitarily equivalent.

The correspondence between irreducible unitary representations and infinitesimally unitary irreducible admissible representations is completed by the following theorem.

Theorem 9.3. If π is an infinitesimally unitary irreducible admissible representation, then there exists an irreducible unitary representation $\tilde{\pi}$ that is infinitesimally equivalent with π.

Proof omitted.

§2. A Third Way of Treating Admissible Representations

We have treated admissible representations globally by using G, and we have treated them infinitesimally by using $U(\mathfrak{g}^{\mathbb{C}})$ on the K-finite vectors. A third way of proceeding combines some of the advantages of each approach and is useful for several purposes. Also it is the approach that has been useful in studying representations of groups defined over non-archimedean fields.

Let

$$C_K = \{f \in C^{\infty}_{\text{com}}(G) \mid f \text{ is right and left } K\text{-finite}\}. \tag{9.1}$$

The condition means that the two-sided translates of f by K span a finite-dimensional space. Then C_K is an algebra under convolution.

Lemma 9.4. C_K is stable under the left-invariant differential operators on G and also under the right-invariant differential operators.

Proof. Let X be in $\mathfrak{g}$ and regard X as a left-invariant differential operator. If f is in C_K, then

$$Xf(x) = \frac{d}{dt} f(x \exp tX)\bigg|_{t=0}$$

makes it clear that Xf is left K-finite. Let the right K translates of f lie in the span of $f_1, \ldots, f_n$. Then

$$Xf(xk) = \frac{d}{dt} f(x(\exp t\,\mathrm{Ad}(k)X)k)\Big|_{t=0} \in \operatorname{span}_j \frac{d}{dt} f_j(x \exp t\,\mathrm{Ad}(k)X)\Big|_{t=0}$$
$$\subseteq \operatorname{span}_j \{\mathfrak{g} f_j\},$$

and Xf is right K-finite. The conclusions about right-invariant differential operators are proved similarly.

Since $C_K \subseteq C_{\mathrm{com}}^{\infty}(G)$, we already have a definition of $\pi(f)$ for f in C_K and π a representation. (See Chapter I.)

Proposition 9.5. If π is a representation on V, then $\pi(C_K)V$ consists of K-finite vectors. If π is admissible and v_0 is K-finite, then

$$\pi(C_K)v_0 = \pi(U(\mathfrak{g}^{\mathbb{C}}))v_0.$$

Remark. The proposition gives a sense in which C_K can be used as a substitute for $U(\mathfrak{g}^{\mathbb{C}})$. It is sometimes helpful to use C_K in place of $U(\mathfrak{g}^{\mathbb{C}})$ when differentiation leads to difficult analytic problems and integration does not.

Proof. The first statement is trivial. For the second statement, let V_0 denote the space of K-finite vectors. Then Corollary 8.10 gives

$$\pi(C_K)v_0 \subseteq V_0 \cap \overline{\operatorname{span} \pi(G)v_0} = \pi(U(\mathfrak{g}^{\mathbb{C}}))v_0.$$

In the reverse direction let v_0 lie in $\sum_{\tau \in \mathscr{S}} n_\tau \tau$, where $\mathscr{S}$ is a finite set of K types, and let e be the function on K given by $e = \sum_{\tau \in \mathscr{S}} d_\tau \bar{\chi}_\tau$. Then

$$E = \pi(e\,dk) = \int_K e(k)\pi(k)\,dk$$

is the orthogonal projection of V on $\sum_{\tau \in \mathscr{S}} n_\tau \tau$. For f in C_K let us write $e *_K f$ for $(e\,dk) * f$, i.e.,

$$e *_K f(x) = \int_K e(k) f(k^{-1}x)\,dk. \tag{9.2}$$

Then $e *_K f$ is in C_K and

$$\pi(e *_K f) = E\pi(f). \tag{9.3}$$

Now $E\pi(C_K)v_0$ is a linear subspace of $\sum_{\tau \in \mathscr{S}} n_\tau \tau$, and an argument with an approximate identity shows v_0 is in its closure. (See the proof of Theorem 3.15.) Since $\sum_{\tau \in \mathscr{S}} n_\tau \tau$ is finite-dimensional, we conclude that there exists f_0 in C_K with $E\pi(f)v_0 = v_0$. Then $\pi(e *_K f_0)v_0 = v_0$ by (9.3). If X is in $\mathfrak{g}$, then $\pi(X)\pi(f_0)$ is π of a right-invariant derivative of f_0. Writing $U(\mathfrak{g}^{\mathbb{C}})_R$ for the action of $U(\mathfrak{g}^{\mathbb{C}})$ by right-invariant derivatives and taking into account Lemma 9.4, we see that

$$\pi(U(\mathfrak{g}^{\mathbb{C}}))v_0 = \pi(U(\mathfrak{g}^{\mathbb{C}}))\pi(e *_K f_0)v_0 \subseteq \pi(U(\mathfrak{g}^{\mathbb{C}})_R(e *_K f_0))v_0 \subseteq \pi(C_K)v_0.$$

This completes the proof.

§3. Equivalent Definitions of Discrete Series

The proposition below has been alluded to in §2.5 and used in §6.4 and §8.14.

Proposition 9.6. For an irreducible unitary representation π on V, the following three conditions are equivalent:

(i) Some nonzero K-finite matrix coefficient is in $L^2(G)$.
(ii) All matrix coefficients are in $L^2(G)$.
(iii) π is equivalent with a direct summand of the right regular representation R of G on $L^2(G)$.

When these conditions are satisfied, there exists a positive number d_π such that

$$\int_G (\pi(x)u_1, v_1)\overline{(\pi(x)u_2, v_2)}\, dx = d_\pi^{-1}(u_1, u_2)\overline{(v_1, v_2)} \tag{9.4}$$

for all u_1, u_2, v_1, and v_2 in V.

Terminology. When these conditions are satisfied, we say π is in the **discrete series** of G, and we call d_π the **formal degree** of π.

Remarks.

(1) For G compact, every irreducible representation is in the discrete series. If Haar measure has total mass 1, then the formal degree is the degree, by Schur orthogonality.

(2) For $G = SL(2, \mathbb{R})$ and in the case of holomorphic discrete series, our terminology here is consistent with terminology used earlier. The terminology is consistent also with that in §8.14.

(3) We give a proof that shows how to use the algebra C_K, even though a shorter proof is possible by using some functional analysis.

Proof. Let V_0 be the set of K-finite vectors in V, and let us introduce the condition

(ii′) All K-finite matrix coefficients are in $L^2(G)$.

It is clear that (ii) $\Rightarrow$ (ii′) $\Rightarrow$ (i).

Let us prove that (i) $\Rightarrow$ (ii′). Thus let $(\pi(x)u_0, v_0)$ be a nonzero K-finite matrix coefficient in $L^2(G)$. For f in C_K, let $f^*(y) = \overline{f(y^{-1})}$. If f and h are in C_K and if f^* and h have support in a compact set E, then we have

$$\begin{aligned}
&\int_G |(\pi(f^*)\pi(x)\pi(h)u_0, v_0)|^2\,dx \\
&\quad = \int_G \left|\int_{E\times E} f^*(y)h(y')(\pi(yxy')u_0, v_0)\,dy\,dy'\right|^2 dx \\
&\quad \leq \int_G \left[\int_{E\times E} |f^*(y)h(y')|^2\,dy\,dy'\right]\left[\int_{E\times E} |(\pi(yxy')u_0, v_0)|^2\,dy\,dy'\right] dx \\
&\qquad\qquad\qquad\qquad\qquad\qquad \text{by the Schwarz inequality} \\
&\quad = \|f\|_2^2\,\|h\|_2^2 \int_{G\times E\times E} |(\pi(yxy')u_0, v_0)|^2\,dy\,dy'\,dx \\
&\quad = \|f\|_2^2\,\|h\|_2^2 |E|^2 \int_G |(\pi(x)u_0, v_0)|^2\,dx \qquad \text{by } yxy' \to x \\
&\quad < \infty.
\end{aligned}$$

Thus the function

$$(\pi(f^*)\pi(x)\pi(h)u_0, v_0) = (\pi(x)\pi(h)u_0, \pi(f)v_0)$$

is in $L^2(G)$. By Proposition 9.5 and the irreducibility of π,

$$\{\pi(h)u_0 \,|\, h \in C_K\} = \{\pi(f)v_0 \,|\, f \in C_K\} = V_0,$$

and thus (ii′) follows.

Next we prove that (ii′) implies (ii), (iii), and (9.4). Fix $v_0 \neq 0$ in V_0, and define $B: V_0 \to L^2(G)$ by

$$B(u) = (\pi(\cdot)u, v_0). \tag{9.5}$$

For X in $\mathfrak{g}$, (8.10) implies

$$B(\pi(X)u) = (\pi(\cdot)\pi(X)u, v_0) = R(X)(\pi(\cdot)u, v_0) = R(X)B(u).$$

Hence B is an infinitesimal equivalence between V_0 and a subspace S_0 of the right K-finite members of $L^2(G)$. The members of S_0 are real analytic on G, just as in Theorem 8.7, and the closure S of S_0 is G-invariant and irreducible, just as in Theorem 8.9 and its corollaries. Hence B is an infinitesimal equivalence between two irreducible unitary representations and must be a scalar multiple of a unitary operator, by Corollary 9.2 and Schur's Lemma (Proposition 1.5).

Since B is a multiple of a unitary operator,

$$\int_G |(\pi(x)u, v_0)|^2\,dx = c_{v_0}\|u\|^2 \tag{9.6}$$

for all u in V and v_0 in V_0. In (9.6) if we move $\pi(x)$ over to v_0 as $\pi(x^{-1})$ and then change notation, we obtain

$$\int_G |(\pi(x)u_0, v_0)|^2\,dx = c_{u_0}\|v_0\|^2 \tag{9.7}$$

for all u_0 and v_0 in V_0. Comparing (9.6) and (9.7), we see that

$$\int_G |(\pi(x)u_0, v_0)|^2 \, dx = c\|u_0\|^2\|v_0\|^2 \tag{9.8}$$

for all u_0 and v_0 in V_0. Passing to the limit with Fatou's Lemma, we obtain

$$\int_G |(\pi(x)u, v)|^2 \, dx \leq c\|u\|^2\|v\|^2 \tag{9.9}$$

for all u and v in V. This proves (ii).

According to (9.8), the linear map $u \to (\pi(x)u, v_0)$ of V into $L^2(G)$ is isometric on V_0, and (9.9) says it is bounded everywhere. Hence it is isometric everywhere. Repeating this argument with the linear map $v \to (v, \pi(x)u)$, we see that equality holds in (9.9). Then polarization completes the proof of (9.4).

Finally we prove that (iii) implies (i). Let E be the orthogonal projection of $L^2(G)$ on a closed subspace invariant under R. Then $R(x)E = ER(x)$ for all x in G implies

$$R(f)E = ER(f) \qquad \text{for all } f \text{ in } L^1(G). \tag{9.10}$$

Let us observe that f in L^1 and h in L^2 implies

$$\begin{aligned} R(f)h(x) &= \int_G f(y)R(y)h(x)\, dy = \int_G f(y)h(xy)\, dy \\ &= \int_G h(xy^{-1})f(y^{-1})\, dy = h * \overline{f^*}(x). \end{aligned} \tag{9.11}$$

Choose h in $C_{\text{com}}(G)$ so that $Eh \neq 0$. We may take h to be right K-finite, and then Eh will be right K-finite. If f is in $C_{\text{com}}(G)$, then (9.10) and (9.11) give

$$Eh * f = R(\overline{f^*})Eh = ER(\overline{f^*})h = E(h * f).$$

Hence

$$\|Eh * f\|_2 \leq \|h * f\|_2 \leq \|h\|_1\|f\|_2.$$

Left convolution by $(Eh)^*$ is the adjoint on $C_{\text{com}}(G)$ of left convolution by Eh, and thus

$$\|(Eh)^* * f\|_2 \leq \|h\|_1\|f\|_2.$$

For general f in L^2, choose f_n in $C_{\text{com}}(G)$ with $f_n \to f$. Then $(Eh)^* * f_n$ tends to some element in L^2 and it tends to $(Eh)^* * f$ in L^∞. Hence $(Eh)^* * f$ is in L^2 for any f in L^2.

We apply this conclusion to $f = Eh$. Since f_1 and f_2 in L^2 implies

$$(R(x)f_1, f_2) = \int_G f_1(yx)\overline{f_2(y)}\, dy = \int_G \overline{f_2(y^{-1})}f_1(y^{-1}x)\, dy = f_2^* * f_1(x),$$

we have

$$(R(x)Eh, Eh) = Eh^* * Eh(x).$$

The previous paragraph shows this is in L^2, and we have arranged that Eh is K-finite and nonzero in the given invariant subspace. This proves (i).

§4. Motivation in General and the Construction in SU(1,1)

We turn now to the main business of this chapter, the actual construction of discrete series. The approach we shall use is completely different from all the ones used earlier. In this section we shall indicate how one might be led to this approach, and we shall carry out the construction for SU(1, 1).

There are three important underlying ideas. The first comes from the theory of compact groups, particularly the Peter-Weyl Theorem. For an irreducible unitary matrix representation π of a compact group G, each row of matrix coefficients of π gives an irreducible subspace of the right regular representation of $L^2(G)$ of type π, and these d_π subspaces are orthogonal (Corollary 1.10) and together exhaust the subspace of $L^2(G)$ of members transforming under the right regular representation according to π. Abstractly these spaces are indistinguishable, and their noncanonical nature contributes to the difficulty in realizing π concretely.

Matters become canonical if we bring in the left regular representation also, which mixes around these spaces. In fact, each column of matrix coefficients of π gives an irreducible subspace of the left regular representation of $L^2(G)$ of type $\bar{\pi}$ (the complex conjugate representation), and the two-sided action on the d_π^2-dimensional space of matrix coefficients is irreducible. The two-sided action is actually an action of $G \times G$: the group $G \times G$ has a representation in $L^2(G)$ in which the first factor acts by the left regular representation L of G and the second factor acts by the right regular representation R. The group $G \times G$ acts irreducibly on our d_π^2-dimensional space under the representation $\bar{\pi} \otimes \pi$. Moreover, $\bar{\pi} \otimes \pi$ occurs with multiplicity one in $L^2(G)$, and so the space is canonical.

We can view matters in terms of group actions as follows: $G \times G$ acts transitively on G by

$$(x, y)g = xgy^{-1}$$

with isotropy subgroup diag $G = \{(x, x) | x \in G\}$ at $g = 1$. So

$$G \cong (G \times G)/\text{diag } G.$$

Haar measure on G corresponds to the normalized invariant measure on $(G \times G)/\text{diag } G$. We form the left regular representation $\mathscr{L}$ of $G \times G$ on G corresponding to our group action:

$$\mathscr{L}(x, y)f(g) = f((x, y)^{-1}g) = f(x^{-1}gy). \tag{9.12}$$

What we are looking for is the invariant subspace of type $\bar{\pi} \otimes \pi$: we know it has multiplicity one and so is canonical.

The first important underlying idea is to adopt a similar philosophy for noncompact G when π is in the discrete series. The justification for such a philosophy comes from Proposition 9.6. Specifically let π be in the discrete series of G. The proposition gives us a concrete way of realizing π as a subrepresentation of the right regular representation R of G on $L^2(G)$, as follows. Let $\{\phi_i\}$ be an orthonormal basis of the space on which π acts, put $\pi(x)_{ij} = (\pi(x)\phi_j, \phi_i)$, and let d_π be the formal degree of π.

First fix i. Then the functions $d_\pi^{1/2}\pi(x)_{ij}$, as j varies, are orthonormal in $L^2(G)$, by Proposition 9.6. Using the definitions, we readily check rigorously that

$$R(g)(\pi(x)_{ij}) = \sum_k \pi(x)_{ik}\pi(g)_{kj} \tag{9.13}$$

with the series absolutely convergent. Hence the closed span of our orthonormal set is invariant under R. Define

$$R(g)_{i'j'} = (R(g)d_\pi^{1/2}\pi(x)_{ij'}, d_\pi^{1/2}\pi(x)_{ii'})_{L^2(G)}.$$

Substituting from (9.13), we find that $R(g)_{i'j'} = \pi(g)_{i'j'}$. That is, the linear extension of the map $\phi_j \to d_\pi^{1/2}\pi(x)_{ij}$ is a unitary equivalence.

Now consider all $d_\pi^{1/2}\pi(x)_{ij}$, with both i and j varying. The proposition says these are orthonormal, and a similar computation to the one above shows that closed linear span is left invariant, as well as right invariant. We regard this closed linear span in $L^2(G)$ as a representation space of $G \times G$ under $\mathscr{L}$ defined as in (9.12), and we want to see that $\mathscr{L}$ is of type $\bar{\pi} \otimes \pi$. To do so, we compute the "matrix" of $\mathscr{L}(x, y)$ relative to the orthonormal basis $d_\pi^{1/2}\pi(x)_{ij}$. Since

$$\mathscr{L}(x, y)(\pi(g)_{ij}) = \pi(x^{-1}gy)_{ij} = \sum_{k,l} \overline{\pi(x)_{ki}}\pi(y)_{lj}\pi(g)_{kl},$$

we obtain

$$\mathscr{L}(x, y)_{(kl)(ij)} = \bar{\pi}(x)_{ki}\pi(y)_{lj} = (\bar{\pi}(x) \otimes \pi(y))_{(kl)(ij)} = (\bar{\pi} \otimes \pi)(x, y)_{(kl)(ij)}.$$

Thus $\bar{\pi} \otimes \pi$ occurs in $L^2((G \times G)/\text{diag } G)$. Moreover, it has multiplicity one. In fact, within any subspace where it occurs we can find an orthonormal basis $\{f_{ij}\}$ such that

$$f_{ij}(x^{-1}gy) = (\bar{\pi} \otimes \pi)(x, y)(f_{ij}(g)) = \sum_{k,l} \bar{\pi}(x)_{ki}\pi(y)_{lj}f_{kl}(g)$$

for all x, g, y. Putting $g = x = 1$, we obtain

$$f_{ij}(y) = \sum_l f_{il}(1)\pi(y)_{lj}.$$

So each f_{ij} is in the space we have already dealt with, and the multiplicity is one. Thus the irreducible subspace of type $\bar{\pi} \otimes \pi$ is canonical.

The functions in the above subspaces are generated by the matrix coefficients of π. In the case of a K-finite matrix coefficient, we are led by our isomorphism to a left $(K \times K)$-finite function on $(G \times G)/\text{diag } G$. As a function on G, this matrix coefficient is an eigenfunction of $Z(\mathfrak{g}^{\mathbb{C}})$. Let us show that the isomorphism $G \cong (G \times G)/\text{diag } G$ carries $Z(\mathfrak{g}^{\mathbb{C}})$, regarded as an algebra of differential operators, isomorphically onto the algebra $D((G \times G)/\text{diag } G)$ of (linear) left $(G \times G)$-invariant differential operators on $(G \times G)/\text{diag } G$. The correspondence $G \leftrightarrow (G \times G)/\text{diag } G$ leads to a correspondence of functions $f \leftrightarrow f^*$ given by

$$f(g) = f^*((g, 1) \text{ diag } G).$$

We readily check that our group actions on functions are related by

$$\begin{aligned} (L(x)f)^* &= \mathscr{L}(x, 1)f^* \\ (R(y)f)^* &= \mathscr{L}(1, y)f^*. \end{aligned} \tag{9.14}$$

Moreover, the correspondence $D \leftrightarrow D^*$ given by

$$Df(g) = D^* f^*((g, 1) \text{ diag } G)$$

makes differential operators correspond to differential operators. Now (9.14) gives

$$\begin{aligned} L(x)Df(g) &= Df(x^{-1}g) = D^*f^*((x^{-1}g, 1) \text{ diag } G) \\ &= \mathscr{L}(x, 1)D^*f^*((g, 1) \text{ diag } G) \\ D^*(\mathscr{L}(x, 1)f^*)((g, 1) \text{ diag } G) &= DL(x)f(g) \\ R(y)Df(g) &= Df(gy) = D^*f^*((gy, 1) \text{ diag } G) \\ &= D^*f^*((g, y^{-1}) \text{ diag } G) \\ &= \mathscr{L}(1, y)D^*f^*((g, 1) \text{ diag } G) \\ D^*(\mathscr{L}(1, y)f^*)((g, 1) \text{ diag } G) &= DR(y)f(g). \end{aligned}$$

These four relations, taken in pairs, show that D^* is left $(G \times G)$-invariant if and only if D is left and right G-invariant. By Proposition 3.8, we conclude that $G \cong (G \times G)/\text{diag } G$ leads to an isomorphism

$$Z(\mathfrak{g}^{\mathbb{C}}) \cong D((G \times G)/\text{diag } G). \tag{9.15}$$

In summary, we now know that we can look for a discrete series π by looking for left $(K \times K)$-finite functions on $(G \times G)/\text{diag } G$ that are eigenfunctions of all left $(G \times G)$-invariant differential operators on $(G \times G)/\text{diag } G$.

The second important underlying idea is more subtle: It is that we can solve the above problem by first solving a dual problem. Let us describe this duality for $G = SU(1, 1)$. For simplicity we write $G^* = G \times G$ and $K^* = K \times K$. It will be vital to distinguish among isomorphic copies of G

that occur; therefore let us write everything as block matrices:

$$G^* = \left\{ \begin{pmatrix} x & 0 \\ 0 & y \end{pmatrix} \middle| x, y \in G \right\}$$

$$K^* = \left\{ \begin{pmatrix} k & 0 \\ 0 & l \end{pmatrix} \middle| k, l \in K \right\}$$

$$\operatorname{diag} G = \left\{ \begin{pmatrix} x & 0 \\ 0 & x \end{pmatrix} \middle| x \in G \right\}.$$

In this notation we are looking for a left K^*-finite function on $G^*/\operatorname{diag} G$ that is an eigenfunction of $D(G^*/\operatorname{diag} G)$. We may interpret $\operatorname{diag} G$ as the set of points fixed under the involution $\imath$ of G^* given by

$$\imath \begin{pmatrix} x & 0 \\ 0 & y \end{pmatrix} = \begin{pmatrix} y & 0 \\ 0 & x \end{pmatrix}.$$

The idea of the duality is to replace the triple $(K^*, G^*, \operatorname{diag} G)$ by another triple in such a way that $\imath$ gets replaced by the Cartan involution Θ.

The replacement is done in a fashion analogous to what is done in §5.1 for the unitary trick. To carry it out, we work with the Lie algebras. Let θ be the Cartan involution of $\mathfrak{g}$ or of $\mathfrak{g}^* = \begin{pmatrix} \mathfrak{g} & \\ & \mathfrak{g} \end{pmatrix}$. Then $\theta\imath$ is another involution of $\mathfrak{g}^*$:

$$\theta\imath \begin{pmatrix} X & \\ & Y \end{pmatrix} = \theta \begin{pmatrix} Y & \\ & X \end{pmatrix} = \begin{pmatrix} \theta Y & \\ & \theta X \end{pmatrix}.$$

So $\mathfrak{g}^*$ is the sum of the $+1$ eigenspace and the -1 eigenspace. For the dual, we use the sum of the $+1$ eigenspace and i times the -1 eigenspace. The eigenspaces of $\theta\imath$ in $\mathfrak{g}^*$ are

$$+1: \left\{ \begin{pmatrix} X & \\ & \theta X \end{pmatrix} \middle| X \in \mathfrak{g} \right\}$$

$$-1: \left\{ \begin{pmatrix} X & \\ & -\theta X \end{pmatrix} \middle| X \in \mathfrak{g} \right\}.$$

Hence the dual of $\mathfrak{g}^*$ is $\left\{ \begin{pmatrix} X & \\ & \theta X \end{pmatrix} + i \begin{pmatrix} Y & \\ & -\theta Y \end{pmatrix} \right\}$. If as usual we extend θ to all of $\mathfrak{sl}(2, \mathbb{C})$ so as to be conjugate linear, then the dual of $\mathfrak{g}^*$ is

$$\mathfrak{g}_{\mathbb{C}} = \left\{ \begin{pmatrix} Z & \\ & \theta Z \end{pmatrix} \middle| Z \in \mathfrak{sl}(2, \mathbb{C}) = \mathfrak{g}^{\mathbb{C}} \right\}.$$

Carrying out the same computation for $\mathfrak{k}^*$ and $\operatorname{diag} \mathfrak{g}$, we find that the dual of $\mathfrak{k}^*$ is

$$\mathfrak{k}_{\mathbb{C}} = \left\{ \begin{pmatrix} X+iY & \\ & X-iY \end{pmatrix} \middle| X, Y \in \mathfrak{k} \right\} = \left\{ \begin{pmatrix} Z & \\ & \theta Z \end{pmatrix} Z \in \mathfrak{k}^{\mathbb{C}} \right\}$$

and the dual of diag $\mathfrak{g}$ is

$$\mathfrak{u} = \left\{ \begin{pmatrix} X & \\ & X \end{pmatrix} \middle| X \in \mathfrak{su}(2) = \mathfrak{k} \oplus i\mathfrak{p} \right\}.$$

Passing to groups with obvious notation, we thus take the dual of the triple $(K^*, G^*, \operatorname{diag} G)$ to be $(K_{\mathbb{C}}, G_{\mathbb{C}}, U)$, with realizations of the Lie algebras as above. In the dual triple, U is the set of fixed points of the Cartan involution of $G_{\mathbb{C}}$.

One point about the duality is that the K^*-finite function we are seeking on $G^*/\operatorname{diag} G$ can be sought as a $K_{\mathbb{C}}$-finite function on $G_{\mathbb{C}}/U$. The key to this correspondence is Proposition 9.7 below. It uses antidiag x to mean $\begin{pmatrix} x & 0 \\ 0 & x^{-1} \end{pmatrix}$.

Proposition 9.7. For $G = \mathrm{SU}(1, 1)$ the decompositions

$$G^* = (K \times 1)(\text{antidiag} \exp \mathfrak{p}(\mathfrak{g}))(\operatorname{diag} G)$$

$$G_{\mathbb{C}} = (\exp \mathfrak{p}(\mathfrak{k}_{\mathbb{C}}))(\text{antidiag} \exp \mathfrak{p}(\mathfrak{g}))U$$

are valid in the sense that multiplication in each case is a real analytic diffeomorphism onto.

Proof. Existence for G^* follows from the Cartan decomposition $G = K \exp \mathfrak{p}(\mathfrak{g})$ of G and the possibility of taking square roots in $\exp \mathfrak{p}(\mathfrak{g})$ in diffeomorphic fashion:

$$\begin{pmatrix} g & \\ & g' \end{pmatrix} = \begin{pmatrix} gg'^{-1} & \\ & 1 \end{pmatrix}\begin{pmatrix} g' & \\ & g' \end{pmatrix} = \begin{pmatrix} k & \\ & 1 \end{pmatrix}\begin{pmatrix} p & \\ & p^{-1} \end{pmatrix}\begin{pmatrix} p & \\ & p \end{pmatrix}\begin{pmatrix} g' & \\ & g' \end{pmatrix}.$$

We lump the last two factors on the right, and we have the required decomposition. For uniqueness, let $g = kph$ and $g' = p^{-1}h$. Then $gg'^{-1} = kp^2$, and so k and p are uniquely determined. Then h is uniquely determined also.

The existence question for $G_{\mathbb{C}}$ is whether an equality

$$\begin{pmatrix} z & \\ & \Theta z \end{pmatrix} = \begin{pmatrix} d & \\ & \Theta d \end{pmatrix}\begin{pmatrix} x & \\ & \Theta x \end{pmatrix}\begin{pmatrix} u & \\ & \Theta u \end{pmatrix}$$

is valid with d in $\exp \mathfrak{p}(\mathfrak{k}^{\mathbb{C}})$, x in $\exp \mathfrak{p}(\mathfrak{g})$, and u in the compact form. Here Θ as usual is adjoint inverse. Thus we want to decompose $z = dxu$. Given z, we note that $z(\Theta z)^{-1}$ is in $\exp \mathfrak{p}(\mathfrak{g}^{\mathbb{C}})$. We shall use the easily verified (for SU(1, 1)) fact that every positive definite element p can be decomposed uniquely as

$$p = dx^2 d \qquad \text{with} \qquad d \in \exp \mathfrak{p}(\mathfrak{k}^{\mathbb{C}}), \qquad x \in \exp \mathfrak{p}(\mathfrak{g}); \tag{9.16}$$

moreover, the decomposition takes place diffeomorphically. Take $p = z(\Theta z)^{-1}$ and define u by $z = dxu$. Then u is in $G^{\mathbb{C}}$ and

$$\begin{aligned} u(\Theta u)^{-1} &= x^{-1}d^{-1}z(\Theta z)^{-1}(\Theta d)(\Theta x) = x^{-1}d^{-1}(z\Theta z^{-1})d^{-1}x^{-1} \\ &= x^{-1}d^{-1}(dx^2 d)d^{-1}x^{-1} = 1. \end{aligned}$$

Hence u is in the compact form, and existence follows. For uniqueness we note that $z = dxu$ forces $z(\Theta z)^{-1} = dx^2 d$, and so d and x are uniquely determined by (9.16). Then u is uniquely determined also.

Let us combine Proposition 9.7 with the fact that K^* and $K_{\mathbb{C}}$ have the same complexification $\left(\text{with Lie algebra } \begin{pmatrix} \mathfrak{k}^{\mathbb{C}} & \\ & \mathfrak{k}^{\mathbb{C}} \end{pmatrix}\right)$. To pass from our function on G^* to one on $G_{\mathbb{C}}$, we restrict to antidiag exp $\mathfrak{p}(\mathfrak{g})$ and extend trivially on the right under U. On the left we take the representation of K^* that gives the left K^* dependence, extend holomorphically to $(K^*)^{\mathbb{C}} = (K_{\mathbb{C}})^{\mathbb{C}}$, and restrict to exp $\mathfrak{p}(\mathfrak{k}_{\mathbb{C}})$. The result is a law for how the transformed function on $G_{\mathbb{C}}$ is to transform on the left under $K_{\mathbb{C}}$ and in particular under exp $\mathfrak{p}(\mathfrak{k}_{\mathbb{C}})$. There are some technical consistency questions to be handled, such as why the transformation law on $G_{\mathbb{C}}$ works correctly under all of $K_{\mathbb{C}}$, but we shall postpone them until we consider general G.

At any rate, the result is that we now seek a function on $G_{\mathbb{C}}/U$ that transforms within a finite-dimensional space under left translation by $K_{\mathbb{C}}$. It is reasonable to expect that the condition on our original function of being an eigenfunction of $D(G^*/\text{diag } G)$ should transform into being an eigenfunction of $D(G_{\mathbb{C}}/U)$. How can we write down such a function?

The third important underlying idea gives a way of generating functions on any G/K that are eigenfunctions of $D(G/K)$. To guess at the answer, let us consider the case that $G = \text{SL}(2, \mathbb{R})$ and the eigenvalue is 0. Here G/K is the upper half plane, and the invariant Laplacian is one of the operators in $D(G/K)$. Since the invariant Laplacian is the product of a function and the usual Euclidean Laplacian, we are trying to generate harmonic functions. One way of doing so is by Poisson integrals, which we can regard as integrated (in t) combinations of the Poisson kernel

$$P(x + iy, t) = \frac{y}{(x - t)^2 + y^2}. \tag{9.17}$$

What we need, then, is a group-theoretic expression for the Poisson kernel. We can write down such an expression easily if we allow the action of

SL(2, $\mathbb{R}$) on the upper half plane to be given by the inverse transpose of the usual action. Thus

$$\begin{pmatrix} a & b \\ c & d \end{pmatrix}(z) = \frac{dz - c}{-b + az}.$$

The correspondence of the half plane to G/K is given explicitly by $g(i) \leftrightarrow gK$. Because of the Iwasawa decomposition $G = \bar{N}AK$, we can coordinatize G/K in group-theoretic notation by $\bar{N} \times A$: If

$$\bar{n} = \begin{pmatrix} 1 & 0 \\ -x & 1 \end{pmatrix} \quad \text{and} \quad a = \begin{pmatrix} y^{-1/2} & \\ & y^{1/2} \end{pmatrix}, \tag{9.18}$$

then $x + iy \leftrightarrow \bar{n}aK$. Relative to $G = KAN$, we readily compute that

$$\exp\left\{-2\rho H\begin{pmatrix} a & b \\ c & d \end{pmatrix}\right\} = (a^2 + c^2)^{-1}. \tag{9.19}$$

With $\bar{n}$ and a as in (9.18) and with $\bar{n}' = \begin{pmatrix} 1 & 0 \\ -t & 1 \end{pmatrix}$, we have

$$a^{-1}\bar{n}^{-1}\bar{n}' = \begin{pmatrix} y^{1/2} & 0 \\ y^{-1/2}(x - t) & y^{-1/2} \end{pmatrix},$$

and thus (9.19) shows that the Poisson kernel (9.17) is given by $\exp\{-2\rho H(a^{-1}\bar{n}^{-1}\bar{n}')\}$. Dropping our coordinates $\bar{N} \times A$, we see that the value of the Poisson kernel at gK in G/K and $\bar{n}'$ on the boundary is

$$P(gK, \bar{n}') = e^{-2\rho H(g^{-1}\bar{n}')}.$$

We can ask whether this formula can be perturbed to give eigenfunctions for other eigenvalues, and the affirmative answer is given in Proposition 9.9 below. The proposition requires our being able to make computations with $D(G/K)$, and we first prove a lemma that relates $D(G/K)$ to $U(\mathfrak{g}^{\mathbb{C}})$ and allows us to make such computations.

Lemma 9.8. Let G be a linear connected reductive group. Then the algebra $D(G/K)$ is canonically isomorphic with the algebra of restrictions of Ad(K)-invariant members of $U(\mathfrak{g}^{\mathbb{C}})$ to right K-invariant functions on G.

Proof. For f defined on G/K, let $\tilde{f}$ be the right K-invariant lift to G defined by $\tilde{f}(x) = f(xK)$. Let $U(\mathfrak{g}^{\mathbb{C}})^K$ denote the set of Ad(K)-invariant members of $U(\mathfrak{g}^{\mathbb{C}})$. Given D in $U(\mathfrak{g}^{\mathbb{C}})^K$, we define a linear transformation E by $(Ef)^{\sim} = D\tilde{f}$ for f in $C^\infty(G/K)$. We check readily that E carries $C^\infty(G/K)$ into itself and that $D \to E$ is an algebra homomorphism carrying 1 to 1. As a map on restrictions of operators D to right K-invariant functions on G, the map is clearly one-one.

To see that E is a differential operator, we use canonical coordinates around g in G of the form

$$(x_1, \ldots, x_n) \leftrightarrow g \exp(x_1 X_1 + \ldots + x_m X_m) \exp(x_{m+1} X_{m+1} + \ldots + x_n X_n), \tag{9.20}$$

where $X_1, \ldots, X_m$ is a basis of $\mathfrak{p}$ and $X_{m+1}, \ldots, X_n$ is a basis of $\mathfrak{k}$. Then

$$(x_1, \ldots, x_m) \leftrightarrow g \exp(x_1 X_1 + \ldots + x_m X_m)K \tag{9.21}$$

provides coordinates around gK in G/K. For f in $C^\infty(G/K)$, we can write f in the coordinates (9.21). Then the lift $\tilde{f}$, in the coordinates (9.20), is independent of $x_{m+1}, \ldots, x_n$. Since D commutes with right translation by K, $D\tilde{f}$ is independent of $x_{m+1}, \ldots, x_n$, too. If we write D out in terms of partial derivatives in the coordinates (9.20), then the result is that D gives the same answer on $\tilde{f}$ as the operator in which all terms are dropped that involve partials with respect to $x_{m+1}, \ldots, x_n$. The remaining partial derivatives have a sense in the coordinates (9.21) on G/K, and the equality $(Ef)^\sim = D\tilde{f}$ says this expression is an expression for E in terms of partial derivatives. Thus E is a differential operator and hence is in $D(G/K)$.

To complete the proof, we must show $D \to E$ is onto $D(G/K)$. For this part of the argument, we shall use the compactness of K. Given E in $D(G/K)$, write E out in the coordinates (9.21) about the identity coset as

$$Ef(xK) = \sum a_{j_1 \cdots j_m}(xK) \frac{\partial^{j_1 + \ldots + j_m} f}{\partial x_1^{j_1} \cdots \partial x_m^{j_m}} (\exp(x_1 X_1 + \ldots + x_m X_m)K)$$

and take $x = 1$. By left invariance we see that the expression for E at gK in the coordinates (9.21) is

$$Ef(gK) = \sum a_{j_1 \cdots j_m}(1) \frac{\partial^{j_1 + \ldots + j_m} f}{\partial x_1^{j_1} \cdots \partial x_m^{j_m}} (g \exp(x_1 X_1 + \ldots + x_m X_m)K)_{\text{all } x_j = 0}. \tag{9.22}$$

Define $D_1 F$ for F in $C^\infty(G)$ by

$$D_1 F(g) = \sum a_{j_1 \cdots j_m}(1) \frac{\partial^{j_1 + \ldots + j_m} F}{\partial x_1^{j_1} \cdots \partial x_m^{j_m}} (g \exp(x_1 X_1 + \ldots + x_m X_m) \times \exp(x_{m+1} X_{m+1} + \ldots + x_n X_n))_{\text{all } x_j = 0}$$

Then $D_1 \tilde{f} = (Ef)^\sim$, and D_1 is a member of $U(\mathfrak{g}^{\mathbb{C}})$. Setting $D = \int_K \mathrm{Ad}(k) D_1 \, dk$ and using the invariance of (9.22) under $g \to gk$, we obtain the required operator mapping onto E.

Proposition 9.9. Let $G = KA_\mathfrak{p}N_\mathfrak{p}$ be an Iwasawa decomposition of the linear connected reductive group G. For ν in $(\mathfrak{a}_\mathfrak{p}')^{\mathbb{C}}$ and x in G, the function

$$e^{-\nu H(g^{-1}x)}, \tag{9.23}$$

as g varies, is an eigenfunction of $D(G/K)$. The eigenvalue is computed as follows: For D in $D(G/K)$, let $\tilde{D}$ be any lift to an $\mathrm{Ad}(K)$-invariant member of $U(\mathfrak{g}^{\mathbb{C}})$, and let $\tilde{D}_{\mathfrak{a}}$ be the first component of $\tilde{D}$ relative to the decomposition

$$U(\mathfrak{g}^{\mathbb{C}}) = U(\mathfrak{a}_{\mathfrak{p}}^{\mathbb{C}}) \oplus (\mathfrak{n}_{\mathfrak{p}}^{\mathbb{C}} U(\mathfrak{g}^{\mathbb{C}}) + U(\mathfrak{g}^{\mathbb{C}})\mathfrak{k}^{\mathbb{C}}). \tag{9.24}$$

Then the eigenvalue of (9.23) under D is $\nu(\tilde{D}_{\mathfrak{a}})$.

Remark. In defining $\nu(\tilde{D}_{\mathfrak{a}})$, we are making use of a familiar argument based on Proposition 3.1 that linear functionals on an abelian Lie algebra extend to homomorphisms of the universal enveloping algebra into $\mathbb{C}$ carrying 1 to 1.

Proof. Fix a lift $\tilde{D}$ of D. Let H' denote the log of the $A_{\mathfrak{p}}$ component relative to $G = N_{\mathfrak{p}} A_{\mathfrak{p}} K$, and let $\kappa'(\cdot)$ denote the K component. Then

$$e^{-\nu H(g^{-1}x)} = e^{\nu H'(x^{-1}g)}. \tag{9.25}$$

When we apply $\tilde{D}$, the effect appears on the right side of g in the right-hand side of (9.25). Thus we have

$$\begin{aligned}
\tilde{D}(e^{-\nu H(g^{-1}x)})_{g=1} &= \tilde{D}(e^{\nu H'(x^{-1}\cdot)})_{\cdot=g} \\
&= \tilde{D}(e^{\nu H'(x^{-1}g\cdot)})_{\cdot=1} \\
&= e^{\nu H'(x^{-1}g)}\tilde{D}(e^{\nu H'(\kappa'(x^{-1}g)\cdot)})_{\cdot=1} \\
&= e^{\nu H'(x^{-1}g)}(\mathrm{Ad}(\kappa'(x^{-1}g))\tilde{D})(e^{\nu H'(\cdot)})_{\cdot=1} \\
&= e^{\nu H'(x^{-1}g)}\tilde{D}(e^{\nu H'(\cdot)})_{\cdot=1},
\end{aligned}$$

the last step holding by the $\mathrm{Ad}(K)$ invariance of $\tilde{D}$. Thus (9.23) is an eigenfunction, and the eigenvalue is $\tilde{D}(e^{\nu H'(\cdot)})_{\cdot=1}$.

Since $\mathfrak{g} = \mathfrak{n}_{\mathfrak{p}} \oplus \mathfrak{a}_{\mathfrak{p}} \oplus \mathfrak{k}$, the Birkhoff-Witt Theorem gives us (9.24). The term in parentheses on the right of (9.24) annihilates $e^{\nu H'(\cdot)}$ at the identity, and hence the eigenvalue is $\nu(\tilde{D}_{\mathfrak{a}})$.

Now let us return to our setting with $G = SU(1, 1)$. We seek a left $K_{\mathbb{C}}$-finite function on $G_{\mathbb{C}}/U$ that is an eigenfunction of $D(G_{\mathbb{C}}/U)$. We expect from the above discussion that we obtain a fairly general eigenfunction by using a combination as u varies of functions $e^{-\nu H_{\mathbb{C}}(g^{-1}u)}$ with ν fixed, where $H_{\mathbb{C}}(\cdot)$ is the $H(\cdot)$ function for $G_{\mathbb{C}}$. (If we replace u by a general x in $G_{\mathbb{C}}$, we obtain no more general a function because of the Iwasawa decomposition in $G_{\mathbb{C}}$.) Thus

$$\psi_{\mathbb{C}}(g_{\mathbb{C}}) = \int_U e^{-\nu H_{\mathbb{C}}(g_{\mathbb{C}}^{-1}u)} h(u)\, du, \qquad g_{\mathbb{C}} \in G_{\mathbb{C}}, \tag{9.26}$$

is such an eigenfunction; here we write the combination as an integration with respect to a function, but integration with respect to a measure or a more general object will produce eigenfunctions, too.

If $\psi_{\mathbb{C}}$ in (9.26) is the function we are seeking, then its transformation law on the left under $K_{\mathbb{C}}$ ought to produce some restriction on h if that kind of integration formula is close to being one-one, and we proceed as if the formula is one-one. If l is in $K_{\mathbb{C}} \cap U = \operatorname{diag} K$, then the relationship between $\psi_{\mathbb{C}}(lg_{\mathbb{C}})$ and $\psi_{\mathbb{C}}(g_{\mathbb{C}})$ will lead to a relationship between $h(lu)$ and $h(u)$.

Thus we need to pin down the transformation law under diag K on the left. Since diag K is common to G^* and $G_{\mathbb{C}}$, left multiplication by a member of diag K interprets directly as conjugation of our original combination of matrix coefficients by a member of K. There is no loss of generality in assuming this conjugation acts trivially, since our function on G can be

$$g \to \operatorname{Tr}(E_\tau \pi(g) E_\tau),$$

where E_τ is the projection to the K type τ, and then it will be invariant under conjugation by K.

We expect then that we lose no generality by requiring that h in (9.26) be left invariant under diag K. The most primitive case to achieve this invariance occurs when h is replaced by Haar measure on diag K, and we are led to

$$\psi_{\mathbb{C}}(g_{\mathbb{C}}) = \int_{\operatorname{diag} K} e^{-\nu H_{\mathbb{C}}(g_{\mathbb{C}}^{-1} l)}\, dl. \tag{9.27}$$

The question is what conditions on ν ensure that $\psi_{\mathbb{C}}$ is left $K_{\mathbb{C}}$-finite. To obtain a sufficient condition, we use the following lemma.

Lemma 9.10. With $\psi_{\mathbb{C}}$ as in (9.27),

$$\psi_{\mathbb{C}}(k_{\mathbb{C}} g_{\mathbb{C}}) = \int_{\operatorname{diag} K} e^{(\nu - 2\rho_{K_{\mathbb{C}}}) H_{\mathbb{C}}(k_{\mathbb{C}} l)} e^{-\nu H_{\mathbb{C}}(g_{\mathbb{C}}^{-1} l)}\, dl \tag{9.28}$$

for $k_{\mathbb{C}} \in K_{\mathbb{C}}$ and $g_{\mathbb{C}} \in G_{\mathbb{C}}$, provided the Iwasawa decompositions of $K_{\mathbb{C}}$ and $G_{\mathbb{C}}$ are chosen compatibly.

Proof. Let the Iwasawa decomposition of x in $G_{\mathbb{C}}$ be written as $x = \kappa(x) e^{H_{\mathbb{C}}(x)} n$. Replacing $g_{\mathbb{C}}$ by $k_{\mathbb{C}} g_{\mathbb{C}}$ in (9.27) and expanding the resulting exponential, we obtain

$$\psi_{\mathbb{C}}(k_{\mathbb{C}} g_{\mathbb{C}}) = \int_{\operatorname{diag} K} e^{-\nu H_{\mathbb{C}}(g_{\mathbb{C}}^{-1} \kappa(k_{\mathbb{C}}^{-1} l))} e^{-\nu H_{\mathbb{C}}(k_{\mathbb{C}}^{-1} l)}\, dl.$$

Since diag K is the maximal compact subgroup of $K_{\mathbb{C}}$, we can handle the change of variables $l' = \kappa(k_{\mathbb{C}}^{-1} l)$. This is the same change of variables one makes to see that $k_{\mathbb{C}}$ acts in unitary fashion in the compact picture of the principal series representation of $K_{\mathbb{C}}$ induced with the trivial parameter on MA, and the relevant formulas are

$$l = \kappa(k_{\mathbb{C}} l') \qquad \text{and} \qquad \int_K f(l)\, dl = \int_K f(\kappa(k_{\mathbb{C}} l')) e^{-2\rho_{K_{\mathbb{C}}} H_{\mathbb{C}}(k_{\mathbb{C}} l')}\, dl'.$$

Substituting and changing l' back to l, we obtain

$$\begin{aligned}\psi_{\mathbb{C}}(k_{\mathbb{C}}g_{\mathbb{C}}) &= \int_{\operatorname{diag} K} e^{-\nu H_{\mathbb{C}}(g_{\mathbb{C}}^{-1}l)}e^{-\nu H_{\mathbb{C}}(k_{\mathbb{C}}^{-1}\kappa(k_{\mathbb{C}}l))}e^{-2\rho_{K_{\mathbb{C}}}H_{\mathbb{C}}(k_{\mathbb{C}}l)}\,dl \\ &= \int_{\operatorname{diag} K} e^{(\nu-2\rho_{K_{\mathbb{C}}})H_{\mathbb{C}}(k_{\mathbb{C}}l)}e^{-\nu H_{\mathbb{C}}(g_{\mathbb{C}}^{-1}l)}\,dl,\end{aligned}$$

as required.

For the case of SU(1, 1), l commutes with $k_{\mathbb{C}}$ and drops out of the formula (9.28), and thus ψ is automatically left $K_{\mathbb{C}}$-finite. For a more general G, the question is a little more subtle. The natural question is whether the space of functions of l spanned by $\{\exp(\nu - 2\rho_{K_{\mathbb{C}}})H_{\mathbb{C}}(k_{\mathbb{C}}l) \mid k_{\mathbb{C}} \in K_{\mathbb{C}}\}$ is finite-dimensional. This span is the representation space of $K_{\mathbb{C}}$ for a representation with a diag K fixed vector (the function with $k_{\mathbb{C}} = 1$), and we want to know a sufficient condition for it to be finite-dimensional; we shall give such a condition in §5.

At any rate, for $SU(1, 1)$ the left $K_{\mathbb{C}}$ finiteness imposes no restriction on ν. However, there is an integrality condition on ν for us to be able to reverse our duality and pass from $\psi_{\mathbb{C}}$ back to a function ψ^* on G^*. Our transformation law for $\psi_{\mathbb{C}}$ when $G = \mathrm{SU}(1, 1)$ is

$$\psi_{\mathbb{C}}(k_{\mathbb{C}}g_{\mathbb{C}}) = e^{\nu H_{\mathbb{C}}(k_{\mathbb{C}})}\psi_{\mathbb{C}}(g_{\mathbb{C}}), \qquad g_{\mathbb{C}} \in G_{\mathbb{C}}.$$

If we write elements of $K_{\mathbb{C}}$ explicitly as

$$K_{\mathbb{C}} = \left\{\begin{pmatrix} e^{t+i\theta} & & & \\ & e^{-t-i\theta} & & \\ & & e^{-t+i\theta} & \\ & & & e^{t-i\theta}\end{pmatrix}\right\}$$

and define a complex number r by

$$e^{\nu H_{\mathbb{C}}(\text{above})} = e^{2rt}, \tag{9.29}$$

then we want to know what condition on r allows e^{2rt} to extend holomorphically to the complexification $(K_{\mathbb{C}})^{\mathbb{C}}$. In this situation $\mathfrak{k}_{\mathbb{C}}$ is a Cartan subalgebra of $\mathfrak{g}_{\mathbb{C}}$, and the analytic group that corresponds to $(\mathfrak{g}_{\mathbb{C}})^{\mathbb{C}}$ is $(G_{\mathbb{C}})^{\mathbb{C}} = \begin{pmatrix} G^{\mathbb{C}} & \\ & G^{\mathbb{C}}\end{pmatrix}$, which is simply connected. Therefore a necessary and sufficient condition is that the linear functional

$$\begin{pmatrix} t+i\theta & & & \\ & -t-i\theta & & \\ & & -t+i\theta & \\ & & & t-i\theta\end{pmatrix} \to 2rt$$

be algebraically integral relative to $\mathfrak{g}_{\mathbb{C}}$, i.e., that

$$\begin{pmatrix} t+i\theta & \\ & -t-i\theta \end{pmatrix} \to 2rt \tag{9.30}$$

be algebraically integral relative to $\mathfrak{g}^{\mathbb{C}}$. The roots for the real Lie algebra $\mathfrak{g}^{\mathbb{C}}$ relative to the Cartan subalgebra are $\pm 2(t+i\theta)$ and $\pm 2(t-i\theta)$. We readily check that algebraic integrality of (9.30) means that r is an integer. Then the holomorphic extension of (9.29) to $(K_{\mathbb{C}})^{\mathbb{C}}$ is

$$\begin{pmatrix} a & & & \\ & a^{-1} & & \\ & & b & \\ & & & b^{-1} \end{pmatrix} \to (ab^{-1})^r. \tag{9.31}$$

Let ψ^* be the function on G^* that corresponds to $\psi_{\mathbb{C}}$. The transformation law for $\psi^*(k^*g^*)$ in terms of $\psi^*(g^*)$ is given by specializing (9.31) to k^*. If $k_\theta = \begin{pmatrix} e^{i\theta} & \\ & e^{-i\theta} \end{pmatrix}$ and $k_\theta^* = \begin{pmatrix} k_\theta & \\ & 1 \end{pmatrix}$, then (9.31) gives

$$\psi^*(k_\theta^* g^*) = e^{ir\theta}\psi^*(g^*).$$

What we are really interested in is not the function ψ^* on $G^*/\text{diag}\, G$ but the corresponding function ψ on G given by

$$\psi(g) = \psi^*\begin{pmatrix} g & \\ & 1 \end{pmatrix}.$$

To evaluate this, we return to our decomposition in Proposition 9.7. We can write

$$g = k_\theta k_\varphi \begin{pmatrix} \cosh s & \sinh s \\ \sinh s & \cosh s \end{pmatrix} k_\varphi^{-1} \tag{9.32}$$

with $s \geq 0$. With

$$p = k_\varphi \begin{pmatrix} \cosh s/2 & \sinh s/2 \\ \sinh s/2 & \cosh s/2 \end{pmatrix} k_\varphi^{-1},$$

we have

$$\begin{pmatrix} g & \\ & 1 \end{pmatrix} = \begin{pmatrix} k_\theta & \\ & 1 \end{pmatrix} \begin{pmatrix} p & \\ & p^{-1} \end{pmatrix} \begin{pmatrix} p & \\ & p \end{pmatrix}$$

and hence

$$\psi(g) = e^{ir\theta}\psi^*\left(\begin{pmatrix} p & \\ & p^{-1} \end{pmatrix}\begin{pmatrix} p & \\ & p \end{pmatrix}\right) = e^{ir\theta}\psi\begin{pmatrix} p & \\ & p^{-1} \end{pmatrix} = e^{ir\theta}\psi_{\mathbb{C}}\begin{pmatrix} p & \\ & p^{-1} \end{pmatrix}.$$

If we define

$$a_s = \begin{pmatrix} \cosh s & \sinh s \\ \sinh s & \cosh s \end{pmatrix},$$

then

$$\begin{aligned} \psi(g) &= e^{ir\theta}\psi_{\mathbb{C}}\left((\operatorname{diag} k_\varphi)\begin{pmatrix} a_{s/2} & 0 \\ 0 & a_{-s/2} \end{pmatrix}(\operatorname{diag} k_\varphi)^{-1}\right) \\ &= e^{ir\theta}\psi_{\mathbb{C}}\begin{pmatrix} a_{s/2} & 0 \\ 0 & a_{-s/2} \end{pmatrix}. \end{aligned} \tag{9.33}$$

Formula (9.33) gives us a function on SU(1, 1) that is K-finite and $Z(\mathfrak{g}^{\mathbb{C}})$-finite, in view of our construction. Corollary 8.42 says ψ comes from discrete series if it is in $L^2(\mathrm{SU}(1, 1))$. To see when ψ is in L^2, we note that formula (9.32) sets up matters so that we can use the $KA_{\mathfrak{p}}^+K$ integration formula (Proposition 5.28 and the example following it). All we need to do is calculate $\psi_{\mathbb{C}}\begin{pmatrix} a_{s/2} & 0 \\ 0 & a_{-s/2} \end{pmatrix}$ from (9.27). In this calculation we have to use an Iwasawa decomposition of $G_{\mathbb{C}}$ compatible with that of $K_{\mathbb{C}}$; hence the A component is diagonal. We readily compute

$$\begin{pmatrix} a & b \\ c & d \end{pmatrix} = \kappa\begin{pmatrix} a & b \\ c & d \end{pmatrix}\begin{pmatrix} (|a|^2 + |c|^2)^{1/2} & \text{—} \\ 0 & (|a|^2 + |c|^2)^{-1/2} \end{pmatrix}. \tag{9.34}$$

To use (9.27), we require the A component of

$$a_{s/2}^{-1}k_\theta = \begin{pmatrix} e^{i\theta}\cosh s/2 & \text{—} \\ -e^{i\theta}\sinh s/2 & \text{—} \end{pmatrix}.$$

Formula (9.34) says the A component is

$$\begin{pmatrix} (\cosh s)^{1/2} & 0 \\ 0 & (\cosh s)^{-1/2} \end{pmatrix},$$

independently of θ. The integration in (9.27) drops out, and (9.31) gives

$$\psi(g) = e^{ir\theta}[(\cosh s)^{1/2}[(\cosh s)^{-1/2}]^{-1}]^r = e^{ir\theta}(\cosh s)^r. \tag{9.35}$$

With a suitable normalization of Haar measure, we obtain

$$\int_{\mathrm{SU}(1,1)} |\psi(g)|^2\, dg = \int_0^\infty (\cosh s)^{2r}\sinh 2s\, ds,$$

which is finite for $r < -1$. If we check over the example following Proposition 5.28 and compare the K dependence here with that in §2.6, we see that ψ is a multiple of the complex conjugate of $(\mathscr{D}_n^+(\cdot)1, 1)$ for $n = -r$. Thus ψ is a multiple of $(\mathscr{D}_n^-(\cdot)1, 1)$ for $n = -r$.

Finally if we use the lower triangular N in (9.34) in place of the upper triangular N, we can trace through the calculation again, obtaining finiteness of the square integral for $r > 1$. This time ψ is a multiple of $(\mathscr{D}_n^+(\cdot)1, 1)$ for $n = +r$.

§5. Finite-Dimensional Spherical Representations

Before taking up the construction of discrete series in the general case, we digress to obtain results that we can combine with Lemma 9.10 to get left $K_{\mathbb{C}}$ finiteness for the function $\psi_{\mathbb{C}}$. The connection between the left $K_{\mathbb{C}}$-finiteness problem and finite-dimensional representations is clarified by Proposition 9.11 below.

In this section we work with a linear connected semisimple group G, and we fix an Iwasawa decomposition $G = KA_{\mathfrak{p}}N_{\mathfrak{p}}$. If $\mathfrak{b}_{\mathfrak{p}}$ denotes a maximal abelian subspace of $\mathfrak{m}_{\mathfrak{p}}$, then $\mathfrak{a}_{\mathfrak{p}} \oplus \mathfrak{b}_{\mathfrak{p}}$ is a Cartan subalgebra of $\mathfrak{g}$. We shall analyze finite-dimensional representations of G using this Cartan subalgebra, just as we did in §7.9. Weights are real on $\mathfrak{a}_{\mathfrak{p}}$ and imaginary on $\mathfrak{b}_{\mathfrak{p}}$. The restriction to $\mathfrak{a}_{\mathfrak{p}}$ of a weight will be called a **restricted weight.** We introduce positive systems Δ^+ and Σ^+ of roots and restricted roots so that the nonzero restrictions to $\mathfrak{a}_{\mathfrak{p}}$ of the members of Δ^+ comprise Σ^+, and then the highest restricted weight is the restriction of the highest weight.

Our application later in this chapter will be in the situation that G is complex (namely G will be what has been called $K_{\mathbb{C}}$). The proofs in this section simplify in that situation, but we retain the greater generality for applications later.

Proposition 9.11. Suppose ν in $\mathfrak{a}_{\mathfrak{p}}'$ is the highest restricted weight for an irreducible finite-dimensional representation π of G having a nonzero K-fixed vector. Then the set of functions of $k \in K$ given by $\{k \to e^{\nu H(xk)},\ x \in G\}$ spans a finite-dimensional space, and the left regular representation of G acts on this space irreducibly as the **contragredient** (= transpose inverse) $\pi^{\vee}$ of π.

Remark. The action of the left regular representation of G is understood as follows: g carries $(k \to e^{\nu H(xk)})$ to $(k \to e^{\nu H(g^{-1}xk)})$.

Proof. Introduce an inner product $(\cdot, \cdot)$ that makes $\pi(K)$ unitary, let ϕ_ν be a nonzero highest restricted weight vector, and let ϕ_K be a nonzero K-fixed vector. We then have

$$(\pi(kan)\phi_\nu, \phi_K) = (\pi(a)\phi_\nu, \pi(k)^{-1}\phi_K) = e^{\nu \log a}(\phi_\nu, \phi_K).$$

By irreducibility of π, the left side cannot be identically 0, and hence (ϕ_ν, ϕ_K) on the right side is nonzero. Consequently

$$e^{\nu H(x)} = (\pi(x)\phi_\nu, \phi_K) \tag{9.36}$$

except for a nonzero constant factor. Let $\langle\cdot,\cdot\rangle$ refer to the pairing with linear functionals and let ϕ'_K be the linear functional $(\cdot, \phi_K)$. Let $\{\phi_j\}$ be a basis of the space on which π acts, and let ϕ'_j be the dual basis. Then

$$e^{\nu H(xk)} = \langle \pi(xk)\phi_\nu, \phi'_K\rangle = \sum_j \langle \pi(x)\phi_j, \phi'_K\rangle\langle \pi(k)\phi_\nu, \phi'_j\rangle,$$

and so the functions in question lie in the finite-dimensional space $V = \sum_j \mathbb{C}\langle \pi(k)\phi_\nu, \phi'_j\rangle$.

Now $\pi^\vee$ acts on the space of (ϕ')'s in the dual space, and we let B be the linear map of this dual space into V given by $B(\phi') = \langle \pi(k)\phi_\nu, \phi'\rangle$. Then

$$\begin{aligned} B(\pi^\vee(g)\phi') &= \langle \pi(k)\phi_\nu, \pi^\vee(g)\phi'\rangle = \langle \pi(g^{-1}k)\phi_\nu, \phi'\rangle \\ &= L(g)\langle \pi(k)\phi_\nu, \phi'\rangle, \end{aligned}$$

and B is equivariant. Since $\pi^\vee$ is irreducible, B is an isomorphism. Thus the left regular representation acts by $\pi^\vee$ on V.

Lemma 9.12. If π is an irreducible finite-dimensional representation of G, then the identity component of $M_\mathfrak{p}$ acts irreducibly in the highest restricted weight space of π.

Proof. Let $v \neq 0$ be in the highest restricted weight space. Using the decomposition $\mathfrak{g} = \bar{\mathfrak{n}}_\mathfrak{p} \oplus \mathfrak{m}_\mathfrak{p} \oplus \mathfrak{a}_\mathfrak{p} \oplus \mathfrak{n}_\mathfrak{p}$ and applying the Birkhoff-Witt Theorem, let $U(\mathfrak{g}^\mathbb{C})$ act on v. Then $\mathfrak{n}_\mathfrak{p}$ pushes restricted weights up, giving 0. Also $\mathfrak{a}_\mathfrak{p}$ acts as scalars in each restricted weight space, $\mathfrak{m}_\mathfrak{p}$ leaves each restricted weight space stable, and $\bar{\mathfrak{n}}_\mathfrak{p}$ pushes restricted weights down. Hence only $U(\mathfrak{m}^\mathbb{C})$ can be responsible for moving from v to other vectors in the highest restricted weight space. Since π is irreducible, we conclude that $U(\mathfrak{m}^\mathbb{C})$ carries v onto the highest restricted weight space. Since v is arbitrary, $\mathfrak{m}$ acts irreducibly.

Lemma 9.13. For each restricted root β, let $\gamma_\beta = \exp 2\pi i|\beta|^{-2}H_\beta$. Then $M_\mathfrak{p}$ is generated by its identity component and the elements γ_β.

Proof omitted.

Remark. Let $X \neq 0$ be in the restricted root space $\mathfrak{g}_\beta$. Then X, θX, and $[X, \theta X]$ span a subalgebra isomorphic to $\mathfrak{sl}(2, \mathbb{R})$, and the corresponding analytic subgroup is a homomorphic image of $SL(2, \mathbb{R})$. Under this homomorphism, $\begin{pmatrix} -1 & 0 \\ 0 & -1 \end{pmatrix}$ maps to γ_β, as one easily checks.

Theorem 9.14. For an irreducible finite-dimensional representation π of G, the following statements are equivalent:

(1) π has a nonzero K-fixed vector.

(2) $M_\mathfrak{p}$ acts by the trivial representation in the highest restricted weight space of π.

(3) The highest weight $\tilde{\nu}$ of π vanishes on $\mathfrak{b}_\mathfrak{p}$, and the restriction ν of $\tilde{\nu}$ to $\mathfrak{a}_\mathfrak{p}$ is such that $\langle \nu, \beta \rangle / |\beta|^2$ is an integer for every restricted root β.

Proof.

(1) $\Rightarrow$ (2). Decompose a K-fixed vector v into its components in the various restricted weight spaces. By (9.36) the component in the space for ν is nonzero. Each of the components of v has to be $M_\mathfrak{p}$-fixed, since the $\mathfrak{a}_\mathfrak{p}$ projections commute with $M_\mathfrak{p}$, and so $M_\mathfrak{p}$ acts trivially on some nonzero vector in the highest restricted weight space. Then (2) follows from Lemma 9.12.

(2) $\Rightarrow$ (1). Let $v \neq 0$ be in the highest restricted weight space, with restricted weight ν. Then $\int_K \pi(k) v \, dk$ is obviously fixed by K, and the problem is to see it is not 0. Since v is assumed to be fixed by $M_\mathfrak{p}$, $\pi(k)v$ is a function on $K/M_\mathfrak{p}$. By (5.25)

$$\int_K \pi(k) v \, dk = \int_{\bar{N}_\mathfrak{p}} \pi(\kappa(\bar{n})) v e^{-2\rho H(\bar{n})} \, d\bar{n} = \int_{\bar{N}_\mathfrak{p}} \pi(\bar{n}) v e^{(-\nu - 2\rho) H(\bar{n})} \, d\bar{n}.$$

Here $e^{(-\nu-2\rho)H(\bar{n})}$ is everywhere positive, and $(\pi(\bar{n})v, v)$ is equal to $|v|^2 \neq 0$, since $\bar{\mathfrak{n}}_\mathfrak{p}$ decreases restricted weights and exp carries $\bar{\mathfrak{n}}_\mathfrak{p}$ onto $\bar{N}_\mathfrak{p}$ by (A.127). Hence $\int_K \pi(k) v \, dk$ is not 0.

(2) $\Rightarrow$ (3). Since $(M_\mathfrak{p})_0$ acts trivially, it follows immediately that $\tilde{\nu}$ vanishes on $\mathfrak{b}_\mathfrak{p}$. For each restricted root β, $\pi(\gamma_\beta)$ is 1 on the highest restricted weight space, by assumption. (Here γ_β is as in Lemma 9.13.) Possibly by passing to a (linear) cover of G, we may assume by the unitary trick that π extends holomorphically to $G^\mathbb{C}$. Then we can compute $\pi(\gamma_\beta)$ on a vector v of restricted weight ν as

$$\pi(\gamma_\beta) v = \exp(2\pi i |\beta|^{-2} \operatorname{ad} H_\beta) v = e^{2\pi i \langle \nu, \beta \rangle / |\beta|^2} v.$$

Since the left side equals v, $\langle \nu, \beta \rangle / |\beta|^2$ must be an integer.

(3) $\Rightarrow$ (2). Since ν vanishes on $\mathfrak{b}_\mathfrak{p}$, the highest weight of the action of $(M_\mathfrak{p})_0$ on the highest restricted weight space is 0. (The action is irreducible by Lemma 9.12.) Thus $(M_\mathfrak{p})_0$ acts trivially. The calculation of $\pi(\gamma_\beta)v$ above shows that each γ_β acts trivially on the highest restricted weight space. Then (2) follows from Lemma 9.13.

Corollary 9.15. If ν is a member of $\mathfrak{a}_\mathfrak{p}'$ such that $\langle \nu, \beta \rangle / |\beta|^2$ is an integer ≥ 0 for every β in Σ^+, then ν is the highest restricted weight of an irreducible finite-dimensional representation of G with a nonzero K-fixed vector.

Proof. We may assume that G has a simply connected complexification $G^\mathbb{C}$, since the resulting representation will have to be scalar on Z_G (because $Z_G \subseteq K$) and will therefore descend to every group covered by G. Let $\tilde{\nu}$ be the extension of ν to $\mathfrak{a}_\mathfrak{p} \oplus \mathfrak{b}_\mathfrak{p}$ that is 0 on $\mathfrak{b}_\mathfrak{p}$. Then $\tilde{\nu}$ is dominant. If

$\tilde{\nu}$ is algebraically integral, then it is the highest weight of some π, and the implication (3) $\Rightarrow$ (1) in Theorem 9.14 says π has a nonzero K-fixed vector.

Thus we are to prove $\tilde{\nu}$ is algebraically integral. Let α be a root, let β be its restriction, and let $|\alpha|^2 = C|\beta|^2$. Then

$$\frac{2\langle \nu, \alpha\rangle}{|\alpha|^2} = \frac{2\langle \nu, \beta\rangle}{C|\beta|^2},$$

and it is enough to show either that $2/C$ is an integer or that $|2/C| = 1/2$ and $2\langle \nu, \beta\rangle/|\beta|^2$ is even. Write $\alpha = \beta + \delta$, where δ is the restriction to $\mathfrak{b}_{\mathfrak{p}}$. Let us agree for this proof to extend θ to $\mathfrak{g}^{\mathbb{C}}$ in complex linear fashion (usually we extend it in conjugate linear fashion). If X_α is a root vector, then θX_α is a root vector for $\theta\alpha = -\beta + \delta$. In particular, $\theta\alpha$ is a root. Thus $-\theta\alpha = \beta - \delta$ is a root with the same length as α.

If α and $-\theta\alpha$ are multiples of one another, then $\delta = 0$ and $C = 1$, so that $2/C$ is an integer. If α and $-\theta\alpha$ are not multiples of one another, then the Schwarz inequality gives

$$\begin{aligned}(-1 \text{ or } 0 \text{ or } 1) &= \frac{2\langle \alpha, -\theta\alpha\rangle}{|\alpha|^2} = \frac{2\langle \beta+\delta, \beta-\delta\rangle}{|\alpha|^2} \\ &= \frac{2(|\beta|^2 - |\delta|^2)}{|\alpha|^2} = \frac{2(2|\beta|^2 - |\alpha|^2)}{|\alpha|^2} = \frac{4}{C} - 2. \qquad (9.37)\end{aligned}$$

If the left side of (9.37) is -1, then $(\beta + \delta) + (\beta - \delta) = 2\beta$ is a root, hence also a restricted root. By assumption $\langle \nu, 2\beta\rangle/|2\beta|^2$ is an integer; hence $\langle \nu, \beta\rangle/|\beta|^2$ is even. If the left side of (9.37) is 0, then (9.37) shows $4/C = 2$; thus $2/C = 1$ is an integer.

Finally we show that the left side of (9.37) cannot be 1. If it is 1, then $(\beta + \delta) - (\beta - \delta) = 2\delta$ is a root vanishing on $\mathfrak{a}_{\mathfrak{p}}$, and hence the root vector is in $\mathfrak{m}_{\mathfrak{p}}^{\mathbb{C}} \subseteq \mathfrak{k}^{\mathbb{C}}$. However, this root is also equal to $\alpha + \theta\alpha$, and a root vector is $[X_\alpha, \theta X_\alpha]$, which is sent to its negative by the extended θ. Hence the root vector is in $\mathfrak{p}^{\mathbb{C}}$. Thus the root vector is in $\mathfrak{k}^{\mathbb{C}} \cap \mathfrak{p}^{\mathbb{C}} = 0$, and we have a contradiction.

§6. Duality in the General Case

Before we can construct discrete series for groups G other than SU(1, 1), we make a second digression, this time to establish the duality of §4 with more care.

Let G be linear connected semisimple. Without loss of generality we may assume $\mathfrak{k} \cap i\mathfrak{p} = 0$ (since G can be rewritten as a group of real matrices if necessary). It will be convenient to assume for now that $G \subseteq G^{\mathbb{C}}$ with $G^{\mathbb{C}}$ simply connected.

Pursuing the duality as in §4, we take

$$G^* = \left\{\begin{pmatrix} x & 0 \\ 0 & y \end{pmatrix} \middle| x, y \in G\right\} \leftrightarrow G_{\mathbb{C}} = \left\{\begin{pmatrix} z & 0 \\ 0 & \Theta z \end{pmatrix} \middle| z \in G^{\mathbb{C}}\right\}$$

$$K^* = \left\{\begin{pmatrix} k & 0 \\ 0 & l \end{pmatrix} \middle| k, l \in K\right\} \leftrightarrow K_{\mathbb{C}} = \left\{\begin{pmatrix} k & 0 \\ 0 & \Theta k \end{pmatrix} \middle| k \in K^{\mathbb{C}}\right\}$$

$$\operatorname{diag} G = \left\{\begin{pmatrix} x & 0 \\ 0 & x \end{pmatrix} \middle| x \in G\right\} \leftrightarrow U = \left\{\begin{pmatrix} u & 0 \\ 0 & u \end{pmatrix} \middle| u \in \exp(\mathfrak{k} + i\mathfrak{p})\right\}$$

$$\operatorname{diag} K = \left\{\begin{pmatrix} k & 0 \\ 0 & k \end{pmatrix} \middle| k \in K\right\} \leftrightarrow \operatorname{diag} K,$$

and we use the obvious notation for the Lie algebras. Also we let

$$G_0 = \left\{\begin{pmatrix} x & 0 \\ 0 & \Theta x \end{pmatrix} \middle| x \in G\right\}.$$

All the groups in question lie inside

$$(G^*)^{\mathbb{C}} = (G_{\mathbb{C}})^{\mathbb{C}} = \begin{pmatrix} G^{\mathbb{C}} & 0 \\ 0 & G^{\mathbb{C}} \end{pmatrix} \cong G^{\mathbb{C}} \times G^{\mathbb{C}},$$

which is simply connected. The groups K^* and $K_{\mathbb{C}}$ satisfy

$$(K^*)^{\mathbb{C}} = (K_{\mathbb{C}})^{\mathbb{C}} = \begin{pmatrix} K^{\mathbb{C}} & 0 \\ 0 & K^{\mathbb{C}} \end{pmatrix} \cong K^{\mathbb{C}} \times K^{\mathbb{C}}.$$

This group need not be simply connected, nor need it be semisimple. The group G_0 is the intersection of G^* and $G_{\mathbb{C}}$.

Proposition 9.16. The decompositions

$$G^* = (K \times 1)(\text{antidiag} \exp \mathfrak{p}(\mathfrak{g}))(\operatorname{diag} G)$$

$$G_{\mathbb{C}} = (\exp \mathfrak{p}(\mathfrak{k}_{\mathbb{C}}))(\text{antidiag} \exp \mathfrak{p}(\mathfrak{g}))U$$

are valid in the sense that multiplication in each case is a real analytic diffeomorphism onto.

Proof. This is the same as for Proposition 9.7, except that Lemma 9.17 below replaces formula (9.16). In the lemma G is taken as $G^{\mathbb{C}}$, and σ is taken as the conjugation of $G^{\mathbb{C}}$ with respect to G.

Lemma 9.17. Let σ be an automorphism of order 2 of a linear connected reductive group G such that $\sigma\Theta = \Theta\sigma$, and let $\mathfrak{g} = \mathfrak{h} \oplus \mathfrak{q}$ be the decomposition of $\mathfrak{g}$ into $+1$ and -1 eigenspaces. Then the map $(X, Y) \to (\exp X)(\exp Y)(\exp X)$ is a diffeomorphism of $(\mathfrak{q} \cap \mathfrak{p}) \times (\mathfrak{h} \cap \mathfrak{p})$ onto $\exp \mathfrak{p}$.

Proof. It is clear that the map is real analytic into $\exp \mathfrak{p}$. For the remaining properties, let $p \in \exp \mathfrak{p}$ be given. The motivation is as follows: We think of connecting p and $\sigma(p)$ by the version of a line appropriate to positive definite matrices, namely $(\exp tX)p(\exp tX)$, $0 \le t \le 1$. The midpoint of the line ought to be fixed by σ and will be taken as $\exp Y$. So before we come to the proof, let us solve

$$(\exp X)p(\exp X) = \sigma(p) \tag{9.38}$$

for X. We write $p = p_1^2$ and $\sigma(p) = p_2^2$ with p_1 and p_2 in $\exp \mathfrak{p}$. Then $p_2 = \sigma(p_1)$. So (9.38) leads to

$$\begin{aligned} p_1(\exp X)p_1p_1(\exp X)p_1 &= p_1p_2^2p_1 \\ (p_1(\exp X)p_1)^2 &= p_1p_2^2p_1 \\ \exp X &= p_1^{-1}(p_1p_2^2p_1)^{1/2}p_1^{-1}. \end{aligned} \tag{9.39}$$

Now we come to the proof. Define $p_1 = p^{1/2}$ and $p_2 = \sigma(p)^{1/2}$, and let $\exp X$ be as in (9.39). Then

$$\exp \sigma(X) = \sigma(\exp X) = p_2^{-1}(p_2p_1^2p_2)^{1/2}p_2^{-1}.$$

Unwinding the calculation in the preceding paragraph, we see this is $\exp(-X)$. Thus $\sigma(X) = -X$. Define

$$p_0 = (\exp \tfrac{1}{2}X)p_1^2(\exp \tfrac{1}{2}X).$$

Then

$$p = p_1^2 = \exp(-\tfrac{1}{2}X)p_0(\exp(-\tfrac{1}{2}X), \tag{9.40}$$

and we have an onto map if we show $\sigma(p_0) = p_0$. For this, we compute that

$$\sigma(p_0) = \exp(\tfrac{1}{2}\sigma(X))p_2^2 \exp(\tfrac{1}{2}\sigma(X)) = \exp(-\tfrac{1}{2}X)p_2^2 \exp(-\tfrac{1}{2}X),$$

and (9.38) shows this is

$$\begin{aligned} &= \exp(-\tfrac{1}{2}X)(\exp X)p_1^2(\exp X)\exp(-\tfrac{1}{2}X) = (\exp \tfrac{1}{2}X)p_1^2(\exp \tfrac{1}{2}X) \\ &= p_0. \end{aligned}$$

Thus (9.40) shows the map is onto, and it is easy to go over this argument to verify that the map is one-one and the inverse is smooth, hence real analytic.

Lemma 9.18.

(a) The real analytic right (diag G)-invariant functions f on G^* stand in one-one correspondence with the real analytic functions f_1 on $K^* \times G_0$ satisfying

$$f_1(k^*l, l^{-1}xl_1) = f_1(k^*, x) \qquad \text{for } k^* \in K^*, x \in G_0, \text{ and } l, l_1 \in \text{diag } K. \tag{9.41a}$$

The correspondence is $f_1(k^*, x) = f(k^*x(\text{diag } G))$.

(b) The real analytic right U-invariant functions f on $G_{\mathbb{C}}$ stand in one-one correspondence with the real analytic functions f_1 on $K_{\mathbb{C}} \times G_0$ satisfying

$$f_1(k_{\mathbb{C}}l, l^{-1}xl_1) = f_1(k_{\mathbb{C}}, x) \qquad \text{for } k_{\mathbb{C}} \in K_{\mathbb{C}}, x \in G_0, \text{ and } l, l_1 \in \text{diag } K. \tag{9.41b}$$

The correspondence is $f_1(k_{\mathbb{C}}, x) = f(k_{\mathbb{C}}xU)$.

Proof. For both (i) and (ii) it is clear that the map $f \to f_1$ leads to the functional equation (9.41) for f_1. In the reverse direction if we are given f_1, we can define f by means of the respective product decompositions of Proposition 9.16:

$$f\left(\begin{pmatrix} k & 0 \\ 0 & 1 \end{pmatrix}\begin{pmatrix} p & 0 \\ 0 & p^{-1} \end{pmatrix} \text{diag } G\right) = f_1\left(\begin{pmatrix} k & 0 \\ 0 & 1 \end{pmatrix}, \begin{pmatrix} p & 0 \\ 0 & p^{-1} \end{pmatrix}\right)$$
$$\text{with } k \in K, p \in \exp \mathfrak{p}(\mathfrak{g}) \tag{9.42a}$$

$$f\left(\begin{pmatrix} q & 0 \\ 0 & \Theta q \end{pmatrix}\begin{pmatrix} p & 0 \\ 0 & \Theta p \end{pmatrix} U\right) = f_1\left(\begin{pmatrix} q & 0 \\ 0 & \Theta q \end{pmatrix}, \begin{pmatrix} p & 0 \\ 0 & \Theta p \end{pmatrix}\right)$$
$$\text{with } q \in \exp \mathfrak{p}(\mathfrak{k}^{\mathbb{C}}), p \in \exp \mathfrak{p}(\mathfrak{g}). \tag{9.42b}$$

The problem is to show that the respective relations between f_1 and f hold globally:

$$f_1(k^*, x) = f(k^*x(\text{diag } G)) \qquad \text{with } k^* \in K^*, x \in G_0 \tag{9.43a}$$

$$f_1(k_{\mathbb{C}}, x) = f(k_{\mathbb{C}}xU) \qquad \text{with } k_{\mathbb{C}} \in K_{\mathbb{C}}, x \in G_0. \tag{9.43b}$$

For (a), let $\begin{pmatrix} g & 0 \\ 0 & g' \end{pmatrix} = k^*x$ with

$$k^* = \begin{pmatrix} k_1 & 0 \\ 0 & k_2 \end{pmatrix} \qquad \text{and} \qquad x = \begin{pmatrix} x_0 & 0 \\ 0 & \Theta x_0 \end{pmatrix}.$$

Define l and then l_1 by

$$l = \begin{pmatrix} k_2^{-1} & 0 \\ 0 & k_2^{-1} \end{pmatrix}, \qquad l_1 = \begin{pmatrix} k_0 & 0 \\ 0 & k_0 \end{pmatrix} \qquad \text{such that } (k_2x_0)k_0 \in \exp \mathfrak{p}(\mathfrak{g}).$$

The functional equation (9.41a) for f_1 and the definition (9.42a) of f yield

$$f_1(k^*, x) = f_1\left(\begin{pmatrix} k_1k_2^{-1} & 0 \\ 0 & 1 \end{pmatrix}, \begin{pmatrix} k_2x_0k_0 & 0 \\ 0 & k_2(\Theta x_0)k_0 \end{pmatrix}\right) = f\begin{pmatrix} g & 0 \\ 0 & g' \end{pmatrix}$$

since $k_2(\Theta x_0)k_0 = \Theta(k_2x_0k_0) = (k_2x_0k_0)^{-1}$. For (b) let $g_{\mathbb{C}} = k_{\mathbb{C}}x$ with

$$k_{\mathbb{C}} = \begin{pmatrix} k^{\mathbb{C}} & 0 \\ 0 & \Theta k^{\mathbb{C}} \end{pmatrix} \qquad \text{and} \qquad x = \begin{pmatrix} x_0 & 0 \\ 0 & \Theta x_0 \end{pmatrix}.$$

Let the Cartan decomposition of $k^{\mathbb{C}}$ in $K^{\mathbb{C}}$ be $k^{\mathbb{C}} = qk_2$ with $q \in \exp \mathfrak{p}(\mathfrak{k}^{\mathbb{C}})$ and $k_2 \in K$, and choose $k_0 \in K$ so that $(k_2 x_0)k_0$ is in $\exp \mathfrak{p}(\mathfrak{g})$. Then the functional equation (9.41b) of f_1 and the definition (9.42b) of f give

$$\begin{aligned} f_1(k_{\mathbb{C}}, x) &= f_1\left(\begin{pmatrix} qk_2 & 0 \\ 0 & q^{-1}k_2 \end{pmatrix}, \begin{pmatrix} x_0 & 0 \\ 0 & \Theta x_0 \end{pmatrix}\right) \\ &= f_1\left(\begin{pmatrix} q & 0 \\ 0 & q^{-1} \end{pmatrix}, \begin{pmatrix} k_2 x_0 k_0 & 0 \\ 0 & k_2(\Theta x_0)k_0 \end{pmatrix}\right) \\ &= f\left(\begin{pmatrix} q & 0 \\ 0 & q^{-1} \end{pmatrix}\begin{pmatrix} k_2 x_0 k_0 & 0 \\ 0 & \Theta(k_2 x_0 k_0) \end{pmatrix} U\right) = f(k_{\mathbb{C}} x U). \end{aligned}$$

This completes the proof.

Fix a holomorphic irreducible finite-dimensional representation τ of $(K^*)^{\mathbb{C}} = (K_{\mathbb{C}})^{\mathbb{C}} \cong K^{\mathbb{C}} \times K^{\mathbb{C}}$. Then τ defines irreducible representations of K^* and $K_{\mathbb{C}}$ on the same space by restriction of the action; we denote these restrictions by τ also. Let us be given a real analytic function f^* on G^* that is fixed under right translation by diag G and transforms according to τ under the left regular representation $\mathscr{L}$ of K^*.

[The latter condition means that f^* lies in an $\mathscr{L}(K^*)$-invariant finite-dimensional space of functions on G^* that is equivalent with τ. Since $\mathscr{L}(K^*)f^*$ generates this space, the functions in the space are real analytic and are fixed under right translation by diag G. It is useful to have the condition on f^* written in terms of a basis: If $\{f_i\}$ is a basis of the space, then we are requiring that

$$f^* = \sum c_i f_i \tag{9.44a}$$

and

$$\mathscr{L}(k^*)f_j = \sum_i \tau_{ij}(k^*) f_i \tag{9.44b}$$

with τ_{ij} the matrix of τ in some basis. Notice that these equations imply

$$\mathscr{L}(k^*)f^*(g_0) = \sum_{i,j} c_j \tau_{ij}(k^*) f_i(g_0) \tag{9.44c}$$

for g_0 in G_0.]

Starting from f^*, we can use Lemma 9.18a to pass to the function f_1^* on $K^* \times G_0$ given by

$$f_1^*(k^*, g_0) = f^*(k^*, g_0) = \mathscr{L}(k^*)^{-1} f^*(g_0) = \sum_{i,j} c_j \tau_{ij}(k^{*-1}) f_i(g_0).$$

Since τ is holomorphic on $(K^*)^{\mathbb{C}}$, the right side here shows we can use analytic continuation to extend f_1^* to a function on $(K^*)^{\mathbb{C}} \times G_0 = (K_{\mathbb{C}})^{\mathbb{C}} \times G_0$. (Any two analytic continuations to this set will coincide, and so the construction is independent of basis.) The extended function satisfies the functional equation (9.41a) even with k^* in $(K^*)^{\mathbb{C}}$, by uniqueness of analytic continuation. Hence the restriction $(f_{\mathbb{C}})_1$ to $K_{\mathbb{C}} \times G_0$ satisfies (9.41b). Using Lemma 9.18b, we obtain a real analytic function $f_{\mathbb{C}}$ on $G_{\mathbb{C}}$ that is fixed under right translation by U and transforms according to τ under $\mathscr{L}(K_{\mathbb{C}})$.

In this way we have defined a linear map

$$f^* \to f_{\mathbb{C}}. \tag{9.45}$$

The domain is real analytic functions on G^* fixed under right translation by diag G and transforming according to τ under $\mathscr{L}(K^*)$. The range is real analytic functions on $G_{\mathbb{C}}$ fixed under right translation by U and transforming according to τ under $\mathscr{L}(K_{\mathbb{C}})$. Clearly we can reverse the construction, and thus (9.45) is invertible.

If we are given two functions f^* and $f_{\mathbb{C}}$ as above, then the above reasoning shows they are matched under (9.45) if there exists a function F on $(K^*)^{\mathbb{C}} \times G_0 = (K_{\mathbb{C}})^{\mathbb{C}} \times G_0$ that is holomorphic in the first variable and restricts to $f^*(k^*g_0)$ and $f_{\mathbb{C}}(k_{\mathbb{C}}g_0)$ on $K^* \times G_0$ and $K_{\mathbb{C}} \times G_0$, respectively.

We give next a result telling how (9.45) respects differential operators. The Ad(diag G)-invariant members $U((\mathfrak{g}^*)^{\mathbb{C}})^{\text{diag } G}$ of $U((\mathfrak{g}^*)^{\mathbb{C}})$ act on smooth functions on G^* by left invariant differentiation, and in particular they act on our functions f^* above, yielding functions of the same type. Similarly $U((\mathfrak{g}_{\mathbb{C}})^{\mathbb{C}})^U$ acts on the functions $f_{\mathbb{C}}$ above, yielding functions of the same type. Since $(\mathfrak{g}^*)^{\mathbb{C}} = (\mathfrak{g}_{\mathbb{C}})^{\mathbb{C}}$, we have $U((\mathfrak{g}^*)^{\mathbb{C}}) = U((\mathfrak{g}_{\mathbb{C}})^{\mathbb{C}})$. Moreover, diag G and U both have diag $G^{\mathbb{C}}$ as complexification, and Ad extends to act holomorphically on the universal enveloping algebra. Thus we have

$$U((\mathfrak{g}^*)^{\mathbb{C}})^{\text{diag } G} = U((\mathfrak{g}_{\mathbb{C}})^{\mathbb{C}})^U. \tag{9.46}$$

Now the definition of left-invariant differentiation depends upon the real group involved, not just the complexification, since a first-order element $X + iY$ is defined to act as $\tilde{X} + i\tilde{Y}$, where $\tilde{X}$ and $\tilde{Y}$ are the left invariant vector fields obtained from X and Y. Thus there is no reason in general why corresponding operators on the two sides of (9.46) should be closely related, even on G_0, which is the intersection of the two groups. Nevertheless, we have the following result for our functions f^* and $f_{\mathbb{C}}$.

Proposition 9.19. Let D be a member of (9.46). Under the map $f^* \to f_{\mathbb{C}}$ of (9.45), Df^* maps to $Df_{\mathbb{C}}$.

Proof. First we show that any D in $U((\mathfrak{g}^*)^{\mathbb{C}}) = U((\mathfrak{g}_{\mathbb{C}})^{\mathbb{C}})$ satisfies

$$Df^*(x_0) = Df_{\mathbb{C}}(x_0) \text{ for } x_0 = \begin{pmatrix} p_0 & 0 \\ 0 & p_0^{-1} \end{pmatrix} \in \text{antidiag} \exp \mathfrak{p}(\mathfrak{g}). \tag{9.47}$$

Since f^* and $f_{\mathbb{C}}$ are real analytic, they both extend to be holomorphic on a sufficiently small connected neighborhood E of x_0 in $(G^*)^{\mathbb{C}} = (G_{\mathbb{C}})^{\mathbb{C}}$.

Now f^* and $f_{\mathbb{C}}$ agree everywhere on antidiag exp $\mathfrak{p}(\mathfrak{g})$, and it follows from Proposition 9.16 and the transformation laws of f^* and $f_{\mathbb{C}}$ that the holomorphic extensions of f^* and $f_{\mathbb{C}}$ to E are equal. Denote by F the common extension.

For any first order term $Z = X + iY$ in $U((\mathfrak{g}^*)^{\mathbb{C}})$, the fact that F is holomorphic means that $i(YF) = (iY)F$ locally and hence that ZF can be computed as a complex derivative:

$$ZF(x_0) = \frac{d}{dz} F(x_0 \exp zZ)\bigg|_{z=0}. \tag{9.48}$$

The same considerations apply to $Z = X' + iY'$ in $U((\mathfrak{g}_{\mathbb{C}})^{\mathbb{C}})$, and again we are led to (9.48) as the formula. Thus first-order elements give the same result, and by iteration we obtain (9.47).

Next Df^* and $Df_{\mathbb{C}}$ are both fixed under right translation by diag K, and it follows that they agree on G_0. To complete the proof, let us introduce notation as in (9.44). Since f^* maps to $f_{\mathbb{C}}$, we have

$$\mathscr{L}(k^*)f^*(g_0) = \sum_{i,j} c_i \tau_{ij}(k^*) f_i^*(g_0)$$

and

$$\mathscr{L}(k_{\mathbb{C}})f_{\mathbb{C}}(g_0) = \sum_{i,j} c_i \tau_{ij}(k_{\mathbb{C}})(f_i)_{\mathbb{C}}(g_0),$$

where f_i^* and $(f_i)_{\mathbb{C}}$ correspond. Applying D and using its left invariance, we obtain

$$Df^*(k^* g_0) = \sum_{i,j} c_i \tau_{ij}(k^{*-1}) Df_i^*(g_0)$$

and

$$Df_{\mathbb{C}}(k_{\mathbb{C}} g_0) = \sum_{i,j} c_i \tau_{ij}(k_{\mathbb{C}}^{-1}) D(f_i)_{\mathbb{C}}(g_0).$$

Since f_i^* and $(f_i)_{\mathbb{C}}$ correspond, Df_i^* equals $D(f_i)_{\mathbb{C}}$ on G_0. Thus the two right sides here have an obvious common analytic continuation to $(K^*)^{\mathbb{C}} \times G_0 = (K_{\mathbb{C}})^{\mathbb{C}} \times G_0$, and hence Df^* corresponds to $Df_{\mathbb{C}}$ under (9.45).

§7. Construction of Discrete Series

Now we are set to define some representations that will turn out to be the full discrete series of G. In the construction we shall assume that rank G = rank K, i.e., that G has a compact Cartan subgroup. In Chapter XII we shall see that this equal-rank condition is necessary for the existence of discrete series.

Thus let G be linear connected semisimple with rank G = rank K, and let $\mathfrak{b} \subseteq \mathfrak{k} \subseteq \mathfrak{g}$ be a Cartan subalgebra. A first effect of the equal-rank condition is that we can define roots of $\mathfrak{g}^{\mathbb{C}}$ and $\mathfrak{k}^{\mathbb{C}}$ with respect to the same

Cartan subalgebra $\mathfrak{b}^{\mathbb{C}}$. Thus let

$$\Delta = \text{roots of } (\mathfrak{g}^{\mathbb{C}}, \mathfrak{b}^{\mathbb{C}})$$
$$\Delta_K = \text{roots of } (\mathfrak{k}^{\mathbb{C}}, \mathfrak{b}^{\mathbb{C}}).$$

Since root spaces are one-dimensional and we have the bracket relations

$$[\mathfrak{b}^{\mathbb{C}}, \mathfrak{k}^{\mathbb{C}}] \subseteq \mathfrak{k}^{\mathbb{C}} \quad \text{and} \quad [\mathfrak{b}^{\mathbb{C}}, \mathfrak{p}^{\mathbb{C}}] \subseteq \mathfrak{p}^{\mathbb{C}},$$

each root space lies either in $\mathfrak{k}^{\mathbb{C}}$ or $\mathfrak{p}^{\mathbb{C}}$. We call the roots in Δ **compact** or **noncompact** accordingly. It is clear that Δ_K is exactly the set of compact roots; let Δ_n be the set of noncompact roots ($= \Delta - \Delta_K$). Let W_G and W_K be the Weyl groups of Δ and Δ_K, respectively. An important identity is

$$W_K = W(B:G),$$

where B is the analytic subgroup with Lie algebra $\mathfrak{b}$; in fact, $W_K = W(B:K)$ by Theorem 4.41, and we know that $W(B:K)$ coincides with $W(B:G)$. Relative to any choice of positive system Δ^+ we might make, we shall take $\Delta_K^+ = \Delta^+ \cap \Delta_K$, and we let δ_G and δ_K be the respective half-sums of positive roots.

Theorem 9.20. Let G be linear connected semisimple with rank $G =$ rank K. Suppose that λ in $(i\mathfrak{b})'$ is **nonsingular** relative to Δ (i.e., $\langle \lambda, \alpha \rangle \neq 0$ for all α in Δ) and that Δ^+ is defined as

$$\Delta^+ = \{\alpha \in \Delta \mid \langle \lambda, \alpha \rangle > 0\}. \tag{9.49}$$

If $\lambda + \delta_G$ is analytically integral, then there exists a discrete series representation π_λ of G with the following properties:

(a) π_λ has infinitesimal character χ_λ.
(b) $\pi_\lambda|_K$ contains with multiplicity one the K type with highest weight

$$\Lambda = \lambda + \delta_G - 2\delta_K.$$

(c) If Λ' is the highest weight of a K type in $\pi_\lambda|_K$, then Λ' is of the form

$$\Lambda' = \Lambda + \sum_{\alpha \in \Delta^+} n_\alpha \alpha \qquad \text{for integers } n_\alpha \geq 0.$$

Two such constructed representations π_λ are equivalent if and only if their parameters λ are conjugate under W_K.

Terminology. λ is called the **Harish-Chandra parameter** of the discrete series π_λ, and the K type parameter Λ is called the **Blattner parameter.** The origins of these names are discussed in the bibliographical notes.

Remarks.

(1) For holomorphic discrete series (Chapter VI), the parameter λ we used coincides with our present Blattner parameter, but the positive system in Chapter VI is not that in (9.49). If $(\Delta^+)'$ and δ_G' refer to the pos-

itive system in Chapter VI, the requirement was

$$\langle \lambda + \delta'_G, \alpha \rangle \begin{cases} \text{positive for } \alpha > 0 \text{ compact} \\ \text{negative for } \alpha > 0 \text{ noncompact.} \end{cases}$$

The relationship is $\Delta^+ = -w_K(\Delta^+)'$, where w_K is the element of W_K carrying Δ_K^+ to $-\Delta_K^+$. Unwinding matters, we can check that $\lambda + \delta'_G$ is the Harish-Chandra parameter. It is not too hard to verify directly that the representation in Chapter VI satisfies (a), (b), and (c) here. At any rate the representations in Chapter VI arise as the special case of Theorem 9.20 in which $-w_K\Delta^+$ comes from a good ordering, in the sense of Chapter VI.

(2) All the parameters $w\lambda$ for $w \in W_G$ give the same infinitesimal character. The final statement of the theorem says that exactly $|W_G|/|W_K|$ of the discrete series $\pi_{w\lambda}$ are mutually inequivalent. When G is noncompact simple and satisfies the conditions of Chapter VI, only two of these arise as holomorphic discrete series.

(3) According to the theorem, there exist groups possessing discrete series but no holomorphic discrete series. The group $SO(2n, 1)$ for $n > 1$ is an example.

The proof of Theorem 9.20 will occupy the remainder of this chapter. We shall implement the duality in this section and then get at the detailed properties of π_λ in the next section and the one after it.

Let us note that the assumption rank G = rank K implies $\mathfrak{k} \cap i\mathfrak{p} = 0$, so that $G^{\mathbb{C}}$ is obtained easily by the construction of §5.1. [In fact, assume on the contrary that $\mathfrak{k} \cap i\mathfrak{p} \neq 0$, and construct $\mathfrak{b}$ by letting $\mathfrak{b}_0$ be a maximal abelian subalgebra of $\mathfrak{k} \cap i\mathfrak{p}$ and by insisting that $\mathfrak{b} \supseteq \mathfrak{b}_0$; then $\mathfrak{b}$ is not maximal abelian in $\mathfrak{g}$ because $i\mathfrak{b}_0$ (which is in $\mathfrak{p}$) can be adjoined.]

In proving the theorem, there is no loss in generality in assuming $G \subseteq G^{\mathbb{C}}$ with $G^{\mathbb{C}}$ simply connected. In fact, if we are given G with a cover $\tilde{G}$ contained in a simply connected complexification, then we first construct π_λ for $\tilde{G}$. Since $\lambda + \delta_G$ is analytically integral for G, so is Λ. Since $Z_{\tilde{G}} \subseteq \exp \mathfrak{b}$, every element in the kernel of the covering homomorphism acts trivially on the Λ $\tilde{K}$ type of π_λ, and by irreducibility every element in the kernel acts as a global scalar. Hence π_λ descends to a representation of G with the required properties.

In this setting we begin the proof by relating the situation in G to a situation in the group $G_{\mathbb{C}}$ defined through the duality of §6. Let $\mathfrak{a}$ and $\mathfrak{m}$ be defined by

$$\mathfrak{a} = \left\{ \begin{pmatrix} H & 0 \\ 0 & -H \end{pmatrix} \middle| H \in i\mathfrak{b} \right\}$$

$$\mathfrak{m} = \left\{ \begin{pmatrix} H & 0 \\ 0 & H \end{pmatrix} \middle| H \in \mathfrak{b} \right\}.$$

Since $\mathfrak{b}^{\mathbb{C}}$ is a Cartan subalgebra of $\mathfrak{g}^{\mathbb{C}}$, $\mathfrak{a} \oplus \mathfrak{m}$ is a Cartan subalgebra of

$$\mathfrak{g}_{\mathbb{C}} = \left\{ \begin{pmatrix} X & 0 \\ 0 & \theta X \end{pmatrix} \middle| X \in \mathfrak{g}^{\mathbb{C}} \right\},$$

and $\mathfrak{m}$ is maximal abelian in diag($\mathfrak{k} \oplus i\mathfrak{p}$), i.e., $\mathfrak{m} = Z_{\mathfrak{u}}(\mathfrak{a})$. Together these facts imply that $\mathfrak{a}$ is an Iwasawa $\mathfrak{a}_{\mathfrak{p}}$ subalgebra of $\mathfrak{g}_{\mathbb{C}}$ and that the corresponding algebra $\mathfrak{m}_{\mathfrak{p}}$ is exactly $\mathfrak{m}$. Now $\mathfrak{a} \oplus \mathfrak{m}$ lies in the smaller subalgebra $\mathfrak{k}_{\mathbb{C}}$, and hence $\mathfrak{a}$ is an Iwasawa $\mathfrak{a}_{\mathfrak{p}}$ subalgebra also for $\mathfrak{k}_{\mathbb{C}}$ and the corresponding $\mathfrak{m}_{\mathfrak{p}}$ is still $\mathfrak{m}$. Let $\Sigma(\mathfrak{g}_{\mathbb{C}})$ and $\Sigma(\mathfrak{k}_{\mathbb{C}})$ denote the respective systems of restricted roots.

If α is a root of $(\mathfrak{g}^{\mathbb{C}}, \mathfrak{b}^{\mathbb{C}})$, the root space has complex dimension 1. Let X be in the root space, and form $\begin{pmatrix} X & \\ & \theta X \end{pmatrix}$ in $\mathfrak{g}_{\mathbb{C}}$. For $\begin{pmatrix} H & \\ & -H \end{pmatrix}$ in $\mathfrak{a}$, we have $\theta H = -H$ and hence

$$\begin{aligned} \left[\begin{pmatrix} H & \\ & -H \end{pmatrix}, \begin{pmatrix} X & \\ & \theta X \end{pmatrix} \right] &= \begin{pmatrix} [H, X] & \\ & -[H, \theta X] \end{pmatrix} \\ &= \begin{pmatrix} [H, X] & \\ & \theta[H, X] \end{pmatrix} \\ &= \alpha(H) \begin{pmatrix} X & \\ & \theta X \end{pmatrix}. \end{aligned}$$

Thus $\begin{pmatrix} X & \\ & \theta X \end{pmatrix}$ is a restricted root vector for the restricted root

$$\begin{pmatrix} H & \\ & -H \end{pmatrix} \to \alpha(H). \tag{9.50}$$

This construction yields a two-dimensional real subspace (spanned by X and iX) of the restricted root space for (9.50). Since Δ gives us a full root space decomposition of $\mathfrak{g}^{\mathbb{C}}$, the image of (9.50) gives us a full restricted root space decomposition of $\mathfrak{g}_{\mathbb{C}}$. Thus (9.50) accounts for all restricted roots of $\mathfrak{g}_{\mathbb{C}}$ with their correct multiplicities 2. Clearly Δ_K corresponds to $\Sigma(\mathfrak{k}_{\mathbb{C}})$.

From Δ^+ defined in the theorem and from (9.50), we can define $\Sigma^+(\mathfrak{g}_{\mathbb{C}})$ and the subset $\Sigma^+(\mathfrak{k}_{\mathbb{C}})$. If $\mathfrak{n}^{\mathbb{C}}$ and $\theta(\mathfrak{n}^{\mathbb{C}}) = \bar{\mathfrak{n}}^{\mathbb{C}}$ denote the sums of the positive and negative root spaces for $\mathfrak{g}^{\mathbb{C}}$, then

$$\mathfrak{n}_{\mathbb{C}} = \left\{ \begin{pmatrix} X & \\ & \theta X \end{pmatrix} \middle| X \in \mathfrak{n}^{\mathbb{C}} \right\} \quad \text{and} \quad \bar{\mathfrak{n}}_{\mathbb{C}} = \left\{ \begin{pmatrix} X & \\ & \theta X \end{pmatrix} \middle| X \in \bar{\mathfrak{n}}^{\mathbb{C}} \right\}$$

are the sums of the positive and negative restricted root spaces of $\mathfrak{g}_{\mathbb{C}}$.

Let $H_{\mathbb{C}}$ be the log of the A component of the Iwasawa decomposition of $G_{\mathbb{C}}$, and let $\rho_{G_{\mathbb{C}}}$ and $\rho_{K_{\mathbb{C}}}$ be the half sums of the members of $\Sigma^+(\mathfrak{g}_{\mathbb{C}})$ and $\Sigma^+(\mathfrak{k}_{\mathbb{C}})$, with multiplicities counted.

With λ defined on $\mathfrak{b}^{\mathbb{C}}$ we define $\lambda_{\mathbb{C}}$ on $\mathfrak{a} \oplus \mathfrak{m}$ by

$$\lambda_{\mathbb{C}}\begin{pmatrix} H & \\ & -H \end{pmatrix} = 2\lambda(H) \qquad \text{for } H \in i\mathfrak{b} \quad \text{(definition on } \mathfrak{a})$$

$$\lambda_{\mathbb{C}}\begin{pmatrix} H & \\ & H \end{pmatrix} = 0 \qquad \text{for } H \in \mathfrak{b} \quad \text{(definition on } \mathfrak{m}). \tag{9.51}$$

We immediately verify the following properties of this definition:

$$\rho_{G_{\mathbb{C}}} = (\delta_G)_{\mathbb{C}} \tag{9.52a}$$

$$\rho_{K_{\mathbb{C}}} = (\delta_K)_{\mathbb{C}} \tag{9.52b}$$

$$\text{the restricted root (9.50) is } \tfrac{1}{2}\alpha_{\mathbb{C}}. \tag{9.53}$$

Guided by (9.27), we define a real analytic right U-invariant function $\psi_{\mathbb{C}}$ on $G_{\mathbb{C}}$ by

$$\psi_{\mathbb{C}}(g_{\mathbb{C}}) = \int_{\operatorname{diag} K} e^{-(\lambda_{\mathbb{C}} + \rho_{G_{\mathbb{C}}})H_{\mathbb{C}}(g_{\mathbb{C}}^{-1}l)}\, dl.$$

Then Lemma 9.10 gives

$$\psi_{\mathbb{C}}(k_{\mathbb{C}}g_{\mathbb{C}}) = \int_{\operatorname{diag} K} e^{(\lambda_{\mathbb{C}} + \rho_{G_{\mathbb{C}}} - 2\rho_{K_{\mathbb{C}}})H_{\mathbb{C}}(k_{\mathbb{C}}l)} e^{-(\lambda_{\mathbb{C}} + \rho_{G_{\mathbb{C}}})H_{\mathbb{C}}(g_{\mathbb{C}}^{-1}l)}\, dl. \tag{9.54}$$

The function $\psi_{\mathbb{C}}$ will certainly be left $K_{\mathbb{C}}$-finite if the first factor of the integrand lies in a finite-dimensional space as $k_{\mathbb{C}}$ varies. To deal with this property, we use Proposition 9.11 and Corollary 9.15.

To make use of Corollary 9.15, let us first observe from (9.52) that $\Lambda_{\mathbb{C}} = \lambda_{\mathbb{C}} + \rho_{G_{\mathbb{C}}} - 2\rho_{K_{\mathbb{C}}}$. We shall use $\nu = \Lambda_{\mathbb{C}}$ in the corollary. If β denotes the restricted root (9.50) obtained from $\alpha \in \Delta$, then (9.53) gives

$$\frac{\langle \Lambda_{\mathbb{C}}, \beta\rangle}{|\beta|^2} = \frac{\langle \Lambda_{\mathbb{C}}, \frac{1}{2}\alpha_{\mathbb{C}}\rangle}{|\frac{1}{2}\alpha_{\mathbb{C}}|^2} = \frac{2\langle \Lambda, \alpha\rangle}{|\alpha|^2}. \tag{9.55}$$

The assumed integrality of Λ implies the right side of (9.55) is an integer. Moreover, if α is simple for Δ_K^+, then

$$\frac{2\langle \lambda, \alpha\rangle}{|\alpha|^2} > 0, \qquad \frac{2\langle \delta_G, \alpha\rangle}{|\alpha|^2} \geq 1, \qquad \text{and} \qquad \frac{2\langle \delta_K, \alpha\rangle}{|\alpha|^2} = 1$$

imply the right side of (9.55) is > -1. Hence it is ≥ 0. Thus the corollary says $\Lambda_{\mathbb{C}}$ is the highest restricted weight of an irreducible finite-dimensional representation τ of $K_{\mathbb{C}}$ with a nonzero vector fixed by diag K.

Proposition 9.11 now applies. It says that the functions of l given by

$$e^{(\lambda_{\mathbb{C}} + \rho_{G_{\mathbb{C}}} - 2\rho_{K_{\mathbb{C}}})H_{\mathbb{C}}(k_{\mathbb{C}}l)} = e^{\Lambda_{\mathbb{C}}H_{\mathbb{C}}(k_{\mathbb{C}}l)}$$

span a finite-dimensional space as $k_{\mathbb{C}}$ varies and that the left regular representation of $K_{\mathbb{C}}$ acts irreducibly on this space by the contragredient $\tau^{\vee}$ of τ. Thus $\psi_{\mathbb{C}}$ is left $K_{\mathbb{C}}$-finite, transforming according to $\tau^{\vee}$.

To apply the duality and carry $\psi_{\mathbb{C}}$ over to ψ^* on G^* by means of Lemma 9.18, we have to show $\tau^\vee$ has a single-valued holomorphic continuation to $(K_{\mathbb{C}})^{\mathbb{C}}$. The lowest weight of $\tau^\vee$ on $\mathfrak{a} \oplus \mathfrak{m}$ is $-\nu$, and it is enough to show that the complex-linear extension of $-\nu$ to $(\mathfrak{a} \oplus \mathfrak{m})^{\mathbb{C}}$ is analytically integral. Since $G^{\mathbb{C}} \times G^{\mathbb{C}}$ is simply connected, this integrality is the same as algebraic integrality relative to the roots of $(\mathfrak{g}^{\mathbb{C}} \times \mathfrak{g}^{\mathbb{C}}, (\mathfrak{a} \oplus \mathfrak{m})^{\mathbb{C}})$, i.e., the roots of $(\mathfrak{g}^{\mathbb{C}} \times \mathfrak{g}^{\mathbb{C}}, \mathfrak{b}^{\mathbb{C}} \times \mathfrak{b}^{\mathbb{C}})$.

The roots of $(\mathfrak{g}^{\mathbb{C}} \times \mathfrak{g}^{\mathbb{C}}, \mathfrak{b}^{\mathbb{C}} \times \mathfrak{b}^{\mathbb{C}})$ are of the form

$$\{(\alpha, 0) \mid \alpha \in \Delta\} \cup \{(0, \alpha) \mid \alpha \in \Delta\}.$$

Let us consider the first type, the second type being similar. Then

$$\frac{2\langle -\nu, (\alpha, 0)\rangle}{|(\alpha, 0)|^2} = -\nu\begin{pmatrix} 2H_\alpha/|\alpha|^2 & \\ & 0 \end{pmatrix}. \tag{9.56a}$$

Here

$$\begin{pmatrix} 2H_\alpha/|\alpha|^2 & \\ & 0 \end{pmatrix} = \begin{pmatrix} H_\alpha/|\alpha|^2 & \\ & -H_\alpha/|\alpha|^2 \end{pmatrix} + i\begin{pmatrix} H_\alpha/i|\alpha|^2 & \\ & H_\alpha/|\alpha|^2 \end{pmatrix} \tag{9.56b}$$

is a decomposition into $\mathfrak{a} \oplus i\mathfrak{m}$, and hence the right side of (9.56a) is

$$\begin{aligned} &= -\nu\begin{pmatrix} H_\alpha/|\alpha|^2 & \\ & -H_\alpha/|\alpha|^2 \end{pmatrix} = -(\lambda_{\mathbb{C}} + \rho_{G_{\mathbb{C}}} - 2\rho_{K_{\mathbb{C}}})\begin{pmatrix} H_\alpha/|\alpha|^2 & \\ & -H_\alpha/|\alpha|^2 \end{pmatrix} \\ &= -2(\lambda + \delta_G - 2\delta_K)(H_\alpha/|\alpha|^2) \qquad \text{by (9.51) and (9.52)} \\ &= -2\langle \lambda + \delta_G - 2\delta_K, \alpha\rangle/|\alpha|^2. \end{aligned}$$

Since Λ is assumed integral, this quantity is an integer. We conclude that τ extends holomorphically to $(K_{\mathbb{C}})^{\mathbb{C}} = (K^*)^{\mathbb{C}}$.

Applying Lemma 9.18, we obtain a corresponding real analytic function ψ^* on G^* that is right (diag G)-invariant and transforms according to $\tau^\vee$ under the left regular representation of K^*. The definition

$$\psi(g) = \psi^*\begin{pmatrix} g & 0 \\ 0 & 1 \end{pmatrix}$$

gives us a function on G.

Lemma 9.21. Under the assumptions of Theorem 9.20, the function ψ on G has the following properties:

(a) Under the right regular representation R, $R(Z)\psi = \chi_\lambda(Z)\psi$ for all $Z \in Z(\mathfrak{g}^{\mathbb{C}})$.
(b) Under the "two-sided" representation (L, R) of $K \times K$, ψ transforms according to $\tau_\Lambda^\vee \otimes \tau_\Lambda$, where τ_Λ is the representation of K with highest weight Λ.
(c) ψ is in $L^2(G)$.

Proof of (a). There are natural ways of imbedding $U(\mathfrak{g}^{\mathbb{C}})$ into $U(\mathfrak{g}^{*\mathbb{C}})$, one of which is

$$X \to X_{\mathbb{C}} = X \otimes 1.$$

Under this imbedding, $Z(\mathfrak{g}^{\mathbb{C}})$ maps into $Z(\mathfrak{g}^{*\mathbb{C}})$. Let Z be in $Z(\mathfrak{g}^{\mathbb{C}})$, and let $Z_{\mathbb{C}}$ act as a left-invariant differential operator on G^*. Since

$$\psi(g) = \psi^*\left(\begin{pmatrix} g & 0 \\ 0 & 1 \end{pmatrix} \operatorname{diag} G\right),$$

it is clear that

$$R(Z)\psi(g) = Z_{\mathbb{C}}\psi^*\left(\begin{pmatrix} g & 0 \\ 0 & 1 \end{pmatrix} \operatorname{diag} G\right). \tag{9.57}$$

Next recall from §8.5 that $\gamma'_{\Delta^-}(Z)$ yields the terms of Z containing only $\mathfrak{b}^{\mathbb{C}}$ factors when Z is expanded relative to the basis given by the Birkhoff-Witt Theorem from $\mathfrak{n}^{\mathbb{C}} \oplus \mathfrak{b}^{\mathbb{C}} \oplus \bar{\mathfrak{n}}^{\mathbb{C}}$. Meanwhile $\mu'_{\Sigma^-(\mathfrak{g}_{\mathbb{C}})}(Z_{\mathbb{C}})$, introduced in §8.6, yields the terms of $Z_{\mathbb{C}}$ containing only $(\mathfrak{a} \oplus \mathfrak{m})^{\mathbb{C}}$ factors when $Z_{\mathbb{C}}$ is expanded relative to

$$(\mathfrak{n}_{\mathbb{C}})^{\mathbb{C}} \oplus (\mathfrak{a} \oplus \mathfrak{m})^{\mathbb{C}} \oplus (\bar{\mathfrak{n}}_{\mathbb{C}})^{\mathbb{C}}. \tag{9.58}$$

Let us write X_α for a typical root vector of $(\mathfrak{g}^{\mathbb{C}}, \mathfrak{b}^{\mathbb{C}})$ and H for a typical element of $i\mathfrak{b}$. The corresponding members of $\mathfrak{g}_{\mathbb{C}}$ are the restricted root vector $\begin{pmatrix} X_\alpha & \\ & \theta X_\alpha \end{pmatrix}$ and the member $\begin{pmatrix} H & \\ & -H \end{pmatrix}$ of $\mathfrak{a}$.

Expand Z, letting a typical term be a complex multiple of

$$X_{\alpha_1}^{p_1} \cdots X_{\alpha_n}^{p_n} H_1^{q_1} \cdots H_l^{q_l} X_{-\alpha_1}^{r_1} \cdots X_{-\alpha_n}^{r_n}.$$

If α is in Δ^+, we can write

$$X_\alpha \otimes 1 = \frac{1}{2}(X_\alpha \otimes 1 + 1 \otimes \theta X_\alpha) + i\left(\left(\left(\frac{1}{2i} X_\alpha\right) \otimes 1\right) + 1 \otimes \theta\left(\frac{1}{2i} X_\alpha\right)\right)$$

to exhibit $X_\alpha \otimes 1$ as in $(\mathfrak{n}_{\mathbb{C}})^{\mathbb{C}}$. Similarly $X_{-\alpha} \otimes 1$ is in $(\bar{\mathfrak{n}}_{\mathbb{C}})^{\mathbb{C}}$. Also

$$H \otimes 1 = \frac{1}{2}(H \otimes 1 - 1 \otimes H) + i\left(\left(\frac{1}{2i} H\right) \otimes 1 + 1 \otimes \left(\frac{1}{2i} H\right)\right) \tag{9.59}$$

exhibits $H \otimes 1$ as in $\mathfrak{a} \otimes i\mathfrak{m} \subseteq (\mathfrak{a} \otimes \mathfrak{m})^{\mathbb{C}}$. Thus, under our map $X \to X_{\mathbb{C}}$, the same terms drop out in computing $\gamma'_{\Delta^-}(Z)$ as in computing $\mu'_{\Sigma^-(\mathfrak{g}_{\mathbb{C}})}(Z_{\mathbb{C}})$. Moreover, if ν and $\nu_{\mathbb{C}}$ are related as in (9.51), then (9.59) shows that

$$\nu_{\mathbb{C}}(H \otimes 1) = \frac{1}{2}\nu_{\mathbb{C}}\begin{pmatrix} H & \\ & -H \end{pmatrix} = \frac{1}{2} 2\nu(H) = \nu(H).$$

Consequently

$$\nu(\gamma'_{\Delta^-}(Z)) = \nu_{\mathbb{C}}(\mu'_{\Sigma^-(\mathfrak{g}_{\mathbb{C}})}(Z_{\mathbb{C}})). \tag{9.60}$$

Now $Z_{\mathbb{C}}$ gives a member of $D(G_{\mathbb{C}}/U)$ via Lemma 9.8, and Proposition 9.9 says that the eigenvalue of $Z_{\mathbb{C}}$ on $\exp\{-\nu_{\mathbb{C}} H_{\mathbb{C}}(g^{-1}l)\}$ is $\nu_{\mathbb{C}}((Z_{\mathbb{C}})_{\mathfrak{a}})$. In the computation of $\mu'_{\Sigma^-(\mathfrak{g}_{\mathbb{C}})}(Z_{\mathbb{C}})$ from (9.58), any term with an $(\bar{\mathfrak{n}}_{\mathbb{C}})^{\mathbb{C}}$ factor has also some $(\mathfrak{n}_{\mathbb{C}})^{\mathbb{C}}$ factor and hence drops out in the computation of $(Z_{\mathbb{C}})_{\mathfrak{a}}$ in (9.24). Also any term with an $\mathfrak{m}^{\mathbb{C}}$ factor drops out when (9.24) is used since $\mathfrak{m} \subseteq \mathfrak{u}$. Since $\nu_{\mathbb{C}}$ vanishes on $\mathfrak{m}$, we conclude

$$\nu_{\mathbb{C}}(\mu'_{\Sigma^-(\mathfrak{g}_{\mathbb{C}})}(Z_{\mathbb{C}})) = \nu_{\mathbb{C}}((Z_{\mathbb{C}})_{\mathfrak{a}}). \tag{9.61}$$

Putting (9.61) and Proposition 9.9 together, we see that the eigenvalue of $Z_{\mathbb{C}}$ on $\psi_{\mathbb{C}}$ is

$$\begin{aligned}(\lambda_{\mathbb{C}} + \rho_{G_{\mathbb{C}}})(\mu'_{\Sigma^-(\mathfrak{g}_{\mathbb{C}})}(Z_{\mathbb{C}})) &= (\lambda + \delta_G)_{\mathbb{C}}(\mu'_{\Sigma^-(\mathfrak{g}_{\mathbb{C}})}(Z_{\mathbb{C}})) && \text{by (9.52a)}\\ &= (\lambda + \delta_G)\gamma'_{\Delta^-}(Z) && \text{by (9.60)}\\ &= \lambda(\gamma(Z)) = \chi_\lambda(Z) && \text{by (8.25).}\end{aligned}$$

According to Proposition 9.19, $Z_{\mathbb{C}}$ acts on $\psi_{\mathbb{C}}$ in the same way it acts on ψ^*. By (9.57), $R(Z)$ acts on ψ with eigenvalue $\chi_\lambda(Z)$.

Proof of (b). We know that ψ^* transforms under $\mathscr{L}(K \times K)$ according to $\tau^\vee$. Hence ψ transforms under (L, R) according to $\tau^\vee$. Since the weights of $\tau^\vee$ are the negatives of the weights of τ and since τ has highest weight $\Lambda_{\mathbb{C}}$ relative to $(\mathfrak{a} \oplus \mathfrak{m})^{\mathbb{C}}$ and to the positive system for $\mathfrak{k}^{\mathbb{C}} \oplus \mathfrak{k}^{\mathbb{C}}$ compatible with $\Sigma^+(\mathfrak{k}_{\mathbb{C}})$, it follows that $\tau^\vee$ has lowest weight $-\Lambda_{\mathbb{C}}$. In this positive system the positive roots are $(\alpha, 0)$ with $\alpha \in \Delta_K^+$ and $(0, \alpha)$ with $-\alpha \in \Delta_K^+$. In fact, the restriction to $\mathfrak{a}$ of a root $(\alpha, 0)$ is $\frac{1}{2}\alpha_{\mathbb{C}}$, while the restriction of $(0, \alpha)$ is $-\frac{1}{2}\alpha_{\mathbb{C}}$, because

$$(\alpha, 0)\begin{pmatrix} H & \\ & -H \end{pmatrix} = \alpha(H) = \tfrac{1}{2}\alpha_{\mathbb{C}}\begin{pmatrix} H & \\ & -H \end{pmatrix}$$

$$(0, \alpha)\begin{pmatrix} H & \\ & -H \end{pmatrix} = \alpha(-H) = -\tfrac{1}{2}\alpha_{\mathbb{C}}\begin{pmatrix} H & \\ & -H \end{pmatrix}.$$

Writing

$$\begin{pmatrix} H_1 & \\ & H_2 \end{pmatrix} = \begin{pmatrix} \frac{1}{2}(H_1 - H_2) & \\ & -\frac{1}{2}(H_1 - H_2) \end{pmatrix} + \begin{pmatrix} \frac{1}{2}(H_1 + H_2) & \\ & \frac{1}{2}(H_1 + H_2) \end{pmatrix} \in \mathfrak{a}^{\mathbb{C}} \oplus \mathfrak{m}^{\mathbb{C}},$$

we have

$$-\Lambda_{\mathbb{C}}\begin{pmatrix} H_1 & \\ & H_2 \end{pmatrix} = -2\Lambda(\tfrac{1}{2}(H_1 - H_2) = \Lambda(H_2) - \Lambda(H_1).$$

The other weights of $\tau^\vee$ will thus be of the form

$$(\Lambda - \textstyle\sum \alpha_j)\,(H_2) - (\Lambda - \sum \beta_j)\,(H_1)$$

with all α_j and β_j in Δ_K^+. Consequently $-\Lambda$ is the lowest weight on the H_1 part, and Λ is the highest weight on the H_2 part. Thus ψ transforms under (L, R) according to $\tau_\Lambda^\sim \otimes \tau_\Lambda$.

Proof of (c) when $\langle \lambda, \alpha \rangle$ is sufficiently large for all $\alpha \in \Delta^+$. Fix an Iwasawa $A_\mathfrak{p}$ for G. (This is quite different from the Iwasawa A of $G^\mathbb{C}$: Recall the situation when $G = \mathrm{SU}(1, 1)$.) We shall use the $KA_\mathfrak{p}^+K$ integration formula given in Proposition 5.28. Letting $g = k_1 a k_2$, write

$$\begin{pmatrix} g & \\ & 1 \end{pmatrix} = \begin{pmatrix} k_1 & \\ & k_2^{-1} \end{pmatrix} \begin{pmatrix} a^{1/2} & \\ & a^{-1/2} \end{pmatrix} \begin{pmatrix} a^{1/2}k_2 & \\ & a^{1/2}k_2 \end{pmatrix}.$$

Then

$$\psi(g) = \psi^*\left(\begin{pmatrix} g & \\ & 1 \end{pmatrix} \operatorname{diag} G\right) = \psi^*\left(\begin{pmatrix} k_1 & \\ & k_2^{-1} \end{pmatrix} \begin{pmatrix} a^{1/2} & \\ & a^{-1/2} \end{pmatrix} \operatorname{diag} G\right).$$

To calculate the effect of the K^* action on the left, we rewrite (9.54) by means of (9.36) as

$$\psi_\mathbb{C}(k_\mathbb{C} g_\mathbb{C}) = c \int_{\operatorname{diag} K} (\tau(k_\mathbb{C} l)\phi_{\Lambda_\mathbb{C}}, \phi_K) e^{-(\lambda_\mathbb{C} + \rho_{G_\mathbb{C}}) H_\mathbb{C}(g_\mathbb{C}^{-1} l)}\, dl$$

with $c \neq 0$. This formula makes it clear how to obtain the analytic continuation of the $K_\mathbb{C}$ dependence. Since

$$a_1 = \begin{pmatrix} a^{1/2} & \\ & a^{-1/2} \end{pmatrix}$$

is in $G^* \cap G_\mathbb{C} = G_0$, we can use the Schwarz inequality to obtain

$$|\psi(g)| \leq c |\phi_{\Lambda_\mathbb{C}}|\, |\phi_K| \int_{\operatorname{diag} K} e^{-(\lambda_\mathbb{C} + \rho_{G_\mathbb{C}}) H_\mathbb{C}(a_1^{-1} l)}\, dl.$$

To complete the proof, it is enough by Proposition 5.28 to prove that

$$\int_{\operatorname{diag} K} e^{-(\lambda_\mathbb{C} + \rho_{G_\mathbb{C}}) H_\mathbb{C}(a_1^{-1} l)}\, dl \leq C e^{-(1+\varepsilon)\rho_G \log a} \tag{9.62}$$

for some $\varepsilon > 0$ and all $a \in A_\mathfrak{p}^+$ when $\langle \lambda, \alpha \rangle$ is sufficiently large for all $\alpha \in \Delta^+$.

Let $\beta_1, \ldots, \beta_n$ be the simple restricted roots of $G_\mathbb{C}$, and define "dual" elements ω_i by $2\langle \omega_i, \beta_j \rangle / |\beta_j|^2 = 4\delta_{ij}$. By Proposition 7.19, there exists an irreducible finite-dimensional representation Φ_i of $G_\mathbb{C}$ with highest restricted weight ω_i, $1 \leq i \leq n$. The assumed dominance of λ means that $\lambda_\mathbb{C} + \rho_{G_\mathbb{C}}$ is a nonnegative combination of the ω_i, say $\lambda_\mathbb{C} + \rho_{G_\mathbb{C}} = \sum c_i \omega_i$. Now Proposition 7.17 says that

$$e^{2\omega_i H_\mathbb{C}(g_\mathbb{C})} = |\Phi_i(g_\mathbb{C})\phi_0^{(i)}|^2$$

if $\phi_0^{(i)}$ is any unit highest restricted weight vector of Φ_i.

We shall prove shortly that

$$H_{\mathbb{C}}(a_1^{-1}l) \neq 0 \qquad \text{for } a_1 \neq 1 \tag{9.63}$$

and that

$$H_{\mathbb{C}}(a_1^{-1}l) = H_{\mathbb{C}}(a_1 l). \tag{9.64}$$

Let us see that (9.63) and (9.64) allow us to complete the proof. Let $\{v_j^{(i)}\}$ be an orthonormal basis of the space for Φ_i that diagonalizes all $\Phi_i(a_1)$, and let $a_j^{(i)}$ be the eigenvalue of $\Phi_i(a_1)$ on $v_j^{(i)}$. If we write

$$\Phi_i(l)\phi_0^{(i)} = \sum b_j^{(i)} v_j^{(i)},$$

then

$$e^{2\omega_i H_{\mathbb{C}}(a_1^{-1}l)} = |\Phi_i(a_1^{-1}l)\phi_0^{(i)}|^2 = \sum_j (b_j^{(i)}/a_j^{(i)})^2. \tag{9.65}$$

In view of (9.64), we have also

$$e^{2\omega_i H_{\mathbb{C}}(a_1^{-1}l)} = \sum_j (b_j^{(i)} a_j^{(i)})^2. \tag{9.66}$$

Adding (9.65) and (9.66) and using the inequality $x + x^{-1} \geq 2$, we obtain

$$e^{2\omega_i H_{\mathbb{C}}(a_1^{-1}l)} \geq \sum_j (b_j^{(i)})^2 = 1. \tag{9.67}$$

Consequently (9.63) says that for each $a_1 \neq 1$ and for each l there is some i such that

$$e^{2\omega_i H_{\mathbb{C}}(a_1^{-1}l)} > 1.$$

Then (9.65) and (9.66) say that $a_j^{(i)} \neq 1$ for some j for which $b_j^{(i)} \neq 0$. For the element $a_t = \exp t(\log a_1)$, the same j gives

$$e^{2\omega_i H_{\mathbb{C}}(a_t^{-1}l)} \geq (b_j^{(i)})^2 \cosh(2t \log a_j^{(i)}). \tag{9.68}$$

An easy compactness argument allows us to get uniform constants on the right if we start with a_1 in the exponential of the unit sphere of $\overline{\mathfrak{a}_{\mathfrak{p}}^+}$. So each $a_1^{-1}l$ gives us the estimate (9.68) for one i and the estimate (9.67) for the remaining i's. Since $\lambda_{\mathbb{C}} + \rho_{G_{\mathbb{C}}} = \sum c_i\omega_i$, we get an exponential lower bound

$$e^{(\lambda_{\mathbb{C}} + \rho_{G_{\mathbb{C}}})H_{\mathbb{C}}(a_1^{-1}l)} \geq e^{\eta\rho_G \log a} \qquad (\eta > 0)$$

uniformly in a_1 and l. The number η can be taken as a fixed multiple of the smallest c_i. Thus (9.62) follows if $\langle\lambda, \alpha\rangle$ is sufficiently large for all α in Δ^+.

To complete the proof of the lemma, we thus need to prove (9.63) and (9.64). For (9.64), introduce the automorphism $\tilde{\theta}$ of $\mathfrak{g}_{\mathbb{C}}$ given by

$$\tilde{\theta}\begin{pmatrix} X + iY & \\ & \theta X - i\theta Y \end{pmatrix} = \begin{pmatrix} \theta X + i\theta Y & \\ & X - iY \end{pmatrix} \qquad \text{for } X, Y \in \mathfrak{g},$$

and let be the corresponding automorphism of $G_{\mathbb{C}}$. By inspection, $\tilde{\theta} = 1$

on $\mathfrak{a}$. It follows that $\tilde{\theta}$ carries each restricted root space to itself and hence satisfies $\tilde{\theta}(\mathfrak{n}_{\mathbb{C}}) = \mathfrak{n}_{\mathbb{C}}$. Also by inspection, $\tilde{\theta}(\mathfrak{u}) = \mathfrak{u}$. Therefore

$$H_{\mathbb{C}}(\tilde{\Theta} g_{\mathbb{C}}) = H_{\mathbb{C}}(g_{\mathbb{C}}) \tag{9.69}$$

for all $g_{\mathbb{C}}$ in $G_{\mathbb{C}}$. For X in $\mathfrak{p}(\mathfrak{g})$ we have

$$\tilde{\theta}\begin{pmatrix} X & \\ & -X \end{pmatrix} = \begin{pmatrix} -X & \\ & X \end{pmatrix};$$

thus $\tilde{\Theta}(a_1) = a_1^{-1}$. Also $\tilde{\theta}$ is the identity on diag $\mathfrak{k}$ and hence $\tilde{\Theta}(l) = l$ for l in diag K. Substituting $g_{\mathbb{C}} = a_1 l$ in (9.69), we obtain (9.64).

For (9.63), suppose $H_{\mathbb{C}}(a_1^{-1}l) = 0$. Then $H_{\mathbb{C}}(l^{-1}a_1^{-1}l) = 0$, and $x = l^{-1}a_1^{-1}l$ is a member of $G_{\mathbb{C}}$ of the form

$$x = \exp\begin{pmatrix} X & \\ & -X \end{pmatrix} \qquad \text{with } X \in \mathfrak{p}(\mathfrak{g}) \tag{9.70}$$

such that $H_{\mathbb{C}}(x) = 0$. Since $\tilde{\Theta} x = x^{-1}$, (9.69) says that the Iwasawa decompositions of x and x^{-1} are

$$x = un \quad \text{and} \quad x^{-1} = u'n' \qquad \text{with } u, u' \in U \text{ and } n, n' \in N_{\mathbb{C}}.$$

Since $\Theta x = x^{-1}$, we obtain

$$u(\Theta n) = \Theta x = x^{-1} = u'n'$$

and hence
$$\Theta n = u^{-1}u'n'.$$

In view of Proposition 7.17, $|\Phi(\theta n)v| = |v|$ whenever Φ is an irreducible finite-dimensional representation of $G_{\mathbb{C}}$ and v is a highest restricted weight vector. By (A.127), the exponential mapping carries $\bar{\mathfrak{n}}_{\mathbb{C}}$ onto $\bar{N}_{\mathbb{C}}$. Letting $\Theta n = \exp Y$ and taking into account the orthogonality of vectors belonging to different restricted weights, we see that $\Phi(\exp Y) - 1$ annihilates v. We expand Y according to restricted root spaces and let Y_1 be the component belonging to the restricted root $-\beta$ closest to 0. Then we see that the differential of Φ on Y_1 annihilates v, hence that β is orthogonal to the weight of v. Since Φ is arbitrary, we conclude that $Y_1 = 0$, hence that $Y = 0$, $\Theta n = 1$, $n = 1$, and x is in U. Since x is in U and x also is of the form (9.70), x is 1. Thus $a_1 = 1$. This completes the proof of the lemma.

Remarks. For the proof of the Lemma 9.21c without the restriction on λ, see the bibliographical notes. A way of avoiding this extra part of the proof is outlined in the Problems for Chapter XII.

From the lemma we see that there exist nonzero $Z(\mathfrak{g}^{\mathbb{C}})$-finite K-finite functions in $L^2(G)$ whenever rank G = rank K. In view of Corollary 8.42

this existence result is enough to prove that G possesses discrete series representations. The lemma does not by itself give us the fine detail of Theorem 9.20, however. To get at the fine detail, we shall study K types more closely.

§8. Limitations on K Types

We continue with the proof of Theorem 9.20, retaining the notation and conventions of §7. The next step is to get control of the K types that can occur in the closed linear span of the left and right translates of ψ.

Briefly the idea is as follows. We think in terms of the setting $G^*/\text{diag}\, G$, starting with a function φ in the space generated by ψ and assuming that φ transforms on both left and right according to the same irreducible representation of K. In the $G^*/\text{diag}\, G$ setting, φ^* transforms according to a representation of K^* with a vector fixed by diag K, and we arrange that φ^* (just like ψ^*) is fixed by diag K. We pass to $\varphi_{\mathbb{C}}$ in the setting of $G_{\mathbb{C}}/U$ and consider $\mathscr{L}(U(\mathfrak{g}_{\mathbb{C}}^{\mathbb{C}}))$ acting on $\varphi_{\mathbb{C}}$. Here any $\mathscr{L}(D)$ acts by *right-invariant* differentiation. We shall see that $\varphi_{\mathbb{C}}$ is an eigenfunction of the subset of operators $\mathscr{L}(Z(\mathfrak{k}_{\mathbb{C}}^{\mathbb{C}}))$, and the point of the argument will be to calculate the eigenvalue in two ways. One way comes directly from the $K_{\mathbb{C}}$ transformation law of $\varphi_{\mathbb{C}}$, and the other comes from realizing $\varphi_{\mathbb{C}}$ as the result of applying a differential operator to $\psi_{\mathbb{C}}$, which in turn is given by an explicit integral.

The important technical point of the argument will be to show that certain functions on $G_{\mathbb{C}}$ are linearly independent, so that a comparison of the eigenvalues is possible. We take up the question of proving this linear independence only after we see explicitly the need for it.

We begin by giving a few facts about right-invariant differential operators on any G. If X is in the Lie algebra $\mathfrak{g}$, the corresponding **right-invariant differential operator** is

$$X_R f(x) = \frac{d}{dt} f((\exp tX)^{-1}x)\bigg|_{t=0}$$

We can rewrite X_R in terms of the left-invariant operator X as follows: If $f^{\sim}(x) = f(x^{-1})$, then

$$X_R f = (X(f^{\sim}))^{\sim}. \tag{9.71}$$

We extend (9.71) to a definition of D_R for any D in $U(\mathfrak{g}^{\mathbb{C}})$ by letting

$$D_R f = (D(f^{\sim}))^{\sim}.$$

The operator D_R gives the effect $\mathscr{L}(D)$ of D in the left regular representation of G on G, and we have $(DD')_R = D_R D'_R$.

Proposition 9.22. Let f^* and $f_{\mathbb{C}}$ be real analytic functions that correspond under the map (9.45) and that transform according to an irreducible holomorphic representation τ of $(K^*)^{\mathbb{C}} = (K_{\mathbb{C}})^{\mathbb{C}}$. If D is a member of $U((\mathfrak{g}^*)^{\mathbb{C}}) = U((\mathfrak{g}_{\mathbb{C}})^{\mathbb{C}})$ such that $D_R f^*$ transforms according to an irreducible holomorphic representation τ' of $(K^*)^{\mathbb{C}}$, then $D_R f_{\mathbb{C}}$ transforms according to τ' and $D_R f^*$ corresponds to $D_R f_{\mathbb{C}}$ under the map (9.45).

Proof. As in Proposition 9.19, $D_R f^*$ and $D_R f_{\mathbb{C}}$ agree on G_0. Since D_R acts on the left of the group variable, $D_R f^*$ and $D_R f_{\mathbb{C}}$ are invariant under diag G and U, respectively, on the right. We shall produce a holomorphic function F on $(K^*)^{\mathbb{C}} \times G_0 = (K_{\mathbb{C}})^{\mathbb{C}} \times G_0$ that restricts to $D_R f^*(k^* g_0)$ and $D_R f_{\mathbb{C}}(k_{\mathbb{C}} g_0)$ on $K^* \times G_0$ and $K_{\mathbb{C}} \times G_0$, respectively. Uniqueness for analytic extensions then completes the proof. If $\mathscr{L}$ denotes the left action by $(K^*)^{\mathbb{C}}$ or $(K_{\mathbb{C}})^{\mathbb{C}}$, the function F is

$$F(k_{\mathbb{C}}^{\mathbb{C}}, g_0) = (\mathrm{Ad}(k_{\mathbb{C}}^{\mathbb{C}})D)_R(\mathscr{L}(k_{\mathbb{C}}^{\mathbb{C}})f^*)(g_0),$$

and the proposition follows.

Let V be the closed linear span in $L^2(G^*/\mathrm{diag}\, G)$ of the left G^* translates of ψ^*, with all the functions viewed as defined on G^*. (The function ψ^* is in $L^2(G^*/\mathrm{diag}\, G)$ by Lemma 9.21c.) Then $\mathscr{L}$ defines a unitary representation of G^* on V.

Lemma 9.23. The representation $\mathscr{L}$ of G^* on V is admissible.

Proof. Let V_0 be the vector space span of $\mathscr{L}(G^*)\psi^*$, let $\tau_{\lambda'} \otimes \tau_{\lambda''}$ be an irreducible representation of K^*, and let P be the orthogonal projection of V on the K^* type $\tau_{\lambda'} \otimes \tau_{\lambda''}$. We can write $P = P_{\lambda'}P_{\lambda''}$, where

$$P_{\lambda'} = d_{\lambda'} \int_K \overline{\chi_{\lambda'}(k, 1)}\mathscr{L}(k, 1)\, dk$$

and $P_{\lambda''}$ is defined similarly relative to the second coordinate. Since P is a bounded projection operator, $P(V)$ equals the closure of $P(V_0)$. Let $F = \sum_{i,j} c_{ij}\mathscr{L}(g_i, g_j)\psi^*$ be in V_0. Then we can write

$$P(F) = P_{\lambda'}\left(\sum_i \mathscr{L}(g_i, 1)\, P_{\lambda''}\left(\sum_j c_{ij}\mathscr{L}(1, g_j)\psi^*\right)\right). \tag{9.72}$$

Let us interpret this equation on G rather than on $G^*/\mathrm{diag}\, G$. By Lemma 9.21, ψ is $Z(\mathfrak{g}^{\mathbb{C}})$-finite K-finite (on both sides) and lies in $L^2(G)$. By Corollary 8.42, ψ lies in a subspace that is the finite sum of discrete series under the right regular representation R of G. The functions

$$P_{\lambda''}\left(\sum_j c_{ij}\mathscr{L}(1, g_j)\psi^*\right), \qquad \text{realized on } G, \tag{9.73}$$

lie in the $\tau_{\chi''}$ K type for this subrepresentation of R, and hence the space of all functions (9.73) is finite-dimensional. Moreover, these functions are $Z(\mathfrak{g}^{\mathbb{C}})$-finite K-finite in $L^2(G)$, and hence each such lies in the finite direct sum of discrete series under the left regular representation, again by Corollary 8.42. Hence $P(V_0)$, which is the space of all functions (9.72), is finite-dimensional. Then $P(V)$ is finite-dimensional, and admissibility is proved.

Now suppose that the contragredient of a K^* type of the form $\tau' = \tau_{\Lambda'}^{\vee} \otimes \tau_{\Lambda'}$ occurs in V. By a computation similar to that in Proposition 8.6, we see that τ' has a (diag K)-fixed vector. Therefore we can find a function φ^* in V that transforms irreducibly according to τ' under $\mathscr{L}(K^*)$ and is invariant under the subgroup diag K. Because of Lemma 9.23 and because of the identity $\mathscr{L}(D) = D_R$, we can write

$$\varphi^* = \mathscr{L}(D)\psi^* = D_R\psi^*$$

for some member D of $U(\mathfrak{g}^{*\mathbb{C}})$. Averaging D by Ad(diag K) and using the left (diag K)-invariance of φ^* and ψ^*, we see that we may assume D is fixed by Ad(diag K).

As a consequence of Schur's Lemma, $\tau'^{\vee}|_{\text{diag } K}$ and $\tau'|_{\text{diag } K}$ each contain the trivial representation of diag K with multiplicity one. Also since K^* is compact, $\tau'^{\vee}$ and τ' extend to holomorphic representations of $K^{*\mathbb{C}}$ (by the Borel-Weil Theorem, for example). The highest weight of τ' relative to $(\mathfrak{a} \oplus \mathfrak{m})^{\mathbb{C}}$ is $\Lambda'_{\mathbb{C}}$, where $\Lambda'_{\mathbb{C}}$ and Λ' are related as in (9.51).

Taking into account Proposition 9.22, we can therefore pass to the dual situation with $G_{\mathbb{C}}/U$ by the usual mapping (9.45). The function $\varphi_{\mathbb{C}}$ is right U-invariant, transforms according to $\tau'^{\vee}$ on the left under $K_{\mathbb{C}}$, is fixed under left translation by diag K, and satisfies

$$\varphi_{\mathbb{C}} = \mathscr{L}(D)\psi_{\mathbb{C}} = D_R\psi_{\mathbb{C}}. \tag{9.74}$$

We immediately obtain our first formula for the desired eigenvalue.

Lemma 9.24. For each $g_{\mathbb{C}}$ in $G_{\mathbb{C}}$, the function $\varphi_{\mathbb{C}}$ on $G_{\mathbb{C}}$ is an eigenfunction of $\mathscr{L}(Z(\mathfrak{k}_{\mathbb{C}}^{\mathbb{C}}))$. For an element $Z_{\mathbb{C}} = Z \otimes 1$ with Z in $Z(\mathfrak{k}^{\mathbb{C}})$, the eigenvalue is $\chi_{-(\Lambda'+\delta_K)}(Z)$.

Remark. Here the infinitesimal character $\chi_{-\Lambda'-\delta_K}$ is understood to be relative to $\mathfrak{k}_{\mathbb{C}}$, rather than all of $\mathfrak{g}_{\mathbb{C}}$.

Proof. Let $Z_{\mathbb{C}}$ be arbitrary in $Z(\mathfrak{k}_{\mathbb{C}}^{\mathbb{C}})$. In view of (9.44), we have

$$= \sum_{i,j} c_j \tau_{ij}'^{\vee}(k_{\mathbb{C}})\varphi_i(g_{\mathbb{C}}).$$

By (8.10) and the irreducibility of $\tau'^{\vee}$, $\mathscr{L}(Z_{\mathbb{C}})$ therefore acts by the scalar $\tau'^{\vee}(Z_{\mathbb{C}})$. To compute $\tau'^{\vee}(Z_{\mathbb{C}})$, we decompose $\mathfrak{k}_{\mathbb{C}}$ as

$$\mathfrak{k}_{\mathbb{C}} = (\mathfrak{n}_{\mathbb{C}} \cap \mathfrak{k}_{\mathbb{C}}) \oplus \mathfrak{a} \oplus \mathfrak{m} \oplus (\bar{\mathfrak{n}}_{\mathbb{C}} \cap \mathfrak{k}_{\mathbb{C}}),$$

expand $Z_{\mathbb{C}}$ out in corresponding fashion via the Birkhoff-Witt Theorem, and apply the resulting expansion to a lowest restricted weight vector of $\tau'^{\vee}$. Then we see that the eigenvalue is $(-\Lambda'_{\mathbb{C}})(\mu'_{\Sigma^-(\mathfrak{k}_{\mathbb{C}})}(Z_{\mathbb{C}}))$.

Finally suppose that $Z_{\mathbb{C}}$ is of the special form $Z \oplus 1$. Then we can argue as in the proof of Lemma 9.21a to see that the expression for the eigenvalue simplifies to $\chi_{-\Lambda'-\delta_K}(Z)$. This completes the proof.

Now we shall compute the desired eigenvalue a second way. Starting from the identity

$$\psi_{\mathbb{C}}(g_{\mathbb{C}}) = \int_{\operatorname{diag} K} e^{-(\Lambda_{\mathbb{C}} + 2\rho_{K\mathbb{C}})H_{\mathbb{C}}(g_{\mathbb{C}}^{-1}l)}\, dl.$$

we apply (9.74) and the Ad(diag K) invariance of D to write

$$\varphi_{\mathbb{C}}(g_{\mathbb{C}}) = \int_{\operatorname{diag} K} [De^{-(\Lambda_{\mathbb{C}} + 2\rho_{K_{\mathbb{C}}})H_{\mathbb{C}}(x)}]_{x = g_{\mathbb{C}}^{-1}l}\, dl. \tag{9.75}$$

Let $\beta_1, \ldots, \beta_n$ be the noncompact positive roots of $(\mathfrak{g}^{\mathbb{C}}, \mathfrak{b}^{\mathbb{C}})$, and let $W_{-\beta_1}, \ldots, W_{-\beta_n}$ be root vectors for their negatives. Then

$$X_{-\beta_j} = \begin{pmatrix} W_{-\beta_j} & \\ & \theta W_{-\beta_j} \end{pmatrix} \quad \text{and} \quad Y_{-\beta_j} = \begin{pmatrix} iW_{-\beta_j} & \\ & \theta(iW_{-\beta_j}) \end{pmatrix}$$

are restricted root vectors. Let $\mathfrak{s}$ denote the linear span in $\mathfrak{g}_{\mathbb{C}}$ of

$$X_{-\beta_1}, Y_{-\beta_1}, \ldots, X_{-\beta_n}, Y_{-\beta_n}. \tag{9.76}$$

Then we easily verify that $\mathfrak{g}_{\mathbb{C}}$ has a direct sum decomposition

$$\mathfrak{g}_{\mathbb{C}} = \operatorname{diag} \mathfrak{k} \oplus \mathfrak{s} \oplus \mathfrak{a} \oplus \mathfrak{n}_{\mathbb{C}}. \tag{9.77}$$

We choose an ordered basis for $\mathfrak{g}_{\mathbb{C}}$ compatible with the decomposition (9.77), taking (9.76) as the ordered basis of the $\mathfrak{s}$ part. Then we expand members of $U((\mathfrak{g}_{\mathbb{C}})^{\mathbb{C}})$ by the Birkhoff-Witt Theorem in the basis of monomials given by that theorem. Specializing to the member D, we group terms containing like monomials

$$X_{-\beta_1}^{p_1} Y_{-\beta_1}^{q_1} \cdots X_{-\beta_n}^{p_n} Y_{-\beta_n}^{q_n}$$

from $\mathfrak{s}$, and then, with obvious notation, we can view D as a member of

$$\sum_{(p,q)} U((\operatorname{diag} \mathfrak{k})^{\mathbb{C}})(XY)_{-\beta}^{(p,q)} U(\mathfrak{a}^{\mathbb{C}}) U((\mathfrak{n}_{\mathbb{C}})^{\mathbb{C}}).$$

When we substitute for D in (9.75), any term containing a nonconstant member of $U(\text{diag}\, \mathfrak{k}^{\mathbb{C}})$ drops out because of the integration in the l variable, and any term containing a nonconstant member of $U(\mathfrak{n}_{\mathbb{C}}^{\mathbb{C}})$ drops out because $H_{\mathbb{C}}(\cdot)$ is right $N_{\mathbb{C}}$-invariant. Consequently there are members $H_{(p,q)}$ of $U(\mathfrak{a}^{\mathbb{C}})$ such that

$$\varphi_{\mathbb{C}}(g_{\mathbb{C}}) = \sum_{(p,q)} \int_{\text{diag}\, K} [(XY)_{-\beta}^{(p,q)} H_{(p,q)} e^{-(\Lambda_{\mathbb{C}} + 2\rho_{K_{\mathbb{C}}})H_{\mathbb{C}}(x)}]_{x = g_{\mathbb{C}}^{-1} l}\, dl. \tag{9.78}$$

It is necessary to take into account the role of $M = Z_U(A) = \exp(\text{diag}\, \mathfrak{b})$. The function $H_{\mathbb{C}}(\cdot)$ is right M-invariant, and also the integration over l in diag K in (9.78) enables us to insert an m on the right of l. Averaging over m in M, we see we may as well replace our differential operators by Ad(M)-invariant operators. Each $H_{(p,q)}$ is fixed by Ad(M). In the case of $(XY)_{-\beta}^{(p,q)}$, both $X_{-\beta_j}$ and $Y_{-\beta_j}$ are mapped by any Ad(m) into linear combinations of $X_{-\beta_j}$ and $Y_{-\beta_j}$, and $X_{-\beta_j}$ commutes with $Y_{-\beta_j}$. Hence the Ad(M) average of $(XY)_{-\beta}^{(p,q)}$ is a linear combination of various $(XY)_{-\beta}^{(p',q')}$. Moreover, every pair (p', q') that results from the (p, q) term has $p_j' + q_j' = p_j + q_j$ for all j. Consequently $\sum (p_j' + q_j')\beta_j$ is the same for all the terms that result from averaging $(XY)_{-\beta}^{(p,q)}$.

Let us now think of evaluating $H_{p,q}$ on the exponential in (9.78), recognizing that it gives a scalar, and let us group the averaged versions of the various $(XY)_{-\beta}^{(p,q)}$ according to their homogeneity

$$\sum \beta = \sum (p_j + q_j)\beta_j.$$

The result is that there exist nonzero Ad(M)-invariant homogeneous polynomials $P_{\Sigma\beta}(X, Y)$ such that

$$\varphi_{\mathbb{C}}(g_{\mathbb{C}}) = \sum_{\Sigma\beta} \int_{\text{diag}\, K} [P_{\Sigma\beta}(X, Y) e^{-(\Lambda_{\mathbb{C}} + 2\rho_{K_{\mathbb{C}}})H_{\mathbb{C}}(x)}]_{x = g_{\mathbb{C}}^{-1} l}\, dl. \tag{9.79}$$

The second result that we need about nonvanishing coefficients is the following lemma.

Lemma 9.25. The functions on $G_{\mathbb{C}}$ given by

$$\int_{\text{diag}\, K} [P_{\Sigma\beta}(X, Y) e^{-(\Lambda_{\mathbb{C}} + 2\rho_{K_{\mathbb{C}}})H_{\mathbb{C}}(x)}]_{x = g_{\mathbb{C}}^{-1} l}\, dl \tag{9.80}$$

are linearly independent.

We postpone the proof to §9. From this lemma we can obtain our second formula for the desired eigenvalue.

Lemma 9.26. Suppose that the term of homogeneity $\sum \beta$ is present in (9.79). If $Z_{\mathbb{C}} = Z \otimes 1$ with Z in $Z(\mathfrak{k}^{\mathbb{C}})$, then $\mathscr{L}(Z_{\mathbb{C}})$ acts on $\varphi_{\mathbb{C}}$ by the scalar $\chi_{-(\Lambda + \delta_K + \frac{1}{2}\Sigma\beta)}$.

Remark. Again the infinitesimal character is relative to $\mathfrak{k}_{\mathbb{C}}$, not $\mathfrak{g}_{\mathbb{C}}$.

Proof. To calculate $\mathscr{L}(Z_{\mathbb{C}})\varphi_{\mathbb{C}}$, we first form

$$\begin{aligned}\mathscr{L}(k_{\mathbb{C}})\varphi(g_{\mathbb{C}}) &= \varphi_{\mathbb{C}}(k_{\mathbb{C}}^{-1}g_{\mathbb{C}}) \\ &= \sum_{\Sigma\beta} \int_{\operatorname{diag} K} [P_{\Sigma\beta}(X, Y)e^{-(\Lambda_{\mathbb{C}}+2\rho_{K_{\mathbb{C}}})H_{\mathbb{C}}(x)}]_{x=g_{\mathbb{C}}^{-1}k_{\mathbb{C}}l}\, dl.\end{aligned}$$

Then $\mathscr{L}(Z_{\mathbb{C}})\varphi_{\mathbb{C}}(g_{\mathbb{C}})$ has the same formal appearance, with $Z_{\mathbb{C}}$ replacing $k_{\mathbb{C}}$ on the right. However, $Z_{\mathbb{C}}$ is Ad(diag K)-invariant and thus commutes with the l. Thus.

$$\mathscr{L}(Z_{\mathbb{C}})\varphi_{\mathbb{C}}(g_{\mathbb{C}}) = \sum_{\Sigma\beta} \int_{\operatorname{diag} K} [Z_{\mathbb{C}}P_{\Sigma\beta}(X, Y)e^{-(\Lambda_{\mathbb{C}}+2\rho_{K_{\mathbb{C}}})H_{\mathbb{C}}(x)}]_{x=g_{\mathbb{C}}^{-1}l}\, dl. \tag{9.81}$$

Write
$$\mathfrak{k}_{\mathbb{C}} = \operatorname{diag}\mathfrak{k} \oplus \mathfrak{a} \oplus (\mathfrak{n}_{\mathbb{C}} \cap \mathfrak{k}_{\mathbb{C}})$$

and expand $Z_{\mathbb{C}}$ by the Birkhoff-Witt Theorem into a sum of members of

$$U(\operatorname{diag}\mathfrak{k}^{\mathbb{C}})U(\mathfrak{a}^{\mathbb{C}})U((\mathfrak{n}_{\mathbb{C}} \cap \mathrm{k}_{\mathbb{C}})^{\mathbb{C}}).$$

Any term with a nonconstant member of $U(\operatorname{diag}\mathfrak{k}^{\mathbb{C}})$ drops out in (9.81) because of the integration in the l variable, and any term with a nonconstant member of $U((\mathfrak{n}_{\mathbb{C}} \cap \mathfrak{k}_{\mathbb{C}})^{\mathbb{C}})$ must have a nonconstant member of $U(\operatorname{diag}\mathfrak{k}^{\mathbb{C}})$ present and therefore drops out. Hence $Z_{\mathbb{C}}$ can be replaced on the right side of (9.81) by the $U(\mathfrak{a}^{\mathbb{C}})$ part of $\mu'_{\Sigma^-(\mathfrak{k}_{\mathbb{C}})}(Z_{\mathbb{C}})$.

For any H in $U(\mathfrak{a}^{\mathbb{C}})$, (9.53) gives

$$HP_{\Sigma\beta}(X, Y) = P_{\Sigma\beta}(X, Y)\{H - (\tfrac{1}{2}\textstyle\sum\beta)_{\mathbb{C}}(H)1\}.$$

Thus the effect of $Z_{\mathbb{C}}$ on the $\sum\beta$ term in (9.81) is to multiply it by

$$-(\Lambda_{\mathbb{C}} + 2\rho_{K_{\mathbb{C}}})(\mu'_{\Sigma^+(\mathfrak{k}_{\mathbb{C}})}(Z_{\mathbb{C}})) - (\tfrac{1}{2}\textstyle\sum\beta)_{\mathbb{C}}(\mu'_{\Sigma^+(\mathfrak{k}_{\mathbb{C}})}(Z_{\mathbb{C}})).$$

The asserted independence in Lemma 9.25 therefore says that $\mathscr{L}(Z_{\mathbb{C}})$ acts on $\varphi_{\mathbb{C}}$ with the eigenvalue

$$(-(\Lambda_{\mathbb{C}} + 2\rho_{K_{\mathbb{C}}} + (\tfrac{1}{2}\textstyle\sum\beta)_{\mathbb{C}}))(\mu'_{\Sigma^+(\mathfrak{k}_{\mathbb{C}})}(Z_{\mathbb{C}})).$$

By (9.52a) and the kind of calculation yielding (9.60), this expression equals

$$(-(\Lambda + 2\delta_K + \tfrac{1}{2}\textstyle\sum\beta))(\gamma'_{\Delta_K^+}(Z)).$$

Formula (8.25) identifies this expression as $\chi_{-(\Lambda+\delta_K+\frac{1}{2}\Sigma\beta)}(Z)$ and completes the proof of the lemma.

Comparing Lemma 9.24 and Lemma 9.26 and bringing in Proposition 8.20, we obtain

$$\Lambda + \delta_K + \tfrac{1}{2}\textstyle\sum\beta = w_K(\Lambda' + \delta_K) \tag{9.82}$$

for some w_K in W_K. Since $w_K(\Lambda' + \delta_K)$ is a weight of the representation of $\mathfrak{k}$ with highest weight $\Lambda' + \delta_K$, the right side of (9.82) is the difference of $\Lambda' + \delta_K$ and a sum of positive compact roots. Thus

$$\Lambda' = \Lambda + \sum_{\alpha \in \Delta^+} n_\alpha \alpha \tag{9.83}$$

with each n_α an integer or half integer ≥ 0.

Lemma 9.27. The K^* type $\tau^\vee$ has multiplicity one in the representation $\mathscr{L}$ on the space V generated by $\mathscr{L}(G^*)\psi^*$, and $\mathscr{L}(G^*)$ acts irreducibly on V.

Proof. If $\tau^\vee$ were to occur a second time, we could arrange in the above argument for $\varphi_{\mathbb{C}}$ to be a function genuinely distinct from $\psi_{\mathbb{C}}$ but still transforming by $\tau^\vee$. Tracing over the above argument, we see that the member D of $U((\mathfrak{g}_{\mathbb{C}})^{\mathbb{C}})$ implementing (9.74) has no $\mathfrak{s}$ part. From (9.78) we then can read off that $\varphi_{\mathbb{C}}$ is a multiple of $\psi_{\mathbb{C}}$, contradiction.

Thus $\tau^\vee$ has multiplicity one. Now suppose P is an orthogonal projection in V commuting with $\mathscr{L}(G^*)$. Since $\tau^\vee$ has multiplicity one, $P\psi^* = c\psi^*$ for some constant c. If $\varphi^* = \mathscr{L}(D)\psi^*$ is any K^*-finite member of V, we then have

$$P\varphi^* = P\mathscr{L}(D)\psi^* = \mathscr{L}(D)P\psi^* = c\mathscr{L}(D)\psi^* = c\varphi^*.$$

Hence P is scalar, and $\mathscr{L}(G^*)$ acts irreducibly on V.

Proof of Theorem 9.20. We pass from $G^*/\text{diag}\, G$ to G. Think of V as the closed linear span in $L^2(G)$ of the left and right G translates of ψ. According to Lemmas 9.21b and 9.27, the representation (L, R) of $G \times G$ on V is irreducible and the $K \times K$ type $\tau_\Lambda^\vee \otimes \tau_\Lambda$ occurs with multiplicity one. Then we can choose a function f on K so that $L(f)$ acts as the orthogonal projection of the $\tau_\Lambda^\vee \otimes \tau_\Lambda$ space E to a subspace E_1 that is irreducible under $R(K)$. Put $\psi_1 = L(f)\psi$, and let V_1 be the closed linear span of $R(G)\psi_1$.

Lemma 9.21a shows that V_1 has infinitesimal character χ_λ, and Corollary 8.42 shows that V_1 is admissible and is a finite direct sum of discrete series. Let us see that V_1 is irreducible.

First we show that $E \cap V_1 = E_1$. In fact, $E_1 \subseteq E$ trivially, and $E_1 \subseteq V_1$ since the irreducibility of E_1 under $R(K)$ implies $E_1 = \text{span}\{R(K)\psi_1\} \subseteq V_1$. For the reverse inclusion, let v_1 in $E \cap V_1$ be given. Since v_1 is in V_1, we can choose D in $U(\mathfrak{g}^{\mathbb{C}})$ with $R(D)\psi_1 = v_1$. Since v_1 is in E, $L(f)$ is defined on v_1. Then

$$L(f)v_1 = L(f)R(D)\psi_1 = R(D)L(f)\psi_1 = R(D)\psi_1 = v_1,$$

and v_1 is in E_1.

Now ψ_1 is $L(K)$-finite of K type $\tau_\Lambda^\vee$. Since $L(K)$ commutes with $R(K)$, every member of V_1 is $L(K)$-finite of K type $\tau_\Lambda^\vee$. In particular, this statement is true of any member of V_1 that is $R(K)$-finite of K type τ_Λ. Consequently any member of V_1 in the τ_Λ subspace transforms by $\tau_\Lambda^\vee \otimes \tau_\Lambda$ under $(L(K), R(K))$ and hence lies in E. From the previous paragraph we conclude τ_Λ occurs with multiplicity one in V_1. The same kind of argument as in Lemma 9.27 allows us to deduce that $R(G)$ acts irreducibly on V_1.

Let π_λ be the restriction of R to V_1. So far we have proved that π_λ is a discrete series representation of G and that (a) and (b) hold in the theorem. It follows from §4 that $\bar{\pi}_\lambda \otimes \pi_\lambda$ imbeds in V. (Actually Lemma 9.27 then forces $\bar{\pi}_\lambda \otimes \pi_\lambda$ to exhaust V.) If $\tau_{\Lambda'}$ occurs as a K type in π_λ, then $\tau_{\Lambda'}^\vee \otimes \tau_{\Lambda'}$ must occur as a $K \times K$ type in $\bar{\pi}_\lambda \otimes \pi_\lambda$, and it follows from (9.83) that (c) holds in the theorem, except possibly with half integers for some of the n_α's.

Let us see that the n_α's are actually integers. The highest weight vector of the τ_Λ K type of π_λ generates the whole representation. Thus by applying members of $\mathfrak{b}^{\mathbb{C}}$ and root vectors, we can generate a spanning set for the K-finite vectors. All such vectors are weight vectors, and their weights differ from Λ by an integral combination of roots. It therefore follows that the numbers n_α in (c) are integers. This proves (c).

If λ and λ' are conjugate under W_K, it is a routine matter to track down an equivalence of π_λ and $\pi_{\lambda'}$ by using a representative in K of the element of W_K. Conversely suppose λ and λ' are not conjugate under W_K. If π_λ and $\pi_{\lambda'}$ are equivalent, then their infinitesimal characters must agree, and (a) of the theorem implies λ and λ' are conjugate via W_G. We may assume λ is Δ^+ dominant and λ' is Δ_K^+ dominant. Let

$$\Lambda = \lambda + \delta_G - 2\delta_K \qquad \text{and} \qquad \Lambda' = \lambda' + \delta'_G - 2\delta_K \tag{9.84}$$

be the K types given by (b) of the theorem. Applying (c), we have

$$\Lambda' + 2\delta_K = \Lambda + 2\delta_K + \sum_{\alpha\in\Delta^+} n_\alpha\alpha = \lambda + \delta_G + \sum_{\alpha\in\Delta^+} n_\alpha\alpha$$

Since $\lambda + \delta_G$ is Δ^+ dominant,

$$\begin{aligned}|\Lambda' + 2\delta_K|^2 &= |\lambda + \delta_G|^2 + 2\langle\lambda + \delta_G, \textstyle\sum n_\alpha\alpha\rangle + |\textstyle\sum n_\alpha\alpha|^2 \\ &\geq |\Lambda + 2\delta_K|^2 + |\textstyle\sum n_\alpha\alpha|^2.\end{aligned} \tag{9.85}$$

Hence $|\Lambda' + 2\delta_K|^2 \geq |\Lambda + 2\delta_K|^2$ with equality only if $\sum n_\alpha\alpha = 0$ (and hence $\Lambda' = \Lambda$). Reversing the roles of Λ and Λ', we obtain $|\Lambda + 2\delta_K|^2 \geq |\Lambda' + 2\delta_K|^2$. We are forced to conclude $\Lambda' = \Lambda$. From (9.84), we find $\lambda + \delta_G = \lambda' + \delta'_G$. Since $\lambda + \delta_G$ and $\lambda' + \delta'_G$ are nonsingular and are dominant for their respective positive systems in Δ, we conclude that their respective positive systems are the same. Thus λ and λ' are both Δ^+ dominant and are conjugate via W_G. Consequently $\lambda' = \lambda$.

This completes the proof of Theorem 9.20, except for one question about linear independence addressed in Lemma 9.25. We take up this final detail in the next section.

§9. Lemma on Linear Independence

We have one point to dispose of in order to complete the proof of Theorem 9.20. This is the matter of proving the linear independence asserted in Lemma 9.25, and we shall take up this question now.

We begin with a lemma fundamental to the subject of analysis on symmetric spaces. It generalizes the radial convergence of Poisson integrals in the unit disc to their boundary values when the boundary values are continuous. The data for the special case of Poisson integrals are $G = SU(1, 1)$ and $\nu = \rho$.

Lemma 9.28. Let G be linear connected reductive, and let $S_\mathfrak{p} = M_\mathfrak{p} A_\mathfrak{p} N_\mathfrak{p}$ be a minimal parabolic subgroup. If ν is any member of $(\mathfrak{a}_\mathfrak{p}')^\mathbb{C}$ with $\operatorname{Re}\nu$ in the open positive Weyl chamber and if f is any continuous function on K that is right invariant under $M_\mathfrak{p}$, then

$$\lim_{\substack{a\to\infty\\ S_\mathfrak{p}}} e^{-(\nu-\rho)\log a}\int_K e^{-(\nu+\rho)H(a^{-1}k)}f(k)\,dk = f(1)\int_{\bar{N}_\mathfrak{p}} e^{-(\nu+\rho)H(\bar{n})}\,d\bar{n}.$$

Proof. By (5.25) and (5.15) we have

$$\begin{aligned}
&\int_K e^{-(\nu+\rho)H(a^{-1}k)}f(k)\,dk\\
&\quad= \int_{\bar{N}_\mathfrak{p}} e^{-(\nu+\rho)H(a^{-1}\kappa(\bar{n}))}e^{-2\rho H(\bar{n})}f(\kappa(\bar{n}))\,d\bar{n}\\
&\quad= \int_{\bar{N}_\mathfrak{p}} e^{(\nu-\rho)H(\bar{n})}e^{-(\nu+\rho)H(a^{-1}\bar{n})}f(\kappa(\bar{n}))\,d\bar{n}\\
&\quad= e^{(\nu+\rho)\log a}\int_{\bar{N}_\mathfrak{p}} e^{(\nu-\rho)H(\bar{n})}e^{-(\nu+\rho)H(a^{-1}\bar{n}a)}f(\kappa(\bar{n}))\,d\bar{n}\\
&\quad= e^{(\nu-\rho)\log a}\int_{\bar{N}_\mathfrak{p}} e^{(\nu-\rho)H(a\bar{n}a^{-1})}e^{-(\nu+\rho)H(\bar{n})}f(\kappa(a\bar{n}a^{-1}))\,d\bar{n}.
\end{aligned}$$

The last paragraph of the proof of Lemma 7.23 shows that the integrand is dominated by a fixed integrable function for a in $A_\mathfrak{p}^+$. Moreover, $a\bar{n}a^{-1} \to 1$ since $\operatorname{Ad}(a)X \to 0$ for X in $\bar{\mathfrak{n}}_\mathfrak{p}$ and since exp carries $\bar{\mathfrak{n}}_\mathfrak{p}$ onto $\bar{N}_\mathfrak{p}$. (See (A.127).) Thus we have dominated convergence, and the lemma follows.

Proposition 9.29. Let G be linear connected reductive, and let $S_\mathfrak{p} = M_\mathfrak{p} A_\mathfrak{p} N_\mathfrak{p}$ be a minimal parabolic subgroup. Suppose that ν is a member of the open positive Weyl chamber of $\mathfrak{a}_\mathfrak{p}'$, and suppose that T is a distribution on K. If T annihilates the function $k \to e^{-(\nu+\rho)H(g^{-1}k)}$ for every g in G, then T annihilates every right $M_\mathfrak{p}$-invariant function in $C^\infty(K)$.

Remark. For $G = \mathrm{SU}(1, 1)$ and $\nu = \rho$, the proposition says that the Poisson integral is a one-one mapping on distributions.

Proof. Let F be a right $M_\mathfrak{p}$-invariant function in $C^\infty(K)$, and let f be the smooth right $M_\mathfrak{p}$-invariant function given by

$$f(k) = \int_{M_\mathfrak{p}} T\{R(m^{-1}k^{-1})F\}\, dm,$$

where R denotes the right regular representation. Then we have

$$\begin{aligned}\int_K e^{-(\nu+\rho)H(g^{-1}k)}f(k)\, dk &= \int_{K\times M_\mathfrak{p}} e^{-(\nu+\rho)H(g^{-1}k)}T\{R(m^{-1}k^{-1}F\}\, dm\, dk \\ &= \int_K e^{-(\nu+\rho)H(g^{-1}k)}T\{k' \to F(k'k^{-1})\}\, dk \\ &= T\left\{k' \to \int_K e^{-(\nu+\rho)H(g^{-1}k)}F(k'k^{-1})\, dk\right\} \\ &= T\left\{k' \to \int_K e^{-(\nu+\rho)H(g^{-1}k^{-1}k')}F(k)\, dk\right\} \\ &= \int_K T\{k' \to e^{-(\nu+\rho)H(g^{-1}k^{-1}k')}\}F(k)\, dk \\ &= 0.\end{aligned}$$

Since $\int_{\bar{N}_\mathfrak{p}} e^{-(\nu+\rho)H(n)}\, d\bar{n}$ cannot be 0 for ν real-valued, Lemma 9.28 applies and gives $f(1) = 0$. Since F is right $M_\mathfrak{p}$-invariant, $f(1) = T(F)$. Thus $T(F) = 0$, as was to be proved.

Let us return to the situation in Lemma 9.25. Arguing by contradiction, suppose an M-invariant linear combination $P(X, Y)$ of monomials $(XY)^{(p,q)}_{-\beta}$ leads to 0 in (9.80). We are to show $P = 0$. Our assumption is

$$\int_{\mathrm{diag}\, K} P(X, Y)[e^{-(\lambda_\mathbb{C}+\rho_{G_\mathbb{C}})H_\mathbb{C}(x)}]_{x=g_\mathbb{C}^{-1}l}\, dl = 0 \text{ for all } g_\mathbb{C} \in G_\mathbb{C}. \quad (9.86)$$

If f is any smooth function on U, we extend f to a function f^* on $G_\mathbb{C}$ by means of the Iwasawa decomposition of $G_\mathbb{C}$, letting

$$f^*(uan_\mathbb{C}) = e^{-(\lambda_\mathbb{C}+\rho_{G_\mathbb{C}})\log a}f(u).$$

Then we define a distribution T on U by

$$Tf = \int_{\mathrm{diag}\, K} (P(X, Y)f^*)(l)\, dl. \quad (9.87)$$

The assumption (9.86) is exactly that

$$T\{u \to e^{-(\lambda_\mathbb{C}+\rho_{G_\mathbb{C}})H_\mathbb{C}(g_\mathbb{C}^{-1}u)}\} = 0 \qquad \text{for all } g_\mathbb{C} \in G_\mathbb{C}. \quad (9.88)$$

We apply Proposition 9.29, using $G_\mathbb{C}$ as the linear connected reductive group and U as maximal compact subgroup. The $M_\mathfrak{p}$ group is just our M. According to the proposition, (9.88) implies that $T(f) = 0$ for every right M-invariant smooth function f on U. We shall complete the proof of

Lemma 9.25, and hence of Theorem 9.20, by sketching the construction of a right M-invariant smooth function f for which (9.87) is not 0.

Let $\mathfrak{l}'$ be the orthogonal complement of $\mathfrak{m}$ in diag $\mathfrak{k}$; $\mathfrak{l}'$ is stable under $\operatorname{Ad}(M)$. We let $Q_1, \ldots, Q_k$ be a basis of $\mathfrak{l}'$. Then

$$(\exp(q_1Q_1 + \ldots + q_kQ_k))(\exp t_1X_1)(\exp t_1'Y_1)\cdots(\exp t_n'Y_n)$$
$$\times (\exp H_M)(\exp H_A)(\exp X)$$

with $H_M \in \mathfrak{m}$, $H_A \in \mathfrak{a}$, and $X \in \mathfrak{n}_{\mathbb{C}}$, describes a coordinate system near the identity in $G_{\mathbb{C}}$. We can regard $\exp(q_1Q_1 + \ldots + q_kQ_k)$ as a coordinate system near the identity coset in $(\operatorname{diag} K)/M$, and we can write down in these coordinates some nonnegative left M-invariant function supported near the identity and not identically 0. Call this function $f_1(q_1, \ldots, q_k)$. Next we choose a function $f_2(t_1, t_1', \ldots, t_n')$ such that the $\operatorname{Ad}(M)$-invariant Euclidean differential operator on $\mathfrak{s}$ made from $P(X, Y)$ and applied to f_2 is nonvanishing at the identity. Finally we choose $f_3(\exp H_M \exp H_A \exp X)$ to be locally $e^{-(\lambda_{\mathbb{C}}+\rho_{G_{\mathbb{C}}})(H_A)}$. Then we obtain a function f_0 defined near the identity of $G_{\mathbb{C}}$ as the product $f_1f_2f_3$ and having support near the identity. The function f obtained by restricting f_0 to U has the required properties.

§10. Problems

1. When rank $G =$ rank K, show that the identity $W_K = W(B{:}G)$ is a special case of the result of Problem 6 in Chapter V.
2. In $SL(2, \mathbb{R})$, the Casimir operator $\Omega = \frac{1}{2}h^2 + ef + fe$ is an operator $\tilde{D}$ to which Proposition 9.9 applies. What is the eigenvalue of Ω on $e^{-\nu H(g^{-1}x)}$ when $\nu\begin{pmatrix} t & \\ & -t\end{pmatrix} = zt$? Verify that $\nu = 2\rho$ leads to eigenvalue 0.
3. Let G be linear connected semisimple, and fix a minimal parabolic subgroup $S_\mathfrak{p} = M_\mathfrak{p}A_\mathfrak{p}N_\mathfrak{p}$. Let π be a finite-dimensional irreducible representation of G acting in a space V^π, let ν be the highest restricted weight, let σ be the (irreducible) representation of $M_\mathfrak{p}$ on the highest restricted weight space, and let E be the projection of V^π onto the highest restricted weight space. For v in V^π, define $f_v(g) = E(\pi(g)^{-1}v)$.
   (a) Prove that the map $v \to f_v$ is G-equivariant from π into $U(\bar{S}_\mathfrak{p}, \sigma, \nu + \rho_\mathfrak{p})$, where $\rho_\mathfrak{p}$ corresponds to $S_\mathfrak{p}$.
   (b) Conclude that the Langlands parameters of π are $(S_\mathfrak{p}, \sigma, \nu + \rho_\mathfrak{p})$.
4. Prove the following special case of Lemma 9.13: $M_\mathfrak{p}$ is connected if G is complex.
5. Give a simpler proof of Corollary 9.15 in the case that G is complex, using the relationships in §1 of Appendix C.

Problems 6 to 14 address the form of $\psi(g)$ for $G = \mathrm{SU}(n, 1)$. We take B to be the diagonal subgroup and define $a_s = \exp s(E_{1,n+1} + E_{n+1,1})$.

6. As in the proof of Lemma 9.20c, obtain the following formula for $\psi(g)$: If $g = k_1 a_s k_2$, then
$$\psi(g) = \psi^*\left(\begin{pmatrix} k_1 & \\ & k_2^{-1} \end{pmatrix}\begin{pmatrix} a_{s/2} & \\ & a_{-s/2} \end{pmatrix} \operatorname{diag} G\right).$$
Here the first matrix on the right side is in K^*, and the second is in G_0.

7. Combining (9.54) and (9.36), obtain the following formula for $\psi_{\mathbb{C}}(k_{\mathbb{C}} g_{\mathbb{C}})$:
$$\psi_{\mathbb{C}}(k_{\mathbb{C}} g_{\mathbb{C}}) = c \int_{\operatorname{diag} K} (\tau(k_{\mathbb{C}} l)\phi_{\Lambda_{\mathbb{C}}}, \phi_K) e^{-(\lambda_{\mathbb{C}} + \rho_{G_{\mathbb{C}}})H_{\mathbb{C}}(g_{\mathbb{C}}^{-1} l)}\, dl.$$

8. Use the results of Problems 6 and 7 and the correspondence (9.45) to prove
$$\psi(g) = c \int_{\operatorname{diag} K} \left(\tau\left(\begin{pmatrix} k_1 & \\ & k_2^{-1} \end{pmatrix} l\right)\phi_{\Lambda_{\mathbb{C}}}, \phi_K\right)$$
$$\times \exp\left\{-(\lambda_{\mathbb{C}} + \rho_{G_{\mathbb{C}}})H_{\mathbb{C}}\left(\begin{pmatrix} a_{-s/2} & \\ & a_{s/2} \end{pmatrix} l\right)\right\} dl.$$

9. Writing out $|\psi(g)|^2$ as a double integral from Problem 8 and using Schur orthogonality, prove that
$$\int_{K \times K} |\psi(k_1 a_s k_2)|^2\, dk_1\, dk_2 = c' \int_{(\operatorname{diag} K) \times (\operatorname{diag} K)} (\tau(l)\phi_{\Lambda_{\mathbb{C}}}, \tau(l')\phi_{\Lambda_{\mathbb{C}}})$$
$$\times \exp\left\{-(\lambda_{\mathbb{C}} + \rho_{G_{\mathbb{C}}})H_{\mathbb{C}}\left(\begin{pmatrix} a_{-s/2} & \\ & a_{s/2} \end{pmatrix} l\right)\right\}$$
$$\times \exp\left\{-(\lambda_{\mathbb{C}} + \rho_{G_{\mathbb{C}}})H_{\mathbb{C}}\left(\begin{pmatrix} a_{-s/2} & \\ & a_{s/2} \end{pmatrix} l'\right)\right\}$$
$$\times\, dl\, dl'.$$

10. To compute $\int_G |\psi(g)|^2\, dg$, we are to integrate the left side in Problem 9 against a certain function of s. What is the function of s? [Hint: See Proposition 5.28.]

11. We can treat the right side of Problem 9 crudely by dropping the bounded factor $(\tau(l)\phi_{\Lambda_{\mathbb{C}}}, \tau(l')\phi_{\Lambda_{\mathbb{C}}})$. Using (9.51), show that
$$\int_{\operatorname{diag} K} \exp\left\{-(\lambda_{\mathbb{C}} + \rho_{G_{\mathbb{C}}})H_{\mathbb{C}}\left(\begin{pmatrix} a_{-s/2} & \\ & a_{s/2} \end{pmatrix} l\right)\right\} dl$$
$$= \int_K e^{-2(\lambda + \delta_G)H^{\mathbb{C}}(a_{-s/2}k)}\, dk,$$
where $H^{\mathbb{C}}$ is the Iwasawa $H(\cdot)$ function for $\mathrm{SL}(n + 1, \mathbb{C})$.

12. In SU(2, 1), write $k = (k_{ij})$ with $k_{ij} = 0$ when exactly one of i and j is 3. Show, using the techniques of Problems 2–3 of Chapter VII, that the integrand on the right side in Problem 11 is

$$(|k_{11}|^2 \cosh s + |k_{21}|^2)^{-p}(\cosh s)^{-q},$$

where p and q depend on λ. Find p and q in terms of an explicit parametrization of λ.

13. Deduce explicitly the square integrability of ψ for SU(2, 1) when λ is sufficiently large by observing that λ large forces q to be large. The first factor (involving the exponent p) can be dropped in the calculation.

14. What are the explicit formulas in SU(n, 1) that correspond to those for SU(2, 1) in Problems 12 and 13?

CHAPTER X

Global Characters

§1. Existence

Chapter VIII established that an important step in understanding all irreducible admissible representations is to understand the discrete series. We constructed some discrete series in Chapter IX (see Theorem 9.20) and now need tools to address the question of completeness. Several approaches to this question are possible, and we follow an analytic one: Motivated by Corollary 8.42, we shall show ultimately that there are enough discrete series representations to handle the Fourier analysis of $Z(\mathfrak{g}^{\mathbb{C}})$-finite K-finite functions in $L^2(G)$. Our Fourier analysis will be based on global characters of representations—suitable generalizations of the characters of irreducible representations of compact groups.

For an infinite-dimensional representation π, the series defining the trace

$$\sum_j (\pi(x)v_j, v_j), \qquad \text{with } \{v_j\} \text{ an orthonormal basis,}$$

does not converge, and we are led to consider a summability method applied to this series. A method to try for a Lie group is to average each term by a smooth function of compact support and then compute the sum.

We say that a linear operator L on the Hilbert space V is of **trace class** if $\sum |(B^{-1}LBv_i, v_i)| < \infty$ for every orthonormal basis $\{v_i\}$ and every bounded linear B with a bounded linear inverse. In this case, $\sum (B^{-1}LBv_i, v_i)$ is independent of $\{v_i\}$ and of B and is called the **trace** of L.

Lemma 10.1. If L satisfies $\sum_{i,j} |(Lv_i, v_j)| < \infty$ in some orthonormal basis $\{v_i\}$, then L is of trace class. Moreover, if B_1 and B_2 are any bounded linear operators, then B_1LB_2 and LB_2B_1 and B_2B_1L are all of trace class, and their traces are equal.

Proof straightforward.

We say an admissible representation π of a linear connected reductive group G has a **global character** Θ if $\pi(f)$ is of trace class for all f in $C_{\text{com}}^\infty(G)$ and if $f \to \operatorname{Tr} \pi(f) = \Theta(f)$ is a **distribution** (i.e., a continuous linear functional on $C^\infty_{\text{com}}(G)$). In this case, it is clear that the distribution Θ has to be **invariant** in the sense of agreeing on $f(x)$ and any conjugate $f(gxg^{-1})$.

Theorem 10.2. Every admissible representation π of a linear connected reductive group G whose decomposition $\pi|_K = \sum_{\tau \in K} n_\tau \tau$ has $n_\tau \leq C \dim \tau$ has a global character. In particular, every irreducible unitary representation of G has a character, and even every irreducible admissible representation of G has a character.

Remarks. The condition $n_\tau \leq C \dim \tau$ is applicable to irreducible unitary representations by Theorem 8.1. If we apply the Subrepresentation Theorem and the first remark after Proposition 8.4, we see that it is applicable to all irreducible admissible representations.

First part of proof. Fix z in $Z(\mathfrak{k}^\mathbb{C})$, and let f be in $C^\infty_{\text{com}}(G)$. Fix temporarily a K type τ_λ, and let E_λ be the orthogonal projection on the subspace V_λ that transforms under K according to τ_λ. The value of $\tau_\lambda(z)$ is given by $\chi_{\lambda+\delta_K}(z)$, with the notation referring to the Harish-Chandra homomorphism of $Z(\mathfrak{k}^\mathbb{C})$. If $\{v_j\}$ is an orthonormal basis of V_λ, then

$$\begin{aligned}
\sum_{i,j} |(\pi(f)v_i, v_j)| &= \sum_{i,j} |(E_\lambda \pi(f)v_i, v_j)| \\
&= |\chi_{\lambda+\delta_K}(z)|^{-1} \sum_{i,j} |(\pi(z)E_\lambda \pi(f)v_i, v_j)| \\
&= |\chi_{\lambda+\delta_K}(z)|^{-1} \sum_{i,j} |(E_\lambda \pi(z)\pi(f)v_i, v_j)| \\
&\leq |\chi_{\lambda+\delta_K}(z)|^{-1} (\dim V_\lambda)^2 \|\pi(zf)\| \qquad \text{if } zf \text{ is defined by } \textit{right}\text{-invariant differentiation} \\
&\leq C^2 d_\lambda^4 |\chi_{\lambda+\delta_K}(z)|^{-1} \|\pi(zf)\|. \qquad (10.1)
\end{aligned}$$

Summing both sides of (10.1) on λ, we see from Lemma 10.1 that $\pi(f)$ is of trace class if z can be chosen so that

$$\sum_\lambda d_\lambda^4 |\chi_{\lambda+\delta_K}(z)|^{-1} \qquad (10.2)$$

is finite. In this case we see by dropping the off-diagonal terms that

$$|\text{Tr}\, \pi(f)| \leq C^2 \left(\sum_\lambda d_\lambda^4 |\chi_{\lambda+\delta_K}(z)|^{-1} \right) \|\pi(zf)\|$$

and therefore that $f \to \text{Tr}\, \pi(f)$ is a distribution. Thus the proof will be complete if we produce z in $Z(\mathfrak{k}^\mathbb{C})$ with (10.2) finite.

Lemma 10.3. Let K_1 be a compact connected semisimple group. For τ in $\hat{K}_1$, let d_τ denote the degree of τ. Then

$$\sum_{\tau \in \hat{K}_1} d_\tau^{-2} < \infty.$$

Proof. Fix a Cartan subalgebra $\mathfrak{h}$ of $\mathfrak{k}_1$ and a positive system of roots Δ^+, and let $\alpha_1, \ldots, \alpha_l$ be the simple roots. For λ dominant and algebraically integral, let

$$d_\lambda = \frac{\prod_{\alpha \in \Delta^+} \langle \lambda + \delta, \alpha \rangle}{\prod_{\alpha \in \Delta^+} \langle \delta, \alpha \rangle}.$$

It is enough to prove that $\sum d_\lambda^{-2} < \infty$, by the Weyl dimension formula. Dropping all factors except those for the simple roots and applying Proposition 4.33, we have

$$d_\lambda \geq \prod_{j=1}^{l} \frac{\langle \lambda + \delta, \alpha_j \rangle}{\langle \delta, \alpha_j \rangle} = \prod_{j=1}^{l} \frac{2\langle \lambda + \delta, \alpha_j \rangle / |\alpha_j|^2}{2\langle \delta, \alpha_j \rangle / |\alpha_j|^2} = \prod_{j=1}^{l} \frac{2\langle \lambda + \delta, \alpha_j \rangle}{|\alpha_j|^2}.$$

The dominant algebraically integral λ's are parametrized by $(\mathbb{Z}^l)^+$, with $(n_1, \ldots, n_l) \leftrightarrow \sum n_i \lambda_i$ and with λ_i defined by

$$\frac{2\langle \lambda_i, \alpha_j \rangle}{|\alpha_j|^2} = \delta_{ij}.$$

Hence our estimate is

$$d_\lambda \geq \prod_{j=1}^{l} (n_j + 1),$$

and the lemma follows from the fact that

$$\sum_{n_1=0}^{\infty} \cdots \sum_{n_l=0}^{\infty} (n_1 + 1)^{-2} \cdots (n_l + 1)^{-2} < \infty.$$

Lemma 10.4. Let K be a compact connected Lie group. Then there exists z_1 in $Z(\mathfrak{k}^{\mathbb{C}})$ such that

$$\chi_{\lambda + \delta_K}(z_1) = d_\lambda^2(1 + \|\lambda|_{Z_{\mathfrak{k}}}\|^2)I$$

for every irreducible representation τ_λ.

Proof. Let $\mathfrak{h} \subseteq \mathfrak{k}$ be a Cartan subalgebra, let Δ be the roots, and let $\mathscr{H} = U(\mathfrak{h}^{\mathbb{C}})$. In view of Theorem 8.18, we are to produce a member X of $\mathscr{H}^W$ with $(\lambda + \delta_K)(X) = d_\lambda^2(1 + \|\lambda|_{Z_{\mathfrak{k}}}\|^2)I$ for all dominant integral λ. The Weyl dimension formula shows that

$$X = \left(\prod_{\alpha \in \Delta^+} \langle \delta_K, \alpha \rangle^{-1} H_\alpha \right)^2 (1 + \sum H_j^2)$$

has the required properties if $\{H_j\}$ is an orthonormal basis of the center $Z_{\mathfrak{k}}$.

Second part of proof of Theorem 10.2. Let $n \geq \max\{3, \dim Z_{\mathfrak{k}}\}$, and put $z = z_1^n$, where z_1 is as in Lemma 10.4. Then Lemma 10.3 (applied with K_1 the semisimple part of K) shows that the sum (10.2) is finite. The theorem follows.

Proposition 10.5. If π and π' are infinitesimally equivalent admissible representations of a linear connected reductive group and both have global characters, then the characters of π and π' are equal.

Proof. Let E_τ be the orthogonal projection of the space V for π onto the subspace transforming by the K type τ. Choosing an orthonormal basis of V compatible with the decomposition of V into K types, we see that

$$\operatorname{Tr} \pi(f) = \sum_{\tau \in \hat{K}} \int_G f(x) \operatorname{Tr}(E_\tau \pi(x) E_\tau)\, dx.$$

Now $\operatorname{Tr}(E_\tau \pi(x) E_\tau)$ is built from K-finite matrix coefficients and is unchanged by infinitesimal equivalence. Hence it follows that $\operatorname{Tr} \pi(f) = \operatorname{Tr} \pi'(f)$.

Theorem 10.6. Let $\pi_1, \ldots, \pi_n$ be infinitesimally inequivalent, irreducible admissible representations with global characters $\Theta_1, \ldots, \Theta_n$. Then $\Theta_1, \ldots, \Theta_n$ are linearly independent.

Proof. We shall use the convolution algebra C_K defined by (9.1). Let $\mathscr{S}$ be a finite set of K types to be specified, and let $e = \sum_{\tau \in \mathscr{S}} d_\tau \bar{\chi}_\tau$ be as in the proof of Proposition 9.5. Then $e *_K C_K *_K e$ is a subalgebra of C_K. (As in the proof of Proposition 9.5, we are abbreviating $e\, dk$ as e in writing convolutions.) Let V_i be the Hilbert space on which π_i acts, and define $E_i = \pi_i(e\, dk)$. Each space $E_i(V_i)$ is finite-dimensional, and π_i gives a representation of $e *_K C_K *_K e$ on $E_i(V_i)$. The proof now proceeds in seven steps.

(1) The representation π_i of $e *_K C_K *_K e$ on $E_i(V_i)$ is irreducible if nonzero. [In fact, let $u \neq 0$ and v be in $E_i(V_i)$. By irreducibility of $\pi_i(U(\mathfrak{g}^{\mathbb{C}}))$ on the K-finite vectors of π_i and by Proposition 9.5, we can find f in C_K with $\pi_i(f)u = v$. Then

$$\pi_i(e *_K f *_K e)u = E_i \pi_i(f) E_i u = v,$$

and the irreducibility follows.]

(2) For each i, $\pi_i(e *_K C_K *_K e) = \operatorname{End}(E_i(V_i))$. [In fact, this is a consequence of Step (1).]

(3) If $i \neq j$ and if both $E_i(V_i)$ and $E_j(V_j)$ are nonzero, then the representations π_i and π_j of $e *_K C_K *_K e$ on $E_i(V_i)$ and $E_j(V_j)$ are inequivalent. [In fact, let $L: E_i(V_i) \to E_j(V_j)$ be an equivalence. We shall use L to construct an infinitesimal equivalence of π_i and π_j, in contradiction to the assumed inequivalence. For this purpose, fix v_i in $E_i(V_i)$, and let $v_j = L(v_i)$. We attempt to map $\pi_i(f)v_i$ to $\pi_j(f)v_j$ for f in C_K. If such a map is well defined,

then it certainly gives an infinitesimal equivalence. It will be well defined if

$$\pi_i(f)v_i = 0 \quad \text{implies} \quad \pi_j(f)v_j = 0. \tag{10.3}$$

We show that the existence of L forces (10.3) to hold. Thus let $\pi_i(f)v_i = 0$. For any h in C_K we then have

$$\begin{aligned} E_j\pi_j(h)(\pi_j(f)v_j) &= \pi_j(e *_K h * f *_K e)v_j \\ &= \pi_j(e *_K h * f *_K e)Lv_i \\ &= L\pi_i(e *_K h * f *_K e)v_i \\ &= L\pi_i(e *_K h)\pi_i(f)v_i \\ &= 0. \end{aligned}$$

Since $\pi_j(U(\mathfrak{g}^{\mathbb{C}}))$ is irreducible on the K-finite vectors and since h is arbitrary in C_K, we conclude from Proposition 9.5 that $\pi_j(f)v_j = 0$.]

(4) If $i \neq j$ and if $E_j(V_j) \neq 0$, then there exists h in $e *_K C_K *_K e$ such that $\pi_j(h)$ is nonsingular on $E_j(V_j)$ and $\pi_i(h)$ is 0 on $E_i(V_i)$. [In fact, Step (3) and its proof show that for any $v_j \neq 0$ in $E_j(V_j)$ and for any v_i in $E_i(V_i)$, we can find f in C_K with

$$\pi_i(f)v_i = 0 \quad \text{and} \quad \pi_j(f)v_j \neq 0. \tag{10.4}$$

Let v be in $E_j(V_j)$ and let $u_1, u_2, \ldots$ be a basis of $E_i(V_i)$. Apply (10.4) with $v_i = u_1$ and $v_j = v$ to obtain f_1 with

$$\pi_i(f_1)u_1 = 0 \quad \text{and} \quad \pi_j(f_1)v \neq 0.$$

Next apply (10.4) with $v_i = \pi_i(f_1)u_2$ and $v_j = \pi_j(f_1)v$ to obtain f_2, and then

$$\pi_i(f_2 * f_1)u_1 = \pi_i(f_2 * f_1)u_2 = 0 \quad \text{and} \quad \pi_j(f_2 * f_1)v \neq 0.$$

Continuing in this way, we ultimately obtain g in C_K with $\pi_i(g) = 0$ on $E_i(V_i)$ and with $\pi_j(g)v \neq 0$. Taking into account the irreducibility of π_j, we may assume $\pi_j(g)v$ is in $E_j(V_j)$. Then $\pi_i(e *_K g *_K e) = 0$ on $E_i(V_i)$ and $\pi_j(e *_K g *_K e)v \neq 0$. By Step (2) we may convolve with a further member of C_K and assume that $\pi_j(e *_K g *_K e)$ is of rank one on $E_j(V_j)$ and has a prescribed image $\mathbb{C}v'$. Since v and v' are arbitrary, we can add such functions to obtain h as asserted.]

(5) If h_i, $1 \leq i \leq p$, are members of $e *_K C_K *_K e$ such that $\sum_{i=1}^{p} \operatorname{Tr} \pi_i(h_i * f) = 0$ for all f in $e *_K C_K *_K e$, then $\pi_i(h_i) = 0$ for $1 \leq i \leq p$. [In fact, we proceed by induction. The case $p = 1$ follows immediately from Step (2). Assume the result for $p - 1$, and suppose

$$\sum_{i=1}^{p} \operatorname{Tr} \pi_i(h_i * f) = 0 \tag{10.5}$$

for all f in $e *_K C_K *_K e$. By inductive hypothesis we may assume $\pi_1(h_1) \neq 0$, hence that $E_1(V_1) \neq 0$. By Step (4) we can find h in $e *_K C_K *_K e$ with $\pi_1(h)$

nonsingular on $E_1(V_1)$ and with $\pi_2(h) = 0$. Replacing f by $h * f$ in (10.5), we have

$$\sum_{i=1}^{p} \operatorname{Tr} \pi_i((h_i * h) * f) = 0.$$

The term for $i = 2$ is 0 since $\pi_2(h) = 0$. Hence our inductive hypothesis applies and says that $\pi_i(h_i * h) = 0$ for all i. For $i = 1$, this gives $\pi_1(h_1) = 0$ since $\pi_1(h)$ is nonsingular on $E_1(V_1)$. Thus the term for $i = 1$ can be dropped in (10.5), and we can once again apply our inductive hypothesis. We obtain $\pi_i(h_i) = 0$ for all i, as asserted.]

(6) If $\mathscr{S} \subseteq K$ is chosen so large that $E_i(V_i) \neq 0$ for all i, then the linear functionals $f \to \operatorname{Tr} \pi_i(f)$ on $e *_K C_K *_K e$ are linearly independent. [In fact, suppose that

$$\sum_{i=1}^{n} c_i \operatorname{Tr} \pi_i(f) = 0 \tag{10.6}$$

for all f in $e *_K C_K *_K e$. By Step (2), choose h_i in $e *_K C_K *_K e$ with $\pi_i(h_i) = c_i I$ on $E_i(V_i)$. Then (10.6) leads to (10.5) for $p = n$, and Step (5) gives us $\pi_i(h_i) = 0$ for all i. So all c_i are 0.]

(7) $\Theta_1, \ldots, \Theta_n$ are linearly independent. [In fact, if $\mathscr{S}$ is chosen as in Step (6), then the restrictions of $\Theta_1, \ldots, \Theta_n$ to $e *_K C_K *_K e$ are linearly independent.]

§2. Character Formulas for SL(2, $\mathbb{R}$)

In this section we shall derive formulas for some of the characters of the group $G = SL(2, \mathbb{R})$. The characters, as distributions, depend upon a particular normalization of Haar measure of G. It will turn out that the characters of irreducible representations are given as functions times the Haar measure, and then it is clear from the definition that these functions do not depend upon the normalization of Haar measure. However, we do not know that characters are of this form at this stage, and thus we shall fix normalizations now. We use the Iwasawa decomposition $G = KN_\mathfrak{p}A_\mathfrak{p}$ to determine the Haar measure of G, writing

$$x = k_\theta n_s a_t = \begin{pmatrix} \cos\theta & \sin\theta \\ -\sin\theta & \cos\theta \end{pmatrix} \begin{pmatrix} 1 & s \\ 0 & 1 \end{pmatrix} \begin{pmatrix} e^t & 0 \\ 0 & e^{-t} \end{pmatrix}$$

and taking

$$dx = \frac{d\theta}{2\pi}\, ds\, dt. \tag{10.7}$$

The group $M_\mathfrak{p}$ is $\{\pm I\}$, and we take its Haar measure to have total mass one. These are two nonconjugate Cartan subgroups $T = M_\mathfrak{p} A_\mathfrak{p}$ and

$B = K$ in G, and we take their Haar measures to be $d(\pm)\,da$ and $d\theta/2\pi$, respectively. With these choices fixed, we then have well-defined quotient measures dx^* on G/T and G/B.

To recognize an expression as the character of a representation and also to make use of the resulting formula, we use the Weyl integration formula (Proposition 5.27), which says

$$\int_G f(x)\,dx = \int_B (e^{-i\theta} - e^{i\theta}) F_f^B(\theta) \frac{d\theta}{2\pi} \pm \frac{1}{2} \int_T |e^t - e^{-t}| F_f^T(\pm a_t)\, d(\pm)\, dt, \tag{10.8}$$

where

$$F_f^B(\theta) = (e^{i\theta} - e^{-i\theta}) \int_{G/B} f(x k_\theta x^{-1})\, dx^* \qquad \text{for } \sin\theta \neq 0 \tag{10.9a}$$

and

$$F_f^T(\pm a_t) = \pm |e^t - e^{-t}| \int_{G/T} f(x(\pm a_t) x^{-1})\, dx^* \qquad \text{for } t \neq 0. \tag{10.9b}$$

(One reason for using notation that isolates the particular expressions in (10.9) is given in Lemma 10.7 below; another reason is the role that F_f^B and F_f^T play in the next chapter.) If a representation π has a global character Θ given as a locally integrable function, then

$$\begin{aligned} \operatorname{Tr} \pi(f) &= \int_G \Theta(x) f(x)\, dx \\ &= \int_B (e^{-i\theta} - e^{i\theta}) \Theta(\theta) F_f^B(\theta) \frac{d\theta}{2\pi} \\ &\quad \pm \frac{1}{2} \int_T |e^t - e^{-t}| \Theta(\pm a_t) F_f^T(\pm a_t)\, d(\pm)\, dt. \end{aligned} \tag{10.10}$$

Conversely we seek conditions under which we can pass from a formula like (10.10) to the conclusion that Θ defines a locally integrable function on G that gives the character of π. We give such a result in Proposition 10.8. Control of the right side of (10.10) is possible because of the following lemma, whose proof is given after Proposition 10.8.

Lemma 10.7. If f is an L^∞ function of compact support on SL(2, ℝ), then F_f^B and F_f^T are L^∞ functions of compact support.

Proposition 10.8. If $\Theta(\theta)$ and $\Theta(\pm a_t)$ are functions on B and T such that

(1) $(e^{i\theta} - e^{-i\theta})\Theta(\theta)$ and $\pm|e^t - e^{-t}|\Theta(\pm a_t)$ are locally integrable on B and T, and
(2) $\Theta(\pm a_t)$ is invariant under $t \to -t$,

then the function

$$\Theta(x) = \begin{cases} \Theta(\theta) & \text{if } x \text{ is conjugate to } k_\theta \\ \Theta(\pm a_t) & \text{if } x \text{ is conjugate to } \pm a_t \end{cases}$$

is a well-defined locally integrable function on G. Moreover, if also Tr $\pi(f)$ is given by the right side of (10.10), then the character of π is given by the function $\Theta(x)$.

Proof. It is easy to see that $\Theta(x)$ is well defined. Replacing $\Theta(x)$ by $|\Theta(x)|$ and using (1) and Lemma 10.7, we see that $\Theta(x)$ is locally integrable. The two integrals in (10.10) are then equal if f is in $C_{\text{com}}^\infty(G)$ by the Weyl integration formula, and hence $\Theta(x)\,dx$ is the character of π.

The next two lemmas together prove Lemma 10.7. The second lemma also gives us a start at calculating the characters of the nonunitary principal series.

Lemma 10.9. If E is a bounded subset of SL(2, $\mathbb{R}$), then there exists a constant $C = C(E)$ such that the Haar measure of the set of x in SL(2, $\mathbb{R}$) with $xk_\theta x^{-1}$ in E is $\le C/|\sin\theta|$.

Proof. We decompose x according to the $K\overline{A_{\mathfrak{p}}^+}K$ decomposition and use the corresponding integration formula (Proposition 5.28). Only the $A_{\mathfrak{p}}$ part is significant for this lemma. If $a_t k_\theta a_t^{-1}$ lies in a bounded set, then $e^{2t}|\sin\theta|$ is bounded. Hence the measure in question is bounded by a multiple of

$$\int_{\{t \mid e^{2t}|\sin\theta| \le c\}} \sinh 2t\,dt \le C/|\sin\theta|.$$

Lemma 10.10. For f in $C_{\text{com}}^\infty(G)$ and for $t \ne 0$,

$$F_f^T(\pm a_t) = \pm e^t \int_{K\times N} f(k(\pm a_t)nk^{-1})\,dk\,dn.$$

Proof. We may assume $f \ge 0$. For any $h \ge 0$ on $N_{\mathfrak{p}}$, we have

$$\int_{-\infty}^{\infty} h\begin{pmatrix} 1 & s \\ 0 & 1 \end{pmatrix} ds = |1 - e^{-2t}| \int_{-\infty}^{\infty} h\begin{pmatrix} 1 & (e^{-2t}-1)s \\ 0 & 1 \end{pmatrix} ds,$$

which we can rewrite as

$$\int_N h(n)\,dn = |1 - e^{-2t}| \int_N h(a_t^{-1} n a_t n^{-1})\,dn. \tag{10.11}$$

Put $h(n) = f(k(\pm a_t)nk^{-1})$ and integrate over K to obtain

$$\int_{K\times N} f(k(\pm a_t)nk^{-1})\,dk\,dn = |1 - e^{-2t}| \int_{K\times N} f(kn(\pm a_t)n^{-1}k^{-1})\,dk\,dn.$$

Our normalizations of Haar measures make dx^* on G/T equal to $dk\,dn$, and thus the right side is

$$= |1 - e^{-2t}| \int_{G/T} f(x(\pm a_t)x^{-1})\,dx^*.$$

Multiplying through by $\pm e^t$, we obtain the formula in the lemma.

Lemmas 10.9 and 10.10 together complete the proof of Lemma 10.7. We turn our attention now to computing specific characters. Computations of traces are easiest for operators that are given by integration on all of L^2 of a compact manifold, because of the formula in Lemma 10.11 below. Proposition 10.12 will use this formula to calculate characters for the nonunitary principal series.

Lemma 10.11. Let X be a compact C^∞ manifold, and let dx be a measure on X that is a smooth function times Lebesgue measure in each coordinate neighborhood. Let L be in $C^\infty(X \times X)$, and define a bounded operator T on $L^2(X, dx)$ by

$$Tf(x) = \int_X L(x, y) f(y)\,dy.$$

Then T is of trace class, and its trace is

$$\operatorname{Tr} T = \int_X L(x, x)\,dx.$$

Proof. Use of a smooth partition of unity reduces the proof to the case that L has support in a coordinate neighborhood, hence to the case that X is a cube centered at the origin of side 2π and L is smooth and has support near the origin. We give the proof in the one-dimensional case, the general case being similar. Thus we take X to be the interval $[-\pi, \pi]$ and assume that $L(x, y)$ is smooth in (x, y) and vanishes off $\left[-\frac{\pi}{3}, \frac{\pi}{3}\right]$; we use $\frac{1}{2\pi}\,dx$ as the measure.

We have

$$T(e^{iN\cdot})(x) = \frac{1}{2\pi} \int_{-\pi}^{\pi} L(x, y) e^{iNy}\,dy.$$

Hence

$$\langle T(e^{iN\cdot}), e^{iM\cdot} \rangle = \left(\frac{1}{2\pi}\right)^2 \int_{-\pi}^{\pi} \int_{-\pi}^{\pi} L(x, y) e^{iNy} e^{-iMx}\,dy\,dx. \tag{10.12}$$

The right side is a Fourier coefficient of L, and the sum of the absolute values of the Fourier coefficients is finite since L is smooth. By Lemma 10.1, T is of trace class.

To compute the trace, we start from (10.12) with $M = N$. We change variables, letting $u = y - x$ and $v = y + x$, and the right side of (10.12) becomes

$$\left(\frac{1}{2\pi}\right)^2 \int_{-\pi}^{\pi} \int_{-\pi}^{\pi} \tfrac{1}{2} L(\tfrac{1}{2}(v - u), \tfrac{1}{2}(v + u))\, e^{iNu}\, du\, dv$$

because of the small support of L. We sum on N, moving the sum under the first integral and recognizing the sum inside as the sum of Fourier coefficients in the u variable. Thus

$$\operatorname{Tr} T = \frac{1}{4\pi} \int_{-\pi}^{\pi} L(\tfrac{1}{2}v, \tfrac{1}{2}v)\, dv.$$

Replacing $\frac{1}{2}v$ by v and again taking into account the small support of L, we obtain the formula asserted.

Proposition 10.12. The character of the nonunitary principal series representation $U(S_{\mathfrak{p}}, \sigma, \nu)$ of $\mathrm{SL}(2, \mathbb{R})$ is given by the locally integrable function

$$\Theta_{\sigma,\nu}(x) = \begin{cases} 0 & \text{if } x \text{ conjugate to } k_\theta \\ \dfrac{\sigma(\pm)(e^{\nu \log a_t} + e^{-\nu \log a_t})}{|e^t - e^{-t}|} & \text{if } x \text{ conjugate to } \pm a_t. \end{cases}$$

Proof. The idea is to express $U(S_{\mathfrak{p}}, \sigma, \nu, f)$ as a kernel operator

$$U(S_{\mathfrak{p}}, \sigma, \nu, f)\varphi(k) = \int_K L(k, k')\varphi(k')\, dk'$$

on K with L in $C^\infty(K \times K)$, and then the trace is given by $\int_K L(k, k)\, dk$, according to Lemma 10.11. If φ is in the subspace of $L^2(K)$ transforming according to σ on the right, we have

$$U(S_{\mathfrak{p}}, \sigma, \nu, x)\varphi(k) = e^{-(\nu+\rho)H(x^{-1}k)}\varphi(\kappa(x^{-1}k)) \tag{10.13}$$

and hence

$$\begin{aligned} U(S_{\mathfrak{p}}, \sigma, \nu, f)\varphi(k) &= \int_G e^{-(\nu+\rho)H(x^{-1}k)} f(x)\varphi(\kappa(x^{-1}k))\, dx \\ &= \int_G e^{-(\nu+\rho)H(x)} f(kx^{-1})\varphi(\kappa(x))\, dx \qquad \text{under } x^{-1}k \to x \\ &= \int_{KN_{\mathfrak{p}}A_{\mathfrak{p}}} e^{-(\nu+\rho)\log a} f(ka^{-1}n^{-1}k'^{-1})\varphi(k')\, dk'\, dn\, da. \end{aligned}$$

The trace of $U(S_{\mathfrak{p}}, \sigma, \nu, f)$ on the subspace that transforms according to σ is the same as the trace of $U(S_{\mathfrak{p}}, \sigma, \nu, f)E$ on all of $L^2(K)$, where E is the orthogonal projection

$$E\varphi(k) = \int_{M_{\mathfrak{p}}} \sigma(m)\varphi(km)\, dm. \tag{10.14}$$

Thus φ in $L^2(K)$ implies

$$\begin{aligned} U(S_{\mathfrak{p}}, \sigma, \nu, f)E\varphi(k) &= \int_{KN_{\mathfrak{p}}A_{\mathfrak{p}} \times M_{\mathfrak{p}}} e^{-(\nu+\rho)\log a}\sigma(m) f(ka^{-1}n^{-1}mk'^{-1})\varphi(k')\, dm\, dk'\, dn\, da \\ &= \int_K \left[\int_{M_{\mathfrak{p}}A_{\mathfrak{p}}N_{\mathfrak{p}}} e^{(\nu+\rho)\log a}\sigma(m) f(kmank'^{-1})\, dm\, da\, dn \right] \varphi(k')\, dk'. \end{aligned}$$

We can take the expression in brackets as our kernel L, and thus

$$\begin{aligned} \operatorname{Tr} U(S_{\mathfrak{p}}, \sigma, \nu, f) &= \int_{K \times M_{\mathfrak{p}}A_{\mathfrak{p}}N_{\mathfrak{p}}} e^{(\nu+\rho)\log a}\sigma(m) f(kmank^{-1})\, dm\, da\, dn\, dk \\ &= \int_T e^{(\nu+\rho)\log a_t}\sigma(\pm)\left[\int_{K \times N_{\mathfrak{p}}} f(k(\pm a_t)nk^{-1})\, dk\, dn \right] d(\pm)\, dt \\ &= \int_T e^{\nu \log a_t}\sigma(\pm)\left[|e^t - e^{-t}| \int_{G/T} f(x(\pm a_t)x^{-1})\, dx^* \right] d(\pm)\, dt \end{aligned} \tag{10.15}$$

by Lemma 10.10. On the right side we can change x to xw, with $w = \begin{pmatrix} 0 & 1 \\ -1 & 0 \end{pmatrix}$, without affecting any integrals, and then we can change variables on T to bring w out so that it conjugates a_t. Averaging the changed formula and the unchanged formula, we obtain

$$\operatorname{Tr} U(S_{\mathfrak{p}}, \sigma, \nu, f) = \pm\tfrac{1}{2} \int_T \sigma(\pm)(e^{\nu \log a_t} + e^{-\nu \log a_t}) F_f^T(\pm a_t)\, d(\pm)\, dt. \tag{10.16}$$

Appealing to Proposition 10.8, we see that the character is as asserted.

Corollary 10.13. The character of the sum $\mathscr{D}_n^+ \oplus \mathscr{D}_n^-$, for $n \geq 1$ in SL(2, ℝ), is given by the locally integrable function

$$\Theta_n(x) = \begin{cases} \dfrac{-e^{i(n-1)\theta} + e^{-i(n-1)\theta}}{e^{i\theta} - e^{-i\theta}} & \text{if } x \text{ is conjugate to } k_\theta \\[2ex] (\pm)^n \dfrac{e^{(n-1)t}(1 - \operatorname{sgn} t) + e^{-(n-1)t}(1 + \operatorname{sgn} t)}{|e^t - e^{-t}|} & \text{if } x \text{ is conjugate to } \pm a_t. \end{cases}$$

Proof. The idea is that $\mathscr{D}_n^+ \oplus \mathscr{D}_n^-$ is a subrepresentation of a nonunitary principal series representation and the quotient is an identifiable finite-dimensional representation. The relevant inclusion is

$$\mathscr{D}_n^+ \oplus \mathscr{D}_n^- \subseteq \begin{cases} U(S_{\mathfrak{p}}, +, (n-1)\rho) & \text{if } n \text{ even} \\ U(S_{\mathfrak{p}}, -, (n-1)\rho) & \text{if } n \text{ odd,} \end{cases} \tag{10.17}$$

with quotient isomorphic to the finite-dimensional representation Φ_{n-2} defined in (2.1).

An elementary proof of (10.17) and the quotient relation is possible by calculating the Lie algebra action, but we prefer to give a conversational proof by means of the Subrepresentation Theorem and the Langlands classification. Thus we first ask what the possible candidates are for the Langlands parameters of Φ_{n-2}. Since Φ_{n-2} is not tempered, the parabolic subgroup parameter must be $S_{\mathfrak{p}}$. Since the infinitesimal character and the value on the central element $\begin{pmatrix} -1 & 0 \\ 0 & -1 \end{pmatrix}$ must match properly, we see that the only possibility is that Φ_{n-2} is the Langlands quotient for the right side of (10.17).

Meanwhile the Subrepresentation Theorem says that $\mathscr{D}_n^+$ and $\mathscr{D}_n^-$ must imbed in the nonunitary principal series. The infinitesimal character and value on $\begin{pmatrix} -1 & 0 \\ 0 & -1 \end{pmatrix}$ say that the only possible parameters for the imbedding are the ones in (10.14) or the same parameters with the $\mathfrak{a}_{\mathfrak{p}}$ parameter negated. The latter is ruled out since Φ_{n-2} is the unique irreducible subrepresentation there. So the inclusion (10.17) follows.

Referring to the realizations in Chapter II, we see that $\mathscr{D}_n^+ \oplus \mathscr{D}_n^-$ has K types $\pm n, \pm(n+2), \pm(n+4), \ldots,$ where the K type l refers to $\begin{pmatrix} e^{i\theta} & 0 \\ 0 & e^{-i\theta} \end{pmatrix} \to e^{il\theta}$. Also Φ_{n-2} has K types $\pm(n-2), \pm(n-4), \ldots$. Thus $\mathscr{D}_n^+ \oplus \mathscr{D}_n^-$ and Φ_{n-2} exhaust the K types of the nonunitary principal series representation. Since the K types of the nonunitary principal series have multiplicity one, we conclude that the quotient of the nonunitary principal series by $\mathscr{D}_n^+ \oplus \mathscr{D}_n^-$ is Φ_{n-2}.

Consequently the characters satisfy

$$\Theta(\mathscr{D}_n^+ \oplus \mathscr{D}_n^-) = \Theta(U(S_{\mathfrak{p}}, (-)^n, (n-1)\rho)) - \Theta(\Phi_{n-2}). \qquad (10.18)$$

The first term on the right is given by Proposition 10.12, and the second term is given by direct calculation or the Weyl character formula. The corollary follows.

When using characters of SL(2, $\mathbb{R}$), it is often enough to know the character of $\mathscr{D}_n^+ \oplus \mathscr{D}_n^-$ and unnecessary to know the characters of $\mathscr{D}_n^+$ and $\mathscr{D}_n^-$ separately. This is the case, for example, in the analysis of $L^2(G)$ that we shall do in Chapter XI. In such cases Corollary 10.13 suffices. When we need the individual characters, we must appeal to Proposition 10.14, which lies much deeper.

Proposition 10.14. For $n \geq 1$ in SL(2, ℝ), the character of $\mathscr{D}_n^+$ is given by the locally integrable function

$$\Theta_n^+(x) = \begin{cases} \dfrac{-e^{i(n-1)\theta}}{e^{i\theta} - e^{-i\theta}} & \text{if } x \text{ conjugate to } k_\theta \\ (\pm)^n \dfrac{e^{(n-1)t}(1 - \operatorname{sgn} t) + e^{-(n-1)t}(1 + \operatorname{sgn} t)}{2|e^t - e^{-t}|} & \text{if } x \text{ conjugate to } \pm a_t. \end{cases}$$

The character of $\mathscr{D}_n^-$ is given by the locally integrable function

$$\Theta_n^-(x) = \begin{cases} \dfrac{e^{-i(n-1)\theta}}{e^{i\theta} - e^{-i\theta}} & \text{if } x \text{ conjugate to } k_\theta \\ (\pm)^n \dfrac{e^{(n-1)t}(1 - \operatorname{sgn} t) + e^{-(n-1)t}(1 + \operatorname{sgn} t)}{2|e^t - e^{-t}|} & \text{if } x \text{ conjugate to } \pm a_t. \end{cases}$$

Remarks. We give part of the proof now and part later. Now we show that if Θ_n^+ and Θ_n^- are functions, their values are as indicated. Theorem 10.25 will imply that Θ_n^+ and Θ_n^- are, at any rate, functions on the set $\{x \in \mathrm{SL}(2, \mathbb{R}) \mid |\operatorname{Tr} x| \neq 2\}$, and Theorem 10.36 will imply that they are locally integrable functions on the full group. Notice that $\Theta_n^+(x) = \Theta_n^-(x)$ if $|\operatorname{Tr} x| > 2$, i.e., if x is conjugate to $\pm a_t$ with $t \neq 0$. A mnemonic for the values of Θ_n^+ and Θ_n^- on k_θ is that they give the formal sum of the geometric series of eigenvalues for the action of k_θ; in fact, manipulations with this geometric series lie at the heart of the proof.

Proof of formulas. The formula on $\pm a_t$ is the easier one. In fact, it is easy to see from the definition of $\mathscr{D}_n^+$ and $\mathscr{D}_n^-$ that conjugation by $\begin{pmatrix} 1 & 0 \\ 0 & -1 \end{pmatrix}$ carries $\mathscr{D}_n^+$ to $\mathscr{D}_n^-$. Hence

$$\Theta_n^+\left(\begin{pmatrix} 1 & 0 \\ 0 & -1 \end{pmatrix}(\pm a_t)\begin{pmatrix} 1 & 0 \\ 0 & -1 \end{pmatrix}\right) = \Theta_n^-(\pm a_t).$$

Since $\begin{pmatrix} 1 & 0 \\ 0 & -1 \end{pmatrix}$ commutes with $\pm a_t$, $\Theta_n^+(\pm a_t) = \Theta_n^-(\pm a_t)$. The sum of Θ_n^+ and Θ_n^- is given by Corollary 10.13, and hence the formulas for $\Theta_n^+(\pm a_t)$ and $\Theta_n^-(\pm a_t)$ follow.

Now we derive the formula for $\Theta_n^+(k_\theta)$. Recall from §2.6 that the K types of $\mathscr{D}_n^+$ are $k_\theta \to e^{i(n+2l)\theta}$, $l \geq 0$, all with multiplicity one. We choose an orthonormal basis ϕ_l of the representation space compatible with this decomposition. (When $\mathscr{D}_n^+$ is transformed to a representation of SU(1, 1), the vector ϕ_l transforms to a multiple of the function z^l in the unit disc.)

Let f be a C^∞ function with compact support in SL(2, $\mathbb{R}$) within the set $\{x \mid |\operatorname{Tr} x| < 2\}$. We make the following formal computation, justifying the steps afterward:

$$\operatorname{Tr} \mathscr{D}_n^+(f) = \operatorname{Tr} \int_G f(x)\mathscr{D}_n^+(x)\, dx \tag{10.19a}$$

$$= 4 \operatorname{Tr} \int_G \int_B f(xk_\theta x^{-1})\, \mathscr{D}_n^+(xk_\theta x^{-1}) \sin^2\theta\, d\theta\, dx \tag{10.19b}$$

$$= 4 \operatorname{Tr} \int_G \mathscr{D}_n^+(x)\left[\int_B f(xk_\theta x^{-1})\mathscr{D}_n^+(k_\theta) \sin^2\theta\, d\theta\right]\mathscr{D}_n^+(x)^{-1}\, dx \tag{10.19c}$$

$$= 4 \int_G \operatorname{Tr}\left\{\mathscr{D}_n^+(x)\left[\int_B f(xk_\theta x^{-1})\mathscr{D}_n^+(k_\theta) \sin^2\theta\, d\theta\right]\mathscr{D}_n^+(x)^{-1}\right\} dx \tag{10.19d}$$

$$= 4 \int_G \operatorname{Tr}\left\{\int_B f(xk_\theta x^{-1})\mathscr{D}_n^+(k_\theta) \sin^2\theta\, d\theta\right\} dx \tag{10.19e}$$

$$= 4 \int_G \left[\sum_{l \geq 0} \int_B f(xk_\theta x^{-1})(\mathscr{D}_n^+(k_\theta)\phi_l, \phi_l) \sin^2\theta\, d\theta\right] dx \tag{10.19f}$$

$$= 4 \int_G \left[\sum_{l \geq 0} \int_B f(xk_\theta x^{-1}) e^{i(n+2l)\theta} \sin^2\theta\, d\theta\right] dx \tag{10.19g}$$

$$= 4 \int_G \left[\lim_{r\uparrow 1} \sum_{l \geq 0} \int_B f(xk_\theta x^{-1})(re^{i\theta})^{n+2l} \sin^2\theta\, d\theta\right] dx \tag{10.19h}$$

$$= 4 \int_G \lim_{r\uparrow 1}\left[\int_B f(xk_\theta x^{-1})\left(\frac{-(re^{i\theta})^{n-1}}{re^{i\theta} - (re^{i\theta})^{-1}}\right) \sin^2\theta\, d\theta\right] dx \tag{10.19i}$$

$$= 4 \int_G \int_B f(xk_\theta x^{-1})\left(\frac{-e^{i(n-1)\theta}}{e^{i\theta} - e^{-i\theta}}\right) \sin^2\theta\, d\theta\, dx. \tag{10.19j}$$

As soon as we have justified the steps, the formula for $\Theta_n^+(k_\theta)$ follows as in Proposition 10.8, and the formula for $\Theta_n^-(k_\theta)$ is proved similarly.

To justify the steps, we proceed as follows. For each trace, we substitute the explicit value computed in terms of the basis $\{\phi_l\}$. Thus, for example, the right side of (10.19a) is written as

$$\sum_{l \geq 0} \int_G f(x)(\mathscr{D}_n^+(x)\phi_l, \phi_l)\, dx.$$

Formula (10.19b) is from the Weyl integration formula, as written out in (10.8) and (10.9); the integral over T vanishes because f has support in the set $\{x \mid |\operatorname{Tr} x| < 2\}$. To get from (10.19b) to (10.19c), we expand out

$$(\mathscr{D}_n^+(xk_\theta x^{-1})\phi_l, \phi_l) = \sum_{p,q} (\mathscr{D}_n^+(x)\phi_p, \phi_l)(\mathscr{D}_n^+(k_\theta)\phi_q, \phi_p)\overline{(\mathscr{D}_n^+(x)\phi_q, \phi_l)}.$$

The terms with $p \neq q$ vanish. The remaining sum (over p alone) can be interchanged with the integral over B because the Fourier coefficients of the C^∞ function $k_\theta \to f(xk_\theta x^{-1}) \sin^2 \theta$ decrease rapidly.

Next we use the estimate on the Fourier coefficient

$$\int_B f(xk_\theta x^{-1})e^{i(n+2p)\theta} \sin^2 \theta \, d\theta,$$

together with the compact support of this integral as a function of x to justify interchanging $\sum_{l \geq 0}$ and $\int_G$. This gets us essentially to (10.19d). Interchanging $\sum_{l \geq 0}$ and $\sum_{p \geq 0}$ by Fubini's Theorem, we compute the sum over l. Since

$$\sum_{l \geq 0} |(\mathscr{D}_n^+(x)\phi_p, \phi_l)|^2 = |\mathscr{D}_n^+(x)\phi_p|^2 = 1,$$

we arrive at (10.19e), written out explicitly as in (10.19f).

Formula (10.19g) comes by evaluating the inner product, and (10.19h) is by Abel's Theorem. In (10.19h) we can trivially interchange $\sum_{l \geq 0}$ and $\int_B$ and then sum the geometric series to arrive at (10.19i). The presence of $\sin^2 \theta$ allows us to use dominated convergence to pass to (10.19j). The full computation is then justified, and the proof is complete.

§3. Induced Characters

$SL(2, \mathbb{R})$ provides a good model for a substantial part of the general theory of global characters. An instance of this fact is the part of the theory dealing with characters of induced representations. We shall see in this section how the computation of characters of nonunitary principal series (Proposition 10.12) generalizes and how each step of the argument has a natural generalization.

The situation is as follows: G is linear connected semisimple, $S = MAN$ is a parabolic subgroup, σ is an irreducible unitary representation of M, and ν is in $(\mathfrak{a}')^{\mathbb{C}}$. To avoid some integrality problems, we shall assume $G \subseteq G^{\mathbb{C}}$ with $G^{\mathbb{C}}$ simply connected. Theorem 10.2 assures us that σ has a global character χ_σ, which is a distribution on M, and we shall suppose that this distribution is a locally integrable function times Haar measure on M. By abuse of notation, we agree to write $\chi_\sigma(m)$ for this locally integrable function. It is easy to see that $\chi_\sigma(m)$ is invariant under conjugation on M. We seek an expression for the character $\Theta_{\sigma,\nu}$ of the induced representation $U(S, \sigma, \nu)$. This character exists by Theorem 10.2.

We begin by imitating the argument, given in the proof of Proposition 10.12, that $U(S, \sigma, \nu, f)$ is given by a kernel operator if f is in $C_{\text{com}}^\infty(G)$. In place of the Iwasawa decomposition, we use the decomposition $G = KMAN$ of (5.11), writing the decomposition of an element x of G

non-uniquely as

$$x = \kappa(x)\mu(x)(\exp H(x))n.$$

Formula (10.13) becomes

$$U(S, \sigma, \nu, x)\varphi(k) = e^{-(\nu+\rho)H(x^{-1}k)}\sigma(\mu(x^{-1}k))^{-1}\varphi(\kappa(x^{-1}k)),$$

with φ now vector-valued, taking its values in the Hilbert space V^σ on which σ operates. The orthogonal projection E of (10.14) is replaced by

$$E\varphi(k) = \int_{K\cap M} \sigma(k_M)\varphi(kk_M)\,dk_M,$$

and we have

$$U(S, \sigma, \nu, f)E\varphi(k)$$

$$= \int_G \left[\int_{MAN} f(kmank'^{-1})e^{(\nu+\rho)\log a}\sigma(m)\,dm\,da\,dn\right]\varphi(k')\,dk'. \quad (10.20)$$

To compute the trace of $U(S, \sigma, \nu, f)E$ on $L^2(K, V^\sigma)$, we need a vector-valued analog of Lemma 10.11, which is as follows.

Lemma 10.15. Let X be a compact C^∞ manifold, and let dx be a measure on X that is a smooth function times Lebesgue measure in each coordinate neighborhood. Let L be a smooth function on $X \times X$ with values in trace-class operators on a separable Hilbert space V, and define a bounded operator T on $L^2(X, V)$ by $Tf(x) = \int_X L(x, y)f(y)\,dy$. If T is of trace class, then its trace is

$$\operatorname{Tr} T = \int_X \operatorname{Tr}(L(x, x))\,dx.$$

Sketch of proof. Lemma 10.15 follows easily from Lemma 10.11 by computing the trace of T relative to an orthonormal basis $\varphi_i(x)\psi_j$, where $\{\varphi_i\}$ is an orthonormal basis of $L^2(X, dx)$ and $\{\psi_j\}$ is an orthonormal basis of V.

In (10.20) the expression in brackets is a smooth compact average of a trace class operator, since χ_σ exists, and we know that $U(S, \sigma, \nu, f)E$ is of trace class. From Lemma 10.15 we conclude

$$\begin{aligned}\Theta_{\sigma,\nu}(f) &= \operatorname{Tr} U(S, \sigma, \nu, f)\\ &= \int_{KMAN} f(kmank^{-1})e^{(\nu+\rho)\log a}\chi_\sigma(m)\,dm\,da\,dn\,dk. \quad (10.21)\end{aligned}$$

Shortly we shall use the notation

$$f^{(S)}(ma) = e^{\rho\log a}\int_{K\times N} f(kmank^{-1})\,dk\,dn. \quad (10.22)$$

In this notation we can rewrite (10.21) as

$$\Theta_{\sigma,\nu}(f) = (e^{\nu} \otimes \chi_\sigma)(f^{(S)}). \tag{10.23}$$

First we introduce suitable generalizations of the functions F_f^B and F_f^T defined for $\mathrm{SL}(2, \mathbb{R})$ in (10.9). There will be one such function for each Θ-stable Cartan subgroup of G, although in the end we shall concentrate on only one Cartan subgroup from each conjugacy class. Thus let T be a Cartan subgroup, and let $\mathfrak{t}$ be its Lie algebra. The Cartan involution Θ determines a product decomposition

$$T = BA \qquad \text{with } B \subseteq K \text{ and } A \subseteq \exp \mathfrak{p}; \tag{10.24}$$

this group A need not be related to the A in MAN above. Let $\mathfrak{t} = \mathfrak{b} \oplus \mathfrak{a}$ be the corresponding decomposition of Lie algebras. Every root of $\Delta(\mathfrak{t}^{\mathbb{C}}:\mathfrak{g}^{\mathbb{C}})$ is real on $\mathfrak{a} \oplus i\mathfrak{b}$. We say that a root of $\Delta(\mathfrak{t}^{\mathbb{C}}:\mathfrak{g}^{\mathbb{C}})$ is **real, imaginary,** or **complex** according as its values on $\mathfrak{t}$ are real, imaginary, or neither. Thus the real roots are those that vanish on $\mathfrak{b}$, the imaginary roots are those that vanish on $\mathfrak{a}$, and the complex roots are those that vanish identically on neither $\mathfrak{a}$ nor $\mathfrak{b}$. For a real root α, the exponential ξ_α is real-valued on all of T. (See the Problems at the end of the chapter.) Let Δ_R and Δ_I refer to the subsets of real and imaginary roots. Fix a positive system $\Delta^+(\mathfrak{t}^{\mathbb{C}}:\mathfrak{g}^{\mathbb{C}})$, and define a Weyl denominator as in Proposition 5.27 by

$$D(h) = \xi_\delta(h) \prod_{\alpha \in \Delta^+(\mathfrak{t}^{\mathbb{C}}:\mathfrak{g}^{\mathbb{C}})} [1 - \xi_\alpha(h)^{-1}] \qquad \text{for } h \in T.$$

(In this formula the half sum δ of positive roots is analytically integral because $G^{\mathbb{C}}$ is simply connected.) Also define

$$D'_R(h) = \prod_{\alpha \in \Delta_R^+(\mathfrak{t}^{\mathbb{C}}:\mathfrak{g}^{\mathbb{C}})} [1 - \xi_\alpha(h)^{-1}] \qquad \text{for } h \in T.$$
$$\varepsilon_R(h) = \operatorname{sgn} D'_R(h).$$

If we want to emphasize T, we include T as a superscript or subscript. If we want to emphasize G, we use G/T as superscript.

For f in $C^\infty_{\mathrm{com}}(G)$, we can now define

$$F_f^{G/T}(h) = \varepsilon_R^T(h) D^{G/T}(h) \int_{G/T} f(xhx^{-1})\, dx^* \qquad \text{for } D^{G/T}(h) \neq 0.$$

If we compare this formula with (10.9), we see that the $\varepsilon_R^T(h)$ accounts in $\mathrm{SL}(2, \mathbb{R})$ for the absolute value signs around $e^t - e^{-t}$ and the lack of absolute value signs around $e^{i\theta} - e^{-i\theta}$. The $\pm$ in front of the formula for $F_f^{G/T}$ in $\mathrm{SL}(2, \mathbb{R})$ comes from $\xi_\delta(h)$.

The Weyl integration formula (Proposition 5.27) says that the contribution to $\int_G f(x)\, dx$ from the conjugates of T is

$$|W(T:G)|^{-1} \int_T F_f^{G/T}(h) \varepsilon_R^T(h) \overline{D^{G/T}(h)}\, dh.$$

It will be more convenient to eliminate the complex conjugation here. Thus let us verify the identity

$$|D(h)|^2 = (-1)^{|\Delta_I^+|} D(h)^2, \tag{10.25a}$$

where $|\Delta_I^+|$ is the number of positive imaginary roots in any ordering. For the proof, let us observe that the nonreal roots come in complex conjugate pairs. Thus

$$\prod_{\alpha \notin \Delta_R} (1 - \xi_\alpha(h)^{-1}) \text{ is real and } \geq 0.$$

On the other hand, if α is a real root, then $1 - \xi_\alpha(h)^{-1}$ is ≥ 0 if and only if $1 - \xi_{-\alpha}(h)^{-1}$ is ≤ 0. Hence $(1 - \xi_\alpha(h)^{-1})(1 - \xi_{-\alpha}(h)^{-1})$ is ≤ 0. Thus

$$\begin{aligned}
0 &\leq (-1)^{|\Delta_R^+|} \prod_{\alpha \in \Delta} (1 - \xi_\alpha(h)^{-1}) \\
&= (-1)^{|\Delta_R^+|} \prod_{\alpha \in \Delta^+} (1 - \xi_\alpha(h)^{-1})(1 - \xi_\alpha(h)) \\
&= (-1)^{|\Delta^+| - |\Delta_R^+|} \prod_{\alpha \in \Delta^+} (1 - \xi_\alpha(h)^{-1})(\xi_\alpha(h) - 1) \\
&= (-1)^{|\Delta^+| - |\Delta_R^+|} \xi_{2\delta}(h) \prod_{\alpha \in \Delta^+} (1 - \xi_\alpha(h)^{-1})^2 \\
&= (-1)^{|\Delta^+| - |\Delta_R^+|} D(h)^2,
\end{aligned}$$

and (10.25a) follows.

Let us write

$$s^{G/T} = (-1)^{|\Delta_I^+|}. \tag{10.25b}$$

Then (10.25a) shows that the contribution to $\int_G f(x)\,dx$ from the conjugates of T is

$$s^{G/T} |W(T:G)|^{-1} \int_T F_f^{G/T}(h) \varepsilon_R^T(h) D^{G/T}(h)\,dh. \tag{10.25c}$$

Let $H_1, \ldots, H_r$ be a complete set of nonconjugate Θ-stable Cartan subgroups of the linear reductive group MA. (As usual, we shall overlook the possible disconnectedness of M.) We rewrite (10.21) or (10.23), using the Weyl integration formula for MA. We use our F_f notation, too; although $(MA)^{\mathbb{C}}$ need not be simply connected, the effect of any nonintegrality cancels in any formulas we shall use. Thus (10.25) gives

$$\begin{aligned}
\Theta_{\sigma,\nu}(f) = \sum_{j=1}^{r} s^{MA/H_j} |W(H_j:MA)|^{-1} \\
\times \int_{H_j} (e^\nu \otimes \chi_\sigma)(h) \varepsilon_R^{MA/H_j}(h) F_{f^{(S)}}^{MA/H_j}(h) D^{MA/H_j}(h)\,dh.
\end{aligned} \tag{10.26}$$

Next we bring in the generalization of (10.11) given in the following lemma.

Lemma 10.16. If g is ≥ 0 on N and h is in MA with $\det(\mathrm{Ad}(h)^{-1} - 1)|_{\mathfrak{n}} \neq 0$, then

$$\int_N g(n)\,dn = |\det(\mathrm{Ad}(h)^{-1} - 1)|_{\mathfrak{n}}| \int_N g(h^{-1}nhn^{-1})\,dn.$$

Sketch of proof. Let $\varphi: N \to N$ be the map $n' = \varphi(n) = (h^{-1}nh)n^{-1}$. Then

$$\varphi(n \exp tX) = n' \exp(t\mathrm{Ad}(nh^{-1})X) \exp(-t\mathrm{Ad}(n)X).$$

Identifying tangent spaces via left translation, we therefore have

$$(d\varphi)_n(X) = (\mathrm{Ad}(nh^{-1}) - \mathrm{Ad}(n))X = \mathrm{Ad}(n)(\mathrm{Ad}(h)^{-1} - 1)X.$$

Hence $\det(d\varphi)_n = \det(\mathrm{Ad}(h)^{-1} - 1)|_{\mathfrak{n}} \neq 0$. An inductive argument with N then shows that φ is one-one onto, and the lemma follows.

We can use Lemma 10.16 to relate F_f for MA to F_f for G. To match things properly, we need to assume some compatibility of positive systems—that $\Delta^+(\mathfrak{h}_j^{\mathbb{C}}:(\mathfrak{m} \oplus \mathfrak{a})^{\mathbb{C}}) \subseteq \Delta^+(\mathfrak{h}_j^{\mathbb{C}}:\mathfrak{g}^{\mathbb{C}})$ and that the nonzero restrictions to $\mathfrak{a}$ of the members of $\Delta^+(\mathfrak{h}_j^{\mathbb{C}}:\mathfrak{g}^{\mathbb{C}})$ are the positive restricted roots that define $\mathfrak{n}$. Under these compatibility conditions, we obtain the following lemma.

Lemma 10.17. $\xi_{-\delta_M}(h)F_{f^{(S)}}^{MA/H_j}(h) = e^{\rho H(h)}\xi_{-\delta}(h)F_f^{G/H_j}(h)$ for h regular.

Remark. The factor $\xi_\delta(h)e^{-\rho H(h)}\xi_{-\delta_M}(h)$ is real analytic and its square is everywhere ± 1. In particular, it drops out if H_j is connected. For qualitative use of the lemma, the factor should be ignored.

Sketch of proof. We apply Lemma 10.16 with $g(n) = f(kmhm^{-1}nk^{-1})$ and integrate the result over K. Then we find

$$e^{-\rho H(h)}f^{(S)}(mhm^{-1}) = \left|\det(\mathrm{Ad}(h)^{-1} - 1)|_{\mathfrak{n}}\right| \int_{K\times N} f(knmhm^{-1}n^{-1}k^{-1})\,dk\,dn$$

for m in MA. Integration over $MA/H_j \cong M/(H_j \cap M)$ gives

$$\begin{aligned} F_{f^{(S)}}^{MA/H_j}(h) = {} & [e^{\rho H(h)}D^{MA/H_j}(h)|\det(\mathrm{Ad}(h)^{-1} - 1)|_{\mathfrak{n}}|] \\ & \times \varepsilon_R^{MA/H_j}(h)\varepsilon_R^{G/H_j}(h)D^{G/H_j}(h)^{-1}F_f^{G/H_j}(h). \end{aligned}$$

We easily verify the identities

$$\varepsilon_R^{G/H_j}(h) = \mathrm{sgn}[\det(1 - \mathrm{Ad}(h)^{-1})|_{\mathfrak{n}}]\varepsilon_R^{MA/H_j}(h)$$

and

$$\xi_{-\delta_M}(h)D^{MA/H_j}(h)\det(1 - \mathrm{Ad}(h)^{-1})|_{\mathfrak{n}} = \xi_{-\delta}(h)D^{G/H_j}(h),$$

and then the lemma follows.

Let us substitute from Lemma 10.17 into (10.26). The quantity $F_f^{G/H_j}(\varepsilon_R D)^{G/H_j}$ is invariant under conjugation by members of $W(H_j\colon MA)$, and so is $e^\nu \otimes \chi_\sigma$. Moreover the quotient

$$(\varepsilon_R D)_j^{G/MA} = e^{\rho H(\cdot)}\xi_{-\delta}\xi_{\delta_M}(\varepsilon_R D)^{G/H_j}/(\varepsilon_R D)^{MA/H_j}$$

can be seen to be invariant under conjugation by all of $W(H_j\colon G)$. Therefore we can rewrite (10.26) as

$$\Theta_{\sigma,\nu}(f) = \sum_{j=1}^{r} s^{G/H_j}|W(H_j\colon G)|^{-1}$$
$$\times \int_{H_j} F_f^{G/H_j}(h)\left\{\frac{s^{MA/H_j}s^{G/H_j}\displaystyle\sum_{w\in W(H_j:G)/W(H_j:MA)}(e^{w\nu}\otimes\chi_{w\sigma})(h)}{(\varepsilon_R D)_j^{G/MA}(h)}\right\}$$
$$\times (\varepsilon_R D)^{G/H_j}(h)\,dh \tag{10.27}$$

Except for the question of local integrability analogous to Proposition 10.8, (10.27) allows us to read off induced global characters quite generally, with nice formulas in a number of situations. For example, suppose that $S = MAN$ is a minimal parabolic subgroup. If B is a Cartan subgroup of the compact group M, then BA is the only Cartan subgroup of MA up to conjugacy. So $r = 1$ in (10.27). In this case the signs s within the braces cancel. Moreover, $W(H_j\colon MA)$ is simply $W(B\colon M)$, and $W(H_j\colon G)$ can be seen to be a semidirect product of $W(B\colon M)$ with $W(A\colon G)$. (See the Problems for Chapter V.) Modulo the local integrability question, we have obtained the following formula for the character of a nonunitary principal series representation.

Proposition 10.18. Let $S = MAN$ be a *minimal* parabolic subgroup, and let B be a Cartan subgroup of M. Then the character of the nonunitary principal series representation $U(S, \sigma, \nu)$ of G is given by the locally integrable function

$$\Theta_{\sigma,\nu}(x) = \begin{cases} \dfrac{\displaystyle\sum_{w\in W(A:G)}(e^{w\nu}\otimes\chi_{w\sigma})(h)}{(\varepsilon_R D)^{G/MA}(h)} & \text{if } h \in BA \\ 0 & \text{if } h \text{ nonconjugate to a member of } BA. \end{cases}$$

Remark. To understand this formula, it is convenient to ignore various plus and minus signs. With this convention, we can write the $D^{G/MA}$ in the denominator as $D^{G/BA}/D^{MA/BA}$ and group the $D^{M/B}$ with $\chi_{w\sigma}$. The product $D^{M/B}\chi_{w\sigma}$ is just the numerator in the Weyl character formula. Thus, apart from jumps arising from minus signs, the singularity in the formula for $\Theta_{\sigma,\nu}$ is given by $D^{G/BA}$ in the denominator.

In using (10.27) to get character formulas in the general case, we have to take into account one thing more beyond the local integrability question. Namely if $T_1, \ldots, T_s$ are a complete set of nonconjugate Θ-stable Cartan subgroups of G, we are trying to match (10.27) with a formula

$$\Theta_{\sigma,\nu}(f) = \sum_{k=1}^{s} s^{G/T_k}|W(T_k:G)|^{-1} \int_{T_k} F_f^{G/T_k}(h)\Theta_{\sigma,\nu}(h)(\varepsilon_R D)^{G/T_k}(h)\, dh. \tag{10.28}$$

When the H_j's can be matched with a subset of T_k's, there is no problem. But the contrary occurs when two nonconjugate Cartan subgroups of MA are conjugate in G. This happens, for example, in $G = \mathrm{SL}(4, \mathbb{R})$ if MAN is the block upper-triangular subgroup with 2-by-2 diagonal blocks. Then MA has four nonconjugate Cartan subgroups, but G has only three. For such a situation, we have to sum the expressions in braces in (10.27) corresponding to all H_j leading to the same T_k. Let us not write an explicit formula, being content with the following qualitative conclusion. Again the proof is complete modulo the local integrability question.

Proposition 10.19. If $S = MAN$ is a parabolic subgroup and if σ is an irreducible unitary representation of M with character of the form $\chi_\sigma(m)\, dm$, then the global character of the induced representation $U(S, \sigma, \nu)$ is a locally integrable function that is nonvanishing only on Cartan subgroups of G that are G conjugate to Cartan subgroups of MA.

The local integrability question is subtle, and we do not yet have the tools to handle it. For the time being, we shall state Lemma 10.20, which generalizes Lemma 10.7, without proof, and we shall obtain Proposition 10.21, which generalizes Proposition 10.8, as an immediate consequence. We shall return to Lemma 10.20 in §11.6, showing at that time that our development has not been circular.

Lemma 10.20. If f is an L^∞ function of compact support on G, then $F_f^{T_k}$ is an L^∞ function of compact support on T_k for $1 \leq k \leq s$.

Remarks. We shall sketch the proof in §11.6. Let us observe now that the lemma reduces to the case that T_k is compact. In fact, if $T_k = B_k A_k$ is the decomposition (10.24), then we can form a parabolic subgroup $S = MAN$ in the standard way such that $A = A_k$ and such that B_k is a compact Cartan subgroup of M_k. The reduction comes about by applying Lemma 10.17.

Proposition 10.21. Suppose for each k with $1 \leq k \leq s$ that $\Theta_k(h)$ is a function on T_k such that $D^{G/T_k}(h)\Theta_k(h)$ is locally integrable on T_k and $\Theta_k(h)$ is invariant under $W(T_k:G)$. Then the definition $\Theta(x) = \Theta_k(h)$ for x conjugate to $h \in T_k$ creates an almost-everywhere defined locally integrable function on G. Moreover, if Tr $\pi(f)$ is given by the right side of (10.28) with $\Theta(x)$ in place of $\Theta_{\sigma,\nu}(x)$, then the character of π is given by the function $\Theta(x)$.

Proof. Same as for Proposition 10.8.

§4. Differential Equations

The previous two sections gave formulas for characters in specific situations. Starting in this section, we shall study characters of general irreducible admissible representations. As with our study of matrix coefficients in Chapter VIII, we shall use differential equations as a tool to limit the possibilities for how characters can behave. If an irreducible admissible representation π is such that $\pi(z) = \chi(z)I$ for z in $Z(\mathfrak{g}^{\mathbb{C}})$, then we shall see that the global character Θ of π satisfies $z\Theta = \chi(z)\Theta$ in a suitable sense. In Chapter VIII we combined the analogous equation for matrix coefficients with transformation laws under K to get detailed information about matrix coefficients. Here we do not have useful transformation laws under K, but we can instead take advantage of the invariance of Θ under conjugation.

If D is any linear differential operator on a connected open subset U of Euclidean space, then there is a linear differential operator D^{tr} such that

$$\int_U (Df_1)(x)f_2(x)\,dx = \int_U f_1(x)(D^{\text{tr}}f_2)(x)\,dx$$

whenever f_1 and f_2 are smooth and at least one of them has compact support in U. The map $D \to D^{\text{tr}}$ is determined by the following facts: It is complex linear, it reverses the order of composition, it sends $a(x)I$ into itself, and it sends $\partial/\partial x_j$ into $-\partial/\partial x_j$.

On our group G in any local coordinate system, Haar measure dx is a nonvanishing smooth function times Lebesgue measure, and it follows by using a partition of unity that there is an analogous map $D \to D^{\text{tr}}$ for G with the property that

$$\int_G (Df_1)(x)f_2(x)\,dx = \int_G f_1(x)(D^{\text{tr}}f_2)(x)\,dx \tag{10.29}$$

whenever f_1 and f_2 are smooth and at least one of them has compact support.

We use equation (10.29) to motivate the definition of $D\Theta$ when Θ is a distribution. [If Θ is of the form $f_1(x)\,dx$, we require that $D\Theta$ be $Df_1(x)\,dx$, and (10.29) gives us a defining relation.] Thus if Θ is a distribution on G, we define $D\Theta$ to be the distribution given by

$$(D\Theta)(f) = \Theta(D^{\text{tr}}f) \qquad \text{for } f \in C_{\text{com}}^{\infty}(G). \tag{10.30}$$

For a left-invariant differential operator D, we can compute D^{tr} by starting with the following lemma.

Lemma 10.22. Let X in $\mathfrak{g}$ act by left invariant differentiation, and let f_1 and f_2 be smooth functions on G with at least one of them of compact support. Then

$$\int_G (Xf_1)(x)f_2(x)\,dx = -\int_G f_1(x)(Xf_2)(x)\,dx.$$

Remark. Therefore $X^{\mathrm{tr}} = -X$ for the left invariant vector field X.

Proof.
$$\begin{aligned}\int_G (Xf_1)(x)f_2(x)\,dx &= \int_G \frac{d}{dt}[f_1(x\exp tX)]_{t=0} f_2(x)\,dx \\ &= \frac{d}{dt}\left[\int_G f_1(x\exp tX)f_2(x)\,dx\right]_{t=0} \\ &= \frac{d}{dt}\left[\int_G f_1(x)f_2(x(\exp tX)^{-1})\,dx\right]_{t=0} \\ &= \int_G f_1(x)(-Xf_2)(x)\,dx.\end{aligned}$$

Corollary 10.23. The restriction of the map $D \to D^{\mathrm{tr}}$ to $U(\mathfrak{g}^{\mathbb{C}})$ is an associative algebra antiautomorphism of $U(\mathfrak{g}^{\mathbb{C}})$ that extends the Lie algebra antiautomorphism $X \to -X$ of $\mathfrak{g}$. The map $D \to D^{\mathrm{tr}}$ carries $Z(g^{\mathbb{C}})$ to itself.

Proposition 10.24. Let π be an irreducible admissible representation such that $\pi(z) = \chi(z)I$ for z in $Z(\mathfrak{g}^{\mathbb{C}})$, and let Θ be the global character of π. If z is considered as a left-invariant differential operator, then $z\Theta = \chi(z)\Theta$.

Proof. If f is in $C^\infty_{\mathrm{com}}(G)$, then $(z\Theta)(f) = \Theta(z^{\mathrm{tr}}f)$. Let $\{v_i\}$ be an orthonormal basis of K-finite vectors for the space on which π acts. Then

$$\begin{aligned}(\pi(z^{\mathrm{tr}}f)v_i, v_i) &= \int_G (\pi(x)v_i, v_i)(z^{\mathrm{tr}}f)(x)\,dx \\ &= \int_G z(\pi(x)v_i, v_i)f(x)\,dx && \text{by (10.29)} \\ &= \int_G (\pi(x)\pi(z)v_i, v_i)f(x)\,dx && \text{by (8.10)} \\ &= \chi(z)(\pi(f)v_i, v_i).\end{aligned}$$

Summing on i, we obtain $(z\Theta)(f) = \chi(z)\Theta(f)$, as required.

§5. Analyticity on the Regular Set, Overview and Example

The fact that all the examples of global characters in §§2–3 are given as locally integrable functions (times Haar measure) raises the question: To what extent are general characters given as functions? The deep answer is that irreducible characters are always given in this fashion and moreover the function is real analytic on the regular set of G. In this section we shall address the part of this answer that deals with the regular set.

We recall the discussion of the regular set in §5.4. If $\{T_i\}$ is a complete set of (necessarily abelian) nonconjugate Θ-stable Cartan subgroups of G, then the set $(T_i')^G$ of G conjugates of the regular elements in T_i is open in G, and these sets, as i varies, form a partition of the set of regular elements of G. Since global characters are invariant under conjugation, we expect the analysis of characters on the regular set to be closely tied to Cartan subgroups.

Theorem 10.25. The restriction of any irreducible global character of G to (smooth functions compactly supported in) the regular set of G is a real analytic function invariant under conjugation.

The idea of the proof is to use the invariance and the differential equations of §4 to express the character in terms of its "restriction" to each T_i' and to show that the restrictions each satisfy an analytic elliptic differential equation. The execution of this idea is more subtle than this simple description, however. There is a certain very general part of the argument at the start, and then some calculations are necessary. Our procedure will be to describe this general part briefly, to do the calculations for $SL(2, \mathbb{R})$, to justify most of the general part, and then to do the calculations for all G.

Fix a Cartan subgroup T. One can show that any member of G that conjugates a regular element of T into T normalizes the whole group T. Consequently the map $p: G/T \times T' \to (T')^G$ given by

$$p(gT, t) = gtg^{-1}$$

is everywhere $|W(T:G)|$ to one. Its differential is nonsingular, by Proposition 5.27, and it is thus a local diffeomorphism. If μ is a distribution on $(T')^G$, we can lift μ to a distribution $\tilde{\mu}$ on $G/T \times T'$ by the formula

$$\tilde{\mu}(F) = \mu(\varphi), \qquad \text{where } \varphi(y) = |W(T:G)|^{-1} \sum_{x \in p^{-1}(y)} F(x). \tag{10.31}$$

If μ is invariant under conjugation, then it is easy to see that $\tilde{\mu}$ is invariant under left translation of F in the first variable. In this generality we shall see in Lemma 10.28 that $\tilde{\mu}$ splits as a product

$$\tilde{\mu} = d\dot{g} \times \mu_T, \tag{10.32a}$$

where $d\dot{g}$ is the left-invariant measure on G/T and μ_T is a distribution on T'. Equation (10.32a) is to be interpreted in the following sense:

$$\text{If } F_2(t') = \int_{G/T} F(gT, t')\, d\dot{g}, \qquad \text{then } \tilde{\mu}(F) = \mu_T(F_2). \tag{10.32b}$$

For now, let us ignore the fact that the first variable of the space $G/T \times T'$ on which $\tilde{\mu}$ lives is G/T, not simply G. If we apply a member D of $U(\mathfrak{g}^{\mathbb{C}})$ to $\tilde{\mu}$ in the first variable and use (10.30), then Lemma 10.22 (with $f_2 = 1$) says that the only surviving part is the constant term of D, applied to $\tilde{\mu}$. In terms of μ on G', this means that the derivatives at gtg^{-1} in the direction of g will go away, provided the constant term is absent. Since the only complementary directions are those along T, we expect to be able to reduce any differential operator to an action along T. Comparing with $\tilde{\mu}$ on $G/T \times T'$, we expect to be able to reduce the action of any member of $U(\mathfrak{g}^{\mathbb{C}})$ to an action of operators on T, applied to μ_T.

When μ is the character of an irreducible admissible representation, Proposition 10.24 says that μ is an eigendistribution of $Z(\mathfrak{g}^{\mathbb{C}})$. The resulting equations ought therefore to be reflected in equations satisfied by μ_T. In fact, we shall see that μ_T does satisfy such a system and that the system contains elliptic operators. Theorem 10.25 will follow from these facts.

In more detail let φ be given in $C^\infty_{\text{com}}((T')^G)$ and let $\tilde{F}$ be the function on $G \times T'$ given by

$$\tilde{F}(g, t) = \varphi(gtg^{-1}).$$

The function $\tilde{F}$ is right-invariant under T and hence descends to a function F on $G/T \times T'$. If we construct a function on $(T')^G$ from F by (10.31), we return to our given φ. Whenever D_1 and D_2 are left-invariant differential operators on G and T, respectively, we denote by $(D_1, D_2)\tilde{F}(g, t)$ the effect of applying D_1 in the first variable and D_2 in the second variable. Let $X_1, \ldots, X_r$ be in $\mathfrak{g}^{\mathbb{C}}$, and let L and R denote "left by" and "right by," respectively, in $U(\mathfrak{g}^{\mathbb{C}})$. Then we shall see in Lemma 10.29 that

$$(X_1 \cdots X_r, H_1 \cdots H_s)\tilde{F}(g, t) = ((\mathrm{Ad}(g)D)\varphi)(gtg^{-1}), \tag{10.33}$$

where D is the variable coefficient differential operator

$$D = (L_{\mathrm{Ad}(t)^{-1}X_1} - R_{X_1}) \cdots (L_{\mathrm{Ad}(t)^{-1}X_r} - R_{X_r})(H_1 \cdots H_s). \tag{10.34}$$

Apart from the fact that the left side of (10.33) is defined on $G \times T'$, not $G/T \times T'$, application of $\tilde{\mu}$ to the left side of (10.33) should give 0 whenever $r > 0$.

Now we can examine how to proceed in $SL(2, \mathbb{R})$. Let $T = \{t\}_{t \in T}$ be the diagonal subgroup, and let $\{h, e, f\}$ be the usual basis of $\mathfrak{sl}(2, \mathbb{R})$. The Casimir operator is given by

$$\Omega = \tfrac{1}{2}h^2 + ef + fe.$$

We shall write Ω as a linear combination (with functions of t as coefficients) of operators D as in (10.34). Then we shall substitute into (10.33), use the fact that $\operatorname{Ad}(g)\Omega = \Omega$, and essentially apply an irreducible character to both sides of (10.33).

Let $\xi_\alpha : T \to \mathbb{C}^\times$ be the exponentiated root given by $\xi_\alpha\begin{pmatrix} x & 0 \\ 0 & x^{-1}\end{pmatrix} = x^2$, $x \in \mathbb{R}^\times$. Define $\Gamma_t(X_1 \cdots X_r, H_1 \cdots H_s)$ to be the right side of (10.34). Since $\operatorname{Ad}(t)^{-1}e = \xi_\alpha(t)^{-1}e$ and $\operatorname{Ad}(t)^{-1}f = \xi_\alpha(t)f$, we readily compute that

$$\begin{aligned}\Gamma_t(ef, 1) = \Gamma_t(fe, 1) &= ef + fe - \xi_\alpha(t)fe - \xi_\alpha(t)^{-1}ef \\ &= \tfrac{1}{2}(2 - \xi_\alpha(t)^{-1} - \xi_\alpha(t))(ef + fe) + \tfrac{1}{2}(\xi_\alpha(t) - \xi_\alpha(t)^{-1})h.\end{aligned}$$

Thus

$$\Gamma_t(ef, 1) - \tfrac{1}{2}(\xi_\alpha(t) - \xi_\alpha(t)^{-1})\Gamma_t(1, h) = \tfrac{1}{2}(2 - \xi_\alpha(t)^{-1} - \xi_\alpha(t))(ef + fe).$$

Dividing by the coefficient on the right and adding $\frac{1}{2}h^2$, we obtain

$$\Omega = \Gamma_t(1, \tfrac{1}{2}h^2) + \frac{\xi_\alpha(t) - \xi_\alpha(t)^{-1}}{\xi_\alpha(t) - 2 + \xi_\alpha(t)^{-1}}\Gamma_t(1, h) - \frac{2}{\xi_\alpha(t) - 2 + \xi_\alpha(t)^{-1}}\Gamma_t(ef, 1).$$

Returning to (10.33), we therefore have

$$\begin{aligned}\Omega\varphi(gtg^{-1}) = (1, \tfrac{1}{2}h^2)\tilde{F}(g, t) &+ \frac{\xi_\alpha(t) - \xi_\alpha(t)^{-1}}{\xi_\alpha(t) - 2 + \xi_\alpha(t)^{-1}}(1, h)\tilde{F}(g, t) \\ &- \frac{2}{\xi_\alpha(t) - 2 + \xi_\alpha(t)^{-1}}(ef, 1)\tilde{F}(g, t).\end{aligned}$$

The various operators on the right that act in the first coordinate are all $\operatorname{Ad}(T)$-invariant, and thus we can rewrite the formula on $G/T \times T'$. With $F(gT, t)$ defined as $\tilde{F}(g, t)$, we have

$$\begin{aligned}\Omega\varphi(gtg^{-1}) = (1, \tfrac{1}{2}h^2)F(gH, t) &+ \frac{\xi_\alpha(t) - \xi_\alpha(t)^{-1}}{\xi_\alpha(t) - 2 + \xi_\alpha(t)^{-1}}(1, h)F(gT, t) \\ &- \frac{2}{\xi_\alpha(t) - 2 + \xi_\alpha(t)^{-1}}(ef, 1)F(gT, t).\end{aligned} \tag{10.35}$$

Let $w = \begin{pmatrix} 0 & 1 \\ -1 & 0\end{pmatrix}$ represent the nontrivial element of $W(T{:}G)$. On the right side of (10.35) if we replace (g, t) by $(gw, w^{-1}tw)$, all three terms remain unchanged; to see this invariance for the last term, we use the identity $\Gamma_t(ef, 1) = \Gamma_t(fe, 1)$.

Consequently (10.35) says that the function on $G/T \times T'$ on the right side of (10.35) is related to $\Omega\varphi$ on $(T')^G$ as in (10.31). Since μ is invariant, we obtain

$$\mu(\Omega\varphi) = \tilde{\mu}((1, \tfrac{1}{2}h^2)F) + \tilde{\mu}\left(\left(1, \frac{\xi_\alpha(t) - \xi_\alpha(t)^{-1}}{\xi_\alpha(t) - 2 + \xi_\alpha(t)^{-1}}h\right)F\right).$$

If F_2 is defined on T' in terms of F by (10.32b), we obtain

$$\mu(\Omega\varphi) = \mu_T(\tfrac{1}{2}h^2F_2) + \mu_T\left(\frac{\xi_\alpha(t) - \xi_\alpha(t)^{-1}}{\xi_\alpha(t) - 2 + \xi_\alpha(t)^{-1}}\, hF_2\right). \tag{10.36}$$

Now suppose that μ is an irreducible character of $\mathrm{SL}(2, \mathbb{R})$, so that $\Omega\mu = \chi(\Omega)\mu$. Since $\Omega = \Omega^{\mathrm{tr}}$ and $\mu(\varphi) = \mu_T(F_2)$, (10.36) gives

$$\mu_T\left(\left(\tfrac{1}{2}h^2 + \frac{\xi_\alpha(t) - \xi_\alpha(t)^{-1}}{\xi_\alpha(t) - 2 + \xi_\alpha(t)^{-1}}\, h - \chi(\Omega)\right)F_2\right) = 0. \tag{10.37}$$

Let us write $\beta_t(\Omega)$ for the operator

$$\beta_t(\Omega) = \tfrac{1}{2}h^2 + \frac{\xi_\alpha(t) - \xi_\alpha(t)^{-1}}{\xi_\alpha(t) - 2 + \xi_\alpha(t)^{-1}}\, h$$

in (10.37). This operator becomes more transparent by using the Harish-Chandra homomorphism γ. Recall that $\gamma(\Omega) = \frac{1}{2}(h^2 - 1)$. Let $\xi_\delta\begin{pmatrix} x & 0 \\ 0 & x^{-1}\end{pmatrix} = x$, and form the Weyl denominator

$$D_T(t) = \xi_\delta(t)(1 - \xi_\alpha(t)^{-1}) = \xi_\delta(t) - \xi_\delta(t)^{-1}.$$

(See Proposition 5.27.) It is straightforward to use the relations $h\xi_\delta = \xi_\delta$ and $h\xi_\delta^{-1} = -\xi_\delta^{-1}$ to calculate

$$D_T(t)^{-1}\tfrac{1}{2}h^2(D_TF_2)(t) = \tfrac{1}{2}h^2F_2(t) + \frac{\xi_\delta(t) + \xi_\delta(t)^{-1}}{\xi_\delta(t) + \xi_\delta(t)^{-1}}\, hF_2(t) + \tfrac{1}{2}F_2(t).$$

Consequently

$$D_T(t)^{-1}\gamma(\Omega)(D_TF_2)(t) = \beta_t(\Omega)F_2(t). \tag{10.38}$$

Equation (10.37) thus says

$$\mu_T(D_T(t)^{-1}(\gamma(\Omega) - \chi(\Omega))(D_TF_2)) = 0.$$

Since D_T is nonvanishing on T', (10.30) gives

$$(\gamma(\Omega) - \chi(\Omega))(D_T^{-1}\mu_T) = 0. \tag{10.39}$$

Therefore $D_T^{-1}\mu_T$ is a solution of a well-understood second order ordinary differential equation with constant coefficients and hence is a real analytic function. We could write down the function explicitly but will not do so now. Write

$$D_T^{-1}\mu_T = s^{G/T}|W(T{:}G)|^{-1}\tau_T(t)\, dt. \tag{10.40}$$

(Here $s^{G/T}$ is given in general by (10.25) but reduces to 1 for T the noncompact Cartan subgroup of $\mathrm{SL}(2, \mathbb{R})$.)

The definition of $\tilde{\mu}$ by means of (10.31) makes $\tilde{\mu}$ invariant under $(gH, t) \to (gwH, w^{-1}tw)$, and this property is reflected in μ_T as invariance under $t \to w^{-1}tw = t^{-1}$. Then it follows that τ_T is odd in the sense that $\tau_T(t^{-1}) = -\tau_T(t)$. Moreover, formulas (10.25), (10.31), (10.32), and (10.40) combine to give

$$\begin{aligned}\mu(\varphi) = \tilde{\mu}(F) &= s^{G/T}|W(T:G)|^{-1}\int_{G/T\times T'} D_T(t)\tau_T(t)F(gT, t)\, d\dot{g}\, dt\\ &= s^{G/T}|W(T:G)|^{-1}\int_{T'}(\tau_T(t)/D_T(t))\left[\int_{G/T}\varphi(gtg^{-1})\, d\dot{g}\right]D_T(t)^2\, dt\\ &= s^{G/T}|W(T:G)|^{-1}\int_{T'}(\tau_T(t)/D_T(t))F_f^T(t)\varepsilon_R^T(t)D_T(t)\, dt.\end{aligned}$$

By Proposition 10.8, the character is given by the locally integrable function $\tau_T(t)/D_T(t)$ on the conjugates of the element t of T'.

For the compact Cartan subgroup of SL(2, $\mathbb{R}$), essentially the same calculation is valid, provided we reinterpret h, e, and f suitably. For h we take the matrix $\begin{pmatrix} 0 & i \\ -i & 0 \end{pmatrix}$, and we let e and f be appropriately normalized root vectors. This time the Weyl group $W(T:G)$ is trivial and the sign $s^{G/T}$ is -1. Thus the constants have different values, but the proof goes through. Hence any irreducible character for SL(2, $\mathbb{R}$) is a real analytic function on the regular set, and the numerator $\tau(t)$ of the character on either Cartan subgroup satisfies the differential equation

$$\tfrac{1}{2}(h^2 - 1)\tau = \chi(\Omega)\tau. \tag{10.41}$$

§6. Analyticity on the Regular Set, General Case

In this section we shall give some of the details for the proof of Theorem 10.25. As in §5, let T be a Cartan subgroup of G, let $\mathfrak{t}$ be the Lie algebra of T, and let $p: G/T \times T' \to (T')^G$ be the map

$$p(gT, t) = gtg^{-1}.$$

Let μ be an invariant distribution on $(T')^G$, and let $\tilde{\mu}$ be the distribution on $G/T \times T'$ given by

$$\tilde{\mu}(F) = \mu(\varphi), \qquad \text{where } \varphi(y) = |W(T:G)|^{-1}\sum_{x\in p^{-1}(y)} F(x). \tag{10.42}$$

Lemma 10.26. Each left-invariant distribution on an analytic group is a multiple of left Haar measure.

Proof. If v is such a distribution and h is a smooth function of compact support, then the distribution $v * h$ is a left-invariant function times left

Haar measure. This function must be constant. Passing to the limit as h passes through an approximate identity, we easily obtain the lemma.

Lemma 10.27. Let H be a closed subgroup of an analytic group G such that G/H has a G-invariant measure $d\dot{g}$. If M is a smooth manifold and if $\tilde{v}$ is a G-invariant distribution on $G/H \times M$, then there exists a (unique) distribution v_M on M such that

$$\tilde{v} = dg \times v_M$$

in the sense that

$$F_2(m) = \int_{G/H} F(gH, m)\, d\dot{g} \qquad \text{implies} \qquad \tilde{v}(F) = v_M(F_2). \quad (10.43)$$

Sketch of proof. By means of a partition of unity, one sees that the map $\tilde{f} \to f$ of $C_{\text{com}}^{\infty}(G)$ to $C_{\text{com}}^{\infty}(G/H)$ given by $f(gH) = \int_H \tilde{f}(gh)\, dh$ is actually onto. Then it is easy to lift $\tilde{v}$ canonically to a left G-invariant distribution $\tilde{v}^*$ on $G \times M$. Fix f_2 in $C_{\text{com}}^{\infty}(M)$. The map $\tilde{f}_1 \to \tilde{v}^*(\tilde{f}_1 \times f_2)$ is a left G-invariant distribution on G and must be a multiple of Haar measure, by Lemma 10.26. Let us write

$$\tilde{v}^*(\tilde{f}_1 \times f_2) = c(f_2) \int_G \tilde{f}_1(g)\, dg.$$

Passing from $\tilde{f}_1$ on G to f_1 on G/H, we obtain

$$\tilde{v}(f_1 \times f_2) = c(f_2) \int_G f_1(gH)\, d\dot{g}$$

The functional $f_2 \to c(f_2)$ is a distribution on M, and we take v_M to be this distribution. Then (10.43) follows when F is of the form $f_1 \times f_2$. Since linear combinations of such functions are dense in $C_{\text{com}}^{\infty}(G/H \times M)$, the lemma follows.

Lemma 10.28. The distribution $\tilde{\mu}$ of (10.42) is of the form $d\dot{g} \times \mu_T$ on $G/T \times T'$ for some distribution μ_T on T'. Moreover, μ_T is invariant under conjugation by the normalizer of T in G.

Proof. The first statement is a special case of Lemma 10.27. The distribution $\tilde{\mu}$ has a further invariance property: If w represents a member of $W(T{:}G)$, then $\tilde{\mu}$ is invariant under $(gT, t) \to (gwT, w^{-1}tw)$. Then it follows that μ_T is invariant under conjugation by w^{-1}.

Lemma 10.29. If $\varphi \in C^{\infty}((T')^G)$ and $\tilde{F} \in C^{\infty}(G \times T')$ are related by $\tilde{F}(g, t) = \varphi(gtg^{-1})$ and if $X_1, \ldots, X_r$ are in $\mathfrak{g}^{\mathbb{C}}$ and $H_1, \ldots, H_s$ are in $\mathfrak{t}^{\mathbb{C}}$, then

$$(X_1 \cdots X_r, H_1 \cdots H_s)\tilde{F}(g, t) = ((\text{Ad}(g)D)\varphi)(gtg^{-1}),$$

where D is the variable-coefficient differential operator

$$D = (L_{\mathrm{Ad}(t)^{-1}X_1} - R_{X_1}) \cdots (L_{\mathrm{Ad}(t)^{-1}X_r} - R_{X_r})(H_1 \cdots H_s).$$

Remark. As in §5, L and R refer to "left by" and "right by," respectively.

Proof. We induct on r, the case $r = 0$ being clear. Assume the identity is valid for $r - 1$. Then

$$(X_2 \cdots X_r, H_1 \cdots H_s)\tilde{F}(g, t) = ((\mathrm{Ad}(g)b)\varphi)(gtg^{-1})$$

for a suitable variable-coefficient operator b. Then

$$\begin{aligned}
&(X_1 \cdots X_r, H_1 \cdots H_s)F(g, t) \\
&\quad = \frac{d}{du}(\mathrm{Ad}(g \exp uX_1)b)F(g(\exp uX_1)t(\exp uX_1)^{-1}g^{-1})_{u=0} \\
&\quad = \frac{d}{du}\{(\mathrm{Ad}(g \exp uX_1)b)F(gtg^{-1}) + (\mathrm{Ad}(g)b)F(g(\exp uX_1)tg^{-1}) \\
&\qquad - (\mathrm{Ad}(g)b)F(gt(\exp uX_1)g^{-1})\}_{u=0} \\
&\quad = (\mathrm{Ad}(g)\,\mathrm{ad}(X_1)b)F(gtg^{-1}) + (\mathrm{Ad}(gt^{-1})X_1)(\mathrm{Ad}(g)b)F(gtg^{-1}) \\
&\qquad - (\mathrm{Ad}(g)X_1)(\mathrm{Ad}(g)b)F(gtg^{-1}) \\
&\quad = \mathrm{Ad}(g)\{X_1 b - bX_1 + (\mathrm{Ad}(t^{-1})X_1)b - X_1 b\}F(gtg^{-1}) \\
&\quad = (\mathrm{Ad}(g)\{(L_{\mathrm{Ad}(t)^{-1}X_1} - R_{X_1})b\})F(gtg^{-1}).
\end{aligned}$$

The induction is complete, and the lemma follows.

Lemma 10.30. For each t in T', there exists a unique linear mapping $\Gamma_t : U(\mathfrak{g}^{\mathbb{C}}) \otimes_{\mathbb{C}} U(\mathfrak{t}^{\mathbb{C}}) \to U(\mathfrak{g}^{\mathbb{C}})$ such that $\Gamma_t(1 \otimes H) = H$ for $H \in U(\mathfrak{t}^{\mathbb{C}})$ and

$$\Gamma_t(X_1 \cdots X_r \otimes H) = (L_{\mathrm{Ad}(t)^{-1}X_1} - R_{X_1}) \cdots (L_{\mathrm{Ad}(t)^{-1}X_r} - R_{X_r})H$$

for all $X_1, \ldots, X_r$ in $\mathfrak{g}^{\mathbb{C}}$ and H in $U(\mathfrak{t}^{\mathbb{C}})$. Moreover, if an ordered basis of $\mathfrak{t}^{\mathbb{C}}$ is extended to an ordered basis of $\mathfrak{g}^{\mathbb{C}}$ by adjoining a basis of root vectors $Y_1, \ldots, Y_s$, then Γ_t maps the space

$$\{\textstyle\sum c_{i_1 \cdots i_s} Y_1^{i_1} \cdots Y_s^{i_s}\} \otimes U(\mathfrak{t}^{\mathbb{C}}) \tag{10.44}$$

one-one onto $U(\mathfrak{g}^{\mathbb{C}})$.

Proof. The uniqueness of Γ_t is obvious; let us prove existence. For X in $\mathfrak{g}$, let $\sigma_t(X)$ be the endomorphism of $U(\mathfrak{g}^{\mathbb{C}})$ given by $\sigma_t(X)D = (L_{\mathrm{Ad}(t)^{-1}X} - R_X)D$. Then it is easy to check that σ_t is a representation of $\mathfrak{g}$ on $U(\mathfrak{g}^{\mathbb{C}})$. Hence σ_t extends to a representation of $U(\mathfrak{g}^{\mathbb{C}})$ on $U(\mathfrak{g}^{\mathbb{C}})$. That is, we have a well-defined bilinear map of $U(\mathfrak{g}^{\mathbb{C}}) \times U(\mathfrak{g}^{\mathbb{C}})$ into $U(\mathfrak{g}^{\mathbb{C}})$ given by

$$(D_1, D_2) \to \sigma_t(D_1)D_2.$$

We restrict this bilinear map to D_2 in $U(\mathfrak{t}^{\mathbb{C}})$, and then Γ_t is the associated linear map on $U(\mathfrak{g}^{\mathbb{C}}) \otimes_{\mathbb{C}} U(\mathfrak{t}^{\mathbb{C}})$. This proves existence.

Let $\mathfrak{q}$ be the complex span of Y_i, and let

$$\mathfrak{q}^d = \sum_{i_1 + \ldots + i_s \le d} c_{i_1 \cdots i_s} Y_1^{i_1} \cdots Y_s^{i_s}.$$

We shall prove by induction on r that

$$\sum_{d+e \le r} \Gamma_t(\mathfrak{q}^d \otimes U^e(\mathfrak{t}^{\mathbb{C}})) = U^r(\mathfrak{g}^{\mathbb{C}}). \tag{10.45}$$

It is clear that the left side is contained in the right side. Let

$$A(t) = (\mathrm{Ad}(t)^{-1} - I)|_{\mathfrak{q}}.$$

Since t is regular, $A(t)$ is nonsingular on $\mathfrak{q}$ and is diagonalized by the Y_i. Recall that $\mathrm{ad}\, X = L_X - R_X$ for X in $\mathfrak{g}^{\mathbb{C}}$. If $H_1, \ldots, H_e$ are in $\mathfrak{t}^{\mathbb{C}}$, then

$$\begin{aligned}
&\Gamma_t(Y_{j_1} \cdots Y_{j_d} \otimes H_1 \cdots H_e) \\
&\quad = (L_{\mathrm{Ad}(t)^{-1} Y_{j_1}} - R_{Y_{j_1}}) \cdots (L_{\mathrm{Ad}(t)^{-1} Y_{j_d}} - R_{Y_{j_d}}) H_1 \cdots H_e \\
&\quad = (L_{A(t) Y_{j_1}} + \mathrm{ad}\, Y_{j_1}) \cdots (L_{A(t) Y_{j_d}} + \mathrm{ad}\, Y_{j_d}) H_1 \cdots H_e \\
&\quad \equiv (A(t) Y_{j_1}) \cdots (A(t) Y_{j_d}) H_1 \cdots H_e \bmod U^{d+e-1}(\mathfrak{g}^{\mathbb{C}}).
\end{aligned} \tag{10.47}$$

The expression on the right is a nonzero multiple of $Y_{j_1} \cdots Y_{j_d} H_1 \cdots H_e$, and hence every monomial $Y_1^{i_1} \cdots Y_s^{i_s} H_1 \cdots H_e$ with $i_1 + \ldots + i_s = d$ is in the image of $\mathfrak{q}^d \otimes U^e(\mathfrak{t}^{\mathbb{C}})$. In view of the Birkhoff-Witt Theorem, (10.45) follows. This proves Γ_t maps (10.44) onto $U(\mathfrak{g}^{\mathbb{C}})$.

Finally we have $\mathfrak{g}^{\mathbb{C}} = \mathfrak{q} \oplus \mathfrak{t}^{\mathbb{C}}$, and thus

$$\sum_{d+e \le r} (\mathfrak{q}^d \otimes U^e(\mathfrak{t}^{\mathbb{C}}))$$

and $U^r(\mathfrak{g}^{\mathbb{C}})$ have the same dimension, by the Birkhoff-Witt Theorem. Thus Γ_t is one-one on (10.44). This completes the proof.

Fix an enumeration of $\Delta(\mathfrak{t}^{\mathbb{C}} : \mathfrak{g}^{\mathbb{C}})$, and let $Y_1, \ldots, Y_s$ be an ordered basis of corresponding root vectors. Define

$$\mathcal{Q} = \{\textstyle\sum c_{i_1 \cdots i_s} Y_1^{i_1} \cdots Y_s^{i_s}\}$$

and

$$\mathcal{Q}' = \left\{ \sum_{i_1 + \ldots + i_s > 0} c_{i_1 \cdots i_s} Y_1^{i_1} \cdots Y_s^{i_s} \right\}.$$

Fix t in T'. Then it follows from Lemma 10.30 that to each D in $U(\mathfrak{g}^{\mathbb{C}})$ corresponds a unique member $\beta_t(D)$ of $U(\mathfrak{t}^{\mathbb{C}})$ such that

$$D - \beta_t(D) \qquad \text{is in} \qquad \Gamma_t(\mathcal{Q}' \otimes U(\mathfrak{t}^{\mathbb{C}})).$$

Going over the proof of Lemma 10.30, we see that $\beta_t(D)$ depends in a real analytic fashion on the parameter t, for t in T'. The element $\beta_t(D)$ does not depend on the enumeration of roots used in its definition, in view of the following lemma.

Lemma 10.31. The members of $\mathfrak{g}^{\mathbb{C}}$, as a subset of $U(\mathfrak{g}^{\mathbb{C}})$, generate an associative subalgebra $U(\mathfrak{g}^{\mathbb{C}})'$ of $U(\mathfrak{g}^{\mathbb{C}})$ of codimension one. This subalgebra has the property that

$$\Gamma_t(\mathscr{Q}' \otimes U(\mathfrak{t}^{\mathbb{C}})) = \Gamma_t(U(\mathfrak{g}^{\mathbb{C}})' \otimes U(\mathfrak{t}^{\mathbb{C}})). \tag{10.48}$$

Proof. In the tensor algebra $T(\mathfrak{g}^{\mathbb{C}})$, the direct sum of the spaces $\bigotimes^n \mathfrak{g}^{\mathbb{C}}$ for $n \geq 1$ is an associative subalgebra $T(\mathfrak{g}^{\mathbb{C}})'$, and $T(\mathfrak{g}^{\mathbb{C}})'$ maps onto the subalgebra $U(\mathfrak{g}^{\mathbb{C}})'$ generated by $\mathfrak{g}^{\mathbb{C}}$. The kernel of the canonical map of $T(\mathfrak{g}^{\mathbb{C}})$ onto $U(\mathfrak{g}^{\mathbb{C}})$ is contained in $T(\mathfrak{g}^{\mathbb{C}})'$, and hence the multiples of the identity of $U(\mathfrak{g}^{\mathbb{C}})$ are not in $U(\mathfrak{g}^{\mathbb{C}})'$. Since it is clear from the definition that $U(\mathfrak{g}^{\mathbb{C}})'$ has codimension at most one, the codimension is exactly one.

From the Birkhoff-Witt Theorem, we then see that $U(\mathfrak{g}^{\mathbb{C}})'$ has a basis consisting of all nontrivial monomials

$$Y_1^{i_1} \cdots Y_s^{i_s} H_1^{j_1} \cdots H_r^{j_r},$$

where $Y_1, \ldots, Y_s$ is as in Lemma 10.30 and $H_1, \ldots, H_r$ is a basis of $\mathfrak{t}^{\mathbb{C}}$. Since $L_{\mathrm{Ad}(t)^{-1}H_k} - R_{H_k}$ is 0 on $U(\mathfrak{t}^{\mathbb{C}})$ for $1 \leq k \leq r$,

we have
$$\Gamma(Y_1^{i_1} \cdots Y_s^{i_s} H_1^{j_1} \cdots H_r^{j_r} \otimes U(\mathfrak{t}^{\mathbb{C}})) = 0$$

if $j_1 + \ldots + j_r > 0$. Thus (10.48) follows.

Lemma 10.32. Let $\varphi \in C^{\infty}_{\mathrm{com}}((T')^G)$ and $F \in C^{\infty}_{\mathrm{com}}(G/T \times T')$ be related by $F(gT, t) = \varphi(gtg^{-1})$, and define $F_2(t) = \int_{G/T} F(gT, t)\, d\dot{g}$. If z is in $Z(\mathfrak{g}^{\mathbb{C}})$, then the distribution μ_T on T' (given in Lemma 10.28) satisfies

$$\mu(z\varphi) = \mu_T(\beta_t(z) F_2(t)).$$

Proof. By Lemma 10.30 we can write

$$z = \beta_t(z) + \sum_i \Gamma_t(q'_{i,t} \otimes h_{i,t})$$

with $q'_{i,t}$ in $\mathscr{Q}'$ and $h_{i,t}$ in $U(\mathfrak{t}^{\mathbb{C}})$. Let t_0 be in T. Applying $\mathrm{Ad}(t_0)$ to both sides, we see that

$$z = \beta_t(z) + \sum_i \Gamma_t(\mathrm{Ad}(t_0) q'_{i,t} \otimes h_{i,t}).$$

The uniqueness in Lemma 10.30 therefore says that $\mathrm{Ad}(T)$ fixes each $q'_{i,t}$. Hence $q'_{i,t}$ descends to a left-invariant differential operator on G/T that we denote by $q'_{i,t}$ also.

With F and $\tilde{F}$ denoting the lifts of φ to $G/T \times T'$ and $G \times T'$, respectively, we use Lemma 10.29 to write

$$\begin{aligned}(z\varphi)(gtg^{-1}) &= ((\mathrm{Ad}(g)z)\varphi)(gtg^{-1}) \\ &= (1, \beta_t(z))\tilde{F}(g, t) + \sum_i (q'_{i,t}, h_{i,t})\tilde{F}(g, t).\end{aligned} \tag{10.49}$$

Let $F_z(gT, t) = (z\varphi)(gtg^{-1})$. Substituting on the left in (10.49) and replacing each term on the right of (10.49) by its version on $G/T \times T'$, we have

$$F_z(gT, t) = (1, \beta_t(z))F(gT, t) + \sum_i (q'_{i,t}, h_{i,t})F(gT, t).$$

Application of $\tilde{\mu}$ to both sides gives

$$\begin{aligned}\mu(z\varphi) &= \tilde{\mu}(F_z) \\ &= \mu_T\left(\int_{G/T} (1, \beta_t(z))F(gT, t)\, d\dot{g}\right) + \sum \mu_T\left(\int_{G/T} (q'_{i,t}, h_{i,t})F(gT, t)\, d\dot{g}\right) \\ &= \mu_T(\beta_t(z)F_2(t)) + \sum \mu_T\left(\int_{G/T} (q'_{i,t}, h_{i,t})F(gT, t)\, d\dot{g}\right).\end{aligned}$$

To complete the proof, it is enough to show that

$$\int_{G/T} (q'_{i,t}, h_{i,t})F(gT, t)\, d\dot{g} = 0. \tag{10.50}$$

A straightforward argument with a partition of unity shows that there exists F^* in $C^\infty_{\text{com}}(G \times T')$ such that

$$F(gT, t) = \int_T F^*(gt_0, t)\, dt_0.$$

Substituting, we see that the left side of (10.50) is

$$= \int_{G/T \times T} (q'_{i,t}, h_{i,t})F^*(gt_0, t)\, dt_0\, d\dot{g} = \int_G (q'_{i,t}, h_{i,t})F^*(g, t)\, dg.$$

By Lemma 10.22 (with $f_1(x) = F^*(x, t)$ and $f_2(x) = 1$), we see that the right side here is 0. Thus (10.50) is proved, and the lemma follows.

In order to pass from Lemma 10.32 to a proof of Theorem 10.25, we need a formula for $\beta_t(z)$. The expected generalization of (10.38) from SL(2, $\mathbb{R}$) to G provides such a formula. It uses the Weyl denominator

$$D_T(t) = \xi_\delta(t) \prod_{\alpha > 0} (1 - \xi_\alpha(t)^{-1})$$

defined in Proposition 5.27 and the remark following it. Implicit in the formula for D_T is a choice of a positive system of roots for $(\mathfrak{g}^{\mathbb{C}}, \mathfrak{t}^{\mathbb{C}})$, and then the product is over all the positive roots. A change of positive system affects $D_T(t)$ at most by a minus sign.

The linear functional δ is algebraically integral (Proposition 4.33) but not necessarily analytically integral in a suitable sense. But without loss of generality, we may assume that $\mathfrak{k} \cap i\mathfrak{p} = 0$, so that $G^{\mathbb{C}}$ is a well-defined group of matrices, and we may assume that $G^{\mathbb{C}}$ is simply connected. Then ξ_δ is well defined on $\exp \mathfrak{t}^{\mathbb{C}} \subseteq G^{\mathbb{C}}$ and restricts to a well-defined function on T (cf. Theorem 5.22c). With this understanding, we have the following formula for $\beta_t(z)$.

Theorem 10.33. If t is in T' and z is in $Z(\mathfrak{g}^{\mathbb{C}})$, then

$$\beta_t(z) = D_T(t)^{-1}(\gamma(z) \circ D_T(t)), \tag{10.51}$$

where γ is the Harish-Chandra homomorphism.

Remark. In more detail $\beta_t(z)$ as a differential operator is obtained by multiplying by $D_T(t)$, applying the differential operator $\gamma(z)$, and then multiplying by $D_T(t)^{-1}$. To prove Theorem 10.33, we require the following lemma.

Lemma 10.34. If D is in $U(\mathfrak{g}^{\mathbb{C}})$, then there exists an integer $l \geq 0$ and there exist members $H_1, \ldots, H_m$ of $U(\mathfrak{t}^{\mathbb{C}})$ and members $\sigma_1, \ldots, \sigma_m$ of the lattice generated by the roots such that

$$\beta_t(D) = D_T(t)^{-2l} \sum_{j=1}^{m} \xi_{\sigma_j}(t) H_j.$$

Proof. We go over the proof of Lemma 10.30, with particular attention to (10.46) and (10.47). The determinant of $A(t)$ is $(-1)^k \xi_{2\delta}(t) D_T(t)^2$, where k is the number of positive roots and the entries are linear combinations of ξ_σ's. To prove the lemma, we proceed by induction on the degree of D, and we may assume D is a monomial:

$$D = Y_1^{i_1} \cdots Y_s^{i_s} H_1^{j_1} \cdots H_r^{j_r}.$$

Then (10.47) shows that

$$D \equiv D_T(t)^{-2u} \xi(t) \Gamma_t(Y_1^{i_1} \cdots Y_s^{i_s} \otimes H_1^{j_1} \cdots H_r^{j_r}) \bmod U^{u+v-1}(\mathfrak{g}^{\mathbb{C}}),$$

where $u = \sum i_m$, $v = \sum j_n$, and ξ is a linear combination of ξ_{σ_j}'s. Since the terms of

$$D - D_T(t)^{-2u} \xi(t) \Gamma_t(Y_1^{i_1} \cdots Y_s^{i_s} \otimes H_1^{j_1} \cdots H_r^{j_r})$$

have lower degree than D, the induction hypothesis applies. Then we can substitute and come to the conclusion of the lemma.

Proof of Theorem 10.33. If H is in $U(\mathfrak{t}^{\mathbb{C}})$ and ξ_λ is defined on T, then $H\xi_\lambda = \lambda(H)\xi_\lambda$. Then it follows that $\gamma(z) \circ D_T(t)$ is of the form $\xi_\delta(t) \sum \xi_{\sigma_j}(t) H_j$ for suitable H_j in $U(\mathfrak{t}^{\mathbb{C}})$ and σ_j in the lattice generated by the roots. By Lemma 10.34, the difference of the two sides of (10.51) is of the form

$$\beta_t(z) - D_T(t)^{-1}(\gamma(z) D_T(t)) = D_T(t)^{-r} \sum \xi_{\sigma_j}(t) H_j \tag{10.52}$$

for some (possibly different) H_j's and σ_j's.

Let λ be the highest weight of an irreducible finite-dimensional representation, and let Θ_λ be the character as a function. Define $\tilde{F}_\lambda(g, t) =$

$\Theta_\lambda(gtg^{-1})$. Since Θ_λ is invariant, $\tilde{F}(g, t) = \Theta_\lambda(t)$ is independent of g. Hence (10.49) gives

$$(\beta_t(z)\Theta_\lambda)(t) = (1, \beta_t(z))\tilde{F}_\lambda(g, t) = (z\Theta_\lambda)(gtg^{-1}). \tag{10.53}$$

Since Θ_λ is a finite sum of matrix coefficients, (8.10) gives $z\Theta_\lambda = \chi_{\lambda+\delta}(z)\Theta_\lambda$. Thus (10.53) simplifies to

$$\beta_t(z)\Theta_\lambda(t) = \chi_{\lambda+\delta}(z)\Theta_\lambda(t).$$

By the Weyl character formula, write

$$\Theta_\lambda(t) = D_T(t)^{-1} \sum_{s \in W(\mathfrak{t}^{\mathbb{C}}:\mathfrak{g}^{\mathbb{C}})} (\det s)\xi_{s(\lambda+\delta)}(t).$$

The effect of $\gamma(z)D_T(t)$ on $\Theta_\lambda(t)$ is therefore

$$\begin{aligned} &= \gamma(z)(\textstyle\sum (\det s)\xi_{s(\lambda+\delta)}(t)) \\ &= \textstyle\sum (\det s)s(\lambda + \delta)(\gamma(z))\xi_{s(\lambda+\delta)}(t) \\ &= \chi_{\lambda+\delta}(z)(D_T(t)\Theta_\lambda(t)), \end{aligned}$$

the last equality holding by (8.33). Hence

$$(D_T^{-1}\gamma(z)D_T)\Theta_\lambda = \chi_{\lambda+\delta}(z)\Theta_\lambda. \tag{10.54}$$

Combining (10.53) and (10.54), we see that $\beta_t(z) - D_T(t)^{-1}(\gamma(z) \circ D_T(t))$ annihilates Θ_λ for each choice of λ.

Now we rewrite Θ_λ in terms of weights as

$$\Theta_\lambda = \xi_\lambda + \sum m_{\lambda'}\xi_{\lambda'},$$

with each λ' the difference of λ and a positive sum of positive roots, and we use (10.52). If σ_1 is the largest of the σ_j's (in the lexicographic order) such that the corresponding H_j is not 0, we obtain

$$\begin{aligned} 0 &= D_T(t)^{-r}\textstyle\sum \xi_{\sigma_j}(t)H_j\Theta_\lambda(t) \\ &= D_T(t)^{-r}(\xi_{\sigma_1}(t)H_1\xi_\lambda(t) + \displaystyle\sum_{j>1} \xi_{\sigma_j}(t)H_j\xi_\lambda(t) + \sum_j \sum_{\lambda'<\lambda} m_{\lambda'}\xi_{\sigma_j}(t)H_j\xi_{\lambda'}(t)). \end{aligned}$$

Hence

$$0 = \lambda(H_1)\xi_{\sigma_1+\lambda}(t) + \sum_{j>1} \lambda(H_j)\xi_{\sigma_j+\lambda}(t) + \sum_j \sum_{\lambda'<\lambda} m_{\lambda'}\lambda'(H_j)\xi_{\sigma_j+\lambda'}(t).$$

The only occurrence of $\xi_{\sigma_1+\lambda}$ on the right is in the first term. Since distinct exponentials are linearly independent, we conclude $\lambda(H_1) = 0$. Since λ is an arbitrary highest weight, the polynomial function $\lambda \to \lambda(H_1)$ vanishes at the lattice points in an "octant" and must vanish identically. Thus $H_1 = 0$, in contradiction to the choice of σ_1, and the theorem follows.

Proof of Theorem 10.25. Let μ be an irreducible character. Then μ satisfies $z\mu = \chi_\lambda(z)\mu$ for all z in $Z(\mathfrak{g}^{\mathbb{C}})$ and some λ, by Corollary 8.14, Proposition

8.21, and Proposition 10.24. If z^{tr} is as in Corollary 10.23 and if φ and F_2 are as in Lemma 10.32, then we have

$$\begin{aligned}(\gamma(z)\mu_T D_T^{-1})(D_T F_2) &= \mu_T(D_T^{-1}\gamma(z)^{\text{tr}} D_T F_2) && \text{by definition}\\ &= \mu_T(D_T^{-1}\gamma(z^{\text{tr}})D_T F_2) && \text{by easy computation}\\ &= \mu_T(\beta_t(z^{\text{tr}})F_2(t)) && \text{by Theorem 10.33}\\ &= \mu(z^{\text{tr}}\varphi) && \text{by Lemma 10.32}\\ &= (z\mu)(\varphi) && \text{by definition}\\ &= \chi_\lambda(z)\mu(\varphi) && \text{by definition of } \lambda\\ &= \chi_\lambda(z)\mu_T(F_2) && \text{by Lemma 10.32}\\ &= (\chi_\lambda(z)\mu_T D_T^{-1})(D_T F_2).\end{aligned}$$

Since D_T is nonvanishing on T', $D_T F_2$ is an arbitrary function in $C_{\text{com}}^\infty(T')$. Thus

$$\gamma(z)(\mu_T D_T^{-1}) = \chi_\lambda(z)(\mu_T D_T^{-1}) \tag{10.55}$$

for all z in $Z(\mathfrak{g}^{\mathbb{C}})$.

By Theorem 8.19, $U(\mathfrak{t}^{\mathbb{C}})$ is a finitely generated free module over the subalgebra $U(\mathfrak{t}^{\mathbb{C}})^W$ of $W(\mathfrak{t}^{\mathbb{C}}:\mathfrak{g}^{\mathbb{C}})$-invariant elements, which in turn is the image of γ. Since $U(\mathfrak{t}^{\mathbb{C}})^W$ by Theorem 8.19 is a polynomial algebra, it is Noetherian, by the Hilbert Basis Theorem. Hence the finitely generated $U(\mathfrak{t}^{\mathbb{C}})^W$ module $U(\mathfrak{t}^{\mathbb{C}})$ satisfies the ascending chain condition for $U(\mathfrak{t}^{\mathbb{C}})^W$ submodules. If $\square$ is any member of $U(\mathfrak{t}^{\mathbb{C}})$, $\square$ therefore satisfies an equation

$$\square^n + \gamma(z_{n-1})\square^{n-1} + \ldots + \gamma(z_1)\square + \gamma(z_0) = 0.$$

Applying both sides to $\mu_T D_T^{-1}$ and using (10.55), we obtain

$$(\square^n + \chi_\lambda(z_{n-1})\square^{n-1} + \ldots + \chi_\lambda(z_1)\square + \chi_\lambda(z_0))(\mu_T D_T^{-1}) = 0. \tag{10.56}$$

Let us choose $\square = \sum H_j^2$, where $\{H_j\}$ is a basis of $\mathfrak{t}$. Then the operator on the left side of (10.56) is elliptic and has real analytic coefficients, and (10.56) says this elliptic operator annihilates the distribution $\mu_T D_T^{-1}$ on T'. Hence $\mu_T D_T^{-1}$ is given by a real analytic function (times Haar measure on T). Unwinding matters by means of Lemma 10.28 and equation (10.42), we see that μ is given on $(T')^G$ by a real analytic function (times Haar measure on G).

§7. Formula on the Regular Set

We continue with notations as in §6. We have seen that an irreducible character μ is given by a function on the regular set, and we now seek to solve the system (10.55) of differential equations to find a formula for the function. On the subset $(T')^G$, let us write $\mu = \mu(x)\,dx$. With the same understandings that precede the statement of Theorem 10.33, we can write

$$\mu(gtg^{-1}) = \frac{\tau_T(t)}{D_T(t)}, \qquad g \in G \text{ and } t \in T', \tag{10.57}$$

where $\tau_T(t)$ is a real analytic function on T' that is odd under the action of $W(T:G)$. Arguing as with $\mathrm{SL}(2, \mathbb{R})$ at the end of §5, we see that

$$D_T^{-1}\mu_T = s^{G/T}|W(T:G)|^{-1}\tau_T(t)\,dt,$$

and it is the left side of this expression that satisfies the system (10.55) of differential equations. Thus if the original character μ satisfies $z\mu = \chi_\lambda(z)\mu$ for z in $Z(\mathfrak{g}^{\mathbb{C}})$, then the **numerator** $\tau_T(t)$ satisfies

$$\gamma(z)\tau_T = \chi_\lambda(z)\tau_T \qquad \text{for } z \text{ in } Z(\mathfrak{g}^{\mathbb{C}}). \tag{10.58}$$

We can solve this system.

Theorem 10.35. Suppose that the irreducible character μ satisfies $z\mu = \chi_\lambda(z)\mu$ for all z in $Z(\mathfrak{g}^{\mathbb{C}})$. Let W_λ be the subgroup of elements s of $W(\mathfrak{t}^{\mathbb{C}}:\mathfrak{g}^{\mathbb{C}})$ for which $s\lambda = \lambda$. Fix t in T and let $\mathfrak{t}_1$ be a connected component of the set of all H in $\mathfrak{t}$ such that $D_T(t \exp H) \neq 0$. Then there exist uniquely determined polynomial functions p_w on $\mathfrak{t}$ ($w \in W(\mathfrak{t}^{\mathbb{C}}:\mathfrak{g}^{\mathbb{C}})$) such that $p_{ws} = p_w$ for s in W_λ and such that the numerator τ_T of μ, as given in (10.57), satisfies

$$\tau_T(t \exp H) = \sum_{w \in W(\mathfrak{t}^{\mathbb{C}}:\mathfrak{g}^{\mathbb{C}})} p_w(H)e^{w\lambda(H)} \qquad \text{for all } H \text{ in } \mathfrak{t}_1.$$

Moreover, the degrees of the polynomials p_w are all less than $|W_\lambda|$.

Remark. The element t should be regarded as a representative of a component of T. It has to be included in the formula because T can be disconnected and the character formula can change abruptly from one component to another. (Recall Propositions 10.12 and 10.14 for $\mathrm{SL}(2, \mathbb{R})$.)

Proof. Let $W = W(\mathfrak{t}^{\mathbb{C}}:\mathfrak{g}^{\mathbb{C}})$. In view of Theorem 8.18, we can rewrite the system (10.58) as

$$v\tau_T = \lambda(v)\tau_T \tag{10.59}$$

for all v in the W invariants $U(\mathfrak{t}^{\mathbb{C}})^W$ of $U(\mathfrak{t}^{\mathbb{C}})$. For u in $U(\mathfrak{t}^{\mathbb{C}})$, consider the polynomial

$$\prod_{w \in W}(\zeta - u^w) = \zeta^{|W|} + v_1\zeta^{|W|-1} + \ldots + v_{|W|}, \tag{10.60}$$

where u^w is shorthand for $\mathrm{Ad}(w)u$ and ζ is an indeterminate. The coefficients on the right side of (10.60) are members of $U(\mathfrak{t}^{\mathbb{C}})^W$. Replacing ζ by u, we obtain

$$u^{|W|} + v_1u^{|W|-1} + \ldots + v_{|W|} = 0.$$

The result of applying both sides to τ_T is

$$u^{|W|}\tau_T + \lambda(v_1)u^{|W|-1}\tau_T + \ldots + \lambda(v_{|W|})\tau_T = 0. \tag{10.61}$$

Since λ is a homomorphism of $U(\mathfrak{t}^{\mathbb{C}})$ into $\mathbb{C}$, (10.60) gives

$$\prod_{w\in W} (\zeta - \lambda(u^w)) = \zeta^{|W|} + \lambda(v_1)\zeta^{|W|-1} + \ldots + \lambda(v_{|W|}).$$

Therefore (10.61) can be rewritten as

$$\prod_{w\in W} (u - \lambda(u^w))\tau_T = 0, \qquad u \in U(\mathfrak{t}^{\mathbb{C}}). \tag{10.62}$$

Let us introduce a basis $H_1, \ldots, H_l$ of $\mathfrak{t}$ and coordinatize $\mathfrak{t}$ by writing $H = x_1H_1 + \ldots + x_lH_l$. Then we can define a coordinatized version τ_T^* of τ_T by

$$\tau_T^*(x_1, \ldots, x_l) = \tau_T(t \exp(x_1H_1 + \ldots + x_lH_l)) = \tau_T(t \exp H).$$

This function allows us to compute the effect of $U(\mathfrak{t}^{\mathbb{C}})$ on τ_T:

$$(H_j^k\tau_T)(t \exp H) = \frac{d^k}{dr^k}\tau_T(t \exp H \exp rH_j)_{r=0} = \frac{\partial^k}{\partial x_j^k}\tau_T^*(x_1, \ldots, x_l).$$

For the special case $u = H_j$, (10.62) therefore becomes

$$\prod_{w\in W} (H_j - \lambda(H_j^w))\tau_T^* = 0. \tag{10.63}$$

Next let us enumerate the Weyl group elements as $w_1, w_2, \ldots, w_{|W|}$ and define

$$f_{k_1,\ldots,k_l} = \left\{\prod_{1\le j\le k_1}\left(\frac{\partial}{\partial x_1} - \lambda(H_1^{w_j})\right)\right\}\cdots\left\{\prod_{1\le j\le k_l}\left(\frac{\partial}{\partial x_l} - \lambda(H^{w_j})\right)\right\}\tau_T^* \tag{10.64}$$

for $0 \le k_1 < |W|, \ldots, 0 \le k_l < |W|$. Then (10.63) says that the $\mathbb{C}^{|W|^l}$-valued function $f = \{f_{k_1,\ldots,k_l}\}$ is a solution of an obvious system

$$\frac{\partial f}{\partial x_j} = A_j f, \tag{10.65}$$

where A_j is a matrix of constants of size $|W|^l$-by-$|W|^l$. Theorems B.8 and B.9 imply that every local solution of (10.65) on the set of x's corresponding to $\mathfrak{t}_1$ is global and that the dimension of the space of solutions is $\le |W|^l$.

Any solution f of (10.65) is determined by its entry $f_{0,\ldots,0}$, in view of (10.64), and the following functions provide $|W|^l$ independent $f_{0,\ldots,0}$'s:

$$x_1^{m_1}\cdots x_l^{m_l}e^{v_1x_1+\ldots+v_lx_l},$$

where $m_1, \ldots, m_l$ are integers ≥ 0, v_i is one of the numbers $\lambda(H_i^w)$, and m_i

is less than the number of w's in W for which $\lambda(H_i^w) = v_i$. Thus we have a basis of solutions. Unwinding matters, we see that τ_T must be of the form

$$\tau_T(t \exp H) = p_1(H)e^{\lambda_1(H)} + \ldots + p_n(H)e^{\lambda_n(H)}, \; H \in \mathfrak{t}_1, \tag{10.66}$$

for suitable linear functionals λ_i and polynomials p_i.

Now we bring in the system (10.59). Any H_0 in $\mathfrak{t}^{\mathbb{C}}$ certainly gives $H_0 e^{\lambda_j} = \lambda_j(H_0)e^{\lambda_j}$. Hence v in $U(\mathfrak{t}^{\mathbb{C}})^W$ implies

$$0 = (v - \lambda(v))\tau_T = \sum_{i=1}^{l} p_i' e^{\lambda_i}, \tag{10.67}$$

in view of (10.59) and (10.66). Here $p_i' = (\lambda_i(v) - \lambda(v))p_i + q_i$ for some polynomial q_i of degree lower than that of p_i. By linear independence we can conclude from (10.67) that $p_i' = 0$ for each i. Thus $\lambda_i(v) = \lambda(v)$ for all v in $U(\mathfrak{t}^{\mathbb{C}})^W$, and Proposition 8.20 implies $\lambda_i = s_i\lambda$ for some s_i in W. In other words,

$$\tau_T(t \exp H) = \sum_{w \in W} p_w(H)e^{w\lambda(H)} \qquad \text{for } H \in \mathfrak{t}_1. \tag{10.68}$$

Moreover, there is no loss of generality in assuming that $p_{ws} = p_w$ for all w in W and s in W_λ.

Since the asserted uniqueness in the statement of the theorem is obvious, it suffices, in order to complete the proof, to prove that the degree of p_w in (10.68) is less than $|W_\lambda|$. To prove this fact, we fix w and suppose by way of argument by contradiction that p_w has degree $n \geq |W_\lambda|$. Let q_w be the part of p_w that is homogeneous of degree n. For each H in $\mathfrak{t}_1$, the function of x given by $x \to q_w(xH)$ is a multiple of x^n. If the multiple is 0 for all H, then q_w is 0 as a function and hence 0 as a polynomial, contradiction. Thus there exists H such that $x \to q_w(xH)$ is of degree n, and then $x \to p_w(xH)$ must also be of degree n. The set of such H is evidently open. Meanwhile an open dense subset of $\mathfrak{t}_1$ is such that $w\lambda(H) = \lambda(H)$ only for w in W_λ. Therefore we can choose H_0 in $\mathfrak{t}_1$ such that $p_w(xH_0)$ is of degree n and $w\lambda(H_0) \neq \lambda(H_0)$ for all $w \notin W_\lambda$. Letting u be xH_0 and substituting the solution (10.68) into the equation (10.62), we obtain a contradiction. This completes the proof of the theorem.

§8. Behavior on the Singular Set

We have seen that the global character of an irreducible admissible representation is given as a distribution on the regular set of G by a real analytic function, and Theorem 10.35 gives a formula for this function. In fact, the proofs did not use representation theory but used only two properties of the distribution: its invariance under conjugation and the fact that $Z(\mathfrak{g}^{\mathbb{C}})$ acts as scalars. We call any such distribution on G an

invariant eigendistribution. It is a deep theorem of Harish-Chandra that the singular set of G cannot make an additional contribution to an invariant eigendistribution.

Theorem 10.36. Any invariant eigendistribution on G is given on all of G by a locally integrable function (whose restriction to the regular set is real analytic).

Since there are several expositions of this theorem in the literature, we omit all discussion of the proof. For the moment we shall indicate the power of the theorem with three easy corollaries.

Corollary 10.37. Up to infinitesimal equivalence, there are only finitely many irreducible admissible representations with a given infinitesimal character.

Remark. It is implicit in the results of Chapter VIII (especially Theorem 8.39) that there are only finitely many irreducible admissible representations with a given infinitesimal character that contain a given K type. The corollary strengthens that conclusion.

Proof. Because of Theorem 10.36, the character is completely determined by the formula on the regular set, which is given in Theorem 10.35. That formula implies that the space of invariant eigendistributions with a given infinitesimal character is finite-dimensional. Thus the corollary follows from the linear independence asserted in Theorem 10.6.

Corollary 10.38. If π is an admissible representation possessing an infinitesimal character, then π has a finite composition series in the following sense: There exist subrepresentations

$$\pi = \pi_1 \supset \pi_2 \supset \ldots \supset \pi_n = 0$$

such that each quotient representation π_i/π_i is irreducible. Moreover, the set of quotients and their multiplicities are the same (apart from infinitesimal equivalence) for all composition series of π.

Proof. An argument with Zorn's Lemma given in the proof of Corollary 8.42 shows that there exist subrepresentations π_i and π_{i+1} of π with $\pi_i \supset \pi_{i+1}$ and π_i/π_{i+1} irreducible. If $\pi \neq \pi_i$, we repeat the argument with π replaced by π/π_i and lift the result back to obtain

$$\pi \supseteq \pi_j \supset \pi_{j+1} \supseteq \pi_i$$

with π_j/π_{j+1} irreducible. Also if $\pi_{i+1} \neq 0$, we repeat the argument with π replaced by π_{i+1} and obtain

$$\pi_{i+1} \supseteq \pi_k \supset \pi_{k+1} \supseteq 0$$

with π_k/π_{k+1} irreducible. And we continue in this fashion.

The process must stop. In fact, there are only finitely many possible distinct irreducible representations that can occur (up to equivalence), by Corollary 10.37. And each of these can occur only finitely many times because of the finite multiplicity in π of any of its K types. Thus we obtain a composition series.

If

$$\pi = \pi'_1 \supset \pi'_2 \supset \ldots \supset \pi'_m = 0$$

is a second composition series, then it is easy to see that the sum of the characters of the π_i/π_{i+1}, for $1 \leq i \leq n-1$, equals the sum of the characters of the π'_j/π'_{j+1}, for $1 \leq j \leq m-1$. The quotients must coincide, apart from order and infinitesimal equivalence, by Theorem 10.6.

Corollary 10.39. Each member of the nonunitary principal series of G has a finite composition series.

Proof. The representation has an infinitesimal character by Proposition 8.22 and is admissible by Proposition 8.4. Thus we can apply Corollary 10.38.

In the course of the proof of Theorem 10.36, Harish-Chandra obtained some information about the behavior, upon approach to the singular set, of the real analytic function that defines an invariant eigendistribution. Before stating his result and an extension of it, we collect the relevant terminology. If μ is an invariant eigendistribution, we write $\mu = \mu(x)\,dx$ on the subset $(T')^G$ of elements in G conjugate to the regular set in T. As in (10.57), we define the numerator $\tau_T(t)$ by

$$\mu(gtg^{-1}) = \tau_T(t)/D_T(t) \qquad \text{for } g \in G \text{ and } t \in T'.$$

Theorem 10.35 shows that $\tau_T(t)$ extends to a real analytic function on the closure of any component of T'. The question is how these extended functions fit together. As in §3, let

$$D'_R(t) = \prod_{\alpha \in \Delta_R^+(\mathfrak{t}^{\mathbb{C}} : \mathfrak{g}^{\mathbb{C}})} (1 - \xi_\alpha(t)^{-1}) \qquad \text{for } t \in T.$$

Define

$$T'_R = \{t \in T \mid D'_R(t) \neq 0\}$$

$$\varepsilon_R^T(t) = \operatorname{sgn} D'_R(t) \qquad \text{for } t \in T'_R.$$

Theorem 10.40. Let $\tau_T(t)$ be the numerator on T' of the function defining an invariant eigendistribution on G. Then $\tau_T(t)$ and $\varepsilon_R^T(t)\tau_T(t)$ extend to real analytic functions on the closure of each component of T'_R. The extensions of $\varepsilon_R^T(t)\tau_T(t)$ are suitably compatible so that their union is a well-defined continuous function on T.

Example. For SL(2, $\mathbb{R}$) let us revert to notation closer to that in §2, writing

$$B = \left\{ \begin{pmatrix} \cos\theta & \sin\theta \\ -\sin\theta & \cos\theta \end{pmatrix} \right\}, \qquad T = \left\{ \pm \begin{pmatrix} e^x & 0 \\ 0 & e^{-x} \end{pmatrix} \right\}.$$

Then B' is the subset of B where $\sin\theta \neq 0$, and T' is the subset of T where $x \neq 0$. The two roots relative to $\mathfrak{b}^{\mathbb{C}}$ are imaginary, not real; hence $B'_R = B$ and $\varepsilon_R^B = 1$. The two roots relative to $\mathfrak{t}^{\mathbb{C}}$ are both real; hence $T'_R = T'$ and ε_R^T is given by sgn x.

Let us compare the assertions in Theorem 10.40 with the formulas in Propositions 10.12 and 10.14 for the characters of the discrete series $\mathscr{D}_n^+$ and the nonunitary principal series $U(S_{\mathfrak{p}}, \sigma, \nu)$. On B' the respective numerators are $-e^{i(n-1)\theta}$ and 0, which are real analytic on all of B. On T' the respective numerators are

$$(\pm)^n \tfrac{1}{2}(\operatorname{sgn} x)[e^{(n-1)x}(1 - \operatorname{sgn} x) + e^{-(n-1)x}(1 + \operatorname{sgn} x)]$$

and

$$\sigma(\pm)(\operatorname{sgn} x)[e^{cx} + e^{-cx}]$$

for a suitable complex constant c that depends on ν. These numerators both extend to be real analytic for $x \geq 0$ or for $x \leq 0$. When we multiply by ε_R^T, which is given by sgn x, both numerators extend to be continuous at $x = 0$. Note in the case of the discrete series that this continuous extension is not differentiable at $x = 0$.

§9. Families of Admissible Representations

In view of Theorems 10.35 and 10.36, we see that the infinitesimal character λ enters the formula for a global character on G only in a relatively harmless-looking way: through the exponentials $e^{w\lambda(H)}$ occurring in the numerator and through the effect of the stability group W_λ in bounding the degrees of the polynomials $p_w(H)$ in the numerator. It is natural to ask whether we can generate other global characters simply by moving the parameter λ while leaving everything else alone. This is in fact the case, and we shall show in this section how to accomplish this movement of parameters.

What is needed is a construction on the level of representations that we can then analyze on the level of global characters. The construction we use will be called **Zuckerman tensoring.** It consists of tensoring with a finite-dimensional representation and projecting according to the effect of $Z(\mathfrak{g}^{\mathbb{C}})$. Before defining matters more precisely, we shall narrow the class of representations that we shall work with.

An admissible representation is said to be **finitely generated** if there exist finitely many K-finite vectors v_j such that $\sum_j U(\mathfrak{g}^{\mathbb{C}})v_j$ exhausts the K-finite vectors. A finitely generated admissible representation is called a

Harish-Chandra module. We shall investigate the behavior of $Z(\mathfrak{g}^{\mathbb{C}})$ on such a representation; to be quite definite in using infinitesimal characters, let us temporarily fix a Cartan subgroup T of G and its Lie algebra $\mathfrak{t}$.

Proposition 10.41. Let (π, V) be a Harish-Chandra module, and let V_0 be the subspace of K-finite vectors. Let W be the Weyl group $W(\mathfrak{t}^{\mathbb{C}}:\mathfrak{g}^{\mathbb{C}})$. Then there exist linear functionals $\lambda_1, \ldots, \lambda_n$ on $\mathfrak{t}^{\mathbb{C}}$, $\mathfrak{g}$-invariant subspaces $V_1, \ldots, V_n$ of V_0, and an integer $d > 0$ such that

(a) $\lambda_1, \ldots, \lambda_n$ are mutually inequivalent under W
(b) $V_0 = V_1 \oplus \ldots \oplus V_n$
(c) $(z - \chi_{\lambda_j}(z))^d$ acts as the 0 operator on V_j for all z in $Z(\mathfrak{g}^{\mathbb{C}})$.

Properties (a), (b), and (c) characterize $V_1, \ldots, V_n$ uniquely and characterize $\lambda_1, \ldots, \lambda_n$ uniquely up to the operation of W. Moreover, for each j with $1 \leq j \leq n$, there exists z_j in $Z(\mathfrak{g}^{\mathbb{C}})$ such that V_j is the image of the action on V_0 by

$$\prod_{i \neq j} (z_j - \chi_{\lambda_i}(z_j))^d.$$

Remarks. It follows from Corollary 8.10 that the closures $\bar{V}_j$ of V_j are each invariant under G, that their sum is direct, and that their sum is dense in V. On $\bar{V}_j$, the restriction of π comes close to having λ_j as infinitesimal character, missing only because the integer d may be greater than one. We say that the representation on $\bar{V}_j$ has **generalized infinitesimal character** λ_j. In proving the proposition, it will be convenient to drop references to π, referring only to the action of the associative algebra $U(\mathfrak{g}^{\mathbb{C}})$ on the module V_0.

Proof. Since V_0 is finitely generated and admissible, we can find a finite-dimensional generating subspace V_0' of V_0 that is the sum of the full subspaces of V_0 corresponding to a certain set of K types. Since $Z(\mathfrak{g}^{\mathbb{C}})$ commutes with K, $Z(\mathfrak{g}^{\mathbb{C}})$ carries V_0' into itself and provides a commuting family of linear maps of the finite-dimensional space V_0' into itself. Applying Lie's Theorem (A.13) to suitable finite-dimensional subspaces of $Z(\mathfrak{g}^{\mathbb{C}})$, we see that we can regard the action of $Z(\mathfrak{g}^{\mathbb{C}})$ on V_0' as simultaneously upper triangular in some basis. The diagonal entries in this action are necessarily homomorphisms of $Z(\mathfrak{g}^{\mathbb{C}})$ into $\mathbb{C}$, and we define $\chi_{\lambda_1}, \ldots, \chi_{\lambda_n}$ to be the distinct ones. Let V_j' be the generalized eigenspace corresponding to χ_{λ_j}; the Jordan normal form says that

$$V_0' = V_1' \oplus \ldots \oplus V_n'. \tag{10.69}$$

Moreover, $(z - \chi_{\lambda_j}(z))^d$ acts as 0 on V_j' if $d = \dim V_0'$ and z is in $Z(\mathfrak{g}^{\mathbb{C}})$, and $(z - \chi_{\lambda_j}(z))^d$ maps the other spaces V_i' into themselves.

Define $V_j = U(\mathfrak{g}^{\mathbb{C}})V_j'$. Applying $U(\mathfrak{g}^{\mathbb{C}})$ to (10.69), we obtain

$$V_0 = V_1 + \ldots + V_n. \tag{10.70}$$

Moreover, $(z - \chi_{\lambda_j}(z))^d$ acts as 0 on V_j if z is in $Z(\mathfrak{g}^{\mathbb{C}})$. By Proposition 8.20, the linear functionals $\lambda_1, \ldots, \lambda_n$ are inequivalent under the Weyl group W. To complete the proof that (a), (b), and (c) are satisfied, we still have to prove that the sum (10.70) is direct. But first let us construct the elements z_j asserted to exist in the statement of the proposition.

Fix j with $1 \leq j \leq n$. Since λ_j is not in $W\lambda_i$ if $i \neq j$, we can find a W-invariant member H_j of $U(\mathfrak{t}^{\mathbb{C}})$ with $\lambda_j(H)$ different from $\lambda_i(H)$ for all $i \neq j$. Choose z_j in $Z(\mathfrak{g}^{\mathbb{C}})$ so that the Harish-Chandra homomorphism maps z_j onto H_j. Then $\chi_{\lambda_j}(z_j)$ is different from $\chi_{\lambda_i}(z_j)$ for all $i \neq j$. By linear algebra the element

$$\prod_{i \neq j} (z_j - \chi_{\lambda_i}(z_j))^d \tag{10.71}$$

carries V_0' onto V_j' and is nonsingular on V_j'. Applying $U(\mathfrak{g}^{\mathbb{C}})$, we see that the element (10.71) carries V_0 onto V_j. Thus z_j has the required property.

We shall use z_j to show that the sum in (10.70) is direct. In fact, the element (10.71) annihilates $\sum_{i \neq j} V_i$ and annihilates no nonzero member of V_j. Thus $V_j \cap (\sum_{i \neq j} V_i) = 0$, and the sum is direct.

Finally we prove the asserted uniqueness. Any $\lambda_1, \ldots, \lambda_n$ satisfying (a), (b), (c) give generalized eigenvalues for the action of $Z(\mathfrak{g}^{\mathbb{C}})$, and $V_1, \ldots, V_n$ are contained in the generalized eigenspaces. Suppose on some finite-dimensional subspace V_0'' of V_0 that $Z(\mathfrak{g}^{\mathbb{C}})$ were to act with some other generalized eigenvalue χ_λ. Arguing as above with $V_0' + V_0''$ in place of V_0', we see from the resulting (b) that the original V_0' could not generate V_0, in contradiction to hypothesis. Thus the χ_{λ_j} are uniquely characterized as the full set of generalized eigenvalues of $Z(\mathfrak{g}^{\mathbb{C}})$. Our argument with the elements z_j showed that the V_j's are the full corresponding generalized eigenspaces. Since χ_{λ_j} determines λ_j up to the action of W, the required uniqueness follows.

Corollary 10.42. Every Harish-Chandra module has a finite composition series. Hence it has a global character that equals the sum of the global characters of its irreducible subquotients, and this global character is given by a locally integrable function that is real analytic on the regular set of G. Moreover, the multiplicities of the various irreducible subquotients of a composition series are independent of the composition series.

Remark. The number of subquotients of a composition series, each counted according to its multiplicity, is called the **length** of the given Harish-Chandra module.

Proof. For existence, Proposition 10.41 reduces matters to the case where $(z - \chi_\lambda(z))^d$ acts as 0 for some λ and d. In this case we can obtain a chain of invariant subspaces as the kernels of the various powers of $z - \chi_\lambda(z)$;

then matters are reduced to the case of a representation with an infinitesimal character. The latter case is handled by Corollary 10.38.

The existence of the global character is immediate, and it is clear from the definition that the global character is the sum of the characters of the irreducible subquotients. The character is a function by Theorem 10.36.

Finally the uniqueness is a consequence of the linear independence of irreducible characters that is given in Theorem 10.6.

If (π, V) is a Harish-Chandra module and λ is an infinitesimal character, then Proposition 10.41 gives us a canonically defined projection p_λ from the space of K-finite vectors of V to the subspace with generalized infinitesimal character λ. This projection depends on the choice of Cartan subgroup T only notationally, since the generalized eigenspaces under $Z(\mathfrak{g}^{\mathbb{C}})$ do not depend on T. To make the correspondence in notation, let $\mathfrak{t}_1$ and $\mathfrak{t}_2$ be two different Cartan subalgebras. Then $\mathfrak{t}_1^{\mathbb{C}}$ and $\mathfrak{t}_2^{\mathbb{C}}$ are conjugate via the complexification of G, and the conjugation carries the one λ parameter to the other.

Instead of thinking of p_λ as a projection operator, we can think about it on a grander level: For one thing it carries Harish-Chandra modules to Harish-Chandra modules; we pass to K-finite vectors, apply p_λ, and take the closure. But in addition, it carries $\mathfrak{g}$ intertwining maps to $\mathfrak{g}$ intertwining maps (by restriction to the image of p_λ), and it does so compatibly with its effect on representations. In the language of category theory, the Harish-Chandra modules and $\mathfrak{g}$ intertwining maps form a category, and p_λ is a functor on this category. The next proposition addresses closure properties of this category and the effect on p_λ.

Proposition 10.43. Any subrepresentation or quotient representation of a Harish-Chandra module is a Harish-Chandra module. Moreover, the functor p_λ is exact in the following sense: If

$$0 \to V_1 \to V \to V/V_1 \to 0$$

is an exact sequence of Harish-Chandra modules, then the sequence

$$0 \to p_\lambda(V_1) \to p_\lambda(V) \to p_\lambda(V/V_1) \to 0,$$

with the maps induced by p_λ, is also exact.

Proof. The first statement follows from Corollary 10.42. The second statement is a consequence of the fact that we can regard p_λ as a projection operator.

If (π, V) is a Harish-Chandra module, we let $\Theta(V)$ denote the global character of V, regarded as a function. It follows from Proposition 10.43

that p_λ yields a well-defined mapping on global characters of Harish-Chandra modules by the rule

$$p_\lambda(\Theta(V)) = \Theta(p_\lambda V).$$

[We have only to show that $\Theta(V) = \Theta(V')$ implies $\Theta(p_\lambda V) = \Theta(p_\lambda V')$, and this we see from Proposition 10.43 by induction on the length of V.]

Another operation that we shall use is the operation of tensor product with a finite-dimensional representation. This operation carries Harish-Chandra modules to Harish-Chandra modules, and we write it as $(\cdot) \otimes F$, where F denotes the finite-dimensional representation. It carries a $\mathfrak{g}$ map L to the $\mathfrak{g}$ map $L \otimes 1$ and again yields a functor.

The functor $(\cdot) \otimes F$ shares with p_λ the property of carrying short exact sequences to short exact sequences. Thus it too yields a well-defined mapping on global characters of Harish-Chandra modules. It is clear from the definition of character that the formula is

$$\Theta(V \otimes F) = \Theta(V)\Theta(F), \tag{10.72}$$

in the sense of pointwise multiplication.

Both p_λ and F have meaning without reference to a Cartan subgroup. However, if we refer them both to a θ-stable Cartan subalgebra or subgroup, we can introduce a compatibility condition on them. Thus, fix a Cartan subgroup T and the corresponding subalgebra $\mathfrak{t}$. We say that λ is **real** if $\lambda|_{\mathfrak{t}\cap\mathfrak{p}}$ is real and $\lambda|_{\mathfrak{t}\cap\mathfrak{k}}$ is imaginary. The condition of reality for an infinitesimal character is independent of $\mathfrak{t}$; in any event, we can define $\operatorname{Re}\lambda$ and $\operatorname{Im}\lambda$ in the obvious way and obtain $\lambda = \operatorname{Re}\lambda + i\operatorname{Im}\lambda$. Choose a positive system $\Delta^+(\mathfrak{t}^{\mathbb{C}}:\mathfrak{g}^{\mathbb{C}})$ that makes $\operatorname{Re}\lambda$ dominant. Let F^ν and $F_{-\nu}$ denote the finite-dimensional irreducible representations of G with highest weight ν and lowest weight $-\nu$, respectively.

Under the assumption that $\operatorname{Re}\lambda$ and ν are both dominant for $\Delta^+(\mathfrak{t}^{\mathbb{C}}:\mathfrak{g}^{\mathbb{C}})$, we define the Zuckerman tensoring functors φ and ψ by

$$\varphi^\lambda_{\lambda+\nu} = p_{\lambda+\nu}[(\cdot) \otimes F^\nu]p_\lambda$$

and

$$\psi^{\lambda+\nu}_\lambda = p_\lambda[(\cdot) \otimes F_{-\nu}]p_{\lambda+\nu}.$$

The functor $\varphi^\lambda_{\lambda+\nu}$ is well defined independently of the choice of T and of $\Delta^+(\mathfrak{t}^{\mathbb{C}}:\mathfrak{g}^{\mathbb{C}})$. The functor $\psi^{\lambda+\nu}_\lambda$ is defined only under a compatibility condition on p_λ and the finite-dimensional representation, but this condition and the result of applying $\psi^{\lambda+\nu}_\lambda$ are well defined independently of the choice of T and of $\Delta^+(\mathfrak{t}^{\mathbb{C}}:\mathfrak{g}^{\mathbb{C}})$. Both $\varphi^\lambda_{\lambda+\nu}$ and $\psi^{\lambda+\nu}_\lambda$ carry short exact sequences to short exact sequences.

In Proposition 10.44 we shall see that φ and ψ have the effect on global characters of moving the infinitesimal character parameter in a simple way. First let us note that if (π, V) has generalized infinitesimal character

λ, then the global character $\Theta(V)$ is given on any component of T'_R by the same formula as in Theorem 10.35:

$$D_T(t_0 \exp H)\Theta(V)(t_0 \exp H) = \sum_{w \in W(\mathfrak{t}^{\mathbb{C}}:\mathfrak{g}^{\mathbb{C}})} p_w(H)e^{w\lambda(H)}, \tag{10.73}$$

where t_0 is a representative of the component in question. The reason is that $\Theta(V)$ is the sum of finitely many irreducible characters each with infinitesimal character λ. [Here, as earlier with explicit formulas for characters, we shall assume that G is contained in a simply connected complexification $G^{\mathbb{C}}$ in order to ensure that D_T is well defined.]

Proposition 10.44. Let G be contained in a simply connected complexification $G^{\mathbb{C}}$. Fix a Cartan subgroup T, a positive system $\Delta^+(\mathfrak{t}^{\mathbb{C}}:\mathfrak{g}^{\mathbb{C}})$, and linear functionals λ and ν on $\mathfrak{t}^{\mathbb{C}}$ such that Re λ and ν are dominant and ν is integral. Let W_λ and $W_{\lambda+\nu}$ be the subgroups of $W = W(\mathfrak{t}^{\mathbb{C}}:\mathfrak{g}^{\mathbb{C}})$ leaving fixed λ and $\lambda + \nu$, respectively.

(a) If V_1 is a Harish-Chandra module with generalized infinitesimal character λ and with global character given by

$$D_T(t_0 \exp H)\Theta(V_1)(t_0 \exp H) = \sum_{w \in W} p_w(H)e^{w\lambda(H)}, \tag{10.74}$$

then the global character of $\varphi^{\lambda}_{\lambda+\nu}(V_1)$ is given by

$$\begin{aligned} &D_T(t_0 \exp H)\Theta(\varphi^{\lambda}_{\lambda+\nu}V_1)(t_0 \exp H) \\ &\quad = |W_\lambda/W_{\lambda+\nu}|^{-1} \sum_{w \in W} \xi_{w\nu}(t_0)p_w(H)e^{w(\lambda+\nu)(H)}. \end{aligned} \tag{10.75}$$

(b) If V_2 is a Harish-Chandra module with generalized infinitesimal character $\lambda + \nu$ and with global character given by

$$D_T(t_0 \exp H)\Theta(V_2)(t_0 \exp H) = \sum_{w \in W} q_w(H)e^{w(\lambda+\nu)(H)}, \tag{10.76}$$

then the global character of $\psi^{\lambda+\nu}_{\lambda}(V_2)$ is given by

$$D_T(t_0 \exp H)\Theta(\psi^{\lambda+\nu}_{\lambda}V_2)(t_0 \exp H) = \sum_{w \in W} \xi_{-w\nu}(t_0)q_w(H)e^{w\lambda(H)}. \tag{10.77}$$

Proof of (a). Let us write the character of F^ν as

$$\Theta(F^\nu)(t_0 \exp H) = \sum_{\sigma \in P(F^\nu)} \xi_\sigma(t_0)e^{\sigma(H)},$$

with $P(F^\nu)$ denoting the weights of F^ν repeated according to their multiplicities. By (10.72) and (10.74) the character of $V_1 \otimes F^\nu$ is given by

$$\begin{aligned} &D_T(t_0 \exp H)\Theta(V_1 \otimes F^\nu)(t_0 \exp H) \\ &\quad = \sum_{w \in W} \sum_{\sigma \in P(F^\nu)} \xi_\sigma(t_0)p_w(H)e^{(w\lambda+\sigma)(H)} \\ &\quad = \sum_{w \in W} \sum_{\sigma \in P(F^\nu)} \xi_{w\sigma}(t_0)p_w(H)e^{w(\lambda+\sigma)(H)}. \end{aligned} \tag{10.78}$$

Since Proposition 10.41 ensures that $V_1 \otimes F^\nu$ splits on the K-finite level as the direct sum of Harish-Chandra modules that possess generalized infinitesimal characters, equation (10.73) makes it clear how to project by $p_{\lambda+\nu}$: We simply retain just those terms in (10.78) for which $w(\lambda + \sigma)$ is W-conjugate to $\lambda + \nu$. Thus we seek those σ for which there exists s in W with $sw(\lambda + \sigma) = \lambda + \nu$.

Since Re λ is dominant and $sw\sigma$ is a weight of F^ν, the real part of this equation forces

$$\operatorname{Re} sw\lambda = \operatorname{Re} \lambda \qquad \text{and} \qquad sw\sigma = \nu.$$

Meanwhile the imaginary part says Im $sw\lambda$ = Im λ. Thus

$$sw\lambda = \lambda \qquad \text{and} \qquad sw\sigma = \nu.$$

Hence σ must be of the form $w^{-1}s^{-1}\nu$ for a suitable s in W, and thus its multiplicity in F^ν must be one. We shall reparametrize the σ's by the s's. This parametrization contains some redundance because if s satisfies $sw(\lambda + \sigma) = \lambda + \nu$, then s can be replaced by any element of $W_{\lambda+\nu}s$ with the same effect. Thus

$$\begin{aligned} &D_T(t_0 \exp H)\Theta(\varphi_{\lambda+\nu}^{\lambda} V_1)(t_0 \exp H) \\ &\quad = |W_{\lambda+\nu}|^{-1} \sum_{w \in W} \left\{ \sum_{\substack{s \in W \\ sw\lambda = \lambda}} \xi_{s^{-1}\nu}(t_0) p_w(H) e^{s^{-1}(\lambda+\nu)(H)} \right\}. \end{aligned}$$

Replacing s by sw^{-1} in the inner sum, we obtain

$$\begin{aligned} &D_T(t_0 \exp H)\Theta(\varphi_{\lambda+\nu}^{\lambda} V_1)(t_0 \exp H) \\ &\quad = |W_{\lambda+\nu}|^{-1} \sum_{w \in W} \sum_{s \in W_\lambda} \xi_{ws^{-1}\nu}(t_0) p_w(H) e^{ws^{-1}(\lambda+\nu)(H)}. \end{aligned}$$

Next we interchange sums and replace w by ws. The polynomial becomes $p_{ws}(H)$, which equals $p_w(H)$ since s is in W_λ. Then (10.75) follows.

Proof of (b). In similar fashion the character of $V_2 \otimes F_{-\nu}$ is given by

$$\begin{aligned} &D_T(t_0 \exp H)\Theta(V_2 \otimes F_{-\nu})(t_0 \exp H) \\ &\quad = \sum_{w \in W} \sum_{\sigma \in P(F_{-\nu})} \xi_\sigma(t_0) q_w(H) e^{[w(\lambda+\nu)+\sigma](H)} \\ &\quad = \sum_{w \in W} \sum_{\sigma \in P(F^\nu)} \xi_{-w\sigma}(t_0) q_w(H) e^{w(\lambda+\nu-\sigma)(H)}. \end{aligned} \tag{10.79}$$

To project by p_λ, we retain those terms for which $w(\lambda + \nu - \sigma)$ is W-conjugate to λ. Thus we seek those σ for which there exists s in W with $sw(\lambda + \nu - \sigma) = \lambda$.

Since $\nu - \sigma$ is a sum of positive roots and Re λ is dominant, we have

$$\begin{aligned} |\operatorname{Re} \lambda + \nu - \sigma|^2 &= |\operatorname{Re} \lambda|^2 + 2\langle \operatorname{Re} \lambda, \nu - \sigma \rangle + |\nu - \sigma|^2 \\ &\geq |\operatorname{Re} \lambda|^2 + |\nu - \sigma|^2. \end{aligned}$$

Thus the existence of s implies $\sigma = \nu$. Thus the inner sum drops out in (10.79) under the projection p_λ, and (b) follows.

Corollary 10.45. Under the assumptions of Proposition 10.44, $\varphi_{\lambda+\nu}^{\lambda}$ maps any nonzero Harish-Chandra module with generalized infinitesimal character λ to a nonzero module.

Proof. This result is evident from (10.75).

Corollary 10.46. Under the assumptions of Proposition 10.44, let V be any Harish-Chandra module with generalized infinitesimal character λ. Then the global character $\Theta(V)$ of V satisfies

$$\Theta(\psi_{\lambda}^{\lambda+\nu}\varphi_{\lambda+\nu}^{\lambda}V) = |W_\lambda/W_{\lambda+\nu}|\Theta(V).$$

Proof. We apply (10.77) with $V_2 = \varphi_{\lambda+\nu}^{\lambda}V$. According to (10.75), the formula for the polynomials $q_w(H)$ is

$$q_w(H) = |W_\lambda/W_{\lambda+\nu}|^{-1}\xi_{w\nu}(t_0)p_w(H).$$

Thus the corollary follows.

Corollary 10.47. Under the assumptions of Proposition 10.44, suppose that $W_\lambda = W_{\lambda+\nu}$. Then

(a) $\varphi_{\lambda+\nu}^{\lambda}$ carries any irreducible admissible representation of infinitesimal character λ to an irreducible admissible representation of infinitesimal character $\lambda + \nu$,

(b) $\psi_{\lambda}^{\lambda+\nu}$ carries any irreducible admissible representation of infinitesimal character $\lambda + \nu$ to an irreducible admissible representation of infinitesimal character λ,

(c) if V_1 is irreducible admissible and has infinitesimal character λ, then $\psi_{\lambda}^{\lambda+\nu}\varphi_{\lambda+\nu}^{\lambda}V_1$ is infinitesimally equivalent with V_1,

(d) if V_2 is irreducible admissible and has infinitesimal character $\lambda + \nu$, then $\varphi_{\lambda+\nu}^{\lambda}\psi_{\lambda}^{\lambda+\nu}V_2$ is infinitesimally equivalent with V_2.

Proof. Conclusion (c) follows from Corollary 10.46 and the irreducibility of V_1. In similar fashion under the assumption $W_\lambda = W_{\lambda+\nu}$ we see from Proposition 10.44 that

$$\Theta(\varphi_{\lambda+\nu}^{\lambda}\psi_{\lambda}^{\lambda+\nu}V_2) = \Theta(V_2).$$

Thus (d) follows if V_2 is irreducible.

Next we combine (d) with Corollary 10.45, and (b) follows. Also (d) and the exactness of $\psi_{\lambda}^{\lambda+\nu}$ imply that $\psi_{\lambda}^{\lambda+\nu}$ is nonzero on nonzero modules of generalized infinitesimal character $\lambda + \nu$. Hence (a) follows from (c).

Remarks. One can show more—that $\psi\varphi$ and $\varphi\psi$ are naturally equivalent with the identity functor (in the sense of category theory). We omit the details.

The fourth corollary is so striking that we state it as a theorem.

Theorem 10.48. In the character formula for the global character of an irreducible admissible representation (see Theorem 10.35), the polynomials are all constants.

Proof. Let V be irreducible, say with infinitesimal character λ, and let δ be half the sum of the positive roots. The character of $\varphi_{\lambda+\delta}^{\lambda}(V)$ is given in terms of the character of V as in Proposition 10.44a. Since $\lambda + \delta$ is regular, Theorem 10.35 says that the polynomials in the expression for $\varphi_{\lambda+\delta}^{\lambda}(V)$, which are uniquely determined, are all constant. By inspection the polynomials in the expression for V must have been constant.

Before coming to one final property of Zuckerman tensoring, we prove the following lemma.

Lemma 10.49. Under the assumptions of Proposition 10.44, let V_1 and V_2 be Harish-Chandra modules with respective generalized infinitesimal characters λ and $\lambda + \nu$. Then the vector space of $\mathfrak{g}$-intertwining operators from $\psi_{\lambda}^{\lambda+\nu}(V_2)$ to V_1 is canonically isomorphic with the vector space of $\mathfrak{g}$-intertwining operators from V_2 to $\varphi_{\lambda+\nu}^{\lambda}(V_1)$.

Proof. It is easy to see that the representation contragredient to F^{ν} is canonically isomorphic to $F_{-\nu}$. From this fact, one exhibits a canonical isomorphism

$$\operatorname{Hom}_{\mathfrak{g}}(F_{-\nu} \otimes V_2, V_1) \cong \operatorname{Hom}_{\mathfrak{g}}(V_2, F^{\nu} \otimes V_1). \tag{10.80}$$

Since $\mathfrak{g}$-intertwining operators preserve generalized infinitesimal characters, we obtain

$$\begin{aligned}
\operatorname{Hom}_{\mathfrak{g}}(\psi_{\lambda}^{\lambda+\nu}V_2, V_1) &= \operatorname{Hom}_{\mathfrak{g}}(p_{\lambda}(F_{-\nu} \otimes V_2), V_1) \\
&\cong \operatorname{Hom}_{\mathfrak{g}}(F_{-\nu} \otimes V_2, V_1) \\
&\cong \operatorname{Hom}_{\mathfrak{g}}(V_2, F^{\nu} \otimes V_1) \qquad \text{by (10.80)} \\
&\cong \operatorname{Hom}_{\mathfrak{g}}(V_2, p_{\lambda+\nu}(F^{\nu} \otimes V_1)) \\
&= \operatorname{Hom}_{\mathfrak{g}}(V_2, \varphi_{\lambda+\nu}^{\lambda}V_1).
\end{aligned}$$

Theorem 10.50. Fix a Cartan subgroup T of G, a positive system $\Delta^+(\mathfrak{t}^{\mathbb{C}}:\mathfrak{g}^{\mathbb{C}})$, and linear functionals λ and ν on $\mathfrak{t}^{\mathbb{C}}$ such that $\operatorname{Re}\lambda$ and ν are dominant and ν is algebraically integral. If (π, V) is an irreducible admissible representation, then the global character of $\psi_{\lambda}^{\lambda+\nu}(V)$ is a multiple of an irreducible character.

Remarks. That is, $\psi_{\lambda}^{\lambda+\nu}(V)$ is **primary,** at least on the level of characters. In fact, one can show that $\psi_{\nu}^{\lambda+\nu}(V)$ is irreducible, but the proof of this fact uses a detailed classification of irreducible tempered representations.

Proof. Let V_1 be an irreducible quotient of $\psi_\lambda^{\lambda+\nu}(V)$, and let j be the quotient mapping. Find the $\mathfrak{g}$-intertwining operator i of V into $\varphi_{\lambda+\nu}^{\lambda}(V_1)$ that corresponds to j under the isomorphism in Lemma 10.49. Then i is not the 0 map since the 0 map must correspond to the 0 map of $\psi_\lambda^{\lambda+\nu}(V)$ into V_1. Since V is irreducible, i must therefore be one-one. Thus we have an exact sequence

$$0 \longrightarrow V \xrightarrow{\ i\ } \varphi_{\lambda+\nu}^{\lambda}(V_1) \longrightarrow \text{quotient} \longrightarrow 0.$$

Since $\psi_\lambda^{\lambda+\nu}$ is an exact functor, the sequence

$$0 \longrightarrow \psi_\lambda^{\lambda+\nu}(V) \xrightarrow{\psi \circ i} \psi_\lambda^{\lambda+\nu}\varphi_{\lambda+\nu}^{\lambda}(V_1) \longrightarrow \psi_\lambda^{\lambda+\nu}(\text{quotient}) \longrightarrow 0$$

is exact. Now we can pass to characters. Corollary 10.46 says that the character of $\psi_\lambda^{\lambda+\nu}\varphi_{\lambda+\nu}^{\lambda}(V_1)$ is a multiple of the character of V_1, which is by assumption irreducible. Therefore the character of the included representation $\psi_\lambda^{\lambda+\nu}(V)$ must be a multiple of the character of V_1.

§10. Problems

1. For $G = SL(4, \mathbb{R})$ and T equal to the diagonal subgroup, calculate $\varepsilon_R^{G/T}$, D_T^G, and $(\varepsilon_R D)^{G/MA}$. Verify that $(\varepsilon_R D)^{G/MA}$ is invariant under $W(T:G)$.
2. Call an admissible representation standard if it is of the form $U(S, \sigma, \nu)$, where $S = MAN$ is a parabolic subgroup and σ is an irreducible tempered representation of M. Prove that the character of any irreducible admissible representation of G is a linear combination with integer coefficients (not necessarily ≥ 0) of characters of standard representations. [Hint: Induct on the Langlands A parameter of the given representation, using Proposition 8.61. Take into account a generalization of Corollary 10.39 and the common infinitesimal character of all subquotients of a standard representation.]

Problems 3 to 8 concern the nature of Zuckerman tensoring with nonunitary principal series. Fix a minimal parabolic subgroup MAN of G. Let (ξ, V^ξ) be a finite-dimensional representation of MAN (not assumed to be the direct sum of irreducible representations) and let $U(\xi)$ be the "induced representation" of G acting in a completion of the space

$$V^{U(\xi)} = \{f: G \to V^\xi \mid f(xman) = e^{-\rho \log a}\xi(man)^{-1}f(x)\}$$

by $U(\xi, g)f(x) = f(g^{-1}x)$.

3. Using Lie's Theorem and the compactness of M, prove that ξ is of the form $\sigma \otimes e^\nu \otimes 1$ if ξ is irreducible.
4. Let $V_1 \subseteq V^\xi$ be an MAN-invariant subspace, and let ξ_1 and ξ_2 be the corresponding subrepresentation and quotient representation on V_1

and V^ξ/V_1, respectively. Let $p: V^\xi \to V^\xi/V_1$ be the quotient map, and regard $V^{U(\xi_1)}$ as a subspace of $V^{U(\xi)}$. For f in $V^{U(\xi)}$ define $p(f)(x) = p(f(x))$. Prove on the level of K-finite vectors that p is a $\mathfrak{g}$-intertwining operator of $V^{U(\xi)}$ onto $V^{U(\xi_2)}$ with kernel $V^{U(\xi_1)}$.

5. Let $V^\xi = V_1 \supset V_2 \supset \ldots \supset V_d \supset 0$ be a composition series of MAN subrepresentations, let ξ_i be the restriction of ξ to V_i, and let the irreducible quotient on V_i/V_{i+1} be denoted $\sigma_i \otimes e^{\nu_i} \otimes 1$ (in accordance with Problem 3). Prove that there is a natural chain of G representations
$$V^{U(\xi)} = V^{U(\xi_1)} \supset V^{U(\xi_2)} \supset \ldots \supset V^{U(\xi_d)} \supset 0$$
and that the representation on $V^{U(\xi_i)}/V^{U(\xi_{i+1})}$ is infinitesimally equivalent with the nonunitary principal series $U(MAN, \sigma_i, \nu_i)$.

6. Let (π, V^π) be a finite-dimensional representation of G, and set $\tau = \pi|_{MAN}$. For f in $V^{U(\xi)}$ and v in V^π, define $Q(f \otimes v)(g) = f(g) \otimes \pi(g)v$. Prove that $Q: V^{U(\xi)} \otimes V^\pi \to V^{U(\xi \otimes \tau)}$ on the level of K-finite vectors is a one-one $\mathfrak{g}$-intertwining operator of $U(\xi) \otimes \pi$ onto $U(\xi \otimes \tau)$.

7. Suppose $\xi = \sigma \otimes e^\nu \otimes 1$ is irreducible and (π, V^π) is a finite-dimensional representation of G. If $\xi \otimes \pi|_{MAN}$ has a composition series
$$V^\xi \otimes V^\pi = V_1 \supset V_2 \supset \ldots \supset V_d \supset 0$$
with irreducible quotients $\sigma_i \otimes e^{\nu_i} \otimes 1$ on V_i/V_{i+1}, prove that $U(\xi) \otimes \pi$ has a corresponding chain and that the respective quotients are infinitesimally equivalent with the $U(MAN, \sigma_i, \nu_i)$ for $1 \leq i \leq d$.

8. Extend the Lie algebra $\mathfrak{a}$ of A to a Cartan subalgebra $\mathfrak{b} \oplus \mathfrak{a}$ of $\mathfrak{g}$, and introduce a positive system for $\Delta((\mathfrak{b} \oplus \mathfrak{a})^{\mathbb{C}} : \mathfrak{g}^{\mathbb{C}})$ in which $i\mathfrak{b}$ comes before $\mathfrak{a}$. Suppose in the notation of Problem 7 that ν is imaginary and π has lowest weight $-\mu$ vanishing on $\mathfrak{a}$. Suppose further that the highest weight of σ on M_0 is of the form $\lambda + \mu$ with λ dominant. Prove that only one quotient $U(MAN, \sigma_i, \nu_i)$ in Problem 7 survives the operation of $\psi_\lambda^{\lambda+\mu}$ on $U(\xi)$, that the surviving ν_i is ν, that the highest weight of σ_i on M_0 is λ, and that σ agrees with σ_i on the group F defined in Lemma 9.13.

CHAPTER XI

Introduction to Plancherel Formula

§1. Constructive Proof for SU(2)

The Plancherel formula for SU(2) reads

$$\int_{\mathrm{SU}(2)} |F(x)|^2\,dx = \sum_{n=0}^{\infty} (n+1)\|\Phi_n(F)\|_{\mathrm{HS}}^2, \qquad F \in L^2(\mathrm{SU}(2)). \quad (11.1)$$

Here Φ_n is the irreducible representation given in (2.1), and dx is normalized to have total mass one. We know from Theorem 2.4 that the Φ_n form a complete set of inequivalent irreducible unitary representations of SU(2), and hence (11.1) is an immediate consequence of the Peter-Weyl Theorem.

In this section we shall give a different proof that has possibilities for generalizing to noncompact groups. First of all, as we noted in §2.7, it is enough to prove the Fourier inversion formula

$$f(1) = \sum_{n=0}^{\infty} (n+1)\,\mathrm{Tr}\,\Phi_n(f), \qquad f \in C^\infty(\mathrm{SU}(2)), \qquad (11.2)$$

since (11.1) follows from (11.2) by taking $f = F^* * F$ and doing a passage to the limit. If Θ_n denotes the character of Φ_n, we can rewrite (11.2) as

$$f(1) = \sum_{n=0}^{\infty} (n+1)\Theta_n(f), \qquad f \in C^\infty(\mathrm{SU}(2)), \qquad (11.3)$$

where

$$\Theta_n(f) = \int_G \Theta_n(x) f(x)\,dx,$$

since

$$\mathrm{Tr}\,\Phi_n(f) = \mathrm{Tr}\int_G f(x)\Phi_n(x)\,dx = \int_G \mathrm{Tr}(f(x)\Phi_n(x))\,dx = \Theta_n(f).$$

Thus we are to prove (11.3). With $t_\theta = \begin{pmatrix} e^{i\theta} & 0 \\ 0 & e^{-i\theta} \end{pmatrix}$, the character Θ_n is given on the maximal torus $T = \{t_\theta\}$ by

$$\Theta_n(t_\theta) = \frac{e^{i(n+1)\theta} - e^{-i(n+1)\theta}}{e^{i\theta} - e^{-i\theta}}, \qquad (11.4)$$

by direct computation or the Weyl character formula. We shall use (11.4) and also the Weyl integration formula, which we specialize from Theorem 4.45 as

$$\int_G f(x)\,dx = \frac{1}{2}\int_T |e^{i\theta} - e^{-i\theta}|^2 \int_G f(xt_\theta x^{-1})\,dx\,\frac{dt_\theta}{2\pi}. \tag{11.5}$$

We introduce

$$F_f(\theta) = (e^{i\theta} - e^{-i\theta})\int_G f(xt_\theta x^{-1})\,dx. \tag{11.6}$$

For the proof of (11.3) we compute

$$\begin{aligned}
\Theta_n(f) &= \int_G \Theta_n(x) f(x)\,dx \\
&= \frac{1}{2}\int_T |e^{i\theta} - e^{-i\theta}|^2 \int_G \Theta_n(t_\theta) f(xt_\theta x^{-1})\,dx\,\frac{dt_\theta}{2\pi} && \text{by (11.5)} \\
&= \frac{1}{2}\int_T (e^{-i\theta} - e^{i\theta})\frac{e^{i(n+1)\theta} - e^{-i(n+1)\theta}}{e^{i\theta} - e^{-i\theta}} F_f(\theta)\frac{dt_\theta}{2\pi} && \text{by (11.4) and (11.6)} \\
&= \int_0^{2\pi} -\frac{1}{2}(e^{i(n+1)\theta} - e^{-i(n+1)\theta})F_f(\theta)\frac{d\theta}{2\pi}.
\end{aligned}$$

Multiplication by $n + 1$ gives

$$(n+1)\Theta_n(f) = \frac{i}{2}\int_0^{2\pi} \frac{d}{d\theta}(e^{i(n+1)\theta} + e^{-i(n+1)\theta})F_f(\theta)\frac{d\theta}{2\pi}.$$

We sum for $n \geq 0$, noting that we can insert a term with $\frac{d}{d\theta}e^{i0\theta}$ on the right since it is 0. We obtain

$$\begin{aligned}
\sum_{n=0}^{\infty}(n+1)\Theta_n(f) &= \sum_{k=-\infty}^{\infty}\frac{i}{2}\int_0^{2\pi}\frac{d}{d\theta}(e^{ik\theta})F_f(\theta)\frac{d\theta}{2\pi} \\
&= \frac{1}{2i}\sum_{k=-\infty}^{\infty}\frac{1}{2\pi}\int_0^{2\pi} e^{ik\theta}\frac{d}{d\theta}F_f(\theta)\,d\theta && \text{by integration by parts} \\
&= \frac{1}{2i}\frac{d}{d\theta}F_f(\theta)\bigg|_{\theta=0} && \text{from the sum of the Fourier coefficients.}
\end{aligned} \tag{11.7}$$

Now

$$\begin{aligned}
\frac{d}{d\theta}F_f(\theta) &= \frac{d}{d\theta}\left(2i\sin\theta\int_G f(xt_\theta x^{-1})\,dx\right) \\
&= 2i(\sin\theta)\frac{d}{d\theta}\int_G f(xt_\theta x^{-1})\,dx + 2i(\cos\theta)\int_G f(xt_\theta x^{-1})\,dx.
\end{aligned}$$

At $\theta = 0$, the first term vanishes, and thus

$$\frac{d}{d\theta} F_f(\theta)\bigg|_{\theta=0} = 2if(1). \tag{11.8}$$

Combining (11.7) and (11.8), we obtain (11.3).

§2. Constructive Proof for SL(2, ℂ)

It is simple matter to generalize the argument of §1 so that it applies to any compact connected Lie group. The factor $(n+1)$ in (11.3) gets replaced by the degree, and the coefficient $(e^{i\theta} - e^{-i\theta})$ in (11.6) gets replaced by the Weyl denominator (4.29), which can be written by Corollary 4.47 as either a sum or a product. Formula (11.8) for $\frac{d}{d\theta} F_f(\theta)\big|_{\theta=0}$ appropriately generalizes to a formula for a higher order derivative of F_f; the derivative is $\prod_{\alpha \in \Delta^+} \partial(\alpha)$, the same one as in the proof of Theorem 4.48.

Our goal in this section is to establish an analog of (11.3) for a noncompact semisimple group. It is tempting to begin with $SL(2, \mathbb{R})$, but a glance at (11.5) points to a serious difficulty: We want to analyze essentially arbitrary L^2 functions on G and can therefore omit only a set of measure 0 in G. The conjugates of the elements of a particular Cartan subgroup contain a nonempty open set, necessarily of positive measure. In $SL(2, \mathbb{R})$ there are two nonconjugate Cartan subgroups, and we must therefore take both of them into account in integration. The Weyl integration formula makes these matters explicit.

Consequently we shall begin by studying instead $SL(2, \mathbb{C})$, in which every Cartan subgroup is conjugate to the diagonal subgroup T. For $G = SL(2, \mathbb{C})$, we look for a concrete space Ω and a measure $d\mu$ on Ω such that

$$f(1) = \int_\Omega \operatorname{Tr} \pi_\omega(f)\, d\mu(\omega) \qquad \text{for all } f \text{ in } C^\infty_{\text{com}}(G). \tag{11.9}$$

We shall refer to an explicit formula of this type as the **Plancherel formula** for G, since the analog of (11.1) follows directly from it. Strictly speaking, (11.9) is the analog of the Fourier inversion formula.

There are three ingredients to our approach, if we are to proceed as in §1.

(1) Normalization of Haar measure. We use the natural normalization of Haar measure for G from $G = KNA$ as follows:

$$A = \left\{ \begin{pmatrix} e^u & 0 \\ 0 & e^{-u} \end{pmatrix} \middle|\, u \in \mathbb{R} \right\}, \qquad \text{Haar measure } du$$

$$N = \left\{ \begin{pmatrix} 1 & x+iy \\ 0 & 1 \end{pmatrix} \middle| x, y \in \mathbb{R} \right\}, \qquad \text{Haar measure } dn = dx\,dy$$

$$K = SU(2), \qquad \text{Haar measure } dk \text{ of total mass 1.}$$

We take Haar measure on G to be

$$dg = dk\,dx\,dy\,du. \tag{11.10}$$

Our Cartan subgroup is

$$T = \left\{ t = a_u m_\theta = \begin{pmatrix} e^{u+i\theta} & 0 \\ 0 & e^{-u-i\theta} \end{pmatrix} \middle| u, \theta \in \mathbb{R} \right\}, \tag{11.11}$$

and we take the Haar measure on T to be

$$dt = \frac{1}{2\pi}\,du\,d\theta. \tag{11.12}$$

(2) The function F_f^T. Since $\mathfrak{g}^{\mathbb{C}} \cong \mathfrak{sl}(2, \mathbb{C}) \oplus \mathfrak{sl}(2, \mathbb{C})$, the root system of $\mathfrak{g}^{\mathbb{C}}$ with respect to $\mathfrak{t}^{\mathbb{C}}$ has four members. The two positive roots can be taken on $\mathfrak{t}$ to be

$$\alpha\begin{pmatrix} u+\theta & 0 \\ 0 & -u-i\theta \end{pmatrix} = 2(u+i\theta) \text{ and } \bar{\alpha}\begin{pmatrix} u+i\theta & 0 \\ 0 & -u-i\theta \end{pmatrix} = 2(u-i\theta).$$

The Weyl denominator on the element t in (11.11) is then

$$D_t(t) = (e^{u+i\theta} - e^{-(u+i\theta)})(e^{u-i\theta} - e^{-u+i\theta}) = 2(\cosh 2u - \cos 2\theta).$$

Hence the singular points are where $u = 0$ and θ is 0 or π. The formula for F_f^T is

$$F_f^T(t) = D_T(t) \int_{G/T} f(gtg^{-1})\,d\dot{g}, \qquad t \in T', \tag{11.13}$$

and (10.22) and Lemma 10.17 say we can rewrite it as

$$F_f^T(t) = e^{2u} \int_{K \times N} f(ktnk^{-1})\,dk\,dn. \tag{11.14}$$

The latter formula makes it clear that F_f^T extends to a member of $C_{\text{com}}^\infty(T)$.

(3) Character formulas. Formula (10.21) identifies the principal series characters. Let us parametrize the principal series more concretely as $U(S, \sigma_n, iv)$, where

$$\sigma_n\begin{pmatrix} e^{i\theta} & 0 \\ 0 & e^{-i\theta} \end{pmatrix} = e^{in\theta}$$

and where we use v for two purposes in writing $iv\begin{pmatrix} u & 0 \\ 0 & -u \end{pmatrix} = ivu$. Then

(10.21) says that the character of $U(S, \sigma_n, iv)$ is given by

$$\begin{aligned}\Theta_{\sigma_n, iv}(f) &= \int_T \int_{K \times N} f(ka_u m_\theta n k^{-1}) e^{(2+iv)u} e^{in\theta}\, dk\, dn \frac{1}{2\pi}\, du\, d\theta \\ &= \int_T e^{i(vu+n\theta)} F_f^T(a_u m_\theta) \frac{1}{2\pi}\, du\, d\theta.\end{aligned}$$

This formula we recognize as an abelian group Fourier transform on T (real Fourier transform taken in u and Fourier series coefficient taken in θ). Since the function F_f^T is in $C_{\text{com}}^\infty(T)$, Fourier inversion holds, and we have

$$F_f^T(a_u m_\theta) = \sum_{n=-\infty}^{\infty} \frac{1}{2\pi} \int_{v=-\infty}^{\infty} \Theta_{\sigma_n, iv}(f) e^{-i(vu+n\theta)}\, dv. \tag{11.15}$$

Beyond the three ingredients, we need to relate $f(1)$ to a suitable derivative of F_f^T, in analogy with (11.8). Again the appropriate derivative is given by the "product of the positive roots" in the following sense: The operator is $\partial(\bar{\alpha})\, \partial(\alpha)$, where

$$\partial(\alpha)h(t) = \frac{d}{ds} h(t \exp(s \operatorname{Re} H_\alpha))\bigg|_{s=0} + i \frac{d}{ds} h(t \exp(s \operatorname{Im} H_\alpha))\bigg|_{s=0}$$

and where $\partial(\bar{\alpha})$ is defined similarly. In this expression, H_α is the member $\operatorname{Re} H_\alpha + i \operatorname{Im} H_\alpha$ of $\mathfrak{t} \oplus i\mathfrak{t}$ corresponding to α under the complexification of the bilinear form obtained from the real part of the trace. Then $\operatorname{Re} H_\alpha = \begin{pmatrix} 1 & 0 \\ 0 & -1 \end{pmatrix}$ and $\operatorname{Im} H_\alpha = \begin{pmatrix} -i & 0 \\ 0 & i \end{pmatrix}$, and it follows that

$$\partial(\alpha)h(a_u m_\theta) = \frac{\partial}{\partial u} h(a_u m_\theta) - i \frac{\partial}{\partial \theta} h(a_u m_\theta)$$

$$\partial(\bar{\alpha})h(a_u m_\theta) = \frac{\partial}{\partial u} h(a_u m_\theta) + i \frac{\partial}{\partial \theta} h(a_u m_\theta).$$

Consequently

$$\partial(\bar{\alpha})\, \partial(\alpha) h(a_u m_\theta) = \left(\frac{\partial^2}{\partial u^2} + \frac{\partial^2}{\partial \theta^2} \right) h(a_u m_\theta). \tag{11.16}$$

The formula relating F_f^T and $f(1)$ is given in the following key lemma.

Lemma 11.1. In SL(2, ℂ),

$$(2\pi)^2 f(1) = -\tfrac{1}{2} \partial(\bar{\alpha})\, \partial(\alpha) F_f^T(1) \tag{11.17}$$

for every f in $C_{\text{com}}^\infty(G)$.

Let us postpone the proof of the lemma temporarily. In (11.15) we can differentiate under the sum and integral because $\Theta_{\sigma_n,iv}(f)$ is a Schwartz function, being the Fourier transform of the member F_f^T of $C_{\text{com}}^\infty(T)$. Using (11.16), we therefore have

$$\partial(\bar{\alpha})\,\partial(\alpha)F_f^T(a_u m_\theta) = -\sum_{n=-\infty}^{\infty} \frac{1}{2\pi}\int_{-\infty}^{\infty} \Theta_{\sigma_n,iv}(f)(n^2+v^2)e^{-i(vu+n\theta)}\,dv.$$

Applying the lemma, we obtain the desired Plancherel formula, as follows.

Theorem 11.2. For f in $C_{\text{com}}^\infty(\text{SL}(2,\mathbb{C}))$ and Haar measure normalized as in (11.10),

$$(2\pi)^3 f(1) = \frac{1}{2}\sum_{n=-\infty}^{\infty}\int_{-\infty}^{\infty} \Theta_{\sigma_n,iv}(f)(n^2+v^2)\,dv. \tag{11.18}$$

Thus the space Ω is the parameter space $\{(\sigma_n, iv)\}$ of the principal series, and the Plancherel measure is a multiple of $(n^2+v^2)\,dv$. Technically we should lump together equivalent representations in (11.18). We have $U(S,\sigma_n,iv) \cong U(S,\sigma_{-n},-iv)$, and there are no other equivalences (by inspection of the character formula in Proposition 10.18). So it is a simple matter to rewrite (11.18) to take the equivalences into account.

Now let us take up the proof of Lemma 11.1. There are two really new ingredients beyond what we have—a formula for the Jacobian of the exponential map in a Lie group and a version of the Weyl integration formula applicable to the Lie algebra $\mathfrak{g}$. We isolate the formula for the Jacobian as Lemma 11.3 and the version of the Weyl integration formula as Lemma 11.4, offering some remarks but no proofs. Then we pass to the proof of Lemma 11.1 itself.

Lemma 11.3. Let G be a connected Lie group with Lie algebra $\mathfrak{g}$. About X in $\mathfrak{g}$ and $\exp X$ in G, identify the respective tangent spaces to $\mathfrak{g}$ and G with the vector space $\mathfrak{g}$ by means of Euclidean translation in $\mathfrak{g}$ and left group translation in G, respectively. Under these identifications the differential of the exponential map at X is given by

$$\frac{(1-\exp(-\operatorname{ad} X))}{\operatorname{ad} X} = \sum_{n=1}^{\infty}(-1)^{n-1}\frac{(\operatorname{ad} X)^{n-1}}{n!}.$$

Lemma 11.3 has an implication for changing variables in integrals: Let dX denote a multiple of Lebesgue measure on $\mathfrak{g}$, and let $d_l x$ be a left Haar measure on G. Then there is a constant c_G with the following property. Suppose that F is a continuous function on $\mathfrak{g}$ with compact support in an open set U on which the exponential map is one-one and regular. We define a function f on G by

$$f(x) = \begin{cases} F(\exp^{-1} x) & \text{on } \exp U \\ 0 & \text{off } \exp U. \end{cases}$$

Then

$$\int_G f(x)\,dx = c_G \int_{\mathfrak{g}} F(X)|\det\{(1 - \exp(-\operatorname{ad} X))/\operatorname{ad} X\}|\,dX. \tag{11.19}$$

For any linear connected reductive group G, we give a **Weyl integration formula** on $\mathfrak{g}$ that rewrites integrals in terms of the elements of Cartan subalgebras and their transforms by $\operatorname{Ad}(G)$. In view of Lemma 11.3, such a formula near 0 in $\mathfrak{g}$ is equivalent with the corresponding formula in G near 1 (cf. Proposition 5.27). However, on the global level it is preferable to give analogous derivations of the formulas rather than to obtain one as a consequence of the other.

Lemma 11.4. Let G be linear connected reductive, let $\mathfrak{h}_1, \ldots, \mathfrak{h}_r$ be a maximal set of nonconjugate θ-stable Cartan subalgebras, and let $H_1, \ldots, H_r$ be the corresponding Cartan subgroups. Let the invariant measures on each H_j and G/H_j be normalized so that

$$\int_G f(x)\,dx = \int_{G/H_j}\left[\int_{H_j} f(gh)\,dh\right]d\dot{g} \qquad \text{for all } f \in C_{\text{com}}(G).$$

Normalize Lebesgue measure dH on each $\mathfrak{h}_j$ so that $c_{H_j} = 1$ in (11.19), and let dX be a Lebesgue measure on $\mathfrak{g}$. Then every F in $C_{\text{com}}(\mathfrak{g})$ satisfies

$$c_G \int_{\mathfrak{g}} F(X)\,dX = \sum_j \frac{1}{|W(H_j : G)|} \int_{G/H_j \times \mathfrak{h}_j} F(\operatorname{Ad}(g)H)|\omega_{\mathfrak{h}_j}(H)|^2\,dH\,d\dot{g},$$

where

$$\omega_{\mathfrak{h}_j}(H) = \prod_{\alpha \in \Delta^+(\mathfrak{h}_j^{\mathbb{C}} : \mathfrak{g}^{\mathbb{C}})} \alpha(H).$$

Passing to the proof of Lemma 11.1, let us specialize Lemma 11.4 to the case $G = SL(2, \mathbb{C})$. Then we have

$$c_G \int_{\mathfrak{sl}(2,\mathbb{C})} F(X)\,dX = \tfrac{1}{2} \int_{G/T \times \mathfrak{t}} F(\operatorname{Ad}(g)H)|\alpha(H)\bar{\alpha}(H)|^2\,d\dot{g}\,dH.$$

Writing $G = K\bar{N}A$ and taking into account the normalizations of Haar measure in (11.10) and (11.12), we see that we can rewrite this equation as

$$c_G \int_{\mathfrak{sl}(2,\mathbb{C})} F(X)\,dX = \tfrac{1}{2} \int_{K \times \bar{N} \times \mathfrak{t}} F(\operatorname{Ad}(k\bar{n})H)|\alpha(H)\bar{\alpha}(H)|^2\,dk\,d\bar{n}\,dH.$$

Now let

$$\bar{n} = \begin{pmatrix} 1 & 0 \\ z' & 1 \end{pmatrix} = \exp Z'$$

and define Lebesgue measure on $\theta\mathfrak{n}$ by $dZ' \leftrightarrow d\bar{n}$. Multiplying the matrices, we see that

$$\mathrm{Ad}(\bar{n})H = H + \alpha(H)Z'.$$

Thus our integration formula becomes

$$c_G \int_{\mathfrak{sl}(2,\mathbb{C})} F(X)\, dX$$

$$= \tfrac{1}{2} \int_{K\times(\mathfrak{t}\oplus\theta\mathfrak{n})} F(\mathrm{Ad}(k)(H + \alpha(H)Z'))|\alpha(H)\bar{\alpha}(H)|^2\, dk\, dH\, dZ'.$$

Replacing $\alpha(H)Z'$ by Z', we obtain

$$c_G \int_{\mathfrak{sl}(2,\mathbb{C})} F(X)\, dX$$

$$= \tfrac{1}{2} \int_{K\times(\mathfrak{t}\oplus\theta\mathfrak{n})} F(\mathrm{Ad}(k)(H + Z'))|\alpha(H)|^2\, dk\, dH\, dZ'. \tag{11.20}$$

This remains valid for integrable F on $\mathfrak{sl}(2, \mathbb{C})$.

This formula is the real starting point for the proof of Lemma 11.1, which takes place in two steps. The first step is to use (11.20) and Euclidean Fourier inversion to obtain an analog of Lemma 11.1 on $\mathfrak{g}$, and the second step is to pass to G by means of the consequence (11.19) of Lemma 11.3.

To fix the notation, let us choose a particular normalization of Lebesgue measure on $\mathfrak{g}$. A general member of $\mathfrak{g}$ is of the form

$$X = \begin{pmatrix} u + i\theta & x + iy \\ x' + iy' & -(u + i\theta) \end{pmatrix},$$

and we choose

$$dX = du\, d\theta\, dx\, dy\, dx'\, dy'. \tag{11.21}$$

The analog of Lemma 11.1 on $\mathfrak{g}$ is the following: If F is in $C^\infty_{\mathrm{com}}(\mathfrak{sl}(2, \mathbb{C}))$ and if $\square_H$ denotes the Laplacian $\dfrac{\partial^2}{\partial u^2} + \dfrac{\partial^2}{\partial \theta^2}$ on $\mathfrak{t}$, then

$$(2\pi)^2 c_G F(0) = -\frac{1}{2}\left[\square_H \int_{K\times\mathfrak{n}} F(\mathrm{Ad}(k)(H + Z))\, dZ\, dk\right]_{H=0}. \tag{11.22}$$

To prove (11.22), we apply (11.20) not to F but to a Euclidean Fourier transform $\mathscr{F}_{\mathfrak{g}}F$ of F. To pin down this Fourier transform, we use the inner product

$$\langle X, Y\rangle = -\mathrm{Re}\,\mathrm{Tr}(X(\theta Y)) = \mathrm{Re}\,\mathrm{Tr}(XY^*)$$

on $\mathfrak{g}$ and take

$$\mathscr{F}_{\mathfrak{g}}F(X_2) = \int_{\mathfrak{g}} F(X_1)e^{-i\langle X_1, X_2\rangle}\, dX_1.$$

The Lie algebra $\mathfrak{g}$ is the orthogonal sum of $\theta\mathfrak{n}$, $\mathfrak{t}$, and $\mathfrak{n}$, and these subalgebras have individual Fourier transforms given by

$$\mathscr{F}_{\theta\mathfrak{n}}F(Z_2', H, Z) = \int_{\theta\mathfrak{n}} F(Z_1', H, Z)e^{-\langle Z_1', Z_2'\rangle}\, dx_1'\, dy_1'$$

$$\mathscr{F}_{\mathfrak{t}}F(Z', H_2, Z) = \int_{\mathfrak{t}} F(Z', H_1, Z)e^{-\langle H_1, H_2\rangle}\, du_1\, d\theta_1$$

$$\mathscr{F}_{\mathfrak{n}}F(Z', H, Z_2) = \int_{\mathfrak{n}} F(Z', H, Z_1)e^{-\langle Z_1, Z_2\rangle}\, dx_1\, dy_1.$$

We have

$$\mathscr{F}_{\mathfrak{g}} = \mathscr{F}_{\theta\mathfrak{n}}\mathscr{F}_{\mathfrak{t}}\mathscr{F}_{\mathfrak{n}}, \tag{11.23}$$

with the three operators on the right commuting. Since the inner product on $\mathfrak{g}$ is $\mathrm{Ad}(K)$-invariant, $\mathscr{F}_{\mathfrak{g}}$ commutes with the action of $\mathrm{Ad}(K)$. Tracking down the various normalizations of Lebesgue measure we have used, we find the following inversion formulas:

$$F(Z_1', H, Z) = (2\pi)^{-2} \int_{\theta\mathfrak{n}} \mathscr{F}_{\theta\mathfrak{n}}F(Z_2', H, Z)e^{\langle Z_1', Z_2'\rangle}\, dx_2'\, dy_2' \tag{11.24a}$$

$$F(Z', H_1, Z) = \pi^{-2} \int_{\mathfrak{t}} \mathscr{F}_{\mathfrak{t}}F(Z', H_2, Z)e^{\langle H_1, H_2\rangle}\, du_2\, d\theta_2 \tag{11.24b}$$

$$F(Z', H, Z_1) = (2\pi)^{-2} \int_{\mathfrak{n}} \mathscr{F}_{\mathfrak{n}}F(Z', H, Z_2)e^{\langle Z_1, Z_2\rangle}\, dx_2\, dy_2. \tag{11.24c}$$

Hence (11.21) and (11.23) give

$$F(0) = 2^{-4}\pi^{-6} \int_{\mathfrak{g}} \mathscr{F}_{\mathfrak{g}}F(X)\, dX. \tag{11.25}$$

Combining (11.25) and (11.20) and writing F_k for $F(\mathrm{Ad}(k)\cdot)$, we have

$$\begin{aligned}
2^4\pi^6 c_G F(0) &= \tfrac{1}{2} \int_{K\times(\mathfrak{t}\oplus\theta\mathfrak{n})} \mathscr{F}_{\mathfrak{g}}F_k(H + Z')|\alpha(H)|^2\, du\, d\theta\, dZ'\, dk \\
&= \tfrac{1}{2} \int_{K\times(\mathfrak{t}\oplus\theta\mathfrak{n})} \mathscr{F}_{\theta\mathfrak{n}}\mathscr{F}_{\mathfrak{t}}\mathscr{F}_{\mathfrak{n}}F_k(H + Z')|\alpha(H)|^2\, dZ'\, du\, d\theta\, dk \\
&\qquad\qquad\qquad\qquad\qquad\qquad\qquad\qquad\qquad \text{by (11.23)} \\
&= \tfrac{1}{2}(2\pi)^2 \int_{K\times\mathfrak{t}} \mathscr{F}_{\mathfrak{t}}\mathscr{F}_{\mathfrak{n}}F_k(H)|\alpha(H)|^2\, du\, d\theta\, dk \qquad \text{by (11.24a)} \\
&= -\tfrac{1}{2}(2\pi)^2 \int_{K\times\mathfrak{t}} \mathscr{F}_{\mathfrak{t}}(\Box_H \mathscr{F}_{\mathfrak{n}}F_k)(H_2)\, du_2\, d\theta_2\, dk \\
&= -\tfrac{1}{2}(2\pi)^2 \int_K (\Box_H \mathscr{F}_{\mathfrak{n}}F_k)(H)\Big|_{H=0}\, dk \qquad \text{by (11.24b)}.
\end{aligned}$$

Putting in the definition of $\mathscr{F}_{\mathfrak{n}}$, we obtain (11.22).

The second step in the proof of Lemma 11.1 is to pass from $\mathfrak{g}$ to G. The computation of $f(1)$ and $\partial(\bar{\alpha})\,\partial(\alpha)F_f^T(1)$ involve only the values of f on a set

$$\bigcup_{k\in K} k(UN)k^{-1}, \tag{11.26}$$

where U is an arbitrarily small neighborhood of 1 in T. One such U consists of those members of T on which the eigenvalues have positive real part. For this U there is a neighborhood E of 0 in $\mathfrak{g}$ that maps one-one onto U under the exponential map and on which the exponential map is everywhere regular (by Lemma 11.3). Without loss of generality we can then adjust f so as to be supported in the open set (11.26). Then the integration formula (11.19) is applicable, and we can track down the effect of carrying the formula (11.22) to G. The result is the identity in Lemma 11.1. This completes the proof of the Plancherel formula for $SL(2, \mathbb{C})$.

All Cartan subgroups are conjugate within any complex semisimple group, and the above style of argument is applicable. With only a few technical improvements, the argument leads to the Plancherel formula for all such groups. The formula is qualitatively similar to (11.18).

§3. Constructive Proof for SL(2, ℝ)

Our goal in this section is to establish an analog of (11.3) for $G = SL(2, \mathbb{R})$. We expect complications to arise in the derivation because of the presence of nonconjugate Cartan subgroups. Nevertheless, we still look to include the same three ingredients in our approach as in §1 and §2:

(1) Normalization of Haar measure. We use the normalization of Haar measure for G from $G = KNA$, as written explicitly in (10.7). We shall need to know what normalization is forced on Haar measure in the $G = KA^+K$ decomposition. If we write

$$x = \begin{pmatrix} \cos\varphi_1 & \sin\varphi_1 \\ -\sin\varphi_1 & \cos\varphi_1 \end{pmatrix} \begin{pmatrix} e^r & 0 \\ 0 & e^{-r} \end{pmatrix} \begin{pmatrix} \cos\varphi_2 & \sin\varphi_2 \\ -\sin\varphi_2 & \cos\varphi_2 \end{pmatrix}$$

with $r \geq 0$, then we can calculate a Jacobian determinant and see that

$$dx = 4\pi \sinh 2r\, dr \frac{d\varphi_1}{2\pi}\frac{d\varphi_2}{2\pi}$$

locally. However, when we integrate over all x in G, we integrate once in the KNA coordinates and twice in the KA^+K coordinates. Thus the global formula is

$$dx = 2\pi \sinh 2r\, dr \frac{d\varphi_1}{2\pi}\frac{d\varphi_2}{2\pi}. \tag{11.27}$$

(2) The functions F_f. The Weyl integration formula is given as (10.8) in terms of two functions F_f, one for each conjugacy class of Cartan subgroups. The functions are denoted $F_f^B(\theta)$ and $F_f^T(\pm a_t)$ and are defined in (10.9).

(3) Character formulas. From Corollary 10.13 the character of the sum $\mathscr{D}_{n+1}^+ \oplus \mathscr{D}_{n+1}^-$ of discrete series is given by

$$\begin{aligned} \Theta_{n+1}(\theta) &= \frac{-e^{in\theta} + e^{-in\theta}}{e^{i\theta} - e^{-i\theta}} \\ \Theta_{n+1}(\pm a_t) &= (\pm)^{n+1}\frac{e^{nt}(1 - \operatorname{sgn} t) + e^{-nt}(1 + \operatorname{sgn} t)}{|e^t - e^{-t}|}. \end{aligned} \tag{11.28}$$

For the principal series, let us use ν for two purposes, writing

$$i\nu\begin{pmatrix} t & 0 \\ 0 & -t\end{pmatrix} = i\nu t.$$

Then formula (10.16) in Proposition 10.12 gives

$$\begin{aligned} \Theta_{\sigma,i\nu}(f) &= \pm\tfrac{1}{2}\int_T \sigma(\pm)(e^{i\nu t} + e^{-i\nu t})F_f^T(\pm a_t)\, d(\pm)\, dt \\ &= \pm\int_T \sigma(\pm)e^{i\nu t}F_f^T(\pm a_t)\, d(\pm)\, dt \qquad \text{since } F_f^T(\pm a_t) = F_f^T(\pm a_{-t}) \\ &= \tfrac{1}{2}\int_T [F_f^T(+a_t) - \sigma(-)F_f^T(-a_t)]e^{i\nu t}\, dt. \end{aligned} \tag{11.29}$$

We can view this formula as identifying $\Theta_{\sigma,i\nu}(f)$ as a Fourier transform.

Before continuing, we make an observation that will simplify the notation. Both F_f^B and F_f^T involve conjugation by members of G, hence by members at least of K as soon as we use a decomposition of G. Our observation will allow us to drop the conjugation over K. The observation is that if the Plancherel formula (11.9) holds for all f in $C_{\text{com}}^\infty(G)$ with $f(kxk^{-1}) = f(x)$ for $k \in K$ and $x \in G$, then it holds for all f in $C_{\text{com}}^\infty(G)$. [In fact, given f in $C_{\text{com}}^\infty(G)$, we can define

$$f^\#(x) = \int_K f(kxk^{-1})\, dk.$$

The two sides of (11.9) are the same for $f^\#$ as they are for f, and thus the identity (11.9) for $f^\#$ implies the identity (11.9) for f. Thus we have

Assumption made without loss of generality: $f(x) = f(kxk^{-1})$ for all $k \in K$ and $x \in G$.

Now let us pursue the analogy with SU(2). For that purpose, we investigate the behavior of $F_f^B(\theta)$ near the singular points $\theta = 0$ and $\theta = \pi$. Remembering that $f(x) = f(kxk^{-1})$, we evaluate $F_f^B(\theta)$ by using the $G = KA^+K$ decomposition of Haar measure given in (11.27). Then we have

$$\begin{aligned} F_f^B(\theta) &= 4\pi i \sin\theta \int_0^\infty f\left(\begin{pmatrix} e^r & 0 \\ 0 & e^{-r} \end{pmatrix}\begin{pmatrix} \cos\theta & \sin\theta \\ -\sin\theta & \cos\theta \end{pmatrix}\begin{pmatrix} e^{-r} & 0 \\ 0 & e^r \end{pmatrix}\right) \sinh 2r\, dr \\ &= 2\pi i \sin\theta \int_0^\infty f\begin{pmatrix} \cos\theta & e^{2r}\sin\theta \\ -e^{-2r}\sin\theta & \cos\theta \end{pmatrix}(e^{2r} - e^{-2r})\, dr \qquad (11.30) \\ &= 2\pi i s \int_0^\infty F(se^{2r}, se^{-2r})(e^{2r} - e^{-2r})\, dr, \end{aligned}$$

where s denotes $\sin\theta$ and F is the C^∞ function on the subset $\{(u, v) \in \mathbb{R}^2 \mid |uv| < 1\}$ given by

$$F(u, v) = f\begin{pmatrix} \pm\sqrt{1 - uv} & u \\ -v & \pm\sqrt{1 - uv} \end{pmatrix}. \qquad (11.31)$$

The ambiguous sign is chosen as plus if we are examining $\theta = 0$ and minus if we are examining $\theta = \pi$.

Lemma 11.5. Let F be a C^∞ function of compact support in

$$\{(u, v) \in \mathbb{R}^2 \mid |uv| < 1\},$$

and put

$$I(s) = \pi s \int_0^\infty F(se^{2r}, se^{-2r})(e^{2r} - e^{-2r})\, dr.$$

Then

(a) $\displaystyle\lim_{s \downarrow 0} I(s) = \frac{\pi}{2} \int_0^\infty F(u, 0)\, du$

(b) $\displaystyle\lim_{s \downarrow 0} \frac{d}{ds} I(s) = -\pi F(0, 0)$

(c) there are constants c_1 and c_2 such that

$$\left|\frac{d}{ds} I(s) + \pi F(0, 0)\right| \le s(c_1 + c_2 \log s^{-1})$$

for $0 < s \le 1/2$.

Remark. One can get a quick intuitive idea of (a) and (b) by computing $I(s)$ when F is the characteristic function of a square centered at the origin.

Proof. $I(s)$ is the sum of

$$\pi s \int_0^\infty F(se^{2r}, se^{-2r})e^{2r}\,dr = \frac{\pi s}{2}\int_1^\infty F(sx, sx^{-1})\,dx$$

$$= \frac{\pi}{2}\int_s^\infty F(y, s^2y^{-1})\,dy \qquad (11.32a)$$

and

$$\pi s \int_0^\infty F(se^{2r}, se^{-2r})(-e^{-2r})\,dr = -\frac{\pi s}{2}\int_0^1 F(sx^{-1}, sx)\,dx$$

$$= -\frac{\pi}{2}\int_0^s F(s^2y^{-1}, y)\,dy. \qquad (11.32b)$$

As $s \downarrow 0$, (11.32a) tends to $\frac{\pi}{2}\int_0^\infty F(u, 0)\,du$ because F is bounded continuous on the set in question and has bounded support. Also (11.32b) tends to 0. This proves (a).

For (b) and (c), we compute $\frac{d}{ds}$ of the right side of (11.32a) to be

$$-\frac{1}{2}\pi F(s, s) + \pi s \int_s^\infty F_2(y, s^2y^{-1})\frac{dy}{y},$$

with F_2 referring to the derivative in the second variable. For the second term, we use the boundedness of F_2 and the boundedness of the support to bound the term by

$$\pi s \int_s^a C\frac{dy}{y} = s(\pi C \log a + \pi C \log s^{-1}),$$

which is of the form required by (c). We compute and estimate $\frac{d}{ds}$ of (11.32b) similarly, and then (b) and (c) follow.

Applying the lemma with F as in (11.31) and the signs chosen as $+$, we obtain

$$F_f^B(0^+) = \pi i \int_0^\infty f\begin{pmatrix} 1 & u \\ 0 & 1 \end{pmatrix} du \qquad (11.33a)$$

$$\left(\frac{d}{d\theta}F_f^B\right)(0^+) = -2\pi i f(1). \qquad (11.33b)$$

[As matters stand, a general f in $C_{\text{com}}^\infty(G)$ with $f(x) = f(kxk^{-1})$ does not necessarily lead to F satisfying the support condition in Lemma 11.1. However, we see from (11.30) that we can adjust f to get the support condition satisfied without changing any of the quantities in (11.33).] Arguing similarly with 0^-, π^+, and π^- and putting the results together, we obtain

$$F_f^B(0^+) - F_f^B(0^-) = \pi i \int_{-\infty}^{\infty} f\begin{pmatrix} 1 & s \\ 0 & 1 \end{pmatrix} ds \tag{11.34a}$$

$$F_f^B(\pi^+) - F_f^B(\pi^-) = -\pi i \int_{-\infty}^{\infty} f\left(-\begin{pmatrix} 1 & s \\ 0 & 1 \end{pmatrix}\right) ds \tag{11.34b}$$

$$\frac{d}{d\theta} F_f^B(0^+) = \frac{d}{d\theta} F_f^B(0^-) = -2\pi i f(1). \tag{11.35}$$

Since $f(kxk^{-1}) = f(x)$, we can rewrite (11.34) by Lemma 10.10 as

$$F_f^B(0^+) - F_f^B(0^-) = \pi i F_f^T(a_0) \tag{11.36a}$$

$$F_f^B(\pi^+) - F_f^B(\pi^-) = \pi i F_f^T(-a_0). \tag{11.36b}$$

Although F_f^B has jump discontinuities at $\theta = 0$ and π, we still have the derivative formula (11.35) analogous to (11.8). By contrast the function F_f^T extends to a member of $C_{\text{com}}^\infty(T)$, by Lemma 10.10; moreover, it is an even function and thus $\dfrac{d}{dt} F_f^T(0) = 0$.

With this much information about F_f^B and F_f^T, we can see how to adjust the argument for SU(2) and SL(2, $\mathbb{C}$) so as to handle SL(2, $\mathbb{R}$). Formula (11.35) says we can recover $f(1)$ from $\dfrac{d}{d\theta} F_f^B$. From (10.10) and (11.28), we can write the discrete series sum of characters as

$$\begin{aligned} \Theta_{n+1}(f) &= \int_B (e^{in\theta} - e^{-in\theta}) F_f^B(\theta) \frac{d\theta}{2\pi} \\ &= +\tfrac{1}{2} \int_T \{e^{nt}(1 - \operatorname{sgn} t) + e^{-nt}(1 + \operatorname{sgn} t)\} F_f^T(\pm a_t)(\pm)^n \, d(\pm)\, dt. \end{aligned} \tag{11.37}$$

We can attempt to get $\dfrac{d}{d\theta} F_f^B$ by summing a Fourier series after doing an integration by parts in the first term on the right. However, we should remember that F_f^B has jump discontinuities, and the jumps will enter the formula when we integrate by parts. These jumps are given in terms of

F_f^T, in view of (11.36). Hence (11.37) allows us to express $f(1)$ in terms of discrete series characters and F_f^T.

But also we can express F_f^T in terms of principal series characters by applying Fourier inversion to (11.29). Thus in principle we can express $f(1)$ in terms of discrete series characters and principal series characters. This is the formula we seek. We require the details.

First we write out the formula for F_f^T. Formula (11.29) and Fourier inversion together give

$$F_f^T(+a_t) - F_f^T(-a_t) = \frac{1}{2\pi}\int_{-\infty}^{\infty} 2\Theta_{+,iv}(f)e^{-ivt}\,dv$$

and $$F_f^T(+a_t) + F_f^T(-a_t) = \frac{1}{2\pi}\int_{-\infty}^{\infty} 2\Theta_{-,iv}(f)e^{-ivt}\,dv.$$

Thus $$F_f^T(+a_t) = \frac{1}{2\pi}\int_{-\infty}^{\infty} [\Theta_{-,iv}(f) + \Theta_{+,iv}(f)]e^{-ivt}\,dv$$

$$F_f^T(-a_t) = \frac{1}{2\pi}\int_{-\infty}^{\infty} [\Theta_{-,iv}(f) - \Theta_{+,iv}(f)]e^{-ivt}\,dv. \tag{11.38}$$

Now we are in a position to derive the Plancherel formula explicitly. Multiplying (11.37) through by n with $n > 0$, we expand out $d(\pm)$ and then integrate by parts on intervals of continuity:

$$\begin{aligned} n\Theta_{n+1}(f) &= \frac{n}{2\pi}\int_0^{2\pi} (e^{in\theta} - e^{-in\theta})F_f^B(\theta)\,d\theta \\ &\quad + \frac{n}{4}\int_{-\infty}^{\infty} \{e^{nt}(1-\operatorname{sgn} t) + e^{-nt}(1+\operatorname{sgn} t)\}F_f^T(+a_t)\,dt \\ &\quad + \frac{n}{4}(-1)^n\int_{-\infty}^{\infty} \{e^{nt}(1-\operatorname{sgn} t) + e^{-nt}(1+\operatorname{sgn} t)\}F_f^T(-a_t)\,dt \\ &= \frac{1}{\pi i}[F_f^B(0^-) - (-1)^nF_f^B(\pi^+) + (-1)^nF_f^B(\pi^-) - F_f^B(0^+)] \\ &\quad - \frac{1}{2\pi i}\int_0^{2\pi} (e^{in\theta} + e^{-in\theta})\frac{d}{d\theta}F_f^B(\theta)\,d\theta \\ &\quad + F_f^T(a_0) + \frac{1}{2}\int_{-\infty}^{\infty} e^{-n|t|}(\operatorname{sgn} t)\frac{d}{dt}F_f^T(a_t)\,dt \\ &\quad + (-1)^nF_f^T(-a_0) + \frac{1}{2}(-1)^n\int_{-\infty}^{\infty} e^{-n|t|}(\operatorname{sgn} t)\frac{d}{dt}F_f^T(-a_t)\,dt. \end{aligned}$$

By (11.36), the jumps cancel exactly. Also we need a term for $n = 0$, and we use

$$-\frac{1}{2\pi i}\int_0^{2\pi}\frac{d}{d\theta}F_f^B(\theta)\,d\theta = -\frac{1}{2\pi i}\left(\int_\pi^{2\pi}+\int_0^{\pi}\right)$$
$$= -\frac{1}{2\pi i}[F_f^B(0^-)-F_f^B(\pi^+)+F_f^B(\pi^-)-F_f^B(0^+)]$$
$$= \tfrac{1}{2}[F_f^T(a_0)+F_f^T(-a_0)] \qquad \text{by (11.36).}$$

Summing over n, we have

$$\sum_{n=2}^{\infty}(n-1)\Theta_n(f) = i\sum_{k=-\infty}^{\infty}\frac{1}{2\pi}\int_0^{2\pi}e^{ik\theta}\frac{d}{d\theta}F_f^B(\theta)\,d\theta$$
$$-\frac{1}{2}[F_f^T(-a_0)+F_f^T(-a_0)]$$
$$+\frac{1}{2}\sum_{n=1}^{\infty}\int_{-\infty}^{\infty}e^{-n|t|}(\operatorname{sgn} t)\frac{d}{dt}F_f^T(a_t)\,dt$$
$$+\frac{1}{2}\sum_{n=1}^{\infty}(-1)^n\int_{-\infty}^{\infty}e^{-n|t|}(\operatorname{sgn} t)\frac{d}{dt}F_f^T(-a_t)\,dt.$$

Lemma 11.5c allows us to use Dini's Test to see that the Fourier series of $\frac{d}{d\theta}F_f^B$ sums to $\frac{d}{d\theta}F_f^B$ at $\theta = 0$. Hence (11.35) gives

$$2\pi f(1) = \sum_{n=2}^{\infty}(n-1)\Theta_n(f)+\frac{1}{2}F_f^T(a_0)$$
$$-\frac{1}{2}\sum_{n=1}^{\infty}\int_{-\infty}^{\infty}e^{-n|t|}(\operatorname{sgn} t)\frac{d}{dt}F_f^T(a_t)\,dt$$
$$+\frac{1}{2}F_f^T(-a_0)-\frac{1}{2}\sum_{n=1}^{\infty}(-1)^n\int_{-\infty}^{\infty}e^{-n|t|}(\operatorname{sgn} t)\frac{d}{dt}F_f^T(-a_t)\,dt.$$

We define a real Fourier transform by $\hat{h}(\xi) = \int_{-\infty}^{\infty}h(x)e^{-i\xi x}\,dx$. If g and h are in $L^1(\mathbb{R})$, then it is a one-line consequence of Fubini's Theorem that $\int_{-\infty}^{\infty}g\hat{h}\,dx = \int_{-\infty}^{\infty}\hat{g}h\,dx$. We apply this identity with

$$g(t) = e^{-n|t|}\operatorname{sgn} t$$
$$h(v) = -\frac{iv}{2\pi}[\Theta_{-,iv}(f)\pm\Theta_{+,iv}(f)],$$

using the Fourier transform formulas

$$\hat{g}(v) = \frac{-2iv}{n^2+v^2}$$

$$\hat{h}(t) = \frac{d}{dt}F_f^T(\pm a_t),$$

which follow, respectively, by direct calculation and by differentiation of (11.38). Then we find that

$$2\pi f(1) = \sum_{n=2}^{\infty} (n-1)\Theta_n(f) + \sum_{\substack{n=-\infty \\ n \text{ odd}}}^{\infty} \frac{1}{2\pi} \int_{-\infty}^{\infty} \frac{v^2}{n^2+v^2} \Theta_{+,iv}(f)\, dv$$

$$+ \sum_{\substack{n=-\infty \\ n \text{ even}}}^{\infty} \frac{1}{2\pi} \int_{-\infty}^{\infty} \frac{v^2}{n^2+v^2} \Theta_{-,iv}(f)\, dv. \tag{11.39}$$

To get our formula, we have only to evaluate

$$\sum_{\substack{n=-\infty \\ n \text{ odd}}}^{\infty} \frac{v^2}{n^2+v^2} \qquad \text{and} \qquad \sum_{\substack{n=-\infty \\ n \text{ even}}}^{\infty} \frac{v^2}{n^2+v^2}.$$

We use the identity

$$\pi \cot \pi z = \frac{1}{z} + \sum_{n=1}^{\infty} \left(\frac{1}{z-n} + \frac{1}{z+n} \right) = \sum_{n=-\infty}^{\infty} \frac{z}{z^2-n^2}$$

and find that

$$\sum_{\substack{n=-\infty \\ n \text{ even}}}^{\infty} \frac{v^2}{n^2+v^2} = \frac{\pi i v}{2} \cot \frac{\pi i v}{2} = \frac{\pi v}{2} \coth \pi v/2$$

and

$$\sum_{\substack{n=-\infty \\ n \text{ odd}}}^{\infty} \frac{v^2}{n^2+v^2} = \pi i v \cot \pi i v - \frac{\pi i v}{2} \cot \frac{\pi i v}{2} = \frac{\pi v}{2} \tanh \frac{\pi v}{2}.$$

Substituting into (11.39), we obtain the desired Plancherel formula for $SL(2, \mathbb{R})$, as follows.

Theorem 11.6. For f in $C_{\text{com}}^{\infty}(SL(2, \mathbb{R})$ and Haar measure normalized as in (10.7),

$$2\pi f(1) = \sum_{n=2}^{\infty} (n-1)\Theta_n(f) + \frac{1}{4} \int_{-\infty}^{\infty} \Theta_{+,iv}(f) v \tanh(\pi v/2)\, dv$$

$$+ \frac{1}{4} \int_{-\infty}^{\infty} \Theta_{-,iv}(f) v \coth(\pi v/2)\, dv. \tag{10.40}$$

§4. Ingredients of Proof for General Case

To arrive at a Plancherel formula for an arbitrary linear connected reductive G, we should expect to incorporate aspects of all three proofs of the previous three sections. Let us begin by examining the three ingredients that were singled out in §§2–3:

Normalization of Haar measure. The Plancherel measure $d\mu(\omega)$ in (11.9) depends on the normalization of Haar measure for G, which enters the definition of $\pi_\omega(f)$. We shall not attempt to fix choices of Haar measures here but shall simply indicate that concrete normalizations are available. For compact subgroups of G, such as K and $M_\mathfrak{p}$, we can require that Haar measure have total mass one. In view of (5.25), it is then natural to fix Haar measure on the various subgroups $\bar{N}$ of G by pre-assignment of the integral of $e^{-2\rho H(\bar{n})}$. Haar measure on the corresponding subgroup N can be taken as Θ of Haar measure on $\bar{N}$.

The Lie algebra $\mathfrak{a}_\mathfrak{p}$ inherits an inner product from the real part of the trace form and gets identified isometrically with Euclidean space. Therefore we can carry Lebesgue measure from the Euclidean space and its subspaces to $\mathfrak{a}_\mathfrak{p}$ and its subspaces. Exponentiating, we obtain natural normalizations of Haar measure on each of the various subgroups A. Finally we can pass to G and to the various subgroups M so as to respect the normalization (5.17) suggested by the Iwasawa decomposition.

The functions F_f. For simplicity we assume that G is contained in a simply connected complexification $G^\mathbb{C}$ or at least that G is such that half the sum of the positive roots is analytically integral on $G^\mathbb{C}$. As in §10.3, let $T_1, \ldots, T_s$ be a complete set of nonconjugate Θ-stable Cartan subgroups of G. For each $T = T_j$, we define

$$F_f^{G/T}(h) = \varepsilon_R^T(h) D_T(h) \int_{G/T} f(xhx^{-1})\, d\dot{x} \qquad \text{for } h \in C_{\text{com}}^\infty(G)$$

just as in §10.3. The Weyl integration formula (Proposition 5.27) allows us to write the integral of f over G as a sum of suitable integrals (10.25c) of the various F_f^{G/T_j}, $1 \le j \le s$. To make maximum use of the formula, we need some properties of the functions F_f. We list them in qualitative form now and return to consider them in more detail in §6. When precisely formulated, they clarify the roles of the coefficients $\varepsilon_R^T(h)$ and $D_T(h)$.

(1) Smoothness properties. We shall see that F_f^T is smooth on the regular set T' and has reasonable limiting behavior on the singular set. (We have seen from the examples that good control of the singularities is needed because we have to do Euclidean Fourier inversion with F_f^T and also handle an integration by parts.)

(2) Reduction formulas for F_f^T. If T is noncompact, Lemma 10.17 gives a formula for F_f^T in terms of an F_f for a compact Cartan subgroup in a lower-dimensional group G.

(3) Control of jumps. Whenever the two-Cartan situation of $SL(2, \mathbb{R})$ is imbedded in a pair of Cartan subgroups of G, then we shall see that the jump formulas (11.36) imbed also. Specifically if we have two Θ-stable Cartan subalgebras

$$\mathfrak{t}_1 = \mathfrak{t}^\perp \oplus \mathfrak{b}_{SL} \qquad \text{and} \qquad \mathfrak{t}_2 = \mathfrak{t}^\perp \oplus \mathfrak{a}_{SL}$$

with $\mathfrak{b}_{SL} \subseteq \mathfrak{k}$, $\mathfrak{a}_{SL} \subseteq \mathfrak{p}$, and $\mathfrak{b}_{SL}$ and $\mathfrak{a}_{SL}$ contained in a common $\mathfrak{sl}(2, \mathbb{R})$, then the jump discontinuities of $F_f^{T_1}$ at generic points of $\exp \mathfrak{t}^\perp$ will be given by the limiting values of $F_f^{T_2}$ on $\exp \mathfrak{t}^\perp$. These relationships will allow us to carry out integrations by parts even though the integrand may be discontinuous.

(4) Connection with $f(1)$. Let $\mathfrak{h}_j$ be the Lie algebra of T_j, and choose a positive system within the root system $\Delta(\mathfrak{h}_j^{\mathbb{C}}:\mathfrak{g}^{\mathbb{C}})$. As in §2, we have a first-order differential operator $\partial(\alpha)$ on T_j for each α in $\Delta^+(\mathfrak{h}_j^{\mathbb{C}}:\mathfrak{g}^{\mathbb{C}})$; $\partial(\alpha)$ is identified with the element H_α of $U(\mathfrak{h}_j^{\mathbb{C}})$. Define

$$\partial(\varpi_{T_j}) = \prod_{\alpha \in \Delta^+(\mathfrak{h}_j^{\mathbb{C}}:\mathfrak{g}^{\mathbb{C}})} \partial(\alpha).$$

Then we shall have a theorem to the effect that

$$\partial(\varpi_{T_j}) F_f^{T_j}(1) = c_{T_j} f(1)$$

for all f in $C_{\mathrm{com}}^\infty(G)$, that $c_{T_j} = 0$ unless T_j is as compact as possible, and that $c_{T_j} \neq 0$ in the maximally compact case.

Character formulas. Fix attention on one of the Cartan subgroups $T = T_j$. Following the construction of §5.5, we can build a parabolic subgroup $S = MAN$ from T such that $T \subseteq MA$ and $T \cap M$ is compact. The groups M and A are unique, but N is determined uniquely only after fixing a positive system for the roots of $\mathfrak{g}$ with respect to $\mathfrak{a}$. Since M has a compact Cartan subgroup, namely $T \cap M$, M has discrete series, by an easy argument based on Theorem 9.20 and given in Chapter XII. To each discrete series σ of M and imaginary-valued linear functional ν on $\mathfrak{a}$, we can form the induced unitary representation $U(S, \sigma, \nu)$. In §10.3 the global character $\Theta_{\sigma,\nu}$ of $U(S, \sigma, \nu)$ is computed in terms of the global character of σ.

It is the characters $\Theta_{\sigma,\nu}$, as σ, ν, and T_j vary, that enter the Plancherel formula of G. [We shall see that changing N has no effect on the equivalence class of $U(S, \sigma, \nu)$. In principle, this fact could be sorted out from the formula for $\Theta_{\sigma,\nu}$, but we can see it without any computation from Theorem 8.38a and a normalization argument to be given in Chapter XIV.]

A key property of the character formulas, from the point of view of the Plancherel formula, is given in Proposition 10.19: The character $\Theta_{\sigma,\nu}$ is nonvanishing only on Cartan subgroups that are conjugate within G to Cartan subgroups of MA. Nevertheless, we do not yet have available a completely explicit formula for $\Theta_{\sigma,\nu}$ because we do not yet have character formulas for discrete series.

All that we know so far about discrete series characters is what we know about general irreducible characters. They are given by locally integrable functions analytic on the regular set (Theorem 10.36). Relative to the Cartan subgroup T if the infinitesimal character is λ, the numerator of the character is of the form

$$\tau_T(t \exp H) = \sum_{w \in W(\mathfrak{t}^{\mathbb{C}}:\mathfrak{g}^{\mathbb{C}})} c_w(t) e^{w\lambda(H)}, \; t \in T, \, t \exp H \text{ in component of } T',$$

with $c_w(t)$ locally constant in H (Theorems 10.35 and 10.48). The denominator is the usual Weyl denominator $D_T(t \exp H)$. The product $\varepsilon_R^T(t)\tau_T(t)$ extends to be real analytic on the set $T'_R \supseteq T'$ (Theorem 10.40) and to be continuous on T.

§5. Scheme of Proof for General Case

Despite not yet having explicit formulas for discrete series characters, we can set up a rough scheme of proof of the Plancherel formula in general on the basis of the examples in §§1–3 and the facts assembled in §4.

Let us work within $\mathrm{Sp}(2, \mathbb{R})$. This group has four nonconjugate Cartan subalgebras, and they are listed explicitly in §5.4. The one with parameters θ_1 and θ_2 we denote $\mathfrak{b}$, the one on the lower left with parameters θ and t we denote $\mathfrak{t}_1$, the one on the upper right with parameters θ and t we denote $\mathfrak{t}_2$, and the one with parameters s and t we denote $\mathfrak{t}$. Let B, T_1, T_2, and T be the corresponding Cartan subgroups. We can order these subgroups by saying that one is $\leq$ another if the first is G-conjugate to a subgroup of the MA associated to the second. The result is a partial ordering that is described by Figure 11.1.

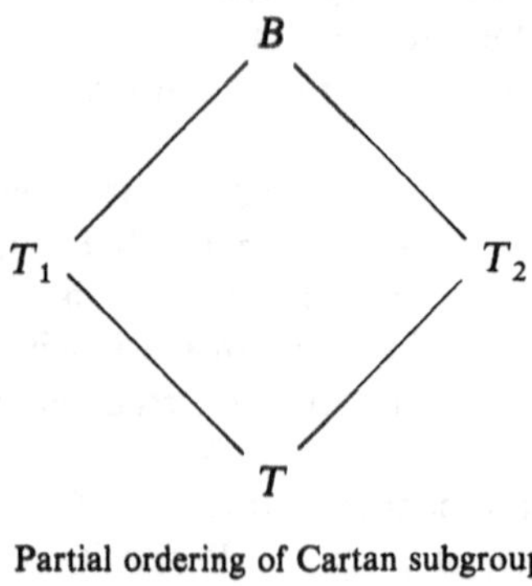

FIGURE 11.1. Partial ordering of Cartan subgroups of $\mathrm{Sp}(2, \mathbb{R})$

The root system relative to any of the Cartan subalgebras is of type C_2. Let us examine $\Delta(\mathfrak{b}^{\mathbb{C}}:\mathfrak{g}^{\mathbb{C}})$ more closely. Write e_1 and e_2 for the linear functionals on $\mathfrak{b}^{\mathbb{C}}$ giving $i\theta_1$ and $i\theta_2$, respectively. The roots are $\pm e_1 \pm e_2$, $\pm 2e_1$, and $\pm 2e_2$. The defining matrix J of the symplectic group (see

§1.1) is in the center of $\mathfrak{k}$, and the compact roots are characterized in Lemma 6.1 as the ones that vanish on the center of $\mathfrak{k}$. Thus $\pm(e_1 - e_2)$ are compact, and the other roots are noncompact. The smoothness properties of F_f^B alluded to in the previous section will say that F_f^B is smooth across the locus in B where $\xi_\alpha(b) = 1$ only for compact roots (in analogy with what happens in SL(2, $\mathbb{R}$)) but that jumps should be expected whenever $\xi_\alpha(b) = 1$ for some noncompact root. Since B is just a torus, parametrized by $|\theta_1| \leq \pi$ and $|\theta_2| \leq \pi$, the possible singularities of F_f^B are as in Figure 11.2.

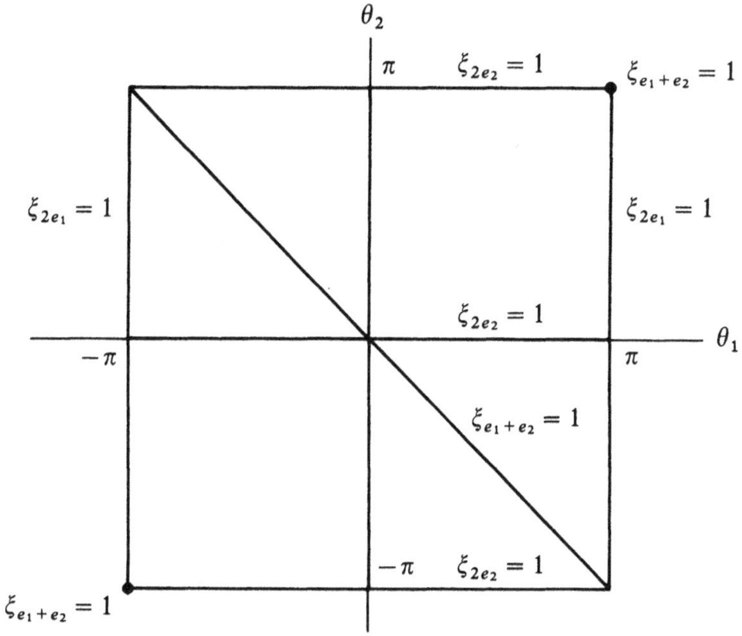

FIGURE 11.2. Singular set for F_f^B in Sp(2, $\mathbb{R}$)

The jump of F_f^B across the locus where $\xi_{2e_1} = 1$ or $\xi_{2e_2} = 1$ is given in terms of $F_f^{T_1}$ (where the t parameter is 0), and the jump across the locus where $\xi_{e_1+e_2} = 1$ is given in terms of $F_f^{T_2}$. When the θ parameter of $F_f^{T_1}$ or $F_f^{T_2}$ is 0, there may be a jump also, and the size of the jump is given in terms of F_f^T.

Reviewing the proof of the Plancherel formula for SL(2, $\mathbb{R}$), we see that we should expect to work with the (double) Fourier series of F_f^B. Since F_f^B is usually not continuous, the series does not converge absolutely, and the manner in which the terms are summed may be relevant. Let us pay only a little attention to this problem for now, taking it up further in Chapter XIII.

The terms (10.25c) of the Weyl integration formula involve certain constants, and we collect them for the Cartan subgroups H of Sp(2, $\mathbb{R}$) in Table 11.1.

TABLE 11.1 Constants in Weyl integration formula of Sp(2, $\mathbb{R}$)

	$\lvert W(H:G)\rvert$	$\lvert\Delta_I^+\rvert$	$\varepsilon^{G/H} = (-1)^{\lvert\Delta_I^+\rvert}$
B	2	4	+1
T_1	2	1	−1
T_2	4	1	−1
T	8	0	+1

Let U be one of the representations discussed in §4, and let H be one of the four Cartan subgroups. Let Θ_H be the character of U written as a function and restricted to H, and let τ_H be the numerator of the character on H. For f in $C_{\text{com}}^\infty(\text{Sp}(2, \mathbb{R}))$, (10.25) gives

$$\begin{aligned}\operatorname{Tr} U(f) &= \int_G f(x)\Theta(x)\,dx \\ &= \tfrac{1}{2}\int_{B\times G/B} \Theta_B(h)|D_B(h)|^2 f(ghg^{-1})\,d\dot{g}\,dh + \tfrac{1}{2}\int_{T_1\times G/T_1}[\text{——}] \\ &\quad + \tfrac{1}{4}\int_{T_2\times G/T_2}[\text{——}] + \tfrac{1}{8}\int_{T\times G/T}[\text{——}] \\ &= \tfrac{1}{2}\int_B \tau_B(h)F_f^B(h)\,dh - \tfrac{1}{2}\int_{T_1} \varepsilon_R^{T_1}(h)\tau_{T_1}(h)F_f^{T_1}(h)\,dh \\ &\quad - \tfrac{1}{4}\int_{T_2} \varepsilon_R^{T_2}(h)\tau_{T_2}(h)F_f^{T_2}(h)\,dh + \tfrac{1}{8}\int_T \varepsilon_R^T(h)\tau_T(h)F_f^T(h)\,dh.\end{aligned}$$

Let us now put a subscript B, T_1, T_2, or T on U to denote the specific series of representations to which U belongs (see §4), and let us take into account the vanishing properties of the character that are forced by Figure 11.1. Then we can write our equations symbolically as

$$\begin{aligned}\operatorname{Tr} U_B(f) &= \tfrac{1}{2}\int_B - \tfrac{1}{2}\int_{T_1} - \tfrac{1}{4}\int_{T_2} + \tfrac{1}{8}\int_T \\ \operatorname{Tr} U_{T_1}(f) &= \ 0 \ - \tfrac{1}{2}\int_{T_1} + \ 0 \ + \tfrac{1}{8}\int_T \\ \operatorname{Tr} U_{T_2}(f) &= \ 0 \ + \ 0 \ - \tfrac{1}{4}\int_{T_2} + \tfrac{1}{8}\int_T \\ \operatorname{Tr} U_T(f) &= \ 0 \ + \ 0 \ + \ 0 \ + \tfrac{1}{8}\int_T.\end{aligned}$$

The triangular nature of these equations makes it possible to imitate the argument for SL(2, $\mathbb{R}$). We start with $\operatorname{Tr} U_B(f)$. If the infinitesimal character of U_B is λ (relative to one of our Cartan subgroups H), then

$$\partial(\varpi_H)\varepsilon_R^H\tau_H = \left\{\prod_{\alpha\in\Delta^+(\mathfrak{h}^{\mathbb{C}}:\mathfrak{g}^{\mathbb{C}})} \langle\lambda, \alpha\rangle\right\}\varepsilon_R^H\tau_H \tag{11.41}$$

because of the explicit formula for τ_H. We multiply the equation for $\operatorname{Tr} U_B(f)$ through by $\prod \langle \lambda, \alpha \rangle$ and introduce the differential operator of (11.41) in each term on the right side of the equation. Then we integrate by parts, taking into account the discontinuities of F_f^H; the factor $\varepsilon_R^H \tau_R$ has no discontinuities, by Theorem 10.40. Our jump formulas for F_f^H should make the total effect of the discontinuities drop out, and we should have

$$\left\{\prod_{\alpha \in \Delta^+(\mathfrak{b}^{\mathbb{C}}:\mathfrak{g}^{\mathbb{C}})} \langle \lambda, \alpha \rangle\right\} \operatorname{Tr} U_B(f) = \tfrac{1}{2} \int_B \tau_B\, \partial(\varpi_B) F_f^B \, dh$$
$$- \tfrac{1}{2} \int_{T_1} \varepsilon_R^{T_1} \tau_{T_1}\, \partial(\varpi_{T_1}) F_f^{T_1} \, dh$$
$$- \tfrac{1}{4} \int_{T_2} \varepsilon_R^{T_2} \tau_{T_2}\, \partial(\varpi_{T_2}) F_f^{T_2} \, dh$$
$$+ \tfrac{1}{8} \int_{T} \varepsilon_R^{T} \tau_{T}\, \partial(\varpi_{T}) F_f^{T} \, dh.$$

Each $\partial(\varpi_H) F_f^H$ is free of discontinuities, with derivatives that have good enough limiting behavior to allow summing over λ.

However, in $SL(2, \mathbb{R})$ the $n = 0$ term was missing at this stage, and we should expect missing λ's here. For them we should introduce canceling jump terms as in $SL(2, \mathbb{R})$. Then we can sum on λ, recognize the sum of the first term on the right side as adding to a multiple of $\partial(\varpi_B) F_f^B(1)$, and substitute a multiple of $f(1)$. Then F_f^B has been eliminated.

Next we substitute for $F_f^{T_1}$, $F_f^{T_2}$, and F_f^T from Lemma 10.17 in order to simplify the remaining F_f's. Then we expect to eliminate $F_f^{T_1}$ in a fashion similar to the elimination of F_f^B, continue by handling $F_f^{T_2}$, and finish up with F_f^T. The details appear hazier as we look farther and farther down the path, but there is some hope of success. The one thing that is certain is that we cannot push through the argument without more explicit character formulas.

§6. Properties of F_f

In this section we discuss in more detail the properties of F_f outlined in §4. As in §4, we assume that G is such that half the sum of the positive roots is analytically integral on $G^{\mathbb{C}}$. We denote a typical Θ-stable Cartan subgroup by T, and we let T' be the subset of regular elements. Further notation is as in §10.3.

The results in this section allow us incidentally to complete the proof of Lemma 10.20. After giving the proof, we will note a route to follow through our development in order to avoid any circularity.

The first property of F_f is its smoothness on the regular set.

Proposition 11.7. For each Θ-stable Cartan subgroup T, F_f^T is in $C^\infty(T')$ if f is in $C_{\text{com}}^\infty(G)$. Moreover, F_f^T vanishes outside a bounded subset of T.

Proof of smoothness. Fix h_0 in T', and let E be a compact neighborhood of h_0 in T'. The Weyl integration formula reflects the fact that the map of $G/T \times T'$ into $(T')^G$ given by $(g, h) \to ghg^{-1}$ has Jacobian determinant $|D_T(h)|^2$. Thus the map is locally a diffeomorphism and is necessarily an open map. Moreover, the fiber over the image point ghg^{-1} consists of $|W(T:G)|$ pre-images, namely the elements $(gwT, w^{-1}hw)$ as w ranges through a system of representatives of $W(T:G)$. Hence $(g, h) \to ghg^{-1}$ is a closed map into $(T')^G$. Thus

$$E^G = \{ghg^{-1} \mid g \in G \text{ and } h \in E\}$$

is closed in $(T')^G$. Referring to the definition of regular element, we see that the closure of E^G in G contains no singular elements. Thus Proposition 5.22d allows us to conclude that E^G is closed in G. Hence its intersection with the support of f is compact. It follows that there exists f_1 in $C_{\text{com}}^\infty((T')^G)$ such that f_1 agrees with f on E^G, i.e., such that $F_{f_1}^T$ agrees with F_f^T on E.

Thus the asserted smoothness in the proposition is reduced to the case of the function f_1 supported in $(T')^G$. The properties of the map $(g, h) \to ghg^{-1}$ imply that the inverse image of a compact set is compact. Thus $(g, h) \to f_1(ghg^{-1})$ is in $C_{\text{com}}^\infty(G/T \times T')$. Integration in the first variable yields a C_{com}^∞ function in the second variable. Hence $F_{f_1}^T$ is in $C_{\text{com}}^\infty(T')$.

The second assertion in the proposition is proved by reduction to the case that T is compact. Such a reduction is a frequently used technique with F_f and is based on Lemma 10.17. Let us review the technique in the context of proving the second assertion of the proposition. As in §10.3, we decompose the Lie algebra $\mathfrak{t}$ of T as $\mathfrak{t} = \mathfrak{a} \oplus \mathfrak{b}$ with $\mathfrak{a} \subseteq \mathfrak{p}$ and $\mathfrak{b} \subseteq \mathfrak{k}$. In the usual way (see §5.5) we construct $\mathfrak{m}$ so that $Z_{\mathfrak{g}}(\mathfrak{a}) = \mathfrak{m} \oplus \mathfrak{a}$. Then $\mathfrak{b}$ is a compact Cartan subalgebra of $\mathfrak{m}$.

Choose a positive system of roots of $(\mathfrak{g}, \mathfrak{a})$, and form the corresponding subalgebra $\mathfrak{n}$ and the parabolic subalgebra $S = MAN$. Except for the question of orderings, the fact that $T \subseteq MA$ means that Lemma 10.17 applies and gives

$$\xi_{-\delta_M}(h) F_{f^{(S)}}^{MA/T}(h) = e^{\rho H(h)} \xi_{-\delta}(h) F_f^{G/T}(h) \tag{11.42}$$

whenever h is G-regular; if the orderings are not compatible, this equality may be off by a minus sign independent of h. Now $f^{(S)}$ is clearly in $C_{\text{com}}^\infty(MA)$, and the closure of the set

$$\{a \in A \mid f^{(S)}(ma) \neq 0 \text{ for some } m \in M\}$$

is compact and contains the closure of

$$\{a \in A \mid F_{f(s)}^{MA/T}(ba) \neq 0 \text{ for some } b \text{ in } T \cap K\}.$$

The support of $F_{f(s)}^{MA/T}$ is contained in the product of $T \cap K$ with the latter set and therefore is a bounded subset of T. From (11.42) we conclude that F_f^T vanishes outside a bounded subset of T.

Euclidean Fourier inversion of F_f played an important role in the previous three sections. Since singularities affect Fourier inversion, it is important to identify the extent to which F_f^T extends to be smooth on T. Let us again write $\mathfrak{t} = \mathfrak{a} \oplus \mathfrak{b}$ and form $\mathfrak{m}$ as above. The imaginary roots of $\Delta(\mathfrak{t}^{\mathbb{C}}:\mathfrak{g}^{\mathbb{C}})$, in the sense of §10.3, are identified via restriction to $\mathfrak{b}^{\mathbb{C}}$ with the roots of $\Delta(\mathfrak{b}^{\mathbb{C}}:\mathfrak{m}^{\mathbb{C}})$. We say such an imaginary root of $\Delta(\mathfrak{t}^{\mathbb{C}}:\mathfrak{g}^{\mathbb{C}})$ is **compact** or **noncompact** according as the corresponding member of $\Delta(\mathfrak{b}^{\mathbb{C}}:\mathfrak{m}^{\mathbb{C}})$ is compact or noncompact. A compact imaginary root α is one such that the intersection of $\mathfrak{g}$ with the three-dimensional subalgebra of $\mathfrak{g}^{\mathbb{C}}$ built from α is isomorphic to $\mathfrak{su}(2)$; a noncompact imaginary root α is one for which this intersection is isomorphic to $\mathfrak{sl}(2, \mathbb{R})$.

Proposition 11.8. Let f be in $C_{\text{com}}^\infty(G)$. Then $F_f^{G/T}(h)$, initially defined as a C^∞ function for h in T', extends to a C^∞ function on the set of all h in T such that $\xi_\alpha(h) \neq 1$ for any noncompact imaginary root in $\Delta(\mathfrak{t}^{\mathbb{C}}:\mathfrak{g}^{\mathbb{C}})$.

Notation. We denote by T'' the set where $\xi_\alpha(h)$ equals one for no noncompact imaginary root α. T'' is open dense in T and has finitely many components.

Proof. The usual reduction argument by means of Lemma 10.17 immediately reduces matters to the case that the Cartan subgroup is compact (and all roots are imaginary). Although the reduction might not lead to a connected group, any disconnectedness is easily handled, and we shall ignore this complication. Thus let B be a compact (connected) Cartan subgroup. We factor the conjugation mapping of $G/B \times B$ to G as

$$G/B \times B \to G/K \times K/B \times B \to G/K \times K \to G; \qquad (11.43)$$

these maps are respectively the Cartan decomposition $G = (\exp \mathfrak{p})K$, the conjugation map $(k, b) \to kbk^{-1}$ on K, and the map $(g, k) \to gkg^{-1}$.

We shall need to know the regularity properties of the third map. Under the natural identifications of tangent spaces, the composition, by the Weyl integration formula, has Jacobian determinant $|D_B^{G/B}(b)|^2$, except possibly for a sign. Noting that

$$\det(1 - \operatorname{Ad}(b)^{-1})_{\mathfrak{g}/\mathfrak{b}} = \prod_{\alpha \in \Delta} (1 - \xi_\alpha(b)^{-1})$$

and referring to (10.25a) and its proof, we see that

$$|D_B^{G/B}(b)|^2 = \det(1 - \mathrm{Ad}(b)^{-1})_{\mathfrak{g}/\mathfrak{b}}. \tag{11.44}$$

Meanwhile the first map in (11.43) is a diffeomorphism, and the second map has Jacobian determinant

$$|D_B^{K/B}(b)|^2 = \det(1 - \mathrm{Ad}(b)^{-1})_{\mathfrak{k}/\mathfrak{b}}$$

by a calculation similar to the one for (11.44). It follows that the third map, carrying (g, k) in $G/K \times K$ to gkg^{-1} in G, has Jacobian determinant equal to the product of a nonvanishing function and $\det(1 - \mathrm{Ad}(k)^{-1})_{\mathfrak{p}}$.

Let

$$K'' = \{k \in K \mid \det(1 - \mathrm{Ad}(k)^{-1})_{\mathfrak{p}} \neq 0\}$$
$$B'' = B \cap K''.$$

The set B'' is the set where $\xi_\alpha(b)$ is not one for any noncompact root, and we are to control F_f^B on this set. What we have just seen is that the map $(g, k) \to gkg^{-1}$ of $G/K \times K''$ to G is a local diffeomorphism onto an open set. We show this map is one-one also. Let

$$g_1 k_1 g_1^{-1} = g_2 k_2 g_2^{-1}$$

with k_1 and k_2 in K''. We rewrite this equation as $gk_1g^{-1} = k_2$ and then as

$$(\exp X)k(\exp X)^{-1} = k_2 \qquad \text{with } X \in \mathfrak{p},\ k \in K,\ k_2 \in K.$$

This equation implies

$$k^{-1}(\exp X)k = k^{-1}k_2 \exp X.$$

The left side is in $\exp \mathfrak{p}$, and the uniqueness of the Cartan decomposition implies $\mathrm{Ad}(k)^{-1}X = X$ and $k = k_2$. Since k is in K'', X must be 0. Therefore the map is one-one.

It follows that whenever f is a C^∞ function compactly supported in the open image of $G/K \times K''$, then the function

$$f_1(k) = \int_G f(gkg^{-1})\, dg$$

is a C^∞ function on K''. Arguing as in the proof of Proposition 11.7, we can extend this conclusion to the case that f is arbitrary in $C_{\mathrm{com}}^\infty(G)$. Restricting to B'' and multiplying by $D_B^{G/B}(b)$, we obtain the desired result.

Deeper properties of F_f depend upon how F_f interacts with differential equations. The fundamental property is the one in Proposition 11.9 below.

Proposition 11.9. Let f be in $C_{\mathrm{com}}^\infty(G)$, and let z be in $Z(\mathfrak{g}^{\mathbb{C}})$. For h in T',

$$F_{zf}^T(h) = (\gamma(z)F_f^T)(h),$$

where γ denotes the Harish-Chandra homomorphism relative to T.

Proof. Checking over the estimates in Proposition 11.7, we see that $\mu(f) = F_f^T(h)$ is an invariant distribution on G. To see the effect of $Z(\mathfrak{g}^{\mathbb{C}})$ in terms of T, we apply the results of §10.6. In the notation of that section, $\tilde{\mu}$ is the distribution on $G/T \times T'$ given by

$$\tilde{\mu}(\tilde{f}) = \varepsilon_R(h)D_T(h)\int_{G/T}|W(T:G)|^{-1}\sum_{w\in W(T:G)}\tilde{f}(gwT, w^{-1}hw)\,d\dot{g},$$

and μ_T in Lemma 10.28 is evidently therefore

$$\mu_T = |W(T:G)|^{-1}\varepsilon_R(h)D_T(h)\sum_{w\in W(T:G)}\delta_{w^{-1}hw},$$

where δ_p denotes the Dirac distribution at p. Lemma 10.32 and Theorem 10.33 say that

$$F_{zf}^T(h) = \mu_T(D_T^{-1}\gamma(z)(D_T F_2)),$$

where $F_2(h') = \int_{G/T} f(gh'g^{-1})\,d\dot{g}$. Sorting matters out, we obtain the conclusion of the proposition.

Remarks. In the case of $SL(2, \mathbb{R})$, this proposition gives us new information about $F_f^B(\theta)$. Let the usual basis $\{h, e, f\}$ of $\mathfrak{sl}(2, \mathbb{C})$ be computed relative to the compact Cartan subgroup B, with h corresponding to $i\dfrac{d}{d\theta}$ when we parametrize B by $\begin{pmatrix}\cos\theta & \sin\theta\\ -\sin\theta & \cos\theta\end{pmatrix}$. A suitable linear combination z of the Casimir operator and 1 has $\gamma(z) = -h^2$, and thus Proposition 11.9 says

$$\frac{d^2}{d\theta^2}F_f^B(\theta) = F_{zf}^B(\theta).$$

If we apply to zf our known properties of F_f^B from §3, we see that $\dfrac{d^2}{d\theta^2}F_f^B(\theta)$ has at worst a jump discontinuity at $\theta = 0$ and $\dfrac{d^3}{d\theta^3}F_f^B(\theta)$ has equal right and left limits at $\theta = 0$. Iterating this calculation by using $z^k f$ in place of zf, we see that all even order derivatives at $\theta = 0$ have at worst jump discontinuities, while the odd order derivatives at $\theta = 0$ have equal right and left limits. The formulas are

$$\frac{d^{2k}}{d\theta^{2k}}F_f^B(0^+) - \frac{d^{2k}}{d\theta^{2k}}F_f^B(0^-) = (-1)^k\pi i\left[\frac{d^{2k}}{dt^{2k}}F_f^T(a_t)\right]_{t=0} \tag{11.45a}$$

$$\frac{d^{2k}}{d\theta^{2k}}F_f^B(\pi^+) - \frac{d^{2k}}{d\theta^{2k}}F_f^B(\pi^-) = (-1)^k\pi i\left[\frac{d^{2k}}{dt^{2k}}F_f^T(-a_t)\right]_{t=0} \tag{11.45b}$$

$$\frac{d^{2k+1}}{d\theta^{2k+1}}F_f^B(0^+) = \frac{d^{2k+1}}{d\theta^{2k+1}}F_f^B(0^-) = 2\pi i(z^k f)(1).$$

Since $F_f^T(\pm a_t)$ is an even function of t, its odd order derivatives are 0. Thus the last equation implies

$$\frac{d^{2k+1}}{d\theta^{2k+1}} F_f^B(0^+) - \frac{d^{2k+1}}{d\theta^{2k+1}} F_f^B(0^-) = c\left[\frac{d^{2k+1}}{dt^{2k+1}} F_f^T(a_t)\right]_{t=0} \tag{11.46a}$$

for any choice of c we might want. Similarly

$$\frac{d^{2k+1}}{d\theta^{2k+1}} F_f^B(\pi^+) - \frac{d^{2k+1}}{d\theta^{2k+1}} F_f^B(\pi^-) = c\left[\frac{d^{2k+1}}{dt^{2k+1}} F_f^T(-a_t)\right]_{t=0}. \tag{11.46b}$$

Equations (11.46) are an odd way to formulate the equivalent condition

$$\frac{d^{2k+1}}{d\theta^{2k+1}} F_f^B(\theta) \text{ is continuous for all } \theta. \tag{11.47}$$

However, (11.46) is the statement that generalizes nicely.

Proposition 11.9 is a remarkably powerful tool. The proof of the next proposition will show how to use this tool to obtain control of derivatives of all orders of F_f^T in the general case. We recall the notation T'' introduced with Proposition 11.8.

Proposition 11.10. If f is in $C_{\text{com}}^\infty(G)$, then the restriction of F_f^T to any component of T'' extends to a C^∞ function on the closure of that component in T.

Remarks. For $SL(2, \mathbb{R})$ we know this result from the above remarks. For the noncompact T, T'' is all of T. For the compact Cartan B, $F_f^B(\theta)$ is a C^∞ function on $0 < \theta < \pi$ and $\pi < \theta < 2\pi$, and its derivatives of all orders have limits at the endpoints of these intervals, as we have just seen.

First part of proof. The first step is to strip away the group theory to expose the heart of the matter as a question in Euclidean analysis. Applying Lemma 10.17, we easily reduce matters to the case that T is compact, say $T = B$. Again the reduction might not lead to a connected group, but the disconnectedness is easily handled. Thus we shall assume B is connected.

From the definition of F_f^B, it is clear that $|F_f^B| \le |F_{|f|}^B|$. Consequently the Weyl integration formula gives

$$\int_B |D_B(b)|\,|F_f^B(b)|\,db \le |W(B:G)|\,\|f\|_1. \tag{11.48}$$

Since B is compact, the Laplacian element Δ of $U(\mathfrak{b}^{\mathbb{C}})$ is invariant under $W(\mathfrak{b}^{\mathbb{C}}:\mathfrak{g}^{\mathbb{C}})$ and therefore satisfies $\Delta = \gamma(z_0)$ for some element z_0 in $Z(\mathfrak{g}^{\mathbb{C}})$.

Combining (11.48) and Proposition 11.9, we see that

$$\int_B |D_B(b)|\,|\Delta^m F_f^B(b)|\,db \le |W(B:G)|\,\|z_0^m f\|_1 \tag{11.49}$$

for every $m \ge 0$.

To reduce from the torus B to Euclidean space, it is necessary to understand the nature of the intersection of B'' with a neighborhood in B. Thus fix b_0 in B and choose a ball E_0 centered at 0 in $\mathfrak{b}$ so small that

$$H \in E_0 \subseteq \mathfrak{b} \to b_0 \exp H$$

gives local coordinates about b_0 and so that every root α with $\xi_\alpha(b_0) \ne 1$ also has $\xi_\alpha(b_0 \exp H) \ne 1$ for all H in E_0. Then the members H of E_0 such that $b_0 \exp H$ is in B'' are those with $\alpha(H) \ne 0$ for every noncompact root α satisfying $\xi_\alpha(b_0) = 1$, i.e., the complement of finitely many hyperplanes. A component of this set is the intersection of a cone C_0 with E_0, and it follows easily from the Fundamental Theorem of Calculus that F_f^B extends as a C^∞ function to the closure of $b_0 \exp(C_0 \cap E_0)$ provided all $U(\mathfrak{b}^{\mathbb{C}})$ derivatives of F_f^B are bounded on $b_0 \exp(C_0 \cap E_0)$. In fact, since the regular set B' is dense in B'', it is enough to prove the boundedness on B'.

Thus we can work directly with B'. The members H of E_0 such that $b_0 \exp H$ is in B' are those with $\alpha(H) \ne 0$ for every root α satisfying $\xi_\alpha(b_0) = 1$. Because we are interested in an arbitrarily small neighborhood of b_0, it is enough to prove the boundedness one component at a time just on the set in B' corresponding to a ball E_1 of half the radius of E_0.

Since B is assumed connected, we can write $b_0 = \exp H_0$ with H_0 in $\mathfrak{b}$. The roots α for which $\xi_\alpha(b_0) = 1$ are those for which $\alpha(H_0)$ is in $2\pi i\mathbb{Z}$. If α and β have this property, so does $s_\alpha(\beta)$, and it follows that the roots with this property form an abstract root system in a subspace of $i\mathfrak{b}$. Using our component of $B' \cap b_0 \exp E_1$ to determine positivity, we let $\alpha_1, \ldots, \alpha_k$ be the simple roots for this system. Our component is then parametrized by the subset $C_1 \cap E_1$ of E_1 where $\alpha_j(iH) > 0$ for $1 \le j \le k$. (Here k may be strictly less than $\dim B$.) The component is $b_0 \exp(C_1 \cap E_1)$, and the pullback $D_B(b_0 \exp \cdot)$ of the Weyl denominator from it is comparable in size with $\prod_{j=1}^k \alpha_j^{n_j}$ for suitable integers $n_j \ge 0$.

The heart of the matter is to prove that any derivative of F_f^B is bounded on the subneighborhood $b_0 \exp(C_1 \cap \frac{1}{2}E_1)$ after multiplication by a suitable monomial in $\alpha_1, \ldots, \alpha_k$. This is a Euclidean problem and is settled by the following lemma, whose proof we defer briefly.

Lemma 11.11. In $\mathbb{R}^n$ let $l_1, \ldots, l_k$ be linearly independent polynomials homogeneous of degree 1, and suppose that the open set C where all $l_j(x)$ are positive is nonempty. Let E_R be the ball of radius R centered at the origin. Let integers $n_j \geq 0$ be given, and let $P(x) = \prod_{j=1}^{k} l_j(x)^{n_j}$. Suppose that F is a C^∞ function on $C \cap E_R$ such that

$$\int_{C \cap E_R} P(x)|\Delta^m F(x)|\, dx = c_m < \infty$$

for all integers $m \geq 0$. Then to each derivative DF of F corresponds a monomial Q_D in $l_1, \ldots, l_k$ such that $Q_D \cdot DF$ is bounded on $C \cap E_{R/2}$, and the bound for $Q_D \cdot DF$ depends on F only through finitely many of the c_m's.

Conclusion of proof of Proposition 11.10. To continue, we shall bring in the effect of other elements of $Z(\mathfrak{g}^{\mathbb{C}})$ besides the powers of z_0. By Theorem 8.19, $U(\mathfrak{b}^{\mathbb{C}})$ is finitely generated over the Weyl group invariants in $U(\mathfrak{b}^{\mathbb{C}})$, say with generators $D_1, \ldots, D_s$ and with $D_1 = 1$. In the notation of Lemma 11.11, put $Q = Q_{D_1} \cdots Q_{D_s}$, and regard F_f^B as defined on $C \cap E_R$. The claim is that $Q \cdot DF_f^B$ is bounded on the set $C \cap E_{R/2}$ in the lemma, for all D in $U(\mathfrak{b}^{\mathbb{C}})$. In fact, we just write $D = \sum_{j=1}^{s} D_j \gamma(z_j)$ for suitable z_j in $Z(\mathfrak{g}^{\mathbb{C}})$. As a result of the lemma, we know that $Q_j \cdot D_j F_f^B$ is bounded on $C \cap E_{R/2}$, and we can apply this conclusion with $z_j f$ in place of f. By Proposition 11.9, $Q_j \cdot D_j \gamma(z_j) F_f^B$ is bounded on $C \cap E_{R/2}$. Hence $Q \cdot DF_f^B$ is bounded on $C \cap E_{R/2}$.

Now let us return to the group notation. The effect of the lemma and the above remarks is that there is a single monomial Q in $\alpha_1, \ldots, \alpha_k$ such that the product of Q with any derivative of F_f^B is bounded on $b_0 \exp(C \cap \frac{1}{2}E_1)$. From this conclusion we want to see that each derivative of F_f^B is itself bounded. This is a simple exercise with the Fundamental Theorem of Calculus applied one variable at a time, since integration in a variable x_j brings the function one power of x_j closer to being bounded (except for logarithmic factors). We omit the easy details.

Proof of Lemma 11.11. We may assume that $k \geq 1$. We shall prove the boundedness of $Q_1 F$ under the assumption that $l_1 = x_1, \ldots, l_k = x_k$. We omit the details for $Q_D \cdot DF$ and for general l's; only technical improvements in the argument are needed to handle these matters. Clearly we may take $R = 1$.

We shall use the existence of a fundamental solution for Δ^m in $\mathbb{R}^n$. The relevant formulas, valid in the sense of distributions on $C_{\text{com}}^\infty(\mathbb{R}^n)$, are

$$\Delta^{l+(n-1)/2}|x|^{2l-1} = c_l \delta_0 \qquad \text{for } n \text{ odd}, l \geq 1 \tag{11.50a}$$

$$\Delta^{l+n/2}(|x|^{2l} \log|x|) = c_l' \delta_0 \qquad \text{for } n \text{ even}, l \geq 0. \tag{11.50b}$$

Here c_l and c_l' are nonzero numbers, and δ_0 denotes the Dirac distribution at the origin.

Let m be an integer large enough so that (11.50) applies to Δ^m, and let $g(x)$ be the fundamental solution of Δ^m indicated by (11.50). Let $\psi(r)$ be a

C^∞ function from $\mathbb{R}^+$ into $[0, 1]$ that is 1 for $0 \le r \le 2$ and is 0 for $r \ge 3$. Define

$$\Psi_\varepsilon(x) = \psi(\varepsilon^{-1}|x|) \qquad \text{for } x \in \mathbb{R}^n,\ \varepsilon > 0$$

and put

$$g_\varepsilon(x) = \Psi_\varepsilon(x) g(x).$$

Then

$$\Delta^m g_\varepsilon = \delta_0 + \beta_\varepsilon, \tag{11.51}$$

where β_ε is in $C^\infty_{\text{com}}(\mathbb{R}^n)$ and satisfies

$$\text{support } \beta_\varepsilon \subseteq \text{support } \Psi_\varepsilon \subseteq \{x \mid 2\varepsilon \le |x| \le 3\varepsilon\}. \tag{11.52}$$

The function β_ε is the sum of products of derivatives of Ψ_ε and derivatives of g away from 0, and thus we easily read off the inequality

$$|\beta_\varepsilon(x)| \le c|x|^{-p(m)}, \qquad |x| \le 1$$

for a constant c and a power $p(m)$ independent of ε. Taking into account (11.52), we see that

$$\sup_{x \in \mathbb{R}^n} |\beta_\varepsilon(x)| \le c\varepsilon^{-p(m)} \tag{11.53}$$

for a possibly different constant c.

Fix x_0 to be studied, writing $x_0 = (x_{1,0}, \ldots, x_{n,0})$, and let

$$\varepsilon = \tfrac{1}{4} \min\{x_{1,0}, \ldots, x_{k,0}\}.$$

Let Φ_ε be a C^∞ function that is 1 on the set

$$\{x \in E_{1/2} \mid x_j \ge \varepsilon \text{ for all } j \text{ with } 1 \le j \le k\}$$

and is 0 off $E_{3/4}$ and on the set

$$\{x \in E_1 \mid x_j \le \tfrac{1}{2}\varepsilon \text{ for some } j \text{ with } 1 \le j \le k\}.$$

and let F_ε be the member $\Phi_\varepsilon F$ of $C^\infty_{\text{com}}(\mathbb{R}^n)$. Applying the two sides of (11.51) to F_ε, we obtain

$$\begin{aligned}
F(x_0) &= F_\varepsilon(x_0) \\
&= \int_{\mathbb{R}_n} F_\varepsilon(x) \Delta^m g_\varepsilon(x_0 - x)\, dx - \int_{\mathbb{R}_n} F_\varepsilon(x) \beta_\varepsilon(x_0 - x)\, dx \\
&= \int_{\mathbb{R}_n} \Delta^m F_\varepsilon(x) g_\varepsilon(x_0 - x)\, dx - \int_{\mathbb{R}_n} F_\varepsilon(x) \beta_\varepsilon(x_0 - x)\, dx \\
&= \int_{|x_0 - x| \le 3\varepsilon} \Delta^m F_\varepsilon(x) g_\varepsilon(x_0 - x)\, dx - \int_{|x_0 - x| \le 3\varepsilon} F_\varepsilon(x) \beta_\varepsilon(x_0 - x)\, dx.
\end{aligned}$$

On the set where $|x_0 - x| \le 3\varepsilon$, F_ε equals F. Also β_ε satisfies (11.53). Thus the above expression gives us

$$|F(x_0)| \le \int_{|x_0 - x| \le 3\varepsilon} |\Delta^m F(x)|\, dx + c\varepsilon^{-p(m)} \int_{|x_0 - x| \le 3\varepsilon} |F(x)|\, dx. \tag{11.54}$$

With c denoting a new constant at each occurrence, we have

$$P(x_0) \le c \min_{\substack{|x_0 - x| \le 3\varepsilon \\ x \in E_1}} P(x).$$

Hence (11.54) implies

$$\begin{aligned} P(x_0)|F(x_0)| & \\ &\le c \int_{|x_0 - x| \le 3\varepsilon} P(x)|\Delta^m F(x)|\,dx + c\varepsilon^{-p(m)} \int_{|x_0 - x| \le 3\varepsilon} P(x)|F(x)|\,dx \\ &\le c \int_{C \cap E_1} P(x)|\Delta^m F(x)|\,dx + c\varepsilon^{-p(m)} \int_{C \cap E_1} P(x)|F(x)|\,dx. \end{aligned}$$

Let $R(x) = (x_1 x_2 \cdots x_k)^{p(m)}$. Then $R(x) \le c\varepsilon^{p(m)}$, and it follows that the polynomial $Q_1 = PR$ satisfies

$$\begin{aligned} Q_1(x_0)|F(x_0)| &\le c \int_{C \cap E_1} P(x)|\Delta^m F(x)|\,dx + c \int_{C \cap E_1} P(x)|F(x)|\,dx \\ &= cc_m + cc_0 < \infty. \end{aligned}$$

This estimate is independent of x_0 (and ε) and completes the proof of the lemma.

Now we have the tools to prove Lemma 10.20—that $f \to F_f^T$ carries $L^\infty_{\text{com}}(G)$ into $L^\infty_{\text{com}}(T)$. Apart from the coefficients $\varepsilon_R^T(h)$ and $D_T(h)$, the map $f \to F_f^T$ is positivity preserving. Proposition 11.10 shows that the image of a function in $C^\infty_{\text{com}}(G)$ is bounded and hence the image of a function in $L^\infty_{\text{com}}(G)$ is bounded. The statement about supports follows similarly from Proposition 11.7.

For SL(2, $\mathbb{R}$) and SL(2, $\mathbb{C}$) we did not need Lemma 10.20. In the case of SL(2, $\mathbb{R}$) we gave in §10.2 a self-contained derivation of the character formulas for the principal series and for the sum of a pair of discrete series, and then we gave in §11.3 a self-contained proof of the Plancherel formula. The formulas for the individual discrete series characters used the deeper properties of general characters from §10.6 and §10.8. In the case of SL(2, $\mathbb{C}$), $f \to F_f^T$ carries $C^\infty_{\text{com}}(G)$ into $C^\infty_{\text{com}}(T)$ as an immediate consequence of Lemma 10.17, and again Lemma 10.20 is not needed. The derivation in §11.2 of the Plancherel formula relies on no other outside properties of F_f^T.

For other groups we avoid circularity in our development as follows: After settling existence questions for characters as in §10.1, we skip immediately to §§10.4–10.8 to settle the regularity questions for characters. Then we define F_f as in §10.3, prove its first properties as in Proposition 11.7, and return to §10.3 to prove the reduction formula in Lemma 10.17. Then we can return to the present section to prove the other properties so far and go back to §10.3 to prove Lemma 10.20 and obtain the formulas for induced characters. At this stage all the results so far on F_f and characters have been derived in noncircular fashion. We are thus prepared to continue our development of the properties of F_f.

We resume the development by discussing control of the jumps of F_f^T. By Proposition 11.8, the only possible singularities of F_f^T occur at points h in T where $\xi_\alpha(h) = 1$ for at least one noncompact imaginary root α in $\Delta(\mathfrak{t}^{\mathbb{C}} : \mathfrak{g}^{\mathbb{C}})$. We shall study the simplest such situation. This will involve a relationship between two Cartan subgroups, and we denote them B and A as a reminder that B has one more compact dimension than A does.

Thus fix a Θ-stable Cartan subgroup B of G, and fix an element h_0 in B such that

(i) there exists exactly one β in $\Delta^+(\mathfrak{b}^{\mathbb{C}} : \mathfrak{g}^{\mathbb{C}})$ such that $\xi_\beta(h_0) = 1$, and
(ii) β is imaginary and noncompact.

(Such an element h_0 is called a **semiregular element** of B.) The identity component of the centralizer $Z_G(h_0)$ is a linear connected reductive group, and we can calculate its Lie algebra as

$$Z_{\mathfrak{g}}(h_0) = \mathfrak{g} \cap (\mathfrak{b}^{\mathbb{C}} \oplus \mathfrak{g}_\beta \oplus \mathfrak{g}_{-\beta}) \cong \mathfrak{b}^\perp \oplus \mathfrak{sl}(2, \mathbb{R}), \tag{11.55}$$

where $\mathfrak{b}^\perp$ denotes the orthocomplement of H_β in $\mathfrak{b}$. The idea will be to capture the singularity of F_f^B at h_0 as coming from the $\mathfrak{sl}(2, \mathbb{R})$.

For this purpose we shall introduce a **Cayley transform** $\mathbf{c}_\beta$ that is 1 on $\mathfrak{b}^\perp$ and carries the complexification of one natural Cartan subalgebra of $\mathfrak{sl}(2, \mathbb{R})$ to the other. (See Problem 9 in Chapter VI for a prototype of this mapping.) We work first with the explicit matrices in $\mathfrak{sl}(2, \mathbb{R})$ and then abstract the situation in order to handle our copy of $\mathfrak{sl}(2, \mathbb{R})$ imbedded by means of (11.55). We start with the standard basis $\{h, e, f\}$ of $\mathfrak{sl}(2, \mathbb{C})$, but computed relative to the rotation subgroup rather than the diagonal subgroup as starting point:

$$h_B = \begin{pmatrix} 0 & i \\ -i & 0 \end{pmatrix}, \qquad e_B = \frac{1}{2}\begin{pmatrix} 1 & -i \\ -i & -1 \end{pmatrix}, \qquad f_B = \frac{1}{2}\begin{pmatrix} 1 & i \\ i & -1 \end{pmatrix}.$$

These satisfy the usual bracket relations

$$[h_B, e_B] = 2e_B, \qquad [h_B, f_B] = -2f_B, \qquad [e_B, f_B] = h_B$$

and in addition are such that $e_B + f_B$ and $i(e_B - f_B)$ are in $\mathfrak{sl}(2, \mathbb{R})$. The Cayley transform within $\mathfrak{sl}(2, \mathbb{C})$ is the unitary mapping

$$\operatorname{Ad}\left(\frac{\sqrt{2}}{2}\begin{pmatrix} 1 & i \\ i & 1 \end{pmatrix}\right) = \operatorname{Ad}\left(\exp \frac{\pi}{4}(f_B - e_B)\right).$$

This carries h_B, e_B, f_B, respectively, to complex multiples of h, e, and f. Moreover, it carries $\mathbb{R}\begin{pmatrix} 0 & 1 \\ -1 & 0 \end{pmatrix}$ to $i\mathbb{R}\begin{pmatrix} 1 & 0 \\ 0 & 1 \end{pmatrix}$.

To imbed this construction into G, let us assume that $\mathfrak{g} \cap i\mathfrak{g} = 0$, so that we can regard $\mathfrak{g}^{\mathbb{C}}$ as the naively complexified version of $\mathfrak{g}$. Let $H'_\beta = 2|\beta|^{-2}H_\beta$, and choose root vectors E'_β in $\mathfrak{g}_\beta$ and $E'_{-\beta} = -\theta E_\beta$ in $\mathfrak{g}_{-\beta}$ such that $B_0(E'_\beta, E'_{-\beta}) = 2/|\beta|^2$ and such that $E'_\beta + E'_{-\beta}$ and $i(E'_{-\beta} - E'_{-\beta})$ are in $\mathfrak{g}$. (A little checking shows that this choice is possible and not quite unique. Recall that θ is not complex linear on $\mathfrak{g}^{\mathbb{C}}$.) Then we define $\mathbf{c}_\beta = \operatorname{Ad}\left(\exp \frac{\pi}{4}\left(E'_{-\beta} - E'_\beta\right)\right)$. Here $\mathbf{c}_\beta$ fixes $\mathfrak{b}^\perp$ and behaves like the above Cayley transform within the $\mathfrak{sl}(2, \mathbb{C})$ summand of $Z_{\mathfrak{g}^{\mathbb{C}}}(h_0)$.

In terms of $\mathbf{c}_\beta$, we define a new Cartan subalgebra $\mathfrak{a}$ of $\mathfrak{g}$ by

$$\mathfrak{a} = \mathfrak{g} \cap \mathbf{c}_\beta(\mathfrak{b}^{\mathbb{C}}) = \mathfrak{b}^\perp \oplus \mathbb{R}(E'_\beta + E'_{-\beta}).$$

The corresponding Cartan subgroup we denote A. (The nonuniqueness of the choice of E'_β results in A's not being quite uniquely defined, but our discussion will apply to any such constructed A.) We have $\mathbf{c}_\beta(\mathfrak{b}^{\mathbb{C}}) = \mathfrak{a}^{\mathbb{C}}$, and we extend $\mathbf{c}_\beta$ to carry $U(\mathfrak{b}^{\mathbb{C}})$ isomorphically onto $U(\mathfrak{a}^{\mathbb{C}})$. If α is in $\Delta(\mathfrak{b}^{\mathbb{C}}:\mathfrak{g}^{\mathbb{C}})$, we define $\mathbf{c}_\beta(\alpha)$ as a member of $\Delta(\mathfrak{a}^{\mathbb{C}}:\mathfrak{g}^{\mathbb{C}})$ by $\mathbf{c}_\beta(\alpha)(H) = \alpha(\mathbf{c}_\beta^{-1}(H))$ for $H \in \mathfrak{a}^{\mathbb{C}}$.

Since the semiregular element h_0 centralizes the whole $\mathfrak{sl}(2, \mathbb{C})$ built from β, h_0 centralizes $\mathfrak{a}$ and is therefore in A. Then we see that $\xi_{\mathbf{c}_\beta(\alpha)}(h_0) = \xi_\alpha(h_0)$. Since $\mathbf{c}_\beta(\beta)$ is real, we see that $\xi_\gamma(h_0) \neq 1$ for all noncompact imaginary roots in $\Delta(\mathfrak{a}^{\mathbb{C}}:\mathfrak{g}^{\mathbb{C}})$. By Proposition 11.8, F_f^A is C^∞ at h_0. By contrast F_f^B is normally not C^∞ at h_0, but Proposition 11.10 does provide us with limiting values of DF_f^B for every D in $U(\mathfrak{a}^{\mathbb{C}})$:

$$DF_f^B(h_0)^\pm = \lim_{t \to 0\pm} DF_f^B(h_0 \exp itH'_\beta).$$

Proposition 11.12. Fix a Cartan subgroup B of G and a semiregular element h_0 in B, and let A be the more noncompact Cartan subgroup constructed by a Cayley transform $\mathbf{c}_\beta$ and containing h_0. Then there exists a constant $c \neq 0$ such that

$$DF_f^B(h_0)^+ - DF_f^B(h_0)^- = c\mathbf{c}_\beta(D)F_f^A(h_0)$$

for all f in $C_{\text{com}}^\infty(G)$ and all D in $U(\mathfrak{b}^{\mathbb{C}})$.

Idea of proof. Applying Lemma 10.17, we reduce to the case that B is compact. Let us assume that B is connected. We calculate each F_f in stages, first by integrating the conjugates within the imbedded $SL(2, \mathbb{R})$ and then by integrating conjugates over $G/SL(2, \mathbb{R})$. (The kind of calculation in the proof of Proposition 11.8 controls the second integral.) Then we apply the $SL(2, \mathbb{R})$ result to the inside integral.

Corollary 11.13. Fix a Cartan subgroup B of G and a semiregular element h_0 in B, and let β be the unique root in $\Delta^+(\mathfrak{b}^{\mathbb{C}}:\mathfrak{g}^{\mathbb{C}})$ such that $\xi_\beta(h_0) = 1$.

If D is any member of $U(\mathfrak{b}^{\mathbb{C}})$ such that $s_\beta D = -D$ and if f is in $C^\infty_{\text{com}}(G)$, then F^B_f extends to a continuous function in a neighborhood of h_0.

Sketch of proof. Form $\mathbf{c}_\beta$ and F^A_f as in Proposition 11.12. One checks that reflection in $\mathbf{c}_\beta(\beta)$ carries ε^A_R into its negative near h_0 and therefore that F^A_f is even near h_0 under this reflection. The assumption $s_\beta D = -D$ forces $\mathbf{c}_\beta(D)F^A_f$ to be odd about h_0 in the direction of $E'_\beta + E'_{-\beta}$. Hence $\mathbf{c}_\beta(D)F^A_f(h_0) = 0$, and the result follows by applying Proposition 11.12.

Corollary 11.14. Fix a Cartan subgroup T of G and an element h_0 of T. If f is in $C^\infty_{\text{com}}(G)$ and if D is any member of $U(\mathfrak{t}^{\mathbb{C}})$ such that $s_\beta D = -D$ for every noncompact imaginary root β in $\Delta(\mathfrak{t}^{\mathbb{C}}:\mathfrak{g}^{\mathbb{C}})$ with $\xi_\beta(h_0) = 1$, then DF^T_f extends to a continuous function in a neighborhood of h_0.

Sketch of proof. Corollary 11.13 implies that DF^T_f is continuous at every semiregular element in a sufficiently small neighborhood of h_0, and Proposition 11.10 implies that F^T_f extends to a C^∞ function on the closure of each component of T''. If we are given two distinct components with h_0 in both their closures, we can connect points arbitrarily close to h_0 in one component to points arbitrarily close to h_0 in the other component by curves passing through only regular elements and semiregular elements. Along such curves, DF^T_f is continuous. Hence the limiting values of DF^T_f at h_0 from the two components must be equal.

Corollary 11.15. Let f be in $C^\infty_{\text{com}}(G)$. For each Cartan subgroup T, $\partial(\varpi_T)F^T_f$ is continuous in a neighborhood of 1. Moreover, $\partial(\varpi_T)F^T_f(1) = 0$ unless the compact part of T has the maximum possible dimension.

The first statement is a special case of Corollary 11.14. The second statement follows from Proposition 11.12 as soon as it is known that any T other than the maximally compact one can be taken as A in the proposition.

To this end, let us discuss Cayley transforms a bit more systematically. What we have really seen so far is the following: Whenever B is a Cartan subgroup such that $\Delta(\mathfrak{b}^{\mathbb{C}}:\mathfrak{g}^{\mathbb{C}})$ has a noncompact imaginary root β, then we can construct a **Cayley transform** $\mathbf{c}_\beta$ and a more noncompact Cartan subgroup A as follows: Let $H'_\beta = 2|\beta|^{-2}H_\beta$, and choose root vectors E'_β in $\mathfrak{g}_\beta$ and $E'_{-\beta} = -\theta E'_\beta$ in $\mathfrak{g}_{-\beta}$ such that $B_0(E'_\beta, E'_{-\beta}) = 2/|\beta|^2$ and such that $E'_\beta + E'_{-\beta}$ and $i(E'_\beta - E'_{-\beta})$ are in $\mathfrak{g}$. Then we define

$$\mathbf{c}_\beta = \operatorname{Ad}\left(\exp\frac{\pi}{4}\left(E'_{-\beta} - E'_\beta\right)\right) \tag{11.56a}$$

and
$$\mathfrak{a} = \mathfrak{g} \cap \mathbf{c}_\beta(\mathfrak{b}^{\mathbb{C}}) = \mathfrak{b}^\perp \oplus \mathbb{R}(E'_\beta + E'_{-\beta}). \tag{11.56b}$$

An inverse construction starts from a Cartan subgroup A such that $\Delta(\mathfrak{a}^{\mathbb{C}}:\mathfrak{g}^{\mathbb{C}})$ has a real root α and leads via a Cayley transform $\mathbf{d}_\alpha$ to a more compact Cartan subgroup B. Namely let $H'_\alpha = 2|\alpha|^{-2}H_\alpha$. Since α is real, there is a root vector for α within $\mathfrak{g}$ itself. Choose E'_α in $\mathfrak{g} \cap \mathfrak{g}_\alpha$ so that $E'_{-\alpha} = \theta E'_\alpha$ satisfies Re $B_0(E'_\alpha, E'_{-\alpha}) = -2/|\alpha|^2$, and define

$$\mathbf{d}_\alpha = \mathrm{Ad}\left(\exp\frac{\pi}{4}\left(E'_{-\alpha} - E'_\alpha\right)\right) \tag{11.57a}$$

and

$$\mathfrak{b} = \mathfrak{a}^\perp \oplus \mathbb{R}(E'_\alpha + E'_{-\alpha}), \tag{11.57b}$$

where $\mathfrak{a}^\perp$ is the orthocomplement of H_α in $\mathfrak{a}$.

This construction has consequences for the extreme Cartan subgroups, those that are maximally compact or maximally noncompact, and we isolate these consequences in the following proposition. The proposition supplies the needed fact to complete the proof of Corollary 11.15.

Proposition 11.16.

(a) The maximally compact Cartan subgroup is characterized as the unique Cartan subgroup (up to conjugacy) having no real roots.

(b) The maximally noncompact Cartan subgroup is characterized as the unique Cartan subgroup (up to conjugacy) having no noncompact imaginary roots.

Proof. That the maximally compact and noncompact Cartan subgroups have these properties follows from the constructions above with Cayley transforms. Conversely suppose as in (a) that $\mathfrak{t} = \mathfrak{b} \oplus \mathfrak{a}$ is a θ-stable Cartan subalgebra with no real roots. We show that $\mathfrak{b}$ is maximal abelian in $\mathfrak{k}$ by showing that $Z_{\mathfrak{k}^{\mathbb{C}}}(\mathfrak{b}^{\mathbb{C}}) = \mathfrak{b}^{\mathbb{C}}$. Thus let X in $\mathfrak{k}^{\mathbb{C}}$ commute with $\mathfrak{b}^{\mathbb{C}}$ and expand X according to the root space decomposition of $\mathfrak{g}^{\mathbb{C}}$ as

$$X = H + \sum_{\alpha \in \Delta(\mathfrak{t}^{\mathbb{C}} : \mathfrak{g}^{\mathbb{C}})} E_\alpha. \tag{11.58}$$

Bracketing with an element H_B of $\mathfrak{b}$, we obtain

$$0 = \sum_\alpha \alpha(H_B)E_\alpha.$$

Thus $E_\alpha \neq 0$ in (11.58) implies α vanishes on $\mathfrak{b}$ and so is real. Since there are no real roots, only the H is present in (11.58). Thus X is in $\mathfrak{b}^{\mathbb{C}}$, and (a) follows. The argument for (b) is similar, showing that if $\mathfrak{t} = \mathfrak{b} \oplus \mathfrak{a}$ has no noncompact imaginary roots, then $\mathfrak{a}$ is maximal abelian in $\mathfrak{p}$.

The final property of F_f is the relationship with $f(1)$.

Theorem 11.17. Fix a Θ-stable Cartan subgroup B of G whose compact part has maximum possible dimension. Then there exists a constant $c \neq 0$ such that every f in $C^\infty_{\mathrm{com}}(G)$ satisfies

$$\partial(\varpi_B)F^B_f(1) = cf(1).$$

In §2 we saw this result in microcosm by means of a proof for $G = SL(2, \mathbb{C})$. The hard part there and here is to use Euclidean Fourier analysis on the Lie algebra $\mathfrak{g}$ to obtain an analogous result on $\mathfrak{g}$, and then one lifts the theorem to G. The general case is much more complicated, as one must take into account multiple Cartan subalgebras and the discontinuities of F_f. We omit the details.

§7. Hirai's Patching Conditions

We have made extensive use of the restriction to the regular set G' of G of irreducible global characters and more general invariant eigendistributions. From Chapter X we know that this restriction is a real analytic function on G' that is invariant under conjugation and is an eigenfunction of $Z(\mathfrak{g}^{\mathbb{C}})$ on G'. In order to tie down characters as much as possible, it is of interest to ask the converse question: What additional properties on such a function ensure that it arises by restriction from an invariant eigendistribution?

We shall not attempt a precise formulation of the answer to this question. Roughly speaking, the answer is that the smoothness conditions in §10.8 should be satisfied and also a kind of **patching condition** between "adjacent" Cartan subgroups should be satisfied. The patching condition is given as Theorem 11.18 below.

Theorem 11.18. Let μ be an invariant eigendistribution on G. For each Θ-stable Cartan subgroup T, introduce a positive system $\Delta^+(\mathfrak{t}^{\mathbb{C}}:\mathfrak{g}^{\mathbb{C}})$, and let τ_T be the numerator of μ on $(T')^G$. Then the system $\{\tau_T\}$ satisfies the following conditions: Let B be any Cartan subgroup, let h_0 be any semiregular element of B, and let the Cayley transform $\mathbf{c}_\beta$ and Cartan subgroup A (one dimension more noncompact than B) be constructed from B and h_0. Suppose that $\mathbf{c}_\beta(\Delta^+(\mathfrak{b}^{\mathbb{C}}:\mathfrak{g}^{\mathbb{C}})) = \Delta^+(\mathfrak{a}^{\mathbb{C}}:\mathfrak{g}^{\mathbb{C}})$. Then

$$D(\varepsilon_R^B\tau_B)(h_0) = \tfrac{1}{2}[\mathbf{c}_\beta(D)(\varepsilon_R^A\tau_A)^+(h_0) - \mathbf{c}_\beta(D)(\varepsilon_R^A\tau_A)^-(h_0)] \tag{11.59}$$

for every member D of $U(\mathfrak{b}^{\mathbb{C}})$ such that $s_\beta D = -D$.

Remarks. By Theorem 10.40, $\varepsilon_R^B\tau_B$ is real analytic at h_0, and thus the left side of (11.59) is defined. It is to be understood on the right side that $D(\varepsilon_R^A\tau_A)$ has a jump discontinuity at h_0, with the two limits

$$\lim_{t\to 0\pm} D(\varepsilon_R^A\tau_A)(h_0 \exp tH_{\mathbf{c}_\beta(\beta)}) = D(\varepsilon_R^A\tau_A)^{\pm}(h_0)$$

existing; in fact, use of the element $s_{\mathbf{c}_\beta(\beta)}$ in $W(A:G)$ shows the two limits are negatives of one another.

The reason there has to be some relation like (11.59) is as follows. Let z be in $Z(\mathfrak{g}^{\mathbb{C}})$ and let $z\mu = \chi(z)\mu$. For f in $C_{\text{com}}^\infty(G)$ we can write out the

Weyl integration formula for $(z\mu)(f) = \mu(z^{tr}f)$ and get integrals involving the functions $\varepsilon_R^T\tau_T$ and $\gamma(z^{tr})F_f^T$ for all T's. On the other hand, we can write out the Weyl integration formula for $\chi(z)\mu(f)$ and get integrals involving the functions $\gamma(z)\varepsilon_R^T\tau_T$ and F_f^T. The two formulas appear to be related by integration by parts, but the boundary terms are missing. Hence the boundary terms must give 0, and the resulting formula is (11.59).

The argument for SL(2, $\mathbb{R}$) is sufficiently illustrative, and we shall give it shortly. First let us see what the conditions mean for SL(2, $\mathbb{R}$).

Example 1. Nonunitary principal series $U(S_{\mathfrak{p}}, \sigma, \nu)$ in SL(2, $\mathbb{R}$).

In the notation of Theorem 11.18, B is the compact Cartan subgroup of SL(2, $\mathbb{R}$), and A is the diagonal subgroup, which is called T in §10.2. The element h_0 can be the identity or minus the identity. For $U(S_{\mathfrak{p}}, \sigma, \nu)$, τ_B is 0, and thus the left side of (11.59) is 0. Since $D_A(\pm a_t) = \pm(e^t - e^{-t})$, the other side involves

$$\varepsilon_R(\pm a_t)\tau(\pm a_t) = \pm\sigma(\pm)(e^{\nu \log a_t} + e^{-\nu \log a_t}).$$

This is smooth at $t = 0$, and its derivatives have no jumps. Thus the right side of (11.59) is 0.

Example 2. Discrete series or limit $\mathscr{D}_n^+$, $n \geq 1$, in SL(2, $\mathbb{R}$).

Let

$$h_B = \begin{pmatrix} 0 & -i \\ i & 0 \end{pmatrix}, \qquad h_A = \begin{pmatrix} 1 & 0 \\ 0 & -1 \end{pmatrix}, \qquad \mathbf{c}_\beta = \operatorname{Ad}\left(\frac{\sqrt{2}}{2}\begin{pmatrix} 1 & -i \\ -i & 1 \end{pmatrix}\right).$$

Direct computation gives $\mathbf{c}_\beta^{-1}(h_A) = h_B$. If we take the Weyl denominator D_A as usual to be $D_A(\pm a_t) = \pm(e^t - e^{-t})$, then h_A is in the positive Weyl chamber of $\mathfrak{a}$. For the Cayley transform $\mathbf{c}_\beta$ to preserve positive roots, h_B must be in the positive Weyl chamber of $i\mathfrak{b}$. Thus the positive root β has $\beta(h_B) = 2$. Consequently

$$\xi_\beta(k_\theta) = \xi_\beta(\exp(i\theta h_B)) = e^{2i\theta},$$

and the Weyl denominator D_B is given by the usual formula $D_B(\theta) = e^{i\theta} - e^{-i\theta}$. Thus the use of $\{h_B, h_B, \mathbf{c}_\beta\}$ in Theorem 11.18 is compatible with the positive systems leading to our usual Weyl denominators. Thus Proposition 10.14 gives

$$\varepsilon_R^B(\theta)\tau_B(\theta) = -e^{i(n-1)\theta} \tag{11.60a}$$

$$\varepsilon_R^A(\pm a_t)\tau_A(\pm a_t) = \pm(\pm)^n \tfrac{1}{2}[e^{(n-1)t}(1 - \operatorname{sgn} t) + e^{-(n-1)t}(1 + \operatorname{sgn} t)]. \tag{11.60b}$$

Next let us make the derivatives concrete. We have

$$k_\theta = \exp(i\theta h_B) \qquad \text{and} \qquad a_t = \exp(t h_A).$$

Thus $$\frac{d}{d\theta} \text{ on } B \text{ corresponds to } ih_B$$

$$\frac{d}{dt} \text{ on } A \text{ corresponds to } h_A,$$

and we can regard $\mathbf{c}_\beta$ as carrying $i\dfrac{d}{d\theta}$ to $-\dfrac{d}{dt}$. The operator D in the theorem is a complex linear combination of odd powers of $i\dfrac{d}{d\theta}$, and we examine $i\dfrac{d}{d\theta}$ itself. In view of (11.60), the theorem dictates that

$$\begin{aligned}
&i\frac{d}{d\theta}(-e^{i(n-1)\theta})_{\theta=0}\\
&\quad= -\frac{1}{4}\left[\frac{d}{dt}(e^{(n-1)t}(1-\operatorname{sgn} t) + e^{-(n-1)t}(1+\operatorname{sgn} t)\right]_{t=0^+}\\
&\quad\quad+\frac{1}{4}\left[\frac{d}{dt}(e^{(n-1)t}(1-\operatorname{sgn} t) + e^{-(n-1)t}(1+\operatorname{sgn} t)\right]_{t=0^-}
\end{aligned}$$

and

$$\begin{aligned}
&i\frac{d}{d\theta}(-e^{i(n-1)\theta})_{\theta=\pi}\\
&\quad= -\frac{(-1)^{n-1}}{4}\left[\frac{d}{dt}(e^{(n-1)t}(1-\operatorname{sgn} t) + e^{-(n-1)t}(1+\operatorname{sgn} t)\right]_{t=0^+}\\
&\quad\quad+\frac{(-1)^{n-1}}{4}\left[\frac{d}{dt}(e^{(n-1)t}(1-\operatorname{sgn} t) + e^{-(n-1)t}(1+\operatorname{sgn} t)\right]_{t=0^-}
\end{aligned}$$

We can verify these formulas by direct calculation.

Proof of *Theorem 11.18 for* $G = SL(2, \mathbb{R})$. We continue with the notation in the above examples and shall establish (11.59) for $D = h_B \leftrightarrow i\dfrac{d}{d\theta}$ at $h_0 = 1$ in the group. The computations for odd powers of D and for $h_0 = -1$ are completely analogous. We shall use the jump formulas for F_f^B for $SL(2, \mathbb{R})$, and the proof for general G proceeds along the same lines by means of Proposition 11.12.

In $SL(2, \mathbb{R})$ we start with a function f in $C_{\text{com}}^\infty(G)$ supported in a small enough neighborhood of the identity so that we can use the Weyl integration formula to write

$$\mu(f) = -\frac{1}{2\pi}\int_{-\varepsilon}^{\varepsilon} \tau_B(\theta)F_f^B(\theta)\,d\theta + \frac{1}{4}\int_{-\varepsilon}^{\varepsilon} \varepsilon_R^A(a_t)\tau_A(a_t)F_f^A(a_t)\,dt$$

for some $\varepsilon < \pi/2$. Let us write $z\mu = \chi(z)\mu$ for z in $Z(\mathfrak{g}^{\mathbb{C}})$. Choose z_0 in $Z(\mathfrak{g}^{\mathbb{C}})$ such that the Harish-Chandra homomorphism maps z_0 to $\dfrac{d^2}{d\theta^2}$ and $-\dfrac{d^2}{dt^2}$ relative to the two Cartan subgroups. Then Proposition 11.9 gives

$$\mu(z_0 f) = -\frac{1}{2\pi}\int_{-\varepsilon}^{\varepsilon} \tau_B(\theta)\frac{d^2}{d\theta^2}F_f^B(\theta)\,d\theta - \frac{1}{4}\int_{-\varepsilon}^{\varepsilon} \varepsilon_R^A(a_t)\tau_A(a_t)\frac{d^2}{dt^2}F_f^A(a_t)\,dt.$$

Since $\dfrac{d}{d\theta}F_f^B$ and $\varepsilon_R^A(a_t)\tau_A(a_t)$ are continuous, we can integrate by parts once without a boundary term coming from $\theta = 0$ or $t = 0$. We obtain

$$\begin{aligned}\mu(z_0 f) &= \frac{1}{2\pi}\int_{-\varepsilon}^{\varepsilon}\left\{\frac{d}{d\theta}\tau_B(\theta)\right\}\frac{d}{d\theta}F_f^B(\theta)\,d\theta \\ &\quad + \frac{1}{4}\int_{-\varepsilon}^{\varepsilon}\frac{d}{dt}\{\varepsilon_R^A(a_t)\tau_A(a_t)\}\frac{d}{dt}F_f^A(a_t)\,dt. \qquad (11.61)\end{aligned}$$

Meanwhile (10.58) gives

$$\begin{aligned}\mu(z_0 f) &= z_0^{\mathrm{tr}}\mu(f) = \chi(z_0^{\mathrm{tr}})\mu(f) \\ &= -\frac{1}{2\pi}\int_{-\varepsilon}^{\varepsilon}\{\chi(z_0^{\mathrm{tr}})\tau_B(\theta)\}F_f^B(\theta)\,d\theta \\ &\quad + \frac{1}{4}\int_{-\varepsilon}^{\varepsilon}\{\chi(z_0^{\mathrm{tr}})\varepsilon_R^A(a_t)\tau_A(a_t)\}F_f^A(a_t)\,dt \\ &= -\frac{1}{2\pi}\int_{-\varepsilon}^{\varepsilon}\left\{\frac{d^2}{d\theta^2}\tau_B(\theta)\right\}F_f^B(\theta)\,d\theta \\ &\quad - \frac{1}{4}\int_{-\varepsilon}^{\varepsilon}\frac{d^2}{dt^2}\{\varepsilon_R^A(a_t)\tau_A(a_t)\}F_f^A(a_t)\}\,dt. \qquad (11.62)\end{aligned}$$

Formula (11.61) is what we would obtain from (11.62) by doing an integration by parts and dropping the boundary term (coming from discontinuities at 0). Therefore the boundary term is 0:

$$\begin{aligned}0 &= \frac{1}{2\pi}\left[F_f^B(\theta)\frac{d}{d\theta}\tau_B(\theta)\right]_{\theta=0^+} - \frac{1}{2\pi}\left[F_f^B(\theta)\frac{d}{d\theta}\tau_B(\theta)\right]_{\theta=0^-} \\ &\quad + \frac{1}{4}\left[\frac{d}{dt}\{\varepsilon_R^A(a_t)\tau_A(a_t)\}F_f^A(a_t)\right]_{t=0^+} - \frac{1}{4}\left[\frac{d}{dt}\{\varepsilon_R^A(a_t)\tau_A(a_t)\}F_f^A(a_t)\right]_{t=0^-}.\end{aligned}$$

This identity simplifies to

$$\begin{aligned}&\frac{1}{2\pi}[F_f^B(0^+) - F_f^B(0^-)]\left[\frac{d}{d\theta}\tau_B(\theta)\right]_{\theta=0} \\ &= -\frac{1}{4}F_f^A(a_0)\left[\frac{d}{dt}\{\varepsilon_R^A(a_t)\tau_A(a_t)\}_{t=0^+} - \frac{d}{dt}\{\varepsilon_R^A(a_t)\tau_A(a_t)\}_{t=0^-}\right].\end{aligned}$$

We substitute for the jump of F_f^B from (11.36a) and then multiply by i and put D for $i\dfrac{d}{d\theta}$ and $\mathbf{c}_\beta(D)$ for $-\dfrac{d}{dt}$ to obtain

$$\tfrac{1}{2}F_f^A(a_0)D(\varepsilon_R^B\tau_B)(1) = \tfrac{1}{4}F_f^A(a_0)[\mathbf{c}_\beta(D)(\varepsilon_R^A\tau_A)^+(1) - \mathbf{c}_\beta(D)(\varepsilon_R^A\tau_A)^-(1)].$$

From Lemma 10.10 it is apparent that we can choose f so that $F_f^A(a_0) \neq 0$. With this choice we can cancel $F_f^A(a_0)$ and obtain the conclusion of the theorem.

§8. Problems

1. Derive the Plancherel formula (11.9) for a general compact connected Lie group by using the style of argument in §1.
2. For SL(2, $\mathbb{R}$), the expression $I(s)$ in Lemma 11.5 is related to $F_f^B(\theta)$ by $s = \sin\theta$. However, not all functions F studied in that lemma arise from F_f^B for various f. Show, in fact, that $I''(s)$ does not have a limit as $s \downarrow 0$ unless the partial derivative of F in the second variable vanishes at the origin. By contrast, $\dfrac{d^2}{d\theta^2}F_f^B(\theta)$ always has a limit as $\theta \downarrow 0$ if f is in C_{com}^∞ (SL(2, $\mathbb{R}$)), as a result of Proposition 11.9.

CHAPTER XII

Exhaustion of Discrete Series

§1. Boundedness of Numerators of Characters

In Theorem 9.20 we constructed a family of discrete series representations of a linear connected semisimple group G under the assumption that rank G = rank K. These representations were given in terms of a Harish-Chandra parameter λ described as follows. Let $\mathfrak{b}$ be a compact Cartan subalgebra, and let $\lambda \in (i\mathfrak{b})'$ be nonsingular relative to $\Delta(\mathfrak{b}^{\mathbb{C}}:\mathfrak{g}^{\mathbb{C}})$ and have $\lambda + \delta_G$ analytically integral. The discrete series with parameter λ is denoted π_λ, and $\pi_\lambda \cong \pi_{\lambda'}$ if and only if λ and λ' are conjugate via $W_K = W(\mathfrak{b}^{\mathbb{C}}:\mathfrak{k}^{\mathbb{C}})$.

The main results of this chapter are that these representations exhaust the discrete series when rank G = rank K and that there are no discrete series at all when rank $G \neq$ rank K. In the latter part of the chapter we shall investigate further properties of discrete series and related topics.

In the first three sections we shall assume that rank G = rank K, and we shall compute the global character Θ_λ of π_λ. As noted in §9.7, the assumption rank G = rank K implies $\mathfrak{k} \cap i\mathfrak{p} = 0$. Thus the complexification $G^{\mathbb{C}}$ is obtained easily by the construction of §5.1. To express the character neatly, it will be convenient to assume $G^{\mathbb{C}}$ is simply connected; as usual there is no loss of generality in this assumption. Let $B = \exp \mathfrak{b}$ be the compact Cartan subgroup corresponding to $\mathfrak{b}$.

Let $\mathfrak{t}$ be any θ-stable Cartan subalgebra, and let T be the corresponding Cartan subgroup. Fix a positive system for $\Delta(\mathfrak{t}^{\mathbb{C}}:\mathfrak{g}^{\mathbb{C}})$. Then we know from Theorem 10.35 that the function defining Θ_λ (cf. Theorem 10.36) is given on $(T')^G$ as a quotient $\tau_{\lambda,T}(h)/D_T(h)$, where D_T is the Weyl denominator and where the numerator $\tau_{\lambda,T}$ is locally just a linear combination of exponentials (cf. Theorem 10.48). Our goal in this section is to prove the following theorem.

Theorem 12.1. Let rank G = rank K, and let Θ_λ be the global character of the discrete series with Harish-Chandra parameter λ. On each Cartan subgroup T, the numerator $\tau_{\lambda,T}$ of Θ_λ is a bounded function on T.

Remarks. The proof will not use the explicit form of the discrete series representation, not even the fact that rank G = rank K. It will show that each exponential function in $\tau_{\lambda,T}$ is a bounded function on its domain. For an earlier example, see Proposition 10.14.

Lemma 12.2. Corresponding to each Θ_λ, there is a member z of $Z(\mathfrak{k}^{\mathbb{C}})$ and there is a constant $C < \infty$ such that

$$|\Theta_\lambda(f)| \le C\|zf\|_2$$

for all f in $C^\infty_{\text{com}}(G)$. Here z acts on f by left-invariant differentiation.

Proof. Let π_λ act on the space V. Fix a K type μ, let V_μ be the subspace of V transforming according to μ, and let $\{v_j\}$ be an orthonormal basis of V_μ. For each z in $Z(\mathfrak{k}^{\mathbb{C}})$ and for each j, we have

$$\begin{aligned}\chi^K_{\mu+\delta_K}(z)&\int_G f(x)(\pi_\lambda(x)v_j, v_j)\,dx\\ &= \int_G f(x)(\pi_\lambda(x)\pi_\lambda(z)v_j, v_j)\,dx\\ &= \int_G f(x)z(\pi_\lambda(x)v_j, v_j)\,dx \qquad \text{by (8.10)}\\ &= \int_G z^{\text{tr}}f(x)(\pi_\lambda(x)v_j, v_j)\,dx \qquad \text{since } f \text{ has compact support}\end{aligned}$$

Thus

$$\sum_j (\pi_\lambda(f)v_j, v_j) = \chi^K_{\mu+\delta_K}(z)^{-1}\sum_j \int_G (z^{\text{tr}}f)(x)(\pi_\lambda(x)v_j, v_j)\,dx.$$

If $d(\pi_\lambda)$ denotes the formal degree of π_λ, then

$$\|(\pi_\lambda(x)v_j, v_j)\|_2 = d(\pi_\lambda)^{-1/2}$$

by Proposition 9.6. Thus we can combine the Schwarz inequality and Theorem 8.1 to obtain

$$\left|\sum_j (\pi_\lambda(f)v_j, v_j)\right| \le d(\pi_\lambda)^{-1/2}|\chi^K_{\mu+\delta_K}(z)|^{-1}d^2_\mu\|z^{\text{tr}}f\|_2. \tag{12.1}$$

The main part of the proof of Theorem 10.2 was to show that there exists z in $Z(\mathfrak{k}^{\mathbb{C}})$for which (10.2) is finite. Choosing z in this fashion, we see that

$$C = d(\pi_\lambda)^{-1/2}\sum_\mu |\chi^K_{\mu+\delta_K}(z)|^{-1}d^2_\mu$$

is finite, and (12.1) shows that

$$|\Theta_\lambda(f)| \le C\|z^{\text{tr}}f\|_2.$$

This proves the lemma.

To get a handle on the numerator $\tau_{\lambda,T}$, we shall regard $\tau_{\lambda,T}(h)\,dh$ as a distribution on T and estimate integrals of the form

$$\int_T \tau_{\lambda,T}(h)g(h)\,dh. \tag{12.2}$$

It will be enough to deal with (12.2) when g is in $C_{\text{com}}^{\infty}(T')$. Since $\tau_{\lambda,T}$ is **odd** under $W(T:G)$ (i.e., $\tau_{\lambda,T}(whw^{-1}) = (\det w)\tau_{\lambda,T}(h)$ for w in $W(T:G)$), only the odd part of g, given by

$$g_{\text{odd}}(h) = |W(T:G)|^{-1} \sum_{w \in W(T:G)} (\det w) g(whw^{-1}),$$

contributes to (12.2). We begin by constructing f in $C_{\text{com}}^{\infty}(G)$ with

$$s^{G/T}|W(T:G)|^{-1}\varepsilon_R^T(h)F_f^T(h) = g_{\text{odd}}(h), \tag{12.3}$$

and then the Weyl integration formula gives

$$\Theta_\lambda(f) = \int_T \tau_{\lambda,T}(h) g_{\text{odd}}(h)\, dh = \int_T \tau_{\lambda,T}(h) g(h)\, dh. \tag{12.4}$$

Lemma 12.2 says that we can bound this expression by estimating $\|zf\|_2$.

To construct f from g, fix a right $N_K(T)$ invariant function $\psi(x)$ on G that descends to G/T as a function in $C_{\text{com}}^{\infty}(G/T)$ with total integral one. Choose a compact set E in G so that support$(\psi) \subseteq ET$. Without loss of generality, we can enlarge E so that $E = E^{-1}$. For g in $C_{\text{com}}^{\infty}(T')$, we define φ_g in $C_{\text{com}}^{\infty}((T')^G)$ by

$$\begin{aligned} \varphi_g(xhx^{-1}) &= s^{G/T}|W(T:G)|\psi(x)D_T(h)^{-1}g_{\text{odd}}(h) \\ &= s^{G/T}\psi(x)D_T(h)^{-1} \sum_{w \in W(T:G)} (\det w) g(whw^{-1}). \end{aligned} \tag{12.5}$$

Then (12.3) holds for $f = \varphi_g$.

To estimate $\|z\varphi_g\|_2$, we need to connect decay estimates and information about conjugacy classes. For this purpose, we shall use the spherical function φ_0^G, which has better invariance properties than the more geometrically defined measures of decay. For each α in $\Delta(\mathfrak{t}^{\mathbb{C}}:\mathfrak{g}^{\mathbb{C}})$, let $\eta_\alpha = (1 - \xi_\alpha^{-1})^{-1}$, so that $D_T^{-1} = \xi_\delta^{-1}\prod_{\alpha>0}\eta_\alpha$. Let $\mathscr{R}$ be the algebra of functions on T generated by 1 and the η_α. Notice that $H \in \mathfrak{t}^{\mathbb{C}}$ implies

$$H\eta_\alpha = (1 - \xi_\alpha^{-1})^{-2}H\xi_{-\alpha} = -\alpha(H)\xi_{-\alpha}\eta_\alpha^2 = \alpha(H)\eta_\alpha\eta_{-\alpha}; \tag{12.6}$$

hence $U(\mathfrak{t}^{\mathbb{C}})$ maps $\mathscr{R}$ into itself. The members of $\mathscr{R}$ blow up at the singular set, but they remain bounded at points of T tending to infinity and remaining suitably distant from the singular set.

Lemma 12.3. If D is in $U(\mathfrak{g}^{\mathbb{C}})$ and m is an integer ≥ 0, then there exist finitely many η_j in $\mathscr{R}$ and H_j in $U(\mathfrak{t}^{\mathbb{C}})$ and there exists $C < \infty$ such that

$$\begin{aligned} |D\varphi_g(x)| \leq C \sup_{h \in T'} &\left\{(1 + \|h\|)^m \sum_j \sum_{w \in W(T:G)} |\eta_j(h)H_j g(whw^{-1})|\right\} \\ &\times \varphi_0^G(x)(1 + \|x\|)^{-m} \end{aligned}$$

for all x in G and all g in $C_{\text{com}}^{\infty}(T')$.

Remarks. This lemma makes no mention of representations and will be used again later. Our application of this lemma in this section will be for D in $Z(\mathfrak{k}^{\mathbb{C}})$ and for m large enough so that $\varphi_0^G(x)(1 + \|x\|)^{-m}$ is in $L^2(G)$. The proof of Lemma 12.3 will be preceded by two preparatory lemmas.

Recall the mapping $\Gamma_h: U(\mathfrak{g}^{\mathbb{C}}) \otimes_{\mathbb{C}} U(\mathfrak{t}^{\mathbb{C}}) \to U(\mathfrak{g}^{\mathbb{C}})$ of Lemma 10.30. The first preparatory lemma is a variant of the calculation in Lemma 10.34.

Lemma 12.4. If D is in $U(\mathfrak{g}^{\mathbb{C}})$, then there exist finitely many η_j in $\mathscr{R}$, X_j in $U(\mathfrak{g}^{\mathbb{C}})$, and H_j in $U(\mathfrak{t}^{\mathbb{C}})$ such that

$$D = \sum \eta_j(h)\Gamma_h(X_j \otimes H_j)$$

for all h in T'.

Proof. We go over the proof in Lemma 10.30 that Γ_h maps the space (10.44) one-one onto $U(\mathfrak{g}^{\mathbb{C}})$ if h is in T'. The relevant formulas are (10.46) and (10.47). In (10.47), the vectors $Y_{j_1}, \ldots, Y_{j_d}$ are root vectors, on which $A(h)$ acts with eigenvalue a product of functions η_α^{-1}. Thus we can invert (10.47), solving for $Y_{j_1} \cdots Y_{j_d} H_1 \cdots H_e$ modulo $U^{d+e-1}(\mathfrak{g}^{\mathbb{C}})$, and the coefficient will be in $\mathscr{R}$. Proceeding inductively on the degree, we obtain the lemma.

Lemma 12.5. Fix a compact set E_1 in G. Then there exist constants $c > 0$ and $C < \infty$ such that

$$c\varphi_0^G(x) \le \varphi_0^G(y_1 x y_2) \le C\varphi_0^G(x) \tag{12.7a}$$

and

$$1 + \|y_1 x y_2\| \le C(1 + \|x\|) \tag{12.7b}$$

for all $x \in G$, $y_1 \in E_1$, and $y_2 \in E_1$.

Proof. We have

$$\varphi_0^G(xy) = \int_K e^{-\rho_{\mathfrak{p}} H_{\mathfrak{p}}(y^{-1}x^{-1}k)}\, dk = \int_K e^{-\rho_{\mathfrak{p}} H_{\mathfrak{p}}(y^{-1}\kappa(x^{-1}k))} e^{-\rho_{\mathfrak{p}} H(x^{-1}k)}\, dk$$

and hence

$$\inf_{k \in K} \{e^{-\rho_{\mathfrak{p}} H_{\mathfrak{p}}(y^{-1}k)}\}\varphi_0^G(x) \le \varphi_0^G(xy) \le \sup_{k \in K} \{e^{-\rho_{\mathfrak{p}} H_{\mathfrak{p}}(y^{-1}k)}\}. \tag{12.8}$$

In addition, we know that φ_0^G is a matrix coefficient of the form

$$\varphi_0^G(x) = (U(x)\phi, \phi) \tag{12.9a}$$

for a certain unitary (principal series) representation U. From this equation and the fact that φ_0^G is real-valued, it follows that

$$\varphi_0^G(x^{-1}) = \varphi_0^G(x). \tag{12.9b}$$

Combining (12.8) and (12.9), we obtain (12.7a). Formula (12.7b) follows from (8.62) and the fact that $\|x^{-1}\| = \|x\|$.

Proof of Lemma 12.3. Given D in $U^k(\mathfrak{g}^{\mathbb{C}})$, let $\{D^{(i)}\}$ be a basis for $U^k(\mathfrak{g}^{\mathbb{C}})$. Then we can write

$$\mathrm{Ad}(x)^{-1}D = \sum a^{(i)}(x)D^{(i)}$$

with each $a^{(i)}(x)$ real analytic on G. Applying Lemma 12.4 to each $D^{(i)}$ and adding the results with coefficients $a^{(i)}(x)$, we obtain

$$\mathrm{Ad}(x)^{-1}D = \sum_{i,j} a^{(i)}(x)\eta_j^{(i)}(h)\Gamma_h(X_j^{(i)} \otimes H_j^{(i)}).$$

Using this formula and Lemma 10.29, we compute $D\varphi_g(xhx^{-1})$, where φ_g is given by (12.5). The result is

$$\begin{aligned} D\varphi_g(xhx^{-1}) = s^{G/T} &\sum_{i,j} \{a^{(i)}(x)X_j^{(i)}\psi(x)\}\eta_j^{(i)}(h) \\ &\times H_j^{(i)}\{D_T(h)^{-1} \sum_w (\det w)g(whw^{-1})\}. \end{aligned} \tag{12.10}$$

Except for a locally constant sign, $D\varphi_g$ does not depend on the choice of the positive system $\Delta^+(\mathfrak{t}^{\mathbb{C}}:\mathfrak{g}^{\mathbb{C}})$, and we shall adapt the positive system (and D_T) to the particular element h. Write $T = T^- A$ with $T^- = T \cap K$ and $A = T \cap \exp \mathfrak{p}$, and decompose h accordingly as $h = ba$. We can arrange that $\mathfrak{a}$ comes before $i\mathfrak{t}^-$ in the ordering and that a is in the exponential of the positive Weyl chamber of $\mathfrak{a}$. Defining δ' to be the half sum of the positive roots for this ordering (δ' depends in locally constant fashion on h), we can write $D_T(h)^{-1} = \xi_{\delta'}^{-1}(h)\eta(h)$ with η in $\mathscr{R}$. Then it follows from (12.6) that any $U(\mathfrak{t}^{\mathbb{C}})$ derivative of $D_T(h)^{-1}$ is the product of $\xi_{\delta'}^{-1}(h)$ by a member of $\mathscr{R}$. Putting

$$C_1 = \max_{i,j} \sup_{x\in E} \{a^{(i)}(x)X_j^{(i)}\psi(x)\}$$

and substituting into (12.10), we obtain

$$|D\varphi_g(xhx^{-1})| \leq C_1|\xi_{\delta'}(h)|^{-1} \sum_{m,w} |\eta_m(h)(H_m g)(whw^{-1})|$$

for suitable η_m in $\mathscr{R}$ and H_m in $U(\mathfrak{t}^{\mathbb{C}})$. Now $h = ba$ implies

$$|\xi_{\delta'}(h)|^{-1} = e^{-\rho' \log a},$$

and it is a simple consequence of (7.54) and Lemma 7.16a that

$$\varphi_0^G(a) \geq e^{-\rho' \log a} \qquad \text{for } a \in \exp \mathfrak{a}^+. \tag{12.11}$$

Since $\varphi_0^G(a) = \varphi_0^G(ba) = \varphi_0^G(h)$, we obtain

$$|D\varphi_g(xhx^{-1})| \leq C_1\varphi_0^G(h) \sum_{m,w} |\eta_m(h)(H_m g)(whw^{-1})|. \tag{12.12}$$

The quantities η_m and H_m may depend on the component of h in T', but by including all such η_m and H_m obtained from all components, we can interpret (12.12) as a single inequality valid for all h in T' and all x in E.

With C as in (12.7b) and with x in E, we thus have

$$\begin{aligned}|D\varphi_g(xhx^{-1})\varphi_0^G(xhx^{-1})^{-1}(1 + \|xhx^{-1}\|)^m| &\le C_1 \frac{\varphi_0^G(h)}{\varphi_0^G(xhx^{-1})} C(1 + \|h\|)^m \sum_{m,w} |\eta_m(h)(H_m g)(whw^{-1})| \\ &\le C_1 C c^{-1}(1 + \|h\|)^m \sum_{m,w} |\eta_m(h)(H_m g)(whw^{-1})|.\end{aligned}$$

We take the sup of the right side for h in T'. On the left side, $D\varphi_g$ vanishes at points in G that are not of the form xhx^{-1} with x in E and h in T', and thus Lemma 12.3 follows.

Proof of Theorem 12.1. From (12.4) and Lemma 12.2 we have

$$\left|\int_T \tau_{\lambda,T}(h)g(h)\,dh\right| \le \text{Const}\|z\varphi_g\|_2$$

for a suitable z in $U(\mathfrak{k}^{\mathbb{C}}) \subseteq U(\mathfrak{g}^{\mathbb{C}})$. We apply Lemma 12.3 with $D = z$ and with m chosen large enough so that $\varphi_0^G(x)(1 + \|x\|)^{-m}$ is in $L^2(G)$. (This choice of m is possible by (7.52).) Then we conclude that

$$\left|\int_T \tau_{\lambda,T}(h)g(h)\,dh\right| \le \text{Const} \sup_{h \in T'} \left\{(1 + \|h\|)^m \sum_{j,w} |\eta_j(h)H_j g(whw^{-1})|\right\} \tag{12.13}$$

for all $g \in C_{\text{com}}^\infty(T')$.

Write $T = T^- A$ with $T^- = T \cap K$ and $A = T \cap \exp \mathfrak{p}$, and let $\mathscr{C}$ be a Weyl chamber in $\mathfrak{a}$. In the notation of §10.8, the subset $T^- \exp \mathscr{C}$ of T is contained in T'_R and is connected. Theorem 10.40 thus says that $\tau_T(h)$ has a uniform formula there, say

$$\tau_T(b \exp H) = \sum_{w \in W(\mathfrak{t}^{\mathbb{C}}:\mathfrak{g}^{\mathbb{C}})} c_w \xi_{w\lambda}(b) e^{w\lambda(H)}, \qquad b \in T^- \text{ and } h \in \mathscr{C}.$$

Fix a smooth function g_1 on T^- that is compactly supported in

$$\{b \in T^- \mid \xi_\alpha(b) \neq 1 \text{ for all imaginary roots } \alpha\}.$$

Let g_2 be a fixed smooth function supported in the ball of radius one centered at the origin in $\mathfrak{a}$. For each H_0 in $\mathscr{C}$ such that the ball of radius two and center H_0 is in $\mathscr{C}$, we form the function

$$g_{H_0}(b \exp H) = g_1(b)g_2(H_0 + H), \qquad b \in T^- \text{ and } H \in \mathfrak{a}.$$

In (12.13) the functions η_j are uniformly bounded on the union of the supports of all such g_{H_0}, and thus the right side of (12.13) grows no faster than a polynomial as $H_0 \to \infty$. Meanwhile the left side of (12.13) is

$$\left|\sum_w \left\{c_w \int_{T^-} g_1(b)\xi_{w\lambda}(b)\,db \int_{\mathfrak{a}} g_2(H)e^{w\lambda(H)}\,dH\right\} e^{w\lambda(H_0)}\right|.$$

Since g_1 and g_2 are at our disposal, we see that for each w either $c_w = 0$ or $e^{w\lambda}$ is bounded on $\mathscr{C}$. Since $\mathscr{C}$ is arbitrary and the boundedness in the T^- direction is automatic, the theorem follows.

§2. Use of Patching Conditions

The patching conditions of §11.7 give information relating the values of a character on different Cartan subgroups. In this section we shall combine this information with Theorem 12.1 to show that Θ_λ is in principle determined by its value on the compact Cartan subgroup.

Theorem 12.6. Let rank G = rank K, and let $\lambda \in (i\mathfrak{b})'$ be nonsingular relative to $\Delta(\mathfrak{b}^{\mathbb{C}}:\mathfrak{g}^{\mathbb{C}})$ and have $\lambda + \delta_G$ analytically integral. If Θ is an invariant eigendistribution on G with infinitesimal character λ such that the numerators τ_T of Θ are bounded on every Cartan subgroup T, then Θ is uniquely determined by its numerator τ_B on the compact Cartan subgroup B.

Example. SU(2, 1).

The diagonal subgroup is a compact Cartan subgroup B, and we can use the usual notation for roots for $\mathfrak{su}(2,1)^{\mathbb{C}} = \mathfrak{sl}(3,\mathbb{C})$ as in §4.1. The compact roots are $\pm(e_1 - e_2)$, and $\pm(e_2 - e_3)$ and $\pm(e_1 - e_3)$ are noncompact. Every integral form is given by $ke_1 + le_2 + me_3$, but this realization is redundant since $e_1 + e_2 + e_3$ acts as 0. The form is nonsingular when the integers k, l, and m are distinct; and $W_G = W(\mathfrak{b}^{\mathbb{C}}:\mathfrak{g}^{\mathbb{C}})$ acts by permutation of the indices $\{1, 2, 3\}$. Thus every nonsingular integral form is given by $w\lambda$ for some $w \in W(\mathfrak{b}^{\mathbb{C}}:\mathfrak{g}^{\mathbb{C}})$ and for

$$\lambda = ke_1 + le_2 \qquad \text{with } k > l > 0.$$

Here λ may be regarded as the infinitesimal character under study. Then τ_B, which is real analytic on all of B (by Theorem 10.40), is necessarily of the form

$$\tau_B(b) = \sum_{w \in W_G} c_w \xi_{w\lambda}(b).$$

Actually the fact that τ_B is an odd function under $W_K = W(B:G) = W(\mathfrak{b}^{\mathbb{C}}:\mathfrak{k}^{\mathbb{C}})$ forces restrictions on the coefficients c_w, but these restrictions play no role in applying the patching conditions.

Let us form the Cayley transform relative to $e_2 - e_3$, obtaining a noncompact Cartan subgroup

$$T = \left\{ \begin{pmatrix} e^{-2i\theta} & & \\ & e^{i\theta}\cosh t & e^{i\theta}\sinh t \\ & e^{i\theta}\sinh t & e^{i\theta}\cosh t \end{pmatrix} \right\}.$$

It is convenient (and customary when unambiguous) to use the same notation for roots of $\Delta(\mathfrak{t}^{\mathbb{C}}:\mathfrak{g}^{\mathbb{C}})$ as for those of $\Delta(\mathfrak{b}^{\mathbb{C}}:\mathfrak{g}^{\mathbb{C}})$. Then we can parametrize T by its logarithms

$$H = xH_{e_2-e_3} + iyH_{e_2+e_3-2e_1}, \qquad x \in \mathbb{R}, y \in [-\pi, \pi).$$

The subset T'_R is given by $x \neq 0$ and has two components $\{x > 0\}$ and $\{x < 0\}$. Let us denote the values of τ_T on these components by τ_T^+ and τ_T^-. Then we have

$$\tau_T^+(x, y) = \sum_{w \in W(\mathfrak{t}^{\mathbb{C}}:\mathfrak{g}^{\mathbb{C}})} d_w^+ \xi_{w\lambda}(x, y) = \sum_w d_w^+ e^{x\langle w\lambda, e_2-e_3\rangle} e^{iy\langle w\lambda, e_2+e_3-2e_1\rangle} \tag{12.13'}$$

and similarly for $\tau_T^-(x, y)$ with coefficients d_w^-.

In (12.13′), x is arbitrarily large positively. Thus the boundedness of τ_T^+ for $x > 0$ forces $\langle w\lambda, e_2 - e_3\rangle < 0$ whenever $d_w^+ \neq 0$. Table 12.1 shows that

$$\tau_T^+(x, y) = d_{13}^+ e^{-x(k-l)+iy(k+l)} + d_{23}^+ e^{-xl+iy(l-2k)} + d_{132}^+ e^{-xk+iy(k-2l)}$$

and similarly

$$\tau_T^-(x, y) = d_1^- e^{xl+iy(l-2k)} + d_{12}^- e^{xk+iy(k-2l)} + d_{123}^- e^{x(k-l)+iy(k+l)}. \tag{12.14}$$

The function $\varepsilon_R^T(x, y)$ is just sgn x. The function $\varepsilon_R^T\tau_T$, according to Theorem 10.40, extends to be continuous at $x = 0$, and the result is that the d^-'s are determined by the d^+'s. The patching condition of Theorem 11.18 is that

$$H_{e_2-e_3}(\varepsilon_R^T\tau_T^+)(0^+, iy) = H_{e_2-e_3}(\tau_B)(\exp iyH_{e_2+e_3-2e_1}). \tag{12.15}$$

On the left side, $H_{e_2-e_3}$ acts on the wth term of (12.13′) by multiplication by $\langle w\lambda, e_2 - e_3\rangle$. Thus (12.15) says

$$\begin{aligned} -(k-l)d_{13}^+ e^{iy(k+l)} - ld_{23}^+ e^{iy(l-2k)} - kd_{132}^+ e^{iy(k-2l)} \\ = l(c_1 - c_{23})e^{iy(l-2k)} + (k-l)(c_{123} - c_{13})e^{iy(k+l)} \\ + k(c_{12} - c_{132})e^{iy(k-2l)}. \end{aligned}$$

TABLE 12.1. Eligible w in τ_T^+ for SU(2, 1)

w	$w\lambda$	$\langle w\lambda, e_2 - e_3\rangle$
1	$ke_1 + le_2$	+
(12)	$ke_2 + le_1$	+
(13)	$ke_3 + le_2$	−
(23)	$ke_1 + le_3$	−
(123)	$ke_2 + le_3$	+
(132)	$ke_3 + le_1$	−

The three exponentials are linearly independent, and we can equate coefficients to conclude

$$d_{13}^+ = -(c_{123} - c_{13}), \qquad d_{23}^+ = -(c_1 - c_{23}), \qquad d_{132}^+ = -(c_{12} - c_{132}). \tag{12.16a}$$

Therefore also

$$d_1^- = (c_1 - c_{23}), \qquad d_{12}^- = (c_{12} - c_{132}), \qquad d_{123}^- = (c_{123} - c_{13}). \tag{12.16b}$$

Substituting in (12.14), we see that τ_B indeed determines τ_T. We shall use (12.16) in the next section to write down the characters Θ_λ for SU(2, 1) explicitly.

Proof of Theorem 12.6. We may assume $G^{\mathbb{C}}$ is simply connected. We show by induction on the dimension of the noncompact part of T that τ_T is completely determined, the case of dimension 0 being the trivial case. Thus let a noncompact T be given, and let $t_1 \in T$ be regular. On the connected component C of t_1 within T'_R, τ_T is of the form

$$\tau_T(t_1 \exp H) = \sum c_w^C(t_1) e^{w\lambda(H)},$$

and we are to show that each $c_w^C(t_1)$ is completely determined by the data for the Cartan subgroups that are more compact than T.

First we shall normalize t_1, replacing it by an element t_0 that is easier to handle. Relative to the set Δ_R of real roots, let

$$\begin{aligned} \Delta_1 &= \{\alpha \in \Delta_R \,|\, \xi_\alpha(t_1) > 0\} \\ \mathfrak{a}_1 &= \sum_{\alpha \in \Delta_1} \mathbb{R} H_\alpha. \end{aligned}$$

For the time being, let us suppose that Δ_1 is nonempty. Then Δ_1 is a root system in $\mathfrak{a}_1'$, and we determine a positive system Δ_1^+ by

$$\Delta_1^+ = \{\alpha \in \Delta_1 \,|\, \xi_\alpha(t_1) > 1\}.$$

Let $\{\alpha_1, \ldots, \alpha_d\}$ be the corresponding simple system for Δ_1, and define $H_0 \in \mathfrak{a}_1$ by

$$e^{\alpha_j(H_0)} = \xi_{\alpha_j}(t_1).$$

Put $t_0 = t_1 \exp(-H_0)$, so that $\xi_\alpha(t_0) = 1$ for $\alpha \in \Delta_1$. Then t_0 is in $\bar{C}$ because $\alpha \in \Delta_R - \Delta_1$ and $H \in \mathfrak{a}$ imply $\xi_\alpha(t_1 \exp H) \neq 1$ (since ξ_α is negative at all such points). Hence $C \supseteq t_0 \exp \mathscr{C}$, where

$$\mathscr{C} = \{H \in \mathfrak{t} \,|\, \alpha_j(H) > 0 \text{ for } 1 \leq j \leq d\},$$

and $H \in \mathscr{C}$ implies

$$\tau_T(t_0 \exp H) = \sum_w c_w^C(t_0) e^{w\lambda(H)} = \sum_w c_w^C(t_1) e^{-w\lambda(H_0)} e^{w\lambda(H)}.$$

It is enough to show that the numbers $c_w^C(t_0)$ are completely determined.

Since λ is nonsingular and $\mathscr{C}$ is open in $\mathfrak{t}$, the functions $e^{w\lambda}$ are linearly independent on $\mathscr{C}$. Thus the assumption of boundedness of τ_T implies that each $e^{w\lambda}$ with $c_w^C(t_0) \neq 0$ is itself bounded on C. Let us analyze this restriction. Let P be the orthogonal projection of $\mathfrak{t}'$ on $\operatorname{span}\{\alpha_1, \ldots, \alpha_d\}$, and define Λ_i to be the member of $\operatorname{span}\{\alpha_1, \ldots, \alpha_d\}$ with $\langle \Lambda_i, \alpha_j \rangle = \delta_{ij}$. Then $tH_{\Lambda i}$ is in $\bar{\mathscr{C}}$ for $t \geq 0$. If $e^{w\lambda}$ is to be bounded on $\mathscr{C}$ (and hence also on $\bar{\mathscr{C}}$, we must have $\langle w\lambda, \Lambda_i \rangle \leq 0$. Thus

$$\langle w\lambda, \Lambda_i \rangle > 0 \text{ for some } i \text{ implies } c_w^C(t_0) = 0. \tag{12.17}$$

We shall apply the Cayley transform $\mathbf{d}_{\alpha_j}$ and use the corresponding patching condition of Theorem 11.18 for the differential operator H_{α_j}. Let

$$\mathscr{C}_j^* = \{H \in \bar{\mathscr{C}} \mid \alpha_j(H) = 0,\ \alpha_i(H) > 0 \text{ for } i \neq j\}.$$

For H in $\mathscr{C}_j^*$ we have

$$H_{\alpha_j}\tau_T(t_0 \exp H)^+ = \sum c_w^C(t_0)\langle w\lambda, \alpha_j \rangle e^{w\lambda(H)}. \tag{12.18a}$$

By means of $\mathbf{d}_{\alpha_j}$, we pass to a more compact Cartan subgroup T^*, and the set $t_0 \exp \mathscr{C}_j^*$ is contained in a component C^* of $T_R^{*\prime}$. On C^* we have a formula for τ_{T^*} that we are taking as known inductively, say

$$\tau_{T^*}(t_0 \exp H) = \sum_w c_w^{C^*}(t_0) e^{w\lambda(H)}.$$

The patching condition implies that

$$\varepsilon_R^T(C) H_{\alpha_j}\tau_T(t_0 \exp H)^+ = \varepsilon_R^{T^*}(C^*) H_{\alpha_j}\tau_{T^*}(t_0 \exp H) \qquad \text{for } H \in \mathscr{C}_j^*.$$

Here

$$H_{\alpha_j}\tau_{T^*}(t_0 \exp H) = \sum_w c_w^{C^*}(t_0)\langle w\lambda, \alpha_j \rangle e^{w\lambda(H)} \qquad \text{for } H \in \mathscr{C}_j^*. \tag{12.18b}$$

Distinct exponentials on $\mathscr{C}_j^*$ are linearly independent. Comparing (12.18a) and (12.18b), we see that we want to know which $w\lambda$'s restrict to the same linear functional on the kernel of α_j. Thus suppose

$$w_2\lambda = w_1\lambda + c\alpha_j.$$

Computing the length squared of both sides and using the fact that $|w_1\lambda| = |w_2\lambda|$, we obtain $2c\langle w_1\lambda, \alpha_j \rangle + c^2|\alpha_j|^2 = 0$. Hence

$$c = 0 \qquad \text{or} \qquad c = -\frac{2\langle w_1\lambda, \alpha_j \rangle}{|\alpha_j|^2}.$$

Thus $w_2\lambda = w_1\lambda$ or $w_2\lambda = s_{\alpha_j}w_1\lambda$.

Applying the patching condition to (12.18a), we see that $\langle w\lambda, \alpha_j \rangle \neq 0$ implies that $c_w^C(t_0) - c_{s_{\alpha_j}w}^C(t_0)$ is known. But $\langle w\lambda, \alpha_j \rangle \neq 0$ is automatic since λ is nonsingular. Thus

$$c_w^C(t_0) - c_{s_{\alpha_j}w}^C(t_0) \text{ is known for all } w \text{ and all } j. \tag{12.19}$$

Let W_1 be the Weyl group of Δ_1. Given w, find w_0 in W_1 such that $w_0 w\lambda$ is dominant for Δ_1^+. Then we can write $Pw_0 w\lambda = \sum c_i \Lambda_i$ with all $c_i \geq 0$. Taking the inner product with Λ_i and using Lemma 8.57, we see that $\langle w_0 w\lambda, \Lambda_i \rangle \geq c_i |\Lambda_i|^2$. Hence $\langle w_0 w\lambda, \Lambda_i \rangle > 0$ for some i unless $Pw_0 w\lambda = 0$. If $Pw_0 w\lambda = 0$, then $w_0 w\lambda$ is orthogonal to $\alpha_1, \ldots, \alpha_d$, in contradiction with the nonsingularity of λ. Thus we conclude from (12.17) that $c_{w_0 w}^C(t_0) = 0$. Now w_0^{-1} can be expanded as a noncommuting product of the elements s_{α_j}, and we can apply (12.19) recursively to see ultimately that $c_w^C(t_0) = c_{w_0^{-1} w_0 w}^C(t_0)$ is known. This completes the argument if Δ_1 is nonempty.

If Δ_1 is empty, then all of $t_1 \exp \mathfrak{t}$ is contained in the component C. Consequently we can deal with the coefficients $c_w^C(t_1)$ directly. Suppose some $c_w^C(t_1)$ is not 0. Then $e^{w\lambda}$ must be bounded on all of $\mathfrak{t}$, in particular on the line uH_α if α is a real root. Thus if there is a real root α, then $\langle w\lambda, \alpha \rangle = 0$, and this equality cannot hold since λ is nonsingular. We conclude there are no real roots. By Proposition 11.16a, T is compact, contradiction. Therefore Δ_1 empty (and T noncompact) implies $c_w^C(t_1) = 0$ for all w.

§3. Formula for Discrete Series Characters

Now we come to the final step in determining the discrete series characters Θ_λ, the computation of the numerator $\tau_{\lambda,B}$ on the compact Cartan subgroup. We assume, without actually including it as a condition, that δ_G is analytically integral, as when $G^{\mathbb{C}}$ is simply connected.

Theorem 12.7. Let rank G = rank K, and let $\lambda \in (i\mathfrak{b})'$ be nonsingular relative to $\Delta(\mathfrak{b}^{\mathbb{C}} : \mathfrak{g}^{\mathbb{C}})$ and have $\lambda + \delta_G$ analytically integral. Fix $\Delta^+(\mathfrak{b}^{\mathbb{C}} : \mathfrak{g}^{\mathbb{C}})$ so that λ is dominant. Then the discrete series character Θ_λ has the following properties:

(a) On the compact Cartan subgroup B, the numerator $\tau_{\lambda,B}$ of Θ_λ is given by

$$\tau_{\lambda,B}(b) = (-1)^q \sum_{w \in W_K} (\det w)\xi_{w\lambda}(b),$$

where $q = \frac{1}{2}\dim(G/K)$.

(b) $z\Theta_\lambda = \chi_\lambda(z)\Theta_\lambda$ for every z in $Z(\mathfrak{g}^{\mathbb{C}})$.

(c) On every Cartan subgroup T, the numerator $\tau_{\lambda,T}$ of Θ_λ is bounded.

Moreover, Θ_λ is the only invariant eigendistribution on G having these three properties.

Example. SU(2, 1).

We continue with the example of §2, referring everything to the simple system $\Pi = \{e_1 - e_2, e_2 - e_3\}$. We fix an infinitesimal character

$$\lambda = ke_1 + le_2 \qquad \text{with } k > l > 0.$$

The distinct discrete series with this infinitesimal character are parametrized by $W_K \backslash W_G$ applied to this λ, and we select Δ_K^+ dominant representatives of the parameters as in Table 12.2. The positive system of the theorem may be different from the one determined by Π, and that difference affects the sign of the Weyl denominator. Using Π as simple system for $\Delta(\mathfrak{b}^{\mathbb{C}}:\mathfrak{g}^{\mathbb{C}})$ and the Cayley transform by $e_2 - e_3$ as simple system for $\Delta(\mathfrak{t}^{\mathbb{C}}:\mathfrak{g}^{\mathbb{C}})$, we obtain the character formulas in Table 12.3 from Theorem 12.7a and from (12.16). The number q is 2 in Theorem 12.7a.

TABLE 12.2. Harish-Chandra parameters for SU(2, 1)

PARAMETER	FORM	RELATIVE SIGN FOR Δ_B FOR TWO ORDERINGS
$ke_1 + le_2$	λ	+
$ke_1 + le_3$	$(23)\lambda$	−
$ke_3 + le_1$	$(132)\lambda$	+

TABLE 12.3. Discrete series characters for SU(2, 1)

FORM	τ_B	τ_T^+	τ_T^-
λ	$\xi_\lambda - \xi_{(12)\lambda}$	$\xi_{(132)\lambda} - \xi_{(23)\lambda}$	$\xi_\lambda - \xi_{(12)\lambda}$
$(23)\lambda$	$\xi_{(123)\lambda} - \xi_{(23)\lambda}$	$-\xi_{(13)\lambda} - \xi_{(23)\lambda}$	$\xi_{(123)\lambda} + \xi_\lambda$
$(132)\lambda$	$\xi_{(132)\lambda} - \xi_{(13)\lambda}$	$\xi_{(132)\lambda} - \xi_{(13)\lambda}$	$\xi_{(123)\lambda} - \xi_{(12)\lambda}$

Now let us turn to the proof of Theorem 12.7. We know that Θ_λ satisfies (b), and Theorem 12.1 proves that Θ_λ satisfies (c). The uniqueness follows from Theorem 12.6. Thus it remains to prove that Θ_λ satisfies (a). Our only way so far of distinguishing among the discrete series $\pi_{w\lambda}$ for $w \in W_G$ is by means of the information about K types given in Theorem 9.20, and the problem is to show how this information affects global characters. Before giving the proof in detail, we give a brief outline, ignoring some technicalities.

The first step is to introduce the K character of a representation of G. If π is the representation, the K character will be the distribution on K

given by

$$\Theta_K(f) = \operatorname{Tr} \int_K f(k)\pi(k)\, dk, \qquad f \in C^\infty(K). \tag{12.20}$$

We prove that the K character exists and is given on the G-regular set of K by a function $\Theta_K(k)$. Moreover Θ_K coincides on B with the function that realizes the global G character of π. We shall apply this fact to Θ_λ. First we observe that

$$\frac{\tau_{\lambda,B}}{D_B^G}\left(\prod_{\alpha \in \Delta_n^+} (\xi_{\alpha/2} - \xi_{-\alpha/2})\right) = \sum_{\substack{w \in W_G \\ w\lambda \text{ dominant} \\ \text{for } \Delta_K^+}} c_w \chi_{w\lambda - \delta_K} \text{ on } B', \tag{12.21}$$

where $\chi_{w\lambda-\delta_K}$ is the character of the finite-dimensional irreducible representation of K with highest weight $w\lambda - \delta_K$.

Next we interpret $\prod_{\alpha \in \Delta_n^+} (\xi_{\alpha/2} - \xi_{-\alpha/2})$ as the difference of two specific finite-dimensional characters of K, say $\chi^+ - \chi^-$. So (12.21) and our fact about K characters give

$$\Theta_{\lambda,K}(\chi^+ - \chi^-) = \sum_{\substack{w \in W_G \\ w\lambda \text{ dominant} \\ \text{for } \Delta_K^+}} c_w \chi_{w\lambda - \delta_K} \text{ on } B'. \tag{12.22}$$

The left side here should be viewed as the character of the formal difference of two tensor products, and we show that this difference is finite-dimensional. Then (12.22) identifies the difference. A consideration of weights shows that the only K type of π_λ that can give a net contribution to the left side is the one with highest weight $\lambda - \delta_K + \delta_n$, and the theorem follows easily.

Lemma 12.8. Let rank G = rank K, let π be an irreducible admissible representation of G, and let Θ be the global character of π, regarded as a function $\Theta(x)$. Then the K character of π, defined as in (12.20), exists as a distribution on K. Moreover it is given on the regular set $(B')^G \cap K$ by a real analytic function $\Theta_K(k)$, and Θ_K coincides with Θ on $(B')^G \cap K$.

Remark. This result in the case of $SL(2, \mathbb{R})$ is essentially what is proved in lines (10.19a) to (10.19f) in the course of the calculation of the character of $\mathscr{D}_n^+$.

Proof. Let $\{v_i\}$ be an orthonormal basis of the representation space of π. The same computation as in (10.1) shows for $f \in C^\infty(K)$ that

$$\sum_{i,j} \left| \int_K f(k)(\pi(k)v_i, v_j) \right| dk = \sum_\mu d_\mu^4 |\chi_{\mu+\delta_K}(z)|^{-1} \|\pi(zf)\| \tag{12.23}$$

for suitably defined zf whenever z is in $Z(\mathfrak{k}^{\mathbb{C}})$. The finiteness of (10.2) for some z allows us to conclude that Θ_K exists as a distribution.

We let $\psi(b)$ be the following quotient of squares of Weyl denominators: $\psi(b) = |D_B^G(b)|^2/|D_B^K(b)|^2$. Then ψ is invariant under conjugation by W_K, and we can extend ψ to a member of $C^\infty(K)$ invariant under conjugation. For $h \in C_{\text{com}}^\infty((B')^G)$, we make a computation of $\Theta(h)$, justifying the steps afterward. We have

$$\begin{aligned}
&\int_G F(x)\Theta(x)\,dx \\
&= \operatorname{Tr} \int_G F(x)\pi(x)\,dx && (12.24\text{a}) \\
&= |W_K|^{-1} \operatorname{Tr} \int_{G\times B} F(xbx^{-1})\pi(xbx^{-1})|D_B^G(b)|^2\,db\,dx \\
&= |W_K|^{-1} \operatorname{Tr} \int_{G\times K\times B} F(xkbk^{-1}x^{-1})\pi(xkbk^{-1}x^{-1})\psi(kbk^{-1})|D_B^K(b)|^2 \\
&\qquad \times db\,dk\,dx \\
&= \operatorname{Tr} \int_G \int_K F(xkx^{-1})\pi(xkx^{-1})\psi(k)\,dk\,dx \\
&= \int_G \left[\operatorname{Tr} \int_K F(xkx^{-1})\pi(x)\pi(k)\pi(x)^{-1}\psi(k)\,dk \right] dx && (12.24\text{b}) \\
&= \int_G \left[\operatorname{Tr} \int_K F(xkx^{-1})\pi(k)\psi(k)\,dk \right] dx && (12.24\text{c}) \\
&= \operatorname{Tr} \int_G \int_K F(xkx^{-1})\pi(k)\psi(k)\,dk\,dx. && (12.24\text{d})
\end{aligned}$$

The first few steps are by the Weyl integration formula. To pass to (12.24b), we use dominated convergence; the support is compact in the x variable since F is in $C_{\text{com}}^\infty((B')^G)$, and everything is uniformly bounded as a result of (12.23). A similar argument allows us to pass from (12.24c) to (12.24d). The equality of (12.24b) and (12.24c) follows by the invariance of traces under conjugation.

To apply (12.24) we need the following formula for how Haar measure for G decomposes under the Cartan decomposition:

$$\int_G g(x)\,dx = \int_{\mathfrak{p}\times K} g((\exp X)k)J(X)\,dk\,dX \tag{12.25}$$

for a positive smooth function J on $\mathfrak{p}$. [In fact, there is certainly such a formula with $J(k, X)$ in place of $J(X)$. If we replace $f(x)$ by $f(xk_0)$ and unwind matters, then we find that $J(k, X)$ is constant in the first variable. Thus we have (12.25).]

Fix a right K-invariant smooth function $h \geq 0$ on G with compact support and with total integral 1. Now let $f \in C_{\text{com}}^\infty((B')^G \cap K)$ be given. We define $F \in C_{\text{com}}^\infty((B')^G)$ by

$$F(xkx^{-1}) = h(x)f(k_x k k_x^{-1})\psi(k_x k k_x^{-1})^{-1},$$

where k_x is given by the Cartan decomposition: $x = (\exp X)k_x$. For this F the left side of (12.24a) is

$$\begin{aligned}
&\int_G F(x)\Theta(x)\,dx \\
&= |W_K|^{-1} \int_{G\times B} F(xbx^{-1})\Theta(b)|D_B^G(b)|^2\,db\,dx \\
&= |W_K|^{-1} \int_{\mathfrak{p}\times K\times B} F((\exp X)kbk^{-1}(\exp X)^{-1})\Theta(b)|D_B^G(b)|^2\,db\,J(X)\,dk\,dX \\
&= |W_K|^{-1} \left(\int_{\mathfrak{p}} h(\exp X)J(X)\,dX\right) \int_{K\times B} f(kbk^{-1})\psi(kbk^{-1})^{-1}\Theta(b)|D_B^G(b)|^2 \\
&\quad \times db\,dk \\
&= |W_K|^{-1} \left(\int_{\mathfrak{p}} h(\exp X)J(X)\,dX\right) \int_{K\times B} f(kbk^{-1})\Theta(b)|D_B^K(b)|^2\,db\,dk.
\end{aligned} \tag{12.26}$$

Meanwhile the (equal) right side of (12.24d) is

$$\begin{aligned}
&\operatorname{Tr} \int_{G\times K} F(xkx^{-1})\pi(k)\psi(k)\,dk\,dx \\
&\qquad = \operatorname{Tr} \int_{\mathfrak{p}\times K\times K} h(\exp X) f(k_0kk_0^{-1})\psi(k_0kk_0^{-1})^{-1}\pi(k)\psi(k)J(X)\,dX\,dk_0\,dk \\
&\qquad = \left(\int_{\mathfrak{p}} h(\exp X)J(X)\,dX\right) \operatorname{Tr} \int_{K\times K} f(k_0kk_0^{-1})\pi(k)\,dk_0\,dk \\
&\qquad = \left(\int_{\mathfrak{p}} h(\exp X)J(X)\,dX\right) \operatorname{Tr} \int_{K\times K} f(k)\pi(k_0^{-1}kk_0)\,dk_0\,dk \\
&\qquad = \left(\int_{\mathfrak{p}} h(\exp X)J(X)\,dX\right) \operatorname{Tr} \int_K f(k)\pi(k)\,dk.
\end{aligned} \tag{12.27}$$

Equating (12.26) and (12.27), we obtain the remaining conclusions of the lemma.

The next step in the proof of Theorem 12.7 is to recognize

$$\prod_{\alpha\in\Delta_n^+} (\xi_{\alpha/2} - \xi_{-\alpha/2}) = \xi_{\delta_n} \prod_{\alpha\in\Delta_n^+} (1 - \xi_\alpha^{-1})$$

as the difference of two finite-dimensional characters for K. For this purpose we begin by studying the "spin representations" of $SO(2m)$, working explicitly with coordinates for concreteness.

Thus let $u_1, \ldots, u_{2m}$ be the standard orthonormal basis of $\mathbb{R}^{2m}$. The **Clifford algebra** $\operatorname{Cliff}(\mathbb{R}^{2m})$ is an associative algebra over $\mathbb{R}$ of dimension 2^{2m} with a basis parametrized by subsets of $\{1, \ldots, 2m\}$ and given by

$$\{u_{i_1}u_{i_2}\cdots u_{i_k} \mid i_1 < i_2 < \ldots < i_k\}.$$

The generators multiply by the rules

$$u_i^2 = -1, \qquad u_i u_j = -u_j u_i \quad \text{if} \quad i \neq j.$$

The Clifford algebra, like any associative algebra, becomes a Lie algebra under the bracket operation $[x, y] = xy - yx$, and

$$\mathfrak{q} = \sum_{i \neq j} \mathbb{R} u_i u_j$$

is a Lie subalgebra isomorphic to $\mathfrak{so}(2m)$, the isomorphism being

$$\tfrac{1}{2} u_i u_j \leftrightarrow E_{ji} - E_{ij}. \tag{12.28}$$

Let us call this isomorphism $\varphi : \mathfrak{so}(2m) \to \mathfrak{q}$. If we continue (as we have so far done) to identify members of $\mathbb{R}^{2m}$ with their images as first-degree elements of $\mathrm{Cliff}(\mathbb{R}^{2m})$, then we can check that

$$[\varphi(x), u_j] = x u_j \qquad \text{for all} \quad x \in \mathfrak{so}(2m). \tag{12.29}$$

Here the left side is a bracket in $\mathrm{Cliff}(\mathbb{R}^{2m})$, and the right side is the product of the matrix x by the column vector u_j, all reinterpreted in $\mathrm{Cliff}(\mathbb{R}^{2m})$.

Now let us pass to the complexification $\mathrm{Cliff}^{\mathbb{C}}(\mathbb{R}^{2m})$ and denote left multiplication by c, putting $c(x)y = xy$. Then c is a representation of the associative algebra $\mathrm{Cliff}^{\mathbb{C}}(\mathbb{R}^{2m})$ on itself, hence also of the Lie algebra. For $1 \leq j \leq m$, let

$$z_j = u_{2j-1} + i u_{2j} \qquad \text{and} \qquad \bar{z}_j = u_{2j-1} - i u_{2j}.$$

For each subset S of $\{1, \ldots, m\}$, define

$$z_S = \Bigl(\prod_{j \in S} z_j\Bigr)\Bigl(\prod_{j=1}^{m} \bar{z}_j\Bigr),$$

with each product arranged so that the indices are in increasing order. Finally put

$$\mathscr{S} = \sum_{S \subseteq \{1, \ldots, m\}} \mathbb{C} z_S.$$

Lemma 12.9. Under the representation c of the Lie algebra $\mathrm{Cliff}^{\mathbb{C}}(\mathbb{R}^{2m})$ on itself, $\mathscr{S}$ is an invariant subspace of dimension 2^m. If c is restricted to the Lie subalgebra $\mathfrak{q} \cong \mathfrak{so}(2m)$, then $\mathscr{S}$ splits as the direct sum of two invariant subspaces

$$\mathscr{S}^+ = \sum_{|S| \text{ even}} \mathbb{C} z_S \qquad \text{and} \qquad \mathscr{S}^- = \sum_{|S| \text{ odd}} \mathbb{C} z_S,$$

each of dimension 2^{m-1}. With the isomorphism (12.28) and the notation of §4.1 in effect, the weights of $\mathscr{S}^+$ are

$$\pm \tfrac{1}{2} e_1 \pm \tfrac{1}{2} e_2 \pm \ldots \pm \tfrac{1}{2} e_m \tag{12.30}$$

with an even number of minus signs, and the weights of $\mathscr{S}^-$ are (12.30) with an odd number of minus signs. The weight vectors are the z_S's.

Remarks. The representations $(c \circ \varphi, \mathscr{S}^+)$ and $(c \circ \varphi, \mathscr{S}^-)$ are called the **spin representations** of $\mathfrak{so}(2m)$. From the Weyl dimension formula we can check that they are irreducible on $\mathfrak{so}(2m)$, but we shall not use this fact.

Proof. We check directly that

$$z_j^2 = \bar{z}_j^2 = 0 \qquad \text{and} \qquad \bar{z}_j z_j \bar{z}_j = -4z_j,$$

and then it follows that

$$c(z_j)z_S = \begin{cases} \pm z_{S\cup\{j\}} & \text{if } j \notin S \\ 0 & \text{if } j \in S \end{cases}$$

$$c(\bar{z}_j)z_S = \begin{cases} 0 & \text{if } j \notin S \\ \pm 4z_{S-\{j\}} & \text{if } j \in S. \end{cases}$$

Since the z_j's and $\bar{z}_j$'s together generate $\mathrm{Cliff}^{\mathbb{C}}(\mathbb{R}^{2m})$, $\mathscr{S}$ is an invariant subspace. The parity of the number of elements of S changes under each $c(z_j)$ or $c(\bar{z}_j)$, hence under each $c(u_{2j-1})$ or $c(u_{2j})$. Hence $c(\mathfrak{q})$ leaves $\mathscr{S}^+$ and $\mathscr{S}^-$ stable.

Under the isomorphism φ in (12.28) and the notation of §4.1, the Cartan subalgebra of $\mathfrak{q}$ is $\sum \mathbb{R}u_{2j}u_{2j-1}$, and $\frac{1}{2}iu_{2j}u_{2j-1}$ is φ of the element of $\mathfrak{q}^{\mathbb{C}}$ on which e_j is 1 and e_i is 0 for $i \neq j$. Since

$$\begin{aligned} c(iu_{2j}u_{2j-1})z_j &= -z_j \\ c(iu_{2j}u_{2j-1})\bar{z}_j &= +\bar{z}_j, \end{aligned} \tag{12.31}$$

we obtain

$$c(\varphi(h))z_S = \frac{1}{2}\left(\sum_{j \notin S} e_j - \sum_{j \in S} e_j\right)(h)z_S$$

for all h in the complexification of the Cartan subalgebra of $\mathfrak{so}(2m)$. The lemma follows.

Complementing (12.31) are the identities

$$z_j(iu_{2j}u_{2j-1}) = +z_j \qquad \text{and} \qquad \bar{z}_j(iu_{2j}u_{2j-1}) = -\bar{z}_j.$$

Combining these identities with (12.31), we obtain

$$[iu_{2j}u_{2j-1}, z_j] = -2z_j \qquad \text{and} \qquad [iu_{2j}u_{2j-1}, \bar{z}_j] = +2\bar{z}_j.$$

Hence h in the complexification of the Cartan subalgebra of $\mathfrak{so}(2m)$ implies

$$[\varphi(h), z_j] = -e_j(h)z_j \qquad \text{and} \qquad [\varphi(h), \bar{z}_j] = +e_j(h)\bar{z}_j. \tag{12.32}$$

Now let us pass to our arbitrary linear connected semisimple group G with rank $G =$ rank K. The real part of the trace form gives an inner product on $\mathfrak{p}$. We fix a positive system $\Delta^+(\mathfrak{b}^{\mathbb{C}}:\mathfrak{g}^{\mathbb{C}})$ and let $\alpha_1, \ldots, \alpha_m$ be an enumeration of the positive noncompact roots. From Problem 5 in Chapter VI, we know that we can choose root vectors E_{α_j} and $E_{-\alpha_j}$ for α_j and $-\alpha_j$ such that

$$u_{2j-1} = E_{\alpha_j} + E_{-\alpha_j} \qquad \text{and} \qquad u_{2j} = i(E_{\alpha_j} - E_{-\alpha_j}) \tag{12.33}$$

are in $\mathfrak{g}$ (hence in $\mathfrak{p}$), and we normalize them so that u_{2j-1} and u_{2j} have norm one. Then $\{u_1, \ldots, u_{2m}\}$ is an orthonormal basis of $\mathfrak{p}$, and we can apply the above theory, regarding $\mathfrak{p}$ as $\mathbb{R}^{2m}$. Then we have

$$z_j = 2E_{-\alpha_j} \qquad \text{and} \qquad \bar{z}_j = 2E_{\alpha_j}.$$

If Y is in $\mathfrak{k}$, then ad Y is in $\mathfrak{so}(\mathfrak{p}) = \mathfrak{so}(\mathbb{R}^{2m})$. Thus, for X in $\mathfrak{p}^{\mathbb{C}}$, (12.29) gives

$$[\varphi(\text{ad } Y), X] = (\text{ad } Y)X \tag{12.34}$$

within $\text{Cliff}^{\mathbb{C}}(\mathfrak{p})$.

For H in $\mathfrak{b}$, we compute directly that

$$\begin{aligned}(\text{ad } H)u_{2j-1} &= -i\alpha_j(H)u_{2j}\\ (\text{ad } H)u_{2j} &= +i\alpha_j(H)u_{2j-1}.\end{aligned}$$

Using these formulas to write out the matrix of ad H, we see that ad H is in the Cartan subalgebra of $\mathfrak{so}(2m)$. Thus (12.32) applies with $h =$ ad H, giving

$$\begin{aligned}[\varphi(\text{ad } H), E_{-\alpha_j}] &= -e_j(\text{ad } H)E_{-\alpha_j}\\ [\varphi(\text{ad } H), E_{\alpha_j}] &= e_j(\text{ad } H)E_{\alpha_j}.\end{aligned}$$

Substituting from (12.34), we see that

$$\alpha_j(H) = e_j(\text{ad } H) \qquad \text{for } H \in \mathfrak{b}^{\mathbb{C}}. \tag{12.35}$$

Referring to Lemma 12.9, we see that we have constructed two (possibly reducible) representations $(c \circ \varphi \circ \text{ad}, \mathcal{S}^+)$ and $(c \circ \varphi \circ \text{ad}, \mathcal{S}^-)$ of $\mathfrak{k}$, both of dimension 2^{m-1}. We refer to these briefly as $(c, \mathcal{S}^+)$ and $(c, \mathcal{S}^-)$. The weights, according to the lemma and (12.35), are

$$\pm\tfrac{1}{2}\alpha_1 \pm \tfrac{1}{2}\alpha_2 \pm \cdots \pm \tfrac{1}{2}\alpha_m, \tag{12.36}$$

with an even number of minus signs in the case of $\mathcal{S}^+$ and an odd number in the case of $\mathcal{S}^-$. All of them have multiplicity one.

Lifting c to a representation of K involves a technical problem in that the weights (12.36) may not be analytically integral. We get around this difficulty as follows: Let Spin$(2m)$ denote the double cover of SO$(2m)$;

we regard the Lie algebra of Spin(2*m*) as just $\mathfrak{so}(2m)$. For $m > 1$, Spin(2*m*) is simply connected, and c therefore lifts to give representations of Spin(2*m*) on $\mathscr{S}^+$ and $\mathscr{S}^-$; for $m = 1$, we check directly that Spin(2) acts on $\mathscr{S}^+$ and $\mathscr{S}^-$. Let $\tilde{K}$ be the analytic subgroup of Spin(2*m*) with Lie algebra $\mathfrak{k} \subseteq \mathfrak{so}(2m)$. Either $\tilde{K}$ is isomorphic to K, or else $\tilde{K}$ is a double cover of K. In any case the restriction of the spin representations gives us representations of $\tilde{K}$ on $\mathscr{S}^+$ and $\mathscr{S}^-$. We denote their characters by χ^+ and χ^-.

Lemma 12.10. The characters χ^+ and χ^- of the representations $\mathscr{S}^+$ and $\mathscr{S}^-$ of $\tilde{K}$ satisfy

$$\chi^+ - \chi^- = \prod_{\alpha \in \Delta_n^+} (\xi_{\alpha/2} - \xi_{-\alpha/2}) = \xi_{\delta_n} \prod_{\alpha \in \Delta_n^+} (1 - \xi_\alpha^{-1}). \tag{12.37}$$

Remark. The middle expression in (12.37) is the formal one that is easy to work with; its rigorous form is the expression on the right side in (12.37).

Proof. The middle expression in (12.37), when multiplied out, is

$$= \sum_{\substack{S \subseteq \Delta_n^+ \\ |S| \text{ even}}} \xi_{(1/2)(\sum_{\alpha \notin S}\alpha - \sum_{\alpha \in S}\alpha)} - \sum_{\substack{S \subseteq \Delta_n^+ \\ |S| \text{ odd}}} \xi_{(1/2)(\sum_{\alpha \notin S}\alpha - \sum_{\alpha \in S}\alpha)},$$

which is $\chi^+ - \chi^-$ by (12.36).

Let π be an admissible representation of G, and let V_0 be the space of K-finite vectors. The **formal Dirac operator** for π is the operator $D: V_0 \otimes \mathscr{S} \to V_0 \otimes \mathscr{S}$ given by

$$D = \sum_i \pi(X_i) \otimes c(X_i), \tag{12.38}$$

where $\{X_i\}$ is any orthonormal basis of $\mathfrak{p}$ and $c(X_i)$ denotes left multiplication in the Clifford algebra. This definition is independent of the basis $\{X_i\}$. [In fact, if $\{Y_j\}$ is another orthonormal basis, write $Y_j = \sum_i a_{ij} X_i$ for an orthogonal matrix (a_{ij}). If $(a^{-1})_{jk}$ denotes a typical entry of the inverse matrix, we have

$$\begin{aligned}\sum_j \pi(Y_j) \otimes c(Y_j) &= \sum_{i,j,k} a_{ij} a_{kj} \pi(X_i) \otimes c(X_k) \\ &= \sum_{i,j,k} a_{ij} (a^{-1})_{jk} \pi(X_i) \otimes c(X_k) = \sum_i \pi(X_i) \otimes c(X_i).\end{aligned}$$

Hence (12.38) is independent of basis.]

The group $\tilde{K}$ acts in $V_0 \otimes \mathscr{S}$ by $\pi \otimes c$, and each $\tilde{K}$ type has finite multiplicity. The above independence of basis shows that D commutes with this action by $\tilde{K}$. Hence D maps each $\tilde{K}$ type into itself. Moreover, it is clear from (12.38) that D maps $V_0 \otimes \mathscr{S}^+$ to $V_0 \otimes \mathscr{S}^-$ and vice versa. We shall use D to prove the following lemma.

Lemma 12.11. Let $\pi|_K = \sum_{\tau \in K} n_\tau \tau$ be the K space decomposition of an irreducible admissible representation π, and expand

$$\left(\sum_{\tau \in K} n_\tau \tau\right)(\chi^+ - \chi^-)$$

as a formal (possibly infinite) linear combination of irreducible characters of $\tilde{K}$. Then the resulting expansion is in fact a finite linear combination of irreducible characters of $\tilde{K}$.

To prove Lemma 12.11, we use the formula for D^2 given in Lemma 12.12 below.

Lemma 12.12. Let π be an admissible representation of G, and let the Casimir operators of G and K be given by

$$\Omega_K = -(Y_1^2 + \ldots + Y_n^2)$$
$$\Omega = (u_1^2 + \ldots + u_m^2) - (Y_1^2 + \ldots + Y_n^2),$$

where $Y_1, \ldots, Y_n$ is an orthonormal basis of $\mathfrak{k}$ (with respect to $-\text{Re } B_0$) and $u_1, \ldots, u_m$ is the orthonormal basis of $\mathfrak{p}$ (with respect to $\text{Re } B_0$) given in (12.33). Then the square of the formal Dirac operator is given by

$$D^2 = (\pi \otimes c)(\Omega_K) - \pi(\Omega) \otimes 1 - 1 \otimes c(\Omega_K).$$

Proof. For Y in $\mathfrak{k}$, we need a formula for $(\varphi \text{ ad } Y)(Y)$ as a member of $\mathfrak{q} \subseteq \text{Cliff } \mathfrak{p}$. We must have $(\varphi \text{ ad } Y)(Y) = \sum_{i<j} c_{ij} u_i u_j$ for suitable constants c_{ij}. Bracketing both sides with u_k and expanding out the results by means of (12.34), we readily find

$$\varphi(\text{ad } Y) = \frac{1}{2} \sum_{i<j} \text{Re } B_0([Y, u_i], u_j) u_i u_j. \tag{12.39}$$

Now

$$\begin{aligned} D^2 &= \sum_{i,j} \pi(u_i u_j) \otimes c(u_i u_j) \\ &= \sum_i \pi(u_i^2) \otimes c(-1) + \sum_{i<j} \pi[u_i, u_j] \otimes c(u_i u_j). \end{aligned} \tag{12.40}$$

Since $\{Y_q\}$ is an orthonormal basis of $\mathfrak{k}$ relative to $-\text{Re B}_0$, we have

$$[u_i, u_j] = -\sum_q \text{Re } B_0([u_i, u_j], Y_q) Y_q = -\sum_q \text{Re } B_0([Y_q, u_i], u_j) Y_q.$$

Thus the second term on the right of (12.40) is

$$\begin{aligned} &= -\sum_q \pi(Y_q) \otimes c\left(\sum_{i<j} \text{Re } B_0([Y_q, u_i], u_j) u_i u_j\right) \\ &= -2 \sum_q \pi(Y_q) \otimes c(\varphi(\text{ad } Y_q)) \qquad \text{by (12.39).} \end{aligned}$$

In this expression, $c \circ \varphi \circ \text{ad}$ is the representation of $\mathfrak{k}$ that we have abbreviated c. Thus this sum is

$$= -\sum_q \{(\pi(Y_q) \otimes 1 + 1 \otimes c(Y_q))^2 - \pi(Y_q^2) \otimes 1 - 1 \otimes c(Y_q)^2\}$$
$$= (\pi \otimes c)(\Omega_K) + \sum_q (\pi(Y_q^2) \otimes 1) - 1 \otimes c(\Omega_K).$$

Substituting into (12.40), we obtain the conclusion of Lemma 12.12.

Proof of Lemma 12.11. Let V_0 be the space of K-finite vectors of π. What is to be shown is that all but finitely many irreducible representations of $\tilde{K}$ have the same multiplicity in $V_0 \otimes \mathscr{S}^+$ as in $V_0 \otimes \mathscr{S}^-$. The operator D maps $V_0 \otimes \mathscr{S}^+$ to $V_0 \otimes \mathscr{S}^-$ and $V_0 \otimes \mathscr{S}^-$ to $V_0 \otimes \mathscr{S}^+$ while preserving $\tilde{K}$ types. Thus it is enough to know that D^2 is one-one onto for all but finitely many $\tilde{K}$ types. Since the subspace of $V_0 \otimes \mathscr{S}$ of a given $\tilde{K}$ type is finite-dimensional, it is enough to show that D^2 is one-one on all but finitely many $\tilde{K}$ types.

We use the formula of Lemma 12.12. The operator $\pi(\Omega) \otimes 1$ is scalar since π is irreducible, and $1 \otimes c(\Omega_K)$ is diagonable with bounded eigenvalues. The operator $(\pi \otimes c)(\Omega_K)$ is scalar on each $\tilde{K}$ type, and for any positive number C, the scalar $|(\pi \otimes c)(\Omega_K)|$ is $\leq C$ on only finitely many $\tilde{K}$ types. Thus D^2 can have a kernel within only finitely many $\tilde{K}$ types, and Lemma 12.11 is proved.

Proof of Theorem 12.7. We are to prove property (a). Let

$$W_1 = \{w \in W_G \mid w\lambda \text{ is dominant for } \Delta_K^+\}.$$

Since $\tau_{\lambda,B}$ is an odd function under the action of W_K, we can wrtie

$$\tau_{\lambda,B}(b) = \sum_{w \in W_1} c_w \sum_{s \in W_K} (\det s)\xi_{sw\lambda}(b).$$

Lifting $\tau_{\lambda,B}$ to $\tilde{K}$, we can rewrite $\tau_{\lambda,B}$ in terms of finite-dimensional characters as in (12.21) on the regular set B'.

We form the $\tilde{K}$ representations $\pi_\lambda \otimes \mathscr{S}^+$ and $\pi_\lambda \otimes \mathscr{S}^-$. Their distribution characters (on $\tilde{K}$) exist by Lemma 12.8 and are given as functions on $\tilde{B}'$, namely $\Theta_\lambda(b)\chi^+(b)$ and $\Theta_\lambda(b)\chi^-(b)$. Lemma 12.11 implies that the difference of these two distribution characters is a function on $\tilde{K}$, and (12.21) identifies this function. We can rewrite (12.21) as (12.22), valid on B. Obviously the numbers c_w must now be integers.

Now suppose $c_w \neq 0$. Then the representation of $\tilde{K}$ of highest weight $w\lambda - \delta_K$ occurs in the tensor product of some $\tilde{K}$ type of $\pi|_K$, say the one of type μ, and either $\mathscr{S}^+$ or $\mathscr{S}^-$. By Problem 13 in Chapter IV, $w\lambda - \delta_K$ must be the sum of μ and some weight of $\mathscr{S}^+$ or $\mathscr{S}^-$, say

$$w\lambda - \delta_K = \mu + \left(-\delta_n + \sum_{\alpha \in S \subseteq \Delta_n^+} \alpha\right). \qquad (12.41a)$$

Theorem 9.20 says that μ is of the form $\lambda - \delta_K + \delta_n + \sum_{\beta \in \Delta^+} \beta$, and thus

$$w\lambda = \lambda + \sum_{\beta \in \Delta^+} n_\beta \beta + \sum_{\alpha \in S \subseteq \Delta_n^+} \alpha. \tag{12.41b}$$

Since λ is Δ^+ dominant, this can happen only if all of the n_β's are 0 and the set $S \subseteq \Delta_n^+$ is empty. Since λ is nonsingular, $w = 1$. Thus $c_w = 0$ for $w \neq 1$. Moreover this identity of weights occurs in only one way, and thus $|c_1| \leq 1$, by Problem 14 of Chapter IV. If $c_1 = 0$, then $\Theta_\lambda = 0$ (by the already proved uniqueness), which is not the case. Thus $c_1 = +1$ or $c_1 = -1$.

To identify the sign, we return to (12.41) to decide whether the weight of $\mathscr{S}$ is in $\mathscr{S}^+$ or $\mathscr{S}^-$. The relevant weight is $-\delta_n$. To achieve this weight in (12.36), we need $m = |\Delta_n^+| = \frac{1}{2}\dim(G/K) = q$ minus signs. Thus the weight is in $\mathscr{S}^+$ if q is even, in $\mathscr{S}^-$ if q is odd. When the weight is in $\mathscr{S}^+$, we want the right side of (12.22) to include a character with positive coefficient, and when it is in $\mathscr{S}^-$, we want the right side of (12.22) to include a character with negative coefficient. Thus $c_1 = (-1)^q$, and the proof is complete.

§4. Schwartz Space

In this section and the next, we shall show that the only discrete series representations of a linear connected semisimple G are those given in Theorem 9.20. One tool for this purpose will be the character formulas just obtained. Another is the analytic tool F_f. If we visualize the Plancherel formula for G as applied to a matrix coefficient of a discrete series representation, only that one discrete series ought to contribute to the formula. So it would be useful to study F_f when f is such a matrix coefficient. Unfortunately such a function f is not in $C_{\text{com}}^\infty(G)$. Thus we want to expand the domain of $f \to F_f$ to include a wider class of functions. We begin with motivation that identifies this wider class of functions, which will be called the "Schwartz space" of G.

Lemma 12.3 used $\varphi_0^G(x)(1 + \|x\|)^{-m}$ to measure a decay rate on G. This rate of decay is a natural generalization of polynomial decrease in $\mathbb{R}^n$, as the next two propositions show. We shall sketch a proof of the first proposition and indicate what changes are needed for the second.

As we observed in §9.8, members of $U(\mathfrak{g}^{\mathbb{C}})$ can act as left-invariant differential operators or as right-invariant differential operators, and we denote these two actions by the element D of $U(\mathfrak{g}^{\mathbb{C}})$ by the symbols D and D_R, respectively.

Proposition 12.13. Let G be linear connected reductive. For each pair (D, E) in $U(\mathfrak{g}^{\mathbb{C}}) \times U(\mathfrak{g}^{\mathbb{C}})$, let $\nu_{D,E}$ be the seminorm on $C^\infty(G)$ given by the

L^2 norm of the mixed left and right derivative:

$$\nu_{D,E}(F) = \|DE_R F\|_{2,G}.$$

Then there exist finitely many pairs (D_i, E_i) such that

$$|F(x)| \le \varphi_0^G(x) \sum \nu_{D_i,E_i}(F)$$

for all F in $C^\infty(G)$ and x in G.

Proposition 12.14. Let G be linear connected reductive. For each pair (D, E) in $U(\mathfrak{g}^{\mathbb{C}}) \times U(\mathfrak{g}^{\mathbb{C}})$ and integer $m \ge 0$, let $\nu_{D,E}^{(m)}$ be the seminorm on $C^\infty(G)$ given by

$$\nu_{D,E}^{(m)}(F) = \|(1 + \|\cdot\|)^m DE_R F(\cdot)\|_{2,G}$$

For any $m \ge 0$, there exist finitely many triples (D_i, E_i, m_i) such that

$$|F(x)| \le \frac{\varphi_0^G(x)}{(1 + \|x\|)^m} \sum \nu_{D_i,E_i}^{(m_i)}(F)$$

for all F in $C^\infty(G)$ and x in G.

Lemma 12.15. Let Σ be the set of restricted roots of G. For a in $\exp \mathfrak{a}_\mathfrak{p}^+$, let

$$w(a) = \prod_{\beta \in \Sigma^+} (1 - e^{-2\beta \log a})^{\dim \mathfrak{g}_\beta}.$$

For each $H \in U(\mathfrak{a}_\mathfrak{p}^{\mathbb{C}})$ and F in $C^\infty(\exp \mathfrak{a}_\mathfrak{p}^+)$, let $\|HF\|_{2,w}$ denote the L^2 norm of HF with respect to $w(a)\,da$. Then there exists a polynomial Q on $\mathfrak{a}_\mathfrak{p}^+$ and there exist finitely many H_i in $U(\mathfrak{a}_\mathfrak{p}^{\mathbb{C}})$ such that

$$\sup_{a \in \exp \mathfrak{a}_\mathfrak{p}^+} |Q(\log a)F(a)| \le \sum_i \|H_i F\|_{2,w}$$

for all F in $C^\infty(\exp \mathfrak{a}_\mathfrak{p}^+)$.

Remarks. The proof of this lemma is very similar to that for Lemma 11.11 and will be omitted. The function $w(a)$ here plays the same role as $P(x)$ in Lemma 11.11. If we apply Lemma 12.15 to a suitable finite number of derivatives of F and integrate the estimates on the derivatives in the same way as in using Lemma 11.11, then we see that we can take $Q = 1$ in Lemma 12.15 if only we use enough H_i's.

Proof of Proposition 12.13. Let g be in $C^\infty(\exp \mathfrak{a}_\mathfrak{p}^+)$, and apply Lemma 12.15 (strengthened as in the remarks) to $e^{\rho \log a} g(a)$ to obtain

$$\sup_{a \in \exp \mathfrak{a}_\mathfrak{p}^+} |e^{\rho \log a} g(a)| \le \sum_i \|H_i(e^{\rho \log a} g)\|_{2,w}$$
$$= \sum_i \|e^{-\rho \log a} H_i(e^{\rho \log a} g)\|_{2,D_A},$$

where $D_A(a) = e^{2\rho \log a} w(a)$ is the factor that appears when Haar measure for G is decomposed according to the $KA_{\mathfrak{p}}K$ decomposition. Applying the Leibniz rule to each $H_i(e^{\rho \log a} g)$ and redefining the H_i's suitably, we find

$$\sup_a |e^{\rho \log a} g(a)| \leq \sum_i \|H_i g\|_{2,D_A}. \tag{12.42}$$

Given F in $C^\infty(G)$, define $F_{k_1,k_2}(a) = F(k_1 a k_2)$ for k_1 and k_2 in K and a in $\exp \mathfrak{a}_{\mathfrak{p}}^+$. If H is in $U(\mathfrak{a}_{\mathfrak{p}}^{\mathbb{C}})$, then

$$HF_{k_1,k_2}(a) = (\mathrm{Ad}(k_2)^{-1}H)F(k_1 a k_2) = ((\mathrm{Ad}(k_2)^{-1}H)F)_{k_1,k_2}(a).$$

On the right side, $\mathrm{Ad}(k_2)^{-1}H$ can be expanded as a linear combination of members of $U(\mathfrak{g}^{\mathbb{C}})$ with coefficients that are C^∞ functions of k_2. Hence there are members X_j of $U(\mathfrak{g}^{\mathbb{C}})$ depending on H such that

$$|HF_{k_1,k_2}(a)| \leq \sum_j |(X_j F)_{k_1,k_2}(a)|.$$

We apply this fact to each H_i in (12.42) and let X_j run through the union of the necessary sets of members of $U(\mathfrak{g}^{\mathbb{C}})$. Then we see that

$$\sup_a |e^{\rho \log a} F_{k_1,k_2}(a)| \leq \sum_j \|(X_j F)_{k_1,k_2}\|_{2,D_A}.$$

By (12.11) this inequality implies

$$|F(k_1 a k_2)| \leq \varphi_0^G(a) \sum_j \|(X_j F)_{k_1,k_2}\|_{2,D_A}.$$

Let T and T' be members of $U(\mathfrak{k}^{\mathbb{C}})$. Replacing F by $TT'_R F$ in the above equation, squaring, and integrating with $dk_1\, dk_2$, we obtain

$$\begin{aligned}\int_{K\times K} |TT'_R F(k_1\, a k_2)|^2\, dk_1\, dk_2 &\leq C\varphi_0^G(a)^2 \sum_j \|X_j TT'_R F\|_{2,G}^2 \\ &\leq C\varphi_0^G(a)^2 \Big(\sum_j \|X_j TT'_R F\|_{2,G}\Big)^2,\end{aligned} \tag{12.43}$$

where C is the number of terms in the sum on j. By Sobolev's Theorem applied to $K \times K$, there are finitely many pairs (T_l, T'_l) in $U(\mathfrak{k}^{\mathbb{C}}) \times U(\mathfrak{k}^{\mathbb{C}})$ such that

$$\sup_{K\times K} |F(k_1 a k_2)| \leq \sum_l \|T_l (T'_l)_R F\|_{2,K\times K}. \tag{12.44}$$

The proposition follows by combining (12.43) and (12.44).

Remarks about Proposition 12.14. In the above proof we apply (12.42) with $(1 + \|a\|^2)^{m/2} g(a)$ in place of $g(a)$. Using the Leibniz rule and adjusting notation, we obtain

$$\sup_a |(1 + \|a\|)^r e^{\rho \log a} g(a)| \leq \sum_j \|(1 + \|a\|)^{r_i} H_i g\|_{2,D_A}.$$

Then we trace through the remainder of the proof, making the obvious adjustments, and the proposition follows.

The two propositions, in combination with the square integrability of $\varphi_0^G(x)(1 + \|x\|)^{-m}$ for large m, suggest that the natural generalization of the Schwartz space from $\mathbb{R}^n$ to G for dealing with analysis of $L^2(G)$ should incorporate the function φ_0^G. Accordingly we define the **Schwartz space** $\mathscr{C}(G)$ for a linear connected reductive group G to be the set of all F in $C^\infty(G)$ for which every mixed right and left derivative, when multiplied by any polynomial in $\|x\|$, remains bounded by a multiple of $\varphi_0^G(x)$. We make $\mathscr{C}(G)$ into a linear topological space by using all seminorms

$$\nu_{D,E,m}(F) = \sup_{x \in G} |(1 + \|x\|)^m \varphi_0^G(x)^{-1} D E_R F(x)|.$$

Here D and E are arbitrary in $U(\mathfrak{g}^{\mathbb{C}})$, and m is any integer ≥ 0.

Example. Any K-finite matrix coefficient of a discrete series representation is in $\mathscr{C}(G)$. For the function itself, the appropriate decay estimate is given in Corollary 8.52. The derivatives of the function are other matrix coefficients and thus satisfy the same kind of estimate. More generally, any $Z(\mathfrak{g}^{\mathbb{C}})$-finite K-finite function in $L^2(G)$ is in $\mathscr{C}(G)$, by the above remarks and Corollary 8.42.

It is clear that $\mathscr{C}(G)$ is complete as a linear topological space. Formula (7.52) shows that $\mathscr{C}(G) \subseteq L^2(G)$ and that the inclusion is continuous. Further elementary properties of $\mathscr{C}(G)$ are given in the following proposition.

Proposition 12.16. Let G be linear connected reductive.

(a) $C_{\text{com}}^\infty(G)$ is dense in $\mathscr{C}(G)$.
(b) Convolution carries $\mathscr{C}(G) \times \mathscr{C}(G)$ into $\mathscr{C}(G)$ and is continuous.

Proof.
(a) For $t > 0$, let χ_t be the characteristic function of $\{x \in G \mid \|x\| \leq t\}$. Fix a function h in $C_{\text{com}}^\infty(G)$ with integral 1, and define $h_t = h * \chi_t * h$. Then h_t is in $C_{\text{com}}^\infty(G)$. Let $c = \max\{\|y\| \mid y \in \text{support}(h)\}$. If we write

$$h_t(x) = \int_{G \times G} h(y)\chi_t(y^{-1}xz^{-1})h(z)\, dy\, dz$$

and use (8.62), then we see that

$$h_t(x) = 1 \qquad \text{if} \quad \|x\| \leq t - 2c. \tag{12.45}$$

If we compute a mixed derivative $D'E'_R h_t$, then the derivatives can be applied in the factors h in the convolution, and it follows that

$$\sup_{t>0} \sup_{x \in G} |D'E'_R h_t| < \infty. \tag{12.46}$$

For f in $\mathscr{C}(G)$ the claim is that $\lim_{t\to\infty} h_t f = f$ in the topology of $\mathscr{C}(G)$; this will prove (a). Thus let (D, E, m) be given and consider

$$\sup_{x\in G} \left|(1 + \|x\|)^m DE_R(f(x)(1 - h_t(x)))\right|.$$

Computing the effect of DE_R on the product by the Leibniz rule, we see that we are to estimate a finite sum of terms

$$\sup_{x\in G}\left|(1 + \|x\|)^{m+1}(D'E'_R f)(x)D''E''_R(1 - h_t)(x)\right|. \tag{12.47}$$

Now $\sup_{x\in G} \left|(1 + \|x\|)^{m+1}(D'E'_R f)(x)\right|$ is finite since f is in $\mathscr{C}(G)$, and

$$\lim_{t\to\infty} \sup_{x\in G}\left|(1 + \|x\|)^{-1}D''E''_R(1 - h_t)(x)\right| = 0$$

by (12.45) and (12.46). Thus (12.47) tends to 0, and (a) follows.

(b) Let $\varphi_m(x) = (1 + \|x\|)^{-m}\varphi_0^G(x)$, and choose r such that φ_r is in $L^2(G)$. We first prove that $m_1 \geq m + r$ implies

$$\left|\varphi_m * \varphi_{m_1}(x)\right| \leq c\left|\varphi_m(x)\right| \qquad \text{for all } x \text{ in } G. \tag{12.48}$$

Writing $xk = (xky^{-1})y$ and applying (8.62), we have

$$(1 + \|x\|) \leq (1 + \|xky^{-1}\|)(1 + \|y\|). \tag{12.49}$$

Thus we compute

$$\begin{aligned}
\varphi_m * \varphi_{m_1}(x) &= \int_G \varphi_m(xy^{-1})\varphi_{m_1}(y)\,dy \\
&= \int_{G\times K} \varphi_m(xky^{-1})\varphi_{m_1}(y)\,dk\,dy && \text{by a change of variables in } y \\
&\leq \int_{G\times K} \varphi_0^G(xky^{-1})\frac{(1 + \|y\|)^{m-m_1}}{(1 + \|x\|)^m}\varphi_0^G(y)\,dk\,dy && \text{by (12.49)} \\
&= \varphi_m(x)\int_G \frac{\varphi_0^G(y^{-1})\varphi_0^G(y)\,dy}{(1 + \|y\|)^{m_1-m}} && \text{by (7.45)} \\
&= c\varphi_m(x)
\end{aligned}$$

by (12.9) and the fact that $m_1 - m \geq r$. This proves (12.48).

If f and g are in $\mathscr{C}(G)$, then

$$DE_R(f * g) = E_R f * Dg.$$

For any m, both $|E_R f|$ and $|Dg|$ are dominated by multiples of φ_{m+r}. Thus (12.48) implies $DE_R(f * g)$ is in $\mathscr{C}(G)$ and that $(f, g) \to f * g$ is continuous.

Theorem 12.17. For each Θ-stable Cartan subgroup T and each h in T', $F_f^T(h)$ remains convergent for f in $\mathscr{C}(G)$. This extension of $f \to F_f$ from

$C^\infty_{\text{com}}(G)$ to the Schwartz space has the following properties:

(a) The mapping $f \to f^{(S)}$ given in (10.22) carries $\mathscr{C}(G)$ continuously into $\mathscr{C}(MA)$.

(b) The formula

$$\xi_{-\delta_M}(h)F^{MA/T}_{f^{(S)}}(h) = e^{\rho H(h)}\xi_{-\delta}(h)F^{G/T}_f(h), \qquad h \in T',$$

of Lemma 10.17 persists for f in $\mathscr{C}(G)$.

(c) The formula

$$F^T_{zf}(h) = (\gamma(z)F^T_f)(h), \qquad z \in Z(\mathfrak{g}^{\mathbb{C}}),\ h \in T',$$

of Proposition 11.9 persists for f in $\mathscr{C}(G)$.

(d) The results of Proposition 11.8 and 11.10 extend to $\mathscr{C}(G)$ as follows: If f is in $\mathscr{C}(G)$, then the restriction of F^T_f to the intersection of T' with any component of

$$T'' = \{h \in T \mid \xi_\alpha(h) \neq 1 \text{ for all noncompact imaginary } \alpha\}$$

extends to a C^∞ function on the closure of this component. On such a component, each derivative of F^T_f remains bounded when multiplied by any polynomial. The resulting seminorms on F^T_f depend continuously on f for f in $\mathscr{C}(G)$.

(e) The jump conditions at semiregular points (given in Proposition 11.12) persist for f in $\mathscr{C}(G)$.

(f) Corollary 11.14, concerning the continuity of certain derivatives of F^T_f at points of $T - T''$, persists for f in $\mathscr{C}(G)$.

(g) Corollary 11.15 and Theorem 11.17, concerning $\partial(\varpi_T)F^T_f(1)$, persist for f in $\mathscr{C}(G)$.

The proof of this theorem involves difficult analysis that cannot be easily summarized. Moreover, there are several expository accounts of it in the literature. Accordingly we omit all discussion of the proof.

§5. Exhaustion of Discrete Series

We come to the main results of this chapter. Together these results say that the only discrete series for a linear connected semisimple group are those in Theorem 9.20. We begin with a simple lemma, apply the facts about F_f in a second lemma, and state the main results as two theorems.

Lemma 12.18. Let G be linear connected reductive, let T be a Θ-stable Cartan subgroup that is noncompact, and let χ_λ be any infinitesimal character. Then the only global square integrable C^∞ solution on T to the system

$$\gamma(z)F = \chi_\lambda(z)F \qquad \text{for all } z \text{ in } Z(\mathfrak{g}^{\mathbb{C}}) \tag{12.50}$$

is the 0 solution.

Proof. Let $\mathfrak{t}$ be the Lie algebra of T. Going over the proof of Theorem 10.35, we see that a global solution of (12.50) must be of the form

$$F(t \exp H) = \sum_{w \in W(\mathfrak{t}^{\mathbb{C}}:\mathfrak{g}^{\mathbb{C}})} p_{w,t}(H)e^{w\lambda(H)} \tag{12.51}$$

for $H \in \mathfrak{t}$ and $t \in T$. We regard t as fixed and H as varying in this formula, and we are to see that the result cannot be in L^2 of the component of t in T unless it is identically 0. Applying Fubini's Theorem, we see that it is sufficient to handle the one-variable case.

Thus suppose we have a finite sum

$$f(t) = \sum_{s,q} c_{s,q}e^{st}t^q$$

in $L^2(\mathbb{R}^1)$. We want to prove that all $c_{s,q}$ are 0. Put

$$h(T) = \int_0^T |f(t)|^2\, dt = \sum_{s,q} d_{s,q}e^{sT}T^q.$$

Since f is in L^2, $h(T)$ is bounded for $T \geq 0$. By Lemma B.24, $d_{s,q} \neq 0$ implies $\operatorname{Re} s \leq 0$. Also $d_{s,q} \neq 0$ with $\operatorname{Re} s = 0$ implies $q = 0$. Computing $|f(t)|^2$ from $h(t)$ explicitly by differentiation, we see that $|f(t)|^2$ is bounded for $t \geq 0$. Hence $f(t)$ is bounded for $t \geq 0$. By Lemma B.24, $c_{s,q} \neq 0$ implies $\operatorname{Re} s \leq 0$. Also $c_{s,q} \neq 0$ with $\operatorname{Re} s = 0$ implies $q = 0$. Replacing $f(t)$ by $f(-t)$ and applying these facts, we are led to conclude that $c_{s,q} = 0$ whenever $\operatorname{Re} s \neq 0$ or $q \neq 0$. Thus we can rewrite f as

$$f(t) = \sum_{\theta} c_\theta e^{i\theta t}.$$

By direct calculation we see that $\int_0^T |f(t)|^2\, dt = T\sum |c_\theta|^2 + O(1)$. Since f is in L^2, $\sum |c_\theta|^2 = 0$. Thus $f = 0$.

Lemma 12.19. Let G be linear connected reductive, and let f be a $Z(\mathfrak{g}^{\mathbb{C}})$-finite K-finite function in $L^2(G)$. Then $F_f^T = 0$ for every Θ-stable noncompact Cartan subgroup T of G. If B is compact, then F_f^B is in $C^\infty(B)$.

Proof. By Corollary 8.42 we may assume that f is a K-finite matrix coefficient for a discrete series representation. The example in the previous section shows that f is in $\mathscr{C}(G)$. Let T be a Θ-stable Cartan subgroup whose noncompact part has the largest possible dimension among all Cartan subgroups on which F_f is not identically 0. Theorem 12.17e shows that every derivative of F_f^T is continuous at every semiregular point of

T. Using the same reasoning as in the sketch of proof of Corollary 11.14, we conclude that F_f^T extends to a C^∞ function on all of T. According to Theorem 12.17d, F_f^T is then actually a Schwartz function on all of T, hence is in $L^2(T)$. On the other hand, since f is a matrix coefficient of a discrete series representation, (8.10) shows that f satisfies

$$zf = \chi_\lambda(f) \qquad \text{for all } z \text{ in } Z(\mathfrak{g}^{\mathbb{C}}),$$

where χ_λ is the infinitesimal character of the discrete series representation. Theorem 12.17c implies that F_f^T satisfies (12.50). Therefore Lemma 12.18 shows that either T is compact or $F_f^T = 0$. This proves both conclusions of the lemma.

Theorem 12.20. A linear connected semisimple group G has discrete series representations if and only if rank G = rank K.

Proof. We may assume that $\mathfrak{k} \cap i\mathfrak{p} = 0$ and $G^{\mathbb{C}}$ is simply connected. If rank G = rank K, then Theorem 9.20 exhibits discrete series. In the converse direction let f be a nonzero K-finite matrix coefficient of a discrete series representation, and let B be a Cartan subgroup of G that is as compact as possible. Arguing by contradiction, suppose B is noncompact. Then Lemma 12.19 gives $F_f^B = 0$. From Theorem 12.17g it follows that $f(1) = 0$. Because of (8.10) this argument applies equally well to all left-invariant derivatives of f, showing that they vanish at the identity. Since f is real analytic, f is identically 0, contradiction. Thus B must be compact, and the theorem follows.

Theorem 12.21. Let G be linear connected semisimple with rank G = rank K. Then the only discrete series representations, up to equivalence, are the representations π_λ of Theorem 9.20.

Proof. We may assume that $G^{\mathbb{C}}$ is simply connected. Let B be a compact Cartan subgroup, and let f be a nonzero K-finite matrix coefficient of some discrete series representation π. Arguing as in Theorem 12.20 and passing to a left invariant derivative f if necessary, we may assume that $\partial(\varpi_B)F_f^B(1) \neq 0$, in particular that F_f^B is not identically 0. If λ is the infinitesimal character of π, then F_f^B must be of the form in (12.51). We may assume, by lumping terms if necessary, that the linear functionals $w\lambda$ for which $p_{w,t} \not\equiv 0$ are linearly independent. Since F_f^B is smooth according to Lemma 12.19, it follows that the polynomial terms are absent and the linear functionals $w\lambda$ are analytically integral.

Let us prove that λ is nonsingular. We have just shown that

$$F_f^B(\exp H) = \sum_{w \in W_G} c_w e^{w\lambda(H)}.$$

Therefore

$$\partial(\varpi_B)F_f^B(\exp H) = \sum_{w \in W_G} c_w \left(\prod_{\alpha \in \Delta^+} \langle w\lambda, \alpha \rangle \right) e^{w\lambda(H)}. \tag{12.52}$$

If λ is singular, then every term on the right side of (12.52) is 0. Evaluating both sides at $H = 0$, we obtain a contradiction to the choice of f. We conclude λ is nonsingular.

With λ nonsingular and analytically integral (so that $\lambda + \delta_G$ is analytically integral), we can appeal to Theorem 12.6: The character of π is determined by its numerator τ_B on the Cartan subgroup B. This numerator is odd under the action of W_K, and the dimension of the linear span of all discrete series with infinitesimal character λ is therefore $\leq |W_G/W_K|$. On the other hand, Theorem 9.20 does produce $|W_G/W_K|$ inequivalent discrete series representations with infinitesimal character λ. Since global characters of inequivalent irreducible admissible representations are linearly independent (Theorem 10.6), π must be equivalent with one of the representations given by Theorem 9.20.

Corollary 12.22. Let G be linear connected semisimple with rank $G =$ rank K. Then

(a) any given K type μ occurs in only finitely many discrete series.
(b) the trivial K type appears in no discrete series unless G is compact.

Proof.

(a) Fix Δ_K^+, and let μ be dominant relative to Δ_K^+. Let $\pi = \pi_\lambda$ be a discrete series representation. Possibly by replacing λ by $w\lambda$ for some w in W_K, we may assume λ is dominant for Δ_K^+. Then we can introduce Δ^+ as in (9.49). If μ occurs in π_λ, then Theorem 9.20c says that

$$\mu = \lambda + \delta_G - 2\delta_K + \sum_{\alpha \in \Delta^+} n_\alpha \alpha.$$

Since $\lambda + \delta_G$ is dominant for Δ^+, we have

$$\|\mu + 2\delta_K\| > \|\lambda + \delta_G\| \tag{12.53}$$

with equality only if $\mu = \lambda + \delta_K - 2\delta_K$. Only finitely many integral λ have this property, and (a) follows.

(b) We use the same notation as in (a), assuming now that π_λ contains the trivial K type. Then

$$2\delta_K = \lambda + \delta_G + \sum_{\alpha \in \Delta^+} n_\alpha \alpha. \tag{12.54}$$

Since λ is integral and dominant nonsingular, every simple root β for Δ^+ satisfies

$$2\langle \lambda, \beta \rangle / |\beta|^2 \geq 1 = 2\langle \delta_G, \beta \rangle / |\beta|^2.$$

Hence $\lambda + \delta_G = 2\delta_G + \lambda'$, where the form $\lambda' = \lambda - \delta_G$ is dominant. We rewrite (12.54) as

$$\lambda' + 2\delta_n + \sum_{\alpha \in \Delta^+} n_\alpha \alpha = 0.$$

By Lemma 8.58, λ' is a nonnegative combination of positive roots. Thus this relation gives us a contradiction unless $\delta_n = 0$, in which case G is compact.

§6. Tempered Distributions

Historically the name "tempered representation," introduced in Chapter VII, arose from the connection between the Schwartz space and the characters of representations. We say that a distribution on a linear connected reductive G is **tempered** if it extends continuously from the dense subset $C_{\text{com}}^\infty(G)$ to all of $\mathscr{C}(G)$. The main result concerning tempered distributions is the following theorem, which plays a role in the analysis associated with the Plancherel Theorem.

Theorem 12.23. If Θ is the global character of an irreducible admissible representation π of the linear connected reductive group G, then the following conditions are equivalent:

(a) Θ is a tempered distribution.
(b) On every Cartan subgroup of G, the numerator of Θ is bounded.
(c) π is a tempered representation.

We shall show now that (a) and (b) are equivalent and imply (c). After Theorem 14.76 we shall complete the proof by showing that (c) implies (b). Theorem 12.23 has a counterpart for invariant eigendistributions—that the tempered condition is equivalent with at-most-polynomial growth for the numerators of Θ. There is no analog of (c) in this case.

Proof that (b) ⇒ (a). Let $\{T_i\}$ be a complete system of nonconjugate Cartan subgroups, and let $\{\tau_{T_i}\}$ be the corresponding numerators of Θ. By (b), let C be a bound for all the τ_{T_i}. By the Weyl integration formula and (10.25), f in $C_{\text{com}}^\infty(G)$ implies

$$\Theta(f) = \sum_i s^{G/T_i}|W(T_i:G)|^{-1} \int_{T_i} \varepsilon_R^{T_i}(h)\tau_{T_i}(h)F_f^{T_i}(h)\,dh,$$

and hence

$$|\Theta(f)| \leq C \sum_i \int_{T_i} |F_f^{T_i}(h)|\,dh.$$

This inequality, combined with Theorem 12.17d, shows Θ is tempered. This proves (a).

Lemma 12.24. If Θ is an invariant distribution on G that is tempered, then Θ remains continuous on $\mathscr{C}(G)$ when $\mathscr{C}(G)$ is given the weaker topology defined by only those seminorms that do not involve right invariant derivatives.

Proof. Fix h in $C_{\text{com}}^{\infty}(G)$ with integral one, and let S be the support of h. For f in $C_{\text{com}}^{\infty}(G)$, define

$$f_*(x) = \int_G h(y) f(yxy^{-1})\, dy.$$

Then

$$f_*(z_1 x z_2) = \int_G h(yz_1^{-1}) f(yxy^{-1}(yz_2y^{-1})(yz_1y^{-1}))\, dy.$$

Using the Leibniz rule, we see that to any D and E in $U(\mathfrak{g}^{\mathbb{C}})$ correspond members E_j of $U(\mathfrak{g}^{\mathbb{C}})$ such that

$$|DE_R f_*(x)| \le \sum_j \sup_{y \in S} |(\text{Ad}(y)D)(\text{Ad}(y)E_j) f(yxy^{-1})|.$$

Expanding out $\text{Ad}(y)D$ and $\text{Ad}(y)E_j$ in terms of a suitable basis of $U(\mathfrak{g}^{\mathbb{C}})$ and using the compactness of S, we see that

$$|DE_R f_*(x)| \le \sum_i \sup_{y \in S} |D_i f(yxy^{-1})|.$$

Thus Lemma 12.5 implies that

$$\nu_{D,E,m}(f_*) \le c \sum_i \nu_{D_i,1,m}(f).$$

Since Θ is invariant and h has integral one, we have

$$\Theta(f) = \Theta(f_*).$$

Since Θ is tempered, there exist D_k, E_k in $U(\mathfrak{g}^{\mathbb{C}})$ and $m_k \ge 0$ with

$$|\Theta(f_*)| \le \sum_k \nu_{D_k,E_k,m_k}(f_*).$$

Combining the last three displays above, we obtain the conclusion of the lemma.

Proof that (a) $\Rightarrow$ (b) in Theorem 12.23. Suppose Θ is a tempered irreducible character. Let τ_T be the numerator of Θ on T. Given g in $C_{\text{com}}^{\infty}(T')$, construct φ_g as in (12.5). Then φ_g is in $C_{\text{com}}^{\infty}(G)$ and

$$\Theta(\varphi_g) = \int_T \tau_T(h) g(h)\, dh \tag{12.55}$$

as in §1. Since Θ is tempered, Lemma 12.24 implies

$$|\Theta(\varphi_g)| \le \sum \nu_{D_j,1,m_j}(\varphi_g).$$

Bounding the right side here by means of Lemma 12.3 and substituting from (12.55), we see that

$$\left|\int_T \tau_T(h)g(h)\,dh\right| \le c \sup_{h\in T'} \left\{(1+\|h\|)^m \sum_{j,w} |\eta_j(h)H_j g(whw^{-1})|\right\}$$

for suitable $m \ge 0$, η_j in $\mathscr{R}$, and H_j in $U(\mathfrak{t}^{\mathbb{C}})$. This is the same kind of inequality as in (12.13), which occurs in the proof of Theorem 12.1. Following through the rest of that proof, we see that τ_T is bounded.

Lemma 12.25. Let $F \in C^\infty(G)$ be $Z(\mathfrak{g}^{\mathbb{C}})$-finite and be left and right K-finite. If $F(x)\,dx$ is a tempered distribution, then

$$|F(x)| \le C(1+\|x\|)^m \varphi_0^G(x)$$

for suitable constants C and m.

Proof. Since $F(x)\,dx$ is tempered, we can write

$$\left|\int_G Ff\,dx\right| \le \sum_i \nu_{D_i,E_i,m_i}(f) \tag{12.56}$$

for all f in $C^\infty_{\text{com}}(G)$. Since F is $Z(\mathfrak{g}^{\mathbb{C}})$-finite and K-finite, Corollary 8.41 implies that there exist h and h' in $C^\infty(G)$ with $h' * F * h = F * h = F$. Put $g(x) = h(x^{-1})$ and $g'(x) = h(x^{-1})$. Then

$$\int_G Ff\,dx = \int_G (h' * F * h)f\,dx = \int_G (g' * f * g)F\,dx.$$

Replacing f by $g' * f * g$ in (12.56), we therefore obtain

$$\left|\int_G Ff\,dx\right| \le \sum_i \nu_{D_i,E_i,m_i}(g' * f * g) = \sum_i \nu_{1,1,m_i}((E_i)_R g' * f * D_i g).$$

From Lemma 12.5 and this inequality, we see that

$$\left|\int_G Ff\,dx\right| \le C\nu_{1,1,m}(f)$$

for suitable constants C and m. Replacing $f(x)$ by $f(x)(1+\|x\|)^m \varphi_0^G(x)^{-1}$, we can rewrite this result as

$$\left|\int_G \{F(x)(1+\|x\|)^{-m}\varphi_0^G(x)\}f(x)\,dx\right| \le C \sup_{x\in G} |f(x)|$$

for all f in $C^\infty_{\text{com}}(G)$. Therefore

$$F(x)(1+\|x\|)^{-m}\varphi_0^G(x) \text{ is in } L^1(G). \tag{12.57}$$

Since g is in $C^\infty_{\text{com}}(G)$, $|g(x)|$ is dominated by $C_1(1 + \|x\|)^{-m}\varphi_0^G(x)$ for some C_1. Therefore $F = F * h$ implies

$$|F(x)| \leq \int_G |F(xy)|\,|g(y)|\,dy \leq C_1 \int_G |F(xy)|(1 + \|y\|)^{-m}\varphi_0^G(y)\,\text{dy}. \quad (12.58)$$

Since F is K-finite on the left, there is a K-finite function β on K with $F = \beta *_K F$. Therefore

$$\begin{aligned}
|F(x)| &\leq \|\beta\|_\infty \int_K |F(k^{-1}x)|\,dk \\
&\leq C_1\|\beta\|_\infty \int_{G\times K} |F(k^{-1}xy)|(1 + \|y\|)^{-m}\varphi_0^G(y)\,dk\,dy \qquad \text{by (12.58)} \\
&= C_1\|\beta\|_\infty \int_{G\times K} |F(y)|(1 + \|x^{-1}ky\|)^{-m}\varphi_0^G(x^{-1}ky)\,dk\,dy \\
&\leq C_1\|\beta\|_\infty(1 + \|x\|)^m \int_{G\times K} |F(y)|(1 + \|y\|)^{-m}\varphi_0^G(x^{-1}ky)\,dk\,dy \\
&\qquad\qquad \text{by (12.49)} \\
&= C_1\|\beta\|_\infty(1 + \|x\|)^m\varphi_0^G(x) \int |F(y)|(1 + \|y\|)^{-m}\varphi_0^G(y)\,dy \\
&\qquad\qquad \text{by (7.45) and (12.9)} \\
&= \text{Const}(1 + \|x\|)^m\varphi_0^G(x) \qquad \text{by (12.57)}.
\end{aligned}$$

This proves the lemma.

Proof that (a) ⇒ (c) in Theorem 12.23. Let Θ be a given tempered irreducible character, say for a representation π. Let $\tau_1, \ldots, \tau_n$ be distinct irreducible representations of K, and let χ be the sum of the product of their degrees by their characters, so that convolution by χ is the projection into the subspace of $L^2(K)$ spanned by the matrix coefficients of $\tau_1, \ldots, \tau_n$. Then $\chi *_K \Theta *_K \chi$ is a tempered distribution on G that is $Z(\mathfrak{g}^{\mathbb{C}})$-finite and left and right K-finite. By the same proof as for Theorem 8.7, $\chi *_K \Theta *_K \chi$ must be given by a real analytic function $F(x)$ on G. In fact, it is easy to see that

$$F(x) = \text{Tr}(E\pi(x)E),$$

where E is the orthogonal projection on the sum of the K types $\bar\tau_1, \ldots, \bar\tau_n$.

According to Lemma 12.25, $F(x)$ satisfies a decay estimate

$$|F(x)| \leq C(1 + \|x\|)^m\varphi_0^G(x)$$

for some C and m. Using translations and integrations within K, we see that any K-finite matrix coefficient of π corresponding to the K types $\bar\tau_1, \ldots, \bar\tau_n$ satisfies a similar estimate. Therefore every K-finite matrix coefficient of π is in $L^{2+\varepsilon}(G)$ for all $\varepsilon > 0$. According to the equivalent definitions in Theorem 8.53, π is tempered.

§7. Limits of Discrete Series

Our objective in this section is to apply the Zuckerman tensoring functor ψ of §10.9 to discrete series in order to obtain some new representations. The prototype for such a construction is SL(2, $\mathbb{R}$), where application of ψ to the discrete series $\mathscr{D}_n^+$ and $\mathscr{D}_n^-$ with $n \geq 2$ leads to the limits of discrete series $\mathscr{D}_1^+$ and $\mathscr{D}_1^-$, by Propositions 10.14 and 10.44b.

Thus suppose rank G = rank K, and for now assume $G^{\mathbb{C}}$ is simply connected. Let B be a compact Cartan subgroup, and let $\mathfrak{b}$ be its Lie algebra. We shall work with an arbitrary integral form λ on $\mathfrak{b}$. If λ is nonsingular, we shall recover the discrete series π_λ with Harish-Chandra parameter λ and with global character Θ_λ. Thus our chief interest will be in the case that λ is singular.

Fix a positive system Δ^+ for $\Delta = \Delta(\mathfrak{b}^{\mathbb{C}} : \mathfrak{g}^{\mathbb{C}})$ that makes λ dominant; Δ^+ will not be unique when λ is singular. Then choose any dominant integral form μ such that $\lambda + \mu$ is nonsingular (e.g., $\mu = \delta_G$). Let $\pi(\lambda, \Delta^+)$ be the admissible representation $\psi_\lambda^{\lambda+\mu}(\pi_{\lambda+\mu})$. We call $\pi(\lambda, \Delta^+)$ a **limit of discrete series** if λ is singular. (If λ is nonsingular, $\pi(\lambda, \Delta^+)$ is just the discrete series π_λ up to infinitesimal equivalence, in view of the results in §§12.1–12.3.)

For any admissible π with infinitesimal character $\lambda + \mu$, it is not hard to see that

$$\psi_\lambda^{\lambda+\mu'}\psi_{\lambda+\mu'}^{\lambda+\mu}\pi \text{ is infinitesimally equivalent with } \psi_\lambda^{\lambda+\mu}\pi, \qquad (12.59)$$

provided λ, μ', and $\mu - \mu'$ are all dominant. Consequently distinct choices of μ lead to infinitesimally equivalent versions of $\pi(\lambda, \Delta^+)$, and also $\psi_\lambda^{\lambda'}\pi(\lambda', \Delta^+)$ is infinitesimally equivalent with $\pi(\lambda, \Delta^+)$ if $\lambda' - \lambda$ is dominant. Let $\Theta(\lambda, \Delta^+)$ be the global character of $\pi(\lambda, \Delta^+)$.

Theorem 12.26. Let G be linear connected reductive with rank G = rank K and with $G^{\mathbb{C}}$ simply connected. The limits of discrete series representations of G and their global characters have the following properties:

(a) $\Theta(\lambda, \Delta^+)$ is a tempered character and either is 0 or is an irreducible character.

(b) $\Theta(\lambda, \Delta^+) = 0$ if and only if $\langle \lambda, \alpha \rangle = 0$ for some Δ^+ simple root α that is compact.

(c) The K types of $\pi(\lambda, \Delta^+)$ are all of the form

$$\lambda - \delta_K + \delta_n + \sum_{\alpha \in \Delta^+} n_\alpha \alpha$$

with all n_α integers ≥ 0. The K type $\lambda - \delta_K + \delta_n$ has multiplicity one if $\pi(\lambda, \Delta^+)$ is not 0.

(d) If $\Theta(\lambda, \Delta^+) \neq 0$, then $\Theta(\lambda', (\Delta^+)') = \Theta(\lambda, \Delta^+)$ if and only if there is some $w \in W_K$ with $\lambda' = w\lambda$ and $(\Delta^+)' = w\Delta^+$.

Remarks.

(1) In the proof we shall dispose of the easy parts first, isolating two statements to treat separately. These two statements are essentially the second half of (c) and the sufficiency of (b), and for each one the proof will be preceded by a lemma.

(2) When $G^{\mathbb{C}}$ is not necessarily simply connected, we obtain a version of Theorem 12.26 by passing to the quotient from the case where $G^{\mathbb{C}}$ is simply connected. To pass to G from $\tilde{G}$, where $\tilde{G}^{\mathbb{C}}$ is simply connected, we want to know which representations $\pi(\lambda, \Delta^+)$ are trivial on the kernel of the covering homomorphism $\varphi: \tilde{G} \to G$. Since $\ker \varphi \subseteq Z_{\tilde{G}}$, we can read of $\pi(\lambda, \Delta^+)$ on an element z of $Z_{\tilde{G}}$ as the scalar value of $\xi_{\lambda - \delta_K + \delta_n}(z)$, by (c). Since $\xi_{2\delta_K}(z) = 1$, we see that the representations $\pi(\lambda, \Delta^+)$ that descend to G are exactly those for which $\lambda + \delta$ is analytically integral.

Before beginning the proof of Theorem 12.26, let us underline the significance of (a) and (b). Recall that the Langlands classification reduced classification of irreducible admissible representations to an understanding of the irreducible tempered unitary representations. The following corollary is an immediate consequence of Theorems 12.26 and 8.53.

Corollary 12.27. Let G be linear connected reductive with rank $G =$ rank K. Let λ be a linear form on $\mathfrak{b}$, and let Δ^+ be a positive system for which λ is dominant. If $\lambda + \delta$ is analytically integral and $\langle \lambda, \alpha \rangle \neq 0$ for every Δ^+ simple root α that is compact, then $\pi(\lambda, \Delta^+)$ is an irreducible tempered unitary representation of G.

Proof of easy parts of Theorem 12.26. Let $\Lambda = \lambda - \delta_K + \delta_n$. We assume for now the following statements:

(b′) $\Theta(\lambda, \Delta^+) = 0$ if $\langle \lambda, \alpha \rangle = 0$ for some Δ^+ simple root α that is compact.

(c′) If Λ is Δ_K^+ dominant, then the K type Λ occurs in $\pi(\lambda, \Delta^+)$.

Let us prove the remaining parts of the theorem.

(b) If Λ is Δ_K^+ dominant, then (c′) shows that $\Theta(\lambda, \Delta^+) \neq 0$. Thus assume that Λ is not Δ_K^+ dominant. Let α be a Δ_K^+ simple root such that $\langle \Lambda, \alpha \rangle < 0$. Then

$$0 > \frac{2\langle \Lambda, \alpha \rangle}{|\alpha|^2} = \frac{2\langle \lambda + \delta - 2\delta_K, \alpha \rangle}{|\alpha|^2} \geq \frac{2\langle \delta, \alpha \rangle}{|\alpha|^2} - 2\frac{2\langle \delta_K, \alpha \rangle}{|\alpha|^2}$$

$$= \frac{2\langle \delta, \alpha \rangle}{|\alpha|^2} - 2.$$

Hence $2\langle \delta, \alpha \rangle / |\alpha|^2 = 1$, and α is Δ^+ simple. In this case, (b′) shows $\Theta(\lambda, \Delta^+) = 0$.

(c) Let $F_{-\mu}$ be an irreducible finite-dimensional representation of G with lowest weight $-\mu$. Decomposing $\pi_{\lambda+\mu}$ and $F_{-\mu}$ under K and taking the tensor product, we see from Problem 13 of Chapter IV that the K types of $\pi_{\lambda+\mu} \otimes F_{-\mu}$ have highest weight $\lambda' + \nu$, where λ' is a K type of $\pi_{\lambda+\mu}$ and ν is a weight of $F_{-\mu}$. Since $\lambda' = \lambda + \mu + \sum_{\alpha \in \Delta^+} m_\alpha \alpha$ and $\nu = -\mu + \sum_{\alpha \in \Delta^+} n_\alpha \alpha$, the first statement in (c) follows. If Λ is Δ_K^+ dominant, then Problem 17 in Chapter IV shows that τ_Λ occurs in the tensor product of $\tau_{\Lambda+\mu}$ and the irreducible representation of K with lowest weight $-\mu$, and thus Λ occurs whenever it is Δ_K^+ dominant. Moreover, Problem 17 shows also that τ_Λ occurs just once. If, on the other hand, Λ is not Δ_K^+ dominant, we have just seen that $\Theta(\lambda, \Delta^+) = 0$. Thus (c) follows.

(d) If w is in W_K, then $\Theta(w(\lambda + \mu), w\Delta^+) = \Theta(\lambda + \mu, \Delta^+)$ as a result of Theorem 9.20. Applying $\psi_\lambda^{\lambda+\mu}$, we obtain $\Theta(w\lambda, w\Delta^+) = \Theta(\lambda, \Delta^+)$. Conversely suppose $\Theta(\lambda', (\Delta^+)') = \Theta(\lambda, \Delta^+) \neq 0$. Choose $w \in W_G$ with $(\Delta^+)' = w\Delta^+$. Since the two infinitesimal characters must match, we have $\chi_{\lambda'} = \chi_\lambda$, and then the dominance forces $\lambda' = w\lambda$. Adjusting matters by a member of W_K, we may assume that

$$\Theta(w\lambda, w\Delta^+) = \Theta(\lambda, \Delta^+) \neq 0$$

with $(w\Delta^+)_K = \Delta_K^+$. Except for the K type Λ, all of the K types Λ' of $\pi(\lambda, \Delta^+)$ have

$$|\Lambda' + 2\delta_K|^2 > |\Lambda + 2\delta_K|^2 = |\lambda + \delta|^2$$

by (c) and the Δ^+ dominance of $\lambda + \delta$. One K type of $\pi(\lambda, \Delta^+)$ is the one given to us by (c′) for $\pi(w\lambda, w\Delta^+)$, namely

$$\Lambda' = w\lambda - \delta_K + \delta_n' = w\lambda - 2\delta_K + w\delta,$$

and it has

$$|\Lambda' + 2\delta_K|^2 = |w\lambda + w\delta|^2 = |\lambda + \delta|^2 = |\Lambda + 2\delta_K|^2.$$

Therefore $\Lambda' = \Lambda$, and we obtain $w(\lambda + \delta) = \lambda + \delta$. Since $\lambda + \delta$ is nonsingular, w must be 1, by Chevalley's Lemma.

(a) By Theorem 10.50, $\Theta(\lambda, \Delta^+)$ either is 0 or is a multiple of an irreducible character. If it is nonzero, (c) identifies a K type of multiplicity one. Hence it must be irreducible. Let us see that it is tempered. Theorem 12.6, especially formulas (12.17) and (12.18), ties down the character formula of a discrete series $\pi_{\lambda+\mu}$, taking into account the boundedness of the numerators. The boundedness of each term is decided by the inner product of $\lambda + \mu$ with some roots. When we move the parameter via ψ, the inner products do not change sign. Thus all the terms in the numerators remain bounded, and $\Theta(\lambda, \Delta^+)$ is tempered by Theorem 12.23. This completes the proof of the easy parts of the theorem.

Now we prepare to prove (c′) in Theorem 12.26. The key observation will be that the projection according to the infinitesimal character in the definition of ψ can be replaced by projection according to the Casimir operator Ω. Recall from §8.3 that Ω is defined relative to the real part of the trace form. Its relation to the infinitesimal character is given in the following lemma.

Lemma 12.28. If ν is in the real linear span of the integral forms, then $\chi_\nu(\Omega) = |\nu|^2 - |\delta|^2$, where χ_ν refers to the infinitesimal character and where δ is the half-sum of the positive roots in some order.

Proof. Let $\mathfrak{t} = \mathfrak{b} \oplus \mathfrak{a}$ be the Cartan subalgebra on which ν is defined, with $\mathfrak{b} = \mathfrak{t} \cap \mathfrak{k}$ and $\mathfrak{a} = \mathfrak{t} \cap \mathfrak{p}$, and let $\{H_i\}$ be an orthonormal basis of $\mathfrak{a} \oplus i\mathfrak{b}$. Normalize root vectors so that $[E_\alpha, E_{-\alpha}] = H_\alpha$ for all roots α, and fix a positive system for $\Delta(\mathfrak{t}^{\mathbb{C}}:\mathfrak{g}^{\mathbb{C}})$. Then we easily check that

$$\begin{aligned}\Omega &= \sum H_i^2 + \sum_{\alpha>0} (E_\alpha E_{-\alpha} + E_{-\alpha}E_\alpha)\\ &= \sum H_i^2 + 2H_\delta + \sum_{\alpha>0} 2E_{-\alpha}E_\alpha.\end{aligned} \tag{12.60}$$

Hence
$$\gamma'(\Omega) = \sum H_i^2 + 2H_\delta$$
and
$$\gamma(\Omega) = \sum (H_i - \delta(H_i))^2 + 2(H_\delta - \delta(H_\delta)).$$

Since $\{H_i\}$ is an orthonormal basis, this expression simplifies to

$$\gamma(\Omega) = \sum H_i^2 - |\delta|^2. \tag{12.61}$$

To compute $\chi_\nu(\Omega)$, we apply ν to both sides of this equation, and the lemma follows.

Proof of (c′) in Theorem 12.26. According to the definition, $\pi(\lambda, \Delta^+)$ is obtained by projecting $\pi_{\lambda+\mu} \otimes F_{-\mu}$ according to the infinitesimal character λ. The value of the Casimir operator on the image is $|\lambda|^2 - |\delta|^2$, by Lemma 12.28. If we refer to (10.79) and Theorem 10.35, we see that the only possible (generalized) infinitesimal characters in $\pi_{\lambda+\mu} \otimes F_{-\mu}$ are $\chi_{\lambda+\mu-\sigma}$, where σ is a weight of the irreducible finite-dimensional representation of G with highest weight μ. Here $\sigma = \mu - Q$ with Q a nonnegative sum of positive roots. The corresponding value of the Casimir operator for $\chi_{\lambda+\mu-\sigma}$ is $|\lambda + Q|^2 - |\delta|^2$, by Lemma 12.28. If $Q \neq 0$, then the dominance of λ forces this quantity to be greater than $|\lambda|^2 - |\delta|^2$. Thus $\pi(\lambda, \Delta^+)$ is obtained from $\pi_{\lambda+\mu} \otimes F_{-\mu}$ simply by taking all vectors on which the Casimir operator Ω acts with eigenvalue $|\lambda|^2 - |\delta|^2$.

In the course of the proof so far of Theorem 12.26, we have already observed that the K type Λ occurs in $\pi_{\lambda+\mu} \otimes F_{-\mu}$ with multiplicity one, specifically within the tensor product of $\tau_{\Lambda+\mu} \subseteq \pi_{\lambda+\mu}|_K$ and the representation $\tau_{-\mu}$ of K within $F_{-\mu}|_K$, where $\tau_{-\mu}$ has lowest weight $-\mu$. Since

Ω commutes with K, Ω acts as a scalar on this K type in $\pi_{\lambda+\mu} \otimes F_{-\mu}$. It is enough to identify this scalar as $|\lambda|^2 - |\delta|^2$.

Thus let $v = \sum_j v_j \otimes f_j$ be a nonzero vector in the K type Λ, decomposed with each v_j a weight vector of $\tau_{\Lambda+\mu}$ and each f_j a weight vector of $\tau_{-\mu}$. Write

$$\Omega = \sum Y_i^2 - \sum X_i^2 = \sum Y_i^2 + \Omega_K,$$

where $\{X_i\}$ is an orthonormal basis of $\mathfrak{k}$ relative to $-\operatorname{Re} B_0$ and $\{Y_i\}$ is an orthonormal basis of $\mathfrak{p}$ relative to $\operatorname{Re} B_0$. Then

$$(\Omega - \Omega_K)v = \sum_{i,j} Y_i^2 v_j \otimes f_j + \sum_{i,j} v_j \otimes Y_i^2 f_j + \sum_{i,j} Y_i v_j \otimes Y_i f_j.$$

The vectors $Y_i f_j$ in the last term on the right side have weight differing by a noncompact root from a weight of $\tau_{-\mu}$. Since a noncompact root cannot be an integral combination of compact roots (see Problem 13 at the end of the chapter), this term cannot contribute to our answer. Thus

$$(\Omega - \Omega_K)v = \sum_j (\Omega - \Omega_K)v_j \otimes f_j + \sum_j v_j \otimes (\Omega - \Omega_K)f_j. \quad (12.62)$$

Now we note from Lemma 12.28 that

$$\Omega_K v = \chi^K_{\Lambda+\delta_K}(\Omega_K)v = (|\Lambda + \delta_K|^2 - |\delta_K|^2)v,$$

$$\Omega v_j = (|\lambda + \mu|^2 - |\delta|^2)v_j \qquad \text{and} \qquad \Omega_K v_j = (|\Lambda + \mu + \delta_K|^2 - |\delta_K|^2)v_j,$$

$$\Omega f_j = (|\mu + \delta|^2 - |\delta|^2)f_j \qquad \text{and} \qquad \Omega_K f_j = (|\mu + \delta_K|^2 - |\delta|_K|^2)f_j.$$

Substituting all this information into (12.62) and simplifying, we obtain $\Omega v = (|\lambda|^2 - |\delta|^2)v$, and (c′) in Theorem 12.26 follows.

The last step in the proof of Theorem 12.26 is the proof of (b′), which uses the following lemma. The lemma will be used again in §9.

Lemma 12.29. Let rank G = rank K, let $G^{\mathbb{C}}$ be simply connected, and let $\lambda \in (i\mathfrak{b})'$ be integral. Suppose that λ is singular with respect to just one pair of roots $\pm\alpha$ of $\Delta(\mathfrak{b}^{\mathbb{C}}:\mathfrak{g}^{\mathbb{C}})$, i.e., $\langle\lambda, \beta\rangle \neq 0$ for $\beta \neq \pm\alpha$. If Θ is an invariant eigendistribution on G with infinitesimal character λ such that the numerators τ_T of Θ are bounded on every Cartan subgroup T and contain no nonconstant polynomial terms, then Θ is uniquely determined by its numerator τ_T on the various Cartan subgroups whose noncompact parts have dimension ≤ 1.

Proof. We imitate the proof of Theorem 12.6. We show by induction on the dimension of the noncompact part of T that τ_T is completely determined, the cases of dimension 0 or 1 being the trivial cases for the induction. Thus let a noncompact T and a regular $t_1 \in T$ be given. Let $\pm\alpha$ be the roots with $\langle\lambda, \pm\alpha\rangle = 0$. We construct Δ_1 as in the proof of Theorem 12.6.

First suppose Δ_1 is empty. Then all of $t_1 \exp \mathfrak{t}$ is contained in the component C. If $c_w^C(t_1) \neq 0$, then $e^{w\lambda}$ must be bounded on all of $\mathfrak{t}$, in particular on the line uH_β if β is a real root. Then it follows that $\langle w\lambda, \beta\rangle = 0$, $\langle \lambda, w^{-1}\beta\rangle = 0$, $w^{-1}\beta = \pm\alpha$, and $\beta = \pm w\alpha$. Consequently there is at most one pair $\pm\beta$ of real roots. A single Cayley transform of T then leads to a Cartan subgroup with no real roots, which must then be compact by Proposition 11.16a. Hence the noncompact part of T has dimension ≤ 1, and τ_T is assumed known.

Thus we may assume Δ_1 is nonempty. With t_0 as in the proof of Theorem 12.6, we are to prove that the numbers $c_w = c_w^C(t_0)$ are completely determined. Since $s_\alpha\lambda = \lambda$, we may eliminate any ambiguity in the definitions of the numbers c_w by assuming

$$c_w = c_{ws_\alpha} \qquad \text{for all } w \in W = W(\mathfrak{t}^{\mathbb{C}}:\mathfrak{g}^{\mathbb{C}}). \tag{12.63}$$

The arguments that establish (12.17) and (12.19) prove here that

$$\langle w\lambda, \Lambda_i\rangle > 0 \text{ for some } i \text{ implies } c_w = 0 \tag{12.64}$$

$$\langle w\lambda, \alpha_j\rangle \neq 0 \text{ implies } c_w - c_{s_{\alpha_j}w} \text{ known.} \tag{12.65}$$

Suppose in (12.65) that $\langle w\lambda, \alpha_j\rangle = 0$. Then $\langle \lambda, w^{-1}\alpha_j\rangle = 0$, $w^{-1}\alpha_j = \pm\alpha$, $w^{-1}s_{\alpha_j}w = s_\alpha$, and $s_{\alpha_j}w = ws_\alpha$. Thus (12.63) gives $c_w = c_{s_{\alpha_j}w}$. When combined with (12.65), this fact gives

$$c_w \text{ known implies } c_{s_{\alpha_j}w} \text{ known for all } j. \tag{12.66}$$

Given w, find w_0 in the Weyl group of Δ_1 such that $w_0 w\lambda$ is dominant for Δ_1^+. As in the proof of Theorem 12.6, either $\langle w_0 w\lambda, \Lambda_i\rangle > 0$ for some i or $w_0 w\lambda$ is orthogonal to $\alpha_1, \ldots, \alpha_d$. In the first case we can use (12.64) and (12.66) in place of (12.17) and (12.19) to see that c_w is completely determined.

Thus suppose $w_0 w\lambda$ is orthogonal to $\alpha_1, \ldots, \alpha_d$. Then certainly $d = 1$ since λ is orthogonal only to $\pm\alpha$. So $\langle w_0 w\lambda, \alpha_1\rangle = 0$. With $d = 1$, we have $\Delta_1 = \{\pm\alpha_1\}$, and the Weyl group of Δ_1 is $\{1, s_{\alpha_1}\}$. Thus w_0 is 1 or s_{α_1}, by construction. Consequently

$$\langle w\lambda, \alpha_1\rangle = 0. \tag{12.67}$$

The root α_1 is real. Suppose there is another real root $\beta \neq \pm\alpha_1$. Let

$$H = H_\beta - \frac{\langle \beta, \alpha_1\rangle}{|\alpha_1|^2}\alpha_1.$$

The elements $t_1 \exp uH$ must lie in the component C for all real u since $\xi_{\alpha_1}(t_1 \exp uH) = \xi_{\alpha_1}(t_1)$, and hence $e^{w\lambda(uH)}$ must be bounded, $-\infty < u < \infty$. Thus $w\lambda(H) = 0$, and we see from (12.67) that

$$\langle w\lambda, \beta\rangle = 0.$$

Since $\beta \neq \pm\alpha_1$, this equation contradicts (12.67). Hence $\pm\alpha_1$ are the only real roots. Again a single Cayley transform must lead to a compact Cartan subgroup, and thus this T has noncompact part of dimension ≤ 1.

Proof of (b′) in Theorem 12.26. Taking (12.59) into account, we see that we may assume $\langle\lambda, \alpha\rangle = 0$ for some compact root α and that $\langle\lambda, \gamma\rangle \neq 0$ for all roots $\gamma \neq \pm\alpha$. Since $\Theta(\lambda, \Delta^+)$ is tempered (Theorem 12.26a), Lemma 12.29 shows it is enough to show that the numerators τ_T of $\Theta(\lambda, \Delta^+)$ vanish when the noncompact part of T has dimension ≤ 1. On B we have

$$\tau_B = (-1)^q \sum_{w \in W_K} (\det w)\xi_{w\lambda},$$

and this expression reduces to 0 if we compute the sum in pairs, grouping the w term and the ws_α term. (Here we use that α is compact, so that s_α is in W_K.)

Thus suppose that T has noncompact part of dimension 1. We may assume that T is obtained from B by Cayley transform with respect to some noncompact root β, and we shall use the same notation for roots in $\Delta(\mathfrak{b}^{\mathbb{C}}:\mathfrak{g}^{\mathbb{C}})$ as for their Cayley transforms in $\Delta(\mathfrak{t}^{\mathbb{C}}:\mathfrak{g}^{\mathbb{C}})$. We seek to compute the coefficients $c_w^{\mathbb{C}}(t_1)$, where t_1 is regular in T. These coefficients do not change as we move within a fixed Weyl chamber under the ψ functor, by Proposition 10.44b. But when we arrive at λ, the two exponentials $e^{w\lambda}$ and $e^{ws_\alpha\lambda}$ coincide, and we can lump the coefficients. Thus we want to prove that

$$c_w^{\mathbb{C}}(t_1) + c_{ws_\alpha}^{\mathbb{C}}(t_1) = 0. \tag{12.68}$$

If $\xi_\beta(t_1) < 0$, then the proof of Theorem 12.6 shows that each term on the left side of (12.68) is 0 before application of ψ and hence is 0 after application of ψ. Thus we may assume that $\xi_\beta(t_1) > 0$, and we are led, in the notation of the proof of Theorem 12.6, to the situation where $\Delta_1 = \{\pm\beta\}$ and where the identity to prove is

$$c_w^{\mathbb{C}}(t_0) + c_{ws_\alpha}^{\mathbb{C}}(t_0) = 0. \tag{12.69}$$

Suppose $\langle w\lambda, \beta\rangle \neq 0$. If we add the patching conditions (12.19) for w and ws_α, we obtain

$$c_w^{\mathbb{C}}(t_0) + c_{ws_\alpha}^{\mathbb{C}}(t_0) - c_{s_\beta w}^{\mathbb{C}}(t_0) - c_{s_\beta ws_\alpha}^{\mathbb{C}}(t_0) = 0. \tag{12.70}$$

Say $\langle w\lambda, \beta\rangle > 0$. Then also $\langle ws_\alpha\lambda, \beta\rangle > 0$. If λ' is dominant integral nonsingular, then (12.17) shows that the coefficients $c_w^{\mathbb{C}}(t_0)$ and $c_{ws_\alpha}^{\mathbb{C}}(t_0)$ are both 0 for $\pi_{\lambda'}$ and hence are 0 after application of ψ. Thus (12.69) follows if $\langle w\lambda, \beta\rangle > 0$. If, on the other hand, $\langle w\lambda, \beta\rangle$ is <0, then we see similarly that the third and fourth terms in (12.70) are 0, and thus (12.70) proves (12.69).

Suppose $\langle w\lambda, \beta\rangle = 0$. Then $\langle \lambda, w^{-1}\beta\rangle = 0$, $w^{-1}\beta = \pm\alpha$, $w^{-1}s_\beta w = s_\alpha$, and $ws_\alpha = s_\beta w$. Thus in this case the identity (12.69) that we want to prove is

$$c_w^C(t_0) + c_{s_\beta w}^C(t_0) = 0. \tag{12.71}$$

The patching condition (12.19) says that the left side of (12.71), up to a sign, is the sum of the coefficients of τ_B for the terms w and ws_α. Since these terms have opposite sign (α being compact), the left side of (12.71) is 0. This completes the proof of Theorem 12.26.

In Theorem 12.26, the representations $\mathcal{D}_1^+$ and $\mathcal{D}_1^-$ of $SL(2, \mathbb{R})$ arise as limits of discrete series in which the parameter λ is 0 and is orthogonal to a noncompact root. But to see the effect of singularities with respect to compact roots, we have to go to a more complicated example.

Example. SU(2, 1) with $\lambda = 0$.

If we fix $\Delta_K^+ = \{e_1 - e_2\}$, there are three possible choices for Δ^+, corresponding to how we view the approach to the limiting value λ. The three possibilities are given in Table 12.2 in §3, with $k \to 0$ and $l \to 0$, $k > l > 0$. If we refer to Table 12.3, we see the effect of this passage to the limit on the characters. The cases where Δ^+ is obtained from simple roots by

$$\Delta^+ \leftrightarrow \{e_1 - e_2, e_2 - e_3\} \qquad \text{and} \qquad \Delta^+ \leftrightarrow \{e_3 - e_1, e_1 - e_2\}$$

correspond to the first and third entries in Table 12.3. In these cases, $\lambda = 0$ is orthogonal to a simple root that is compact, namely $e_1 - e_2$. And we see from the table that τ_B, τ_T^+, and τ_T^- are all 0. The middle entry in Table 12.3 comes from

$$\Delta^+ \leftrightarrow \{e_1 - e_3, e_3 - e_2\}. \tag{12.72}$$

Here there are no simple roots that are compact, and indeed the table shows that τ_T^+ and τ_T^- are not 0, in accord with Theorem 12.26b. In fact, with Δ^+ as in (12.72), one can check with Proposition 10.18 that $\Theta(0, \Delta^+)$ coincides with the character of the unitary principal series with σ trivial and $\nu = 0$. We shall pursue this example further in Chapter XIV.

§8. Discrete Series of M

We were led to construct and analyze discrete series as a result of their role in the Langlands classification. And we now have definitive results for the discrete series of G, which is assumed to be linear connected semisimple. However, the Langlands classification and Theorem 8.53 together require that we understand also the discrete series of M for each parabolic subgroup MAN of G, and M can fail to be semisimple, even fail to be connected.

For the first time in our theory, the possible disconnectedness of M becomes significant, as we shall see a little here and more in Chapter XIV. Thus we must have a careful statement of what happens. The qualitative facts are simple enough. We first extend our theory to linear connected reductive groups with compact center, thereby handling M_0. Next we identify the discrete series of $M^\# = M_0 Z_M$. Finally we induce these to M, and they remain irreducible. The only equivalences are the obvious ones. Moreover, we shall see that $M^\#$ has a simple constructive description.

Let us consider matters in a little more detail. We continue to assume that G is linear connected semisimple and that $\mathfrak{k} \cap i\mathfrak{p} = 0$. Fix an Iwasawa $A_\mathfrak{p}$. In Lemma 9.13 we noted that

$$M_\mathfrak{p} = (M_\mathfrak{p})_0 F, \tag{12.73}$$

where F is the group generated by all elements γ_β for $\beta \in \Sigma$. The group F is finite abelian and may be characterized also as

$$F = K \cap \exp i\mathfrak{a}_\mathfrak{p}. \tag{12.74}$$

In particular, F lies in the center of $M_\mathfrak{p}$.

If MAN is a parabolic subgroup containing a fixed minimal $M_\mathfrak{p} A_\mathfrak{p} N_\mathfrak{p}$, then we defined M in such a way that $M = M_0 M_\mathfrak{p}$, and it follows that

$$M = M_0 F. \tag{12.75}$$

In this generality, F need not lie in the center of M. [For example, in $SL(n, \mathbb{R})$, F is the diagonal subgroup with entries ± 1, and one choice of M is $M = G$.] We define a subgroup $M^\#$ of M by

$$M^\# = M_0 Z_M.$$

Suppose that rank $\mathfrak{m}$ = rank$(\mathfrak{k} \cap \mathfrak{m})$. Let $\mathfrak{b}^-$ be a maximal abelian subspace of $\mathfrak{k} \cap \mathfrak{m}$, and let $B^- = \exp \mathfrak{b}^-$. Then $\mathfrak{a} \oplus \mathfrak{b}^-$ is a Cartan subalgebra of $\mathfrak{g}$. If we have a real root α in $\Delta((\mathfrak{a} \oplus \mathfrak{b}^-)^\mathbb{C} : \mathfrak{g}^\mathbb{C})$, then we can form H_α as a member of $\mathfrak{a}$ and consider the element

$$\exp 2\pi i |\alpha|^{-2} H_\alpha. \tag{12.76}$$

Meanwhile we can restrict α to $\mathfrak{a}$ and extend it by 0 on $\mathfrak{a}_M$ to obtain a member of Σ, which we write temporarily as α_*. It is clear that $H_{\alpha_*} = H_\alpha$ and hence that γ_{α_*} is given by (12.76). To take advantage of this fact, it will help to use the same letter for the real root and the member of Σ. Our conclusion is: If α is a real root, then we can define γ_α without reference to $\mathfrak{a}_\mathfrak{p}$ as given by (12.76), and γ_α is a member of the center of M.

[It is in M by (12.75) and is central because it is in $\exp i\mathfrak{a}$.] We define $F(B^-)$ to be the subgroup of F generated by all γ_α with α a real root of $\Delta((\mathfrak{a} \oplus \mathfrak{b}^-)^{\mathbb{C}} : \mathfrak{g}^{\mathbb{C}})$.

Lemma 12.30. The group $F(B^-)$ is in the center of M and satisfies

(a) $M^\# = M_0 F(B^-)$
(b) $Z_M = (Z_M \cap B^-)F(B^-) = Z_{M_0}F(B^-)$
(c) $Z_G(\mathfrak{a} \oplus \mathfrak{b}^-) = AB^-F(B^-)$
(d) $Z_M(\mathfrak{b}^-) = B^-F(B^-) \subseteq M^\#$.

Proof. We have seen that $F(B^-) \subseteq Z_M$. A little argument with the Cartan decomposition, which we omit, shows that

$$Z_G(\mathfrak{b}^-) = Z_K(\mathfrak{b}^-) \exp(\mathfrak{p} \cap Z_{\mathfrak{g}}(\mathfrak{b}^-)).$$

Each factor on the right is connected, by Corollary 4.22, and thus $Z_G(\mathfrak{b}^-)$ is connected. Consequently $Z_G(\mathfrak{b}^-)$ is the analytic subgroup of G with Lie algebra

$$Z_{\mathfrak{g}}(\mathfrak{b}^-) = \mathfrak{a} \oplus \mathfrak{b}^- \oplus \sum_{\alpha \text{ real}} \mathfrak{g}_\alpha = (\mathfrak{a} + \mathfrak{b}^-) + \left(\sum_{\alpha \text{ real}} \mathbb{R}H_\alpha + \sum \mathfrak{g}_\alpha \right).$$

The second grouped term on the right is a semisimple subalgebra whose corresponding analytic group G_r is closed (cf. the proof of Proposition 5.5), so that our theory applies. Its $\mathfrak{m}_{\mathfrak{p}}$ subalgebra is 0, and (12.73) shows that its $M_{\mathfrak{p}}$ subgroup equals $F(B^-)$. Thus the centralizer of $\sum_{\alpha \text{ real}} \mathbb{R}H_\alpha$ in G_r is $F(B^-)\exp(\sum_{\alpha \text{ real}} \mathbb{R}H_\alpha)$, and we have

$$Z_G(\mathfrak{a} \oplus \mathfrak{b}^-) \subseteq Z_G(\mathfrak{b}^-) \cap Z_G(\exp \sum_{\alpha \text{ real}} \mathbb{R}H_\alpha) \subseteq AB^-F(B^-).$$

The reverse inclusion is obvious, and (c) follows. Intersecting with M, we obtain (d).

For (a), we have $M^\# \supseteq M_0F(B^-)$ and $M_0 \subseteq M_0F(B^-)$. Since $Z_M \subseteq Z_M(\mathfrak{b}^-)$, (d) shows $Z_M \subseteq B^-F(B^-) \subseteq M_0F(B^-)$. Thus (a) follows.

For (b), the first equality follows from the inclusions $F(B^-) \subseteq Z_M \subseteq B^-F(B^-)$. The second follows because Z_{M_0} lies in every maximal torus of $K \cap M_0$, in particular B^-, by Corollary 4.24. This proves the lemma.

The first step in parametrizing the discrete series of M is to handle M_0. We can write $M_0 = M_{ss}(Z_M)_0$ by Proposition 5.5. Since $(Z_M)_0$ is compact and central in M_0, a discrete series of M_0 will be scalar when restricted to $(Z_M)_0$ and will be in the discrete series of M_{ss} when restricted to M_{ss}. We can reverse matters and pass from M_{ss} and $(Z_M)_0$ back to M_0 as long as the scalar on $(Z_M)_0$ agrees with the value of the discrete series on

$M_{ss} \cap (Z_M)_0$. The value can be read off from any $K \cap M_{ss}$ type of the discrete series, and we are led to the following parametrization: Let $\lambda \in (i\mathfrak{b}^-)'$ be such that $\lambda - \delta_M$ is analytically integral and λ is nonsingular. Then there exists a discrete series $\pi_\lambda^{M_0}$ of M_0 whose restriction to M_{ss} is the discrete series with Harish-Chandra parameter $\lambda|_{\mathfrak{b}^- \cap \mathfrak{m}_{ss}}$ and whose restriction to $(Z_M)_0$ is scalar with differential $\lambda|_{\mathfrak{b}^- \cap \mathfrak{z}_\mathfrak{m}}$. All discrete series of M_0 are obtained in this way, and two such are equivalent if and only if their parameters are conjugate via $W(B^-:M_0)$.

The next step is to pass to $M^\#$. Lemma 12.30 gives $M^\# = M_0F(B^-)$. Since $F(B^-)$ is central in M, we can proceed in the same way, obtaining a parametrization by pairs (λ, χ) as follows. Let $\lambda \in (i\mathfrak{b}^-)'$ be such that $\lambda - \delta_M$ is analytically integral and λ is nonsingular. Let χ be a character of $F(B^-)$ that agrees with $\xi_{\lambda-\delta_M}$ on $B^- \cap F(B^-)$. Then there exists a discrete series $\pi^{M^\#}(\lambda, \chi)$ of $M^\#$ whose restriction to M_0 is $\pi_\lambda^{M_0}$ and whose restriction to $F(B^-)$ is the scalar χ. All discrete series of $M^\#$ are obtained this way, and two such are equivalent if and only if their χ's are equal and their parameters λ are conjugate by $W(B^-:M_0)$.

Finally we pass to M. The result will be given as Proposition 12.32 after the following lemma.

Lemma 12.31. Let $\pi = \pi^{M^\#}(\lambda, \chi)$ be a discrete series representation of $M^\#$ If w is in M and the representation $w\pi$ defined by $w\pi(m) = \pi(w^{-1}mw)$ is equivalent with π, then w is in $M^\#$.

Proof. Since $M = M_0F$, we may assume that w is in F, hence in $K \cap M$. Since $\mathfrak{b}^-$ and $\text{Ad}(w)\mathfrak{b}^-$ are maximal abelian subspaces of $\mathfrak{k} \cap \mathfrak{m}$, we can use Theorem 4.42 to adjust w by a member of $K \cap M_0$ so that $\text{Ad}(w)\mathfrak{b}^- = \mathfrak{b}^-$. Then w represents a member of $W(B^-:M)$. Since $M = M_0F$ and $F \subseteq \exp i\mathfrak{a}_\mathfrak{p}$, it follows that MA is contained in the analytic subgroup of $G^\mathbb{C}$ with Lie algebra $(\mathfrak{m} \oplus \mathfrak{a})^\mathbb{C}$. Consequently we have an inclusion

$$W(B^-:M) \subseteq W((\mathfrak{b}^-)^\mathbb{C}:\mathfrak{m}^\mathbb{C}). \tag{12.77}$$

Even without the assumption $w\pi \cong \pi$, a look at global characters shows that $(w\pi)|_{M_0}$ must be $\pi^{M_0}(w\lambda)$. The assumption $w\pi \cong \pi$ forces $\pi^{M_0}(w\lambda) \cong \pi^{M_0}(\lambda)$, and Theorem 9.20 shows that w fixes λ, except for the action by a representative of a member of $W(B^-:M_0)$.

So we may assume $w\lambda = \lambda$. Using Chevalley's Lemma, we see from (12.77) and the nonsingularity of λ that w centralizes $\mathfrak{b}^-$. By Lemma 12.30d, w is in $M^\#$.

Proposition 12.32. The discrete series representations of M are exactly the representations $\pi^M(\lambda, \chi) = \text{ind}_{M^\#}^M \pi^{M^\#}(\lambda, \chi)$. Two such representations $\pi^M(\lambda', \chi')$ and $\pi^M(\lambda, \chi)$ are equivalent if and only if $\chi' = \chi$ and $\lambda' = w\lambda$ for some w in $W(B^-:M)$. The restriction to $M^\#$ of $\pi(\lambda, \chi)$ is given by

$$\pi^M(\lambda, \chi)|_{M^\#} = \sum_{w \in W(B^-:M)/W(B^-:M_0)} \pi^{M^\#}(w\lambda, \chi). \tag{12.78}$$

Remarks. The appropriate notion of **induced representation** here is the one for compact groups used in Chapter I, with the assumption of compactness dropped. It applies since $M^\#$ has finite index in M.

Proof. Let π_0 be a discrete series representation of $M^\#$. Tracking down the definitions, we see that the induced representation π splits under $M^\#$ as

$$\pi|_{M^\#} = (\operatorname{ind}_{M^\#}^M \pi_0)|_{M^\#} = \sum_{w \in M/M^\#} w\pi_0. \tag{12.79}$$

Lemma 12.31 says that the terms on the right side are inequivalent. Therefore any bounded linear operator commuting with $\pi|_{M^\#}$ must be scalar on the space of each $w\pi_0$. If the operator commutes with all of M, these scalars must match. Therefore π is irreducible. It is then clear that π is in the discrete series of M. Lemma 12.31 and equation (12.79) establish (12.78).

Conversely suppose π is in the discrete series of M. Then π is admissible, and so is $\pi|_{M^\#}$. Since π is also unitary, $\pi|_{M^\#}$ splits as the orthogonal direct sum of irreducible unitary representations of $M^\#$. Let π_0 be one such representation. Then it is clear that π_0 is in the discrete series of $M^\#$. To complete the proof, we show that

$$\pi \cong \operatorname{ind}_{M^\#}^M \pi_0. \tag{12.80}$$

Thus let E be the orthogonal projection on the space on which π_0 acts. We map π to the induced space in equivariant fashion by the mapping $v \to f_v$ given by

$$f_v(m) = E(\pi(m)^{-1}v).$$

This map is nonzero since $f_v(1) = v$ for v in the space of π_0. Since the induced representation was shown to the irreducible, the map is an equivalence. Thus (12.80) follows, and the proof is complete.

Example. In $SL(3, \mathbb{R})$, let MAN be block upper triangular with a 2-by-2 block and then a 1-by-1 block. In the upper left 2-by-2 block, a member of M is a 2-by-2 real matrix of determinant ± 1. In the lower right 1-by-1 block, it is ± 1, and the total determinant is one. So $M \cong SL^\pm(2, \mathbb{R})$, where

$$SL^\pm(2, \mathbb{R}) = \{g \in GL(2, \mathbb{R}) | \det g = \pm 1\}. \tag{12.81}$$

We take B^- to be the usual rotation subgroup in the 2-by-2 block of M, and then the matrix $\operatorname{diag}(1, -1, -1)$ operates on B^- as a nontrivial member of $W(B^-:M)$. In this situation, $M^\# = M_0 \cong SL(2, \mathbb{R})$. The discrete series of $SL^\pm(2, \mathbb{R})$ are the representations $\mathscr{D}_n$, $n \geq 2$, given by

$$\mathscr{D}_n \cong \operatorname{ind}_{SL}^{SL^\pm} \mathscr{D}_n^+ \cong \operatorname{ind}_{SL}^{SL^\pm} \mathscr{D}_n^-;$$

the lumping of the parameters is caused by the nontrivial element of $W(B^-:M)$. When restricted to $SL(2, \mathbb{R})$, $\mathscr{D}_n$ splits as $\mathscr{D}_n^+ \oplus \mathscr{D}_n^-$.

Later we shall need a notion of **limit of discrete series** for M. Such representations have no intrinsic defining property, and we simply define them constructively by analogy with limits of discrete series for G. Again we handle M_0 first. Let $\lambda \in (i\mathfrak{b}^-)'$ be such that $\lambda - \delta_M$ is analytically integral and λ is nonsingular. Fix a positive system Δ_M^+ for $\Delta((\mathfrak{b}^-)^{\mathbb{C}}:\mathfrak{m}^{\mathbb{C}})$ such that λ is dominant. Then there is a unique $\pi^{M_0}(\lambda, \Delta_M^+)$ whose restriction to M_{ss} is the limit of discrete series with parameter $(\lambda|_{\mathfrak{b}^-\cap\mathfrak{m}_{ss}}, \Delta_M^+)$ and whose restriction to $(Z_M)_0$ is scalar with differential $\lambda|_{\mathfrak{b}^-\cap Z_{\mathfrak{m}}}$. Next we construct $\pi^{M^\#}(\lambda, \Delta_M^+, \chi)$ whenever (λ, Δ_M^+) is as above and χ is a character of $F(B^-)$ that agrees with $\xi_{\lambda-\delta_M}$ on $B^- \cap F(B^-)$. Finally we define limits of discrete series for M by

$$\pi^M(\lambda, \Delta_M^+, \chi) = \operatorname{ind}_{M^\#}^M \pi^{M^\#}(\lambda, \Delta_M^+, \chi); \tag{12.82}$$

the character of this representation is denoted $\Theta^M(\lambda, \Delta_M^+, \chi)$.

Proposition 12.33. The limits of discrete series on M and their global characters have the following properties:

(a) $\Theta^M(\lambda, \Delta_M^+, \chi)$ is tempered and either is 0 or is irreducible.
(b) $\Theta^M(\lambda, \Delta_M^+, \chi)$ is 0 if and only if $\langle\lambda, \alpha\rangle = 0$ for some Δ_M^+ simple root α that is compact.
(c) If $\Theta^M(\lambda, \Delta_M^+, \chi) \neq 0$, then $\Theta^M(\lambda', (\Delta_M^+)', \chi') = \Theta^M(\lambda, \Delta_M^+, \chi)$ if and only if $\chi' = \chi$ and $(\lambda', (\Delta_M^+)')$ is conjugate to (λ, Δ_M^+) via $W(B^-:M)$.
(d) The restriction to $M^\#$ of $\pi^M(\lambda, \Delta_M^+, \chi)$ is given by

$$\pi^M(\lambda, \Delta_M^+, \chi)|_{M^\#} = \sum_{w \in W(B^-:M)/W(B^-:M_0)} \pi^{M^\#}(w\lambda, w\Delta_M^+, \chi).$$

Proof. Conclusion (b) is immediate from Theorem 12.26. To prove (a), we have to check that the induction (12.82) leads to an irreducible representation; the proof is the same as for Lemma 12.31 and Proposition 12.32, except that we now use (λ, Δ_M^+) in the argument in place of λ. This argument leads us to the formula in (d) also. Result (d) then implies (c).

Concluding remark. Our use of discrete series and limits of discrete series for M will largely be in the context of induced representations from a parabolic subgroup M. Since we have seen that the representations of M are induced, we can apply the double induction formula. Thus every standard induced representation with σ a discrete series or limit of discrete series is of the form

$$U(MAN, \sigma, \nu) \cong \operatorname{ind}_{M^\# AN}^G(\sigma^\# \otimes e^\nu \otimes 1). \tag{12.83}$$

§9. Schmid's Identity

Theorem 12.26 says that the nonzero limits of discrete series of G are irreducible tempered unitary representations and are not discrete series. According to Theorem 8.53, such a representation must either coincide with a standard induced representation or else exhibit reducibility of such a representation; the standard induced representation may be taken to be induced from discrete series on some M.

In the case of SL(2, $\mathbb{R}$), it is $\mathscr{D}_1^+$ and $\mathscr{D}_1^-$ that are addressed by Theorem 12.26, and in their case we know that the relevant result is the identity $\mathscr{P}^{-,0} \cong \mathscr{D}_1^+ \oplus \mathscr{D}_1^-$. Our goal in this section is to establish a generalization of this result.

Theorem 12.34. Let G be linear connected reductive with rank $G =$ rank K, and let $\mathfrak{b} \subseteq \mathfrak{k}$ be a compact Cartan subalgebra. Fix a positive system Δ^+ for $\Delta(\mathfrak{b}^{\mathbb{C}}:\mathfrak{g}^{\mathbb{C}})$, and suppose α is a simple noncompact root. Suppose that the Cayley transform $\mathbf{c}_\alpha$ leads from $\mathfrak{b}$ to a Cartan subalgebra $\mathfrak{b}^- \oplus \mathfrak{a}$, and let $S = MAN$ be a parabolic subgroup constructed from $\mathfrak{b}^- \oplus \mathfrak{a}$. Let Δ_M^+ be the positive system of $\Delta((\mathfrak{b}^-)^{\mathbb{C}}:\mathfrak{m}^{\mathbb{C}})$ given by

$$\Delta_M^+ = \{\mathbf{c}_\alpha(\gamma) \mid \gamma \in \Delta^+ \text{ and } \gamma \perp \alpha\}.$$

If λ is a member of $(i\mathfrak{b})'$ such that

(i) $\lambda - \delta$ is analytically integral
(ii) λ is Δ^+ dominant
(iii) $\langle \lambda, \alpha \rangle = 0$, and $\langle \lambda, \beta \rangle \neq 0$ for $\beta \neq \pm\alpha$,

then
$$U(S, \pi^M(\lambda|_{\mathfrak{b}^-}, \Delta_M^+, \zeta), 0) \cong \pi(\lambda, \Delta^+) \oplus \pi(\lambda, s_\alpha \Delta^+),$$

where ζ is the character of $F(B^-) = \{1, \gamma_{\mathbf{c}_\alpha(\alpha)}\}$ given by

$$\zeta(\gamma_{\mathbf{c}_\alpha(\alpha)}) = (-1)^{2\langle \rho_\alpha, \alpha \rangle / |\alpha|^2}$$

and where ρ_α is half the sum of the roots of $\Delta(\mathfrak{b}^{\mathbb{C}}:\mathfrak{g}^{\mathbb{C}})$ whose inner product with α is positive.

Remarks. The Lie algebra $\mathfrak{a}$ in the statement is one-dimensional, and $\pm\mathbf{c}_\alpha(\alpha)$ are the only real roots. Thus $F(B^-) = \{1, \gamma_\alpha\}$. By (iii), $\lambda|_{\mathfrak{b}^-}$ is nonsingular. It is implicit in the statement of the theorem that $\pi^M(\lambda|_{\mathfrak{b}^-}, \Delta_M^+, \zeta)$ is a well-defined discrete series representation of M, i.e., that $\lambda|_{\mathfrak{b}^-} - \delta_M$ is analytically integral for B^- and that ζ agrees with $\xi_{\lambda|_{\mathfrak{b}^-} - \delta_M}$ on $B^- \cap F(B^-)$.

First part of proof. First let us check that $\lambda|_{\mathfrak{b}^-} - \delta_M$ is analytically integral for B^- and that ζ agrees with $\xi|_{\mathfrak{b}^-} - \delta_M$ on $B^- \cap F(B^-)$. It will be convenient to identify forms on $\mathfrak{b}$ with their Cayley transforms, writing everything relative to $\mathfrak{b}$. Introduce a new positive system $(\Delta^+)'$ by using the ordering obtained from a basis consisting of H_α and a basis of $i\mathfrak{b}^-$ that leads to Δ_M^+. The half sum δ' for this ordering satisfies $\delta' = \rho_\alpha + \delta_M$. Now δ' is a Weyl group transform of δ and hence differs from δ by a sum of roots. Thus $\lambda - \delta'$ is analytically integral for B, by (i). We write

$$\lambda - \delta' = (\lambda - \delta_M) - \rho_\alpha. \tag{12.84}$$

Now ρ_α vanishes on $\mathfrak{b}^-$, and thus we can see by restricting (12.84) to $\mathfrak{b}^-$ that $\lambda|_{\mathfrak{b}^-} - \delta_M$ is analytically integral for B^-. Next, $\xi_{\lambda-\delta'}(\gamma_\alpha)$ is well defined and is given by

$$\begin{aligned}\xi_{\lambda-\delta'}(\gamma_\alpha) = e^{2\pi i\langle\lambda-\delta',\alpha\rangle/|\alpha|^2} &= (-1)^{2\langle\lambda-\delta',\alpha\rangle/|\alpha|^2} \\ &= (-1)^{2\langle\rho_\alpha,\alpha\rangle/|\alpha|^2} = \zeta(\gamma_\alpha),\end{aligned} \tag{12.85}$$

by (12.84). If γ_α is in $\exp \mathfrak{b}^-$, write $\gamma_\alpha = \exp H$ with $H \in \mathfrak{b}^-$. Then

$$\xi_{\lambda-\delta'}(\gamma_\alpha) = \xi_{\lambda|_{\mathfrak{b}^-}-\delta_M}(\exp H)$$

since $\rho_\alpha(H) = 0$. Hence ζ agrees with $\xi_{\lambda|_{\mathfrak{b}^-}-\delta_M}$ on $B^- \cap F(B^-)$.

The theorem will be proved as a global character identity. This will be sufficient by Theorem 10.6, since the induced representation is unitary and the limits of discrete series are irreducible. Referring to (10.27) and Theorem 12.1, we see that all the characters in question are tempered. By Lemma 12.29 it suffices to prove the character identity just on Cartan subgroups whose noncompact parts have dimension ≤ 1.

On B, the numerators for the two limit of discrete series characters are the same, and the denominators are of opposite sign because of s_α. Thus the sum is 0. The character of the induced representation matches this 0 on B, by Proposition 10.19.

Next consider a Cartan subgroup T whose noncompact part has dimension 1. The Cartan subgroup corresponding to $\mathfrak{b}^- \oplus \mathfrak{a}$ is $B^-AF(B^-)$ by Lemma 12.30, and we exclude for now the case that T is conjugate to this. Then again the induced character vanishes on T, by Proposition 10.19. Let Θ be the sum of the two limit of discrete series characters. Let us see how close the patching conditions come to proving that Θ vanishes on T. Without loss of generality we may assume that T is obtained from B by Cayley transform relative to some noncompact root β.

To use the patching conditions, let us refer to the proof of Lemma 12.29. In the notation of that proof, we will get all coefficients 0 in any situation where Δ_1 is empty, since such coefficients are 0 for discrete series. Thus we may assume that $\Delta_1 = \{\pm\beta\}$. Let $\{c_w\}$ be the unknown coefficients in the numerator of Θ, arranged so that $c_w = c_{ws_\alpha}$. By (12.64), c_w is 0 if $\langle w\lambda, \beta\rangle > 0$, and by (12.66) $c_w = 0$ if $c_{s_\beta w} = 0$. Thus c_w is 0 if $\langle w\lambda, \beta\rangle \neq 0$. In other words, the only way we can have a nonzero coefficient is if $\langle w\lambda, \beta\rangle = 0$ for some w. In this case, $\langle \lambda, w^{-1}\beta\rangle = 0$ and so $w^{-1}\beta = \pm\alpha$ by hypothesis (iii). That is, β must be conjugate to $\pm\alpha$ via

$W(\mathfrak{b}^{\mathbb{C}}:\mathfrak{g}^{\mathbb{C}})$. The Problems at the end of the chapter show that this conjugacy implies that β is conjugate to $\pm\alpha$ by W_K. Hence T is conjugate to $B^-AF(B^-)$. Therefore Θ vanishes on a Cartan subgroup T with noncompact part of dimension 1 unless T is conjugate to $B^-AF(B^-)$. And, of course, in the nonconjugate case, Θ then matches the induced character on T.

Second part of proof. To complete the proof, it therefore suffices to show that Θ matches the induced character on $B^-AF(B^-)$. We shall carry out this step only under some simplifying assumptions. For a while, we shall assume that M is connected, and we shall obtain a formula for the sum of the two limit of discrete series characters. Then we shall specialize to SU(2, 1) to see that we have a match with the induced character.

We shall refer all positive systems to the one Δ^+ in the statement of the theorem. Let us compute the numerator of the character of $\pi(\lambda, \Delta^+)$ on $B^-AF(B^-)$. We use the patching conditions to determine the character of $\pi(\delta, \Delta^+)$ and then move the parameter. The notation is as in the proof of (b′) for Theorem 12.26. Since M is connected, ξ_α is positive on $B^-AF(B^-)$. Thus $\Delta_1 = \{\pm\alpha\}$. At a point where $\xi_\alpha(t_0) = 1$, we study the coefficients $c_w = c_w^{\mathbb{C}}(t_0)$. We know that

$$\langle w\delta, \alpha\rangle > 0 \quad \text{implies} \quad c_w = 0$$

from the boundedness of the numerator and that

$$c_w - c_{s_\alpha w} = \begin{cases} (-1)^q(\det w) & \text{if } w \in W_K \text{ or } s_\alpha w \in W_K \\ 0 & \text{otherwise} \end{cases}$$

from the patching condition. Thus the numerator of $\Theta(\lambda, \Delta^+)$ on the + component of $(B^-AF(B^-))'_R$ is

$$\sum_{\substack{w \in W_K \\ \langle w\delta,\alpha\rangle > 0}} (-1)^{q+1}(\det w)\xi_{s_\alpha w\lambda} + \sum_{\substack{s_\alpha w \in W_K \\ \langle w\delta,\alpha\rangle > 0}} (-1)^{q+1}(\det w)\xi_{s_\alpha w\lambda}. \tag{12.86}$$

We need also the numerator $\Theta(\lambda, s_\alpha\Delta^+)$, again relative to Δ^+. A similar computation shows it to be

$$\sum_{\substack{w \in W_K \\ \langle ws_\alpha\delta,\alpha\rangle > 0}} (-1)^{q}(\det w)\xi_{s_\alpha w\lambda} + \sum_{\substack{s_\alpha w \in W_K \\ \langle ws_\alpha\delta,\alpha\rangle > 0}} (-1)^{q}(\det w)\xi_{s_\alpha w\lambda}. \tag{12.87}$$

The sum of (12.86) and (12.87) gives us the numerator τ^+ of the character of the sum of the two limits of discrete series. After a change of variables in two of the four sums, we can write the result as

$$\tau^+ = (-1)^{q+1}\left\{ \sum_{\substack{w \in W_K \\ w^{-1}\alpha > 0}} - \sum_{\substack{w \in W_K \\ w^{-1}\alpha < 0}} - \sum_{\substack{w \in W_K \\ s_\alpha w^{-1}\alpha > 0}} + \sum_{\substack{w \in W_K \\ s_\alpha w^{-1}\alpha < 0}} \right\}(\det w)\xi_{s_\alpha w\lambda}. \tag{12.88}$$

Fix attention on the contribution to (12.88) from a single element w. Every w contributes twice. If $w^{-1}\alpha$ and $s_\alpha w^{-1}\alpha$ have the same sign, the two contributions cancel, and conversely. Now the only way $w^{-1}\alpha$ and $s_\alpha w^{-1}\alpha$ can have opposite signs is if $w^{-1}\alpha = \pm\alpha$, since α is simple. In these cases $w^{-1}s_\alpha w = s_\alpha$, so that $s_\alpha w\lambda = ws_\alpha\lambda = w\lambda$. Thus we obtain

$$\tau^+ = \sum_{\substack{w\in W_K\\ w\alpha=\alpha}} (-1)^{q+1}2(\det w)\xi_{w\lambda} + \sum_{\substack{w\in W_K\\ w\alpha=-\alpha}} (-1)^q 2(\det w)\xi_{w\lambda}. \quad (12.89)$$

The corresponding formula for τ^- is completely determined by (12.89) and Theorem 10.40.

Now we specialize to SU(2, 1). Then $q = 2$ and $W_K = \{1, s_{e_1-e_2}\}$. For α noncompact and w in W_K, the only contribution to (12.89) is from $w = 1$ in the first term. Thus $\tau^+ = -2\xi_{w\lambda}$. To complete the proof, we need only compare with Proposition 10.18. Before making the comparison, we have to change the positive system Δ^+ to match the one in Proposition 10.18, where $\mathfrak{a}$ comes before $\mathfrak{m}$, so that the Weyl denominators are the same. This change introduces a minus sign, and thus the characters match as asserted.

§10. Problems

1. Using the patching conditions, show that the coefficients $c_w^C(t_1)$ in the numerators of discrete series characters are necessarily integers.

Problems 2 to 4 establish absolute convergence of the sum that will be seen in Chapter XIII to be the contribution of the discrete series to the Plancherel formula of G.

2. Let $\{T_i\}$ be a complete system of nonconjugate Θ-stable Cartan subgroups. Using Theorem 12.17d, prove that

$$f \to \sum_i \int_{T_i} |F_f^{G/T_i}(t)|\, dt$$

is a continuous complex-valued function on the Schwartz space $\mathscr{C}(G)$.

3. Starting from $\chi_\lambda(z)\Theta_\lambda(f) = \Theta_\lambda(z^{\text{tr}}f)$ for f in $\mathscr{C}(G)$, bound $\Theta_\lambda(f)\varpi(\lambda)$, where $\varpi = \prod_{\alpha>0}\alpha$, by the product of $|\varpi(\lambda)/\chi_\lambda(z)|$ with a multiple of $\sum_i \int_{T_i} |F_{z^{\text{tr}}f}^{G/T_i}(t)|\, dt$. [Hint: Show that the numerators of discrete series for G are bounded on each T_i in a fashion independent of λ.]
4. Combine Problems 2 and 3 with the selection of a suitable z to prove that $\sum \Theta_\lambda(f)\varpi(\lambda)$ is absolutely convergent for f in $\mathscr{C}(G)$ and the mapping of f to this sum is continuous on $\mathscr{C}(G)$.

Problems 5 to 11 give an environment for realizing discrete series concretely. Let π_λ be a discrete series, let Λ be its Blattner parameter, and let τ_Λ be a representation of K with highest weight Λ, acting irreducibly in a space V_Λ. According to Problems 11–14 in Chapter IV,

$$\tau_\Lambda \otimes \mathrm{Ad}(K)|_{\mathfrak{p}^{\mathbb{C}}} = \sum_{\beta \in \Delta_n} m_\beta \tau_{\Lambda+\beta}$$

with each m_β equal to 0 or 1. Let τ_Λ^- be the subrepresentation of this tensor product given by

$$\tau_\Lambda^- = \sum_{\beta \in \Delta_n^+} m_{-\beta}\tau_{\Lambda-\beta}.$$

Let V_Λ^- be the subspace of $V_\Lambda \otimes \mathfrak{p}^{\mathbb{C}}$ in which τ_Λ^- operates, and let $P: V_\Lambda \otimes \mathfrak{p}^{\mathbb{C}} \to V_\Lambda^-$ be the orthogonal projection. Define

$$C^\infty(G, \tau_\Lambda) = \{f: G \to V_\Lambda \text{ smooth} \mid f(kg) = \tau_\Lambda(k)f(g) \text{ for } g \in G, k \in K\}$$

and define $C^\infty(G, \tau_\Lambda^-)$ similarly. The **Schmid $\mathscr{D}$ operator** carries $C^\infty(G, \tau_\Lambda)$ into $C^\infty(G, \tau_\Lambda^-)$ and is given by

$$\mathscr{D}f(g) = \sum_{i=1}^{2n} P((X_i)_R f(g) \otimes X_i),$$

where $\{X_i\}$ is an orthonormal basis of $\mathfrak{p}$ and $(X_i)_R$ refers to right invariant differentiation.

5. Verify that $\tau_\Lambda^-(k)P(v) = P((\tau_\Lambda(k) \otimes \mathrm{Ad}(k))v)$ for all v in $V_\Lambda \otimes \mathfrak{p}^{\mathbb{C}}$ and that $P(v) = 0$ if v is a weight vector whose weight is not of the form $\Lambda - \sum_{\alpha \in \Delta^+} n_\alpha \alpha$ with all $n_\alpha \geq 0$.
6. Show that $\mathscr{D}$ is independent of the basis and is equivariant with respect to right translation by G. Formulate and prove a property of equivariance under left translation by K.
7. Let V^λ be the K-finite subspace for π_λ, and let E_Λ be the orthogonal projection on the K type Λ. For any φ in V^λ, define f_φ by

$$f_\varphi(x) = E_\Lambda \pi_\lambda(x)\varphi.$$

   Prove that $\varphi \to f_\varphi$ is a $\mathfrak{g}$-equivariant map of V^λ into $C^\infty(G, \tau_\Lambda)$ and that f_φ is in $L^2(G)$.
8. With f_φ as in Problem 7, prove that $\mathscr{D}f_\varphi$ is in L^2 and that $z(\mathscr{D}f_\varphi) = \chi_\lambda(z)\mathscr{D}f_\varphi$ for z in $Z(\mathfrak{g}^{\mathbb{C}})$. Conclude that $\mathscr{D}f_\varphi$ is in the finite sum of discrete series under the *left* regular representation. Using knowledge of the K types of π_λ, prove that $\mathscr{D}f_\varphi = 0$. Hence π_λ imbeds infinitesimally into the space of square integrable members of ker $\mathscr{D}$ within $C^\infty(G, \tau_\Lambda)$.
9. Let $\mathscr{S}$ be the space of the spin representations as in §3. Introduce a Dirac operator in this context by the definition

$$Df(g) = \sum_{i=1}^{2n} (I \otimes c(X_i))X_i f(g)$$

for f in $C^\infty(G, \tau_\Lambda) \subseteq C^\infty(G, \tau_{\Lambda-\delta_n} \otimes c)$ under the additional assumption that $\Lambda - \delta_n$ is Δ_K^+ dominant. Prove that $\ker \mathscr{D} \subseteq \ker D$. [Hint: Let

$$D_1 f(g) = \sum X_i f(g) \otimes X_i$$

$$D_2 = I \otimes c \qquad \text{(as a map from } V_{\Lambda-\delta_n} \otimes (\mathscr{S} \otimes \mathfrak{p}^{\mathbb{C}}) \text{ to } V_{\Lambda-\delta_n} \otimes \mathscr{S}),$$

and show that $\mathscr{D} = PD_1$ and $D = D_2 D_1$. Obtain the conclusion $\ker \mathscr{D} \subseteq \ker D$ by showing that $D_2(I - P) = 0$ follows from Problem 5.]

10. Appealing to a suitable variant of Lemma 12.12, show that the Casimir operator of G is scalar on $\ker \mathscr{D}$, under the additional assumption in Problem 9.
11. In SU(2, 1) for a suitable Δ^+, the parameters $\lambda = 0$ and $\Lambda = 0$ lead to a nonzero limit of discrete series. Show that $\mathscr{D}$ is the 0 operator in this case. Hence the Casimir operator of G does not act as a scalar on $\ker \mathscr{D}$.

Problems 12 to 13 compare the integer span of the roots with the integer span of the compact roots. It is assumed that rank G = rank K.

12. Fix a positive system Δ^+. Attach to each simple noncompact root the integer 1 and to each simple compact root the integer 0; extend additively to the group generated by the roots, obtaining a function $\gamma \to n(\gamma)$. Using an inductive argument and taking into account the bracket relations for $\mathfrak{k}$ and $\mathfrak{p}$, prove that $n(\gamma)$ is odd when γ is a positive noncompact root and is even when γ is a positive compact root.
13. Making use of the function $\gamma \to (-1)^{n(\gamma)}$, prove that a noncompact root can never be an integer combination of compact roots.

Problems 14 to 18 relate the condition rank G = rank K to properties of restricted roots.

14. Suppose rank G = rank K and $\mathfrak{b}$ is a compact Cartan subalgebra. Define H_0 to be the member of $i\mathfrak{b}$ with $\gamma(H_0) = n(\gamma)$ for each simple root γ, where $n(\gamma)$ is as in Problem 12. Let w be the member of K given by $w = \exp i\pi H_0$. Prove that $\mathrm{Ad}(w)$ is $+1$ on $\mathfrak{k}^{\mathbb{C}}$ and is -1 on $\mathfrak{p}^{\mathbb{C}}$. Hence $\mathrm{Ad}(w) = \theta$ on $\mathfrak{g}$, where θ is the Cartan involution.
15. When rank G = rank K, conclude from Problem 14 the following strong form of the statement that -1 is in $W(A_\mathfrak{p}:G)$: If the Cartan subgroup of G built from $\mathfrak{a}_\mathfrak{p}$ is denoted $A_\mathfrak{p} B_\mathfrak{p} F(B_\mathfrak{p})$, then the transformation that is -1 on $\mathfrak{a}_\mathfrak{p}$ and is $+1$ on $\mathfrak{b}_\mathfrak{p}$ is in $W(A_\mathfrak{p} B_\mathfrak{p} F(B_\mathfrak{p}):G)$.
16. Using Chevalley's Lemma and Problem 15, prove when rank G = rank K that there exists an orthogonal basis of restricted roots of $\mathfrak{a}_\mathfrak{p}$

and that these restricted roots may be taken to be the restrictions of real roots relative to $\mathfrak{a}_\mathfrak{p} \oplus \mathfrak{b}_\mathfrak{p}$.

17. In the situation of Problem 16, let $\alpha_1, \ldots, \alpha_l$ be the orthogonal basis.
    (a) Show that if $\alpha_i \pm \alpha_j$ are roots, then $\alpha_i \pm \alpha_k$ cannot be roots when $k \neq j$.
    (b) If α_i has the property that $\alpha_i \pm \alpha_j$ are roots for some (unique) j, replace α_i, α_j in the basis by $\alpha_i + \alpha_j, \alpha_i - \alpha_j$. Repeat for each pair of roots among $\alpha_1, \ldots, \alpha_l$ for which this condition holds. Show that the new orthogonal basis has the property that no sum or difference of the basis members can be a root. [*Terminology.* The members of the new basis are **strongly orthogonal.**]
18. Suppose G has the property that $\mathrm{Ad}(w) = \theta$ on $\mathfrak{g}$ for some w in K. Tracing through the constructions of Problems 15–17, prove that rank G = rank K. [Hint: Use Cayley transforms, starting from the strongly orthogonal set in Problem 17b.]

Problems 19 to 22 relate conjugacy of noncompact roots via W_G to conjugacy via W_K. It is assumed that rank G = rank K, and references are to a Cartan subalgebra $\mathfrak{b} \subseteq \mathfrak{k}$. The final result is that if β and γ are noncompact roots that are conjugate by W_G, then β is conjugate by W_K to γ or $-\gamma$. Without loss of generality, we assume that $\mathfrak{g}$ is simple. Problem 15 in Chapter VI provides background for this situation.

19. Suppose all noncompact roots have the same length. First assume that $\mathrm{Ad}(K)$ acts irreducibly on $\mathfrak{p}^{\mathbb{C}}$. Fix a positive system Δ^+, and let α be the highest weight of $\mathfrak{p}^{\mathbb{C}}$ (the largest noncompact root). Given β noncompact, use W_K to move β so as to be Δ_K^+ dominant. Using that β is a dominant weight and α is the highest weight and $|\alpha| = |\beta|$, prove that $\beta = \alpha$. Conclude that all noncompact roots are conjugate via W_K.
20. Suppose all noncompact roots have the same length. Adapt the argument of Problem 19 to allow for reducibility $\mathfrak{p}^{\mathbb{C}} = \mathfrak{p}^+ \oplus \mathfrak{p}^-$, showing that one of $\pm\beta$ must be conjugate to the largest noncompact root for $\mathfrak{p}^+$ and the other must be conjugate to the largest noncompact root for $\mathfrak{p}^-$. Conclude that if β and γ are noncompact, then γ is conjugate to one of β or $-\beta$ via W_K.
21. Suppose the noncompact roots are of two different lengths. (There cannot be more than two.) We shall assume the relation between the two lengths is given by $|\alpha_1|^2 = 2|\alpha_2|^2$, since the only other relation $|\alpha_1|^2 = 3|\alpha_2|^2$ occurs in only one group. Adapt the argument in Problems 19 and 20 to show that if β and γ are long and noncompact, then γ is conjugate to one of β or $-\beta$ via W_K.
22. Suppose the noncompact roots are of two different lengths, related as in Problem 21. Let α be the largest noncompact root for $\mathfrak{p}^{\mathbb{C}}$ or $\mathfrak{p}^+$,

and let β be a short dominant noncompact root for $\mathfrak{p}^{\mathbb{C}}$ or $\mathfrak{p}^+$. Write $\alpha = \beta + \sum n_\gamma \gamma$ with the γ in Δ_K^+ and all $n_\gamma > 0$.

(a) Prove that β is orthogonal to all γ that occur in the above decomposition.

(b) Using (a), prove that $\langle \alpha, \beta \rangle = |\beta|^2$. Conclude that $\alpha = \beta$ is a short compact root γ and that $\langle \beta, \gamma \rangle = 0$.

(c) Suppose that the situation in (b) occurs twice, with $\alpha - \beta_1 = \gamma_1$ and $\alpha - \beta_2 = \gamma_2$. If $\langle \beta_1, \beta_2 \rangle \neq 0$, show that β_1 and one of $\pm\beta_2$ are conjugate via the member $s_{\beta_1 \pm \beta_2}$ of W_K. If $\langle \beta_1, \beta_2 \rangle = 0$, use Parseval's equality to show that $\alpha = \pm\beta_1 \pm \beta_2$, in contradiction to Problem 13.

(d) Conclude that if β and γ are short noncompact roots, then γ is conjugate to one of β or $-\beta$ via W_K.

Problems 23 to 24 give a painless construction of the discrete series for which the square integrability is difficult to prove in Chapter IX. It is assumed that rank G = rank K. One takes as known the uniqueness results in §§12.1–12.3 and the construction in Chapter IX for very nonsingular λ.

23. Let π be an irreducible tempered representation of G with nonsingular integral infinitesimal character. Prove that π is in the discrete series. [Hint: Use Theorem 8.53 and show that the $\mathfrak{a}$ parameter is 0. Use nonsingularity to conclude that there are no real roots.]

24. Starting from π_λ for λ very nonsingular, construct π_λ for all nonsingular λ by Zuckerman tensoring. Going over the proof of Theorem 12.26 and applying Problem 23, show how this construction enables one to deduce Theorem 9.20 for all nonsingular λ from the theorem for λ very nonsingular.

Problems 25 to 29 show how some limits of discrete series of a suitable M can exhibit reducibility of induced representations. Let G be the double cover of the split group $SO_0(4, 4)$, and regard G as contained in $G^{\mathbb{C}}$ with $G^{\mathbb{C}}$ simply connected. Say that one Cartan subgroup T_1 of G is $\geq$ another T_2 if T_2 is conjugate in G to a subgroup of $M_1 A_1$.

25. Classify the Θ-stable Cartan subgroups of G, apart from conjugacy. There is one each of compact dimension 0, 1, 3, 4, and there are three of compact dimension 2.

26. Let T be a Cartan subgroup of compact dimension 3. Show that the associated M is isomorphic to $SL(2, \mathbb{R}) \times SL(2, \mathbb{R}) \times SL(2, \mathbb{R})$ with the element $\begin{pmatrix} 1 & 0 \\ 0 & -1 \end{pmatrix} \times \begin{pmatrix} 1 & 0 \\ 0 & -1 \end{pmatrix} \times \begin{pmatrix} 1 & 0 \\ 0 & -1 \end{pmatrix}$ adjoined.

27. Show that the $M_{\mathfrak{p}}$ group of the M in Problem 26 is of order 16. Let σ be a character of $M_{\mathfrak{p}}$ that is nontrivial on each $SL(2, \mathbb{R})$ factor. Show that the principal series representation of M corresponding to this σ and to $\mathfrak{a}_{\mathfrak{p}}$ parameter 0 is reducible and has four inequivalent irreducible constituents.

28. Show that M has four limits of discrete series, and identify them as the constituents in Problem 27.

29. Inducing everything to G, show that there is a principal series representation $U(S_{\mathfrak{p}}, \sigma, 0)$ of G that reduces into at least four pieces. What can be said about the value of σ on the elements γ_α of the $M_{\mathfrak{p}}$ subgroup of G?

CHAPTER XIII

Plancherel Formula

§1. Ideas and Ingredients

We discussed a scheme of proof of the Plancherel formula in §§11.4 and 11.5. Since then, we have obtained more precise information about F_f and the discrete series characters, and we can now proceed with the derivation of the Plancherel formula.

In §11.5 we saw that the derivation might be so overwhelmingly complicated as to be impractical in general. A few extra devices reduce the complications significantly:

(1) For the most part, we shall work with an "averaged" version $\mathscr{F}_f$ of F_f to take the maximum possible symmetry into account.

(2) We shall replace discrete series characters by their "averaged" versions.

(3) We shall observe that the "averaged" discrete series characters satisfy reduction identities that make their formulas manageable.

(4) We shall invert F_f (or actually $\mathscr{F}_f$) before taking the derivative with respect to the product of the positive roots and applying Theorem 11.17.

We shall do the real-rank-one case in some detail as a start. In §2, we use only device (4), saving introduction of (1) and (2) for §3. In §5, we take up the case of $\mathrm{Sp}(2, \mathbb{R})$, which is fairly typical of the general case except that the reduction identities in device (3) do not come into play. The remainder of the chapter describes (2) and (3) in detail and their role in the general case, and it addresses the resulting formula.

§2. Real-Rank-One Groups, Part I

Let G be linear connected semisimple with real rank one. Without loss of generality, we shall assume that $\mathfrak{k} \cap i\mathfrak{p} = 0$ and that $G^{\mathbb{C}}$ is simply connected. Since G itself is the only nonminimal parabolic subgroup, we drop the subscripts "$\mathfrak{p}$" used in connection with minimal parabolic subgroups. Fix a minimal parabolic subgroup MAN. Unless G is essentially $\mathrm{SL}(2, \mathbb{R})$, which is already understood from Chapter XI, the group M is connected, by Problems 21–22 of Chapter V. Thus we shall assume M is connected. Let B^- be a Cartan subgroup of M; then $T = B^-A$ is a Cartan subgroup of G. We choose a positive system for $\Delta(\mathfrak{t}^{\mathbb{C}}:\mathfrak{g}^{\mathbb{C}})$ such that the nonzero restrictions to $\mathfrak{a}$ of the positive roots are the restricted roots defining N. Note that $|W(A:G)| = 2$.

Our first objective is to express $F_f^T = F_f^{G/T}$ in terms of characters. When T is the only Cartan subgroup of G up to conjugacy (as, for example, in $SL(2, \mathbb{C})$), we will be able to differentiate this formula to obtain the Plancherel formula.

To get started, we need to know that δ_M is analytically integral for B^-. In fact, in our positive system, δ_M is the restriction to $\mathfrak{b}^-$ of δ_G, which is analytically integral since $G^{\mathbb{C}}$ is simply connected.

Let σ_Λ be the irreducible finite-dimensional representation of M with highest weight Λ. Combining (10.25c) and Proposition 10.18, we see that the nonunitary principal series characters are given by

$$\Theta_{\sigma_\Lambda,\nu}(f) = s^{G/T}|W(T:G)|^{-1} \int_T \left(\sum_{w \in W(A:G)} D^{M/B^-} \chi_{w\sigma_\Lambda} \otimes e^{w\nu} \right)(h) F_f^T(h)\, dh. \tag{13.1}$$

Let us write $w\Lambda$ for the highest weight of $w\sigma_\Lambda$. If w is the nontrivial element of $W(A:G)$, w preserves the weight lattice in $\mathfrak{b}^-$ and preserves positivity, thus must fix δ_M. Using the Weyl character formula, we can therefore rewrite (13.1) as

$$\Theta_{\sigma_\Lambda,\nu}(f) = s^{G/T}|W(T:G)|^{-1} \sum_{\substack{w \in W(A:G) \\ s \in W(B^-:M)}} (\det s) \times \int_T (\xi_{sw(\Lambda+\delta_M)} \otimes e^{w\nu})(h) F_f^T(h)\, dh. \tag{13.2}$$

To solve for F_f^T, we recognize the expression on the right as a linear combination of values of the abelian group Fourier transform of F_f^T. Before applying Fourier inversion, let us be more precise about normalization of Haar measures. The number of positive restricted roots is at most two, and we write β for the unique or larger one. Put $H'_\beta = 2|\beta|^{-2}H_\beta$, parametrize A by $\exp tH'_\beta$, and let Haar measure on A be given by dt. Let Haar measure on B^- have total mass one, and take the product measure as Haar measure dh for T. For the dual, we use the customary discrete measure on the dual of B^- (the lattice of analytically integral forms on $\mathfrak{b}^-$). As with $SL(2, \mathbb{R})$ and $SL(2, \mathbb{C})$, we parametrize the dual $\hat{A}$ of A by using ν to mean two things: an arbitrary linear functional on $\mathfrak{a}$, or the numerical multiple of some standard linear functional ν_0 that yields the given linear functional. Thus $\hat{A} = \{i\nu \mid \nu \in \mathfrak{a}'\}$, and we write

$\nu = v\nu_0$ with $\nu_0(H'_\beta) = 1$. Haar measure $d\nu$ on $\hat{A}$ refers to Lebesgue measure for the numerical multiple. If φ is in $C^\infty_{\text{com}}(T)$, its Fourier transform $\hat{\varphi}$ is

$$\hat{\varphi}(\mu, \nu) = \int_T (\xi_\mu \otimes e^{i\nu})(h)\varphi(h)\, dh, \ \mu \text{ integral on } \mathfrak{b}^-, \ \nu \in \mathfrak{a}',$$

and the associated inversion formula is

$$\varphi(h) = \frac{1}{2\pi} \sum_\mu \int_{\nu \in \mathfrak{a}'} (\xi_{-\mu} \otimes e^{-i\nu})(h)\hat{\varphi}(\mu, \nu)\, d\nu.$$

Theorem 13.1. If G has real rank one with $G^{\mathbb{C}}$ simply connected and M connected, then

$$(F_f^T)\hat{}\,(s(\Lambda + \delta_M), \nu) = s^{G/T}(\det s)\Theta_{\sigma_\Lambda, i\nu}(f) \tag{13.3}$$

and $(F_f^T)\hat{}\,(\mu, \nu) = 0$ if μ is singular with respect to $\Delta((\mathfrak{b}^-)^{\mathbb{C}}:\mathfrak{m}^{\mathbb{C}})$. Moreover,

$$F_f^T(h) = \frac{1}{2\pi} s^{G/T} \sum_{\substack{\Lambda \text{ dominant} \\ \text{integral on } \mathfrak{b}^-}} \int_{\nu \in \mathfrak{a}'} \left(\sum_{s \in W(B^- : M)} (\det s)\xi_{-s(\Lambda + \delta_M)} \otimes e^{-i\nu} \right)(h) \times \Theta_{\sigma_\Lambda, i\nu}(f)\, d\nu. \tag{13.4}$$

Proof. The second conclusion follows immediately from the first by Fourier inversion since $F_f^{G/T}$ is in $C^\infty_{\text{com}}(T)$. For the first conclusion, we note that

$$(F_f^T)\hat{}\,(s\mu, \nu) = (\det s)(F_f^{G/T})\hat{}\,(\mu, \nu) \tag{13.5}$$

for $s \in W(B^- : M)$ since F_f^T is odd under the action of $W(B^- : M)$. In particular, $(F_f^{G/T})\hat{}\,(\mu, \nu)$ vanishes when μ is singular. Moreover,

$$|W(T:G)| = |W(A:G)|\,|W(B^-:M)| \tag{13.6}$$

from Problem 6 in Chapter V. Also if w is in $W(A:G)$, we can choose a representative $\tilde{w}$ in K by Lemma 5.16 that normalizes $\mathfrak{b}^-$. We can further assume that $\tilde{w}$ leaves stable the set of positive roots of M, and then $\tilde{w}\Lambda$ equals what we have called $w\Lambda$. Using this fact and tracking down the change of variables $\tilde{w}^{-1}h\tilde{w} \to h$ on T, we easily see that

$$(F_f^T)\hat{}\,(w(\Lambda + \delta_M), w\nu) = (F_f^T)\hat{}\,(\Lambda + \delta_M, \nu). \tag{13.7}$$

Formulas (13.5) and (13.7) say that all the terms in (13.2), as s and w vary, are equal, and (13.6) allows us to cancel the appropriate coefficients to arrive at (13.3).

If G does not have a compact Cartan subgroup, then T is a Θ-stable Cartan subgroup whose compact part has maximum possible dimension.

Thus we can apply Theorem 11.7 to obtain the Plancherel formula by differentiating (13.4) and setting $h = 1$.

Theorem 13.2. If G has real rank one with M connected and rank $G >$ rank K, then there is a constant $c \neq 0$ depending on the normalization of Haar measure such that

$$f(1) = c \sum_{\substack{\Lambda \text{ dominant} \\ \text{integral on } \mathfrak{b}^-}} \int_{\nu \in \mathfrak{a}'} \Theta_{\sigma_\Lambda, i\nu}(f)\{\prod \langle \Lambda + \delta_M + i\nu, \alpha\rangle\}\, d\nu$$

for all $f \in C_{\text{com}}^\infty(G)$.

Proof. We apply $\partial(\varpi_T)$ to (13.4), and there is no problem in carrying the derivative under the sum and integral. We have

$$\begin{aligned} &\partial(\varpi_T)(\xi_{-s(\Lambda+\delta_M)} \otimes e^{-i\nu}) \\ &= \left\{\prod_{\alpha>0} \langle -s(\Lambda + \delta_M) - i\nu, \alpha\rangle\right\}(\xi_{-s(\Lambda+\delta_M)} \otimes e^{-i\nu}) \\ &= s^{G/T}(\det s)\left\{\prod_{\alpha>0} \langle \Lambda + \delta_M + i\nu, \alpha\rangle\right\}(\xi_{-s(\Lambda+\delta_M)} \otimes e^{-i\nu}), \end{aligned}$$

and the theorem follows.

§3. Real-Rank-One Groups, Part II

We retain the notation and assumptions of §2 except that we now assume also that rank $G =$ rank K. Fix a notion of positivity for the restricted roots, and fix $\Delta^+(\mathfrak{t}^{\mathbb{C}}:\mathfrak{g}^{\mathbb{C}})$ so that any nonzero restriction to $\mathfrak{a}$ of a member of $\Delta^+(\mathfrak{t}^{\mathbb{C}}:\mathfrak{g}^{\mathbb{C}})$ is positive. By Proposition 11.16a, $\Delta(\mathfrak{t}^{\mathbb{C}}:\mathfrak{g}^{\mathbb{C}})$ has a real root, say α, and we may assume α is positive. By means of the Cayley transform $\mathbf{d}_\alpha$, we can pass from $\mathfrak{t}$ to a Cartan subalgebra $\mathfrak{b} \subseteq \mathfrak{k}$ and to the associated compact Cartan subgroup B. Let $\Delta^+(\mathfrak{b}^{\mathbb{C}}:\mathfrak{g}^{\mathbb{C}})$ be the image of $\Delta^+(\mathfrak{t}^{\mathbb{C}}:\mathfrak{g}^{\mathbb{C}})$ under $\mathbf{d}_\alpha$, and let $\tilde{\alpha}$ be the image of α. When there is no danger of confusion, we use the same names for members of $(\mathfrak{b}^{\mathbb{C}})'$ as for the corresponding members of $(\mathfrak{t}^{\mathbb{C}})'$.

Our objectives in this section will be to obtain an inversion formula for an averaged version of F_f^B and to deduce the Plancherel formula for G by differentiation of the inversion formula. The **averaged version** of F_f^B is

$$\mathscr{F}_f^B(b) = \sum_{w \in W_G} (\det w) F_f^B(w^{-1}bw),$$

where W_G is as usual the Weyl group of $\Delta(\mathfrak{b}^{\mathbb{C}}:\mathfrak{g}^{\mathbb{C}})$. By Theorem 11.17 we have

$$\partial(\varpi_B)F_f^B(1) = c_G f(1) \tag{13.8a}$$

for a nonzero constant c and all f in $C_{\text{com}}^{\infty}(G)$. Then it follows that

$$\partial(\varpi_B)\mathscr{F}_f^B(1) = c_G|W_G|f(1). \tag{13.8b}$$

The function $\mathscr{F}_f^B$ has only small advantages over F_f^B when G is of real rank one; $\mathscr{F}_f^B$ makes certain sums over Weyl groups be a little more manageable, and it eliminates some terms corresponding to singular parameters.

The Plancherel formula for G will involve the discrete series and the unitary principal series. Let L be the lattice of integral forms on $\mathfrak{b}$, and let L' be the subset of nonsingular forms. Since $G^{\mathbb{C}}$ is simply connected, L' parametrizes the discrete series of G, and two parameters give the same discrete series if and only if they are conjugate under $W_K = W(\mathfrak{b}^{\mathbb{C}}:\mathfrak{k}^{\mathbb{C}})$. Let Θ_λ be a discrete series character. The Weyl denominator used in Theorem 12.7 to describe Θ_λ differs from the one for $\Delta^+(\mathfrak{b}^{\mathbb{C}}:\mathfrak{g}^{\mathbb{C}})$ by the sign $\varepsilon(\lambda) = \prod_{\alpha>0}\langle\lambda,\alpha\rangle$. Therefore the numerator $\tau_{\lambda,B}$ of Θ_λ relative to the system $\Delta^+(\mathfrak{b}^{\mathbb{C}}:\mathfrak{g}^{\mathbb{C}})$ is given on B by

$$\tau_{\lambda,B}(b) = (-1)^q\varepsilon(\lambda)\sum_{W_K}(\det s)\xi_{s\lambda}(b),$$

where $q = \frac{1}{2}\dim G/K$. On the noncompact (connected) Cartan subgroup $T = B^-A$, the numerator of Θ_λ is given in two parts, one on $T^+ = \{h \mid \xi_\alpha(h) > 1\}$ and one on $T^- = \{h \mid \xi_\alpha(h) < 1\}$, according to the results of Chapter XII. [The easiest place from which to obtain explicit formulas is the proof of Theorem 12.34, particularly formulas (12.85) and (12.87).] Identifying W_K with a subgroup of $W(\mathfrak{t}^{\mathbb{C}}:\mathfrak{g}^{\mathbb{C}})$, we can write the numerator $\tau_{\lambda,T}$ of Θ_λ as

$$\tau_{\lambda,T}(b^-a) = \begin{cases} -(-1)^q\varepsilon(\lambda)\sum_{W_K}(\det s)\operatorname{sgn}\langle s\lambda,\alpha\rangle\xi_{s\lambda}(b^-)e^{-|s\lambda\log a|}, & b^-a \in T^+ \\ +(-1)^q\varepsilon(\lambda)\sum_{W_K}(\det s)\operatorname{sgn}\langle s\lambda,\alpha\rangle\xi_{s\lambda}(b^-)e^{-|s\lambda\log a|}, & b^-a \in T^-. \end{cases}$$

We introduce "characters" of **averaged discrete series** by the formula

$$\Theta_\lambda^* = (-1)^q\varepsilon(\lambda)|W_K|^{-1}\sum_{w\in W_G}\Theta_{w\lambda}.$$

These transform in the parameter λ under the action of W_G according to the rule $\Theta_{w\lambda}^* = (\det w)\Theta_\lambda^*$. The corresponding numerators $\tau_{\lambda,B}^*$ and $\tau_{\lambda,T}^*$ are given by

$$\tau^*_{\lambda,B}(b) = \sum_{w \in W_G} (\det w)\xi_{w\lambda}(b)$$

$$\tau^*_{\lambda,T}(b^- a) = \begin{cases} -\sum_{W_G} (\det w) \operatorname{sgn}\langle w\lambda, \alpha\rangle \xi_{w\lambda}(b^-)e^{-|w\lambda \log a|} & \text{for } b^- a \in T^+ \\ +\sum_{W_G} (\det w) \operatorname{sgn}\langle w\lambda, \alpha\rangle \xi_{w\lambda}(b^-)e^{-|w\lambda \log a|} & \text{for } b^- a \in T^-. \end{cases}$$

In these formulas we have used the same symbol for λ and its Cayley transform.

To handle the unitary principal series, we shall have to relate the lattice of integral forms on $\mathfrak{b}$ to the lattice of integral forms on $\mathfrak{b}^-$. The relevant fact for this purpose is given in the following lemma.

Lemma 13.3. If G is of real rank one, if rank G = rank K, if M is connected, and if the compact Cartan subgroup B is built from a Cayley transform $\mathbf{d}_\alpha$, then $B = B^- B_\alpha$ and $B^- \cap B_\alpha = \{1, \gamma_\alpha\}$, where

$$B_\alpha = \{\exp 2i\theta|\alpha|^{-2}H_\alpha \,|\, \theta \in \mathbb{R}\}.$$

Proof. Clearly $\mathfrak{b} = \mathfrak{b}^- \oplus \mathfrak{b}_\alpha$, where $\mathfrak{b}_\alpha = \mathbb{R}iH_{\tilde{\alpha}}$; thus $B = B^- B_\alpha$. The element γ_α occurs in B_α with $\theta = \pi$. Moreover, since γ_α is in Z_M and M is connected, γ_α lies in every Cartan subgroup of M, hence in B^-, by Corollary 4.24. Thus $B^- \cap B_\alpha \supseteq \{1, \gamma_\alpha\}$. In the reverse direction the only elements of B_α that commute with H_α are the ones with θ a multiple of π, by reference to $SL(2, \mathbb{R})$. Hence $B^- \cap B_\alpha \subseteq M \cap B_\alpha \subseteq \{1, \gamma_\alpha\}$, and the lemma follows.

Remarks. The integral forms λ on $\mathfrak{b}$ divide into two classes, depending on whether $\xi_\lambda(\gamma_\alpha) = +1$ or $\xi_\lambda(\gamma_\alpha) = -1$, i.e., depending on whether $2\langle\lambda, \alpha\rangle/|\alpha|^2$ is even or odd. Accordingly we can decompose $\lambda = \lambda^- + \frac{1}{2}n\alpha$, where λ^- is an analytically integral form on $\mathfrak{b}^-$ and $n = 2\langle\lambda, \alpha\rangle/|\alpha|^2$, and we have

$$\xi_\lambda(\gamma_\alpha) = \xi_{\lambda^-}(\gamma_\alpha) = (-1)^n. \tag{13.9}$$

The lemma implies that we can go backwards: If λ^- is an integral form on $\mathfrak{b}^-$ and n is an integer such that the second equality in (13.9) is valid, then $\lambda = \lambda^- + \frac{1}{2}n\alpha$ is an integral form on $\mathfrak{b}$. Let us say that an integral form λ^- on $\mathfrak{b}^-$ is **even** or **odd** depending on whether $\xi_{\lambda^-}(\gamma_\alpha)$ is $+1$ or -1.

Theorem 13.4. Suppose that G has real rank one, that $G^{\mathbb{C}}$ is simply connected, that M is connected, and that rank G = rank K. Fix $b \in B$ regular, and write, for each $w \in W_G$,

$$w^{-1}bw = b_w^- \exp(2i\theta_w|\alpha|^{-2}H_{\tilde{\alpha}})$$

with $0 < \theta_w < \pi$ as a decomposition according to $B^- B_\alpha$. Then

$$\begin{aligned}
\mathscr{F}_f^B(b) &= (-1)^r |W(B:G)| \sum_{\lambda \in L'} \Theta_\lambda^*(f)\xi_{-\lambda}(b) \\
&+ \frac{i(-1)^r|W(B:G)|}{2|W(T:G)|} \sum_{w \in W_G} \det w \sum_{\substack{\Lambda \text{ dominant} \\ \text{integral} \\ \text{on } ib^-}} \\
&\times \Bigg\{ \int_{-\infty}^{\infty} \Theta_{\sigma_\Lambda, iv}(f) \Bigg[\sum_{s \in W(B^-:M)} (\det s)\xi_{-s(\Lambda+\delta_M)}(b_w^-) \Bigg] \Bigg[\frac{\sinh v(\theta_w - \pi)}{\sinh v\pi} \Bigg] dv \\
&+ \int_{-\infty}^{\infty} \Theta_{\sigma_\Lambda, iv}(f) \Bigg[\sum_{s \in W(B^-:M)} (\det s)\xi_{-s(\Lambda+\delta_M)}(b_w^- \gamma_\alpha) \Bigg] \Bigg[\frac{\sinh v\theta_w}{\sinh v\pi} \Bigg] dv \Bigg\}
\end{aligned}$$

for all f in $C_{\text{com}}^\infty(G)$. Here $r = \frac{1}{2}(\dim G - \text{rank } G)$.

Remarks. The proof requires attention to the manner of summing conditionally convergent series, and we shall normally disregard these questions. For information about how these matters are handled, see the bibliographical notes.

Proof. We begin by computing $\Theta_\lambda^*(f)$ for $\lambda \in L'$. Formula (10.25c) gives

$$\begin{aligned}
\Theta_\lambda^*(f) &= s^{G/B}|W(B:G)|^{-1} \int_B \tau_{\lambda,B}^*(b) F_f^B(b)\, db \\
&\quad + s^{G/T}|W(T:G)|^{-1} \int_T F_f^T(h)\varepsilon_R^T(h)\tau_{\lambda,T}^*(h)\, dh.
\end{aligned}$$

The integral over B is

$$\begin{aligned}
&= \int_B \sum_{w \in W_G} (\det w)\xi_{w\lambda}(b) F_f^B(b)\, db \\
&= \int_B \xi_\lambda(b) \sum_{w \in W_G} (\det w) F_f^B(wbw^{-1})\, db \\
&= \int_B \xi_\lambda(b) \mathscr{F}_f^B(b)\, db,
\end{aligned}$$

while the integral over T is

$$\begin{aligned}
&= -\int_{T^+} (+1) \sum_{w \in W_G} (\det w) \operatorname{sgn}\langle w\lambda, \alpha\rangle \xi_{w\lambda}(b^-) e^{-|w\lambda \log a|} F_f^T(b^- a)\, db^-\, da \\
&\quad + \int_{T^-} (-1) \sum_{w \in W_G} (\det w) \operatorname{sgn}\langle w\lambda, \alpha\rangle \xi_{w\lambda}(b^-) e^{-|w\lambda \log a|} F_f^T(b^- a)\, db^-\, da \\
&= -\int_T \sum_{w \in W_G} (\det w) \operatorname{sgn}\langle w\lambda, \alpha\rangle \xi_{w\lambda}(b^-) e^{-|w\lambda \log a|} F_f^T(b^- a)\, db^-\, da.
\end{aligned}$$

Referring to (10.25b), we see that $s^{G/T} = -s^{G/B}$ and that $s^{G/B} = (-1)^r$.

Putting all these formulas together, we obtain

$$\begin{aligned}\Theta_\lambda^*(f) = (-1)^r|W(B:G)|^{-1} \int_B \xi_\lambda(b)\mathscr{F}_f^B(b)\,db \\ +(-1)^r|W(T:G)|^{-1} \int\sum(\det w)\operatorname{sgn}\langle w\lambda, \alpha\rangle\xi_{w\lambda}(b^-)e^{-|w\lambda \log a|} \\ \times F_f^T(b^- a)\,db^-\,da.\end{aligned} \tag{13.10}$$

We can regard this formula as the computation of the Fourier coefficient $(\mathscr{F}_f^B)\check{}(\lambda)$ for λ nonsingular. For λ singular we have $(\mathscr{F}_f^B)\check{}(\lambda) = 0$ because $\mathscr{F}_f^B$ is odd under W_G and so is $(\mathscr{F}_f^B)\check{}$. Thus we have all Fourier coefficients of $\mathscr{F}_f^B$. The Fourier series of $\mathscr{F}_f^B$ at worst has certain jumps and converges, while $\sum_\lambda \Theta_\lambda^*(f)\xi_{-\lambda}(b)$ converges absolutely, by Problem 4 in Chapter XII. Thus we can solve for $\mathscr{F}_f^B$, obtaining

$$\begin{aligned}\mathscr{F}_f^B(b) = (-1)^r|W(B:G)| \sum_{\lambda\in L'} \Theta_\lambda^*(f)\xi_{-\lambda}(b) - \frac{|W(B:G)|}{|W(T:G)|}\sum_{\lambda\in L'} \\ \times \left\{\sum_{w\in W_G}\int_T (\det w)\operatorname{sgn}\langle w\lambda, \alpha\rangle\xi_{w\lambda}(b^-)e^{-|w\lambda \log a|}\right. \\ \left.\times F_f^T(b^- a)\,db^-\,da\right\}\xi_{-\lambda}(b)\end{aligned} \tag{13.11}$$

for b regular. Delicately changing the parametrization of the sum in the second term on the right, we can rewrite this as

$$\begin{aligned}\mathscr{F}_f^B(b) = (-1)^r|W(B:G)| \sum_{\lambda\in L'} \Theta_\lambda^*(f)\xi_{-\lambda}(b) - \frac{|W(B:G)|}{|W(T:G)|}\sum_{w\in W_G}(\det w) \\ \times \sum_{\lambda\in L'}\left\{\int_T \operatorname{sgn}\langle \lambda, \alpha\rangle\xi_\lambda(b^-)e^{-|\lambda \log a|}F_f^T(b^- a)\,db^-\,da\right\}\xi_{-w\lambda}(b).\end{aligned} \tag{13.12}$$

Since $\sum_{W_G}(\det w)\xi_{-w\lambda} = 0$ if λ is singular, we can extend the sum over the nonsingular subset L' in the second term on the right side to all of L, interpreting $\operatorname{sgn}\langle\lambda, \alpha\rangle$ as 0 when $\langle\lambda, \alpha\rangle = 0$.

For fixed w, we look at everything after "det w" on the right side of (13.12). We write $\lambda = \lambda^- + \frac{1}{2}n\alpha$, where λ^- is the restriction to $\mathfrak{b}^-$. For each w in W_G, we write $w^{-1}bw = b_w^- b_w^\alpha$ by means of the decomposition $B = B^- B_\alpha$ of Lemma 13.3, and we define θ_w by

$$b_w^\alpha = \exp 2i\theta_w|\alpha|^{-2}H_{\bar{\alpha}}. \tag{13.13}$$

In the decomposition of $w^{-1}bw$ into two factors, we can put γ_α into either factor. Thus we can fix matters by insisting that $0 < \theta_w < \pi$. Then

$$\xi_{-w\lambda}(b) = \xi_{-\lambda}(w^{-1}bw) = \xi_{-\lambda^-}(b_w^-)e^{-in\theta_w}.$$

The inner sum in (13.12) we are now taking over all λ in L. We divide this sum into two parts, depending on whether n is even or odd, and we take into account the remarks and notation after Lemma 13.3. With L_e denoting the set of even integral forms on $\mathfrak{b}^-$, the sum over all λ with n even is

$$= \sum_{n\in 2\mathbb{Z}} \sum_{\lambda^- \in L_e} \left\{ \int_{B^- A} (\operatorname{sgn} n) \xi_{\lambda^-}(b^-) e^{-|(1/2)n\alpha \log a|} F_f^T(b^- a)\, db^-\, da \right\}$$

$$\times\ \xi_{-\lambda^-}(b_w^-) e^{-in\theta_w}$$

$$= \frac{1}{2} \sum_{n\in 2\mathbb{Z}} \int_A (\operatorname{sgn} n) e^{-|(1/2)n\alpha \log a|} e^{-in\theta_w}$$

$$\times \left\{ \sum_{\lambda^- \in L_e} \left[\int_{B^-} \xi_{\lambda^-}(b^-)(F_f^T(b^- a) + F_f^T(b^- \gamma_\alpha a))\, db^- \right] \xi_{-\lambda^-}(b_w^-) \right\} da.$$

Since $F_f^T(b^- a) + F_f^T(b^- \gamma_\alpha a)$ is an even function on B^- relative to translation by γ_α, the odd Fourier coefficients on B^- vanish. Then this expression collapses to

$$= \frac{1}{2} \sum_{n\in 2\mathbb{Z}} \int_A (\operatorname{sgn} n) e^{-|(1/2)n\alpha \log a|} e^{-in\theta_w} (F_f^T(b_w^- a) + F_f^T(b_w^- \gamma_\alpha a))\, da.$$

In similar fashion, we treat the sum over all λ with n odd. This time we arrange for F_f^T to provide an odd function, replacing $F_f^T(b^- a)$ by $\frac{1}{2}(F_f^T(b^- a) - F_f^T(b^- \gamma_\alpha a))$ in the integrand. The comparable sum is then

$$= \frac{1}{2} \sum_{n\in 2\mathbb{Z}+1} \int_A (\operatorname{sgn} n) e^{-|(1/2)n\alpha \log a|} e^{-in\theta_w} (F_f^T(b_w^- a) - F_f^T(b_w^- \gamma_\alpha a))\, da.$$

Hence the inner sum in (13.12), taken over all λ in L, is

$$= \frac{1}{2} \sum_{n\in \mathbb{Z}} \int_A (\operatorname{sgn} n) e^{-|(1/2)n\alpha \log a|} \{ e^{-in\theta_w} F_f^T(b_w^- a) + e^{-in(\theta_w + \pi)} F_f^T(b_w^- \gamma_\alpha a)\}\, da. \tag{13.14}$$

In our normalization of da in the previous section, we wrote $da = dt$, where $a = a_t = \exp tH'_\beta$ with β equal to the larger restricted root. In our present case, consideration of lengths forces β to be the restriction to $\mathfrak{a}$ of α. Thus

$$\alpha \log a = \alpha(tH'_\beta) = 2t.$$

Substituting this expression into (13.14), we see that (13.14) is

$$= \frac{1}{2} \sum_{n\in \mathbb{Z}} \int_{-\infty}^{\infty} (\operatorname{sgn} n) e^{-|nt|} \{ e^{-in\theta_w} F_f^T(b_w^- a_t) + e^{-in(\theta_w + \pi)} F_f^T(b_w^- \gamma_\alpha a_t)\}\, dt.$$

There is no difficulty in interchanging sum and integral here since $e^{i\theta_w}$ cannot be ± 1 when b in (13.12) is regular. Summing the series, substituting into (13.12), and using the fact that

$$\frac{e^{-|t|}}{1 - 2e^{-|t|}\cos\theta + e^{-2|t|}} = \frac{e^t}{1 - 2e^t\cos\theta + e^{2t}},$$

we obtain

$$\begin{aligned}\mathscr{F}_f^B(b) &= (-1)^r|W(B:G)| \sum_{\lambda\in L'} \Theta_\lambda^*(f)\xi_{-\lambda}(b) - \frac{|W(B:G)|}{2|W(T:G)|} \\ &\times \sum_{w\in W_G} \det w \left\{\int_{-\infty}^{\infty} F_f^T(b_w^- a_t)\left[\frac{e^t(e^{-i\theta_w} - e^{i\theta_w})}{1 - 2e^t\cos\theta_w + e^{2t}}\right]dt\right. \\ &\left. + \int_{-\infty}^{\infty} F_f^T(b_w^-\gamma_\alpha a_t)\left[\frac{e^t(e^{-i(\theta_w+\pi)} - e^{i(\theta_w+\pi)})}{1 - 2e^t\cos(\theta_w+\pi) + e^{2t}}\right]dt\right\}\end{aligned}$$

Next we substitute for $F_f^T(h)$ from (13.4) and use the parametrization $\nu = v\nu_0$ of §2, which makes $e^{-i\nu\log a_t} = e^{-ivt}$. The result is

$$\begin{aligned}&\mathscr{F}_f^B(b) \\ &= (-1)^r|W(B:G)| \sum_{\lambda\in L'} \Theta_\lambda^*(f)\xi_{-\lambda}(b) + \frac{(-1)^r|W(B:G)|}{4\pi|W(T:G)|} \sum_{w\in W_G} \det w \sum_{\substack{\Lambda \text{ dominant} \\ \text{integral} \\ \text{on } i\mathfrak{b}^-}} \\ &\times \left\{\int_{v=-\infty}^{\infty} \Theta_{\sigma_\Lambda, iv}(f)\left[\sum_{w\in W(B^-:M)} (\det s)\xi_{-s(\Lambda+\delta_M)}(b_w^-)\right] I(v, \theta_w)\, dv\right. \\ &+ \int_{v=-\infty}^{\infty} \Theta_{\sigma_\Lambda, iv}(f)\left[\sum_{w\in W(B^-:M)} (\det s)\xi_{-s(\Lambda+\delta_M)}(b_w^-\gamma_\alpha)\right] \\ &\left.\times\, I(v, \theta_w + \pi)\, dv\right\} \qquad (13.15)\end{aligned}$$

where

$$I(v,\theta) = \int_{t=-\infty}^{\infty} e^{-ivt}\left[\frac{e^t(e^{-i\theta} - e^{i\theta})}{1 - 2e^t\cos\theta + e^{2t}}\right]dt.$$

The integrals $I(v, \theta)$ are Fourier transforms that can be evaluated explicitly. First we change variables:

$$I(v,\theta) = (e^{-i\theta} - e^{i\theta})\int_0^{\infty} \frac{x^{-iv}\,dx}{1 - 2x\cos\theta + x^2}. \qquad (13.16)$$

Next, we evaluate the integral on the right side by a contour integration valid when $0 < |v| < 1$. The contour goes out the positive real axis from the origin, goes counterclockwise once around a large circle, and then comes back to the origin along the positive real axis. The result is that (13.16) is

$$= \frac{2\pi i \sinh \nu(\theta - \pi)}{\sinh \nu\pi} \qquad \text{if } 0 < \theta < 2\pi \text{ and } \theta \neq \pi. \tag{13.17}$$

Finally we observe that (13.16) and (13.17) are both real analytic for $-\infty < \nu < \infty$; therefore they agree for all real ν. Substituting this formula into (13.15) completes the proof of the theorem.

Theorem 13.5. Suppose that G has real rank one, that $G^{\mathbb{C}}$ is simply connected, that M is connected, and that rank $G =$ rank K. Let L_e and L_o be the sets of $\Delta^+((\mathfrak{b}^-)^{\mathbb{C}}:\mathfrak{m}^{\mathbb{C}})$ dominant nonsingular integral forms λ^- on $\mathfrak{b}^-$ for which $\xi_{\lambda^-}(\gamma_\alpha) = +1$ and -1, respectively. Then there is a constant $c \neq 0$ depending on the normalization of Haar measure such that

$$\begin{aligned} f(1) = c \sum_{\lambda \in L'} \left(\varepsilon(\lambda) \prod_{\gamma > 0} \langle \lambda, \gamma\rangle\right) \Theta_\lambda(f) + \frac{1}{2} i(-1)^q c \frac{|W(B:G)|}{|W(T:G)|} \\ \times \left\{ \sum_{\Lambda + \delta_M \in L_e} \int_{-\infty}^{\infty} \Theta_{\sigma_\Lambda, i\nu}(f) \left(\prod_{\gamma > 0} \langle \Lambda + \delta_M + i\nu, \gamma\rangle \right) \coth(\pi\nu/2)\, d\nu \right. \\ \left. + \sum_{\Lambda + \delta_M \in L_o} \int_{-\infty}^{\infty} \Theta_{\sigma_\Lambda, i\nu}(f) \left(\prod_{\gamma > 0} \langle \Lambda + \delta_M + i\nu, \gamma\rangle \right) \tanh(\pi\nu/2)\, d\nu \right\}. \end{aligned} \tag{13.18}$$

Remark. The coefficient of i in the principal series term cancels with the i from $\langle \Lambda + \delta_M + i\nu, \alpha\rangle$, the remaining part of the Plancherel measure being real. The Plancherel measure is nonnegative, and more information about the signs of its parts appears in the Problems at the end of the chapter.

Proof. We apply Theorem 11.17 and (13.8), differentiating $\mathscr{F}_f^B(b)$ in Theorem 13.4 by $\partial(\varpi_B)$ and putting $b = 1$. A full proof should do the interchanges of limits with some care, but we shall omit these details. Also we shall not pay full attention to the limiting behavior of the decomposition $w^{-1}bw = b_w^- b_w^\alpha$ as b tends to the identity. The constant c will be $(-1)^q c_G^{-1}$, with c_G as in (13.8) and with $q = \frac{1}{2} \dim G/K$.

When we apply $\partial(\varpi_B)$ to the first term of $\mathscr{F}_f^B(b)$ in Theorem 13.4, the result is

$$(-1)^{r+q} \sum_{\lambda \in L'} \sum_{w \in W_G} \left(\varepsilon(\lambda) \prod_{\gamma > 0} \langle -\lambda, \gamma\rangle\right) \Theta_{w\lambda}(f) \xi_{-\lambda}(b).$$

At $b = 1$, this reduces to

$$(-1)^q |W_G| \sum_{\lambda \in L'} \left(\varepsilon(\lambda) \prod_{\gamma > 0} \langle \lambda, \gamma\rangle\right) \Theta_\lambda(f).$$

If we take into account (13.8b), we get the asserted first term on the right side of (13.18).

In the principal series terms of $\mathscr{F}_f^B(b)$, let us suppose first that $\Lambda + \delta_M$ is in L_e, i.e., that $\xi_{\Lambda+\delta_M}(\gamma_\alpha) = 1$. Then we can extend $\xi_{\Lambda+\delta_M}$ to a character of B, making it trivial on B_α, and the extension satisfies

$$\xi_{-s(\Lambda+\delta_M)}(b_w^-) = \xi_{-s(\Lambda+\delta_M)}(b_w^- b_w^\alpha) = \xi_{-ws(\Lambda+\delta_M)}(b). \tag{13.19}$$

Let $\tilde{\nu}$ and $\tilde{\nu}_0$ be the Cayley transforms of the linear functionals ν and ν_0. Let us disregard the fact that b_w^α may be near γ_α rather than 1, as b tends to 1. Then we can define and use $\xi_{i\tilde{\nu}}$ locally near 1 in B. From (13.13), we obtain $\xi_{\tilde{\nu}_0}(b_w^\alpha) = e^{i\theta_w}$. Raising this equation to the $i\nu$ power, we have

$$\xi_{iw\tilde{\nu}}(b) = \xi_{i\tilde{\nu}}(b_w^\alpha) = (\xi_{\tilde{\nu}_0}(b_w^\alpha))^{i\nu} = e^{-\nu\theta_w}. \tag{13.20}$$

We pick out the Λ term from each of the principal series integrals in the expression for $\mathscr{F}_f^B$. The sum of the contributions from the terms in brackets, in view of (13.19) and (13.20), is

$$\begin{aligned} &= \sum_{s\in W(B^-:M)} (\det s)\xi_{-ws(\Lambda+\delta_M)}(b)\left[\frac{\sinh \nu(\theta_w - \pi) + \sinh \nu\theta_w}{\sinh \nu\pi}\right] \\ &= \sum_s (\det s)\xi_{-ws(\Lambda+\delta_M)}(b)\left[\frac{\sinh \nu(\theta_w - \pi/2)}{\sinh \nu\pi/2}\right] \\ &= \sum_s (\det s)\xi_{-ws(\Lambda+\delta_M)}(b)\left[\frac{e^{-\nu\pi/2}\xi_{-iw\tilde{\nu}}(b) - e^{\nu\pi/2}\xi_{iw\tilde{\nu}}(b)}{2\sinh \nu\pi/2}\right]. \end{aligned} \tag{13.21}$$

The element s in $W(B^-:M)$ fixes $\tilde{\nu}$, and thus we have

$$\begin{aligned} &\partial(\varpi_B)\xi_{-ws(\Lambda+\delta_M)-iw\tilde{\nu}}(b) \\ &\quad = (-1)^r(\det ws)\left(\prod_{\gamma>0}\langle\Lambda+\delta_M+i\tilde{\nu},\gamma\rangle\right)\xi_{-ws(\Lambda+\delta_M+i\tilde{\nu})}(b). \end{aligned}$$

Since

$$\begin{aligned} \prod_{\gamma>0}\langle\Lambda+\delta_M-i\tilde{\nu},\gamma\rangle &= \prod_{\gamma>0}\langle s_\alpha(\Lambda+\delta_M+i\tilde{\nu},\gamma\rangle \\ &= -\prod_{\gamma>0}\langle\Lambda+\delta_M+i\tilde{\nu},\gamma\rangle, \end{aligned}$$

we have also

$$\begin{aligned} &\partial(\varpi_B)\xi_{-ws(\Lambda+\delta_M)+iw\tilde{\nu}}(b) \\ &\quad = -(-1)^r(\det ws)\left(\prod\langle\Lambda+\delta_M+i\tilde{\nu},\gamma\rangle\right)\xi_{-ws(\Lambda+\delta_M-i\tilde{\nu})}(b). \end{aligned}$$

Thus we have no difficulty applying $\partial(\varpi_B)$ to (13.21). The determinants cancel, and the product factors out. When the various coefficients are taken into account, we obtain the first of the two principal series terms in (13.18).

The contribution from $\Lambda + \delta_M$ in L_o is handled in similar fashion and leads to the other principal series term in (13.18). This completes the proof.

§4. Averaged Discrete Series

In order to proceed more generally, it is necessary to be more systematic in dealing with discrete series characters. What we shall do is reformulate some of the identities in the proof of Theorem 12.6 from a more global point of view and then introduce averaged discrete series and obtain their corresponding identities.

Thus suppose rank G = rank K and $G \subseteq G^{\mathbb{C}}$ with $G^{\mathbb{C}}$ simply connected. Fix a compact Cartan subgroup $B \subseteq K$, and let $T = B^- F(B^-)A$ be another Θ-stable Cartan subgroup (cf. Lemma 12.30c). We wish to study the numerator of the discrete series character Θ_λ near the regular element $t_1 = b^- a$ of T. We know that $\mathfrak{b}^{\mathbb{C}}$ and $\mathfrak{t}^{\mathbb{C}}$ are conjugate within $G^{\mathbb{C}}$, and we temporarily let $\mathbf{c}: \mathfrak{b}^{\mathbb{C}} \to \mathfrak{t}^{\mathbb{C}}$ be such a conjugation.

Fix positive systems $\Delta^+(\mathfrak{b}^{\mathbb{C}}:\mathfrak{g}^{\mathbb{C}})$ and $\Delta^+(\mathfrak{t}^{\mathbb{C}}:\mathfrak{g}^{\mathbb{C}})$ compatible under $\mathbf{c}$, and define Weyl denominators accordingly. Then the numerator of Θ_λ on B is given by

$$D_B \Theta_\lambda = (-1)^q \varepsilon(\lambda) \sum_{s \in W_K} (\det s) \xi_{s\lambda}, \tag{13.22}$$

where $q = \frac{1}{2} \dim G/K$ and $\varepsilon(\lambda) = \operatorname{sgn} \prod_{\beta > 0} \langle \lambda, \beta \rangle$. On the component C of $t_1 = b^- a$ in T'_R, we have described the T numerator of Θ_λ as given by

$$D_T(t_1 \exp H)\Theta_\lambda(t_1 \exp H) = \sum c_w^C(t_1) e^{w\lambda(H)}, \tag{13.23}$$

omitting mention of $\mathbf{c}$ on the right side. We shall start to include $\mathbf{c}$ now in such formulas in order to be quite specific.

We proceed in stages to work toward defining more systematically the constants that appear in the discrete series numerators.

(1) Since we need a numerator formula only on a complete system of nonconjugate Θ-stable Cartan subgroups, we shall conjugate T conveniently. It is clear that the component C of t_1 in T'_R is of the form $C = B_1^- A_1$, where B_1^- is the component of b^- in $B^- F(B^-)$ and where A_1 is a certain subset of A. Problem 8 at the end of the chapter shows that we can conjugate B_1^- into B by an element of K. Changing the definitions of T and t_1 accordingly, we now assume that $B_1^- \subseteq B$. A uniform formula like (13.23) is applicable to the whole component C. We rewrite (13.23) in a way that does not single out t_1, does not make use of $\Delta^+(\mathfrak{t}^{\mathbb{C}}:\mathfrak{g}^{\mathbb{C}})$,

but does include the group-level version of $\mathbf{c}$, which carries $B^{\mathbb{C}}$ to $T^{\mathbb{C}}$:

$$D_B(\mathbf{c}^{-1}t)\Theta_\lambda(t) = \sum_{w \in W_G} (\det w) d(w, \lambda, C)\xi_{w\lambda}(\mathbf{c}^{-1}t), \qquad t \in C \subseteq T. \tag{13.24}$$

(2) Next we pin down a choice of the mapping $\mathbf{c}$. Let $\mathfrak{l}$ be the centralizer in $\mathfrak{g}$ of B_1^-, and let L and $L^{\mathbb{C}}$ be the analytic subgroups corresponding to $\mathfrak{l}$ and $\mathfrak{l}^{\mathbb{C}}$. L is certainly linear connected reductive. Since B_1^- is contained in both B and T, $\mathfrak{b}$ and $\mathfrak{t}$ are both contained in $\mathfrak{l}$ and are evidently Cartan subalgebras of $\mathfrak{l}$. Therefore $\mathfrak{b}^{\mathbb{C}}$ and $\mathfrak{t}^{\mathbb{C}}$ are conjugate within $L^{\mathbb{C}}$. We take as a first approximation to $\mathbf{c}$ such a conjugation $\mathbf{c}'$ carrying $\mathfrak{b}^{\mathbb{C}}$ to $\mathfrak{t}^{\mathbb{C}}$. Since $\mathbf{c}'$ fixes b^- and $b^- \exp \mathfrak{b}^-$, $\mathbf{c}'$ fixes $\mathfrak{b}^-$. The mapping $\mathbf{c}'$ carries $\Delta(\mathfrak{b}^{\mathbb{C}}:\mathfrak{g}^{\mathbb{C}})$ to $\Delta(\mathfrak{t}^{\mathbb{C}}:\mathfrak{g}^{\mathbb{C}})$ under the definition $\mathbf{c}'(\beta)(H) = \beta(\mathbf{c}'^{-1}(H))$. Since $\mathbf{c}'$ fixes $\mathfrak{b}^-$, β and $\mathbf{c}'(\beta)$ agree on $\mathfrak{b}^-$.

(3) As in the proof of Theorem 12.6, we introduce

$$\Delta_1 = \{\alpha \in \Delta(\mathfrak{t}^{\mathbb{C}}:\mathfrak{g}^{\mathbb{C}}) \mid \alpha \text{ is real and } \xi_\alpha(t_1) > 0\}.$$

Then Problem 9 at the end of the chapter shows that

$$\Delta_1 = \{\alpha \in \Delta(\mathfrak{t}^{\mathbb{C}}:\mathfrak{g}^{\mathbb{C}}) \mid \xi_\alpha(B_1^-) \equiv 1\},$$

and Problem 10 shows that

$$\Delta_1 = \Delta(\mathfrak{t}^{\mathbb{C}}:\mathfrak{l}^{\mathbb{C}}) = \mathbf{c}'\Delta(\mathfrak{b}^{\mathbb{C}}:\mathfrak{l}^{\mathbb{C}}).$$

In particular, every member of $\Delta(\mathfrak{t}^{\mathbb{C}}:\mathfrak{l}^{\mathbb{C}})$ is a real root. Thus the semisimple part of L is a group split over $\mathbb{R}$, in the terminology of Appendix C, and we can identify the roots and restricted roots of T. From Theorem 5.17, it follows that the Weyl group $W(T_0:L)$ for the restricted roots equals all of $W(\mathfrak{t}^{\mathbb{C}}:\mathfrak{l}^{\mathbb{C}})$.

(4) As in the proof of Theorem 12.6, the factor a of t_1 (or the factor A_1 of the component C) determines a positive system Δ_1^+ of roots for Δ_1, hence for $\Delta(\mathfrak{t}^{\mathbb{C}}:\mathfrak{l}^{\mathbb{C}})$, by the rule that $\alpha > 0$ if and only if $\xi_\alpha(a) > 1$. We can now define $\mathbf{c}$ to be the composition of $\mathbf{c}'$ followed by the unique member of $W(\mathfrak{t}^{\mathbb{C}}:\mathfrak{l}^{\mathbb{C}})$ such that $\mathbf{c}^{-1}\Delta_1^+ \subseteq \Delta^+(\mathfrak{b}^{\mathbb{C}}:\mathfrak{g}^{\mathbb{C}})$. The restriction of $\mathbf{c}$ to $\mathfrak{b}^{\mathbb{C}}$, carrying $\mathfrak{b}^{\mathbb{C}}$ onto $\mathfrak{t}^{\mathbb{C}}$, then is characterized by the properties

(i) $\mathbf{c}$ is in $\mathrm{Ad}(G^{\mathbb{C}})$
(ii) $\mathbf{c}$ fixes $\mathfrak{b}^-$
(iii) $\mathbf{c}$ preserves positivity of roots vanishing on $\mathfrak{b}^-$ (i.e., carries $\Delta^+(\mathfrak{b}^{\mathbb{C}}:\mathfrak{l}^{\mathbb{C}}) \subseteq \Delta^+(\mathfrak{b}^{\mathbb{C}}:\mathfrak{g}^{\mathbb{C}})$ to $\Delta_1^+ = \Delta^+(\mathfrak{t}^{\mathbb{C}}:\mathfrak{l}^{\mathbb{C}})$).

Since the additional factor used in defining $\mathbf{c}$ is in $\mathrm{Ad}(L^{\mathbb{C}})$ and $L^{\mathbb{C}}$ centralizes B_1^-, $\mathbf{c}$ has the additional property

(iv) $\mathbf{c}$ fixes B_1^-.

(5) In our first step, we conjugated B_1^- into B via K. The remainder of the construction involves an arbitrary choice that needs to be sorted out; namely T can still be conjugated while B_1^- is held fixed. The most general such conjugation from within K is by $L \cap K$. Thus let us look at the effect on our formulas of replacing T by lTl^{-1}, $l \in L \cap K$. Then $\mathfrak{t}$ is replaced by $\mathrm{Ad}(l)\mathfrak{t}$, and $\mathbf{c}$ is replaced by $\mathrm{Ad}(l)\mathbf{c}$ on the Lie algebra level (or $l\mathbf{c}(\cdot)l^{-1}$ on the group level), in view of the uniqueness in Step 4. Referring to (13.24), we see that

$$d(w, \lambda, lCl^{-1}) = d(w, \lambda, C). \tag{13.25}$$

(6) Let $\alpha_1, \ldots, \alpha_d$ be the simple roots in $\Delta_1^+ = \Delta^+(\mathfrak{t}^{\mathbb{C}}:\mathfrak{l}^{\mathbb{C}})$, and let $\Lambda_1, \ldots, \Lambda_d$ be members of $\sum \mathbb{R}\alpha_j$ with $\langle \Lambda_i, \alpha_j \rangle = \delta_{ij}$. Then (12.17), in terms of our new notation (13.24), says

$$w \in W_G \text{ and } \langle w\lambda, \mathbf{c}^{-1}\Lambda_i \rangle > 0 \text{ for some } i \Rightarrow d(w, \lambda, C) = 0. \tag{13.26}$$

(7) Next let us translate the patching condition in the proof of Theorem 12.6 into the notation of (13.24). Let α be a simple root of $\Delta^+(\mathfrak{t}^{\mathbb{C}}:\mathfrak{l}^{\mathbb{C}})$. In order to get a clean condition, it will be necessary to conjugate T as in Step 5. By means of the Cayley transform $\mathbf{d}_\alpha$, we can pass to a candidate for a more compact Cartan subgroup T^*. The difficulty is that the connected component of $B^{-*} = T^* \cap K$ containing B_1^- might not lie in B, as our normalization in Step 1 requires. Thus, starting from α, we immediately observe that $\mathbf{d}_\alpha(iH_\alpha)$ is in $\mathfrak{l} \cap \mathfrak{k}$, and we choose l in $L \cap K$ so that $\mathrm{Ad}(l)\mathbf{d}_\alpha(iH_\alpha)$ is in $\mathfrak{b}$. We replace T by lTl^{-1}, $\mathfrak{t}$ by $\mathrm{Ad}(l)\mathfrak{t}$, $\mathbf{c}$ by $\mathrm{Ad}(l)\mathbf{c}$, and α by $\mathrm{Ad}(l)\alpha$. This change is not harmful because of (13.25). Changing notation so that the new objects are called T, $\mathfrak{t}$, $\mathbf{c}$, and α, we construct $\mathbf{d}_\alpha$ and T^*, and we see from Problem 11 that the connected component of B^{-*} containing B_1^- lies in B.

(8) To define $\mathbf{c}^*: \mathfrak{b}^{\mathbb{C}} \to \mathfrak{t}^{*\mathbb{C}}$, we need to specify a subset A_1^* of A^*. Thus let

$$\begin{aligned}
\Delta_1^* &= \{\mathbf{d}_\alpha(\beta) \mid \beta \in \Delta_1 \text{ and } \beta \perp \alpha\} \\
\Delta_1^{*+} &= \{\mathbf{d}_\alpha(\beta) \mid \beta \in \Delta_1^+ \text{ and } \beta \perp \alpha\} \\
A_1^* &= \{a \in \overline{A_1} \mid \xi_\alpha(a) = 1 \text{ and } \xi_\beta(a) > 1 \text{ for all } \beta \in \Delta_1^+ \text{ with } \beta \perp \alpha\} \\
C^* &= B_1^{-*}A_1^*.
\end{aligned}$$

The set A_1^* determines $\Delta_1^{*+} = \Delta^+(\mathfrak{t}^{*\mathbb{C}}:\mathfrak{l}^{*\mathbb{C}})$ just as in Step 4, and then it determines $\mathbf{c}^*$. Problem 12 shows that $\mathbf{c}^* = \mathbf{d}_\alpha\mathbf{c}$. Let us write α also for $\mathbf{c}^{-1}(\alpha)$. Then the patching condition in (12.18) and (12.19) reduces to

$$d(w, \lambda, C) + d(s_\alpha w, \lambda, C) = d(w, \lambda, C^*) + d(s_\alpha w, \lambda, C^*). \tag{13.27}$$

(Here we are taking into account the factor $\det w$ in (13.24) and the conclusion of Problem 13 that ε_R^T and $\varepsilon_R^{T^*}$ can be dropped.)

(9) From (13.24) and the fact that $\Theta_{s\lambda} = \Theta_\lambda$ for all s in W_K, we see that

$$d(w, \lambda, C) = (\det s)d(ws^{-1}, s\lambda, C), \qquad s \in W_K. \tag{13.28}$$

(10) Formula (13.22) gives us "initial conditions" for our constants $d(w, \lambda, C)$, namely

$$d(w, \lambda, B) = \begin{cases} (-1)^q \varepsilon(\lambda) & \text{if } w \in W_K \\ 0 & \text{if } w \notin W_K. \end{cases} \tag{13.29}$$

The proof of Theorem 12.6 shows that the constants $d(w, \lambda, C)$ are completely determined by conditions (13.29), (13.25), (13.26), and (13.27). In fact, if we trace through the argument, we can come to a stronger conclusion, namely that the only way that C enters the computation of $d(w, \lambda, C)$ is through the root system $\Delta(\mathfrak{t}^{\mathbb{C}}:\mathfrak{l}^{\mathbb{C}})$ and its positive subsystem $\Delta^+(\mathfrak{t}^{\mathbb{C}}:\mathfrak{l}^{\mathbb{C}})$. This stronger conclusion is clarified by the examples we give after the next (and last) step.

(11) We introduce new notation to eliminate the alternating sign in (13.28) and to stress the way in which the constants depend on the component C. Put

$$\Delta_L^+ = \Delta^+(\mathfrak{t}^{\mathbb{C}}:\mathfrak{l}^{\mathbb{C}})$$

$$c(w, \lambda, \Delta_L^+) = (-1)^q \varepsilon(\lambda) d(w, \lambda, C).$$

In terms of the new constants, the numerator formula (13.24) is

$$(-1)^q \varepsilon(\lambda) D_B(\mathbf{c}^{-1}t)\Theta_\lambda(t) = \sum (\det w) c(w, \lambda, \Delta_L^+) \xi_{w\lambda}(\mathbf{c}^{-1}t). \tag{13.30}$$

The properties of the c's corresponding to (13.29), (13.26), and (13.28) are

$$c(w, \lambda, \varnothing) = \begin{cases} 1 & \text{if } w \in W_K \\ 0 & \text{if } w \notin W_K \end{cases} \tag{13.31}$$

$$w \in W_G \text{ and } \langle w\lambda, \mathbf{c}^{-1}\Lambda_i \rangle > 0 \text{ for some } i \Rightarrow c(w, \lambda, \Delta_L^+) = 0 \tag{13.32}$$

$$c(w, \lambda, \Delta_L^+) = c(ws^{-1}, s\lambda, \Delta_L^+), \qquad s \in W_K. \tag{13.33}$$

Let α be simple for Δ_L^+ and let $\Delta_{L,\alpha}^+$ be the subset of Δ_L^+ orthogonal to α. With the understanding that s_α is the member of W that corresponds to reflection in α under $\mathbf{c}$ or $\mathbf{c}^*$, the patching condition (13.27) becomes

$$c(w, \lambda, \Delta_L^+) + c(s_\alpha w, \lambda, \Delta_L^+) = c(w, \lambda, \Delta_{L,\alpha}^+) + c(s_\alpha w, \lambda, \Delta_{L,\alpha}^+). \tag{13.34}$$

Example 1. Any G; Δ_L of rank one.

This case arises, for example, when the Cartan subgroup T has noncompact dimension 1.

Then $\Delta_L^+ = \{\alpha\}$ and $\Delta_{L,\alpha} = \varnothing$. When we apply $\mathbf{c}^{-1}$ (built from α), the image of α is noncompact. Therefore s_α is not in W_K, and (13.31) shows that the right side of (13.34) cannot be 2. In (13.32), the only $\mathbf{c}^{-1}\Lambda_i$ is $\mathbf{c}^{-1}\alpha$, which we abbreviate as α. Then (13.32) says

$$\langle w\lambda, \alpha\rangle > 0 \Rightarrow c(w, \lambda, \{\alpha\}) = 0.$$

For any w, either $\langle w\lambda, \alpha\rangle > 0$ or $\langle s_\alpha w\lambda, \alpha\rangle > 0$, λ being nonsingular. Thus we conclude from (13.34) that

$$c(w, \lambda, \{\alpha\}) = \begin{cases} 1 & \text{if } \langle w\lambda, \alpha\rangle < 0 \text{ and } w \text{ or } s_\alpha w \text{ is in } W_K \\ 0 & \text{otherwise.} \end{cases} \tag{13.35}$$

Example 2. Any G; Δ_L of type C_2.

This case arises, for example, when $G = Sp(2, \mathbb{R})$ and B_I^- is $\{1\}$ or $\{-1\}$.

We use our standard notation for the root system C_2, taking $e_1 - e_2$ to be compact for $\Delta(\mathfrak{b}^{\mathbb{C}}:\mathfrak{l}^{\mathbb{C}})$. Let {——} denote the positive system whose simple roots are those listed between the braces. We are choosing notation so that $\Delta_L^+ = \{e_1 - e_2, 2e_2\}$. By (13.32), $c(w, \lambda, \{e_1 - e_2, 2e\})$ is 0 if $\langle w\lambda, e_1 + e_2\rangle > 0$ or $\langle w\lambda, 2e_1\rangle > 0$. The chambers for $w\lambda$ where $c(w, \lambda, \{e_1 - e_2, 2e_2\})$ might be nonzero are chambers I, II, and III in Figure 13.1.

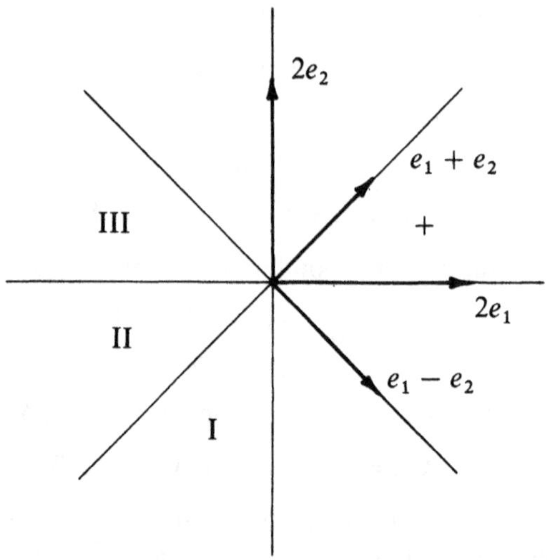

FIGURE 13.1. Chambers for $w\lambda$ in $Sp(2, \mathbb{R})$ where $c(w, \lambda, \Delta_L^+)$ might be nonzero

Suppose $w\lambda$ is in chamber I. Take $\alpha = 2e_2$. Then $s_{2e_2}w$ is not in one of the three numbered chambers, and $c(s_{2e_2}w, \lambda, \{e_1 - e_2, 2e_2\}) = 0$. Since the roots orthogonal to $2e_2$ are $\pm 2e_1$, the patching condition gives

$$c(w, \lambda, \{e_1 - e_2, 2e_2\}) = c(w, \lambda, \{2e_1\} + c(s_{2e_2}w, \lambda, \{2e_1\}). \tag{13.36}$$

To calculate the right side, we use (13.35) for each term. The inner product (in the condition on the right side of (13.35)) does not have to be recalculated; it has to be negative since $w\lambda$ is in chamber I. Thus the first term on the right side of (13.36) is 1 if w or $s_{2e_1}w$ is in $W_K = \{1, s_{e_1-e_2}\}$ and is 0 otherwise. The second term is 1 if $s_{2e_2}w$ or $s_{2e_1}s_{2e_2}w$ is in W_K and is 0 otherwise. One checks for each w that exactly one of w, $s_{2e_1}w$, $s_{2e_2}w$, $s_{2e_1}s_{2e_2}w$ is in W_K. Thus (13.36) gives

$$c(w, \lambda, \{e_1 - e_2, 2e_2\}) = 1 \text{ for all } w \text{ with } w\lambda \text{ in region I.}$$

Suppose $w\lambda$ is in region II. Take $\alpha = e_1 - e_2$. Then $s_{e_1-e_2}w\lambda$ is in region I, and the previous paragraph shows that $c(s_{e_1-e_2}w, \lambda, \{e_1 - e_2, 2e_2\}) = 1$. Thus the patching condition gives

$$c(w, \lambda, \{e_1 - e_2, 2e_2\}) + 1 = c(w, \lambda, \{e_1 + e_2\}) + c(s_{e_1-e_2}w, \lambda, \{e_1 + e_2\}). \tag{13.37}$$

Again we evaluate the right side by means of (13.35). Making the calculation, we find that

$$c(w, \lambda, \{e_1 - e_2, 2e_2\}) = \begin{cases} 1 & \text{if } w\lambda \text{ in region II and } w \text{ in } W_K \cup s_{e_1+e_2}W_K \\ -1 & \text{if } w\lambda \text{ in region II and } w \text{ not in } W_K \cup s_{e_1+e_2}W_K. \end{cases}$$

Suppose $w\lambda$ is in region III. Again take $\alpha = e_1 - e_2$. Then $s_{e_1-e_2}w$ is not in one of the three numbered chambers, and $c(s_{e_1-e_2}w, \lambda, \{e_1 - e_2, 2e_2\}) = 0$. The patching condition is just (13.37) except that the $+1$ is to be dropped from the left side. Thus we find that

$$c(w, \lambda, \{e_1 - e_2, 2e_2\}) = \begin{cases} 2 & \text{if } w\lambda \text{ in region III and } w \text{ in } W_K \cup s_{e_1+e_2}W_K \\ 0 & \text{if } w\lambda \text{ in region III and } w \text{ not in } W_K \cup s_{e_1+e_2}W_K. \end{cases}$$

Following the model of §3, we introduce **averaged discrete series** by the formula

$$\Theta_\lambda^* = (-1)^q \varepsilon(\lambda) |W_K|^{-1} \sum_{w \in W_G} \Theta_{w\lambda}. \tag{13.38}$$

As we shall see, the inclusion of the factor $|W_K|^{-1}$ in this definition effectively eliminates the special role of the system of compact roots in explicit formulas. Using (13.30), we find that the numerator of Θ_λ^* on T is given by

$$D_B(\mathbf{c}^{-1}t)\Theta_\lambda^*(t) = \sum_{w \in W_G} (\det w)\bar{c}(w\lambda, \Delta_L^+)\xi_{w\lambda}(\mathbf{c}^{-1}t), \tag{13.39}$$

where
$$\bar{c}(\lambda, \Delta_L^+) = |W_K|^{-1} \sum_{s \in W_G} c(s, s^{-1}\lambda, \Delta_L^+). \tag{13.40}$$

Properties (13.31) and (13.33) of the c's imply

$$\bar{c}(\lambda, \varnothing) = 1 \text{ for all } \lambda \tag{13.41}$$

$$\langle \lambda, \mathbf{c}^{-1}\Lambda_i \rangle > 0 \text{ for some } i \Rightarrow \bar{c}(\lambda, \Delta_L^+) = 0. \tag{13.42}$$

Applying the patching condition (13.34) to $w^{-1}\lambda$ and summing on w, we obtain

$$\bar{c}(\lambda, \Delta_L^+) + \bar{c}(s_\alpha\lambda, \Delta_L^+) = \bar{c}(\lambda, \Delta_{L,\alpha}^+) + \bar{c}(s_\alpha\lambda, \Delta_{L,\alpha}^+).$$

This condition, in the presence of (13.41) and (13.42), determines the $\bar{c}$'s completely. Tracing through that argument, we see that

$$\bar{c}(\lambda, \Delta_L^+) \text{ depends only on the projection of } \lambda \text{ to the span of } \Delta_L^+. \tag{13.43}$$

Applying this fact to our patching condition for the $\bar{c}$'s, we conclude

$$\bar{c}(\lambda, \Delta_L^+) + \bar{c}(s_\alpha\lambda, \Delta_L^+) = 2\bar{c}(\lambda, \Delta_{L,\alpha}^+). \tag{13.44}$$

Example 1 (continued). Any G; Δ_L of rank one.

We use (13.35), summing $c(w, w^{-1}\lambda, \{\alpha\})$. If $\langle\lambda, \alpha\rangle < 0$, we get a contribution whenever w or $s_\alpha w$ is in W_K, for a total of $2|W_K|$ times. Thus

$$\bar{c}(\lambda, \{\alpha\}) = \begin{cases} 2 & \text{if } \langle\lambda, \alpha\rangle < 0 \\ 0 & \text{otherwise.} \end{cases} \tag{13.45}$$

Example 2 (continued). Any G; Δ_L of type C_2.

Relation (13.42) implies that $c(\lambda, \{e_1 - e_2, 2e_2\}) = 0$ unless λ is in one of the regions I, II, III of Figure 13.1. We follow the earlier argument with the c's, but this time we use (13.45) and the simpler patching condition (13.44). Then we obtain

$$\bar{c}(\lambda, \{e_1 - e_2, 2e_2\}) = \begin{cases} 4 & \text{if } \lambda \text{ in region I} \\ 0 & \text{if } \lambda \text{ in region II} \\ 4 & \text{if } \lambda \text{ in region III.} \end{cases} \tag{13.46}$$

The constants $\bar{c}(\lambda, \Delta_L^+)$ really depend on λ only as abstract quantities in the following sense: Corresponding constants coincide whenever we have two groups G_1 and G_2, two Cartan subgroups T_1 and T_2, two forms λ_1 and λ_2, and two positive systems $\Delta_{L_1}^+$ and $\Delta_{L_2}^+$ such that the systems Δ_{L_1} and Δ_{L_2} are isomorphic in a fashion that preserves positivity and carries the Δ_{L_1} component of λ_1 to the Δ_{L_2} component of λ_2. We see this inductively, using (13.41), (13.42), and (13.44).

A special case of interest arises when $G_1 = G_2 = G$, $T_1 = T_2$, $L_1 = L_2$, and $\Delta_{L_2}^+ = w\Delta_{L_1}^+$ for some w in W_G. Then we conclude that

$$\bar{c}(w\lambda, w\Delta_L^+) = \bar{c}(\lambda, \Delta_L^+), \qquad w \in W_G. \tag{13.47}$$

To make $\bar{c}$ into an abstract function, we should say what its domain is. The group L is always split over $\mathbb{R}$, and Δ_L is its root system. If G itself is split and T is a maximally noncompact Cartan subgroup, then Δ_L will coincide with the roots of G. Also this situation imposes the least nonsingularity conditions on λ. Thus we have a definition of $\bar{c}(\lambda, \Delta^+)$ whenever

Δ is a root system in a subspace of an inner product space that is abstractly isomorphic to the root system of a split group with rank $G =$ rank K, provided Δ^+ is a positive system for Δ and λ is algebraically integral and nonsingular relative to Δ.

The root systems arising from split groups G with rank $G =$ rank K can be characterized abstractly: They are the ones with an orthogonal basis $\beta_1, \ldots, \beta_l$ of roots such that no $\beta_i \pm \beta_j$ is a root. Equivalently they are the root systems for which -1 is in the Weyl group. See Problems 14–18 in Chapter XII.

Theorem 13.6 below gives an explicit formula for $\bar{c}(\lambda, \Delta^+)$. To understand the theorem, some preparation is necessary. In the first place, the same inductive argument that we have already used several times shows that $\bar{c}(\lambda, \Delta^+)$ splits as a product if Δ is a reducible root system:

$$\bar{c}(\lambda, \Delta^+) = \bar{c}(\lambda, \Delta_1^+)\bar{c}(\lambda, \Delta_2^+) \tag{13.48}$$

if $\Delta = \Delta_1 \cup \Delta_2$, $\Delta_1^+ = \Delta_1 \cap \Delta^+$, and $\Delta_2^+ = \Delta_2 \cap \Delta^+$. In case Δ is empty or is of type A_1 or C_2, $\bar{c}(\lambda, \Delta^+)$ is given by (13.41), (13.45), or (13.46). We shall see that the theorem reduces the general case to these special cases and to the product formula (13.48).

Let Δ be as above and let $\varphi \subseteq \Delta$ be a root subsystem, i.e., a subset closed under its own reflections, and let $\varphi^+ = \varphi \cap \Delta^+$. We say that φ is a **two-system** in Δ if the irreducible components of φ are all of type A_1 or C_2 and if every w in the Weyl group $W(\Delta)$ with $w\varphi^+ = \varphi^+$ has determinant $+1$. (It follows that φ has the same span as Δ.) Let $\mathscr{T}(\Delta)$ be the set of two-systems in Δ.

Example. $G = Sp(3, \mathbb{R})$, Δ of type C_3, notation as in Chapter IV.

Let us identify positive systems by their simple roots. For Δ^+ we can use $\{e_1 - e_2, e_2 - e_3, 2e_3\}$. Then there are three two-systems, all of type $C_2 \times A_1$, corresponding to $\{e_1 - e_2, 2e_2; 2e_3\}$, $\{e_1 - e_3, 2e_3; 2e_2\}$, and $\{e_2 - e_3, 2e_3; 2e_1\}$.

If we fix Δ^+, then we can attach a sign sgn φ to each two-system $\varphi \subseteq \Delta$, as follows: Within each factor C_2 of φ, the choice of φ^+ as $\varphi \cap \Delta^+$ picks out two long positive roots $2e_1$, $2e_2$ whose difference is twice a positive root. Within each factor A_1 of φ the choice of φ^+ picks out the positive root. We enumerate these roots of φ in any fashion, except that we insist that each $2e_1$ immediately precede the corresponding $2e_2$. The lexicographic ordering obtained from this enumeration determines a new positive system $(\Delta^+)'$ for Δ, and we must have $(\Delta^+)' = w\Delta^+$ for some $w \in W(\Delta)$. Then we take sgn φ to be det w. Problems 14–20 at the end of the chapter address the fact that sgn φ is well defined, and they address other properties of two-systems.

Example. $G = \mathrm{Sp}(3, \mathbb{R})$, Δ of type C_3, notation as in Chapter IV.

The ordered basis $2e_1, 2e_2, 2e_3$ leads to the usual Δ^+, and this arises from the above construction if the C_2 factor of φ corresponds to $2e_1, 2e_2$ or to $2e_2, 2e_3$. Thus the two-systems $\{e_1 - e_2, 2e_2; 2e_3\}$ and $\{e_2 - e_3, 2e_3; 2e_1\}$ get a positive sign, and we see easily that $\{e_1 - e_3, 2e_3; 2e_2\}$ gets a negative sign.

Theorem 13.6. The averaged discrete series constants $\bar{c}(\lambda, \Delta^+)$ satisfy

$$\bar{c}(\lambda, \Delta^+) = \sum_{\varphi \in \mathscr{T}(\Delta)} (\operatorname{sgn} \varphi)\bar{c}(\lambda, \varphi^+). \tag{13.49}$$

Remarks. It is trivial that the right side of (13.49) satisfies (13.41), and it is clear that the right side satisfies (13.42). The theorem is proved by verifying inductively on rank Δ that the right side satisfies (13.44).

Proof for $\mathrm{Sp}(3, \mathbb{R})$. Let $k(\lambda, \Delta^+)$ be the right side of (13.49), with $\Delta^+ = \{e_1 - e_2, e_2 - e_3, 2e_3\}$. (We list only the simple roots.) For each simple root α in Δ^+, we are to verify that

$$k(\lambda, \Delta^+) + k(s_\alpha\lambda, \Delta^+) = 2\bar{c}(\lambda, \Delta_\alpha^+). \tag{13.50}$$

For $\alpha = e_1 - e_2$, the left side of (13.50) is

$$\begin{aligned} &\bar{c}(\lambda, \{e_1 - e_2, 2e_2; 2e_3\}) + \bar{c}(s_{e_1 - e_2}\lambda, \{e_1 - e_2, 2e_2; 2e_3\} \\ &\quad - \bar{c}(\lambda, \{e_1 - e_3, 2e_3; 2e_2\}) - \bar{c}(s_{e_1 - e_2}\lambda, \{e_1 - e_3, 2e_3; 2e_2\}) \\ &\quad + \bar{c}(\lambda, \{e_2 - e_3, 2e_3; 2e_1\}) + \bar{c}(s_{e_1 - e_2}\lambda, \{e_2 - e_3, 2e_3; 2e_1\}). \end{aligned} \tag{13.51}$$

The last four terms cancel in pairs, by (13.47), and the first two terms add to $2\bar{c}(\lambda, \{e_1 + e_2, 2e_3\})$, by (13.44). Since $\{e_1 + e_2, 2e_3\}$ is Δ_α^+, (13.50) is proved for $\alpha = e_1 - e_2$.

The argument for $\alpha = e_2 - e_3$ is similar. For $\alpha = 2e_3$, we get an expression similar to (13.51), but with s_{2e_3} in place of $s_{e_1 - e_2}$. In this case we can add each pair of terms, by (13.44), and the total is

$$\begin{aligned} &= 2\bar{c}(\lambda, \{e_1 - e_2, 2e_2\}) - 2\bar{c}(\lambda, \{2e_1; 2e_2\}) + 2\bar{c}(\lambda, \{2e_1; 2e_2\}) \\ &= 2\bar{c}(\lambda, \{e_1 - e_2, 2e_2\}). \end{aligned}$$

The right side here matches the right side of (13.50).

§5. Sp(2, $\mathbb{R}$)

We have all the background we need in order to obtain the Plancherel formula for an arbitrary linear connected reductive group G. The argument is conceptually a little simpler if rank G = rank K, and we shall make this assumption during our computations.

In this section we shall work for a time with a general G satisfying rank G = rank K, and then we shall specialize to Sp(2, ℝ). In the next section we shall draw conclusions about the general case.

Let rank G = rank K, and let B be a compact Cartan subgroup. We introduce averaged discrete series characters Θ_λ^* as in (13.38), and we again define $\mathscr{F}_f^{G/B} = \mathscr{F}_f^B$ by

$$\mathscr{F}_f^{G/B}(b) = \sum_{w \in W_G} (\det w) F_f^{G/B}(w^{-1}bw).$$

This time $\mathscr{F}_f^{G/B}$ provides a great advantage over $F_f^{G/B}$ in leading to manageable formulas. The first step is to obtain an analog of the formula for $\Theta_\lambda^*(f)$ in (13.10), but with an averaged version of F_f^T on the right side in place of F_f^T. Again as we proceed, the proof requires attention to the manner of summing conditionally convergent series, and again we normally disregard these questions. For information about how these matters are handled, see the bibliographical notes.

Let Car′ G be a maximal set of nonconjugate Θ-stable Cartan subgroups of G nonconjugate to B. If H is in Car′ G, we decompose the identity component H_0 according to Θ as $H_0 = B^- A$, and then Lemma 12.30c says that $H = F(B^-)B^-A$. Let $\Gamma(H) \subseteq F(B^-)$ be a set of representatives of H/H_0. If γ is in $\Gamma(H)$, then §4 associates to the component γB^- of $F(B^-)B^-$ a subgroup $L = L(\gamma)$ of G and its root system Δ_L. Each choice of positive system Δ_L^+ determines a Δ_L^+ dominant cone $\mathfrak{a}^+$ in the Lie algebra of A and leads to the system of coefficients $\bar{c}(\lambda, \Delta_L^+)$ of §4 used in expanding the numerator of Θ_λ^* on $\gamma B^- \exp \mathfrak{a}^+$. Let $W(L)$ be the Weyl group of Δ_L.

Proceeding as at the start of the proof of Theorem 13.4, simplifying the integral over B in the same way, and substituting from (13.39), we have

$$\begin{aligned}\Theta_\lambda^*(f) &= s^{G/B}|W(B:G)|^{-1} \int \xi_\lambda(b)\mathscr{F}_f^{G/B}(b)\, db \\ &\quad + \sum_{H \in \mathrm{Car}'G} s^{G/H}|W(H:G)|^{-1} \sum_{\gamma \in \Gamma(H)} \sum_{l \in W(L(\gamma))} \\ &\quad \times \int_{\gamma B^- \exp l\mathfrak{a}^+} F_f^{G/H}(h)\varepsilon_R^{G/H}(h) \sum_{w \in W_G} (\det w)\bar{c}(w\lambda, l\Delta_{L(\gamma)}^+)\xi_{w\lambda}(\mathbf{c}^{-1}h)\, dh\end{aligned}$$

for λ nonsingular. Now each $l \in W(L(\gamma))$ has a representative in L, which was constructed to centralize γB^-. Thus in obvious notation we can rewrite $\gamma B^- \exp l\mathfrak{a}^+$ as $l(\gamma B^- \exp \mathfrak{a}^+)l^{-1}$. Changing variables in the integral over $l(\gamma B^- \exp \mathfrak{a}^+)l^{-1}$, let us replace h by lhl^{-1}, with the new h integrated over $\gamma B^- \exp \mathfrak{a}^+$. The identities

$$F_f^{G/H}(lhl^{-1}) = F_f^{G/H}(h), \qquad \varepsilon_R^{G/H}(lhl^{-1}) = (\det l^{-1})\varepsilon_R^{G/H}(h),$$

$$\bar{c}(w\lambda, l\Delta_{L(\gamma)}^+) = \bar{c}(l^{-1}w, \Delta_{L(\gamma)}^+)$$

$$\xi_{w\lambda}(\mathbf{c}^{-1}(lhl^{-1})) = \xi_{l^{-1}w\lambda}(\mathbf{c}^{-1}h)$$

allow us to rewrite the new integral over $\gamma B^- \exp \mathfrak{a}^+$ as

$$\int_{\gamma B^- \exp \mathfrak{a}^+} F_f^{G/H}(h)\varepsilon_R^{G/H}(h) \sum (\det l^{-1}w)\bar{c}(l^{-1}w\lambda, \Delta_{L(\gamma)}^+)\xi_{l^{-1}w\lambda}(\mathbf{c}^{-1}h)\, dh.$$

The factor $\varepsilon_R^{G/H}(h)$ is now $+1$. If we replace w by lw, the l drops out completely, and we have

$$\begin{aligned}\Theta_\lambda^*(f) &= s^{G/B}|W(B:G)|^{-1} \int_B \xi_\lambda(b)\mathscr{F}_f^{G/B}(b)\, db \\ &\quad + \sum_{H \in \mathrm{Car}'\, G} s^{G/H}|W(H:G)|^{-1} \sum_{\gamma \in \Gamma(H)} |W(L(\gamma))| \\ &\quad \times \sum_{w \in W_G} \int_{\gamma B^- \exp \mathfrak{a}^+} (\det w)\bar{c}(w\lambda, \Delta_{L(\gamma)}^+)\xi_{w\lambda}(\mathbf{c}^{-1}h)F_f^{G/H}(h)\, dh.\end{aligned} \tag{13.52}$$

Before solving for $\mathscr{F}_f^B$, we introduce an averaged version of $F_f^{G/H}$. For this average we do not sum over all of W_G since not all elements of W_G normalize H. Instead we use only the part of W_G corresponding to the imaginary roots in $\Delta(\mathfrak{h}^{\mathbb{C}}:\mathfrak{g}^{\mathbb{C}})$. Thus we construct MA in the standard way as in §5.5 such that the noncompact part of H is A and such that the compact part of H is a compact Cartan subgroup of M. Our sum is over W_M, the Weyl group of $\Delta((\mathfrak{b}^-)^{\mathbb{C}}:\mathfrak{m}^{\mathbb{C}}) \cong \Delta(\mathfrak{h}^{\mathbb{C}}:(\mathfrak{m} \oplus \mathfrak{a})^{\mathbb{C}})$. Let

$$\mathscr{F}_f^{G/H}(h) = \sum_{s \in W_M} (\det s)F_f^{G/H}(s^{-1}hs).$$

Each $s \in W_M$ commutes with $\exp i\mathfrak{a}$, hence with the member γ of $\Gamma \subseteq F(B^-)$. Therefore s conjugates the set $\gamma B^- \exp \mathfrak{a}^+$ into itself. Consequently the change of variables $h \to s^{-1}hs$ makes the third line on the right side of (13.52) become

$$\sum_{w \in W_G} \int_{\gamma B^- \exp \mathfrak{a}^+} (\det sw)\bar{c}(w\lambda, \Delta_{L(\gamma)}^+)\xi_{sw\lambda}(\mathbf{c}^{-1}h)(\det s)F_f^{G/H}(s^{-1}hs)\, dh.$$

Since s fixes $\Delta_{L(\gamma)}$, we can change the $\bar{c}$ factor to $\bar{c}(sw\lambda, \Delta_{L(\gamma)}^+)$. Then we can change sw to w. Averaging over s, we obtain

$$\begin{aligned}\Theta_\lambda^*(f) &= s^{G/B}|W(B:G)|^{-1} \int_B \xi_\lambda(b)\mathscr{F}_f^{G/B}(b)\, db \\ &\quad + \sum_{H \in \mathrm{Car}'\, G} s^{G/H}|W(H:G)|^{-1}\, |W_M|^{-1} \sum_{\gamma \in \Gamma(H)} |W(L(\gamma))| \sum_{w \in W_G} \\ &\quad \times \int_{\gamma B^- \exp l\mathfrak{a}^+} (\det w)\bar{c}(w\lambda, l\Delta_{L(\gamma)}^+)\xi_{w\lambda}(\mathbf{c}^{-1}h)\mathscr{F}_f^{G/H}(h)\, dh\end{aligned}$$

This is our analog of (13.10).

Now we invert. The Fourier coefficients of $\mathscr{F}_f^{G/B}$ at singular values of λ are 0 since $\mathscr{F}_f^{G/B}$ is odd under W_G. Thus we can sum the Fourier series with what we have. Understanding that only nonsingular λ's appear in the sum over λ, we have

$$\begin{aligned}\mathscr{F}_f^{G/B}(b) &= s^{G/B}|W(B:G)|\sum_\lambda \Theta_\lambda^*(f)\xi_{-\lambda}(b)\\ &+ \sum_{H\in \mathrm{Car}'\,G} s^{G/B}s^{G/H}|W(B:G)||W(H:G)|^{-1}|W_M|^{-1}\sum_{\gamma\in\Gamma(H)}|W(L(\gamma))|\\ &\times \sum_{w\in W_G}(\det w)\sum_\lambda\left\{\int_{\gamma B^-\exp\mathfrak{a}^+}\bar{c}(w\lambda,\Delta^+_{L(\gamma)})\xi_{w\lambda}(\mathbf{c}^{-1}h)\mathscr{F}_f^{G/H}(h)\,dh\right\}\xi_{-\lambda}(b).\end{aligned} \tag{13.53}$$

Getting this formula into a usable form involves an induction. The idea is to substitute for $\mathscr{F}_f^{G/H}$ in terms of a suitable average $\mathscr{F}_{f(S)}^{MA/H}$ by means of Lemma 10.17. Then we find an expression for $\mathscr{F}_{f(S)}^{MA/H}$ analogous to the one for $\mathscr{F}_f^{G/B}(b)$ in (13.53); instead of using discrete series characters and their explicit formulas, we use the induced series constructed from MA and its characters as given in (10.26).

Thus we define $\mathscr{F}_{f(S)}^{MA/H}$ by

$$\mathscr{F}_{f(S)}^{MA/H} = \sum_{s\in W_M}(\det s)F_{f(S)}^{MA/H}(s^{-1}hs).$$

Then Lemma 10.17 gives us the following relationship:

$$\mathscr{F}_f^{G/H}(h) = \xi_\delta(h)\xi_{-\delta_M}(h)e^{-\rho_A H(h)}\mathscr{F}_{f(S)}^{MA/H}(h) \qquad \text{for } h \text{ regular.} \tag{13.54}$$

(Here we use that any $s\in W_M$ fixes $\delta-\delta_M$ and fixes ρ_A; thus $\xi_\delta(h)\xi_{-\delta_M}e^{-\rho_A H(h)}$ factors out when we average over the conjugates by W_M.

Let $\pi^M(\lambda_M,\chi)$ be a discrete series representation of M as in Proposition 12.32. Here the pair (λ_M,χ) corresponds to a one-dimensional character of $F(B^-)B^-$ with an appropriate nonsingularity property. To simplify the notation, let us denote the global character of $\pi^M(\lambda_M,\chi)$ by $\Theta_{\lambda_M}^M$, suppressing the χ. When we sum over λ_M, we must remember that the parameter space includes all nonsingular characters on $F(B^-)B^-$ and must take χ into account. Proposition 12.32 implies

$$\Theta_{\lambda_M}^M = |W_{K,M}|^{-1}\sum_{w\in W(H:MA)}\Theta_{w\lambda_M}^{M\#}, \tag{13.55a}$$

where $W_{K,M}$ is the Weyl group of $\Delta((\mathfrak{b}^-)^{\mathbb{C}}:(\mathfrak{k}\cap\mathfrak{m})^{\mathbb{C}})$. We define average discrete series for M by

$$\Theta_{\lambda_M}^{*M} = (-1)^{q_M}\varepsilon(\lambda_M)|W(H:MA)|^{-1}\sum_{w\in W_M}\Theta_{w\lambda_M}^M \tag{13.55b}$$

and corresponding averaged induced characters by

$$\Theta_{\lambda_M,iv}^* = \mathrm{ind}_{MAN}^G(\Theta_{\lambda_M}^{*M}\otimes e^{iv}\otimes 1). \tag{13.56}$$

The formula that allows us to use (13.53) inductively is the following generalization of (13.53).

Lemma 13.7. Let $F(B^-)B^-$ be a compact Cartan subgroup of M, so that $H = F(B^-)B^-A$ is a Cartan subgroup of MA. For h in the subset $\gamma B^- \exp \mathfrak{a}^+$ of H, $\mathscr{F}_{f(S)}^{MA/H}(h)$ is given by

$$\begin{aligned}\mathscr{F}_{f(S)}^{MA/H}(h) &= s^{MA/H}|W(H:MA)| \sum_{\lambda_M} \int_{\nu\in\mathfrak{a}'} \Theta_{\lambda_M,i\nu}^*(f)\xi_{-\lambda_M-i\nu}(h)\,d\nu \\ &\quad + \sum_{T\in\mathrm{Car}'\,MA} s^{MA/H}s^{MA/T}|W(H:MA)||W(T:MA)|^{-1}|W_{M_*}|^{-1} \\ &\quad \times \sum_{\gamma_M\in\Gamma(T)} |W(L(\gamma_M))| \sum_{w\in W_M} \sum_{\lambda_M} \int_\nu \\ &\quad \times \left\{\int_{\gamma_M B_*^- \exp \mathfrak{a}^\pm} \bar{c}(w\lambda_M, \Delta_{L(\gamma_M)}^+)\xi_{w(\lambda_M+i\nu)}(\mathbf{c}^{-1}t)\mathscr{F}_{f(S)}^{MA/T}(t)\,dt\right\} \\ &\quad \times \xi_{-\lambda_M-i\nu}(h)\,d\nu.\end{aligned}$$

In this expression it is understood that M_*A_* is constructed in the standard way from T, that

$$\mathscr{F}_{f(S)}^{MA/T}(t) = \sum_{s\in W_{M_*}} (\det s)F_{f(S)}^{MA/T}(s^{-1}ts),$$

and that $d\nu$ is normalized on $\mathfrak{a}'$ so that the Fourier inversion formula holds for $\mathfrak{a}$ and $\mathfrak{a}'$ with constant 1.

Remarks. The first step is to sort out that $\Theta_{\lambda_M}^{*M} = \Theta_{\lambda_M}^{*M^\#}$, where $\Theta_{\lambda_M}^{*M}$ is defined in (13.55b) and $\Theta_{\lambda_M}^{*M^\#}$ is given essentially as in (13.38). Then $\Theta_{\lambda_M,i\nu}^*$ in (13.56) can be computed by means of (10.26), using $\Theta_{\lambda_M}^{*M}$ for χ_σ. The remainder of the argument is formally identical with the derivation of (13.53), and we therefore take Lemma 13.7 as completely proved.

To get the induction going properly with Lemma 13.7 used recursively, there is one more step, which is to relate $\mathscr{F}_{f(S)}^{MA/T}$ to $\mathscr{F}_{f(S_*)}^{M_*A_*/T}$. For this purpose, we use the following generalization of (13.54).

Lemma 13.8. $\mathscr{F}_{f(S)}^{MA/T}(t) = \xi_{\delta_M}(t)\xi_{-\delta_{M_*}}(t)e^{-\rho_{A_*}^{MA}H_*(t)}\mathscr{F}_{f(S_*)}^{M_*A_*/T}(t)$.

Proof. Formula (13.54) applied to M_*A_* and T gives

$$\mathscr{F}_{f(S_*)}^{M_*A_*/T} = \xi_{\delta_{M_*}}e^{\rho_{A_*}H_*(\cdot)}\xi_{-\delta}\mathscr{F}_f^{G/T}.$$

Lemma 10.17 with $H_j = T$, when summed with alternating signs over W_{M_*}, gives

$$\mathscr{F}_{f(S)}^{MA/T} = \xi_{\delta_M}e^{\rho_A H(\cdot)}\xi_{-\delta}\mathscr{F}_f^{G/T}.$$

Thus Lemma 13.8 follows.

Let us now specialize to $G = Sp(2, \mathbb{R})$. There are four conjugacy classes of Cartan subgroups, discussed in detail in §11.5. As in that section, we use B, T_1, T_2, and T as representatives. Here T_1 and T_2 have noncompact dimension one, T_1 is related to B and T by Cayley transform in a long root, and T_2 is related to B and T be Cayley transform in a short root. We shall see below that T_2 is connected and that T_1 has two components. The associated groups MA for B and T are $G \cdot 1$ and $M_\mathfrak{p} A_\mathfrak{p}$; the groups MA for T_1 and T_2 we denote $M_1 A_1$ and $M_2 A_2$.

We start with (13.53). The sum over H in Car′ G has three terms, corresponding to $H = T_1, T_2$, and T. Consider the term corresponding to $H = T$. In this case the M group, which is $M_\mathfrak{p}$, is abelian and has just four elements: $1, \gamma_{2e_2}, \gamma_{e_1-e_2}, \gamma_{2e_1}$. (Here $\gamma_{e_1-e_2}$ is central and equals $\gamma_{2e_1}\gamma_{2e_2}$ and $\gamma_{e_1+e_2}$.) Hence $M_\mathfrak{p}$ coincides with the compact part of T, and $\Gamma(T)$ has the same four members. For $\gamma = 1$ and $\gamma = \gamma_{e_1-e_2}$, the group $L(\gamma)$ is all of Sp(2, ℝ). For $\gamma = \gamma_{2e_1}$ or γ_{2e_2}, the group $L(\gamma)$ contains the two SL(2, ℝ)'s built separately from $2e_1$ and from $2e_2$. So

$$\Delta^+_{L(\gamma)} = \begin{cases} \{e_1 + e_2, e_1 - e_2, 2e_1, 2e_2\} & \text{for } \gamma = 1 \text{ and } \gamma = \gamma_{e_1-e_2} \\ \{2e_1, 2e_2\} & \text{for } \gamma = \gamma_{2e_1} \text{ and } \gamma = \gamma_{2e_2}. \end{cases} \tag{13.57}$$

The constants $\bar{c}(w\lambda, \Delta^+_{L(\gamma)})$ we can then determine explicitly from the previous section.

To handle the term for $H = T$, we need to substitute for $\mathscr{F}^{G/T}_f(h)$. For this purpose we use (13.54) and Lemma 13.7. The sum over Car′ $M_\mathfrak{p} A_\mathfrak{p}$ is empty, and $s^{M_\mathfrak{p} A_\mathfrak{p}/T}$ and $|W(T{:}M_\mathfrak{p} A_\mathfrak{p})|$ are both 1. Put

$$s(h) = \xi_\delta(h)\xi_{-\delta_{M_\mathfrak{p}}}(h)e^{-\rho_\mathfrak{p} H(h)} = \begin{cases} 1 & \text{on components of } 1 \text{ and } \gamma_{2e_1} \\ -1 & \text{on components of } \gamma_{2e_2} \text{ and } \gamma_{e_1-e_2}. \end{cases}$$

Then

$$\mathscr{F}^{G/T}_f(h) = s(h) \sum_{\lambda_{M_\mathfrak{p}}} \int_{\nu \in \mathfrak{a}'_\mathfrak{p}} \Theta^*_{\lambda_{M_\mathfrak{p}}, i\nu}(f)\xi_{-\lambda_{M_\mathfrak{p}}-i\nu}(h)\, d\nu.$$

Apart from numerical coefficients, the term in (13.53) corresponding to $H = T$ is thus

$$\sum_{\gamma \in \Gamma(T)} |W(L(\gamma))| s(\gamma) \sum_{w \in W_G} (\det w) \sum_\lambda \left\{ \int_{\gamma \exp \mathfrak{a}^+} \bar{c}(w\lambda, \Delta^+_{L(\gamma)})\xi_{w\lambda}(\mathbf{c}^{-1}h) \right.$$
$$\left. \times \sum_{\lambda_{M_\mathfrak{p}}} \int_{\nu \in \mathfrak{a}'_\mathfrak{p}} \Theta^*_{\lambda_{M_\mathfrak{p}}, i\nu}(f)\xi_{-\lambda_{M_\mathfrak{p}}-i\nu}(h)\, d\nu\, dh \right\} \xi_{-\lambda}(b).$$

Interchanging limits, we rewrite this as

$$\sum_{\gamma \in \Gamma(T)} |W(L(\gamma))| s(\gamma) \sum_{w \in W_G} (\det w) \sum_{\lambda_{M_\mathfrak{p}}} \int_{\nu \in \mathfrak{a}'_\mathfrak{p}} \left[\sum_\lambda \int_{\gamma \exp \mathfrak{a}^+} \bar{c}(\lambda, \Delta^+_{L(\gamma)})\xi_\lambda(\mathbf{c}^{-1}h) \right.$$
$$\left. \times \xi_{-\lambda_{M_\mathfrak{p}}-i\nu}(h)\xi_{-\lambda}(w^{-1}bw)\, dh \right] \Theta^*_{\lambda_{M_\mathfrak{p}}, i\nu}(f)\, d\nu. \tag{13.58}$$

Let us see that we can express the terms in brackets as elementary functions, at least insofar as they affect (13.58). Once we have done so, we see that the $H = T$ term in (13.53) leads to a simple integral of principal series characters evaluated on f. The terms for $\gamma = 1$ and $\gamma = \gamma_{e_1 - e_2}$ involve things not previously encountered. In either case $\Delta^+_{L(\gamma)}$ is all of Δ^+, by (13.57). Also, since γ is central, we can rewrite $\xi_\lambda(\mathbf{c}^{-1}\gamma)\xi_{-\lambda}(w^{-1}bw)$ as $\xi_{-\lambda}(w^{-1}\gamma bw)$. Thus for either γ, we want to evaluate

$$\sum_{\lambda} \int_{\exp \mathfrak{a}^+} \bar{c}(\lambda, \Delta^+)\xi_\lambda(\mathbf{c}^{-1}h)\xi_{-i\nu}(h)\xi_{-\lambda}(b)\, dh, \tag{13.59}$$

at least to the extent that it affects (13.58). Let us write "$\equiv$" for "equal insofar as it affects (13.58)." Two operations on (13.59) that lead to $\equiv$ expressions are as follows:

(1) replacing ν by $p\nu$, for p in the Weyl group. [In fact, we use the fact (evident from the explicit character formula) that $\Theta^*_{p(\lambda_{M_p}, i\nu)} = \Theta^*_{\lambda_{M_p}, i\nu}$, and then we shift the p to the λ_{M_p} and ν in (13.58).]
(2) replacing $\xi_{-\lambda}(b)$ by $(\det p)\xi_{-\lambda}(p^{-1}bp)$, for p in the Weyl group. [In fact, we just change variables in the w in (13.58).]

In obvious notation, (13.59) is

$$= \sum_{\substack{(m,n)\\ \text{nonsingular}}} \int_{r>s>0} \bar{c}((m, n), \Delta^+)e^{mr+ns}e^{-i(r\nu_1+s\nu_2)}e^{-i(m\theta_1+n\theta_2)}\, dr\, ds. \tag{13.60}$$

The big advantage of using averaged discrete series and averaged F_f functions is that we can now substitute for $\bar{c}$ from (13.46). Then (13.60) is

$$\begin{aligned} &= 4 \sum_{\substack{n<m<0\\ \text{or}\\ 0<n<-m}} \int_{r>s>0} e^{mr+ns}e^{-i(r\nu_1+s\nu_2)}e^{-i(m\theta_1+n\theta_2)}\, dr\, ds \\ &= 4 \sum_{\substack{n<m<0\\ \text{or}\\ 0<n<-m}} \frac{e^{-i(m\theta_1+n\theta_2)}}{(m - i\nu_1)(m + n - i\nu_1 - i\nu_2)}. \end{aligned} \tag{13.61}$$

We apply operations (1) and (2) above to (13.61) but with coefficient 2, using $p = s_{e_1 - e_2}$ and then interchanging m and n. Then (13.61) is

$$\begin{aligned} &\equiv 2 \sum_{\substack{n<m<0\\ \text{or}\\ 0<n<-m}} \frac{e^{-i(m\theta_1+n\theta_2)}}{(m - i\nu_1)(m + n - i\nu_1 - i\nu_2)} \\ &\quad - 2 \sum_{\substack{m<n<0\\ \text{or}\\ 0<m<-n}} \frac{e^{-i(m\theta_1+n\theta_2)}}{(n - i\nu_2)(m + n - i\nu_1 - i\nu_2)}. \end{aligned} \tag{13.62}$$

Since

$$-\frac{1}{(n-iv_2)(m+n-iv_1-iv_2)} = \frac{1}{(m-iv_1)(m+n-iv_1-iv_2)} - \frac{1}{(m-iv_1)(n-iv_2)},$$

(13.62) is

$$= 2 \sum_{\substack{m+n<0 \\ \text{nonsingular}}} \frac{e^{-i(m\theta_1+n\theta_2)}}{(m-iv_1)(m+n-iv_1-iv_2)} - 2 \sum_{\substack{m<n<0 \\ \text{or} \\ 0<m<-n}} \frac{e^{-i(m\theta_1+n\theta_2)}}{(m-iv_1)(n-iv_2)}. \tag{13.63}$$

We apply operations (1) and (2) above to the first term of (13.63) but with coefficient 1, using $p = s_{2e_1}s_{2e_2}$ and then replacing m by $-m$ and n by $-n$. Then that term is

$$\equiv \sum_{\substack{(m,n) \\ \text{nonsingular}}} \frac{e^{-i(m\theta_1+n\theta_2)}}{(m-iv_1)(m+n-iv_1-iv_2)}. \tag{13.64}$$

To the part of the second term of (13.63) where $0 < m < -n$, we apply operations (1) and (2) with $p = s_{2e_1}$, and then we replace m by $-m$. We find that that term is

$$\equiv -2 \sum_{\substack{m<0,\, n<0 \\ \text{nonsingular}}} \frac{e^{-i(m\theta_1+n\theta_2)}}{(m-iv_1)(n-iv_2)}.$$

Finally we apply operations (1) and (2) to this last expression with $s = s_{e_1-e_2}$ and see that it is $\equiv 0$. Hence (13.59) is $\equiv$ to (13.64). It is easy to check by the same techniques that the sum of all the singular terms of the expression in (13.64) is $\equiv 0$, and thus (13.59) is

$$\equiv \sum_{(m,n)} \frac{e^{-i(m\theta_1+n\theta_2)}}{(m-iv_1)(m+n-iv_1-iv_2)} = \sum_m \frac{e^{-im(\theta_1-\theta_2)}}{m-iv_1} \sum_n \frac{e^{-in\theta_2}}{n-iv_1-iv_2}.$$

This expression is a full Fourier series, and one can check directly that it is the Fourier series of

$$\frac{(\pi i)^2}{\sinh \pi v_1 \sinh \pi(v_1+v_2)} \times \begin{cases} e^{v_1(\theta_1 \mp \pi)} e^{v_2(\theta_2 \mp (1+v_1v_2^{-1})\pi)} & \text{if } 0 < |\theta_2| < |\theta_1| < \pi \\ e^{v_1\theta_1} e^{v_2(\theta_2 \mp \pi)} & \text{if } 0 < |\theta_1| < |\theta_2| < \pi. \end{cases}$$

Therefore the terms for 1 and $\gamma_{e_1-e_2}$ in (13.58) simplify to a sum over λ_{M_p} and integral over ν of a piecewise elementary function times $\Theta^*_{\lambda_{M_p},i\nu}(f)$.

For the terms with $\gamma = \gamma_{2e_1}$ and $\gamma = \gamma_{2e_2}$, $\mathfrak{a}^+$ is a quadrant in the plane, and the integral in brackets splits as a product because of (13.48). The resulting expressions are the same expressions that one encounters in $SL(2, \mathbb{R})$ and lead to piecewise elementary functions.

Now let us turn to the other H's contributing to (13.53). The Cartan subgroup $T_2 = B_2A_2$ is connected since $F(B_2) = \{1, \gamma_{e_1+e_2}\}$ and $\gamma_{e_1+e_2} = \gamma_{e_1-e_2}$ is in the identity component. On the other hand, γ_{2e_1} is not in the identity component of T_1, and T_1 consequently has two components. Despite this difference in T_1 and T_2, there is no major difference in how the corresponding terms in (13.53) are handled. Thus we consider only T_2.

For $H = T_2$ in (13.53), $\Gamma(T_2) = \{1\}$ and $\Delta^+_{L(1)} = \{e_1 + e_2\}$. The constant $\bar{c}$ is thus a rank one constant given as in (13.45), and the integration over $\gamma B^- \exp \mathfrak{a}^+$ is over the product of a circle and a half line. We must substitute for $\mathscr{F}_f^{G/T_2}(h)$ in (13.53) from (13.54) and Lemma 13.7. The substitution leads to two terms, one corresponding to the first term on the right side in Lemma 13.7 and the other corresponding to the term from the Cartan subgroup T on the right side in Lemma 13.7 (since $\mathrm{Car}'\, M_2A_2 = \{T\}$).

For the first term, the idea is that the sum of the component of λ parallel to $e_1 - e_2$ is an operation inverse to integrating in the B^- factor of h, by virtue of summing a Fourier series. The resulting expression is essentially encountered in $SL(2, \mathbb{R})$ and so can be handled. For the second term, the $\mathscr{F}^{M_2A_2/T}$ can be replaced by an expression involving principal series characters by means of Lemma 13.8 and 13.7. The region of integration $\gamma_M B_*^- \exp \mathfrak{a}_*^+$ reduces to a half plane in $\gamma \exp \mathfrak{a}_\mathfrak{p}$; one variable is handled by Euclidean Fourier inversion, and the other leads to the usual sort of expression encountered in $SL(2, \mathbb{R})$. Hence the second terms can be handled. Putting all these facts together, we arrive at the following result.

Theorem 13.9. In $Sp(2, \mathbb{R})$ there exist computable piecewise elementary functions $q^H(b, \lambda, i\nu)$ corresponding to each of the three standard induced series of representations such that $\mathscr{F}_f^B(b)$, for b regular and f in $C^\infty_{\mathrm{com}}(G)$, is given by

$$\begin{aligned}\mathscr{F}_f^B(b) = s^{G/B}|W(B:G)| &\sum_\lambda \Theta^*_\lambda(f)\xi_{-\lambda}(b)\\ &+ \sum_{\lambda_{M_1}} \int_{\nu_1} \Theta^*_{\lambda_{M_1},i\nu_1}(f) q^{T_1}(b, \lambda_{M_1}, i\nu_1)\, d\nu_1\\ &+ \sum_{\lambda_{M_2}} \int_{\nu_2} \Theta_{\lambda_{M_2},i\nu_2}(f) q^{T^2}(b, \lambda_{M_2}, i\nu_2)\, d\nu_2\\ &+ \sum_{\lambda_{M_p}} \int_{\nu} \Theta^*_{\lambda_{M_p},i\nu}(f) q^{T}(b, \lambda_{M_p}, i\nu)\, d\nu.\end{aligned}$$

The sums here extend only over nonsingular values of the discrete parameters. If we follow the details through more carefully, we can differentiate and use Theorem 11.17 to obtain an explicit Plancherel formula, as follows. Again the sums extend only over nonsingular values of the discrete parameters.

Corollary 13.10. In Sp(2, $\mathbb{R}$) there exist computable real analytic elementary functions $p^H(\lambda, i\nu)$ corresponding to each of the three induced series of representations such that $f(1)$, for f in $C^\infty_{\text{com}}(G)$, is given by

$$f(1) = c\sum_{\lambda}\left(\varepsilon(\lambda)\prod_{\gamma>0}\langle\lambda,\gamma\rangle\right)\Theta_\lambda(f) + \sum_{\lambda_{M_1}}\int_{\nu_1}\Theta^*_{\lambda_{M_1},i\nu_1}(f)p^{T_1}(\lambda_{M_1}, i\nu_1)\,d\nu_1$$

$$+ \sum_{\lambda_{M_2}}\int_{\nu_2}\Theta^*_{\lambda_{M_2},i\nu_2}(f)p^{T_2}(\lambda_{M_2}, i\nu_2)\,d\nu_2$$

$$+ \sum_{\lambda_{M_p}}\int_{\nu}\Theta_{\lambda_{M_p},i\nu}(f)p^{T}(\lambda_{M_p}, i\nu)\,d\nu.$$

§6. General Case

For general G with rank G = rank K, we proceed in the same inductive fashion as was done for Sp(2, $\mathbb{R}$) in the previous section. If rank $G \neq$ rank K, the argument changes a little at the beginning but is otherwise the same. Namely we let H be a maximally compact Cartan subgroup of G, we apply (13.54) immediately, and then we go to Lemma 13.7.

Although it requires patient checking of the details, no new ideas are needed to obtain the Plancherel formula in the general case. The reason is that the reduction formulas for $\bar{c}$ in (13.48) and Theorem 13.6 reduce all the integrals that arise to integrals encountered in SL(2, $\mathbb{R}$) and to integrals like the one (13.59) encountered in Sp(2, $\mathbb{R}$).

We shall be content to state the following qualitative version of the Plancherel formula. It is possible to give a complete formula with all details.

Theorem 13.11 (Plancherel formula). Let G be linear connected reductive, and let $H_1, \ldots, H_s$ be a complete set of nonconjugate Θ-stable Cartan subgroups. Then there exist computable real analytic elementary functions $p^{H_j}(\lambda, i\nu)$ corresponding to the standard induced series of representations built from H_j such that

$$f(1) = \sum_{j=1}^{s}\left\{\sum_{\lambda}\int_{\nu}\Theta^*_{\lambda,i\nu}(f)p^{H_j}(\lambda, i\nu)\,d\nu\right\}$$

for all f in $C^\infty_{\text{com}}(G)$. Moreover, if H_1 is maximally compact, then $p^{H_1}(\lambda, i\nu)$

is given by

$$p^{H_1}(\lambda, i\nu) = c \prod_{\alpha \in \Delta^+} \langle \lambda + i\nu, \alpha \rangle$$

for a nonzero constant c.

Remarks. When H_1 is compact, the ν parameter is trivial and drops out. Thus the formula for the H_1 term specializes to what one expects from $Sp(2, \mathbb{R})$.

§7. Problems

1. Prove that the discrete series constant $c(w, \lambda, \Delta_L^+)$ can be nonzero only if $w = us$ with $u \in W(\mathfrak{b}^{\mathbb{C}}:\mathfrak{l}^{\mathbb{C}})$ and $s \in W_K$.

Problems 2 to 7 address aspects and generalizations of the proofs in §§2–3.

2. Suppose in the setting of Theorem 13.1 that rank G = rank K. Prove directly that $\partial(\varpi_T)$ of the right side of (13.4) vanishes at $h = 1$.
3. With β as in §2, suppose that $c\beta + \gamma$ is in $\Delta(\mathfrak{t}^{\mathbb{C}}:\mathfrak{g}^{\mathbb{C}})$ with $\gamma \perp \beta$.
   (a) Using the complex-linear extension of θ to $\mathfrak{g}^{\mathbb{C}}$, prove that $c\beta - \gamma$ is in $\Delta(\mathfrak{t}^{\mathbb{C}}:\mathfrak{g}^{\mathbb{C}})$.
   (b) Prove that 2γ is not in $\Delta(\mathfrak{t}^{\mathbb{C}}:\mathfrak{g}^{\mathbb{C}})$. [Hint: Show that a root vector would have to be in $\mathfrak{m}^{\mathbb{C}} \cap \mathfrak{p}^{\mathbb{C}}$.]
   (c) Prove that $c \neq 0$ and $\gamma \neq 0$ implies $2\langle c\beta + \gamma, c\beta - \gamma\rangle/|c\beta + \gamma|^2$ is -1 or 0.
4. With β as in §2, suppose $\frac{1}{2}\beta$ is a restricted root. Using Problem 3c, show that there exists a real root α with $\beta = \alpha|_{\mathfrak{a}}$. Hence rank G = rank K, and also $\frac{1}{2}\beta$ is not the restriction of a real root.
5. With β as in §2, compute $\mathrm{Ad}(\gamma_\beta)$ on each restricted root space. Deduce that γ_β is central if and only if $\frac{1}{2}\beta$ is not a restricted root.
6. Go over the proofs of Theorems 13.4 and 13.5 in the case of $G = SL(2, \mathbb{R})$, modifying them appropriately to get a formula for $\mathscr{F}_f^B(b)$ and to get the Plancherel formula.
7. Following through the arguments of §§2–3 for a general G, derive the contribution to the Plancherel formula of G from the representations induced from MAN when $\dim A = 1$.

Problems 8 to 13 fill in details in the 11-step derivation in §4 of properties of discrete series constants.

8. In the notation of Step 1, prove that $kB_1^- k^{-1} \subseteq B$ for some k in K. [Hint: Find k_0 with $k_0 b^- k_0^{-1}$ in B. Call this element b. Find z in $K \cap Z_G(b)_0$ with $\mathrm{Ad}(z)\mathfrak{b}^- \subseteq \mathfrak{b}$. Put $k = zk_0$.]

9. In the notation of Step 3, prove that

$$\Delta_1 = \{\alpha \in \Delta(\mathfrak{t}^{\mathbb{C}}:\mathfrak{g}^{\mathbb{C}}) \mid \xi_\alpha(B_1^-) \equiv 1\}.$$

10. In the notation of Step 3, prove that $\Delta_1 = \Delta(\mathfrak{t}^{\mathbb{C}}:\mathfrak{l}^{\mathbb{C}})$.
11. In the notation of Step 7, prove that the connected component of B^{-*} containing B_1^- lies in B.
12. In the notation of Step 8, prove that $\mathbf{c}^* = \mathbf{d}_\alpha \mathbf{c}$.
13. Show why the factors ε_R^T and $\varepsilon_R^{T^*}$ can be dropped in passing from the patching condition of Chapter XII to formula (13.27).

Problems 14 to 20 derive properties of two-systems and their signs as defined in §4. Let Δ be the root system of a split group G with rank $G =$ rank K, and fix Δ^+. According to Problems 14–18 in Chapter XII, there exists a basis of strongly orthogonal roots. Assume in solving these problems that either all members of Δ have the same length or else that there are two lengths, related by $|\gamma_1|^2 = 2|\gamma_2|^2$. (There is only one simple G that fails to satisfy this condition, and any assertions about its root system can be tested directly.)

14. Let $\alpha_1, \ldots, \alpha_l$ be a basis of strongly orthogonal roots. Let γ be a root, and write $\gamma = \sum c_i \alpha_i$.
    (a) Prove that the possible values of c_i are limited to 0, $\pm\frac{1}{2}$, and ± 1.
    (b) Prove that changing the sign of some c_i leads to another root.
    (c) Let $(\Delta^+)_1$ be the positive system obtained from the lexicographic ordering relative to $\alpha_1, \ldots, \alpha_l$, and let $(\Delta^+)_2$ be the positive system obtained similarly when α_i and α_{i+1} are interchanged. Prove that the number of $(\Delta^+)_1$ positive roots that are $(\Delta^+)_2$ negative is even unless $\frac{1}{2}(\alpha_i - \alpha_{i+1})$ is a root.
15. Let φ be a two-system. Show that φ contains a unique unordered strongly orthogonal *basis* of positive roots $\alpha_1, \ldots, \alpha_l$ such that the two α's contributed by a C_2 factor are the two long positive roots.
16. Let φ be a two-system. Deduce from Problems 14c and 15 that sgn φ, given as in §4, is well defined.
17. Let φ be a two-system, obtain the basis $\alpha_1, \ldots, \alpha_l$ as in Problem 15, and use it to determine a positive system $(\Delta^+)_\varphi$ as in the definition of sgn φ. Let φ' be another two-system, let $\alpha_1', \ldots, \alpha_l'$ be the corresponding basis, and suppose that this basis leads to the same positive system $(\Delta^+)_\varphi$. Assume that the α_j's and α_k''s all have the same length. Prove that $\alpha_1', \ldots, \alpha_l'$ is a permutation of $\alpha_1, \ldots, \alpha_l$. [Hint: Choose i as small as possible so that $\langle \alpha_1, \alpha_i' \rangle \neq 0$. Show that $\alpha_1 - \alpha_i'$ is a root or is 0. If it is a positive root, show that the coefficient of α_i' in the expansion of α_1 must be 1 to allow $\alpha_1 - \alpha_i'$ to have a positive leading

coefficient, and then the length assumption gives a contradiction. If it is a negative root, reverse the roles of α_i' and α_1.]

18. Suppose that φ and φ' are two-systems whose associated bases (Problem 15) have all members of the same length (as, for example, when all members of Δ have the same length). Show that φ' and φ as sets are conjugate by the Weyl group. [Hint: Choose w with $w(\Delta^+)_{\varphi'} = (\Delta^+)_\varphi$, and apply Problem 17 to $w\varphi'$ and φ.]

19. Among the Dynkin diagrams A_n, B_n, C_n, D_n, show that the ones that are allowable as Δ are A_1, B_n, C_n, and D_{2n}. Show that all of the allowable ones except B_n satisfy the assumption of Problem 18. Exhibit distinct two-systems φ and φ' in B_3 with $(\Delta^+)_\varphi = (\Delta^+)_{\varphi'}$, and show that φ and φ' are nevertheless conjugate by the Weyl group.

20. Prove that a two-system exists for each Δ under discussion. [Hint: Start with a strongly orthogonal basis of roots. Group as many disjoint pairs as possible that can correspond to factors C_2. Take φ to be the resulting collection of systems C_2 and A_1. To check the value of det w when $w\varphi^+ = \varphi^+$, use simple roots to see that the C_2's are permuted and so are the A_1's. Find the effect on the strongly orthogonal basis, and apply Problem 14c.]

CHAPTER XIV

Irreducible Tempered Representations

§1. SL(2, $\mathbb{R}$) from a More General Point of View

To make maximum use of the Langlands classification (Theorem 8.54), we need a better understanding of the irreducible tempered representations. Theorem 8.53 already tells us that each irreducible tempered representation is a constituent of some unitary representation induced from a parabolic subgroup MAN with a discrete series on M and a unitary character on A. Thus the heart of the problem is to account fully for whatever reducibility is exhibited by such induced representations.

It is not surprising that the standard intertwining operators of Chapter VII should play a pivotal role in this study. Lemma 8.53 is a first hint that the intertwining operators are intimately connected with the asymptotics of matrix coefficients. For the moment, let us think of the matrix coefficients as solutions of a system of differential equations ($Z(\mathfrak{g}^{\mathbb{C}})$ acting by the infinitesimal character). Part of the classical theory of differential equations associates to certain systems a spectral expansion of the solutions in terms of a measure computable from the dominant terms of the asymptotics. The analog in the group case is a relationship that we shall find between the Plancherel measure and the dominant terms in the asymptotic expansion of matrix coefficients. Thus the pivotal role of the intertwining operators is to connect questions about reducibility with properties of the Plancherel measure.

In this section we shall illustrate these matters in terms of the principal series for SL(2, $\mathbb{R}$). Let $S = MAN$ be the upper triangular group, and let $\bar{S} = MA\bar{N}$ be the lower triangular group. Form $U(S, \sigma, \nu)$. We shall identify ν with $z\rho$, for $z \in \mathbb{C}$, and then a phrase like "ν in the right half plane" means Re $z > 0$.

Fix a nonzero imaginary ν, and let f and g be eigenfunctions of K in the space on which $U(S, \sigma, \nu)$ acts. Then $(U(S, \sigma, \nu)f, g)$ is τ-spherical for an appropriate τ, and Theorems 8.32 and 8.33 give

$$(U(S, \sigma, \nu, ma)f, g) = e^{-\rho \log a}(c_\nu(m)e^{\nu \log a} + c_{-\nu}(m)e^{-\nu \log a} + \text{Error}(m, a)) \tag{14.1}$$

as $a \to +\infty$, with Error(m, a) a term that decreases almost as fast as $e^{-\alpha \log a}$, where α is the positive restricted root.

To use (14.1) effectively, we need to identify c_ν and $c_{-\nu}$ and to know how the error term depends on ν. Thus we start over, allowing ν to be complex. Lemma 7.23 gives us the explicit limit formula

$$\lim_{a \to +\infty} e^{(-\nu+\rho)\log a}(U(S, \sigma, \nu, ma)f, g) = (\sigma(m)A(\bar{S}:S:\sigma:\nu)f(1), g(1))$$

when $\operatorname{Re} \nu > 0$. Letting e denote evaluation at 1 and e^* denote the adjoint (on a suitable finite-dimensional space), we can rewrite this identity as

$$\lim_{a \to +\infty} e^{(-\nu+\rho)\log a}(U(S, \sigma, \nu, ma)f, g) = (e^*\sigma(m)eA(\bar{S}:S:\sigma:\nu)f, g).$$

Going over the proof of Lemma 7.23 carefully, we can bound the rate of convergence when $\nu = z\rho$ and $0 < \operatorname{Re} \nu \leq \nu_0 < \alpha$. The result is

$$\begin{aligned} &|(U(S, \sigma, \nu, ma)f, g) - e^{(\nu-\rho)\log a}(e^*\sigma(m)eA(\bar{S}:S:\sigma:\nu)f, g)| \\ &\qquad \leq \operatorname{Const}_{f,g}[|1 - z| + (\operatorname{Re} z)^{-1}]e^{-(1+\operatorname{Re} z)\rho \log a}. \end{aligned} \tag{14.2}$$

Let $w = \begin{pmatrix} 0 & 1 \\ -1 & 0 \end{pmatrix}$. If we replace k by wk in the first line of the proof of Lemma 7.23 and track down what happens, we are led to an analog of (14.2) valid when $0 < -\operatorname{Re} \nu \leq \nu_0 < \alpha$. The argument uses one new identity that we shall note in §4, namely

$$A(\bar{S}:S:\sigma:\nu)^* = A(S:\bar{S}:\sigma:-\bar{\nu}) \tag{14.3}$$

with the adjoint taken K space by K space. The resulting limit formula is

$$\begin{aligned} &\lim_{a \to +\infty} e^{(\nu+\rho)\log a}(U(S, \sigma, \nu, ma)f, g) \\ &\qquad = (U(w)A(S:\bar{S}:\sigma:\nu)e^*\sigma(w^{-1}mw)eU(w)^{-1}f, g) \end{aligned}$$

for $\operatorname{Re} \nu < 0$, where $U(w)$ is an abbreviation for $U(S, \sigma, \nu, w)$. The more precise estimate, valid when $\nu = z\rho$ and $0 < -\operatorname{Re} \nu \leq \nu_0 < \alpha$ is

$$\begin{aligned} &|(U(S, \sigma, \nu, ma)f, g) \\ &\quad - e^{(-\nu-\rho)\log a}(U(w)A(S:\bar{S}:\sigma:\nu)e^*\sigma(w^{-1}mw)eU(w)^{-1}f, g)| \\ &\qquad \leq \operatorname{Const}_{f,g}[|1 - z| + (\operatorname{Re} z)^{-1}]e^{-(1+\operatorname{Re} z)\rho \log a}. \end{aligned} \tag{14.4}$$

Fix a. With $\nu = z\rho$, let

$$\begin{aligned} F(z) = \frac{z}{(z-1)^2} &[(U(S, \sigma, \nu, ma)f, g) - e^{(\nu-\rho)\log a}(e^*\sigma(m)eA(\bar{S}:S:\sigma:\nu)f, g) \\ &- e^{(-\nu-\rho)\log a}(U(w)A(S:\bar{S}:\sigma:\nu)e^*\sigma(w^{-1}mw)eU(w)^{-1}f, g). \end{aligned} \tag{14.5}$$

Examination of the proof of Theorem 7.12 shows that $z(z-1)^{-2}A(\bar{S}:S:\sigma:\nu)$ is bounded as a function of ν when applied to a single function. Consequently $F(z)$ is bounded in the vertical strip $|\mathrm{Re}\, v| \leq v_0$. This same control of the intertwining operator, together with the estimates (14.2) and (14.4), shows that

$$|F(v)| \leq \mathrm{Const}_{f,g}\, e^{-(\rho + v_0)\log a} \tag{14.6}$$

when $|\mathrm{Re}\, v| = v_0$. By the Maximum Modulus Theorem (applied to $e^{\varepsilon z^2}F(z)$), the bound (14.6) is valid in the strip $|\mathrm{Re}\, v| \leq v_0$.

We can make a significant improvement in this bound. We restrict v to a narrower strip with $|\mathrm{Re}\, v| \leq v_0 < \rho$ and consider, for fixed a, the function $G(z) = z^{-1}(z-1)F(z)$. Then (14.6) implies

$$|G(v)| \leq \mathrm{Const}_{f,g}\, e^{-(\rho + v_0)\log a} \tag{14.7}$$

when $|\mathrm{Re}\, v| = v_0$. The claim is that G is bounded in the strip. Since F is bounded, the only possible problem is from $z = 0$, and it is enough to see that G has no pole at $z = 0$, i.e., that $F(0) = 0$. Thus we pass to the limit $z = 0$ in (14.5) and (14.6), multiply by $e^{-\rho \log a}$, and obtain

$$\begin{aligned}
&|(e^*\sigma(m)e[vA(\bar{S}:S:\sigma:v)]_{v=0}f, g)\\
&\quad + (U(w)[vA(S:\bar{S}:\sigma:v)]_{v=0}e^*\sigma(w^{-1}mw)eU(w^{-1})f, g)|\\
&\quad \leq \mathrm{Const}_{f,g}\, e^{-v_0 \log a}.
\end{aligned}$$

The left side is independent of a, and the right side shows that the left side must be 0. Hence G is bounded near 0 and thus in the whole strip. Applying the Maximum Modulus Theorem, we obtain (14.7) in the whole strip. That is,

$$\begin{aligned}
&|(U(S, \sigma, v, ma)f, g) - e^{(v-\rho)\log a}(e^*\sigma(m)eA(\bar{S}:S:\sigma:v)f, g)\\
&\quad - e^{(-v-\rho)\log a}(U(w)A(S:\bar{S}:\sigma:v)e^*\sigma(w^{-1}mw)eU(w)^{-1}f, g)|\\
&\quad \leq \mathrm{Const}_{f,g}(1 + |\mathrm{Im}\, z|)e^{-(\rho + v_0)\log a}.
\end{aligned} \tag{14.8}$$

This estimate identifies c_v and c_{-v} in (14.1) and bounds the error term in a way that shows the dependence on v.

In this context, let us now investigate the reducibility of $U(S, \sigma, v)$ for v imaginary. We know from either Proposition 2.7 or Theorem 7.2 that $U(S, \sigma, v)$ is irreducible when v is *nonzero* imaginary, and we know from Theorem 7.12 that $A(\bar{S}:S:\sigma:v)$ and $A(S:\bar{S}:\sigma:v)$ are well defined for such v. The composition

$$A(S:\bar{S}:\sigma:v)A(\bar{S}:S:\sigma:v)$$

is necessarily a self-intertwining operator for $U(S, \sigma, v)$ and so must be scalar. We shall check in §4 that this scalar, which we denote $\eta(\bar{S}:S:\sigma:v)$, is meromorphic for v in $\mathbb{C}$ and is analytic and ≥ 0 for v nonzero imaginary. Let us see that the function η controls the irreducibility/reducibility at $v = 0$. First suppose $\eta(\bar{S}:S:\sigma:v)$ has a pole at $v = 0$. Then the pole is at least double since $\eta \geq 0$ for ν imaginary, and $A(\bar{S}:S:\sigma:\nu)$ and $A(S:\bar{S}:\sigma:\nu)$, which have at most simple poles, must each have a genuine pole on every nonzero function in order to produce the double pole in η. We shall use this fact and (14.8) to prove $U(S, \sigma, 0)$ is irreducible.

If, on the contrary, $U(S, \sigma, 0)$ is reducible, then we can choose f and g in (14.1) so as to be in orthogonal irreducible subspaces. Without loss of generality we may take $f(1) = g(1) = 1$. We apply (14.8) with $m = 1$, writing

$$A(\bar{S}:S:\sigma:z\rho)f = z^{-1}a(z)f$$
$$\overline{(f(w), g(w))A(S:\bar{S}:\sigma:z\rho)^*g} = \overline{z^{-1}b(z)g}$$
$$a_t = \begin{pmatrix} e^t & 0 \\ 0 & e^{-t} \end{pmatrix},$$

and we obtain

$$\begin{aligned} &|(U(S, \sigma, z\rho, a_t)f, g) - z^{-1}a(z)e^{(-1+z)t} - z^{-1}b(z)e^{(-1-z)t}| \\ &\quad \leq C(1 + |\text{Im } z|)e^{-(1+\varepsilon)t} \end{aligned}$$

for all imaginary $z \neq 0$ and all $t > 0$. Passing to the limit $z = 0$ with t fixed, we find

$$\left|\lim_{z\to 0} (z^{-1}a(z)e^{zt} + z^{-1}b(z)e^{-zt})\right| \leq Ce^{-\varepsilon t}. \tag{14.9}$$

Write $a(z) = a_0 + a_1 z + \ldots$ and $b(z) = b_0 + b_1 z + \ldots$. We already know that $a_0 + b_0 = 0$; this conclusion is a restatement of the earlier boundedness of $G(z)$ near $z = 0$. With the understanding that $a_0 + b_0 = 0$, (14.9) says that

$$|(a_1 + b_1) + t(a_0 - b_0)| \leq Ce^{-\varepsilon t}.$$

Therefore $a_0 - b_0 = 0$, and so $a_0 = b_0 = 0$. Hence the intertwining operators have no poles, contradiction. We conclude $U(S, \sigma, 0)$ is irreducible when $\eta(\bar{S}:S:\sigma:\nu)$ has a pole at $\nu = 0$.

Now suppose conversely that $\eta(\bar{S}:S:\sigma:\nu)$ is regular and nonvanishing at $\nu = 0$. We shall use the following elementary lemma.

Lemma 14.1. If $\eta(z)$ is meromorphic in $\mathbb{C}$, is not identically 0, and is such that $\eta(z) \geq 0$ on the imaginary axis, then there exists a meromorphic function $\gamma(z)$ in $\mathbb{C}$ such that

$$\eta(z) = \gamma(z)\overline{\gamma(-\bar{z})}. \tag{14.10}$$

The function $\gamma(z)$ can be chosen to be regular and nonvanishing whenever $\eta(z)$ is regular and nonvanishing. If also $\eta(z)$ is even, then $\gamma(z)$ can be chosen to be real for real z.

Proof. Since $\eta(z) \geq 0$ on the imaginary axis, the identity

$$\eta(z) = \overline{\eta(-\bar{z})} \tag{14.11}$$

holds for all imaginary z and therefore, by analytic continuation, for all z in $\mathbb{C}$.

To prove the lemma, first suppose η is entire and nonvanishing. Then we can take $\gamma(z)$ to be one of the two versions of $\eta(z)^{1/2}$ and obtain

$$\gamma(z)\overline{\gamma(-\bar{z})} = \eta(z)^{1/2}\overline{\eta(-\bar{z})^{1/2}} = \pm\eta(z)^{1/2}[\overline{\eta(-\bar{z})}]^{1/2} = \pm[\eta(z)^{1/2}]^2 = \pm\eta(z)$$

with the sign continuous for all z and positive for $z = 0$. Thus (14.10) holds in this case. If η is even, then (14.11) shows that η is real on the real axis. Since $\eta(0) > 0$, η is positive on the real axis. Thus $\gamma = \eta^{1/2}$ is real on the real axis.

In the general case, the zeros and poles of $\eta(z)$ are symmetric about the imaginary axis by (14.11), and the ones on the axis occur with even multiplicity since $\eta \geq 0$ there. Construct $\gamma_0(z)$ as the quotient of two Weierstrass canonical products, so that $\gamma_0(z)$ has the zeros of $\eta(z)$ that lie in the open right half plane, $\gamma_0(z)$ has the poles of $\eta(z)$ that lie in the open left half plane, $\gamma_0(z)$ has zeros and poles of half the orders of those for $\eta(z)$ on the imaginary axis, $\gamma_0(z)$ has no other zeros and poles, and $\gamma_0(z)$ is real on the real axis in case $\eta(z)$ is even (in which case $\eta(z)$ is real on the real axis, by (14.11)). Then we can apply the special case to $\eta(z)\gamma_0(z)^{-1}\overline{\gamma_0(-\bar{z})}^{-1}$ and obtain a function $\gamma_1(z)$. If we put $\gamma(z) = \gamma_0(z)\gamma_1(z)$, then $\gamma(z)$ has the required properties. This proves the lemma.

Let us apply the lemma to $\eta(z) = \eta(\bar{S}:S:\sigma:\nu)$, obtaining $\gamma(z) = \gamma(\bar{S}:S:\sigma:\nu)$. Then let us define $\gamma(S:\bar{S}:\sigma:\nu) = \overline{\gamma(-\bar{z})}$ and

$$\mathscr{A}(\bar{S}:S:\sigma:\nu) = \gamma(\bar{S}:S:\sigma:\nu)^{-1}A(\bar{S}:S:\sigma:\nu)$$

$$\mathscr{A}(S:\bar{S}:\sigma:\nu) = \gamma(S:\bar{S}:\sigma:\nu)^{-1}A(S:\bar{S}:\sigma:\nu).$$

The adjoint relation (14.3) and the formula (14.10) show that $\mathscr{A}(\bar{S}:S:\sigma:\nu)$ is unitary for ν imaginary and has inverse $\mathscr{A}(S:\bar{S}:\sigma:\nu)$ for all ν. In analogy with (7.20), we let

$$\mathscr{A}_S(w, \sigma, \nu) = R(w)\mathscr{A}(\bar{S}:S:\sigma:\nu).$$

This operator intertwines $U(S, \sigma, \nu)$ with $U(S, w\sigma, w\nu) = U(S, \sigma, -\nu)$, and it is unitary for ν imaginary. Thus $\mathscr{A}_S(w, \sigma, 0)$ is a self-intertwining operator for $U(S, \sigma, 0)$.

If $\eta(\bar{S}:S:\sigma:\nu)$ is regular and nonvanishing at $\nu = 0$, then so is $\gamma(\bar{S}:S:\sigma:\nu)$, and $\mathscr{A}_S(w, \sigma, 0)$ is just a multiple of $A_S(w, \sigma, 0)$. From the proof of Theorem 7.12, we can see that $A_S(w, \sigma, 0)$ is not scalar. Therefore $U(S, \sigma, 0)$ is reducible.

Thus for SL(2, $\mathbb{R}$) we see that $U(S, \sigma, 0)$ is irreducible when $\eta(\bar{S}:S:\sigma:\nu)$ has a pole at $\nu = 0$, and it is reducible when $\eta(\bar{S}:S:\sigma:\nu)$ is regular and nonvanishing at $\nu = 0$. The function $\eta(\bar{S}:S:\sigma:\nu)$ cannot vanish at $\nu = 0$ since otherwise $A_S(w, \sigma, 0)^* A_S(w, \sigma, 0)$ would be 0 and $A_S(w, \sigma, 0)$ would be 0, hence scalar. This is a contradiction since we can see from the proof of Theorem 7.12 that $A_S(w, \sigma, 0)$ is not scalar. Thus it remains to identify η and to account constructively for the reducibility that it predicts.

In §7 we shall prove that

$$\eta(\bar{S}:S:\sigma:\nu) = c_\sigma p_\sigma(\nu)^{-1},$$

where $p_\sigma(\nu)$ is the density function in the Plancherel measure for the series of representations $U(S, \sigma, \cdot)$ and where c_σ is a positive constant. Therefore $U(S, \sigma, 0)$ is irreducible in SL(2, $\mathbb{R}$) if and only if $p_\sigma(0) = 0$. Since $p_\sigma(z)$ is given in Theorem 11.6 as a multiple of

$$\begin{cases} z\tan(\pi z/2) & \text{if } \sigma \text{ trivial} \\ z\cot(\pi z/2) & \text{if } \sigma \text{ nontrivial,} \end{cases}$$

this result is consistent with the irreducibility/reducibility results in Proposition 2.7.

In the reducible case we know that the principal series representation splits as $\mathscr{D}_1^+ \oplus \mathscr{D}_1^-$ and that this splitting is a special case of Schmid's character identity given in Theorem 12.34. But let us see an indication that these relationships are causal by referring to the Plancherel formula for the other real-rank-one groups (Theorem 13.4). The Plancherel density for the principal series is the product of the infinitesimal character by a tangent or cotangent. The density is nonvanishing at the origin if and only if we are in a cotangent case and $\Lambda + \delta_M$ is orthogonal only to the real roots $\pm\alpha$. The condition for the cotangent case is that $\xi_{\Lambda+\delta_M}(\gamma_\alpha) = +1$, which is equivalent with integrality of $\Lambda + \delta_M$ for the compact Cartan subgroup B, by Lemma 13.3. In turn, this integrality implies that $\xi_\Lambda(\gamma_\alpha) = (-1)^{2\langle\rho_\alpha, \alpha\rangle/|\alpha|^2}$, by (12.85). This last condition says that the induced representation in Schmid's identity is well defined, and we shall see that the nonorthogonality condition for $\Lambda + \delta_M$ is equivalent with the nonvanishing of both of the limits of discrete series that enter Schmid's identity. Thus if we extrapolate our conclusions of this section to the other real-rank-one groups, then we should expect that all reducibility is accounted for by Schmid's identity.

§2. Eisenstein Integrals

We begin now to consider groups more general than SL(2, $\mathbb{R}$). The first step is to introduce a way of working with operators on the space of a standard induced representation that does not depend on having a particular realization of the inducing representation. In the resulting dictionary, the space of such operators within the sum of a finite number of K types gets replaced by a space ${}^0\mathscr{C}_\sigma(M, \tau_M)$ that depends only on the class of σ, and matrix coefficients of the induced representation get replaced by "Eisenstein integrals." The operator $e^*\sigma(m)e$ in (14.8), which is intimately connected with Frobenius reciprocity, gets absorbed into the correspondence and disappears.

Fix a parabolic subgroup MAN and a discrete series representation σ of M on a Hilbert space V^σ. Let F be a finite set of irreducible representations of K, and put

$$\alpha_F = \sum_{\tau \in F} d_\tau \chi_\tau.$$

Let $\mathscr{H}_F$ be the subspace of the space of $U(MAN, \sigma, \nu)$ transforming under K according to F and realized in the compact picture; elements of $\mathscr{H}_F$ are functions $f: K \to V^\sigma$ such that $f(km) = \sigma(m)^{-1}f(k)$ for $m \in K \cap M$ and such that $\alpha_F *_K f = f$. Let E_F be the orthogonal projection of the induced space on this subspace. We shall be interested in members of $\operatorname{End}(\mathscr{H}_F)$; these operators correspond to matrices operating on the induced space and vanishing outside the entries corresponding to F.

Let V_τ be the finite-dimensional subspace of $C^\infty(K)$ where $\alpha_F *_K f = f$, and let τ be the left regular representation of $C^\infty(K)$ on this space.

Let V_F be the (finite-dimensional) subspace of $C^\infty(K \times K)$ transforming in the first variable according to members of F and in the second variable according to the complex conjugates of members of F. Specifically, if φ is in V_F, it is assumed that

$$\alpha_F *_K \varphi(\cdot, k_2) = \varphi(\cdot, k_2) \qquad \text{for all } k_2 \in K$$

$$\overline{\alpha_F} *_K \varphi(k_1, \cdot) = \varphi(k_1, \cdot) \qquad \text{for all } k_1 \in K.$$

It may be helpful to think of V_F as operating on V_τ as kernels:

$$f \in V_\tau \to \int_K \varphi(\cdot, k_2) f(k_2)\, dk_2 \in V_\tau.$$

The representation τ on V_τ gives rise to a representation of $K \times K$ on V_F, with action by τ in the first variable and action by $\bar{\tau}$ in the second variable. However, it will be more convenient to write the action in the second variable on the right side: Thus we define

$$\tau_1(k_1)\varphi\tau_2(k_2)(k, k') = \varphi(k_1^{-1}k, k_2k') \qquad \text{for } \varphi \in V_F,$$

and τ_1 and τ_2 are both representations equivalent with τ.

The space ${}^0\mathscr{C}_\sigma(M, \tau_M)$ is the set of all functions $\psi: M \to V_F$ such that

$$\psi(m_1mm_2) = \tau_1(m_1)\psi(m)\tau_2(m_2) \quad \text{for } m \in M \text{ and } m_1, m_2 \in K \cap M \tag{14.12}$$

and such that the coordinate entries of ψ (which is regarded as a vector-valued function on M) are finite linear combinations of matrix coefficients of σ. The space ${}^0\mathscr{C}_\sigma(M, \tau_M)$ depends only on the equivalence class of σ.

Proposition 14.2. For T in $\operatorname{End}(\mathscr{H}_F)$, define ψ_T by

$$\psi_T(m)(k_1, k_2) = d_\sigma \operatorname{Tr}(e^*\sigma(m)eU(k_2)TU(k_1)^{-1}),$$

where $e:\mathscr{H}_F \to V^\sigma$ is evaluation at 1, $e^*:V^\sigma \to \mathscr{H}_F$ is the adjoint, $U(k)$ refers to the induced representation (left regular on K), and d_σ is the formal degree of σ. Then ψ_T is in ${}^0\mathscr{C}_\sigma(M, \tau_M)$, and the mapping $T \to \psi_T$ is a linear isomorphism of $\operatorname{End}(\mathscr{H}_F)$ onto ${}^0\mathscr{C}_\sigma(M, \tau_M)$.

Remarks.

(1) Let $\{h_i\}$ be an orthonormal basis of the finite-dimensional space $\mathscr{H}_F$. Then the formula for ψ_T is

$$\psi_T(m)(k_1, k_2) = d_\sigma \sum_i (\sigma(m)Th_i(k_2^{-1}), h_i(k_1^{-1}))_{V^\sigma}. \tag{14.13}$$

(2) The space ${}^0\mathscr{C}_\sigma(M, \tau_M)$ is a subspace of square integrable functions on M with values in $L^2(K \times K)$ and thereby inherits a Hilbert space norm. The proof will show that $T \to \psi_T$ is isometric when $\operatorname{End}(\mathscr{H}_F)$ is given the Hilbert-Schmidt norm.

(3) The space V_F has a natural multiplication (corresponding to its interpretation as kernels) given by

$$\varphi_1 \cdot \varphi_2(k_1, k_2) = \int_K \varphi_1(k_1, k)\varphi_2(k, k_2)\, dk.$$

With this multiplication on the values of ψ, we can define convolution over M. Using Proposition 9.6, one can show that

$$\psi_T * \psi_S = \psi_{ST}. \tag{14.14}$$

Proof. Using (14.13), we check readily that ψ_T is in V_F. If m_1 and m_2 are in $K \cap M$, then

$$\begin{aligned}\psi_T(m_1mm_2)(k_1, k_2) &= d_\sigma \sum_i (\sigma(m_1mm_2)Th_i(k_2^{-1}), h_i(k_1^{-1}))_{V_\sigma}\\ &= d_\sigma \sum_i (\sigma(m)Th_i(k_2^{-1}m_2^{-1}), h_i(k_1^{-1}m_1))_{V_\sigma}\\ &= \psi_T(m)(m_1^{-1}k_1, m_2k_2)\\ &= \tau_1(m_1)\psi_T(m)\tau_2(m_2)(k_1, k_2).\end{aligned}$$

To see that the coordinate entries of ψ_T are linear combinations of matrix coefficients of σ, we take the inner product relative to $L^2(K \times K)$ of both sides of (14.13) with a member of V_F.

Next we check norms in order to see that $T \to \psi_T$ is one-one. We have

$$\begin{aligned}\|\psi_T\|^2 &= d_\sigma \int_{K\times K\times M} \left| \sum_i (\sigma(m)Th_i(k_2^{-1}), h_i(k_1^{-1}))_{V^\sigma} \right|^2 dm\, dk_1\, dk_2 \\ &= d_\sigma \sum_{i,j} \int_{K\times K\times M} (\sigma(m)Th_i(k_2), h_i(k_1))_{V^\sigma} \\ &\qquad \times \overline{(\sigma(m)Th_j(k_2), h_j(k_1))_{V^\sigma}}\, dm\, dk_1\, dk_2 \\ &= \sum_{i,j} \int_{K\times K} (Th_i(k_2), Th_j(k_2))_{V^\sigma}(h_j(k_1), h_i(k_1))_{V^\sigma}\, dk_1\, dk_2 \\ &= \sum_{i,j} (Th_i, Th_j)_{\mathscr{H}_F}(h_j, h_i)_{\mathscr{H}_F} = \|T\|_{\mathrm{HS}}^2.\end{aligned}$$

Hence $T \to \psi_T$ is one-one.

Next we calculate $\dim \mathscr{H}_F$. Let $\tau = \sum l_i\varphi_i$ be a decomposition of τ into irreducible representations of K, and let $\sigma|_{K\cap M} = \sum n_s\rho_s$ be a decomposition of $\sigma|_{K\cap M}$. Schur orthogonality gives $l_i = d_{\varphi_i}$, and Frobenius reciprocity gives

$$\begin{aligned}\sum_s n_s[\tau_{K\cap M}:\rho_s] &= \sum_s n_s \sum_i d_{\varphi_i}[\varphi_i|_{K\cap M}:\rho_s] \\ &= \sum_{s,i} n_s d_{\varphi_i}[\operatorname{ind}_{K\cap M}^K \rho_s:\varphi_i] \\ &= \sum_i d_{\varphi_i}[\operatorname{ind}_{K\cap M}^K \sigma:\varphi_i] \\ &= \sum_i d_{\varphi_i}[U|_K:\varphi_i] = \dim \mathscr{H}_F. \qquad (14.15)\end{aligned}$$

Consequently $\dim \operatorname{End}(\mathscr{H}_F)$ is the square of the left side of (14.15), and the proof will be complete if we show that $\dim {}^0\mathscr{C}(M, \tau_M)$ is no greater than

$$\sum_{s,t} n_s n_t [\tau|_{K\cap M}:\rho_s][\tau|_{K\cap M}:\rho_t]. \qquad (14.16)$$

For this purpose, fix an orthonormal basis $\{f_i\}$ of V_τ, and define

$$\tau(k)_{ij} = (\tau(k)f_j, f_i)_{L^2(K)}.$$

The functions $f_i \otimes \overline{f_j}$ defined by

$$f_i \otimes \overline{f_j}(k_1, k_2) = f_i(k_1)\overline{f_j(k_2)}$$

are an orthonormal basis for V_F. If $v(\cdot, \cdot)$ is in V_F, we let

$$v_{ij} = (v, f_i \otimes \overline{f_j})_{V_F},$$

so that

$$v = \sum v_{ij} f_i \otimes \overline{f_j}. \qquad (14.17)$$

Direct calculation gives

$$(\tau_1(k_1)v\tau_2(k_2))_{ij} = \sum_{k,l} \tau(k_1)_{ik}v_{kl}\tau(k_2)_{lj}. \tag{14.18}$$

Let $\{u_{i'}\}$ be an orthonormal basis of V^σ, and define

$$\sigma(m)_{i'j'} = (\sigma(m)u_{j'}, u_{i'})_{V^\sigma}.$$

Let $\psi \in {}^0\mathscr{C}_\sigma(M, \tau_M)$ be given. Since ψ takes its values in V_F, (14.17) shows that ψ is determined by the numbers

$$\psi(m)_{ij} = (\psi(m), f_i \otimes \bar{f}_j)_{V_F}.$$

By assumption this scalar-valued function is a linear combination of matrix coefficients of σ. Thus let us write

$$\psi(m)_{ij} = \sum_{i',j'} c_{(ij)(i'j')}\sigma(m)_{i'j'}. \tag{14.19}$$

Let m_1 and m_2 be in $K \cap M$. Formula (14.19) gives

$$\begin{aligned}\psi(m_1mm_2)_{ij} &= \sum_{k',l'} c_{(ij)(k'l')}(\sigma(m)\sigma(m_2)u_{l'}, \sigma(m_1)^{-1}u_{k'}) \\ &= \sum_{i',j',k',l'} c_{(ij)(k'l')}(\sigma(m)u_{j'}, u_{i'})(\sigma(m_2)u_{l'}, u_{j'})\overline{(\sigma(m_1)^{-1}u_{k'}, u_{i'})}.\end{aligned}$$

On the other hand, (14.18) and (14.19) together give

$$\begin{aligned}(\tau_1(m_1)\psi(m)\tau_2(m_2))_{ij} &= \sum_{k,l} \tau(m_1)_{ik}\psi(m)_{kl}\tau(m_2)_{lj} \\ &= \sum_{i',j',k',l'} \tau(m_1)_{ik}c_{(kl)(i'j')}(\sigma(m)u_{j'}, u_{i'})\tau(m_2)_{lj}.\end{aligned}$$

Hence (14.12) and the irreducibility of σ imply that the respective coefficients of $(\sigma(m)u_{j'}, u_{i'})$ in the above two expressions are equal:

$$\sum_{k',l'} c_{(ij)(k'l')}\sigma(m_2)_{j'l'}\overline{\sigma(m_1^{-1})_{i'k'}} = \sum_{k,l} \tau(m_1)_{ik}\tau(m_2)_{lj}c_{(kl)(i'j')}.$$

If we replace m_2 by m_2^{-1}, then we can rewrite this equation in obvious notation as

$$\sum_{k',l'} c_{(ij)(k'l')}(\sigma \otimes \bar{\sigma})(m_1, m_2)_{(k'l')(i'j')} = \sum_{k,l} (\tau \otimes \bar{\tau})(m_1, m_2)_{(ij)(kl)}c_{(kl)(i'j')}.$$

This equation says that the matrix $(c_{(ij)(k'l')})$ defines an intertwining operator from the representation $\sigma \otimes \bar{\sigma}$ of $(K \cap M) \times (K \cap M)$ to the representation $\tau \otimes \bar{\tau}$. The dimension of the space of such operators is exactly (14.16). Hence the dimension of ${}^0\mathscr{C}(M, \tau_M)$ is bounded as asserted, and the proof is complete.

We define **Eisenstein integrals** by

$$E(S:\psi:\nu:x) = \int_K \psi(xk)\tau_2(k)^{-1}e^{(\nu-\rho_S)H_S(xk)}\,dk \tag{14.20}$$

for ψ in ${}^0\mathscr{C}_\sigma(M, \tau_M)$, under the convention $\psi(kman) = \tau_1(k)\psi(m)$. Here ρ_S refers to ρ_A with positivity of roots defined by N. The function $E(S:\psi:v:\cdot)$ has domain G and takes values in V_F. For the next proposition, recall the projection E_F whose image is $\mathscr{H}_F$.

Proposition 14.3. Eisenstein integrals are related to matrix coefficients by

$$E(S:\psi_T:v:x)(k_1, k_2) = d_\sigma \operatorname{Tr}(E_F U(S, \sigma, v, k_1^{-1}xk_2)TE_F)$$

for ψ_T in ${}^0\mathscr{C}_\sigma(M, \tau_M)$.

Remarks. Since Proposition 14.2 says that every ψ is of the form ψ_T, the present proposition says explicitly how to obtain all Eisenstein integrals from matrix coefficients. If we specialize the formula to $k_1 = k_2 = 1$ and to T defined by $Th = (h, g)f$, we obtain

$$d_\sigma(U(S, \sigma, v, x)f, g) = E(S:\psi_T:v:x)(1, 1). \tag{14.21}$$

Thus all matrix coefficients can be recovered from Eisenstein integrals.

Proof. We calculate with $\rho = \rho_S$ that

$$\begin{aligned}
&E(S:\psi_T:v:x)(k_1, k_2)\\
&\quad= \int_K (\tau_1(\kappa(xk))\psi_T(\mu(xk))\tau_2(k)^{-1})(k_1, k_2)e^{(v-\rho)H(xk)}\,dk\\
&\quad= \int_K \psi_T(\mu(xk))(\kappa(xk)^{-1}k_1, k^{-1}k_2)e^{(v-\rho)H(xk)}\,dk\\
&\quad= \int_K d_\sigma \sum_i (\sigma(\mu(xk))Th_i(k_2^{-1}k), h_i(k_1^{-1}\kappa(xk)))_{V\sigma}e^{(v-\rho)H(xk)}\,dk\\
&\qquad\qquad\qquad\qquad\qquad\qquad\qquad\qquad\qquad\qquad\qquad\qquad\text{by (14.13)}\\
&\quad= d_\sigma \int_K \sum_i (Th_i(k_2^{-1}k), h_i(k_1^{-1}\kappa(xk)\mu(xk)e^{(\bar{v}-\rho)H(xk)})_{V\sigma}\,dk\\
&\quad= d_\sigma \sum_i \int_K (U(S, \sigma, v, k_2)Th_i(k), U(S, \sigma, -\bar{v}, x^{-1}k_1)h_i(k))_{V\sigma}\,dk\\
&\quad= d_\sigma \sum_i (U(S, \sigma, v, k_2)Th_i, U(S, \sigma, -\bar{v}, x^{-1}k_1)h_i)_{\mathscr{H}_F}\\
&\quad= d_\sigma \sum_i (U(S, \sigma, v, k_1^{-1}xk_2)Th_i\ h_i)_{\mathscr{H}_F} \qquad \text{as explained below}\\
&\quad= d_\sigma \operatorname{Tr}(E_F U(S, \sigma, v, k_1^{-1}xk_2)TE_F),
\end{aligned}$$

as required. One step above uses the adjoint formula

$$U(S, \sigma, v, x)^* = U(S, \sigma, -\bar{v}, x^{-1}). \tag{14.22}$$

This follows by analytic continuation from the fact that $U(S, \sigma, v, x)$ is unitary for v imaginary, or else it can be proved at the same time (cf. §7.2).

Eisenstein integrals satisfy some functional equations that reflect known identities for intertwining operators. For some of these, we must pass outside of ${}^0\mathscr{C}_\sigma(M, \tau_M)$, working with both σ and $w\sigma$, where $w\sigma(m) = \sigma(w^{-1}mw)$

for $w \in N_K(\mathfrak{a})$. When we want to emphasize the space ${}^0\mathscr{C}$ to which ψ belongs, we shall write ψ^σ or $\psi^{w\sigma}$ as needed.

Proposition 14.4. If T is in $\text{End}(\mathscr{H}_F)$ and w is in $N_K(\mathfrak{a})$, then

(a) $E(S_2:\psi^\sigma_{A(S_2:S_1:\sigma:\nu)T}:\nu:x) = E(S_1:\psi^\sigma_{TA(S_2:S_1:\sigma:\nu)}:\nu:x)$ and
(b) $E(S:\psi^\sigma_T:\nu:x) = E(wSw^{-1}:\psi^{w\sigma}_{R(w)TR(w)^{-1}}:w\nu:x)$,

where $R(w)$ denotes right translation of functions by w.

Proof. (a) By Proposition 14.3, we have the following meromorphic identity in ν:

$$\begin{aligned}
E(S_2:\psi^\sigma_{A(S_2:S_1:\sigma:\nu)T}:\nu:x)(k_1,k_2)
&= d_\sigma \operatorname{Tr}(E_F U(S_2,\sigma,\nu,k_1^{-1}xk_2)A(S_2:S_1:\sigma:\nu)TE_F)\\
&= d_\sigma \operatorname{Tr}(A(S_2:S_1:\sigma:\nu)E_F U(S_1,\sigma,\nu,k_1^{-1}xk_2)TE_F) \qquad \text{by Theorem 8.3a}\\
&= d_\sigma \operatorname{Tr}(E_F U(S_1,\sigma,\nu,k_1^{-1}xk_2)TE_F A(S_2:S_1:\sigma:\nu)) \qquad \text{since } \operatorname{Tr}(AB)=\operatorname{Tr}(BA)\\
&= d_\sigma \operatorname{Tr}(E_F U(S_1,\sigma,\nu,k_1^{-1}xk_2)TA(S_2:S_1:\sigma:\nu)E_F)\\
&= E(S_1:\psi^\sigma_{TA(S_2:S_1:\sigma:\nu)}:\nu:x)(k_1,k_2).
\end{aligned}$$

(b) The operator $T' = R(w)TR(w)^{-1}$ is in $\text{End}\,\mathscr{H}'_F$ for the associated space of functions transforming according to $w\sigma$ under $K \cap M$ on the right, and it is straightforward to check that

$$U(S,\sigma,\nu,x)T = R(w)^{-1}U(wSw^{-1},w\sigma,w\nu,x)T'R(w).$$

Hence

$$\begin{aligned}
E(S:\psi^\sigma_T:\nu:x)(k_1,k_2)
&= d_\sigma \operatorname{Tr}(E_F R(w)^{-1}U(wSw^{-1},w\sigma,w\nu,k_1^{-1}xk_2)T'R(w)E_F)\\
&= d_\sigma \operatorname{Tr}(R(w)^{-1}E'_F U(wSw^{-1},w\sigma,w\nu,k_1^{-1}xk_2)T'E'_F R(w))\\
&= d_\sigma \operatorname{Tr}(E'_F U(wSw^{-1},w\sigma,w\nu,k_1^{-1}xk_2)T'E'_F)\\
&= E(wSw^{-1}:\psi^{w\sigma}_{T'}:w\nu:x)(k_1,k_2)
\end{aligned}$$

§3. Asymptotics of Eisenstein Integrals

It is apparent from the definition that Eisenstein integrals are (τ_1,τ_2)-spherical, and Proposition 14.3 shows that $E(S:\psi_T:\nu:x)$ is an eigenfunction of $Z(\mathfrak{g}^{\mathbb{C}})$ with eigenvalue $\chi_{\Lambda+\nu}$, where Λ is the infinitesimal character of σ. Therefore the theory of Chapter VIII gives an asymptotic expansion for $E(S:\psi_T:\nu:x)$. Motivated by §1 of the present chapter, we shall be interested in having explicit formulas for the dominant terms of this expansion and in controlling the error term as a function of ν.

The complex-variables methods of Chapter VIII are ill-suited for addressing these questions, and it is not known how to generalize the method of §1 beyond the real-rank-one case. Instead one uses a real-variables approach. We shall indicate aspects of this approach, omitting a number of details, and we shall state the final results precisely and interpret them.

Let us start in the generality of a τ-spherical eigenfunction F of $Z(\mathfrak{g}^{\mathbb{C}})$, with eigenvalue χ_λ, and let MAN be a parabolic subgroup. We study the asymptotic expansion appropriate to $L = MA$. Let $\mathfrak{t} \supseteq \mathfrak{a}$ be a Cartan subalgebra of $\mathfrak{l} = \mathfrak{m} \oplus \mathfrak{a}$.

Recall from Chapter VIII and Appendix B how we generated a first-order system of equations to study. Let Γ^+ denote the positive roots of $(\mathfrak{g}, \mathfrak{a})$. We use the decomposition $G = KM\overline{A^+}K$, passing to τ-radial components of functions and operators far out along A^+, with $ma \in L$ carried along as a parameter restricted to a compact set. For each $Z \in Z(\mathfrak{g}^{\mathbb{C}})$, Theorem 8.26 or its generalization to the nonminimal case tells us that the τ-radial component $D_\tau(Z)$ of Z is such that

$$D_\tau(Z) - D_\tau(\mu'_{\Gamma^+}(Z))$$

has coefficients that "vanish at infinity" in a suitable sense. Namely if the coefficients are transformed to the polydisc, they vanish at the origin. In terms of $\exp \mathfrak{a}^+$, this means that they vanish at least at fast as $\exp\{-\min \alpha_j(H)\}$ as $H \to \infty$ in $\mathfrak{a}^+$; here the minimum is taken over $1 \leq j \leq l'$ in the notation of §8.12. The operator $D_\tau(\mu'_{\Gamma^+}(Z))$ has constant coefficients, and $D_\tau(Z)$ of course acts on F as $\chi_\lambda(Z)$. We regard our high-order system as

$$\mu'_{\Gamma^+}(Z)F(ma) = \chi_\lambda(Z)F(ma) + \text{error term}.$$

For the present theory it will be necessary to pass from this high-order system to the first-order system with a little more care than in the case of Chapter VIII. The kind of calculation in Lemma 8.29 has to be done more tightly. Put

$$\mathscr{Z}_G = Z(\mathfrak{g}^{\mathbb{C}}) \qquad \text{and} \qquad \mathscr{Z}_L = Z(\mathfrak{l}^{\mathbb{C}}).$$

The one-one map $\mu\colon \mathscr{Z}_G \to \mathscr{Z}_L$ is the composition of the Harish-Chandra isomorphism from $\mathscr{Z}_G$ to the $W(\mathfrak{t}^{\mathbb{C}}:\mathfrak{g}^{\mathbb{C}})$ invariants in $U(\mathfrak{t}^{\mathbb{C}})$, followed by the inverse of the Harish-Chandra isomorphism for $\mathscr{Z}_L$. We use the following sharpened version of Theorem 8.19: There exist $D_1 = 1, D_2, \ldots, D_N$ in $\mathscr{Z}_L$ that are linearly independent over the quotient field of $\mu(\mathscr{Z}_G)$ such that

$$\mathscr{Z}_L = \sum_{j=1}^{N} \mu(\mathscr{Z}_G) D_j; \tag{14.23}$$

moreover, $N = |W(\mathfrak{t}^{\mathbb{C}}:\mathfrak{g}^{\mathbb{C}})/W(\mathfrak{t}^{\mathbb{C}}:\mathfrak{l}^{\mathbb{C}})|$. Put

$$\mathscr{Z}_{0,L} = \mathscr{Z}_L \mu(\ker \chi_\lambda)$$
$$\mathscr{Z}_L^* = \mathscr{Z}_L/\mathscr{Z}_{0,L}.$$

Then $\mathscr{Z}_L^*$ is a left $\mathscr{Z}_L$ module, and as a vector space it has the images $D_1^*, \ldots, D_N^*$ of $D_1, \ldots, D_N$ as a basis. [In fact, if Z_L is in $\mathscr{Z}_L$, we use (14.23) to write

$$Z_L = \sum_{j=1}^{N} \mu(\chi_\lambda(Z_j))D_j + \sum_{j=1}^{N} \mu(Z_j - \chi_\lambda(Z_j))D_j \in \sum_{j=1}^{N} \mu(\chi_\lambda(Z_j))D_j + \mathscr{Z}_{0,L},$$

and this proves spanning. If $\sum c_j D_j = Z_0$ is in $\mathscr{Z}_{0,L}$, write $Z_0 = \sum_i Z_L^{(i)} \mu(Z^{(i)})$ with $Z_L^{(i)} \in \mathscr{Z}_L$ and $Z^{(i)} \in \ker \chi_\lambda$. Expanding $Z_L^{(i)}$ by (14.23) as $\sum_j \mu(Z_j^{(i)})D_j$, we obtain

$$\sum_j \left(c_j - \sum_i \mu(Z^{(i)} Z_j^{(i)}) \right) D_j = 0.$$

The independence in (14.23) forces $c_j = \mu(\sum_i Z^{(i)} Z_j^{(i)})$ for all j and hence c_j is in $\mu(\ker \chi_\lambda)$, mapping to 0. Thus $\{D_1^*, \ldots, D_N^*\}$ is linearly independent.]

For each $Z \in \mathscr{Z}_L$, we can therefore write

$$ZD_j \equiv \sum c_{jk}^{(Z)} D_k \bmod \mathscr{Z}_{0,L}$$

for uniquely determined numbers $c_{jk}^{(Z)}$. These are the relations, as Z runs through a basis of $\mathfrak{a}$, that lead to the first-order system by the construction in the proof of Theorem B.16. Fix $Z = H$ in $\mathfrak{a}$, and consider the effect of the corresponding equation on $F(ma \exp tH)$. As k varies, we make the $D_k F(ma)$'s into a vector-valued unknown function, and it will simplify the notation if we incorporate $e^{\rho \log a}$ into $D_k F$; thus we put

$$\Phi(ma) = e^{\rho \log a}\{D_k F(ma)\}.$$

(More formally the function Φ has its values in the tensor product of the image space of F by the dual of $\mathscr{Z}_L^*$.) Then the equation obtained from $H \in \mathfrak{a}$ is

$$\frac{d}{dt} \Phi(ma \exp tH) = \Gamma\Phi(ma \exp tH) + Q(t)\Phi(ma \exp tH). \quad (14.24)$$

Here we regard Γ and $Q(t)$ as N-by-N matrices with entries in the image space of F. The matrix Γ is a matrix of constants that captures the action of H on $\mathscr{Z}_L^*$, and $Q(t)$ is a matrix vanishing exponentially fast at $+\infty$.

We bring to bear one additional piece of information, a bound on the behavior of F and its derivatives as $t \to +\infty$. Since our interest will be in the case that ν is imaginary and $U(S, \sigma, \nu)$ is tempered, let us assume for concreteness that F and its derivatives are dominated by multiples of the spherical function φ_0^G. This condition translates to mean that $\Phi(ma \exp tH)$

grows no faster than a polynomial in t and that the behavior in m is controlled suitably by φ_0^M. The nature of the real-variables method will be clearer if we state and prove a result about vector-valued functions on $\mathbb{R}^1$ in which the group theory is absent. Regard a column of $\Phi(ma \exp tH)$ as the function $\varphi(t)$ in the following lemma.

Lemma 14.5. Let $\varphi(t)$ be a vector-valued smooth function on $[0, +\infty)$ that is at most of polynomial growth and that satisfies

$$\frac{d}{dt}\varphi(t) = \Gamma\varphi(t) + Q(t)\varphi(t)$$

for a matrix of constants Γ and a smooth matrix $Q(t)$ with $|Q(t)| \le Ce^{-\varepsilon t}$ for some $\varepsilon > 0$. Then

$$\varphi(t) - e^{t\Gamma}\Theta(0)$$

tends to 0 exponentially fast as $t \to +\infty$, where $\Theta(0)$ is the vector given by the absolutely convergent integral

$$\Theta(0) = E_0\varphi(0) + \int_0^\infty e^{-t\Gamma}E_0Q(t)\varphi(t)\,dt.$$

Here E_0 is the projection operator whose image is the sum of the generalized eigenspaces of Γ for eigenvalues of real part 0 and whose kernel is the sum of the remaining generalized eigenspaces.

Remarks. So the dominant terms of $\varphi(t)$ are $e^{t\Gamma}$ times a vector of constants, and the error term is exponentially small. Moreover, we have an explicit expression for the vector of constants.

Proof. Since $e^{-t\Gamma}(\varphi'(t) - \Gamma\varphi(t)) = \dfrac{d}{dt}(e^{-t\Gamma}\varphi(t))$, we can multiply the equation by $e^{-t\Gamma}$ and integrate from 0 to T to obtain

$$\varphi(T) = e^{T\Gamma}\varphi(0) + e^{T\Gamma}\int_0^T e^{-t\Gamma}Q(t)\varphi(t)\,dt. \tag{14.25}$$

Let E_+ and E_- be the projections on the sum of the generalized eigenspaces of Γ for eigenvalues of real part positive and negative, respectively. Then there is a positive number ε_0 less than ε such that

$$|e^{-t\Gamma}E_+| \le ce^{-\varepsilon_0 t} \quad\text{and}\quad |e^{+t\Gamma}E_-| \le ce^{-\varepsilon_0 t} \qquad \text{for} \quad t \ge 0 \tag{14.26a}$$

$$|e^{t\Gamma}E_0| \le ce^{\varepsilon_0|t|} \qquad \text{for} \quad t \in \mathbb{R}. \tag{14.26b}$$

First we show that $E_+\varphi(T)$ and $E_-\varphi(T)$ are exponentially small.

For $E_-\varphi(T)$, we apply E_- to both sides of (14.25) and immediately obtain

$$|E_-\varphi(T)| \le c|E_-\varphi(0)|e^{-\varepsilon_0 T} + c\int_0^T e^{-\varepsilon_0(T-t)}|E_-Q(t)\varphi(t)|\,dt$$

from (14.26a). The first term on the right side is harmless, and the second term is dominated by a multiple of

$$\int_0^T e^{-\varepsilon_0(T-t)}e^{-\varepsilon t}(1+t^n)\,dt \le (1+T^n)e^{-\varepsilon_0 T}\int_0^T e^{-t(\varepsilon-\varepsilon_0)}\,dt$$
$$\le (\varepsilon-\varepsilon_0)^{-1}(1+T^n)e^{-\varepsilon_0 T},$$

which is exponentially small.

For $E_+\varphi(T)$, we apply (14.25) at T and $2T$, subtract $e^{-T\Gamma}$ times the second from the first, and obtain

$$\varphi(T) = e^{-T\Gamma}\varphi(2T) - e^{T\Gamma}\int_T^{2T} e^{-t\Gamma}Q(t)\varphi(t)\,dt.$$

If we apply E_+, then (14.26a) gives

$$|E_+\varphi(T)| \le ce^{-\varepsilon_0 T}|\varphi(2T)| + c\int_T^{2T} e^{-\varepsilon_0(t-T)}|E_+Q(t)\varphi(t)|\,dt.$$

Since φ has at most polynomial growth, the first term is harmless, and the second term is dominated by a multiple of

$$\int_T^{2T} e^{-\varepsilon_0(t-T)}e^{-\varepsilon t}(1+t^n)\,dt \le (1+(2T)^n)e^{-\varepsilon T}\int_T^{2T} e^{-(\varepsilon_0+\varepsilon)(t-T)}\,dt$$
$$\le (\varepsilon_0+\varepsilon)^{-1}(1+(2T)^n)e^{-\varepsilon T},$$

which is exponentially small.

Now let us consider E_0. By (14.26b), the integral

$$\int_0^\infty e^{-t\Gamma}E_0Q(t+s)\varphi(t+s)\,dt$$

is dominated in magnitude by a multiple of

$$\int_0^\infty e^{\varepsilon_0|t|}e^{-\varepsilon(t+s)}(1+(s+t)^n)\,dt \le e^{-\varepsilon s}\int_0^\infty e^{-(\varepsilon-\varepsilon_0)t}(1+(s+t)^n)\,dt$$

and is therefore absolutely convergent. Moreover, the right side of the latter expression tends to 0 exponentially fast in s. Therefore

$$\Theta(s) = E_0\varphi(s) + \int_0^\infty e^{-t\Gamma}E_0Q(t+s)\varphi(t+s)\,dt \tag{14.27}$$

is well defined and $E_0\varphi(s) - \Theta(s)$ tends to 0 exponentially fast in s. Since $E_+\varphi(s)$ and $E_-\varphi(s)$ have been shown to tend to 0 exponentially fast, so does $\varphi(s) - \Theta(s)$.

This is the conclusion of the lemma except for the explicit expression $\Theta(s) = e^{s\Gamma}\Theta(0)$ for $\Theta(s)$. Computing directly from (14.27), we have

$$\Theta(s) - e^{s\Gamma}\Theta(0) = E_0\varphi(s) - e^{s\Gamma}E_0\varphi(0) - e^{s\Gamma}\int_0^s e^{-t\Gamma}E_0Q(t)\varphi(t)\,dt,$$

and the right side is 0 by (14.25). This completes the proof of the lemma.

The proof of Lemma 14.5 is so explicit that one can keep track of the dependence on ma when applying the lemma to our function $\Phi(ma \exp tH)$.

Let us state the final result. A finite-dimensional-vector-valued function F on G is said to satisfy the **weak inequality** if there is some integer $r \geq 0$ such that

$$\sup_{x \in G} |F(x)| \varphi_0^G(x)^{-1}(1 + \|x\|)^{-r} < \infty. \tag{14.28}$$

(This condition is satisfied by K-finite matrix coefficients of representations of G induced from a discrete series on M and a unitary character on A, and it is satisfied also by all derivatives of such functions. Cf. Proposition 7.14 and §8.11.) We define $\mathscr{A}(G, \tau)$ to be the space of all τ-spherical smooth functions on G that are $Z(\mathfrak{g}^{\mathbb{C}})$-finite and satisfy the weak inequality. What we have just seen is the heart of the proof of the following theorem.

Theorem 14.6. For each F in $\mathscr{A}(G, \tau)$ there exists a unique element F_S in $\mathscr{A}(MA, \tau_M)$ such that

$$\lim_{\substack{a \to \infty \\ S}} \{e^{\rho \log a} F(ma) - F_S(ma)\} = 0$$

for all m in MA, the limit occurring exponentially fast along any one-parameter group in A.

Terminology. The function F_S is called the **constant term** of F along S.

However, it is not just the result that is important but also the method, because the method allows us to track down the effect on F_S of moving parameters. Parameters are moving in the case of interest to us, which is Eisenstein integrals. Consider $E(S:\psi:\nu:x)$ with $\psi \in {}^0\mathscr{C}_\sigma(M, \tau_M)$. For ν imaginary, this function is in $\mathscr{A}(G, \tau)$, and the theorem applies. We saw in §§1–2 the utility in having explicit formulas and in being able to handle the dependence of the constant term and the error term as functions of ν. These are complicated matters to which we return in a moment.

First let us address the M dependence of the constant term of an Eisenstein integral. Even in the generality of Theorem 14.6, if F satisfies $ZF = \chi_{\Lambda+\nu}(Z)F$ for $Z \in Z(\mathfrak{g}^{\mathbb{C}})$, then F_S satisfies $Z_L F_S = \chi_{\Lambda+\nu}(Z_L) F_S$ for $Z_L \in \mu(Z(\mathfrak{g}^{\mathbb{C}})) \subseteq Z(\mathfrak{l}^{\mathbb{C}})$. Hence F_S is a linear combination of functions that are right and left K-finite on $L = MA$, satisfy the weak inequality, and have generalized infinitesimal character $w\Lambda + \nu$ for some $w \in W(\mathfrak{t}^{\mathbb{C}}:\mathfrak{l}^{\mathbb{C}})$. In the case that F is the Eisenstein integral considered above, Λ is M-nonsingular and integral. Under these conditions one can show that the conditions on F_S imply that each entry of the vector-valued function F_S is a linear combination of matrix coefficients of discrete series representations (possibly discrete series other than σ); this conclusion should be regarded as a generalization of Problem 23 of Chapter XII.

Now let us consider the dependence on ν. The first step is to separate the M dependence from the A dependence, and for this step one assumes that ν is imaginary and is **regular** (i.e., is not orthogonal to any root of $(\mathfrak{g}, \mathfrak{a})$). The argument giving the separation is not trivial, and the details will be omitted; it is necessary to understand the eigenvalues of the particular matrix Γ of Lemma 14.5 that enters the proof of Theorem 14.6. When one combines this separation with the considerations in the previous paragraph, one arrives at the following theorem.

Theorem 14.7. Fix ν imaginary and regular, and let $S_1 = MAN_1$ and $S_2 = MAN_2$. Then there exist unique elements

$$c_{S_2|S_1}(s:\nu) \text{ in } \operatorname{Hom}({}^0\mathscr{C}_\sigma(M, \tau_M), {}^0\mathscr{C}_{w\sigma}(M, \tau_M)),$$

for $s \in W(A:G)$ and w a representative of s in $N_K(\mathfrak{a})$, such that

$$E_{S_2}(S_1:\psi:\nu:ma) = \sum_{s \in W(A:G)} (c_{S_2|S_1}(s:\nu)\psi)(m)e^{s\nu \log a}$$

for all ψ in ${}^0\mathscr{C}_\sigma(M, \tau_M)$, $m \in M$, and $a \in A$.

To get at the dependence on ν of the constant term and the error term, closer study of the eigenvalues of Γ is again necessary. We shall be content with the following theorem.

Theorem 14.8. Let $S_1 = MAN_1$ and $S_2 = MAN_2$.

(a) All mixed left-invariant and right-invariant derivatives of $E_{S_2}(S_1:\psi:\nu:ma)$ are continuous functions of the pair (ν, ma), for ν imaginary.

(b) If ν is imaginary and regular, then $c_{S_2|S_1}(s:\nu)$ extends to be holomorphic in a neighborhood of ν in $(\mathfrak{a}')^{\mathbb{C}}$, and the corresponding limit formula of Theorem 14.6 persists to such a neighborhood. The rate of convergence depends at most polynomially on $\operatorname{Im} \nu$.

Part (b) of Theorem 14.8 allows us to connect c functions with intertwining operators. In this way we can see explicitly how Theorems 14.7 and 14.8 generalize our result (14.8) for $SL(2, \mathbb{R})$. To obtain these relationships, we first obtain an integral formula for $c_{S|S}(1:\nu)$. We start from the definition (14.20) of Eisenstein integrals and run through the same calculation as in Lemma 7.23 to obtain

$$\lim_{\substack{a \to \infty \\ S}} e^{-(\nu-\rho)\log a}E(S:\psi:\nu:ma) = \int_{\bar{N}} \psi(m\mu(\bar{n})^{-1})\tau_2(\kappa(\bar{n}))^{-1}e^{-(\rho+\nu)H(\bar{n})}\, d\bar{n}$$

if Haar measure on K and $\bar{N}$ are related as in (5.25) and if $\operatorname{Re} \nu$ is in the open positive Weyl chamber. Meanwhile Theorems 14.7 and 14.8 say that

$$\lim_{\substack{a \to \infty \\ S}} \{e^{\rho \log a}E(S:\psi:\nu:ma) - \sum_{s \in W(A:G)} (c_{S|S}(s:\nu)\psi)(m)e^{s\nu \log a}\} = 0$$

if $\operatorname{Re} \nu$ is sufficiently close to 0 with ν in one of the neighborhoods in Theorem 14.8b. If $\operatorname{Re} \nu$ is in the open positive Weyl chamber and is sufficiently close to 0, then we can multiply this relation by $e^{-\nu \log a}$ and use the relation $\operatorname{Re} \nu \log a > \operatorname{Re} s\nu \log a$ for $s \neq 1$ to obtain

$$\lim_{\substack{a \to \infty \\ S}} e^{-(\nu-\rho)\log a} E(S:\psi:\nu:ma) = (c_{S|S}(1:\nu)\psi)(m).$$

Consequently we obtain

$$c_{S|S}(1:\nu)\psi(m) = \int_{\bar{N}} \psi(m\mu(\bar{n})^{-1})\tau_2(\kappa(\bar{n}))^{-1} e^{-(\rho+\nu)H(\bar{n})}\, d\bar{n}, \tag{14.29}$$

which is the desired integral formula.

Corollary 14.9. Each $c_{S_2|S_1}(s:\nu)$ extends to a meromorphic function for ν in $(\mathfrak{a}')^{\mathbb{C}}$. Moreover if T is in $\operatorname{End}(\mathscr{H}_F)$ and s is in $W(A:G)$ with w in $N_K(\mathfrak{a})$ as a representative, then

(a) $c_{S_2|S_2}(s:\nu)\psi^{\sigma}_{A(S_2:S_1:\sigma:\nu)T} = c_{S_2|S_1}(s:\nu)\psi^{\sigma}_{TA(S_2:S_1:\sigma:\nu)}$
(b) $c_{S|S}(s:\nu)\psi^{\sigma}_{T} = c_{S|wSw^{-1}}(1:s\nu)\psi^{w\sigma}_{R(w)TR(w)^{-1}}$
(c) $c_{S|S}(1:\nu)\psi^{\sigma}_{T} = \psi^{\sigma}_{A(\bar{S}:S:\sigma:\nu)T}$ if Haar measure on $\bar{N}$ is normalized as in (5.25).

Proof. For ν in the region in Theorem 14.8b, result (a) follows from Proposition 14.4a by taking the S_2 limit and equating coefficients, and (b) follows similarly from Proposition 14.4b by taking the S limit. For result (c) we prove the formula under the additional assumption that $\operatorname{Re} \nu$ is in the open positive Weyl chamber, by computing from (14.29) and applying (14.13) twice:

$$\begin{aligned}
&c_{S|S}(1:\nu)\psi^{\sigma}_{T}(m)(k_1, k_2) \\
&\quad = \int_{\bar{N}} \psi^{\sigma}_{T}(m\mu(\bar{n})^{-1})\tau_2(\kappa(\bar{n}))^{-1}(k_1, k_2) e^{-(\rho+\nu)H(\bar{n})}\, d\bar{n} \\
&\quad = \int_{\bar{N}} \psi^{\sigma}_{T}(m\mu(\bar{n})^{-1})(k_1, \kappa(\bar{n})^{-1}k_2) e^{-(\rho+\nu)H(\bar{n})}\, d\bar{n} \\
&\quad = d_\sigma \sum_i \int_{\bar{N}} (\sigma(m)\sigma(\mu(\bar{n}))^{-1} T h_i(k_2^{-1}\kappa(\bar{n})), h_i(k_1^{-1})) e^{-(\rho+\nu)H(\bar{n})} \\
&\quad = d_\sigma \sum_i (\sigma(m)A(\bar{S}:S:\sigma:\nu)T h_i(k_2^{-1}), h_i(k_1^{-1})) \\
&\quad = \psi^{\sigma}_{A(\bar{S}:S:\sigma:\nu)T}(m)(k_1, k_2).
\end{aligned}$$

The integral defining the intertwining operator is convergent by Theorem 7.22.

Then (c) extends by analytic continuation to be valid for ν in the region of Theorem 14.8b. Now let us see that $c_{S_2|S_1}(s:\nu)$ can be defined in this region in terms of intertwining operators. Result (c) shows how $c_{S|S}(1:\nu)$

is so defined, and then (a) with $s = 1$ shows how $c_{S_2|S_1}(1:\nu)$ is so defined. Using (b), we see how $c_{S|S}(s:\nu)$ is so defined, and then a second application of (a) shows how $c_{S_2|S_1}(s:\nu)$ is so defined. Since the intertwining operators are meromorphic for ν in $(\mathfrak{a}^{\mathbb{C}})'$, $c_{S_2|S_1}(s:\nu)$ has a meromorphic continuation to $(\mathfrak{a}^{\mathbb{C}})'$. Results (a), (b), and (c) then extend to be valid as meromorphic identities for ν in $(\mathfrak{a}^{\mathbb{C}})'$.

A special case of the formula for $c_{S_2|S_1}(s:\nu)$ given descriptively above is the formula for $c_{S|S}(s:\nu)$, which is worth writing down explicitly. It is

$$c_{S|S}(s:\nu)\psi_T^\sigma = \psi_{T'}^{w\sigma}, \tag{14.30a}$$

where

$$T' = R(w)A(w^{-1}\bar{S}w:S:\sigma:\nu)TA(S:w^{-1}Sw:\sigma:\nu)R(w)^{-1} \tag{14.30b}$$

and Haar measure on $\bar{N}$ is normalized as in (5.25). This formula says that each $c_{S|S}$ is given by a pair of complementary intertwining operators, one operating on the left (of T) and one operating on the right.

Let us see that (14.30) generalizes the limit formula (14.8) for $SL(2, \mathbb{R})$. In this case, we can choose $w = 1$ and $w = \begin{pmatrix} 0 & 1 \\ -1 & 0 \end{pmatrix}$ for the two values of s, and the corresponding operators T' in (14.30b) are

$$T' = A(\bar{S}:S:\sigma:\nu)T \tag{14.31a}$$

and

$$T' = R(w)TA(S:\bar{S}:\sigma:\nu)R(w)^{-1}. \tag{14.31b}$$

We take T to be as in the remarks after Proposition 14.3, and we apply Proposition 14.2 with $k_1 = k_2 = 1$. Then (14.31a) gives

$$\begin{aligned} c_{S|S}(1:\nu)\psi_T^\sigma(m)(1, 1) &= \operatorname{Tr}(e^*\sigma(m)eA(\bar{S}:S:\sigma:\nu)T) \\ &= (e^*\sigma(m)eA(\bar{S}:S:\sigma:\nu)f, g), \end{aligned}$$

and (14.31b) gives

$$\begin{aligned} c_{S|S}(s:\nu)\psi_T^{w\sigma}(m)(1, 1) &= \operatorname{Tr}(e^*w\sigma(m)eR(w)TA(S:\bar{S}:\sigma:\nu)R(w)^{-1}) \\ &= \operatorname{Tr}(A(S:\bar{S}:\sigma:\nu)R(w)^{-1}e^*w\sigma(m)eR(w)T) \\ &= \operatorname{Tr}(A(S:\bar{S}:\sigma:\nu)U(w)e^*\sigma(w^{-1}mw)eU(w)^{-1}T) \\ &\qquad\qquad \text{since } eR(w) = eU(w)^{-1} \\ &= \operatorname{Tr}(U(w)A(S:\bar{S}:\sigma:\nu)e^*\sigma(w^{-1}mw)eU(w)^{-1}T) \\ &= (U(w)A(S:\bar{S}:\sigma:\nu)e^*\sigma(w^{-1}mw)eU(w)^{-1}f, g). \end{aligned}$$

Taking into account (14.21) and applying Theorems 14.6 and 14.7, we obtain

$$\begin{aligned} \lim_{a \to +\infty} \{ & e^{\rho \log a}(U(S, \sigma, \nu, ma)f, g) - e^{\nu \log a}(e^*\sigma(m)eA(\bar{S}:S:\sigma:\nu)f, g) \\ & - e^{-\nu \log a}(U(w)A(S:\bar{S}:\sigma:\nu)e^*\sigma(w^{-1}mw)eU(w)^{-1}f, g)\} = 0, \end{aligned}$$

the limit occurring exponentially fast. So indeed Theorems 14.6 and 14.7 specialize to (14.8). Theorem 14.8b includes the estimate of polynomial growth in Im ν for the remainder term.

§4. The η Functions for Intertwining Operators

Let G be linear connected reductive with compact center. If MAN is a parabolic subgroup, then M_0 is linear connected reductive with compact center. Let σ be an irreducible unitary representation of M, and let $\mathfrak{t}$ be a θ-stable Cartan subalgebra of $\mathfrak{m}$. We say that σ has **real infinitesimal character** if the infinitesimal character of σ is real on $(\mathfrak{t} \cap \mathfrak{p}) \oplus i(\mathfrak{t} \cap \mathfrak{k})$. This property is independent of $\mathfrak{t}$, and all discrete series and limits of discrete series have this property.

Our objective in this section is to show that certain compositions of standard intertwining operators are scalar and to identify properties of such scalars (the η functions). We shall be especially interested in the case that the inducing representation on M has real infinitesimal character. In §1 we saw how the η functions arose and were used in the context of $SL(2, \mathbb{R})$.

Some of the properties of these scalars depend on the formula for the adjoint of an intertwining operator that was promised in (14.3).

Proposition 14.10. Let $S_1 = MAN_1$ and $S_2 = MAN_2$ be parabolic subgroups. If σ is an irreducible unitary representation of M and if adjoints are computed K type by K type in the compact picture, then

$$A(S_2:S_1:\sigma:\nu) = A(S_1:S_2:\sigma:-\bar{\nu})^* \tag{14.32}$$

as an identity of meromorphic functions.

Idea of proof. The proof when the left side is given by a convergent integral is a straightforward but lengthy computation using the techniques of §7.2 and the integral formulas of §5.6. The general case follows by analytic continuation (Theorem 8.38).

Remarks. It follows from this proposition and Theorem 8.38c that

$$A_S(w, \sigma, \nu)^* = A_S(w^{-1}, w\sigma, -w\bar{\nu}) \tag{14.33}$$

if w is in $N_K(\mathfrak{a})$ and if adjoints are interpreted as in the proposition.

We now introduce the η functions in stages, first the real-rank-one minimal case, then the general minimal case, and finally the general not necessarily minimal case.

Proposition 14.11. Suppose G has real rank one and $S_\mathfrak{p} = M_\mathfrak{p}A_\mathfrak{p}N_\mathfrak{p}$ is a minimal parabolic subgroup. If σ is an irreducible unitary representation of M, then there exists a complex-valued meromorphic function

$\eta(\bar{S}_\mathfrak{p}:S_\mathfrak{p}:\sigma:\nu)$ on $(\mathfrak{a}'_\mathfrak{p})^\mathbb{C}$ such that

$$A(S_\mathfrak{p}:\bar{S}_\mathfrak{p}:\sigma:\nu)A(\bar{S}_\mathfrak{p}:S_\mathfrak{p}:\sigma:\nu) = \eta(\bar{S}_\mathfrak{p}:S_\mathfrak{p}:\sigma:\nu)I \tag{14.34}$$

as an identity of meromorphic functions when applied to any smooth f. The function η has no poles for nonreal ν, and η satisfies

(a) $\eta(\bar{S}_\mathfrak{p}:S_\mathfrak{p}:\sigma:\nu) = \eta(S_\mathfrak{p}:\bar{S}_\mathfrak{p}:\sigma:\nu)$
(b) $\eta(\bar{S}_\mathfrak{p}:S_\mathfrak{p}:\sigma:\nu) \geq 0$ for ν imaginary
(c) $\eta(\bar{S}_\mathfrak{p}:S_\mathfrak{p}:\sigma:\nu) = \overline{\eta(\bar{S}_\mathfrak{p}:S_\mathfrak{p}:\sigma:-\bar{\nu})}$
(d) $\eta(\bar{S}_\mathfrak{p}:S_\mathfrak{p}:w\sigma:w\nu) = \eta(\bar{S}_\mathfrak{p}:S_\mathfrak{p}:\sigma:\nu)$ for w in $N_K(\mathfrak{a}_\mathfrak{p})$
(e) $\eta(\bar{S}_\mathfrak{p}:S_\mathfrak{p}:\sigma^\varphi:\nu) = \eta(\bar{S}_\mathfrak{p}:S_\mathfrak{p}:\sigma:\nu)$ if φ is an automorphism of G leaving K stable and the positive chamber of $A_\mathfrak{p}$ fixed and if $\sigma^\varphi(m) = \sigma(\varphi^{-1}(m))$.
(f) $\eta(\bar{S}_\mathfrak{p}:S_\mathfrak{p}:\sigma:\nu) = \eta(\bar{S}_\mathfrak{p}:S_\mathfrak{p}:\sigma:-\nu)$.

Proof. Theorem 7.12 shows that the left side of (14.34) is well defined for ν nonzero imaginary, and Corollary 7.13 identifies the left side as a self-intertwining operator for $U(S_\mathfrak{p}, \sigma, \nu)$, at least on the level of K-finite functions. By Theorem 7.2, the left side is scalar for such ν and f.

Now consider a K-finite block of the left side of (14.34) for all ν. The result is a meromorphic matrix-valued function of ν that is scalar for ν imaginary. Hence it must be scalar for all ν. We conclude that a meromorphic η exists such that (14.34) is valid when applied to any K-finite f. The continuity noted in Theorem 7.12 allows us to pass to general smooth f. Theorem 7.12 implies that poles of η can occur only for real ν.

Conclusion (b) follows from (14.32), and (c) is a consequence of (b). At nonzero imaginary values of ν where η is not zero, (14.34) implies (a) since a left inverse of a square matrix is a right inverse. Thus (a) follows by analytic continuation unless η is identically 0 for imaginary ν. In this case, (14.32) implies that $A(\bar{S}_\mathfrak{p}:S_\mathfrak{p}:\sigma:\nu)$ is 0 for imaginary ν; then both sides of (a) are 0 for ν imaginary and hence for all ν, by analytic continuation. Conclusion (d) follows from (a) and from (7.31).

For (e) we check easily that if f on K transforms by σ under $M_\mathfrak{p}$, then $f \circ \varphi^{-1}$ transforms by σ^φ under $\varphi(M_\mathfrak{p})$. Moreover, $\varphi(M_\mathfrak{p}) = M_\mathfrak{p}$, so that $\varphi(S_\mathfrak{p}) = S_\mathfrak{p}$ and $\varphi(\bar{S}_\mathfrak{p}) = \bar{S}_\mathfrak{p}$. Using the integral formula for intertwining operators and then applying analytic continuation, we therefore find that

$$A(\bar{S}_\mathfrak{p}:S_\mathfrak{p}:\sigma^\varphi:\nu)(f \circ \varphi^{-1})(k) = (A(\bar{S}_\mathfrak{p}:S_\mathfrak{p}:\sigma:\nu)f)(\varphi^{-1}(k)).$$

From this identity we obtain (e). Finally (f) follows by combining (d) and (e) when $\varphi(x) = \Theta(w^{-1}xw)$ for w representing the nontrivial member of $W(A_\mathfrak{p}:G)$.

For general G we recall from §7.5 that we can associate to any reduced positive restricted root β a group $G^{(\beta)}M_\mathfrak{p}$ of real rank one whose $M_\mathfrak{p}$ is the original $M_\mathfrak{p}$ and whose $\mathfrak{a}_\mathfrak{p}$ is $\mathbb{R}H_\beta$. (The group $G^{(\beta)}M_\mathfrak{p}$ may be disconnected, but this difficulty is minor, and we shall ignore it.)

Proposition 14.12. Let G be linear connected reductive with compact center, and suppose that $S_{1,\mathfrak{p}} = M_\mathfrak{p}A_\mathfrak{p}N_{1,\mathfrak{p}}$ and $S_{2,\mathfrak{p}} = M_\mathfrak{p}A_\mathfrak{p}N_{2,\mathfrak{p}}$ are minimal parabolic subgroups. For each irreducible unitary representation σ of $M_\mathfrak{p}$, define

$$\eta(S_{2,\mathfrak{p}}:S_{1,\mathfrak{p}}:\sigma:\nu) = \prod_\beta \eta^{(\beta)}(\bar{S}^{(\beta)}M_\mathfrak{p}:S^{(\beta)}M_\mathfrak{p}:\sigma:\nu|_{\mathbb{R}H_\beta}),$$

where the product on the right side is a product of real-rank-one η factors for the groups $G^{(\beta)}M_\mathfrak{p}$ and is taken over all reduced restricted roots β that are positive for $N_{1,\mathfrak{p}}$ and negative for $N_{2,\mathfrak{p}}$. Then η is meromorphic on $(\mathfrak{a}_\mathfrak{p}')^\mathbb{C}$ and is holomorphic at all points ν where $\operatorname{Im}\nu$ is orthogonal to no $\beta\in\Sigma$ that is positive for $N_{1,\mathfrak{p}}$ and negative for $N_{2,\mathfrak{p}}$. Moreover, η satisfies

(a) $A(S_{1,\mathfrak{p}}:S_{2,\mathfrak{p}}:\sigma:\nu)A(S_{2,\mathfrak{p}}:S_{1,\mathfrak{p}}:\sigma:\nu) = \eta(S_{2,\mathfrak{p}}:S_{1,\mathfrak{p}}:\sigma:\nu)$
(b) $\eta(S_{1,\mathfrak{p}}:S_{2,\mathfrak{p}}:\sigma:\nu) = \eta(S_{2,\mathfrak{p}}:S_{1,\mathfrak{p}}:\sigma:\nu)$.

Sketch of proof. It is immediate from Proposition 14.11 that η is globally meromorphic and is holomorphic where asserted. Conclusion (b) follows from the definition and from conclusions (a) and (f) of Proposition 14.11. For (a), the operator $A(S_{2,\mathfrak{p}}:S_{1,\mathfrak{p}}:\sigma:\nu)$ can be decomposed by Proposition 7.9, by means of the remarks after Proposition 7.10, into a product of operators $A(S':S:\sigma:\nu)$ with the property that $\bar{\mathfrak{n}}\cap\mathfrak{n}' = \bar{\mathfrak{n}}^{(\beta)}$ for some N-positive reduced restricted root β. Then $A(S_{1,\mathfrak{p}}:S_{2,\mathfrak{p}}:\sigma:\nu)$ decomposes into the product of the corresponding operators $A(S:S':\sigma:\nu)$, but in the reverse order. In the resulting decomposition of

$$A(S_{1,\mathfrak{p}}:S_{2,\mathfrak{p}}:\sigma:\nu)A(S_{2,\mathfrak{p}}:S_{1,\mathfrak{p}}:\sigma:\nu),$$

the middle two operators are of the form

$$A(S:S':\sigma:\nu)A(S':S:\sigma:\nu),$$

which behaves, according to Proposition 7.11, like

$$A(S^{(\beta)}M_\mathfrak{p}:\bar{S}^{(\beta)}M_\mathfrak{p}:\sigma:\nu|_{\mathbb{R}H_\beta})A(\bar{S}^{(\beta)}M_\mathfrak{p}:S^{(\beta)}M_\mathfrak{p}:\sigma:\nu|_{\mathbb{R}H_\beta})$$

and therefore collapses to the scalar

$$\eta^{(\beta)}(\bar{S}^{(\beta)}M_\mathfrak{p}:S^{(\beta)}M_\mathfrak{p}:\sigma:\nu|_{\mathbb{R}H_\beta}).$$

The scalar factors out, and the next pair of operators in the middle yields a scalar, and so on. The result is conclusion (a).

Now we pass to the intertwining operators $A(S':S:\xi:\nu)$ of §8.10 for general parabolic subgroups $S = MAN$ and $S' = MAN'$. Since $W(A:G)$ is not necessarily transitive on the Weyl chambers, we cannot necessarily relate this operator to some $A_S(w, \xi, \nu)$. In particular, we cannot take advantage of the Weyl group in decomposing $A(S':S:\xi:\nu)$ into some kind of elementary product.

The elementary operators, as is suggested by the proof of Proposition 14.12, are the ones $A(S':S:\xi:\nu)$ for which $\bar{\mathfrak{n}} \cap \mathfrak{n}' = \bar{\mathfrak{n}}^{(\gamma)}$ for some N-positive reduced root γ of $(\mathfrak{g}, \mathfrak{a})$; here $\bar{\mathfrak{n}}^{(\gamma)}$ is the sum over $c > 0$ of the root spaces $\mathfrak{g}_{-c\gamma}$. We mentioned in §8.10 that Proposition 7.11 (given for minimal parabolics) has a generalization, and we state the generalization now. To the root γ of $(\mathfrak{g}, \mathfrak{a})$, we can associate the group $G^{(\gamma)}M$ in which $G^{(\gamma)}$ is the analytic subgroup of G whose Lie algebra is generated by $\bar{\mathfrak{n}}^{(\gamma)}$ and $\mathfrak{n}^{(\gamma)} = \sum_{c>0} \mathfrak{g}_{c\gamma}$. Let $N^{(\gamma)} = \exp \mathfrak{n}^{(\gamma)}$ and $\bar{N}^{(\gamma)} = \exp \bar{\mathfrak{n}}^{(\gamma)}$. The group $G^{(\gamma)}M$ has two maximal parabolic subgroups of special interest: $S^{(\gamma)}M = M(\exp \mathbb{R}H_\gamma)N^{(\gamma)}$ and $\bar{S}^{(\gamma)}M = M(\exp \mathbb{R}H_\gamma)\bar{N}^{(\gamma)}$. In obvious notation the formula that generalizes Proposition 7.11 is

$$A(S':S:\xi:\nu)F(k) = A(\bar{S}^{(\gamma)}M:S^{(\gamma)}M:\xi:\nu|_{\mathbb{R}H_\gamma})F_k(1) \tag{14.35}$$

if $\bar{\mathfrak{n}} \cap \mathfrak{n}' = \bar{\mathfrak{n}}^{(\gamma)}$.

A general intertwining operator decomposes as a composition of such operators. To see this, let S and S' be given. A sequence $S_i = MAN_i$, $0 \le i \le r$, is called a **string** if there are N-positive reduced roots γ_i, $1 \le i \le r$, of $(\mathfrak{g}, \mathfrak{a})$ such that

$$\bar{N}_{i-1} \cap N_i = \bar{N}^{(\gamma_i)} \quad \text{or} \quad N^{(\gamma_i)} \qquad \text{for} \quad 1 \le i \le r,$$

$$S_0 = S, \qquad \text{and} \qquad S_r = S'.$$

The string is called a **minimal string** from S to S' if in the above equalities $\bar{N}_{i-1} \cap N_i$ is always $\bar{N}^{(\gamma_i)}$. In the latter case, one can show that $\{\gamma_1, \ldots, \gamma_r\}$ is the set of all N-positive reduced roots that are N'-negative. Whenever we have a minimal string, we can iterate Theorem 8.38d and decompose the given operator as a composition

$$A(S':S:\xi:\nu) = A(S_r:S_{r-1}:\xi:\nu) \cdots A(S_2:S_1:\xi:\nu)A(S_1:S_0:\xi:\nu). \tag{14.36}$$

A minimal string always exists from S to S': Namely we choose H and H' in the respective open positive Weyl chambers of $\mathfrak{a}$ for S and S' so that no point on the line segment

$$H(t) = (1 - t)H + tH', \qquad 0 \le t \le 1,$$

is annihilated by more than one N-positive reduced root of $(\mathfrak{g}, \mathfrak{a})$. Let $0 < t_1 < \ldots < t_r < 1$ be the values of t where $H(t)$ is annihilated by such a root, and let $S_i = MAN_i$ be associated to the Weyl chamber in which (t_i, t_{i+1}) lies (with $t_0 = 0$ and $t_{r+1} = 1$). Then S_i gives a minimal string.

Proposition 14.13. Let G be linear connected reductive with compact center, and suppose $S_1 = MAN_1$ and $S_2 = MAN_2$ are parabolic subgroups.

If ξ is an irreducible unitary representation of M, then there exists a complex-valued meromorphic function $\eta(S_2:S_1:\xi:\nu)$ on $(\mathfrak{a}')^{\mathbb{C}}$ such that

$$A(S_1:S_2:\xi:\nu)A(S_2:S_1:\xi:\nu) = \eta(S_2:S_1:\xi:\nu)I \tag{14.37}$$

as an identity of meromorphic functions when applied to any K-finite function. The function η has the following properties:

(a) $\eta(S_1:S_2:\xi:\nu) = \eta(S_2:S_1:\xi:\nu)$,
(b) $\eta(S_2:S_1:\xi:\nu) = \prod_\gamma \eta^{(\gamma)}(\bar{S}^{(\gamma)}M:S^{(\gamma)}M:\xi:\nu|_{\mathbb{R}H_\gamma})$, where the product on the right side is a product of η functions for maximal parabolic subgroups of $G^{(\gamma)}M$ and is taken over all reduced roots γ of $(\mathfrak{g}, \mathfrak{a})$ that are positive for N_1 and negative for N_2,
(c) $\eta(S_2:S_1:\xi:\nu) = \eta(S_{2,\mathfrak{p}}:S_{1,\mathfrak{p}}:\sigma:\nu \oplus \nu_M)$ if ξ imbeds as a subrepresentation of the nonunitary principal series $U(S_M, \sigma, \nu_M)$ of M and if $S_{1,\mathfrak{p}} = M_\mathfrak{p}(AA_M)(N_1N_M)$ and $S_{2,\mathfrak{p}} = M_\mathfrak{p}(AA_M)(N_2N_M)$,
(d) $\eta(S_2:S_1:\xi:\nu)$ is holomorphic at ν if ξ has real infinitesimal character and if Im ν is not orthogonal to any root of $(\mathfrak{g}, \mathfrak{a})$ that is positive for N_1 and negative for N_2,
(e) if dim $A = 1$ then $\eta(\bar{S}:S:\xi:\nu)$ satisfies
   (i) $\eta(\bar{S}:S:\xi:\nu) \geq 0$ for ν imaginary
   (ii) $\eta(\bar{S}:S:\xi:\nu) = \overline{\eta(\bar{S}:S:\xi:-\bar{\nu})}$
   (iii) $\eta(\bar{S}:S:w\xi:w\nu) = \eta(\bar{S}:S:\xi:\nu)$ for $w \in N_K(\mathfrak{a})$
   (iv) $\eta(\bar{S}:S:\xi^\varphi:\nu) = \eta(\bar{S}:S:\xi:\nu)$ if φ is an automorphism of G leaving K stable and the positive chamber of A fixed and if $\xi^\varphi(m) = \xi(\varphi^{-1}(m))$
   (v) $\eta(\bar{S}:S:\xi:-\nu) = \eta(\bar{S}:S:\xi:\nu)$ and η is holomorphic for nonzero imaginary ν provided ξ has real infinitesimal character.

Proof. Using Theorem 8.37, we imbed ξ as a subrepresentation of the nonunitary principal series of M as in (c), and the intertwining operators imbed, too. Thus (14.37) and (c) follow from Proposition 14.12b. Substituting in (14.37) from (14.35) and (14.36), we obtain (b). Conclusion (a) follows from (14.32) by the same argument as for Proposition 14.11a.

Let ξ have real infinitesimal character. In (c) the nonunitary principal series $U(S_M, \sigma, \nu_M)$ must have the same infinitesimal character as ξ, and thus ν_M must be real, by Proposition 8.22. The function $\eta(S_{2,\mathfrak{p}}:S_{1,\mathfrak{p}}:\sigma:\nu \oplus \nu_M)$ is holomorphic at ν if $\text{Im}\langle \nu + \nu_M, \beta\rangle \neq 0$ for every restricted root β that is positive for $N_{1,\mathfrak{p}}$ and negative for $N_{2,\mathfrak{p}}$, by Proposition 14.12. Since ν_M is real, let us consider an $N_{1,\mathfrak{p}}$-positive β with $\text{Im}\langle \nu, \beta\rangle = 0$. If $\beta|_\mathfrak{a} \neq 0$, then $\text{Im}\langle \nu, \beta|_\mathfrak{a}\rangle = 0$ and we have a contradiction to the hypothesis in (d). If $\beta|_\mathfrak{a} = 0$, then β is a positive restricted root of M. Since N_M is contained in both $N_{1,\mathfrak{p}}$ and $N_{2,\mathfrak{p}}$, β is not $N_{2,\mathfrak{p}}$-negative. Hence η is holomorphic as asserted (and is globally meromorphic).

The first four conclusions in (e) are proved as in Proposition 14.11. In (v), the conclusion that η is holomorphic for ν nonzero imaginary is a special case of (d). For the formula we apply (e, ii) and then (c) to obtain

$$\begin{aligned}
\eta(\bar{S}:S:\xi:\nu) &= \bar{\eta}(\bar{S}:S:\xi:-\bar{\nu}) = \bar{\eta}(S_{2,\mathfrak{p}}:S_{1,\mathfrak{p}}:\sigma:-\bar{\nu}\oplus\nu_M) \\
&= \eta(S_{2,\mathfrak{p}}:S_{1,\mathfrak{p}}:\sigma:\nu\oplus(-\bar{\nu}_M)) && \text{by Proposition 14.11c} \\
&= \eta(S_{2,\mathfrak{p}}:S_{1,\mathfrak{p}}:\sigma:\nu\oplus -\nu_M) && \text{since } \nu_M \text{ is real} \\
&= \eta(S_{2,\mathfrak{p}}:S_{1,\mathfrak{p}}:\sigma:-\nu\oplus\nu_M) && \text{by Proposition 14.11f} \\
&= \eta(\bar{S}:S:\xi:-\nu) && \text{by (c).}
\end{aligned}$$

Remarks. No η function vanishes for any imaginary value of ν. In fact, if η were to vanish at ν_0, (14.32) would force the whole intertwining operator to vanish at ν_0 (on all functions for which (14.37) is valid). It is possible to construct a $(K \cap M)$-finite-valued function f in the compact picture of the induced space on which (14.37) is valid and on which the usual integral formula for the intertwining operator is convergent at the identity and gives the true value; by peaking such a function sufficiently, we can force the integral to be nonzero.

§5. First Irreducibility Results

For G linear connected reductive we continue to pursue generalizations of the arguments in §1. To get at irreducibility, we use Eisenstein integrals and their asymptotic expansions.

Lemma 14.14. Let $S = MAN$ be a parabolic subgroup, let σ be a discrete series representation of M, and let $\nu \in (\mathfrak{a}')^{\mathbb{C}}$ be imaginary. If $U(S, \sigma, \nu)$ is reducible, then there exists a sufficiently large finite set F of K types and there exists $T \neq 0$ in End $\mathscr{H}_F$ such that

$$\operatorname{Tr}(E_F U(S, \sigma, \nu, x) T E_F) = 0$$

for all x in G.

Proof. Suppose the induced space splits nontrivially as $V_1 \oplus V_2$ into the orthogonal sum of closed invariant subspaces. Let F contain a K type of V_1 and a K type of V_2. Then $\mathscr{H}_F$ splits nontrivially as $(\mathscr{H}_F \cap V_1) \oplus (\mathscr{H}_F \cap V_2)$. If we let T be any nonzero linear map that is 0 on $\mathscr{H}_F \cap V_2$ and carries $\mathscr{H}_F \cap V_1$ into $\mathscr{H}_F \cap V_2$, then T has the required properties.

Theorem 14.15. Let $S = MAN$ be a parabolic subgroup, and let σ be a discrete series representation of M. Then $U(S, \sigma, \nu)$ is irreducible for all regular imaginary ν.

Remark. This result extends Bruhat's theorem, Theorem 7.2, from the case of a minimal parabolic S.

Proof. Suppose on the contrary that $U(S, \sigma, \nu)$ is reducible. Choose F and $T \neq 0$ as in Lemma 14.14. By Proposition 14.3 the Eisenstein integral $E(S:\psi_T:\nu:x)$ is 0 for all x. Hence the constant term along S of this Eisenstein integral is 0 (see Theorem 14.6). By Theorem 14.7, $c_{S|S}(1:\nu)\psi_T = 0$. Corollary 14.9 implies that $\psi_{T'} = 0$ for $T' = A(\bar{S}:S:\sigma:\nu)T$, and Proposition 14.2 implies that T' itself is 0. Since ν is regular and imaginary, $A(\bar{S}:S:\sigma:\nu)$ and $A(S:\bar{S}:\sigma:\nu)$ are both well defined (with no singularity at ν) by the same argument as for Proposition 14.13d. Therefore

$$0 = A(S:\bar{S}:\sigma:\nu)A(\bar{S}:S:\sigma:\nu)T = \eta(\bar{S}:S:\sigma:\nu)T.$$

The remarks following Proposition 14.13 say that η is nonvanishing, and thus $T = 0$, contradiction.

Theorem 14.16. Let $S = MAN$ be a parabolic subgroup with $\dim A = 1$, and let σ be a discrete series representation of M. Then $U(S, \sigma, 0)$ is irreducible unless

(a) $|W(A:G)| = 2$,
(b) $w\sigma$ is equivalent with σ for all w in $N_K(\mathfrak{a})$, and
(c) $\eta(\bar{S}:S:\sigma:\nu)$ has no pole at $\nu = 0$.

Remarks. For ν nonzero imaginary, $U(S, \sigma, \nu)$ is irreducible by the previous theorem. The proof of the present theorem is similar to an argument in §1, but it is complicated by the fact that we do not know in advance that the intertwining operator has at worst a simple pole at $\nu = 0$.

Proof. Fix H in $\mathfrak{a}^+$. For any finite set F of K types and for any T in $\operatorname{End} \mathscr{H}_F$, we know from Theorems 14.6, 14.7, and 14.8 that

$$\Big|e^{\rho(tH)}E(S:\psi_T:\nu:m \exp tH) - \sum_{s \in W(A:G)} c_{S|S}(s:\nu)\psi_T(m)e^{s\nu(tH)}\Big| \leq Ce^{-\varepsilon t} \tag{14.38}$$

for all imaginary ν in a neighborhood of 0 and for all $t > 0$; here C depends on m and T. Letting $\nu \to 0$, we see that the meromorphic function of ν given by

$$\sum_{s \in W(A:G)} c_{S|S}(s:\nu)\psi_T(m)e^{s\nu(tH)} \tag{14.39}$$

has no pole at $\nu = 0$.

Suppose $U(S, \sigma, 0)$ is reducible. Choose F and $T \neq 0$ as in Lemma 14.14. In this case the first term on the left side of (14.38) vanishes, and we conclude that

$$\Big|\lim_{\nu \to 0} \sum_{s \in W(A:G)} c_{S|S}(s:\nu)\psi_T(m)e^{s\nu(tH)}\Big| \leq Ce^{-\varepsilon t}. \tag{14.40}$$

First suppose that $W(A:G) = \{1\}$. Then $|c_{S|S}(1:0)\psi_T(m)| \le Ce^{-\varepsilon t}$ for all $t > 0$ says $c_{S|S}(1:0)\psi_T = 0$. Corollary 14.9 and Proposition 14.2 imply that

$$A(\bar{S}:S:\sigma:0)T = 0, \tag{14.41}$$

and we shall see below that this relation is impossible if $T \neq 0$. Thus $|W(A:G)| = 2$.

Let s be the nontrivial element of $W(A:G)$, and let us use v to denote both the linear functional and its value on H. Then we can write

$$c_{S|S}(1:v)\psi_T = v^{-k}(a_0 + a_1 v + a_2 v^2 + \ldots)$$
$$c_{S|S}(s:v)\psi_T = v^{-l}(b_0 + b_1 v + b_2 v^2 + \ldots)$$

with each a_j in ${}^0\mathscr{C}_\sigma(M, \tau_M)$ and each b_j in ${}^0\mathscr{C}_{w\sigma}(M, \tau_M)$, where w is a representative of s in $N_K(\mathfrak{a})$. We may assume a_0 and b_0 are not 0. Evaluating at a point m where at least one of $a_0(m)$ and $b_0(m)$ is nonzero, we see from the regularity of (14.39) that $k = l$ and $a_0 + b_0 = 0$. Multiplying the sum in (14.40) by v^{k-1} and letting $v \to 0$, we obtain

$$\left| \lim_{v \to 0} \frac{(a_0 + a_1 v + \ldots)e^{vt} + (b_0 + b_1 v + \ldots)e^{-vt}}{v} \right| \le Ce^{-\varepsilon t}$$

if $k \ge 1$. The limit is given by a derivative since $a_0 + b_0 = 0$, and the inequality thus says

$$|(a_1 + b_1) + t(a_0 - b_0)| \le Ce^{-\varepsilon t}$$

if $k \ge 1$. This inequality forces $a_0 - b_0 = 0$ and hence $a_0 = b_0 = 0$, since $a_0 + b_0 = 0$. Thus we have a contradiction unless $k \le 0$.

We conclude that $c_{S|S}(1:v)\psi_T$ and $c_{S|S}(s:v)\psi_T$ have no pole at $v = 0$. Referring to (14.40), we see that

$$c_{S|S}(1:0)\psi_T + c_{S|S}(s:0)\psi_T = 0.$$

The first term is in ${}^0\mathscr{C}_\sigma(M, \tau_M)$ and the second is in ${}^0\mathscr{C}_{w\sigma}(M, \tau_M)$. These spaces are disjoint unless $w\sigma$ is equivalent with σ, since the Plancherel formula for M implies that inequivalent discrete series have orthogonal matrix coefficients. Thus either both terms are 0 and (14.41) holds, or $w\sigma$ is equivalent with σ. And in the latter case,

$$A(\bar{S}:S:\sigma:v)T \tag{14.42}$$

has no pole at $v = 0$.

We shall now show that the regularity of (14.42) at $v = 0$ forces $\eta(\bar{S}:S:\sigma:v)$ to be regular at $v = 0$ and that if (14.42) tends to 0 at $v = 0$, then η vanishes at $v = 0$. In view of the remarks following Proposition 14.13, this will show that (14.41) is impossible and that the reducibility of $U(S, \sigma, 0)$ forces (c). Hence the proof will be complete.

Thus let us expand $A(\bar{S}:S:\sigma:\nu)$ on $\mathscr{H}_F$ in series as

$$A(\bar{S}:S:\sigma:\nu) = \nu^{-k}(A_0 + A_1\nu + A_2\nu^2 + \ldots)$$

with $A_0 \neq 0$. Then we have

$$A(S:\bar{S}:\sigma:\nu) = A(\bar{S}:S:\sigma:-\bar{\nu})^* = (-\nu)^{-k}(A_0^* - A_1^*\nu + A_2^*\nu^2 - \ldots)$$

and

$$\begin{aligned}\eta(\bar{S}:S:\sigma:\nu)I &= A(S:\bar{S}:\sigma:\nu)A(\bar{S}:S:\sigma:\nu)\\ &= (-1)^k\nu^{-2k}(A_0^*A_0 + \text{higher order}).\end{aligned} \tag{14.43}$$

Since $A_0^*A_0 \neq 0$, $A_0^*A_0$ must be scalar. Thus A_0 is invertible as a member of End $\mathscr{H}_F$. Consequently when we expand (14.42) in series, the first term $\nu^{-k}A_0T$ is not 0. Hence the regularity of (14.42) at $\nu = 0$ forces $k \leq 0$, and (14.43) shows that η is regular at $\nu = 0$. Moreover, if (14.42) vanishes at $\nu = 0$, then $k \leq -1$, and (14.43) shows that η vanishes at $\nu = 0$. These are the conclusions that we observed would complete the proof.

§6. Normalization of Intertwining Operators and Reducibility

To get at reducibility by means of intertwining operators, we follow the model of §1 and introduce normalizing factors γ. We continue to assume that G is linear connected reductive with compact center and that $S = MAN$ is a parabolic subgroup, but now we assume in addition that σ is an irreducible unitary representation of M with real infinitesimal character.

By Lemma 14.1, we select simultaneously, for each scalar-valued meromorphic function $\eta(z)$ of one complex variable that is ≥ 0 on the imaginary axis and is even, a meromorphic $\gamma(z)$ that is real for real z and satisfies $\eta(z) = \gamma(z)\overline{\gamma(-\bar{z})}$.

To normalize the operators $A(\bar{S}:S:\sigma:\nu)$ for which dim $A = 1$, we take $\eta(z) = \eta(\bar{S}:S:\sigma:z\rho_A)$, where ρ_A is defined relative to S. Parts (e, i) and (e, v) of Proposition 14.13 show that $\eta(z)$ satisfies the conditions in the previous paragraph, and we can therefore define $\gamma(\bar{S}:S:\sigma:\nu)$ to be the corresponding function $\gamma(z)$. It is easy to verify that this function has the properties in the lemma below.

Lemma 14.17. If dim $A = 1$, then

(a) $\gamma(S:\bar{S}:\sigma:\nu) = \gamma(\bar{S}:S:\sigma:-\nu)$
(b) $\gamma(S:\bar{S}:\sigma:\nu)\gamma(\bar{S}:S:\sigma:\nu) = \eta(\bar{S}:S:\sigma:\nu)$
(c) $\gamma(\bar{S}:S:\sigma^\varphi:\nu) = \gamma(\bar{S}:S:\sigma:\nu)$ if φ is as in Proposition 14.13 (e, iv).

Next we pass to the elementary operators $A(S':S:\sigma:\nu)$ of §4, for which $\bar{\mathfrak{n}} \cap \mathfrak{n}' = \bar{\mathfrak{n}}^{(\beta)}$ for a single N-positive reduced root β of $(\mathfrak{g}, \mathfrak{a})$. By (14.35) this operator behaves as if dim $A = 1$, and accordingly we use the corresponding normalizing factor

$$\gamma(S':S:\sigma:\nu) = \gamma^{(\beta)}(\bar{S}^{(\beta)}M:S^{(\beta)}M:\sigma:\nu|_{\mathbb{R}H_\beta}).$$

Finally in the general case, we are guided by Proposition 14.13b to define

$$\gamma(S_2:S_1:\sigma:\nu) = \prod_{\substack{\beta \text{ reduced} \\ \beta > 0 \text{ for } N_1 \\ \beta < 0 \text{ for } N_2}} \gamma^{(\beta)}(\bar{S}^{(\beta)}M:S^{(\beta)}M:\sigma:\nu|_{\mathbb{R}H_\beta}).$$

These factors are canonical, once the initial choice of functions $\gamma(z)$ has been made, and do not depend on any particular choices of isomorphisms between groups.

Normalized intertwining operators can now be defined by

$$\mathscr{A}(S_2:S_1:\sigma:\nu) = \gamma(S_2:S_1:\sigma:\nu)^{-1}A(S_2:S_1:\sigma:\nu)$$
$$\mathscr{A}_S(w, \sigma, \nu) = \gamma(w^{-1}Sw:S:\sigma:\nu)^{-1}A_S(w, \sigma, \nu).$$

These formulas are to be interpreted as identities of meromorphic functions. Hence poles of γ may cancel out poles of A at particular points. In any case division by the γ factor is meaningful since no η factor is identically 0. Combining Lemma 14.17a and Proposition 14.13b, we immediately obtain the following lemma.

Lemma 14.18. Let $S = MAN$ and $S' = MAN'$ be parabolic subgroups. Then

(a) $\mathscr{A}(S:S':\sigma:\nu)\mathscr{A}(S':S:\sigma:\nu) = I$

(b) For any minimal string $S_i = MAN_i$, $0 \le i \le r$, from S to S', $\mathscr{A}(S':S:\sigma:\nu) = \mathscr{A}(S_r:S_{r-1}:\sigma:\nu) \cdots \mathscr{A}(S_1:S_0:\sigma:\nu)$.

Theorem 14.19. For any three parabolic subgroups of the form $S_1 = MAN_1$, $S_2 = MAN_2$, and $S_3 = MAN_3$, and for any σ with real infinitesimal character,

$$\mathscr{A}(S_3:S_1:\sigma:\nu) = \mathscr{A}(S_3:S_2:\sigma:\nu)\mathscr{A}(S_2:S_1:\sigma:\nu).$$

Sketch of proof. We suppress σ and ν in the notation. By Lemma 14.18a, we are to show that

$$\mathscr{A}(S_1:S_3)\mathscr{A}(S_3:S_2)\mathscr{A}(S_2:S_1) = I. \tag{14.44}$$

We form minimal strings between each pair of parabolic subgroups and substitute into the left side of (14.44) from Lemma 14.18b. Changing notation, we see that we are to prove the following: Whenever $S_0, S_1, \ldots, S_n$ is a string with $S_0 = S_n$, then

$$\mathscr{A}(S_n:S_{n-1}) \cdots \mathscr{A}(S_2:S_1)\mathscr{A}(S_1:S_0) = I.$$

We argue by induction on n. We may assume that no two consecutive members of the string are the same since $\mathscr{A}(S:S) = I$.

We check that the number of reduced N_0-positive roots of $(\mathfrak{g}, \mathfrak{a})$ that are N_i-positive changes by 1 in passing from i to $i + 1$, and all such changes for a string are increases by 1 if and only if the string is minimal. The first conclusion implies that n is even, and the second implies that a minimal string from S to S' is strictly shorter than a nonminimal one.

For $n = 2$, the result in question follows from Lemma 14.18a. For general (even) n, suppose that the result holds for all shorter circular strings. We examine the strings

$$S_0, S_1, \ldots, S_{n/2} \qquad \text{and} \qquad S_{n/2}, S_{n/2+1}, \ldots, S_n.$$

If both are minimal, we collapse the corresponding operators by Lemma 14.18b and obtain our result from Lemma 14.18a. Otherwise we replace a nonminimal one by a minimal one between the ends, using Lemma 14.18a and our inductive hypothesis. The new total string from S_0 to S_n is shorter, and we finish off the proof by applying our inductive hypothesis a second time.

Theorem 14.19 is one effect of normalization. Others are given below. The first proposition formulates results for the operators $A(S_2:S_1:\sigma:\nu)$, and the second reformulates them for $A_S(w, \sigma, \nu)$. Both results are easy consequences of properties of the η and γ functions, as well as the intertwining operators, and we omit the proofs.

Proposition 14.20. Normalized intertwining operators have the following properties when applied to K-finite members of the appropriate induced space, provided Haar measures are normalized as in Theorem 8.38:

(a) $\mathscr{A}(S_2:S_1:\sigma:\nu)U(S_1, \sigma, \nu, X) = U(S_2, \sigma, \nu, X)\mathscr{A}(S_2:S_1:\sigma:\nu)$ for X in $\mathfrak{g}$

(b) $\mathscr{A}(S_2:S_1:\sigma:\nu) = R(w)^{-1}\mathscr{A}(wS_2w^{-1}:wS_1w^{-1}:w\sigma:w\nu)R(w)$ for w in $N_K(\mathfrak{a})$

(c) $\mathscr{A}(S_2:S_1:\sigma:\nu)^* = \mathscr{A}(S_1:S_2:\sigma:-\bar{\nu})$, with the adjoint defined K type by K type

(d) $\mathscr{A}(S_2:S_1:\sigma:\nu)$ extends to a holomorphic function of ν for ν imaginary, is unitary for every imaginary value of ν, and, for such ν, exhibits $U(S_1, \sigma, \nu)$ and $U(S_2, \sigma, \nu)$ as unitarily equivalent.

Proposition 14.21. Normalized intertwining operators have the following properties when applied to K-finite members of the appropriate induced space, provided Haar measures are normalized as in Theorem 8.38:

(a) $\mathscr{A}_S(w, \sigma, \nu)U(S, \sigma, \nu, X) = U(S, w\sigma, w\nu, X)\mathscr{A}_S(w, \sigma, \nu)$ for all X in $\mathfrak{g}$ and w in $N_K(\mathfrak{a})$

(b) $\mathscr{A}_S(w, E\sigma E^{-1}, \nu) = E\mathscr{A}_S(w, \sigma, \nu)E^{-1}$ if E is a unitary operator on V^σ and w is in $N_K(\mathfrak{a})$
(c) $\mathscr{A}_S(w_1 w_2, \sigma, \nu) = \mathscr{A}_S(w_1, w_2\sigma, w_2\nu)\mathscr{A}_S(w_2, \sigma, \nu)$ for all w_1, w_2 in $N_K(\mathfrak{a})$
(d) $\mathscr{A}_S(w, \sigma, \nu)^* = \mathscr{A}_S(w^{-1}, w\sigma, -w\bar{\nu})$, with the adjoint defined K type by K type
(e) $\mathscr{A}_S(w, \sigma, \nu)$ extends to a holomorphic function of ν for ν imaginary, is unitary for every imaginary value of ν, and, for such ν, exhibits $U(S, \sigma, \nu)$ and $U(S, w\sigma, w\nu)$ as unitarily equivalent.

If ν is imaginary and if $w\sigma = \sigma$ and $w\nu = \nu$, then Proposition 14.21e provides us with unitary self-intertwining operators for $U(S, \sigma, \nu)$. Any nonscalar such operator will exhibit reducibility, by Schur's Lemma. We can relax this condition, insisting only that $w\sigma$ be equivalent with σ, if we are willing to adjust our operators a little. The adjustment is constructed in the lemma below, and properties of the adjusted operators are given in the proposition after the lemma.

Lemma 14.22. Let $H \subseteq H'$ be locally compact groups with H closed and normal and with H'/H cyclic of order n, let w be an element of H' whose powers meet all cosets of H'/H, and let L be an irreducible unitary representation of H on a Hilbert space V such that L and wL are unitarily equivalent. Then it is possible to define $L(w)$ as an operator on V in exactly n ways, differing only by an nth root of unity as a factor, such that L extends to a unitary representation of H' on V.

Remarks. We omit the elementary proof. The operator $L(w)$ is a multiple of an operator exhibiting L and wL as unitarily equivalent.

Proposition 14.23. If w is in $N_K(\mathfrak{a})$ and if $w\sigma$ is equivalent with σ, then the normalized operators $\sigma(w)\mathscr{A}_S(w, \sigma, \nu)$, when applied to K-finite members of the induced space for $U(S, \sigma, \nu)$, have the following properties:

(a) $\sigma(w)\mathscr{A}_S(w, \sigma, \nu)U(S, \sigma, \nu, X) = U(S, \sigma, w\nu, X)\sigma(w)\mathscr{A}_S(w, \sigma, \nu)$ for all X in $\mathfrak{g}$
(b) $[\sigma(w)\mathscr{A}_S(w, \sigma, \nu)]^* = \sigma(w)^{-1}\mathscr{A}_S(w^{-1}, \sigma, -w\bar{\nu})$, with the adjoint defined K type by K type.

Proof. We apply the lemma with $H = M$, $L = \sigma$, and H' equal to the group generated by H and w. Using Proposition 14.21b with $E = \sigma(w)^{\pm 1}$, we readily obtain the results from conclusions (a) and (d) in Proposition 14.21.

If s is in $W(A{:}G)$ and $s[\sigma] = [\sigma]$, then Proposition 14.21b shows that the operator $\sigma(w)\mathscr{A}_S(w, \sigma, \nu)$ is independent of the choice of a representative w in $N_K(\mathfrak{a})$, although it does depend on how the selection is made in Lemma 14.22. So we write $\sigma(s)\mathscr{A}_S(s, \sigma, \nu)$ for the operator. Motivated by

Proposition 14.23, we define, for ν imaginary,

$$W_{\sigma,\nu} = \{s \in W(A:G) \mid s[\sigma] = [\sigma] \text{ and } s\nu = \nu\}.$$

During the normalization process, those operators corresponding to W_{σ,ν_0} for which $\eta(s^{-1}Ss:S:\sigma:\nu)$ was regular at $\nu = \nu_0$ were normalized merely by a finite nonzero constant. The theorem below concerns these operators.

Theorem 14.24. For ν_0 imaginary let

$$R'_{\sigma,\nu_0} = \{r \in W_{\sigma,\nu_0} \mid \eta(r^{-1}Sr:S:\sigma:\nu) \text{ is regular at } \nu_0\}.$$

Then the operators $\sigma(r)\mathscr{A}_S(r, \sigma, \nu_0)$, for r in R'_{σ,ν_0}, are linearly independent.

Example. In SL(2, $\mathbb{R}$), the nontrivial representation σ of M gives rise to an intertwining operator that in the noncompact picture is given by

$$A_S(w, \sigma, z\rho)f(x) = \int_{-\infty}^{\infty} \frac{f(y)\,\mathrm{sgn}(x-y)\,dy}{|x-y|^{1-z}}.$$

Our interest is in $z = 0$, and we study $x = 0$. If f is in $C_{\mathrm{com}}^{\infty}(\mathbb{R} - \{0\})$, then the integral is convergent at $z = 0$. Such an f vanishes at 0, and it is easy to arrange that $A_S(w, \sigma, 0)f$ does not vanish there. Therefore $A_S(w, \sigma, 0)$ is not scalar, and the two operators corresponding to $R'_{\sigma,0}$ are linearly independent.

Idea of proof. The operators in question are essentially given by their original integral formulas, since their normalizing factors at ν_0 are finite nonzero scalars. The integral formulas are still divergent in general, and one works only with functions for which the integrals converge at the identity. The supports of the integrands, when sorted out properly, form a system of sets (of varying dimension) filtered by inclusion, and one constructs functions on these sets that exhibit the linear independence.

Corollary 14.25. Let $S = MAN$ be a parabolic subgroup with $\dim A = 1$, and let σ be a discrete series representation of M. Then $U(S, \sigma, 0)$ is reducible if

(a) $|W(A:G)| = 2$,
(b) $w\sigma$ is equivalent with σ if w represents the nontrivial element of $W(A:G)$, and
(c) $\eta(S:S:\sigma:\nu)$ has no pole at $\nu = 0$.

Remark. This result is a converse to Theorem 14.16.

§7. Connection with Plancherel Formula When $\dim A = 1$

We mentioned in §1 that the η function in SL(2, $\mathbb{R}$) would be essentially the reciprocal of the density in the Plancherel formula. We now take up this result. Our assumptions for this section are that G is linear connected reductive with compact center, that $S = MAN$ is a parabolic subgroup with $\dim A = 1$, and that σ is a discrete series representation of M. In the Plancherel formula for G, there is a term coming from the induced series $U(S, \sigma, iv, \cdot)$, v real-valued. Let us say that it fits into the formula as

$$f(1) = \ldots + \int_{v \in \mathfrak{a}'} \operatorname{Tr} U(S, \sigma, iv, f) p_\sigma(iv)\, dv + \ldots.$$

One can compute $p_\sigma(iv)$ explicitly by the methods of Chapter XIII. From §§5–6 in that chapter, we see that the computation is essentially the same as for the real-rank-one case. The results in the real-rank-one case are given in §§2–3 of Chapter XIII and are reinterpreted in the last paragraph of §14.1. We shall give the general formula without further proof.

Proposition 14.26. Let $S = MAN$ with $\dim A = 1$. Let $\mathfrak{b}^-$ be a compact Cartan subalgebra of $\mathfrak{m}$, let $\sigma = \pi^M(\lambda, \chi)$ be a discrete series representation of M, and let β be a certain positive root of $(\mathfrak{g}, \mathfrak{a})$—the unique one if there is just one, or 2α if α is reduced and 2α is a root of $(\mathfrak{g}, \mathfrak{a})$. Then the Plancherel density $p_\sigma(iv)$ is given by two cases, each involving a product over roots ε in $\Delta((\mathfrak{a} \oplus \mathfrak{b}^-)^{\mathbb{C}} : \mathfrak{g}^{\mathbb{C}})$ and a nonzero constant c_σ depending on σ but not v:

(a) If rank $G >$ rank K, then

$$p_\sigma(iv) = c_\sigma \prod_{\varepsilon|_{\mathfrak{a}} = \beta} \langle \lambda + iv, \varepsilon \rangle. \tag{14.45a}$$

(b) If rank $G =$ rank K, then β extends to a real root of $\Delta((\mathfrak{a} \oplus \mathfrak{b}^-)^{\mathbb{C}} : \mathfrak{g}^{\mathbb{C}})$ and

$$p_\sigma(iv) = c_\sigma \prod_{\substack{\varepsilon|_{\mathfrak{a}} = c\beta \\ c > 0}} \langle \lambda + iv, \varepsilon \rangle \; f_{\sigma,\beta}(iv), \tag{14.45b}$$

where

$$f_{\sigma,\beta}(iv) = \begin{cases} \tanh\left(\dfrac{\pi\langle v, \beta\rangle}{|\beta|^2}\right) & \text{if } \chi(\gamma_\beta) = -(-1)^{2\langle \rho_\beta, \beta\rangle/|\beta|^2} \\[2ex] \coth\left(\dfrac{\pi\langle v, \beta\rangle}{|\beta|^2}\right) & \text{if } \chi(\gamma_\beta) = +(-1)^{2\langle \rho_\beta, \beta\rangle/|\beta|^2} \end{cases} \tag{14.45c}$$

Here ρ_β is half the sum of the members of $\Delta((\mathfrak{a} \oplus \mathfrak{b}^-)^{\mathbb{C}} : \mathfrak{g}^{\mathbb{C}})$ having positive inner product with β.

The main result of this section is the theorem below. In view of Propositions 14.13b and 14.26, it yields explicit formulas, up to constant factors, for all η functions arising from discrete series on M, no matter what $\dim A$ is.

Theorem 14.27. When $S = MAN$ with $\dim A = 1$ and when σ is in the discrete series of M, $\eta(\bar{S}:S:\sigma:i\nu)^{-1}$ is given by a multiple of the Plancherel density $p_\sigma(i\nu)$ for ν real-valued, the multiple depending on σ.

We shall sketch the proof in the real-rank-one case. We adopt the same notational conventions for A and $\mathfrak{a}'$ as in §13.2. Namely we parametrize A by $\exp tH'_\beta$, we define ν_0 in $\mathfrak{a}'$ by $\nu_0(H'_\beta) = 1$, and we use ν to mean two things by writing $\nu = \nu\nu_0$. The proof will use some knowledge of the image of the group Fourier transform $f \to \{U(S, \sigma, \nu, f)\}$; the following lemma will suffice.

Lemma 14.28. With σ fixed, let $\nu \to T_\nu$ be a measurable function from $(0, \infty)$ to the Hilbert-Schmidt operators on the induced space from σ such that $\int_0^\infty \|T_\nu\|_{\mathrm{HS}}^2 p_\sigma(i\nu)\, d\nu < \infty$. Then there exists f in $L^2(G)$ such that

$$\int_G f(x)\overline{h(x)}\, dx = \int_0^\infty \operatorname{Tr}(T_\nu U(S, \sigma, i\nu, h)^*) p_\sigma(i\nu)\, d\nu$$

for all h in $C_{\mathrm{com}}^\infty(G)$ and such that

$$\int_G |f(x)|^2\, dx = \int_0^\infty \|T_\nu\|_{\mathrm{HS}}^2 p_\sigma(i\nu)\, d\nu. \tag{14.46}$$

We save all comments on this lemma for the bibliographical notes and turn to the proof of Theorem 14.27. Fix a finite subset F of $\hat{K}$, and let $\mathscr{H}_F$ as in §2 be the subset of the induced space from σ that transforms by F. We may assume $\mathscr{H}_F \neq 0$. Let

$$\Phi(i\nu, x) = E_F U(S, \sigma, i\nu, x) E_F.$$

The explicit formula for p_σ shows that $p_\sigma(i\nu)$ is nonvanishing for ν nonzero real. Thus any φ in $C_{\mathrm{com}}^\infty((0, \infty))$ has the property that

$$\int_0^\infty |\varphi(\nu)|^2 p_\sigma(i\nu)^{-1}\, d\nu < \infty. \tag{14.47}$$

Define
$$\Phi(\varphi, x) = \int_0^\infty \varphi(\nu)\Phi(i\nu, x)^*\, d\nu.$$

The point of the argument is to compute $\int_G \|\Phi(\varphi, x)\|_{\mathrm{HS}}^2\, dx$ in two ways, once by Lemma 14.28 and once in terms of asymptotic expansions. We isolate the first computation as a lemma.

Lemma 14.29. $\int_G \|\Phi(\varphi, x)\|_{\mathrm{HS}}^2\, dx = (\dim \mathscr{H}_F)^2 \int_0^\infty |\varphi(\nu)|^2 p_\sigma(i\nu)^{-1}\, d\nu.$

Proof. Let $\{\psi_i\}$ be an orthonormal basis of $\mathscr{H}_F$, and define

$$\Phi^{ij}(\varphi, x) = (\Phi(\varphi, x)\psi_i, \psi_j). \tag{14.48}$$

Let T_ν^{ij} be the Hilbert-Schmidt operator on the induced space defined by

$$T_\nu^{ij}\xi = \varphi(\nu)p_\sigma(i\nu)^{-1}(\xi, \psi_j)\psi_i. \tag{14.49}$$

Since $p_\sigma(iv)$ is real, we have

$$(T^{ij})^*\xi = \overline{\varphi(v)}p_\sigma(iv)^{-1}(\xi, \psi_i)\psi_j$$

and
$$T_v^{ij}(T_v^{ij})^*\xi = |\varphi(v)|^2 p_\sigma(iv)^{-2}(\xi, \psi_i)\psi_i.$$

Thus
$$\|T_v^{ij}\|_{\text{HS}}^2 = |\varphi(v)|^2 p_\sigma(iv)^{-2}. \tag{14.50}$$

For h in $C_{\text{com}}^\infty(G)$, (14.48) gives

$$(T_v^{ij}U(S, \sigma, iv, h)\psi_k, \psi_l) = \varphi(v)p_\sigma(iv)^{-1}(U(S, \sigma, iv, h)\psi_k, \psi_j)(\psi_i, \psi_l).$$

Hence

$$\operatorname{Tr}(T_v^{ij}U(S, \sigma, iv, h)) = \varphi(v)p_\sigma(iv)^{-1}(U(S, \sigma, iv, h)\psi_i, \psi_j). \tag{14.51}$$

By (14.47) and (14.50), T_v^{ij} satisfies the assumptions of Lemma 14.28. Choosing $f = f^{ij}$ as in Lemma 14.28, we obtain

$$\begin{aligned}\int_G f^{ij}(x)\overline{h(x)}\,dx &= \int_0^\infty \operatorname{Tr}(T_v^{ij}U(S, \sigma, iv, h)^*)p_\sigma(iv)\,dv \\ &= \int_0^\infty \int_G \varphi(v)\overline{h(x)}(U(S, \sigma, iv, x)^*\psi_i, \psi_j)\,dv \qquad \text{by (14.51)} \\ &= \int_G \Phi^{ij}(\varphi, x)\overline{h(x)}\,dx \qquad \text{by (14.48).}\end{aligned}$$

Since h is arbitrary, $f^{ij}(x) = \Phi^{ij}(\varphi, x)$ almost everywhere, and (14.46) and (14.50) give

$$\int_G |\Phi^{ij}(\varphi, x)|^2\,dx = \int_0^\infty \|T_v^{ij}\|_{\text{HS}}^2 p_\sigma(iv)\,dv = \int_0^\infty |\varphi(v)|^2 p_\sigma(iv)^{-1}\,dv.$$

Summing on i and j, we obtain the lemma.

To prove Theorem 14.27, we compute $\int_G \|\Phi(\varphi, x)\|_{\text{HS}}^2\,dx$ a second time. We use the KAK decomposition and the asymptotic expansion. Proposition 5.28 shows that Haar measure on G is a multiple of

$$D_A(a)\,dk_1\,da\,dk_2 \qquad \text{if} \quad x = k_1ak_2,$$

where $D_A(a)$ is a linear combination of exponentials on A, the largest of which is $e^{2\rho \log a}$. To simplify notation, let us think of Haar measure on G as normalized so that $e^{2\rho \log a}\,dk_1\,da\,dk_2$, with coefficient one, is the dominant term in dx. If we write $\rho = c\nu_0$, then our parametrizations are such that this dominant term becomes $e^{2ct}\,dk_1\,dt\,dk_2$.

From the end of §3, we can write our asymptotic expansion with $m = 1$ as

$$\begin{aligned}|e^{ct}\Phi(iv, \exp tH'_\beta) &- e^{ivt}e^*eA(\bar{S}:S:\sigma:iv) \\ &- e^{-ivt}U(w)A(S:\bar{S}:\sigma:iv)e^*eU(w)^{-1}| \leq P(v)e^{-\varepsilon t}.\end{aligned}$$

Put
$$\hat{\varphi}(t) = \int_{-\infty}^\infty e^{ivt}\varphi(v)\,dv,$$

so that $\|\hat{\varphi}\|_2^2 = 2\pi\|\varphi\|_2^2$. Fix $\nu_1 > 0$ and let E_1, E_2, E_3 denote error terms to be computed separately. Since $\Phi(\varphi, k)$ is a unitary operator that factors out, we have

$$\begin{aligned}
&\int_G \|\Phi(\varphi, x)\|_{\mathrm{HS}}^2\,dx \\
&\quad = \int_0^\infty \|\Phi(\varphi, \exp tH'_\beta)\|_{\mathrm{HS}}^2 e^{2ct}(1 + \text{lower})\,dt \\
&\quad = \int_0^\infty (1+\text{lower})\bigg\|\int_{-\infty}^{\infty} [e^{i\nu t}e^*eA(\bar{S}:S:\sigma:i\nu) \\
&\qquad + e^{-i\nu t}U(w)A(S:\bar{S}:\sigma:i\nu)e^*eU(w)^{-1}]\varphi(\nu)\,d\nu\bigg\|_{\mathrm{HS}}^2 dt + E_1 \\
&\quad = \int_0^\infty \bigg\|\int_{-\infty}^{\infty} [e^{i\nu t}e^*eA(\bar{S}:S:\sigma:i\nu) \\
&\qquad + e^{-i\nu t}U(w)A(S:\bar{S}:\sigma:i\nu)e^*eU(w)^{-1}]\varphi(\nu)\,d\nu\bigg\|_{\mathrm{HS}}^2 dt + E_1 + E_2 \\
&\quad = \int_0^\infty \bigg\|\int_{-\infty}^{\infty} [e^{i\nu t}e^*eA(\bar{S}:S:\sigma:i\nu_1) \\
&\qquad + e^{-i\nu t}U(w)A(S:\bar{S}:\sigma:i\nu_1)e^*eU(w)^{-1}]\varphi(\nu)\,d\nu\bigg\|_{\mathrm{HS}}^2 dt + E_1 + E_2 + E_3 \\
&\quad = \int_0^\infty \|\hat{\varphi}(t)e^*eA(\bar{S}:S:\sigma:i\nu_1) \\
&\qquad + \hat{\varphi}(-t)U(w)A(S:\bar{S}:\sigma:i\nu_1)e^*eU(w)^{-1}\|_{\mathrm{HS}}^2\,dt + E_1 + E_2 + E_3.
\end{aligned} \tag{14.52}$$

Put $\eta(i\nu_1) = \eta(\bar{S}:S:\sigma:i\nu_1)$. Since

$$A(\bar{S}:S:\sigma:i\nu_1)^* = A(S:\bar{S}:\sigma:i\nu_1),$$

we have
$$\|e^*eA(\bar{S}:S:\sigma:i\nu_1)\|_{\mathrm{HS}}^2 = \eta(i\nu_1)\|e^*e\|_{\mathrm{HS}}^2$$
and
$$\|U(w)A(S:\bar{S}:\sigma:i\nu_1)e^*eU(w)^{-1}\|_{\mathrm{HS}}^2 = \eta(i\nu_1)\|e^*e\|_{\mathrm{HS}}^2.$$

Denoting the cross terms of (14.52) as E_4, we can rewrite (14.52) as

$$\begin{aligned}
&= \eta(i\nu_1)\|e^*e\|_{\mathrm{HS}}^2 \int_{-\infty}^{\infty} |\hat{\varphi}(t)|^2\,dt + E_1 + E_2 + E_3 + E_4 \\
&= \eta(i\nu_1)\|e^*e\|_{\mathrm{HS}}^2 2\pi \int_{-\infty}^{\infty} |\varphi(\nu)|^2\,d\nu + E_1 + E_2 + E_3 + E_4.
\end{aligned}$$

This expression equals the right side of the equation in Lemma 14.29. Let φ run through a sequence φ_n of functions with L^2 norm equal to 1 and with φ vanishing off the set $|\nu - \nu_1| \leq 1/n$. In the limit we obtain

$$(\dim \mathscr{H}_F)^2 p_\sigma(i\nu_1)^{-1} = \eta(i\nu_1)\|e^*e\|_{\mathrm{HS}}^2 2\pi + \lim_n (E_1 + E_2 + E_3 + E_4).$$

This is the desired equality for $\nu_1 > 0$ if it is shown that

$$\lim_n (E_1 + E_2 + E_3 + E_4) = 0.$$

Then the equality for other values of ν_1 follows by analytic continuation.

The terms E_1 and E_2 are easily seen to tend to 0 since $\|\varphi_n\|_1 \to 0$ by the Schwarz inequality. Term E_3 is seen to tend to 0 by using the Plancherel theorem for the line, the continuity in ν of the intertwining operators, and the small support of φ_n. With E_4, the norms are harmless and the problem is to show that

$$\lim_n \int_0^\infty \hat{\varphi}_n(t)\overline{\hat{\varphi}_n(-t)}\, dt = 0,$$

which is an exercise in the use of the Hilbert transform. See the bibliographical notes. This completes the proof of Theorem 14.27.

Corollary 14.30. Let $S = MAN$ be a parabolic subgroup with $\dim A = 1$, and let σ be a discrete series representation of M. Then $U(S, \sigma, 0)$ is reducible if and only if

(a) $|W(A:G)| = 2$,
(b) $w\sigma \cong \sigma$ if w represents the nontrivial element of $W(A:G)$, and
(c) $p_\sigma(0) \neq 0$.

Proof. We combine Theorem 14.27 with the results on irreducibility and reducibility, Theorem 14.15 and Corollary 14.25.

Ultimately we shall use the Plancherel densities from the case $\dim A = 1$ to describe reducibility in the general case. For now, we introduce appropriate notation. Let $S = MAN$ be a general parabolic subgroup for which $\mathfrak{m}$ has a compact Cartan subalgebra $\mathfrak{b}^-$, and let $\sigma = \pi^M(\lambda, \chi)$ be a discrete series representation of M. For each reduced N-positive root α of $(\mathfrak{g}, \mathfrak{a})$, we form the group $G^{(\alpha)}M$ built from the root spaces corresponding to multiples of α. For ν in $(\mathfrak{a}')^{\mathbb{C}}$ we define

$$\mu_{\sigma,\alpha}(\nu) = c_\sigma^{(\alpha)-1} p_\sigma^{(\alpha)}(\nu|_{\mathbb{R}H_\alpha}),$$

where $c_\sigma^{(\alpha)}$ and $p_\sigma^{(\alpha)}$ are the constant and the Plancherel density of (14.45) for the group $G^{(\alpha)}M$. Thus $\mu_{\sigma,\alpha}(\nu)$ is given by the product of the polynomial

$$\prod_{\substack{\varepsilon|_{\mathfrak{a}} = c\alpha \\ c > 0}} \langle \lambda + \nu, \varepsilon \rangle$$

by either 1 or a tanh-coth factor. The root β in (14.45c) is to be α if 2α is not a root of $(\mathfrak{g}, \mathfrak{a})$ and is 2α otherwise, and we can use (14.45c) without change to describe the tanh-coth factor provided we interpret ρ_β as half the sum of all members of $\Delta((\mathfrak{a} \oplus \mathfrak{b}^-)^{\mathbb{C}} : \mathfrak{g}^{\mathbb{C}})$ whose restriction to $\mathfrak{a}$ is a positive multiple of β. We call $\mu_{\sigma,\alpha}$ the **Plancherel factor** associated to σ and the reduced root α. For α not reduced, we define $\mu_{\sigma,\alpha}$ to be $\mu_{\sigma,t\alpha}$, where t is the unique nonzero real number such that $t\alpha$ is reduced.

If ν_0 is imaginary, the subset R'_{σ,ν_0} of W_{σ,ν_0} defined in Theorem 14.24 can be redefined in terms of Plancherel factors as follows:

$$R'_{\sigma,\nu_0} = \{r \in W_{\sigma,\nu_0} | \mu_{\sigma,\alpha}(\nu_0) \neq 0 \text{ for all reduced } \alpha > 0 \text{ with } r\alpha < 0\}. \tag{14.53}$$

This formula follows from Proposition 14.13b and Theorem 14.27.

§8. Harish-Chandra's Completeness Theorem

The main result of this section is the completeness of the standard self-intertwining operators for a representation unitarily induced from discrete series.

Theorem 14.31. Let G be linear connected reductive with compact center, let $S = MAN$ be a parabolic subgroup, let σ be a discrete series representation of M, and let ν be an imaginary-valued member of $(\mathfrak{a}')^{\mathbb{C}}$. Then the linear span of the unitary operators $\sigma(w)\mathscr{A}_S(w, \sigma, \nu)$ for $w \in W_{\sigma,\nu}$ coincides with the algebra of bounded operators commuting with $U(S, \sigma, \nu)$.

Remarks. Consequently the dimension of the commuting algebra is $\leq |W_{\sigma,\nu}|$. This dimension is one if and only if $U(S, \sigma, \nu)$ is irreducible.

The proof involves several ingredients. The first is a suitable formulation, in the language of Proposition 14.2, of the element ψ that corresponds to an averaged version of $U(S, \sigma, \nu, x)$. The second is information about the image of the Schwartz space of G under the group Fourier transform. We obtain a particular ψ of interest by using an averaged version of $U(S, \sigma, \nu, x)$ that comes from Fourier inversion. The third ingredient is a formula for this particular element ψ in terms of the c functions that arise in the asymptotic expansions of Eisenstein integrals; this expression shows that we can shape ψ's as freely as possible. And the fourth is an unraveling step that shows that our ability to shape ψ's as freely as possible is equivalent with the asserted completeness. The most difficult step is the third; for that, we shall indicate why one expects some formula and we shall give the formula, but we give none of the derivation. For the other steps we shall give sketches of proof.

The first step is to work with averaged versions of $U(S, \sigma, \nu, x)$. We use notation as in §2. For f in $C^\infty_{\text{com}}(G)$, let $\tilde{f}(x) = f(x^{-1})$. We start with any f in $C^\infty_{\text{com}}(G)$ such that

$$\alpha_F *_K \tilde{f} = \tilde{f} = \tilde{f} *_K \alpha_F. \tag{14.54}$$

If we define

$$\mathbf{f}(x)(k, k') = f(k^{-1}xk') \qquad \text{for} \quad x \in G, k \in K, k' \in K,$$

then we check easily that $\mathbf{f}(x)$ is in V_F for each x and is τ-spherical:

$$\mathbf{f}(k_1xk_2) = \tau_1(k_1)\mathbf{f}(x)\tau_2(k_2). \tag{14.55}$$

Conversely if $\mathbf{f}$ is a smooth function of compact support that has values in V_F and is τ-spherical, then f, defined by $f(x) = \mathbf{f}(x)(1, 1)$, is such that $\tilde{f}$ satisfies (14.54).

Given f or $\mathbf{f}$ as above, we define $\mathbf{f}^{(S)}$ by means of the convergent integral

$$\mathbf{f}^{(S)}(ma) = e^{\rho \log a} \int_N \mathbf{f}(man)\, dn.$$

This definition is suggestive of (10.22), but it is different in that no average over K takes place here. Finally we put

$$\mathbf{f}_\nu^{(S)}(m) = \int_A e^{\bar{\nu} \log a}\mathbf{f}^{(S)}(ma)\, da = \int_{AN} e^{(\bar{\nu}+\rho) \log a}\mathbf{f}(man)\, da\, dn.$$

For each m in M, $\mathbf{f}_\nu^{(S)}(m)$ is a member of V_F. We shall convolve the two functions $\mathbf{f}_\nu^{(S)}$ and ψ_T on M; in the convolution the values at each m are to be multiplied as kernels, in the manner of Remark 3 after Proposition 14.2.

Lemma 14.32. If f in $C_{\text{com}}^\infty(G)$ satisfies (14.54) and if ψ_T is in ${}^0\mathscr{C}_\sigma(M, \tau_M)$, then

$$\mathbf{f}_\nu^{(S)} * \psi_T = \psi_{TU(S,\sigma,\nu,\tilde{f})}.$$

Proof. Using (14.13) and putting $U(x) = U(S, \sigma, \nu, x)$, we have

$$\begin{aligned}
&\mathbf{f}_\nu^{(S)} * \psi_T(m)(k_1, k_2)\\
&\quad = d_\sigma \int_{KMAN} f(k_1^{-1}mank)e^{(\bar{\nu}+\rho)\log a}\\
&\qquad \times \sum_i (\sigma(m^{-1}m_0)Th_i(k_2^{-1}), h_i(k^{-1}))_{V^\sigma}\, dk\, dm\, da\, dn\\
&\quad = d_\sigma \int_{KMAN} f(k_1^{-1}mank)\\
&\qquad \times \sum_i (\sigma(m_0)Th_i(k_2^{-1}), h_i(k^{-1}n^{-1}a^{-1}m^{-1}))_{V^\sigma}\, dk\, dm\, da\, dn\\
&\quad = d_\sigma \int_G f(k_1^{-1}x) \sum_i (\sigma(m_0)Th_i(k_2^{-1}), h_i(x^{-1}))_{V^\sigma}\, dx\\
&\quad = d_\sigma \int_G f(x) \sum_i (\sigma(m_0)Th_i(k_2^{-1}), U(x)h_i(k_1^{-1}))_{V^\sigma}\, dx\\
&\quad = d_\sigma \sum_i (\sigma(m_0)Th_i(k_2^{-1}), U(\tilde{f})h_i(k_1^{-1}))_{V^\sigma}.
\end{aligned}$$

Equation (14.54) implies that $U(\bar{f})$ maps $\mathscr{H}_F$ to itself, and hence the above expression is

$$= d_\sigma \sum_i (\sigma(m_0)eU(k_2)Th_i, eU(k_1)U(\bar{f})h_i)_{V^\sigma}$$
$$= d_\sigma \operatorname{Tr}(U(\bar{f})^* U(k_1)^{-1} e^* \sigma(m_0)eU(k_2)T)$$
$$= d_\sigma \operatorname{Tr}(e^*\sigma(m_0)eU(k_2)TU(\tilde{f})U(k_1)^{-1})$$
$$= \psi_{TU(\tilde{f})},$$

the last step holding by Proposition 14.2.

Now consider the space $\mathscr{C}(M, \tau_M)$ of all Schwartz functions from M to V_F such that (14.12) holds. The space ${}^0\mathscr{C}_\sigma(M, \tau_M)$ is a finite-dimensional subspace, and we let E_σ be the orthogonal projection relative to L^2.

Lemma 14.33. E_σ is a continuous projection of $\mathscr{C}(M, \tau_M)$ onto ${}^0\mathscr{C}_\sigma(M, \tau_M)$ and is given by convolution with $\Theta_{\sigma,F_M}(m)$, where F_M is any finite set of $K \cap M$ types containing all $K \cap M$ types occurring in the reduction of members of F and where

$$\Theta_{\sigma,F_M}(m) = \operatorname{Tr} E_{F_M}\sigma(m)E_{F_M}.$$

Proof. Let $\{v_k\}$ be an orthonormal basis of image E_{F_M}, so that

$$\Theta_{\sigma,F_M}(m) = \sum_k (\sigma(m)v_k, v_k),$$

and let $\tilde{E}_\sigma$ be the operation of convolution by $\Theta_{\sigma,F_M}(m)$. Calculation with Proposition 9.6 gives $\tilde{E}_\sigma(\sigma(m)v_i, v_j) = (\sigma(m)v_i, v_j)$. In particular, $\tilde{E}_\sigma(\Theta_{\sigma,F_M}) = \Theta_{\sigma,F_M}$, and so $\tilde{E}_\sigma$ is a projection. The image of $\tilde{E}_\sigma$ on $\mathscr{C}(M, \tau_M)$ certainly contains ${}^0\mathscr{C}_\sigma(M, \tau_M)$. If we apply Corollary 8.42 and Proposition 12.16b, we see that the image of $\tilde{E}_\sigma$ is contained in ${}^0\mathscr{C}_{\Sigma\sigma_i}(M, \tau_M)$, in which matrix coefficients are allowed from a sum of finitely many discrete series. However, distinct discrete series have orthogonal matrix coefficients (by the Plancherel formula for M, which sets up an isometric mapping that must necessarily preserve orthogonality). Thus the image is just ${}^0\mathscr{C}_\sigma(M, \tau_M)$. The L^2 adjoint of convolution with $h(m)$ is convolution with $\overline{h(m^{-1})}$; since $\Theta_{\sigma,F_M}(m)$ is fixed under this operation, $\tilde{E}_\sigma^* = \tilde{E}_\sigma$. Thus $\tilde{E}_\sigma$ coincides with E_σ.

Let us apply the operator E_σ of Lemma 14.33 to the function $\mathbf{f}_\nu^{(S)}$ of Lemma 14.32. Then we have

$$E_\sigma(\mathbf{f}_\nu^{(S)}) * \psi_T = E_\sigma(\mathbf{f}_\nu^{(S)} * \psi_T) = E_\sigma\psi_{TU(S,\sigma,\nu,\tilde{f})} = \psi_{TU(S,\sigma,\nu,\tilde{f})}$$

since E_σ is given by convolution and since $E_\sigma\psi = \psi$. Now $E_\sigma(\mathbf{f}_\nu^{(S)})$ is $\psi_{T'}$ for some T', by Proposition 14.2. If we take $T = I$, we conclude from

(14.14) that

$$E_\sigma(\mathbf{f}_\nu^{(S)}) = \psi_{U(S,\sigma,\nu,\check{f})}, \tag{14.56}$$

which is the formula we seek that incorporates an averaged version of $U(S, \sigma, \nu, x)$.

The second step is to construct a specific function $\mathbf{f}$ by Fourier inversion.

Lemma 14.34. If φ is in $C_{\text{com}}^\infty(i\mathfrak{a}')$ and ψ is in ${}^0\mathscr{C}_\sigma(M, \tau_M)$, then the function

$$\mathbf{f}_\varphi(x) = \int_{i\mathfrak{a}'} \eta(\bar{S}:S:\sigma:\nu)^{-1}\varphi(\nu)E(S:\psi:\nu:x)\,d\nu$$

is a τ-spherical Schwartz function.

Idea of proof. If φ is supported in the regular set, then the Eisenstein integral is the sum of two terms, the leading term of the asymptotic expansion and the error terms. The leading term contributes the Fourier transform of φ in the A variable, which is rapidly decreasing, and the error terms contribute rapidly decreasing terms. There may be singularities in the components of the leading term (the c functions) if the support of φ meets the singular set, but the factor η^{-1} cancels these out, and the same argument applies. Finally $\mathbf{f}$ is τ-spherical because the Eisenstein integral is τ-spherical.

The third step is to obtain a formula for $\mathbf{f}_{\varphi,\nu}^{(S)}$ when $\mathbf{f}_\varphi$ is as in Lemma 14.34. An important property of the Schwartz space that is suggested by Theorem 12.17 is that $\mathbf{f} \to \mathbf{f}^{(S)}$ carries Schwartz functions on G to Schwartz functions on MA. The Fourier transform in the A variable then carries Schwartz functions of the pair (m, a) to Schwartz functions of (m, ν), ν imaginary. Consequently for any fixed imaginary ν_0,

$$T(\varphi) = E_\sigma(\mathbf{f}_\varphi)_{\nu_0}^{(S)} \tag{14.57}$$

is a distribution on $C_{\text{com}}^\infty(i\mathfrak{a}')$ with values in ${}^0\mathscr{C}_\sigma(M, \tau_M)$.

This distribution measures the image of the group Fourier transform $f \to \{U(S, \sigma, \nu, f)\}$. To see this, let us write the Plancherel formula of Chapter XIII crudely as

$$f(1) = \sum_{S',\sigma'} \int_\nu \operatorname{Tr}(U(S', \sigma', \nu, f))\,d\mu_{S',\sigma'}(\nu).$$

Taking f to be the function $y \to f_\varphi(x^{-1}y)$, we obtain

$$\check{f}_\varphi(x) = f_\varphi(x^{-1}) = \sum_{S',\sigma'\nu} \int_\nu \operatorname{Tr}(U(S', \sigma', \nu, x)U(S', \sigma', \nu, f_\varphi))\,d\mu_{S',\sigma'}(\nu).$$

We can write the function ψ in the definition of $\mathbf{f}_\varphi$ as $\psi = \psi_X$ for some X in End $\mathscr{H}_F$, by Proposition 14.2. Then Proposition 14.3 shows that f_φ

is given by

$$f_\varphi(x) = \mathbf{f}_\varphi(x)(1, 1) = d_\sigma \int_{i\mathfrak{a}'} \eta(\bar{S}:S:\sigma:\nu)^{-1}\varphi(\nu) \operatorname{Tr}(E_F U(S, \sigma, \nu, x) X E_F)\, d\nu. \tag{14.58}$$

From §7 it is reasonable to expect that

$$d\mu_{S,\sigma}(\nu) = d_\sigma \eta(\bar{S}:S:\sigma:\nu)^{-1}\, d\nu, \qquad (\nu \in i\mathfrak{a}')$$

except for a constant factor depending on σ. If $\{\varphi(\nu)X\}$ is in the image of the group Fourier transform in the sense that there is a function $g(x)$ with

$$U(S', \sigma', \nu, g) = \begin{cases} \varphi(\nu)E_F X E_F & \text{if } S' = S, \sigma' = \sigma \\ 0 & \text{otherwise,} \end{cases}$$

then the inversion formula says that $\tilde{g}(x)$ must be given by the right side of (14.58), and thus we must have $\tilde{g} = f_\varphi$ and $g = \tilde{f}_\varphi$. Substituting into (14.56) and (14.57), we obtain

$$T(\varphi) = E_\sigma(\mathbf{f}_\varphi)\nu_0^{(S)} = \psi_{U(S,\sigma,\nu_0,\tilde{f}_\varphi)} = \psi_{U(S,\sigma,\nu_0,g)} = \varphi(\nu_0)\psi_X. \tag{14.59}$$

Under our assumption on the image of the group Fourier transform, T thus has one-point support and is given by (14.59).

This assumption about the image of the group Fourier transform is not quite right, because it does not take into account the known equivalences among representations $U(S', \sigma', \nu)$. The following lemma allows us to control matters without this assumption.

Lemma 14.35. If T is as in (14.57), then the support of T is contained in the finite set $\{s\nu_0 \mid s \in W(A:G)\}$.

Proof. By continuity of T as a function of ν_0, we may assume ν_0 is regular. Let $\mathfrak{b}^-$ be a compact Cartan subalgebra of $\mathfrak{m}$, let $\mathfrak{t} = \mathfrak{a} \oplus \mathfrak{b}^-$, let $\gamma = \gamma_G$ be the Harish-Chandra homomorphism relative to $\mathfrak{t}$, and let $\lambda \in (i\mathfrak{b}^-)'$ be the infinitesimal character of σ. For z in $Z(\mathfrak{g}^{\mathbb{C}})$, let p_z be the polynomial of ν in $i\mathfrak{a}'$ given by

$$p_z(\nu) = \gamma_{\lambda+\nu}(z).$$

As a consequence of Proposition 14.3, we know that

$$zE(S:\psi:\nu:\cdot) = p_z(\nu)E(S:\psi:\nu:\cdot).$$

If we integrate both sides with $\eta^{-1}\varphi\, d\nu$ and group φ with p_z, we see that

$$z\mathbf{f}_\varphi = \mathbf{f}_{p_z\varphi}.$$

Applying the remaining operators that define T, we obtain

$$T(p_z\varphi) = E_\sigma(z\mathbf{f}_\varphi)_{\nu_0}^{(S)}. \tag{14.60}$$

On the other hand, if $\mu: Z(\mathfrak{g}^{\mathbb{C}}) \to Z((\mathfrak{m} \oplus \mathfrak{a})^{\mathbb{C}})$ is the natural map $\gamma_{MA}^{-1} \circ \gamma_G$, then it is clear from the definition that

$$(z\mathbf{f})^{(S)} = \mu(z)\mathbf{f}^{(S)}.$$

Substituting into (14.60), we obtain

$$T(p_z\varphi) = E_\sigma(\mu(z)\mathbf{f}_\varphi^{(S)})_{\nu_0} = p_z(\nu_0)E_\sigma(\mathbf{f}_\varphi^{(S)})_{\nu_0} = p_z(\nu_0)T(\varphi).$$

That is,

$$T((p_z - p_z(\nu_0))\varphi) = 0$$

for all $\varphi \in C_{\text{com}}^\infty(i\mathfrak{a}')$. Allowing φ to be an arbitrary function supported where $p_z - p_z(\nu_0)$ is nonvanishing, we see that

$$\operatorname{supp} T \subseteq \{\nu \,|\, p_z(\nu) = p_z(\nu_0)\}.$$

By Proposition 8.20,

$$\operatorname{supp} T \subseteq \{\nu \,|\, w(\lambda + \nu) = \lambda + \nu_0 \text{ for some } w \in W(\mathfrak{t}^{\mathbb{C}}:\mathfrak{g}^{\mathbb{C}})\}.$$

This equation implies that $w\lambda = \lambda$ and $w\nu = \nu_0$, since ν and ν_0 are imaginary. The remainder of the argument consists in showing that $w\nu = \nu_0$ with ν_0 regular imaginary implies ν and ν_0 are conjugate via $W(A:G)$. See Problem 7 at the end of the chapter.

With the aid of Lemma 14.35, one can derive an explicit formula for the distribution T. The result is given in the lemma below. The derivation of the formula is complicated and will be omitted.

Lemma 14.36. There is a nonzero constant c such that

$$E_\sigma(\mathbf{f}_\varphi)_\nu^{(S)} = cE_\sigma\left(\sum_{s \in W(A:G)} \varphi(s^{-1}\nu)c_{S|S}(1:\nu)^{-1}c_{S|S}(s:s^{-1}\nu)\psi\right) \tag{14.61}$$

for all φ in $C_{\text{com}}^\infty(i\mathfrak{a}')$ and all imaginary ν.

Remarks.

(1) Each term on the right side is well defined for all imaginary ν, as will follow from Lemma 14.37a.

(2) Actually the formula remains valid even if E_σ is dropped from both sides. The effect of E_σ on the right is to drop all terms except those for which $s[\sigma] = [\sigma]$, as will be apparent below.

Now we come to the fourth step, which is to see how general a member of ${}^0\mathscr{C}_\sigma(M, \tau_M)$ can occur on the right side of (14.61) and to unwind matters in terms of intertwining operators. We begin with more identities relating c functions and intertwining operators.

Lemma 14.37. If s in $W(A:G)$ has w in $N_K(\mathfrak{a})$ as a representative, then

(a) $c_{S|S}(1:\nu)^{-1}c_{S|S}(s:s^{-1}\nu)\psi_T^\sigma = \psi_{T'}^{w\sigma}$ with

$$T' = \mathscr{A}_S(w, \sigma, w^{-1}\nu)T\mathscr{A}_S(w, \sigma, w^{-1}\nu)^{-1}.$$

(b) $\psi_{T'}^{w\sigma} = \psi_{T''}^\sigma$ if $w\sigma$ is equivalent with σ and $T'' = \sigma(w)T'\sigma(w)^{-1}$, where $\sigma(w)$ is defined as in Lemma 14.22.

Remarks. To prove (a), we use (14.30) and sort out the η functions that result. To prove (b), we argue directly from Proposition 14.2. Lemma 14.37a makes it clear that the terms in (14.61) with $s[\sigma] \neq [\sigma]$ drop out.

Proof of Theorem 14.31. Let $T \in \operatorname{End} \mathscr{H}_F$ be an operator that commutes with all operators $\sigma(w)\mathscr{A}_S(w, \sigma, \nu)$, $[w] \in W_{\sigma,\nu}$. Lemma 14.37 shows that

$$c_{S|S}(1:\nu)^{-1}c_{S|S}(s:s^{-1}\nu)\psi_T = \psi_T$$

for all s in $W_{\sigma,\nu}$. Choose φ in $C_{\text{com}}^\infty(i\mathfrak{a}')$ so that $\varphi(\nu) = c^{-1}$ and so that $\varphi(s\nu) = 0$ for $s \in W_{\sigma,\nu}$; here c is the constant in (14.59). Then Lemma 14.36 shows that

$$E_\sigma(\mathbf{f}_\varphi)_\nu^{(S)} = \psi_T.$$

Since $C_{\text{com}}^\infty(G)$ is dense in $\mathscr{C}(G)$, we can choose τ-spherical smooth functions $\mathbf{f}_n$ of compact support with $\mathbf{f}_n \to \mathbf{f}_\varphi$. In the notation of (14.56), we have

$$E_\sigma(\mathbf{f}_n)_\nu^{(S)} = \psi_{U(S,\sigma,\nu,\tilde{f}_n)}$$

for all n. By continuity of the map $\mathbf{f} \to E_\sigma(\mathbf{f})_\nu^{(S)}$, we conclude that

$$\psi_T = \lim \psi_{U(S,\sigma,\nu,\tilde{f}_n)},$$

hence that

$$T = \lim U(S, \sigma, \nu, \tilde{f}_n).$$

Thus T is in the closure of $E_F U(S, \sigma, \nu, C_{\text{com}}^\infty(G))E_F$. Since $\operatorname{End} \mathscr{H}_F$ is finite-dimensional, T is in $E_F U(S, \sigma, \nu, C_{\text{com}}^\infty(G))E_F$.

Now we take centralizers, applying the Double Commutant Theorem: Let $\mathscr{A}$ be the self-adjoint algebra of operators generated by all $\sigma(w)\mathscr{A}_S(w, \sigma, \nu)$ for $[w] \in W_{\sigma,\nu}$, and let $\mathscr{A}_F = E_F\mathscr{A}E_F$. Since

$$E_F U(S, \sigma, \nu, C_{\text{com}}^\infty(G))E_F$$

is the algebra of operators on $\mathscr{H}_F$ commuting with $\mathscr{A}_F$, $\mathscr{A}_F$ is the algebra of operators on $\mathscr{H}_F$ commuting with $E_F U(S, \sigma, \nu, C_{\text{com}}^\infty(G))E_F$. Since $\mathscr{A}$ is finite-dimensional, it follows readily that $\mathscr{A}$ is the algebra of bounded operators on the induced space commuting with $U(S, \sigma, \nu, C_{\text{com}}^\infty(G))$, hence commuting with $U(S, \sigma, \nu, G)$.

To complete the proof, it is necessary only to observe that the linear span of all $\sigma(w)\mathscr{A}_S(w, \sigma, \nu)$ for $[w] \in W_{\sigma,\nu}$ is already a self-adjoint algebra. The closure under adjoints follows from Proposition 14.21. The closure under products follows from the following lemma and thereby completes the proof of Theorem 14.31.

Lemma 14.38. For ν imaginary, let w_1 and w_2 be representatives in $N_K(\mathfrak{a})$ of members of $W_{\sigma,\nu}$.

(a) If w_1 and w_2 are in a cyclic extension of $K \cap M$ and if $\sigma(w_1)$ and $\sigma(w_2)$ are compatibly defined, then

$$\sigma(w_1)\mathscr{A}_S(w_1, \sigma, \nu)\sigma(w_2)\mathscr{A}_S(w_2, \sigma, \nu) = \sigma(w_1 w_2)\mathscr{A}_S(w_1 w_2, \sigma, \nu).$$

(b) Whether or not w_1 and w_2 are in a cyclic extension of $K \cap M$,

$$\sigma(w_1)\mathscr{A}_S(w_1, \sigma, \nu)\sigma(w_2)\mathscr{A}_S(w_2, \sigma, \nu) = c\sigma(w_1 w_2)\mathscr{A}_S(w_1 w_2, \sigma, \nu)$$

with c a constant satisfying $|c| = 1$ and given by

$$\sigma(w_1 w_2)^{-1}\sigma(w_1)\sigma(w_2) = cI.$$

Remarks. Lemma 14.22 allows us to define $\sigma(w_1)$ or $\sigma(w_2)$ or $\sigma(w_1 w_2)$, but it does not say that we can define the three operators compatibly.

Proof. We drop the S and the ν in the notation, since they are unaffected throughout. Proposition 14.21b gives

$$\begin{aligned}\mathscr{A}(w_1 w_2, \sigma)^{-1}&\sigma(w_1 w_2)^{-1}\sigma(w_1)\mathscr{A}(w_1, \sigma)\sigma(w_2)\mathscr{A}(w_2, \sigma)\\ &= \mathscr{A}(w_1 w_2, \sigma)^{-1}[\sigma(w_1 w_2)^{-1}\sigma(w_1)\sigma(w_2)]\\ &\quad\times \mathscr{A}(w_1, \sigma(w_2)^{-1}\sigma\sigma(w_2))\mathscr{A}(w_2, \sigma).\end{aligned}$$

It is easy to show that the expression in brackets commutes with $\sigma(m)$ for all m and hence is a scalar cI, by Schur's Lemma. Moreover, the constant c is 1 under the assumptions of (a). In any event, the right side collapses to

$$c\mathscr{A}(w_1 w_2, \sigma)^{-1}\mathscr{A}(w_1, w_2\sigma)\mathscr{A}(w_2, \sigma)$$

and equals cI by Proposition 14.21c.

§9. *R* Group

Theorem 14.31 says that the commuting algebra of $U(S, \sigma, \nu)$ is spanned by the operators $\sigma(w)\mathscr{A}_S(w, \sigma, \nu)$ for $w \in W_{\sigma,\nu}$ when σ is in the discrete series and ν is imaginary. Therefore the dimension of the commuting algebra is $\leq |W_{\sigma,\nu}|$. Theorem 14.24 identifies a lower bound for this dimension as

$|R'_{\sigma,\nu}|$, and formula (14.53) redefines the set $R'_{\sigma,\nu}$ in terms of Plancherel factors. In this section we determine the dimension exactly.

The answer will be $|R'_{\sigma,\nu}|$ for a very strong reason: that the members of $R'_{\sigma,\nu}$ provide all the distinct operators, apart from scalar factors, corresponding to members of $W_{\sigma,\nu}$. In fact, every member of $W_{\sigma,\nu}$ will be expressible as a product of a member of $R'_{\sigma,\nu}$ by a member of $W_{\sigma,\nu}$ whose corresponding operator is scalar; then Lemma 14.38 tells us that the operators multiply in corresponding fashion, except for scalar factors. This decomposition suggests that $R'_{\sigma,\nu}$ might be a group, and we shall in fact characterize $R'_{\sigma,\nu}$ in a way that exhibits it as a group.

Example. G = double covering of $SO_0(4, 4)$, $S = MAN$ minimal parabolic.

This group is split over $\mathbb{R}$, in the sense that $\mathfrak{a}$ is a Cartan subalgebra. Its roots (= restricted roots) form a system Δ of type D_4, and we use the standard notation of Chapter IV in referring to this system. Then M is the 16-element abelian group generated by the elements γ_α for α simple, each of which has order 2. We can define a (one-dimensional) representation σ by deciding, for each simple α, whether $\sigma(\gamma_\alpha) = +1$ or $\sigma(\gamma_\alpha) = -1$. We shall take $\nu = 0$, and we consider $U(S, \sigma, 0)$.

Let us take $\sigma(\gamma_\alpha) = -1$ for each simple root α, and let us determine $W_{\sigma,0}$. If α, β, and $\alpha + \beta$ are all roots, then we readily see from the equality of their lengths that $\sigma(\gamma_\alpha)\sigma(\gamma_\beta) = \sigma(\gamma_{\alpha+\beta})$. Consequently if β in Δ is the sum of n simple roots, then $\sigma(\gamma_\beta) = (-1)^n$. If w is in $W(A:G)$, then

$$w\sigma(\gamma_\beta) = \sigma(w^{-1}\gamma_\beta w) = \sigma(\gamma_{w\beta}). \tag{14.62}$$

So

$$w \text{ is in } W_{\sigma,0} \Leftrightarrow \sigma(\gamma_{w\beta}) = \sigma(\gamma_\beta) \text{ for all } \beta \text{ in } \Delta. \tag{14.63}$$

First we apply this criterion to a reflection s_ε. From the definition of γ_β, we see that

$$\gamma_{s_\varepsilon\beta} = \gamma_\beta\gamma_\varepsilon^{2\langle\varepsilon,\beta\rangle/|\beta|^2}. \tag{14.64}$$

Thus s_ε is in $W_{\sigma,0}$ if and only if

$$\sigma(\gamma_\varepsilon)^{2\langle\varepsilon,\beta\rangle/|\beta|^2} = +1$$

for all β. Since we can find β for which the exponent is 1, the condition is that $\sigma(\gamma_\varepsilon) = +1$. The roots ε in question form the set $\Delta'_{\sigma,0}$ given by

$$\Delta'_{\sigma,0} = \{\pm(e_1 - e_3), \pm(e_1 + e_3), \pm(e_2 - e_4), \pm(e_2 + e_4)\}, \tag{14.65}$$

which is a root system of type $A_1 \times A_1 \times A_1 \times A_1$.

Let us calculate $\sigma(s_\varepsilon)\mathscr{A}_S(s_\varepsilon, \sigma, 0)$ for some root ε in $\Delta'_{\sigma,0}$, say $\varepsilon = e_1 - e_3$, without bothering to choose representatives in K for Weyl group elements.

Using the rules for manipulating intertwining operators, we have

$$\begin{aligned}
&\sigma(s_\varepsilon)\mathscr{A}_S(s_\varepsilon, \sigma, 0)\\
&= \sigma(s_{e_1-e_3})\mathscr{A}_S(s_{e_1-e_2}, s_{e_2-e_3}s_{e_1-e_2}\sigma, 0)\mathscr{A}_S(s_{e_2-e_3}, s_{e_1-e_2}\sigma, 0)\mathscr{A}_S(s_{e_1-e_2}, \sigma, 0)\\
&= \mathscr{A}_S(s_{e_1-e_2}, s_{e_1-e_2}\sigma, 0)[\sigma(s_{e_1-e_3})\mathscr{A}_S(s_{e_2-e_3}, s_{e_1-e_2}\sigma, 0)]\mathscr{A}_S(s_{e_1-e_2}, \sigma, 0)\\
&= \mathscr{A}_S(s_{e_1-e_2}, s_{e_1-e_2}\sigma, 0)[(s_{e_1-e_2}\sigma)(s_{e_2-e_3})\mathscr{A}_S(s_{e_2-e_3}, s_{e_1-e_2}\sigma, 0)]\\
&\quad\times \mathscr{A}_S(s_{e_1-e_2}, \sigma, 0).
\end{aligned}$$

The operators on the ends of the right side are inverse to one another, and the operator in brackets can be regarded as an operator for SL(2, $\mathbb{R}$) by virtue of Proposition 7.11. Since $s_{e_1-e_2}\sigma$ is trivial on $\gamma_{e_2-e_3}$, the representation of the M in SL(2, $\mathbb{R}$) is trivial; then the induced representation of SL(2, $\mathbb{R}$) is irreducible, and the operator in brackets is scalar. Thus the whole operator corresponding to $s_{e_1-e_3}$ is scalar. The same thing happens for the other reflections in members of $\Delta'_{\sigma,0}$.

Now a general element of $W_{\sigma,0}$ must map $\Delta'_{\sigma,0}$ into itself, by (14.63). Composing it with a member of the Weyl group of $\Delta'_{\sigma,0}$, we may assume it leaves stable the set of positive roots of $\Delta'_{\sigma,0}$. Thus it must carry $e_1 - e_3$ to one of

$$e_1 - e_3, e_1 + e_3, e_2 - e_4, e_2 + e_4.$$

We check directly that its effect on $e_1 - e_3$ determines it completely as one of the elements

$$1, s_{e_3-e_4}s_{e_3+e_4}, s_{e_1-e_2}s_{e_3-e_4}, s_{e_1-e_2}s_{e_3+e_4}. \tag{14.66}$$

These elements are all in $R'_{\sigma,0}$. In fact, let us consider $s_{e_3-e_4}s_{e_3+e_4}$. It contributes two SL(2, $\mathbb{R}$) Plancherel factors, one corresponding to σ on $\{1, \gamma_{e_3+e_4}\}$ and one corresponding to σ on $\{1, \gamma_{e_3-e_4}\}$. In each case, the restriction of σ is nontrivial, and the Plancherel factor for SL(2, $\mathbb{R}$) is nonvanishing at 0.

Since (14.66) yields the only possible nonscalar operators from $W_{\sigma,0}$ and since $R'_{\sigma,0}$ leads to linearly independent operators, it follows that the dimension of the span of our operators is 4 and that $R'_{\sigma,0}$ consists of the four elements in (14.66). Notice that $R'_{\sigma,0}$ is in this case a group; we obtained it here as the subgroup of $W_{\sigma,0}$ preserving $\Delta'_{\sigma,0} \cap \Delta^+$.

This example is rather typical of the general case. We shall merely sketch the steps in general, correlating them with the example. There is one new ingredient, which arises because the roots of $(\mathfrak{g}, \mathfrak{a})$, when MAN is not necessarily minimal parabolic, do not necessarily form a root system. At the end of this section we shall define certain roots of $(\mathfrak{g}, \mathfrak{a})$ to be "useful." The following theorem, which we state without proof, allows us to apply all of our knowledge of abstract Weyl groups to $W(A:G)$.

Theorem 14.39. Let G be linear connected reductive with compact center, and let MAN be a parabolic subgroup with rank $M = \operatorname{rank}(K \cap M)$. If α is a root of $(\mathfrak{g}, \mathfrak{a})$, then the reflection s_α is in $W(A:G)$ if and only if α is a useful root or $\frac{1}{3}\alpha$ exists as a root of $(\mathfrak{g}, \mathfrak{a})$. The useful roots form an abstract root system in a subspace of $\mathfrak{a}'$, and $W(A:G)$ is exactly the Weyl group of this root system.

Remarks. If MAN is minimal, then every root of $(\mathfrak{g}, \mathfrak{a})$ is useful, and this theorem reduces to Theorem 5.17. In the general case, the positive multiples of a reduced positive root are $\{1\}$ or $\{1, 2\}$ or $\{1, 2, 3\}$. The last of these happens in only one simple group, and the group is not classical; in it, A has dimension one, and the clause about $\frac{1}{3}\alpha$ is included to handle only this one exceptional situation.

The main result of this section will describe a decomposition of $W_{\sigma,\nu}$ as a semidirect product $W_{\sigma,\nu} = W'_{\sigma,\nu}R_{\sigma,\nu}$, where $W'_{\sigma,\nu}$ is normal and is a Weyl group. The group $R_{\sigma,\nu}$ will turn out to coincide with the set $R'_{\sigma,\nu}$ of (14.53), and the operators $\sigma(r)\mathscr{A}_S(r, \sigma, \nu)$ for r in $R_{\sigma,\nu}$ will turn out to be a linear basis for the commuting algebra of $U(S, \sigma, \nu)$. The full result will be stated as Theorem 14.43.

We begin with the appropriate definitions. In terms of the Plancherel factors $\mu_{\sigma,\alpha}$ of §7, let

$$\Delta'_{\sigma,\nu} = \{\alpha = \text{useful root of } (\mathfrak{g}, \mathfrak{a}) \mid \mu_{\sigma,\alpha}(\nu) = 0\}.$$

Theorem 14.39 tells us that s_α is in $W(A:G)$ whenever α is in $\Delta'_{\sigma,\nu}$. Actually s_α is in the smaller group $W_{\sigma,\nu}$, as the following lemma says.

Lemma 14.40. Suppose α is a root of $(\mathfrak{g}, \mathfrak{a})$ such that $\mu_{\sigma,\alpha}(\nu) = 0$, where σ is in the discrete series of M and ν is imaginary. Then the root reflection s_α exists in $W(A:G)$ and is such that $s_\alpha\sigma \cong \sigma$, $s_\alpha\nu = \nu$, and $\sigma(s_\alpha)\mathscr{A}_S(s_\alpha, \sigma, \nu)$ is scalar for every choice of S.

Proof. We may assume α is reduced. Since $\mu_{\sigma,\alpha}(\nu)$ is a nonzero multiple of the Plancherel density $p_\sigma^{(\alpha)}(\nu|_{\mathbb{R}H_\alpha})$ for the group $G^{(\alpha)}M$, $p_\sigma^{(\alpha)}(\nu|_{\mathbb{R}H_\alpha})$ must be 0. Combining Theorem 14.27 and Proposition 14.13(e, v), we see that $\nu|_{\mathbb{R}H_\alpha} = 0$.

Let us see that s_α exists in $W(A:G)$. Assuming the contrary, let $S_\alpha = S \cap G^{(\alpha)}M$. Then we see from (14.39) for the group $G^{(\alpha)}M$ that the function $c_{S_\alpha|S_\alpha}(1:\nu|_{\mathbb{R}H_\alpha})$ for the group $G^{(\alpha)}M$ has no pole at $\nu = 0$. By Corollary 14.9c, $A(\bar{S}_\alpha:S_\alpha:\sigma:\nu|_{\mathbb{R}H_\alpha})$ has no pole at $\nu = 0$. Then it follows that $\eta^{(\alpha)}(\bar{S}_\alpha:S_\alpha:\sigma:\nu|_{\mathbb{R}H_\alpha})$ has no pole at $\nu = 0$. However, this situation contradicts Theorem 14.27 and the identity $p_\sigma^{(\alpha)}(\nu|_{\mathbb{R}H_\alpha}) = 0$.

Thus s_α exists in $W(A:G)$. Since $\nu|_{\mathbb{R}H_\alpha} = 0$, we have $s_\alpha\nu = \nu$. To see that $s_\alpha\sigma \cong \sigma$, we refer again to (14.39). If $s_\alpha\sigma$ and σ are inequivalent, then the two terms that contribute to the sum (14.39) are in orthogonal spaces, relative to $L^2(M)$. Since the sum is regular at $\nu = 0$, each must be regular. But then the regularity of $c_{S_\alpha|S_\alpha}(1:\nu|_{\mathbb{R}H_\alpha})$ at $\nu = 0$ leads to a contradiction, just as in the previous paragraph. We conclude that $s_\alpha\sigma \cong \sigma$.

Applying Corollary 14.30, we see that the representation $U(S_\alpha, \sigma, 0)$ of $G^{(\alpha)}M$ is irreducible. Therefore the self-intertwining operator $\sigma(s_\alpha)\mathscr{A}_{S_\alpha}(s_\alpha, \sigma, 0)$ is scalar. To complete the proof, we want to apply (14.35) to conclude $\sigma(s_\alpha)\mathscr{A}_S(s_\alpha, \sigma, \nu)$ is scalar. However, the assumption $\bar{\mathfrak{n}} \cap \mathrm{Ad}(s_\alpha)^{-1}\mathfrak{n} = \bar{\mathfrak{n}}^{(\alpha)}$ need not be satisfied. (This assumption is essentially that α is a simple root of $(\mathfrak{g}, \mathfrak{a})$.) So we need to adjust matters.

Thus let $S' = s_\alpha^{-1} S s_\alpha$, and choose a minimal string (in the sense of §4) from S to S', say $S = S_0, S_1, \ldots, S_r = S'$. By Lemma 14.18b,

$$\mathscr{A}(S':S:\sigma:\nu) = \mathscr{A}(S_r:S_{r-1}:\sigma:\nu) \cdots \mathscr{A}(S_1:S_0:\sigma:\nu). \tag{14.67}$$

It was noted with the definition of minimal string that the reduced N-positive roots γ_i such that $\bar{\mathfrak{n}}_{i-1} \cap \mathfrak{n}_i = \bar{\mathfrak{n}}^{(\gamma_i)}$ are exactly those that are N'-negative, and α satisfies this condition. Hence one factor on the right in (14.67), say $\mathscr{A}(S_i:S_{i-1}:\sigma:\nu)$, has $\bar{\mathfrak{n}}_{i-1} \cap \mathfrak{n}_i = \bar{\mathfrak{n}}^{(\alpha)}$. Problem 9 at the end of the chapter shows as a consequence that $\mathfrak{n}_i = \mathrm{Ad}(s_\alpha)^{-1}\mathfrak{n}_{i-1}$. Hence we can at least conclude from (14.35) that $\sigma(s_\alpha)\mathscr{A}_{S_{i-1}}(s_\alpha, \sigma, \nu)$ is a scalar operator, say cI. To be quite precise, let us write

$$\sigma(w)\mathscr{A}_{S_{i-1}}(w, \sigma, \nu) = cI \tag{14.68}$$

with w denoting a representative in $N_K(\mathfrak{a})$ of s_α. For our given parabolic subgroup S, we obtain

$$\begin{aligned}
&\sigma(w)\mathscr{A}_S(w, \sigma, \nu)\\
&\quad = \sigma(w)R(w)\mathscr{A}(w^{-1}Sw:S:\sigma:\nu)\\
&\quad = \sigma(w)R(w)\mathscr{A}(w^{-1}Sw:w^{-1}S_{i-1}w:\sigma:\nu)R(w)^{-1}\sigma(w)^{-1}\\
&\qquad \times \sigma(w)R(w)\mathscr{A}(w^{-1}S_{i-1}w:S_{i-1}:\sigma:\nu)\\
&\qquad \times \mathscr{A}(S_{i-1}:S:\sigma:\nu) \qquad \text{by Lemma 14.18b}\\
&\quad = c\sigma(w)\mathscr{A}(S:S_{i-1}:w\sigma:w\nu)\sigma(w)^{-1}\mathscr{A}(S_{i-1}:S:\sigma:\nu) \qquad \text{by (14.68) and Proposition 14.20b}\\
&\quad = c\mathscr{A}(S:S_{i-1}:\sigma:w\nu)\mathscr{A}(S_{i-1}:S:\sigma:\nu) \qquad \text{by Theorem 8.38e}\\
&\quad = cI,
\end{aligned}$$

the last step holding by Theorem 14.19 since $w\nu = \nu$. This proves the lemma.

Lemma 14.40 admits a converse, as follows.

Lemma 14.41. Suppose α is a root of $(\mathfrak{g}, \mathfrak{a})$, σ is in the discrete series of M, and ν is imaginary. If the root reflection s_α exists in $W(A:G)$ and is such that $s_\alpha\sigma \cong \sigma$, $s_\alpha\nu = \nu$, and $\sigma(s_\alpha)\mathscr{A}_S(s_\alpha, \sigma, \nu)$ is scalar for some S, then $\mu_{\sigma,\alpha}(\nu) = 0$.

Proof. Let $S' = s_\alpha^{-1}Ss_\alpha$, and choose a minimal string from S to S' as in Lemma 14.40. Again some factor $\mathscr{A}(S_i:S_{i-1}:\sigma:\nu)$ on the right side of (14.67) has $\bar{\mathfrak{n}}_{i-1} \cap \mathfrak{n}_i = \bar{\mathfrak{n}}^{(\alpha)}$ and hence $\mathfrak{n}_i = \operatorname{Ad}(s_\alpha)^{-1}\mathfrak{n}_{i-1}$. Let w be a representative in $N_K(\mathfrak{a})$ of s_α. Unwinding the computation at the end of the proof of Lemma 14.40, we see that $\sigma(w)\mathscr{A}_S(w, \sigma, \nu)$ scalar implies $\sigma(w)\mathscr{A}_{S_{i-1}}(w, \sigma, \nu)$ scalar. By (14.35) the operator in $G^{(\alpha)}M$ given as $\sigma(s_\alpha)\mathscr{A}_{S_{i-1} \cap G^{(\alpha)}M}(s_\alpha, \sigma, \nu|_{\mathbb{R}H_\alpha})$ is scalar. The η function for this operator must therefore have a pole at 0 (since the unnormalized operator is not scalar) and hence $\mu_{\sigma,\alpha}(\nu) = 0$ by Theorem 14.27.

Corollary 14.42. For σ in the discrete series of M and ν imaginary, $\Delta'_{\sigma,\nu}$ is given by

$$\Delta'_{\sigma,\nu} = \{\alpha = \text{useful root of } (\mathfrak{g}, \mathfrak{a}) \mid s_\alpha \in W_{\sigma,\nu} \text{ and } \sigma(s_\alpha)\mathscr{A}_S(s_\alpha, \sigma, \nu) = cI\}. \tag{14.69}$$

Consequently $\Delta'_{\sigma,\nu}$ (when nonempty) is a (possibly nonreduced) root system in a subspace of $\mathfrak{a}'$, and $W_{\sigma,\nu}$ carries $\Delta'_{\sigma,\nu}$ into itself.

Proof. Lemma 14.40 implies that $\Delta'_{\sigma,\nu}$ is contained in the right side of (14.69), and Lemma 14.41 implies that $\Delta'_{\sigma,\nu}$ contains the right side. This proves (14.69). The useful roots of $(\mathfrak{g}, \mathfrak{a})$ form a root system in a subspace of $\mathfrak{a}'$, and $\Delta'_{\sigma,\nu}$ is a subset of the useful roots. Therefore $\Delta'_{\sigma,\nu}$ is a root system (abstractly) if and only if it is closed under its own reflections. Consequently the corollary follows if it is shown that $W_{\sigma,\nu}(\Delta'_{\sigma,\nu}) \subseteq \Delta'_{\sigma,\nu}$.

Thus let α be in $\Delta'_{\sigma,\nu}$ and let p be in $W_{\sigma,\nu}$. Then $s_{p\alpha} = ps_\alpha p^{-1}$, and it follows from (14.69) and Lemma 14.38b that the operator corresponding to $s_{p\alpha}$ is scalar. Since $p\alpha$ is certainly useful, $p\alpha$ is in $\Delta'_{\sigma,\nu}$, again by (14.69). This proves the corollary.

Since $\Delta'_{\sigma,\nu}$ is a root system, we now define

$$W'_{\sigma,\nu} = \text{Weyl group of } \Delta'_{\sigma,\nu} \subseteq W(A:G)$$

$$R_{\sigma,\nu} = \{r \in W_{\sigma,\nu} \mid r\alpha > 0 \text{ for every } \alpha > 0 \text{ in } \Delta'_{\sigma,\nu}\}.$$

Corollary 14.42 says that $R_{\sigma,\nu}$ carries $\Delta'_{\sigma,\nu}$ into itself, and it follows that $R_{\sigma,\nu}$ is a subgroup of $W_{\sigma,\nu}$.

Theorem 14.43. Let G be linear connected reductive with compact center. Suppose that σ is in the discrete series of M and ν is imaginary. Then

$W_{\sigma,\nu}$ is the semidirect product $W_{\sigma,\nu} = W'_{\sigma,\nu}R_{\sigma,\nu}$ with $W'_{\sigma,\nu}$ normal. This decomposition has the following properties:

(a) $W'_{\sigma,\nu}$ is the set of s in $W_{\sigma,\nu}$ for which $\sigma(s)\mathscr{A}_S(s, \sigma, \nu)$ is scalar.

(b) $R_{\sigma,\nu}$ is the set of r in $W_{\sigma,\nu}$ for which $\mu_{\sigma,\alpha}(\nu) \neq 0$ for all reduced positive roots α of $(\mathfrak{g}, \mathfrak{a})$ for which $r\alpha$ is negative.

(c) The unitary operators $\sigma(r)\mathscr{A}_S(r, \sigma, \nu)$ for r in $R_{\sigma,\nu}$ are linearly independent and span the commuting algebra of $U(S, \sigma, \nu)$.

(d) The dimension of the commuting algebra of $U(S, \sigma, \nu)$ is $|R_{\sigma,\nu}|$.

Remarks. Conclusion (b) says that $R_{\sigma,\nu} = R'_{\sigma,\nu}$, where $R'_{\sigma,\nu}$ is given in (14.53). By Schur's Lemma, (d) implies that $U(S, \sigma, \nu)$ is irreducible if and only if $R_{\sigma,\nu}$ is trivial.

Proof. Corollary 14.42 says that $W_{\sigma,\nu}(\Delta'_{\sigma,\nu}) \subseteq \Delta'_{\sigma,\nu}$. Consequently $W_{\sigma,\nu}$ conjugates reflections in members of $\Delta'_{\sigma,\nu}$ into reflections in members of $\Delta'_{\sigma,\nu}$. Since these reflections generate $W'_{\sigma,\nu}$, $W'_{\sigma,\nu}$ is normal in $W_{\sigma,\nu}$.

Theorem 4.10 implies that $W'_{\sigma,\nu} \cap R_{\sigma,\nu} = \{1\}$. On the other hand, if $s \in W_{\sigma,\nu}$ is given, then $s(\Delta'_{\sigma,\nu})^+$ is a positive system for $\Delta'_{\sigma,\nu}$, and Theorem 4.10 allows us to choose $s' \in W'_{\sigma,\nu}$ with $(s')^{-1}s(\Delta'_{\sigma,\nu})^+ = (\Delta'_{\sigma,\nu})^+$. Then $r = (s')^{-1}s$ is in $R_{\sigma,\nu}$, and $s = s'r$ decomposes s within $W'_{\sigma,\nu}R_{\sigma,\nu}$. Hence $W_{\sigma,\nu}$ is a semidirect product $W_{\sigma,\nu} = W'_{\sigma,\nu}R_{\sigma,\nu}$ as asserted.

The operators $\sigma(s)\mathscr{A}_S(s, \sigma, \nu)$ corresponding to $W'_{\sigma,\nu}$ are scalar by Lemma 14.38b. Our semidirect product decomposition and Lemma 14.38b therefore say that the dimension of the span of all operators corresponding to $W_{\sigma,\nu}$ is $\leq |R_{\sigma,\nu}|$. On the other hand, the operators corresponding to the set $R'_{\sigma,\nu}$ in (14.53) are linearly independent. If we can prove that $R_{\sigma,\nu} \subseteq R'_{\sigma,\nu}$, then it follows that $R_{\sigma,\nu} = R'_{\sigma,\nu}$ and that the only scalar operators are those from $W'_{\sigma,\nu}$. That is, (a) and (b) hold. Since the span of the operators corresponding to $W_{\sigma,\nu}$ is the same as the span corresponding to $R_{\sigma,\nu}$, Theorem 14.31 gives us (c), and then (d) is an immediate consequence.

Thus we are to prove that $R_{\sigma,\nu} \subseteq R'_{\sigma,\nu}$. Arguing by contradiction, suppose the conclusion fails. Then we can choose r in $R_{\sigma,\nu}$ and a reduced positive root α of $(\mathfrak{g}, \mathfrak{a})$ such that $r\alpha < 0$ and $\mu_{\sigma,\alpha}(\nu) = 0$. Since $r\alpha < 0$, the definition of $R_{\sigma,\nu}$ implies that α is not in $\Delta'_{\sigma,\nu}$. On the other hand, since $\mu_{\sigma,\alpha}(\nu) = 0$, Lemma 14.40 implies that s_α is in $W_{\sigma,\nu}$ and $\sigma(s_\alpha)\mathscr{A}_S(s_\alpha, \sigma, \nu)$ is scalar. By Corollary 14.42, α must be not useful. But then Theorem 14.39 implies that $\frac{1}{2}\alpha$ exists as a root of $(\mathfrak{g}, \mathfrak{a})$, since s_α is in $W(A{:}G)$. Since α was assumed to be reduced, we have a contradiction. This completes the proof of the theorem.

Let us return to the notion of "useful roots," which we have so far left undefined. If $\mathfrak{a}_M$ denotes a maximal abelian subspace of $\mathfrak{m} \cap \mathfrak{p}$, then $\mathfrak{a}_\mathfrak{p} = \mathfrak{a} \oplus \mathfrak{a}_M$ is a maximal abelian subspace of $\mathfrak{p}$. If $\lambda = \lambda_R + \lambda_I$ is the corresponding decomposition of a member of $\mathfrak{a}_\mathfrak{p}'$, then we define the **conjugate** $\bar{\lambda}$ of λ relative to $\mathfrak{a}$ by $\bar{\lambda} = \lambda_R - \lambda_I$.

The roots of $(\mathfrak{g}, \mathfrak{a})$ are the nonzero restrictions to $\mathfrak{a}$ of the roots of $(\mathfrak{g}, \mathfrak{a}_\mathfrak{p})$, and we know that the roots of $(\mathfrak{g}, \mathfrak{a}_\mathfrak{p})$ form a root system.

Lemma 14.44. Under the assumption that rank $M = \operatorname{rank}(K \cap M)$, the roots of $(\mathfrak{g}, \mathfrak{a}_\mathfrak{p})$ are closed under conjugation relative to $\mathfrak{a}$.

Proof. Applying Problem 14 of Chapter XII to M_0, we see that there is an element w in $K \cap M_0$ with $\operatorname{Ad}(w) = -1$ on $\mathfrak{a}_M$. Since w is in M, $\operatorname{Ad}(w)$ is $+1$ on $\mathfrak{a}$. Thus w represents a member of $W(A_\mathfrak{p}:G)$ that is 1 on $\mathfrak{a}$ and -1 on $\mathfrak{a}_M$. These properties say that conjugation is implemented by a member of $W(A_\mathfrak{p}:G)$, and the lemma follows.

Example. $G = SL(6, \mathbb{R})$ with M block diagonal, blocks of sizes 2, 2, 1, 1.

As usual we can take $\mathfrak{a}_\mathfrak{p}$ to be the diagonal subalgebra. The roots of $(\mathfrak{g}, \mathfrak{a}_\mathfrak{p})$ are all $e_i - e_j$, $i \neq j$, with notation as in §4.1. A typical decomposition according to $\mathfrak{a}_\mathfrak{p} = \mathfrak{a} \oplus \mathfrak{a}_M$ is

$$e_1 - e_3 = \tfrac{1}{2}(e_1 + e_2 - e_3 - e_4) + \tfrac{1}{2}(e_1 - e_2 - e_3 + e_4),$$

and the conjugate of this root is

$$e_2 - e_4 = \tfrac{1}{2}(e_1 + e_2 - e_3 - e_4) - \tfrac{1}{2}(e_1 - e_2 - e_3 + e_4).$$

Since the roots of $(\mathfrak{g}, \mathfrak{a}_\mathfrak{p})$ form a root system, it follows that $2\langle\alpha, \bar{\alpha}\rangle/|\alpha|^2$ is an integer. Since $|\alpha| = |\bar{\alpha}|$, this integer can fail to be 0 or $+1$ or -1 only if $\bar{\alpha} = \pm\alpha$. Let us write $\alpha = \alpha_R + \alpha_I$ and examine the cases:

(i) $2\langle\alpha, \bar{\alpha}\rangle/|\alpha|^2 = 2$. Then $\alpha_R = \alpha$ and the reflection s_α normalizes $\mathfrak{a}$ and acts as s_{α_R} on $\mathfrak{a}$.

(ii) $2\langle\alpha, \bar{\alpha}\rangle/|\alpha|^2 = -2$. Then $\alpha_R = 0$.

(iii) $2\langle\alpha, \bar{\alpha}\rangle/|\alpha|^2 = -1$. Then $\alpha + \bar{\alpha}$ is a root of $(\mathfrak{g}, \mathfrak{a}_\mathfrak{p})$ and equals $2\alpha_R$. The reflection $s_{\alpha+\bar{\alpha}}$ normalizes $\mathfrak{a}$ and acts as s_{α_R} on $\mathfrak{a}$.

(iv) $2\langle\alpha, \bar{\alpha}\rangle/|\alpha|^2 = 0$. Then α and $\bar{\alpha}$ are orthogonal, and $s_\alpha s_{\bar{\alpha}} = s_{\alpha_R} s_{\alpha_I}$. Hence $s_\alpha s_{\bar{\alpha}}$ normalizes $\mathfrak{a}$ and acts as s_{α_R} on $\mathfrak{a}$.

(v) $2\langle\alpha, \bar{\alpha}\rangle/|\alpha|^2 = +1$. There is no obvious way of obtaining s_{α_R} from $W(A_\mathfrak{p}:G)$.

Let β be a root of $(\mathfrak{g}, \mathfrak{a})$. We say that β is a **useful root** of $(\mathfrak{g}, \mathfrak{a})$ if $\beta = \alpha_R$ for some root $\alpha = \alpha_R + \alpha_I$ of $(\mathfrak{g}, \mathfrak{a}_\mathfrak{p})$ such that $2\langle\alpha, \bar{\alpha}\rangle/|\alpha|^2 \neq +1$. In this case s_β is certainly in $W(A:G)$. Theorem 14.39 shows the role of useful roots in identifying $W(A:G)$.

Example. $G = SL(6, \mathbb{R})$ with M block diagonal, blocks of sizes 2, 2, 1, 1.

The roots $\alpha = e_i - e_j$ with $2\langle\alpha, \bar{\alpha}\rangle/|\alpha|^2 \neq +1$ are those with i and j both less than 5 or both greater than 4. Their restrictions to $\mathfrak{a}$ are 0 and the useful roots

$$\pm\tfrac{1}{2}(e_1 + e_2 - e_3 - e_4),\ \pm(e_5 - e_6),$$

which form a root system of type $A_1 \times A_1$ in a two-dimensional subspace of the three-dimensional space $\mathfrak{a}'$. The restrictions to $\mathfrak{a}$ of $e_1 - e_5$ and $e_1 - e_6$ are

$$\beta_1 = \tfrac{1}{2}(e_1 + e_2) - e_5 \qquad \text{and} \qquad \beta_2 = \tfrac{1}{2}(e_1 + e_2) - e_6.$$

Here $2\langle\beta_1, \beta_2\rangle/|\beta_1|^2 = -2/3$, and so β_1 and β_2 cannot be part of a root system. Members of $W(A\!:\!G)$ must permute the blocks of M; since 2-by-2 blocks must go into 2-by-2 blocks, the two sets of indices $\{1, 2, 3, 4\}$ and $\{5, 6\}$ do not mix with each other under this action.

§10. Action by Weyl Group on Representations of M

From Theorem 14.43 and Lemma 14.38b, we know that the group $R_{\sigma,\nu}$ controls the reducibility of the standard induced representation $U(S, \sigma, \nu)$, where σ is in the discrete series and ν is imaginary. Since the irreducible constituents of $U(S, \sigma, \nu)$ provide us with examples of irreducible tempered representations and since irreducible tempered representations are one of the ingredients in the Langlands classification, we need to know the size and structure of $R_{\sigma,\nu}$. We address these questions in this section.

The difficulty in understanding $R_{\sigma,\nu}$ is in dealing with the condition $w\sigma \cong \sigma$ for w in $N_K(\mathfrak{a})$. The examples of $SL(n, \mathbb{R})$ and $SL(n, \mathbb{C})$, with MAN the upper-triangular minimal parabolic subgroup in each case, illustrate opposite extremes. For $SL(n, \mathbb{R})$, M is the subgroup of diagonal matrices with diagonal entries ± 1. Although the Weyl group does act by permutations on M, it is not so obvious when $w\sigma = \sigma$. (Think of what happens already in $SL(2, \mathbb{R})$.) The case of $SL(n, \mathbb{C})$ is nicer. To study the effect of the Weyl group, we can work with the action of $\mathfrak{m}$. We have $\mathfrak{m} = i\mathfrak{a}$, and the multiplication-by-i map carries the Weyl group action from $\mathfrak{a}$ to $\mathfrak{m}$. Since Chevalley's Lemma is applicable on $\mathfrak{a}$, it is applicable on $\mathfrak{m}$. Thus the subgroup of the Weyl group leaving σ fixed is generated by reflections for $SL(n, \mathbb{C})$; it need not be generated by reflections for $SL(n, \mathbb{R})$.

Guided by these examples, we study first the effect of the Weyl group $W(A\!:\!G)$ on the identity component M_0. We shall arrange that it operate on Harish-Chandra parameters, and we shall see that the action on the discrete series representations of M is compatible with the action on Harish-Chandra parameters.

Let $\mathfrak{b} \subseteq \mathfrak{k} \cap \mathfrak{m}$ be a compact Cartan subalgebra of $\mathfrak{m}$, so that $\mathfrak{a} \oplus \mathfrak{b}$ is a Cartan subalgebra of $\mathfrak{g}$. We can regard the roots of $(\mathfrak{g}, \mathfrak{a})$ as the nonzero restrictions to $\mathfrak{a}$ of the members of $\Delta((\mathfrak{a} \oplus \mathfrak{b})^{\mathbb{C}}\!:\!\mathfrak{g}^{\mathbb{C}})$. If $\lambda = \lambda_{\mathfrak{a}} + \lambda_{\mathfrak{b}}$ is the decomposition of a linear functional on $(\mathfrak{a} \oplus \mathfrak{b})^{\mathbb{C}}$ according to $\mathfrak{a} \oplus \mathfrak{b}$ and if λ is real-valued on $\mathfrak{a} \oplus i\mathfrak{b}$, then $\lambda_{\mathfrak{a}} - \lambda_{\mathfrak{b}}$ is the complex-linear extension from $\mathfrak{a} \oplus \mathfrak{b}$ to $(\mathfrak{a} \oplus \mathfrak{b})^{\mathbb{C}}$ of the complex conjugate of λ. On $\mathfrak{a} \oplus \mathfrak{b}$, $\lambda_{\mathfrak{a}} - \lambda_{\mathfrak{b}}$ is given as $-\theta\lambda$. Therefore the map $\lambda_{\mathfrak{a}} + \lambda_{\mathfrak{b}} \to \lambda_{\mathfrak{a}} - \lambda_{\mathfrak{b}}$ carries $\Delta((\mathfrak{a} \oplus \mathfrak{b})^{\mathbb{C}}:\mathfrak{g}^{\mathbb{C}})$ into itself, fixing only the real roots.

We say that a root α_R of $(\mathfrak{g}, \mathfrak{a})$ is **odd** if the dimension of $\sum_{c>0} \mathfrak{g}_{c\alpha_R}$ is odd; otherwise we say that α_R is **even.** An odd root is necessarily useful unless 1/3 of it is also a root.

Lemma 14.45. The following three conditions on a root α_R of $(\mathfrak{g}, \mathfrak{a})$ are equivalent:

(a) α_R is odd.
(b) Some multiple of α_R is in $\Delta((\mathfrak{a} \oplus \mathfrak{b})^{\mathbb{C}}:\mathfrak{g}^{\mathbb{C}})$ when extended by 0 on $\mathfrak{b}$.
(c) The Lie algebra $\mathfrak{g}^{(\alpha_R)}$ has a compact Cartan subalgebra.

Proof.

(a) $\Rightarrow$ (b). If α_R is odd, some multiple $c\alpha_R$ with $c > 0$ has odd multiplicity as a root of $(\mathfrak{g}, \mathfrak{a})$. Taking into account the mapping $\lambda \to -\theta\lambda$ above, we see that $c\alpha_R$ is a real root when extended by 0.

(b) $\Rightarrow$ (c). In $\mathfrak{g}^{(\alpha_R)}$, $(\mathbb{R}H_{\alpha_R} \oplus \mathfrak{b}) \cap \mathfrak{g}^{(\alpha_R)}$ is a Cartan subalgebra whose noncompact part has dimension one. If some $c\alpha_R$ is a real root, then we can do a Cayley transform to obtain a compact Cartan subalgebra in $\mathfrak{g}^{(\alpha_R)}$.

(c) $\Rightarrow$ (a). Proposition 11.16a implies that there is a real root in $\Delta((\mathbb{R}H_{\alpha_R} \oplus \mathfrak{b})^{\mathbb{C}} \cap (\mathfrak{g}^{(\alpha_R)})^{\mathbb{C}}:(\mathfrak{g}^{(\alpha_R)})^{\mathbb{C}})$, which is necessarily a multiple $c\alpha_R$ of α_R. There can be no other real roots besides $\pm c\alpha_R$ for $(\mathfrak{g}^{(\alpha_R)})^{\mathbb{C}}$, and thus $c'\alpha_R$ has even multiplicity as a root of $(\mathfrak{g}, \mathfrak{a})$ for $c' \neq \pm c$ and odd multiplicity for $c' = \pm c$. Summing for $c' > 0$, we see that α_R is odd.

Lemma 14.46.

(a) Let α_R be an odd root of $(\mathfrak{g}, \mathfrak{a})$. Then there is a representative w in K of the reflection s_{α_R} on $\mathfrak{a}$ such that w is in the analytic subgroup corresponding to $\mathfrak{g}^{(\alpha_R)}$ and such that $\operatorname{Ad}(w)$ is the identity on $\mathfrak{m}$.

(b) Let α_R be an even useful root of $(\mathfrak{g}, \mathfrak{a})$, and let $\alpha_R \pm \beta$ be roots of $(\mathfrak{g}^{\mathbb{C}}:(\mathfrak{a} \oplus \mathfrak{b})^{\mathbb{C}})$ restricting to α_R. Then there is a representative w in K of the reflection s_{α_R} on $\mathfrak{a}$ such that w is in the analytic subgroup corresponding to $\mathfrak{g}^{(\alpha_R)}$ and such that $\operatorname{Ad}(w)$ is -1 on $\mathbb{R}H_\beta$ and is $+1$ on the orthocomplement of H_β in $i\mathfrak{b}$.

Proof.

(a) Since α_R is odd, Lemma 14.45 says that $\mathfrak{g}^{(\alpha_R)}$ has a compact Cartan subalgebra. By Problem 14 of Chapter XII, there exists w_1 in the analytic subgroup with Lie algebra $\mathfrak{g}^{(\alpha_R)} \cap \mathfrak{k}$ such that $\operatorname{Ad}(w_1)$ is $+1$ on $\mathfrak{g}^{(\alpha_R)} \cap \mathfrak{k}$ and is -1 on $\mathfrak{g}^{(\alpha_R)} \cap \mathfrak{p}$. The same problem shows there is w_2 in $K^{(\alpha_R)} \cap M_0$

such that $\mathrm{Ad}(w_2)$ is $+1$ on $\mathfrak{m}^{(\alpha_R)} \cap \mathfrak{k}$ and is -1 on $\mathfrak{m}^{(\alpha_R)} \cap \mathfrak{p}$. Then $w = w_1 w_2$ has the required properties.

(b) Let $\alpha = \alpha_R + \beta$, and let X_α be a root vector for α. If we write $\bar{\alpha} = \alpha_R - \beta$ and define a conjugation mapping of $\mathfrak{g}^{\mathbb{C}}$ with respect to $\mathfrak{g}$, then $\bar{X}_\alpha$ is a root vector for $\bar{\alpha}$. If we extend θ to $\mathfrak{g}^{\mathbb{C}}$ in complex-linear fashion (not our usual convention), then $\theta\bar{X}_\alpha$ is a root vector for $\theta\bar{\alpha} = -\alpha$. Now the Killing form B has $B(X, -\theta\bar{X}) > 0$ for all $X \neq 0$ in (the semisimple part of) $\mathfrak{g}^{\mathbb{C}}$, and thus we can normalize X_α so that $B(X_\alpha, -\theta\bar{X}_\alpha) = 2/|\alpha|^2$. Then we can set $H = 2|\alpha|^{-2}H_\alpha$ and $X_{-\alpha} = -\theta\bar{X}_\alpha$ and see that

$$[H, X_\alpha] = 2X_\alpha, \qquad [H, X_{-\alpha}] = -2X_{-\alpha}, \qquad [X_\alpha, X_{-\alpha}] = H.$$

That is, $\{H, X_\alpha, X_{-\alpha}\}$ can be taken as the basis $\{h, e, f\}$ within a copy of $\mathfrak{sl}(2, \mathbb{C})$. From the theory of $SL(2, \mathbb{C})$, it follows that

$$w^+ = \exp\frac{\pi}{2}(X_\alpha - X_{-\alpha}) = \exp\frac{\pi}{2}(X_\alpha + \theta\bar{X}_\alpha)$$

represents the Weyl reflection s_α on $\mathfrak{a} \oplus i\mathfrak{b}$. Similarly

$$w^- = \exp\frac{\pi}{2}(\bar{X}_\alpha + \theta X_\alpha)$$

represents the Weyl reflection $s_{\bar{\alpha}}$. The two members $X_\alpha + \theta\bar{X}_\alpha$ and $\bar{X}_\alpha + \theta X_\alpha$ of $\mathfrak{g}^{\mathbb{C}}$ commute since α is even, and thus

$$w^+ w^- = \exp\frac{\pi}{2}(X_\alpha + \bar{X}_\alpha + \theta X_\alpha + \overline{\theta X_\alpha}).$$

Consequently $w = w^+ w^-$ is exhibited as in the analytic subgroup with Lie algebra $\mathfrak{g}^{(\alpha_R)}$. This element w has the required properties.

Lemma 14.46 gives further evidence that it is a manageable problem to decide how $W(A{:}G)$ acts on the restriction of σ to M_0. In particular, reflections in odd roots do nothing, and only the reflections in even useful roots need to be understood. Let us examine the effect on Harish-Chandra parameters more closely, first looking at some examples. Examples in which MAN is minimal parabolic illustrate enough of the general case; since M is then compact, we are looking at the effect on highest weights.

Example 1. $G = SU(p, q)$, $p > q$.

We follow the notation of Appendix C, in which this example is of type AIII. The restricted roots form a system of type $(BC)_q$. Among them, $\alpha_R = \pm f_i \pm f_j$ is even if $i \neq j$ because $\mathfrak{g}^{(\alpha_R)} \cong \mathfrak{sl}(2, \mathbb{C})$. On the other hand, $\alpha_R = \pm f_i$ and $\alpha_R = \pm 2f_i$ are odd since $2f_i$ extends to a real root. Since

$$s_{f_i + f_j} = s_{2f_j} s_{f_i - f_j} s_{2f_j}$$

and since s_{2f_j} acts trivially on highest weights for M (by Lemma 14.46a), it is enough to understand the action of $s_{f_i-f_j}$. In the notation of Appendix C,

$$2f_j = e_j - e_{p+q+1-j},$$

and it is easy to check that $f_i - f_j$ arises as the restriction only of $e_i - e_j$ and $e_{p+q+1-j} - e_{p+q+1-i}$:

$$e_i - e_j = (f_i - f_j) + [\tfrac{1}{2}(e_i + e_{p+q+1-i}) - \tfrac{1}{2}(e_j + e_{p+q+1-j})]$$

$$e_{p+q+1-j} - e_{p+q+1-i} = (f_i - f_j) - [\tfrac{1}{2}(e_i + e_{p+q+1-i}) - \tfrac{1}{2}(e_j + e_{p+q+1-j})].$$

Using Lemma 14.46b, we see that $s_{f_i-f_j}$ can be realized in $i\mathfrak{b}'$ by interchanging the coefficient of $\frac{1}{2}(e_i + e_{p+q+1-i})$ and the coefficient of $\frac{1}{2}(e_j + e_{p+q+1-j})$. Since this action takes place in the center of $\mathfrak{m}$, it preserves $\mathfrak{m}$ dominance and thus gives the correct mapping on highest weights. Thus the group W_e generated by the reflections $s_{f_i-f_j}$ acts on highest weights as the permutation group on q letters with its natural action.

Let us restate matters. There is an isometric imbedding of the restricted roots $f_i - f_j$ into $i\mathfrak{b}'$, namely

$$f_i - f_j \to \tfrac{1}{2}(e_i + e_{p+q+1-i}) - \tfrac{1}{2}(e_j + e_{p+q+1-j}),$$

and the induced action by reflections in $i\mathfrak{b}'$ is the correct action on highest weights. The full Weyl group $W(A:G)$ is a semidirect product $W(A:G) = W_eS$, where S is the normal subgroup of sign changes. The sign changes are generated by the reflections s_{2f_i}, which act trivially on $i\mathfrak{b}'$. Thus the action of $W(A:G)$ on highest weights is explained completely by the decomposition $W(A:G) = W_eS$, the trivial action by S on $i\mathfrak{b}'$, and the action by W_e according to the isometric imbedding of the $f_i - f_j$'s into $i\mathfrak{b}'$.

Example 2. $G = SO_0(p, q)$, $p > q$.

We follow the notation of Appendix C, in which this example is of type BD I. The restricted roots form a system of type B_q. Among them, $\alpha_R = \pm f_i \pm f_j$ is odd if $i \neq j$ because $\mathfrak{g}^{(\alpha_R)} \cong \mathfrak{sl}(2, \mathbb{R})$. The restricted roots $\alpha_R = \pm f_i$ have $\mathfrak{g}^{(\alpha_R)} \cong \mathfrak{so}(p-q+1, 1)$ and so are odd if $p - q$ is odd, even if $p - q$ is even. When $p - q$ is odd, all restricted roots therefore are odd, and so the action by $W(A:G)$ on highest weights is trivial. Thus let $p - q$ be even. The restricted root f_i arises by restriction from

$$e_i \pm e_j = f_i + (\pm e_j), \qquad q + 1 \leq j \leq \tfrac{1}{2}(p+q)$$

Using Lemma 14.46b, we see that s_{f_i} can be realized in $i\mathfrak{b}'$ as reflection in e_j for any j with $q + 1 \leq j \leq \frac{1}{2}(p+q)$. If $j < \frac{1}{2}(p+q)$, then $e_j - e_{(1/2)(p+q)}$ is a root of $\mathfrak{m}$; consequently we check readily that dominance of highest weights is preserved only if s_{f_i} is realized as reflection in $e_{(1/2)(p+q)}$.

Let us restate matters for the case that $p-q$ is even. Let W_e be the group $\{1, s_{f_q}\}$. There is an isometric imbedding of the restricted roots $\pm f_q$ into $i\mathfrak{b}'$, namely

$$f_q \to e_{(1/2)(p+q)},$$

and the induced action by s_{f_q} in $i\mathfrak{b}'$ is the correct action on highest weights. The full Weyl group $W(A:G)$ is a semidirect product $W(A:G) = W_e S$, where S is the normal subgroup of permutations and even sign changes (the Weyl group of the odd roots). The action of $W(A:G)$ on highest weights is explained completely by the decomposition $W(A:G) = W_e S$, the trivial action by S on $i\mathfrak{b}'$, and the action by W_e according to the isometric imbedding of $\pm f_q$ into $i\mathfrak{b}'$.

Now we abstract facts from these examples. Let Γ_u be the root system of useful roots of $(\mathfrak{g}, \mathfrak{a})$. Let

$$\Pi_e = \{\text{those simple roots of } \Gamma_u \text{ that are even}\}$$

$$\mathfrak{a}_e = \sum_{\alpha_R \in \Pi_e} \mathbb{R}H_{\alpha_R}$$

$$W_e = \text{subgroup of } W(A:G) \text{ (or equivalently } W(\Gamma_u)) \text{ generated by reflections in the members of } \Pi_e.$$

Let a positive system $\Delta^+(\mathfrak{b}^{\mathbb{C}}:\mathfrak{m}^{\mathbb{C}})$ of roots of $\mathfrak{m}$ be specified. Introduce a lexicographic ordering on $i\mathfrak{b}'$ compatible with this positive system.

Lemma 14.47. For each α_R in Π_e, let α_I be the smallest positive member of $(i\mathfrak{b})'$ (in the lexicographic ordering) such that $\alpha_R \pm \alpha_I$ are roots in $\Delta((\mathfrak{a} \oplus \mathfrak{b})^{\mathbb{C}}:\mathfrak{g}^{\mathbb{C}})$. Then s_{α_I} preserves $\Delta^+(\mathfrak{b}^{\mathbb{C}}:\mathfrak{m}^{\mathbb{C}})$. Moreover, it is possible to choose a sign $\varepsilon = \pm 1$ for each α_R in Π_e so that the linear extension of the mapping given by $\alpha_R \to J(\alpha_R) = \varepsilon\alpha_I$ is an isometry of $\mathfrak{a}_e'$ into $(i\mathfrak{b})'$.

Remarks. We omit the proof, part of which involves understanding the property "useful" in the context of roots of $\Delta((\mathfrak{a} \oplus \mathfrak{b})^{\mathbb{C}}:\mathfrak{g}^{\mathbb{C}})$. Ultimately the signs are constructed recursively, in any sequence so that each member of Π_e is adjacent in the Dynkin diagram of Π_e to only one previous member.

Fix a mapping $J:\mathfrak{a}_e' \to (i\mathfrak{b})'$ as in the lemma. Let W_Π be the set of simple reflections relative to Γ_u, so that $W_\Pi \subseteq W(A:G)$. Then J defines a map of W_Π into the orthogonal group $O((i\mathfrak{b})')$ as follows: If α_R is in Π_e, map s_{α_R} into $s_{J\alpha_R}$. If α_R is simple for Γ_u but is not in Π_e, map s_{α_R} into the identity.

Theorem 14.48. Let G be linear connected reductive with compact center, and let MAN be a parabolic subgroup with rank $M = \operatorname{rank}(K \cap M)$. Then the mapping of W_Π into $O((i\mathfrak{b})')$ defined by J extends to a group homomorphism of $W(\mathfrak{a})$ into $O((i\mathfrak{b})')$. The resulting action of $W(A:G)$ on $(i\mathfrak{b})'$ has the properties that

(a) each p in $W(A:G)$ has a representative w in $N_K(\mathfrak{a})$ such that $\operatorname{Ad}(w)$ agrees on $(i\mathfrak{b})'$ with the action of p,
(b) for w in W_e, $Jw = wJ$ on $\mathfrak{a}'_e$, and
(c) each w in $W(A:G)$ preserves $\Delta^+(\mathfrak{b}^{\mathbb{C}}:\mathfrak{m}^{\mathbb{C}})$.

Sketch of proof. In the presence of Lemmas 14.46 and 14.47, the crux of the matter is to see that a group action results. Certainly J extends from W_Π to the free group on W_Π, and the problem is to show that the relations act as the identity. It turns out that the relations can be taken to be all of the form $(s_i s_j)^n = 1$, where n is 2, 3, 4, or 6. If s_i and s_j both come from Π_e, then $(J(s_i)J(s_j))^n$ is 1 because of the isometry in Lemma 14.47. If s_j comes from an odd root of $(\mathfrak{g}, \mathfrak{a})$, then we are to show $J(s_i)^n = 1$. Since $J(s_i)$ is a reflection, the only difficulty is if $n = 3$. In this case one shows that the roots of $(\mathfrak{g}, \mathfrak{a})$ corresponding to s_i and s_j are conjugate via $W(A:G)$; this implies s_i is odd and $J(s_i) = 1$.

Corollary 14.49. Let α_R be any useful root of $(\mathfrak{g}, \mathfrak{a})$. Then the action of s_{α_R} on $(i\mathfrak{b})'$ in Theorem 14.48 is given as follows:

(a) it is s_{α_I} if α_R is even and α_I is the smallest positive member of $(i\mathfrak{b})'$ such that $\alpha_R + \alpha_I$ is in $\Delta((\mathfrak{a} \oplus \mathfrak{b})^{\mathbb{C}}:\mathfrak{g}^{\mathbb{C}})$
(b) it is 1 if α_R is odd.

Proof.
(a) Let q be the action of s_{α_R} on $(i\mathfrak{b})'$ given by Theorem 14.48. Then it is clear from Lemma 14.46b that $s_{\alpha_R}q$ and $s_{\alpha_R}s_{\alpha_I}$ are both in $W((\mathfrak{a} \oplus \mathfrak{b})^{\mathbb{C}}:\mathfrak{g}^{\mathbb{C}})$. Hence so is qs_{α_I}. But qs_{α_I} fixes $\mathfrak{a}$ and by Chevalley's Lemma must be in $W(\mathfrak{b}^{\mathbb{C}}:\mathfrak{m}^{\mathbb{C}})$. Using Lemma 14.47 carefully, we see that qs_{α_I} preserves $\Delta^+(\mathfrak{b}^{\mathbb{C}}:\mathfrak{m}^{\mathbb{C}})$. Hence $qs_{\alpha_I} = 1$ and $q = s_{\alpha_I}$.
(b) We repeat the argument in (a) except that we use 1 in place of s_{α_I}. The element s_{α_R} is in $W((\mathfrak{a} \oplus \mathfrak{b})^{\mathbb{C}}:\mathfrak{g}^{\mathbb{C}})$ by Lemma 14.46a.

Let S be the subgroup of elements of $W(A:G)$ that act as the identity on $(i\mathfrak{b})'$ in the action of Theorem 14.48.

Corollary 14.50. S is normal in $W(A:G)$, and $W(A:G)$ is the semidirect product $W(A:G) = W_e S$.

Proof. S is normal because it is the kernel of a homomorphism. If w is in $W_e \cap S$, then w is in W_e and $w = 1$ on $J(\mathfrak{a}'_e)$. By Theorem 14.48, $w = 1$ on $\mathfrak{a}'_e$ and so $w = 1$. To see that $W(A:G) = W_e S$, let w be given, and write w as the product of simple reflections. Strike out the reflections in odd roots, leaving an element w_1 in W_e that has the same action on $(i\mathfrak{b})'$ as

w. Then $w_1^{-1}w$ acts as the identity on $(i\mathfrak{b})'$ and hence is in S. Therefore $w = w_1(w_1^{-1}w)$ is the required decomposition of w.

Now define

$$\Delta_0 = \{\alpha \in \Gamma_u | \alpha \text{ has odd multiplicity}\}.$$

Corollary 14.51. Δ_0 is a reduced root system (if nonempty), and S is its Weyl group.

Proof. It is clear that we can identify Δ_0 with the set of real roots in $\Delta((\mathfrak{a} \oplus \mathfrak{b})^{\mathbb{C}} : \mathfrak{g}^{\mathbb{C}})$. This identification makes it clear that Δ_0 is closed under its own reflections; since it is a subset of a root system, it is a root system (if nonempty). It is reduced since a nontrivial multiple of a real root cannot be a root.

The Weyl group of Δ_0 is contained in S, by Corollary 14.49. To prove the equality $W(\Delta_0) = S$, it is therefore enough to prove that the only w in S that leaves stable the set of positive members of Δ_0 is $w = 1$. By Corollary 14.50 it is enough to prove that the only elements of $W(A:G)$ leaving stable the set of positive members of Δ_0 are the members of W_e. We do so by induction on the length $l(w)$ of w within $W(A:G) = W(\Gamma_u)$, the case $l(w) = 0$ being trivial. Suppose $l(w) > 0$ and suppose w leaves stable the set of positive members of Δ_0. There must be some member β of Π_e such that $w\beta < 0$, since otherwise w permutes the positive roots in Γ_u. Then $l(ws_\beta) < l(w)$, s_β is in W_e, and we claim $ws_\beta\gamma > 0$ for every $\gamma > 0$ in Δ_0. [In fact, if $\gamma > 0$ is in Δ_0, then so is $s_\beta\gamma$ since the only positive roots in Γ_u mapped by s_β into negative roots are the multiples of β, and γ is not a multiple of β. Then $ws_\beta\gamma$ is positive since we are assuming w carries positive roots of Δ_0 into positive roots.] These facts reduce the proof to showing that ws_β is in W_e, and the induction is complete.

The final step is to connect the action of $W(A:G)$ on $(i\mathfrak{b})'$ with the effect on representations. This connection is given in the following theorem and its corollary.

Theorem 14.52. Let G be linear connected reductive with compact center, and let σ be a discrete series or limit of discrete series representation of M with global character $\Theta^M(\lambda, \Delta_M^+, \chi)$. Define the action of $W(A:G)$ on $(i\mathfrak{b})'$ by Theorem 14.48 so as to preserve Δ_M^+. If p is in $W(A:G)$ and w is a representative of p in $N_K(\mathfrak{a})$, then $w\sigma$ is a discrete series or limit of discrete series representation, respectively, and its character is $\Theta^M(p\lambda, \Delta_M^+, p\chi)$.

Remarks. Here $p\chi$ is well defined since the action of $W(A:G)$ on Z_M is well defined.

Proof. First suppose σ is in the discrete series. Integrating the absolute value squared of a matrix coefficient, we see that $w\sigma$ is in the discrete series. Clearly the value of $w\sigma$ on Z_M is given by $w\chi$ (which we can write as $p\chi$). To prove the character formula for $w\sigma$, it is therefore enough to deal with an irreducible constituent σ_0 of $\sigma|_{M_0}$ and with $w\sigma_0$, by Proposition 12.32. We may assume that w is the special representative of p given in Theorem 14.48a.

Let $B = \exp \mathfrak{b}$, and let $\Theta(b) = \tau_B(b)/D_B(b)$ be the global character of σ_0 on the regular set B' of B. (See (10.57) and Theorem 12.7.) Let f be a function on M_0 in the space $C^\infty_{\text{com}}((B')^{M_0})$. Then (10.25c) gives

$$\begin{aligned}\operatorname{Tr} \sigma_0(f) = \Theta(f) &= \int_G f(x)\Theta(x)\,dx \\ &= s^{M_0/B}|W(B:M_0)|^{-1} \int_B F_f^{M_0/B}(b)\tau_B(b)\,db. \qquad (14.69)\end{aligned}$$

If $g(m) = f(wmw^{-1})$, then

$$\begin{aligned}F_g^{M_0/B}(b) &= D^{M_0/B}(b) \int_{M_0} g(mbm^{-1})\,dm \\ &= D^{M_0/B}(b) \int_{M_0} f(wmbm^{-1}w^{-1})\,dm \\ &= D^{M_0/B}(b) \int_{M_0} f(m(wbw^{-1})m^{-1})\,dm \qquad \text{after } wmw^{-1} \to m \\ &= D^{M_0/B}(wbw^{-1}) \int_{M_0} f(m(wbw^{-1})m^{-1}\,dm \\ &= F_f^{M_0/B}(wbw^{-1}), \qquad (14.70)\end{aligned}$$

the next to last equality holding since w leaves the set $\Delta^+(\mathfrak{b}^{\mathbb{C}}:\mathfrak{m}^{\mathbb{C}})$ stable. Hence

$$\begin{aligned}\operatorname{Tr} w\sigma_0(f) &= \operatorname{Tr} \int_{M_0} f(m)\sigma_0(w^{-1}mw)\,dm \\ &= \operatorname{Tr} \int_{M_0} g(m)\sigma_0(m)\,dm = \operatorname{Tr} \sigma_0(g) \\ &= s^{M_0/B}|W(B:M_0)|^{-1} \int_B F_g^{M_0/B}(b)\tau_B(b)\,db \qquad \text{by (14.69)} \\ &= s^{M_0/B}|W(B:M_0)|^{-1} \int_B F_f^{M_0/B}(wbw^{-1})\tau_B(b)\,db \qquad \text{by (14.70)} \\ &= s^{M_0/B}|W(B:M_0)|^{-1} \int_B F_f^{M_0/B}(b)\tau_B(w^{-1}bw)\,db \\ &= s^{M_0/B}|W(B:M_0)|^{-1} \int_B F_f^{M_0/B}(b)w\tau_B(b)\,db.\end{aligned}$$

Thus the w translates of the exponentials in the numerator of the character of σ_0 on B appear in the numerator of the character of $w\sigma_0$. This completes the proof of Theorem 14.52 in the case of a discrete series representation of M.

Applying this result to our given data in the case of a limit of discrete series, we have

$$w\Theta^{M_0}(\lambda + \mu, \Delta_M^+) = \Theta^{M_0}(p(\lambda + \mu), \Delta_M^+) \tag{14.71}$$

whenever μ is Δ_M^+ dominant, M_0 integral, and M_0 nonsingular. Fix such a μ and apply the Zuckerman tensoring functor $\psi_{p\lambda}^{p\lambda+p\mu}$ to both sides of (14.71). We shall show that

$$\psi_{p\lambda}^{p\lambda+p\mu}(w\Theta) = w\psi_{\lambda}^{\lambda+\mu}\Theta, \tag{14.72}$$

and then the result will be

$$\begin{aligned} \Theta^{M_0}(p\lambda, \Delta_M^+) &= \psi_{p\lambda}^{p\lambda+p\mu} w\Theta^{M_0}(\lambda + \mu, \Delta_M^+) && \text{by (14.71)} \\ &= w\psi_{\lambda}^{\lambda+\mu}\Theta^{M_0}(\lambda + \mu, \Delta_M^+) && \text{by (14.72)} \\ &= w\Theta^{M_0}(\lambda, \Delta_M^+). \end{aligned}$$

Extending both sides to $M^\#$ by $p\chi$ and inducing to M, we obtain the conclusion of the theorem in the general case.

Thus we are to prove (14.72). Let π be a representation with global character Θ and with infinitesimal character $\chi_{\lambda+\mu}$. We claim that

$$w\pi \text{ has infinitesimal character } p(\lambda + \mu) \tag{14.73}$$

and

$$wF_{-\mu} = F_{-p\mu}. \tag{14.74}$$

If so, then

$$\begin{aligned} w\psi_{\lambda}^{\lambda+\mu}\pi &= w(p_\lambda(\pi \otimes F_{-\mu})) && \text{by definition} \\ &= p_{p\lambda}(w(\pi \otimes F_{-\mu})) && \text{by (14.73) with } \lambda \text{ in place of } \lambda + \mu \\ &= p_{p\lambda}(w\pi \otimes wF_{-\mu}) \\ &= p_{p\lambda}(w\pi \otimes F_{-p\mu}) && \text{by (14.74)} \\ &= \psi_{p\lambda}^{p\lambda+p\mu}(w\pi) && \text{by (14.73) and definition.} \end{aligned}$$

Thus (14.72) follows if we prove (14.73) and (14.74). Equation (14.74) follows by the same computation as for the discrete series earlier. For (14.73) we need to observe that each z in $Z(\mathfrak{m}^{\mathbb{C}})$ satisfies

$$\chi_{\lambda+\mu}(\mathrm{Ad}(w)^{-1}z) = \chi_{p(\lambda+\mu)}(z).$$

[This identity can be phrased more generally as a statement about automorphisms of $\mathfrak{m}^{\mathbb{C}}$ that map $\mathfrak{b}^{\mathbb{C}}$ into itself and preserve Δ_M^+ and can then be proved readily by using Verma modules.] Then

$$w\pi(z) = \pi(\mathrm{Ad}(w)^{-1}z) = \chi_{\lambda+\mu}(\mathrm{Ad}(w)^{-1}z)I = \chi_{p(\lambda+\mu)}(z)I,$$

and (14.73) follows. This completes the proof of the theorem.

Corollary 14.53. Let σ be a discrete series or nonzero limit of discrete series representation of M with global character $\Theta^M(\lambda, \Delta_M^+, \chi)$. Define the action of $W(A:G)$ on $(i\mathfrak{b})'$ by Theorem 14.48 so as to preserve Δ_M^+. If p is in $W(A:G)$ and w is a representative of p in $N_K(\mathfrak{a})$, then

(a) $w\sigma$ is equivalent with σ if and only if $p\lambda = \lambda$ and $p\chi = \chi$, and
(b) $w\sigma|_{M_0}$ is equivalent with $\sigma|_{M_0}$ if and only if $p\lambda = \lambda$.

Proof. Result (a) is immediate from Proposition 12.33c and Theorem 14.52. For (b) let $B = \exp \mathfrak{b}$. We use Proposition 12.33d and Theorem 14.52 to see that

$$\sigma|_{M_0} \text{ has character } \sum_{s \in W(B:M)/W(B:M_0)} \Theta^{M_0}(s\lambda, s\Delta_M^+) \tag{14.75a}$$

and

$$w\sigma|_{M_0} \text{ has character } \sum_{s \in W(B:M)/W(B:M_0)} \Theta^{M_0}(sp\lambda, s\Delta_M^+). \tag{14.75b}$$

Each of $w\sigma|_{M_0}$ and $\sigma|_{M_0}$ is fully reducible, and they are equivalent if and only if their characters are equal. If $p\lambda = \lambda$, then (14.75) shows $w\sigma|_{M_0} \cong \sigma|_{M_0}$. Conversely if $w\sigma|_{M_0} \cong \sigma|_{M_0}$, then the only term in (14.75b) that can match the term in (14.75a) for $s = 1$ is the $s = 1$ term in (14.75b), by Proposition 12.33c for M_0. That proposition shows also that we must have $p\lambda = \lambda$. Thus (b) follows.

§11. Multiplicity One Theorem

Corollary 14.53 allows us to get at the structure of the R group of §9. We fix now a *discrete series representation* σ of M and an imaginary-valued parameter ν on $\mathfrak{a}$, writing the character of σ as $\Theta^M(\lambda, \Delta_M^+, \chi)$. As earlier, define $W_{\sigma,\nu}$ to be the subgroup of $W(A:G)$ fixing ν and the class of σ, and define $W'_{\sigma,\nu}$ and $R_{\sigma,\nu}$ as in §9. Since our choice for how to write the character of σ involves a choice of Δ_M^+, we define the action of $W(A:G)$ on $(i\mathfrak{b})'$ once and for all by Theorem 14.48 so as to preserve Δ_M^+.

A key step in our analysis will be to show that $R_{\sigma,\nu}$ is contained in the Weyl group S of the odd roots (recall Corollary 14.51). To do so, we shall work both with Weyl group elements fixing $\sigma|_{M_0}$ and with elements fixing σ, and we shall have to relate the two.

Lemma 14.54. If α is an even useful root of $(\mathfrak{g}, \mathfrak{a})$ such that $s_\alpha\lambda = \lambda$, then $s_\alpha[\sigma] = [\sigma]$. In this case the Plancherel factor $\mu_{\sigma,\alpha}(\nu)$ is 0 whenever $\langle \nu, \alpha \rangle = 0$.

Proof. Let β be the smallest positive element of $(i\mathfrak{b})'$ such that $\alpha + \beta$ is a root in $\Delta = \Delta((\mathfrak{a} \oplus \mathfrak{b})_c : \mathfrak{g}^{\mathbb{C}})$. By Corollary 14.49a, $s_\alpha\lambda = \lambda$ implies $\langle \lambda, \beta \rangle = 0$. Then

$$\sum_{\substack{\varepsilon \in \Delta \\ \varepsilon|_{\mathfrak{a}} = c\alpha,\, c > 0}} \langle \lambda, \varepsilon \rangle = 0 \tag{14.76}$$

because the factor with $\varepsilon = \alpha + \beta$ is 0. The left side of (14.76) we recognize as the Plancherel factor $\mu_{\sigma,\alpha}(0)$ defined in §7. (Here the fact that α is even implies that $\mathfrak{g}^{(\alpha)}$ has no compact Cartan subalgebra, by Lemma 14.45, and Proposition 14.26a says there is no additional tanh-coth factor in $\mu_{\sigma,\alpha}$.) Since $\mu_{\sigma,\alpha}(0) = 0$, Lemma 14.40 implies $s_\alpha[\sigma] = [\sigma]$.

For general ν, $\mu_{\sigma,\alpha}(\nu)$ is given as in (14.76) but with factors $\langle \lambda + \nu, \varepsilon \rangle$. If $\langle \nu, \alpha \rangle = 0$, then the factor with $\varepsilon = \alpha + \beta$ still gives us 0. Hence $\mu_{\sigma,\alpha}(\nu) = 0$.

We now define

$$\begin{aligned} W_\lambda &= \{p \in W(A:G) \mid p\lambda = \lambda \text{ in action of } W(A:G) \text{ on } (i\mathfrak{b})'\} \\ W_{e,\lambda} &= W_e \cap W_\lambda \\ \Delta_{e,\lambda} &= \{\beta \in \Gamma_u \mid s_\beta \in W_{e,\lambda}\} \\ \Delta_\lambda &= \Delta_0 \cup S\Delta_{e,\lambda} \\ W_{\lambda,\nu} &= \{p \in W_\lambda \mid p\nu = \nu\} \\ \Delta_{\lambda,\nu} &= \{\alpha \in \Delta_\lambda \mid \langle \nu, \alpha \rangle = 0\}. \end{aligned}$$

In the course of the next four lemmas, we shall identify each of these Δ's as a root system (with the convention that the empty set is a root system) and the corresponding W as its Weyl group. Then we shall prove that $R_{\sigma,\nu} \subseteq S$ in Theorem 14.59. Our proofs use the obvious fact that $W(A:G)$ carries even roots to even roots and odd roots to odd roots.

Lemma 14.55. The group $W_{e,\lambda}$ is generated by the reflections that it contains. Hence $\Delta_{e,\lambda}$ is a root system and $W_{e,\lambda}$ is its Weyl group.

Proof. W_e is a Weyl group, and Theorem 14.48 says that its action on $(i\mathfrak{b})'$ is isometric with its standard action on $\mathfrak{a}'$. Thus Chevalley's Lemma applies, and the subgroup of W_e that fixes λ is a Weyl group, hence is generated by its reflections. Then $\Delta_{e,\lambda}$ is a subset of Γ_u closed under its own reflections and so is a root system (possibly not reduced). Its Weyl group is the group generated by its reflections, namely $W_{e,\lambda}$.

Lemma 14.56. $W_\lambda = W_{e,\lambda}S$. Moreover,

$$W_{e,\lambda} \subseteq W'_{\sigma,0} \subseteq W_{\sigma,0} \subseteq W_\lambda.$$

Proof. Corollary 14.50 gives $\Gamma_u = W_e S$, and S by definition is contained in W_λ. Thus we can intersect with W_λ to obtain $W_\lambda = W_{e,\lambda}S$. For the second statement, $W_{e,\lambda} \subseteq W'_{\sigma,0}$ by Lemma 14.54, $W'_{\sigma,0} \subseteq W_{\sigma,0}$ trivially, and $W_{\sigma,0} \subseteq W_\lambda$ by Corollary 14.53.

Lemma 14.57. Δ_λ is a reduced root system on a subspace of $\mathfrak{a}$, and W_λ is its Weyl group.

Proof. Since $\Delta_\lambda \subseteq \Gamma_u$, Δ_λ will be an abstract root system if it is shown that Δ_λ is closed under its own reflections. Since $W(A:G)$ carries odd roots to odd roots, Δ_0 is closed under all reflections. Next, let α be in Δ_0, p be in S, and β be in $\Delta_{e,\lambda}$. Then $s_\alpha(p\beta) = ps_{p^{-1}\alpha}\beta$ and $p^{-1}\alpha$ is in Δ_0. By Corollary 14.51, $ps_{p^{-1}\alpha}$ is in S. Thus $s_\alpha(p\beta)$ is in $S\Delta_{e,\lambda}$. Finally, let p and q be in S and let α and β be in $\Delta_{e,\lambda}$. Then

$$s_{p\alpha}(q\beta) = ps_\alpha p^{-1}q\beta = pq's_\alpha\beta$$

with $q' = s_\alpha(p^{-1}q)s_\alpha^{-1}$ in S since S is normal in $W(A:G)$. Since pq' is then in S and $s_\alpha\beta$ is in $\Delta_{e,\lambda}$ (Lemma 14.55), $s_{p\alpha}(q\beta)$ is in $S\Delta_{e,\lambda}$. Thus Δ_λ is closed under its own reflections and is a root system.

We know that Δ_0 is reduced (because of its identification with the set of real roots). Let β be in $S\Delta_{e,\lambda}$. Since β is even, Problem 19 at the end of the chapter shows that 2β is not a root of $(\mathfrak{g}, \mathfrak{a})$. Then it follows that Δ_λ is reduced.

Let $W(\Delta_\lambda)$ be the Weyl group of Δ_λ. This group is generated by the reflections s_α for α in Δ_0 and $S\Delta_{e,\lambda}$. If α is in Δ_0, then s_α is in S by Corollary 14.49b; if α is in $S\Delta_{e,\lambda}$, say with $\alpha = p\beta$, then

$$s_\alpha = s_{p\beta} = ps_\beta p^{-1} \in pW_{e,\lambda}p^{-1} \subseteq W_\lambda.$$

In either case s_α is in W_λ; thus we have $W(\Delta_\lambda) \subseteq W_\lambda$. For the reverse inclusion we use the equality $W_\lambda = W_{e,\lambda}S$ of Lemma 14.56. The group S is generated by reflections in members of Δ_0, by Corollary 14.51, and $W_{e,\lambda}$ is the Weyl group of $\Delta_{e,\lambda}$ by Lemma 14.55. Thus S and $W_{e,\lambda}$ are contained in $W(\Delta_\lambda)$, and we must have the desired inclusion $W_\lambda \subseteq W(\Delta_\lambda)$. This proves the lemma.

Lemma 14.58. $\Delta_{\lambda,\nu}$ is a reduced root system on a subspace of $\mathfrak{a}$, and its Weyl group is $W_{\lambda,\nu}$. Moreover,

$$W'_{\sigma,\nu} \subseteq W_{\sigma,\nu} \subseteq W_{\lambda,\nu}.$$

Proof. The first statement follows immediately from Lemma 14.57 and Chevalley's Lemma. The inclusion $W'_{\sigma,\nu} \subseteq W_{\sigma,\nu}$ is trivial, and the inclusion $W_{\sigma,\nu} \subseteq W_{\lambda,\nu}$ follows from Corollary 14.53.

Theorem 14.59. Let G be linear connected reductive with compact center, and let MAN be a parabolic subgroup. For σ in the discrete series of M and ν imaginary on $\mathfrak{a}$, $R_{\sigma,\nu}$ is contained in $S \cap W_{\lambda,\nu}$.

Proof. We know that $R_{\sigma,\nu} \subseteq W_{\sigma,\nu} \subseteq W_{\lambda,\nu}$ by Lemma 14.58, and we need to see that $R_{\sigma,\nu} \subseteq S$. We use the inclusion $R_{\sigma,\nu} \subseteq W_{\lambda,\nu}$ to get a handle on $R_{\sigma,\nu}$ for this purpose. Here $W_{\lambda,\nu}$ is the Weyl group of

$$\Delta_{\lambda,\nu} = \nu^\perp \cap (\Delta_0 \cup S\Delta_{e,\lambda}) = (\nu^\perp \cap \Delta_0) \cup (\nu^\perp \cap S\Delta_{e,\lambda}).$$

First we prove that $\nu^\perp \cap S\Delta_{e,\lambda} \subseteq \Delta'_{\sigma,\nu}$. If α is in $\nu^\perp \cap S\Delta_{e,\lambda}$, then α is an even useful root with $\langle \nu, \alpha \rangle = 0$. Also α is of the form $\alpha = p\beta$ with p in S and β in $\Delta_{e,\lambda}$, so that

$$s_\alpha \lambda = s_{p\beta}\lambda = ps_\beta p^{-1}\lambda = ps_\beta\lambda = p\lambda = \lambda.$$

By Lemma 14.54, s_α is in $W_{\sigma,\nu}$ and $\mu_{\sigma,\alpha}(\nu) = 0$. According to the definition, α is therefore in $\Delta'_{\sigma,\nu}$.

Let us notice that the roots in $\nu^\perp \cap \Delta_0$ are odd while those in $\nu^\perp \cap S\Delta_{e,\lambda}$ are even. Thus we have just showed that the even roots in $\Delta_{\lambda,\nu}$ are all in $\Delta'_{\sigma,\nu}$.

Now let r be in $R_{\sigma,\nu}$, and decompose $r = s_{\alpha_1} \cdots s_{\alpha_n}$ as a minimal product of simple reflections relative to $\Delta_{\lambda,\nu}$. This decomposition is possible since $R_{\sigma,\nu} \subseteq W_{\lambda,\nu}$ and $W_{\lambda,\nu}$ is the Weyl group of the root system $\Delta_{\lambda,\nu}$. Arguing by contradiction, suppose that some α_j is even. Since the decomposition is minimal, Proposition 4.11 implies that $\gamma = s_{\alpha_n} \cdots s_{\alpha_{j+1}}(\alpha_j)$ is a positive root in $\Delta_{\lambda,\nu}$ such that $r\gamma < 0$. Since α_j is even, so is γ. Thus γ, being an even root in $\Delta_{\lambda,\nu}$, must be in $\Delta'_{\sigma,\nu}$. Consequently we obtain the conclusion that r maps the positive member γ of $\Delta'_{\sigma,\nu}$ into a negative root, in contradiction to the definition of $R_{\sigma,\nu}$. Thus $\alpha_1, \ldots, \alpha_n$ are all odd, and r is the product of reflections in members of $\nu^\perp \cap \Delta_0$. By Corollary 14.51, r is in S.

Corollary 14.60. If the Cartan subgroup corresponding to $\mathfrak{a} \oplus \mathfrak{b}$ is maximally compact, then every representation $U(S, \sigma, \nu)$ with σ in the discrete series and ν imaginary is irreducible.

Proof. By Proposition 11.16a there are no real roots. Thus Corollary 14.51 says $S = \{1\}$, and Theorem 14.59 forces $R_{\sigma,\nu} = \{1\}$. By Theorem 14.43c, the identity operator spans the commuting algebra of $U(S, \sigma, \nu)$. Since $U(S, \sigma, \nu)$ is unitary, Schur's Lemma (Proposition 1.5) says $U(S, \sigma, \nu)$ is irreducible.

Example. In the case of complex semisimple Lie groups, there is only one conjugacy class of Cartan subgroups (Theorem 5.22e), and the corresponding parabolic subgroup S is minimal parabolic. The condition of the corollary is satisfied, and the corollary implies that all members of the unitary principal series of a complex semisimple group are irreducible.

The path is short from Theorem 14.59 to the structure of $R_{\sigma,\nu}$ in general, and we follow it now.

Lemma 14.61. Suppose α and β are nonproportional nonorthogonal roots of $(\mathfrak{g}, \mathfrak{a})$ in $\Delta_{\lambda,\nu}$ with $|\alpha| \geq |\beta|$. If α and β are not in $\Delta'_{\sigma,\nu}$, then $s_\beta\alpha$ is in $\Delta'_{\sigma,\nu}$.

Remark. The proof will use a fact that we did not prove in Chapter IV—that the only nonreduced irreducible root systems are those of type $(BC)_n$.

Proof. Lemma 14.54 shows that the even roots in $\Delta_{\lambda,\nu}$ are in $\Delta'_{\sigma,\nu}$. Since α and β are assumed not to be in $\Delta'_{\sigma,\nu}$, they must both be odd. By Lemma 14.40, $\mu_{\sigma,\alpha}(\nu) = 0$ would imply $s_\alpha \in W_{\sigma,\nu}$ and hence $\alpha \in \Delta'_{\sigma,\nu}$. Thus $\mu_{\sigma,\alpha}(\nu)$ must be nonzero, and similarly $\mu_{\sigma,\beta}(\nu)$ is nonzero.

On the other hand, α and β in $\Delta_{\lambda,\nu}$ implies $\langle \nu, \alpha \rangle = \langle \nu, \beta \rangle = 0$. Thus one of the polynomial factors in $\mu_{\sigma,\alpha}(\nu)$ vanishes, and similarly one of the polynomial factors in $\mu_{\sigma,\beta}(\nu)$ vanishes. We conclude that α and β correspond to coth cases in Proposition 14.26. Since the odd members of $\Delta_{\lambda,\nu}$ have already been scaled so as to be of odd multiplicity, (14.45c) gives

$$\chi(\gamma_\alpha) = (-1)^{2\langle \rho_\alpha, \alpha \rangle / |\alpha|^2} \qquad \text{and} \qquad \chi(\gamma_\beta) = (-1)^{2\langle \rho_\beta, \beta \rangle / |\beta|^2}.$$

Now an easy computation gives

$$\chi(\gamma_{s_\beta \alpha}) = \chi(\gamma_\alpha)\chi(\gamma_\beta)^{2\langle \alpha, \beta \rangle / |\alpha|^2} = \chi(\gamma_\alpha)\chi(\gamma_\beta).$$

We shall use this formula to show that $s_\beta \alpha$ corresponds to a tanh case. Then necessarily $\mu_{\sigma, s_\beta \alpha}(\nu) = 0$ since $\langle \nu, s_\beta \alpha \rangle = 0$, and Lemma 14.40 finishes the proof that $s_\beta \alpha$ is in $\Delta'_{\sigma,\nu}$.

In view of the above formulas and (14.45c), we are to show that

$$\frac{2\langle \rho_{s_\beta \alpha}, s_\beta \alpha \rangle}{|s_\beta \alpha|^2} \not\equiv \frac{2\langle \rho_\alpha, \alpha \rangle}{|\alpha|^2} + \frac{2\langle \rho_\beta, \beta \rangle}{|\beta|^2} \mod 2$$

Here ρ_α, for example, is half the sum of the positive multiples of α, counting multiplicities. The term on the left equals the first term on the right. Thus we are to show that

$$2\langle \rho_\beta, \beta \rangle / |\beta|^2 \equiv 1 \mod 2. \tag{14.77}$$

Since β has odd multiplicity, (14.77) will follow if we show that $\frac{1}{2}\beta$ is not a root of $(\mathfrak{g}, \mathfrak{a})$, in view of the limitation on constants c for which $c\beta$ is a root of $(\mathfrak{g}, \mathfrak{a})$ (see the Problems at the end of the chapter.) If $\frac{1}{2}\beta$ is a root of $(\mathfrak{g}, \mathfrak{a})$, then it is useful (Problem 22), and the simple component of Γ_u to which α and β belong is a nonreduced irreducible root system, necessarily of type $(BC)_n$ for some n. Here β is already a root of maximal length, and $|\alpha| \geq |\beta|$. So $|\alpha| = |\beta|$. In $(BC)_n$ any two nonproportional roots of maximal length are orthogonal. Since α and β have been assumed nonorthogonal, we get a contradiction unless $\frac{1}{2}\beta$ is not a root of $(\mathfrak{g}, \mathfrak{a})$. This proves (14.77), and the lemma follows.

Lemma 14.62. Let q be the linear transformation on $\mathfrak{a}'$ given by

$$q = \frac{1}{|R_{\sigma,\nu}|} \sum_{r \in R_{\sigma,\nu}} r. \tag{14.78}$$

Then ν is orthogonal to $(1 - q)\mathfrak{a}'$, and no root of $(\mathfrak{g}, \mathfrak{a})$ in $(1 - q)\mathfrak{a}'$ lies in $\Delta'_{\sigma,\nu}$.

Proof. The operator q is the orthogonal projection on the simultaneous $+1$ eigenspace of the members of $R_{\sigma,\nu}$. Thus the image of $1 - q$ is the same as the kernel of q. If α is in $\Delta'_{\sigma,\nu}$, then $r\alpha$ has the same sign as α, for each r in $R_{\sigma,\nu}$, and hence $q\alpha$ cannot be 0. Thus α cannot be in the image of $1 - q$. Moreover, $q\nu = \nu$, and hence ν is orthogonal to $(1 - q)\mathfrak{a}'$.

A subset $\{\alpha_1, \ldots, \alpha_n\}$ of $\Delta_{\lambda,\nu}$ will be said to be **superorthogonal** (relative to $\Delta_{\lambda,\nu}$) if the only way that $\sum c_j\alpha_j$ can be in $\Delta_{\lambda,\nu}$ is for all but one of the c_j's to be 0.

Lemma 14.63. With q as in (14.78), the positive members of $\Delta_{\lambda,\nu}$ that lie in the space $(1 - q)\mathfrak{a}'$ are superorthogonal.

Proof. Otherwise we can find nonproportional nonorthogonal roots α and β in $(1 - q)\mathfrak{a}' \cap \Delta_{\lambda,\nu}$ with $|\alpha| \geq |\beta|$. By Lemma 14.62, neither α nor β is in $\Delta'_{\sigma,\nu}$. Lemma 14.61 therefore shows that $s_\beta\alpha$ is in $\Delta'_{\sigma,\nu}$. But $s_\beta\alpha$ is also in $(1 - q)\mathfrak{a}'$, and we have a contradiction to Lemma 14.62.

Theorem 14.64. Let G be linear connected reductive with compact center, and let MAN be a parabolic subgroup. For σ in the discrete series of M and ν imaginary on $\mathfrak{a}$, $R_{\sigma,\nu}$ is of the form $\sum \mathbb{Z}_2$. In fact, with q as in (14.78), the set $\mathscr{H} = \{\alpha_1, \ldots, \alpha_k\}$ of positive roots of $(\mathfrak{g}, \mathfrak{a})$ having odd multiplicity and lying in $(1 - q)\mathfrak{a}'$ is superorthogonal relative to $\Delta_{\lambda,\nu}$ and spans $(1 - q)\mathfrak{a}'$; therefore

(a) each r in $R_{\sigma,\nu}$ is of the form $s_{\alpha_{j_1}} \cdots s_{\alpha_{j_n}}$ with $\{\alpha_{j_1}, \ldots, \alpha_{j_n}\} \subseteq \mathscr{H}$
(b) each α_j in $\mathscr{H}$ satisfies $\langle \nu, \alpha_j \rangle = 0$
(c) each α_j occurs in the decomposition of some r in $R_{\sigma,\nu}$.

Proof. The roots in $(1 - q)\mathfrak{a}'$ are orthogonal to ν by Lemma 14.62, and the odd roots are in Δ_λ by definition. Thus $\mathscr{H}$ is contained in $\Delta_{\lambda,\nu}$. By Lemma 14.63, $\mathscr{H}$ is a superorthogonal set. If r is in $R_{\sigma,\nu}$, then r fixes $q\mathfrak{a}'$. By Theorem 14.59, we have $R_{\sigma,\nu} \subseteq S$, and we know that S is the Weyl group of Δ_0. By Chevalley's Lemma, r is the product of reflections in members of Δ_0 fixing $q\mathfrak{a}'$. Such roots must lie in $(1 - q)\mathfrak{a}'$, hence are members of $\mathscr{H}$ up to sign. Thus $\mathscr{H}$ satisfies (a). Conclusion (b) follows from Lemma 14.62.

To prove conclusion (c), we show that the roots needed for (a), as r varies, span $(1 - q)\mathfrak{a}'$. Assume the contrary, and let v be a member of $(1 - q)\mathfrak{a}'$ orthogonal to all such roots. On the one hand, v must be fixed by every element of $R_{\sigma,\nu}$, hence by q, so that v is in the image of q. On the other hand, v is in the image of $(1 - q)$ and hence the kernel of q. Since the image and kernel of q are disjoint, $v = 0$. This proves property (c) and the spanning of $(1 - q)\mathfrak{a}'$ by $\mathscr{H}$. Finally every element of $R_{\sigma,\nu}$ is of order ≤ 2 by (a) and the superorthogonality of $\mathscr{H}$, and it follows that $R_{\sigma,\nu} = \sum \mathbb{Z}_2$.

Remarks. In particular, the cardinality of $R_{\sigma,\nu}$ is a power of 2, say 2^m. The number m is bounded by the cardinality k of the superorthogonal set $\mathscr{H}$, which in turn is bounded by dim $\mathfrak{a}$. The theorem says more than just $R_{\sigma,\nu} = \sum \mathbb{Z}_2$ with this bound, however: The following corollary uses the conclusion about superorthogonality.

Corollary 14.65. For each r in $R_{\sigma,\nu}$ it is possible to select a representative w_r in $N_K(\mathfrak{a})$ such that the members of $\{w_r \mid r \in R_{\sigma,\nu}\}$ commute with one another and centralize M_0.

Proof. Regard each α in $\mathscr{H}$ as a real root in $\Delta((\mathfrak{a} \oplus \mathfrak{b})^{\mathbb{C}} : \mathfrak{g}^{\mathbb{C}})$. Since α is real, we can choose a root vector E_α for it that lies in $\mathfrak{g}$. Following the proof of Proposition 5.15c, we can normalize E_α so that Re $B_0(E_\alpha, \theta E_\alpha) = -2|\alpha|^{-2}$, where B_0 is the trace form. Then $w_\alpha = \exp \frac{\pi}{2}(E_\alpha + \theta E_\alpha)$ is a member of the normalizer $N_K(\mathfrak{a})$, and $\mathrm{Ad}(w_\alpha)$ acts as the reflection s_α on $\mathfrak{a}$.

The various elements $w_{\alpha_1}, \ldots, w_{\alpha_k}$ constructed in this way commute since the superorthogonality implies that the Lie algebra elements $E_{\alpha_1} + \theta E_{\alpha_1}, \ldots, E_{\alpha_k} + \theta E_{\alpha_k}$ commute. Moreover, each $\mathrm{Ad}(w_{\alpha_j})$ is 1 on $\mathfrak{b}$ since the roots α_j are real.

Next we apply Lemma 14.46a to construct $u_{\alpha_1}, \ldots, u_{\alpha_k}$ in $N_K(\mathfrak{a})$ such that each $\mathrm{Ad}(u_{\alpha_j})$ is s_{α_j} on $\mathfrak{a}$ and is 1 on $\mathfrak{m}$. Then the element $w_{\alpha_j}^{-1}u_{\alpha_j}$ is in K, is 1 on $\mathfrak{a}$, and is 1 on $\mathfrak{b}$. By Lemma 12.30, $w_{\alpha_j}^{-1}u_{\alpha_j}$ is of the form $b_{\alpha_j}z_{\alpha_j}$ with b_{α_j} in $B = \exp \mathfrak{b}$ and with z_{α_j} in $F(B) \subseteq Z_M$. Then the element v_{α_j} given by

$$v_{\alpha_j} = u_{\alpha_j}z_{\alpha_j}^{-1} = w_{\alpha_j}b_{\alpha_j}$$

is in $N_K(\mathfrak{a})$, represents s_{α_j} in $W(A:G)$, and centralizes M_0. Since $w_{\alpha_1}, \ldots, w_{\alpha_k}$ commute and centralize B and since B is abelian, the elements v_{α_j} commute with each other. We can choose each w_r to be a product of the appropriate elements $v_{\alpha_1}, \ldots, v_{\alpha_k}$ by Theorem 14.64, and then the various elements w_r commute.

Corollary 14.66 (Multiplicity One Theorem). Let G be linear connected reductive with compact center. For σ in the discrete series of M and ν imaginary on $\mathfrak{a}$, the commuting algebra of $U(S, \sigma, \nu)$ is commutative and its dimension is a power of 2, the power being $\leq$ dim $\mathfrak{a}$. Consequently the irreducible constituents of $U(S, \sigma, \nu)$ all occur with multiplicity one.

Proof. Theorem 14.43c says that the operators $\sigma(r)\mathscr{A}_S(r, \sigma, \nu)$ for r in $R_{\sigma,\nu}$ are a basis of the commuting algebra. By Theorem 14.64, the dimension of the commuting algebra is a power of 2. We shall show that the members of this basis commute.

Let r and s be members of $R_{\sigma,\nu}$, and let w_r and w_s be their representatives in $N_K(\mathfrak{a})$ constructed in Corollary 14.65. From Lemma 14.38b, we know that

$$\sigma(w_r)\mathscr{A}_S(w_r, \sigma, \nu)\sigma(w_s)\mathscr{A}_S(w_s, \sigma, \nu) = c\sigma(w_r w_s)\mathscr{A}_S(w_r w_s, \sigma, \nu),$$

where c is given by

$$\sigma(w_r w_s)^{-1}\sigma(w_r)\sigma(w_s) = cI.$$

An analogous formula holds for $w_s w_r$, with a constant d in place of c. Now $w_r w_s$ and $w_s w_r$ are representatives of the same member of $R_{\sigma,\nu}$ since Theorem 14.64 implies $R_{\sigma,\nu}$ is abelian, and thus the corresponding operators are the same. Thus the operators for w_r and w_s commute if and only if $c = d$, i.e., if and only if

$$\sigma(w_r w_s)^{-1}\sigma(w_r)\sigma(w_s) = \sigma(w_s w_r)^{-1}\sigma(w_s)\sigma(w_r).$$

Since $w_r w_s = w_s w_r$, the condition for the operators to commute is that

$$\sigma(w_r)\sigma(w_s) = \sigma(w_s)\sigma(w_r). \tag{14.79}$$

Now w_r and w_s both commute with M_0, by Corollary 14.65, and thus $\sigma(w_r)$ and $\sigma(w_s)$ are in the commuting algebra of $\sigma|_{M_0}$. Proposition 12.32 implies that the irreducible constituents of $\sigma|_{M_0}$ are inequivalent, and hence the commuting algebra of $\sigma|_{M_0}$ is commutative. Therefore the members $\sigma(w_r)$ and $\sigma(w_s)$ of this commuting algebra commute with each other. This proves (14.79), and we have seen that this implies that the operators $\sigma(r)\mathscr{A}_S(r, \sigma, \nu)$ commute. Since the commuting algebra of $U(S, \sigma, \nu)$ is then commutative, the irreducible constituents of $U(S, \sigma, \nu)$ are inequivalent.

Example. G = double covering of $SO_0(4, 4)$, $S = MAN$ minimal parabolic.

This example was discussed at the start of §9. For a suitable σ and for $\nu = 0$, we determined that $R_{\sigma,\nu}$ was the four-element group (14.66), isomorphic to $\mathbb{Z}_2 \oplus \mathbb{Z}_2$. In this case, $\Delta_{\lambda,\nu}$ is the full root system D_4, and the set $\mathscr{H}$ of superorthogonal roots is given by

$$\mathscr{H} = \{e_1 - e_2, e_3 - e_4, e_3 + e_4\}.$$

§12. Zuckerman Tensoring of Induced Representations

It turns out that the R group points to the natural setting where reducibility occurs. In the example at the end of §11, the setting is a parabolic subgroup associated to the Cartan subgroup built by Cayley transform from the three roots of $\mathscr{H}$. For this parabolic subgroup, M_0 is the sum of three copies of $SL(2, \mathbb{R})$, and the relevant principal series representation of M_0 splits into eight irreducible pieces. When the full effect of the disconnectedness of M is taken into account, the reducibility is into just four irreducible pieces. To get the reducibility for the given principal series representation of G, we simply induce to G the reducibility on the M level.

The reducibility in this example on the level of M_0 comes from the reducibility of $\mathscr{P}^{-,0}$ in $SL(2, \mathbb{R})$, and this is a special case of Schmid's identity (Theorem 12.34). In fact, Schmid's identity is the mechanism in the general case, except that its hypotheses have to be weakened a bit. The task of generalizing Schmid's identity will occupy us in this section and the next. We shall first allow more singularities for the infinitesimal character in question, and then we shall deal with the kind of disconnectedness that M exhibits. The tool for handling more singularities is Zuckerman tensoring, and we have to see that it is compatible with induction.

We continue to let MAN be a parabolic subgroup of G, but the condition that rank $M = \operatorname{rank}(K \cap M)$ will play no role in this section.

We work with **Harish-Chandra modules** of MA, i.e., finitely generated representations that are **admissible** in the sense that each $K \cap M$ type has finite multiplicity. The general facts about Harish-Chandra modules for the connected group G (Proposition 10.41, Corollary 10.42, Proposition 10.43) are equally applicable for the possibly disconnected group MA.

Let $\mathfrak{t} = \mathfrak{a} \oplus \mathfrak{b}$ be a Cartan subalgebra of $\mathfrak{g}$; the condition that the group corresponding to $\mathfrak{b}$ be compact will play no role. Let

$$\Delta_M = \Delta(\mathfrak{b}^{\mathbb{C}}:\mathfrak{m}^{\mathbb{C}}) \qquad \text{and} \qquad \Delta_G = \Delta(\mathfrak{t}^{\mathbb{C}}:\mathfrak{g}^{\mathbb{C}})$$

be the root systems for M and G, and regard Δ_M as a subset of Δ_G. Suppose that positive systems Δ_M^+ and Δ_G^+, not necessarily related to N, have been specified such that

$$\Delta_M^+ \subseteq \Delta_G^+. \tag{14.80}$$

We may assume as usual in our concrete realization of G that $\mathfrak{k} \cap i\mathfrak{p} = 0$. Then let $(MA)^{\mathbb{C}}$ denote the analytic subgroup of $G^{\mathbb{C}}$ with Lie algebra $(\mathfrak{m} \oplus \mathfrak{a})^{\mathbb{C}}$. Since the group M is generated by M_0 and elements γ_α in $\exp \mathfrak{a}_{\mathfrak{p}}^{\mathbb{C}}$ (Lemma 9.13), we have $M \subseteq (MA)^{\mathbb{C}}$.

A linear form μ on $\mathfrak{t}$ or $\mathfrak{t}^{\mathbb{C}}$ will be called **integral** if it exponentiates to the analytic subgroup of $G^{\mathbb{C}}$ with Lie algebra $\mathfrak{t}^{\mathbb{C}}$. If μ is an integral form that is Δ_G^+ dominant, then μ is also Δ_M^+ dominant by (14.80), and we can form an irreducible finite-dimensional representation of $(MA)^{\mathbb{C}}$ with lowest weight $-\mu$. Restricting to M, we obtain a well-defined representation ${}^M F_{-\mu}$ of the possibly disconnected group MA. We may regard ${}^M F_{-\mu}$ also as a representation of MAN with N acting trivially.

Let λ be a linear form on $\mathfrak{t}^{\mathbb{C}}$ such that $\operatorname{Re} \lambda$ is Δ_G^+ dominant. (Here we define $\operatorname{Re} \lambda$ as in §10.9.) If μ is integral and Δ_G^+ dominant and if V is a

Harish-Chandra module of MA, we define

$$ {}^{M}\psi_{\lambda}^{\lambda+\mu}(V) = p_{\lambda}[V \otimes {}^{M}F_{-\mu}] $$

in analogy with the definition of §10.9. Here p_λ is projection according to the infinitesimal character λ.

For emphasis, we write ${}^{G}\psi_{\lambda}^{\lambda+\mu}$ for the usual Zuckerman ψ functor for G.

Theorem 14.67. Let G be linear connected reductive, and let MAN be a parabolic subgroup. For any Harish-Chandra module V of MA, the functors ${}^{M}\psi_{\lambda}^{\lambda+\mu}$ and ${}^{G}\psi_{\lambda}^{\lambda+\mu}$ are related by

$$ {}^{G}\psi_{\lambda}^{\lambda+\mu}(\mathrm{ind}_{MAN}^{G} V) \cong \mathrm{ind}_{MAN}^{G}({}^{M}\psi_{\lambda}^{\lambda+\mu}V). \tag{14.81} $$

Moreover, if V_1 and V_2 are two such Harish-Chandra modules of MA and T is an $\mathfrak{m} \oplus \mathfrak{a}$ map from the $(K \cap M)$-finite vectors of V_1 to those of V_2, then

$$ {}^{G}\psi_{\lambda}^{\lambda+\mu}(\mathrm{ind}_{MAN}^{G} T) \cong \mathrm{ind}_{MAN}^{G}({}^{M}\psi_{\lambda}^{\lambda+\mu}T). \tag{14.82} $$

Proof. One checks that

$$ F_{-\mu} \otimes \mathrm{ind}_{MAN}^{G} V \cong \mathrm{ind}_{MAN}^{G}(F_{-\mu}|_{MAN} \otimes V) \tag{14.83} $$

in the same manner as in Problems 3–8 of Chapter X. Since p_λ is exact, an irreducible MAN subquotient F_i of $F_{-\mu}|_{MAN}$ contributes the subquotient

$$ p_{\lambda}[\mathrm{ind}_{MAN}^{G}(F_i \otimes V)] \tag{14.84} $$

to p_λ of the right side of (14.83). The discussion of §10.9 tells us the generalized infinitesimal characters that can occur in $F_i \otimes V$; they can be read off from the global character and are therefore all of the form $-\nu + t(\lambda + \mu)$, where $-\nu$ is a weght of F_i and t is in $W(\mathfrak{b}^{\mathbb{C}}:\mathfrak{m}^{\mathbb{C}})$. According to (10.27), the exponentials that occur in the numerators of induced global characters are at worst the $W(\mathfrak{t}^{\mathbb{C}}:\mathfrak{g}^{\mathbb{C}})$ translates of the original exponentials. Thus (14.84) can be nonzero only if there exist a weight $-\nu$ of F_i, a member t of $W(\mathfrak{b}^{\mathbb{C}}:\mathfrak{m}^{\mathbb{C}})$, and a member s of $W(\mathfrak{t}^{\mathbb{C}}:\mathfrak{g}^{\mathbb{C}})$ such that

$$ -\nu + t(\lambda + \mu) = s\lambda. $$

In the irreducible representation F_i of MAN, N must act trivially and $F_i|_{MA}$ must be irreducible. Thus $-\nu' = -t^{-1}\nu$ is a weight of F_i, and the member $s' = t^{-1}s$ of $W(\mathfrak{t}^{\mathbb{C}}:\mathfrak{g}^{\mathbb{C}})$ satisfies

$$ \lambda + \mu = s'\lambda + \nu'. $$

Since $\mathrm{Re}\,\lambda$ is dominant and μ is G-highest, we have $s'\mathrm{Re}\,\lambda \leq \mathrm{Re}\,\lambda$ and $\nu' \leq \mu$. Therefore

$$ s'\lambda = \lambda \qquad \text{and} \qquad \nu' = \mu. $$

Thus (14.84) can be nonzero only if F_ι contains the weight $-\mu$, i.e., only if $F_\iota = {}^M F_{-\mu}$. From (14.83) we therefore obtain

$$\begin{aligned}{}^G\psi_\lambda^{\lambda+\mu}(\operatorname{ind}_{MAN}^G V) &\cong p_\lambda[\operatorname{ind}_{MAN}^G({}^M F_{-\mu} \otimes V) \\ &\cong \operatorname{ind}_{MAN}^G[p_\lambda({}^M F_{-\mu} \otimes V)] = \operatorname{ind}_{MAN}^G({}^M\psi_\lambda^{\lambda+\mu} V).\end{aligned}$$

This proves (14.81), and (14.82) follows similarly by tracking down the effect on maps.

§13. Generalized Schmid Identities

Schmid's identity in Theorem 12.34 expresses the sum of two limits of discrete series with certain properties as an induced representation from a maximal parabolic subgroup. One of the assumed properties is that the Harish-Chandra parameters are orthogonal to only one positive root. We begin by dropping this assumption.

Theorem 14.68. Let G be linear connected reductive with rank $G =$ rank K, and let $\mathfrak{b} \subseteq \mathfrak{k}$ be a compact Cartan subalgebra. Fix a positive system Δ^+ for $\Delta(\mathfrak{b}^{\mathbb{C}}:\mathfrak{g}^{\mathbb{C}})$, and suppose α is a simple noncompact root. Suppose the Cayley transform $\mathbf{c}_\alpha$ leads from $\mathfrak{b}$ to a Cartan subalgebra $\mathfrak{b}^- \oplus \mathfrak{a}$, and let $S = MAN$ be a parabolic subgroup constructed from $\mathfrak{b}^- \oplus \mathfrak{a}$. Let Δ_M^+ be the positive system of $\Delta((\mathfrak{b}^-)^{\mathbb{C}}:\mathfrak{m}^{\mathbb{C}})$ given by

$$\Delta_M^+ = \{\mathbf{c}_\alpha(\gamma) \mid \gamma \in \Delta^+ \text{ and } \gamma \perp \alpha\}. \tag{14.85}$$

If λ is a member of $(i\mathfrak{b})'$ such that

(i) $\lambda - \delta$ is analytically integral
(ii) λ is Δ^+ dominant
(iii) $\langle \lambda, \alpha \rangle = 0$,

then

$$U(S, \pi^M(\lambda|_{\mathfrak{b}^-}, \Delta_M^+, \zeta), 0) \cong \pi(\lambda, \Delta^+) \oplus \pi(\lambda, s_\alpha \Delta^+), \tag{14.86}$$

where ζ is the character of $F(B^-) = \{1, \gamma_{\mathbf{c}_\alpha(\alpha)}\}$ given by

$$\zeta(\gamma_{\mathbf{c}_\alpha(\alpha)}) = (-1)^{2\langle \rho_\alpha, \alpha \rangle / |\alpha|^2}$$

and where ρ_α is half the sum of the roots of $\Delta(\mathfrak{b}^{\mathbb{C}}:\mathfrak{g}^{\mathbb{C}})$ whose inner product with α is positive.

Example. $G = SU(2, 1)$, $\lambda = 0$.

We take the Cartan subalgebra $\mathfrak{b}$ to be diagonal as usual. The Iwasawa $\mathfrak{a}$ can be taken to be $\mathbb{R}(E_{13} + E_{31})$; in this case, it is obtained from the Cayley transform $\mathbf{c}_\alpha$ with $\alpha = e_1 - e_3$. To meet the conditions of Theorem 14.68, we introduce a positive system Δ^+ for which α is simple, say

$$\Delta^+ \leftrightarrow \text{simple system } \{e_1 - e_3, e_3 - e_2\}.$$

Then $$s_\alpha\Delta^+ \leftrightarrow \{e_3 - e_1, e_1 - e_2\}.$$

The root $e_1 - e_2$ is compact and is simple for $s_\alpha\Delta^+$, and $\lambda = 0$ is orthogonal to it. Thus $\pi(\lambda, s_\alpha\Delta^+) = 0$. On the other hand, Δ^+ has no simple roots that are compact, and so $\pi(\lambda, \Delta^+) \neq 0$. The group $M = M_\mathfrak{p}$ for SU(2, 1) is a circle and is connected; thus the character ζ in the statement of the theorem can be ignored. The representation $\pi^M(\lambda|_{\mathfrak{b}^-}, \Delta_M^+, \zeta)$ is then simply the trivial character of the circle. Thus the theorem gives us the identity

$$U(S_\mathfrak{p}, 1, 0) \cong \pi(0, \Delta^+), \tag{14.87}$$

which says that the 0th unitary principal series is isomorphic with a certain limit of discrete series.

The proof of Theorem 14.68 is based on Schmid's identity and the following lemma.

Lemma 14.69. Let G be linear connected reductive, and let MAN be a parabolic subgroup built from a Cartan subalgebra $\mathfrak{a} \oplus \mathfrak{b}$ with $\mathfrak{a} \subseteq \mathfrak{p}$ and $\mathfrak{b} \subseteq \mathfrak{k}$. Let a positive system Δ_M^+ for $\Delta((\mathfrak{a} \oplus \mathfrak{b})^\mathbb{C} : (\mathfrak{m} \oplus \mathfrak{a})^\mathbb{C})$ be specified, and let λ be a Δ_M^+ dominant linear form on $(\mathfrak{a} \oplus \mathfrak{b})^\mathbb{C}$ vanishing on $\mathfrak{a}^\mathbb{C}$. Then there exists a positive system Δ_G^+ for $\Delta((\mathfrak{a} \oplus \mathfrak{b})^\mathbb{C} : \mathfrak{g}^\mathbb{C})$ such that

(i) $\Delta_M^+ \subseteq \Delta_G^+$
(ii) λ is Δ_G^+ dominant
(iii) every nonreal α in Δ_G^+ is such that $\theta\alpha$ ($= \alpha(H)$ on $\mathfrak{b}$ and $-\alpha(H)$ on $\mathfrak{a}$) is in Δ_G^+.

Under these conditions, suppose that $\Theta^M(\lambda, \Delta_M^+, \chi)$ is a discrete series or limit of discrete series character, and suppose that ν is imaginary-valued on $\mathfrak{a}$. Then

$$\Theta^M(\lambda, \Delta_M^+, \chi) \otimes e^\nu = {}^M\psi_{\lambda+\nu}^{\lambda+\nu+\mu}(\Theta^M(\lambda + \mu, \Delta_M^+, \chi e^\mu) \otimes e^\nu)$$

and, in obvious notation,

$$\operatorname{ind}_{MAN}^G[\Theta^M(\lambda, \Delta_M^+, \chi) \otimes e^\nu] = \psi_{\lambda+\nu}^{\lambda+\nu+\mu} \operatorname{ind}_{MAN}^G[\Theta^M(\lambda + \mu, \Delta_M^+, \chi e^\mu) \otimes e^\nu]$$

for every integral Δ_G^+ dominant linear form μ on $(\mathfrak{a} \oplus \mathfrak{b})^\mathbb{C}$ such that μ vanishes on $\mathfrak{a}$. Moreover, there exists such a μ satisfying

$$\langle \mu, \gamma \rangle > 0 \qquad \text{for } \gamma \text{ in } \Delta_M^+ \tag{14.88}$$

and even the stronger condition

$$\langle \mu, \gamma \rangle = 0 \qquad \text{for } \gamma \text{ in } \Delta_G^+ \text{ only if } \gamma|_\mathfrak{b} = 0. \tag{14.89}$$

Remarks. Less precisely the lemma says that we can back away in the M direction from the M parameter λ to an M parameter $\lambda + \mu$ that is nonsingular with respect to the nonreal roots of G, and then Zuckerman tensoring on the G level will move the induced representations in the expected way if the $\mathfrak{a}$ parameter ν is imaginary. The condition that the backing away by μ is in the M direction is expressed by the requirement that μ vanish on $\mathfrak{a}$. For μ to be Δ_G^+ dominant while vanishing on $\mathfrak{a}$, the positive system Δ_G^+ must take M before A in a certain way, and this way is expressed by condition (iii) in the statement.

Proof. The proof consists of three geometric constructions and an application of Theorem 14.67. For the geometric constructions, we shall just state the results. But we shall give the details of the application of Theorem 14.67.

The first geometric step is to show that the intersection of $(i\mathfrak{b})'$ with the integral points of $(\mathfrak{a} \oplus i\mathfrak{b})'$ forms a lattice in $(i\mathfrak{b})'$, i.e., a discrete subgroup with compact quotient. The second geometric step is to produce Δ_G^+ satisfying (i), (ii), and (iii), and the third geometric step is to prove the existence of μ satisfying the asserted conditions.

Now suppose that we are given Δ_G^+ satisfying (i), (ii), and (iii). We shall prove the character identities. Thus let μ be given. For the first identity, we notice that the group A acts as a scalar in ${}^M F_{-\mu}$ since ${}^M F_{-\mu}$ is irreducible on MA, and then

$$ {}^M F_{-\mu}(a) = 1 \qquad \text{for } a \text{ in } A \tag{14.90}$$

since μ vanishes on $\mathfrak{a}$. Therefore

$$\begin{aligned}
&{}^M\psi_{\lambda+\nu}^{\lambda+\nu+\mu}(\pi^M(\lambda+\mu, \Delta_M^+, \chi e^{\mu}) \otimes e^{\nu}) \\
&\quad = p_{\lambda+\nu}[\{\operatorname{ind}_{M^\#}^M(\pi^{M_0}(\lambda+\mu, \Delta_M^+) \otimes \chi e^{\mu}) \otimes e^{\nu}\} \otimes {}^M F_{-\mu}] \\
&\quad = p_{\lambda+\nu}[\operatorname{ind}_{M^\#}^M\{(\pi^{M_0}(\lambda+\mu, \Delta_M^+) \otimes {}^M F_{-\mu}|_{M_0}) \otimes \chi\} \otimes e^{\nu}] \\
&\quad = p_{\lambda} \operatorname{ind}_{M^\#}^M[(\pi^{M_0}(\lambda+\mu, \Delta_M^+) \otimes {}^M F_{-\mu}|_{M_0}) \otimes \chi] \otimes e^{\nu} \\
&\quad = \operatorname{ind}_{M^\#}^M p_{\lambda}[\pi^{M_0}(\lambda+\mu, \Delta_M^+) \otimes {}^M F_{-\mu}|_{M_0} \otimes \chi] \otimes e^{\nu},
\end{aligned}$$

the last step holding since $\operatorname{Ad}(M)$ is the identity on $Z(\mathfrak{m}^{\mathbb{C}})$. The last expression here is

$$\begin{aligned}
&= \operatorname{ind}_{M^\#}^M[\pi^{M_0}(\lambda, \Delta_M^+) \otimes \chi] \otimes e^{\nu} \\
&= \pi^M(\lambda, \Delta_M^+, \chi) \otimes e^{\nu}.
\end{aligned}$$

Taking the character of both sides, we obtain the first character identity of the lemma. Applying Theorem 14.67 to this first character identity, we obtain the second character identity of the lemma.

Proof of Theorem 14.68. We shall apply Lemma 14.69 to the decomposition $\mathfrak{g} = \mathfrak{a} \oplus \mathfrak{b}^-$. For Δ_G^+ in that lemma we use $\mathbf{c}_\alpha(\Delta^+)$, and for Δ_M^+ we use the definition (14.85). Then (i) holds. The form λ in the lemma is to be defined on $(\mathfrak{a} \oplus \mathfrak{b}^-)^{\mathbb{C}}$, and we therefore take it to be the $\mathbf{c}_\alpha$ Cayley transform of the λ in the statement of the theorem. Then (ii) holds.

Let $\alpha' = \mathbf{c}_\alpha(\alpha)$. If β in Δ_G^+ is not real, then β is not $\pm\alpha'$, and so $s_\alpha\beta$ is in Δ_G^+ because α is simple. However, $\theta\beta = s_{\alpha'}\beta$, and thus $\theta\beta$ is in Δ_G^+. This proves (iii).

Applying Lemma 3.1, we can find a Δ_G^+ dominant integral form $\mathbf{c}_\alpha(\mu)$ on $(\mathfrak{a} \oplus \mathfrak{b}^-)^{\mathbb{C}}$ that vanishes on $\mathfrak{a}$ and satisfies (14.89). The lemma applied with $v = 0$ says, in obvious notation for induced characters, that

$$\begin{aligned}&\operatorname{ind}_{MAN}^G[\Theta^M|_{\mathfrak{b}^-}, \Delta_M^+, \zeta) \otimes 1] \\ &\quad = \psi_{\mathbf{c}_\alpha(\lambda)}^{\mathbf{c}_\alpha(\lambda+\mu)} \operatorname{ind}_{MAN}^G[\Theta^M((\lambda + \mu)|_{\mathfrak{b}^-}, \Delta_M^+, \zeta e^{\mathbf{c}_\alpha(\mu)}) \otimes 1],\end{aligned}$$

and $e^{\mathbf{c}_\alpha(\mu)}$ is 1 on the element $\gamma_{\alpha'}$ since $\langle \mu, \alpha \rangle = 0$. Thus the right side is

$$= \psi_{\mathbf{c}_\alpha(\lambda)}^{\mathbf{c}_\alpha(\lambda+\mu)} \operatorname{ind}_{MAN}^G[\Theta^M((\lambda + \mu)|_{\mathfrak{b}^-}, \Delta_M^+, \zeta) \otimes 1].$$

Condition (14.89) says that $\langle \lambda + \mu, \beta \rangle = 0$ only for $\beta = \pm\alpha$, hence that the extra hypothesis in (iii) of Theorem 12.34 is satisfied. Since the ψ functor does not depend on a choice of Cartan subgroup or positive system of roots, Theorem 12.34 says that the right side of the above expression is

$$\begin{aligned}&= \psi_\lambda^{\lambda+\mu}(\Theta^G(\lambda + \mu, \Delta^+) + \Theta^G(\lambda + \mu, s_\alpha\Delta^+)) \\ &= \Theta^G(\lambda, \Delta^+) + \Theta^G(\lambda, s_\alpha\Delta^+).\end{aligned}$$

Since the representations in (14.86) are fully reducible, this character identity proves the theorem.

Lemma 14.70. Let $M^*A^*N^*$ be a parabolic subgroup of G with rank $M^* = \operatorname{rank}(K \cap M^*)$, let $\mathfrak{b}^* \subseteq \mathfrak{k} \cap \mathfrak{m}^*$ be a compact Cartan subalgebra of $\mathfrak{m}^*$, and let α be a noncompact root of $\Delta(\mathfrak{b}^{*\mathbb{C}} : \mathfrak{m}^{*\mathbb{C}})$. Suppose that the Cayley transform $\mathbf{c}_\alpha$ leads from the Cartan subalgebra $\mathfrak{a}^* \oplus \mathfrak{b}^*$ to a more noncompact Cartan subalgebra $\mathfrak{a} \oplus \mathfrak{b}$, and suppose that MAN is a parabolic subgroup built from $\mathfrak{a} \oplus \mathfrak{b}$ in the usual way. Let $\alpha' = \mathbf{c}_\alpha(\alpha)$. Then the inclusion mapping of Z_M into $Z_M(M \cap M^{*\#})$ yields an isomorphism

$$Z_M/\{1, \gamma_{\alpha'}\}Z_{M_0}Z_{M^*} \cong Z_M(M \cap M^{*\#})/(M \cap M^{*\#}). \tag{14.91}$$

The groups on either side of (14.91) have order at most 2. They have order exactly 2 if and only if the reflection s_α is present in $W(B^* : M^*)$, where $B^* = \exp \mathfrak{b}^*$.

Examples.

(1) Let $G = M^* = SL(2, \mathbb{R})$, and let MAN be the upper triangular group. Then $Z_M = \{1, \gamma_{\alpha'}\} = Z_{M^*} = M$, and $Z_{M_0} = \{1\}$. Hence both sides of (14.91) collapse to the identity. The Weyl group $W(B^* : M^*) = W(B^* : G)$ contains only the reflections in the compact roots when G is connected, and here there are no compact roots. So $W(B^* : G)$ is trivial.

(2) Let $G = M^* = SL^\pm(2, \mathbb{R})$, and let MAN be the upper triangular group. This example has G disconnected and is not directly covered by the

lemma, but it can be imbedded in a connected example, such as $Sp(2, \mathbb{R})$. In any event, M here is the four-element group of diagonal entries with ± 1 in each diagonal entry, and $Z_M = M$. Moreover, the other subgroups in question are related by $\{1, \gamma_{\alpha'}\} = Z_{M^*} = M \cap M^{*\#}$. So the groups on the two sides of (14.91) have order 2. The reflection s_α in $W(B^*:G)$ has $\begin{pmatrix} 1 & 0 \\ 0 & -1 \end{pmatrix}$ as a representative.

Proof of lemma. First we prove the isomorphism. Clearly Z_M maps onto the right side of (14.91). We have

$$\{1, \gamma_{\alpha'}\} \subseteq M_\mathfrak{p} \cap M_0^*, \qquad Z_{M_0} \subseteq M_0, \qquad Z_{M^*} \subseteq M_\mathfrak{p} \cap M^{*\#},$$

and all three right sides are contained in $M \cap M^{*\#}$. Hence

$$\{1, \gamma_{\alpha'}\} Z_{M_0} Z_{M^*} \subseteq M \cap M^{*\#},$$

and our map yields a quotient homomorphism of the left side of (14.91) onto the right side. The right side of (14.91) is isomorphic with

$$Z_M/(Z_M \cap M \cap M^{*\#}) = Z_M/(Z_M \cap M^{*\#})$$

by standard group theory, and the isomorphism will follow if we show

$$Z_M \cap M^{*\#} \subseteq \{1, \gamma_{\alpha'}\} Z_{M_0} Z_{M^*},$$

hence if we show

$$Z_M \cap M_0^* \subseteq \{1, \gamma_{\alpha'}\} Z_{M_0}.$$

But this is just the special case of Lemma 12.30b in which we replace G by M_0^* and M by $M \cap M_0^*$. The isomorphism follows.

As in (11.57), we choose root vectors $E_{\alpha'}$ and $E_{-\alpha'} = \theta E_{\alpha'}$ in $\mathfrak{g}$ so that $\operatorname{Re} B_0(E_{\alpha'}, E_{-\alpha'}) = -2/|\alpha'|^2$, where B_0 is the trace form. The vectors $E_{\alpha'}$, $E_{-\alpha'}$, $H_{\alpha'}$ span an $\mathfrak{sl}(2, \mathbb{R})$ subalgebra $\mathfrak{s} \subseteq \mathfrak{g}$. Let us see that each $\operatorname{Ad}(z)$ for $z \in Z_M$ carries $\mathfrak{s}$ into itself. In fact, if H is in $\mathfrak{a} \oplus \mathfrak{b}$, then $\operatorname{Ad}(z)E_{\alpha'}$ is in $\mathfrak{g}$ and we have

$$[H, \operatorname{Ad}(z)E_{\alpha'}] = \operatorname{Ad}(z)[\operatorname{Ad}(z)^{-1}H, E_{\alpha'}] = \operatorname{Ad}(z)[H, E_{\alpha'}] = \alpha'(H) \operatorname{Ad}(z)E_{\alpha'}.$$

Thus $\operatorname{Ad}(z)E_{\alpha'} = \pm E_{\alpha'}$. Applying θ, we obtain $\operatorname{Ad}(z)E_{-\alpha'} = \pm E_{-\alpha'}$ with the same sign. Since $E_{\alpha'}$ and $E_{-\alpha'}$ generate $\mathfrak{s}$, $\operatorname{Ad}(z)$ carries $\mathfrak{s}$ into itself and is determined by its effect on $E_{\alpha'}$, which must be ± 1.

Now Ad of each factor of $\{1, \gamma_\alpha\} Z_{M_0} Z_{M^*}$ acts trivially on $\mathfrak{s}$. Conversely if z in Z_M is such that $\operatorname{Ad}(z)$ is 1 on $\mathfrak{s}$, then the fact that $\mathfrak{b}^* = \mathfrak{b} \oplus \mathbb{R}(E_{\alpha'} + E_{-\alpha'})$ means that $\operatorname{Ad}(z) = 1$ on $\mathfrak{b}^*$. By Lemma 12.30d, z is in $M^{*\#}$. Hence the right side of (14.91) exhibits z as in the trivial coset of (14.91). We conclude that the coset of $z \in Z_M$ in (14.91) is determined by the effect of $\operatorname{Ad}(z)$ on $E_{\alpha'}$, which is multiplication by ± 1. Hence each side of (14.91) has order at most 2.

Suppose each side of (14.91) has order 2. Then we have just seen that there exists z in Z_M with $\mathrm{Ad}(z)E_{\alpha'} = -E_{\alpha'}$. This z has

$$\mathrm{Ad}(z)(E_{\alpha'} + \theta E_{\alpha'}) = -(E_{\alpha'} + \theta E_{\alpha'})$$

and $\mathrm{Ad}(z)|_{\mathfrak{b}} = 1$. Since $E_{\alpha'} + \theta E_{\alpha'}$ is a multiple of H_α, $\mathrm{Ad}(z)$ acts as the reflection s_α on $\mathfrak{b}^*$. Thus s_α exists in $W(B^*:M^*)$.

Conversely if s_α exists in $W(B^*:M^*)$, let u be a representative. Let E_α and $E_{-\alpha}$ be root vectors for α. Since u carries α to $-\alpha$,

$$\mathrm{Ad}(u)E_\alpha = e^{i\theta}E_{-\alpha} \qquad \text{and} \qquad \mathrm{Ad}(u)E_{-\alpha} = e^{i\varphi}E_\alpha.$$

Bracketing these equations and using the relation $\mathrm{Ad}(u)H_\alpha = -H_\alpha$, we see that $\varphi = -\theta$. Thus

$$\mathrm{Ad}(u)(E_\alpha + E_{-\alpha}) = e^{i\theta}E_\alpha + e^{-i\theta}E_{-\alpha}.$$

The element $b^* = \exp(i\theta|\alpha|^{-2}H_\alpha)$ of B^* satisfies

$$\mathrm{Ad}(b^*)(E_\alpha + E_{-\alpha}) = e^{i\theta}E_\alpha + e^{-i\theta}E_{-\alpha},$$

and therefore Ad of $w = b^{*-1}u$ fixes $E_\alpha + E_{-\alpha}$. It also fixes $\mathfrak{b}$ and thus centralizes $\mathfrak{a} \oplus \mathfrak{b}$. Since it is in K, Lemma 12.30c says $w = bz$ with b in $B = \exp \mathfrak{b}$ and Z in $F(B) \subseteq Z_M$. Then z is a member of Z_M for which $\mathrm{Ad}(z)$ is not 1 on $\mathfrak{s}$, and it follows that the two sides of (14.91) have order 2.

Theorem 14.71 (generalized Schmid identities). Let G be linear connected reductive with compact center, let $M^*A^*N^*$ be a parabolic subgroup of G with rank $M^* = \mathrm{rank}(K \cap M^*)$, let $\mathfrak{b}^* \subseteq \mathfrak{k} \cap \mathfrak{m}^*$ be a compact Cartan subalgebra of $\mathfrak{m}^*$, and let $\Delta_{M^*}^+$ be a positive system for $\Delta(\mathfrak{b}^{*\mathbb{C}}:\mathfrak{m}^{*\mathbb{C}})$. Let α be a simple noncompact root of $\Delta_{M^*}^+$. Suppose that the Cayley transform $\mathbf{c}_\alpha$ leads from the Cartan subalgebra $\mathfrak{a}^* \oplus \mathfrak{b}^*$ to a more noncompact Cartan subalgebra $\mathfrak{a} \oplus \mathfrak{b}$, and suppose that $S = MAN$ is a parabolic subgroup built from $\mathfrak{a} \oplus \mathfrak{b}$ in the usual way. Put $\Delta_M^+ = \mathbf{c}_\alpha(\alpha^\perp \cap \Delta_{M^*}^+)$. Finally suppose that $\Theta^{M^*}(\lambda, \Delta_{M^*}^+, \chi)$ is a well-defined limit of discrete series character of M^* with $\langle \lambda, \alpha \rangle = 0$, and suppose ν is an imaginary-valued linear functional on $\mathfrak{a}^*$. With $\alpha' = \mathbf{c}_\alpha(\alpha)$ and $B^* = \exp \mathfrak{b}^*$, there are the following cases:

(a) If s_α does not exist in $W(B^*:M^*)$, then $Z_M = \{1, \gamma_{\alpha'}\}Z_{M_0}Z_{M^*}$ and

$$\begin{aligned}\Theta^{M^*}(\lambda, \Delta_{M^*}^+, \chi) \otimes e^\nu + \Theta^{M^*}(\lambda, s_\alpha\Delta_{M^*}^+, \chi) \otimes e^\nu \\ = \mathrm{ind}_{S \cap M^*A^*}^{M^*A^*}[\Theta^M(\lambda|_{\mathfrak{b}}, \Delta_M^+, \zeta \otimes \chi) \otimes e^{\nu \oplus 0}], \quad (14.92\mathrm{a})\end{aligned}$$

where ζ is the character of $\{1, \gamma_{\alpha'}\}$ defined in Theorem 14.68.

(b) If s_α exists in $W(B^*:M^*)$, then $|Z_M/\{1, \gamma_{\alpha'}\}Z_{M_0}Z_{M^*}| = 2$. Let $(\zeta \otimes \chi)^+$ and $(\zeta \otimes \chi)^-$ denote the two extensions of $\zeta \otimes e^{(\lambda - \delta_M)|_{\mathfrak{b}}} \otimes \chi$ from $\{1, \gamma_{\alpha'}\}Z_{M_0}Z_{M^*}$ to Z_M, with ζ as in Theorem

14.68. Then

$$\begin{aligned}\Theta^{M^*}(\lambda, \Delta^+_{M^*}, \chi) \otimes e^\nu &= \operatorname{ind}^{M^*A^*}_{S \cap M^*A^*}[\Theta^M(\lambda|_{\mathfrak{b}}, \Delta^+_M, (\zeta \otimes \chi)^+) \otimes e^{\nu \oplus 0}] \\ &= \operatorname{ind}^{M^*A^*}_{S \cap M^*A^*}[\Theta^M(\lambda|_{\mathfrak{b}}, \Delta^+_M, (\zeta \otimes \chi)^-) \otimes e^{\nu \oplus 0}]. \quad (14.92\text{b})\end{aligned}$$

Remarks. In the context of the examples with Lemma 14.70, identity (a) applies to Example 1, and identity (b) applies to Example 2. Identity (a) is just the familiar reducibility relation for the unitary principal series $\mathscr{P}^{-,0}$ of SL(2, $\mathbb{R}$). On the other hand, there is only one limit of discrete series $\mathscr{D}_1$ in SL$^\pm$(2, $\mathbb{R}$), and identity (b) shows it can be viewed as a unitary principal series representation in two ways.

Proof. We shall suppress some of the variables to simplify the notation. We start by applying Theorem 14.68 to M_0^*, obtaining

$$\Theta^{M_0^*}(\Delta^+_{M^*}) + \Theta^{M_0^*}(s_\alpha \Delta^+_{M^*}) = \operatorname{ind}^{M_0^*}_{S \cap M_0^*}(\Theta^{M \cap M_0^*} \otimes 1^{A \cap M^*}).$$

Next we extend this relation to $M^{*\#}$ by χ; the required consistency condition for this extension is satisfied on the left side since $\Theta^{M^*}(\lambda, \Delta^+_{M^*}, \chi)$ is well defined, and it is therefore satisfied on the right side by (14.86). Inducing to M^* and sorting out matters on the right side by application of the double induction formula, we obtain

$$\begin{aligned}\Theta^{M^*}(\Delta^+_{M^*}) + \Theta^{M^*}(s_\alpha \Delta^+_{M^*}) &= \operatorname{ind}^{M^*}_{S \cap M^{*\#}}(\Theta^{M \cap M^{*\#}} \otimes 1^{A \cap M^*}) \\ &= \operatorname{ind}^{M^*}_{S \cap M^{*\#}}[(\operatorname{ind}^{M}_{Z_M(M \cap M^{*\#})} \operatorname{ind}^{Z_M(M \cap M^{*\#})}_{M \cap M^{*\#}} \Theta^{M \cap M^{*\#}}) \otimes 1^{A \cap M^*}]. \quad (14.93)\end{aligned}$$

(a) If s_α does not exist in $W(B^*:M^*)$, Lemma 14.70 says that $Z_M \subseteq M \cap M^{*\#}$. Therefore $M^\# \subseteq M \cap M^{*\#}$, and the expression in parentheses equals Θ^M. Putting e^ν in place on both sides of (14.93), we obtain (14.92a).

(b) If s_α does exist in $W(B^*:M^*)$, then Lemma 14.70 says that $|Z_M(M \cap M^{*\#})/(M \cap M^{*\#})| = 2$. Since the extension of $M \cap M^{*\#}$ to $Z_M(M \cap M^{*\#})$ is by central elements, we have

$$\operatorname{ind}^{Z_M(M \cap M^{*\#})}_{M \cap M^{*\#}} \Theta^{M \cap M^{*\#}}) = \Theta^{M \cap M^{*\#}} \otimes (\zeta \otimes \chi)^+ + \Theta^{M \cap M^{*\#}} \otimes (\zeta \otimes \chi)^-.$$

Substituting into (14.93) and using the fact that $M^\# \subseteq Z_M(M \cap M^{*\#})$, we obtain

$$\begin{aligned}\Theta^{M^*}(\Delta^+_{M^*}) + \Theta^{M^*}(s_\alpha \Delta^+_{M^*}) &= \operatorname{ind}^{M^*}_{S \cap M^*}[\Theta^M((\zeta \otimes \chi)^+) \otimes 1^{A \cap M^*}] \\ &\quad + \operatorname{ind}^{M^*}_{S \cap M^*}[\Theta^M((\zeta \otimes \chi)^-) \otimes 1^{A \cap M^*}].\end{aligned}$$

Since s_α is in $W(B^*:M^*)$, Proposition 12.33c says the two characters on the left side are equal. To complete the proof, it is enough to show that the two induced characters on the right side are equal, since we can put e^ν in place to get (14.92b).

The equivalences produced by standard intertwining operators show that it is enough to prove that $\pi^M(\lambda|_{\mathfrak{b}}, \Delta_M^+, (\zeta \otimes \chi)^+)$ and $\pi^M(\lambda|_{\mathfrak{b}}, \Delta_M^+, (\zeta \otimes \chi)^-)$ are conjugate by some element of the Weyl group $W((A \cap M^*): M^*)$. The element we use is $s_{\alpha'}$, which exists in $W((A \cap M^*): M^*)$ since α' is a real root. Since α' as a root of $(\mathfrak{m}^*, (\mathfrak{a} \cap \mathfrak{m}^*))$ is odd, $s_{\alpha'}$ fixes $\lambda|_{\mathfrak{b}}$. Thus we want to see that $s_{\alpha'}((\zeta \otimes \chi)^+) = (\zeta \otimes \chi)^-$. The two central characters $(\zeta \otimes \chi)^+$ and $(\zeta \otimes \chi)^-$ are equal on the subgroup $\{1, \gamma_{\alpha'}\}Z_{M_0}Z_{M^*}$ of Z_M, and $s_{\alpha'}$ fixes this subgroup. The difficulty is with a representative z of the unique nontrivial coset of Z_M modulo $\{1, \gamma_{\alpha'}\}Z_{M_0}Z_{M^*}$ (cf. Lemma 14.70). In this case we have

$$(\zeta \otimes \chi)^+(z) = -(\zeta \otimes \chi)^-(z)$$

and thus we are to prove that

$$(\zeta \otimes \chi)^+((w^{-1}zw)z^{-1}) = -1$$

if w is a representative of $s_{\alpha'}$. To do so, we prove

$$w^{-1}zw = \gamma_{\alpha'}z \tag{14.94a}$$

$$(\zeta \otimes \chi)^+(\gamma_{\alpha'}) = -1. \tag{14.94b}$$

Formula (14.94b) says that ζ and χ agree on $\gamma_{\alpha'}$ and that their common value is -1. Their agreement on $\gamma_{\alpha'}$ is implicit in (14.86) and the fact that Θ^M is well defined. Now

$$\zeta(\gamma_{\alpha'}) = (-1)^{2\langle \rho_\alpha, \alpha\rangle/|\alpha|^2},$$

and this will be -1 if and only if the number of pairs $\{\beta, -s_\alpha\beta\}$ of roots with $2\langle\beta, \alpha\rangle/|\alpha|^2 = 1$ is even. In fact, the number of such pairs is 0, because the existence of s_α in $W(B^*: M^*)$ means that s_α has to preserve type—compact or noncompact. Since in such a pair $\{\beta, -s_\alpha\beta\}$, one root will be compact and the other noncompact, there can be no such pairs. This proves (14.94b).

Finally we prove (14.94a). Since z is in Z_M, Lemma 12.30b says that z is in $\exp(\mathfrak{a} \oplus \mathfrak{b})^{\mathbb{C}}$. Thus let $z = \exp H$ with H in $(\mathfrak{a} \oplus \mathfrak{b})^{\mathbb{C}}$, and assume without loss of generality that w representing s_α is chosen as in Lemma 14.46a. Then

$$(w^{-1}zw)z^{-1} = \exp(s_\alpha H - H) = \exp(-2|\alpha|^{-2}\alpha(H)H_\alpha). \tag{14.95}$$

The proof of Lemma 14.70 showed that $\mathrm{Ad}(z)E_{\alpha'} = -E_{\alpha'}$, and it follows that $\alpha(H)$ is an odd multiple of πi. Therefore the right side of (14.95) is an odd power of $\gamma_{\alpha'}$, and (14.94a) follows. This completes the proof of the theorem.

A representation of G will be called a **basic representation** if it is a full induced representation from some parabolic subgroup $S = MAN$ with a discrete series or limit of discrete series on M and a unitary character on A (imaginary parameter on $\mathfrak{a}$). Its global character we may then write as

$$\operatorname{ind}_S^G \Theta^{MA}(\lambda, \Delta_M^+, \chi, \nu) \tag{14.96}$$

by combining $\Theta^M(\lambda, \Delta_M^+, \chi)$ and e^ν as indicated. A character written in this way we call a **basic character.** A basic character written in the form (14.96) in such a way that $\Theta^M(\lambda, \Delta_M^+, \chi)$ is a discrete series character will be called a **basic character induced from discrete series.** These definitions intentionally carry along the parametrization of the character, not just the character itself; we have seen in SU(2, 1) that the 0th principal series character equals a limit of discrete series character, and our definition distinguishes between these two ways of writing the character.

Corollary 14.72. Every basic character is contained in a basic character induced from discrete series (in the sense that the basic character induced from discrete series is the sum of the given basic character and some other global character).

Proof. For a basic character written as in (14.96), the proof proceeds by induction downward on dim A, with G fixed. The first step of the induction is when S is a minimal parabolic subgroup. In this case M is compact. Thus λ nonsingular relative to $\Delta(\mathfrak{b}^{\mathbb{C}}:\mathfrak{m}^{\mathbb{C}})$ gives a finite-dimensional (discrete series) representation of M, while λ singular gives 0.

Now suppose $\Theta^{M^*A^*}(\lambda, \Delta_{M^*}^+, \chi, \nu)$ is given and the result is known when dim $A >$ dim A^*. If λ is nonsingular relative to all simple roots in $\Delta_{M^*}^+$, then $\Theta^{M^*}(\lambda, \Delta_{M^*}^+, \chi)$ is in the discrete series, and there is nothing to prove. If λ is singular with respect to some $\Delta_{M^*}^+$ simple compact root, then Theorem 12.26b says $\Theta^{M^*}(\lambda, \Delta_{M^*}^+, \chi)$ is 0. Finally if λ is singular with respect to some $\Delta_{M^*}^+$ simple noncompact root α, then the conditions of Theorem 14.71 are satisfied and either $\Theta^{M^*A^*}(\lambda, \Delta_{M^*}^+, \chi, \nu)$ or

$$\Theta^{M^*A^*}(\lambda, \Delta_{M^*}^+, \chi, \nu) + \Theta^{M^*A^*}(\lambda, s_\alpha\Delta_{M^*}^+, \chi, \nu)$$

is the left side of a generalized Schmid identity (14.92a) or (14.92b). Inducing the identity from S^* to G, we can apply our inductive hypothesis to the right side to complete the proof.

§14. Inversion of Generalized Schmid Identities

At the end of §13, we used generalized Schmid identities to imbed basic characters in basic characters attached to smaller parabolic subgroups (and built from more noncompact Cartan subgroups). Now we turn matters around, in order to exhibit reducibility.

Theorem 14.73. Let G be linear connected reductive with compact center, let $S = MAN$ be a parabolic subgroup of G with rank $M = \operatorname{rank}(K \cap M)$, let $\mathfrak{b} \subseteq \mathfrak{k} \cap \mathfrak{m}$ be a compact Cartan subalgebra of $\mathfrak{m}$, and let Δ_M^+ be a positive system for $\Delta(\mathfrak{b}^{\mathbb{C}}:\mathfrak{m}^{\mathbb{C}})$. Let α be a real root of $\Delta((\mathfrak{a} \oplus \mathfrak{b})^{\mathbb{C}}:\mathfrak{g}^{\mathbb{C}})$. Suppose that the Cayley transform $\mathbf{d}_\alpha$ leads from the Cartan subalgebra $\mathfrak{a} \oplus \mathfrak{b}$ to a more noncompact Cartan subalgebra $\mathfrak{a}^* \oplus \mathfrak{b}^*$, and suppose that $M^*A^*N^*$ is a parabolic subgroup built from $\mathfrak{a}^* \oplus \mathfrak{b}^*$ in the usual way. Let $\tilde{\alpha} = \mathbf{d}_\alpha(\alpha)$. Then the character

$$\operatorname{ind}_{S \cap M^*A^*}^{M^*A^*} \Theta^{MA}(\lambda_M, \Delta_M^+, \chi_M, \nu_M)$$

is the right side of a generalized Schmid identity (14.92a) or (14.92b) obtained from $\tilde{\alpha}$ if and only if $\langle \nu_M, \alpha \rangle = 0$ and

$$\chi_M(\gamma_\alpha) = (-1)^{2\langle \rho_\alpha, \alpha \rangle / |\alpha|^2}, \tag{14.97}$$

where ρ_α is half the sum of all members of $\Delta((\mathfrak{a} \oplus \mathfrak{b})^{\mathbb{C}}:\mathfrak{g}^{\mathbb{C}})$ whose restriction to $\mathfrak{a}$ is a positive multiple of α. When this condition is satisfied, the positive system $\Delta_{M^*}^+$ can be taken as any positive system for $\Delta(\mathfrak{b}^{*\mathbb{C}}:\mathfrak{m}^{*\mathbb{C}})$ such that

(i) $\mathbf{d}_\alpha \Delta_M^+ \cup \{\tilde{\alpha}\} \subseteq \Delta_{M^*}^+$
(ii) $\lambda_M \oplus 0$ is $\Delta_{M^*}^+$ dominant
(iii) every nonreal β in $\Delta_{M^*}^+$ is such that $\theta\beta$ is in $\Delta_{M^*}^+$.

Proof. The necessity of $\langle \nu_M, \alpha \rangle = 0$ and condition (14.97) are obvious from Theorem 14.71. For the sufficiency we are to construct $\Theta^{M^*A^*}(\lambda, \Delta_{M^*}^+, \chi, \nu)$ with the properties in Theorem 14.71. The definitions of λ, χ, and ν have to be

$$\lambda = \begin{cases} \lambda_M & \text{on } \mathfrak{b} \\ 0 & \text{on } H_{\tilde{\alpha}} \end{cases}$$

$$\chi = \chi_M|_{Z_{M^*}}, \qquad \text{and} \qquad \nu = \nu_M|_{\mathfrak{a}^*}. \tag{14.98}$$

We construct $\Delta_{M^*}^+$ by applying Lemma 14.69 to M_0^*; if $\tilde{\alpha}$ ends up negative, then we replace $\Delta_{M^*}^+$ by $s_{\tilde{\alpha}}\Delta_{M^*}^+$. Then (i), (ii), and (iii) hold above. Problem 30 at the end of the chapter shows that (i), (ii), and (iii) imply that $\tilde{\alpha}$ is simple. The theorem will therefore be proved if we show

$$\lambda - \delta_{M^*} \text{ is } \mathfrak{b}^* \text{ integral, and} \tag{14.99a}$$

$$e^{\lambda - \delta_{M^*}} \text{ agrees with } \chi \text{ on } Z_{M^*} \cap B^*. \tag{14.99b}$$

We can prove (14.99) for the δ_{M^*} obtained from any $\Delta_{M^*}^+$, not necessarily one satisfying (i), (ii), and (iii) above.

Thus form $\Delta((\mathbb{R}H_\alpha \oplus \mathfrak{b})^{\mathbb{C}}:\mathfrak{m}^{*\mathbb{C}})$, order $\mathbb{R}H_\alpha$ before $i\mathfrak{b}$, and transfer the resulting positive system to $\Delta(\mathfrak{b}^{*\mathbb{C}}:\mathfrak{m}^{*\mathbb{C}})$ by the Cayley transform $\mathbf{d}_\alpha$. Then we obtain

$$\delta_{M^*} = \rho_{\tilde{\alpha}} + \delta_M,$$

where $\rho_{\tilde{\alpha}} = \mathbf{d}_\alpha(\rho_\alpha)$.

Let $B_{\tilde{\alpha}} = \exp i\mathbb{R}H_{\tilde{\alpha}}$; this is a circle group in the usual $SL(2, \mathbb{R})$ subgroup built from the real root α. We have $B^* = BB_{\tilde{\alpha}}$ on the level of connected groups and

$$\lambda - \delta_{M^*} = (\lambda_M - \delta_M) + (-\rho_{\tilde{\alpha}}) \tag{14.100}$$

as linear functionals on $\mathfrak{b}^*$. Since Θ^{MA} is given as well defined, $\lambda_M - \delta_M$ exponentiates to B. Thus $\lambda - \delta_{M^*}$ is $\mathfrak{b}^*$ integral if and only if

$$-\rho_{\tilde{\alpha}} \text{ exponentiates to } B_{\tilde{\alpha}}, \text{ and} \tag{14.101a}$$

$$e^{\lambda_M - \delta_M} \text{ and } e^{-\rho_{\tilde{\alpha}}} \text{ agree on } B \cap B_{\tilde{\alpha}}. \tag{14.101b}$$

Let us prove these relations.

For (14.101a), $\rho_{\tilde{\alpha}}$ is a half-integral multiple of $\tilde{\alpha}$, and $\tilde{\alpha}$ exponentiates to B^*, hence to $B_{\tilde{\alpha}}$, since $\tilde{\alpha}$ is a root. Thus the only way ρ_α can fail to exponentiate to $B_{\tilde{\alpha}}$ is if both

$$\rho_{\tilde{\alpha}} = (n + \tfrac{1}{2})\tilde{\alpha} \text{ for some integer } n$$

and

$$\gamma_\alpha = 1,$$

and these two equations together are incompatible with (14.97). This proves (14.101a).

For (14.101b), the most general element of $B_{\tilde{\alpha}}$ is $\exp(i\theta|\tilde{\alpha}|^{-2}H_{\tilde{\alpha}})$, and a root vector E_α for $\tilde{\alpha}$ satisfies

$$\mathrm{Ad}(\exp i\theta|\tilde{\alpha}|^{-2}H_{\tilde{\alpha}})E_{\tilde{\alpha}} = e^{i\theta}E_{\tilde{\alpha}}.$$

Since $\mathrm{Ad}(b)$ fixes $E_{\tilde{\alpha}}$ for b in B, the only elements of B_α that can be in B have θ a multiple of 2π. That is, $B \cap B_\alpha \subseteq \{1, \gamma_\alpha\}$. If γ_α lies in $B \cap B_{\tilde{\alpha}}$, then $e^{\lambda_M - \delta_M}$ and $e^{-\rho_{\tilde{\alpha}}}$ agree on γ_α by (14.97). This proves (14.101b). Thus (14.99a) is proved.

For (14.99b), we use Lemma 12.30b to write

$$Z_{M^*} \cap B^* \subseteq Z_M \cap M_0^* = \{1, \gamma_\alpha\}(Z_M \cap B).$$

Since Θ^{MA} is given as well defined, $e^{\lambda_M - \delta_M}$ agrees with χ_M on $Z_M \cap B$. By (14.100), $e^{\lambda - \delta}$ agrees with χ on $Z_M \cap B$. If γ_α lies in Z_{M^*}, then

$$\chi(\gamma_\alpha) = \chi_M(\gamma_\alpha) = (-1)^{2\langle\rho_\alpha,\alpha\rangle/|\alpha|^2} = (e^{-\rho_{\tilde{\alpha}}})(\gamma_\alpha) = (e^{\lambda - \delta_{M^*}})(\gamma_\alpha)$$

by (14.98), (14.97), and (14.100). Thus (14.99b) is proved, and the theorem follows.

A basic character (with its built-in parametrization) is **final** if the MA character from which it is induced is not the right side of any generalized Schmid identity (14.92a) or (14.92b). According to Theorem 14.73, $\text{ind}_S^G\, \Theta^{MA}(\lambda, \Delta_M^+, \chi, \nu)$ is final if there is no real root α of $\Delta((\mathfrak{a} \oplus \mathfrak{b})^{\mathbb{C}} : \mathfrak{g}^{\mathbb{C}})$ such that $\langle \nu, \alpha \rangle = 0$ and $\chi(\gamma_\alpha) = (-1)^{2\langle \rho_\alpha, \alpha \rangle / |\alpha|^2}$.

Corollary 14.74. Every final basic character is irreducible (or zero).

Proof. Let $\text{ind}_S^G\, \Theta^{MA}(\lambda, \Delta_M^+, \chi, \nu)$ be final and nonzero. Choose a positive system Δ_G^+ for $\Delta((\mathfrak{a} \oplus \mathfrak{b})^{\mathbb{C}} : \mathfrak{g}^{\mathbb{C}})$ satisfying the conditions (i), (ii), and (iii) of Lemma 14.69, and choose μ on $(\mathfrak{a} \oplus \mathfrak{b})^{\mathbb{C}}$ in that lemma such that (14.88) holds. The lemma says that

$$\text{ind}_S^G\, \Theta^{MA}(\lambda, \Delta_M^+, \chi, \nu) = \psi_{\lambda+\nu}^{\lambda+\nu+\mu}\, \text{ind}_S^G\, \Theta^{MA}(\lambda + \mu, \Delta_M^+, \chi e^\mu, \nu). \quad (14.102)$$

We show that the induced representation on the right side of (14.102) is irreducible. Arguing by contradiction, suppose it is reducible. Then its R group is not $\{1\}$, by Theorem 14.43c, and the superorthogonal set $\mathscr{H}$ in Theorem 14.64 must be nonempty. If α is in the set $\mathscr{H}$, then $\langle \nu, \alpha \rangle = 0$ by Theorem 14.64b and

$$(\chi e^\mu)(\gamma_\alpha) = (-1)^{2\langle \rho_\alpha, \alpha \rangle / |\alpha|^2}$$

from the proof of Lemma 14.61. Since $\langle \mu, \mathbf{c}_\alpha(\alpha) \rangle = 0$, $e^\mu(\gamma_\alpha) = 1$. Thus $\chi(\gamma_\alpha)$ satisfies (14.97), and Theorem 14.73 says that the induced character on the right side of (14.102) is not final. Therefore the left side is not final. This is a contradiction, and we conclude that the induced representation is irreducible.

By Theorem 10.50 the right side of (14.102) must be a multiple of an irreducible character. Combining Corollary 14.72 and the Multiplicity One Theorem (Corollary 14.66), we see that the multiple must be one. Hence the given character is irreducible.

Corollary 14.75. A nonzero basic character $\text{ind}_S^G\, \Theta^{MA}(\lambda, \Delta_M^+, \chi, \nu)$ is irreducible if $\langle \nu, \alpha \rangle \neq 0$ for every root α of $(\mathfrak{g}, \mathfrak{a})$ with odd multiplicity.

Theorem 14.76. Every irreducible tempered character is equal to a basic character. The basic character can be taken to be final.

Proof. By Theorem 8.53c, we can imbed such a character in a basic character induced from discrete series. The latter we can attempt to use as the right side of an induced generalized Schmid identity. We can do so unless it is final, in which case Corollary 14.75 says it is irreducible. If it is the right side of an induced generalized Schmid identity, only one term of the left side can be nonzero. We attempt to use this nonzero term as the right side of an induced generalized Schmid identity, and we continue in this way. Since the dimension of A is decreasing by one at each step, we ultimately obtain a final basic character.

Remarks. Theorem 14.76 allows us to complete the proof of Theorem 12.23, which relates temperedness of characters to temperedness of representations. In fact, if Θ is the character of an irreducible tempered representation, we now know that Θ is induced from a limit of discrete series on M and a unitary character on A. The numerators of characters of limits of discrete series are known to be bounded, and induction with a unitary character on A preserves this property. Thus Θ has bounded numerators. These remarks show that (c) implies (b) in Theorem 12.23, and the proof of Theorem 12.23 is now complete.

§15. Complete Reduction of Induced Representations

We come now to the actual reduction into irreducible representations of $U(S, \sigma, \nu)$, where σ is in the discrete series of M and ν is imaginary. Theorem 14.43 indicates that the R group controls the amount of reducibility, and the Multiplicity One Theorem says that R even controls the commuting algebra. But more is so: The superorthogonal set $\mathscr{H}$ of Theorem 14.64 directly points to the irreducible basic characters whose sum is a given basic character induced from discrete series.

To be more explicit, let $\mathscr{H} = \{\alpha_1, \ldots, \alpha_k\}$ be the superorthogonal set of real roots in $\Delta((\mathfrak{a} \oplus \mathfrak{b})^{\mathbb{C}}:\mathfrak{g}^{\mathbb{C}})$ that are obtained from $R_{\sigma,\nu}$. Form the Cayley transforms $\mathbf{d}_{\alpha_1}, \ldots, \mathbf{d}_{\alpha_k}$. The product $\mathbf{d}_{\mathscr{H}} = \prod_{j=1}^k \mathbf{d}_{\alpha_j}$ does not depend on what order the roots α_j are written in, since no $\alpha_i \pm \alpha_j$ is a root, and $\mathbf{d}_{\mathscr{H}}$ (regarded as a finite iterated sequence of Cayley transforms) leads us to the data

$$\mathfrak{b} \oplus \sum i\mathbb{R}H_{\tilde{\alpha}_j} = \mathfrak{b}^*, \qquad \text{where} \quad \tilde{\alpha}_j = \mathbf{d}_{\mathscr{H}}(\alpha_j)$$
$$\mathfrak{a}^* \oplus \sum \mathbb{R}(E_{\tilde{\alpha}_j} + E_{-\tilde{\alpha}_j}) = \mathfrak{a}.$$

Let $\sigma = \pi^M(\lambda_M, \Delta_M^+, \chi)$, and define

$$\lambda = \left.\begin{cases} \lambda_M & \text{on } \mathfrak{b} \\ 0 & \text{on } \sum i\mathbb{R}H_{\tilde{\alpha}_j} \end{cases}\right\} \text{on } \mathfrak{b}^*$$
$$\chi^* = \chi|_{Z_{M^*}} \qquad \text{on } Z_{M^*}$$
$$\nu^* = \nu|_{\mathfrak{a}^*} \qquad \text{on } \mathfrak{a}^*.$$

Form a parabolic subgroup $S^* = M^*A^*N^*$ in the usual way, using any choice of N^*. These definitions give us all the ingredients of some characters

$$\operatorname{ind}_S^G \Theta^{M^*A^*}(\lambda, \Delta_{M^*}^+, \chi^*, \nu^*) \tag{14.103}$$

except for the positive system $\Delta_{M^*}^+$. (We do have to check integrality of λ and compatibility with χ^*, also.) Lemma 14.77 below will say that the number of choices of $\Delta_{M^*}^+$ compatible with Lemma 14.69 is exactly 2^k, and Lemma 14.78 will note that some choices lead to equal characters, even on the level of M^*A^*. We shall see in Theorem 14.79 that the complete decomposition of $\operatorname{ind}_S^G \Theta^{MA}(\lambda_M, \Delta_M^+, \chi, \nu)$ is into the sum of the characters (14.103), but with each one taken only once.

Let $W_{\mathscr{H}}$ be the abelian group of order 2^k in $W(\mathfrak{b}^{*\mathbb{C}}:\mathfrak{m}^{*\mathbb{C}})$ generated by the reflections $s_{\tilde{\alpha}_j}$, $1 \leq j \leq k$, and let $E_{\mathscr{H}}$ be the subgroup of members of $W_{\mathscr{H}}$ realizable in M^*:

$$E_{\mathscr{H}} = W_{\mathscr{H}} \cap W(B^*:M^*).$$

Lemma 14.77. The parameter λ satisfies $\langle \lambda, \tilde{\beta} \rangle \neq 0$ for all roots $\tilde{\beta}$ in $\Delta(\mathfrak{b}^{*\mathbb{C}}:\mathfrak{m}^{*\mathbb{C}})$ other than $\pm\tilde{\alpha}_j$, $1 \leq j \leq k$. Consequently there exist exactly 2^k positive systems $\Delta_{M^*}^+$ for $\Delta(\mathfrak{b}^{*\mathbb{C}}:\mathfrak{m}^{*\mathbb{C}})$ such that

(i) $\Delta_M^+ \subseteq \Delta_{M^*}^+$ (after identification by $\mathbf{d}_{\mathscr{H}}$)
(ii) λ is $\Delta_{M^*}^+$ dominant
(iii) every $\tilde{\beta}$ in $\Delta_{M^*}^+$ such that $\mathbf{d}_{\mathscr{H}}^{-1}\tilde{\beta}$ is nonreal is such that $\mathbf{d}_{\mathscr{H}}(\theta\mathbf{d}_{\mathscr{H}}^{-1}\tilde{\beta})$ is in $\Delta_{M^*}^+$.

All such positive systems for M^* are obtained from one of them $\Delta_{M^*}^+$ as $\{w\Delta_{M^*}^+ \mid w \in W_{\mathscr{H}}\}$. For any such positive system the positive roots from among the $\pm\tilde{\alpha}_j$, $1 \leq j \leq k$, are all simple.

Proof. First we show $\langle \lambda, \tilde{\beta} \rangle \neq 0$ under the circumstances stated. Regard $\tilde{\beta}$ as a member of $\Delta((\mathfrak{a}^* \oplus \mathfrak{b}^*)^{\mathbb{C}}:\mathfrak{g}^{\mathbb{C}})$ that vanishes on $\mathfrak{a}^*$, and form $\mathbf{d}_{\mathscr{H}}^{-1}(\tilde{\beta})$. We may assume that this has nonzero $\mathfrak{a}$ component since λ_M is $\mathfrak{m}$-nonsingular. Let β_R be the root of $(\mathfrak{g}, \mathfrak{a})$ given by $\beta_R = \mathbf{d}_{\mathscr{H}}^{-1}(\tilde{\beta})|_{\mathfrak{a}}$. Then β_R is a linear combination of the roots α_j of $(\mathfrak{g}, \mathfrak{a})$ and hence is in $(1-q)\mathfrak{a}'$, in the notation of (14.78). By Lemma 14.62, no multiple of β_R is in $\Delta'_{\sigma,\nu}$.

By Proposition 14.26 and the definition at the end of §7, we have

$$\mu_{\sigma,\beta_R}(\nu) = \left(\prod_{\substack{\varepsilon|_{\mathfrak{a}} = c\beta_R \\ c>0}} \langle \lambda_M + \nu, \varepsilon \rangle \right) f_{\sigma,\beta'_R} \tag{14.104}$$

where f_{σ,β'_R} is 1 or is given by (14.45c). One of the factors in the polynomial part of this expression comes from $\varepsilon = \mathbf{d}_{\mathscr{H}}^{-1}(\tilde{\beta})$, which we write as $\beta_R + \beta_I$.

Assume $\langle \lambda, \tilde{\beta} \rangle = 0$; we show $\tilde{\beta} = \pm\tilde{\alpha}_j$ for some j. We first show $\beta_I = 0$. If $\beta_I \neq 0$, then $\beta_R - \beta_I$ contributes a second factor to $\mu_{\sigma,\beta_R}(\nu)$, and we have

$$\begin{aligned} \langle \lambda_M + \nu, \beta_R \pm \beta_I \rangle &= \pm\langle \lambda_M, \beta_I \rangle + \langle \nu, \beta_R \rangle \\ &= \pm\langle \lambda, \tilde{\beta} \rangle + \langle \nu, \beta_R \rangle = \langle \nu, \beta_R \rangle. \end{aligned}$$

The right side here is 0 since β_R is a linear combination of the α_j's and Theorem 14.64b gives $\langle \nu, \alpha_j \rangle = 0$ for all j. Thus $\beta_I \neq 0$ means that two factors in the polynomial part of $\mu_{\sigma,\beta_R}(\nu)$ vanish. Since f_{σ,β'_R} has at most simple poles, $\mu_{\sigma,\beta_R}(\nu)$ must vanish. By Lemma 14.40, s_{β_R} exists in $W(A:G)$ and $s_{\beta_R}\sigma \cong \sigma$. Hence β_R (or at least some useful multiple of β_R) is in $\Delta'_{\sigma,\nu}$, contradiction. Thus $\beta_I = 0$.

Consequently β_R is in Δ_0, and the superorthogonality implies $\beta_R = \pm\alpha_j$ for some j and therefore also $\tilde{\beta} = \pm\tilde{\alpha}_j$. That is, $\langle \lambda, \tilde{\beta} \rangle = 0$ implies $\tilde{\beta} = \pm\tilde{\alpha}_j$ for some j.

Lemma 14.69 (applied with G replaced by M_0^*) implies that there exists $\Delta^+_{M^*}$ satisfying (i), (ii), (iii). For any such $\Delta^+_{M^*}$, Problem 30 at the end of the chapter says that the real roots in $\Delta((\mathfrak{a} \oplus \mathfrak{b})^{\mathbb{C}}:\mathfrak{m}^{*\mathbb{C}})$ are generated by the real roots that are simple for $\mathbf{d}_{\mathscr{H}}^{-1}\Delta^+_{M^*}$. Since $\alpha_1, \ldots, \alpha_k$ are superorthogonal among the real roots, $\pm\alpha_j$ must be simple for each j. Then it is clear that we can generate at least 2^k systems $\Delta^+_{M^*}$ by applying $W_{\mathscr{H}}$ to one of them. On the other hand, if $\tilde{\beta}$ is a nonreal root, then we have seen that $\langle \lambda, \tilde{\beta} \rangle \neq 0$. Thus condition (ii) forces the sign of $\tilde{\beta}$ to be determined by the sign of $\langle \lambda, \tilde{\beta} \rangle$. Hence there are only 2^k systems $\Delta^+_{M^*}$.

Fix a reference positive system $\Delta^+_{M^*}$ as in Lemma 14.77. The case $i = k$ in Lemma 14.80 below will show that λ and χ^* satisfy the appropriate conditions so that $\Theta^{M^*A^*}(\lambda, \Delta^+_{M^*}, \chi^*, \nu^*)$ is well defined. Assuming this fact, we have the following lemma.

Lemma 14.78. With $\Delta^+_{M^*}$ as in Lemma 14.77, let w be in $W_{\mathscr{H}}$ and e be in $E_{\mathscr{H}}$. Then

$$\Theta^{M^*A^*}(\lambda, we\Delta^+_{M^*}, \chi^*, \nu^*) = \Theta^{M^*A^*}(\lambda, w\Delta^+_{M^*}, \chi^*, \nu^*).$$

Proof. We have $we = ew$, and λ is $w\Delta^+_{M^*}$ dominant. Also $e\lambda = \lambda$. By Proposition 12.33c,

$$\begin{aligned}\Theta^{M^*A^*}(\lambda, w\Delta^+_{M^*}, \chi^*, \nu^*) &= \Theta^{M^*A^*}(e\lambda, ew\Delta^+_{M^*}, \chi^*, \nu^*)\\ &= \Theta^{M^*A^*}(\lambda, we\Delta^+_{M^*}, \chi^*, \nu^*).\end{aligned}$$

Theorem 14.79. For G a linear connected reductive group with compact center, let $\operatorname{ind}_S^G \Theta^{MA}(\lambda_M, \Delta^+_M, \chi, \nu)$ be a basic character induced from discrete series, let $R_{\sigma,\nu}$ be the R group, and let $\mathscr{H} = \{\alpha_1, \ldots, \alpha_k\}$ be the corresponding superorthogonal set. Then

$$E_{\mathscr{H}} = \left\{ w \in W_{\mathscr{H}} \;\middle|\; \begin{array}{l}\text{for each } r \text{ in } R_{\sigma,\nu},\ w \text{ and } r\\ \text{have an even number of factors } s_{\alpha_j} \text{ in}\\ \text{common, after identification of } s_{\alpha_j} \text{ with}\\ s_{\tilde{\alpha}_j} \text{ by Cayley transform}\end{array} \right\}, \qquad (14.105)$$

and formula (14.105) sets up a canonical isomorphism of $W_{\mathscr{H}}/E_{\mathscr{H}}$ onto the dual group $\hat{R}_{\sigma,\nu}$. Moreover,

$$\operatorname{ind}_S^G \Theta^{MA}(\lambda_M, \Delta_M^+, \chi, \nu) = \sum_{w \in W_{\mathscr{H}}/E_{\mathscr{H}} \cong \hat{R}_{\sigma,\nu}} \operatorname{ind}_{S^*}^G \Theta^{M^*A^*}(\lambda, w\Delta_{M^*}^+, \chi^*, \nu^*) \tag{14.106}$$

for any choice of positive system $\Delta_{M^*}^+$ as in Lemma 14.77. The characters on the right side of (14.106) are all nonzero and irreducible.

Remarks. The critical case for understanding the theorem is the example at the beginning of §9 and the end of §11, in which G is the double covering of $SO_0(4, 4)$. We give much of the proof in the context of that example, first proving two lemmas in general.

We regard the members $\alpha_1, \ldots, \alpha_k$ of $\mathscr{H}$ as listed in some particular enumeration, and we build intermediate data $(\mathfrak{m}_i, \mathfrak{a}_i, \mathfrak{b}_i)$ and subgroups

$$M = M_{(0)} \subseteq M_1 \subseteq M_2 \subseteq \cdots \subseteq M_k = M^* \tag{14.107}$$

by means of the successive Cayley transforms $\mathbf{d}_{\alpha_i}$. For each i, let $\Delta_{M_i,j}^+$ denote the positive systems for $\Delta(\mathfrak{b}_i^{\mathbb{C}}:\mathfrak{m}_i^{\mathbb{C}})$ obtained by applying Lemma 14.69 with G replaced by $(M_i)_0$. The number of such systems for fixed i is exactly 2^i, by Lemma 14.77, the only choice being the signs of the Cayley transforms of $\alpha_1, \ldots, \alpha_i$. Define

$$\begin{aligned} \lambda_i &= \text{extension of } \lambda_M \text{ by } 0 \text{ to } \mathfrak{b}_i \\ \chi_i &= \text{restriction of } \chi \text{ to } Z_{M_i} \text{ from } Z_M \\ \nu_i &= \text{restriction of } \nu \text{ to } \mathfrak{a}_i \text{ from } \mathfrak{a}. \end{aligned} \tag{14.108}$$

Lemma 14.80. For each i with $0 \le i \le k$,

$$\operatorname{ind}_{S \cap M_i A_i}^{M_i A_i} \Theta^{MA}(\lambda_M, \Delta_M^+, \chi, \nu) = \sum_{\text{certain } j\text{'s}} \Theta^{M_i A_i}(\lambda_i, \Delta_{M_i,j}^+, \chi_i, \nu_i) \tag{14.109}$$

with λ_i satisfying the appropriate conditions so that the right side makes sense.

Remark. As we observed before Lemma 14.78, the case $i = k$ of this lemma implies that the characters in Lemma 14.78 and Theorem 14.79 are meaningful.

Proof. The proof is by induction on i by means of generalized Schmid identities and double induction. To use generalized Schmid identities, we apply Theorem 14.73. In view of the statement of that theorem, we have to check successive ρ_α conditions like (14.97), and we have to see that the positive systems are the ones asserted. (The orthogonality condition on v_i follows from the fact that $\langle v, \alpha_j \rangle = 0$ for all j, which is given in Theorem 14.64b.)

Let us first check the ρ_α condition. Let $\rho^{(i)}_{\alpha_{i+1}}$ be half the sum of the Cayley transforms of those members of $\Delta((\mathfrak{a}_i \oplus \mathfrak{b}_i)^{\mathbb{C}} : (\mathfrak{m}_{i+1} \oplus \mathfrak{a}_{i+1})^{\mathbb{C}})$ whose inner product with α_{i+1} is positive. These roots are the members of $\Delta((\mathfrak{a} \oplus \mathfrak{b})^{\mathbb{C}} : \mathfrak{g}^{\mathbb{C}})$ orthogonal to $\alpha_{i+2}, \ldots, \alpha_k$ and having positive inner product with α_{i+1}. We are to check that

$$\chi(\gamma_{\alpha_{i+1}}) = (-1)^{2\langle \rho^{(i)}_{\alpha_{i+1}}, \alpha_{i+1} \rangle / |\alpha_{i+1}|^2}. \tag{14.110}$$

We know from Lemma 14.62 that α_{i+1} is not in $\Delta'_{\sigma,v}$. Since $\langle \lambda + v, \alpha_{i+1} \rangle = 0$, we see from Proposition 14.26b and Lemma 14.40 that

$$\chi(\gamma_{\alpha_{i+1}}) = (-1)^{2\langle \rho_{\alpha_{i+1}}, \alpha_{i+1} \rangle / |\alpha_{i+1}|^2},$$

where $\rho_{\alpha_{i+1}}$ is half the sum of the members of $\Delta((\mathfrak{a} \oplus \mathfrak{b})^{\mathbb{C}} : \mathfrak{g}^{\mathbb{C}})$ orthogonal to all α_j with $j \neq i+1$ and having positive inner product with α_{i+1}. Then (14.110) will follow if we show that

$$2\langle \rho^{(i)}_{\alpha_{i+1}} - \rho_{\alpha_{i+1}}, \alpha_{i+1} \rangle / |\alpha_{i+1}|^2 \tag{14.111}$$

is even. That is, we are to show that the sum (with multiplicities) of all coefficients $c_{i+1} > 0$ of all roots β in $\Delta((\mathfrak{a} \oplus \mathfrak{b})^{\mathbb{C}} : \mathfrak{g}^{\mathbb{C}})$ of the form

$$\beta \equiv c_{i+1}\alpha_{i+1} + c_i\alpha_i + \ldots + c_1\alpha_1 \bmod ((\mathfrak{a}^* \oplus \mathfrak{b})^{\mathbb{C}})'$$

with $c_i, \ldots, c_1$ not all zero is an even integer.

There are two situations. If β has a nonzero component β_I in $((\mathfrak{a}^* \oplus \mathfrak{b})^{\mathbb{C}})'$, then we use the fact that c_{i+1} is an integer or a half-integer. If j is an index $\leq i$ with $c_j \neq 0$, then we can produce new roots by changing the sign of c_j and by changing the sign of β_I. Then β occurs in a set of $4n$ mates for some n, and $4nc_{i+1}$ is an even integer.

If β has 0 component in $((\mathfrak{a}^* \oplus \mathfrak{b})^{\mathbb{C}})'$, then the superorthogonality shows that $c_i = \ldots = c_1 = 0$, and there is no contribution to (14.111). Thus (14.111) is even, and (14.110) follows.

We are left with showing that the positive systems in the lemma are those that result from iterating Theorem 14.73. We induct on i, the case $i = 0$ being trivial. Thus suppose inductively that $\Delta^+_{M_i,j}$ is one of the positive systems that occur in the identity at stage i. By definition $\Delta^+_{M_i,j}$ is one of the 2^i positive systems for $\Delta(\mathfrak{b}_i^{\mathbb{C}} : \mathfrak{m}_i^{\mathbb{C}})$ such that

(i) $\text{Cayley}(\Delta^+_M) \subseteq \Delta^+_{M_i,j}$
(ii) λ_i is $\Delta^+_{M_i,j}$ dominant
(iii) β nonreal in $\Delta^+_{M_i,j}$ implies $\theta\beta$ is in $\Delta^+_{M_i,j}$.

The positive systems $\Delta^+_{M_{i+1}}$ for $\Delta((\mathfrak{a}_{i+1} \oplus \mathfrak{b}_{i+1})^{\mathbb{C}}:(\mathfrak{m}_{i+1} \oplus \mathfrak{a}_{i+1})^{\mathbb{C}})$ that arise by applying Theorem 14.73 to

$$\operatorname{ind}_{S_i \cap M_{i+1}A_{i+1}}^{M_{i+1}A_{i+1}} \Theta^{M_iA_i}(\lambda_i, \Delta^+_{M_i,j}, \chi_i, \nu_i)$$

have to satisfy

(i′) Cayley$(\Delta^+_{M_i,j}) \cup \{\tilde{\alpha}_{i+1}\} \subseteq \Delta^+_{M_{i+1}}$ or this inclusion with $-\tilde{\alpha}_{i+1}$ instead of $\tilde{\alpha}_{i+1}$
(ii′) λ_{i+1} is $\Delta^+_{M_{i+1}}$ dominant
(iii′) β nonreal in $\Delta^+_{M_{i+1}}$ implies $\theta\beta$ is in $\Delta^+_{M_{i+1}}$.

We are to prove that $\Delta^+_{M_{i+1}}$ satisfies conditions (i), (ii), and (iii) for $i + 1$, namely

(i″) Cayley$(\Delta^+_M) \subseteq \Delta^+_{M_{i+1}}$
(ii″) λ_{i+1} is $\Delta^+_{M_{i+1}}$ dominant
(iii″) β nonreal in $\Delta^+_{M_{i+1}}$ implies $\theta\beta$ is in $\Delta^+_{M_{i+1}}$.

But (ii″) and (iii″) are just restatements of (ii′) and (iii′), and (i″) follows by combining (i) and (i′). The induction is complete, and the lemma is proved.

Let $n(i)$, $0 \le i \le k$, denote the number of terms that occur on the right side of (14.109) at stage i. Note that the terms are all nonzero. In fact, Lemma 14.77 shows that the only roots of $\Delta(\mathfrak{b}_i^{\mathbb{C}}:\mathfrak{m}_i^{\mathbb{C}})$ orthogonal to λ_i are the Cayley transforms of $\alpha_1, \ldots, \alpha_i$, and these are noncompact. By Theorem 12.26b the characters that occur on the right side of (14.109) at stage i are nonzero.

In passing from stage $i - 1$ to stage i, we use a generalized Schmid identity of one of the two types (14.92a) or (14.92b), and which one we use depends only on whether, in obvious notation, s_{α_i} is in $W(B_i:M_i)$. It does not depend on the particular positive system. Therefore

$n(0) = 1$

$n(i) = 2n(i - 1)$ if (14.92a) is used to pass from stage $i - 1$ to stage i

$n(i) = n(i - 1)$ if (14.92b) is used to pass from stage $i - 1$ to stage i.

The decisive property of the R group is that it controls the jumps of this function: Jumps must occur as elements of the R group get used up.

Lemma 14.81. The function $n(i)$ satisfies $n(i) = 2n(i - 1)$ if there exists an element r in $R_{\sigma,\nu}$ of the form

$$r = \Bigg(\prod_{\substack{\text{certain} \\ j<i}} s_{\alpha_j}\Bigg) s_{\alpha_i}. \tag{14.112}$$

Proof. Assume the contrary, so that identity (14.92b) is used in passing from stage $i - 1$ to stage i. From (14.94) we know that there is an element

z in $Z_{M_{i-1}}$ such that

$$s_{\alpha_i}\chi_{i-1}(z) \neq \chi_{i-1}(z).$$

The other factor $\prod s_{\alpha_j}$ of r in (14.112) has a representative in M_{i-1}, and hence it fixes χ_{i-1}. Thus we obtain $r\chi_{i-1} \neq \chi_{i-1}$, and we cannot have $r\chi = \chi$, a contradiction.

Proof of Theorem 14.79 in example with $SO_0(4, 4)$. We treat the example at the start of §9, which is continued at the end of §11. In it, G is the double cover of $SO_0(4, 4)$, $S = MAN$ is a minimal parabolic subgroup, and $R_{\sigma,0}$ is given by

$$R_{\sigma,0} = \{1, s_{e_1-e_2}s_{e_3-e_4}, s_{e_1-e_2}s_{e_3+e_4}, s_{e_3-e_4}s_{e_3+e_4}\}.$$

Thus $\mathscr{H} = \{e_1 - e_2, e_3 - e_4, e_3 + e_4\}$, and we take these roots to be α_1, α_2, and α_3, respectively. We form the intermediate data as in (14.107) and (14.108). There are elements of $R_{\sigma,0}$ that end with $e_3 - e_4$ and with $e_3 + e_4$. Thus $n(i)$ jumps in passing from $i = 1$ to $i = 2$ and in passing from $i = 2$ to $i = 3$. Since $|R_{\sigma,0}|$ bounds the number of constituents of our induced representation, $n(i)$ cannot jump in passing from $i = 0$ to $i = 1$ (or else there would be eight constituents). We have found four nonzero constituents, and they must be all be irreducible.

Let us track down the sequence of generalized Schmid identities. We shall identify positive systems by giving the simple roots, and we drop any mention of Cayley transforms and irrelevant parameters. Then we have

$$\begin{aligned}
\operatorname{ind}_S^G \Theta^{MA}(\varnothing) &= \operatorname{ind}_{S_1}^G \Theta^{M_1A_1}(\{e_1 - e_2\}) \qquad \text{since } n(1) = n(0)\\
&= \operatorname{ind}_{S_2}^G[\Theta^{M_2A_2}(\{e_1 - e_2, e_3 - e_4\})\\
&\quad + \Theta^{M_2A_2}(\{e_1 - e_2, -e_3 + e_4\})] \qquad \text{since } n(2) = 2n(1)\\
&= \operatorname{ind}_{S_3}^G[\Theta^{M^*A^*}(\{e_1 - e_2, e_3 - e_4, e_3 + e_4\})\\
&\quad + \Theta^{M^*A^*}(\{e_1 - e_2, e_3 - e_4, -e_3 - e_4\})\\
&\quad + \Theta^{M^*A^*}(\{e_1 - e_2, -e_3 + e_4, e_3 + e_4\})\\
&\quad + \Theta^{M^*A^*}(\{e_1 - e_2, -e_3 + e_4, -e_3 - e_4\})]. \qquad (14.113)
\end{aligned}$$

To prove the theorem, we are to see that the positive systems on the right side are all related to one of them by $W_{\mathscr{H}}/E_{\mathscr{H}}$, we are to see that $E_{\mathscr{H}}$ is given by (14.105), and we are to show that (14.105) sets up a canonical isomorphism of $W_{\mathscr{H}}/E_{\mathscr{H}}$ with $\hat{R}_{\sigma,0}$.

A little computation shows that $\gamma_{e_2-e_3}$, which is in the finite group M and is therefore in M^*, normalizes B^* and acts as (the Cayley transform of) $s_{e_1-e_2}s_{e_3-e_4}s_{e_3+e_4}$. Therefore

$$E_{\mathscr{H}} \supseteq \{1, s_{e_1-e_2}s_{e_3-e_4}s_{e_3+e_4}\}. \qquad (14.114)$$

The two-element set on the right side of (14.114) allows us to pass from the four positive systems on the right side of (14.113) to all eight possible positive systems. Since the characters on the right side of (14.113) are unchanged by the action of $E_{\mathscr{H}}$ (Lemma 14.78), the Multiplicity One Theorem implies that there cannot be further elements of $E_{\mathscr{H}}$. Thus

$$E_{\mathscr{H}} = \{1, s_{e_1 - e_2} s_{e_3 - e_4} s_{e_3 + e_4}\},$$

and the positive systems on the right side on (14.113) are parametrized by $W_{\mathscr{H}}/E_{\mathscr{H}}$.

It is clear in this example that $E_{\mathscr{H}}$ is given by (14.105). We define a map of $W_{\mathscr{H}}$ into $\hat{R}_{\sigma,0}$ by

$$w \in W_{\mathscr{H}} \to \chi_w(r) = \begin{cases} +1 & \text{if } w \text{ and } r \text{ have an even number of factors in common} \\ -1 & \text{if not.} \end{cases}$$

This map is a homomorphism, and its kernel is exactly $E_{\mathscr{H}}$, in view of (14.105). Thus it descends to a one-one homomorphism of $W_{\mathscr{H}}/E_{\mathscr{H}}$ into $\hat{R}_{\sigma,0}$. Since $|W_{\mathscr{H}}/E_{\mathscr{H}}| = 4 = |R_{\sigma,0}| = |\hat{R}_{\sigma,0}|$, the map of $W_{\mathscr{H}}/E_{\mathscr{H}}$ into $\hat{R}_{\sigma,0}$ is an isomorphism. This completes the proof of Theorem 14.79.

§16. Classification

By Theorem 14.76, every irreducible tempered character is basic. Conversely every basic character is tempered. Thus to have a classification of irreducible tempered representations, we have to know how to recognize irreducibility, and we have to be able decide when two basic characters are equal.

For the recognition of irreducibility, again the tool is an appropriately defined R group. But the definitions cannot all remain the same, as the following example shows.

Example. $G = \mathrm{Sp}(2, \mathbb{R})$. Let $S_{\mathfrak{p}} = M_{\mathfrak{p}} A_{\mathfrak{p}} N_{\mathfrak{p}}$ be a minimal parabolic subgroup, and denote the simple roots by $e_1 - e_2$ and $2e_2$. Define σ on $M_{\mathfrak{p}}$ by

$$\sigma(\gamma_{e_1 - e_2}) = -1 \quad \text{and} \quad \sigma(\gamma_{2e_2}) = -1, \tag{14.115}$$

and let $\nu = 0$. Condition (14.115) forces

$$\sigma(\gamma_{e_1 + e_2}) = -1 \quad \text{and} \quad \sigma(\gamma_{2e_1}) = +1.$$

Then we can compute that $\Delta'_{\sigma,0} = \{\pm 2e_1\}$ and that $R_{\sigma,0} = \{1, s_{2e_2}\}$. Thus

$$U = \mathrm{ind}_{S_{\mathfrak{p}}}^{G}(\sigma \otimes 1 \otimes 1)$$

is reducible, with two irreducible constituents. The theory of §15 suggests that we pass to the parabolic subgroup that incorporates $2e_2$ into M. Instead we pass to the parabolic subgroup that incorporates $e_1 - e_2$ into

M. Write $S = MAN$ for this parabolic subgroup; here M is isomorphic with $\mathrm{SL}^{\pm}(2, \mathbb{R})$. Since $\sigma(\gamma_{e_1 - e_2}) = -1$ and $\mathfrak{m} \cong \mathfrak{sl}(2, \mathbb{R})$, the root $e_1 - e_2$ corresponds to a cotangent case in Proposition 14.26. By Theorem 14.73, the character of U is the right side of a generalized Schmid identity for $\alpha = e_1 - e_2$. The presence of the element $\begin{pmatrix} 1 & 0 \\ 0 & -1 \end{pmatrix}$ in $\mathrm{SL}^{\pm}(2, \mathbb{R}) \cong M$ forces this identity to be as in (14.92b). Thus we have

$$U \cong \operatorname{ind}_S^G(\xi \otimes 1 \otimes 1) \qquad \text{for } \xi = \pi^M(0, \Delta_M^+, \chi), \tag{14.116}$$

where χ is defined by $\chi(\gamma_{e_1 + e_2}) = -1$ and Δ_M^+ is just $\{e_1 - e_2\}$. Assuming we should keep the usual definitions of $\Delta'_{\xi,0}$ and $R_{\xi,0}$, let us compute $\mu_{\xi,\, e_1 + e_2}(0)$. There are three polynomial factors, all of them 0 at $\nu = 0$, and a tangent or cotangent. In any case, $\mu_{\xi,\, e_1 + e_2}(0) = 0$. Thus under our usual definitions, $\Delta'_{\xi,0}$ would be $\{\pm(e_1 + e_2)\}$ and $R_{\xi,0}$ would be $\{1\}$. However, we know that U is reducible into two pieces. Thus our usual definitions will not lead to the desired theorem.

It turns out, for reasons that emerge from the proof of Proposition 14.83a below and that will be mentioned explicitly when the above example is continued later in this section, that one should exclude the effect of $2e_1$ and $2e_2$ in the computation of the polynomial part of whatever is to play the role of $\mu_{\xi, e_1 + e_2}(0)$. Then this quantity ends up being nonzero, and $R_{\xi,0}$ ends up as $\{1, s_{e_1 + e_2}\}$.

Thus in the general case with $\sigma = \pi^M(\lambda, \Delta_M^+, \chi)$ a discrete series or limit of discrete series, we define a **modified Plancherel factor** $\mu'_{\sigma,\alpha}(\nu)$ to be equal to $\mu_{\sigma,\alpha}(\nu)$ except that certain polynomial factors $\langle \lambda + \nu, \varepsilon \rangle$ are to be dropped. These factors are of two kinds:

(1) We drop $\langle \lambda + \nu, \varepsilon \rangle$ if $\bar{\varepsilon} = -\theta\varepsilon$ has $2\langle \varepsilon, \bar{\varepsilon} \rangle / |\varepsilon|^2 = +1$.

(2) If α is odd and α' is the positive multiple of α that extends to be a real root, we drop $\langle \lambda + \nu, \varepsilon \rangle$ if ε is of the form $\alpha' + \beta$ with $\beta|_{\mathfrak{a}} = 0$.

Operationally for even α, this means that we define $\mu'_{\sigma,\alpha} = \mu_{\sigma,\alpha}$ if α is useful and $\mu'_{\sigma,\alpha} = 1$ if α is not useful. For α odd, normally we drop only the factors described in (2); however, there is one exceptional group with α odd where (1) dictates that a pair of factors be dropped.

For σ in the discrete series of M, $\mu_{\sigma,\alpha}(\nu)$ and $\mu'_{\sigma,\alpha}(\nu)$ are both zero or both nonzero. The reason is that the parameter λ, which is nonsingular with respect to $\Delta(\mathfrak{b}^{\mathbb{C}} : \mathfrak{m}^{\mathbb{C}})$, is nonorthogonal to the root $\varepsilon - \bar{\varepsilon}$ in (1) and the root β in (2).

Lemma 14.82. If α is a root of $(\mathfrak{g}, \mathfrak{a})$ and ν is imaginary and σ is a discrete series or limit of discrete series representation of M such that $\mu'_{\sigma,\alpha}(\nu) = 0$, then s_α exists in $W(A : G)$ and satisfies $s_\alpha \sigma \cong \sigma$ and $s_\alpha \nu = \nu$.

Remarks. We omit the detailed proof. Let $\sigma = \pi^M(\lambda, \Delta_M^+, \chi)$. If $\mu'_{\sigma,\alpha}(\nu) = 0$, then the definition of μ' forces a multiple of α to be useful and $s_\alpha \nu$ to be ν. It forces also $s_\alpha \lambda = \lambda$ by an argument similar to the one following (14.104). Corollary 14.53 shows that the crux of the matter is thus to prove $s_\alpha \chi = \chi$. To do this, one makes an algebraic computation; Lemma 14.40 is no longer available when σ is not in the discrete series.

We now define

$$\Delta'_{\sigma,\nu} = \{\alpha = \text{useful root of } (\mathfrak{g}, \mathfrak{a}) \,|\, \mu'_{\sigma,\alpha}(\nu) = 0\}.$$

It is a simple matter to see that $W_{\sigma,\nu}$ carries $\Delta'_{\sigma,\nu}$ into itself. Since Lemma 14.82 implies that reflections in members of $\Delta'_{\sigma,\nu}$ are in $W_{\sigma,\nu}$, $\Delta'_{\sigma,\nu}$ is closed under its own reflections and is therefore a root system. Let $W'_{\sigma,\nu}$ be its Weyl group, and define $R_{\sigma,\nu}$ again to be the set of members of $W_{\sigma,\nu}$ carrying $(\Delta'_{\sigma,\nu})^+$ to itself. As in §9, we have a semidirect product decomposition

$$W_{\sigma,\nu} = W'_{\sigma,\nu} R_{\sigma,\nu} \tag{14.117}$$

with $W'_{\sigma,\nu}$ normal. What is lacking is any information about intertwining operators that would allow us to interpret these groups. We return to this point shortly. Anyway, we can trace through §11, seeing what proofs go through. We make the same definitions as there and find, even for σ a limit of discrete series, that

$$R_{\sigma,\nu} \subseteq S \cap W_{\lambda,\nu} \qquad \text{as in Theorem 14.59} \tag{14.118}$$

and $R_{\sigma,\nu}$ is given by a superorthogonal set $\mathscr{H} = \{\alpha_1, \ldots, \alpha_k\}$ as in Theorem 14.64. (14.119)

Let us come back to the question about intertwining operators. The result that extends Theorem 14.43 is as follows.

Proposition 14.83. Let G be linear connected reductive with compact center. Suppose that σ is a discrete series or nonzero limit of discrete series representation of M and that ν is imaginary. Then

(a) the operators $\sigma(s)\mathscr{A}_S(s, \sigma, \nu)$ corresponding to s in $W'_{\sigma,\nu}$ are all scalar
(b) the operators $\sigma(s)\mathscr{A}_S(s, \sigma, \nu)$ corresponding to s in $W_{\sigma,\nu}$ span the commuting algebra of $U(S, \sigma, \nu)$
(c) the dimension of the commuting algebra of $U(S, \sigma, \nu)$ is $\leq |R_{\sigma,\nu}|$.

Remarks. We give parts of the proof. We shall assume that no two roots β and γ have $2\langle\beta, \gamma\rangle/|\beta|^2 = 3$; this condition is satisfied in all classical groups. In any case, notice that (c) follows by combining (a), (b), and (14.117), since the operators $\sigma(s)\mathscr{A}_S(s, \sigma, \nu)$ multiply according to the group law in $W_{\sigma,\nu}$, except for scalar factors.

Proof of (a). We know that $W'_{\sigma,\nu}$ is generated by reflections s_α with α in $\Delta'_{\sigma,\nu}$. Thus we are to show that $\mu'_{\sigma,\alpha}(\nu) = 0$ implies $\sigma(s_\alpha)\mathscr{A}_S(s_\alpha, \sigma, \nu)$ is scalar. In the orthogonal complement of α in $i\mathfrak{a}'$, there is a dense set of members $\nu^\perp$ such that $\langle \nu + \nu^\perp, \beta\rangle \neq 0$ for all roots β of $(\mathfrak{g}, \mathfrak{a})$ that are not multiples of α. Fix such an element $\nu^\perp$. Then $s_\alpha(\nu + \nu^\perp) = \nu + \nu^\perp$, and

$$\mu'_{\sigma,\alpha}(\nu + \nu^\perp) = \mu'_{\sigma,\alpha}(\nu) = 0, \tag{14.120}$$

since $\mu'_{\sigma,\alpha}$ depends only on the projection of the variable in the α direction. We shall prove that $U(S, \sigma, \nu + \nu^\perp)$ is irreducible for such $\nu^\perp$. Then $\sigma(s_\alpha)\mathscr{A}_S(s, \sigma, \nu + \nu^\perp)$ has to be scalar, and a passage to the limit shows $\sigma(s_\alpha)\mathscr{A}_S(s, \sigma, \nu)$ is scalar.

Let σ have character $\Theta^M(\lambda_M, \Delta_M^+, \chi)$. If

$$\operatorname{ind}_S^G \Theta^{MA}(\lambda_M, \Delta_M^+, \chi, \nu + \nu^\perp) \tag{14.121}$$

is a final basic character, then Corollary 14.74 says it is irreducible. Since $\langle \nu + \nu^\perp, \beta\rangle \neq 0$ for $\beta \neq c\alpha$, the only way it can fail to be final is if α is odd and we are in a cotangent case:

$$\chi(\gamma_{\alpha'}) = (-1)^{2\langle \rho_\alpha, \alpha'\rangle/|\alpha'|^2}, \tag{14.122}$$

where α' is the unique positive multiple of α that extends to a real root. Let us normalize matters so that $\alpha' = \alpha$.

Equations (14.120) and (14.122) together imply that there is some complex root β for G such that

$$\beta|_{\mathfrak{a}} = c\alpha \qquad \text{with} \quad c \neq 0 \tag{14.123a}$$

$$\langle \lambda_M, \beta\rangle = 0 \tag{14.123b}$$

$$|\beta| \leq |\alpha|, \tag{14.123c}$$

the last condition coming from the use of μ' in place of μ.

We use the real root α in Theorem 14.73 to form a generalized Schmid identity with a character induced from $\Theta^{MA}(\lambda_M, \Delta_M^+, \chi, \nu + \nu^\perp)$ on the right side. Whatever characters are on the left side of this identity will remain irreducible upon induction to G, by Corollary 14.74. Thus (14.121) will be irreducible (and we will be done) if the generalized Schmid identity is of type (14.92b). So we may assume that the left side of the identity is of the form

$$\Theta^{M^*A^*}(\lambda, \Delta_{M^*}^+, \chi^*, (\nu + \nu^\perp)|_{\mathfrak{a}^*}) + \Theta^{M^*A^*}(\lambda, s_{\tilde\alpha}\Delta_{M^*}^+, \chi^*, (\nu + \nu^\perp)|_{\mathfrak{a}^*}) \tag{14.124}$$

with $\tilde\alpha = \mathbf{d}_\alpha(\alpha)$.

In this case we prove that one of these characters is 0. In (14.123) we may replace β by $-\beta$ if necessary to make $\tilde\beta = \mathbf{d}_\alpha(\beta)$ be positive. We may assume in addition that $\tilde\beta$ is as small as possible so that (14.123) holds. We shall prove in this case that $\tilde\beta$, which is a root of M^*, is simple for $\Delta_{M^*}^+$,

$s_{\tilde{\alpha}}\tilde{\beta}$ is simple for $s_{\tilde{\alpha}}\Delta^+_{M^*}$, and one of $\tilde{\beta}$ and $s_{\tilde{\alpha}}\tilde{\beta}$ is compact. Then Theorem 12.26b shows that the character in (14.124) corresponding to the compact simple root vanishes, and we are done.

Assuming on the contrary that $\tilde{\beta}$ is not simple, let $\tilde{\beta} = \tilde{\gamma} + \tilde{\varepsilon}$ with $\tilde{\gamma}$ and $\tilde{\varepsilon}$ in $\Delta^+_{M^*}$. Set $\gamma = \mathbf{d}_\alpha^{-1}(\tilde{\gamma})$ and $\varepsilon = \mathbf{d}_\alpha^{-1}(\tilde{\varepsilon})$. Then (14.123b) holds for both γ and ε in place of β since it holds for β. By (14.123a) and (14.123c) for β, $\beta|_{\mathfrak{a}} = \pm\frac{1}{2}\alpha$. Since we have assumed that no two roots have normalized inner product 3, it follows that

$$\gamma|_{\mathfrak{a}} = c'\alpha \qquad \text{and} \qquad \varepsilon|_{\mathfrak{a}} = c''\alpha \qquad \text{with} \quad c, c' = 0, \pm\tfrac{1}{2}, \text{or } \pm 1.$$

Thus one of c' and c'' is $\pm\frac{1}{2}$. Say $\gamma|_{\mathfrak{a}} = \pm\frac{1}{2}\alpha$. Then γ satisfies (14.123a). Since $s_\alpha\gamma|_{\mathfrak{a}} = \mp\frac{1}{2}\alpha$, we have $s_\alpha\gamma = \gamma \pm \alpha$. Thus γ satisfies (14.123c), and the inequality $0 < \tilde{\gamma} < \tilde{\beta}$ contradicts the minimality of $\tilde{\beta}$. Hence $\tilde{\beta}$ is simple.

Since $|\tilde{\beta}| \leq |\tilde{\alpha}|$, $s_{\tilde{\alpha}}\tilde{\beta} = \tilde{\beta} \pm \tilde{\alpha}$. Thus the difference of $\tilde{\beta}$ and $s_{\tilde{\alpha}}\tilde{\beta}$ is the noncompact root $\pm\tilde{\alpha}$ of Δ_{M^*}, and it follows that one of $\tilde{\beta}$ and $s_{\tilde{\alpha}}\tilde{\beta}$ is compact. This completes the proof of (a).

Example, continued. In the $Sp(2, \mathbb{R})$ example given in (14.116), we do have vanishing for $\mu_{\xi, e_1+e_2}(0)$. Since the representation (14.116) is actually reducible, we can trace through the above proof to see that there is some complex root β of G (namely the Cayley transform of $2e_1$ or $2e_2$) such that (14.123a) and (14.123b) hold, but (14.123c) does not hold. The minimal $\tilde{\beta}$ is the Cayley transform of $-2e_2$, which is noncompact and simple for Δ^+_G. Its reflection by $e_1 + e_2$ is the Cayley transform of $2e_1$, which is noncompact and simple for $s_{e_1+e_2}\Delta^+_G$. Thus neither term on the left side of the generalized Schmid identity in the proof is 0, and the proof fails. This example gives the reason for the peculiar definition of $\mu'_{\xi,\alpha}$.

Let us turn to the proof of (b). Let σ have character $\Theta^M(\lambda, \Delta^+_M, \chi)$, and write, by Lemma 14.69,

$$\operatorname{ind}_S^G \Theta^{MA}(\lambda, \Delta^+_M, \chi, \nu) = \psi^{\lambda+\nu+\mu}_{\lambda+\nu} \operatorname{ind}_S^G \Theta^{MA}(\lambda + \mu, \Delta^+_M, \chi e^\mu, \nu), \qquad (14.125)$$

with $\sigma' = \pi^M(\lambda + \mu, \Delta^+_M, \chi e^\mu)$ in the discrete series of M. The following lemma is easy to believe, and we shall omit its proof.

Lemma 14.84. In the above notation if w is in $N_K(\mathfrak{a})$ and $w\sigma'$ is equivalent with σ', then $w\sigma$ is equivalent with σ and

$$\psi^{\lambda+\nu+\mu}_{\lambda+\nu}(\sigma'(w)\mathscr{A}_S(w, \sigma', \nu)) = c\sigma(w)\mathscr{A}_S(w, \sigma, \nu)$$

for a nonzero constant $c = c(\nu)$.

Proof of (b) in Proposition 14.83. By the Multiplicity One Theorem,

$$\operatorname{ind}_S^G \Theta^{MA}(\lambda + \mu, \Delta^+_M, \chi e^\mu, \nu) = \Theta_1 + \ldots + \Theta_n \qquad (14.126)$$

for distinct irreducible characters. Since $\psi\Theta_j$ is a multiple of an irreducible character and since the left side of (14.125) is contained in a basic character induced from discrete series (Corollary 14.72), a second application of the Multiplicity One Theorem shows that each $\psi\Theta_j$ is irreducible or 0 and the nonzero $\psi\Theta_j$'s are distinct. All of the representations in question are unitary. Therefore the commuting algebra for the left side of (14.126) is of dimension n and is generated by the projections corresponding to each Θ_j, while the commuting algebra for the left side of (14.125) is of dimension $\#\{j \mid \psi\Theta_j \neq 0\}$ and is generated by the projections corresponding to each nonzero $\psi\Theta_j$. When $\psi\Theta_j \neq 0$, ψ carries the projection corresponding to Θ_j to the projection corresponding to $\psi\Theta_j$, since ψ carries the identity morphism to the identity morphism. We conclude that ψ carries the commuting algebra corresponding to (14.126) *onto* the commuting algebra corresponding to (14.125). Thus (b) follows by applying Theorem 14.31 and Lemma 14.84.

The next step in getting an irreducibility criterion for basic characters is to try to extend to general basic characters the theory in §15 for basic characters induced from discrete series. We immediately run into a problem in trying to generalize Lemma 14.77. It is reasonable that we do not want to allow arbitrary singularities for the Harish-Chandra parameter of σ. Accordingly we say that a basic character

$$\operatorname{ind}_S^G \Theta^{MA}(\lambda_M, \Delta_M^+, \chi, \nu) \tag{14.127}$$

is given by **nondegenerate data** if, for each root $\tilde{\alpha}$ in $\Delta(\mathfrak{b}^{\mathbb{C}}:\mathfrak{m}^{\mathbb{C}})$ with $\langle \lambda_M, \tilde{\alpha}\rangle = 0$, the reflection $s_{\tilde{\alpha}}$ is not in $W(B:M)$.

A basic character induced from discrete series is given with nondegenerate data, since there are no roots of M orthogonal to λ_M. Moreover, the irreducible characters on the right side of (14.106) in the complete reduction theorem are nondegenerate. In fact, Lemma 14.77 shows that it is necessary only to check that $s_{\tilde{\alpha}_j}$ is not in $W(B^*:M^*)$, a result that we can see from Lemma 14.81 by enumerating the roots $\alpha_1, \ldots, \alpha_k$ with α_j at the end.

A little argument gives the following proposition, whose proof we omit. The proposition gives us examples of nondegenerate cases and degenerate cases.

Proposition 14.85. A nonzero basic character (14.127) is given by nondegenerate data if and only if it is not the nonzero part of the left side of some generalized Schmid identity (14.92a) or (14.92b).

Remark. That is, the character is not to be the left side of an identity (14.92b), and it is not to be paired with a 0 character so as to be the left side of an identity (14.92a).

Corollary 14.86. Every nonzero basic character can be rewritten with nondegenerate data.

Proof. We use generalized Schmid identities repeatedly until we can proceed no further. The process must stop since dim A increases at each stage.

From Proposition 14.85 we can see that the exceptional roots in defining $\mu'_{\sigma,\alpha}$ play no role in the nondegenerate case. That is,

$$\text{Nondegenerate} \Rightarrow \mu_{\sigma,\alpha}(\nu) = 0 \text{ only when } \mu'_{\sigma,\alpha}(\nu) = 0. \tag{14.128}$$

Now we can imitate §15. We start from a basic character (14.127) and assume it has nondegenerate data. The R group gives us a superorthogonal set $\mathscr{H} = \{\alpha_1, \ldots, \alpha_k\}$, and we define Cayley transforms and $\mathfrak{b}^*$, $\mathfrak{a}^*$, λ, χ^*, ν^*, $W_{\mathscr{H}}$, and $E_{\mathscr{H}}$ as in §15.

Lemma 14.87. Under the assumption that (14.127) is given by nondegenerate data, the parameter λ satisfies $\langle \lambda, \tilde{\beta} \rangle \neq 0$ for all roots $\tilde{\beta}$ in $\Delta(\mathfrak{b}^{*\mathbb{C}} : \mathfrak{m}^{*\mathbb{C}})$ other than (1) $\pm \tilde{\alpha}_j$, $1 \le j \le k$, and (2) certain roots that are orthogonal to $\tilde{\alpha}_1, \ldots, \tilde{\alpha}_k$. Consequently there exist exactly 2^k positive systems $\Delta^+_{M^*}$ for $\Delta(\mathfrak{b}^{*\mathbb{C}} : \mathfrak{m}^{*\mathbb{C}})$ such that

(i) $\Delta^+_M \subseteq \Delta^+_{M^*}$ (after identification by $\mathbf{d}_{\mathscr{H}}$)
(ii) λ is $\Delta^+_{M^*}$ dominant
(iii) every $\tilde{\beta}$ in $\Delta^+_{M^*}$ such that $\mathbf{d}_{\mathscr{H}}^{-1}\tilde{\beta}$ is nonreal is such that $\mathbf{d}_{\mathscr{H}}(\theta \mathbf{d}_{\mathscr{H}}^{-1}\tilde{\beta})$ is in $\Delta^+_{M^*}$.

All such positive systems for M^* are obtained from one of them $\Delta^+_{M^*}$ as $\{w\Delta^+_{M^*} \mid w \in W_{\mathscr{H}}\}$. For any such positive system the positive roots from among the $\pm\tilde{\alpha}_j$, $1 \le j \le k$, are all simple.

Remarks. The proof is similar to that for Lemma 14.77. That proof showed and used the fact that $\mu_{\sigma,\beta_{\mathbb{R}}}(\nu) \neq 0$. Here we obtain $\mu'_{\sigma,\beta_{\mathbb{R}}}(\nu) \neq 0$, and we use (14.128) to conclude $\mu_{\sigma,\beta_{\mathbb{R}}}(\nu) \neq 0$. Then the proof goes through.

As in §15, we fix a reference positive system $\Delta^+_{M^*}$, show that λ and χ^* satisfy the appropriate conditions so that

$$\Theta^{M^*A^*}(\lambda, \Delta^+_{M^*}, \chi^*, \nu^*)$$

is well defined, and show that application of e in $E_{\mathscr{H}}$ to $\Delta^+_{M^*}$ does not change this character. The result is as follows.

Theorem 14.88. For G a linear connected reductive group with compact center, let $\operatorname{ind}_S^G \Theta^{MA}(\lambda_M, \Delta^+_M, \chi, \nu)$ be a basic character given with nondegenerate data, let $R_{\sigma,\nu}$ be the R group, and let $\mathscr{H} = \{\alpha_1, \ldots, \alpha_k\}$ be the corresponding superorthogonal set. Then

$$E_{\mathscr{H}} = \left\{ w \in W_{\mathscr{H}} \,\middle|\, \begin{array}{l} \text{for each } r \text{ in } R_{\sigma,\nu},\ w \text{ and } r \text{ have an even number} \\ \text{of factors } s_{\alpha_j} \text{ in common, after identification} \\ \text{of } s_{\alpha_j} \text{ with } s_{\tilde{\alpha}_j} \text{ by Cayley transform} \end{array} \right\}, \tag{14.129}$$

and formula (14.129) sets up a canonical isomorphism of $W_{\mathscr{H}}/E_{\mathscr{H}}$ onto the dual group $\hat{R}_{\sigma,\nu}$. Moreover,

$$\operatorname{ind}_S^G \Theta^{MA}(\lambda_M, \Delta_M^+, \chi, \nu) = \sum_{w \in W_{\mathscr{H}}/E_{\mathscr{H}} \cong \hat{R}_{\sigma,\nu}} \operatorname{ind}_{S^*}^G \Theta^{M^*A^*}(\lambda, w\Delta_{M^*}^+, \chi^*, \nu^*). \tag{14.130}$$

for any choice of positive system $\Delta_{M^*}^+$ as in Lemma 14.87. The characters on the right side of (14.130) are all nonzero irreducible and are given with nondegenerate data.

Remarks. The style of proof is the same as in §15. Proposition 14.83c limits the reducibility that can occur, and nondegeneracy forces the appropriate characters to be nonzero. We obtain our irreducibility criterion as a corollary.

Corollary 14.89. A basic character given with nondegenerate data is irreducible if and only if its R group is trivial.

In view of this corollary and Theorem 14.76 and Corollary 14.86, all irreducible tempered characters are basic characters written with nondegenerate data and having trivial R group, and conversely. To complete the classification, we need a way of deciding when two such characters are equal. The additional tool is the following.

Theorem 14.90 (Langlands Disjointness Theorem). If

$$\operatorname{ind}_S^G \Theta^{MA}(\lambda, \Delta_M^+, \chi, \nu) \qquad \text{and} \qquad \operatorname{ind}_{S'}^G \Theta^{M'A'}(\lambda', \Delta_{M'}^{+\prime}, \chi', \nu')$$

are basic characters induced from discrete series that have an irreducible constituent in common, then the two induced characters are equal and there is an element w in K such that $M' = wMw^{-1}$, $A' = wAw^{-1}$, $B' = wBw^{-1}$, $\lambda' = w\lambda$, $\Delta_M^{+\prime} = w\Delta_M^+$, $\chi' = w\chi$, and $\nu' = w\nu$.

An analytic proof is possible starting from Eisenstein integrals. The two representations have a nonzero matrix coefficient in common, hence a nonzero Eisenstein integral in common. Using the theory of the constant term, we can recapture MA and then σ, up to a member of $W(A:G)$, and finally ν. We shall not pursue this matter for now but shall indicate in §15.3 a proof that is largely algebraic, using Eisenstein integrals only at the end.

Theorem 14.91 (Classification theorem). Let G be a linear connected reductive group with compact center. Every irreducible tempered character is basic and can be written with nondegenerate data; when the character is written this way, its R group is trivial. Conversely every basic character with nondegenerate data and trivial R group is an irreducible tempered character. For two irreducible basic characters with nondegenerate data, an equality

$$\operatorname{ind}_S^G \Theta^{MA}(\lambda, \Delta_M^+, \chi, \nu) = \operatorname{ind}_{S'}^G \Theta^{M'A'}(\lambda', \Delta_M^{+\prime}, \chi', \nu') \tag{14.131}$$

holds if and only if there is an element w in K with

$$M' = wMw^{-1}, \qquad A' = wAw^{-1}, \qquad B' = wBw^{-1}, \qquad \lambda' = w\lambda, \tag{14.132}$$
$$\Delta_M^{+\prime} = w\Delta_M^+, \qquad \chi' = w\chi, \qquad \text{and} \qquad \nu' = w\nu.$$

Sketch of proof. We are left with showing that (14.131) implies (14.132). We start from the left side of (14.131) and give a canonical imbedding of the character in a basic character induced from discrete series. Namely we define

$$\mathscr{K} = \{\tilde{\alpha} \in \Delta_M^+ \mid \langle \lambda, \tilde{\alpha} \rangle = 0\}.$$

Nondegeneracy implies $\mathscr{K}$ contains only noncompact roots and the members of $\mathscr{K}$ are simultaneously simple. It follows that the members of $\mathscr{K}$ are superorthogonal. Using generalized Schmid identities obtained from the members of $\mathscr{K}$, we get the canonical imbedding. Moreover, we can recover our original character as one of the explicit constituents given in (14.106).

The same argument applies to the right side of (14.131). Applying Theorem 14.90, we see that the data for the two basic characters induced from discrete series must be conjugate. Unwinding matters, we see that $M'A'$ is conjugate to MA. Thus we may assume that $M'A' = MA$ and that the data for the two basic characters induced from discrete series are conjugate via K. It is then a simple matter to track down the conjugacy of B', λ'_M, χ', and ν' to B, λ_M, χ, and ν.

Thus we may assume everything on the two sides of (14.131) is the same except possibly for the positive systems. Then we refer to (14.106), which shows that two positive systems not conjugate by $E_{\mathscr{K}}$ lead to different terms on the right side of (14.106). The Multiplicity One Theorem implies that distinct terms are distinct characters, and the theorem follows.

§17. Revised Langlands Classification

We can now substitute the classification of irreducible tempered representations (Theorem 14.91) into the Langlands classification (Theorem 8.54) to obtain a new listing of the irreducible admissible representations of a linear connected semisimple group G. After all, Theorem 8.54 gives a classification in terms of induction from irreducible tempered representations, and Theorem 14.91 (even Theorem 14.76) shows that an irreducible tempered representation is itself induced. By the double induction formula, one expects a classification of irreducible admissible representations in terms of induction from a **cuspidal** parabolic subgroup (a parabolic subgroup MAN for which rank $M = \text{rank}(K \cap M)$), with a limit of discrete series on M and a parameter ν on $\mathfrak{a}$ with Re ν in the *closure* of the positive Weyl chamber.

We shall formulate such a result more precisely as Theorem 14.92. It has two features worth noting now: (1) Under the isomorphism given by the double induction formula, the kernels of the appropriate intertwining operators correspond, so that the Langlands quotient representation can be defined without reference to the intermediate parabolic subgroup. (2) The equivalences in Theorem 14.91 with irreducible tempered representations come from mapping MA to $M'A'$, whereas (implicitly through the use in the theorem of a representative from each G conjugacy class of parabolic subgroups) the equivalences in Theorem 8.54 come from mapping MAN to $M'A'N'$.

As a result of (2), we should expect the equivalence condition to become messy when the two stages of induction are combined. In fact, we shall not write down a combined equivalence theorem in general, contenting ourselves with completeness and irreducibility in the general case and with the knowledge that in principle we could unwind the double induction to sort out the equivalences.

In our definition in §16 of "basic character with nondegenerate data," we should note that the nondegeneracy condition can be checked already within M, i.e., it is really a property of the (discrete series or) limit of discrete series on M. Accordingly we define a limit of discrete series $\pi^M(\lambda, \Delta_M^+, \chi)$ of M to be **nondegenerate** if, for each root $\tilde{\alpha}$ in $\Delta(\mathfrak{b}^{\mathbb{C}}:\mathfrak{m}^{\mathbb{C}})$ with $\langle \lambda, \tilde{\alpha} \rangle = 0$, the reflection $s_{\tilde{\alpha}}$ is not in $W(B:M)$. This definition depends only on the representation itself, not on its parameters within M, since $\pi^M(\lambda', \Delta_M^{+'}, \chi') \cong \pi^M(\lambda, \Delta_M^+, \chi)$ implies $\chi' = \chi$, $\lambda' = s\lambda$ and $\Delta_M^{+'} = s\Delta_M^+$ for some $s \in W(B:M)$ if the representation is nonzero. (See Proposition 12.33c.)

Let MAN be a cuspidal parabolic subgroup of G, and let σ be a discrete series or nondegenerate limit of discrete series representation of M. For each root α of $(\mathfrak{g}, \mathfrak{a})$, the Plancherel factor $\mu_{\sigma,\alpha}(\nu)$ is still defined even when ν is not imaginary. Define

$$\begin{aligned} \Delta'_{\sigma,\nu} &= \{\alpha = \text{useful root of } (\mathfrak{g}, \mathfrak{a}) \,|\, s_\alpha \nu = \nu \text{ and } \mu_{\sigma,\alpha}(\nu) = 0\} \\ &= (\text{Re } \nu)^\perp \cap (\text{Im } \nu)^\perp \cap \Delta'_{\sigma,0}. \end{aligned} \tag{14.133a}$$

The right-hand equality shows that $\Delta'_{\sigma,\nu}$ is a root system. Thus it makes sense to define

$$W'_{\sigma,\nu} = \text{Weyl group of } \Delta'_{\sigma,\nu}. \tag{14.133b}$$

By Lemma 14.82, $W'_{\sigma,\nu}$ is a subgroup of

$$W_{\sigma,\nu} = \{s \in W(A:G) \mid s[\sigma] = [\sigma] \text{ and } s\nu = \nu\}. \tag{14.133c}$$

We can then reformulate the completeness of the Langlands classification as follows. The idea is that the R group of the concealed tempered representation is isomorphic to $W_{\sigma,\nu}/W'_{\sigma,\nu}$. Fix a minimal parabolic subgroup $S_\mathfrak{p}$. Recall that a parabolic subgroup S is **standard** if $S \supseteq S_\mathfrak{p}$.

Theorem 14.92 (Langlands classification). Let $S = MAN$ be a cuspidal standard parabolic subgroup of G, let σ be a discrete series or nondegenerate limit of discrete series of M, and let ν be a complex-valued linear functional on $\mathfrak{a}$ with $\operatorname{Re} \nu$ in the closed positive Weyl chamber. Suppose that $W_{\sigma,\nu} = W'_{\sigma,\nu}$. Then the induced representation $U(S, \sigma, \nu)$ has a unique irreducible quotient $J(S, \sigma, \nu)$, and every irreducible admissible representation of G is of the form $J(S, \sigma, \nu)$ for some such triple (S, σ, ν).

Remarks. We call $J(S, \sigma, \nu)$ the **Langlands quotient** of $U(S, \sigma, \nu)$. If $\operatorname{Re} \nu$ is in the open positive Weyl chamber, then no element of $W(A:G)$ fixes ν, and hence $W_{\sigma,\nu} = W'_{\sigma,\nu} = \{1\}$. Thus the key hypothesis for the uniqueness of the irreducible quotient is satisfied. In this case, the notation $J(S, \sigma, \nu)$ and the terminology "Langlands quotient" are consistent with Theorem 8.54.

Proof. First we show the existence of the unique irreducible quotient. Let us regard the roots of $(\mathfrak{g}, \mathfrak{a})$ as the nonzero restrictions to $\mathfrak{a}$ of the restricted roots. Define

$$\Gamma_1 = \{\alpha \in \Delta(\mathfrak{a}:\mathfrak{g}) \mid \langle \operatorname{Re} \nu, \alpha\rangle = 0\}.$$

The set of restricted roots whose restriction to $\mathfrak{a}$ is 0 or is in Γ_1 is a root system generated by simple restricted roots, since $\operatorname{Re} \nu$ is in the closed positive Weyl chamber of $\mathfrak{a}$. Proposition 5.23 associates to the set of simple roots of this root system a parabolic subgroup $S_1 = M_1A_1N_1 \supseteq S$, and we easily see that

M_1 is built from M and members of Γ_1

$\mathfrak{a}_1 =$ common kernel of all $\alpha \in \Delta(\mathfrak{a}:\mathfrak{g})$ with $\langle \operatorname{Re} \nu, \alpha\rangle = 0$

$\mathfrak{n}_1 =$ sum of root spaces $\mathfrak{g}_\alpha$, $\alpha \in \Delta(\mathfrak{a}:\mathfrak{g})$ with $\langle \operatorname{Re} \nu, \alpha\rangle > 0$.

Let $\mathfrak{a}_{M_1} = \mathfrak{a} \cap \mathfrak{m}_1$ and

$\mathfrak{n}_{M_1} =$ sum of root spaces for positive members of Γ_1, and let A_{M_1} and N_{M_1} be the analytic subgroups corresponding to $\mathfrak{a}_{M_1}$ and $\mathfrak{n}_{M_1}$. Then $MA_{M_1}N_{M_1}$ is a parabolic subgroup of M_1, and $\operatorname{Re}\nu$ satisfies

$$\operatorname{Re}\nu|_{\mathfrak{a}_1} \text{ is in open positive Weyl chamber of } \mathfrak{a}_1'$$
$$\operatorname{Re}\nu|_{\mathfrak{a}_{M_1}} = 0.$$

Let
$$\pi = \operatorname{ind}_{S \cap M_1}^{M_1}(\sigma \otimes \exp \nu|_{\mathfrak{a}_{M_1}} \otimes 1).$$

Then π is a basic (tempered unitary) representation of M_1. Since σ is nondegenerate, we can test π for irreducibility by computing its R group. Suppose $s \in W(A:G)$ fixes ν. Then s is the product of reflections fixing ν since $W(A:G)$ is a Weyl group (the Weyl group of the useful roots). If s_α is any such reflection, then $s_\alpha(\operatorname{Re}\nu) = \operatorname{Re}\nu$, and it follows that $\langle \operatorname{Re}\nu, \alpha\rangle = 0$. Hence α is in Γ_1 and contributes to M_1. Therefore the subgroup of $W(A \cap M_1 : M_1)$ fixing σ and ν is the group $W_{\sigma,\nu}$ in (14.133c), and the Δ' system within M_1 is the system $\Delta'_{\sigma,\nu}$ in (14.133a). Since we have assumed $W_{\sigma,\nu} = W'_{\sigma,\nu}$, the R group of π is trivial.

By double induction

$$U(S, \sigma, \nu) \cong U(S_1, \pi, \nu|_{\mathfrak{a}_1}),$$

and the right side has a unique irreducible quotient by Theorem 7.24.

Conversely we want to show that every irreducible admissible representation arises in this way. By Theorem 8.54, we are to identify

$$J(S_1, \pi, \nu_1) = U(S_1, \pi, \nu_1)/\ker A(\bar{S}_1 : S_1 : \pi : \nu_1),$$

where π is irreducible tempered. By Theorem 14.91 (extended to disconnected groups like M_1), we can write π as

$$\pi = \operatorname{ind}_{S'}^{M_1}(\sigma \otimes e^{\nu_2} \otimes 1),$$

where $S' = MA_{M_1}N_{M_1}$ is cuspidal parabolic in M_1, σ is a discrete series or nondegenerate limit of discrete series, and ν_2 is imaginary. Then $S' = S \cap M_1$, where $S = MAN$ is the parabolic subgroup of G with $A = A_{M_1}A_1$ and $N = N_{M_1}N_1$. Put $\nu = \nu_1 \oplus \nu_2$, so that $\operatorname{Re}\nu$ is in the closed positive Weyl chamber. Then

$$U(S, \sigma, \nu) \cong U(S_1, \pi, \nu_1), \tag{14.134}$$

and our given $J(S_1, \pi, \nu_1)$ is the unique irreducible quotient. Our computations in the first part of the proof led us from $U(S, \sigma, \nu)$ to π, and we saw that the R group of π is isomorphic to the quotient $W_{\sigma,\nu}/W'_{\sigma,\nu}$ for G. Since π is irreducible, the R group is trivial; thus $W_{\sigma,\nu} = W'_{\sigma,\nu}$. This proves the theorem.

As we remarked before the proof, the Langlands quotient J can be defined explicitly without reference to the intermediate tempered representation.

In fact, we check directly from the defining integral formula that, under the isomorphism (14.134), the operator that corresponds to $A(\bar{S}_1:S_1:\pi:\nu_1)$ is $A(MAN_{M_1}\bar{N}_1:MAN_{M_1}N:\sigma:\nu)$. Hence

$$J(S, \sigma, \nu) \cong U(S, \sigma, \nu)/\ker A(MAN_{M_1}\bar{N}_1:MAN_{M_1}N:\sigma:\nu). \tag{14.135}$$

Example. $G = \mathrm{SL}(4, \mathbb{R})$. Let $S_\mathfrak{p}$ be the upper triangular group, and let

$$S_1 = \begin{pmatrix} \cdot & \cdot & \cdot & \cdot \\ \cdot & \cdot & \cdot & \cdot \\ \cdot & \cdot & \cdot & \cdot \\ 0 & 0 & 0 & \cdot \end{pmatrix},$$

in the notation of the proof of the theorem. Then $M_1 \cong \mathrm{SL}^{\pm}(3, \mathbb{R})$. We can arrange that an irreducible tempered π on M_1 is induced in nondegenerate fashion from a maximal parabolic subgroup S' of M_1, and we have our choice for this group of either

$$\begin{pmatrix} \cdot & \cdot & \cdot & 0 \\ \cdot & \cdot & \cdot & 0 \\ 0 & 0 & \cdot & 0 \\ 0 & 0 & 0 & \cdot \end{pmatrix} \quad \text{or} \quad \begin{pmatrix} \cdot & \cdot & \cdot & 0 \\ 0 & \cdot & \cdot & 0 \\ 0 & \cdot & \cdot & 0 \\ 0 & 0 & 0 & \cdot \end{pmatrix},$$

since the corresponding groups MA_{M_1} are conjugate within M_1. The corresponding groups $S = MAN$ are

$$\begin{pmatrix} \cdot & \cdot & \cdot & \cdot \\ \cdot & \cdot & \cdot & \cdot \\ 0 & 0 & \cdot & \cdot \\ 0 & 0 & 0 & \cdot \end{pmatrix} \quad \text{and} \quad \begin{pmatrix} \cdot & \cdot & \cdot & \cdot \\ 0 & \cdot & \cdot & \cdot \\ 0 & \cdot & \cdot & \cdot \\ 0 & 0 & 0 & \cdot \end{pmatrix},$$

which are not conjugate in G. Thus in the revised Langlands classification, a representation can arise in two different ways that are not obviously leading to the same representation.

The same circle of ideas as in the proof of Theorem 14.92 can give information about induction from nontempered representations. Here is an example.

Theorem 14.93. Let $S = MAN$ be any parabolic subgroup of G, let σ be an irreducible unitary representation of M with real infinitesimal character, and let ν be an imaginary-valued linear functional on $\mathfrak{a}$ that is nonorthogonal to all roots of $(\mathfrak{g}, \mathfrak{a})$. Then the unitary representation $\mathrm{ind}_S^G(\sigma \otimes e^\nu \otimes 1)$ is irreducible.

Proof when σ is tempered. Taking for granted that Theorem 14.92 extends to the possibly disconnected group M, we can write

$$\sigma = \operatorname{ind}_{M_*A_*N_*}^{M}(\sigma_* \otimes e^{\nu_*} \otimes 1)$$

with $M_*A_*N_*$ cuspidal parabolic in M, σ_* a discrete series or nondegenerate limit of discrete series of M_*, ν_* imaginary on $\mathfrak{a}_*$, and the R group trivial. Since ν_* is imaginary and the infinitesimal character of σ is real, ν_* must be 0 by Proposition 8.22. We can therefore write the trivial R group for σ as $R_{\sigma_*,0}^M$.

By double induction

$$\operatorname{ind}_S^G(\sigma \otimes e^\nu \otimes 1) = \operatorname{ind}_{M_*(AA_*)(NN_*)}^G(\sigma_* \otimes e^{\nu\oplus 0} \otimes 1). \tag{14.136}$$

Thus Theorem 14.92 tells us that we can decide the irreducibility of (14.136) in terms of the R group $R_{\sigma_*,\,\nu\oplus 0}^G$.

Since $R_{\sigma_*,0}^M = \{1\}$, we have

$$W_{\sigma_*,0}^M = W'^M_{\sigma_*,0} \tag{14.137}$$

in obvious notation. Any member of $W(AA_*:G)$ that fixes $\nu \oplus 0$ is the product of reflections fixing $\nu \oplus 0$, and the corresponding roots of $(\mathfrak{g}, \mathfrak{a} \oplus \mathfrak{a}_*)$ must be orthogonal to ν. Since ν is regular, the restriction to $\mathfrak{a}$ of any such root is 0. Thus all such roots are roots of $(\mathfrak{m}, \mathfrak{a}_*)$, and it follows that

$$W_{\sigma_*,\,\nu\oplus 0}^G = W_{\sigma_*,0}^M. \tag{14.138}$$

Let us check that $\Delta'^G_{\sigma_*,\,\nu\oplus 0} = \Delta'^M_{\sigma_*,0}$. Since $\alpha \in \Delta'^G_{\sigma_*,\,\nu\oplus 0}$ implies $s_\alpha \nu = \nu$, we see that $\Delta'^G_{\sigma_*,\,\nu\oplus 0}$ is contained in the set of roots of $(\mathfrak{m}, \mathfrak{a}_*)$. For such roots α, the factors comprising $\mu_{\sigma_*,\alpha}^G(\nu \oplus 0)$ match those comprising $\mu_{\sigma_*,\alpha}^M(0)$, and thus $\Delta'^G_{\sigma_*,\,\nu\oplus 0} = \Delta'^M_{\sigma_*,0}$.

Consequently

$$W'^G_{\sigma_*,\,\nu\oplus 0} = W'^M_{\sigma_*,0}. \tag{14.139}$$

Putting (14.137), (14.138), and (14.139) together, we see that $W_{\sigma_*,\,\nu\oplus 0}^G = W'^G_{\sigma_*,\,\nu\oplus 0}$. Hence $R_{\sigma_*,\,\nu\oplus 0}^G$ is trivial, and (14.136) is irreducible.

Proof for general σ. Let us abbreviate intertwining operators by writing only the N parameters within the parabolic subgroups, the other parameters being clear from the context. (E.g., we write $A(\bar{N}:N)$ in place of $A(\bar{S}:S:\sigma:\nu)$.)

By Theorem 8.54 (extended so as to apply to the disconnected group M), there is a parabolic subgroup $M_*A_*N_*$ in M such that

$$\sigma \cong \operatorname{ind}_{M_*A_*N_*}^M(\sigma_* \otimes e^{\nu_*} \otimes 1)/\ker A(\bar{N}_*:N_*)$$

for some irreducible tempered σ_* on M_* and some ν_* on $\mathfrak{a}_*$ with $\langle \text{Re } \nu_*, \alpha \rangle > 0$ for all positive roots α of $(\mathfrak{m}, \mathfrak{a}_*)$. Since σ has real infinitesimal character, ν_* is real. Since induction is an exact functor,

$$\text{ind}_{MAN}^G(\sigma \otimes e^\nu \otimes 1) \cong \text{ind}_{M_*(AA_*)(NN_*)}^G(\sigma_* \otimes e^{\nu \oplus \nu_*} \otimes 1)/\ker A(N\bar{N}_*:NN_*). \tag{14.140}$$

We shall prove that the "numerator" on the right side has a unique irreducible quotient; hence so does the left side. Since the left side is unitary, we shall be able to conclude that the left side is irreducible, and the proof will be complete.

Choose a different nilpotent subgroup N_0 in place of NN_* in the parabolic subgroup $M_*(AA_*)(NN_*)$ in such a way that $\text{Re}(\nu \oplus \nu_*)$, which equals $0 \oplus \nu_*$, is dominant for the positive roots defined by N_0. Let us see that the intertwining operator $A(NN_*:N_0)$ is an isomorphism at $\nu \oplus \nu_*$. It is enough to show that there are no roots α of $(\mathfrak{g}, \mathfrak{a} \oplus \mathfrak{a}_*)$ such that α is N_0 positive, α is NN_* negative, and $\langle \nu \oplus \nu_*, \alpha \rangle$ is real. Suppose on the contrary that α is such a root. Since $\langle \nu, \alpha \rangle$ is imaginary and $\langle \nu_*, \alpha \rangle$ is real, $\langle \nu, \alpha \rangle$ must be 0. Thus ν is orthogonal to $\alpha|_{\mathfrak{a}}$, which is a root of $(\mathfrak{g}, \mathfrak{a})$ if it is nonzero. Since ν has been assumed regular, we conclude $\alpha|_{\mathfrak{a}} = 0$. Thus α may be regarded as a root of $(\mathfrak{m}, \mathfrak{a}_*)$. Since α as a root of $(\mathfrak{g}, \mathfrak{a})$ is NN_* negative, α as a root of $(\mathfrak{m}, \mathfrak{a}_*)$ is N_* negative, and thus $\langle \nu_*, \alpha \rangle < 0$. On the other hand, the construction of N_0 makes $\langle \nu_*, \alpha \rangle \geq 0$. Thus we have a contradiction, and we must have an isomorphism. Consequently the "numerator" of (14.140) is isomorphic with

$$\text{ind}_{M_*(AA_*)N_0}^G(\sigma_* \otimes e^{\nu \oplus \nu_*} \otimes 1). \tag{14.141}$$

Next we generate a parabolic subgroup $S = M^*A^*N^*$ containing $M_*(AA_*)N_0$. The M^* is to be built from all roots α of $(\mathfrak{g}, \mathfrak{a} \oplus \mathfrak{a}_*)$ such that $\langle \text{Re}(\nu + \nu_*), \alpha \rangle = 0$. These roots are generated by "simple roots" in the obvious sense, since $\text{Re}(\nu + \nu_*)$ is dominant. Hence we do obtain a parabolic subgroup. Define $A_\perp = M^* \cap AA_*$ and $N_\perp = M^* \cap N_0$. By double induction (14.141) is isomorphic to

$$\text{ind}_{M^*A^*N^*}^G\{[\text{ind}_{M_*A_\perp N_\perp}^{M^*}(\sigma_* \otimes \exp(\nu \oplus \nu_*)|_{\mathfrak{a}_\perp} \otimes 1)] \otimes \exp(\nu \oplus \nu_*)|_{\mathfrak{a}^*} \otimes 1\}. \tag{14.142}$$

For the representation in brackets, σ_* is irreducible tempered. The construction of M^* makes $(\nu \oplus \nu_*)|_{\mathfrak{a}_\perp}$ imaginary. We claim there is no root α of $(\mathfrak{m}^*, \mathfrak{a}_\perp)$ with $\langle (\nu \oplus \nu_*)|_{\mathfrak{a}_\perp}, \alpha \rangle = 0$. [In fact, otherwise $\langle \nu, \alpha \rangle = \langle \nu_*, \alpha \rangle = 0$ and α is a root of $(\mathfrak{g}, \mathfrak{a} \oplus \mathfrak{a}_*)$. Once again $\langle \nu, \alpha \rangle = 0$ implies $\alpha|_{\mathfrak{a}} = 0$, and α is a root of $(\mathfrak{m}, \mathfrak{a}_*)$. Since ν_* is nonsingular on $\mathfrak{a}_*$, we have a contradiction.] Therefore our proof in the special case applies and shows that the representation in brackets is irreducible. Since σ_* is basic, we could rewrite the expression in brackets by double induction to see that it is basic, hence tempered. Our construction has made $(\nu \oplus \nu_*)|_{\mathfrak{a}^*}$ have real part in the open positive Weyl chamber of $\mathfrak{a}^{*\prime}$. By Theorem 7.24, the representation (14.142) has a unique irreducible quotient. The theorem follows.

§18. Problems

1. Prove the estimate (14.2) for the rate of convergence of a matrix coefficient of a nonunitary principal series of $SL(2, \mathbb{R})$.
2. Prove the convolution formula $\psi_{ST} = \psi_T * \psi_S$ given as (14.14), using Proposition 9.6 to handle the integrals.
3. Relate the matrix Γ in the differential equation (14.24) to the matrix of constants $c_{jk}^{(H)}$ defined just before it.
4. Using Corollary 8.41, show that all mixed left and right invariant derivatives of members of $\mathscr{A}(G, \tau)$ satisfy the weak inequality.
5. The limit formula in Theorem 14.6 is interpreted for $SL(2, \mathbb{R})$ at the end of §3. Using the result and interpretation for $SL(2, \mathbb{R})$, derive the result and an interpretation for $SL(2, \mathbb{R}) \oplus SL(2, \mathbb{R})$.
6. Derive the adjoint formula (14.33) from the adjoint formula (14.32).
7. Let $\mathfrak{a} \oplus \mathfrak{b}$ be a Cartan subalgebra of $\mathfrak{g}$. Suppose that λ, ν, and ν_0 are in $(\mathfrak{a} \oplus \mathfrak{b})'$, that λ vanishes on $\mathfrak{a}$ and is the infinitesimal character of a discrete series of M, and that ν and ν_0 vanish on $\mathfrak{b}$ and are imaginary. Suppose further that ν_0 is regular for the roots of $(\mathfrak{g}, \mathfrak{a})$ and that $w\lambda = \lambda$ and $w\nu = \nu_0$ for some w in $W((\mathfrak{a} \oplus \mathfrak{b})^{\mathbb{C}} : \mathfrak{g}^{\mathbb{C}})$. Prove that ν and ν_0 are conjugate via $W(A:G)$. [Hint: First pass to the centralizer of λ. There define a conjugate element $\bar{w}$ with $\bar{w}\nu = \nu_0$. Show that $\bar{w}w^{-1}$ is trivial.]
8. Prove the linear independence of operators from $R'_{\sigma,0}$ (Theorem 14.24) when G has real rank one.
9. Suppose s_α exists in $W(A:G)$ and $\bar{\mathfrak{n}} \cap \mathfrak{n}' = \bar{\mathfrak{n}}^{(\alpha)}$. Prove that $\mathfrak{n}' = \mathrm{Ad}(s_\alpha)^{-1}\mathfrak{n}$.
10. Calculation of R groups:
    (a) For $SU(n, 1)$, let MAN be minimal parabolic with

$$A = \left\{\begin{pmatrix} \cosh t & 0 & \sinh t \\ 0 & I & 0 \\ \sinh t & 0 & \cosh t \end{pmatrix}\right\} \quad \text{and} \quad M = \left\{\begin{pmatrix} e^{i\theta} & & \\ & u & \\ & & e^{i\theta} \end{pmatrix}\right\}.$$

Let σ_k for $k \in \mathbb{Z}$ be the one-dimensional representation of M whose value on the above matrix is $e^{ik\theta}$. Prove that $|R_{\sigma_k,0}| = 2$ if $k \equiv n \bmod 2$ and $|k| \geq n$ and that $|R_{\sigma_k,0}| = 1$ otherwise.

(b) For SU(3, 2), determine $R_{\sigma,0}$ for all σ's for the M of a minimal parabolic subgroup.

11. Determine all nonzero final basic characters for SU(2, 1).
12. In SL($2n$, $\mathbb{R}$) for the minimal parabolic subgroup S, let

$$\sigma(\gamma_{e_{2k-1}-e_{2k}}) = -1 \qquad \text{for} \quad 1 \leq k \leq n$$

$$\sigma(\gamma_{e_{2k}-e_{2k+1}}) = +1 \qquad \text{for} \quad 1 \leq k \leq n-1.$$

Determine the two-element group $R_{\sigma,0}$, and describe the two irreducible constituents of $U(S, \sigma, 0)$.

13. The proof of Lemma 14.80 implicitly uses the fact that the domains of the characters χ_i are getting smaller. Why are they getting smaller?
14. Show that Lemma 14.87 fails for the degenerate basic character in Sp(2, $\mathbb{R}$) carried along as an example in §16. How many positive systems $\Delta_{M^*}^+$ satisfy conditions (i), (ii), and (iii) in that lemma? Draw a picture of the Weyl chambers in question.
15. For $G = \text{Sp}(3, \mathbb{R})$, let the simple restricted roots be $e_1 - e_2$, $e_2 - e_3$, $2e_3$. Form the standard parabolic subgroup $S = S_{\{e_1-e_2\}}$, in which $M \cong \text{SL}^{\pm}(2, \mathbb{R}) \oplus \mathbb{Z}_2$. Let $\sigma = \pi^M(0, \{e_1 - e_2\}, \chi)$, where

$$\chi(\gamma_{e_1+e_2}) = -1 \qquad \text{and} \qquad \chi(\gamma_{2e_3}) = 1,$$

and consider the reducibility of $U(S, \sigma, 0)$.

(a) Calculate $R_{\sigma,0}$, showing it has order 4.

(b) Exhibit the reducibility of $U(S, \sigma, 0)$ explicitly by using generalized Schmid identities, showing that $U(S, \sigma, 0)$ is the sum of two irreducible constituents.

16. Suppose $\text{ind}_{MA}^G\, \Theta^{MA}(\lambda, \Delta_M^+, \chi, \nu)$ is a basic character with nondegenerate data.

(a) Prove that the positive roots of Δ_M^+ orthogonal to λ are superorthogonal in Δ_M.

(b) If G is connected, which limits of discrete series of G are nondegenerate?

Problems 17 to 22 deal with relationships among different kinds of roots. They explain, for example, why we do not need multiples of β in (14.45a). Let $\mathfrak{t} = \mathfrak{a}_\mathfrak{p} \oplus \mathfrak{b}_\mathfrak{p} = \mathfrak{a} \oplus \mathfrak{a}_M \oplus \mathfrak{b}_\mathfrak{p}$ be a Cartan subalgebra. To simplify the terminology, we speak of $\mathfrak{t}$ roots, $\mathfrak{a}_\mathfrak{p}$ roots, and $\mathfrak{a}$ roots: A general $\mathfrak{t}$ root we can write as $\alpha_R + \alpha_I + \gamma$, and $\alpha_R + \alpha_I$ (if nonzero) is the associated $\mathfrak{a}_\mathfrak{p}$

root while α_R (if nonzero) is the associated $\mathfrak{a}$ root. For $\mathfrak{t}$ roots α and β, let $\langle \alpha, \beta^\vee \rangle = 2\langle \alpha, \beta \rangle / |\beta|^2$.

17. Prove that $\langle \alpha_R + \alpha_I + \gamma, (\alpha_R + \alpha_I - \gamma)^\vee \rangle \neq +1$. [Hint: Show that a root vector for 2γ is in $\mathfrak{m}_\mathfrak{p} \cap \mathfrak{p} = 0$.]

18. For the $\mathfrak{t}$ root $\alpha_R + \alpha_I + \gamma$, prove that if $2\alpha_R$ is not a $\mathfrak{t}$ root, then $\alpha_I = 0$ or $\gamma = 0$. If $\gamma \neq 0$, then conclude $|\alpha_R| = |\gamma|$. [Hint: Consider $\langle \alpha_R + \alpha_I + \gamma, (\alpha_R - \alpha_I - \gamma)^\vee \rangle$. Then recalculate with the sign of α_I changed and compare with Problem 17.]

19. Prove that α_R even (as an $\mathfrak{a}$ root) implies $2\alpha_R$ is not an $\mathfrak{a}_\mathfrak{p}$ root. [Hint: If $2\alpha_R + \gamma'$ is a $\mathfrak{t}$ root, compare with $\alpha_R + \alpha_I$ or $\alpha_R + \gamma$. In the first case, use Problem 18, and in the second case, use length relations to obtain $2\alpha_R + \gamma' = 2(\alpha_R + \gamma)$.]

20. If $\alpha_R + \alpha_I$ is an $\mathfrak{a}_\mathfrak{p}$ root with $\alpha_I \neq 0$ and if α_R is an even useful $\mathfrak{a}$ root, prove that $|\alpha_R| = |\alpha_I|$.

21. If α_R is an even useful $\mathfrak{a}$ root, prove that $2\alpha_R$ is not an $\mathfrak{a}$ root. [Hint: Argue as in Problem 19, using the length result from Problem 20.]

22. Suppose that no two $\mathfrak{a}_\mathfrak{p}$ roots β_1 and β_2 have $|\beta_1|^2 = 3|\beta_2|^2$.
    (a) Prove that if α_R is a not useful $\mathfrak{a}$ root, then $2\alpha_R$ is not an $\mathfrak{a}_\mathfrak{p}$ root.
    (b) Prove that if α_R is an $\mathfrak{a}$ root of odd multiplicity, then $c\alpha_R$ is not an $\mathfrak{a}$ root for $c > 1$. [Hint: Consider $\langle c\alpha_R + \beta, (c\alpha_R - \beta)^\vee \rangle$. Check separately the three cases where it is 0, $+1$, -1, examining root lengths, to obtain a contradiction.]

Problems 23 to 29 examine the reducibility of a unitary induced representation in which $\dim A = 1$ and the number of irreducible constituents is greater than 2 (which is the order of the Weyl group). Let $G = SU(2, 2)$. Section 5.5 identifies the standard parabolic subgroups of G, one of them being $S = S_{\{f_1 - f_2\}} = MAN$. This subgroup has $\dim A = 1$, and N is abelian.

23. Using the isomorphism $SU(1, 1) \cong SL(2, \mathbb{R})$ in Chapter II as a guide, conjugate SU(2, 2) so that MA is block diagonal and N is block upper triangular. The group N should be naturally identifiable with the additive group of 2-by-2 Hermitian matrices.

24. Show that the action of MA on N by conjugation preserves the signature of a Hermitian matrix (under the identification of N in Problem 23).

25. Motivated by what happens in $SL(2, \mathbb{R})$, show that the unitary representation $\mathrm{ind}_S^G(1 \otimes 1 \otimes 1)$ can be realized in L^2 (2-by-2 Hermitian matrices) with an action given by linear fractional transformations.

26. The set of 2-by-2 Hermitian matrices x may be regarded as a boundary of the complex manifold
$$\left\{ z = x + iy \,\middle|\, \begin{array}{l} x \text{ is 2-by-2 Hermitian and} \\ y \text{ is 2-by-2 Hermitian positive definite} \end{array} \right\}.$$
Show that this space is naturally identified with G/K, the action being by linear fractional transformations.
27. Identify invariant subspaces of "holomorphic" and "antiholomorphic" functions in the representation in Problem 25.
28. [Uses Fourier transform] Show for each of the three possible nonsingular signatures that the space of L^2 functions in Problem 25 whose Fourier transforms are supported on matrices of a particular signature is invariant under G. Conclude that the representation splits into at least three irreducible pieces, even though $\dim A = 1$.
29. If the $\mathfrak{a}$ parameter in the induction in Problem 25 is replaced by something nonzero imaginary, why must the induced representation be irreducible?

Problems 30 to 37 deal with the conditions in Lemma 14.69 for extending a positive system from M to G appropriately. Let a Cartan subalgebra $\mathfrak{a} \oplus \mathfrak{b}$ define $\mathfrak{m}$, let $\Delta_M = \Delta((\mathfrak{a} \oplus \mathfrak{b})^{\mathbb{C}} : (\mathfrak{m} \oplus \mathfrak{a})^{\mathbb{C}})$, and let $\Delta_G = \Delta((\mathfrak{a} \oplus \mathfrak{b})^{\mathbb{C}} : \mathfrak{g}^{\mathbb{C}})$. Fix positive systems Δ_M^+ and Δ_G^+ with $\Delta_M^+ \subseteq \Delta_G^+$. Let (iii) and (iii′) be the following conditions on Δ_G^+:

(iii) Every nonreal α in Δ_G^+ is such that $\bar{\alpha} = -\theta\alpha$ ($= \alpha$ on $\mathfrak{a}$ and $-\alpha$ on $\mathfrak{b}$).

(iii′) If C is the positive Weyl chamber in $\mathfrak{a} \oplus i\mathfrak{b}$ relative to Δ_G^+, then $i\mathfrak{b} \cap \bar{C}$ has nonempty interior in $i\mathfrak{b}$.

30. Prove that (iii) implies that the real roots of Δ_G that are Δ_G^+ simple generate the real roots of Δ_G. [Hint: Expand a real positive α in terms of simple roots as $\sum n_j\alpha_j$, letting the first k simple roots be real, the others nonreal. Apply the conjugation and calculate $\alpha - \sum_{j=1}^k n_j\bar{\alpha}_j$ in two ways.]
31. Prove that (iii′) implies (iii). [Hint: If α nonreal is given, use (iii′) to produce a dominant λ in $(i\mathfrak{b})'$ with $\langle \lambda, \alpha \rangle \neq 0$.]
32. Suppose that (iii) holds. Let w_0 be the Weyl group element $w_0 = w_G w_r$, where w_G is the long element of the Weyl group of Δ_G and w_r is the long element of the Weyl group of the real roots. Prove that
(a) α simple implies $-w_0\bar{\alpha}$ is simple,
(b) the map $\alpha \to -w_0\bar{\alpha}$ has order 2.
33. Suppose that (iii) holds. Using Problem 32, lump the simple roots $\{\alpha_j\}$ into pairs and singletons, nonreal and real, under the action

$\alpha \to -w_0\bar{\alpha}$. After first giving a lower bound for dim $\mathfrak{a}$, prove the bound

$$\dim \mathfrak{b} \leq \#\{\text{nonreal singletons}\} + \#\{\text{nonreal pairs}\}.$$

34. In the setting of Problems 32–33, let $\{\alpha_j\}$ be the set of simple roots, and define H_i in $\mathfrak{a} \oplus i\mathfrak{b}$ by $\alpha_j(H_i) = \delta_{ij}$. Show that all elements

$$\sum_{\{\alpha_m\}\text{ nonreal singleton}} c_m H_m + \sum_{\substack{\{\alpha_i,\alpha_{j(i)}\}\\ \text{nonreal pair}}} c_i(H_i + H_{j(i)}), \quad \text{all coefficients} \geq 0 \tag{14.143}$$

are in $\bar{C}$ and that the dimension of the above set is the sum of the number of nonreal singletons and the number of nonreal pairs, hence at least dim $\mathfrak{b}$.

35. In the setting of Problems 32–34, prove that any element (14.143) is annihilated by all real roots and by all $\alpha_j + w_0\bar{\alpha}_j$ with α_j simple and nonreal.

36. In the setting of Problems 32–35, prove that the linear functionals in Problem 35 span $\mathfrak{a}'$. [Hint: The $\mathfrak{a}'$ component of any nonreal singleton and of $\alpha_j - w_0\bar{\alpha}_j$ for any nonreal pair are in the -1 eigenspace of w_0, hence in the span of the real roots.]

37. Put together the conclusions of Problems 34–36 to prove that (iii) implies (iii′).

CHAPTER XV

Minimal K Types

§1. Definition and Formula

For any linear connected semisimple group G, the Langlands classification, as revised in §14.17, turns out to carry with it information about the K type structure of irreducible admissible representations. The present chapter will introduce for admissible representations a notion of "minimal K type" that will essentially determine the S and σ parameters for Langlands quotients $J(S, \sigma, \nu)$. This development will have several important consequences:

(1) it will exhibit a K type that occurs in $J(S, \sigma, \nu)$, and the multiplicity of that K type will be one
(2) it will provide a new useful normalization of a single intertwining operator
(3) it will relate reducibility of standard induced representations to the structure of the K types
(4) it will provide a largely algebraic proof of the Langlands Disjointness Theorem.

Fix a Cartan subalgebra $\mathfrak{b}$ of $\mathfrak{k}$ and a positive system for $\Delta(\mathfrak{b}^{\mathbb{C}}:\mathfrak{k}^{\mathbb{C}})$. Let π be an admissible representation of G. Among all K types $\tau_{\Lambda'}$ occurring in π, the **minimal** K types of π are the ones τ_Λ for which $|\Lambda' + 2\delta_K|^2$ is minimized by $\Lambda' = \Lambda$. It is trivial that π has at least one but only finitely many minimal K types; π may have more than one even if π is irreducible. The minimal K types do not depend on the choice of positive system for $\Delta(\mathfrak{b}^{\mathbb{C}}:\mathfrak{k}^{\mathbb{C}})$.

Examples.

(1) Discrete series or limit $\pi^G(\lambda_0, \Delta^+)$.

By Theorem 9.20, every K type $\tau_{\Lambda'}$ in the representation is of the form

$$\Lambda' = \Lambda + \sum n_j\alpha_j$$

with $n_j \geq 0$ and $\alpha_j \in \Delta^+ = \Delta^+(\mathfrak{b}^{\mathbb{C}}:\mathfrak{g}^{\mathbb{C}})$; here Λ is the Blattner parameter

$$\Lambda = \lambda_0 + \delta - 2\delta_K.$$

Thus

$$\Lambda' + 2\delta_K = \lambda_0 + \delta + \sum n_j\alpha_j.$$

Taking the norm squared of both sides, we see that $|\Lambda' + 2\delta_K|^2$ is minimized uniquely by $\Lambda' = \Lambda$. Therefore the Blattner parameter gives the unique minimal K type.

(2) Principal series $U(S, \sigma, \nu)$ of $SL(2, \mathbb{R})$.

Let the nth K type be

$$\tau_n\begin{pmatrix} \cos\theta & \sin\theta \\ -\sin\theta & \cos\theta \end{pmatrix} = e^{in\theta}. \tag{15.1}$$

The K types in $U(S, \sigma, \nu)$ are given by all even n if σ is trivial, all odd n if σ is nontrivial. In this group, $2\delta_K$ is 0. Thus the minimal K types are given by $n = 0$ if σ is trivial and $n = \pm 1$ if σ is nontrivial.

For general G the first step is to give a formula for all the minimal K types of any standard induced representation $U(S, \sigma, \nu)$, where σ is a discrete series or nondegenerate limit of discrete series. The decomposition of $U(S, \sigma, \nu)$ under K is independent of ν, by Frobenius reciprocity, and thus the minimal K types do not depend on ν.

Although one can give such a formula in general, we shall limit ourselves to the case rank G = rank K. Essentially all the ideas are present in this case, and the general case involves only technical refinements in statements and proofs.

Thus let G be linear connected semisimple with rank G = rank K, and let $S = MAN$ be a parabolic subgroup built from a Cartan subalgebra $\mathfrak{a} \oplus \mathfrak{b}_-$ with $\mathfrak{a} \subseteq \mathfrak{p}$ and $\mathfrak{b}_- \subseteq \mathfrak{k}$. The assumption rank G = rank K implies that we may assume that there is a Cartan subalgebra $\mathfrak{b}$ of $\mathfrak{g}$ with $\mathfrak{b}_- \subseteq \mathfrak{b} \subseteq \mathfrak{k}$ and that $\mathfrak{b}$ is obtained by Cayley transform from $\mathfrak{a} \oplus \mathfrak{b}_-$. Turning matters around, we may assume that $\mathfrak{b}$ is given and that there is a strongly orthogonal sequence $\alpha_1, \ldots, \alpha_l$ of noncompact roots in $\Delta = \Delta(\mathfrak{b}^{\mathbb{C}}:\mathfrak{g}^{\mathbb{C}})$ such that $\mathfrak{a} \oplus \mathfrak{b}_-$ is obtained from $\mathfrak{b}$ by application of the Cayley transform $\mathbf{c} = \mathbf{c}_{\alpha_1} \cdots \mathbf{c}_{\alpha_l}$.

We shall usually write formal results in terms of $\Delta(\mathfrak{b}^{\mathbb{C}}:\mathfrak{g}^{\mathbb{C}})$ and often think in terms of the Cayley transform $\Delta((\mathfrak{a} \oplus \mathfrak{b}_-)^{\mathbb{C}}:\mathfrak{g}^{\mathbb{C}})$. Taking into account the genuine definitions in the latter root system, let us say that

$$\text{a root in } \Delta \text{ is } \begin{cases} \textbf{real} & \text{if in } \sum \mathbb{R}\alpha_j \\ \textbf{imaginary} & \text{if orthogonal to } \sum \mathbb{R}\alpha_j \\ \textbf{complex} & \text{otherwise.} \end{cases}$$

Let

$$\Delta_r = \{\text{real roots in } \Delta\}$$

$$\mathfrak{b}_r = \sum \mathbb{R}iH_{\alpha_j}.$$

Then $\mathfrak{b} = \mathfrak{b}_- \oplus \mathfrak{b}_r$. Let E be the orthogonal projection of $(\mathfrak{b}^{\mathbb{C}})'$ onto $(\mathfrak{b}_r^{\mathbb{C}})'$ given by

$$E(\gamma) = \sum_{j=1}^{l} \frac{\langle \gamma, \alpha_j \rangle}{|\alpha_j|^2} \alpha_j. \tag{15.2}$$

The subalgebra

$$\mathfrak{g}_r = \mathfrak{g} \cap \left(\mathfrak{b}_r^{\mathbb{C}} \oplus \sum_{\beta \in \Delta_r} \mathbb{C} E_\beta \right)$$

is a θ-stable semisimple subalgebra of $\mathfrak{g}$ that is split over $\mathbb{R}$. (Note that $\mathfrak{a}$ is a Cartan subalgebra of $\mathfrak{g}_r$.) Let G_r be the analytic subgroup of G with Lie algebra $\mathfrak{g}_r$; G_r is linear connected semisimple. The group $K_r = K \cap G_r$ is a maximal compact subgroup of G_r, and its Lie algebra is $\mathfrak{k}_r = \mathfrak{k} \cap \mathfrak{g}_r$. Moreover, rank G_r = rank K_r since $\mathfrak{b}_r$ is a Cartan subalgebra of $\mathfrak{g}_r$ contained in $\mathfrak{k}_r$. The root system of $(\mathfrak{g}_r^{\mathbb{C}}, \mathfrak{b}_r^{\mathbb{C}})$ may be identified with Δ_r.

Example. $G = SU(p, q)$, $p > q$, minimal parabolic subgroup.

In the notation of Appendix C, where this example is of type A III, we can take

$$\alpha_j = e_j - e_{p+q+1-j} \qquad \text{for } 1 \leq j \leq q.$$

The root system Δ_r consists only of the roots $\pm\alpha_j$ and is therefore a q-fold product of systems of type A_1. The group G_r is the direct sum of q copies of $SL(2, \mathbb{R})$, one copy coming from each α_j.

The Lie algebra $\mathfrak{b}_-$ is a compact Cartan subalgebra of $\mathfrak{m}$, and $\mathfrak{m}^{\mathbb{C}}$ is given in terms of the root system $\Delta((\mathfrak{a} \oplus \mathfrak{b}_-)^{\mathbb{C}} : \mathfrak{g}^{\mathbb{C}})$ by the roots that restrict to 0 on $\mathfrak{a}$:

$$\mathfrak{m}^{\mathbb{C}} = \mathfrak{b}_-^{\mathbb{C}} \oplus \sum_{\substack{\beta \in \Delta \\ \beta|_{\mathfrak{b}_r} = 0}} \mathbb{C}\, \mathbf{c}(E_\beta).$$

Thus if we write

$$\Delta_- = \{\beta \in \Delta \,|\, \beta|_{\mathfrak{b}_r} = 0\},$$

then we can identify Δ_- with $\Delta(\mathfrak{b}_-^{\mathbb{C}} : \mathfrak{m}^{\mathbb{C}})$.

Each root vector $\mathbf{c}(E_\beta)$ for $\mathfrak{m}^{\mathbb{C}}$ is either in $\mathfrak{k}^{\mathbb{C}}$ or in $\mathfrak{p}^{\mathbb{C}}$, and we call β ***M*-compact** or ***M*-noncompact** accordingly. Under the identification of Δ_- with $\Delta(\mathfrak{b}_-^{\mathbb{C}} : \mathfrak{m}^{\mathbb{C}})$, let

$$\Delta_{-,c} = \{M\text{-compact roots in } \Delta_-\}.$$

Discrete series and nondegenerate limits of discrete series of M are induced from $M^\# = M_0 Z_M$, and we know from Lemma 12.30 that $M^\# = M_0 F(B_-)$, where $F(B_-)$ is generated by the elements $\gamma_{\mathbf{c}(\beta)}$ with $\beta \in \Delta_r$. Under our definition of G_r, the group $F(B_-)$ is exactly the M group for a minimal parabolic subgroup of G_r; we denote it by M_r. Thus discrete series and nondegenerate limits of discrete series of M are given in terms of a pair $(\lambda_0, (\Delta_-)^+)$ and a character χ of M_r.

Let $\sigma = \pi^M(\lambda_0, (\Delta_-)^+, \chi)$ be a discrete series or nondegenerate limit of discrete series of M. The choice of positive system $(\Delta_-)^+$ for Δ_- determines half sums $\delta_{-,c}$ and δ_- of the positive members of $\Delta_{-,c}$ and Δ_-, respectively. In terms of $\delta_{-,c}$ and δ_-, the Blattner parameter λ of $\pi^{M_0}(\lambda_0, (\Delta_-)^+)$ is given by

$$\lambda = \lambda_0 + \delta_- - 2\delta_{-,c}.$$

According to Lemma 14.69, we can find a positive system Δ^+ for $\Delta = \Delta(\mathfrak{b}^{\mathbb{C}}:\mathfrak{g}^{\mathbb{C}})$ such that

(i) $(\Delta_-)^+ \subseteq \Delta^+$
(ii) λ_0 (extended by 0 on $\mathfrak{b}_r$) is Δ^+ dominant
(iii) β in $\Delta^+ - \Delta_r$ implies $s_{\alpha_1} \cdots s_{\alpha_l}\beta$ is positive.

Problem 30 in Chapter XIV shows that Δ_r is generated by the simple roots that it contains. Therefore we can change the system Δ^+ by any element in the Weyl group $W(\Delta_r)$ of Δ_r, and properties (i), (ii), and (iii) will remain valid. Thus we apply whatever element of $W(\Delta_r)$ is appropriate so that positivity of roots in Δ_r is determined by lexicographic ordering relative to the ordered basis. (That is, $\sum c_j\alpha_j$ is > 0 if $c_1 = \ldots = c_{i-1} = 0$ and $c_i > 0$ for some i.) After Δ^+ is redefined by application of such a Weyl group element, Δ^+ satisfies (i), (ii), (iii), and the additional property

(iv) $\beta = \sum c_j\alpha_j$ in Δ_r is in Δ^+ if the first nonzero c_i is positive.

Any positive system Δ^+ satisfying (i), (ii), (iii), and (iv) we shall say is **compatible** with λ_0 and $(\Delta_-)^+$.

Theorem 15.1 (Minimal K type formula). Let G be linear connected semisimple with rank G = rank K, let $\mathfrak{b} \subseteq \mathfrak{k}$ be a compact Cartan subalgebra, let $S = MAN$ be a cuspidal parabolic subgroup built by Cayley transform from a sequence $\alpha_1, \ldots, \alpha_l$ of strongly orthogonal noncompact roots, and let G_r be the split group built from all roots in the span of the α_j's. Let $\sigma = \pi^M(\lambda_0, (\Delta_-)^+, \chi)$ be a discrete series or nondegenerate limit of discrete series of M, and let Δ^+ be a positive system for $\Delta = \Delta(\mathfrak{b}^{\mathbb{C}}:\mathfrak{g}^{\mathbb{C}})$ compatible with λ_0 and $(\Delta_-)^+$. Then every minimal K type τ_Λ of $U(S, \sigma, \nu)$ is of multiplicity one in $U(S, \sigma, \nu)$ and has Λ of the form

$$\Lambda = \lambda - E(2\delta_K) + 2\delta_{K_r} + \mu, \tag{15.3}$$

where λ is the Blattner parameter for λ_0, E is the projection in (15.2), and τ_μ is a minimal K_r type for the principal series representations of G_r with

M_r parameter

$$\sigma_r = \chi \cdot \exp(E(2\delta_K) - 2\delta_{K_r})|_{M_r}; \tag{15.4}$$

here $\exp(E(2\delta_K) - 2\delta_{K_r})$ is a well-defined one-dimensional representation of $K_r \supseteq M_r$. Conversely every minimal K_r type τ_μ for this σ_r is such that Λ in (15.3) is analytically integral; if Λ is also Δ_K^+ dominant, then τ_Λ is a minimal K type of $U(S, \sigma, v)$.

Example. $G = SU(3, 1)$, minimal parabolic, $\alpha = \alpha_1 = e_3 - e_4$.

In the usual notation for $G^{\mathbb{C}} = SL(4, \mathbb{C})$ as given in §4.1, Δ_- is given by $\Delta_- = \{\pm(e_1 - e_2)\}$. The group M is connected, and we suppose that

$$\sigma = \pi^M(\lambda_0, \{e_1 - e_2\}),$$

where $\lambda_0 = e_1 - e_2 + 2(e_3 + e_4)$. There is only one compatible Δ^+ in this case, and its simple roots are

$$e_3 - e_4, e_4 - e_1, e_1 - e_2.$$

Then $2\delta_K = 2(e_3 - e_2)$ and $E(2\delta_K) = |\alpha|^{-2}\langle 2\delta_K, \alpha\rangle\alpha = e_3 - e_4$. Since $\delta_- = \delta_{-,c} = \frac{1}{2}(e_1 - e_2)$, the Blattner parameter (highest weight) for σ is

$$\lambda = \tfrac{1}{2}(e_1 - e_2) + 2(e_3 + e_4).$$

The group G_r is the $SU(1, 1)$ built from α, K_r is the (diagonal) circle subgroup of G_r, and $2\delta_{K_r} = 0$. The group M_r contains two elements, 1 and $\gamma_{e_3 - e_4} = \operatorname{diag}(1, 1, -1, -1)$, and

$$\exp(E(2\delta_K) - 2\delta_{K_r})(\gamma_{e_3 - e_4}) = [\exp(e_3 - e_4)](\gamma_{e_3 - e_4}) = +1.$$

We can rewrite $\gamma_{e_3 - e_4}$ as the exponential of a member of $\mathfrak{b}_-$, namely

$$\gamma_{e_3 - e_4} = \exp i\pi \operatorname{diag}(0, -2, 1, 1),$$

and then we can see that the character corresponding to λ takes the value -1 on $\gamma_{e_3 - e_4}$. Hence $\chi(\gamma_{e_3 - e_4}) = -1$ and $\sigma_r(\gamma_{e_3 - e_4}) = -1$. Hence μ is to be a minimal K_r type for the principal series of $SU(1, 1)$ with the nontrivial character of M_r. Thus $\mu = \pm\frac{1}{2}\alpha$. Putting everything together, we obtain from the theorem the formula

$$\Lambda = \lambda - E(2\delta_K) + 2\delta_{K_r} + \mu = \tfrac{1}{2}e_1 + \tfrac{1}{2}e_2 + e_3 + 3e_4 \pm \tfrac{1}{2}(e_3 - e_4).$$

The difference of any two coefficients is integral, and thus Λ is integral. For either choice of sign, the inner product of Λ with the simple roots of Δ_K^+, namely $e_3 - e_1$ and $e_1 - e_2$, is ≥ 0. Thus there are two minimal K types,

$$\Lambda = \tfrac{1}{2}(e_1 + e_2 + 3e_3 + 5e_4)$$

and

$$\Lambda = \tfrac{1}{2}(e_1 + e_2 + e_3 + 7e_4).$$

Remarks.

(1) In general the theorem reduces matters to minimal parabolics in split groups. This case will be discussed further in the next section, especially in Theorem 15.4. Often the reduction is to a sum of copies of $SL(2, \mathbb{R})$.

(2) The proof of integrality of Λ proceeds by an inductive construction similar to that in §14.14 and contains no really new ideas. The assertions about dominance can be verified by inspection in any particular example and require an exhaustive analysis of roots in general. Thus the parts of the proof of most immediate interest are the occurrence of τ_Λ in the induced representation, the reduction of the multiplicity assertion to minimal parabolics in split groups, and the fact that every minimal K type is of the form (15.3), under the assumptions that (15.3) is automatically integral and some expression (15.3) is Δ_K^+ dominant.

Proof of occurrence of τ_Λ. Under the assumption that Λ is integral and Δ_K^+ dominant, let v_Λ be a nonzero highest weight vector of τ_Λ, and let

$$\mathscr{S} = \text{span } \tau_\Lambda(K_r)v_\Lambda. \tag{15.5}$$

Since $\Delta_{K_r}^+ \subseteq \Delta_K^+$, $\mathscr{S}$ is irreducible under $\tau_\Lambda|_{K_r}$, and its highest weight is

$$\Lambda|_{\mathfrak{b}_r} = -E(2\delta_K) + 2\delta_{K_r} + \mu. \tag{15.6}$$

Since $\tau_\mu|_{M_r}$ contains σ_r, it follows from (15.6) and (15.4) that $(\tau_\Lambda|_{K_r}, \mathscr{S})|_{M_r}$ contains χ. Since M_r is abelian, there is a one-dimensional subspace $\mathbb{C}v_0$ of $\mathscr{S}$ in which $\tau_\Lambda|_{M_r}$ acts by χ.

We shall show that every vector in $\mathscr{S}$ (and, in particular, the vector v_0) is a highest weight vector under $\tau_\Lambda|_{K \cap M_0}$, relative to the Cartan subalgebra $\mathfrak{b}_-$. First we show from (15.3) that

$$\tau_\Lambda(H_-)v = \lambda(H_-)v \qquad \text{for all } H_- \text{ in } \mathfrak{b}_-. \tag{15.7}$$

[In fact, this equality holds for $v = v_\Lambda$. If X is in $U(\mathfrak{k}_r^{\mathbb{C}})$, then X commutes with H_-, and

$$\tau_\Lambda(H_-)\tau_\Lambda(X)v = \tau_\Lambda(X)\tau_\Lambda(H_-)v = \lambda(H_-)\tau_\Lambda(X)v.$$

So (15.7) holds for $\tau_\Lambda(U(\mathfrak{k}_r^{\mathbb{C}}))v_\Lambda = \mathscr{S}$.]

Now we consider root vectors relative to $\Delta_{-,c}^+$. Thus suppose $\tilde{E}_\beta = \mathbf{c}(E_\beta)$ with β in $\Delta_{-,c}^+$. From (15.7), we know that $\tau_\Lambda(\mathfrak{b})\mathscr{S} \subseteq \mathscr{S}$. Thus $\mathscr{S}$ is spanned by weight vectors under $\tau_\Lambda(\mathfrak{b})$. So we are to show that

$$\tau_\Lambda(\tilde{E}_\beta)v_{\Lambda'} = 0 \tag{15.8}$$

for every weight vector $v_{\Lambda'}$ in $\mathscr{S}$. According to (15.7), $\Lambda'|_{\mathfrak{b}_-} = \lambda$.

There are two cases. One is that β is strongly orthogonal to all α_j. In this case, $\tilde{E}_\beta = E_\beta$ and β is in Δ_K^+. Then $\tau_\Lambda(\tilde{E}_\beta)v_{\Lambda'} = \tau_\Lambda(E_\beta)v_{\Lambda'}$ is a weight vector of weight $\Lambda' + \beta$. If the vector is not zero, then $\Lambda - (\Lambda' + \beta)$ has to be a sum of members of Δ_K^+ and has to be $-\beta|_{\mathfrak{b}_-}$ on $\mathfrak{b}_-$. Since β is positive, this condition contradicts property (iii) of Δ^+. Therefore (15.8) holds for this β.

The other case is that β fails to be strongly orthogonal to some α_j. In this case Problem 2 shows that there is some index j such that $\tilde{E}_\beta$ is a linear combination of $E_{\beta+\alpha_j}$ and $E_{\beta-\alpha_j}$. Moreover, β must be noncompact, and $\beta \pm \alpha_j$ are compact. Then $\tau_\Lambda(\tilde{E}_\beta)v_{\Lambda'}$ is a linear combination of vectors of weights $\Lambda' + \beta + \alpha_j$ and $\Lambda' + \beta - \alpha_j$. Using property (iii) of Δ^+ as in the previous paragraph, we see that both vectors are 0. Therefore (15.8) holds for this β.

Thus we have proved that v_0 is a highest weight vector under $\tau_\Lambda|_{K \cap M_0}$ with $\mathfrak{b}_-$ weight λ. Consequently it generates an irreducible representation of $K \cap M_0$ of type λ. Since M_r acts by χ on v_0 and M_r commutes with M_0, the restriction of τ_Λ to $K \cap M^\#$ contains the irreducible representation with highest weight λ that is χ on M_r. By Frobenius reciprocity, the restriction to K of

$$\operatorname{ind}_{M^\# AN}^G(\pi^{M^\#}(\lambda_0, (\Delta_-)^+, \chi) \otimes e^\nu \otimes 1)$$

contains τ_Λ. But this representation is just $U(S, \sigma, \nu)$, and the result follows.

The proof that every minimal K type is of the form (15.3) uses the following identity, whose proof we omit.

Proposition 15.2. Under the assumption that Δ^+ is a positive system compatible with $(\Delta_-)^+$ and some λ_0,

$$2(\delta_K - \delta_{-,c}) = \delta - \delta_- - \delta_r + E(2\delta_K).$$

Remark. As a result of this proposition, we can rewrite the minimal K type formula (15.3) as

$$\Lambda + 2\delta_K = \lambda_0 + (\delta - \delta_r) + (\mu + 2\delta_{K_r}), \tag{15.9}$$

which is the formula we shall apply in the next section. For now we shall be content with the following consequence of the proposition.

Corollary 15.3. $\langle \delta_K - \delta_{-,c}, \gamma \rangle \geq 0$ for every γ in $(\Delta_-)^+$.

Proof. We may assume γ is simple for $(\Delta_-)^+$. Applying the proposition, we have

$$\frac{4\langle \delta_K - \delta_{-,c}, \gamma \rangle}{|\gamma|^2} = \frac{2\langle \delta - \delta_-, \gamma \rangle}{|\gamma|^2} = \frac{2\langle \delta, \gamma \rangle}{|\gamma|^2} - 1 \geq 0.$$

Proof that minimal K types are as in (15.3). Under the assumption that Λ in (15.3) is integral and Δ_K^+ dominant, let τ_{Λ_0} be a minimal K type. By

Frobenius reciprocity, $\tau_{\Lambda_0}|_{K\cap M^\#}$ contains some $K \cap M^\#$ type of σ, say $\sigma_{\lambda'}$ with

$$\sigma_{\lambda'} = \begin{cases} \text{irred. rep. of } K \cap M_0 \text{ with highest weight } \lambda' \text{ on } \mathfrak{b}_- \\ \chi \text{ on } M_r. \end{cases} \tag{15.10}$$

A little argument similar to the first part of the proof that τ_Λ occurs in $U(S, \sigma, \nu)$ shows that τ_{Λ_0} has a weight $\lambda' + \omega$ such that ω is $\Delta^+_{K_r}$ dominant and

$$\tau_\omega|_{M_r} \text{ contains } \chi. \tag{15.11}$$

We omit the details of this step.

Let us write down some consequences of these facts. Since $\lambda' + \omega$ is a weight of τ_{Λ_0}, we have

$$\lambda' + \omega = \Lambda_0 - \sum n_i\beta_i, \qquad n_i \geq 0, \qquad \beta_i \in \Delta^+_K \tag{15.12}$$

$$|\lambda' + \omega|^2 \leq |\Lambda_0|^2. \tag{15.13}$$

Since the Blattner weight of $\sigma|_{M_0}$ is minimal,

$$|\lambda' + 2\delta_{-,c}|^2 \geq |\lambda + 2\delta_{-,c}|^2. \tag{15.14}$$

Since τ_Λ has been shown to occur in the induced representation and τ_{Λ_0} is a minimal K type,

$$|\Lambda_0 + 2\delta_K|^2 \leq |\Lambda + 2\delta_K|^2. \tag{15.15}$$

Then we have

$$\begin{aligned}
&|\lambda|^2 + |\omega|^2 \\
&\quad = |\lambda + 2\delta_{-,c}|^2 - 4\langle\lambda, \delta_{-,c}\rangle - 4|\delta_{-,c}|^2 + |\omega|^2 \\
&\quad \leq |\lambda' + 2\delta_{-,c}|^2 - 4\langle\lambda, \delta_{-,c}\rangle - 4|\delta_{-,c}|^2 + |\omega|^2 && \text{by (15.14)} \\
&\quad = |\lambda'|^2 + 4\langle\lambda' - \lambda, \delta_{-,c}\rangle + |\omega|^2 \\
&\quad = |\lambda' + \omega|^2 + 4\langle\lambda' - \lambda, \delta_{-,c}\rangle \\
&\quad \leq |\Lambda_0|^2 + 4\langle\lambda' - \lambda, \delta_{-,c}\rangle && \text{by (15.13)} \\
&\quad = |\Lambda_0 + 2\delta_K|^2 - 4\langle\Lambda_0, \delta_K\rangle - 4|\delta_K|^2 + 4\langle\lambda' - \lambda, \delta_{-,c}\rangle \\
&\quad \leq |\Lambda + 2\delta_K|^2 - 4\langle\Lambda_0, \delta_K\rangle - 4|\delta_K|^2 + 4\langle\lambda' - \lambda, \delta_{-,c}\rangle && \text{by (15.15)} \\
&\quad = |\Lambda|^2 + 4\langle\Lambda - \Lambda_0, \delta_K\rangle + 4\langle\lambda' - \lambda, \delta_{-,c}\rangle \\
&\quad = |\lambda|^2 + |2\delta_{K_r} - E(2\delta_K) + \mu|^2 + 4\langle\Lambda - \Lambda_0, \delta_K\rangle + 4\langle\lambda' - \lambda, \delta_{-,c}\rangle \\
&\quad = |\lambda|^2 + |2\delta_{K_r} - E(2\delta_K) + \mu|^2 + 4\langle\lambda' - \lambda, \delta_{-,c}\rangle \\
&\qquad + 4\langle(\lambda + 2\delta_{K_r} - E(2\delta_K) + \mu) - (\lambda' + \omega + \textstyle\sum n_i\beta_i), \delta_K\rangle && \text{by (15.12)} \\
&\quad = |\lambda|^2 + |2\delta_{K_r} - E(2\delta_K) + \mu|^2 - 4\langle\lambda' - \lambda, \delta_K - \delta_{-,c}\rangle \\
&\qquad - 4\langle E(2\delta_K) - 2\delta_{K_r} - \mu + \omega, \delta_K\rangle - 4\langle\textstyle\sum n_i\beta_i, \delta_K\rangle \\
&\quad \leq |\lambda|^2 + |2\delta_{K_r} - E(2\delta_K) + \mu|^2 - 4\langle\lambda' - \lambda, \delta_K - \delta_{-,c}\rangle \\
&\qquad - 4\langle E(2\delta_K) - 2\delta_{K_r} - \mu + \omega, \delta_K\rangle && \text{by (15.12).}
\end{aligned}$$

By Theorem 9.20, $\lambda' - \lambda$ is a sum of members of $(\Delta_-)^+$. Thus Corollary 15.3 shows that

$$4\langle \lambda' - \lambda, \delta_K - \delta_{-,c}\rangle \geq 0.$$

Thus we have

$$|\lambda|^2 + |\omega|^2 \leq |\lambda|^2 + |2\delta_{K_r} - E(2\delta_K) + \mu|^2 - 4\langle E(2\delta_K) - 2\delta_{K_r} - \mu + \omega, \delta_K\rangle. \tag{15.16}$$

Going over the argument, we note for later reference that equality in (15.16) would force

$$\lambda = \lambda' \quad \text{by uniqueness of minimal } K \text{ types for discrete series and limits} \tag{15.17a}$$

$$\Lambda \text{ is a minimal } K \text{ type} \quad \text{since } |\Lambda_0 + 2\delta_K|^2 = |\Lambda + 2\delta_K|^2 \tag{15.17b}$$

$$\lambda' + \omega = \Lambda_0 \quad \text{since all } n_i \text{ equal } 0. \tag{15.17c}$$

Using (15.17a) and (15.11), we can rewrite (15.17c) as

$$\Lambda_0 = \lambda + \omega, \text{ where } \tau_\omega|_{M_r} \text{ contains } \chi. \tag{15.17d}$$

Subtracting $|\lambda|^2$ from both sides of (15.16) and rearranging the terms on the right side, we obtain

$$|\omega|^2 \leq |\mu + 2\delta_{K_r}|^2 - |E(2\delta_K)|^2 - 2\langle \omega, E(2\delta_K)\rangle.$$

Thus

$$|\omega + E(2\delta_K)|^2 \leq |\mu + 2\delta_{K_r}|^2$$

or

$$|(\omega + E(2\delta_K) - 2\delta_{K_r}) + 2\delta_{K_r}|^2 \leq |\mu + 2\delta_{K_r}|^2. \tag{15.18}$$

Equality in (15.18) will force (15.17).

Now (15.11) says that $\tau_\omega|_{M_r}$ contains χ. Therefore

$$\tau_{\omega + E(2\delta_K) - 2\delta_{K_r}}|_{M_r}$$

contains the character σ_r defined in (15.4). Since τ_μ is a minimal K_r type for the principal series of G_r corresponding to σ_r, Frobenius reciprocity forces equality to hold in (15.18) and says that $\tau_{\mu'}$ with

$$\mu' = \omega + E(2\delta_K) - 2\delta_{K_r}$$

is another minimal K_r type. Since equality holds in (15.18), (15.17d) is valid. This proves the formula.

Proof of multiplicity one, given result in G_r. By Frobenius reciprocity, we are to count occurrences of $K \cap M^\#$ types of σ within $\tau_\Lambda|_{K \cap M^\#}$. According to the above proof, the only one that can occur is the one σ_λ given by (15.10) and corresponding to the Blattner parameter.

Let v_λ be a highest weight vector for an occurrence of σ_λ in $\tau_\Lambda|_{K\cap M^\#}$. Applying $\tau_\Lambda(b)$, we easily see that v_λ is a sum of weight vectors v_i with weights $\lambda + \omega_i$, where $\omega_i|_{\mathfrak{b}_-} = 0$. These weights agree with Λ on $\mathfrak{b}_-$. Since Λ is highest and Δ_r is spanned by simple roots, these weights are all obtained from Λ by subtracting sums of members of $\Delta_{K_r}^+$. Thus the weight vectors are in the space $\mathscr{S}$ defined in (15.5), and v_λ is in $\mathscr{S}$.

Now $\mathscr{S}$ is irreducible under $\tau_\Lambda|_{K_r}$. Since $E(2\delta_K) - 2\delta_{K_r}$ is the highest weight of a one-dimensional representation of K_r, (15.6) shows that τ_μ acts irreducibly in $\mathscr{S}$. By our assumption that the multiplicity-one result is valid in G_r, $\tau_\mu|_{M_r}$ contains σ_r just once. Unwinding matters, we see that $(\tau_\Lambda|_{K_r}, \mathscr{S})|_{M_r}$ contains χ just once. Therefore the vector v_λ is uniquely determined up to a scalar.

§2. Inversion Problem

Now we take up the problem of recovering the Cartan subalgebra $\mathfrak{a} \oplus \mathfrak{b}_-$ and the parameter λ_0 from a minimal K type τ_Λ. Doing so requires knowing something more about minimal K types for the principal series of a split group.

Example. $G = \mathrm{Sp}(n, \mathbb{R})$, minimal parabolic subgroup.

In the notation of §4.1, the simple roots are $e_1 - e_2, \ldots, e_{n-1} - e_n, 2e_n$. So the elements $\gamma_{e_1 - e_2}, \ldots, \gamma_{e_{n-1}-e_n}, \gamma_{2e_n}$ generate $M = M_\mathfrak{p}$ independently. Since $\gamma_{e_i - e_j} = \gamma_{2e_i}\gamma_{2e_j}$, the elements

$$\gamma_{2e_1}, \gamma_{2e_2}, \ldots, \gamma_{2e_n} \tag{15.19}$$

generate M independently. Thus a character σ of M is specified by $\sigma(\gamma_{2e_j})$, $1 \leq j \leq n$.

To study K types containing σ, it is enough to consider all conjugates of σ by $W(A:G)$ at once, since $W(A:G)$ conjugates any representation of K into itself. Thus it is enough to study the $n + 1$ choices of σ given by

$$\sigma_k(\gamma_{2e_j}) = \begin{cases} -1 & \text{for } 1 \leq j \leq k \\ +1 & \text{otherwise,} \end{cases}$$

$0 \leq k \leq n$.

Using a Cayley transform $\mathbf{d}_{2e_1} \cdots \mathbf{d}_{2e_n}$ leads us to compact Cartan subgroup B containing (15.19). Using the notation e_j also in connection with B, let $\mu' = \sum c_j e_j$ be a weight of a representation τ_μ of K. Then $\xi_{\mu'}(\gamma_{2e_j}) = (-1)^{c_j}$. Taking conjugacy into account, we see that $\tau_\mu|_M$ contains σ_k if and only if τ_μ has a dominant weight $\sum c_j e_j$ with exactly k of the c_j's odd, the remainder even.

One τ_μ for which this condition is satisfied has $\mu' = \mu = \sum_{j=1}^k e_j$. It is the natural representation of unitary matrices on alternating tensors of rank k and has dimension $\binom{n}{k}$. Its reduction under M contains each conjugate of σ_k exactly once and nothing else. A second obvious τ_μ containing σ_k has $\mu' = \mu = -\sum_{j=1}^k e_{n+1-j}$, and it is different from the previous one if $k \neq 0$. This τ_μ is the complex conjugate of the first and again contains each conjugate of σ_k exactly once, nothing else.

Problems 13 to 18 at the end of the chapter outline a combinatorial argument that these two representations of K are the only minimal K types of the σ_k principal series of $Sp(n, \mathbb{R})$. They have two further properties of interest, in that they are "fine" and "small." They are "fine" in the following sense: For each root β of $(\mathfrak{g}, \mathfrak{a})$, the $SL(2, \mathbb{R})$ subgroup corresponding to β has its K_β acting only by K_β types with $n = 0$ and $n = \pm 1$, in the notation of (15.1). To see this, we extend the Lie algebra $\mathfrak{k}_\beta$ of K_β to a Cartan subalgebra of K, notice that each weight string has length at most 2, and conclude that K_β acts in the fashion asserted.

For the sense in which our two representations of K are "small," let us introduce a new positive system Δ^+ with simple roots

$$e_1 + e_n, -e_n - e_2, e_2 + e_{n-1}, -e_{n-1} - e_3, e_3 + e_{n-2}, \ldots,$$

the last simple root being $2e_{(n+1)/2}$ if n is odd, $-2e_{(n+2)/2}$ if n is even. Then we can check that

$$2\delta_K = \sum_{j=1}^{[(n+1)/2]} (n - 2j + 1)(e_j - e_{n+1-j})$$

$$\sigma = \sum_{j=1}^{[(n+1)/2]} [(n + 2 - 2j)e_j - (n + 1 - 2j)e_{n+1-j}]$$

$\mu + 2\delta_K$ is Δ^+ dominant (no matter what k is).

Hence

$$\mu + 2\delta_K - \delta = \mu - \sum_{j=1}^{[(n+1)/2]} e_j,$$

which, in the system Δ^+, is a linear combination of positive roots with all coefficients negative. This is the sense in which τ_μ is "small."

Let G be a split group with rank G = rank K and with $\mathfrak{b}$ a compact Cartan subalgebra. Let τ_μ be a K type occurring in the σ principal series of G, i.e., having $\tau_\mu|_{M_\mathfrak{p}} \supseteq \sigma$. We say that τ_μ is **fine** if, for each root β of $(\mathfrak{g}, \mathfrak{a}_\mathfrak{p})$, the $SL(2, \mathbb{R})$ subgroup corresponding to β has its K_β acting only by K_β types with $n = 0$ and $n = \pm 1$, in the notation of (15.1). We say that τ_μ is **small** if, in some positive system Δ^+ for $\Delta(\mathfrak{b}^\mathbb{C}:\mathfrak{g}^\mathbb{C})$ such that $\mu + 2\delta_K$ is Δ^+ dominant, $\mu + 2\delta_K - \delta$ is a linear combination of positive roots with all coefficients negative.

Theorem 15.4. Let G be a linear connected semisimple group that is split over $\mathbb{R}$ and has rank G = rank K, and let τ_μ be a K type occurring in the σ principal series of G. Then the following conditions are equivalent:

(a) τ_μ is minimal
(b) τ_μ is fine
(c) τ_μ is small.

In this case, $\tau_\mu|_{M_\mathfrak{p}}$ contains each $W(A_\mathfrak{p}:G)$ conjugate of σ exactly once, and it contains no other characters of $M_\mathfrak{p}$.

Operationally in the theorem, the minimal K types are the ones whose existence is trivial, the fine K types are the ones that are easy to construct in concrete cases, and the small K types are the ones that we shall use for the inversion problem. The last statement of the theorem, by Frobenius reciprocity, implies that τ_μ occurs with multiplicity one in the σ principal series. As remarked in connection with Theorem 15.1, the proof of Theorem 15.1 really only reduces the multiplicity one assertion to the case considered in Theorem 15.4, and Theorem 15.4 is needed to finish off the multiplicity question.

The example with $\mathrm{Sp}(n, \mathbb{R})$ at the start of this section produced fine K types and observed (1) that they are minimal and small, (2) that there are no other minimal K types, and (3) that the ones produced do decompose under $M_\mathfrak{p}$ as asserted. Therefore it showed for $\mathrm{Sp}(n, \mathbb{R})$ that minimal implies fine and small and that minimal implies the asserted decomposition under $M_\mathfrak{p}$.

The proof in general that fine and small are equivalent is not too difficult. However, the proof that minimal is equivalent with fine is a complicated combinatorial argument; the only known proof of this equivalence organizes the combinatorics as a proof in cohomology theory. We omit the details.

Let us return to the situation of §1: G has rank G = rank K, MAN is cuspidal parabolic, σ is a discrete series or nondegenerate limit of discrete series of M, and Δ^+ is compatible with the positive system used to write down parameters for σ. By (15.9), we can write the minimal K type formula as

$$\Lambda + 2\delta_K = \lambda_0 + (\delta - \delta_r) + (\mu + 2\delta_{K_r}). \tag{15.20}$$

We introduce a new positive system $(\Delta^+)'$ by changing the notion of positivity on Δ_r (and only there) so that τ_μ is exhibited as small (since τ_μ minimal is equivalent with τ_μ small, by Theorem 15.4). This means in particular that $\mu + 2\delta_{K_r}$ is to be $(\Delta_r^+)'$ dominant. With "primes" referring to objects in the new ordering, we have

$$\begin{aligned} \Delta_{K,r}^+ &= (\Delta_{K,r}^+)' \qquad \text{and} \qquad \Delta_K^+ = (\Delta_K^+)' \\ \delta_{K_r} &= \delta'_{K_r} \qquad \text{and} \qquad \delta_K = \delta'_K \\ \delta - \delta_r &= \delta' - \delta'_r. \end{aligned} \tag{15.21}$$

Since τ_μ is small, we can write

$$\mu + 2\delta_{K_r} - \delta'_r = -\sum_{\beta_i \in (\Delta_r^+)'} c_i\beta_i, \qquad c_i \geq 0. \tag{15.22}$$

Substituting from (15.21) and (15.22) into (15.20), we obtain

$$\Lambda + 2\delta_K - \delta' = \lambda_0 - \sum_{\beta_i \in (\Delta_r^+)'} c_i\beta_i, \qquad c_i \geq 0. \tag{15.23}$$

We shall use (15.23) to recover λ_0 from Λ. But we need one more fact: that $\Lambda + 2\delta_K$ is $(\Delta^+)'$ dominant. To see this, we let β be $(\Delta^+)'$ simple and we compute $\langle \Lambda + 2\delta_K, \beta\rangle$, distinguishing two cases:

(1) $\beta|_{\mathfrak{b}_-} \neq 0$. Then β is in Δ^+ and thus has $\langle \lambda_0, \beta\rangle \geq 0$. Also each β_i is the sum of simple roots other than β, and so $\langle -\sum c_i\beta_i, \beta\rangle \geq 0$. Since also $\langle \delta', \beta\rangle \geq 0$, we conclude $\langle \Lambda + 2\delta_K, \beta\rangle \geq 0$ from (15.23).

(2) $\beta|_{\mathfrak{b}_-} = 0$, so that β is in Δ_r. Then $\langle \lambda_0, \beta\rangle = \langle \delta - \delta_r, \beta\rangle = 0$, and $\langle \Lambda + 2\delta_K, \beta\rangle \geq 0$ by (15.20).

Proposition 15.5. Under the assumption rank G = rank K, Λ determines λ_0 in formula (15.23) in the following sense. Suppose Δ_K^+ is fixed and Λ is any Δ_K^+ dominant form. Let $(\Delta^+)'$ and $(\Delta^+)''$ be positive systems such that $\Lambda + 2\delta_K$ is dominant for each and such that

(a) $\Lambda + 2\delta_K - \delta' = \lambda'_0 - \sum c'_i\beta'_i, \qquad c'_i \geq 0$
$\Lambda + 2\delta_K - \delta'' = \lambda''_0 - \sum c''_i\beta''_i, \qquad c''_i \geq 0$
(b) λ'_0 and λ''_0 are dominant for $(\Delta^+)'$ and $(\Delta^+)''$, respectively
(c) β'_i and β''_i are positive for $(\Delta^+)'$ and $(\Delta^+)''$, respectively, for all i
(d) $\langle \lambda'_0, \beta'_i\rangle = 0$ and $\langle \lambda''_0, \beta''_i\rangle = 0$ for all i.

Then $\lambda'_0 = \lambda''_0$.

Proof. By Lemma 8.56, the properties in the $(\Delta^+)'$ system characterize λ'_0 as the unique nearest $(\Delta^+)'$ dominant form to $\Lambda + 2\delta_K - \delta'$. (The lemma is to be applied to an abstract root system, not to the system in $\mathfrak{a}'_{\mathfrak{p}}$.) In more detail the lemma says that the set $\mathscr{F} = \mathscr{F}(\Lambda + 2\delta_K - \delta')$ of indices has

$$\Lambda + 2\delta_K - \delta' = \sum_{j \notin \mathscr{F}} b_j\omega_j - \sum_{j \in \mathscr{F}} a_j\gamma_j \qquad (\gamma_j \text{ simple for } (\Delta^+)')$$

$$\lambda'_0 = \sum_{j \notin \mathscr{F}} b_j\omega_j$$

$$b_j > 0 \text{ for } j \notin \mathscr{F} \text{ and } a_j \geq 0 \text{ for } j \in \mathscr{F}.$$

Let $\mathscr{F}' = \{j \mid \langle \Lambda + 2\delta_K, \gamma_j\rangle = 0\}$, and let W' be the subgroup of the Weyl group $W(\Delta)$ fixing $\Lambda + 2\delta_K$. Certainly $\mathscr{F}' \subseteq \mathscr{F}$. Also Chevalley's Lemma and the $(\Delta^+)'$ dominance of $\Lambda + 2\delta_K$ imply that W' is generated by the reflections s_{γ_j} with j in $\mathscr{F}'$. Then $s \in W'$ implies $s\lambda_0' = \lambda_0'$ because

$$s_{\gamma_i}(\sum b_j\omega_j) = \sum b_j\omega_j - \frac{2\sum b_j\langle\omega_j, \gamma_i\rangle}{|\gamma_i|^2}\gamma_i = \sum b_j\omega_j.$$

Let s be the unique element of $W(\Delta)$ such that $s(\Delta^+)' = (\Delta^+)''$. We shall show in the lemma below that $s(\Lambda + 2\delta_K) = \Lambda + 2\delta_K$. Then s is in W' and $s\lambda_0 = \lambda_0$. Moreover $s\delta' = \delta''$. Hence application of s to the identity

$$\Lambda + 2\delta_K - \delta' = \lambda_0' - \sum c_i'\beta_i'$$

gives

$$\Lambda + 2\delta_K - \delta'' = \lambda_0' - \sum c_i' s\beta_i'.$$

Thus Lemma 8.56 characterizes both λ_0' and λ_0'' as the unique nearest $(\Delta^+)''$ dominant form to $\Lambda + 2\delta_K - \delta''$, and we conclude $\lambda_0' = \lambda_0''$, as required.

Thus the proof of Proposition 15.5 is completed by applying the following lemma with $v = \Lambda + 2\delta_K$.

Lemma 15.6. Let Δ be an abstract root system, let $(\Delta^+)'$ and $(\Delta^+)''$ be two positive systems for Δ, and let s be the member of the Weyl group $W(\Delta)$ with $s(\Delta^+)' = (\Delta^+)''$. If v is a vector that is dominant for both $(\Delta^+)'$ and $(\Delta^+)''$, then $sv = v$.

This lemma in turn relies on the following result.

Lemma 15.7. Let Δ^+ be a positive system for an abstract root system Δ. Suppose v and v' are dominant, α is simple in Δ^+, and w in $W(\Delta)$ satisfies $l(s_\alpha w) = l(w) + 1$. Then $\langle s_\alpha wv, v'\rangle \le \langle wv, v'\rangle$.

Proof. The condition $l(s_\alpha w) = l(w) + 1$ means that $w^{-1}\alpha > 0$. (Otherwise $-w^{-1}\alpha$ is a positive root whose image under w is negative and is sent to a positive root by s_α.) Now

$$wv - s_\alpha wv = \frac{2\langle wv, \alpha\rangle}{|\alpha|^2}\alpha = \frac{2\langle v, w^{-1}\alpha\rangle}{|\alpha|^2}\alpha$$

and hence

$$\langle wv, v'\rangle - \langle s_\alpha wv, v'\rangle = \frac{2\langle v, w^{-1}\alpha\rangle\langle\alpha, v'\rangle}{|\alpha|^2},$$

from which Lemma 15.7 follows.

Proof of Lemma 15.6. Let $v' = sv$, and write s as a minimal product $s_{\gamma_n}\cdots s_{\gamma_1}$ of simple reflections in $(\Delta^+)'$. Applying Lemma 15.7 recursively, we see that

$$\langle v', v'\rangle \le \cdots \le \langle s_{\gamma_2}s_{\gamma_1}v, v'\rangle \le \langle s_{\gamma_1}v, v'\rangle \le \langle v, v'\rangle.$$

Since $|v| = |v'|$, we have equality in the Schwarz inequality and must have $v' = v$.

Theorem 15.8. Under the assumption that rank $G =$ rank K, let $S = MAN$ and $S' = M'A'N'$ be two cuspidal parabolic subgroups. Suppose that σ and σ' are discrete series of M and M', respectively. If $U(S, \sigma, 0)$ and $U(S', \sigma', 0)$ have a minimal K type in common, then $M'A'$ is conjugate to MA via K in a way that carries σ' to σ.

Proof. Let $\sigma = \pi^M(\lambda_0, \Delta_M^+, \chi)$, and let $\{\alpha_1, \ldots, \alpha_l\}$, Δ^+, and μ be the data in Theorem 15.1 leading to the highest weight Λ of the common minimal K type. Let $\sigma' = \pi^{M'}(\lambda_0', (\Delta_M^+)' \chi')$, and let $\{\alpha_1', \ldots, \alpha_{l'}'\}$, $(\Delta^+)'$, and μ' be the data leading to the weight for the same K type that is highest for $(\Delta_K^+)'$. Since $(\Delta_K^+)'$ may be different from Δ_K^+, we can be sure only that $w\Lambda$ results from (15.3), where w is in $N_K(\mathfrak{b})$ and $w\Delta_K^+ = (\Delta_K^+)'$.

By (15.23), we can write in obvious notation

$$\Lambda + 2\delta_K - \delta = \lambda_0 - \sum c_i\beta_i$$

$$w(\Lambda + 2\delta_K) - \delta' = \lambda_0' - \sum c_i'\beta_i'.$$

The second of these relations implies

$$\Lambda + 2\delta_K - w^{-1}\delta' = w^{-1}\lambda_0' - \sum c_i' w^{-1}\beta_i',$$

and Proposition 15.5 says $w^{-1}\lambda_0' = \lambda_0$, i.e., $\lambda_0' = w\lambda_0$. Thus we may conjugate matters from the outset so that $U(S, \sigma, 0)$ and $U(S', \sigma', 0)$ are defined from the same $\mathfrak{b}$ and the same λ_0.

With this normalization, let $L = Z_G(H_{\lambda_0})$. Then L is linear connected reductive (by an application of Corollary 4.22), and its Lie algebra $\mathfrak{l}$ has

$$\mathfrak{l}^{\mathbb{C}} = \mathfrak{b}^{\mathbb{C}} + \sum_{\beta \perp \lambda_0} \mathfrak{g}_\beta .$$

We shall show that the Lie algebra $\mathfrak{a}$ associated to $U(S, \pi^M(\lambda_0, \Delta_M^+, \chi), 0)$ is an Iwasawa $\mathfrak{a}$ for $\mathfrak{l}$.

The Lie algebra $\mathfrak{a}$ for $U(S, \pi^M(\lambda_0, \Delta_M^+, \chi), 0)$ is built by Cayley transform from $\alpha_1, \ldots, \alpha_l$ and is contained in $\mathfrak{l} \cap \mathfrak{p}$. Let $\alpha_1'', \ldots, \alpha_l''$ be the corresponding transformed roots, and extend $\mathfrak{a}$ to an Iwasawa $\mathfrak{a}_\mathfrak{p}^L$ for L. Suppose this extension is nontrivial . Since rank $L =$ rank $L \cap K$, it follows from Problems 14–17 of Chapter XII that a successive product of reflections in real roots is -1 on $\mathfrak{a}_\mathfrak{p}^L$. Composing with $s_{\alpha_1''} \cdots s_{\alpha_l''}$ and applying Chevalley's Lemma, we obtain a real root orthogonal to $\alpha_1'', \ldots \alpha_l''$. Transforming back to roots relative to $\mathfrak{b}$, we obtain a member of Δ_- orthogonal to λ_0. Since our given representation σ is in the discrete series, this is a contradiction. Thus $\mathfrak{a} = \mathfrak{a}_\mathfrak{p}^L$.

The Lie algebra $\mathfrak{a}$ for $U(S', \pi^{M'}(\lambda_0, (\Delta_M^+)', \chi'), 0)$ is similarly an Iwasawa $\mathfrak{a}$ for $\mathfrak{l}$, and the two must be conjugate by a member of $L \cap K$, by Theorem 5.13. Sorting matters out, we see that we can conjugate $M'A'$ to MA while leaving λ_0 fixed as a parameter on $\mathfrak{b}_-$. Moreover, the nonsingularity of λ_0

means that $(\Delta_M^+)'$ will map to Δ_M^+. Thus from the start we may assume that we are working with

$$U(S, \pi^M(\lambda_0, \Delta_M^+, \chi'), 0) \qquad \text{and} \qquad U(S, \pi^M(\lambda_0, \Delta_M^+, \chi), 0)$$

and that they have a minimal K type in common.

In the minimal K type formula, we can now fix common data $\alpha_1, \ldots, \alpha_l$ and Δ^+ for both representations, and thus we can conclude $\mu' = \mu$. By Theorem 15.4, χ' is conjugate to χ via $W(A:G_r)$. The roots leading to this Weyl group are odd and thus $W(A:G_r)$ acts trivially on M_0. In particular the Weyl group element conjugating χ' to χ fixes (λ_0, Δ_M^+). This completes the proof of the theorem.

§3. Connection with Intertwining Operators

For the reducible unitary principal series $\mathscr{P}^{-,0}$ of SL(2, $\mathbb{R}$), there are two minimal K types, and one occurs in each irreducible constituent. The next theorem indicates that this phenomenon is a general one.

Theorem 15.9. Let G be linear connected semisimple (with rank $G =$ rank K), and let $U(S, \sigma, \nu)$ be a basic representation induced from discrete series. Then each minimal K type of an irreducible constituent of $U(S, \sigma, \nu)$ is minimal for $U(S, \sigma, \nu)$.

Proof. By Theorem 14.79, let the irreducible constituent have global character

$$\operatorname{ind}_{MAN}^G \Theta^{MA}(\lambda_0, \Delta_M^+, \chi, \nu)$$

in nondegenerate form, and let $\{\alpha_1, \ldots, \alpha_l\}$, Δ^+, and μ be the data to use in the minimal K type formula to obtain a parameter Λ. By nondegeneracy the members β of Δ_M^+ for which $\langle \lambda_0, \beta \rangle = 0$ are all noncompact and Δ_M^+ simple; let us enumerate them as $\beta_1, \ldots, \beta_k$. They are strongly orthogonal.

In Theorem 14.79, we see that the corresponding basic representation induced from discrete series has character

$$\operatorname{ind}_{M_*A_*N_*}^G \Theta^{M_*A_*}(\lambda_0|_{\mathfrak{b}_*}, \Delta_M^+ \cap \beta_1^\perp \cap \ldots \cap \beta_k^\perp, \chi^*, \nu \oplus 0),$$

where $\mathfrak{a}_* \oplus \mathfrak{b}_*$ is obtained from $\mathfrak{a} \oplus \mathfrak{b}_-$ by a Cayley transform $\mathbf{c}_{\beta_1} \cdots \mathbf{c}_{\beta_k}$ and where χ^* is some extension of χ. If we use $\{\beta_1, \ldots, \beta_k, \alpha_1, \ldots, \alpha_l\}$ as an ordered basis, we easily check that Δ^+ is a compatible positive system to use in the minimal K type formula with the representation induced from discrete series.

Let Δ_r be the system of "real" roots for the irreducible constituent, and let Δ_s be the system of "real" roots for the representation induced from discrete series. Let G_r and G_s be the respective split groups. The claim is that Λ arises from the minimal K type formula for the representation induced from discrete series when we use the parameter

$$\mu' = \mu + 2\delta_{K_r} - 2\delta_{K_s} - \delta_r + \delta_s \tag{15.24}$$

as minimal K_s type for the group G_s.

To verify this claim, we use the version of the minimal K type formula given in (15.9) as

$$\Lambda = \lambda_0 - 2\delta_K + (\delta - \delta_r) + (\mu + 2\delta_{K_r}).$$

Substituting, we have

$$\Lambda = \lambda_0 - 2\delta_K + (\delta - \delta_s) + (\mu' + 2\delta_{K_s}).$$

If β is in Δ_s, then β is orthogonal to λ_0 and $\delta - \delta_s$, and hence

$$\langle \Lambda + 2\delta_K, \beta \rangle = \langle \mu' + 2\delta_{K_s}, \beta \rangle. \tag{15.25}$$

Since Δ_s is generated by the Δ^+ simple roots that it contains, this relation implies that μ' is $\Delta_{K_s}^+$ dominant.

Following the derivation in §2, let us change the positive system of Δ_r to make $\mu + 2\delta_{K_r}$ dominant for $(\Delta_r^+)'$. We have seen that $\Lambda + 2\delta_K$ is then $(\Delta^+)'$ dominant in the new positive system for Δ. Referring to (15.25), we see that $\mu' + 2\delta_{K_s}$ is $(\Delta_s^+)'$ dominant in the new positive system for Δ_s. Since the new positive system does not change positivity for roots outside Δ_r, we have

$$\delta_s - \delta_r = \delta_s' - \delta_r'.$$

Substituting into (15.24), we see that

$$\mu' + 2\delta_{K_s} - \delta_s' = \mu + 2\delta_{K_r} - \delta_r,$$

and it follows that $\tau_{\mu'}$ is small for G_s. By Theorem 15.4, $\tau_{\mu'}$ is minimal for the σ_s principal series of G_s if $\tau_{\mu'}|_{M_s} \supseteq \sigma_s$. Thus to see that Λ arises from the minimal K type formula applied to the representation induced from discrete series (and thereby to complete the proof), we have only to show that the particular character

$$\sigma_s = \chi^* \cdot \exp(E_s(2\delta_K) - 2\delta_{K_s})|_{M_s}$$

of M_s occurs in $\tau_{\mu'}|_{M_s}$. Since

$$\Lambda|_{\mathfrak{b}_s} = -E_s(2\delta_K) + 2\delta_{K_s} + \mu'$$

as in (15.6), we have only to show that the restriction of $\tau_\Lambda|_{K_s}$ to M_s contains χ^*. This result is proved inductively by examining the effect of adjoining one β_j at a time to Δ_r, and we omit the details.

Now we can indicate how to prove the Langlands Disjointness Theorem (Theorem 14.90). (We are working in this chapter under the assumption that rank G = rank K, but we mentioned that this assumption is not really necessary, as long as the results are phrased carefully enough.) We shall skip one step at the end, supplying references for it in the Notes.

Proof of Theorem 14.90. Suppose we have two basic representations induced from discrete series that have an irreducible constituent in common. A minimal K type of the irreducible constituent will be common to the two basic representations induced from discrete series. By Theorem 15.9 this K type is minimal in the two basic representations induced from discrete series. By Theorem 15.8 we may assume the two basic representations induced from discrete series are $U(S, \sigma, \nu_0)$ and $U(S, \sigma, \nu_0')$, the N factors of the parabolic subgroups being irrelevant since ν_0 and ν_0' are assumed imaginary.

Let us see that $\nu_0' = s_0\nu_0$ for some $s_0 \in W(A:G)$. In fact, if $\sigma = \pi^M(\lambda_0, \Delta_M^+, \chi)$ with λ_0 defined on $\mathfrak{b}_-$, then consideration of infinitesimal characters shows that $\lambda_0 + \nu_0' = w(\lambda_0 + \nu_0)$ for some w in $W((\mathfrak{a} \oplus \mathfrak{b}_-)^{\mathbb{C}}:\mathfrak{g}^{\mathbb{C}})$. Since λ_0 is real while ν_0' and ν_0 are imaginary, we have $w\lambda_0 = \lambda_0$ and $\nu_0' = w\nu_0$. Referring to Problem 5 at the end of the chapter, we conclude that $\nu_0' = s_0\nu_0$ for some $s_0 \in W(A:G)$. (Moreover, s_0 fixes λ_0, but it may *a priori* move χ.)

Now $U(S, \sigma, \nu_0)$ and $U(S, \sigma, s_0\nu_0)$ have a nonzero K-finite matrix coefficient in common. In terms of Eisenstein integrals this means that there are ψ and ψ' in ${}^0\mathscr{C}_\sigma(M, \tau_M)$ such that $E(S:\psi:\nu_0:x)$ is nonzero and

$$E(S:\psi:\nu_0:x) = E(S:\psi':s_0\nu_0:x). \tag{15.26}$$

For ν regular and imaginary, we know from Theorem 14.7 that

$$E_S(S:\psi:\nu:ma) = \sum_{s \in W(A:G)} (c_{S|S}(s:\nu)\psi)(m)e^{s\nu \log a}.$$

Projecting by E_σ to ${}^0\mathscr{C}_\sigma(M, \tau_M)$, we have

$$E_\sigma[E_S(S:\psi:\nu:ma)] = \sum_{s \in W_{\sigma,0}} (c_{S|S}(s:\nu)\psi)(m)e^{s\nu \log a}.$$

Passing to the limit $\nu \to \nu_0$ and using the continuity asserted in Theorem 14.8, we obtain

$$E_\sigma[E_S(S:\psi:\nu_0:ma)] = \lim_{\nu \to \nu_0} \sum_{s \in W_{\sigma,0}} (c_{S|S}(s:\nu)\psi)(m)e^{s\nu \log a}. \tag{15.27}$$

By (15.26) we have also

$$E_\sigma[E_S(S:\psi':s_0\nu_0:ma)] = \lim_{\nu \to \nu_0} \sum_{s \in W_{\sigma,0}} (c_{S|S}(s:\nu)\psi')(m)e^{ss_0\nu \log a}. \tag{15.28}$$

The a dependence in (15.27) involves derivatives (in ν) of $e^{s\nu \log a}$, $s \in W_{\sigma,0}$, at $\nu = \nu_0$, while the a dependence in (15.28) involves derivatives of $e^{ss_0\nu \log a}$, $s \in W_{\sigma,0}$, at $\nu = \nu_0$. Assume $s_0\nu_0$ is not in $W_{\sigma,0}\nu_0$. Then the a behaviors are independent of one another, and we conclude that

$$\lim_{\nu \to \nu_0} \sum_{s \in W_{\sigma,0}} (c_{S|S}(s:\nu)\psi)(m)e^{s\nu \log a} = 0.$$

Arguing similarly with $E_{w\sigma}$ for all $w \in W(A:G)$, we conclude that $E_S(S:\psi:\nu_0:ma) = 0$.

Next we shall prove that $E_{S_1}(S:\psi:\nu_0:ma) = 0$ for all parabolic subgroups $S_1 = MAN_1$ with the same MA as for S. Referring to the Notes for Chapter XV, we can see from this conclusion that $E(S:\psi:\nu_0:x)$ itself must be 0, in contradiction to its construction. This contradiction shows that $s_0\nu_0$ is in $W_{\sigma,0}\nu_0$, hence that our representations are $U(S, \sigma, \nu_0)$ and $U(S, \sigma, s_0\nu_0)$ with $s_0\nu_0$ in $W_{\sigma,0}\nu_0$. Thus to complete the proof, we need only show that $E_{S_1}(S:\psi:\nu_0:ma)$ is 0 for all S_1.

To prove the asserted vanishing of constant terms, it is enough to show that $E_{S'}(S:\psi:\nu_0:\cdot) = 0$ implies $E_{S''}(S:\psi:\nu_0:\cdot) = 0$ whenever $S' = MAN'$, $S'' = MAN''$, and $\bar{\mathfrak{n}}' \cap \mathfrak{n}'' = \bar{\mathfrak{n}}^{(\alpha)}$ for some root α of $(\mathfrak{g}, \mathfrak{a})$. The argument will depend on whether s_α exists in $W(A:G)$.

First suppose s_α exists in $W(A:G)$. Let w be a representative in K. The function $F(x) = E(S:\psi:\nu_0:x)$ satisfies

$$F(k_1xk_2) = \tau_1(k_1)F(x)\tau_2(k_2).$$

Taking $k_1 = w$ and $k_2 = w^{-1}$ and inserting this relation into Theorem 14.6, we see that

$$F_{S''}(ma) = \tau_1(w)^{-1}F_{S'}(w(ma)w^{-1})\tau_2(w),$$

and our assertion about the vanishing of the constant term therefore follows in this case.

Now suppose s_α does not exist in $W(A:G)$. By Lemma 14.40, $\mu_{w\sigma,\alpha}(\nu)$ is holomorphic and nowhere vanishing for ν imaginary and for w in $W(A:G)$. Hence the normalizing factors $\gamma(S'':S':w\sigma:\nu)$, $\gamma(S':S'':w\sigma:\nu)$, $\gamma(\bar{S}'':\bar{S}':w\sigma:\nu)$, and $\gamma(\bar{S}':\bar{S}'':w\sigma:\nu)$ are holomorphic and nowhere vanishing for such ν and w.

Let us write $\psi = \psi_T$ by Proposition 14.2. Fix w_0 in $W(A:G)$, and let ν be regular and imaginary. By our assumption on $E_{S'}$, in combination with Theorems 14.7 and 14.8 and the kind of independence argument earlier in the proof, we see that the expression

$$\sum_{w \in w_0 W_{\sigma,\nu_0}} c_{S'|S}(w:\nu)\psi_T^\sigma(m)e^{w\nu \log a}$$

tends to 0 (i.e., is $o(1)$) as ν tends to ν_0. Since $\bar{\mathfrak{n}}' \cap \mathfrak{n}'' = \bar{\mathfrak{n}}^{(\alpha)}$ implies

$$\begin{aligned}\frac{\gamma(\bar{S}'':S'':w_0\sigma:w_0\nu)}{\gamma(\bar{S}':S':w_0\sigma:w_0\nu)} &= \frac{\gamma(\bar{S}'':\bar{S}':w_0\sigma:w_0\nu)\gamma(\bar{S}':S'':w_0\sigma:w_0\nu)}{\gamma(S'':S':w_0\sigma:w_0\nu)\gamma(\bar{S}':S'':w_0\sigma:w_0\nu)} \\ &= \frac{\gamma(\bar{S}'':\bar{S}':w_0\sigma:w_0\nu)}{\gamma(S'':S':w_0\sigma:w_0\nu)}\end{aligned}$$

and since the right side is holomorphic and nonvanishing for ν imaginary, we can multiply through by it and then replace $w_0\nu$ by $w\nu$ with an error that is $o(1)$. Thus we obtain

$$\sum_{w \in w_0 W_{\sigma,\nu_0}} \frac{\gamma(\bar{S}'':S'':w_0\sigma:w\nu)}{\gamma(\bar{S}':S':w_0\sigma:w\nu)} c_{S'|S}(w:\nu)\psi_T^\sigma(m)e^{w\nu \log a} = o(1)$$

as $\nu \to \nu_0$. By Corollary 14.9, we can rewrite this relation as

$$o(1) = \sum_{w \in w_0 W_{\sigma,\nu_0}} \frac{\gamma(\bar{S}'':S'':w_0\sigma:w\nu)}{\gamma(\bar{S}':S':w_0\sigma:w\nu)} e^{w\nu \log a}$$
$$\times \psi^{w\sigma}_{R(w)A(w^{-1}\bar{S}'w:S')A(S':S)TA(S':S)^{-1}A(S':w^{-1}S'w)R(w)^{-1}}(m)$$

with (σ, ν) suppressed in all four intertwining operators. Since

$$\gamma(w^{-1}\bar{S}'w:w^{-1}S'w:\sigma:\nu) = \gamma(\bar{S}':S':w\sigma:w\nu) = \gamma(\bar{S}':S':w_0\sigma:w\nu),$$

we can collapse our relation to

$$o(1) = \sum_{w \in w_0 W_{\sigma,\nu_0}} \gamma(\bar{S}'':S'':w_0\sigma:w\nu)e^{w\nu \log a}$$
$$\times \psi^{w\sigma}_{R(w)\mathscr{A}(w^{-1}\bar{S}'w:S)T\mathscr{A}(S:w^{-1}S'w)R(w)^{-1}}(m)$$
$$= \sum_{w \in w_0 W_{\sigma,\nu_0}} \gamma(\bar{S}'':S'':w_0\sigma:w\nu)e^{w\nu \log a}$$
$$\times \psi^{w\sigma}_{\mathscr{A}(\bar{S}':wSw^{-1}:w\sigma:w\nu)R(w)TR(w)^{-1}\mathscr{A}(wSw^{-1}:S':w\sigma:w\nu)}(m).$$

If $w = w_0 s$, the factor $\psi^{w\sigma}_{[\quad]}$ is

$$= \psi^{w_0\sigma}_{(w_0\sigma)(w_0 s w_0^{-1})[\quad](w_0\sigma)(w_0 s w_0^{-1})^{-1}}(m).$$

Letting L refer to "left by" and R refer to "right by," we obtain from our relation

$$o(1) = L_{\mathscr{A}(\bar{S}'':\bar{S}':w_0\sigma:w_0\nu_0)}R_{\mathscr{A}(S':S'':w_0\sigma:w_0\nu_0)}$$
$$\times \left(\sum_{s \in W_{\sigma,\nu_0}} \gamma(\bar{S}'':S'':w_0\sigma:w_0 s\nu)e^{w_0 s\nu \log a} \right.$$
$$\left. \times \psi^{w_0\sigma}_{(w_0\sigma)(w_0 s w_0^{-1})[\quad](w_0\sigma)(w_0 s w_0^{-1})^{-1}}(m) \right)$$
$$= \sum_{w \in w_0 W_{\sigma,\nu_0}} \gamma(\bar{S}'':S'':w\sigma:w\nu)e^{w\nu \log a}$$
$$\times \psi^{w\sigma}_{\mathscr{A}(\bar{S}'':\bar{S}':w\sigma:w\nu_0)[\quad]\mathscr{A}(S':S'':w\sigma:w\nu_0)}(m).$$

Replacing $w\nu_0$ by $w\nu$ in the intertwining operators, we introduce an error that is $o(1)$. Then our relation simplifies to

$$o(1) = \sum_{w \in w_0 W_{\sigma,\nu_0}} \gamma(\bar{S}'':S'':w\sigma:w\nu)e^{w\nu \log a}$$
$$\times \psi^{w\sigma}_{\mathscr{A}(\bar{S}'':wSw^{-1}:w\sigma:w\nu)R(w)TR(w)^{-1}\mathscr{A}(wSw^{-1}:S'':w\sigma:w\nu)}(m)$$
$$= \sum_{w \in w_0 W_{\sigma,\nu_0}} c_{S''|S}(w:\nu)\psi_T^\sigma(m).$$

Summing on coset representatives w_0 and using Theorems 14.7 and 14.8, we obtain $E_{S''}(S:\psi:\nu_0:ma) = 0$ as asserted. This completes the proof of the Langlands Disjointness Theorem.

Theorem 15.10. Let G be linear connected semisimple, and let $U(S, \sigma, \nu)$ be a standard induced representation in which $S = MAN$ is a cuspidal parabolic subgroup, σ is a discrete series or nondegenerate limit of discrete series, and Re ν is in the closed positive Weyl chamber. Suppose in the notation of (14.133) that $W_{\sigma,\nu} = W'_{\sigma,\nu}$, so that $U(S, \sigma, \nu)$ has a unique irreducible quotient $J(S, \sigma, \nu)$. Then $J(S, \sigma, \nu)$ contains all the minimal K types of $U(S, \sigma, \nu)$.

Proof. Assume the contrary. Among all counterexamples $(J(S, \sigma, \nu), \tau_\Lambda)$, choose one with $|\Lambda + 2\delta_K|^2$ as small as possible; we can do so since the range of $|\Lambda' + 2\delta_K|^2$ is a discrete set of positive numbers.

For our chosen counterexample, τ_Λ lies in $U(S, \sigma, \nu)$ but not in $J(S, \sigma, \nu)$. Therefore it lies in some other subquotient of $U(S, \sigma, \nu)$. By Theorem 14.92 this other subquotient is necessarily of the form $J(S', \sigma', \nu')$. By Proposition 8.61,

$$|\mathrm{Re}\, \nu'| < |\mathrm{Re}\, \nu|. \tag{15.29}$$

However, $J(S', \sigma', \nu')$ has the same infinitesimal character as $U(S, \sigma, \nu)$ and $J(S, \sigma, \nu)$. Thus if we let λ_0 and λ_0' be the infinitesimal characters of σ and σ', respectively, then $\lambda_0' + \nu'$ is conjugate to $\lambda_0 + \nu$ and necessarily $\lambda_0' + \mathrm{Re}\, \nu'$ is conjugate to $\lambda_0 + \mathrm{Re}\, \nu$. From (15.29) we conclude

$$|\lambda_0'| > |\lambda_0|. \tag{15.30}$$

Now τ_Λ is a minimal K type for $U(S, \sigma, \nu)$, and Proposition 15.5 shows that we can recover λ_0 (possibly up to some conjugacy) from τ_Λ. Then it follows from (15.30) that τ_Λ is not a minimal K type of $U(S', \sigma', \nu')$. Since τ_Λ does occur in $U(S', \sigma', \nu')$, it follows that a minimal K type $\tau_{\Lambda'}$ of $U(S', \sigma', \nu')$ has $|\Lambda' + 2\delta_K|^2 < |\Lambda + 2\delta_K|^2$. Since τ_Λ is minimal for $J(S', \sigma', \nu')$, $\tau_{\Lambda'}$ cannot occur in $J(S', \sigma', \nu')$. Thus $(J(S', \sigma', \nu'), \tau_{\Lambda'})$ provides a smaller counterexample, contradiction.

The results of this chapter have important implications for intertwining operators. Theorem 15.9 implies that the minimal K types of a basic representation induced from discrete series distinguish the irreducible constituents from one another. The standard intertwining operators $\sigma(r)\mathscr{A}_S(r, \sigma, \nu)$ corresponding to the R group are unitary and have order 2 and thus are $+1$ on some constituents, -1 on others. Moreover, these operators determine all the reducibility. Consequently all the reducibility is determined by knowing the signs of the intertwining operators on the minimal K types.

Theorem 15.10 is related to intertwining operators because the Langlands quotient J of U is the image of an intertwining operator corresponding to $S \to \bar{S}$. Theorem 15.10 says that the minimal K types survive the quotient mapping from U to J. Thus the operator has its largest possible behavior on the minimal K types. For Re ν in the open positive Weyl chamber, the operator is nonvanishing (and scalar!) on each minimal K type, and for Re ν in the closed positive Weyl chamber, the operator can anyway be normalized by its behavior on some minimal K type and the resulting operator will be everywhere regular for Re ν in the closed positive Weyl chamber.

§4. Problems

1. With $\alpha_1, \ldots, \alpha_l$ as in §1, suppose $\beta \in \Delta$ is strongly orthogonal to all α_j.
   (a) Using the explicit form of the Cayley transform $\mathbf{c}$ of §1, show that $\mathbf{c}(E_\beta) = E_\beta$.
   (b) Conclude that β is M-compact if and only if β is G-compact.
2. With $\alpha_1, \ldots, \alpha_l$ as in §1, suppose $\beta \in \Delta$ is orthogonal to all α_j but that $\beta \pm \alpha_j$ are roots for some j.
   (a) Prove that the index j is unique, say $j = j_0$.
   (b) Prove that $\beta \pm \alpha_{j_0} \pm \alpha_i$ are not roots if $i \neq j_0$.
   (c) Using the explicit form of the Cayley transform $\mathbf{c}$ of §1, make a calculation that shows that $\mathbf{c}(E_\beta)$ is a linear combination of $E_{\beta+\alpha_{j_0}}$ and $E_{\beta-\alpha_{j_0}}$.
   (d) Conclude that β is M-compact if and only if β is G-noncompact.

Problems 3 to 5 improve on the conclusion of Problem 7 in Chapter XIV. Let $\mathfrak{a} \oplus \mathfrak{b}$ be a Cartan subalgebra of $\mathfrak{g}$ with $\mathfrak{a} \subseteq \mathfrak{p}$ and $\mathfrak{b} \subseteq \mathfrak{k}$. Suppose that λ, ν, and ν_0 are in $(\mathfrak{a} \oplus \mathfrak{b})'$, that λ vanishes on $\mathfrak{a}$ and is the infinitesimal character of a discrete series of M, and that ν and ν_0 vanish on $\mathfrak{b}$ and are imaginary. Suppose further that $w\lambda = \lambda$ and $w\nu = \nu_0$ for some w in $W((\mathfrak{a} \oplus \mathfrak{b})^{\mathbb{C}} : \mathfrak{g}^{\mathbb{C}})$. It is to be proved that ν and ν_0 are conjugate via $W(A:G)$.

3. Show that there is no loss of generality in assuming that there are no imaginary roots and that $\lambda = 0$. [Hint: Consider $\mathfrak{z} = Z_{\mathfrak{g}}(H_\lambda)$ and the corresponding analytic subgroup.]
4. In the situation in Problem 3:
   (a) Show that every root of $(\mathfrak{g}, \mathfrak{a})$ is useful.
   (b) Show that there exist positive systems for the roots of $(\mathfrak{g}^{\mathbb{C}}, (\mathfrak{a} \oplus \mathfrak{b})^{\mathbb{C}})$ and $(\mathfrak{g}, \mathfrak{a})$ such that $i\nu$ is dominant and restriction to $\mathfrak{a}$ carries positive roots to positive roots.
5. In the situation of Problem 3, use Problem 4 to show in addition that $i\nu_0$ may be assumed dominant for the roots of $(\mathfrak{g}, \mathfrak{a})$. Conclude that

$i\nu_0$ is dominant for the roots of $(\mathfrak{g}^{\mathbb{C}}, (\mathfrak{a} \oplus \mathfrak{b})^{\mathbb{C}})$, and apply Lemma 15.6 to complete the proof of conjugacy via $W(A:G)$.

Problems 6 to 8 provide restrictions on the K_r type τ_μ that occurs in the minimal K type formula. They use the fact given in Theorem 15.4 that τ_μ is fine. It is assumed that rank G = rank K.

6. Prove that $|2\langle\mu, \gamma\rangle/|\gamma|^2| \le 1$ for every noncompact root γ in Δ_r that can be imbedded in a strongly orthogonal basis of $\sum_{j=1}^{l} \mathbb{R}\alpha_j$ consisting of noncompact roots.
7. Deduce from Problem 6 that $|2\langle\mu, \gamma\rangle/|\gamma|^2| \le 1$ for every noncompact long root γ in Δ_r.
8. Prove that $\mu = \sum c_j\alpha_j$ has $c_j = 0$ or $+\frac{1}{2}$ or $-\frac{1}{2}$ for each j. [Hint: μ is analytically integral on $\mathfrak{b}_r$, hence is algebraically integral for the roots of $\mathfrak{g}_r$. Apply Problem 6.]

Problems 9 to 12 calculate minimal K types for nonunitary principal series of real-rank-one groups G with rank G = rank K. Let the strongly orthogonal sequence of noncompact roots be just α, and let ρ_α be half the sum of the members of Δ whose inner product with α is positive. The K_r parameter μ is just 0 or $\pm\frac{1}{2}\alpha$.

9. Prove that
$$\rho_\alpha - \frac{\langle 2\delta_K, \alpha\rangle}{|\alpha|^2}\alpha + \tfrac{1}{2}\alpha$$
is an integral multiple of α. [Hint: For $c > 0$, group 4-tuples of roots $\pm c\alpha \pm \varepsilon$, where $\varepsilon \perp \alpha$, considering which values of c are possible, which roots are compact or noncompact, and which roots are positive. Take for granted that no relation $|\beta|^2 = 3|\gamma|^2$ occurs for β and γ in Δ if one of β and γ is nonorthogonal to α.]
10. Deduce from Problem 9 that
$$\frac{2\langle E(2\delta_K) - 2\delta_{K_r}, \alpha\rangle}{|\alpha|^2} \equiv 1 + \frac{2\langle \rho_\alpha, \alpha\rangle}{|\alpha|^2} \bmod 2.$$
11. Deduce from Problem 10 that μ assumes one or both of the values $\pm\frac{1}{2}\alpha$ when
$$\chi(\gamma_{c(\alpha)}) = (-1)^{2\langle\rho_\alpha, \alpha\rangle/|\alpha|^2} \qquad \text{(cotangent case)}$$
and is 0 when
$$\chi(\gamma_{c(\alpha)}) = -(-1)^{2\langle\rho_\alpha, \alpha\rangle/|\alpha|^2} \qquad \text{(tangent case).}$$

12. Let $G = \mathrm{SU}(n, 1)$, and let σ_k be the one-dimensional representation of M defined in Problem 10 of Chapter XIV. Find the minimal K types of $U(S, \sigma_k, 0)$. Verify that the number of minimal K types equals $|R_{\sigma_k,0}|$.

Problems 13 to 18 outline calculations for $\mathrm{Sp}(n, \mathbb{R})$ that identify the minimal K types for the principal series, showing they are fine. Fix k with $0 \le k \le n$, and let σ_k be the character of $M_\mathfrak{p}$ defined in the discussion of $\mathrm{Sp}(n, \mathbb{R})$ in §2. Let $\mu = \sum_{j=1}^k e_j$; we know that τ_μ is a fine K type with $\tau_\mu|_{M_\mathfrak{p}}$ containing σ_k. Let τ_{μ_0} be a minimal K type of $U(S_\mathfrak{p}, \sigma_k, 0)$.

13. Show that τ_{μ_0} contains a weight $\sum a_j e_j$ with a_j odd for $j \le k$ and even for $j > k$. Conclude that τ_{μ_0} contains a Δ_K^+ dominant weight $\mu_1 = \sum c_j e_j$ (i.e., $c_1 \ge c_2 \ge \ldots \ge c_n$) with k of the c_j odd and the others even.
14. Prove that μ_1 in Problem 13 satisfies $|\mu_1 + 2\delta_K|^2 \le |\mu_0 + 2\delta_K|^2$.
15. Assume that $k \le [n/2]$, and calculate $|\mu_1 + 2\delta_K|^2 - |\mu + 2\delta_K|^2$. Show that if it can be proved that

$$\sum_{j=1}^{k} (n - 2j + 1)(c_j - c_{n+1-j} - 1) \ge 0, \tag{15.31}$$

then it follows that μ_0 is either μ or $\sum_{j=1}^k (-e_{n+1-j})$.
16. Assume that $k \le [n/2]$. Show that (15.31) holds if $c_k - c_{n+1-k} \ge 1$.
17. Assume that $k \le [n/2]$ and that $c_k = c_{n+1-k}$. Let $j = j_0$ (≥ 1) be the largest integer such that $c_{k+1-j} = c_{n-k+j}$. Prove that $c_{j_0} - c_{n-j_0+1} \ge 2$, treating separately the cases that c_{k+1-j_0} is even and c_{k+1-j_0} is odd. Lumping terms appropriately, deduce (15.31) from this fact.
18. For $k > [n/2]$, prove that μ_0 is either μ or $\sum_{j=1}^k (-e_{n+1-j})$ by giving an argument analogous to that in Problems 15 to 17.

CHAPTER XVI

Unitary Representations

§1. SL(2, ℝ) and SL(2, ℂ)

For $G = SL(2, \mathbb{C})$, there is no discrete series, by Theorem 12.20. Thus Problems 9–14 of Chapter VIII identify the irreducible admissible representations as the finite-dimensional ones and certain nonunitary principal series. These are the only candidates for irreducible unitary representations, by Theorem 8.1, and we can determine the irreducible unitary representations simply by deciding which irreducible admissible representations are infinitesimally unitary, according to Theorem 9.3.

The finite-dimensional irreducible representations are not infinitesimally unitary, apart from the trivial representation, because of the unitary trick (see Proposition 5.7 and Corollary 2.3). Thus suppose we are given a nonunitary principal series $\mathscr{P}^{k,z}$ induced from the kth character of the circle group $M = M_{\mathfrak{p}}$ and the parameter $\frac{1}{2}z\rho$ on $\mathfrak{a} = \mathfrak{a}_{\mathfrak{p}}$. Proposition 10.18 gives the global character of $\mathscr{P}^{k,z}$ on the element

$$a_t m_\theta = \begin{pmatrix} e^{t+i\theta} & 0 \\ 0 & e^{-t-i\theta} \end{pmatrix}$$

as

$$\Theta(a_t m_\theta) = \frac{e^{zt+ik\theta} + e^{-zt-ik\theta}}{2(\cosh 2t + \cos 2\theta)}. \tag{16.1}$$

Now a unitary representation π has

$$\pi(x)^* = \pi(x^{-1}) \qquad \text{for all } x, \tag{16.2}$$

and this property forces the global character to satisfy

$$\overline{\Theta(x)} = \Theta(x^{-1}) \tag{16.3}$$

on each Cartan subgroup. Referring to (16.1), we see that $\mathscr{P}^{k,z}$ can be infinitesimally unitary only if z is imaginary or if z is real and $k = 0$. The case that z is imaginary is the case that $\mathscr{P}^{k,z}$ is in the unitary principal series, and these representations are already unitary.

Thus the only candidates for infinitesimally unitary irreducible representations besides the trivial representation and the unitary principal series are $\mathscr{P}^{0,x}$ with x real and positive. (Positivity is given to us in Problems 9–14 of Chapter VIII as a result of the Langlands classification.)

For a unitary representation all matrix coefficients must be bounded. Moreover, the K-finite matrix coefficients as a set are not changed under infinitesimal equivalence, as we see easily from Theorem 8.7 and Equation (8.10). Since we can read off asymptotic behavior of K-finite matrix coefficients from the Langlands classification, we see that $\mathscr{P}^{0,x}$ cannot be unitary for $x > 2$. At $x = 2$, $\mathscr{P}^{0,x}$ is reducible, and the trivial representation of G is the Langlands quotient.

Thus we are to consider $\mathscr{P}^{0,x}$ for $0 < x < 2$. Problems 9–14 of Chapter VIII show that these representations are irreducible. We shall show that they are infinitesimally unitary. Proposition 9.1 shows that there is at most one candidate for an inner product, and we begin by determining it. If $\mathscr{P}^{0,x}$ has been made unitary by introduction of a suitable inner product, then (16.2) must hold. Let this inner product be denoted by $\langle\cdot,\cdot\rangle$, and let $(\cdot,\cdot)_{L^2(K)}$ be the usual inner product on K. For a suitable operator L on K-finite vectors, we have

$$\langle f, g\rangle = (Lf, g)_{L^2(K)}. \tag{16.4}$$

Condition (16.2) says that $\mathscr{P}^{0,x}(X)$ is skew-Hermitian relative to $\langle\cdot,\cdot\rangle$ for every X in $\mathfrak{sl}(2,\mathbb{C})$. Thus X in $\mathfrak{sl}(2,\mathbb{C})$ implies

$$\begin{aligned}(L\mathscr{P}^{0,x}(X)f, g) &= \langle \mathscr{P}^{0,x}(X)f, g\rangle = -\langle f, \mathscr{P}^{0,x}(X)g\rangle \\ &= -(Lf, \mathscr{P}^{0,x}(X)g) = -(\mathscr{P}^{0,x}(X)^*Lf, g).\end{aligned}$$

Hence $\quad L\mathscr{P}^{0,x}(X) = -\mathscr{P}^{0,x}(X)^*L.$

Formula (14.22) relates the adjoint of a standard induced representation to another standard induced representation, and we conclude

$$L\mathscr{P}^{0,x}(X) = \mathscr{P}^{0,-x}(X)L.$$

From Proposition 14.23a, we see that L must be the standard intertwining operator

$$\sigma(w)A_S(w,\sigma,\nu) \qquad \text{with} \quad w = \begin{pmatrix} 0 & 1 \\ -1 & 0 \end{pmatrix}, \quad \sigma = 1, \quad \nu = \tfrac{1}{2}x\rho, \tag{16.5}$$

except for normalization. In view of (16.4), the question is whether (16.5) is a multiple of a semidefinite operator. We use the following trivial lemma.

Lemma 16.1. Let $F(x)$ be a continuous function from a topological space X to the vector space of n-by-n Hermitian matrices such that $F(x_0)$ is positive definite for some x_0 and such that $\det F(x)$ is nonvanishing for x in a dense connected subset Y. Then $F(x)$ is positive definite for x in Y, and $F(x)$ is positive semidefinite for all x in X.

We use the interval $0 \le x \le 2$ as our topological space. Since $\mathscr{P}^{0,x}$ is irreducible for $0 < x < 2$, the (well-defined) operator $A_S(w, 1, \frac{1}{2}x\rho)$ has 0 kernel. Also the operator commutes with K and thus carries each K type to itself. Hence we shall apply the lemma to the restriction of the operator to a finite sum of K types.

The operator is Hermitian by Proposition 14.23b since w represents an element of order 2 in $W(A_\mathfrak{p}:G)$. To deal with the fact that $F(x) = \sigma(w)A_S(w, 1, \frac{1}{2}x\rho)|_{\text{subspace}}$ may blow up at $x = 0$, we simply normalize the operator. Multiplication of the operator by x gives an adequate normalization in this case. The normalized operator has to be scalar at $x = 0$ since $\mathscr{P}^{0,0}$ is irreducible. Thus the lemma applies and shows the operator is definite for $0 < x < 2$.

As a result, the representations $\mathscr{P}^{0,x}$ are infinitesimally unitary for $0 < x < 2$. The unitary versions of these representations are called the **complementary series,** which we denote $\mathscr{C}^x$, $0 < x < 2$. The inner product for the complementary series is given by (2.12) in the noncompact picture. Our calculations have thus established the completeness in the following theorem. The statement about equivalences follows by inspection of the formula for global characters—or alternatively by use of Theorems 8.54 and 14.91.

Theorem 16.2. In $G = \mathrm{SL}(2, \mathbb{C})$, the only irreducible unitary representations up to unitary equivalence are

(a) the trivial representation
(b) the unitary principal series $\mathscr{P}^{k,iy}$ with $k \in \mathbb{Z}$ and $y \in \mathbb{R}$
(c) the complementary series $\mathscr{C}^x$ with $0 < x < 2$.

Moreover, the only equivalences among these representations are $\mathscr{P}^{k,iy} \cong \mathscr{P}^{-k,-iy}$.

Now let us consider $G = \mathrm{SL}(2, \mathbb{R})$. The discussion proceeds in similar fashion. By Theorem 12.21, the only discrete series representations are $\mathscr{D}_n^{\pm}$, $n \ge 2$. By Problems 2–8 in Chapter VIII, the remaining irreducible tempered representations are $\mathscr{P}^{+,iy}$ for y in $\mathbb{R}$, $\mathscr{P}^{-,iy}$ for y nonzero in $\mathbb{R}$, and $\mathscr{D}_1^{\pm}$. Moreover, the only further candidates for irreducible unitary representations are the trivial representation and $\mathscr{P}^{\pm,z}$ for certain z with $\operatorname{Re} z > 0$.

Consideration of global characters again shows z must be real, and the necessity of bounded matrix coefficients forces $z = x$ with $0 < x \le 1$. Since $\mathscr{P}^{+,1}$ is reducible (with Langlands quotient the trivial representation), we need only consider $\mathscr{P}^{+,x}$ for $0 < x < 1$ and $\mathscr{P}^{-,x}$ for $0 < x \le 1$. Problems 2–8 of Chapter VIII show these are irreducible.

Again we argue from (16.4) that the only candidate for an invariant inner product is given by the operator $\sigma(w)A_S(w, \pm, x\rho)$, apart from a scalar factor. Here w is $\begin{pmatrix} 0 & 1 \\ -1 & 0 \end{pmatrix}$. For $\mathscr{P}^{+,x}$ we argue as with $SL(2, \mathbb{C})$ to see that the representations are infinitesimally unitary. The unitary versions of these representations we again call **complementary series,** denoting them $\mathscr{C}^x$, $0 < x < 1$; the inner product is given in §2.5 in the noncompact picture.

For $\mathscr{P}^{-,x}$, we use Lemma 16.1 to argue that the presence of a single unitary representation would imply that $\sigma(w)A_S(w, -, 0)$ is definite. Since this operator is unitary, it would have to be scalar. But we know it is not scalar. Hence there is no x with $0 < x \leq 1$ for which $\mathscr{P}^{-,x}$ is infinitesimally unitary. Our calculations have thus established the completeness in the following theorem. Again we can decide all questions about equivalences either by inspecting the formula for the global characters or by using Theorems 8.54 and 14.91.

Theorem 16.3. In $G = SL(2, \mathbb{R})$, the only irreducible unitary representations up to unitary equivalence are

(a) the trivial representation
(b) the discrete series $\mathscr{D}_n^{\pm}$, $n \geq 2$, and the limits of discrete series $\mathscr{D}_1^{\pm}$
(c) the irreducible members of the unitary principal series, $\mathscr{P}^{+,iy}$ with y real and $\mathscr{P}^{-,iy}$ with y nonzero real
(d) the complementary series $\mathscr{C}^x$ with $0 < x < 1$.

Moreover, the only equivalences among these representations are $\mathscr{P}^{+,iy} \cong \mathscr{P}^{+,-iy}$ and $\mathscr{P}^{-,iy} \cong \mathscr{P}^{-,-iy}$.

§2. Continuity Arguments and Complementary Series

Some of the arguments in §1 extend to give information about irreducible unitary representations for general G. Again the problem is to decide which irreducible admissible representations are infinitesimally unitary. By Theorem 14.92 it is enough to consider the Langlands quotient $J(S, \sigma, \nu)$, where $S = MAN$ is a cuspidal standard parabolic subgroup, σ is a discrete series or nondegenerate limit of discrete series of M, and ν is in $(\mathfrak{a}')^{\mathbb{C}}$ with $\operatorname{Re}\nu$ in the closed positive Weyl chamber and with $W_{\sigma,\nu} = W'_{\sigma,\nu}$.

It is too difficult to examine the global character of J to check (16.3). Instead we shall derive a necessary reality condition on the parameters by using our classification theorems. This result will be given as Theorem 16.6. In the meantime we look for simpler conditions.

Again a necessary condition for unitarity is boundedness of matrix coefficients. Problems 5–7 at the end of the chapter show that this condition forces Re ν to be in the convex hull of all values of ρ_A as N varies.

Taking a cue from (16.4) and (16.5), we can write down a sufficient condition for unitarity that generalizes the construction of the complementary series in SL(2, $\mathbb{C}$) and SL(2, $\mathbb{R}$).

Proposition 16.4. Under the assumption that σ is a discrete series or nondegenerate limit of discrete series and that $J(S, \sigma, \nu)$ is defined, suppose that s is in $W'_{\sigma,0}$, that $s^2 = 1$, that $s\nu = -\bar{\nu}$, and that $\sigma(s)\mathscr{A}_S(s, \sigma, t\nu)$ is invertible (on each K type) for $0 < t < 1$. Then $J(S, \sigma, \nu)$ is infinitesimally unitary.

Proof. If w is in $N_K(\mathfrak{a})$, then we recall that $\sigma(w)\mathscr{A}_S(w, \sigma, t\nu)$ depends only on the element that w represents in $W(A:G)$. Since $s^2 = 1$ and $s\nu = -\bar{\nu}$ and $s[\sigma] = [\sigma]$, it follows from Proposition 14.23b that $\sigma(s)\mathscr{A}_S(s, \sigma, t\nu)$ is Hermitian. By assumption it is nonsingular for $0 < t < 1$, and the fact that s is in $W'_{\sigma,0}$ means that it is scalar at $t = 0$. By Lemma 16.1, it is semidefinite at $t = 1$. Therefore we obtain a semidefinite inner product on $U(S, \sigma, \nu)$ by the definition

$$\langle f, g \rangle = (Lf, g)_{L^2(K)} \qquad \text{with} \quad L = \sigma(s)\mathscr{A}_S(s, \sigma, \nu).$$

Because $s\nu = -\bar{\nu}$, the intertwining property of Proposition 14.23a implies that

$$LU(S, \sigma, \nu, X) = U(S, \sigma, -\bar{\nu}, X)L, \qquad X \in \mathfrak{g}.$$

From (14.22) we obtain

$$LU(S, \sigma, \nu, X) = -U(S, \sigma, \nu, X)^*L, \qquad X \in \mathfrak{g},$$

and it follows that

$$\langle U(S, \sigma, \nu, X)f, g \rangle = -\langle f, U(S, \sigma, \nu, X)g \rangle, \qquad X \in \mathfrak{g}.$$

Thus $U(S, \sigma, \nu)$ is infinitesimally unitary, except for the requirement that the form $\langle \cdot, \cdot \rangle$ be definite. However, the kernel of L is the kernel of the quotient map $U \to J$, as we shall see during the proof of Theorem 16.6, and thus $\langle \cdot, \cdot \rangle$ descends to a definite form on J. Therefore $J(S, \sigma, \nu)$ is infinitesimally unitary.

In Proposition 16.4 we do have some control over the invertibility of $\sigma(s)\mathscr{A}_S(s, \sigma, t\nu)$, provided we normalize the operator carefully. For this purpose we shall use an *ad hoc* normalization of intertwining operators, rather than the systematic procedure of §14.6.

Changing notation, let us consider an operator $A(\bar{S}:S:\sigma:\nu)$ with $\dim A = 1$. By Theorem 7.22 this operator is finite-valued for Re $\nu > 0$.

At $v = 0$ it has at most a simple pole. To see this, we go over the proof of Theorem 14.16 carefully: Formula (14.38) remains valid, we still have $k = l$, we can still multiply (14.38) by v^{k-1}, and we get a contradiction if $k > 1$ by letting $v \to 0$. Therefore, in obvious notation, either $A(\bar{S}:S:\sigma:v)$ or $vA(\bar{S}:S:\sigma:v)$ is nonvanishing and regular for Re $v \geq 0$.

Then we can use (14.35) to pass to a normalization of the operators $A(S':S:\sigma:v)$ with $\bar{\mathfrak{n}} \cap \mathfrak{n}' = \bar{\mathfrak{n}}^{(\beta)}$, taking the normalization to be the same as for the operators with dim $\mathfrak{a} = 1$. Finally we want to normalize our given operator, which is essentially $A(s^{-1}Ss:S:\sigma:tv)$. Fix a minimal string $S = S_0, S_1, \ldots, S_n = s^{-1}Ss$, decompose the operator accordingly, and normalize each factor of the operator as above. Use the product of the normalizations for the whole operator. This normalization will sometimes not result in the usual identities among intertwining operators, but it has the following nice properties:

(i) the normalized operators are regular and nonvanishing for Re v in the closure of the positive Weyl chamber
(ii) $\mathscr{A}(s^{-1}Ss:S:\sigma:tv) = \mathscr{A}(S_n:S_{n-1}:\sigma:tv) \cdots \mathscr{A}(S_1:S_0:\sigma:tv)$.

Each operator on the right is essentially an operator in a case where dim $A = 1$, and it is essentially the Langlands operator in that case. Thus $\mathscr{A}(s^{-1}Ss:S:\sigma:tv)$ has a nonzero kernel if and only if $\mathscr{A}(S_k:S_{k-1}:\sigma:tv)$ has a kernel for some k, which happens if and only if a suitable induced representation is reducible in a case where dim $A = 1$ and the A parameter has real part strictly positive.

A simple example occurs with G split over $\mathbb{R}$ and with S a minimal parabolic subgroup. Each operator on the right side of (ii) is essentially an operator for $SL(2, \mathbb{R})$, and we can check whether it has a kernel by looking at where the nonunitary principal series of $SL(2, \mathbb{R})$ is reducible. The result of applying Proposition 16.4 is as follows.

Corollary 16.5. Let G be split over $\mathbb{R}$, let $S = MAN$ be a minimal parabolic subgroup, let σ be a character of the finite abelian group M, and suppose that $J(S, \sigma, v)$ is defined. Suppose that there is an element $s \in W'_{\sigma,0}$ with $s^2 = 1$ such that $sv = -\bar{v}$. If

$$\frac{|\langle \text{Re } v, \alpha \rangle|}{|\alpha|^2} \leq \frac{1}{2}$$

for every restricted root α, then $J(S, \sigma, v)$ is infinitesimally unitary.

§3. Criterion for Unitary Representations

Our substitute for using formulas for global characters in limiting unitarity is the following theorem. It reduces the problem of determining the unitary dual of G to decisions about positivity of intertwining operators.

Theorem 16.6. Let G be linear connected semisimple, let $S = MAN$ be a cuspidal standard parabolic subgroup of G, let σ be a discrete series or nondegenerate limit of discrete series representation of M, and let ν be in $(\mathfrak{a}')^{\mathbb{C}}$ with Re ν in the closed positive Weyl chamber and with $W_{\sigma,\nu} = W'_{\sigma,\nu}$. Then $J(S, \sigma, \nu)$ is infinitesimally unitary if and only if

(i) there exists s in $W(A:G)$ such that $s^2 = 1$, $s[\sigma] = [\sigma]$, and $s\nu = -\bar{\nu}$, and

(ii) the standard intertwining operator $\sigma(s)A_S(s, \sigma, \nu)$, when normalized to be pole-free and not identically zero as

$$\sigma(s)\mathscr{A}_S(s, \sigma, \nu) \tag{16.6}$$

is positive or negative semidefinite on the K-finite vectors.

If $J(S, \sigma, \nu)$ is infinitesimally unitary, then every s satisfying (i) is such that the operator (16.6) is positive or negative semidefinite on K-finite vectors.

Remark. In fact, the proof will show that $J(S, \sigma, \nu)$ has a nonzero invariant Hermitian form if and only if (i) holds and that (16.6) gives this form whenever s is as in (i).

Before coming to the proof, we make some observations about adjoints. The representation $J(S, \sigma, \nu)$ we can identify with some $J(S_1, \pi, \nu_1)$ as in the proof of Theorem 14.92; here $S_1 \supseteq S$, π is irreducible tempered, and ν_1 has real part in the open positive Weyl chamber. The representation $J(S_1, \pi, \nu_1)$, as a quotient of $U(S_1, \pi, \nu_1)$, is a representation by bounded operators in the quotient Hilbert space. It is therefore meaningful to speak of the adjoint $J(S_1, \pi, \nu_1, X)^*$ for $X \in \mathfrak{g}$ if we agree to define the adjoint K type by K type. If $J(S, \sigma, \nu)$ is infinitesimally unitary, then so is $J(S_1, \pi, \nu_1)$ and it follows from the same argument as with (16.4) that

$$J(S_1, \pi, \bar{\nu}_1, X) \text{ is infinitesimally equivalent with } J(S_1, \pi, \nu_1, -X)^*. \tag{16.7}$$

We begin with a lemma that identifies this adjoint.

Lemma 16.7. $J(S_1, \pi, \nu_1, -X)^*$ is infinitesimally equivalent with $J(\bar{S}_1, \pi, -\bar{\nu}_1, X)$.

Proof. Since the intertwining operators act separately on orthogonal finite-dimensional spaces, the space

$$\begin{aligned} V &= \ker A(\bar{S}_1 : S_1 : \pi : \nu_1))^{\perp} \\ &= \text{image } A(\bar{S}_1 : S_1 : \pi : \nu_1)^*, \end{aligned}$$

is

which, by Proposition 14.10, is

$$= \text{image } A(S_1:\bar{S}_1:\pi:-\bar{v}_1).$$

Let E be the orthogonal projection of $L^2(K)$ onto V. Then $J(S_1, \pi, v_1, X)$ acts in V as $EU(S_1, \pi, v_1, X)E$. By (14.22), $J(S_1, \pi, v_1, -X)^*$ acts in V as

$$EU(S_1, \pi, v_1, -X)^*E = EU(S_1, \pi, -\bar{v}_1, X)E.$$

Now $A(S_1:\bar{S}_1:\pi:-\bar{v}_1)$ is a linear isomorphism from

$$\tilde{V} = (\ker A(S_1:\bar{S}_1:\pi:\bar{v}_1))^{\perp}$$

onto $V = \text{image } A(S_1:\bar{S}_1:\pi:-\bar{v}_1)$. Therefore $J(S_1, \pi, v_1 - X)^*$ in its action on V pulls back from V to a unique operator $S(X)$ on $\tilde{V}$ satisfying

$$J(S_1, \pi, v_1, -X)^*A(S_1:\bar{S}_1:\pi:-\bar{v}_1) = A(S_1:\bar{S}_1:\pi:-\bar{v}_1)S(X).$$

The left side of this relation is

$$\begin{aligned} &= EU(S_1, \pi, -\bar{v}_1, X)EA(S_1:\bar{S}_1:\pi:-\bar{v}_1) \\ &= EU(S_1, \pi, -\bar{v}_1, X)A(S_1:\bar{S}_1:\pi:-\bar{v}_1)\tilde{E} \qquad (\tilde{E} = \text{projection on } \tilde{V}) \\ &= EA(S_1:\bar{S}_1:\pi:-\bar{v}_1)U(\bar{S}_1, \pi, -\bar{v}_1, X)\tilde{E} \\ &= A(S_1:\bar{S}_1:\pi:-\bar{v}_1)\tilde{E}U(\bar{S}_1, \pi, -\bar{v}_1, X)\tilde{E}. \end{aligned}$$

Hence $\qquad S(X) = \tilde{E}U(\bar{S}_1, \pi, -\bar{v}_1, X)\tilde{E} = J(\bar{S}_1, \pi, -\bar{v}_1, X).$

Thus the linear isomorphism $A(S_1:\bar{S}_1:\pi:-\bar{v}_1)$ from $\tilde{V}$ to V exhibits the required infinitesimal equivalence.

Proof of Theorem 16.6. Suppose $J(S, \sigma, v)$ is infinitesimally unitary. The lemma and formula (16.7) together show that $J(S_1, \pi, v_1)$ is infinitesimally equivalent with $J(\bar{S}_1, \pi, -\bar{v}_1)$. By uniqueness in the Langlands classification (Theorem 8.54), there is an element w_1 in K mapping the data (S_1, π, v_1) to the data $(\bar{S}_1, \pi, -\bar{v}_1)$. Following the notation of §14.17, write $S_1 = M_1A_1N_1$. Then A_1 is the noncompact part of the center of $S_1 \cap \bar{S}_1$, and it follows that w_1 is in $N_K(\mathfrak{a}_1)$. Then we have

$$w_1S_1w_1^{-1} = \bar{S}_1, \qquad w_1\pi \cong \pi, \qquad \text{and} \qquad w_1v_1 = -\bar{v}_1. \tag{16.8}$$

We shall apply Theorem 14.91 to the equivalence of the irreducible tempered representations

$$\pi = \operatorname{ind}_{MA_{M_1}N_{M_1}}^{M_1}(\sigma \otimes \exp v|_{\mathfrak{a}_{M_1}} \otimes 1)$$

and $\qquad w_1\pi \cong \operatorname{ind}_{w_1(MA_{M_1}N_{M_1})w_1^{-1}}^{M_1}(w_1\sigma \otimes \exp w_1v|_{w_1\mathfrak{a}_{M_1}} \otimes 1).$

The theorem says that such an equivalence must be implemented by an element $w_2 \in K \cap M_1$ conjugating all the data but N_{M_1} for π to the corresponding data for $w_1\pi$. Thus

$$w_1 M w_1^{-1} = w_2 M w_2^{-1} \tag{16.9a}$$

$$w_1 A_{M_1} w_1^{-1} = w_2 A_{M_1} w_2^{-1} \tag{16.9b}$$

$$w_1\sigma \cong w_2\sigma \tag{16.9c}$$

$$w_1 v\big|_{w_1\mathfrak{a}_{M_1}} = w_2 v\big|_{w_2\mathfrak{a}_{M_1}}. \tag{16.9d}$$

We shall list some properties of $w_2^{-1}w_1$. Since w_2 is in M_1, (16.8) gives

$$(w_2^{-1}w_1)S_1(w_2^{-1}w_1)^{-1} = \bar{S}_1. \tag{16.10a}$$

Also w_1 in $N_K(\mathfrak{a}_1)$ and w_2 in $Z_K(\mathfrak{a}_1)$ imply that $w_2^{-1}w_1$ is in $N_K(\mathfrak{a}_1)$, and (16.9b) shows that $w_2^{-1}w_1$ is in $N_K(\mathfrak{a}_{M_1})$. Thus

$$w_2^{-1}w_1 \in N_K(\mathfrak{a}_1) \cap N_K(\mathfrak{a}_{M_1}) \subseteq N_K(\mathfrak{a}). \tag{16.10b}$$

From (16.9c) we have

$$w_2^{-1}w_1\sigma \cong \sigma. \tag{16.10c}$$

From (16.9d) and (16.10b), we have

$$w_2^{-1}w_1(v|_{\mathfrak{a}_{M_1}}) = v|_{\mathfrak{a}_{M_1}},$$

which is imaginary. Hence

$$w_2^{-1}w_1(v|_{\mathfrak{a}_{M_1}}) = -(\bar{v}|_{\mathfrak{a}_{M_1}}),$$

and (16.8) gives

$$w_2^{-1}w_1 v = -\bar{v}. \tag{16.10d}$$

Let us observe that $w_2^{-1}w_1$ normalizes the root system $\Delta'_{\sigma,v}$ defined in (14.133). In fact, α in $\Delta'_{\sigma,v}$ implies $w_2^{-1}w_1\alpha$ is useful, and we have

$$s_{w_2^{-1}w_1\alpha}v = (w_2^{-1}w_1)s_\alpha(w_2^{-1}w_1)^{-1}v = -w_2^{-1}w_1 s_\alpha\bar{v} = -w_2^{-1}w_1\bar{v} = v$$

by two applications of (16.10d). Also

$$\begin{aligned}\mu_{\sigma,w_2^{-1}w_1\alpha}(v) &= \mu_{(w_2^{-1}w_1)^{-1}\sigma,\alpha}((w_2^{-1}w_1)^{-1}v)\\ &= \mu_{\sigma,\alpha}(-\bar{v}) \qquad \text{by (16.10c) and (16.10d)}\\ &= \overline{\mu_{\sigma,\alpha}(v)} \qquad \text{by Proposition 14.13e(ii).}\end{aligned}$$

Hence $\mu_{\sigma,w_2^{-1}w_1\alpha}(v)$ and $\mu_{\sigma,\alpha}(v)$ are both zero or both nonzero, and $w_2^{-1}w_1$ normalizes $\Delta'_{\sigma,v}$.

Thus we can choose w_3 in $K \cap M_1$ representing a member of $W'_{\sigma,v}$ such that

$$w_3^{-1}w_2^{-1}w_1\Delta'^{+}_{\sigma,v} = \Delta'^{+}_{\sigma,v}. \tag{16.11}$$

Then it is clear that

$$(w_3^{-1}w_2^{-1}w_1)S_1(w_3^{-1}w_2^{-1}w_1)^{-1} = \bar{S}_1 \tag{16.12a}$$

and

$$w_3^{-1}w_2^{-1}w_1 \in N_K(\mathfrak{a}_1) \cap N_K(\mathfrak{a}_{M_1}). \tag{16.12b}$$

Since $W'_{\sigma,\nu} \subseteq W_{\sigma,\nu}$, (16.10c) and (16.10d) give

$$w_3^{-1}w_2^{-1}w_1\sigma \cong \sigma \tag{16.12c}$$

$$w_3^{-1}w_2^{-1}w_1\nu = -\bar{\nu}. \tag{16.12d}$$

Let s be the class of $w_3^{-1}w_2^{-1}w_1$ in $W(A:G)$. We show that s satisfies (i). In fact, $s[\sigma] = [\sigma]$ and $s\nu = -\bar{\nu}$ by (16.12c) and (16.12d). Hence s^2 fixes $[\sigma]$ and ν and is in $W_{\sigma,\nu}$. Since $W_{\sigma,\nu} = W'_{\sigma,\nu}$, s^2 is in $W'_{\sigma,\nu}$. Then (16.11) shows that s^2 leaves $\Delta'^{+}_{\sigma,\nu}$ stable, and it follows that $s^2 = 1$. This proves that s satisfies (i).

Now let us drop the assumption that $J(S, \sigma, \nu)$ is infinitesimally unitary and assume instead that s is any element of $W(A:G)$ satisfying (i). To complete the proof, it is enough to show that $J(S, \sigma, \nu)$ is infinitesimally unitary if and only if (16.6) is semidefinite.

Since s exists and fixes $[\sigma]$, the operator (16.6) is well defined. By Proposition 14.23, (16.6) gives an invariant Hermitian form on $U(S, \sigma, \nu)$ by the definition

$$\langle f, g \rangle = (\sigma(s)\mathscr{A}_S(s, \sigma, \nu)f, g)_{L^2(K)}. \tag{16.13}$$

We shall check that this form descends to $J(S, \sigma, \nu)$. Once it does, it provides a nonzero invariant Hermitian form for $J(S, \sigma, \nu)$, and such a form is unique apart from scalar multiples, by Proposition 9.1. Therefore $J(S, \sigma, \nu)$ is infinitesimally unitary if and only if this form for J is semidefinite, hence if and only if the corresponding form (16.13) for U is semidefinite, hence if and only if (16.6) is a semidefinite operator on K-finite vectors.

Thus we are to prove that the form (16.13) descends to $J(S, \sigma, \nu)$. To do so, we show that the kernel of the operator (16.6) contains the kernel of the operator

$$A(MAN_{M_1}\bar{N}_1 : MAN_{M_1}N_1 : \sigma : \nu) \tag{16.14}$$

that defines $J(S, \sigma, \nu)$. (See (14.135).) In fact, we shall exhibit (16.6) as the composition of (16.14) followed by a suitably normalized version of

$$\sigma(s)R(s)A(s^{-1}MANs : MAN_{M_1}\bar{N}_1 : \sigma : \nu),$$

and the result will follow.

Such a composition formula follows from Theorem 8.38d if it is shown that

$$\mathrm{Ad}(s)^{-1}\mathfrak{n} \cap \mathfrak{n} \subseteq (\mathfrak{n}_{M_1} \oplus \bar{\mathfrak{n}}_1) \cap \mathfrak{n}. \tag{16.15}$$

The right side of (16.15) is just $\mathfrak{n}_{M_1}$, and thus (16.15) amounts to the assertion that any positive root α of $(\mathfrak{g}, \mathfrak{a})$ such that $s^{-1}\alpha$ is positive is necessarily a root of M_1, i.e., is orthogonal to Re ν.

Thus suppose $\alpha > 0$ and $s^{-1}\alpha > 0$. Since $s\nu = -\bar{\nu}$, $s \operatorname{Re} \nu = -\operatorname{Re} \nu$. Since Re ν is dominant, $\langle \operatorname{Re} \nu, \alpha \rangle \geq 0$ and $\langle \operatorname{Re} \nu, s^{-1}\alpha \rangle \geq 0$. The latter relation implies $0 \leq \langle s \operatorname{Re} \nu, \alpha \rangle = -\langle \operatorname{Re} \nu, \alpha \rangle$. Thus $\langle \operatorname{Re} \nu, \alpha \rangle = 0$. This completes the proof.

If we bring in the theory of minimal K types, then the easy part of Theorem 16.6 has the following corollary, which goes in the direction converse to Proposition 16.4.

Proposition 16.8. Under the assumption that σ is a discrete series or nondegenerate limit of discrete series and that $J(S, \sigma, \nu)$ is defined, suppose that s is in $W_{\sigma,0}$ but not $W'_{\sigma,0}$, that $s^2 = 1$, and that $s\nu = -\bar{\nu}$. Then $J(S, \sigma, \nu)$ is not infinitesimally unitary.

Proof. Theorem 16.6 says that it suffices to prove that $\sigma(s)\mathscr{A}_S(s, \sigma, \nu)$ is not semidefinite. Since s is in $W_{\sigma,0}$ but not $W'_{\sigma,0}$, we can combine Proposition 14.83 and Theorem 14.88 to see that $\sigma(s)\mathscr{A}_S(s, \sigma, 0)$ is not scalar. Since the operator is Hermitian and unitary, it is $+1$ on some irreducible constituents and -1 on some others. By Theorem 15.9 it is $+1$ on some minimal K types of $U(S, \sigma, 0)$ and -1 on some others. The operator $\sigma(s)\mathscr{A}_S(s, \sigma, t\nu)$, $0 \leq t \leq 1$, is real on each K type, and Theorem 15.10 says it is nonvanishing on the minimal ones. Therefore $\sigma(s)\mathscr{A}_S(s, \sigma, \nu)$ is not semidefinite.

§4. Reduction to Real Infinitesimal Character

Although Theorem 16.6 is a kind of classification theorem for irreducible unitary representations, it does not really solve the classification problem. It does not, for example, give an effective way of deciding whether a particular $J(S, \sigma, \nu)$ is unitary. We conclude this chapter with a result that is effective at simplifying the problem (although it does not solve the problem completely): Theorem 16.10 will show that it is enough to solve the classification problem for cases where the infinitesimal character is real, i.e., where the parameter ν is real-valued on $\mathfrak{a}$. We begin with a lemma.

Lemma 16.9. Let $S = MAN$ be a parabolic subgroup, let σ be an irreducible admissible representation of M with a nonzero invariant Hermitian form on the $(K \cap M)$-finite vectors, and let ν be an imaginary parameter on $\mathfrak{a}$. Define

$$U(S, \sigma, \nu) = \operatorname{ind}_S^G(\sigma \otimes e^{\nu} \otimes 1)$$

as an admissible representation of G. If $U(S, \sigma, \nu)$ is irreducible, then it is infinitesimally unitary if and only if σ is infinitesimally unitary.

Proof. Let $\langle \cdot, \cdot \rangle_M$ be the given nonzero invariant Hermitian form for σ. If f and g are K-finite members of the induced space, we define

$$\langle f, g \rangle_G = \int_K \langle f(k), g(k) \rangle_M \, dk. \tag{16.16}$$

The same calculation as in §7.2 (where it is assumed that $\langle \cdot, \cdot \rangle_M$ is definite) shows that $\langle \cdot, \cdot \rangle_G$ is an invariant Hermitian form on the K-finite vectors of $U(S, \sigma, \nu)$. Since σ and $U(S, \sigma, \nu)$ are assumed irreducible, Proposition 9.1 shows that $\langle \cdot, \cdot \rangle_M$ and $\langle \cdot, \cdot \rangle_G$ are unique up to scalar factors as invariant Hermitian forms for σ and $U(S, \sigma, \nu)$, respectively. It is obvious that $\langle \cdot, \cdot \rangle_M$ definite implies $\langle \cdot, \cdot \rangle_G$ definite, i.e., that σ infinitesimally unitary implies $U(S, \sigma, \nu)$ infinitesimally unitary. To complete the proof, we show that $\langle \cdot, \cdot \rangle_M$ indefinite implies $\langle \cdot, \cdot \rangle_G$ not semidefinite.

The trick is now to forget about G, working only with representations of K and $K \cap M$. We have

$$U(S, \sigma, \nu)\big|_K = \operatorname{ind}_{K \cap M}^K(\sigma|_{K \cap M}).$$

Let us show that the $(K \cap M)$-finite subspace for σ splits as the (infinite) direct sum of irreducible $(K \cap M)$-invariant subspaces that are orthogonal with respect to $\langle \cdot, \cdot \rangle_M$. Since $\langle \cdot, \cdot \rangle_M$ is invariant, we have $\langle u, v \rangle_M = 0$ if u and v belong to different $K \cap M$ types of the space for σ. So the problem is to decompose the finite-dimensional subspace V corresponding to a single $K \cap M$ type.

Now σ, being an admissible representation, acts by definition in a Hilbert space V^σ with some inner product $(\cdot, \cdot)_{V^\sigma}$. Moreover, the action of $\sigma(K \cap M)$ is assumed unitary, by our definition of admissibility. Within the finite-dimensional space V, let us write

$$\langle u, v \rangle_M = (Lu, v)_{V^\sigma}$$

for some $L \in \operatorname{End} V$. It is apparent that L is Hermitian and commutes with $\sigma(K \cap M)$. Therefore V is the orthogonal sum (with respect to $(\cdot, \cdot)_{V^\sigma}$) of the eigenspaces of L for the distinct eigenvalues, and each such eigenspace is stable under $K \cap M$. Lumping these spaces according to the sign of the eigenvalue, we can therefore write V as an orthogonal sum

$$V = V_+ \oplus V_0 \oplus V_-,$$

where each summand is stable under $K \cap M$ and where $\langle \cdot, \cdot \rangle_M$ is positive definite on V_+, 0 on V_0, and negative definite on V_-. Then we can decompose V_+, V_0, V_- and obtain the required decomposition into irreducible subspaces that are orthogonal under $\langle \cdot, \cdot \rangle_M$.

On each irreducible summand under $K \cap M$, $\langle \cdot, \cdot \rangle_M$ is positive definite, or 0, or negative definite (by Propositions 9.1 and 1.6). We can induce each summand separately, and these characteristics persist for $\langle \cdot, \cdot \rangle_G$, in view of (16.16). Moreover, (16.16) shows that the induced spaces corresponding to different subspaces on the M level are orthogonal under $\langle \cdot, \cdot \rangle_G$ on the G level. Since induction is an exact functor, $U(S, \sigma, \nu)|_K$ is nothing more than the (orthogonal) sum of these subspaces that are obtained from $K \cap M$. Hence the presence of a positive subspace for $U(S, \sigma, \nu)|_K$ implies the presence of a positive subspace for $\sigma|_{K \cap M}$, and similarly for negative subspaces. This completes the proof.

Theorem 16.10. Let G be linear connected semisimple, let $S = MAN$ be a cuspidal standard parabolic subgroup of G, let σ be a discrete series or nondegenerate limit of discrete series representation of M, and let ν be in $(\mathfrak{a}')^{\mathbb{C}}$ with $\operatorname{Re} \nu$ in the closed positive Weyl chamber and with $W_{\sigma,\nu} = W'_{\sigma,\nu}$, so that $J(S, \sigma, \nu)$ is defined. Suppose that there exists s in $W(A:G)$ such that $s^2 = 1$, $s[\sigma] = [\sigma]$, and $s\nu = -\bar{\nu}$, so that $J(S, \sigma, \nu)$ has a nonzero invariant Hermitian form. Suppose further that ν is not real. Then there exists a proper parabolic subgroup $M_1A_1N_1$ with M_1A_1 canonical and there exists a canonical irreducible admissible representation ξ of M_1 such that $J(S, \sigma, \nu)$ is infinitesimally unitary if and only if ξ is infinitesimally unitary.

Remarks.

(1) The representation ξ of M_1 will have real infinitesimal character and will possess a nonzero invariant Hermitian form. The proof will exhibit the Langlands parameters of ξ explicitly, so that one can stay within the context of Theorem 16.6 in giving a criterion for ξ to be infinitesimally unitary.

(2) Thus if one attacks the problem of classifying irreducible unitary representations by induction on the dimension of G, the only representations $J(S, \sigma, \nu)$ of G for which condition (ii) in Theorem 16.6 needs to be checked for G are those with real infinitesimal character.

Example. $G = SU(2, 2)$, $S = S_{\mathfrak{p}} = M_{\mathfrak{p}}A_{\mathfrak{p}}N_{\mathfrak{p}}$ minimal, $\sigma = 1$.

We follow the notation of Appendix C, in which this example is of type A III. The restricted roots form a system of type C_2, with simple roots $f_1 - f_2$ and $2f_2$. Let

$$\nu = (f_1 + f_2) + it(-f_1 + f_2) = \operatorname{Re} \nu + i \operatorname{Im} \nu$$

with t real and nonzero. We can take $s = s_{f_1+f_2}$. We shall see that the group M_1 is to be built from the restricted roots orthogonal to $\operatorname{Im} \nu$, thus from $\pm(f_1 + f_2)$. Since these restricted roots are not generated by simple roots, some conjugation is needed to obtain a parabolic subgroup containing $S_\mathfrak{p}$, and this conjugation will play a role in the proof.

Anyway, we use $\sigma = 1$ and $f_1 + f_2$ as Langlands parameters for a representation of the group M_1, which is essentially SL(2, $\mathbb{C}$). The corresponding representation of M_1 is trivial, hence unitary. The theorem says that our original $J(S_\mathfrak{p}, 1, \nu)$ is infinitesimally unitary.

Proof of theorem. For the positive system of roots of $(\mathfrak{g}, \mathfrak{a})$ defining N, Re ν is dominant by assumption. Let us introduce the positive system corresponding to some N' such that Im ν is dominant and such that the N'-positive roots orthogonal to Im ν are the same as the N-positive ones. The roots of $(\mathfrak{g}, \mathfrak{a})$ orthogonal to Im ν then define the M group M_1 of a parabolic subgroup $M_1A_1N_1$ containing $S' = MAN'$, by the same kind of construction as in the proof of Theorem 14.92. Since Im ν is assumed to be nonzero, M_1 is not all of G. Let $A_{M_1} = M_1 \cap A$ and $N_{M_1} = M_1 \cap N$, so that

$$S' \cap M_1 = MA_{M_1}N_{M_1}$$

is a parabolic subgroup of M_1. We shall define

$$\xi = J^{M_1}(S' \cap M_1, \sigma, \nu|_{\mathfrak{a}_{M_1}}).$$

(As usual, we ignore the technicality that we have defined Langlands quotients only for connected groups.)

Let us check that this definition of ξ makes sense. Let $\nu' = \nu|_{\mathfrak{a}_{M_1}}$. By Theorem 14.92 we are to check that $W^{M_1}_{\sigma,\nu'} = W'^{M_1}_{\sigma,\nu'}$. If p is an element of $W^G_{\sigma,\nu}$, then $p\nu = \nu$. Consequently $p(\text{Im } \nu) = \text{Im } \nu$. By Chevalley's Lemma, p is the product of reflections fixing Im ν, and the roots for these reflections must be orthogonal to Im ν. By construction these roots contribute to M_1. Therefore $W^G_{\sigma,\nu} = W^{M_1}_{\sigma,\nu'}$. Arguing in the same way as in the proof of Theorem 14.93, we see that $\Delta'^G_{\sigma,\nu} = \Delta'^{M_1}_{\sigma,\nu'}$ and hence that $W'^G_{\sigma,\nu} = W'^{M_1}_{\sigma,\nu'}$. Since the given existence of $J(S, \sigma, \nu)$ implies that $W^G_{\sigma,\nu} = W'^G_{\sigma,\nu}$, we conclude that $W^{M_1}_{\sigma,\nu'} = W'^{M_1}_{\sigma,\nu'}$ and hence that ξ is well defined.

The given Weyl group element s satisfies $s\nu = -\bar{\nu}$. Consequently $s(\text{Im } \nu) = \text{Im } \nu$, and the argument in the previous paragraph shows that s may be regarded as an element of $W(A_{M_1}:M_1)$. Then it follows from the remark attached to Theorem 16.6 that ξ has a nonzero invariant Hermitian form on the $(K \cap M_1)$-finite vectors.

We shall show that

$$J(S, \sigma, \nu) \cong \text{ind}^G_{S_1}(\xi \otimes \exp(\nu|_{\mathfrak{a}_1}) \otimes 1) \tag{16.17a}$$

and that

$$\nu|_{\mathfrak{a}_1} \text{ is imaginary,} \tag{16.17b}$$

and then application of Lemma 16.9 will complete the proof.

The key step in establishing (16.17a) is to show that $A(S':S:\sigma:\nu)$ is invertible on each K type. Proposition 14.13d says that it is enough to show that Im ν is not orthogonal to any root β of $(\mathfrak{g}, \mathfrak{a})$ that is positive for N and negative for N'. Assuming the contrary, suppose that β is a root with $\langle \text{Im } \nu, \beta \rangle = 0$. By construction the N'-positive roots orthogonal to Im ν are the same as the N-positive ones. Hence β cannot be both positive for N and negative for N'. Thus $A(S':S:\sigma:\nu)$ is invertible on each K type.

Consequently we have isomorphisms

$$\begin{aligned} U(S, \sigma, \nu) &\cong U(S', \sigma, \nu) \\ &= \text{ind}_{S'}^{G}(\sigma \otimes e^{\nu} \otimes 1) \\ &\cong \text{ind}_{S_1}^{G}[\text{ind}_{S \cap M_1}^{M_1}(\sigma \otimes \exp(\nu|_{\mathfrak{a}_{M_1}}) \otimes 1) \otimes \exp(\nu|_{\mathfrak{a}_1}) \otimes 1]. \qquad (16.18) \end{aligned}$$

The left side here has $J(S, \sigma, \nu)$ as a unique irreducible quotient, and the right side maps onto

$$\text{ind}_{S_1}^{G}[\xi \otimes \exp(\nu|_{\mathfrak{a}_1}) \otimes 1].$$

It is enough to show that the kernels of the two quotient mappings are isomorphic. The kernel for the quotient map with domain the left side of (16.18) is

$$\ker \mathscr{A}(s^{-1}Ss:S:\sigma:\nu) \qquad (16.19)$$

by the last part of the proof of Theorem 16.6. Similarly the kernel of the quotient map

$$\text{ind}_{S \cap M_1}^{M_1}(\sigma \otimes \exp(\nu|_{\mathfrak{a}_{M_1}}) \otimes 1) \to \xi$$

is $\ker \mathscr{A}(s^{-1}(S' \cap M_1)s:S' \cap M_1:\sigma:(\nu|_{\mathfrak{a}_{M_1}}))$. Since induction is an exact functor, the kernel of the quotient map with domain the right side of (16.18) is

$$\text{ind}[\ker \mathscr{A}(s^{-1}(S' \cap M_1)s:\sigma:(\nu|_{\mathfrak{a}_{M_1}})] = \ker \mathscr{A}(s^{-1}S's:S':\sigma:\nu).$$

In this identity we have used the fact that $s^{-1}N_1 s = N_1$. This kernel is isomorphic with (16.19) since $A(S':S:\sigma:\nu)$ is invertible on each K type. This proves (16.17a).

Finally we prove (16.17b). Since $s^2 = 1$, iterated application of Chevalley's Lemma shows that s can be written as the product $s = \prod s_{\alpha_i}$ of mutually orthogonal root reflections, with the number of reflections equal to the dimension of the -1 eigenspace of s. The roots α_i span the -1 eigenspace, and Re ν is in this space. Thus Re $\nu = \sum c_i \alpha_i$ for suitable constants c_i. Since the $+1$ and -1 eigenspaces are orthogonal, each α_i is orthogonal to the $+1$ eigenspace, which contains Im ν. Thus each α_i contributes to M_1 and vanishes on $\mathfrak{a}_1$. Hence the combination Re ν of the α_i's vanishes on $\mathfrak{a}_1$. Then (16.17b) follows, and the proof is complete.

§5. Problems

1. Section 6 of Chapter II shows how to realize the nonunitary principal series $\mathscr{P}^{+,z}$ for SU(1, 1) (which is isomorphic to SL(2, $\mathbb{R}$)) in L^2 of the circle. Find the complementary series norm in this realization.

Problems 2 to 4 address irreducibility of nonunitary principal series and existence of complementary series for groups of real rank one.

2. Using Theorem 14.92 and an "induction" on the $\mathfrak{a}$ parameter, prove in any group G of real rank one that $U(S, \sigma, \nu)$ can be reducible only when the infinitesimal character is integral. [Hint: See Problems 15 to 19 in Chapter VIII.]
3. If G has real rank one and $U(S, \sigma, \nu)$ has an infinitesimal character that is integral and is orthogonal to two nonproportional roots, prove that $U(S, \sigma, \nu)$ is irreducible. [Hint: For Re $\nu > 0$, prove that $A(\bar{S}:S:\sigma:\nu)$ has 0 kernel by considering $A(S:\bar{S}:\sigma:\nu)A(\bar{S}:S:\sigma:\nu)$.]
4. For $G = SO_0(n, 1)$ and $G = SU(n, 1)$ with $n \geq 2$, prove that the Langlands quotients $J(S, 1, \nu)$ of nonunitary principal series representations are infinitesimally unitary for all real-valued ν between 0 and $\rho_{\mathfrak{p}}$.

Problems 5 to 7 translate into information about Langlands parameters the fact that infinitesimally unitary implies bounded matrix coefficients. Fix a cuspidal parabolic subgroup $S = MAN$, a discrete series or nondegenerate limit of discrete series σ of M, and a member ν of $(\mathfrak{a}')^{\mathbb{C}}$ such that Re ν is in the closed positive Weyl chamber and $J(S, \sigma, \nu)$ is defined.

5. Unwind Theorem 14.92 and use Theorem 8.54 to prove that $J(S, \sigma, \nu)$ has bounded matrix coefficients if and only if $\exp[(\nu - \rho_A)(H)]$ is bounded for H in the closed positive Weyl chamber of $\mathfrak{a}$.
6. Let $S' = MAN'$ range through all possible parabolic subgroups obtained by using different notions of positive roots while keeping MA fixed. In each case let ρ_A' be the corresponding half sum of positive roots, counting multiplicities. Also imbed $\mathfrak{a} \subseteq \mathfrak{a}_{\mathfrak{p}}$ and choose a positive system of restricted roots such that the nonzero restrictions to $\mathfrak{a}$ of the positive restricted roots are N-positive. Let $M_{\mathfrak{p}}A_{\mathfrak{p}}N_{\mathfrak{p}}$ be the corresponding minimal parabolic subgroup, and let $\rho_{\mathfrak{p}}$ be the corresponding half-sum of positive restricted roots, counting multiplicities. Let $\rho_{\mathfrak{p}}''$ range through the corresponding half sums for all the minimal parabolic subgroups $M_{\mathfrak{p}}A_{\mathfrak{p}}N_{\mathfrak{p}}''$, including those not obtained by any kind of compatible ordering relative to some S'.
   (a) Show that Re ν is in the closed positive Weyl chamber of the dual of $\mathfrak{a}_{\mathfrak{p}}$.

(b) Show that if Re ν is in the closed convex hull of all the ρ'_A, then it is also in the closed convex hull of all $\rho''_{\mathfrak{p}}$.

(c) For $s \in W(A_{\mathfrak{p}}:G)$, show that $s\rho_{\mathfrak{p}} = \rho_{\mathfrak{p}} - \sum c_j(s)\alpha_j$, where all $c_j(s)$ are ≥ 0 and where $\{\alpha_j\}$ is the set of simple restricted roots.

(d) Show that Re ν in the closed convex hull of all $\rho''_{\mathfrak{p}}$ implies Re $\nu - \rho_{\mathfrak{p}}$ is a linear combination of simple restricted roots with negative coefficients.

(e) Conclude that if Re ν is in the closed convex hull of all the ρ'_A, then $\exp[(\nu - \rho_A)(H)]$ is bounded for H in the closed positive Weyl chamber of $\mathfrak{a}$.

7. In the notation of Problem 6, suppose that Re $\nu(H) \leq \rho_{\mathfrak{p}}(H)$ for all H in the closed positive Weyl chamber of $\mathfrak{a}_{\mathfrak{p}}$. Let l be a linear functional on $\mathfrak{a}'$ with $l(\rho'_A) \leq 1$ for all choices of ρ'_A, and extend l to $\mathfrak{a}'_{\mathfrak{p}}$ so as to be 0 on the orthogonal complement of $\mathfrak{a}'$.

(a) Write $l = \langle \cdot, \varepsilon_1 \rangle$ with ε_1 in $\mathfrak{a}'$. Choose a closed Weyl chamber of $\mathfrak{a}'$ containing ε_1, and let ρ'_A be the ρ for that chamber of $\mathfrak{a}'$. Extend the notion of positive root of $(\mathfrak{g}, \mathfrak{a})$ to a notion of positive restricted root so that nonzero restrictions to $\mathfrak{a}$ of positive restricted roots are positive roots of $(\mathfrak{g}, \mathfrak{a})$, and let $\rho'_{\mathfrak{p}}$ be the corresponding ρ for that chamber of $\mathfrak{a}'_{\mathfrak{p}}$. Choose s in $W(A_{\mathfrak{p}}:G)$ so that $s\rho'_{\mathfrak{p}} = \rho_{\mathfrak{p}}$. Put $\varepsilon = s\varepsilon_1$. Then justify the following chain of inequalities:

$$l(\text{Re } \nu) \leq \langle \text{Re } \nu, \varepsilon \rangle \leq \langle \rho_{\mathfrak{p}}, \varepsilon \rangle = \langle \rho'_{\mathfrak{p}}, \varepsilon_1 \rangle = \langle \rho'_A, \varepsilon_1 \rangle = l(\rho'_A) \leq 1.$$

(b) Conclude that if $\exp[(\nu - \rho_A)(H)]$ is bounded for H in the closed positive Weyl chamber of $\mathfrak{a}$, then Re ν is in the closed convex hull of all the ρ'_A.

APPENDIX A

Elementary Theory of Lie Groups

§1. Lie Algebras

(A.1) A Lie algebra $\mathfrak{g}$ is a vector space with a bilinear multiplication $[x, y]$ such that $[x, x] = 0$ for all x and such that the **Jacobi identity** holds:

$$[[x, y], z] + [[y, z], x] + [[z, x], y] = 0.$$

(A.2) Left multiplication in $\mathfrak{g}$ by x is denoted ad x. Then ad x is in End $\mathfrak{g}$, and the map ad: $\mathfrak{g} \to$ End $\mathfrak{g}$ is linear. In fact, each $L =$ ad x is a **derivation** in the sense that

$$L[x, y] = [x, Ly] + [Lx, y];$$

in the presence of the other defining properties of a Lie algebra, this fact is equivalent with the Jacobi identity. The set of derivations is denoted Der $\mathfrak{g}$.

(A.3) A **homomorphism** $\varphi: \mathfrak{g} \to \mathfrak{h}$ is a linear map carrying brackets to brackets. It is an **isomorphism** if it is one-one onto. If $\mathfrak{a}$ and $\mathfrak{b}$ are subsets of $\mathfrak{g}$, then $[\mathfrak{a}, \mathfrak{b}]$ denotes the span of the brackets of a member of $\mathfrak{a}$ with a member of $\mathfrak{b}$. A (Lie) **subalgebra** $\mathfrak{h}$ is a subspace with $[\mathfrak{h}, \mathfrak{h}] \subseteq \mathfrak{h}$. An **ideal** $\mathfrak{h}$ is a subspace with $[\mathfrak{h}, \mathfrak{g}] \subseteq \mathfrak{h}$. The kernel of a homomorphism is an ideal. Conversely the quotient vector space of a Lie algebra by an ideal is a Lie algebra in a natural way, and the quotient map is a homomorphism onto.

(A.4) If $\mathfrak{a}$ and $\mathfrak{b}$ are ideals in $\mathfrak{g}$, so are $\mathfrak{a} + \mathfrak{b}$, $\mathfrak{a} \cap \mathfrak{b}$, and $[\mathfrak{a}, \mathfrak{b}]$.

(A.5) The **commutator series** of $\mathfrak{g}$ is the nonincreasing sequence of ideals $\mathfrak{g}^j$ with $\mathfrak{g}^0 = \mathfrak{g}$ and $\mathfrak{g}^{j+1} = [\mathfrak{g}^j, \mathfrak{g}^j]$. The **lower central series** of $\mathfrak{g}$ is the nonincreasing sequence of ideals $\mathfrak{g}_j$ with $\mathfrak{g}_0 = \mathfrak{g}$ and $\mathfrak{g}_{j+1} = [\mathfrak{g}, \mathfrak{g}_j]$. We say $\mathfrak{g}$ is **solvable** if its commutator series ends in 0, **nilpotent** if its lower central series ends in 0, and **abelian** if $[\mathfrak{g}, \mathfrak{g}] = 0$. The center of a Lie algebra is an abelian ideal.

(A.6) Any subalgebra or quotient of a solvable Lie algebra is solvable. Conversely if $\mathfrak{a}$ is a solvable ideal in $\mathfrak{g}$ and if $\mathfrak{g}/\mathfrak{a}$ is solvable, then $\mathfrak{g}$ is solvable. Consequently if $\mathfrak{g}$ is a finite-dimensional Lie algebra, then there exists a unique solvable ideal $\mathfrak{r}$ in $\mathfrak{g}$ containing all solvable ideals of $\mathfrak{g}$; $\mathfrak{r}$ is called the **radical** of $\mathfrak{g}$ and is written $\mathfrak{r} =$ rad $\mathfrak{g}$.

(A.7) A finite-dimensional Lie algebra $\mathfrak{g}$ is **simple** if it is nonabelian and has no proper nonzero ideals. It is **semisimple** if rad $\mathfrak{g} = 0$, i.e., if $\mathfrak{g}$ has no nonzero solvable ideals. Simple implies semisimple, and simple implies $[\mathfrak{g}, \mathfrak{g}] = \mathfrak{g}$. For any finite-dimensional Lie algebra $\mathfrak{g}$, $\mathfrak{g}/\text{rad}\, \mathfrak{g}$ is semisimple.

(A.8) If $\mathfrak{a}$ and $\mathfrak{b}$ are Lie algebras, their constructive (external) direct sum $\mathfrak{g}$ as vector spaces becomes a Lie algebra with consistent multiplication within $\mathfrak{a}$ and $\mathfrak{b}$ and with $[\mathfrak{a}, \mathfrak{b}] = 0$. We say $\mathfrak{g}$ is the **direct sum** of $\mathfrak{a}$ and $\mathfrak{b}$ and write $\mathfrak{g} = \mathfrak{a} \oplus \mathfrak{b}$. Here $\mathfrak{a}$ and $\mathfrak{b}$ are ideals in $\mathfrak{g}$. Conversely if $\mathfrak{g}$ is given and if $\mathfrak{a}$ and $\mathfrak{b}$ are ideals in $\mathfrak{g}$ such that $\mathfrak{g}$ is the (internal) direct sum of $\mathfrak{a}$ and $\mathfrak{b}$ as vector spaces, then $\mathfrak{g}$ is isomorphic with the Lie algebra direct sum of $\mathfrak{a}$ and $\mathfrak{b}$, and it is customary to say that $\mathfrak{g}$ *is* the direct sum of $\mathfrak{a}$ and $\mathfrak{b}$.

(A.9) If $\mathfrak{g}$ is a Lie algebra, then Der $\mathfrak{g}$ is a Lie algebra, and $\text{ad}: \mathfrak{g} \to \text{Der}\, \mathfrak{g} \subseteq \text{End}\, \mathfrak{g}$ is a homomorphism. If $\mathfrak{g} = \mathfrak{a} \oplus \mathfrak{b}$ as vector spaces and $\mathfrak{a}$ is a subalgebra of $\mathfrak{g}$ and $\mathfrak{b}$ is an ideal, then each ad a (for a in $\mathfrak{a}$) leaves $\mathfrak{b}$ stable and ad $a|_{\mathfrak{b}}$ is in Der $\mathfrak{b}$. Hence $\pi(a) = \text{ad}\, a|_{\mathfrak{b}}$ defines a homomorphism from $\mathfrak{a}$ to Der $\mathfrak{b}$. Then $\mathfrak{a}$, $\mathfrak{b}$, and π determine the bracket structure of $\mathfrak{g}$. We say $\mathfrak{g} = \mathfrak{a} \oplus_\pi \mathfrak{b}$ is a **semidirect product.**

(A.10) Conversely if $\mathfrak{a}$ and $\mathfrak{b}$ are Lie algebras and π is a homomorphism from $\mathfrak{a}$ into Der $\mathfrak{b}$, then there exists a unique Lie algebra structure on the vector space direct sum $\mathfrak{g} = \mathfrak{a} \oplus \mathfrak{b}$ retaining the given brackets within $\mathfrak{a}$ and $\mathfrak{b}$ and satisfying $[a, b] = \pi(a)b$ for $a \in \mathfrak{a}$ and $b \in \mathfrak{b}$. In $\mathfrak{g}$, $\mathfrak{a}$ is a subalgebra and $\mathfrak{b}$ is an ideal.

§2. Structure Theory of Lie Algebras

Henceforth $\mathfrak{g}$ will denote a finite-dimensional Lie algebra, and the underlying field will be assumed to have characteristic 0.

(A.11) If $\mathfrak{g}$ is n-dimensional, then $\mathfrak{g}$ is solvable if and only if there exists a sequence of subalgebras

$$\mathfrak{g} = \mathfrak{a}_0 \supset \mathfrak{a}_1 \supset \mathfrak{a}_2 \supset \ldots \supset \mathfrak{a}_n = 0$$

such that, for each i, $\mathfrak{a}_{i+1}$ is an ideal in $\mathfrak{a}_i$ and $\dim(\mathfrak{a}_i/\mathfrak{a}_{i+1}) = 1$.

(A.12) **Lie's Theorem.** Let $\mathfrak{g}$ be solvable, let V be a finite-dimensional vector space, and let $\pi: \mathfrak{g} \to \text{End}\, V$ be a homomorphism. If the underlying field is algebraically closed, then there exists a simultaneous eigenvector $v \neq 0$ for all the members of $\pi(\mathfrak{g})$. More generally even if the field is not algebraically closed, there exists a simultaneous eigenvector if the characteristic polynomials of all $\pi(x)$, for x in $\mathfrak{g}$, split over the given field.

(A.13) Under the assumptions on $\mathfrak{g}$, V, π, and the field in (A.12), there exists a sequence of subspaces

$$V = V_0 \supset V_1 \supset V_2 \supset \ldots \supset V_m = 0$$

such that each V_i is stable under $\pi(\mathfrak{g})$ and $\dim(V_i/V_{i+1}) = 1$. Consequently V has a basis with respect to which all the matrices of $\pi(\mathfrak{g})$ are upper triangular.

(A.14) If $\mathfrak{g}$ is nilpotent and the member $\mathfrak{g}_k$ of the lower central series is 0, then $(\operatorname{ad} x)^k = 0$ for all x in $\mathfrak{g}$. Hence ad x is a nilpotent endomorphism.

(A.15) If ad $\mathfrak{g}$ is a nilpotent Lie algebra, then $\mathfrak{g}$ is a nilpotent Lie algebra.

(A.16) **Engel's Theorem.** Let V be a finite-dimensional vector space, and let $\mathfrak{g}$ be a Lie algebra of nilpotent endomorphisms of V. Then

(1) $\mathfrak{g}$ is a nilpotent Lie algebra
(2) there exists $v \neq 0$ in V with $x(v) = 0$ for all x in $\mathfrak{g}$
(3) in a suitable basis of V, all x are upper triangular with 0 on the diagonal.

(A.17) If $\mathfrak{g}$ is a Lie algebra such that each ad x, for x in $\mathfrak{g}$, is nilpotent, then $\mathfrak{g}$ is a nilpotent Lie algebra.

(A.18) Let $C(u, v)$ be a bilinear form (not necessarily symmetric) on the finite-dimensional vector space V. Define a one-sided radical for C by

$$\operatorname{rad} C = \{v \in V \mid C(v, u) = 0 \text{ for all } u \in V\}.$$

Next define $\varphi: V \to V'$ (with V' the dual of V) by

$$\langle \varphi(v), u \rangle = C(v, u).$$

Then $\ker \varphi = \operatorname{rad} C$, and so φ is an isomorphism (onto) if and only if C is **nondegenerate** (i.e., $\operatorname{rad} C = 0$). If U is a subspace of V, let

$$U^\perp = \{v \in V \mid C(v, u) = 0 \text{ for all } u \in U\}.$$

Then $U \cap U^\perp = \operatorname{rad}(C|_{U \times U})$. Hence even if C is nondegenerate, it can happen that $U \cap U^\perp \neq 0$.

(A.19) Let $C(u, v)$ be a bilinear form on the finite-dimensional vector space V, and suppose C is nondegenerate. If U is a subspace of V, then

(1) $\dim U + \dim U^\perp = \dim V$
(2) $V = U \oplus U^\perp$ if and only if $C|_{U \times U}$ is nondegenerate.

(A.20) The **Killing form** $B(x,y)$ on $\mathfrak{g}$ is the symmetric bilinear form given by $B(x, y) = \operatorname{Tr}(\operatorname{ad} x \operatorname{ad} y)$. Each ad x, for x in $\mathfrak{g}$, is skew-symmetric with respect to B.

(A.21) **Cartan's criterion for solvability.** The Lie algebra $\mathfrak{g}$ is solvable if and only if $B(x, y) = 0$ for all x in $\mathfrak{g}$ and all y in $[\mathfrak{g}, \mathfrak{g}]$.

(A.22) For any $\mathfrak{g}$, $\operatorname{rad} B \subseteq \operatorname{rad} \mathfrak{g}$.

(A.23) **Cartan's criterion for semisimplicity.** The Lie algebra $\mathfrak{g}$ is semisimple if and only if the Killing form for $\mathfrak{g}$ is nondegenerate.

(A.24) The Lie algebra $\mathfrak{g}$ is semisimple if and only if $\mathfrak{g} = \mathfrak{g}_1 \oplus \ldots \oplus \mathfrak{g}_n$ with the $\mathfrak{g}_j$'s ideals that are each simple Lie algebras. In this case, the decomposition is unique apart from permutation of the summands, and the only ideals of $\mathfrak{g}$ are the sums of the various $\mathfrak{g}_j$'s.

(A.25) If $\mathfrak{g}$ is semisimple, then $[\mathfrak{g}, \mathfrak{g}] = \mathfrak{g}$. If $\mathfrak{a}$ is any ideal in $\mathfrak{g}$, then $\mathfrak{a}^\perp$ (relative to the Killing form) is an ideal and $\mathfrak{g} = \mathfrak{a} \oplus \mathfrak{a}^\perp$.

§3. Fundamental Group and Covering Spaces

In this section X will denote a separable metric space, i.e, a regular Hausdorff space with a countable base. Shortly X will be assumed to have certain connectedness properties.

(A.26) A **path** in X is a continuous function $a:[0, \|a\|] \to X$. A **loop** is a path with $a(0) = a(\|a\|)$, and $a(0)$ is the **base point** of the loop. An **identity path** is a path with $\|a\| = 0$. A **constant path** is a path with $a(t) = a(0)$ for all t. The **inverse path** a^{-1} to a is the path with

$$a^{-1}(t) = a(\|a\| - t) \qquad \text{for} \quad 0 \le t \le \|a\|.$$

(A.27) If a and b are paths with $a(\|a\|) = b(0)$, then the **product** $c = a \cdot b$ is defined by

$$c(t) = \begin{cases} a(t) & \text{for} \quad 0 \le t \le \|a\| \\ b(t - \|a\|) & \text{for} \quad \|a\| \le t \le \|a\| + \|b\|. \end{cases}$$

The product has $\|c\| = \|a\| + \|b\|$.

(A.28) If $a \cdot b$ and $b \cdot c$ are defined, then $(a \cdot b) \cdot c$ and $a \cdot (b \cdot c)$ are defined and $(a \cdot b) \cdot c = a \cdot (b \cdot c)$. If i is an identity path, then $i \cdot a = a$ whenever $i \cdot a$ is defined and $b \cdot i = b$ whenever $b \cdot i$ is defined. The products $a \cdot a^{-1}$ and $a^{-1} \cdot a$ are always defined.

(A.29) Two paths a and b with $a(0) = b(0)$ and $a(\|a\|) = b(\|b\|)$ are **equivalent**, written $a \simeq b$, if there are continuous functions $s:[0, 1] \to [0, \infty)$ and

$$h:\{(u, t) \mid u \in [0, 1] \text{ and } t \in [0, s(u)]\} \to X$$

such that

$$h(0, t) = a(t), \qquad h(u, 0) = a(0) = b(0),$$

$$h(1, t) = b(t), \qquad h(u, s(u)) = a(\|a\|) = b(\|b\|).$$

"Equivalent" is an equivalence relation, and a typical class is denoted $[a]$.

(A.30) If $a \simeq a'$ and $b \simeq b'$ and if $a \cdot b$ is defined, then $a' \cdot b'$ is defined and $a \cdot b \simeq a' \cdot b'$. Also $a^{-1} \simeq (a')^{-1}$. Constant paths are equivalent with identity paths. The products $a \cdot a^{-1}$ and $a^{-1} \cdot a$ are equivalent with constant paths. Therefore **class multiplication** is well defined under the rules

$$[a][b] = [ab] \qquad \text{and} \qquad [a]^{-1} = [a^{-1}].$$

(A.31) If $[a][b]$ and $[b][c]$ are defined, then $([a][b])[c]$ and $[a]([b][c])$ are defined and equal. If 1 is the identity path $a(0)$, then $[1][a] = [a]$ and $[a][a]^{-1} = [1]$. If 1 is the identity path $a(\|a\|)$, then $[a][1] = [a]$ and $[a]^{-1}[a] = [1]$.

(A.32) For fixed p in X, the set of classes of loops in X with base point p is a group under class multiplication. This group is called the **fundamental group** of X with base point p and is denoted $\pi(X, p)$.

(A.33) If p and q are in X, then any path c from p to q canonically defines an isomorphism of $\pi(X, p)$ onto $\pi(X, q)$.

(A.34) If X is pathwise connected, then (A.33) implies $\pi(X, p)$ as an abstract group is independent of p. We say a pathwise connected X is **simply connected** if $\pi(X) = \{1\}$.

(A.35) Let $f: X \to Y$ be continuous between two separable metric spaces. Then f induces a well-defined homomorphism

$$f_*: \pi(X, p) \to \pi(Y, f(p))$$

by $f_*([a]) = [f \circ a]$. The induced homomorphism has the property $(f \circ g)_* = f_* g_*$.

For the remainder of this section, we shall assume X and Y are separable metric spaces that are pathwise connected and locally pathwise connected.

(A.36) Let $e: X \to Y$ be continuous, and let V be open in Y. We say V is **evenly covered** by e if each connected component of $e^{-1}(V)$ is mapped by e homeomorphically onto V. We say e is a **covering map** if each y in Y has an open neighborhood V_y that is evenly covered by e; in this case Y is called the **base space** and X is the **covering space.**

(A.37) **Path-lifting theorem.** Suppose $e: X \to Y$ is a covering map. If $y(t)$, $0 \le t \le 1$, is a path in Y and if x_0 is in $e^{-1}(y(0))$, then there exists a unique path $x(t)$, $0 \le t \le 1$, in X with $x(0) = x_0$ and $e(x(t)) = y(t)$.

(A.38) Let $e: X \to Y$ be a covering map and let V be an open subset of Y that is evenly covered. Then the components of $e^{-1}(V)$ are open.

(A.39) **Covering homotopy theorem.** Let $e: X \to Y$ be a covering map, let K be a compact space, and let $f_0: K \to X$ be continuous. If $g: K \times [0, 1] \to Y$ is continuous and satisfies $g(\cdot, 0) = ef_0$, then there is a unique continuous $f: K \times [0, 1] \to X$ such that $f(\cdot, 0) = f_0$ and $g = ef$.

(A.40) If $e: X \to Y$ is a covering map, then a path $x(t)$ in X is a contractible loop (i.e., a loop equivalent with a constant map) if and only if the projected $ex(t)$ is a contractible loop. Consequently e_* is one-one.

(A.41) If $e:X \to Y$ is a covering map, if y_0 is in Y, and if x_0 is in $e^{-1}(y_0)$, then the lift of a loop $y(t)$ based at y_0 is a loop if and only if $[y(t)]$ is in $e_*\pi(X, x_0)$.

(A.42) **Map-lifting theorem.** If $e:X \to Y$ is a covering map, if P is a pathwise connected, locally pathwise connected, separable metric space, if $g:P \to Y$ is continuous, and if p_0 is in $g^{-1}(y_0)$ and x_0 is in $e^{-1}(y_0)$, then there exists a continuous $f:P \to X$ with $f(p_0) = x_0$ and $g = ef$ if and only if $g_*(\pi(P, p_0)) \subseteq e_*(\pi(X, x_0))$. When f exists, it is unique.

(A.43) Let $e:X \to Y$ and $e':X' \to Y$ be coverings. We say e and e' are **equivalent coverings** if there is a homeomorphism ι of X onto X' such that $e'\iota = e$.

(A.44) **Main uniqueness theorem.** Let $e:X \to Y$ and $e':X' \to Y$ be covering maps, and let y_0 be in Y. Then e and e' are equivalent coverings if and only if base points x_0 in X and x_0' in X' can be chosen so that $e(x_0) = e'(x_0') = y$ and $e_*(\pi(X, x_0)) = e_*'(\pi(X', x_0'))$.

(A.45) If e: $X \to Y$ is a covering, then any pathwise connected open subset Q of Y such that any loop in Q is contractible in Y is evenly covered.

(A.46) Y is said to be **locally simply connected** if each y in Y has an open pathwise connected, simply connected neighborhood. In this case, each y in Y has arbitrarily small open pathwise connected neighborhoods Q such that any loop in Q is contractible in Y, and (A.45) applies. Moreover, for such a space Y, $\pi(Y, y_0)$ is countable.

(A.47) **Main existence theorem.** If Y is locally simply connected, if y_0 is in Y, and if H is a subgroup of $\pi(Y, y_0)$, then there exists a covering space X with covering map $e:X \to Y$ and with point x_0 in X such that $e(x_0) = y_0$ and $e_*(\pi(X, x_0)) = H$.

(A.48) By (A.44) and (A.47), if Y is locally simply connected, then Y has a simply connected covering space that is unique up to equivalence. This space is called the **universal covering space** of Y.

The next few items establish formulas for specific fundamental groups.

(A.49) $\mathbb{R}^n$ is simply connected.

(A.50) If X and Y are pathwise connected separable metric spaces with $x_0 \in X$ and $y_0 \in Y$, then $\pi(X \times Y, (x_0, y_0))$ is canonically isomorphic with $\pi(X, x_0) \times \pi(Y, y_0)$.

(A.51) Let S^1 be the circle $S^1 = \{z \in \mathbb{C} \mid |z| = 1\}$. Let $e:\mathbb{R}^1 \to S^1$ be the map $e(x) = e^{ix}$. Then e is a covering map, $\pi(S^1) = \mathbb{Z}$, and $y(t)$, $0 \le t \le 2\pi$, is a generator of $\pi(S^1)$.

(A.52) Let X be pathwise connected and locally pathwise connected. If there exist connected, simply connected open subsets X_1 and X_2 of X with $X = X_1 \cup X_2$ and with $X_1 \cap X_2$ connected, then X is simply connected.

(A.53) Let X be the universal covering space of a pathwise connected, locally pathwise connected, locally simply connceted, separable metric space Y, and let $e: X \to Y$ be the covering map. A **deck transformation** of X is a homeomorphism f of X that satisfies $ef = e$.

(A.54) Let $e: X \to Y$ be a universal covering map, and let $e(x_0) = y_0$. Then

(1) $\pi(Y, y_0)$ is in one-one correspondence with $e^{-1}(y_0)$, the correspondence being that x_1 in $e^{-1}(y_0)$ corresponds to

$$[e(\text{any path from } x_0 \text{ to } x_1)].$$

(2) the group of deck transformations H of X acts simply transitively on $e^{-1}(y_0)$.

(3) the correspondence that associates to a deck transformation f in H the member of $\pi(Y, y_0)$ corresponding to $f(x_0)$ is a group isomorphism of H onto $\pi(Y, y_0)$.

§4. Topological Groups

(A.55) A **topological group** is a Hausdorff space that is a group such that multiplication and inversion are continuous. Such a group G is regular as a topological space, and left and right translations are homeomorphisms.

(A.56) A topological group G has the following properties:

(1) Let H be a closed subgroup, and let G/H have the quotient topology. Then the projection of G onto G/H is open, and G/H is a Hausdorff regular space such that the action of G on G/H is jointly continuous. If G has a countable base, so does G/H.

(2) Let H be a closed subgroup. If H and G/H are connected, then G is connected.

(3) Let H be a closed subgroup. If H and G/H are compact, then G is compact.

(4) If H is a closed normal subgroup, then G/H is a topological group.

(5) Any open subgroup is closed.

(6) The identity component G_0 of 1 in G is a closed normal subgroup.

(7) If G is connected, then any neighborhood of 1 generates G.

(8) If H is a **discrete subgroup** of G (i.e., if every subset of H is relatively open), then H is a closed subgroup.

(9) If G is connected, then any discrete normal subgroup of G lies in the center of G.

For the next nine items, G will denote a pathwise connected, locally pathwise connected, separable metric topological group, and H will denote a closed subgroup of G that is locally pathwise connected but not necessarily connected.

(A.57) The quotient G/H is pathwise connected and locally pathwise connected.

(A.58) If H_0 is the identity component of H, then the natural map of G/H_0 onto G/H given by $gH_0 \to gH$ is a covering map.

(A.59) If G/H is simply connected, then H is connected.

(A.60) If H is discrete, then the quotient map of G onto G/H is a covering map.

(A.61) Suppose G is simply connected and H is discrete, so that the quotient map $G \to G/H$ is a covering map. Then the group of deck transformations of G is exactly the group of right translations in G by members of H. Consequently $\pi(G/H, 1 \cdot H)$ is canonically isomorphic with H.

(A.62) Let G be locally simply connected, let $\tilde{G}$ be the universal covering space with covering map $e:\tilde{G} \to G$, and let $\tilde{1}$ be in $e^{-1}(1)$. Then there exists a unique multiplication on $\tilde{G}$ that makes $\tilde{G}$ into a topological group in such a way that e is a group homomorphism and $\tilde{G}$ has $\tilde{1}$ as identity. The group $\tilde{G}$ is called the **universal covering group** of G.

(A.63) If G/H is simply connected and if G and H are locally simply connected, then $\pi(G, 1)$ is isomorphic to a quotient group of $\pi(H, 1)$.

(A.64) Let $a(t)$ and $b(t)$ be two loops based at 1 with $\|a\| = \|b\|$. Then the loop product $a \cdot b$ (with the loops arranged end to end) is equivalent with the pointwise group product $a(t)b(t)$ (with the multiplication taken in G). Consequently multiplication in $\pi(G, 1)$ coincides with pointwise multiplication of loops in G.

(A.65) $\pi(G, 1)$ is abelian.

(A.66) If G and H are topological groups whose topologies each are locally compact and have a countable base and if $\varphi:G \to H$ is a one-one continuous homomorphism of G onto H, then φ^{-1} is continuous.

§5. Vector Fields and Submanifolds

This section uses standard terminology for manifolds and functions between them. The kinds of manifolds of interest in Lie theory are C^∞, real analytic, and complex analytic. The word "smooth" will be used interchangeably with "C^∞." The notation $C^\infty(U)$ refers to the real-valued smooth functions on U.

(A.67) Let M be a smooth manifold and let p be in M. The space $\mathscr{C}_p(M)$ of **germs** of smooth functions at p is defined as follows: Let $\tilde{\mathscr{C}}_p(M)$ be the union of all $C^\infty(U)$ for U open with $p \in U \subseteq M$. If p lies in $U \cap V$ and if f is in $C^\infty(U)$ and g is in $C^\infty(V)$, then $f + G$ and fg are defined on $U \cap V$. We say f and g are equivalent in $\tilde{\mathscr{C}}_p(M)$ if $f = g$ in a neighborhood of p.

Then $\mathscr{C}_p(M)$ is the set of equivalence classes, and the above definitions of addition and multiplication pass to $\mathscr{C}_p(M)$ and make $\mathscr{C}_p(M)$ an algebra. Evaluation at p is a well-defined multiplicative linear functional on $\mathscr{C}_p(M)$.

(A.68) The **tangent space** $T_p(M)$ of the smooth manifold M at the point p is defined to be the real vector space of all linear functionals L on $\mathscr{C}_p(M)$ satsifying

$$L(\alpha\beta) = \alpha(p)L(\beta) + \beta(p)L(\alpha) \qquad \text{for } \alpha, \beta \in \mathscr{C}_p(M).$$

(A.69) If M is a smooth manifold of dimension n and p is in M, then $\dim_{\mathbb{R}} T_p(M) = n$. Let (U, φ) be a chart with p in U and with φ given by

$$\varphi(q) = (x^1(q), \ldots, x^n(q)), \qquad q \in U.$$

Then the tangent vectors $f \to \left[\dfrac{\partial f}{\partial x^i}\right]_p$ defined as

$$\left[\frac{\partial(f \circ \varphi^{-1})}{\partial u^i}\right]_{\{u^j = x^j(p)\}}$$

form a basis of $T_p(M)$, and

$$L(f) = \sum_{i=1}^n L(x^i)\frac{\partial f}{\partial x^i} \qquad \text{for } L \text{ in } T_p(M).$$

(A.70) For M a smooth manifold, let

$$T(M) = \{(p, L) \,|\, p \in M \text{ and } L \in T_p(M)\},$$

and let γ be the projection of $T(M)$ on M. A **vector field** X on M is a function from M to $T(M)$ such that $\gamma \circ X = \text{identity}$. We denote by X_p the value of X at p, which is in $T_p(M)$. The vector field X is said to be **smooth** if, in acting on $C^\infty(M)$ by $(Xf)(p) = X_p f$, it carries smooth functions to smooth functions. In any case, the expression of X in a chart (U, φ) with $\varphi = (x^1, \ldots, x^n)$ is

$$Xf(p) = \sum \frac{\partial f}{\partial x^i}(p)Xx^i(p) \qquad \text{for } p \in U,$$

and X is smooth if and only if Xx^i is smooth for each coordinate function of each chart on M. If X is smooth and f is in $C^\infty(U)$ with U open in M, then Xf is in $C^\infty(U)$.

(A.71) If M is a smooth manifold and X and Y are smooth vector fields on M, then $[X, Y] = XY - YX$ is also a smooth vector field. The vector space of smooth vector fields becomes a Lie algebra over $\mathbb{R}$ with this definition of bracket. It is also a $C^\infty(M)$ module under the definition $(fX)(g) = f(Xg)$.

(A.72) The **differential** $d\Phi_p$ at p of a smooth mapping $\Phi: M \to N$ is defined by

$$d\Phi_p(L)(g) = L(g \circ \Phi) \qquad \text{for } L \in T_p(M),\, g \in \mathscr{C}_{\Phi(p)}(N).$$

It is a linear map from $T_p(M)$ to $T_{\Phi(p)}(N)$. With respect to the bases $\left[\dfrac{\partial}{\partial x^j}\right]_p$ and $\left[\dfrac{\partial}{\partial y^i}\right]_{\Phi(p)}$, its matrix is

$$\left[\frac{\partial \Phi^i}{\partial u^j}\right]_{\{u^k = x^k(p)\}}.$$

It satisfies the composition rule

$$d(\Psi \circ \Phi)_p = d\Psi_{\Phi(p)} \circ d\Phi_p.$$

(A.73) A smooth mapping $\Phi: M \to N$ is said to be **regular** at the point p of M if $d\Phi_p$ is one-one.

(A.74) Let $\Phi: M \to N$ be a smooth mapping, and let X and Y be smooth vector fields on M and N, respectively. We say X and Y are **Φ-related** if

$$d\Phi_p(X_p) = Y_{\Phi(p)} \qquad \text{for all } p \in M.$$

(A.75) Let $\Phi: M \to N$ be an everywhere regular smooth mapping. For p in M, let $\tilde{T}_p = d\Phi_p(T_p)$. If Y is a smooth vector field on N, then there exists a smooth vector field X on M that is Φ-related to Y if and only if $Y_{\Phi(p)}$ is in $\tilde{T}_p$ for all p in M. In this case, X is unique.

(A.76) Let $\Phi: M \to N$ be a smooth mapping, let X_1 and X_2 be smooth vector fields on M, and let Y_1 and Y_2 be smooth vector fields on N. If X_1 and Y_1 are Φ-related and if X_2 and Y_2 are Φ-related, then $[X_1, X_2]$ and $[Y_1, Y_2]$ are Φ-related.

(A.77) A smooth manifold with a countable base satisfies all the hypotheses in the existence theorem (A.47) for covering spaces.

(A.78) Let N be a smooth manifold of dimension n, and suppose N has a countable base. Let M be a covering space with covering map $e: M \to N$. Then M uniquely admits the structure of a smooth manifold of dimension n in such a way that e is smooth and everywhere regular. Moreover, if P is another smooth mainfold, if $g: P \to N$ is smooth, and if $f: P \to M$ is a continuous lift of g with $ef = g$, then f is smooth.

(A.79) Let M be a smooth mainfold. A manifold S is a **submanifold** of M if

(1) $S \subseteq M$ as a set, and
(2) the inclusion mapping of S into M is smooth and everywhere regular.

In this case the inclusion is necessarily continuous, but S is not necessarily homeomorphic with its image in M.

(A.80) Let M be a smooth manifold of dimension m, and let S be a submanifold of dimension s. Then S has the following properties:

(1) Let (U, φ) be an M chart at a point p of S, say $\varphi = (y^1, \ldots, y^m)$. Then the restrictions to S of some subset of s of the functions y^j define an S chart near p.
(2) Let (U, φ) be an S chart at a point p of S, say $\varphi = (x^1, \ldots, x^s)$. Then there is an M chart $(z^1, \ldots, z^m)$ at p such that x^j near p is the restriction to S of z^j for $1 \leq j \leq s$.
(3) If $U \subseteq M$ is an open set and $f: U \to N$ is smooth, then the restriction of f to $U \cap S$ is smooth into N. If $V \subseteq S$ is open and p is in V and $f: V \to N$ is smooth, then in a small enough S neighborhood of p, f is the restriction of a smooth function from M into N.
(4) If the dimension of S is m, then S is an **open submanifold** (open set with the inherited manifold structure).
(5) Let $\iota: S \to M$ be the inclusion, and identify $S_p = d\iota_p(T_p(S))$ with the tangent space to S at p. Let X be a smooth vector field on M such that X_p is in S_p for all p in S. If Y denotes the smooth vector field on S given by (A.75) such that $d\iota_p(Y_p) = X_p$ for all p in S, then Y is called the **contraction** of X to S. If X_1 and X_2 have contractions Y_1 and Y_2, respectively, then $[Y_1, Y_2]$ is the contraction of $[X_1, X_2]$, by (A.76).

(A.81) Let M be a smooth manifold. A **curve** $c(t)$ on M is a continuous function from an open interval of $\mathbb{R}^1$ into M. Let X be a smooth vector field on M. A curve $c(t)$ on M is an **integral curve** for X if $c(t)$ is smooth and if $X_{c(t)} = dc_t\left(\frac{d}{dt}\right)$ for all t in the domain of c.

(A.82) **Existence theorem for integral curves.** Let M be a smooth manifold of dimension m, and let p be in M.

(1) If X is a smooth vector field on M, then there exists an $\varepsilon > 0$ and there exists an integral curve $c(t)$ defined for $-\varepsilon < t < \varepsilon$ such that $c(0) = p$. The integral curve is unique on any interval about 0 where it exists. For small t, the curve is a submanifold if $X_p \neq 0$.
(2) Let $X_1, \ldots, X_n$ be smooth vector fields on M, and let V be a bounded open subset of 0 in $\mathbb{R}^n$. For λ in V, put $X_\lambda = \sum_{j=1}^n \lambda^j X_j$. Then there exists an $\varepsilon > 0$ and there exists a system of integral curves $c(t, \lambda)$, defined for $-\varepsilon < t < \varepsilon$ and λ in V, such that $c(\cdot, \lambda)$ is an integral curve for X_λ with $c(0, \lambda) = p$. Each curve $c(t, \lambda)$ is unique, and the map $c:(-\varepsilon, \varepsilon) \times V \to M$ is smooth. If $m = n$, if the vectors $(X_1)_p, \ldots, (X_m)_p$ are linearly independent, and if δ is any positive number $< \varepsilon$, then $c(\delta, \cdot)$ is a diffeomorphism from an open subneighborhood of 0 (depending on δ) onto an open subset of M, and its inverse defines a chart at p.

(3) If X is a smooth vector field with $X_p \neq 0$, then there is a cubical neighborhood U of p of side some 2ε (i.e., $\varphi(x^1, \ldots, x^m)$ maps U onto a cube of side 2ε centered at $\varphi(p) = 0$) such that the curves

$$c_u(t) = \varphi^{-1}(t, u^2, \ldots, u^m), \quad -\varepsilon < t < \varepsilon, \quad |u^j| < \varepsilon \text{ for } 2 \leq j \leq m,$$

are integral curves for X.

(A.83) Let M be a smooth manifold. An s-**distribution** $\mathscr{S}$ on M is a function whose value at p in M is $(p, \mathscr{S}_p)$, where $\mathscr{S}_p$ is an s-dimensional subspace of $T_p(M)$. $\mathscr{S}$ is **smooth** if to each p in M correspond an open neighborhood V and a system $X^1, \ldots, X^s$ of smooth vector fields whose values at each q in V form a basis of $\mathscr{S}_q$. The system $X^1, \ldots, X^s$ is called a **local base** for $\mathscr{S}$ around p. A smooth vector field X is said to **belong to** $\mathscr{S}$ if X_p is in $\mathscr{S}_p$ for all p.

(A.84) Let $\mathscr{S}$ be a smooth s-distribution on a smooth manifold M. A submanifold S of M is called an **integral submanifold** for $\mathscr{S}$ if, for each p in S, $\mathscr{S}_p$ coincides with the tangent space to S at p (realized as a subspace of $T_p(M)$). In this case if X_1 and X_2 are smooth vector fields that belong to $\mathscr{S}$, then $[X_1, X_2]$ belongs to $\mathscr{S}$ at points of S, by (A.80.5). A smooth s-distribution $\mathscr{S}$ on M is **involutive** if the set of smooth vector fields that belong to $\mathscr{S}$ is closed under bracket.

(A.85) Let Σ be a set of smooth vector fields on M such that

(1) the span $\mathscr{S}_p$ of the vectors X_p with X in Σ has dimension s for each p in M.
(2) whenever X and Y are in Σ, then $[X, Y]$ is a finite linear combination of the members of Σ with smooth functions as coefficients.

Then $\mathscr{S} = \{(p, \mathscr{S}_p)\}$ is a smooth involutive s-distribution on M.

(A.86) **Local Frobenius theorem.** Let $\mathscr{S}$ be a smooth involutive s-distribution on a smooth manifold M of dimension m. If p is in M, then there exists a cubical neighborhood U of p of side some 2ε such that whenever $\xi^{s+1}, \ldots, \xi^m$ are $m - s$ numbers with $|\xi^{s+h}| < \varepsilon$ for $1 \leq h \leq m - s$, then the slice of U defined by the equations $x^{s+h}(q) = \xi^{s+h}$ for $1 \leq h \leq m - s$ is an integral submanifold of M.

(A.87) Let $\mathscr{S}$ be a smooth involutive s-distribution on the smooth manifold M. Then M decomposes (uniquely) as a disjoint union $M = \bigcup_{\alpha \in \Gamma} S_\alpha$ such that each S_α is an integral submanifold of M and is maximal in the sense of being contained in no larger integral submanifold. If S is any integral submanifold of M such that $S \cap S_\alpha \neq 0$, then $S \subseteq S_\alpha$ and S is an open submanifold of S_α. (Consequently the differentiable structure on S is uniquely determined.) Finally if M has a countable base, then so does each S_α; in this case, if $f: N \to M$ is a smooth map such that $f(N) \subseteq S_\alpha$, then $f: N \to S_\alpha$ is smooth.

§6. Lie Groups

(A.88) An **analytic group** G is a topological group with the structure of a smooth manifold such that multiplication is smooth from $G \times G$ to G and inversion is smooth from G to G. Such a group is connected and locally compact, it is generated by any compact neighborhood of 1, and it has a countable base.

(A.89) A **Lie group** is a locally connected topological group with a countable base such that the identity component (which has to be open) is an analytic group.

(A.90) Let G be an analytic group, and let $L_x: G \to G$ be left translation by $x: L_x(y) = xy$. L_x is a diffeomorphism. A vector field X on G is **left-invariant** if X is L_x-related to itself for all x. Equivalently X, as an operator on functions, commutes with left translations.

(A.91) If G is an analytic group, then the map $X \to X_1$ is an isomorphism of the vector space of left-invariant vector fields on G onto $T_1(G)$, and the inverse map is $Xf(x) = X_1(L_x f)$, where $L_x f(y) = f(xy)$. Every left-invariant vector field on G is smooth, and the bracket of two left-invariant vector fields is left-invariant.

(A.92) If G is an analytic group, set $\mathfrak{g} = T_1(G)$. Then $\mathfrak{g}$ becomes a Lie algebra over $\mathbb{R}$ by virtue of (A.91) and is called the **Lie algebra** of G. The Lie algebra of a Lie group is the Lie algebra of the identity component.

(A.93) Let $\mathfrak{g}$ be the Lie algebra of $GL(n, \mathbb{R})$ and let e_{ij} be the i-j^{th} entry function. Then the map $\varphi: \mathfrak{g} \to \mathfrak{gl}(n, \mathbb{R})$ given by $\varphi(X)_{ij} = Xe_{ij}$ is a Lie algebra isomorphism onto. Similarly the Lie algebra of $GL(n, \mathbb{C})$ is canonically identified with $\mathfrak{gl}(n, \mathbb{C})$.

(A.94) **Exponential mapping.** Let G be an analytic group with Lie algebra $\mathfrak{g}$. For X in $\mathfrak{g}$ let $\tilde{X}$ denote the corresponding left-invariant vector field on G. Then there exists a unique function $\exp: \mathfrak{g} \to G$ such that, for each X in $\mathfrak{g}$, $\exp tX$ for $-\infty < t < \infty$ is an integral curve for $\tilde{X}$ with $\exp 0 = 1$. Moreover, exp is smooth, the identity

$$\exp(s + t)X = \exp sX \exp tX$$

is valid for all real s and t, and exp is a diffeomorphism from any sufficiently small open neighborhood of 0 in $\mathfrak{g}$ onto an open neighborhood of 1 in G. Hence $\exp^{-1}$ gives a chart near 1.

(A.95) The conclusion in (A.94) that $\exp^{-1}$ gives a chart implies the following: Let $X_1, \ldots, X_n$ be a basis of $\mathfrak{g}$. Then

$$x^j(g(\exp(\lambda^1 X_1 + \ldots + \lambda^n X_n)) = \lambda^j$$

is a system of local coordinates near g in G. These are called **canonical coordinates of the first kind.**

(A.96) Every analytic group has canonically the structure of a real analytic manifold, and multiplication and inversion are real analytic mappings. Every left-invariant vector field has real analytic coefficients, and exp is real analytic.

(A.97) Suppose an analytic group G possesses a complex analytic structure such that multiplication and inversion are holomorphic. Then the complex structure induces a multiplication-by-i mapping in the Lie algebra $\mathfrak{g} = T_1(G)$ such that $\mathfrak{g}$ becomes a Lie algebra over $\mathbb{C}$. Every left-invariant vector field has holomorphic coefficients, and exp is a holomorphic mapping.

(A.98) **Taylor's Theorem** for C^∞ functions. Let G be an analytic group with Lie algebra $\mathfrak{g}$. If X is in $\mathfrak{g}$ and f is in $C^\infty(G)$ and if $\tilde{X}$ denotes the left-invariant vector field corresponding to X, then

$$(\tilde{X}^n f)(g \exp tX) = \frac{d^n}{dt^n} f(g \exp tX) \qquad \text{for } g \text{ in } G.$$

Moreover, for X in any bounded set

$$f(\exp X) = \sum_{k=0}^{n} \frac{1}{k!} (\tilde{X}^k f)(1) + R_n(X),$$

where $|R_n(X)| \leq C_n |X|^{n+1}$.

(A.99) Let X be in the Lie algebra $\mathfrak{g}$ of an analytic group G, and let $\tilde{X}$ denote the corresponding left-invariant vector field. Then

$$\tilde{X} f(g) = \frac{d}{dt} f(g \exp tX)\Big|_{t=0} \quad \text{for all } g \in G.$$

(A.100) **Taylor's Theorem** for real analytic functions. Let G be an analytic group with Lie algebra $\mathfrak{g}$. For X in $\mathfrak{g}$, let $\tilde{X}$ denote the corresponding left-invariant vector field. If f is real analytic in a neighborhood of 1 and if X is sufficiently small in $\mathfrak{g}$, then

$$f(\exp X) = \sum_{k=0}^{\infty} \frac{1}{k!} (\tilde{X}^k f)(1).$$

(A.101) In $GL(n, \mathbb{R})$ and $GL(n, \mathbb{C})$, if the Lie algebras are identified with Lie algebras of matrices via (A.93), then $\exp X$ is given by the usual series for e^X when X is a matrix.

(A.102) An **analytic subgroup** H of an analytic group G is an analytic group that is both a submanifold and a subgroup.

(A.103) Let G be an analytic group with Lie algebra $\mathfrak{g}$.

(1) If H is an analytic subgroup of G with Lie algebra $\mathfrak{h}$, then the differential of the inclusion, as a mapping

$$\mathfrak{h} = T_1(H) \to \tilde{T}_1(H) \subseteq \mathfrak{g},$$

is a one-one Lie algebra homomorphism. Thus $\tilde{T}_1(H)$ can be regarded as the Lie algebra of H.

(2) The correspondence $H \to \tilde{T}_1(H) \subseteq \mathfrak{g}$ in (1) of analytic subgroups of G to Lie subalgebras of $\mathfrak{g}$ is one-one onto.

(3) If $\varphi: M \to G$ is a smooth function such that $\varphi(M) \subseteq H$ for an analytic subgroup H, then $\varphi: M \to H$ is smooth.

(4) If H is an analytic subgroup with Lie algebra $\mathfrak{h}$, if $\iota_H: H \to G$ and $\iota_{\mathfrak{h}}: \mathfrak{h} \to \mathfrak{g}$ denote the inclusions, and if $\exp_H$ and $\exp_G$ denote the exponential maps, then

$$\iota_H \circ \exp_H = \exp_G \circ \iota_{\mathfrak{h}}.$$

(5) If H is an analytic subgroup with Lie algebra $\mathfrak{h}$ and if X is in $\mathfrak{g}$, then X is in $\mathfrak{h}$ if and only if $\exp_G tX$ is in H for $-\infty < t < \infty$.

(A.104) Let G be an analytic group with Lie algebra $\mathfrak{g}$, and suppose $\mathfrak{g}$ is a direct sum of subspaces

$$\mathfrak{g} = \mathfrak{g}_1 \oplus \ldots \oplus \mathfrak{g}_k.$$

Then on any sufficiently small open neighborhoods U_j of 0 in the respective $\mathfrak{g}_j$'s, the map

$$\Phi(X_1, \ldots, X_k) = g(\exp X_1) \cdots (\exp X_k)$$

is a diffeomorphism of $U_1 \times \ldots \times U_k$ onto an open neighborhood of g in G. When the $\mathfrak{g}_j$'s are all one-dimensional, the local coordinates given by the inverse map are called **canonical coordinates of the second kind.**

(A.105) Let G be an analytic group with Lie algebra $\mathfrak{g}$. If X and Y are in $\mathfrak{g}$, then

(1) $\exp tX \exp tY = \exp\{t(X + Y) + \frac{1}{2}t^2[X, Y] + O(t^3)\}$
(2) $\exp(-tX) \exp(-tY) \exp tX \exp tY = \exp\{t^2[X, Y] + O(t^3)\}$
(3) $\exp tX \exp tY \exp(-tX) = \exp\{tY + t^2[X, Y] + O(t^3)\}$,

where $O(t^3)$ denotes a function from an interval to $\mathfrak{g}$ such that, for some $\varepsilon > 0$, $t^{-3}O(t^3)$ extends to be (bounded and) smooth for $|t| < \varepsilon$.

(A.106) Let G be an analytic group. A **Lie subgroup** H of G is an abstract subgroup that in some topology is a Lie group and is such that its identity component is an analytic subgroup of G. The Lie algebra of a Lie subgroup is defined as the Lie algebra of its identity component.

(A.107) Let G be an analytic group, and let H be a closed subgroup. Then there exists a (unique) differentiable structure on the identity component of H such that H, in its relative topology, is a Lie subgroup.

(A.108) Let G be an analytic group of dimension n, let H be a closed subgroup of dimension s, and let $\pi: G \to G/H$ be the quotient map. Then there exists a cubical chart (U, φ) of side 2ε around 1 in G, say

$\varphi = (x^1, \ldots, x^n)$, such that

(1) each slice with $x^{s+1} = \xi^{s+1}, \ldots, x^n = \xi^n$ is a relatively open set in some coset gH and these cosets are all distinct
(2) the restriction of π to the slice $x^1 = 0, \ldots, x^s = 0$ is a homeomorphism onto an open set and therefore determines a chart about the identity coset in G/H.

If the translates in G/H of the chart in (2) are used as charts to cover G/H, then G/H becomes a smooth manifold such that π and the action of G are smooth. Moreover, any smooth map $\sigma: G \to M$ that factors continuously through G/H as $\sigma = \bar{\sigma} \circ \pi$ is such that $\bar{\sigma}$ is smooth; if $\bar{\sigma}$ is G-equivariant and one-one and onto, then $\bar{\sigma}$ is a diffeomorphism of G/H onto M.

(A.109) Let $\varphi: G \to H$ be a smooth homomorphism between analytic groups, and let $d\varphi: \mathfrak{g} \to \mathfrak{h}$ be the differential at the identity. Then

(1) $d\varphi$ is a Lie algebra homomorphism
(2) φ is uniquely determined by $d\varphi$
(3) the image of φ is an analytic subgroup H' of H, and $\varphi: G \to H'$ is smooth
(4) the respective exponential mappings satisfy $\exp_H \circ\, d\varphi = \varphi \circ \exp_G$.

(A.110) If G is an analytic group and φ: $G \to GL(n, \mathbb{C})$ is a smooth homomorphism, then $\varphi \circ \exp_G$ can be computed as $e^{d\varphi}$.

(A.111) If G is a Lie group with Lie algebra $\mathfrak{g}$ and if g is in G, let $\mathrm{Ad}(g): \mathfrak{g} \to \mathfrak{g}$ be the differential at the identity of the isomorphism $x \to gxg^{-1}$ of G. Then Ad is a smooth homomorphism of G into the general linear group on $\mathfrak{g}$. Ad is called the **adjoint representation** of G. The differential of Ad at the identity is ad. Consequently $\mathrm{Ad}(\exp X) = e^{\mathrm{ad}\, X}$ for X in $\mathfrak{g}$.

(A.112) Let G be an analytic group. Then the canonical manifold structure on the universal covering group $\tilde{G}$ of G (see (A.62) and (A.78)) makes $\tilde{G}$ into an analytic group such that the covering map is a smooth homomorphism. The groups $\tilde{G}$ and G have canonically isomorphic Lie algebras.

(A.113) If G and H are analytic groups with G simply connected and if $\varphi: \mathfrak{g} \to \mathfrak{h}$ is a homomorphism between their Lie algebras, then there exists a smooth homomorphism of G into H whose differential at the identity is φ.

(A.114) Any two simply connected analytic groups with isomorphic Lie algebras are isomorphic.

(A.115) If G is any analytic group, then $G \cong \tilde{G}/H$, where $\tilde{G}$ is the universal covering group of G and H is a discrete subgroup of the center of $\tilde{G}$. Conversely the most general analytic group with Lie algebra that of G is isomorphic to $\tilde{G}/H'$ for some central discrete subgroup H' of $\tilde{G}$.

(A.116) If G is an analytic group and H is an analytic subgroup, then $H \subseteq \text{center}(G)$ if and only if $\mathfrak{h} \subseteq \text{center}(\mathfrak{g})$.

(A.117) An analytic group is abelian if and only if its Lie algebra is abelian.

(A.118) The most general abelian analytic group is of the form $\mathbb{R}^l \times T^k$, where T^k denotes the k-dimensional **torus** group (the product of k circle groups).

(A.119) The most general compact abelian analytic group is a torus.

(A.120) If G is an analytic group and X and Y are in its Lie algebra, then $[X, Y] = 0$ if and only if $\exp sX$ and $\exp tY$ commute for all real s and t.

(A.121) If G is an analytic group with $G = H_1 \times H_2$ as analytic groups and if $\mathfrak{g}$, $\mathfrak{h}_1$, and $\mathfrak{h}_2$ are the Lie algebras, then $\mathfrak{g} = \mathfrak{h}_1 \oplus \mathfrak{h}_2$. Conversely if H_1 and H_2 are analytic subgroups of G with $\mathfrak{g} = \mathfrak{h}_1 \oplus \mathfrak{h}_2$ and if G is simply connected, then $G = H_1 \times H_2$.

(A.122) Let V be a finite-dimensional algebra over $\mathbb{R}$, not necessarily associative. Let $\text{Aut}\, V \subseteq GL(V)$ be the group of all algebra automorphisms of V. Then $\text{Aut}\, V$ is a Lie subgroup of $GL(V)$ and its Lie algebra is $\text{Der}\, V$, the Lie algebra of derivations of V.

(A.123) Let G and H be analytic groups. We say G **acts on** H **by automorphisms** if a smooth map $\tau: G \times H \to H$ is specified such that $g \to \tau(g, \cdot)$ is a homomorphism into the automorphism group of H. In this case we can form the **semidirect product** $G \ltimes_\tau H$ of the groups as follows: The manifold is $G \times H$, and the multiplication and inversion mappings are given by

$$(g_1, h_1)(g_2, h_2) = (g_1 g_2, \tau(g_2^{-1}, h_1)h_2)$$
$$(g, h)^{-1} = (g^{-1}, \tau(g, h^{-1})).$$

Then $G \ltimes_\tau H$ is an analytic group, G and H are closed subgroups, and H is normal.

(A.124) Let G and H be analytic groups, and let G act on H by automorphisms (by means of τ). If $\bar{\tau}(g)$ denotes the differential of $\tau(g, \cdot)$ at the identity of H, then $\bar{\tau}$ is a smooth homomorphism of G into $\text{Aut}\, \mathfrak{h}$. Hence $d\bar{\tau}$ is a homomorphism of $\mathfrak{g}$ into $\text{Der}\, \mathfrak{h}$, and the semidirect product $\mathfrak{g} \oplus_{d\bar{\tau}} \mathfrak{h}$ is well defined. The Lie algebra of $G \ltimes_\tau H$ is $\mathfrak{g} \oplus_{d\bar{\tau}} \mathfrak{h}$.

(A.125) Conversely let G and H be simply connected analytic groups with Lie algebras $\mathfrak{g}$ and $\mathfrak{h}$, and let $\pi: \mathfrak{g} \to \text{Der}\, \mathfrak{h}$ be a homomorphism. Then there exists a unique action τ of G on H by automorphisms such that $d\bar{\tau} = \pi$, and $G \ltimes_\tau H$ is a simply connected analytic group with Lie algebra $\mathfrak{g} \oplus_\pi \mathfrak{h}$.

(A.126) If $\mathfrak{g}$ is a finite-dimensional real solvable Lie algebra, then there exists a simply connected analytic group G with Lie algebra $\mathfrak{g}$, and G is diffeomorphic to $\mathbb{R}^n$ for some n.

(A.127) If $\mathfrak{g}$ is a finite-dimensional real nilpotent Lie algebra and G is a simply connected analytic group with $\mathfrak{g}$ as Lie algebra, then the exponential map is a diffeomorphism of $\mathfrak{g}$ onto G.

(A.128) If $\mathfrak{g}$ is a finite-dimensional real semisimple Lie algebra and if $\mathfrak{g} = \mathfrak{g}_1 \oplus \ldots \oplus \mathfrak{g}_k$ is its decomposition (A.24) into simple components, then there exists a simply connected analytic group G with Lie algebra $\mathfrak{g}$, and G is of the form $G = G_1 \times \ldots \times G_k$, where the G_j's are the simply connected analytic groups corresponding to the $\mathfrak{g}_j$'s and are given as the universal covering groups of the analytic subgroups of $\mathfrak{gl}(\mathfrak{g}_j)$ with Lie algebra ad $\mathfrak{g}_j$.

(A.129) **Levi decomposition.** If $\mathfrak{g}$ is a finite-dimensional real Lie algebra, then $\mathfrak{g}$ is the semidirect product of a semisimple factor and the (solvable) radical of $\mathfrak{g}$. Consequently (by (A.125), (A.126), and (A.128)) there exists a simply connected analytic group with Lie algebra $\mathfrak{g}$.

APPENDIX B

Regular Singular Points of Partial Differential Equations

This appendix develops that part of the theory of regular singular points for partial differential equations useful in studying semisimple groups. For any irreducible admissible representation (see Chapter VIII), every matrix coefficient is an eigenfunction of the center of the universal enveloping algebra. With the help of the Harish-Chandra homomorphism, we are led to a system of differential equations on the subgroup A in the Iwasawa decomposition. In the initial application, the theory that follows is applied to that system after a change of variables.

§1. Summary of Classical One-Variable Theory

In this section we review, mostly for motivation, the theory of regular singular points for analytic ordinary differential equations. We omit the proofs but will include proofs of some generalizations in later sections.

A **first-order analytic ordinary linear system** in a simply connected domain E of $\mathbb{C}$ is a system of the form

$$w' = A(z)w, \tag{B.1}$$

where $A(z)$ is an n-by-n matrix of analytic functions and

$$w(z) = \begin{pmatrix} w_1(z) \\ \vdots \\ w_n(z) \end{pmatrix}$$

is a vector-valued analytic function to be found.

Theorem B.1. If $w' = A(z)w$ is a first-order analytic ordinary linear system in the simply connected domain $E \subseteq \mathbb{C}$, if z_0 is in E, and if w_0 is in $\mathbb{C}^n$, then there exists a unique analytic solution $w(z)$ satisfying $w(z_0) = w_0$.

Remarks. The proof of existence uses Picard iterations, just as in the real-variable case. In some applications one wants to allow one or more parameters in the function A, say $A(z, \lambda)$. The existence proof then shows that the dependence of the solution on (z, λ) is as good as the dependence of A on (z, λ): If $A(z, \lambda)$ is smooth, the dependence of the solution on (z, λ) is smooth, and the solution is analytic in any parameters that $A(z, \lambda)$ is analytic in.

Fix a first-order analytic ordinary linear system (B.1) in a simply connected domain E. In view of Theorem B.1 the solutions of (B.1) on E form an n-dimensional complex vector space. Let $\varphi_1, \ldots, \varphi_n$ be a basis. The matrix-valued function $\Phi(z)$ whose n columns are the n independent solutions $\varphi_1, \ldots, \varphi_n$ is called a **fundamental matrix** for (B.1). The most general basis is of course obtained by taking a nonsingular linear combination of $\varphi_1, \ldots, \varphi_n$, and the resulting fundamental matrix is $\Phi(z)C$ with $\det C \neq 0$.

Proposition B.2. Let $\Phi(z)$ be a fundamental matrix for (B.1). Then

(a) $\Phi'(z) = A(z)\Phi(z)$
(b) $(\det \Phi)' = (\operatorname{Tr} A(z)) \det \Phi$
(c) $\det \Phi$ is nowhere vanishing and is given by

$$\det \Phi(z) = \left(\exp \int_{z_0}^{z} \operatorname{Tr} A(\zeta)\, d\zeta\right) \det \Phi(z_0).$$

Corollary B.3. If $\Phi(z)$ is a matrix-valued solution of $\Phi'(z) = A(z)\Phi(z)$, then either $\det \Phi(z) \equiv 0$ or $\det \Phi(z)$ is nowhere zero, and $\Phi(z)$ is a fundamental matrix for (B.1) exactly in the latter case.

A more canonical definition of fundamental matrix is possible and involves viewing the fundamental matrix at a point z as a linear transformation on the range $\mathbb{C}^n$ after fixing a point z_0 in the domain. Namely for v in $\mathbb{C}^n$ let φ_v be the solution of (B.1) with $\varphi_v(z_0) = v$. Then $L(z): \mathbb{C}^n \to \mathbb{C}^n$ given by $L(z)v = \varphi_v(z)$ is linear, and its matrix in the standard basis of $\mathbb{C}^n$ is $\Phi(z)$ defined in such a way that $\Phi(z_0) = I$. (We shall adopt this point of view in the several-variable case, starting in the next section.)

If A is an n-by-n matrix, we define $\exp A$ in the usual way by a power series. In analogy with what is done with scalars, we define $z^A = \exp(A \log z)$ for z in $\mathbb{C}$.

Now we shall examine systems in which the coefficient matrix is allowed an isolated singularity. We say the system

$$w' = A(z)w \tag{B.2}$$

is an **analytic system** in the punctured disc

$$0 < |z - z_0| < a \tag{B.3}$$

if $A(z)$ is an n-by-n matrix of (single-valued) analytic functions in (B.3). Since Theorem B.1 ensures the existence of local solutions of (B.2) (in any simply connected subdomain of (B.3)), the problem is how these local solutions are pieced together.

Theorem B.4. If $w' = A(z)w$ is an analytic system in the punctured disc (B.3), then there exists a multiple-valued analytic fundamental matrix $\Phi(z)$ of the form

$$\Phi(z) = S(z)(z - z_0)^P, \qquad 0 < |z - z_0| < a, \tag{B.4}$$

where S is single-valued analytic in (B.3) and where P is a matrix of constants. Moreover, every fundamental matrix has this form for suitable S and P.

Let $w' = A(z)w$ be an analytic system for $0 < |z - z_0| < a$. The form of the fundamental matrix is as in (B.4). The point z_0 is called a **regular singular point** if $S(z)$ can be chosen so as to have at most a pole at z_0. [In this case we can write $S(z) = S_1(z)(z - z_0)^{-m}$, and $\Phi(z) = S_1(z)(z - z_0)^{P-mI}$. That is, the S part can be assumed to be analytic at z_0.]

Theorem B.5. If $w' = A(z)w$ is an analytic system in the punctured disc (B.3) and if $A(z)$ has a simple pole at z_0, then z_0 is a regular singular point.

The condition in Theorem B.5 is sufficient for a regular singular point but not necessary, as we shall observe presently. In the meantime a converse result is as follows.

Proposition B.6. If $w' = A(z)w$ is an analytic system in the punctured disc (B.3) and if z_0 is a regular singular point, then $A(z)$ has only a pole at z_0, and $(z - z_0)\operatorname{Tr} A(z)$ is analytic at z_0.

Suppose $w' = A(z)w$ is an analytic system for $0 < |z| < a$ of the kind in Theorem B.5. We investigate the qualitative nature of a solution $\varphi(z)$ of the system. First we arrange that $\varphi(z)$ is a column of a fundamental matrix $\Phi(z)$, and then we write

$$\Phi(z) = S(z)z^P \tag{B.5}$$

with $S(z)$ analytic for $|z| < a$. We return to (B.5) in a moment.

An example is the **Euler system** $w' = z^{-1}Rw$, where R is a matrix of constants. In this case, z^R is a fundamental matrix. The idea in the general case when $z^{-1}R$ is replaced by

$$A(z) = z^{-1}\left(\sum_{k=0}^{\infty} A_k z^k\right) \tag{B.6}$$

is to relate the system $w' = A(z)w$ to the Euler system $w' = z^{-1}A_0w$. If no two distinct eigenvalues of A_0 differ by an integer, then it is possible to arrange in (B.5) that $P = R$ and that $S(0) = I$. This is not quite true in the general case, but a substitute result (Theorem B.7 below) is available and will be sufficient for our purposes.

In (B.5) we apply the Jordan decomposition theorem to P, writing $P = D + N$ with D diagonable, N nilpotent, and $DN = ND$. The eigenvalues of D are the eigenvalues of P. We have

$$z^P = z^D z^N,$$

and z^D is a matrix whose entries are linear combinations of powers of z, the powers being eigenvalues of D (hence eigenvalues of P). Since N is nilpotent,

$$z^N = \exp(N \log z) = \sum_{m=0}^{n-1} \frac{1}{m!} N^m (\log z)^m.$$

Substituting in (B.5), we see that we can write

$$\varphi(z) = \sum_{j=1}^{r} \sum_{m=0}^{n-1} z^{s_j}(\log z)^m f_{s_j,m}(z)$$

with each $f_{s_j,m}(z)$ analytic for $|z| < a$; here $\{s_j\}$ is the set of distinct eigenvalues of P. When we substitute a power series for each $f_{s_j,m}(z)$, we can rewrite $\varphi(z)$ as

$$\varphi(z) = \sum_{j=1}^{r} \sum_{k=0}^{\infty} \left(\sum_{m=0}^{n-1} c_{s_j+k,m} z^{s_j+k} (\log z)^m \right) = \sum_{j=1}^{r} \sum_{k=0}^{\infty} \varphi_{s_j+k}(z). \qquad \text{(B.7)}$$

If two s_j's satisfy $s_j - s_{j'} = l$ with $l > 0$ an integer, we can drop $s_{j'}$ and regroup (B.7). Thus we may assume in (B.7) that $\{s_j\}$ is a *subset* of the eigenvalues of P and that no two s_j's differ by an integer. Then we say that $s_j + k$ is a **leading exponent** of φ if $\varphi_{s_j+k}(z) \not\equiv 0$ but $\varphi_{s_j+l} \equiv 0$ for all integers l with $0 \le l \le k - 1$.

Theorem B.7. Suppose $w' = A(z)w$ is an analytic system for $0 < |z| < a$ such that $A(z)$ has only a simple pole at 0, and write $A(z)$ as in (B.6). If φ is any solution, then every leading term φ_s of φ is such that s is an eigenvalue of A_0.

The form in which we shall apply the theory of regular singular points to groups will differ in two respects from what has been given above: The system will consist of equations of higher order than the first, and the domain will normally be several variables.

A single n^{th} order equation can be converted to a system: For the equation

$$w^{(n)} + a_{n-1}w^{(n-1)} + \ldots + a_1(z)w' + a_0(z)w = 0, \tag{B.8}$$

let

$$\mathbf{w} = \begin{pmatrix} w_1 \\ w_2 \\ \vdots \\ w_n \end{pmatrix} = \begin{pmatrix} w \\ w' \\ \vdots \\ w^{(n-1)} \end{pmatrix}. \tag{B.9}$$

Then the equation (B.8) is equivalent with the system

$$\mathbf{w}' = \begin{pmatrix} 0 & 1 & & & & \\ & 0 & 1 & & & \\ & & 0 & 1 & & \\ & & & \ddots & & \\ & & & & 0 & 1 \\ -a_0(z) & -a_1(z) & & \cdots & & -a_{n-1}(z) \end{pmatrix} \mathbf{w}.$$

Similar remarks apply to a system of higher order equations.

When a single n^{th} order equation is treated in this fashion, the initial conditions in Theorem B.1 become the values of $w(z_0), w'(z_0), \ldots, w^{(n-1)}(z_0)$. The fundamental matrix takes the form

$$\Phi(z) = \begin{pmatrix} \varphi_1(z) & \varphi_2(z) & \cdots & \varphi_n(z) \\ \varphi_1'(z) & \varphi_2'(z) & \cdots & \varphi_n'(z) \\ & & \ddots & \\ \varphi_1^{(n-1)}(z) & \varphi_2^{(n-1)}(z) & \cdots & \varphi_n^{(n-1)}(z) \end{pmatrix},$$

and det $\Phi(z)$ is called the **Wronskian** of the nth order equation.

To connect n^{th} order equations with Theorem B.4 about regular singular points, we do not use the reduction (B.9), because when applied to the familiar equation

$$w'' + \frac{a(z)}{z}w' + \frac{b(z)}{z^2}w = 0,$$

it yields $\mathbf{w}' = A(z)\mathbf{w}$ with

$$A(z) = \begin{pmatrix} 0 & 1 \\ -\dfrac{b(z)}{z^2} & -\dfrac{a(z)}{z} \end{pmatrix},$$

which is not of the form requested by the theorem. Instead we make a different reduction. Namely with

$$w^{(n)} + \frac{a_{n-1}(z)}{z} w^{(n-1)} + \ldots + \frac{a_1(z)}{z^{n-1}} w' + \frac{a_0(z)}{z^n} w = 0, \tag{B.10}$$

we can define

$$\mathbf{w} = \begin{pmatrix} w_1 \\ \vdots \\ w_n \end{pmatrix} \quad \text{by} \quad w_k = z^{k-1} w^{(k-1)}. \tag{B.11}$$

Then we are led to the system $\mathbf{w}' = A(z)\mathbf{w}$ with

$$A(z) = z^{-1} \begin{pmatrix} 0 & 1 & & & & \\ & 1 & 1 & & & \\ & & 2 & 1 & & \\ & & & 3 & 1 & \\ & & & & \ddots & 1 \\ -a_0(z) & -a_1(z) & & \cdots & & (n-1) - a_{n-1}(z) \end{pmatrix}, \tag{B.12}$$

and the theorem applies. (This example shows also that a simple pole for $A(z)$ is not a necessary condition for a regular singular point.) One can show in fact that the singularity of (B.10) at $z = 0$ is regular if and only if all the $a_j(z)$ are analytic at $z = 0$.

The traditional way of solving (B.10) begins by seeking a solution of the form z^s times a power series, and one is led immediately to the **indicial equation**

$$\begin{aligned} &s(s-1)\cdots(s-n+1) + a_{n-1}(0)s(s-1)\cdots(s-n+2) + \cdots \\ &\quad + a_1(0)s + a_0(0) = 0. \end{aligned} \tag{B.13}$$

Solving for s gives leading exponents of solutions. If instead we use the transformation (B.11), we can apply Theorem B.7. The characteristic polynomial of the relevant matrix (the matrix in (B.12)) is just the left side of (B.13), except possibly for a minus sign, and thus the solutions of the indicial equation are the eigenvalues mentioned in Theorem B.7.

§2. Uniqueness and Analytic Continuation of Solutions in Several Variables

In l complex variables z with $w(z)$ in $\mathbb{C}^n$, we consider the system

$$\frac{\partial w}{\partial z_i} = A_i(z)w. \tag{B.14}$$

Here $A_i(z)$, for $1 \leq i \leq l$, is an n-by-n matrix of holomorphic functions on some connected open set E in $\mathbb{C}^l$. At first we allow E to be general, and later we specialize to the case that E is a product of half planes or a product of punctured discs.

We shall not need an existance theorem since we shall be dealing with solutions (from representation theory) already known to exist. A theory concerned with existence of solutions will normally assume an **integrability condition** motivated as follows: If w satisfies (B.14), then

$$\frac{\partial^2 w}{\partial z_j\, \partial z_i} = \frac{\partial}{\partial z_j}(A_i(z)w) = \frac{\partial A_i}{\partial z_j} + A_i \frac{\partial w}{\partial z_j} = \left(\frac{\partial A_i}{\partial z_j} + A_i A_j\right) w,$$

and this is

$$= \frac{\partial^2 w}{\partial z_i\, \partial z_j} = \left(\frac{\partial A_j}{\partial z_i} + A_j A_i\right) w.$$

Hence

$$\left(\frac{\partial A_i}{\partial z_j} - \frac{\partial A_j}{\partial z_i} + [A_i, A_j]\right) w = 0$$

for all z. For any point $z = z_0$, this condition places a limitation on

$$\dim\{w(z_0) \mid w \text{ is a solution of (B.14)}\}$$

unless the operator in question is 0. The condition

$$\frac{\partial A_i}{\partial z_j} - \frac{\partial A_j}{\partial z_i} + [A_i, A_j] = 0 \tag{B.15}$$

is thus normally assumed as an integrability condition in an existence theorem. *We shall not assume (B.15) since we shall not have to deal with existence of solutions.*

Let

$$H^l = \{z \in \mathbb{C}^l \mid \operatorname{Re} z_i > 0 \text{ for } 1 \leq i \leq l\}$$
$$D^l = \{z \in \mathbb{C}^l \mid |z_i| < 1 \text{ for } 1 \leq i \leq l\}$$
$$(D^\times)^l = \{z \in \mathbb{C}^l \mid 0 < |z_i| < 1 \text{ for } 1 \leq i \leq l\}.$$

Theorem B.8.

(a) If $E \subseteq \mathbb{C}^l$ is open and connected and if $\varphi(z)$ is a holomorphic solution of (B.14) in E with $\varphi(z_0) = 0$, then $\varphi(z) \equiv 0$.

(b) If $E \subseteq H^l$ is open connected nonempty and if (B.14) is defined on H^l, then any holomorphic solution $\varphi(z)$ of (B.14) on E extends to a holomorphic solution on H^l.

Proof.

(a) Let z_0 and z be in a ball contained in E. Then $z_0 + t(z - z_0)$ is in E for $0 \le t \le 1$. We have

$$\begin{aligned}\frac{d}{dt}\varphi(z_0 + t(z - z_0))\\ &= \sum_{i=1}^{l}(z_i - z_{0,i})\frac{\partial\varphi}{\partial z_i}(z_0 + t(z - z_0))\\ &= \left[\sum_{i=1}^{l}(z_i - z_{0,i})A_i(z_0 + t(z - z_0))\right]\varphi(z_0 + t(z - z_0)).\end{aligned}$$

The uniqueness theorem for ordinary differential equations thus implies (a) on the ball. Since φ is holomorphic on E, $\varphi(z) \equiv 0$ on E.

(b) Without loss of generality, E is the product of open discs $E_1 \times \ldots \times E_l$. We prove the result by induction on l. For $l = 1$ the result follows from Theorem B.1. Suppose the result is known for $l - 1$ or fewer variables. Fix $z_{0,l}$ in the lth coordinate in E_l, regard $z_1, \ldots, z_{l-1}$ as parameters, and apply the one-variable theory with parameters (Theorem B.1, the remarks after Theorem B.1, and Proposition B.2c) in the variable z_l to define Φ from H^l into n-by-n matrices with

$$\frac{\partial\Phi}{\partial z_l} = A_l(z)\Phi \qquad \text{and} \qquad \Phi(z_1, \ldots, z_{l-1}, z_{0,l}) = I.$$

Since the dependence of $A_l(z)$ and the initial conditions on the parameters $z_1, \ldots, z_{l-1}$ is holomorphic, Φ is holomorphic.

Now $\varphi(z_1, \ldots, z_l)$ and $\Phi(z_1, \ldots, z_l)\varphi(z_1, \ldots, z_{l-1}, z_{0,l})$ are both functions w on E_l satisfying

$$\frac{\partial w}{\partial z_l} = \mathbf{A}_l(z)w \qquad \text{and} \qquad w(z_1, \ldots, z_{l-1}, z_{0,l}) = \varphi(z_1, \ldots, z_{l-1}, z_{0,l}),$$

and so the one-variable uniqueness theorem implies

$$\varphi(z_1, \ldots, z_l) = \Phi(z_1, \ldots, z_l)\varphi(z_1, \ldots, z_{l-1}, z_{0,l}) \tag{B.16}$$

for z_l in E_l. By inductive hypothesis, the function

$$(z_1, \ldots, z_{l-1}) \to \varphi(z_1, \ldots, z_{l-1}, z_{0,l})$$

extends to H^{l-1} so as to satisfy (B.14) for $1 \le i \le l - 1$ and to be holomorphic in $l - 1$ variables. Equation (B.16) then shows that φ extends to a holomorphic map on H^l. We can regard the extension as taking place in three steps—first in passing from the left side of (B.16) for z in E to the right side, second in extending the right side in $z_1, \ldots, z_{l-1}$, and third in passing back to the left side of (B.16). Thus the extended φ extends the original φ defined on the open set E, and the extended φ must satisfy (B.14).

Theorem B.9. Let the system (B.14) be defined on H^l, let U be a nonempty connected open subset of $(\mathbb{R}^+)^l$, and let φ be a C^∞ function on U satisfying (B.14) on U. Then φ extends to a holomorphic function on H^l satisfying (B.14).

Proof. We may assume U is an open rectangular set. With $x_2, \ldots, x_l$ fixed, φ satisfies the ordinary differential equation

$$\frac{\partial}{\partial z_1} \varphi(z_1, x_2, \ldots, x_l) = A_1(z_1, x_2, \ldots, x_l)\varphi(z_1, x_2, \ldots, x_l) \quad \text{(B.16')}$$

for $z_1 = x_1$. Comparing the uniqueness theorem for real-variable differential equations with the existence part of Theorem B.1, we see that every solution of (B.16′) in one variable is analytic and extends for $z_1 \in x_1 + i\mathbb{R}$. Theorem B.1 allows us allow to carry along C^∞ parameters, and thus

$$\varphi(z_1, x_2, \ldots, x_l) \text{ is globally } C^\infty.$$

Moreover, $\varphi(z_1, x_2, \ldots, x_l)$ satisfies

$$\frac{\partial}{\partial x_i} \varphi(z_1, x_2, \ldots, x_l) = A_i(z_1, x_2, \ldots, x_l)\varphi(z_1, x_2, \ldots, x_l)$$

for $i \geq 1$ since it satisfies this for $z_1 = x_1$ and since both sides are analytic in z_1.

Next we argue with z_2, obtaining an analytic extension from x_2 to z_2 in the same way. The extended function will still be analytic in z_1, by the remarks after Theorem B.1 about analytic dependence on parameters. We proceed through $z_3, \ldots, z_l$ and finally obtain a C^∞ function φ on $U \times i\mathbb{R}^l$ analytic in each variable and satisfying (B.14). Then φ has a complex differential and so is holomorphic. Hence it extends to H^l by Theorem B.8b.

§3. Analog of Fundamental Matrix

Let us start with the system (B.14) in a simply connected open set $E \subseteq \mathbb{C}^l$. The dimension of the space $\mathscr{V}$ of global single-valued solutions is $\leq n$, by Theorem B.8a. Fix z_0 in E and let

$$V = \{\varphi(z_0) \mid \varphi \in \mathscr{V}\} \subseteq \mathbb{C}^n.$$

The map $\varphi \to \varphi(z_0)$ is one-one by Theorem B.8a, and thus $\dim V = \dim \mathscr{V}$. Let $v \to \varphi_v$ be the inverse map carrying V onto $\mathscr{V}$: here φ_v is the unique solution with $\varphi_v(z_0) = v$. Define $\Phi(z): V \to \mathbb{C}^n$ by

$$\Phi(z)v = \varphi_v(z).$$

Then Φ is a holomorphic map from E to $\mathrm{Hom}(V, \mathbb{C}^n)$. Moreover, Φ satisfies

(i) $\Phi(z_0) =$ inclusion of V into $\mathbb{C}^n$

(ii) $\dfrac{\partial \Phi}{\partial z_i} = A_i(z)\Phi$ for $1 \leq i \leq l$.

By Theorem B.8a, Φ is uniquely determined by these conditions.

The map Φ is the analog of the fundamental matrix. For each z in E, $\Phi(z)$ is one-one, again by Theorem B.8a. Every solution of (B.14) is of the form

$$\varphi(z) = \Phi(z)v$$

for some v in V. (In fact, we can take $v = \varphi(z_0)$.)

We shall want to consider the system (B.14) on $(D^\times)^l$. For this purpose we need to allow multiple-valued solutions. Global (multiple-valued) holomorphic functions on $(D^\times)^l$ are defined in the same way as in the theory of one complex variable: One introduces notions of **function element** (f, Ω) (= a single-valued holomorphic f on a connected open set Ω), **direct analytic continuation** (= two function elements with overlapping domains whose f's agree on the common domain), **analytic continuation** (= two function elements related by a finite sequence of direct analytic continuations), **global function** (= a set of function elements that are analytic continuations of one another), and **complete global function** (= a maximal global function).

Define

$$p: H^l \to (D^\times)^l \text{ by } p(\zeta_1, \ldots, \zeta_l) = (e^{-\zeta_1}, \ldots, e^{-\zeta_l}) = (z_1, \ldots, z_l).$$

If we have a holomorphic function $\tilde{F}$ on H^l, we can construct a global function F on $(D^\times)^l$ in such a way that each function element (F, Ω) gives rise to $\tilde{\Omega} \subseteq H^l$ such that $p: \tilde{\Omega} \to \Omega$ is one-one and $\tilde{F}(p^{-1}z) = F(z)$ for $z \in \Omega$, $p^{-1}z \in \tilde{\Omega}$. We call F a **multiple-valued function** on $(D^\times)^l$.

Consider the system

$$\frac{\partial w}{\partial z_i} = A_i(z)w \qquad \text{for } z \in (D^\times)^l. \tag{B.17}$$

If (F, Ω) is a function element w satisfying (B.17), then $\tilde{F}(\zeta) = F(p\zeta)$ satisfies

$$\frac{\partial}{\partial \zeta_i} \tilde{F}(\zeta) = \sum_{j=1}^{l} \frac{\partial F}{\partial z_j} \frac{\partial p_j}{\partial \zeta_i} = -e^{-\zeta_i} \frac{\partial F}{\partial z_i}(p\zeta) = -e^{-\zeta_i} A_i(p\zeta)\tilde{F}(\zeta).$$

Hence $\tilde{F}$ satisfies

$$\frac{\partial \tilde{w}}{\partial \zeta_i} = A_i(\zeta)\tilde{w} \qquad \text{for } \zeta \in H^l, \tag{B.18}$$

where

$$A_i(\zeta) = -e^{-\zeta_i} A_i(p\zeta). \tag{B.19}$$

Proposition B.10. If $\varphi(z)$ is a holomorphic solution of (B.17) on a connected open set in $(D^\times)^l$, then φ extends to a complete global solution of (B.17) on $(D^\times)^l$. This solution arises from the corresponding $\tilde{\varphi}(\zeta)$ solution of (B.18) on H^l.

Proof. Lift $\varphi(z)$ to $\tilde{\varphi}(\zeta)$ locally. Extend to a solution on H^l by Theorem B.8b, and push down to the required multiple-valued extension of φ.

In view of the proposition, we can extend our notion of a several-variable analog $\Phi(z)$ of fundamental matrix to the case that $(D^\times)^l$ is the domain, by allowing $\Phi(z)$ to be multiple-valued.

Lemma B.11. If $B_1, \ldots, B_k$ are nonsingular commuting s-by-s matrices, then there exist commuting $A_1, \ldots, A_k$ with $B_j = \exp A_j$ for $1 \leq j \leq k$.

Proof. In the case $k = 1$, we may assume B_1 is in Jordan form. Then B_1 comes in blocks of the form $\lambda I + N = \lambda(I + \lambda^{-1}N)$ with N strictly upper triangular and with $\lambda \neq 0$. For X strictly upper triangular, define

$$\log(I + X) = \sum_{m=0}^{\infty} (-1)^{m+1} \frac{X^m}{m};$$

this is a polynomial since $X^s = 0$. It is readily checked that

$$\exp \log(I + X) = I + X.$$

Thus a block of B_1 is of the form

$$\begin{aligned}(\lambda I)(I + \lambda^{-1}N) &= \exp[(\log \lambda)I] \exp[\log(I + \lambda^{-1}N)] \\ &= \exp[(\log \lambda)I + \log(I + \lambda^{-1}N)],\end{aligned}$$

and B_1 is an exponential.

For general k, the $B_1, \ldots, B_k$ can be put simultaneously in Jordan form. Then the argument in the previous paragraph works, and the matrices A_j commute.

Theorem B.12. Associated to the system (B.17) on $(D^\times)^l$ are a subspace V of $\mathbb{C}^n$ and a multiple-valued "fundamental matrix" $\Phi(z): V \to \mathbb{C}^n$ of the form

$$\Phi(z) = S(z) \prod_{i=1}^{l} z_i^{R_i}, \tag{B.20}$$

where $S:(D^\times)^l \to \mathrm{End}(V, \mathbb{C}^n)$ is (single-valued) holomorphic and $R_1, \dots, R_l$ are commuting members of $\mathrm{End}(V)$. If $\varphi(z)$ is any multiple-valued solution of (B.17) on $(D^\times)^l$, then $\varphi(z) = \Phi(z)v$ for some v in V.

Proof. Let $\tilde{\Phi}(\zeta)$ be the "fundamental matrix" for (B.18) on H^l, relative to ζ_0, and let $\{e_j\}$ be the standard basis of $\mathbb{C}^l$. Now $A_i(\zeta) = -e^{-\zeta_i}A_i(p\zeta)$ satisfies

$$A_i(\zeta + 2\pi i e_j) = A_i(\zeta)$$

since $A_i(z)$ is single-valued. Then

$$\frac{\partial\tilde{\Phi}}{\partial\zeta_i}(\zeta + 2\pi i e_j) = A_i(\zeta + 2\pi i e_j)\tilde{\Phi}(\zeta + 2\pi i e_j) = A_i(\zeta)\tilde{\Phi}(\zeta + 2\pi i e_j).$$

Hence $\tilde{\Phi}(\Phi + 2\pi i e_j)v$ is a solution of (B.18) for each v in V. Then

$$\tilde{\Phi}(\zeta + 2\pi i e_j)v = \tilde{\Phi}(\zeta)\tilde{\Phi}(\zeta_0 + 2\pi i e_j)v,$$

and we have

$$\tilde{\Phi}(\zeta + 2\pi i e_j) = \tilde{\Phi}(\zeta)\tilde{\Phi}(\zeta_0 + 2\pi i e_j).$$

Write

$$B_j = \tilde{\Phi}(\zeta_0 + 2\pi i e_j).$$

The operator B_j is a nonsingular endomorphism of V because we can invert this construction, using $-2\pi i e_j$ in place of $2\pi i e_j$. Since it satisfies

$$\tilde{\Phi}(\zeta + 2\pi i e_j) = \tilde{\Phi}(\zeta)B_j, \tag{B.21}$$

we obtain

$$\tilde{\Phi}(\zeta + 2\pi i e_i + 2\pi i e_j) = \tilde{\Phi}(\zeta + 2\pi i e_i)B_j = \tilde{\Phi}(\zeta)B_iB_j$$

and symmetrically

$$\tilde{\Phi}(\zeta + 2\pi i e_i + 2\pi i e_j) = \tilde{\Phi}(\zeta)B_jB_i.$$

Thus $B_iB_j = B_jB_i$. By Lemma B.11, choose commuting elements $R_1, \dots, R_l$ of End V such that $B_j = \exp(-2\pi i R_j)$ for $1 \le j \le l$, and define

$$\tilde{S}(\zeta) = \tilde{\Phi}(\zeta)\exp\sum_{j=1}^{l}\zeta_jR_j = \tilde{\Phi}(\zeta)\prod_{j=1}^{l}\exp\zeta_jR_j. \tag{B.22}$$

From (B.22) and (B.21), we obtain

$$\tilde{S}(\zeta + 2\pi i e_j) = \tilde{\Phi}(\zeta + 2\pi i e_j)e^{2\pi i R_j}\prod e^{\zeta_i R_i} = \tilde{\Phi}(\zeta)\prod e^{\zeta_i R_i} = \tilde{S}(\zeta).$$

Hence $S(z)$ defined by $\tilde{S}(\zeta) = S(p\zeta)$ is single-valued, and

$$\Phi(z) = S(z)\prod e^{-\zeta_1 R_i} = S(z)\prod z_i^{R_i}.$$

The theorem follows.

§4. Regular Singularities

In this section we generalize Theorem B.5 and equation (B.7) to the several-variable case.

Lemma B.13. If $\varphi(t) = \begin{pmatrix} \varphi_1(t) \\ \cdots \\ \varphi_n(t) \end{pmatrix}$ is a $\mathbb{C}^n$-valued function of one variable of class C^1, then $\frac{d}{dt}|\varphi(t)| \leq \left|\frac{d}{dt}\varphi(t)\right|$ at points where $\varphi(t) \neq 0$.

Proof. By the Schwarz inequality

$$\frac{d}{dt}|\varphi(t)| = \left|\frac{1}{\varphi(t)}\right| \sum_{j=1}^{n} \operatorname{Re}(\varphi_j'(t)\overline{\varphi_j(t)}) \leq \left|\frac{1}{\varphi(t)}\right| (|\varphi'(t)|\,|\varphi(t)|) = |\varphi'(t)|.$$

Theorem B.14. In the system (B.17), suppose that $A_i(z) = z_i^{-1}H_i(z)$ with H_i holomorphic on all of D^l, $1 \leq i \leq l$, and let the fundamental matrix be

$$\Phi(z) = S(z) \prod_{i=1}^{l} z_i^{R_i}$$

as in (B.20). Then there are positive integers $m_1, \ldots, m_l$ so that $z_1^{m_1} \cdots z_l^{m_l}S(z)$ is holomorphic on D^l. Consequently the decomposition (B.20) can be rewritten so that $S(z)$ is holomorphic on D^l.

Proof. The problem is to show that $z_1^{m_1} \cdots z_l^{m_l}S(z)$ is bounded on compact subsets of D^l. [In this case the Riemann Removable Singularity Theorem applied inductively, one variable at a time, shows $z_1^{m_1} \cdots z_l^{m_l}S(z)$ is analytic in each variable on H^l. Therefore it is holomorphic, by Hartogs's Theorem.]

The system (B.18) in H^l becomes

$$\frac{\partial \tilde{w}}{\partial \zeta_i} = -\tilde{H}_i(\zeta)\tilde{w} \qquad \text{with } \tilde{H}_i(\zeta) = H_i(e^{-\zeta_1}, \ldots, e^{-\zeta_l}).$$

Let $F(\zeta)$ be a solution $\not\equiv 0$ in H^l, and let

$$\varphi(x_1, y_1) = \tilde{F}(x_1 + iy_1, \zeta_2, \ldots, \zeta_l)$$

for fixed $\zeta_2, \ldots, \zeta_l$. Then

$$\frac{\partial}{\partial x_1}\varphi(x_1, y_1) = -\tilde{H}_1(x_1 + iy_1, \zeta_2, \ldots, \zeta_l)\varphi(x_1, y_1).$$

Since $H_1(z)$ is continuous on D^l, it follows that if we fix $c > 0$ then $|H_1(z)| \leq M_1$ for $|z_j| \leq e^{-c}$, i.e., for $x_1 \geq c$ and $\operatorname{Re}\zeta_j \geq c$, $2 \leq j \leq l$. Thus

$$\left|\frac{\partial}{\partial x_1}\varphi(x_1, y_1)\right| \leq M_1|\varphi(x_1, y_1)|.$$

Now $\tilde{F}$ is nowhere 0, by Theorem B.8a, and Lemma B.13 implies

$$\frac{\partial}{\partial x_1}|\varphi(x_1, y_1)| \leq M_1|\varphi(x_1, y_1)|$$

for $x_1 \geq c$ and $\operatorname{Re}\zeta_j \geq c$, $2 \leq j \leq l$. Hence

$$|\varphi(x_1, y_1)| \leq |\varphi(c, y_1)|e^{M_1(x_1-c)} \qquad \text{for } x_1 \geq c, \operatorname{Re}\zeta_j \geq c, 2 \leq j \leq l.$$

Next we can set $\psi(x_2, y_2) = \tilde{F}(\zeta_1, x_2 + iy_2, \ldots, \zeta_l)$ and argue similarly, and we can continue in this way. In the end we get

$$|\tilde{F}(\zeta_1, \ldots, \zeta_l)| \leq |\tilde{F}(c + iy_1, \ldots, c + iy_l)|e^{\Sigma M_j(x_j-c)}.$$

Since $\tilde{F}(c + iy_1, \ldots, c + iy_l)$ is continuous for $0 \leq y_j \leq 2\pi$, we conclude

$$|\tilde{F}(\zeta)| \leq C_0 e^{M\cdot x} \qquad \text{for} \quad x_j \geq c, 0 \leq y_j \leq 2\pi, 1 \leq j \leq l.$$

Since $\tilde{F}$ is a solution, we have $\tilde{F}(\zeta) = \tilde{\Phi}(\zeta)v$ for some v in V. Letting $v_1, \ldots, v_k$ be an orthonormal basis for V and writing $v = \sum c_j v_j$, we have

$$|\tilde{\Phi}(\zeta)v| = |\textstyle\sum c_j\tilde{F}_j(\zeta)| \leq \sum|c_j||\tilde{F}_j(\zeta)| \leq e^{M\cdot x}\sum|c_j||C_j| \leq e^{M\cdot x}(\sum C_j^2)^{1/2}|v|.$$

Hence in the operator norm

$$\|\tilde{\Phi}(\zeta)\| \leq Ce^{M\cdot x} \qquad \text{for some } C \text{ if } x_j \geq c, 0 \leq y_j \leq 2\pi.$$

Then

$$\|\tilde{S}(\zeta)\| \leq \|\tilde{S}(\zeta)e^{-\zeta\cdot R}\|\,\|e^{\zeta\cdot R}\| \leq Ce^{M\cdot x}\|e^{\zeta_1 R_1}\|\cdots\|e^{\zeta_l R_l}\|$$

and so

$$\|S(z)\| \leq C|z_1|^{-M_1}\cdots|z_l|^{-M_l}\|z_1^{-R_1}\|\cdots\|z_l^{-R_l}\|. \tag{B.23}$$

Now $0 \leq y_j \leq 2\pi$ implies

$$\begin{aligned}\|z_j^{-R_j}\| &= \|\exp(-R_j \log z_j)\| \leq \exp\|-R_j \log z_j\| \\ &= \exp(|\log z_j|\,\|R_j\|) \leq a_j|z_j|^{-b_j}.\end{aligned}$$

Substituting this inequality into (B.23), we obtain the desired estimate on $S(z)$, and the theorem follows.

Let

$$(\mathbb{Z}^+)^l = \{k = \{k_1, \ldots, k_l\} \mid \text{each } k_i \text{ is an integer} \geq 0\}.$$

For k in $(\mathbb{Z}^+)^l$, define

$$|k| = |(k_1, \ldots, k_l)| = \textstyle\sum k_i$$
$$(\log z)^k = (\log z_1)^{k_1}\cdots(\log z_l)^{k_l}.$$

If $s = (s^{(1)}, \ldots, s^{(l)})$ is in $\mathbb{C}^l$, let

$$z^s = z_1^{s(1)} \cdots z_l^{s(l)}.$$

Theorem B.15. Suppose $H_1(z), \ldots, H_l(z)$ are holomorphic functions from D^l to End $\mathbb{C}^n$. Then

(a) the system

$$\frac{\partial w}{\partial z_i} = z_i^{-1} H_i(z) w \tag{B.24}$$

in $(D^\times)^l$ has fundamental matrix (relative to z_0) of the form

$$\Phi(z) = S(z)\left(\sum_{i=1}^{r} z^{s_i} p_i(\log z)\right), \tag{B.25}$$

where $S(z)$ is holomorphic from D^l to Hom$(V, \mathbb{C}^n)$, the s_i are in $\mathbb{C}^l$, and the $p_i(\log z)$ are functions with values in End V having polynomial entries in $\log z$ of degree $\leq l(n-1)$.

(b) any global multiple-valued holomorphic solution φ of the system (B.24) in $(D^\times)^l$ has the form

$$\varphi(z) = \sum_{i=1}^{r} \sum_{0 \leq |m| \leq l(n-1)} z^{s_i} (\log z)^m f_{s_i,m}(z) \tag{B.26}$$

with each $f_{s_i,m}(z)$ holomorphic from D^l into $\mathbb{C}^n$.

(c) any C^∞ solution of the system (B.24) on $(0, 1)^l$ extends to a global multiple-valued holomorphic solution on $(D^\times)^l$.

Proof.

(a) By Theorem B.14 we are to show that

$$z_1^{R_1} \cdots z_l^{R_l} = \sum_{i=1}^{r} z^{s_i} p_i(\log z) \tag{B.27}$$

for suitable vectors s_i and polynomials p_i. Since the R_j's commute, they have simultaneous Jordan forms, which in turn decompose into mutually commuting diagonal and upper triangular parts. Thus we can write $R_j = D_j + N_j$ with D_j diagonable, N_j nilpotent, and all the D_j and N_j commuting. Let I_i, $1 \leq i \leq r$, be projections on independent subspaces on which all the D_j are scalar. Say $D_j I_i = s_i^{(j)} I_i$. Let s_i be the tuple $(s_i^{(j)})$. Then

$$z_j^{D_j} = \sum_{i=1}^{r} z_j^{D_j I_i} I_i = \sum_{i=1}^{r} z_j^{s_i^{(j)}} I_i$$

and

$$\prod z_j^{D_j} = \sum_{i=1}^{r} z^{s_i} I_i.$$

Moreover, $\prod_{j=1}^{l} z_j^{N_j}$ is of the form $p(\log z)$ with p a polynomial of degree $\leq l(n-1)$ because

$$z_j^{N_j} = \exp(N_j \log z_j) = \sum_{t=0}^{n-1} (\log z_j)^t N_j^t.$$

Hence

$$z_1^{R_1} \cdots z_l^{R_l} = \left(\prod z_j^{D_j}\right)\left(\prod z_j^{N_j}\right) = \sum_{i=1}^{r} z^{s_i} I_i p(\log z),$$

and (B.27) follows with $p_i(\log z) = I_i p(\log z)$.

(b) We have $\varphi(z) = \Phi(z)v$ for some v in V (namely $v = \varphi(z_0)$), and (b) then follows from (a).

(c) We transform the system (B.24) in $(D^\times)^l$ into a system in H^l of the form (B.18) and (B.19). Then we can apply Theorem B.9 and transform back to $(D^\times)^l$.

§5. Systems of Higher Order

Motivation. In $SL(2, \mathbb{R})$ one encounters the following situation when studying matrix coefficients of irreducible representations: U_1 and U_2 are two finite-dimensional vector spaces, and B_1 and B_2 are known members of End U_1 and End U_2, respectively. An unknown function F on $(0, \infty)$ with values in $U = \text{Hom}(U_2, U_1)$ is to be an eigenfunction, with eigenvalue χ, of the operator Δ given by

$$\frac{1}{2}\Delta F(t) = \frac{d^2F}{dt^2} + (\coth t)\frac{dF}{dt} + (\sinh t)^{-2}(F(t)B_2^2 + B_1^2 F(t)) - \frac{\cosh t}{(\sinh t)^2} B_1 F(t) B_2. \tag{B.28}$$

Here, right by B_2 and left by B_1 may be regarded as endomorphisms of U and can be written as operators applied on the left to $F(t)$. Then the operator on the right side of (B.28) is seen to have all its coefficients of the form $\sum_{m=0}^{\infty} a(m)e^{-mt}$, where $a(m)$ is in End U and the series is absolutely convergent for $t > 0$.

Let $\mathscr{U}$ be the space of all constant coefficient differential operators in $\frac{d}{dt}$ on $\mathbb{R}^+$. One can show that the two-element set $\left\{u_1 = 1, u_2 = \frac{d}{dt}\right\}$ has the property that any member u of $\mathscr{U}$ can be written as

$$u = \sum_{i=1}^{2}\sum_{j=1}^{2} g_{ij}(t) u_i D_j \tag{B.29}$$

where $D_1 = I$ and $D_2 = \Delta$ and where the g_{ij} are functions of the form $\sum_{m=0}^{\infty} a(m)e^{-mt}$ as above. Let $\tilde{F}: \mathbb{R}^+ \to U \oplus U$ be defined by $\tilde{F} = (u_1 F, u_2 F)$, and apply (B.29) to F with u specialized as $u = \frac{d}{dt} u_l$, $l = 1$ or 2. We have

$$\frac{d}{dt} u_l F = \sum_{i,j} g_{ij,l}(t) u_i D_j F = \sum_i \left(\sum_j g_{ij,l}(t) \chi_j \right) u_i F,$$

where $\chi_1 = 1$ and χ_2 is the eigenvalue χ of Δ. If we assemble the results for $l = 1$ and 2 into a vector equation, we see that

$$\frac{d\tilde{F}}{dt} = \Gamma(t)\tilde{F}$$

with $\Gamma(t)$ in $\operatorname{End}(U \oplus U)$. This formula can be transformed into an equation to which Theorem B.15 applies if we make the substitution $z = e^{-t}$. The theorem gives us an expansion for $\tilde{F}(t)$ and in particular for $u_1 F = F$.

We shall establish more general results about such differential equations now, casting everything in terms of variables analogous to $z = e^{-t}$.

Thus let

U = finite-dimensional complex vector space (replacing $\mathbb{C}^n$ earlier)

$\mathscr{H}_U$ = {holomorphic functions from D^l to U}

$\mathscr{H}_{\operatorname{End} U}$ = {holomorphic functions from D^l to End U}.

If k is in $(\mathbb{Z}^+)^l$, we let

$$\partial^k = \left(\frac{\partial}{\partial z_1}\right)^{k_1} \cdots \left(\frac{\partial}{\partial z_l}\right)^{k_l}.$$

Let
$$\mathscr{D} = \left\{ \sum_{\substack{k \in (\mathbb{Z}^+)^l \\ \text{finite sum}}} A_k \partial^k \,\middle|\, A_k \in \mathscr{H}_{\operatorname{End} U} \text{ for all } k \right\}.$$

Members of $\mathscr{D}$ act as differential operators on $\mathscr{H}_U$ with values in $\mathscr{H}_U$. $\mathscr{D}$ is an associative algebra with identity, and $\mathscr{H}_{\operatorname{End} U}$ is the subalgebra of 0th degree operators; we can thus regard $\mathscr{D}$ as a left $\mathscr{H}_{\operatorname{End} U}$ module. Let

$\mathscr{D}^*$ = left $\mathscr{H}_{\operatorname{End} U}$ submodule of $\mathscr{D}$ generated by all monomials in the $z_i \frac{\partial}{\partial z_i}$.

It is easy to check that

$$\mathscr{D}^* = \left\{ \sum_{\substack{k \in (\mathbb{Z}^+)^l \\ \text{finite sum}}} A_k z^k \partial^k \,\middle|\, A_k \in \mathscr{H}_{\operatorname{End} U} \text{ for all } k \right\}$$
$$= \left\{ \sum_{\substack{k \in (\mathbb{Z}^+)^l \\ \text{finite sum}}} B_k (z\partial)^k \,\middle|\, B_k \in \mathscr{H}_{\operatorname{End} U} \text{ for all } k \right\}.$$

The second formula shows that $\mathscr{D}^*$ is an algebra.

Let $\{D_\alpha\}$ be a family of operators from $\mathscr{D}^*$. We say the system $\{D_\alpha w = 0\}$ has a **simple singularity** (along $D^l - (D^\times)^l$) if the left ideal $\mathscr{I}$ in $\mathscr{D}^*$ generated by $\{D_\alpha\}$ has the property that $\mathscr{D}^*/\mathscr{I}$ is finitely generated *as an* $\mathscr{H}_{\text{End }U}$ *module.*

Example 1. Let $l = 1$ and $U = \mathbb{C}$. Let $\{D_\alpha\}$ be the one-element set

$$D = \sum_{k=0}^{M} f_k(z) z^k \frac{d^k}{dz^k}$$

with f_k analytic for $|z| < 1$ and $f_M(z) \neq 0$ for $z \neq 0$. Then one can show that $Dw = 0$ has a simple singularity (at $z = 0$) if and only if $f_M(0) \neq 0$. $\left[\text{If } f_M(0) \neq 0, \text{ the elements } z^j \frac{d^j}{dz^j}, 0 \leq j \leq M - 1, \text{ represent generators of } \mathscr{D}^*/\mathscr{I}.\right]$

Example 2. Let l be general and let $U = \mathbb{C}^n$. Let $\{D_\alpha w = 0\}$ be the system

$$z_i \frac{\partial w}{\partial z_i} = H_i(z) w, \qquad 1 \leq i \leq l,$$

with H_i holomorphic on D^l. This system has a simple singularity.

Theorem B.16. Suppose that the system $\{D_\alpha w = 0\}$, with operators in $\mathscr{D}^*$, has a simple singularity. Then there exist an integer M and a finite subset $S \subseteq \mathbb{C}^l$ such that any global multiple-valued holomorphic solution φ of the system on $(D^\times)^l$ has the form

$$\varphi(z) = \sum_{s \in S} \sum_{0 \leq |m| \leq M} z^s (\log z)^m f_{s,m}(z)$$

with each $f_{s,m}$ in $\mathscr{H}_U$. Moreover, any C^∞ solution of the system on $(0, 1)^l$ extends to a global multiple-valued holomorphic solution on $(D^\times)^l$.

Proof. Choose $D_1, \ldots, D_N$ in $\mathscr{D}^*$ with $D_1 = 1$ such that any member of $\mathscr{D}^*$ can be written in the form

$$h_1 D_1 + \ldots + h_N D_N + \psi \tag{B.30}$$

with h_j in $\mathscr{H}_{\text{End }U}$ and with ψ in the left ideal $\mathscr{I}$ generated by the D_α. Expanding $z_i \dfrac{\partial}{\partial z_i} D_j$ in the form (B.30), we obtain $c_{jk}^{(i)}$ in $\mathscr{H}_{\text{End }U}$ such that

$$z_i \frac{\partial}{\partial z_i} D_j = \sum_{k=1}^{N} c_{jk}^{(i)} D_k \bmod \mathscr{I}. \tag{B.31}$$

Put $\varphi_j = D_j\varphi$ for $1 \le j \le N$, and let $\boldsymbol{\varphi} = \{\varphi_j\}$. Then (B.31) applied to φ_j gives

$$z_i \frac{\partial \varphi_j}{\partial z_i} = \sum_{k=1}^{N} c_{jk}^{(i)}(z)\varphi_k \qquad \text{for } 1 \le i \le l,\ 1 \le j \le N.$$

Hence

$$\frac{\partial \boldsymbol{\varphi}}{\partial z_i} = z_i^{-1} H_i(z)\boldsymbol{\varphi}$$

with $H_i = \{c_{jk}^{(i)}\}$. By Theorem B.15b there exist S and M such that

$$\boldsymbol{\varphi}(z) = \sum_{s \in S} \sum_{0 \le |m| \le M} z^s (\log z)^m F_{s,m}(z) \tag{B.32}$$

with $F_{s,m}$ in $\mathscr{H}_{NU}$, where NU denotes the sum of N copies of U; here S and M are independent of $\boldsymbol{\varphi}$ and depend only on the data in (B.31). Composing (B.32) with evaluation in the first U component, we obtain the required formula for $\varphi_1 = D_1\varphi = \varphi$. The statement in the theorem about C^∞ solutions follows from Theorem B.15c.

§6. Leading Exponents and the Analog of the Indicial Equation

Let us rewrite the form (B.26) of solutions in Theorems B.15 and B.16. So far, we have written

$$\varphi(z) = \sum_{i=1}^{r} \sum_{0 \le |m| \le M} z^{s_i} (\log z)^m f_{s_i,m}(z).$$

Putting

$$f_{s_i,m}(z) = \sum_{k \in (\mathbb{Z}^+)^l} c_{s_i,k,m} z^k$$

we obtain

$$\varphi(z) = \sum_{i=1}^{r} \sum_{k \in (\mathbb{Z}^+)^l} \varphi_{s_i,k}(z) \tag{B.33a}$$

with

$$\varphi_{s_i,k}(z) = \sum_{0 \le |m| \le M} c_{s_i,k,m}\, z^{s_i + k} (\log z)^m. \tag{B.33b}$$

In (B.33) we can have $s_i + k = s_i' + k'$, and we shall eliminate this redundance.

We say s and t in $\mathbb{C}^l$ are **integrally equivalent** if $s - t$ is in $\mathbb{Z}^l$. In this case we say $s \le t$ if $t - s$ is in $(\mathbb{Z}^+)^l$. If $s_i \le s_j$ in (B.33a), we can lump together the contributions from s_i and s_j, eliminating the t terms. More generally if $t_1, \ldots, t_j$ are integrally equivalent, there exists s in $\mathbb{C}^l$ with $s \le t_i$ for $1 \le i \le j$, and all the t_i terms can be lumped into s terms. Thus we obtain, with redefined notation

$$\varphi(z) = \sum_{i=1}^{r} \sum_{k \in (\mathbb{Z}^+)^l} \varphi_{s_i+k}(z) \tag{B.34}$$

with $s_1, \ldots, s_r$ integrally inequivalent.

With $\varphi(z)$ expanded as in (B.34), we define

$$\mathscr{E}(\varphi) = \{t \in \mathbb{C}^l \mid \varphi_t \not\equiv 0\} = \text{set of } \textbf{exponents} \text{ of } \varphi$$

$$\begin{aligned}\mathscr{E}^0(\varphi) &= \{\text{minimal elements of } \mathscr{E}(\varphi) \text{ under } \leq\} \\ &= \text{set of } \textbf{leading exponents} \text{ of } \varphi.\end{aligned}$$

If t is a leading exponent, we call φ_t a **leading term** of φ.

Proposition B.17. With $\varphi(z)$ expanded as in (B.34),

(a) to each $t \in \mathscr{E}(\varphi)$ corresponds some $s \in \mathscr{E}^0(\varphi)$ such that $s \leq t$.
(b) $\mathscr{E}^0(\varphi)$ is a finite set. In fact, any nonempty subset of $(\mathbb{Z}^+)^l$ has only finitely many minimal elements.

Proof.
(a) Otherwise we could find an infinite sequence $\{t_j\}$ in $\mathscr{E}(\varphi)$ with $t = t_1 > t_2 > t_3 > \ldots$. These elements must all be in $s_i + (\mathbb{Z}^+)^l$, in order to be integrally equivalent, and then we obtain a strictly decreasing infinite sequence in $(\mathbb{Z}^+)^l$, contradiction.

(b) We induct on l, the case $l = 1$ being clear. Let $(n_1^{(j)}, \ldots, n_{l-1}^{(j)})$, $1 \leq j \leq J$, be the minimal elements among the projections of a subset of $(\mathbb{Z}^+)^l$ on the first $l - 1$ coordinates, and let $n_l^{(j)}$ be l^{th} coordinates for each. Put $M = \max_{1 \leq j \leq J}(n_l^{(j)})$. For $0 \leq m \leq M$, form the finitely many minimal elements of the form $(k_1, \ldots, k_{l-1}, m)$. The claim is that every minimal element in the given set is of this form. In fact, suppose $(i_1, \ldots, i_l)$ is minimal and $i_l > M$. We have

$$i_1 \geq n_1^{(j)}, \ldots, i_{l-1} \geq n_{l-1}^{(j)}$$

for some j. Then $i_l > M \geq n_l^{(j)}$ contradicts minimality of $(i_1, \ldots, i_l)$.

We wish to identify $\mathscr{E}^0(\varphi)$ as closely as possible. Let Δ be the algebra over End U having l commuting indeterminates $\delta_1, \ldots, \delta_l$ in the center and given by

$$\Delta = (\text{End } U)[\delta_1, \ldots, \delta_l].$$

Lemma B.18. There exists a unique algebra homomorphism $\sigma: \mathscr{D}^* \to \Delta$ such that

$$\sigma\left(z_i \frac{\partial}{\partial z_i}\right) = \delta_i \qquad \text{and} \qquad \sigma(f) = f(0) \text{ for } f \in \mathscr{H}_{\text{End } U}.$$

The homomorphism σ is onto, and its kernel is equal to the left ideal generated by $z_1, \ldots, z_l$ in $\mathscr{D}^*$.

Remarks. When $l = 1$ and $U = \mathbb{C}$, one computes that σ is given by

$$\sigma\left(z^m \frac{d^m}{dz^m}\right) = \delta(\delta - 1) \cdots (\delta - m + 1).$$

If $D = z^M \frac{d^M}{dz^M} + \sum_{m=0}^{M-1} f_m(z) z^m \frac{d^m}{dz^m}$, then $\sigma(D)$ is just the indicial polynomial of D.

Proof. Uniqueness of σ is clear. For existence, we can write each D in $\mathscr{D}^*$ uniquely as

$$D = \sum_{k \in (\mathbb{Z}^+)^l} A_k (z\partial)^k \qquad \text{with} \qquad A_k \in \mathscr{H}_{\operatorname{End} U}.$$

Define
$$\sigma(D) = \sum_k A_k(0) \delta^k. \tag{B.35}$$

Then σ is $\mathbb{C}$-linear and onto, and it satisfies $\sigma\left(z_i \frac{\partial}{\partial z_i}\right) = \delta_i$ and $\delta(f) = f(0)$. To prove that σ is multiplicative, it is enough to show

$$\sigma(f(z\partial)^k D) = f(0)\delta^k \sigma(D).$$

This follows from the easily verified identities

$$\sigma(fD) = f(0)\sigma(D)$$

and
$$\sigma\left(z_i \frac{\partial}{\partial z_i} D\right) = \delta_i \sigma(D).$$

Finally the kernel of σ is clearly the right ideal generated by $z_1, \ldots, z_l$, and we are to prove the right ideal generated by $z_1, \ldots, z_l$ equals the left ideal generated by $z_1, \ldots, z_l$. We have

$$\left(z_i \frac{\partial}{\partial z_i}\right) z_i = z_i \left(1 + z_i \frac{\partial}{\partial z_i}\right).$$

Thus z_i can be commuted past an operator, introducing an error term that has coefficient z_i. The result follows.

Now let $\{D_\alpha w = 0\}$ be a system with a simple singularity, with each D_α in $\mathscr{D}^*$, and let $\mathscr{I}$ be the left ideal in $\mathscr{D}^*$ generated by $\{D_\alpha\}$. By assumption there exist $D_1, \ldots, D_N$ in $\mathscr{D}^*$ such that any D in $\mathscr{D}^*$ can be written as

$$D = h_1 D_1 + \ldots + h_N D_N + \psi$$

with h_j in $\mathscr{H}_{\operatorname{End} U}$ and with ψ in $\mathscr{I}$. Applying σ to all such D, we obtain

$$\Delta = \sigma(\mathscr{D}^*) = (\operatorname{End} U)\sigma(D_1) + \ldots + (\operatorname{End} U)\sigma(D_N) + \sigma(\mathscr{I}).$$

Therefore $\sigma(\mathscr{I})$ is a left ideal in Δ of finite codimension. The finite-dimensional Δ module $\Delta/\sigma(\mathscr{I})$ is called the **indicial module** associated to the system.

Example. Let $l = 1$ and $U = \mathbb{C}^n$, and let the system be

$$\frac{dw}{dz} = z^{-1}H(z)w.$$

We rewrite the system as $Dw = 0$ with $D = z\dfrac{d}{dz} - H(z)$. Then $\sigma(D) = \delta - H(0)$. We can therefore identify $\Delta/\sigma(\mathscr{I})$ as all polynomials in $H(0)$ within End $\mathbb{C}^n$. Under this identification, left multiplication by δ on $\Delta/\sigma(\mathscr{I})$ becomes left multiplication by $H(0)$, whose eigenvalues are the same as the eigenvalues of $H(0)$. These remarks will make it clear that Theorem B.7 is a special case of Theorem B.19 below.

The algebra Δ acts on the finite-dimensional vector space $\Delta/\sigma(\mathscr{I})$, and the transformations "left by δ_i" form a commuting family and have simultaneous generalized eigenspaces. A typical eigenvalue will be denoted $s = (s^{(1)}, \ldots, s^{(l)})$, and the generalized eigenspace will be denoted $(\Delta/\sigma(\mathscr{I}))^{(s)}$.

Theorem B.19. Any leading exponent s of any solution φ of the system $\{D_\alpha w = 0\}$ with a simple singularity is a simultaneous eigenvalue of the operators "left by δ_i" on the indicial module.

The proof will be preceded by three lemmas that relate the system to an "Euler system." For D in $\mathscr{D}^*$ we again write

$$D = \sum_{k \in (\mathbb{Z}^+)^l} A_k (z\partial)^k.$$

Then

$$\sigma(D) = \sum_{k \in (\mathbb{Z}^+)^l} A_k(0)\delta^k.$$

If we compose σ with the one-one homomorphism $\delta^k \to (z\partial)^k$, we arrive at an operator

$$D_0 = \sum_{k \in (\mathbb{Z}^+)^l} A_k(0)(z\partial)^k.$$

Then

$$\sigma(D) = \sigma(D_0). \tag{B.36}$$

If $\{D_\alpha w = 0\}$ is a system with a simple singularity, the system $\{(D_\alpha)_0 w = 0\}$ is called the corresponding **Euler system.**

Lemma B.20. If $\{D_\alpha w = 0\}$ has a simple singularity, so does the corresponding Euler system.

Proof. Let $\mathcal{I}_0$ be the left ideal generated by the $(D_\alpha)_0$. Since $\sigma(D_\alpha) = \sigma((D_\alpha)_0)$, we have $\sigma(\mathcal{I}) = \sigma(\mathcal{I}_0)$. Then the simple singularity of $\{D_\alpha w = 0\}$ implies that $\Delta/\sigma(\mathcal{I}_0)$ is finite-dimensional. Consequently there exists an integer K such that

$$\sum_{|k| \leq K} (\text{End } U)(z\partial)^k + \mathcal{I}_0 \supseteq \sum_{\text{all } k} (\text{End } U)(z\partial)^k.$$

If $|k| > K$, we can thus choose D_k in $\mathcal{I}_0$ with $(z\partial)^k - D_k$ of order $\leq |k| - 1$. For any h in $\mathcal{H}_{\text{End } U}$, we then have

$$h(z)(z\partial)^k - h(z)D_k \text{ of order } \leq |k| - 1.$$

Iterating this procedure, we see that $\mathcal{D}^* = \mathcal{I}_0 + \mathcal{D}^*_K$, where $\mathcal{D}^*_K$ is the set of operators of order $\leq K$, and hence $\mathcal{D}^*/\mathcal{I}_0$ is finitely generated as an $\mathcal{H}_{\text{End } U}$ module.

Lemma B.21. Let φ be a solution of $\{D_\alpha w = 0\}$, written as in (B.34).

(a) For D in $\mathcal{D}^*$, each exponent of $D\varphi$ is $\geq$ some leading exponent of φ.

(b) For D in $\mathcal{D}^*$, $D_0\varphi_{s_i+k} = (D_0\varphi)_{s_i+k}$. Consequently $D_0\varphi = 0$ only if $D_0\varphi_{s_i+k} = 0$ for each $s_i + k$.

Proof. We have

$$\varphi(z) = \sum_{i=1}^{r} \sum_{k \in (\mathbb{Z}^+)^l} \varphi_{s_i+k}$$

with

$$\varphi_{s_i+k} = \sum_{0 \leq |m| \leq M} c_{s_i+k,\, m} z^{s_i+k} (\log z)^m.$$

Then

$$z_1 \frac{\partial}{\partial z_1} \varphi_{s_i+k}(z) = \sum_{0 \leq |m| \leq M} c_{s_i+k,\, m}(s_i + k)_1 z^{s_i+k}(\log z)^m + \sum_{0 \leq |m| \leq M} c_{s_i+k,\, m} m_1 z^{s_i+k}(\log z)^{m-e_1}. \tag{B.37}$$

It is clear therefore that

$$(z\partial)^t \varphi_{s_i+k} = \psi_{s_i+k} \qquad \text{for an appropriate } \psi.$$

Conclusions (a) and (b) follow immediately.

Lemma B.22. If φ is a solution of the system $\{D_\alpha w = 0\}$ with a simple singularity, then any leading term of φ satisfies the corresponding Euler system $\{(D_\alpha)_0 w = 0\}$.

Proof. Write D for an operator D_α. Since $D\varphi = 0$, we have

$$D_0\varphi = -(D - D_0)\varphi.$$

By (B.36), $\sigma(D - D_0) = 0$. Thus $D - D_0$ is in $\ker(\sigma)$, and Lemma B.18 says that

$$-(D - D_0) = Q_1 z_1 + \ldots + Q_l z_l \qquad \text{with } Q_1, \ldots, Q_l \text{ in } \mathscr{D}^*.$$

Hence
$$D_0\varphi = Q_1(z_1\varphi) + \ldots + Q_l(z_l\varphi).$$

Let s be a leading exponent of φ. Then $s + e_j$ is a leading exponent of $z_j\varphi$. If s is an exponent of $Q_j(z_j\varphi)$, Lemma B.21a shows that some leading s_0 for $z_j\varphi$ has $s_0 \leq s < s + e_j$, and thus $s + e_j$ cannot be a leading exponent of $z_j\varphi$, contradiction. So $(Q_j(z_j\varphi))_s = 0$, and Lemma B.21b gives

$$D_0\varphi_s = (D_0\varphi)_s = \sum_{j=1}^{l} (Q_j(z_j\varphi))_s = 0.$$

Proof of Theorem B.19. Let φ be a solution of the system $\{D_\alpha w = 0\}$, and let s be a leading exponent. By Lemma B.22, $(D_\alpha)_0\varphi_s = 0$. Here the system $\{(D_\alpha)_0 w = 0\}$ has a simple singularity by Lemma B.20. In (B.33a) with φ_s on the left side, only $k = 0$ contributes to the right. Thus (B.33b) gives

$$\varphi_s(z) = z^s \sum_{0 \leq |m| \leq M} c_{s,m}(\log z)^m, \qquad c_{s,m} \in U. \tag{B.38}$$

If we identify $\delta^k \to (z\partial)^k$, then Δ gets identified with a certain subalgebra $\mathscr{D}_0^*$ of $\mathscr{D}^*$. The property of φ_s is that each D in the left ideal I of $\mathscr{D}_0^*$ generated by $\{(D_\alpha)_0\}$ satisfies $D\varphi_s = 0$.

For simplicity let us now change notation, writing (B.38) as

$$\varphi(z) = z^s \sum_{0 \leq |m| \leq M} \varphi_m(\log z)^m, \qquad \varphi_m \in U. \tag{B.39}$$

For any function $\psi(z)$ of the form

$$\psi(z) = z^s \sum_{0 \leq |m| \leq M} \psi_m(\log z)^m, \tag{B.40}$$

we see from (B.37) that

$$\psi_m = \frac{1}{m!}((z\partial - s)^m\psi)_0, \tag{B.41}$$

where $m! = m_1! \cdots m_l!$

In the notation of (B.39) and (B.40), put

$$T_\varphi(D) = (D\varphi)_0 \qquad \text{for } D \in \mathscr{D}_0^*.$$

Then T_φ is in $\mathrm{Hom}_{\mathrm{End}\,U}(\mathscr{D}_0^*, U)$. Since $T_\varphi(D) = 0$ for D in I (because $D\varphi = 0$ when D is in I), we can regard T_φ as a member of $\mathrm{Hom}_{\mathrm{End}\,U}(\mathscr{D}_0^*/I, U)$. Moreover, (B.41) gives

$$T_\varphi((z\partial - s)^m) = ((z\partial - s)^m\varphi)_0 = m!\varphi_m,$$

which shows that T_φ is not the 0 member of $\mathrm{Hom}_{\mathrm{End}\,U}(\mathscr{D}_0^*/I, U)$. For $t \neq s$, we shall show that T_φ vanishes on the generalized eigenspace $(\mathscr{D}_0^*/I)^{(t)}$ for the action of $z\partial$, and it will follow that $(\mathscr{D}_0^*/I)^{(s)} \neq 0$. Since

$$\mathscr{D}_0^*/I \cong \Delta/\sigma(\mathscr{I}_0) = \Delta/\sigma(\mathscr{I}),$$

by (B.36), we will have $(\Delta/\sigma(\mathscr{I}))^{(s)} \neq 0$, which is the conclusion of the theorem.

Thus let D be in $(\mathscr{D}_0^*/I)^{(t)}$ and suppose that $0 \neq T_\varphi(D) = (D\varphi)_0$. Then $D\varphi \neq 0$ and we let m be maximal such that $(D\varphi)_m \neq 0$. Form

$$E = (z\partial - s)^m D \text{ in } (\mathscr{D}_0^*/I)^{(t)}.$$

By (B.41)

$$(E\varphi)_0 = ((z\partial - s)^m D\varphi)_0 = m!(D\varphi)_m \neq 0. \tag{B.42}$$

If $m' \neq 0$, then

$$(E\varphi)_{m'} = ((z\partial - s)^{m'} E\varphi)_0 = ((z\partial - s)^{m'+m} D\varphi)_0 = (m'+m)!(D\varphi)_{m'+m} = 0.$$

Therefore

$$E\varphi = z^s(E\varphi)_0. \tag{B.43}$$

Write $t = (t^{(1)}, \ldots, t^{(l)})$. Since E is in $(\Delta/I)^{(t)}$, for each i there is an integer k_i such that $\left(z_i \frac{\partial}{\partial z_i} - t^{(i)}\right)^{k_i} E$ is in I. Thus

$$\left(z_i \frac{\partial}{\partial z_i} - t^{(i)}\right)^{k_i} E\varphi = 0,$$

and (B.43) gives

$$\begin{aligned} 0 &= \left(z_i \frac{\partial}{\partial z_i} - t^{(i)}\right)^{k_i} z^s(E\varphi)_0 \\ &= (s^{(i)} - t^{(i)})\left(z_i \frac{\partial}{\partial z_i} - t^{(i)}\right)^{k_i - 1} z^s(E\varphi)_0 \\ &= \cdots = (s^{(i)} - t^{(i)})^{k_i} z^s(E\varphi)_0. \end{aligned}$$

Choosing i so that $s^{(i)} \neq t^{(i)}$, we see that $(E\varphi)_0 = 0$, in contradiction to (B.42). We conclude that $(\mathscr{D}_0^*/I)^{(t)}$ must have been 0, and the theorem follows.

Corollary B.23. The space of solutions to a system $\{D_\alpha w = 0\}$ with a simple singularity is finite-dimensional.

Proof. If $\varphi \neq 0$ is a solution, Proposition B.17a says that some leading term φ_s is not zero. By Theorem B.19 this index s is a simultaneous eigenvalue for the operators "left by δ_i" on the indicial module. Consequently if S is the (finite) set of simultaneous eigenvalues for the operators "left by δ_i" on the indicial module, then the linear map on solutions given by

$$\varphi = \sum_{s \in \mathbb{C}^l} \varphi_s \to \sum_{s \in S} \varphi_s$$

is one-one. The space of all functions

$$\psi_s = z^s \sum_{0 \leq |m| \leq M} (\psi_s)_m (\log z)^m$$

is finite-dimensional, M being bounded according to Theorem B.16. Hence the space of solutions is finite-dimensional.

§7. Uniqueness of Representation

If we group terms in (B.33) differently, we can rewrite our solutions (B.26) in still another way as

$$\sum_{s \in \mathbb{C}^l} \sum_{0 \leq |m| \leq M} c_{s,m} z^s (\log z)^m, \tag{B.44}$$

with appropriate limitations on the subsets $\{s \in \mathbb{C}^l\}$. This section will discuss the extent to which the sum (B.44) determines properties of the coefficients $c_{s,m}$. The first lemma applies in the setting where z^s is replaced by $e^{s \cdot t}$, and then we return to the notation of (B.44).

Lemma B.24. Suppose that the finite sum $\sum_{s,q} c_{s,q} e^{s \cdot t} t^q$ (in which each s is in $\mathbb{C}^l$, each q is in $\mathbb{R}^l$, and the $c_{s,q}$ are in a common finite-dimensional inner product space) satisfies an inequality

$$\left| \sum_{s,q} c_{s,q} e^{s \cdot t} t^q \right| \leq c e^{s_0 \cdot t} t^{q_0}$$

for all t in $\mathbb{R}^l$ with $t_j \geq r$ for $1 \leq j \leq l$. Then $c_{s,q} \neq 0$ implies $\operatorname{Re} s \leq s_0$ entry by entry. Also if $\operatorname{Re} s = s_0$ and if $c_{s,q} \neq 0$, then $q \leq q_0$.

Proof. Since we can treat one variable of t at a time, we may assume $l = 1$. Dividing, we may assume $s_0 = 0$ and $q_0 = 0$. Let s' be the maximum value of $\operatorname{Re} s$ for which some $c_{s,q}$ is nonzero, and let q' be the maximum value of q for which some $c_{s,q}$ is nonzero with $\operatorname{Re} s = s'$. Then we can rewrite the expression under consideration as

$$\sum_{s,q} c_{s,q} e^{st} t^q = e^{s't} t^{q'} \left(\sum_{s,q} c_{s,q} e^{(s-s')t} t^{q-q'} \right)$$

and take the lim sup of the right side as $t \to +\infty$. Since the left side is by assumption bounded, we will be done if we show that the term in parentheses on the right side does not tend to 0. In the argument it will be enough to sum over those terms with $\operatorname{Re} s = s'$ and $q = q'$, since the others tend to 0. Changing notation, we are to show that

$$\limsup_{t\to+\infty}\left|\sum c_\theta e^{i\theta t}\right| \neq 0 \tag{B.45}$$

unless all coefficients are 0. Putting $f(t) = \sum c_\theta e^{i\theta t}$, we easily compute that

$$\lim_{T\to\infty} T^{-1}\int_0^T |f(t)|^2\,dt = \sum |c_\theta|^2.$$

Since

$$\lim_{T\to\infty} T^{-1}\int_0^T |f(t)|^2\,dt \leq \limsup_{t\to+\infty} |f(t)|^2,$$

(B.45) and the lemma follow.

Theorem B.25. Let $\mathscr{F}$ be a finite subset of $\mathbb{C}^l$, let $\mathscr{M}$ be a finite set of integer tuples ≥ 0. and let $r > 0$. Suppose, for each s in $\mathscr{F}$ and for each integer tuple m in $\mathscr{M}$, that $F_{s,m}(z)$ is a $\mathbb{C}^N$-valued function holomorphic in the polydisc $(D^r)^l$ of radius r in each of its l coordinates. Let $F(z)$ be the function

$$F(z) = \sum_{s\in\mathscr{F}}\sum_{m\in\mathscr{M}} z^s(\log z)^m F_{s,m}(z),$$

and rearrange $F(z)$ as

$$F(z) = \sum_{s\in\mathbb{C}^l}\sum_{m\in\mathscr{M}} c_{s,m} z^s(\log z)^m.$$

If there exist an $\varepsilon > 0$, a real number s_0, and an integer $m_0 \geq 0$ such that

$$|F(x)| \leq Cx^{s_0}|\log x|^{m_0} \tag{B.46}$$

whenever $0 < x_j \leq \varepsilon$ for $1 \leq j \leq l$, then $c_{s,m} \neq 0$ implies $\operatorname{Re} s \geq s_0$ entry by entry. Also if $\operatorname{Re} s = s_0$ and if $c_{s,m} \neq 0$, then $m \leq m_0$ entry by entry.

Proof. We may assume $\varepsilon < r$. Write

$$F(z_1, z_2, \ldots, z_l) = \sum_{\substack{s_1\in\mathbb{C}\\ m_1\in\mathbb{Z}^+}} (c_{s_1,m_1}(z_2,\ldots,z_l)\, z_1^{s_1}\,(\log z_1)^{m_1}),$$

where

$$c_{s_1,m_1}(z_2,\ldots,z_l) = \sum_{\substack{s(1)=s_1\\ m(1)=m_1}} c_{s,m} z_2^{s_2}\cdots z_l^{s_l}\,(\log z_2)^{m_2}\cdots(\log z_l)^{m_l}.$$

Suppose the result is known for 1 variable and $l-1$ variables. If $c_{s,m} \neq 0$ and if we define $s_1 = s(1)$ and $m_1 = m(1)$, then the result for $l-1$ variables shows c_{s_1,m_1} is not identically 0. Then in turn the result for one variable, in the presence of (B.46), says that $\operatorname{Re} s_1$ does not exceed the first coordinate $s_{0,1}$ of s_0. Moreover, if $\operatorname{Re} s_1 = s_{0,1}$, then m_1 is $\leq m_{0,1}$. Since we could have singled out z_j in place of z_1 in this argument, the result follows for l variables.

In other words, the theorem reduces, via induction, to the case $l = 1$. In this case let m_1 be the highest power of $|\log z|$ that appears. Also let us decompose each $F_{s,m}$ as

$$F_{s,m}(z) = P_{s,m}(z) + R_{s,m}(z),$$

where $P_{s,m}(z)$ is a polynomial containing all terms z^n of $F_{s,m}$ for which $\operatorname{Re}(s + n) \le s_0$ and where $R_{s,m}(z)$ contains the remaining terms. Then let $P(z) = \sum z^s(\log z)^m P_{s,m}(z)$ and $R(z) = \sum z^s(\log z)^m R_{s,m}(z)$. Each term of any $z^s(\log z)^m R_{s,m}(z)$ is dominated for $0 < x \le \varepsilon$ by a multiple of $x^{s_0}|\log x|^{m_1}$. By the assumed analyticity

$$|R(x)| \le C'x^{s_0}\,|\log x|^{m_1} \qquad \text{for} \quad 0 < x \le \varepsilon.$$

Therefore (B.46) implies

$$|P(x)| \le C''x^{s_0}|\log x|^{\max(m,m_1)} \qquad \text{for} \quad 0 < x \le \varepsilon.$$

Changing variables and applying Lemma B.24, we see that each power z^s that contributes to $P(z)$ has $\operatorname{Re} s \ge s_0$. Since the only powers z^s that contribute to $R(z)$ have $\operatorname{Re} s > s_0$, it follows for F that $c_{s,m} \ne 0$ implies $\operatorname{Re} s \ge s_0$. This completes the proof of the first assertion of the theorem for the case $l = 1$.

For the second assertion, we review the definition of P and R and observe that in fact $R(x)$ is dominated by a multiple of $x^{s_0+\delta}$ for $0 < x \le \varepsilon$, for some sufficiently small $\delta > 0$. Thus (B.46) implies $P(x)$ is actually dominated by a multiple of $x^{s_0}|\log x|^{m_0}$. Appealing to Lemma B.24, we see that $c_{s,m} \ne 0$ in P implies $m \le m_0$. This completes the proof of the theorem.

Corollary B.26. Let $\mathcal{F}$ be a finite subset of $\mathbb{C}^l$, let $\mathcal{M}$ be a finite set of integer tuples ≥ 0, and let $r > 0$. Suppose, for each s in $\mathcal{F}$ and for each integer tuple m in $\mathcal{M}$, that $F_{s,m}(z)$ is a $\mathbb{C}^N$-valued function holomorphic in the polydisc $(D^r)^l$. Let $F(z)$ be the function

$$F(z) = \sum_{s\in\mathcal{F}} \sum_{m\in\mathcal{M}} z^s(\log z)^m F_{s,m}(z),$$

and rearrange $F(z)$ as

$$F(z) = \sum_{s\in\mathbb{C}^l} \sum_{m\in\mathcal{M}} c_{s,m} z^s(\log z)^m.$$

If there exists an $\varepsilon > 0$ such that $F(x) = 0$ whenever $0 < x_j \le \varepsilon$ for $1 \le j \le l$, then all coefficients $c_{s,m}$ are 0.

APPENDIX C

Roots and Restricted Roots for Classical Groups

§1. Complex Groups

The Dynkin diagrams of the simple complex classical groups appear in §4.3. The distinct such diagrams, together with a group attached to each, are identified in Table C.1.

TABLE C.1. Distinct irreducible classical Dynkin diagrams

DIAGRAM	VALUES OF n	G	K	FUNDAMENTAL GROUP
A_n	$n \geq 1$	$SL(n+1, \mathbb{C})$	$SU(n+1)$	trivial
B_n	$n \geq 2$	$SO(2n+1, \mathbb{C})$	$SO(2n+1)$	$\mathbb{Z}_2$
C_n	$n \geq 3$	$Sp(n, \mathbb{C})$	$Sp(n)$	trivial
D_n	$n \geq 4$	$SO(2n, \mathbb{C})$	$SO(2n)$	$\mathbb{Z}_2$

The subalgebra $\mathfrak{a}_\mathfrak{p}$ is always such that $\mathfrak{m}_\mathfrak{p} = J(\mathfrak{a}_\mathfrak{p})$, where J is the multiplication-by-i map within $\mathfrak{g}$, and $\mathfrak{t} = \mathfrak{a}_\mathfrak{p} \oplus \mathfrak{m}_\mathfrak{p}$ is a Cartan subalgebra of $\mathfrak{g}$. The word "Diagram" in Table C.1 refers to the root system $\Delta(\mathfrak{t}:\mathfrak{g})$.

The Lie algebra $\mathfrak{m}_p$ is a Cartan subalgebra of $\mathfrak{k}$, and we have

$$\dim \mathfrak{t} = 2 \dim \mathfrak{m}_\mathfrak{p} = 2 \dim \mathfrak{a}_p.$$

Since we use the term "rank G" to refer to $\dim_\mathbb{R} \mathfrak{t}$, we have

$$\text{rank } G = 2 \text{ rank } K = 2 \text{ real rank } G.$$

The restricted root system $\Delta(\mathfrak{a}_\mathfrak{p}:\mathfrak{g})$ is naturally identified with $\Delta(\mathfrak{t}:\mathfrak{g})$ by complex-linear extension of the restricted roots from $\mathfrak{a}_\mathfrak{p}$ to $\mathfrak{t}$, and the Dynkin diagrams are as in Table C.1. The diagram of $\Delta(\mathfrak{t}^\mathbb{C}:\mathfrak{g}^\mathbb{C})$ consists of two copies of the diagram of $\Delta(\mathfrak{t}:\mathfrak{g})$ (cf. Proposition 2.5).

§2. Noncompact Real Groups

The other noncompact simple classical groups (up to local isomorphism) are the subject of Table C.2. In certain low-dimensional cases, the groups

TABLE C.2. Root data for noncompact real groups

CARTAN NUMBERING	G	K	RANK G	RANK K	REAL RANK G	$\Delta(\mathfrak{t}^{\mathbb{C}}:\mathfrak{g}^{\mathbb{C}})$	$\Delta(\mathfrak{a}_{\mathfrak{p}}:\mathfrak{g})$
A I	$SL(n, \mathbb{R})$	$SO(n)$	$n-1$	$[\frac{1}{2}n]$	$n-1$	A_{n-1}	A_{n-1}
A II	$SL(n, \mathbb{H})$	$Sp(n)$	$2n-1$	n	$n-1$	A_{2n-1}	A_{n-1}
A III	$SU(p, q)$	$S(U(p) \times U(q))$	$p+q-1$	$p+q-1$	q	A_{p+q-1}	C_q if $p = q$ $(BC)_q$ if $p > q$
BD I	$SO_0(p, q)$	$SO(p) \times SO(q)$	$[\frac{1}{2}(p+q)]$	$[\frac{1}{2}p] + [\frac{1}{2}q]$	q	$B_{(p+q-1)/2}$ if $p+q$ odd $D_{(p+q)/2}$ if $p+q$ even	D_q if $p = q$ B_q if $p > q$
D III	$SO^*(2n)$	$U(n)$	n	n	$[\frac{1}{2}n]$	D_n	$C_{n/2}$ if n even $(BC)_{(n-1)/2}$ if n odd
C I	$Sp(n, \mathbb{R})$	$U(n)$	n	n	n	C_n	C_n
C II	$Sp(p, q)$	$Sp(p) \times Sp(q)$	$p+q$	$p+q$	q	C_{p+q}	C_q if $p = q$ $(BC)_q$ if $p > q$

fail to be simple, and some of the low-dimensional groups are isomorphic with one another. For the isometry groups of quadratic forms, the notation assumes that $p \geq q$.

§3. Roots vs. Restricted Roots in Noncompact Real Groups

Let $\mathfrak{b}_\mathfrak{p}$ be a maximal abelian subspace of $\mathfrak{m}_\mathfrak{p}$, and let $\mathfrak{t} = \mathfrak{a}_\mathfrak{p} \oplus \mathfrak{b}_\mathfrak{p}$; $\mathfrak{t}$ is a Cartan subalgebra of $\mathfrak{g}$ whose noncompact part has dimension as large as possible. For each of the groups in §2, we shall identify (a choice of) $\mathfrak{a}_\mathfrak{p}$ and tell how the roots of $\Delta(\mathfrak{t}^\mathbb{C}:\mathfrak{g}^\mathbb{C})$ and $\Delta(\mathfrak{a}_\mathfrak{p}:\mathfrak{g})$ are related. For $\Delta(\mathfrak{t}^\mathbb{C}:\mathfrak{g}^\mathbb{C})$ we use the concrete notation for root systems in §4.3, e.g., $\pm e_i \pm e_j$, $\pm e_j$, $\pm 2e_j$ as appropriate. But $\mathfrak{g}^\mathbb{C}$ is normally not identical with one of the complex Lie algebras in §4.3, and thus $\Delta(\mathfrak{t}^\mathbb{C}:\mathfrak{g}^\mathbb{C})$ has to be interpreted in terms of an isomorphism, which we do not specify. For $\Delta(\mathfrak{a}_\mathfrak{p}:\mathfrak{g})$ we use analogous notation but with "f" in place of "e"; in this case we shall be very concrete, telling both the explicit definitions of the f's and their relation to the abstract e's. A group is called **split** (over $\mathbb{R}$) if $\mathfrak{a}_\mathfrak{p}$ is a Cartan subalgebra, i.e., if $\mathfrak{b}_\mathfrak{p} = 0$. In this case, $\mathfrak{m}_\mathfrak{p} = 0$, and the f's and e's may be identified with one another.

In §7.5 we attached a connected semisimple subgroup $G^{(\beta)}$ to each restricted root β. Conjugate roots (under $W(A_\mathfrak{p}:G)$) have conjugate subgroups, and we shall identify the Lie algebra $\mathfrak{g}^{(\beta)}$ of $G^{(\beta)}$ for each simple restricted root β. The $M_\mathfrak{p}$ groups of the various $G^{(\beta)}$'s generate $M_\mathfrak{p}$, as a consequence of Lemma 9.13, and we tell the resulting form of $M_\mathfrak{p}$.

Type A I. $G = \mathrm{SL}(n, \mathbb{R})$, split over $\mathbb{R}$.

$\Delta(\mathfrak{t}^\mathbb{C}:\mathfrak{g}^\mathbb{C})$ of type A_{n-1}.
$\mathfrak{a}_\mathfrak{p}$ = diagonal matrices with real entries and trace 0.
$f_j = e_j$ = evaluation of jth diagonal entry $1 \leq j \leq n$.
$\Delta(\mathfrak{a}_\mathfrak{p}:\mathfrak{g})$ of type A_{n-1}:

$f_1 - f_2$ $\quad$ $f_2 - f_3$ $\quad$ $f_{n-1} - f_n$

$\mathfrak{g}^{(\beta)}$: $\quad$ $\mathfrak{sl}(2, \mathbb{R})$ $\quad$ $\mathfrak{sl}(2, \mathbb{R})$ $\quad$ $\mathfrak{sl}(2, \mathbb{R})$

$\mathfrak{m}_\mathfrak{p} = 0$.
$M_\mathfrak{p}$ = diagonal matrices with ± 1 in each diagonal entry
$\cong n - 1$ copies of $\mathbb{Z}_2$.

Type A II. $G = \mathrm{SL}(n, \mathbb{H})$, quaternion realization.

$\Delta(\mathfrak{t}^\mathbb{C}:\mathfrak{g}^\mathbb{C})$ of type A_{2n-1}.
$\mathfrak{a}_\mathfrak{p}$ = diagonal matrices with real entries and trace 0.

$f_j = \frac{1}{2}(e_{2j-1} + e_{2j})$ = evaluation of jth diagonal entry, $1 \le j \le n$.
$\Delta(\mathfrak{a}_\mathfrak{p}:\mathfrak{g})$ of type A_{n-1}:

$f_1 - f_2$ $\quad f_2 - f_3 \quad\quad f_{n-1} - f_n$

$\mathfrak{g}^{(\beta)}$: $\quad \mathfrak{sl}(2, \mathbb{H}) \quad \mathfrak{sl}(2, \mathbb{H}) \quad \mathfrak{sl}(2, \mathbb{H}) \quad (\mathfrak{sl}(2, \mathbb{H}) \cong \mathfrak{so}(5, 1))$

$\mathfrak{m}_\mathfrak{p}$ semisimple and built from $e_1 - e_2, e_3 - e_4, \ldots, e_{2n-1} - e_{2n}$.
$M_\mathfrak{p}$ = diagonal matrices with diagonal entries of modulus one $\cong (\mathrm{SU}(2))^n$.

Type A III. $G = \mathrm{SU}(p, q)$, $p \ge q$.

$\Delta(\mathfrak{t}^{\mathbb{C}}:\mathfrak{g}^{\mathbb{C}})$ of type A_{p+q-1}.
$\mathfrak{a}_\mathfrak{p} = \mathbb{R}(E_{1,p+q} + E_{p+q,1}) + \mathbb{R}(E_{2,p+q-1} + E_{p+q-1,2}) + \cdots + \mathbb{R}(E_{q,p+1} + E_{p+1,q})$.
$f_j = \frac{1}{2}(e_j - e_{p+q+1-j})$ = coefficient of $(E_{j,p+q+1-j} + E_{p+q+1-j,j})$, $1 \le j \le n$.
$\Delta(\mathfrak{a}_\mathfrak{p}:\mathfrak{g})$ of type C_q if $p = q$ and type $(BC)_q$ if $p > q$:

$(C_q$ or $(BC)_q)$

$f_1 - f_2 \quad\quad f_{q-1} - f_q \quad 2f_q$

$\mathfrak{g}^{(\beta)}$: $\quad \mathfrak{sl}(2, \mathbb{C}) \quad\quad \mathfrak{sl}(2, \mathbb{C}) \quad \mathfrak{su}(p - q + 1, 1)$

$\mathfrak{m}_\mathfrak{p}$ of the form

$$\mathfrak{m}_\mathfrak{p} = \begin{cases} \mathbb{R}^{q-1} \oplus \text{semisimple} & \text{if } p = q \\ \mathbb{R}^{q} \oplus \text{semisimple} & \text{if } p > q, \end{cases}$$

with semisimple part built from

$$e_{q+1} - e_{q+2}, e_{q+2} - e_{q+3}, \ldots, e_{p-1} - e_p$$

and with abelian part consisting of the kernel in $\mathfrak{b}_\mathfrak{p}$ of

$$e_1 - e_{p+q}, e_2 - e_{p+q-1}, \ldots, e_q - e_{p+1}, e_{q+1}, e_{q+2}, \ldots, e_p.$$

$M_\mathfrak{p}$ is connected unless $p = q$ and consists of all members of $\mathrm{SU}(p + q)$ whose off-diagonal entries vanish except between coordinates $q + 1$ and p and whose diagonal entries t_j satisfy $t_j = t_{p+q+1-j}$ for $1 \le j \le q$. When $p = q$, $M_\mathfrak{p}$ has two components.

Type BD I. $G = \mathrm{SO}_0(p, q)$, $p \ge q$, split over $\mathbb{R}$ if $p = q + 1$ or $p = q$.

$\Delta(\mathfrak{t}^{\mathbb{C}}:\mathfrak{g}^{\mathbb{C}})$ of type $B_{(p+q-1)/2}$ if $p + q$ odd, of type $D_{(p+q)/2}$ if $p + q$ even.
$\mathfrak{a}_\mathfrak{p} = \mathbb{R}(E_{1,p+q} + E_{p+q,1}) + \mathbb{R}(E_{2,p+q-1} + E_{p+q-1,2}) + \cdots + \mathbb{R}(E_{q,p+1} + E_{p+1,q})$.

$f_j = e_j =$ coefficient of $(E_{j,\,p+q+1-j} + E_{p+q+1-j,\,j})$, $1 \le j \le q$.
$\Delta(\mathfrak{a}_{\mathfrak{p}}:\mathfrak{g})$ of type B_q if $p > q$, of type D_q if $p = q$:

(if $p > q$)

	$f_1 - f_2$	$f_{q-1} - f_q$	f_q
$\mathfrak{g}^{(\beta)}$:	$\mathfrak{sl}(2, \mathbb{R})$	$\mathfrak{sl}(2, \mathbb{R})$	$\mathfrak{so}(p - q + 1, 1)$

$\mathfrak{m}_{\mathfrak{p}}$ built from $e_{q+1}, \ldots, e_{[(p+q)/2]}$
$\mathfrak{m}_{\mathfrak{p}} = 0$ if $p = q + 1$ or $p = q$
$\mathfrak{m}_{\mathfrak{p}}$ abelian if $p = q + 2$
$\mathfrak{m}_{\mathfrak{p}}$ semsimple otherwise.

$M_{\mathfrak{p}} = SO(p - q) \times (\mathbb{Z}_2)^{q-1}$ for $p \ge q + 1$, with
(i) $SO(p - q)$ built from the middle $p - q$ entries
(ii) $(\mathbb{Z}_2)^{q-1}$ generated by even products of diagonal matrices γ_j defined by

$$\gamma_j = \begin{cases} -1 & \text{in diagonal entries } j \text{ and } p + q + 1 - j \\ +1 & \text{in remaining entries.} \end{cases}$$

Type D III. $G = SO^*(2n)$.

$\Delta(\mathfrak{t}^{\mathbb{C}}:\mathfrak{g}^{\mathbb{C}})$ of type D_n.

$$\mathfrak{a}_{\mathfrak{p}} = \sum_{j=1}^{[n/2]} \mathbb{R}(E_{n+2j,\,2j-1} - E_{n+2j-1,2j} + E_{2j-1,\,n+2j} - E_{2j,n+2j-1}).$$

$f_j = \tfrac{1}{2}(e_{2j-1} + e_{2j}) =$ coefficient of jth term above, $1 \le j \le [n/2]$.
$\Delta(\mathfrak{a}_{\mathfrak{p}}:\mathfrak{g})$ of type $C_{n/2}$ if n even, $(BC)_{[n/2]}$ if n odd:

	$f_1 - f_2$	$f_{[n/2]-1} - f_{[n/2]}$	$2f_{[n/2]}$	
$\mathfrak{g}^{(\beta)}$:	$\mathfrak{sl}(2, \mathbb{H})$	$\mathfrak{sl}(2, \mathbb{H})$	$\mathfrak{sl}(2, \mathbb{R})$	if n even
	$\mathfrak{sl}(2, \mathbb{H})$	$\mathfrak{sl}(2, \mathbb{H})$	$\mathfrak{su}(3, 1)$	if n odd

$\mathfrak{m}_{\mathfrak{p}}$ semisimple and built from $e_1 - e_2, e_3 - e_4, \ldots, e_{n-1} - e_n$ if n even.
$\mathfrak{m}_{\mathfrak{p}}$ of the form $\mathbb{R} \oplus$ semisimple if n odd, with the semisimple part built from $e_1 - e_2, e_3 - e_4, \ldots, e_{n-2} - e_{n-1}$, and with the abelian part corresponding to e_n.
$M_{\mathfrak{p}}$ isomorphic to $\{SU(2)^{n/2}/\mathbb{Z}_2\} \times \mathbb{Z}_2$ if n even.
$M_{\mathfrak{p}}$ connected with Lie algebra isomorphic to $\mathbb{R} \oplus \sum_{j=1}^{[n/2]} \mathfrak{su}(2)$ if n odd.

Type C I. $G = Sp(n, \mathbb{R})$, split over $\mathbb{R}$.
$\Delta(\mathfrak{t}^{\mathbb{C}}, \mathfrak{g}^{\mathbb{C}})$ of type C_n.
$\mathfrak{a}_p =$ diagonal matrices with real entries and with j^{th} entry the negative of the $(n + j)^{\text{th}}$.

$f_j = e_j$ = evaluation of jth diagonal entry, $1 \leq j \leq n$.
$\Delta(\mathfrak{a}_\mathfrak{p}:\mathfrak{g})$ of type C_n:

$f_1 - f_2$ $\quad\cdots\quad$ $f_{n-1} - f_n$ $\quad$ $2f_n$

$\mathfrak{g}^{(\beta)}$: $\mathfrak{sl}(2, \mathbb{R})$ $\quad\cdots\quad$ $\mathfrak{sl}(2, \mathbb{R})$ $\quad$ $\mathfrak{sl}(2, \mathbb{R})$

$\mathfrak{m}_\mathfrak{p} = 0$.
$M_\mathfrak{p} \cong (\mathbb{Z}_2)^n$, generated by diagonal matrices γ_j defined by

$$\gamma_j = \begin{cases} -1 & \text{in diagonal entries } j \text{ and } n+j \\ +1 & \text{in remaining diagonal entries.} \end{cases}$$

Type C II. $G = \mathrm{Sp}(p, q)$, $p \geq q$, quaternion realization.
$\Delta(\mathfrak{t}^\mathbb{C}:\mathfrak{g}^\mathbb{C})$ of type C_{p+q}.
$\mathfrak{a}_\mathfrak{p} = \mathbb{R}(E_{1,p+q} + E_{p+q,1}) + \mathbb{R}(E_{2,p+q-1} + E_{p+q-1,2}) + \dots + \mathbb{R}(E_{q,p+1} + E_{p+1,q})$.
$f_j = \frac{1}{2}(e_{2j-1} + e_{2j})$ = coefficient of $(E_{j,p+q+1-j} + E_{p+q+1-j,j})$, $1 \leq j \leq q$.
$\Delta(\mathfrak{a}_\mathfrak{p}, \mathfrak{g})$ of type $(BC)_q$ if $p > q$ and type C_q if $p = q$:

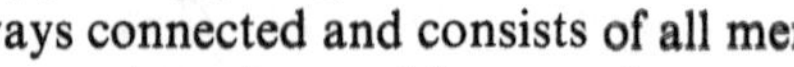

(C_q or $(BC)_q$)

$f_1 - f_2$ $\quad\cdots\quad$ $f_{q-1} - f_q$ $\quad$ $2f_q$

$\mathfrak{g}^{(\beta)}$: $\mathfrak{sl}(2, \mathbb{H})$ $\quad\cdots\quad$ $\mathfrak{sl}(2, \mathbb{H})$ $\quad$ $\mathfrak{sp}(p - q + 1, 1)$

$\mathfrak{m}_\mathfrak{p}$ built from $e_1 - e_2, e_3 - e_4, \dots, e_{2q-1} - e_{2q}$ and all e_j with $j > 2q$.
$M_\mathfrak{p} \cong \mathrm{SU}(2)^q \times \mathrm{Sp}(p - q)$.
$M_\mathfrak{p}$ is always connected and consists of all members of $\mathrm{Sp}(p + q)$ whose off-diagonal entries vanish except between coordinates $q + 1$ and p and whose j^{th} diagonal entry equals its $(p + q + 1 - j)^{\text{th}}$ for $1 \leq j \leq q$.

Notes

Chapter I

The classification and early structure theory for real semisimple Lie groups are due to E. Cartan, the fundamental paper being Cartan [1927]. The approach here, using a definition of linear connected semisimple group that builds in the Cartan involution, is due to Mostow. Propositions 1.1, 1.2, and 1.4 are substantially in Cartan [1927]. See Helgason [1962], Chapter VI, for more detail.

The classical groups are listed in Cartan [1927] and studied in detail in Chevalley [1946]. They are reviewed in Helgason [1962], Chapter IX, in the context of a classification. The compact classical groups are the subject of Weyl [1946]. Helgason gives realizations of all the noncompact classical groups without reference to quaternions and proves the connectivity of the groups; Problem 8 in §7 summarizes this argument. More information beyond the Problems in §7 about the isomorphism between the complex and quaternion realizations of Sp(n) and Sp(p, q) appears in pp. 134–135 of Varadarajan [1974].

Mostow [1949] gave a group-theoretic proof of Proposition 1.2, and the argument here is a specialization of Mostow's argument. For the missing computation of the Jacobian determinant, see p. 216 of Helgason [1962].

For finite groups the theory of representations was begun by Frobenius [1898] and Schur [1905]. The definition of representation that we give for a general topological group is more restrictive than it needs to be; the complex vector space could as well be a complete linear topological space. However, such extra generality is unnecessary for our purposes. The idea of considering $\Phi(L^1(G))$ for Φ a unitary representation is the basis of the theory of C^* and W^* algebras begun by Murray and von Neumann [1936]; some of the many books on the subject are by Dixmier [1957] and [1964] and by Naimark [1964].

The subject of analyzing other representations of noncompact semisimple groups besides the irreducible ones and L^2 of the group is a vast one, and some introductory information about it appears at the end of these Notes.

In the abstract theory of representations of compact groups, one generalizes from the case of finite groups, and the new step is Theorem 1.11, proved by Peter and Weyl [1927]. The proof here follows one of Weyl's original ideas and uses the Hilbert-Schmidt Theorem (see p. 242 of Riesz and Sz. Nagy [1955]); the generality of the argument was underscored by Cartan [1929].

The use of Haar measure is critical in the abstract theory for compact groups. The standard proof of existence and uniqueness of this measure for Lie groups is in Chapter V of Chevalley [1946]. A brief argument for existence in the case of a general compact group is in von Neumann [1934]. For a modern development of Haar measure and its properties for locally compact groups, see Loomis [1953].

Induced representations for finite groups were introduced by Frobenius [1898]. For locally compact groups, see Mackey [1952].

Chapter II

The classification of the finite-dimensional complex-linear irreducible representations of $\mathfrak{sl}(2, \mathbb{C})$ is old and is at least implicit in Cartan's thesis [1894], in which the complex semisimple Lie algebras are classified. The unitary trick appears in §§1–2 of Weyl [1926b] and in Weyl's book [1946], p. 265.

The irreducible unitary representations of SL(2, $\mathbb{R}$) were classified by Bargmann [1947], who cast his results in terms of SU(1, 1). Bargmann announced also the classification of the irreducible unitary representations of SL(2, $\mathbb{C}$) but did not publish the latter classification because the same results were obtained independently by Gelfand and Naimark [1947]. Some other papers published at about the same time purport to give classifications but have mistakes.

The argument in the proof of Proposition 2.6 is due to Gelfand and Naimark [1947], and the proof of Proposition 2.7 for the principal series of SL(2, $\mathbb{R}$) is merely an elaboration of that argument. For a treatment of the Euclidean Fourier analysis used in those proofs, see Stein and Weiss [1971]. The theorem of Privalov appears as Theorem 1.9 on p. 203 of Zygmund [1959]. The approach to the principal series in Problems 10 to 13 in §8 is the one used by Gelfand, Graev, and Pyatetskii-Shapiro [1969].

Gelfand and Naimark [1950] gave a suitable definition of the nonunitary principal series for a wide class of semisimple groups, pointing to the definition in general. Harish-Chandra [1953] and [1954a] used these representations, first in the case of complex groups and later more generally, as a kind of universal representation. Kunze and Stein [1960] were the first to use the nonunitary principal series of SL(2, $\mathbb{R}$) as an analytic tool. But the question of the decomposition of the nonunitary principal series into its constituents does not seem to have been answered explicitly for any particular group before Sally [1967].

Bargmann obtained the ingredients and proof, but not the explicit formulation, of the Plancherel formula for SL(2, $\mathbb{R}$), using results about special functions and differential equations. Segal [1950] gave the first formulation of an abstract Plancherel formula and proved that such a formula exists for any Type I separable unimodular locally compact group. Harish-Chandra [1952] obtained the explicit Plancherel formula for SL(2, $\mathbb{R}$) and gave an elementary proof, quite different from Bargmann's. The Plancherel formula for SL(2, $\mathbb{C}$) had been obtained earlier by Gelfand and Naimark [1950].

A book devoted entirely to the representation theory of SL(2, $\mathbb{R}$) is the one by Lang [1975]. The book by Sugiura [1975] has a chapter on the representation theory of SL(2, $\mathbb{R}$).

Chapter III

The universal enveloping algebra in essence was introduced by Poincàré [1899] and [1900]. The paper [1899] announces a result equivalent with Theorem 3.2, and Poincaré [1900] gives a sketchy proof. Garrett Birkhoff [1937] and Witt [1937] rediscovered the theorem and proved it in more generality.

Theorem 3.6 is stated by Godement [1952] and identified on p. 537 as an unpublished result of L. Schwartz; a published proof is in Harish-Chandra [1956d] as Lemma 13. No generality is gained by adjusting the definition of left-invariant differential operator so as to allow an infinite-order operator that is of finite order on each compact subset of a chart. The analytic proof of Theorem 3.2 that we give is similar to that in Helgason [1962], except that Helgason uses the operators $X(M)$ defined in Problem 7. For an algebraic proof of Theorem 3.2, see Jacobson [1962].

The assumption made in the proof of Theorem 3.2 that the complex Lie algebra $\mathfrak{g}$ is the complexification of a real Lie algebra is not universally valid for complex Lie algebras, but it is valid in the semisimple case. The proof as presented works equally well if $\mathfrak{g}$ imbeds in a larger complex Lie algebra that is a complexification.

For further discussion of the universal enveloping algebra and its properties, see Helgason [1962], pp. 90–92, 97–99, 386, and 391–393, Jacobson [1962], Chapter V, and Dixmier [1974].

The symmetrization mapping in Problems 9 to 16 is due to Gelfand [1950b] and Harish-Chandra [1953]. When a Lie algebra decomposes naturally as a direct sum in which one or both factors fail to be subalgebras (as in the Cartan decomposition $\mathfrak{g} = \mathfrak{k} \oplus \mathfrak{p}$), the symmetrization mapping gives a canonical way of decomposing the universal enveloping algebra in a corresponding fashion. We have chosen never to pursue this canonical decomposition in the text, but we shall point out in these Notes instances of where it can be helpful in later chapters.

Bargmann [1947] assumed when necessary in his work on $SL(2, \mathbb{R})$ that there was a dense set of C^∞ vectors. Gårding [1947] showed by an argument like that for Theorem 3.15 that this assumption was automatically satisfied. Segal [1951] introduced the general notion of a representation of a Lie algebra corresponding to an infinite-dimensional representation of a Lie group, but he used the Gårding subspace as domain, instead of the ostensibly larger C^∞ subspace. Harish-Chandra [1953] pointed out an advantage of using the full C^∞ subspace.

The traditional preparation for Harish-Chandra [1953] involves proving that the center of the universal enveloping algebra acts by scalars on the C^∞ subspace of an irreducible unitary representation. This was proved by Segal [1952] for the Gårding subspace, and an easy supplementary argument extends the result to the C^∞ subspace. However, we obtain in Chapter VIII what we need of this result more directly for linear connected semisimple groups.

For any analytic group, the Gårding subspace is always actually equal to the full C^∞ subspace. This result was obtained by Dixmier and Malliavin [1978].

Chapter IV

Most of the results of this chapter are due either to E. Cartan or H. Weyl, although in some cases streamlined proofs were obtained later by other authors.

Killing [1888–90] introduced roots and announced a classification of complex simple Lie algebras, but his work contained many errors and gaps. Cartan [1894] in his thesis made the classification rigorous, proceeding with such thoroughness that the classification is now attributed to Cartan alone and not to Killing. Nevertheless, Killing must be given credit for identifying this particular classification problem as a manageable one and for calling attention to the five exceptional Lie algebras. The classification was simplified later by Dynkin [1946] by the introduction of simple roots and his diagrams. For a derivation of the classification, see Jacobson [1962]. The notion of an abstract root system is implicit in the work of some authors and is explicit in Bourbaki [1968]. See Helgason [1978], p. 536, for further bibliographical information.

The traditional theory treats complex semisimple Lie algebras as the basic objects of study and ultimately shows that they have compact forms (i.e., they are complexifications of Lie algebras of compact semisimple groups). We have turned matters around, treating the compact groups as the basic objects of study and getting to the detailed theory of roots more quickly; in so doing, we lose the theorem on the existence of the compact form. The fact that Dynkin diagrams do not depend on the choices made is essentially proved here; it follows at once from Theorems 4.41 and 4.42 later in the chapter. The converse result that the Dynkin diagram determines the Lie algebra of the compact group may be found in Helgason [1962], §3.5 and especially p. 148.

Cartan [1894] proved the existence of the five exceptional simple Lie algebras over $\mathbb{C}$ by giving their multiplication tables. Current proofs of the existence of the exceptional Lie algebras are more conceptual and overlap with the theory of Kač-Moody algebras; see Helgason [1978], §10.4, for such a proof. For a somewhat longer, but still conceptual, existence proof that originated with Chevalley [1948] and Harish-Chandra [1951], see Jacobson [1962]. For detailed information about the root systems of the exceptional Lie algebras, see the appendices of Bourbaki [1968].

The exceptional Lie algebras can all be related to the Cayley numbers; the final paper identifying these algebras is by Schafer [1966].

Proposition 4.12 is attributed to Chevalley by Harish-Chandra [1956d].

The central result of the algebraic theory of representations is the theorem of the highest weight (Theorem 4.28), which is due to Cartan [1913]. The proof given here by means of Verma modules is due to Harish-Chandra [1951]. Verma modules are so named because of later work by Verma [1968] in this area.

The Weyl group and much of the analytic theory, including Theorem 4.19 and 4.26 and also the results of §10, are due to Weyl [1925], [1926a], and [1926b]. For a proof of Lemma 4.17, see Jacobson [1953], pp. 79–83. Our proof of Lemma 4.27 is taken from Varadarajan [1974], p. 343, and ultimately from Cartier [1955]. Some of the other material in §6 is taken from Helgason [1962], p. 247, and ultimately from Serre [1955].

Theorem 4.19 is a difficult one to prove in the general case. Serre [1955] discusses proofs by Riemannian geometry and by algebraic topology; the proof by Riemannian geometry is made to look a little more group-theoretic by Hochschild [1965]. Wallach [1973] gives a proof that uses much more group theory than topology. Varadarajan [1974] gives a completely different proof that proceeds by induction on the dimension of G.

Theorem 4.41 is sometimes attributed to Cartan but seems to be due to Weyl [1926b], Satz 1; the quick proof here of that theorem is from Hunt [1956]. For a proof of the Weyl integration formula (Theorem 4.44), see Adams [1969], pp. 142–144. Jacobson [1962] gives an algebraic proof of the Weyl character formula; a newer algebraic proof is a corollary of the work in Part II of Bernstein, Gelfand, and Gelfand [1975] and appears in Dixmier [1974] and Humphreys [1972]. This latter style of proof simultaneously proves the Kostant multiplicity formula (Problems 6 to 9), which originally appeared in Kostant [1959].

There are several books with substantial sections devoted to the representation theory of compact Lie groups and/or complex semisimple Lie algebras. Among these are the ones by Adams [1969], Dixmier [1974], Humphreys [1972], Jacobson [1962], Lichtenberg [1970], Seminaire "Sophus Lie" [1955], Varadarajan [1974], Wallach [1973], Warner [1972a], Weyl [1946], Wigner [1959], and Želobenko [1973]. To this list one can add the books by Helgason [1978] and Hochschild [1965] as including extensive structure theory of compact groups and complex semisimple Lie algebras, though essentially no representation theory.

The books by Humphreys [1972], Lichtenberg [1970], Wigner [1959], and Želobenko [1973] discuss decompositions of tensor products of representations into irreducible constituents. The treatment of such decompositions in Problems 10 to 17 is based in part on results in the book by Humphreys.

Chapter V

Goto [1948] and Mostow [1950] investigated conditions that ensure that an analytic subgroup is closed. The circle of ideas in this direction in §1 is that of Goto. The unitary trick consists of two parts—the existence of the compact form, proved as Satz 6 by Weyl [1926a], and the comparison of $\mathfrak{g}$ and $\mathfrak{u}$, done in §§1–2 of Weyl [1926b].

The Iwasawa, *KAK*, and Bruhat decompositions are discussed in more detail in Warner [1972a], Wallach [1973], and Helgason [1978]. The Iwasawa decomposition (Proposition 5.10 and Theorem 5.12) is due to Iwasawa [1949], but Lemma 5.11 came later and appears as Lemma 26 of Harish-Chandra [1953]. Theorem 5.13 is due to Cartan [1927], and the proof here is a simple variant of the one by Hunt [1956] for Theorem 4.41. For the Weyl group $W(A:G)$, cf. Cartan [1927]. Lemma 5.16 is due to Harish-Chandra. Aspects of the Bruhat decomposition occur in the work of Gelfand and Naimark [1950] in defining the principal series representations of the complex classical groups; Bruhat [1956] recognized the full decomposition (publishing his results in some announcements), and Harish-Chandra [1956c] gave a general proof.

The work on Cartan subalgebras and Cartan subgroups for noncompact groups is taken from Harish-Chandra [1956d], §2. Parabolic subgroups in special cases appear in the work of Gelfand and Naimark [1950] and also in an announcement by Harish-Chandra [1954d]; in his later work Harish-Chandra introduced the term "Langlands decomposition" to refer to the *MAN* decomposition.

For further information about the structure of the roots and Weyl group associated with a parabolic subgroup, see Knapp [1975] and [1982]. The double cosets for a parabolic subgroup are unions of double cosets for a minimal parabolic subgroup; Bruhat [1956] conjectured precisely which double cosets of a minimal parabolic subgroup lump to form a double coset of a given parabolic subgroup, and Borel and Tits [1965] proved the conjecture as their Corollary 5.20.

Helgason [1962], §10.1, relates differential forms to integral formulas and gives derivations of a number of the formulas left unproved in §6 here. A more general version of Proposition 5.27, valid in the context of locally compact groups, appears in Bourbaki [1963], p. 66. The technique of proof of Proposition 5.31 occurs in Lemma 13 of Kunze and Stein [1967]. The Weyl integration formula for noncompact groups is due to Harish-Chandra [1965b], Lemma 41, and [1966b], Lemma 91.

For the Borel-Weil Theorem, see Serre [1954]. The theorem had been discovered earlier by Tits [1955], pp. 112–113, and it becomes a special case of Harish-Chandra's earlier theory of holomorphic discrete series [1955], [1956a], and [1956b] the way we present the theory in Chapter VI. The Borel-Weil Theorem is normally stated in the language of holomorphic vector bundles; see Chapter 6 of Wallach [1973] for such a formulation. We follow Harish-Chandra's approach in using Lemma 5.32 as the heart of the proof; a number of proofs of the theorem rely on a suitable result from algebraic geometry or algebraic topology in place of this lemma.

The connectedness of $M_{\mathfrak{p}}$ for most real-rank-one groups has been known for a long time from case-by-case analysis. The general proof outlined in Problems 22 to 25 appears in Knapp [1975].

Chapter VI

E. Cartan [1935] studied bounded symmetric domains and related them to semisimple groups. He constructed the classical domains and stated without proof (p. 151) the existence of two exceptional ones. For a noncompact simple Lie group, G/K cannot have an invariant complex structure unless the center of K has positive dimension, and in this case the dimension of the center of K is one (cf. Problem 2). The reason is that the complex structure for G/K gives a multiplication-by-i mapping on $\mathfrak{p}$, and one can show that this mapping arises from a member of ad $\mathfrak{k}$. Thus $Z_{\mathfrak{g}}(\mathfrak{c}) = \mathfrak{k}$ is necessary and sufficient in the semisimple case for G/K to be complex. For an exposition of these matters without any use of symmetric spaces, see Knapp [1972].

The geometry and analysis of bounded symmetric domains has been studied extensively. The book by Hua [1963] gives an introduction to this subject. See also Korányi and Wolf [1965]. The work by Korányi and Wolf includes the geometry of Cayley transforms, in the sense of Problems 9 to 11 of §6.

Holomorphic discrete series were introduced by Harish-Chandra [1955], [1956a], and [1956b]. From the point of view of the treatment here, the second and third paper are more relevant than the first. The decomposition in Theorem 6.3 is summarized on p. 591 of [1956b]. The construction of $\mathfrak{a}$, stated here as Lemma 6.4, appears as the Corollary to Lemma 8 of [1956b]; the proof is outlined in Problems 4 to 8 here. The key lemma, stated as Lemma 6.8 here and as Lemma 5.32 in Chapter V, is Lemma 6 of [1956a]. The finiteness of the integral in §5 is proved in §9 of [1956b].

Harish-Chandra's three papers go considerably beyond the results given in this chapter. The papers contain also (1) a classification of irreducible admissible representations with a highest weight vector, (2) a treatment of nonlinear groups as well as linear ones, and (3) a calculation of the formal degree (see Chapter IX here for the definition).

For further information, see the Notes for Chapter XVI.

Chapter VII

Induced representations appear in the noncompact picture in the work of Gelfand and Naimark [1950], then in something like the compact picture in some 1951 announcements by Harish-Chandra and in Harish-Chandra [1952]. Bruhat [1956] systematically used the induced picture and worked with arbitrary Lie groups. Kunze and Stein [1960], [1961], and [1967] exploited the relationships among the three pictures. For induced representations in a more abstract setting, see Mackey [1952].

Each of the early authors who worked with $U(S, \sigma, \nu)$ knew that the representation is unitary if ν is imaginary. We have followed the style of argument of Kunze and Stein [1967]. The double induction formula appears in Gelfand and Naimark [1950], Mackey [1952], and Bruhat [1956]. Proposition 7.1 is due to Gelfand and Naimark [1950], and we have sketched their proof.

Mackey [1951] studied intertwining operators for finite groups, and the "Motivation" in §3 is essentially Theorem 1 of his paper. Bruhat [1956] adapted the Mackey arguments to Lie groups, using distributions, and proved Theorem 7.2, among other things.

Kunze and Stein [1960] studied the equivalences (7.13) for $SL(2, \mathbb{R})$ by Euclidean Fourier analysis. Kunze and Stein [1967] introduced formal intertwining operators (7.20) for the nonunitary principal series of a general G. The use of the operator (7.16) as the more basic object of study was later suggested by Wallach. The argument that (7.16) formally gives an intertwining operator is adapted from Kunze and Stein [1967] and Schiffmann [1971].

Bhanu Murti [1961a] and [1961b] obtained formula (7.25) for $SL(n, \mathbb{R})$ and $Sp(n, \mathbb{R})$ to answer a question raised by Harish-Chandra [1958b], at least for these groups. Gindikin and Karpelevič [1962] generalized the result to arbitrary G; their proof essentially includes Proposition 7.4 as an intermediate step. The generalization (7.26) is due to Helgason [1967]. Proposition 7.6 is in §13 of Harish-Chandra [1958a]; Harish-Chandra obtained it by using properties of the hypergeometric equation satisfied by the restriction to A of a spherical function. Schiffmann [1971] and Helgason [1970] gave more direct evaluations of the integral.

Schiffmann [1971] recognized that the Gindikin-Karpelevič idea was exactly what was needed to obtain a region of convergence for the Kunze-Stein formal intertwining operators for the nonunitary principal series. The results of §6 are all due to him.

Theorem 7.12 is due to Knapp and Stein [1971]. The computation that precedes it is implicit in Kunze and Stein [1967] and is the analytic basis for transferring intertwining operators from one picture to another.

Spherical functions were studied by Godement [1952] and by Harish-Chandra [1958a] and [1958b]. Propositions 7.15 and 7.17 and Lemma 7.16 are due to Harish-Chandra.

Proposition 7.14 and Theorem 7.22 are new and generalize the corresponding results for $\lambda = 0$, which appear in Langlands [1973]. Lemma 7.23 and Theorem 7.24 are due to Langlands and to Miličić [1977]. For the proof of Lemma 7.23 in the general case, see the proof of Lemma 3.12 of Langlands [1973] or Lemma 20.1 on p. 49 of Harish-Chandra [1976a].

Chapter VIII

Harish-Chandra [1953] began an investigation of infinite-dimensional representations of semisimple Lie groups from the point of view of the action of the Lie algebra on K-finite vectors. Although he obtained full results even for nonlinear groups and for representations in Banach spaces, we limit this discussion to linear groups and to representations in Hilbert spaces. He proved that a representation of a semisimple group always has a dense subspace of analytic vectors ("well-behaved" vectors in his terminology), vectors u such that $g \to \pi(g)u$ is real analytic. It follows that the analytic vectors of each K type are dense in the space of all vectors of that K type. This is a stronger result than the present Theorem 8.7, since admissibility is not assumed. For an irreducible unitary representation, Segal [1952] had proved that $Z(\mathfrak{g}^{\mathbb{C}})$ acts by scalars on the Gårding subspace. Harish-Chandra observed that whenever a representation has $Z(\mathfrak{g}^{\mathbb{C}})$ acting by scalars on the Gårding subspace, then the action by scalars persists for the C^∞ vectors and thus can be restricted to the analytic K-finite vectors. He then concentrated on representations for which $Z(\mathfrak{g}^{\mathbb{C}})$ acts as scalars, calling such representations "quasisimple."

He proved that an analytic K-finite vector v always has the property in a quasisimple representation that the closure of $U(\mathfrak{g}^{\mathbb{C}})v$ is G-invariant and has all K types of finite multiplicity. In particular, this is the case for an irreducible unitary representation, and we arrive at Harish-Chandra's original proof of Theorem 8.1, but without the explicit bound for the multiplicities. (An explicit bound came later, in Harish-Chandra [1954b].)

At this stage it is natural to consider admissible representations. Harish-Chandra did so, without assuming (as we have done) that K acts by unitary operators, and obtained the correspondence given here in Corollary 8.10, using the existence of analytic vectors. Combining the two approaches, he proved that a semisimple group is of Type I, settling affirmatively a conjecture of Mautner [1950].

Since that time, some simpler proofs and generalizations have appeared. Nelson [1959] proved that any representation of a connected Lie group has a dense subspace of analytic vectors. Nelson's proof has the flavor of the proof that solutions of analytic elliptic partial differential equations are real analytic (for which, see p. 136 of Schwartz [1950]), and we have given in Theorem 8.7 a proof of a special case that uses this more familiar theorem.

Godement gave a simple proof of Theorem 8.1, and we have followed the variant of Godement's proof given by Dixmier [1964].

Harish-Chandra [1951] proved Theorem 8.18 about the homomorphism γ and also established the properties of infinitesimal characters given in Propositions 8.20 and 8.21. For a proof of the Hilbert Nullstellensatz, see Zariski and Samuel [1960], pp. 164–167. Theorem 8.19 is due to Chevalley [1955]; for an exposition, see Varadarajan [1974], pp. 380–387.

We have given a proof in only a special case that the Harish-Chandra homomorphism γ is onto $\mathscr{H}^W$. A proof in the same style can be pushed through in general, as follows: The construction of z_k in (8.29) and the congruence for $\gamma(z_k)$ in (8.30) work in general, and the standard representation can be replaced by any finite-dimensional representation for the computation of the trace in (8.29). If this is done, the argument goes through without the special notions of elementary symmetric polynomials and substitution of roots of unity, but such a proof is quite messy.

A cleaner proof is obtained by proving a corresponding "onto" statement about the restriction map from Ad(G)-invariant polynomials on $\mathfrak{g}^{\mathbb{C}}$ to W-invariant polynomials on $\mathfrak{h}^{\mathbb{C}}$ and then transferring the result via the symmetrization mapping (cf. the Notes for Chapter III) to the situation here. See Varadarajan [1974], pp. 333–341, for an exposition.

As was mentioned in the Notes for Chapter II, Bargmann [1947] obtained results tantamount to the Plancherel formula for SL(2, $\mathbb{R}$). His method was to use a known technique for the spectral analysis of linear ordinary differential operators, relating the asymptotic expansion of the eigenfunctions of the operator to the spectral measure (see Coddington and Levinson [1955]). The operator was the Casimir operator, viewed as an operator on functions on $A_{\mathfrak{p}}$, and the eigenfunctions were the K-finite matrix coefficients of the irreducible representations.

Motivated by Bargmann's work, Harish-Chandra [1958a] and [1958b] began a similar analysis of the invariant differential operators on G/K in order to get a Plancherel formula for $L^2(G/K)$, i.e., the subspace of $L^2(G)$ of right K-invariant functions. The operators are to be regarded as acting on functions on $A_{\mathfrak{p}}$, and the restrictions to $A_{\mathfrak{p}}$ of the spherical functions are the eigenfunctions. Harish-Chandra obtained asymptotic expansions and proceeded from there to the desired Plancherel formula, except for two conjectures that were proved subsequently. One of these conjectures follows from the work of Gindikin and Karpelevič [1962] and §3 of Helgason [1964], and the other was proved by Harish-Chandra [1966b] in §21.

Harish-Chandra expected that one step toward a full Plancherel formula for $L^2(G)$ would be a similar analysis of the asymptotic expansion of a general K-finite matrix coefficient of an irreducible representation. He carried out such an analysis in two brilliant papers [1960a] and [1960b] that he chose not to publish, and §§7, 8, and 12 of the present chapter are based on this material. (The original papers have been published in Harish-Chandra's collected works.) This material has been simplified somewhat by Casselman [1975b], Miličić [1976], and Casselman and Miličić [1982]. For further information, see the Notes for Appendix B.

Ironically Harish-Chandra then discovered that this analysis was inadequate for his purposes, and this discovery may account for his failure to publish the two papers. We have indicated in §14.3 one way in which the analysis was inadequate.

Nevertheless, the two papers Harish-Chandra [1960a] and [1960b] do have far-reaching applications, as this chapter indicates. The Subrepresentation Theorem, given as Theorem 8.37, is due to Casselman [1975b] and appears in published form in Casselman and Miličić [1982]; it strengthens a subquotient theorem of Harish-Chandra [1954a] that every irreducible admissible representation is infinitesimally equivalent with a subquotient of a nonunitary principal series representation. For further developments along these lines, see Casselman and Osborne [1975], Miličić [1977], and Hecht and Schmid [1983a] and [1983b].

Theorem 8.38 on intertwining operators is due to Wallach [1975a] and Knapp and Stein [1980]. The circle of ideas in Theorem 8.39 and its corollaries is in Harish-Chandra [1966b] and is implicit to some extent in Harish-Chandra [1953]; in particular, Corollary 8.41 appears as Theorem 1 of Harish-Chandra [1966b] and Corollary 8.42 is Lemma 77. The inequality (8.62) is Lemma 10 of that paper.

For Theorems 8.47 and 8.48, see Casselman and Miličić [1982]. Theorem 8.48 for $p = 2$ is part of the central Lemmas 42 and 43 of Harish-Chandra [1966b], and so is Theorem 8.51. For further research on the size of matrix coefficients, see Trombi and Varadarajan [1972] and Hecht [1979].

Theorem 8.53 on irreducible tempered representations is due to Langlands [1973]. Trombi [1977] has a different proof. A published exposition of this theorem appears in Chapter 4 of Borel and Wallach [1980]. The proof we give uses parts of the argument by Langlands and parts of the argument by Borel and Wallach.

The Langlands classification (Theorem 8.54) appeared originally in §§3–4 of Langlands [1973], and Borel and Wallach [1980] give an exposition. The geometric preliminaries (Lemmas 8.55 to 8.60) were simplified to their present form by Carmona [1983]. The reader who wants to review the original proof by Langlands of Theorem 8.54 is well advised to skip directly to §3 of the Langlands paper.

Problems 20 to 25 derive some of the results of Harish-Chandra [1958a] and [1958b] on asymptotic expansions as special cases of the results of the later papers [1960a] and [1960b]. The examples in Problems 26 and 27 were communicated by Vogan.

The "asymptotic expansions" of this chapter are more than ordinary asymptotic expansions: they are convergent series expansions. Wallach [1983] has developed a theory of asymptotics for eigenfunctions of $Z(\mathfrak{g}^{\mathbb{C}})$ that are K-finite on only one side and that satisfy some mild growth conditions. In this case the expansions are asymptotic expansions in the traditional sense.

Chapter IX

Proposition 9.1 and Corollary 9.2 illustrate the power of the results of Harish-Chandra [1953] relating representations of G to representations of $\mathfrak{g}$; Schur's Lemma here becomes an algebraic result and no longer requires any functional analysis, even though the representations are infinite-dimensional. Theorem 9 of the Harish-Chandra paper is what we quote as Theorem 9.3.

When Bargmann [1947] treated SL(2, $\mathbb{R}$), he proved essentially Proposition 9.6 for the representations $\mathcal{D}_n^{\pm}$ with $n \geq 2$. Godement [1947] showed that the result is not special to SL(2, $\mathbb{R}$), giving a proof by functional analysis of a corresponding more general theorem for locally compact unimodular groups. We have chosen instead to give a proof using the methods of semisimple groups, and Y. Benoist provided part of this argument for us.

Early successes with the Plancherel formula by Gelfand and Naimark [1950] for SL(n, $\mathbb{C}$), by Harish-Chandra [1952] for SL(2, $\mathbb{R}$), and by Gelfand and Graev [1953a] and [1953b] for SL(n, $\mathbb{R}$) quickly pointed to the approach for the general case that we discuss in §11.5. This kind of proof insists that some discrete family of representations contribute to the Plancherel formula if and only if G has a compact Cartan subgroup and moreover that the discrete family must consist of discrete series representations. On the other hand, the only known discrete series at that time were the holomorphic ones. Thus Harish-Chandra conjectured at the 1954 International Congress of Mathematicians the existence of discrete series whenever rank G = rank K. The smallest group for which rank G = rank K and for which there are no holomorphic discrete series is SO(4, 1), and Dixmier [1961] showed that indeed there are discrete series for this group.

Harish-Chandra [1965b], [1966a], and [1966b] then tackled the problem from a different point of view, using global character theory, and was able to classify discrete series in general without actually constructing them. We take up this matter further in the Notes for Chapter XII, along with some discussion of constructions that depend on the character theory.

The construction we give in §§4, 6, 7, and 8 is completely different and is due to Flensted-Jensen [1980]. Flensted-Jensen had earlier studied both the work of Berger [1957] on **semisimple symmetric spaces** (G/H, where H is the set of fixed points of an involution) and the explicit realization of zonal functions for representations of compact groups by the use of duality, and he combined these approaches in the 1980 paper to give an explicit realization of "discrete series for semisimple symmetric spaces." Specialization of his theory to $(G \times G)/G$ gives the discrete series of G. The main result in the chapter is Theorem 9.20. Actually Lemma 9.21c, which gives the required square integrability, is proved in the chapter only for sufficiently nonsingular Harish-Chandra parameters. Thus the Flensted-Jensen construction as presented obtains only the discrete series whose Harish-Chandra parameters are sufficiently nonsingular. Problems 23 and 24 of Chapter XII show how to obtain easily the remaining discrete series.

The Harish-Chandra parameter is so named because it is the one that occurs in Harish-Chandra's formula for the global character. If one expands the expression for the global character formally in an infinite series as in the Notes for Chapter XII, one obtains a formula that looks as if it might be the reduction of the discrete series representation under the action of K. The conjecture that this formula actually gives the K-type decomposition was attributed (orally) by Harish-Chandra to R. Blattner; its first complete appearance in print is in Schmid [1968], although part of it is in Okamoto and Ozeki [1967]. The formula has thus come to be known as the **Blattner conjecture** or the **Blattner formula.** (See the Notes for Chapter XII for a discussion of its proof.) One K type stands out in the Blattner formula as special, and its highest weight is called the Blattner parameter. The qualitative description of the other highest weights given as in Theorem 9.20c is motivated by an easy computation with the Blattner formula.

Flensted-Jensen and Okamoto [1984] have found a variant of the second computation of the eigenvalue in §8 that gives a sharper conclusion about K types than that in Theorem 9.20c. The conclusion is that the highest weight of any K type in π_λ is obtained from τ_Λ by adding to Λ a sum of positive noncompact roots. The trick is to start from the formula for $\psi_{\mathbb{C}}(g_{\mathbb{C}})$ just before (9.75), recognize diag K as the maximal compact subgroup of $K_{\mathbb{C}}$, and transform the integral to one over the $\bar{N}$ space for $K_{\mathbb{C}}$. Then we get a comparable formula in place of (9.75) upon differentiation. In place of (9.77), we use a decomposition of $\mathfrak{g}_{\mathbb{C}}$ that involves the

$\bar{\mathfrak{n}}$ space of $\mathfrak{k}_{\mathbb{C}}$, the same space $\mathfrak{s}$, the full Cartan subalgebra $\mathfrak{a} \oplus \mathfrak{m}$, and the same $\mathfrak{n}_{\mathbb{C}}$. Natural adjustments to the rest of the argument lead to the sharper conclusion.

In connection with Lemma 9.8, see Gelfand [1950a]. A useful generalization of this lemma is due to Helgason [1959]; in it, K is replaced by any subgroup H such that Ad(H) leaves stable some vector space complement to the Lie algebra of H. See p. 395 of Helgason [1962]. Proposition 9.9 appears explicitly as Lemma 3 of Harish-Chandra [1958a].

Lemma 9.13 plays an important role in later chapters of the book because it identifies the disconnectedness of parabolic subgroups exactly. The lemma is proved in two stages. First one shows that $M_{\mathfrak{p}}$ is generated by its identity component and $G \cap \exp i\mathfrak{a}_{\mathfrak{p}}$; see Lemma 9 of Satake [1960] and Lemmas 1 and 3 of Moore [1964]. Then one identifies $G \cap \exp i\mathfrak{a}_{\mathfrak{p}}$ as the group generated by the elements γ_β; see Loos [1969b], Theorems 3.4a and 3.6 on pp. 75–77, for the key step that $2\{H \in i\mathfrak{a}_{\mathfrak{p}} \mid \exp H \in K\}$ is contained in the lattice generated by the vectors $4\pi i|\beta|^{-2}H_\beta$.

Parts of Theorem 9.14 are due to Harish-Chandra [1958a], §2, and parts are due to Helgason [1970], §III.3. See also Wallach [1971] and [1972] and Lepowsky and Wallach [1973].

The proof of Lemma 9.17 is adapted from p. 159 of Loos [1969a]. The proof of Lemma 9.24 is made tidier with the symmetrization mapping, which is what Flensted-Jensen uses.

Furstenberg [1963] introduced Poisson integrals for symmetric spaces G/K, and Moore [1964] identified certain insufficiently understood group theory constructs in Furstenberg's work. Helgason [1967] proved Lemma 9.28 as a generalization of the classical result that Poisson integrals of continuous functions tend radially to their continuous boundary values. See §2 of Knapp and Williamson [1971] for an exposition of the elementary parts of this theory.

Flensted-Jensen's theory of discrete series for symmetric spaces has developed into a study in its own right, and more bibliographical information about this subject appears in the section "Further topics" at the end of the Notes.

Chapter X

For a careful summary of properties of trace class operators, see pp. 34–35 of Borel [1972]. Global characters appear in special cases in Gelfand and Naimark [1950], but Theorem 10.2 is due to Harish-Chandra [1954b] and is proved in §4 of that paper. For the situation in which $\pi(f)$ is always of trace class, Erik Thomas has communicated to us a proof using the closed graph theorem that $f \to \pi(f)$ is automatically a distribution.

Theorem 10.6 is Theorem 6 of Harish-Chandra [1954b]; we have followed the style of proof in Lemma 7.1 of Jacquet and Langlands [1970].

Proposition 10.12 and Corollary 10.13 are in Harish-Chandra [1952]. These results are proved in many expository treatments of the representation theory of SL(2, $\mathbb{R}$), but Proposition 10.14 is not. The latter proposition follows from Harish-Chandra's general work on discrete series but does not seem to have been proved separately earlier in print. The proof we give was communicated to us by Wallach.

The technique for computing induced characters is due to Gelfand and Naimark [1950] for complex groups and to Harish-Chandra [1966a], pp. 93–94, in general.

See Harish-Chandra [1954c], Hirai [1968], Lipsman [1971], and Wolf [1974] for results of this kind. See p. 464 of Warner [1972a] and §3 of Trombi [1978] for expositions. Some of these papers and expositions contain unwise conventions and/or technical errors.

The results of §§4–7 concerning analyticity and formula on the regular set are due to Harish-Chandra [1956d]. Harish-Chandra's proof is more complicated than the one here since his main regularity result in that paper is close to the deeper Theorem 10.40, containing some information about behavior on the singular set. Harish-Chandra's proof differs from the one here also in its use of the symmetrization mapping in handling the subspace $\mathfrak{q}$ of $\mathfrak{g}^{\mathbb{C}}$. Yet another proof uses Sato's theory of hyperfunctions, including the result that solutions of an analytic system of differential equations are analytic off the wave front set; see Duflo, Heckman, and Vergne [1984] for details.

Theorem 10.36 is a deep result of Harish-Chandra [1965b], appearing as Theorem 2 of that paper. The proof takes most of five of Harish-Chandra's papers, of which this is the fifth. Thorough expositions of this theorem appear in Warner [1972b] and Varadarajan [1977]. For a different proof, see Atiyah [1976] and Schmid [1975c]. Harish-Chandra's approach is first to prove an analog of the theorem for the Lie algebra and then to lift the result to the group; the main result for the Lie algebra is Theorem 1 of Harish-Chandra [1965a].

Corollary 10.37 appeared originally as Theorem 3 of Harish-Chandra [1954b], proved by completely different methods.

The use of tensor products with finite-dimensional representations in order to move parameters in infinite-dimensional representations is implicit in Wallach [1971] and [1972] and is explicit in Bernstein, Gelfand, and Gelfand [1975], Lepowsky and Wallach [1973], and Schmid [1975b]. Jantzen [1974], in the case of Verma modules, combined this operation with projection according to an infinitesimal character. Zuckerman [1977] adapted the combined operation to admissible representations. A supplement to Zuckerman's work appears as appendices to Knapp and Zuckerman [1982]. Theorem 10.48 is due to Fomin and Shapovalov [1974]; we have given Zuckerman's proof.

The result in Problem 2 is an unpublished theorem of Zuckerman's. Problems 3 to 8 contain results of Lepowsky and Wallach [1973] and relate them to the contents of the chapter.

Vogan [1979c] has combined the theory of characters and some algebraic techniques for a systematic study of the decomposition into subquotients of the standard representations $U(S, \sigma, \nu)$.

Chapter XI

Gelfand and Naimark [1950] formulated and proved the Plancherel formula (Fourier inversion formula) for $SL(n, \mathbb{C})$ as their formula (27.7). Their method was the method of §§1–2. Harish-Chandra [1954c] showed how to extend the proof to all complex semisimple groups; Lemma 11 in that paper is not quite right, but the main line of proof is still valid.

Harish-Chandra announced his result for complex semisimple groups in 1951. Later Harish-Chandra [1952] proved Theorem 11.6 for $SL(2, \mathbb{R})$, overcoming the difficulties arising from the nonconjugate Cartan subgroups. We follow essentially

Harish-Chandra's original proof. Next Gelfand and Graev [1953a] and [1953b] treated SL(n, ℝ), and this result pointed to the approach for the general case that is outlined in §§4–5.

The results in §6 in the general case are due to Harish-Chandra [1957a], [1957b], [1957c], and [1957d]. The paper [1957a] contains results on differential operators in the same vein as in §10.6, the papers [1957b] and [1957c] prove analogs on the Lie algebra of the desired results on the group, and the paper [1957d] carries the results from the Lie algebra to the Lie group. Our Proposition 11.8 is contained in Lemma 14 and Theorem 2 of Harish-Chandra [1957d]. Our Proposition 11.9 is equivalent with Theorem 3 of that paper, and our Proposition 11.10 is Lemma 18 in that paper. The proof of Proposition 11.10 that we give is adapted from §§22–26 of Harish-Chandra [1965b].

Semiregular elements and their role are discussed in §2 of Harish-Chandra [1957c]. See §2 of Schmid [1975a] for a thorough exposition. Corollary 11.14 appears originally as Lemma 40 of Harish-Chandra [1965b]; the note at the end of Harish-Chandra [1957d] suggests that the lemma had been obtained much earlier. Theorem 11.17 is what is asserted in Theorems 4 and 5 of Harish-Chandra [1957d] if we take into account Corollary 11.15.

Thorough expositions of the results of §6 appear in Warner [1972b] and Varadarajan [1977].

Patching conditions make no sense until it has been proved that invariant eigendistributions are given by locally integrable functions. Harish-Chandra [1965a] proved a patching condition on the Lie algebra level as Theorem 3; this is the same paper in which he proved that invariant eigendistributions on the Lie algebra are given by functions. The patching condition on the group level (Theorem 11.18) is due to Hirai [1976]. Hirai actually proves something more: he obtains a necessary and sufficient condition for functions to patch together to form an invariant eigendistribution.

Chapter XII

Armed with the knowledge that global characters of irreducible representations are given by locally integrable functions, Harish-Chandra [1965c], [1966a], and [1966b] classified the discrete series by determining their global characters.

This work may be motivated and described as follows: The scheme of proof in §11.5 for the Plancherel formula dictates that the infinitesimal characters of some, if not all, discrete series representations should be algebraically integral and nonsingular, in order to contribute appropriately to the Fourier analysis on the compact Cartan subgroup. Thus we seek invariant eigendistributions with eigenvalue χ_λ, where λ is integral and nonsingular. The proof of the Plancherel formula for complex groups and the development of the properties of F_f and invariant eigendistributions suggest first treating a corresponding problem on $\mathfrak{g}$: look for Ad(G)-invariant distributions τ on $\mathfrak{g}$ satisfying

$$\partial(p)\tau = (p|_{\mathfrak{b}'})(\lambda)\tau$$

for all p in $I(\mathfrak{g}^{\mathbb{C}})$, the set of Ad($G$)-invariant members of the symmetric algebra of $\mathfrak{g}^{\mathbb{C}}$. (Here, to define $p|_{\mathfrak{b}'}$, we regard p as a member of the symmetric algebra of the double dual of $\mathfrak{g}^{\mathbb{C}}$ and then restrict accordingly.)

For the case that G is compact, Harish-Chandra [1957a] had shown that the Lie algebra analog of F_f commutes with the Euclidean Fourier transform, except for a scalar factor. More specifically, fix a Cartan subalgebra $\mathfrak{b}$, let ϖ be the product of the positive roots (as a member of $S(\mathfrak{g}'^{\mathbb{C}})$), and put

$$\phi_f(H) = \varpi(H) \int_G f(\mathrm{Ad}(x)H)\,dx.$$

Let $\mathscr{F}_\mathfrak{g}$ and $\mathscr{F}_\mathfrak{b}$ denote the Euclidean Fourier transforms on $\mathfrak{g}$ and $\mathfrak{b}$, with the Killing form used to define the inner product. Then

$$\phi_{\mathscr{F}_\mathfrak{g} f} = c\mathscr{F}_\mathfrak{b}(\phi_f).$$

Moreover, the Lie algebra analog of Proposition 11.9 is valid:

$$\phi_{\partial(p)f} = \partial(p|_{\mathfrak{b}'})\phi_f.$$

If we put $\tau(f) = \phi_{\mathscr{F}_\mathfrak{g}}(iH_\lambda)$, then we can use the above identities to see directly that τ has the right eigenvalue under $\partial(p)$. Moreover, it is not hard to see that τ is unique up to a scalar as an invariant eigendistribution with that eigenvalue.

The existence argument above works when G is noncompact, as Harish-Chandra [1957b] had shown earlier, but the uniqueness fails. We get distinct distribution solutions $\tau_s(f) = \phi_{\mathscr{F}_\mathfrak{g}}(iH_{s\lambda})$ for each s in W_K/W_G. Even then, there are more solutions. But Harish-Chandra found that the solutions parametrized by s as above span the space of distribution solutions that are tempered (in the Euclidean sense), because temperedness implies that some patching conditions have to be satisfied. (See §14 of the paper [1965c].) By a dimension argument, there has to be one such solution whose formula on $\mathfrak{b}$ is

$$\varpi(H)^{-1} \sum_{s \in W(B:G)} (\det s) e^{\lambda(sH)}.$$

A delicate lifting argument given in the latter half of the paper [1965c] then allows one to obtain a corresponding invariant eigendistribution Θ on G.

The distribution Θ on G has bounded numerators and so is tempered by the proof that (b) implies (a) in Theorem 12.23. Projecting Θ on both sides by a K type as in the proof that (a) implies (c) in Theorem 12.23, we obtain a real analytic function $F(x)$ that satisfies the conditions of Lemma 12.25. The critical Lemma 67 of the paper [1966b] then derives from the nonsingularity of the infinitesimal character a property of the leading exponents that forces $F(x)$ to be in $L^2(G)$. By our Corollary 8.42, $F(x)$ is in the finite sum of subspaces corresponding to discrete series. Supplementary arguments in Harish-Chandra [1966b] sort out the irreducibility of the representation in question and the exact formula for its character. Completeness is proved by the method we have given in §5. Schiffmann [1968] gives an exposition of Harish-Chandra's approach to discrete series.

Realization of the global character of an irreducible representation as the product of a Jacobian factor by the orbital integral (i.e., F_f) of the invariant measure on an orbit of $\mathrm{Ad}(G)$ in the dual of the Lie algebra of G has become a study in itself. Kirillov [1968] used this approach for nilpotent groups and conjectured its wider applicability, and such formulas have come to bear his name. Rossmann

[1978] began a study of the semisimple case, and further work appears in Vergne [1979], Berline and Vergne [1985], and Duflo, Heckman, and Vergne [1984].

This chapter adopts a completely different approach to discrete series. Chapter IX gives the Flensted-Jensen construction, and Chapter XII needs only to find what the characters must be. Their construction is not an issue. Parts of the argument were communicated to us by Vogan. The results of §1 are distilled from Harish-Chandra [1966b], particularly §19 and §37. See also §8.3.8 of Warner [1972b]. The patching conditions of §2 are a transposition into the setting of Hirai [1976] of the kind of computation in §14 of Harish-Chandra [1965c]. The techniques in §3 appear originally in Parthasarathy [1972] and Schmid [1971] and [1975a]. Lemma 12.8 is in Schmid [1971], and Schmid attributes it to Harish-Chandra [1956d]. The use of the Clifford algebra and the Dirac operator in this situation is due to Parthasarathy [1972], and Lemma 12.12 is in that paper. Lemma 12.11 appeared originally as Lemma 8.26 of Schmid [1975a].

The material in §§4–6 on the Schwartz space, exhaustion of discrete series, and tempered representations is mostly from Harish-Chandra [1966b], but some preliminary estimates are in the second half of Harish-Chandra [1966a]. Varadarajan [1973] gives an excellent exposition of this work. The simple proof of Proposition 12.16 is Varadarajan's, and his 1973 exposition guides the reader through Harish-Chandra's extension of F_f to the Schwartz space (Theorem 12.17). The original development of the results in §5 is in §§32, 33, and 36 of Harish-Chandra [1966b].

Harish-Chandra [1956b] mentions the existence of representations that turn out to be limits of holomorphic discrete series, and Knapp and Okamoto [1972] construct these representations globally and show how they exhibit reducibility of some induced representations. The first appearance of limits of discrete series in any greater generality is in the work of Schmid [1975a], who constructed those whose infinitesimal characters are orthogonal to a noncompact root. Schmid's technique is to start with a discrete series character and to move the parameter (on each component of each Cartan subgroup) to the edge and beyond the edge of the Weyl chamber that makes it dominant; he shows that the patching conditions of Hirai [1976] are satisfied and thus obtains a whole family of invariant eigendistributions. The identity we give as Theorem 12.34, proved on the level of invariant eigendistributions, helps him to show that any of his invariant eigendistributions is a character if its parameter is dominant and is nonsingular apart from one positive noncompact root.

The introduction of Zuckerman tensoring (Zuckerman [1977]) allowed the construction of all limits of discrete series and the identification of their characters with some of Schmid's invariant eigendistributions. For the most general singularities of the infinitesimal character, Theorem 12.26 was proved in the following papers originally: The temperedness is in Schmid [1975a], and the irreducibility of the nonzero characters follows from a theorem of Zuckerman [1977] and the multiplicity-one part of conclusion (c). The direct part of conclusion (b) (isolated as conclusion (b′)) and all of conclusion (c) are in Hecht and Schmid [1975]. Conclusion (d) and the converse half of (b) are due to Knapp and Zuckerman [1982] and go hand in hand with the development of a generalization of Schmid's identity.

We have given a simpler proof in this chapter, avoiding the induction, by proving conclusion (c′) directly.

For Lemma 12.30, see §2 of Schmid [1975a] and §2 of Knapp and Zuckerman [1982]. Lemma 12.31 is due to Lipsman [1971]. Theorem 12.34 is formula (1.6) of Schmid [1975a]. See also Schmid [1977].

Two important aspects of discrete series that we have not addressed in this chapter are their realization and their decomposition under K. The Borel-Weil Theorem for a compact group G realizes the irreducible representations in a space of holomorphic functions on $G^{\mathbb{C}}$ transforming in a particular way under a parabolic subgroup B built from a maximal torus T, i.e., as the space of holomorphic sections of a certain complex line bundle over G/T built as in §5.7 from a dominant integral parameter on the Cartan subalgebra. This space may be regarded as a $\bar{\partial}$ cohomology space in degree 0. Bott [1957] and Kostant [1961] generalized the Borel-Weil Theorem by showing that one could repeat the construction with any integral parameter λ, dominant or not: If $\lambda + \delta$ is nonsingular, the cohomology is nonvanishing in just one degree and in that degree yields an irreducible representation whose infinitesimal character is $\lambda + \delta$.

Soon after Harish-Chandra's work on the discrete series, Griffiths showed a similar vanishing theorem for higher cohomology in connection with Γ-invariant sections in the case that G is noncompact, Γ is discrete, G/Γ is compact, and the parameter has a suitable regularity property. Langlands [1966b] observed that a formal computation of characters suggested that the cohomology in the remaining degree realizes an appropriate discrete series representation. Langlands put all these facts together and made a specific conjecture about how to realize all discrete series representations in an L^2 version of $\bar{\partial}$ cohomology of a line bundle over G/B, where B is a compact Cartan subgroup.

Okamoto and Ozeki [1967] dealt with an analog of this problem in the case that G/K is complex, the analog being for vector bundles over G/K. They showed that a parameter that we have called the "Blattner parameter" plays an important role in the Langlands conjecture. Schmid [1968] showed that a formal expansion of the global character of a discrete series representation (the full "Blattner formula") gives a bound for the multiplicities of the K types that can occur in the cohomology representation.

The Blattner formula comes from the following heuristic calculation. Starting from the character formula on B

$$\Theta_\lambda(b) = (-1)^q D_B(b)^{-1} \sum_{s \in W(B:G)} (\det s)\xi_{s\lambda}(b)$$

with notation as in Theorem 12.17, we write

$$D_B = D_K \xi_{\delta_n} \prod_{\alpha \in \Delta_n^+} (1 - \xi_{-\alpha}) = D_K D_n,$$

where D_K is the Weyl denominator for K. Since D_B and D_K are odd under $W(B:G)$, D_n must be even. We put D_n^{-1} inside the sum over s and rewrite D_n^{-1}, using its evenness, as the reciprocal of

$$D_n = (-1)^q \xi_{-s\delta_n} \prod_{\alpha \in \Delta_n^+} (1 - \xi_{s\alpha}).$$

Expanding each $(1 - \xi_{s\alpha})^{-1}$ as a geometric series, we obtain

$$\Theta_\lambda = D_K^{-1} \sum_{s,\alpha,n_\alpha} (\det s)\xi_{s(\lambda+\delta_n+\sum n_\alpha \alpha)},$$

the sum extending over s in $W(B:G)$, α in Δ_n^+, and $n_\alpha \geq 0$. Because of Lemma 12.8, we expect that this expression formally matches $\sum_{\Lambda'} (\text{mult } \tau_{\Lambda'})\chi_{\Lambda'}$. Taking into account the Weyl character formula, we look for the coefficient of $\xi_{\Lambda'+\delta_K}$ in the formal expression for Θ_λ. Letting $\mathscr{P}(\nu)$ be the number of ways of writing ν as $\sum_{\alpha\in\Delta_n^+} n_\alpha \alpha$ with all $n_\alpha \geq 0$, we easily find that the coefficient of $\xi_{\Lambda'+\delta_K}$ is

$$\sum_{s\in W(B:G)} (\det s)\mathscr{P}(s(\Lambda' + \delta_K) - \lambda - \delta_n).$$

The **Blattner conjecture** is the assertion that this integer is the multiplicity of $\tau_{\Lambda'}$ in the discrete series π_λ.

Narasimhan and Okamoto [1970] proved the G/K analog of the Langlands conjecture for G/K complex under the assumption that the Harish-Chandra parameter is sufficiently regular. Griffiths and Schmid [1969] were able to refine Griffiths' original arguments to obtain a vanishing theorem for the original Langlands conjecture for sufficiently regular parameter, and Schmid [1971] found a justifiable argument with characters that proved the conjecture under the same assumptions. Proof of theorems in Schmid [1968] for the most part appear in the 1971 paper. Parthasarathy [1972] formulated and proved a G/K analog of the Langlands conjecture without the assumption that G/K is complex; his device was to use the Dirac operator in place of $\bar{\partial}$, but he too assumed that the parameter is sufficiently regular. His results can be rephrased for the operator $\mathscr{D}$ discussed in Problems 5 to 11; this operator had been introduced in Schmid [1968]. Problems 5 to 11 give some facts about $\mathscr{D}$ that are easy in retrospect.

Subsequently Schmid [1975a], [1975b], and [1976] proved the full Langlands conjecture, including the original version and the G/K analogs, with an assist from Zuckerman [1977]. During the same interval, Hecht and Schmid [1975] proved the Blattner conjecture. For a constructive analytic realization of discrete series, see Knapp and Wallach [1976] and Blank [1985].

Enright and Varadarajan [1975] approached the question of finding an algebraic construction of the discrete series. They found such a construction and proved that it works for sufficiently regular parameters. Application of Zuckerman tensoring proves that it works for all regular parameters, as Wallach [1976] notes; Wallach also gives a different algebraic construction. Duflo [1979b] gives an exposition of all these matters, starting with Harish-Chandra's results. See Enright and Wallach [1980] and Vogan [1981] for later directions that these investigations take.

Problems 2 to 4 address a result proved in Harish-Chandra [1966a].

Chapter XIII

The method of §11.5 applies directly to prove the Plancherel formula for real-rank-one groups. Okamoto [1965] proved this fact (modulo a conjecture) for those real-rank-one groups with rank G = rank K, and Hirai [1966] noted that there was no difficulty with the one remaining class of groups, the covering groups of

$SO_0(2n+1, 1)$. Harish-Chandra [1966a] gave an independent proof of the Plancherel formula for real-rank-one groups with rank G = rank K; his grouping of terms was a little different from Okamoto's, but he left no unproved conjectures.

The chief difficulty in pushing through a proof for general G was the lack of knowledge about discrete series characters on noncompact Cartan subgroups, and the effect of this problem appears already in $Sp(2, \mathbb{R})$. Harish-Chandra [1970] introduced new methods to get around this difficulty, announcing a proof in the general case. But this paper gives no clear way of writing down an explicit Plancherel measure. In a subsequent announcement, Harish-Chandra [1972] solved this problem, too. The latter paper gives a clear overview of his approach to the Plancherel formula, which is based on an analog with the situation for $L^2(G/\Gamma)$, Γ discrete and G/Γ of finite volume. He called this analogy the "philosophy of cusp forms," and its origins may be seen in the exposition Harish-Chandra [1968] of work of Langlands. The detailed proof of the Plancherel formula is in Harish-Chandra [1975], [1976a], and [1976b].

In the meantime, Sally and Warner [1973] took up the question of inverting F_f^B; their motivation was related to analysis of G/Γ. But a by-product of their work is a proof of the Plancherel formula that is different from the ones mentioned above. This is the proof we give in §§2–3, except that we have ignored some of the convergence difficulties. One should consult the Sally and Warner paper for the full proof.

Herb [1976] and [1979a] undertook the general problem of inverting F_f^B, analyzing what properties of discrete series characters are available and showing what kinds of terms have to be handled. Herb and Sally [1979] identified a relationship between limits of discrete series characters and some singular terms in the inversion problem.

Motivated by work of Shelstad [1979a] on stable orbital integrals, Chao [1977] obtained an inversion formula for the averaged version $\mathscr{F}_f^B$ of F_f^B for the group $Sp(2, \mathbb{R})$. Chao found that the averaged discrete series were manageable and that the singular terms dropped out. He then obtained the Plancherel formula by differentiation.

The break through in extending Chao's work to other groups came from the discovery by Herb [1979b] and [1981a] that the averaged discrete series constants are sums of products given in terms of two-systems. This formula (Theorem 13.6) and the other results in the latter part of §4 are due to Herb. Motivation for such a formula comes partly from work of Hirai [1977] and of Midorikawa [1977].

Problems 14 to 20 derive some properties of two-systems that are given in Herb's work. In some instances the Problems suggest a general proof for a result that she verified case by case. Parts of Problems 8 to 13 are taken from computations of Harish-Chandra [1965a], §§13–14, 21, and 23.

Herb [1982], [1981a], and [1983b] used the formulas for averaged discrete series characters to invert $\mathscr{F}_f$ (for each Cartan subgroup) and obtain the Plancherel formula. Our proof in §5 of the Plancherel formula for $Sp(2, \mathbb{R})$ follows Herb [1982] rather than Chao [1977], except that we have normalized our measures differently.

For an extension of the Plancherel formula to a wide class of nonlinear groups (wider than Harish-Chandra's class), see Herb and Wolf [1983].

Shelstad [1979b], [1981], and [1982] showed that the discrete series constants for G could be given in terms of averaged discrete series constants for G and some smaller groups (not necessarily subgroups). Herb [1981a] and [1983a] gave explicit formulas for implementing Shelstad's work. Herb [1983a] also applied the formulas for the discrete series constants to give an inversion formula for F_f for each Cartan subgroup.

Chapter XIV

Bargmann [1947] addressed the harmonic analysis of $L^2(\mathrm{SL}(2, \mathbb{R}))$ as in §1 by finding the asymptotics of matrix coefficients and relating them to what we now know as the Plancherel formula. In the late 1960's, Harish-Chandra switched his own approach to the Plancherel formula, abandoning the scheme of proof in §11.5 and instead taking advantage of the kind of relationship exploited by Bargmann. See Harish-Chandra [1968] for motivation for the change.

The first connection between asymptotics and intertwining operators is in Knapp and Stein [1970]. Knapp and Stein [1971] derived the inequality (14.6). The strengthened version in (14.8) is due to Arthur [1970], and so is the kind of irreducibility proof that follows (14.8). Lemma 14.1 is from Knapp and Stein [1971].

Eisenstein integrals were introduced by Harish-Chandra [1970] and [1972] in two announcements; motivation for them is in Harish-Chandra [1968]. Their detailed development appears starting with §19 of Harish-Chandra [1975]. Wallach [1975] identified the exact relationship (Proposition 14.3) between Eisenstein integrals and matrix coefficients; see also §7 of Harish-Chandra [1976b]. For Proposition 14.4, see §9 of Knapp and Stein [1980].

The real-variables method by which one obtains asymptotics of Eisenstein integrals appears first (in the context of semisimple groups) in §§27–30 of Harish-Chandra [1966b]. Our Theorem 14.6 is Theorem 21.1 of Harish-Chandra [1975]. Arthur [1970], for groups of real rank one, introduced techniques for allowing parameters to vary, and these techniques were extended to general groups G in §§2–10 of Harish-Chandra [1976a]. Our Theorems 14.7 and 14.8 are Theorem 18.1 of Harish-Chandra [1976a].

Eisenstein integrals, c functions, and the real-variables method of handling asymptotics have been developed further by other authors for a number of purposes. See Cohn [1974] and Wallach [1975a] for properties of c functions, van den Ban [1982] for integral formulas for asymptotics, Duistermaat, Kolk, and Varadarajan [1983] for connections with the subject of oscillatory integrals, Kashiwara and Oshima [1977] for the use of hyperfunctions, and Blank [1981] for application to the problem of finding global realizations of limits of discrete series.

Formula (14.19) is Theorem 19.1 of Harish-Chandra [1976a]. For Corollary 14.9, see §9 of Knapp and Stein [1980]. Formula (14.30) was obtained independently by Arthur [1974] by different methods. It had been observed earlier in unpublished work of Wallach in 1971–72 for the case of minimal parabolic subgroups.

Proposition 14.10 is due to Schiffmann [1971] in the case of minimal parabolic subgroups and to Knapp and Stein [1980] in the general case. The η functions were introduced by Knapp and Stein [1971] and [1980]. The first paper treats only the case of a minimal parabolic subgroup and uses different notation; the second paper treats the general case.

Theorem 14.15 is due to Harish-Chandra; see Lemma 13.3 of Harish-Chandra [1976b]. In the case that S is a minimal parabolic subgroup, Theorem 14.16 was discovered independently by Arthur [1970] and Knapp and Stein [1971]. The result in the general case was announced in Harish-Chandra [1972]; its proof is contained in the proof of Lemma 39.1 of Harish-Chandra [1976b].

The results in §6 are taken from Knapp and Stein [1980]. In the case of a minimal parabolic subgroup, some of the results were already in Knapp and Stein [1971]. Kunze and Stein [1960], [1961], and [1967] earlier had shown the importance of normalizing intertwining operators appropriately. They worked with explicit normalizations, which they were able to construct only in certain circumstances.

Proposition 14.26 is proved in §§24–36 of Harish-Chandra [1976b] by the method using asymptotics. For a proof in the style of Chapter XIII, see Herb [1982].

The heart of the proof of Theorem 14.27 is to relate the Plancherel measure to some of the other objects in this chapter. Harish-Chandra [1976b] essentially defines p_σ (called $\mu(\omega:\nu)$ in that paper) as the reciprocal of our η function (see Lemma 12.1 of that paper). The Plancherel density enters in a roundabout way. For a K-finite $Z(\mathfrak{g}^{\mathbb{C}})$-finite function f, the formula for obtaining $f(1)$ in terms of a derivative of F_f^B allows one to obtain a Fourier expansion of f in terms of discrete series characters (see §23). Direct integration of this formula, applied to M and combined with various identities, gives the part of the Plancherel formula corresponding to a single Cartan subgroup. See especially §§20 and 27.

The proof we give of Theorem 14.27 when S is minimal parabolic is from Knapp and Stein [1971] and does not use the differential equations. See §11 of that paper. Lemma 14.28 is essentially in Naimark [1958].

The completeness theorem and most of the other results in §8 are due to Harish-Chandra [1976b]. The completeness theorem is stated in Harish-Chandra's language as Theorem 38.1 of that paper and is translated into a result about intertwining operators in Corollary 9.8 of Knapp and Stein [1980]. This translation had been made earlier by Wallach in unpublished work in 1971–72 for the case of minimal parabolic subgroups. Lemma 14.38 is contained in Lemma 14.2 of Knapp and Stein [1980].

In connection with §9, useful roots first appear in Knapp [1975], and Theorem 14.39 is proved there. Further properties of useful roots are derived in Knapp [1982]. The R group was introduced by Knapp and Stein [1972], who noted its semidirect product decomposition in the case of minimal parabolic subgroups. Theorem 14.43 and the results that precede it are from Knapp and Stein [1980], except that part of Lemma 14.40 is from Lemma 39.1 of Harish-Chandra [1976b].

Most of the results of §§10–11, including the Multiplicity One Theorem, are from Knapp [1982], but a number of the simplified proofs that we give are taken from Knapp and Zuckerman [1982]. The fact that the R group is the direct sum of two-element groups when G is split and the parabolic subgroup is minimal is Theorem 15.1 of Knapp and Stein [1980]. The original proof of the Multiplicity One Theorem was obtained without the benefit of Theorem 14.64, which appears in Knapp and Zuckerman [1982].

Zuckerman tensoring of induced representations is treated in Appendix B of Knapp and Zuckerman [1982], having been announced in Zuckerman [1977].

Generalized Schmid identities appear in §§1–5 of Knapp and Zuckerman [1982] and are developed inductively along with limits of discrete series. We have been able to separate their development here by giving a direct proof of conclusion (c′) of Theorem 12.26.

Inversion of generalized Schmid identities is in §6 of Knapp and Zuckerman [1982]. The conditions on positive systems in Lemma 14.69 are not identical with the ones in Lemma 3.1 of Knapp and Zuckerman [1982], but Problems 30 to 37 show that they are equivalent.

The complete reduction of representations induced from discrete series is in §8 of Knapp and Zuckerman [1982]. The results of §7 of that paper are not needed here, since we have settled for less than the maximum amount of information about intertwining operators. For the results of §16 through Corollary 14.89, see §§10–13 of the 1982 paper.

The Langlands Disjointness Theorem (Theorem 14.90) is due to Langlands [1973] and is discussed in detail in these Notes in connection with Chapter XV.

The classification theorem, Theorem 14.91, is Theorem 14.2 of Knapp and Zuckerman [1982], having been announced in Knapp and Zuckerman [1977]. Its combination with the original Langlands classification into Theorem 14.92 appeared originally in the 1977 paper.

Theorem 14.93 is an unpublished theorem of Harish-Chandra obtained approximately in 1971. The original proof is quite different, predating all of the classification theorems.

Problems 17 to 22 are based on results in §2 of Knapp [1982].

Chapter XV

The theory of minimal K types is due to Vogan [1979a], who called them "lowest." The notion that there should be a distinguished K type of multiplicity one is an old one and plays a role in Parthasarathy, Rao, and Varadarajan [1967] and in Lepowsky [1971]. But the real motivation for Vogan's definition comes from discrete series and Schmid's sequence of papers that use the Blattner K type as a point of departure for the analysis of the representations (see the Notes for Chapter XII).

Vogan's approach is to start with any highest weight Λ for K and to give an explicit construction of a parameter λ_0 as in (15.23). This construction is his Proposition 4.1, and the proof is simplified in Carmona [1983]. The assumption rank $G =$ rank K makes the proof still simpler, but Vogan and Carmona do not make this assumption.

An extra requirement in the proposition makes the collection of β_i's as large as possible. The parameter λ_0 then allows one to define a subspace $\mathfrak{b}_-$ of $\mathfrak{b}$ as the common kernel of the β_i's. Vogan next uses a cohomological argument to prove in Proposition 5.8 that if τ_Λ is a minimal K type of an irreducible admissible representation π, then some Weyl group conjugate of the infinitesimal character of π has the same restriction to $\mathfrak{b}_-$ as λ_0. In the special case of a split group, it follows easily that minimal implies small for the nonunitary principal series (see our

Theorem 15.4). The remainder of the proof of Theorem 15.4 is in §6 of Vogan's paper, as follows: It is not hard to see that small is equivalent with fine and that all fine K types are suitably conjugate within G. Therefore $|\Lambda + 2\delta_K|^2$ is constant for fine K types in a particular principal series, and so fine implies minimal.

Theorem 16.1 and Proposition 16.2 are announced in Knapp [1983]. A formula related to Theorem 16.1, but with the quantifiers arranged differently, appears in §7 of Vogan's paper.

Theorem 15.8 is a restatement of Lemma 8.8 in Vogan [1979a]. Theorem 15.9 is announced (in a strong form) as Theorem 1.4, but the proof appears only in the unpublished Part II of the paper, Vogan [1979b]. Vogan proves actually that each irreducible constituent of $U(S, \sigma, 0)$ contains only one minimal K type.

The Langlands Disjointness Theorem, whose proof is indicated in §3, is in Langlands [1973], pp. 65 and 76–78. The step in the argument whose proof was left for these Notes is handled as follows: Recall the space $\mathscr{A}(G, \tau)$ we defined in §14.3. Section 25 of Harish-Chandra [1975] defines a notion of "negligibility" relative to S of an f in $\mathscr{A}(G, \tau)$, writing $f_S \sim 0$ for it. If $S' = M'A'N'$ with A' not conjugate to A, then an Eisenstein integral built from S is S'-negligible, by Lemma 18.3 of Harish-Chandra [1976a]. In our proof, what we have seen implies that the Eisenstein integral is S'-negligible when A' is conjugate to A. Therefore it is negligible relative to all cuspidal parabolic subgroups, and Lemma 25.2 of Harish-Chandra [1975] says it is 0.

A version of Theorem 15.10 is announced as Theorem 1.3 of Vogan [1979a], and again the proof is in Part II. The proof we give is a little different and was communicated by Langlands.

Vogan [1981] redoes the theory of minimal K types, with a construction in mind called "cohomological induction." Much of the book revolves around an alternate classification of irreducible admissible representations that grows out of this construction. See the Notes for Chapter XVI for further discussion of this construction. The topics in Vogan's book overlap rather little with the present book, and Vogan's book is one direction that the interested reader can take for further exposition of the representation theory of semisimple Lie groups.

Chapter XVI

For historical information about $SL(2, \mathbb{R})$ and $SL(2, \mathbb{C})$, see the Notes for Chapter II. These classifications do not, of course, need the whole previous fifteen chapters of the book as background. The things that are really needed are the Subrepresentation Theorem, parts of the first form of the Langlands classification (Theorem 8.54), and the material in §9.1. These results identify all the candidates for irreducible unitary representations. To find the discrete series, one can examine the nonunitary principal series for reducibility, working directly with the action by the Lie algebra. One does not need characters in the argument; Lemma 16.5 and the first paragraph of the proof of Theorem 16.6 show that $J(S, \sigma, \nu)$ can be disregarded unless $w\sigma = \sigma$ and $w\nu = -\bar{\nu}$ for some w. The simple arguments in §16.1 then complete the classification.

The use of continuity arguments for establishing that representations are unitary dates from Sally [1967], is more explicit in Kostant [1969] and [1975], and is still more explicit in Knapp and Stein [1971]. Kostant dealt with the spherical principal series (induced representations from a minimal parabolic subgroup with the trivial representation on $M_{\mathfrak{p}}$). First he classified the irreducible quotients of spherical principal series that are infinitesimally unitary in the case that G has real rank one. Then he exhibited unitary cases with small $A_{\mathfrak{p}}$ parameter for general G. Knapp and Stein [1971] worked with the full nonunitary principal series. They indicated a result like Corollary 16.5 and proved explicitly the analog for complex groups. Section 16 of Knapp and Stein [1980] contains further results along the same lines.

It was long ago recognized that Weyl group elements of order 2 have something to do with complementary series. This notion appears in Bruhat [1956] and then in Kunze and Stein [1967] and Kostant [1969].

A version of Theorem 16.6 appropriate to the first Langlands classification (Theorem 8.54) is due to Knapp and Zuckerman [1977], who proved also Lemma 16.7. The version here appears in §1 of Knapp and Speh [1982].

Proposition 16.8 is due to Baldoni Silva [1981], Lemma 16.9 is due to Speh [1981], and Theorem 16.10 is an unpublished result due to Vogan.

The spherical functions with bounded matrix coefficients were identified by Helgason and Johnson [1969]. Problems 5 to 7 represent a generalization of this result from Langlands quotients of spherical principal series to general irreducible admissible representations; this generalization was announced by Knapp and Speh [1982]. Actually in a simple group, the matrix coefficients of an infinite-dimensional irreducible unitary representation must vanish at infinity, according to Howe and Moore [1979].

Considerably more is known about the classification of irreducible unitary representations than is indicated here. Early progress concentrated on particular groups; the first success along these lines was by Dixmier [1961], in connection with SO(4, 1). Later research began to concentrate on techniques. The status of the classification problem as of the early 1980's is given by Knapp and Speh [1982], who include an extensive bibliography.

One technique of interest has concerned representations similar to discrete series. As part of his development of holomorphic discrete series, Harish-Chandra [1955] considered general irreducible admissible representations with a highest weight vector. A systematic study of which ones are unitary was begun by Wallach [1979] and Rossi and Vergne [1976] and has now been completed by Enright, Howe, and Wallach [1983] and by Jakobsen [1983]. The bibliographies of these papers contain further references. These representations may be regarded as obtained from holomorphic discrete series by suitable continuation of the parameter. Enright, Parthasarathy, Wallach, and Wolf [1983] and [1985] have extended this continuation argument somewhat to nonholomorphic discrete series.

Another technique of interest is a construction through the use of "dual reductive pairs" of groups. See Howe [1980], Rallis and Schiffmann [1977], and Adams [1983].

A significant recent development is the discovery that unitary representations often result from a construction known as **cohomological induction.** This construction grew out of the work of Schmid on realizing discrete series (see the Notes for Chapter XII) and the unpublished work of Zuckerman on extending such constructions to handle other representations that appear to be of interest in physics and geometry (see Vogan and Zuckerman [1984]). The construction is due to Vogan [1981], and parts of it appear in Speh and Vogan [1980]. Vogan [1984] gives far-reaching sufficient conditions under which this construction carries unitary representations to unitary representations. See Wallach [1984] for an alternate proof of some of Vogan's theorems and for additional results.

A forward-looking status report that examines this and other constructions is the one by Vogan [1983].

The other side of the classification problem is the elimination of Langlands quotients that are not infinitesimally unitary. One technique for this purpose may be found in Duflo [1979a] and is developed further by Knapp and Speh [1982] and [1983]. Another technique is the use of a Dirac inequality; this technique evolved over a period of years, and final forms of it may be found in Enright [1979] and Baldoni Silva [1981]. A third technique has origins in the work of Klimyk and Gavrilik [1974] and may be found in Baldoni-Silva and Knapp [1984].

Appendix A

Much of this material may be found in the first four chapters of Chevalley [1947]. For the structure theory of Lie algebras, see Jacobson [1962]. For the material on covering spaces, see Chapters 2 and 5 of Massey [1967].

Appendix B

For the one-variable theory in more detail, see Coddington and Levinson [1955], especially Chapter 4.

Harish-Chandra found that an analog for partial differential equations of the classical theory for ordinary differential equations had profound implications for the study of group representations. Harish-Chandra [1960a] developed such a theory, but with a singularity at infinity and with expansions in terms of exponential functions. Wallach [1975b] simplified the theory substantially, in part by applying an idea of Casselman [1975a] of changing variables to bring the singularity to the origin; the proofs become very similar to those for ordinary differential equations. We follow essentially Wallach's treatment of the theory in §§2–4. For Theorem B.9, see the appendix in Warner [1972b]. Deligne [1970] gave a systematic exposition of the whole theory in greater generality than is needed for semisimple groups; the singular locus need not be a product in Deligne's case.

The application of this theory to semisimple groups, including the "Motivation" in §5, is in Harish-Chandra [1960b]. Casselman [1975a] isolated the abstract theory of higher-order systems and introduced the indicial module, using the language of sheaves as Deligne had done. Miličić [1976] simplified this theory for the case of interest in semisimple groups, proceeding in the spirit of Wallach [1975b]. We follow Miličić in §§5–6. The uniqueness results in §7 are in §15 of Harish-Chandra [1958a].

Appendix C

There are several useful sources for explicit information about groups. The tables at the end of Bourbaki [1968] give the Dynkin diagrams, explicit realizations of the root systems, the algebraically integral forms, and information about the Weyl groups. The books by McKay and Patera [1981] and Bremner, Moody, and Patera [1985] give tables of information about finite-dimensional representations.

Section IX.4 of Helgason [1962] includes a table containing much of the information in Table C.2, but for exceptional groups and classical groups both. In addition, that section lists the classical groups with their Cartan decompositions and subgroups $A_{\mathfrak{p}}$.

Sugiura [1959] identifies the conjugacy classes of Cartan subalgebras for all real simple Lie algebras. Araki [1962] gives information about real forms and introduces diagrams for (re)classifying them; see also Tits [1966]. A table of Araki diagrams may be found in pp. 30–32 of Warner [1972a]. This table tells also the multiplicities of the restricted roots but does not tell which diagram goes with which group.

Explicit formulas for composition series and intertwining operators for spherical principal series of real-rank-one groups may be found in Johnson and Wallach [1972] and [1977] and Johnson [1976]. Formulas associated with the full non-unitary principal series of $SO(n, 1)$ and $SU(n, 1)$ may be found in Klimyk and Gavrilik [1976]. For $Sp(n, 1)$, see Holod and Klimyk [1977].

Further Topics

It is mentioned in Chapter I that analysis of some other representations besides $L^2(G)$ is a third large problem in representation theory. This section contains some introductory information about work on this problem.

First, there is L^2 of the noncompact symmetric space G/K. This is normally treated through an analysis of bi-K-invariant functions by means of spherical functions, and the passage to a function f in $L^2(G/K)$ comes by considering the bi-K-invariant function $f^* * f$. An early paper is by Godement [1952]. Harish-Chandra [1958a] and [1958b] did the main analysis; his papers left unresolved two questions, which were later settled in Gindikin and Karpelevič [1962] and §3 of Helgason [1964] and in §21 of Harish-Chandra [1966]. The theory on G/K has been developed further for the study of differential operators on G/K by Helgason [1964] and for the study of multiplier operators on $L^p(G/K)$ by Clerc and Stein [1974] and by Stanton and Tomas [1978]. See also Eguchi and Kumahara [1982].

Also among other representations to analyze, there is L^2 of a "semisimple symmetric space" G/H; here H is the fixed-point set of an involution (in place of the Cartan involution). Fundamental work on the general theory has been done by Flensted-Jensen [1980], Matsuki [1979] Oshima and Matsuki [1983], and Schlichtkrull [1982] and [1983]. Strichartz [1973] and Faraut [1979] analyzed families of particular cases and showed how the results could be applied to study a corresponding family of partial differential equations in Euclidean space. Sekiguchi [1984] has proved theorems about differential equations that generalize both part of Strichartz's and Faraut's work and Harish-Chandra's theorem about regularity of invariant eigendistributions. A readable exposition of the analysis on G/H is the one by Schlichtkrull [1984].

Several authors have considered aspects of $C^\infty_{\text{com}}(G)$ or $L^p(G)$ for $p \neq 2$. See, for example Ehrenpreis and Mautner [1955] and [1957], Kunze and Stein [1960],

Cowling [1978] and [1982], Trombi and Varadarajan [1971] and [1972], Arthur [1983], and Clozel and Delorme [1984].

The notes for Chapter XIII mention that the problem of inverting F_f has been the object of study by a number of people.

Finally there is the huge subject of $L^2(G/\Gamma)$, where Γ is a discrete subgroup of G. This subject is closely related to the Selberg Trace Formula. For orientation one can look at Godement [1966], Langlands [1966a], Harish-Chandra [1968], Arthur [1979] and [1981], and Osborne and Warner [1981].

References

Adams, J. D., Discrete spectrum of the dual reductive pair (O(p, q), Sp($2m$)), *Invent. Math.* 74 (1983), 449–475.

Adams, J. F., *Lectures on Lie Groups*, W. A. Benjamin, New York, 1969.

Araki, S., On root systems and an infinitesimal classification of irreducible symmetric spaces, *J. of Math., Osaka City Univ.* 13 (1962), 1–34.

Arthur, J. G., Harmonic analysis of tempered distributions on semisimple Lie groups of real rank one, Ph.D. Dissertation, Yale University, 1970.

Arthur, J. G., Intertwining integrals for cuspidal parabolic subgroups, mimeographed notes, Yale University, 1974.

Arthur, J., Eisenstein series and the trace formula, *Automorphic Forms, Representations, and L-functions*, Proc. Symp. in Pure Math. 33, Part I (1979), 253–274, American Mathematical Society, Providence, R. I.

Arthur, J., Automorphic representations and number theory, *1980 Seminar on Harmonic Analysis*, Canad. Math. Soc. Conf. Proc. 1 (1981), 3–51, American Mathematical Society, Providence, R. I.

Arthur, J., A Paley-Wiener theorem for real reductive groups, *Acta Math.* 150 (1983), 1–89.

Atiyah, M. F., Characters of semi-simple Lie groups, mimeographed notes, 1976.

Baldoni Silva, M. W., Unitary dual of Sp(n, 1), $n \geq 2$, *Duke Math. J.* 48 (1981), 549–583.

Baldoni-Silva, M. W., and A. W. Knapp, Indefinite intertwining operators, *Proc. Nat. Acad. Sci. USA* 81 (1984), 1272–1275.

Barbasch, D., and D. A. Vogan, The local structure of characters, *J. Func. Anal.* 37 (1980), 27–55.

Bargmann, V., Irreducible unitary representations of the Lorentz group, *Ann. of Math.* 48 (1947), 568–640.

Berger, M., Les espaces symétriques non compacts, *Ann. Sci. Ecole Norm. Sup.* 74 (1957), 85–177.

Berline, N., and M. Vergne, Sur la formule universelle de Kirillov, *Proc. International Congress of Mathematicians (1983, Warsaw, Poland)*, to appear 1985.

Bernstein, I. N., I. M. Gelfand, and S. I. Gelfand, Differential operators on the base affine space and a study of $\mathfrak{g}$-modules, *Lie Groups and Their Representations (Summer School of the Bolyai János Mathematical Society)*, Halsted Press, New York, 1975, 21–64.

Bhanu Murti, T. S., Plancherel's measure for the factor space SL(n; R)/SO(n; R), *Soviet Math. Doklady* 1 (1961a), 860–862.

Bhanu Murti, T. S., The asymptotic behavior of zonal spherical functions on the Siegel upper half-plane, *Soviet Math. Doklady* 1 (1961b), 1325–1329.

Birkhoff, Garrett, Representability of Lie algebras and Lie groups by matrices, *Ann. of Math.* 38 (1937), 526–533.

Blank, B. E., Embedding limits of discrete series of semisimple Lie groups, *1980 Seminar on Harmonic Analysis*, Canad. Math. Soc. Proc. 1 (1981), 55–64, American Mathematical Society, Providence, R. I.

Blank, B. E., Knapp-Wallach Szegö integrals and generalized principal series representations: the parabolic rank one case, *J. Func. Anal.* 60 (1985), 127–145.

Borel, A., *Représentations de Groupes Localement Compacts*, Springer-Verlag Lecture Notes in Math. 276 (1972).

Borel, A., and J. Tits, Groupes réductifs, *I.H.E.S. Publications Mathématiques* 27 (1965), 55–151.

Borel, A., and N. Wallach, *Continuous Cohomology, Discrete Subgroups, and Representations of Reductive Groups*, Princeton University Press, Princeton, 1980.

Bott, R., Homogeneous vector bundles, *Ann. of Math.* 66 (1957), 203–248.

Bourbaki, N., *Livre VI, Intégration: Chaptire 7, Mesur de Haar, Chapitre 8, Convolution et Represéntations*, Hermann, Paris, 1963.

Bourbaki, N., *Groupes et Algèbres de Lie, Chapitres 4, 5, et 6*, Hermann, Paris, 1968.

Bremner, M. R., R. V. Moody, and J. Patera, *Tables of Dominant Weight Multiplicities for Representations of Simple Lie Algebras*, Marcel Dekker, New York, 1985.

Bruhat, F., Sur les représentations induites des groupes de Lie, *Bull. Soc. Math. France* 84 (1956), 97–205.

Carmona, J., Sur la classification des modules admissible irréductibles, *Non Commutative Harmonic Analysis and Lie Groups*, Springer-Verlag Lecture Notes in Math. 1020 (1983), 11–34.

Cartan, E., Sur la structure des groupes de transformations finis et continus, *Thèse*, Nony, Paris, 1894; 2nd ed., Vuibert, 1933.

Cartan, E., Les groupes projectifs qui ne laissent invariante aucune multiplicité plane, *Bull. Soc. Math. France* 41 (1913), 53–96.

Cartan, E., Sur certaines formes riemanniennes remarquables des géométries à groupe fondamental simple, *Ann. Sci. Ecole Norm. Sup.* 44 (1927), 345–467.

Cartan, E., Sur la détermination d'un système orthogonal complet dans un espace de Riemann symétrique clos, *Rend. Circ. Mat. Palermo* 53 (1929), 217–252.

Cartan, E., Sur les domaines bornés homogènes de l'espace de n variables complexes, *Abh. Math. Sem. Univ. Hamburg* 11 (1935), 116–162.

Cartier, P., Structure topologique des groupes de Lie généraux, Exposé 22, Séminaire "Sophus Lie," *Théorie des algèbres de Lie, topologie des groupes de Lie, 1954–55*, Ecole Normale Supérieure, Paris, 1955.

Casselman, W., Systems of analytic partial differential equations of finite codimension, manuscript, 1975a.

Casselman, W., The differential equations satisfied by matrix coefficients, manuscript, 1975b.

Casselman, W., and D. Miličić, Asymptotic behavior of matrix coefficients of admissible representations, *Duke Math. J.* 49 (1982), 869–930.

Casselman, W., and M. S. Osborne, The n-cohomology of representations with an infinitesimal character, *Compositio Math.* 31 (1975), 219–227.

Chao, W.-M., Fourier inversion and Plancherel formula for semisimple Lie groups of real rank two, Ph.D. dissertation, University of Chicago, 1977.

Chevalley, C., *Theory of Lie Groups I*, Princeton University Press, Princeton, 1946.

Chevalley, C., Sur la classification des algèbres de Lie simples et de leurs représentations, *C. R. Acad. Sci. Paris* 227 (1948), 1136–1138.

Chevalley, C., Invariants of finite groups generated by reflections, *Amer. J. Math.* 77 (1955), 778–782.

Clerc, J. L., and E. M. Stein, L^p multipliers for noncompact symmetric spaces, *Proc. Nat. Acad. Sci. USA* 71 (1974), 3911–3912.

Clozel, L., and P. Delorme, Le Théoréme de Paley-Wiener invariant pour les groupes de Lie réductifs, *Invent. Math.* 77 (1984), 427–453.

Coddington, E. A., and N. Levinson, *Theory of Ordinary Differential Equations*, McGraw Hill, New York, 1955.

Cohn, L., *Analytic Theory of the Harish-Chandra C-Function*, Springer-Verlag Lecture Notes in Math. 429 (1974).

Cowling, M., The Kunze-Stein phenomenon, *Ann. of Math.* 107 (1978), 209–234.

Cowling, M., Unitary and uniformly bounded representations of some simple Lie groups, *Harmonic Analysis and Group Representations*, C.I.M.E. Conference at Cortona 1980, Liguori Editore, Naples, Italy, 1982, 49–128.

Deligne, P., *Equations Différentielles à Points Singuliers Reguliers*, Springer-Verlag Lecture Notes in Math. 163 (1970).

Dixmier, J., Représentations intégrables du groupe de De Sitter, *Bull. Soc. Math. France* 89 (1961), 9–41.

Dixmier, J., *Les Algèbres d'opérateurs dans l'Espace Hilbertien* (*Algèbres de Von Neumann*), Gauthiers-Villars, Paris, 1957; 2nd ed., 1969.

Dixmier, J., *Les C*-Algèbres et leurs Représentations*, Gauthier-Villars, Paris, 1964; 2nd ed., 1969.

Dixmier, J., *Algèbres Enveloppantes*, Gauthier-Villars, Paris, 1974.

Dixmier, J., and P. Malliavin, Factorisations de fonctions et de vecteurs indéfiniment différentiables, *Bull. des Sciences Math.*, 102 (1978), 305–330.

Duflo, M., Représentations unitaires irréductibles des groupes simples complexes de rang deux, *Bull. Soc. Math. France* 107 (1979a), 55–96.

Duflo, M., Représentations de carré intégrable des groupes semi-simples réels, Exposé 508, *Séminaire Bourbaki, Volume 1977/78*, Springer-Verlag Lecture Notes in Math. 710 (1979b).

Duflo, M., G. Heckman, and M. Vergne, Projection d'orbites, formule de Kirillov, et formule de Blattner, *Mem. Soc. Math. France, Nouvelle série* 15 (1984), 65–128.

Duistermaat, J. J., J.A.C. Kolk, and V. S. Varadarajan, Functions, flows, and oscillatory integrals on flag manifold and conjugacy classes in real semisimple Lie groups, *Compositio Math.* 49 (1983), 309–398.

Dynkin, E., Classification of the simple Lie groups, *Rec. Math. [Mat. Sbornik] N.S.* 18 (60) (1946), 347–352 (Russian with English summary).

Eguchi, M., and K. Kumahara, An L^p Fourier analysis on symmetric spaces, *J. Func. Anal.* 47 (1982), 230–246.

Ehrenpreis, L., and F. I. Mautner, Some properties of the Fourier transform on semi-simple Lie groups I, *Ann. of Math.* 61 (1955), 406–439.

Ehrenpreis, L., and F. I. Mautner, Some properties of the Fourier transform on semi-simple Lie groups II, *Trans. Amer. Math. Soc.* 84 (1957), 1–55.

Enright, T. J., Relative Lie algebra cohomology and unitary representations of complex Lie groups, *Duke Math. J.* 46 (1979), 513–525.

Enright, T. J., R. Howe, and N. Wallach, A classification of unitary highest weight modules, *Representation Theory of Reductive Groups, Proceedings of the University of Utah Conference 1982*, Birkhäuser, Boston, 1983, 97–143.

Enright, T. J., R. Parthasarathy, N. R. Wallach, and J. A. Wolf, Classes of unitarizable derived functor modules, *Proc. Nat. Acad. Sci. USA* 80 (1983), 7047–7050.

Enright, T. J., R. Parthasarathy, N. R. Wallach, and J. A. Wolf, Unitary derived functor modules with small spectrum, *Acta Math.* 154 (1985), 105–136.

Enright, T. J., and V. S. Varadarajan, On an infinitesimal characterization of the discrete series, *Ann. of Math.* 102 (1975), 1–15.

Enright, T. J., and N. R. Wallach, Notes on homological algebra and representations of Lie algebras, *Duke Math. J.* 47 (1980), 1–15.

Faraut, J., Distributions sphériques sur les espaces hyperboliques, *J. Math. Pures Appl.* 58 (1979), 369–444.

Flensted-Jensen, M., Spherical functions on a real semisimple Lie group. A method of reduction to the complex case, *J. Func. Anal.* 30 (1978), 106–146.

Flensted-Jensen, M., On a fundamental series of representations related to an affine symmetric space, *Non-Commutative Harmonic Analysis*, Springer-Verlag Lecture Notes in Math. 728 (1979), 77–96.

Flensted-Jensen, M., Discrete series for semisimple symmetric spaces, *Ann. of Math.* 111 (1980), 253–311.

Flensted-Jensen, M., and K. Okamoto, An explicit construction of the *K*-finite vectors in the discrete series for an isotropic semisimple symmetric space, *Mem. Soc. Math. France, Nouvelle série* 15 (1984), 157–199.

Fomin, A. I., and N. N. Shapovalov, A property of the characters of real semisimple Lie groups, *Func. Anal. and Its Appl.* 8(3) (1974), 270–271.

Frobenius, G., Sitzungsber. Preuss. Akad. (1898); see A. Speiser, *Théorie der Gruppen von endlicher Ordnung*, 3rd ed., Berlin, 1937.

Furstenberg, H., A Poisson formula for semi-simple Lie groups, *Ann. of Math.* 77 (1963), 335–386.

Gårding, L., Note on continuous representations of Lie groups, *Proc. Nat. Acad. Sci. USA* 33 (1947), 331–332.

Gelbart, S., Holomorphic discrete series for the real symplectic group, *Invent. Math.* 19 (1973), 49–58.

Gelbart, S., Examples of dual reductive pairs, *Automorphic Forms, Representations, and L-functions*, Proc. Symp. in Pure Math. 33, Part I (1979), 287–296, American Mathematical Society, Providence, R. I.

Gelfand, I. M., Spherical functions in symmetric Riemannian spaces, *Doklady Akad. Nauk SSSR (N.S.)* 70 (1950a), 5–8 (Russian).

Gelfand, I. M., The center of an infinitesimal group ring, *Mat. Sbornik N.S.* 26(68) (1950b), 103–112 (Russian).

Gelfand, I. M., and M. I. Graev, On a general method of decomposition of the regular representation of a Lie group into irreducible representations, *Doklady Akad. Nauk SSSR* 92 (1953a), 221–224 (Russian).

Gelfand, I. M., and M. I. Graev, Analog of the Plancherel formula for real semisimple Lie groups, *Doklady Akad. Nauk SSSR* 92 (1953b), 461–464 (Russian).

Gelfand, I. M., M. I. Graev, and I. I. Pyatetskii-Shapiro, *Representation Theory and Automorphic Functions*, W. B. Saunders, Philadelphia, 1969.

Gelfand, I. M., and M. A. Naimark, Unitary representations of the Lorentz group, *Izvestiya Akad. Nauk SSSR, Ser. Mat.* 11 (1947), 411–504 (Russian).

Gelfand, I. M., and M. A. Naimark, *Unitary Representations of the Classical Groups*, Trudy Mat. Inst. Steklov 36, Moskow-Leningrad, 1950 (Russian). German translation: Akademie-Verlag, Berlin, 1957.

Gindikin, S. G., and F. I. Karpelevič, Plancherel measure for Riemann symmetric spaces of nonpositive curvature, *Soviet Math. Doklady* 3 (1962), 962–965.

Godement, R., Sur les relations d'orthogonalité de V. Bargmann, I and II, *C. R. Acad. Sci. Paris* 225 (1947), 521–523 and 657–659.

Godement, R., A theory of spherical functions I, *Trans. Amer. Math. Soc.* 73 (1952), 496–556.

Godement, R., The decomposition of $L^2(G/\Gamma)$ for $\Gamma = SL(2, Z)$, *Algebraic Groups and Discontinuous Subgroups*, Proc. Symp. in Pure Math. 9 (1966), 211–224, American Mathematical Society, Providence, R. I.

Goto, M., Faithful representations of Lie groups I, *Math. Japonicae* 1 (1948), 107–119.

Gross, K. I., and R. A. Kunze, Bessel functions and representation theory I, *J. Func. Anal.* 22 (1976), 73–105.

Gross, K. I., and R. A. Kunze, Bessel functions and representation theory II, *J. Func. Anal.* 25 (1977), 1–49.

Harish-Chandra, On some applications of the universal enveloping algebra of a semisimple Lie algebra, *Trans. Amer. Math. Soc.* 70 (1951), 28–96.

Harish-Chandra, Plancherel formula for the 2×2 real unimodular group, *Proc. Nat. Acad. Sci. USA* 38 (1952), 337–342.

Harish-Chandra, Representations of a semisimple Lie group on a Banach space I, *Trans, Amer. Math. Soc.* 75 (1953), 185–243.

Harish-Chandra, Representations of semisimple Lie groups II, *Trans. Amer. Math. Soc.* 76 (1954a), 26–65.

Harish-Chandra, Representations of semisimple Lie groups III, *Trans. Amer. Math. Soc.* 76 (1954b), 234–253.

Harish-Chandra, The Plancherel formula for complex semisimple Lie groups, *Trans. Amer. Math. Soc.* 76 (1954c), 485–528.

Harish-Chandra, Representations of semisimple Lie groups V, *Proc. Nat. Acad. Sci. USA* 40 (1954d), 1076–1077.

Harish-Chandra, Representations of semisimple Lie groups IV, *Amer. J. Math.* 77 (1955), 743–777.

Harish-Chandra, Representations of semisimple Lie groups V, *Amer. J. Math.* 78 (1956a), 1–41.

Harish-Chandra, Representations of semisimple Lie groups VI, *Amer. J. Math.* 78 (1956b), 564–628.

Harish-Chandra, On a lemma of F. Bruhat, *J. Math. Pures Appl.* 35 (1956c), 203–210.

Harish-Chandra, The characters of semisimple Lie groups, *Trans. Amer. Math. Soc.* 83 (1956d), 98–163.

Harish-Chandra, Differential operators on a semisimple Lie algebra, *Amer. J. Math.* 79 (1957a), 87–120.

Harish-Chandra, Fourier transforms on a semisimple Lie algebra I, *Amer. J. Math.* 79 (1957b), 193–257.

Harish-Chandra, Fourier transforms on a semisimple Lie algebra II, *Amer. J. Math.* 79 (1957c), 653–686.

Harish-Chandra, A formula for semisimple Lie groups, *Amer. J. Math.* 79 (1957d), 733–760.

Harish-Chandra, Spherical functions on a semisimple Lie group I, *Amer. J. Math.* 80 (1958a), 241–310.

Harish-Chandra, Spherical functions on a semisimple Lie group II, *Amer. J. Math.* 80 (1958b), 553–613.

Harish-Chandra, Some results on differential equations, manuscript, 1960a. (See *Collected Papers*, Vol. III, 7–48, Springer-Verlag, New York, 1984.)

Harish-Chandra, Differential equations and semisimple Lie groups, manuscript, 1960b. (See *Collected Papers*, Vol. III, 57–120, Springer-Verlag, New York, 1984.)

Harish-Chandra, Invariant eigendistributions on a semisimple Lie algebra, *I.H.E.S. Publications Mathématiques* 27 (1965a), 5-54.

Harish-Chandra, Invariant eigendistributions on a semisimple Lie group, *Trans. Amer. Math. Soc.* 119 (1965b), 457–508.

Harish-Chandra, Discrete series for semisimple Lie groups I. Construction of invariant eigendistributions, *Acta Math.* 113 (1965c), 241–318.

Harish-Chandra, Two theorems on semi-simple Lie groups, *Ann. of Math.* 83 (1966a), 74–128.

Harish-Chandra, Discrete series for semisimple Lie groups II. Explicit determination of the characters, *Acta Math.* 116 (1966b) 1–111.

Harish-Chandra, *Automorphic Forms on Semisimple Lie Groups*, Springer-Verlag Lecture Notes in Math. 62 (1968).

Harish-Chandra, Harmonic analysis on semisimple Lie groups, *Bull. Amer. Math. Soc.* 76 (1970), 529–551.

Harish-Chandra, On the theory of the Eisenstein integral, *Conference on Harmonic Analysis*, Springer-Verlag Lecture Notes in Math. 266 (1972), 123–149.

Harish-Chandra, Harmonic analysis on real reductive groups I. The Theory of the constant term, *J. Func. Anal.* 19 (1975), 104–204.

Harish-Chandra, Harmonic analysis on real reductive groups II. Wave packets in the Schwartz space, *Invent. Math.* 36 (1976a), 1–55.

Harish-Chandra, Harmonic analysis on real reductive groups III. The Maass-Selberg relations and the Plancherel formula, *Ann. of Math.* 104 (1976b), 117–201.

Hecht, H., On characters and asymptotics of representations of a real reductive Lie group, *Math. Ann.* 242 (1979), 103–126.

Hecht, H., and W. Schmid, A proof of Blattner's conjecture, *Invent. Math.* 31 (1975), 129–154.

Hecht, H., and W. Schmid, Characters, asymptotics, and $\mathfrak{n}$-homology of Harish-Chandra modules, *Acta Math.* 151 (1983a), 49–151.

Hecht, H., and W. Schmid, On the asymptotics of Harish-Chandra modules, *J. Reine Angew. Math.* 343 (1983b), 169–183.

Helgason, S., Differential operators on homogeneous spaces, *Acta Math.* 102 (1959), 239–299.

Helgason, S., *Differential Geometry and Symmetric Spaces*, Academic Press, New York, 1962.

Helgason, S., Fundamental solutions of invariant differential operators on symmetric spaces, *Amer. J. Math.* 86 (1964), 565–601.

Helgason, S., lectures at Massachusetts Institute of Technology, Spring 1967.

Helgason, S., A duality for symmetric spaces with applications to group representations, *Advances in Math.* 5 (1970), 1–154.

Helgason, S., *Differential Geometry, Lie Groups, and Symmetric Spaces*, Academic Press, New York, 1978.

Helgason, S., and K. Johnson, The bounded spherical functions on symmetric spaces, *Advances in Math.* 3 (1969), 586–593.

Herb, R. A., Character formulas for discrete series on semisimple Lie groups, *Nagoya Math. J.* 64 (1976), 47–61.

Herb, R. A., Fourier inversion of invariant integrals on semisimple real Lie groups, *Trans. Amer. Math. Soc.* 249 (1979a), 281–302.

Herb, R. A., Characters of averaged discrete series on semisimple real Lie groups, *Pacific J. Math.* 80 (1979b), 169–177.

Herb, R. A., Fourier inversion and the Plancherel theorem, *Non Commutative Harmonic Analysis and Lie Groups*, Springer-Verlag Lecture Notes in Math. 880 (1981a), 197–210.

Herb, R. A., Discrete series character identities and Fourier inversion, *1980 Seminar on Harmonic Analysis*, Canad. Math. Soc. Conf. Proc. 1 (1981b), 93–98, American Mathematical Society, Providence, R. I.

Herb, R. A., Fourier inversion and the Plancherel theorem for semisimple real Lie groups, *Amer. J. Math.* 104 (1982), 9–58.

Herb, R. A., Discrete series characters and Fourier inversion on semisimple real Lie groups, *Trans. Amer. Math. Soc.* 277 (1983a), 241–262.

Herb, R. A., The Plancherel Theorem for semisimple Lie groups without compact Cartan subgroups, *Non Commutative Harmonic Analysis and Lie Groups*, Springer-Verlag Lecture Notes in Math. 1020 (1983b), 73–79.

Herb, R. A., and P. J. Sally, Singular invariant eigendistributions as characters in the Fourier transform of invariant distributions, *J. Func. Anal.* 33 (1979), 195–210.

Herb, R. A., and J. A. Wolf, The Plancherel theorem for general semisimple groups, preprint, 1983, to appear in *Compositio Math.*

Hirai, T., The Plancherel formula for the Lorentz group of n-th order, *Proc. Japan Acad.* 42 (1966), 323–326.

Hirai, T., The characters of some induced representations of semisimple Lie groups, *J. Math. Kyoto Univ.* 8 (1968), 313–363.

Hirai, T., Invariant eigendistributions of Laplace operators on real simple Lie groups II: General theory for semisimple Lie groups, *Japanese J. Math.* 2 (1976), 27–89.

Hirai, T., Invariant eigendistributions of Laplace operators on real simple Lie groups IV: Explicit forms of the characters of discrete series representations for $Sp(n, R)$, *Japanese J. Math.* 3 (1977), 1–48.

Hochschild, G., *The Structure of Lie Groups*, Holden-Day, San Francisco, 1965.

Holod, P. I., and A. U. Klimyk, Representations of the group $Sp(n, 1)$, I and II, preprints ITP-77-73E and ITP-77-80E, Institute for Theoretical Physics, Kiev, USSR, 1977.

Howe, R., On a notion of rank for unitary representations of the classical groups, duplicated notes, Yale University, 1980.

Howe, R. E., and C. C. Moore, Asymptotic properties of unitary representations, *J. Func. Anal.* 32 (1979), 72–96.

Hua, L.-K., *Harmonic Analysis of Functions of Several Complex Variables in the Classical Domains*, American Mathematical Society, Providence, R. I., 1963; revised 1979.

Humphreys, J. E., *Introduction to Lie Algebras and Representation Theory*, Springer-Verlag, Berlin-Heidelberg-New York, 1972.

Hunt, G. A., A theorem of Élie Cartan, *Proc. Amer. Math. Soc.* 7 (1956), 307–308.

Iwasawa, K., On some types of topological groups, *Ann. of Math.* 50 (1949), 507–558.

Jacobson, N., *Lectures in Abstract Algebra*, Vol. II, D. Van Nostrand, Princeton, 1953.

Jacobson, N., *Lie Algebras*, Interscience, New York, 1962.

Jacquet, H., and R. P. Langlands, *Automorphic Forms on GL(2)*, Springer-Verlag Lecture Notes in Math. 114 (1970).

Jakobsen, H. P., Hermitian symmetric spaces and their unitary highest weight modules, *J. Func. Anal.* 52 (1983), 385–412.

Jantzen, J. C., Zur Charakterformel gewisser Darstellungen halbeinfacher Gruppen und Lie-Algebren, *Math. Zeitschrift* 140 (1974), 127–149.

Jantzen, J. C., *Moduln mit einen höchsten Gewicht*, Springer-Verlag Lecture Notes in Math. 750 (1979).

Johnson, K. D., Composition series and intertwining operators for the spherical principal series II, *Trans. Amer. Math. Soc.* 215 (1976), 269–283.

Johnson, K., and N. R. Wallach, Composition series and intertwining operators for the spherical principal series, *Bull. Amer. Math. Soc.* 78 (1972), 1053–1059.

Johnson, K. D., and N. R. Wallach, Composition series and intertwining operators for the spherical principal series I, *Trans. Amer. Math. Soc.* 229 (1977), 137–173.

Kashiwara, M., and T. Oshima, Systems of differential equations with regular singularities and their boundary value problems. *Ann. of Math.* 106 (1977), 145–200.

Kashiwara, M., and M. Vergne, On the Segal-Shale-Weil representations and harmonic polynomials, *Invent. Math.* 44 (1978), 1–47.

Killing, W., Die Zusammensetzung der stetigen endlichen Transformationsgruppen I, II, III, IV, *Math. Ann.* 31 (1888), 252–290; 33 (1889), 1–48; 34 (1889), 57–122; 36 (1890), 161–189.

Kirillov, A., The characters of unitary representations of Lie groups, *Func. Anal. and Its Appl.* 2 (1968) 133–146.

Klimyk, A. U., and A. M. Gavrilik, The representations of the groups U(n, 1) and $SO_0(n, 1)$, preprint ITP-76-39E, Institute for Theoretical Physics Kiev, USSR, 1976.

Knapp, A. W., Bounded symmetric domains and holomorphic discrete series, *Symmetric Spaces*, W. M. Boothby and G. L. Weiss (ed.), Marcel Dekker, New York, 1972, 211–246.

Knapp, A. W., Weyl group of a cuspidal parabolic, *Ann. Sci. Ecole Norm. Sup.* 8 (1975), 275–294.

Knapp, A. W., Commutativity of intertwining operators for semisimple groups, *Compositio Math.* 46 (1982), 33–84.

Knapp, A. W., Minimal K-type formula, *Non Commutative Harmonic Analysis and Lie Groups*, Springer-Verlag Lecture Notes in Math. 1020 (1983), 107–118.

Knapp, A. W., and K. Okamoto, Limits of holomorphic discrete series, *J. Func. Anal.* 9 (1972), 375–409.

Knapp, A. W., and B. Speh, Status of classification of irreducible unitary representations, *Harmonic Analysis*, Springer-Verlag Lecture Notes in Math. 908 (1982), 1–38.

Knapp, A. W., and B. Speh, The role of basic cases in classification: Theorems about unitary representations applicable to SU(N, 2), *Non Commutative Harmonic Analysis and Lie Groups*, Springer-Verlag Lecture Notes in Math. 1020 (1983), 119–160.

Knapp, A. W., and E. M. Stein, The existence of complementary series, *Problems in Analysis: A Symposium in Honor of Salomon Bochner*, Princeton University Press, Princeton, 1970, 249–259.

Knapp, A. W., and E. M. Stein, Intertwining operators for semisimple groups, *Ann. of Math.* 93 (1971). 489–578.

Knapp, A. W., and E. M. Stein, Irreducibility theorems for the principal series, *Conference on Harmonic Analysis*, Springer-Verlag Lecture Notes in Math. 266 (1972), 197–214.

Knapp, A. W., and and E. M. Stein, Intertwining operators for semisimple groups II, *Invent. Math.* 60 (1980), 9–84.

Knapp, A. W., and N. R. Wallach, Szegö kernels associated with discrete series, *Invent. Math.* 34 (1976), 163–200. (See also 62 (1980), 341–346.)

Knapp, A. W., and R. E. Williamson, Poisson integrals and semisimple groups, *J. d'Analyse Math.* 24 (1971), 53–76.

Knapp, A. W., and G. Zuckerman, Classification theorems for representations of semisimple Lie groups, *Non-Commutative Harmonic Analysis*, Springer-Verlag Lecture Notes in Math. 587 (1977), 138–159.

Knapp, A. W., and G. J. Zuckerman, Classification of irreducible tempered representations of semisimple groups, *Ann. of Math.* 116 (1982), 389–501. (See also 119 (1984), 639.)

Koornwinder, T. H., (ed.), *Representations of Locally Compact Groups with Applications*, Parts I and II, Mathematisch Centrum, Amsterdam, 1979.

Koornwinder, T. H., Jacobi functions and analysis on noncompact semisimple Lie groups, *Special Functions: Group Theoretical Aspects and Applications*, D. Reidel, Hingham, Mass., 1985, 1–85.

Koranyi, A., and J. A. Wolf, Realization of Hermitian symmetric spaces as generalized half-planes, *Ann. of Math.* 81 (1965), 265–288.

Kostant, B., A formula for the multiplicity of a weight, *Trans. Amer. Math. Soc.* 93 (1959), 53–73.

Kostant, B., Lie algebra cohomology and the generalized Borel-Weil theorem, *Ann. of Math.* 74 (1961), 329–387.

Kostant, B., On the existence and irreducibility of certain series of representations, *Bull. Amer. Math. Soc.* 75 (1969), 627–642.

Kostant, B., On the existence and irreducibility of certain series of representations, *Lie Groups and Their Representations (Summer School of the Bolyai János Mathematical Society)*, Halsted Press, New York, 1975, 231–329.

Kunze, R. A., and E. M. Stein, Uniformly bounded representations and harmonic analysis of the 2×2 real unimodular group, *Amer. J. Math.* 82(1960), 1–62.

Kunze, R. A., and E. M. Stein, Uniformly bounded representations II. Analytic continuation of the principal series of representations of the $n \times n$ complex unimodular group, *Amer. J. Math.* 83 (1961), 723–786.

Kunze, R. A., and E. M. Stein, Uniformly bounded representations III. Intertwining operators for the principal series on semisimple groups, *Amer. J. Math.* 89 (1967), 385–442.

Lang, S., $SL_2(R)$, Addison-Wesley, Reading, Mass., 1975.

Langlands, R. P., Eisenstein series, *Algebraic Groups and Discontinuous Subgroups*, Proc. Symp. in Pure Math. 9 (1966a), 235–252, American Mathematical Society, Providence, R. I.

Langlands, R. P., Dimension of spaces of automorphic forms, *Algebraic Groups and Discontinuous Subgroups*, Proc. Symp. in Pure Math. 9 (1966b), 253–257, American Mathematical Society, Providence, R. I.

Langlands, R. P., On the classification of irreducible representations of real algebraic groups, mimeographed notes, Institute for Advanced Study, 1973.

Lepowsky, J., Multiplicity formulas for certain semisimple Lie groups, *Bull. Amer. Math. Soc.* 77 (1971), 601–605.

Lepowsky, J., and N. R. Wallach, Finite- and infinite-dimensional representations of linear semisimple groups, *Trans. Amer. Math. Soc.* 184 (1973), 223–246.

Lichtenberg, D. B., *Unitary Symmetry and Elementary Particles*, Academic Press, New York, 1970; 2nd ed., 1978.

Lipsman, R. L., On the characters and equivalence of continuous series representations, *J. Math. Soc. Japan* 23 (1971), 452–480.

Loomis, L. H., *An Introduction to Abstract Harmonic Analysis*, D. Van Nostrand, Princeton, 1953.

Loos, O., *Symmetric Spaces*, Vol. I, W. A. Benjamin Inc., New York, 1969a.

Loos, O., *Symmetric Spaces*, Vol. II, W. A. Benjamin Inc., New York, 1969b.

Mackey, G. W., On induced representations of groups, *Amer. J. Math.* 73 (1951), 576–592.

Mackey, G. W., Induced representations of locally compact groups I, *Ann. of Math.* 55 (1952), 101–139.

Mackey, G. W., Infinite-dimensional group representations, *Bull. Amer. Math. Soc.* 69 (1963), 628–686.

Mackey, G. W., On the analogy between semisimple Lie groups and certain related semi-direct product groups, *Lie Groups and Their Representations (Summer School of the Bolyai János Mathematical Society)*, Halsted Press, New York, 1975, 339–363.

Mackey, G. W., *The Theory of Unitary Group Representations*, University of Chicago Press, Chicago, 1976.

Mackey, G. W., *Unitary Group Representations in Physics, Probability, and Number Theory*, Benjamin/Cummings Publishing Co., Reading, Mass., 1978.

Massey, W. S., *Algebraic Topology: An Introduction*, Springer-Verlag, New York, 1967.

Matsuki, T., The orbits of affine symmetric spaces under the action of minimal parabolic subgroups, *J. Math. Soc. Japan* 31 (1979), 331–357.

Mautner, F. I., Unitary representations of locally compact groups II, *Ann. of Math.* 52 (1950), 528–556.

McKay, W. G., and J. Patera, *Tables of Dimensions, Indices, and Branching Rules for Representations of Simple Lie Algebras*, Marcel Dekker, New York, 1981.

Midorikawa, H., On the explicit formulae of characters for discrete series representations, *Japanese J. Math.* 3 (1977), 313–368.

Miličić, D., Notes on asymptotics of admissible representations of semisimple Lie groups, manuscript, 1976.

Miličić, D., Asymptotic behaviour of matrix coefficients of the discrete series, *Duke Math. J.* 44 (1977), 59–88.

Moore, C. C., Compactifications of symmetric spaces, *Amer. J. Math.* 86 (1964), 201–218.

Mostow, G. D., A new proof of E. Cartan's theorem on the topology of semi-simple Lie groups, *Bull. Amer. Math. Soc.* 55 (1949), 969–980.

Mostow, G. D., The extensibility of local Lie groups of transformations and groups on surfaces, *Ann. of Math.* 52 (1950), 606–636.

Murray, F. J., and J. von Neumann, On rings of operators, *Ann. of Math.* 37 (1936), 116–229.

Naimark, M. A., A continuous analogue of Schur's Lemma and its application to Plancherel's formula for complex classical groups, *Translations Amer. Math. Soc.* (2) 9 (1958), 217–231.

Naimark, M. A., *Normed Rings*, P. Noordhoff, Groningen, 1964.

Narasimhan, M. S., and K. Okamoto, An analogue of the Borel-Weil-Bott theorem for hermitian symmetric pairs of non-compact type, *Ann. of Math.* 91 (1970), 486–511.

Nelson, E., Analytic vectors, *Ann. of Math.* 70 (1959), 572–615.

Okamoto, K., On the Plancherel formulas for some types of simple Lie groups, *Osaka J. Math.* 2 (1965), 247–282.

Okamoto, K., and H. Ozeki, On square-integrable $\bar{\partial}$-cohomology spaces attached to Hermitian symmetric spaces, *Osaka J. Math.* 4 (1967), 95–110.

Osborne, M. S., and G. Warner, *The Theory of Eisenstein Systems*, Academic Press, New York, 1981.

Oshima, T., Fourier analysis on semisimple symmetric spaces, *Non Commutative Harmonic Analysis and Lie Groups*, Springer-Verlag Lecture Notes in Math. 880 (1981), 357–369.

Oshima, T., and T. Matsuki, A complete description of discrete series for semisimple symmetric spaces, preprint, 1983.

Parthasarathy, R., Dirac operator and the discrete series, *Ann. of Math.* 96 (1972), 1–30.

Parthasarathy, K. R., R. Ranga Rao, and V. S. Varadarajan, Representations of complex semi-simple Lie groups and Lie algebras, *Ann. of Math.* 85 (1967), 383–429.

Peter, F., and H. Weyl, Die Vollständigkeit der primitiven Darstellungen einer geschlossenen kontinuielichen Gruppe, *Math. Ann.* 97 (1927), 737–755.

Poincaré, H., Sur les groupes continus, *C. R. Acad, Sci. Paris* 128 (1899), 1065–1069.

Poincaré, H., Sur les groupes continus, *Trans. Cambridge Philosophical Soc.* 18 (1900), 220–255.

Rallis, S., and G. Schiffmann, Discrete spectrum of the Weil representation, *Bull. Amer. Math. Soc.* 83 (1977), 267–270.

Riesz, F., and B. Sz. Nagy, *Functional Analysis*, Frederick Ungar Publishing Co., New York, 1955.

Rosenberg, J., A quick proof of Harish-Chandra's Plancherel theorem for spherical functions on a semisimple Lie group, *Proc. Amer. Math. Soc.* 63 (1977), 143–149.

Rossi, H., and M. Vergne, Analytic continuation of the holomorphic discrete series of a semisimple Lie group, *Acta Math.* 136 (1976), 1–59.

Rossmann, W., Kirillov's character formula for reductive Lie groups, *Invent. Math.* 48 (1978), 207–220.

Sally, P. J., Analytic continuation of the irreducible unitary representations of the universal covering group of SL(2, *R*), *Memoirs Amer. Math. Soc.* 69 (1967).

Sally, P. J., Intertwining operators and the representations of SL(2, *R*), *J. Func. Anal.* 6 (1970), 441–453.

Sally, P. J., and G. Warner, The Fourier transform on semisimple Lie groups of real rank one, *Acta Math.* 131 (1973), 1–26.

Satake, I., On representations and compactifications of symmetric Riemannian spaces, *Ann. of Math.* 71 (1960), 77–110.

Schafer, R. D., On the simplicity of the Lie algebras E_7 and E_8, *Nederl. Akad. Wetensch. Proc., Ser. A* 69 (1966), 64–69.

Schiffmann, G., Introduction aux travaux d'Harish-Chandra, Exposé 323, *Séminaire Bourbaki*, Volume 1966/67, W. A. Benjamin, New York, 1968.

Schiffmann, G., Intégrales d'entrelacement et fonctions de Whittaker, *Bull. Soc. Math. France* 99 (1971), 3–72.

Schlichtkrull, H., A series of unitary irreducible representations induced from a symmetric subgroup of a semisimple Lie group, *Invent. Math.* 68 (1982), 497–516.

Schlichtkrull, H., The Langlands parameters of Flensted-Jensen's discrete series for semisimple symmetric spaces, *J. Func. Anal.* 50 (1983), 133–150.

Schlichtkrull, H., *Hyperfunctions and Harmonic Analysis on Symmetric Spaces*, Birkhäuser, Boston, 1984.

Schmid, W., Homogeneous complex manifolds and representations of semisimple Lie groups, *Proc. Nat. Acad. Sci. USA* 59 (1968), 56–59.

Schmid, W., On the realization of the discrete series of a semisimple Lie group, *Rice University Studies* 56, No. 2, (1970), 99–108.

Schmid, W., On a conjecture of Langlands, *Ann. of Math.* 93 (1971), 1–42.

Schmid, W., On the characters of discrete series: the Hermitian symmetric case, *Invent. Math.* 30 (1975a), 47–144.

Schmid, W., Some properties of square-integrable representations of semisimple Lie groups, *Ann. of Math.* 102 (1975b), 535–564.

Schmid, W., untitled notes on the regularity theorem for invariant eigendistributions, 1975c.

Schmid, W., L^2-cohomology and the discrete series, *Ann. of Math.* 103 (1976). 375–394.

Schmid, W., Two character identities for semisimple Lie groups, *Non-Commutative Harmonic Analysis*, Springer-Verlag Lecture Notes in Math. 587 (1977), 196–225.

Schmid, W., Poincaré and Lie groups, *Bull. Amer. Math. Soc.* 6 (1982), 175–186.

Schmid, W., Boundary value problems for group invariant differential equations, preprint, 1984, to appear in *Astérisque*.

Schmid, W., Recent developments in representation theory, *Arbeitstagung Bonn 1984*, Springer-Verlag Lecture Notes in Math. 1111 (1985), 135–153.

Schur, I., Neue Begründung der Theorie der Gruppencharaktere, *Sitzungsber. Preuss. Akad.* (1905), 406–513.

Schwartz, L., *Théorie des Distributions I*, Hermann, Paris, 1950.

Segal, I. E., An extension of Plancherel's formula to separable unimodular groups, *Ann. of Math.* 52 (1950), 272–292.

Segal, I. E., A class of operator algebras which are determined by groups, *Duke Math. J.* 18 (1951), 221–265.

Segal, I. E., Hypermaximality of certain operators on Lie groups, *Proc. Amer. Math. Soc.* 3 (1952), 13–15.

Sekiguchi, J., Invariant spherical hyperfunctions on the tangent space of a symmetric space, preprint, 1984.

Séminaire "Sophus Lie," *Théorie des algèbres de Lie, topologie des groupes de Lie, 1954–55*, Ecole Normale Supérieure, Paris, 1955.

Serre, J.-P., Représentations linéaires et espaces homogènes Kählérians des groupes de Lie compacts, Exposé 100, *Séminaire Bourbaki, 6e année, 1953/54*, Inst. Henri Poincaré, Paris, 1954. Reprinted with corrections: 1965.

Serre, J.-P., Tores maximaux des groupes de Lie compacts, Exposé 23, Séminaire "Sophus Lie," *Théorie des algèbres de Lie, topologie des groupes de Lie, 1954–55*, Ecole Normale Supérieure, Paris, 1955.

Shelstad, D., Characters and inner forms of a quasi-split group over *R*, *Compositio Math.* 39 (1979a), 11–45.

Shelstad, D., Orbital integrals and a family of groups attached to a real reductive group, *Ann. Sci. Ecole Norm. Sup.* 12 (1979b), 1–31.

Shelstad, D., Embeddings of *L*-groups, *Canad. J. Math.* 33 (1981), 513–558.

Shelstad, D., *L*-indistinguishability for real groups, *Math. Ann.* 259 (1982), 385–430.

Speh, B., The unitary dual of GL(3, *R*) and GL(4, *R*), *Math. Ann.* 258 (1981), 113–133.

Speh, B., and D. A. Vogan, Reducibility of generalized principal series representations, *Acta Math.* 145 (1980), 227–299.

Stanton, R. J., and P. A. Tomas, Expansions for spherical functions on noncompact symmetric spaces, *Acta Math.* 140 (1978), 251–276.

Stein, E. M., and G. Weiss, *Introduction to Fourier Analysis on Euclidean Spaces*, Princeton University Press, Princeton, 1971.

Strichartz, R. S., Harmonic analysis on hyperboloids, *J. Func. Anal.* 12 (1973), 341–383.

Sugiura, M., Conjugate classes of Cartan subalgebras in real semisimple Lie algebras, *J. Math. Soc. Japan* 11 (1959), 374–434.

Sugiura, M., *Unitary Representations and Harmonic Analysis: An Introduction*, Wiley, New York, 1975.

Tits, J., Sur certaines classes d'espaces homogènes de groupes de Lie, *Acad. Roy. Belg. Cl. Sci. Mém. Coll.* 29 (1955), No. 3.

Tits, J., Classification of algebraic semisimple groups, *Algebraic Groups and Discontinuous Subgroups*, Proc. Symp. in Pure Math. 9 (1966), 33–62, American Mathematical Society, Providence, R. I.

Trombi, P. C., The tempered spectrum of a real semisimple Lie group, *Amer. J. Math.* 99 (1977), 57–75.

Trombi, P. C., *On Harish-Chandra's Theory of the Eisenstein Integral for Real Semisimple Lie Groups*, University of Chicago Lecture Notes in Representation Theory, Chicago, 1978.

Trombi, P. C., and V. S. Varadarajan, Spherical transforms on semisimple Lie groups, *Ann. of Math.* 94 (1971), 246–303.

Trombi, P. C., and V. S. Varadarajan, Asymptotic behaviour of eigen functions on a semisimple Lie group: the discrete spectrum, *Acta Math.* 129 (1972), 237–280.

van den Ban, E. P., Asymptotic expansions and integral formulas for eigenfunctions on a semisimple Lie group, Ph.D. Dissertation, Utrecht, 1982.

Varadarajan, V. S., The theory of characters and the discrete series for semisimple Lie groups, *Harmonic Analysis on Homogeneous Spaces*, Proc. Symp. in Pure Math. 26 (1973), 45–99, American Mathematical Society, Providence, R.I.

Varadarajan, V. S., *Lie Groups, Lie Algebras, and Their Representations*, Prentice-Hall, Englewood Cliffs, N. J., 1974; 2nd edition, Springer-Verlag, New York, 1984.

Varadarajan, V. S., *Harmonic Analysis on Real Reductive Groups*, Springer-Verlag Lecture Notes in Math. 576 (1977).

Vergne, M., On Rossmann's character formula for discrete series, *Invent. Math.* 54 (1979), 11–14.

Vergne, M., A Poisson-Plancherel formula for semi-simple Lie groups, *Ann. of Math.* 115 (1982), 639–666.

Verma, D.-N., Structure of certain induced representations of complex semisimple Lie algebras, *Bull. Amer. Math. Soc.* 74 (1968), 160–166 and 628.

Vogan, D. A., The algebraic structure of the representation of semisimple Lie groups I, *Ann. of Math.* 109 (1979a), 1–60.

Vogan, D. A., The algebraic structure of the representations of semisimple Lie groups II, duplicated notes, Massachusetts Institute of Technology, 1979b.

Vogan, D. A., Irreducible characters of semisimple Lie groups I, *Duke Math. J.* 46 (1979c), 61–108.

Vogan, D. A., *Representations of Real Reductive Lie Groups*, Birkhäuser, Boston, 1981.

Vogan, D. A., Understanding the unitary dual, *Lie Group Representations I*, Springer-Verlag Lecture Notes in Math. 1024 (1983), 264–286.

Vogan, D. A., Unitarizability of certain series of representations, *Ann. of Math.* 120 (1984), 141–187.

Vogan, D. A., and G. J. Zuckerman, Unitary representations with non-zero cohomology, *Compositio Math.* 53 (1984), 51–90.

Wallach, N. R., Cyclic vectors and irreducibility for principal series representations, *Trans. Amer. Math. Soc.* 158 (1971), 107–113.

Wallach, N. R., Cyclic vectors and irreducibility for principal series representations II, *Trans. Amer. Math. Soc.* 164 (1972), 389–396.

Wallach, N. R., *Harmonic Analysis on Homogeneous Spaces*, Marcel Dekker, New York, 1973.

Wallach, N. R., On Harish-Chandra's generalized C-functions, *Amer. J. Math.* 97 (1975a), 386–403.

Wallach, N. R., On regular singularities in several variables, manuscript, 1975b.

Wallach, N. R., On the Enright-Varadarajan modules: a construction of the discrete series, *Ann. Sci. Ecole Norm. Sup.* 9 (1976), 81–101.

Wallach, N. R., The analytic continuation of the discrete series II, *Trans. Amer. Math. Soc.* 251 (1979), 19–37.

Wallach, N. R., Asymptotic expansions of generalized matrix entries of representations of real reductive groups, *Lie Group Representations I*, Springer-Verlag Lecture Notes in Math. 1024 (1983), 287-369.

Wallach, N. R., On the unitarizability of derived functor modules, *Invent. Math.* 78 (1984), 131–141.

Warner, G., *Harmonic Analysis on Semi-Simple Lie Groups*, Vol. I, Springer-Verlag, New York, 1972a.

Warner, G., *Harmonic Analysis on Semi-Simple Lie Groups*, Vol. II, Springer-Verlag, New York, 1972b.

Weyl, H., Theorie der Darstellung kontinuierlicher halb-einfacher Gruppen durch lineare Transformationen I, *Math. Zeitschrift* 23 (1925), 271–309.

Weyl, H., Theorie der Darstellung kontinuierlicher halb-einfachen Gruppen durch lineare Transformationen II, *Math. Zeitschrift* 24 (1926a), 328–376.

Weyl, H., Theorie der Darstellung kontinuierlicher halb-einfacher Gruppen durch lineare Transformationen III, *Math. Zeitschrift* 24 (1926b), 377–395.

Weyl, H., *The Classical Groups, Their Invariants and Representations*, 2nd ed., Princeton University Press, Princeton, 1946.

Wigner, E. P., *Group Theory and Its Application to the Quantum Mechanics of Atomic Spectra*, Academic Press, New York, 1959.

Witt, E., Treue Darstellung Liescher Ringe, *J. Reine Angew. Math.* 177 (1937), 152–160.

Wolf, J. A., Unitary representations on partially holomorphic cohomology spaces, *Memoirs Amer. Math. Soc.* 138 (1974).

Zariski, O., and P. Samuel, *Commutative Algebra*, Vol. II, D. Van Nostrand, Princeton, 1960.

Želobenko, D. P., The analysis of irreducibility in a class of elementary representations of a complex semisimple Lie group, *Math. USSR Izvestija* 2 (1968), 105–128.

Želobenko, D. P., *Compact Lie Groups and Their Representations*, American Mathematical Society, Providence, R. I., 1973.

Zuckerman, G., Tensor products of finite and infinite dimensional representations of semisimple Lie groups, *Ann. of Math.* 106 (1977), 295–308.

Zygmund, A., *Trigonometric Series*, Vol. II, Cambridge Univ. Press, Cambridge, 1959.

Index of Notation

General Notation

$\{x \in A \mid p, q\}$, the set of x in A such that p and q hold
$\varnothing$, empty set
$|E|$ or $\# E$, number of elements in E
$f|_E$, restriction of f to E
$\mathbb{R}$, {real numbers}
$\mathbb{C}$, {complex numbers}
$\mathbb{Z}$, {integers}
$\mathbb{R}^\times$, {nonzero real numbers}
$\mathbb{C}^\times$, {nonzero complex numbers}
$\mathbb{R}^+$, {real numbers ≥ 0}
V', dual of vector space V
End, maps of a vector space to itself
Hom, maps between two vector spaces
I or 1, identity
B^*, adjoint or conjugate transpose of B
B^{tr}, transpose of B
$[w]$, class of w
G_0, identity component of G
$*$, convolution of functions

Specific Notation

In this list, Latin, German, and script symbols appear together and are followed by Greek symbols and non-letters.

Index

GPSR Authorized Representative: Easy Access System Europe - Mustamäe tee 50, 10621 Tallinn, Estonia, gpsr.requests@easproject.com

www.ingramcontent.com/pod-product-compliance
Lightning Source LLC
LaVergne TN
LVHW021001300926
849160LV00007B/212